Selected Titles in This Series

69 **Anatole Katok, Rafael de la Llave, Yakov Pesin, and Howard Weiss, Editors,** Smooth ergodic theory and its applications (University of Washington, Seattle, 1999)

68 **Robert S. Doran and V. S. Varadarajan, Editors,** The mathematical legacy of Harish-Chandra: A celebration of representation theory and harmonic analysis (Baltimore, Maryland, 1998)

67 **Wayne Raskind and Charles Weibel, Editors,** Algebraic K-theory (University of Washington, Seattle, 1997)

66 **Robert S. Doran, Ze-Li Dou, and George T. Gilbert, Editors,** Automorphic forms, automorphic representations, and arithmetic (Texas Christian University, Fort Worth, 1996)

65 **M. Giaquinta, J. Shatah, and S. R. S. Varadhan, Editors,** Differential equations: La Pietra 1996 (Villa La Pietra, Florence, Italy, 1996)

64 **G. Ferreyra, R. Gardner, H. Hermes, and H. Sussmann, Editors,** Differential geometry and control (University of Colorado, Boulder, 1997)

63 **Alejandro Adem, Jon Carlson, Stewart Priddy, and Peter Webb, Editors,** Group representations: Cohomology, group actions and topology (University of Washington, Seattle, 1996)

62 **János Kollár, Robert Lazarsfeld, and David R. Morrison, Editors,** Algebraic geometry—Santa Cruz 1995 (University of California, Santa Cruz, July 1995)

61 **T. N. Bailey and A. W. Knapp, Editors,** Representation theory and automorphic forms (International Centre for Mathematical Sciences, Edinburgh, Scotland, March 1996)

60 **David Jerison, I. M. Singer, and Daniel W. Stroock, Editors,** The legacy of Norbert Wiener: A centennial symposium (Massachusetts Institute of Technology, Cambridge, October 1994)

59 **William Arveson, Thomas Branson, and Irving Segal, Editors,** Quantization, nonlinear partial differential equations, and operator algebra (Massachusetts Institute of Technology, Cambridge, June 1994)

58 **Bill Jacob and Alex Rosenberg, Editors,** K-theory and algebraic geometry: Connections with quadratic forms and division algebras (University of California, Santa Barbara, July 1992)

57 **Michael C. Cranston and Mark A. Pinsky, Editors,** Stochastic analysis (Cornell University, Ithaca, July 1993)

56 **William J. Haboush and Brian J. Parshall, Editors,** Algebraic groups and their generalizations (Pennsylvania State University, University Park, July 1991)

55 **Uwe Jannsen, Steven L. Kleiman, and Jean-Pierre Serre, Editors,** Motives (University of Washington, Seattle, July/August 1991)

54 **Robert Greene and S. T. Yau, Editors,** Differential geometry (University of California, Los Angeles, July 1990)

53 **James A. Carlson, C. Herbert Clemens, and David R. Morrison, Editors,** Complex geometry and Lie theory (Sundance, Utah, May 1989)

52 **Eric Bedford, John P. D'Angelo, Robert E. Greene, and Steven G. Krantz, Editors,** Several complex variables and complex geometry (University of California, Santa Cruz, July 1989)

51 **William B. Arveson and Ronald G. Douglas, Editors,** Operator theory/operator algebras and applications (University of New Hampshire, July 1988)

50 **James Glimm, John Impagliazzo, and Isadore Singer, Editors,** The legacy of John von Neumann (Hofstra University, Hempstead, New York, May/June 1988)

49 **Robert C. Gunning and Leon Ehrenpreis, Editors,** Theta functions – Bowdoin 1987 (Bowdoin College, Brunswick, Maine, July 1987)

48 **R. O. Wells, Jr., Editor,** The mathematical heritage of Hermann Weyl (Duke University, Durham, May 1987)

(Continued in the back of this publication)

Smooth Ergodic Theory and Its Applications

Proceedings of Symposia in
PURE MATHEMATICS

Volume 69

Smooth Ergodic Theory and Its Applications

Proceedings of the AMS Summer Research Institute on Smooth Ergodic Theory and Its Applications
July 26–August 13, 1999,
University of Washington, Seattle

Anatolc Katok
Rafael de la Llave
Yakov Pesin
Howard Weiss
Editors

American Mathematical Society
Providence, Rhode Island

PROCEEDINGS OF THE 1999 AMS SUMMER RESEARCH INSTITUTE ON SMOOTH ERGODIC THEORY AND ITS APPLICATIONS, HELD AT THE UNIVERSITY OF WASHINGTON, SEATTLE, JULY 26–AUGUST 13, 1999

with support from the National Science Foundation, Grant DMS-9813732.

2000 *Mathematics Subject Classification.* Primary 11–XX, 28–XX, 34Cxx, 34Dxx, 37–XX, 53–XX, 70–XX.

Library of Congress Cataloging-in-Publication Data

AMS Summer Research Institute on Smooth Ergodic Theory and Its Applications (1999 : University of Washington)

Smooth ergodic theory and its applications : proceedings of the AMS Summer Research Institute on Smooth Ergodic Theory and Its Applications, July 26–August 13, 1999, University of Washington, Seattle / Anatole Katok... [et al.], editors.

p. cm. — (Proceedings of symposia in pure mathematics, ISSN 0082-0717 ; v. 69)

Includes bibliographical references.

ISBN 0-8218-2682-4 (alk. paper)

1. Ergodic theory—Congresses. I. Katok, A. B. II. Title. III. Series.

QA611.5.A475 1999
515′.42—dc21 2001041216

♾ The paper used in this book is acid-free and falls within the guidelines
established to ensure permanence and durability.
Visit the AMS home page at URL: http://www.ams.org/

10 9 8 7 6 5 4 3 2 1 06 05 04 03 02 01

Contents

Preface

This volume is primarily based on the mini-courses and lectures presented at the American Mathematical Society Summer Research Institute (SRI) on Smooth Ergodic Theory and Its Applications in Seattle, Washington, on July 26–August 13, 1999. Smooth Ergodic Theory studies the statistical properties of differentiable dynamical systems. The origin of this theory can be traced to the seminal works of H. Poincaré, and later many great mathematicians contributed to its development.

One of the most systematically developed areas is the dynamics of systems exhibiting hyperbolic behavior and preserving a natural invariant measure (e.g., absolutely continuous measure, SRB measure, or a more general Gibbs measure). Other major parts of smooth ergodic theory are the study of persistence of stable behavior in certain systems (e.g., KAM-theory) and various rigidity phenomena for dynamical systems and smooth actions of more general groups (e.g., invariant measures for unipotent homogeneous systems, entropies of geodesic flows on negatively curved manifolds, and smooth conjugacy for Anosov systems). Recently, there has been significant progress in a number of other areas including conformal dynamics, coupled map lattices, dimension theory of dynamical systems, and stable ergodicity theory.

Smooth ergodic theory, especially the theory of nonuniformly hyperbolic systems, provides the principle paradigm for the rigorous study of complicated or *chaotic* behavior in deterministic systems. This paradigm asserts that if a non–linear dynamical system exhibits sufficiently pronounced exponential behavior, then global properties of the system can be deduced from studying the linearized system. One can then obtain detailed information on topological properties (such as the growth of periodic orbits, topological entropy, and dimension of invariant sets including attractors), as well as statistical properties (such as the existence of invariant measures, asymptotic behavior of typical orbits, ergodicity, mixing, decay of correlation, and measure-theoretic entropy). Smooth ergodic theory also provides a foundation for numerous applications throughout mathematics (e.g., to Riemannian geometry, number theory, Lie groups, and partial differential equations) as well as other sciences.

During the past decade there have been several major new developments in smooth ergodic theory which have attracted substantial new interest to the field from mathematicians as well as scientists using dynamics in their work. In spite of an impressive literature, it has been extremely difficult for a student, or even an established mathematician who is not an expert in the area, to acquire a working knowledge of smooth ergodic theory and to learn how to use its tools. Accordingly, the Summer Research Institute had a strong educational component including ten mini-courses on various aspects of smooth ergodic theory and its applications which were presented by experts in their fields.

The purpose of this volume is two-fold: first, to serve as a useful gateway to smooth ergodic theory for students and nonspecialists, and, second, to provide a state-of-the-art report on important aspects of the subject at the turn of the millennium. It is a testimony to the vitality of the field that, in spite of the size of the volume, some very important areas had to be omitted from the Summer Research Institute in order to achieve the goal of a sufficiently comprehensive coverage of the core areas.

The structure of the volume reflects this pedagogical aspect and purpose. For reader's convenience the volume is divided into three parts: lecture notes, survey–expository articles, and original research articles.

The first part consists of three long expositions with proofs, each aimed to serve as a comprehensive and largely self-contained introduction to a particular area of smooth ergodic theory. Two of these expositions (by Barreira and Pesin and by de la Llave) represent greatly expanded notes of mini-courses given in Seattle while the third (by Katok and Robinson) is an updated and expanded version of an earlier unpublished manuscript.

The second part is in turn subdivided into thematic sections. All but two of the articles in this part are based either on mini-courses or on survey lectures given in Seattle. The exceptions are the article by Kalinin and Katok written specially for the volume, and by Pöschel which is an update of an earlier unpublished manuscript.

The third part contains a number of original contributions either presented at the research sessions during the Seattle meeting or closely related to the topics discussed there.

All papers in this volume have been refereed by experts whose identities were not disclosed to the authors, and various improvements were made following suggestions of the referees. On behalf of all authors we would like to express our gratitude to the referees for their valuable efforts.

The Institute was funded by the National Science Foundation and sponsored by the American Mathematical Society. The editors/scientific organizers thank Dr. James Maxwell, Associate Executive Director of the AMS, for his considerable efforts in organizing the conference. We also thank the local AMS conference organizer Mr. Wayne Drady who did a superb job. Finally, we would like to thank Ilie Ugarcovici, who very quickly and efficiently prepared files of the papers for publication.

Editors and Scientific Organizers of the Summer Research Institute,
A. Katok, R. de la Llave, Ya. Pesin (managing editor), H. Weiss

Part I

Lecture Notes

Proceedings of Symposia in Pure Mathematics
Volume **69**, 2001

Lectures on Lyapunov Exponents and Smooth Ergodic Theory

L. Barreira and Ya. Pesin

Contents

2000 *Mathematics Subject Classification.* Primary: 37D25, 37C40.

Key words and phrases. Lyapunov exponents, nonuniformly hyperbolic dynamical systems, smooth ergodic theory.

L. Barreira was partially supported by FCT's Funding Program and the NATO grant CRG 970161. Ya. Pesin was partially supported by the National Science Foundation grant #DMS-9704564 and the NATO grant CRG 970161.

Introduction

This manuscript is based on lectures given by Ya. Pesin at the AMS Summer Research Institute (Seattle, Washington, 1999). It presents the core of the nonuniform hyperbolicity theory of smooth dynamical systems. This theory was originated in [**26, 27, 28, 29**] and has since become a mathematical foundation for the paradigm which is widely-known as "deterministic chaos" — the appearance of irregular "chaotic" motions in pure deterministic dynamical systems. We follow the original approach by Ya. Pesin making some improvements and necessary modifications.

The nonuniform hyperbolicity theory is based on the theory of Lyapunov exponents which was originated in the works of Lyapunov [**19**] and Perron [**25**] and was developed further in [**7**]. We provide an extended excursion into this theory. This includes the abstract theory of Lyapunov exponents — that allows one to introduce and study the crucial concept of Lyapunov-Perron regularity (see Section 2) — as well as the advanced stability theory of differential equations (see Sections 1, 3, and 4).

Using the language of the theory of Lyapunov exponents one can view nonuniformly hyperbolic dynamical systems as those where the set of points whose Lyapunov exponents are *all* nonzero is "large", for example, has full measure with respect to an invariant Borel measure (see Sections 6 and 7). In this case the fundamental Multiplicative Ergodic theorem of Oseledets [**24**] implies that almost every point is Lyapunov–Perron regular. Thus, the powerful theory of Lyapunov exponents applies and allows one to carry out a thorough analysis of the local stability of trajectories.

The crucial difference between the classical uniform hyperbolicity and its weakened version of nonuniform hyperbolicity is that the hyperbolicity conditions can get worse when one moves along the trajectory of a nonuniformly hyperbolic point. However, if this point is Lyapunov–Perron regular then the worsening occurs with subexponential rate and the contraction and expansion along stable and unstable directions prevail.

One of the crucial manifestations of this fact is the fundamental Stable Manifold theorem that was established in [**27**] and is a generalization of the classical Hadamard–Perron theorem. In Section 9 we present the proof of the Stable Manifold theorem following the original approach in [**27**] which is essentially an elaboration of the Perron method. In Section 10 we sketch the proof of a slightly more general version of the Stable Manifold theorem (known as the Graph Transform Property) which is due to Hadamard. We also describe several main properties of local stable manifolds of which one of the most important is that their sizes may decrease along trajectories only with subexponential rate (and thus the contraction prevails).

There are several methods for establishing nonuniform hyperbolicity. One of them, which we consider in these lectures, is to show that the Lyapunov exponents of the system are nonzero. One of the first examples was constructed in [**26**]. It is a three dimensional flow on a compact smooth Riemannian manifold which is a reconstruction of an Anosov flow. It reveals some mechanisms for the appearance of zero Lyapunov exponents (see the discussion in Section 8).

Another example are geodesic flows on compact smooth Riemannian manifolds of nonpositive curvature. Let us stress that geodesic flows have always been a good source of examples and have provided inspiration for developing the hyperbolicity theory. In these lectures we consider only geodesic flows on surfaces of nonpositive curvature and show that they are nonuniformly hyperbolic on an open and dense set (see Section 8).

One of the main goals of the nonuniform hyperbolicity theory is to describe the ergodic properties of a smooth dynamical system preserving a smooth invariant measure. This is done in Sections 11, 12, 13, and 15 where we show that such a system has ergodic components of positive measure and also establish the entropy formula that expresses the entropy of the system via its positive Lyapunov exponents. Finally, in Section 16 we apply these results to the geodesic flows on compact surfaces of nonpositive curvature.

Acknowledgment. We want to express our deep gratitude to M. Brin, D. Dolgopyat, C. Pugh, and J. Schmeling for their valuable comments and discussions.

The lectures were used as a handout for a graduate course on Dynamical Systems with Nonzero Lyapunov Exponents that Ya. Pesin taught during the Fall 2000 semester at The Pennsylvania State University. Ya. Pesin would like to thank students for their patience during the course and numerous fruitful remarks which helped improve the text. The authors are specially grateful to students C. Carter, T. Fisher, R. Gunesch, I. Ugarcovici, and A. Windsor for many valuable comments and corrections.

1. Lyapunov Exponents for Differential Equations

Consider a linear differential equation

$$\dot{v} = A(t)v, \tag{1.1}$$

where $v(t) \in \mathbb{C}^n$ and $A(t)$ is a $n \times n$ matrix with complex entries depending continuously on $t \in \mathbb{R}$. We assume that the matrix function $A(t)$ is bounded, i.e.,

$$\sup\{\|A(t)\| : t \in \mathbb{R}\} < \infty. \tag{1.2}$$

It follows that for every $v_0 \in \mathbb{C}^n$ there exists a unique solution $v(t) = v(t, v_0)$ of Equation (1.1) which is defined for every $t \in \mathbb{R}$ and satisfies the initial condition $v(0, v_0) = v_0$.

Consider the trivial solution $v(t) = 0$ for $t \geq 0$. If the matrix function $A(t)$ is constant, i.e., $A(t) = A$ for all $t \geq 0$, then the trivial solution is asymptotically (and indeed, exponentially) stable if and only if the real part of every eigenvalue of the matrix A is negative. A similar result holds in the case when the matrix function $A(t)$ is periodic.

In order to characterize the stability of the trivial solution in the general case we introduce the *Lyapunov exponent* $\chi^+ : \mathbb{C}^n \to \mathbb{R} \cup \{-\infty\}$ of Equation (1.1) by the formula

$$\chi^+(v) = \limsup_{t \to +\infty} \frac{1}{t} \log\|v(t)\|, \tag{1.3}$$

for each $v \in \mathbb{C}^n$, where $v(t)$ is the unique solution of (1.1) satisfying the initial condition $v(0) = v$. It follows immediately from (1.3) that the Lyapunov exponent χ^+ satisfies:

1. $\chi^+(\alpha v) = \chi^+(v)$ for each $v \in \mathbb{C}^n$ and $\alpha \neq 0$;
2. $\chi^+(v + w) \leq \max\{\chi^+(v), \chi^+(w)\}$ for each v, $w \in \mathbb{C}^n$;
3. $\chi^+(0) = -\infty$.

The function χ^+ can take on only finitely many distinct values $\chi_1^+ < \cdots < \chi_s^+$ on $\mathbb{C}^n \setminus \{0\}$, where $s \le n$ (see Section 2 below). Each number χ_i^+ occurs with some multiplicity k_i so that $\sum_{i=1}^s k_i = n$. Note that for every $\varepsilon > 0$ there exists a constant $C_\varepsilon > 0$ such that for every solution $v(t)$ of Equation (1.1) and any $t \ge 0$ we have

$$\|v(t)\| \le C_\varepsilon e^{(\chi_s^+ + \varepsilon)t} \|v(0)\|. \tag{1.4}$$

It follows from (1.4) that if

$$\chi_s^+ < 0 \tag{1.5}$$

then for any sufficiently small $\varepsilon > 0$, every solution $v(t) \to 0$ as $t \to +\infty$ with an exponential rate. In other words the trivial solution $v(t) = 0$ is asymptotically (and indeed, exponentially) stable.

We now consider a nonlinear differential equation

$$\dot{u} = A(t)u + f(t, u), \tag{1.6}$$

which is a *perturbation* of (1.1). We assume that $f(t, 0) = 0$ and hence, $u(t) = 0$ is a solution of (1.6). We also assume that there exists a neighborhood H of 0 in $\mathbb{C}^n$ such that f is continuous on $[0, \infty) \times H$ and that for every u_1, $u_2 \in H$ and $t \ge 0$, we have

$$\|f(t, u_1) - f(t, u_2)\| \le K \|u_1 - u_2\|^q \tag{1.7}$$

for some constants $K > 0$ and $q > 1$. This means that the perturbation $f(t, u)$ is small in H. The number q is called the *order of the perturbation.*

One of the main problems in the Lyapunov stability theory is whether Condition (1.5) implies that the solution $u(t) = 0$ of the perturbed equation (1.6) is asymptotically (and exponentially) stable.

Perron showed that in general the answer is negative (see [**25**]).

EXAMPLE 1.1. Consider the following nonlinear system of differential equations in $\mathbb{R}^2$:

$$\begin{aligned} \dot{u}_1 &= [-\omega - a(\sin \log t + \cos \log t)] u_1, \\ \dot{u}_2 &= [-\omega + a(\sin \log t + \cos \log t)] u_2 + |u_1|^{\lambda+1}, \end{aligned} \tag{1.8}$$

for some positive constants ω, a, and λ. It is a perturbation of the following linear system of differential equations

$$\begin{aligned} \dot{v}_1 &= [-\omega - a(\sin \log t + \cos \log t)] v_1, \\ \dot{v}_2 &= [-\omega + a(\sin \log t + \cos \log t)] v_2. \end{aligned} \tag{1.9}$$

We assume that

$$a < \omega < (2e^{-\pi} + 1)a \quad \text{and} \quad 0 < \lambda < \frac{2a}{\omega - a} - e^{\pi}. \tag{1.10}$$

Notice that the function $f(t, (u_1, u_2)) = (0, |u_1|^{\lambda+1})$ is autonomous, i.e., it does not depend on t, and that it satisfies Condition (1.7) with $q = \lambda + 1 > 1$.

The general solution of (1.8) is given by

$$\begin{aligned} u_1(t) &= c_1 e^{-\omega t - at \sin \log t}, \\ u_2(t) &= c_2 e^{-\omega t + at \sin \log t} + {c_1}^2 e^{-\omega t + at \sin \log t} \int_{t_0}^{t} e^{-(2+\lambda)a\tau \sin \log \tau - \omega \lambda \tau} \, d\tau, \end{aligned} \tag{1.11}$$

while the general solution of (1.9) is given by

$$v_1(t) = d_1 e^{-\omega t - at\sin\log t}, \quad v_2(t) = d_2 e^{-\omega t + at\sin\log t}.$$

Here c_1, c_2, d_1, d_2, and t_0 are arbitrary numbers.

It is easy to check that the values of the Lyapunov exponent associated with Equation (1.9) are $\chi_1^+ = \chi_2^+ = -\omega + a < 0$.

Let $u(t) = (u_1(t), u_2(t))$ be a solution of the nonlinear system of differential equations (1.8). In view of (1.11) it is also a solution of the linear system of differential equations

$$\begin{aligned} \dot{u}_1 &= [-\omega - a(\sin\log t + \cos\log t)]u_1, \\ \dot{u}_2 &= [-\omega + a(\sin\log t + \cos\log t)]u_2 + \delta(t)u_1, \end{aligned} \tag{1.12}$$

where

$$\delta(t) = \operatorname{sgn} c_1 |c_1|^\lambda e^{-\omega\lambda t - a\lambda t\sin\log t}.$$

Note that

$$|\delta(t)| \le |c_1|^\lambda e^{(-\omega+a)\lambda t}$$

and thus by (1.10), Condition (1.2) holds for Equation (1.12). Fix $0 < \varepsilon < \pi/4$ and for each $k \in \mathbb{N}$ set

$$t_k = e^{2k\pi - \frac{1}{2}\pi}, \quad t_k' = e^{2k\pi - \frac{1}{2}\pi - \varepsilon}.$$

Clearly, $t_k \to \infty$ and $t_k' \to \infty$ as $k \to \infty$. One can also see that

$$\int_{t_0}^{t_k} e^{-(2+\lambda)a\tau\sin\log\tau + \omega\lambda\tau}\, d\tau > \int_{t_k'}^{t_k} e^{-(2+\lambda)a\tau\sin\log\tau + \omega\lambda\tau}\, d\tau.$$

For every $\tau \in [t_k', t_k]$ we have

$$2k\pi - \frac{\pi}{2} - \varepsilon \le \log\tau \le 2k\pi - \frac{\pi}{2},$$

$$(2+\lambda)a\tau\cos\varepsilon \le -(2+\lambda)a\tau\sin\log\tau.$$

This implies that

$$\int_{t_k'}^{t_k} e^{-(2+\lambda)a\tau\sin\log\tau + \omega\lambda\tau}\, d\tau \ge \int_{t_k'}^{t_k} e^{(2+\lambda)a\tau\cos\varepsilon - \omega\lambda\tau}\, d\tau.$$

Set $r = (2+\lambda)a\cos\varepsilon - \omega\lambda$. It follows that if $k \in \mathbb{N}$ is sufficiently large then

$$\int_{t_0}^{t_k} e^{-(2+\lambda)a\tau\sin\log\tau + \omega\lambda\tau}\, d\tau > \int_{t_k'}^{t_k} e^{r\tau}\, d\tau > ce^{rt_k},$$

where $c = (1 - e^{-\varepsilon})/r$. Set

$$t_k^* = t_k e^\pi = e^{2k\pi + \frac{1}{2}\pi}.$$

We obtain

$$\begin{aligned} e^{at_k^*\sin\log t_k^*} \int_{t_0}^{t_k^*} e^{-(2+\lambda)a\tau\sin\log\tau + \omega\lambda\tau}\, d\tau &> e^{at_k^*} \int_{t_0}^{t_k} e^{-(2+\lambda)a\tau\sin\log\tau + \omega\lambda\tau}\, d\tau \\ &> ce^{at_k^* + rt_k} = ce^{(a + re^{-\pi})t_k^*}. \end{aligned}$$

It follows from (1.10) that if $c_1 \ne 0$ and ε is sufficiently small, then the Lyapunov exponent of any solution $u(t)$ of (1.8) (which is also a solution of (1.12)) satisfies

$$\chi^+(u) \ge -\omega + a + re^{-\pi} = -\omega + a + [(2+\lambda)a\cos\varepsilon - \omega\lambda]e^{-\pi} > 0.$$

Therefore, the solution $u(t)$ is not asymptotically stable. This completes the construction of the example.

Lyapunov introduced regularity conditions which guarantee asymptotic (and indeed, exponential) stability of the solution $u(t) = 0$ of the perturbed equation (1.6). Although there are many different ways to state the regularity conditions (which we discuss below), for a given differential equation the regularity is often difficult to verify.

We now state the *Lyapunov Stability Theorem* (see [**7**]). It claims that under an additional assumption known as *forward regularity*, Condition (1.5) indeed implies the stability of the trivial solution $u(t) = 0$ of (1.6).

THEOREM 1.2. *Assume that the Lyapunov exponent χ^+ of Equation* (1.1), *with the matrix function $A(t)$ satisfying* (1.2), *is forward regular and satisfies Condition* (1.5). *Then the solution $u(t) = 0$ of the perturbed equation* (1.6) *is asymptotically and exponentially stable.*

The notion of forward regularity will be introduced and discussed in Sections 2 and 3.

2. Abstract Theory of Lyapunov Exponents

Before we proceed with the proof of the Lyapunov Stability Theorem 1.2 we discuss the notions of Lyapunov exponent and regularity. It is more convenient to do this in a formal axiomatic setting using the basic properties 1, 2, and 3 of the Lyapunov exponent χ^+ described in Section 1. This allows one to study Lyapunov exponents in other situations such as sequences of matrices, cocycles, dynamical systems with continuous and discrete time, etc.

Let V be an n-dimensional real vector space. A function $\chi\colon V \to \mathbb{R} \cup \{-\infty\}$ is called a *Lyapunov characteristic exponent* or simply a *Lyapunov exponent* on V if:

1. $\chi(\alpha v) = \chi(v)$ for each $v \in V$ and $\alpha \in \mathbb{R} \setminus \{0\}$;
2. $\chi(v + w) \le \max\{\chi(v), \chi(w)\}$ for each v, $w \in V$;
3. $\chi(0) = -\infty$ (*normalization property*).

We describe some basic properties of Lyapunov exponents.

THEOREM 2.1. *If χ is a Lyapunov exponent, then the following properties hold:*

1. *if v, $w \in V$ are such that $\chi(v) \neq \chi(w)$ then*
$$\chi(v + w) = \max\{\chi(v), \chi(w)\};$$
2. *if $v_1, \dots, v_m \in V$ and $\alpha_1, \dots, \alpha_m \in \mathbb{R} \setminus \{0\}$ then*
$$\chi(\alpha_1 v_1 + \cdots + \alpha_m v_m) \le \max\{\chi(v_i) : 1 \le i \le m\};$$
if, in addition, there exists i such that $\chi(v_i) > \chi(v_j)$ for all $j \neq i$ then
$$\chi(\alpha_1 v_1 + \cdots + \alpha_m v_m) = \chi(v_i);$$
3. *if for some $v_1, \dots, v_m \in V \setminus \{0\}$ the numbers $\chi(v_1), \dots, \chi(v_m)$ are distinct, then the vectors $v_1, \cdots, v_m$ are linearly independent; if, in addition, $m = n$, then the vectors $v_1, \dots, v_n$ form a basis of V;*
4. *the function χ can take no more than n distinct finite values.*

PROOF. Suppose that $\chi(v) < \chi(w)$. We have

$$\chi(v+w) \leq \chi(w) = \chi(v+w-v) \leq \max\{\chi(v+w), \chi(v)\}.$$

It follows that if $\chi(v+w) < \chi(v)$ then $\chi(w) \leq \chi(v)$ which contradicts our assumption. Hence, $\chi(v+w) \geq \chi(v)$, and thus, $\chi(v+w) = \chi(w)$. Statement 1 follows. Statement 2 is an immediate consequence of Statement 1 and Properties 1 and 2 in the definition of Lyapunov exponent.

In order to prove Statement 3 assume on the contrary that the vectors $v_1, \ldots, v_m$ are linearly dependent, i.e., $\alpha_1 v_1 + \cdots + \alpha_m v_m = 0$ with not all constants α_i equal to zero, while $\chi(v_1), \ldots, \chi(v_m)$ are distinct. By Statement 2 and Property 3 in the definition of Lyapunov exponent, we obtain

$$-\infty = \chi(\alpha_1 v_1 + \cdots + \alpha_m v_m) = \max\{\chi(v_i) : 1 \leq i \leq m \text{ and } \alpha_i \neq 0\} \neq -\infty.$$

This contradiction implies Statement 3. Statement 4 follows from Statement 3. □

By Theorem 2.1, the Lyapunov exponent χ can take on only finitely many distinct *values* on $V \setminus \{0\}$. We denote them by

$$\chi_1 < \cdots < \chi_s$$

for some $s \leq n$. In general, χ_1 may be $-\infty$. For each $1 \leq i \leq s$, define

$$V_i = \{v \in V : \chi(v) \leq \chi_i\}. \tag{2.1}$$

Put $V_0 = \{0\}$. It follows from Theorem 2.1 that V_i is a linear subspace of V for each i, and

$$\{0\} = V_0 \subsetneq V_1 \subsetneq \cdots \subsetneq V_s = V. \tag{2.2}$$

We call a collection $\mathcal{V} = \{V_i : i = 0, \ldots, s\}$ of linear subspaces of V satisfying (2.2) a *linear filtration* or simply a *filtration* of V.

The following result gives an equivalent characterization of Lyapunov exponents in terms of filtrations.

THEOREM 2.2. *A function $\chi\colon V \to \mathbb{R} \cup \{-\infty\}$ is a Lyapunov exponent if and only if there exist numbers $\chi_1 < \cdots < \chi_s$ for some $1 \leq s \leq n$, and a filtration $\mathcal{V} = \{V_i : i = 0, \ldots, s\}$ of V such that:*

1. *$\chi(v) \leq \chi_i$ for every $v \in V_i$;*
2. *$\chi(v) = \chi_i$ for every $v \in V_i \setminus V_{i-1}$ and $1 \leq i \leq s$;*
3. *$\chi(0) = -\infty$.*

PROOF. If χ is a Lyapunov exponent then the filtration

$$\mathcal{V} = \{V_i : i = 0, \ldots, s\}$$

defined by (2.1) satisfies Conditions 1 and 3 of Theorem 2.2. Moreover, for any $v \in V_i \setminus V_{i-1}$ we have $\chi_{i-1} < \chi(v) \leq \chi_i$. Since χ takes no value strictly between χ_{i-1} and χ_i, we obtain $\chi(v) = \chi_i$ and Condition 2 follows.

Now suppose that a function χ and a filtration $\mathcal{V}$ satisfy the conditions of the theorem. Observe that $v \in V_i \setminus V_{i-1}$ if and only if $\alpha v \in V_i \setminus V_{i-1}$ for any $\alpha \in \mathbb{R} \setminus \{0\}$. Therefore, by Condition 2, $\chi(\alpha v) = \chi(v)$. Choose now vectors $v_1, v_2 \in V$. Let $\chi(v_j) = \chi_{i_j}$ for $j = 1, 2$. It follows from Conditions 1 and 2 that (2.1) holds. Therefore, $v_j \in V_{i_j}$ for $j = 1, 2$. Without loss of generality we may assume that $i_1 < i_2$. This implies that $v_1 + v_2 \in V_{i_1} \cup V_{i_2} = V_{i_2}$. Hence, by Condition 1, we have

$$\chi(v_1 + v_2) \leq \chi_{i_2} = \max\{\chi(v_1), \chi(v_2)\},$$

and thus, χ is a Lyapunov exponent. □

We refer to the filtration $\mathcal{V} = \{V_i : i = 0, \ldots, s\}$ of V, defined by (2.1), as the *filtration of V associated to χ* and denote it by $\mathcal{V}_\chi$. We call the number

$$k_i = \dim V_i - \dim V_{i-1}$$

the *multiplicity* of the value χ_i, and the collection of pairs

$$\operatorname{Sp} \chi = \{(\chi_i, k_i) : 1 \le i \le s\}$$

the *Lyapunov spectrum* of χ.

Given a filtration $\mathcal{V} = \{V_i : i = 0, \ldots, s\}$ of V and numbers $\chi_1 < \cdots < \chi_s$ define a function $\chi \colon V \to \mathbb{R} \cup \{-\infty\}$ by $\chi(v) = \chi_i$ for every $v \in V_i \setminus V_{i-1}$, $1 \le i \le s$, and $\chi(0) = -\infty$. It is easy to see that the function χ and the filtration $\mathcal{V}$ satisfy the conditions of Theorem 2.2, and thus determine a Lyapunov exponent on V.

We now consider a special class of bases in $\mathbb{R}^n$ which are well-adapted to filtrations in $\mathbb{R}^n$.

A basis $\mathbf{v} = (v_1, \ldots, v_n)$ of V is said to be *normal* with respect to the filtration $\mathcal{V} = \{V_i : i = 0, \ldots, s\}$ if for every $1 \le i \le s$ there exists a basis of V_i composed of n_i vectors from $\{v_1, \ldots, v_n\}$. A normal basis $\mathbf{v}$ is called *ordered* if for every $1 \le i \le s$, the vectors $v_1, \ldots, v_{n_i}$ form a basis of V_i.

Note that for every filtration $\mathcal{V}$ there exists a basis which is normal with respect to $\mathcal{V}$. Moreover, two filtrations coincide if and only if each normal basis with respect to one of them is also normal with respect to the other one. Furthermore, if χ is a Lyapunov exponent and $\mathbf{v}$ is a normal basis with respect to the filtration $\mathcal{V}_\chi$, then among the numbers $\chi(v_1), \ldots, \chi(v_n)$ the value χ_i occurs exactly k_i times, for each $i = 1, \ldots, s$. Hence,

$$\sum_{j=1}^{n} \chi(v_j) = \sum_{i=1}^{s} k_i \chi_i. \tag{2.3}$$

We use this observation to prove the following result.

THEOREM 2.3. *A basis $\mathbf{v}$ of V is normal with respect to the filtration $\mathcal{V}_\chi$ associated to the Lyapunov exponent χ if and only if*

$$\inf \left\{ \sum_{j=1}^{n} \chi(w_j) : \mathbf{w} \text{ is a basis of } V \right\} = \sum_{j=1}^{n} \chi(v_j). \tag{2.4}$$

PROOF. One can easily verify that there exists a normal ordered basis $\mathbf{v}$ adapted to $\mathcal{V}_\chi$.

LEMMA 2.4. *If $\mathbf{w}$ is a basis of V for which $\chi(w_1) \le \cdots \le \chi(w_n)$, then:*

1. *$\chi(w_j) \ge \chi(v_j)$ for every $1 \le j \le n$, and $\chi(w_n) = \chi(v_n)$;*
2. *$\sum_{j=1}^{n} \chi(w_j) \ge \sum_{j=1}^{n} \chi(v_j)$;*
3. *$\mathbf{w}$ is normal if and only if $\chi(w_j) = \chi(v_j)$ for every $1 \le j \le n$;*
4. *$\mathbf{w}$ is normal if and only if $\sum_{j=1}^{n} \chi(w_j) = \sum_{j=1}^{n} \chi(v_j)$.*

PROOF OF THE LEMMA. We notice that since χ_1 is the minimal value of χ on $V \setminus \{0\}$ we have $\chi(w_j) \ge \chi(v_j) = \chi_1$ for every $j = 1, \ldots, n_1$. Assume that $\chi(w_{n_1+1}) = \chi_1$. Then $\chi(w_1) = \cdots = \chi(w_{n_1+1}) = \chi_1$ and

$$n_1 \ge \dim \operatorname{span}\{w_1, \ldots, w_{n_1+1}\} = n_1 + 1,$$

where $\operatorname{span} Z$ denotes the linear space generated by the set of vectors Z. This contradiction implies that $\chi(w_{n_1+1}) \geq \chi_2$ and hence $\chi(w_j) \geq \chi(v_j) = \chi_2$ for every $j = n_1 + 1, \ldots, n_2$.

Repeating the same argument finitely many times we obtain $\chi(w_j) \geq \chi(v_j)$ for every $1 \leq j \leq n$. In particular, $\chi(w_n) = \chi(v_n)$ since $\chi(v_n)$ is the maximum value of χ. Statement 1 follows. Statement 2 is an immediate consequence of Statement 1.

By (2.1), $\chi(w_j) = \chi(v_j)$ for every $1 \leq j \leq n$ if and only if $w_1, \ldots, w_{n_i}$ is a basis of V_i for every $1 \leq i \leq s$ and hence, if and only if the basis $\mathbf{w}$ is normal. This implies Statement 3. The last statement is a consequence of Statements 2 and 3. □

By the lemma, the infimum in (2.4) is equal to

$$\inf\left\{\sum_{j=1}^{n} \chi(w_j) : \mathbf{w} \text{ is a normal basis of } V\right\} = \sum_{i=1}^{s} k_i \chi_i.$$

In view of Statement 4 in the lemma, the basis $\mathbf{v}$ is normal if and only if (2.3) holds, and hence, if and only if (2.4) holds. □

There is a very useful Lyapunov's construction of normal bases which we now describe. This construction defines a sequence of bases

$$\mathbf{w}_i = (w_{i1}, \ldots, w_{in}) = (w_1, \ldots, w_i, v_{i+1}, \ldots, v_n),$$

which makes the sum $\sum_{j=1}^{n} \chi(w_{ij})$ decrease as i increases. Since χ takes on only finitely many values, this process ends up after a finite number of steps.

Set $w_n = v_n$ and let $i < n$. If

$$w \in \operatorname{span}\{v_i, \ldots, v_n\} \setminus \operatorname{span}\{v_{i+1}, \ldots, v_n\}, \tag{2.5}$$

then $w = \alpha(v_i + w_i')$ for some real number $\alpha \neq 0$ and some vector

$$w_i' \in \operatorname{span}\{v_{i+1}, \ldots, v_n\}.$$

Since χ takes on only finitely many values one can choose a vector w_i' such that if $w_i = v_i + w_i'$ then $\chi(w_i)$ is the minimum of $\chi(w)$ for w satisfying (2.5). Observe that there exists a linear transformation $A\colon V \to V$ which has an upper triangular form with respect to the basis $\mathbf{v}$ such that $w_i = Av_i$ for each $i = 1, \ldots, n$. Moreover, since

$$w_i - v_i \in \operatorname{span}\{v_{i+1}, \ldots, v_n\},$$

all entries on the diagonal of A are equal to 1, and hence $\det A \neq 0$ and $\mathbf{w} = (w_1, \ldots, w_n)$ is a basis.

We now show that $\mathbf{w}$ is a normal basis. Assuming the contrary and using Lemma 2.4 we can find a vector v such that $v = \sum_{j=1}^{k} \alpha_j w_{i_j}$ for some $1 \leq k \leq n$, $i_1 < i_2 < \cdots < i_k$, and $\alpha_j \neq 0$ for $j = 1, \ldots, k$, which satisfies

$$\chi(v) < \max\{\chi(w_{i_1}), \ldots, \chi(w_{i_k})\}. \tag{2.6}$$

Since $v \in \operatorname{span}\{w_{i_1}, \ldots, w_{i_n}\}$, we have $\chi(v) \geq \chi(w_{i_1})$ and hence,

$$\chi(v - \alpha_1 w_{i_1}) \leq \max\{\chi(v), \chi(w_{i_1})\} = \chi(v). \tag{2.7}$$

By (2.6) and (2.7),

$$\chi(v - \alpha_1 w_{i_1}) < \max\{\chi(w_{i_2}), \ldots, \chi(w_{i_k})\}.$$

Iterating this procedure we obtain

$$\chi(w_{i_k}) = \chi(v - \sum_{j=1}^{k-1} \alpha_j w_{i_j}) < \chi(w_{i_k}).$$

This contradiction implies that the basis $\mathbf{w}$ is normal.

We now establish the behavior of normal bases under linear transformations. Let $\mathbf{v}$ be a normal ordered basis with respect to a filtration $\mathcal{V} = \{V_i : i = 0, \ldots, s\}$ of a vector space V, and $A\colon V \to V$ an invertible linear transformation.

THEOREM 2.5. *The following properties are equivalent:*

1. $(Av_1, \ldots, Av_n)$ *is a normal ordered basis;*
2. *the transformation* A *preserves the filtration* $\mathcal{V}$, *i.e,* $AV_i = V_i$ *for every* $1 \le i \le s$;
3. *the transformation* A, *with respect to a basis which is a reordering of the basis* $\mathbf{v}$, *has the lower block-triangular form*

$$\begin{pmatrix} A_1 & 0 & \cdots & 0 \\ & A_2 & \ddots & \vdots \\ & & \ddots & 0 \\ & & & A_s \end{pmatrix},$$

where each A_i *is a* $k_i \times k_i$ *matrix with* $\det A_i \ne 0$.

PROOF. The proof is elementary. □

Let $\mathcal{W}$ be another filtration of V. As a consequence of the above statement one can show that *there exists a basis which is normal with respect to both filtrations.* Indeed, Let $\mathbf{v}$ be a normal basis with respect to $\mathcal{W}$. There exists a lower triangular $n \times n$ matrix such that $\mathbf{w} = (Av_1, \ldots, Av_n)$ is a normal basis with respect to $\mathcal{V}$. One can easily show that the basis $\mathbf{w}$ is also normal with respect to $\mathcal{W}$.

Let V be a vector space and W a dual vector space to V. Let also $\mathbf{v} = (v_1, \ldots, v_n)$ be a basis in V and $\mathbf{w} = (w_1, \ldots, w_n)$ a basis in W. We say that $\mathbf{v}$ is *dual* to $\mathbf{w}$ and write $\mathbf{v} \sim \mathbf{w}$ if $\langle v_i, w_j \rangle = \delta_{ij}$ for each i and j.

Let χ be a Lyapunov exponent on V and $\widetilde{\chi}$ a Lyapunov exponent on W. We say that the exponents χ and $\widetilde{\chi}$ are *dual* and write $\chi \sim \widetilde{\chi}$ if for any dual bases $\mathbf{v}$ and $\mathbf{w}$, and every $1 \le i \le n$, we have

$$\chi(v_i) + \widetilde{\chi}(w_i) \ge 0.$$

We denote by $\chi'_1 \le \cdots \le \chi'_n$ the values of χ counted with their multiplicities, i.e., $\chi'_i = \chi(v_i)$ for some normal ordered basis $\mathbf{v}$ of V. Similarly, we denote by $\widetilde{\chi}'_1 \ge \cdots \ge \widetilde{\chi}'_n$ the values of $\widetilde{\chi}$ counted with their multiplicities.

We define the *regularity coefficient* of the dual Lyapunov exponents χ and $\widetilde{\chi}$ by

$$\gamma(\chi, \widetilde{\chi}) = \min \max\{\chi(v_i) + \widetilde{\chi}(w_i) : 1 \le i \le n\},$$

where the minimum is taken over all dual bases $\mathbf{v}$ and $\mathbf{w}$ of V and W. The *Perron coefficient* of χ and $\widetilde{\chi}$ is defined by

$$\pi(\chi, \widetilde{\chi}) = \max\{\chi'_i + \widetilde{\chi}'_i : 1 \le i \le n\}.$$

THEOREM 2.6. *The following statements hold:*

1. $\pi(\chi, \widetilde{\chi}) \le \gamma(\chi, \widetilde{\chi})$;

2. if $\chi \sim \widetilde{\chi}$, *then* $0 \leq \pi(\chi, \widetilde{\chi}) \leq \gamma(\chi, \widetilde{\chi}) \leq n\,\pi(\chi, \widetilde{\chi})$.

PROOF. We begin with the following lemma.

LEMMA 2.7. *Given numbers* $\lambda_1 \leq \cdots \leq \lambda_n$ *and* $\mu_1 \geq \cdots \geq \mu_n$, *and a permutation* σ *of* $\{1, \ldots, n\}$, *we have*

$$\min\{\lambda_i + \mu_{\sigma(i)} : 1 \leq i \leq n\} \leq \min\{\lambda_i + \mu_i : 1 \leq i \leq n\},$$

$$\max\{\lambda_i + \mu_{\sigma(i)} : 1 \leq i \leq n\} \geq \max\{\lambda_i + \mu_i : 1 \leq i \leq n\}.$$

PROOF OF THE LEMMA. Notice that the second inequality follows from the first one in view of the following relations:

$$\begin{aligned}\max\{\lambda_i + \mu_{\sigma(i)} : 1 \leq i \leq n\} &= -\min\{-\mu_{\sigma(i)} - \lambda_i : 1 \leq i \leq n\}\\ &= -\min\{-\mu_i - \lambda_{\sigma^{-1}(i)} : 1 \leq i \leq n\}\\ &\geq -\min\{-\mu_i - \lambda_i : 1 \leq i \leq n\},\\ &= \max\{\lambda_i + \mu_i : 1 \leq i \leq n\}.\end{aligned}$$

We now prove the first inequality. We may assume that σ is not the identity permutation (otherwise the result is trivial). Fix an integer i such that $1 \leq i \leq n$. If $i \leq \sigma(i)$, then $\mu_{\sigma(i)} \leq \mu_i$ and

$$\min\{\lambda_i + \mu_{\sigma(i)} : 1 \leq i \leq n\} \leq \lambda_i + \mu_{\sigma(i)} \leq \lambda_i + \mu_i.$$

If $i > \sigma(i)$, then there exists $k < i$ such that $i \leq \sigma(k)$. Otherwise, we would have $\sigma(1), \ldots, \sigma(i-1) \leq i-1$ and hence, $\sigma(i) \geq i$. It follows that

$$\min\{\lambda_i + \mu_{\sigma(i)} : 1 \leq i \leq n\} \leq \lambda_k + \mu_{\sigma(k)} \leq \lambda_i + \mu_i.$$

The desired result now follows. □

We proceed with the proof of the theorem. Consider dual bases $\mathbf{v}$ and $\mathbf{w}$. Without loss of generality, we may assume that $\chi(v_1) \leq \cdots \leq \chi(v_n)$. Let σ be a permutation of $\{1, \ldots, n\}$ such that the numbers $\mu_{\sigma(i)} = \widetilde{\chi}(w_i)$ satisfy $\mu_1 \geq \cdots \geq \mu_n$. We have $\chi(v_i) \geq \chi_i'$ and $\mu_i \geq \widetilde{\chi}_i'$. By Lemma 2.7 we obtain

$$\begin{aligned}\max\{\chi(v_i) + \widetilde{\chi}(w_i) : 1 \leq i \leq n\} &\geq \max\{\chi(v_i) + \mu_i : 1 \leq i \leq n\}\\ &\geq \max\{\chi_i' + \widetilde{\chi}_i' : 1 \leq i \leq n\}\\ &= \pi(\chi, \widetilde{\chi}).\end{aligned}$$

Therefore, $\gamma(\chi, \widetilde{\chi}) \geq \pi(\chi, \widetilde{\chi})$ and Statement 1 is proven.

We assume now that $\chi \sim \widetilde{\chi}$. It is not difficult to show that one can choose bases $\mathbf{v}$ and $\mathbf{w}$, which are dual and both normal; furthermore, we assume that $\mathbf{v}$ is ordered. It follows that $\chi(v_i) = \chi_i'$ and $\mu_i = \widetilde{\chi}_i'$ for each i. Thus,

$$\begin{aligned}\gamma(\chi, \widetilde{\chi}) &\leq \max\{\chi(v_i) + \widetilde{\chi}(w_i) : 1 \leq i \leq n\}\\ &\leq \sum_{i=1}^{n}(\chi(v_i) + \widetilde{\chi}(w_i)) = \sum_{i=1}^{n}(\chi_i' + \widetilde{\chi}_i')\\ &\leq n \max\{\chi_i' + \widetilde{\chi}_i' : 1 \leq i \leq n\} = n\,\pi(\chi, \widetilde{\chi}).\end{aligned}$$

Finally, since $\chi \sim \widetilde{\chi}$ we have $\gamma(\chi, \widetilde{\chi}) \geq 0$. This implies that $\pi(\chi, \widetilde{\chi}) \geq 0$, and Statement 2 follows. □

We now introduce the crucial concept of regularity of a pair of Lyapunov exponents χ and $\widetilde{\chi}$ in dual vector spaces V and W. Roughly speaking regularity means that the filtrations $\mathcal{V}_\chi$ and $\mathcal{V}_{\widetilde{\chi}}$ are well-adapted to each other (essentially they are orthogonal; see Theorem 2.8 below), and thus implies some special properties which determine its role in the stability theory. On a first glance the regularity requirements seem quite strong and even a bit artificial. However, they hold in "typical" situations.

The pair of Lyapunov exponents $(\chi, \widetilde{\chi})$ is called *regular* if $\chi \sim \widetilde{\chi}$ and $\gamma(\chi, \widetilde{\chi}) = 0$. By Theorem 2.6, this holds if and only if $\pi(\chi, \widetilde{\chi}) = 0$, and also if and only if $\widetilde{\chi}'_i = -\chi'_i$.

THEOREM 2.8. *If the pair $(\chi, \widetilde{\chi})$ is regular, then the filtrations $\mathcal{V}_\chi = \{V_i : i = 0, \ldots, s\}$ and $\mathcal{V}_{\widetilde{\chi}} = \{W_i : i = 0, \ldots, r\}$ are orthogonal, that is, $s = r$, $\dim V_i + \dim W_{s-i} = n$, and $\langle v, w\rangle = 0$ for every $v \in V_i$ and $w \in W_{s-i}$.*

PROOF. Set $m_i = n - \dim W_{s-i} + 1$. Then

$$W_{s-i} = \{w \in W : \widetilde{\chi}(w) \le \widetilde{\chi}'_{m_i}\}$$

and $\widetilde{\chi}'_{m_i} = -\chi'_{m_i}$ in view of Theorem 2.6. Let $\mathbf{v}$ be a normal ordered basis of V and $\mathbf{w}$ the basis of W dual to $\mathbf{v}$. Since $\chi \sim \widetilde{\chi}$ we obtain

$$\begin{aligned} W_{s-i} &= \{w \in W : \chi'_{m_i} + \widetilde{\chi}(w) < 0\} \\ &= \{w \in W : \chi(v_i) + \widetilde{\chi}(w) < 0 \text{ if and only if } i < m_i\} \\ &= \operatorname{span}\{w_{m_i}, \ldots, w_n\}. \end{aligned}$$

This implies that $\mathbf{w}$ is normal and ordered. Now for each i and any normal ordered basis $\widetilde{\mathbf{v}} = (\widetilde{v}_1, \ldots, \widetilde{v}_{n_i}, v_{n_i+1}, \ldots, v_n)$ of V and the dual basis $\widetilde{\mathbf{w}}$ of W the last $n - n_i$ components of $\widetilde{\mathbf{w}}$ have to coincide with those of $\mathbf{w}$. This implies that $r = s$, and

$$W_{s-i} = \operatorname{span}\{w_{n_i+1}, \ldots, w_n\} = V_i^\perp.$$

The desired result now follows. □

With slight changes one can also consider Lyapunov exponents on complex vector spaces.

3. Regularity of Lyapunov Exponents Associated with Differential Equations

We now discuss the regularity properties of the Lyapunov exponent χ^+ defined by (1.3) for Equation (1.1) provided the matrix function $A(t)$ satisfies (1.2). Consider the linear differential equation which is dual to (1.1),

$$\dot{w} = -A(t)^* w, \tag{3.1}$$

where $w(t) \in \mathbb{C}^n$ and $A(t)^*$ denotes the complex-conjugated transpose of $A(t)$. Let $w(t)$ be a unique solution of this equation such that $w(0) = w$. The function $\widetilde{\chi}^+ : \mathbb{C}^n \to \mathbb{R} \cup \{-\infty\}$ given by

$$\widetilde{\chi}^+(w) = \limsup_{t\to+\infty} \frac{1}{t} \log\|w(t)\|$$

defines a Lyapunov exponent associated with Equation (3.1). We note that the exponents χ^+ and $\widetilde{\chi}^+$ are dual. To see that, let $v(t)$ be a solution of the equation

(1.1) and $w(t)$ a solution of the dual equation (3.1). Observe that for every $t \in \mathbb{R}$,

$$\begin{aligned}\frac{d}{dt}\langle v(t), w(t)\rangle &= \langle A(t)v(t), w(t)\rangle + \langle v(t), -A(t)^* w(t)\rangle \\ &= \langle A(t)v(t), w(t)\rangle - \langle A(t)v(t), w(t)\rangle = 0,\end{aligned}$$

where $\langle \cdot, \cdot \rangle$ denote the standard inner product in $\mathbb{C}^n$. Hence

$$\langle v(t), w(t)\rangle = \langle v(0), w(0)\rangle$$

for any $t \in \mathbb{R}$. Choose now dual bases $(v_1, \ldots, v_n)$ and $(w_1, \ldots, w_n)$ of $\mathbb{C}^n$. Let $v_i(t)$ be the unique solution of (1.1) such that $v_i(0) = v_i$, and $w_i(t)$ the unique solution of (3.1) such that $w_i(0) = w_i$, for each i. We obtain

$$\|v_i(t)\| \cdot \|w_i(t)\| \geq 1$$

for every $t \in \mathbb{R}$, and hence, $\chi^+(v_i) + \widetilde{\chi}^+(w_i) \geq 0$ for every i. It follows that the exponents χ^+ and $\widetilde{\chi}^+$ are dual.

We will study the regularity of the pair of exponents $(\chi^+, \widetilde{\chi}^+)$. Let $\mathbf{v} = (v_1, \ldots, v_n)$ be a basis of $\mathbb{C}^n$. We denote by $\Gamma_m(t) = \Gamma_m^{\mathbf{v}}(t)$ the m-volume of the parallelepiped defined by the vectors $v_i(t)$, for $i = 1, \ldots, m$, that are solutions of (1.1) satisfying the initial conditions $v_i(0) = v_i$. Let $V_m(t)$ be the $m \times m$ matrix whose entries are the scalar products $\langle v_i(t), v_j(t)\rangle$ for $i, j = 1, \ldots, m$. Then

$$\Gamma_m(t) = \Gamma_m^{\mathbf{v}}(t) = |\det V_m(t)|^{1/2}.$$

In particular, $\Gamma_1(t) = |v_1(t)|$ and $\Gamma_n(t) = \Gamma_n(0)|\det \mathcal{A}(t)|$, where

$$\mathcal{A}(t) = \exp\left(\int_0^t A(\tau)\, d\tau\right).$$

Note that

$$\det \mathcal{A}(t) = \exp\left(\int_0^t \operatorname{tr} A(\tau)\, d\tau\right). \tag{3.2}$$

The following theorem provides some crucial criteria for the pair $(\chi^+, \widetilde{\chi}^+)$ to be regular.

THEOREM 3.1. *Assume that the matrix function $A(t)$ satisfies (1.2). The following statements are equivalent:*

1. *the pair $(\chi^+, \widetilde{\chi}^+)$ is regular;*
2. $$\lim_{t\to+\infty} \frac{1}{t} \log|\det \mathcal{A}(t)| = \sum_{i=1}^{s} k_i \chi_i^+. \tag{3.3}$$
3. *for any normal ordered basis $\mathbf{v}$ of $\mathbb{C}^n$ and any $1 \leq m \leq n$ the following limit exists:*
$$\lim_{t\to+\infty} \frac{1}{t} \log \Gamma_m^{\mathbf{v}}(t).$$

In addition, if the pair $(\chi^+, \widetilde{\chi}^+)$ is regular then for any normal ordered basis $\mathbf{v}$ of $\mathbb{C}^n$ and any $1 \leq m \leq n$ we have

$$\lim_{t\to+\infty} \frac{1}{t} \log \Gamma_m^{\mathbf{v}}(t) = \sum_{i=1}^{m} \chi^+(v_i). \tag{3.4}$$

PROOF. We adopt the following notations in the proof. Given a function $f\colon (0,\infty)\to\mathbb{R}$ we set

$$\overline{\chi}(f)=\limsup_{t\to+\infty}\frac{1}{t}\log|f(t)| \quad\text{and}\quad \underline{\chi}(f)=\liminf_{t\to+\infty}\frac{1}{t}\log|f(t)|.$$

If, in addition, f is integrable we shall also write

$$\overline{f}=\limsup_{t\to+\infty}\frac{1}{t}\int_0^t f(\tau)\,d\tau \quad\text{and}\quad \underline{f}=\liminf_{t\to+\infty}\frac{1}{t}\int_0^t f(\tau)\,d\tau.$$

We first show that Statement 1 implies Statement 2. We start with an auxiliary result.

LEMMA 3.2. *The following statements hold:*

1. $\underline{\chi}(\det\mathcal{A})=\operatorname{Re}\underline{\operatorname{tr}A}$ *and* $\overline{\chi}(\det\mathcal{A})=\operatorname{Re}\overline{\operatorname{tr}A}$*;*
2. *if* $(v_1,\ldots,v_n)$ *is a basis of* $\mathbb{C}^n$*, then*

$$-\sum_{i=1}^n\widetilde{\chi}^+(v_i)\le\underline{\chi}(\det\mathcal{A})\le\overline{\chi}(\det\mathcal{A})\le\sum_{i=1}^n\chi^+(v_i).$$

PROOF OF THE LEMMA. It follows from (3.2) that

$$\underline{\chi}(\det\mathcal{A})=\liminf_{t\to+\infty}\frac{1}{t}\operatorname{Re}\int_0^t\operatorname{tr}A(\tau)\,d\tau=\operatorname{Re}\underline{\operatorname{tr}A}$$

and

$$\overline{\chi}(\det\mathcal{A})=\limsup_{t\to+\infty}\frac{1}{t}\operatorname{Re}\int_0^t\operatorname{tr}A(\tau)\,d\tau=\operatorname{Re}\overline{\operatorname{tr}A}.$$

This establishes the first statement.

Since $\Gamma_n(0)|\det\mathcal{A}(t)|$ gives the volume of the parallelepiped determined by the vectors $v_1(t),\ldots,v_n(t)$, we have $|\det\mathcal{A}(t)|\le\prod_{i=1}^n\|v_i(t)\|$, and hence,

$$\overline{\chi}(\det\mathcal{A})\le\sum_{i=1}^n\chi^+(v_i).$$

In a similar way,

$$-\underline{\chi}(\det\mathcal{A})=-\operatorname{Re}\underline{\operatorname{tr}A}=\operatorname{Re}\overline{\operatorname{tr}(-A^*)}\le\sum_{i=1}^n\widetilde{\chi}^+(v_i).$$

The lemma follows. □

Let χ_i' and $\widetilde{\chi}_i'$ be the values of the Lyapunov exponents χ^+ and $\widetilde{\chi}^+$, counted with their multiplicities. Choosing a normal basis $(v_1,\ldots,v_n)$ of $\mathbb{C}^n$, it follows from Lemma 3.2 that

$$-\sum_{i=1}^n\widetilde{\chi}_i'\le\underline{\chi}(\det\mathcal{A})\le\overline{\chi}(\det\mathcal{A})\le\sum_{i=1}^n\chi_i'.$$

Therefore,

$$\overline{\chi}(\det\mathcal{A})-\underline{\chi}(\det\mathcal{A})\le\sum_{i=1}^n(\chi_i'+\widetilde{\chi}_i')\le n\,\pi(\chi^+,\widetilde{\chi}^+),$$

This shows that if the pair $(\chi^+,\widetilde{\chi}^+)$ is regular then (3.3) holds.

We now show that Statement 3 implies Statement 1. We split the proof into two steps.

Step 1. For every $t \geq 0$ consider a linear coordinate change in $\mathbb{C}^n$ given by a matrix $U(t)$. We assume that the matrix function $U(t)$ is differentiable. Setting $z(t) = U(t)^{-1}v(t)$ we obtain

$$\dot{v}(t) = \dot{U}(t)z(t) + U(t)\dot{z}(t) = A(t)v(t) = A(t)U(t)z(t).$$

It follows that $\dot{z} = B(t)z$, where the matrix $B(t) = (b_{ij}(t))$ is defined by

$$B(t) = U(t)^{-1}A(t)U(t) - U(t)^{-1}\dot{U}(t). \tag{3.5}$$

We need the following lemma of Perron. Its main manifestation is to show how to reduce Equation (1.1) with a general matrix function $A(t)$ to a linear differential equation with a triangular matrix function.

LEMMA 3.3. *There exists a differentiable matrix function $U(t)$ such that:*

1. *$U(t)$ is unitary for each $t \geq 0$;*
2. *the matrix $B(t)$ is upper triangular for each t;*
3. *$\sup\{|b_{ij}(t)| : t \geq 0, i \neq j\} < \infty$;*
4. *if $k = 1, \ldots, n$ then*

$$\operatorname{Re} b_{kk}(t) = \frac{d}{dt} \log \frac{\Gamma_k^{\mathbf{v}}(t)}{\Gamma_{k-1}^{\mathbf{v}}(t)}.$$

PROOF OF LEMMA 3.3. Given a basis $\mathbf{v} = (v_1, \ldots, v_n)$ we construct the desired matrix function $U(t)$ by applying the Gram–Schmidt orthogonalization procedure to the basis $v_i(t)$, with $i = 1, \ldots, n$, where $v_i(t)$ is the solution of (1.1) satisfying the initial condition $v_i(0) = v_i$. Thus we obtain a collection of functions $u_1(t), \ldots, u_n(t)$ such that $\langle u_i(t), u_j(t)\rangle = \delta_{ij}$ where δ_{ij} is the Kronecker symbol. Let $V(t)$ and $U(t)$ be the matrices with columns $v_1(t), \ldots, v_n(t)$ and $u_1(t), \ldots, u_n(t)$, respectively. The matrix $U(t)$ is unitary for each t. Moreover, the Gram–Schmidt procedure can be effected in such a way that each function $u_k(t)$ is a linear combination of functions $v_1(t), \ldots, v_k(t)$. It follows that the matrix $Z(t) = U(t)^{-1}V(t)$ is upper triangular for each t.

The columns $z_1(t) = U(t)^{-1}v_1(t), \ldots, z_n(t) = U(t)^{-1}v_n(t)$ of the matrix $Z(t)$ form a basis of the space of solutions of the linear differential equation $\dot{z} = B(t)z$. Furthermore, $B(t) = \dot{Z}(t)Z(t)^{-1}$, and as $Z(t)$ is upper triangular so is the matrix $B(t)$.

Since $U(t)$ is unitary, using (3.5) we obtain

$$\begin{aligned} B(t) + B(t)^* &= U(t)^*(A(t) + A(t)^*)U(t) - (U(t)^*\dot{U}(t) + \dot{U}(t)^*U(t)) \\ &= U(t)^*(A(t) + A(t)^*)U(t) - \frac{d}{dt}(U(t)^*U(t)) \\ &= U(t)^*(A(t) + A(t)^*)U(t). \end{aligned}$$

Since $B(t)$ is triangular we conclude that $|b_{ij}(t)| \leq 2\|A(t)\| < \infty$ uniformly over $t \geq 0$ and $i \neq j$, thus establishing the third statement.

In order to prove the last statement of the lemma assume first that all entries of the matrix $Z(t) = (z_{ij}(t))$ are real. Then the entries of the matrix $B(t)$ are real too and

$$b_{kk}(t) = \frac{\dot{z}_{kk}(t)}{z_{kk}(t)} = \frac{d}{dt} \log z_{kk}(t).$$

Observe that

$$v_i(t) = \sum_{1 \leq \ell \leq i} u_\ell(t) z_{\ell i}(t).$$

Therefore,

$$\begin{aligned}\langle v_i(t), v_j(t)\rangle &= \sum_{1\le \ell\le i,\, 1\le m\le j} \delta_{\ell m} z_{\ell i}(t)\overline{z_{mj}(t)}\\ &= \sum_{1\le \ell\le \min\{i,j\}} z_{\ell i}(t)\overline{z_{\ell j}(t)} = \langle z_i(t), z_j(t)\rangle.\end{aligned}$$

Set $\mathbf{z} = (z_1(0), \ldots, z_n(0))$. This implies that $\Gamma_k^{\mathbf{v}}(t) = \Gamma_k^{\mathbf{z}}(t)$ for each k and thus,

$$\frac{\Gamma_k^{\mathbf{v}}(t)}{\Gamma_{k-1}^{\mathbf{v}}(t)} = \frac{\Gamma_k^{\mathbf{z}}(t)}{\Gamma_{k-1}^{\mathbf{z}}(t)} = z_{kk}(t).$$

In the general case (when the entries of $Z(t)$ are not necessarily real) we can write $\Gamma_k^{\mathbf{v}}(t)/\Gamma_{k-1}^{\mathbf{v}}(t) = |z_{kk}(t)|$ and

$$\begin{aligned}\frac{d}{dt}\log|z_{kk}(t)| &= \frac{1}{2}\frac{d}{dt}\log(\overline{z_{kk}(t)}z_{kk}(t))\\ &= \frac{1}{2}\left(\frac{\overline{\dot z_{kk}(t)}}{\overline{z_{kk}(t)}} + \frac{\dot z_{kk}(t)}{z_{kk}(t)}\right)\\ &= \frac{1}{2}(\overline{b_{kk}(t)} + b_{kk}(t)) = \operatorname{Re} b_{kk}(t).\end{aligned}$$

This completes the proof of the lemma. □

We define the $n\times n$ matrix function $Z(t) = (z_{ij}(t))$ as follows: $z_{ij}(t) = 0$ if $j < i$,

$$z_{ij}(t) = e^{\int_0^t b_{ii}(\tau)\,d\tau}$$

if $j = i$, and

$$z_{ij}(t) = \int_{a_{ij}}^t \sum_{k=i+1}^j b_{ik}(s) z_{kj}(s) e^{\int_s^t b_{ii}(\tau)\,d\tau} ds$$

if $j > i$.

LEMMA 3.4. *For any constants a_{ij}, with $1\le i<j\le n$, the columns of the matrix $Z(t)$ form a basis of solutions of the equation $\dot z = B(t)z$.*

PROOF OF LEMMA 3.4. For each i we have $\dot z_{ii}(t) = b_{ii}(t)z_{ii}(t)$, and

$$\dot z_{ij}(t) = \sum_{k=i+1}^j b_{ik}(t)z_{kj}(t) + b_{ii}(t)z_{ij}(t) = \sum_{k=i}^j b_{ik}(t)z_{kj}(t)$$

for each $j > i$. This shows that $\dot Z(t) = B(t)Z(t)$ and hence the columns of $Z(t)$ (i.e., the vectors $\mathbf{z}_i(t) = (z_{1i}(t), \ldots, z_{ni}(t))$) are solutions of the equation $\dot z = B(t)z$. Since $Z(t)$ is upper triangular, we have

$$\det Z(t) = \exp\left(\sum_{i=1}^n \int_0^t b_{ii}(\tau)\,d\tau\right) \ne 0,$$

and hence the vectors $\mathbf{z}_i(t)$ form a basis. □

Step 2. Assume that $\overline{\chi}(\Gamma_m^{\mathbf{v}}) = \underline{\chi}(\Gamma_m^{\mathbf{v}})$ for any normal ordered basis $\mathbf{v}$ and any $1\le m\le n$. We show that the pair $(\chi^+, \widetilde\chi^+)$ is regular. By Lemma 3.3, it is sufficient to consider the equation $\dot z = B(t)z$, where $B(t)$ is a $n\times n$ upper triangular matrix for every t.

LEMMA 3.5. *If* $\overline{\operatorname{Re} b_{ii}} = \underline{\operatorname{Re} b_{ii}} \stackrel{\text{def}}{=} B_i$ *for each* $i = 1, \ldots, n$ *then:*

1. *the pair of Lyapunov exponents corresponding to the equations* $\dot{z} = B(t)z$ *and* $\dot{w} = -B(t)^* w$ *is regular;*
2. *the numbers* $B_1, \ldots, B_n$ *are the values of the Lyapunov exponent* χ^+*;*
3. *the numbers* $-B_1, \ldots, -B_n$ *are the values of the Lyapunov exponent* $\widetilde{\chi}^+$.

PROOF OF LEMMA 3.5. We consider the solutions of the equation $\dot{z} = B(t)z$ described in Lemma 3.4 and show that each column

$$z_i(t) = (z_{1i}(t), \ldots, z_{ni}(t))$$

of $Z(t)$ satisfies

$$\chi^+(z_i) = \limsup_{t \to +\infty} \frac{1}{t} \log \|z_i(t)\| = B_i$$

for some choice of the constants a_{ij}. Since $\operatorname{Re} b_{ii} = B_i$, we clearly have $\chi^+(z_{ii}) = B_i$. Assume now that $\chi^+(z_{kj}) \le B_j$ for each $i+1 \le k \le j$. We show that $\chi^+(z_{ij}) \le B_j$. Observe that for each $\varepsilon > 0$, we have

$$\begin{aligned}\chi^+(z_{ij}) \le \limsup_{t \to +\infty} \frac{1}{t} \Bigg(&\log \left| e^{\int_0^t b_{ii}(\tau)\, d\tau} \right| \\ &+ \log \left| \int_{a_{ij}}^t \sum_{k=i+1}^{j} b_{ik}(s) z_{kj}(s) e^{-\int_0^s b_{ii}(\tau)\, d\tau} ds \right| \Bigg) \\ \le B_i + \limsup_{t \to +\infty} &\frac{1}{t} \log \left| \int_{a_{ij}}^t K n e^{(B_j - B_i + \varepsilon)s}\, ds \right|.\end{aligned}$$

We exploit here the fact that, by Lemma 3.3, $|b_{ij}(t)| \le K$ for some constant K independent of i, j, and t. For each $j > i$ set $a_{ij} = 0$ if $B_j - B_i \ge 0$ and $a_{ij} = +\infty$ if $B_j - B_i < 0$. Then for every sufficiently small $\varepsilon > 0$, we have

$$\chi^+(z_{ij}) \le B_i + \limsup_{t \to +\infty} \frac{1}{t} \log \frac{K n (e^{(B_j - B_i + \varepsilon)t} - 1)}{B_j - B_i + \varepsilon}$$

if $B_j - B_i \ge 0$ and

$$\chi^+(z_{ij}) \le B_i + \limsup_{t \to +\infty} \frac{1}{t} \log \frac{K n e^{(B_j - B_i + \varepsilon)t}}{B_j - B_i + \varepsilon}$$

if $B_j - B_i < 0$. Therefore,

$$\chi^+(z_{ij}) \le B_i + B_j - B_i + \varepsilon = B_j + \varepsilon.$$

Since ε is arbitrary we obtain $\chi^+(z_{ij}) \le B_j$. This shows that $\chi^+(z_i) = B_i$ for each $1 \le i \le n$.

In a similar way, one can show that there exists a lower triangular matrix $W(t)$ such that $\dot{W}(t) = -B(t)^* W(t)$. The entries of the matrix $W(t)$ are defined by $w_{ij}(t) = 0$ if $j > i$,

$$w_{ij}(t) = e^{-\int_0^t \overline{b_{jj}(\tau)}\, d\tau}$$

if $j = i$, and

$$w_{ij}(t) = -\int_{a_{ji}}^t \sum_{k=j}^{i-1} \overline{b_{ki}(s)} w_{kj}(s) e^{-\int_s^t \overline{b_{ii}(\tau)}\, d\tau}\, ds$$

if $j < i$ where the constants a_{ji} are chosen as above. Since $\operatorname{Re} b_{ii} = B_i$, the columns $w_1(t), \ldots, w_n(t)$ of $W(t)$ satisfy

$$\widetilde{\chi}^+(w_i) = \limsup_{t\to+\infty} \frac{1}{t}\log\|w_i(t)\| = -B_i = -\chi^+(z_i).$$

Note that $\chi^+(z_i)+\widetilde{\chi}^+(w_i) = 0$ for each i. In order to prove that the pair $(\chi^+, \widetilde{\chi}^+)$ is regular, it remains to show that the bases $\mathbf{z}$ and $\mathbf{w}$ are dual. Clearly, $\langle z_i(0), w_j(0)\rangle = 0$ for every $i < j$. Moreover, $\langle z_i(0), w_j(0)\rangle = 1$ for each $1 \le i \le n$. Fix $i > j$ and $t > 0$. We have that

$$\langle z_i(t), w_j(t)\rangle = \sum_{k=j}^{i} z_{ki}(t)\overline{w_{kj}(t)}. \tag{3.6}$$

Since $\chi^+(z_{ij}) \le B_j$ and $\widetilde{\chi}^+(w_{ij}) \le -B_j$ for every i, j, and $\varepsilon > 0$, we obtain

$$\begin{aligned}
\overline{\chi}(\langle z_i(0), w_j(0)\rangle) &\le \max_{j+1\le k\le i-1} \overline{\chi}(z_{ki}\overline{w_{kj}}) \\
&\le \max_{j+1\le k\le i-1} \limsup_{t\to+\infty} \frac{1}{t}\left(\log\left|\int_{a_{ki}}^{t} Kne^{(B_i-B_k+\varepsilon)s}\,ds\right|\right. \\
&\quad \left. + \log\left|\int_{a_{jk}}^{t} Kne^{(-B_j+B_k+\varepsilon)s}\,ds\right|\right) \\
&\le \max_{j+1\le k\le i-1} (B_i - B_k - B_j + B_k + 2\varepsilon) = B_i - B_j + 2\varepsilon.
\end{aligned}$$

Since ε is arbitrary, if $B_i - B_j < 0$ we obtain $\overline{\chi}(\langle z_i(0), w_j(0)\rangle) < 0$ and

$$\langle z_i(0), w_j(0)\rangle = \lim_{t\to+\infty} \langle z_i(t), w_j(t)\rangle = 0.$$

If $B_i - B_j \ge 0$, then $a_{ji} = 0$. Moreover, for each k we have $B_i - B_k \ge 0$ or $B_k - B_j \ge 0$, and hence $a_{ki} = 0$ or $a_{jk} = 0$. Letting $t \to 0$ in (3.6) we obtain

$$\langle z_i(0), w_j(0)\rangle = z_{ji}(0)\overline{w_{jj}(0)} + z_{ii}(0)\overline{w_{ij}(0)} + \sum_{k=j+1}^{i-1} z_{ki}(0)\overline{w_{kj}(0)}.$$

Since $i > j$ and $a_{ji} = 0$, we obtain $z_{ji}(0) = w_{ij}(0) = 0$. Moreover, for each k such that $j+1 \le k \le i-1$ we have $a_{ki} = 0$ or $a_{jk} = 0$, and hence, $z_{ki}(0) = 0$ or $w_{kj}(0) = 0$. Therefore each term in the above sum is zero. Thus $\langle z_i(0), w_j(0)\rangle = 0$, and the lemma follows. □

By Lemma 3.3 and Statement 3, we have

$$\begin{aligned}
\frac{1}{t}\int_0^t \operatorname{Re} b_{ii}(\tau)\,d\tau &= \frac{1}{t}\int_0^t \frac{d}{d\tau}\log\frac{\Gamma_i^{\mathbf{v}}(t)}{\Gamma_{i-1}^{\mathbf{v}}(t)}\,d\tau \\
&= \frac{1}{t}\log\frac{\Gamma_i^{\mathbf{v}}(t)/\Gamma_{i-1}^{\mathbf{v}}(t)}{\Gamma_i^{\mathbf{v}}(0)/\Gamma_{i-1}^{\mathbf{v}}(0)} \to \overline{\chi}(\Gamma_i^{\mathbf{v}}) - \overline{\chi}(\Gamma_{i-1}^{\mathbf{v}})
\end{aligned}$$

as $t \to +\infty$. By Lemma 3.5 we conclude that the Lyapunov exponent corresponding to the equation $\dot{z} = B(t)z$ is regular.

We now show that Statement 2 implies Statement 3. By Lemma 3.3 we have

$$\Gamma_m^{\mathbf{v}}(t) = \Gamma_m^{\mathbf{z}}(t) = \prod_{k=1}^{m} z_{kk}(t). \tag{3.7}$$

By Lyapunov's construction of normal bases (see the description after Theorem 2.3) there exists a normal basis $(v_1(0), \ldots, v_n(0))$ of $\mathbb{C}^n$ such that $z_k(0) = e_k + f_k$ for some $f_k \in \operatorname{span}\{e_{k+1}, \ldots, e_n\}$, where $(e_1, \ldots, e_n)$ is the canonical basis of $\mathbb{C}^n$. Since the matrix solution $Z(t)$ is upper triangular for each t, the vectors e_k and $Z(t)f_k$ are orthogonal. Therefore, $\|Z(t)f_k\| \geq |z_{kk}(t)|$, and

$$\chi^+(z_k) \geq \limsup_{t\to\infty} \frac{1}{t} \log|z_{kk}(t)| \stackrel{\text{def}}{=} \lambda_k. \tag{3.8}$$

Without loss of generality we are considering the norm in $\mathbb{C}^n$ given by

$$\|(w_1, \ldots, w_n)\| = |w_1| + \cdots + |w_n|.$$

By (3.7) we obtain

$$\lim_{t\to\infty} \frac{1}{t} \log \Gamma_n^{\mathbf{v}}(t) - \sum_{i=1}^{s} k_i \chi_i^+ \geq \sum_{k=1}^{n} \chi^+(z_k) \geq \sum_{k=1}^{n} \lambda_k. \tag{3.9}$$

Furthermore, by (3.2) and Lemma 3.4 we have

$$\Gamma_n^{\mathbf{z}}(t)/\Gamma_n^{\mathbf{z}}(0) = \exp\left(\int_0^t \operatorname{tr} B(\tau)\, d\tau\right) = \prod_{k=1}^{n} z_{kk}(t),$$

and hence, by (3.7),

$$\lim_{t\to\infty} \frac{1}{t} \log \Gamma_n^{\mathbf{v}}(t) \leq \sum_{k=1}^{n} \lambda_k. \tag{3.10}$$

It follows from (3.8), (3.9), and (3.10) that

$$\chi^+(v_k) = \chi^+(z_k) = \lim_{t\to+\infty} \frac{1}{t} \log\|z_k(t)\| = \lambda_k$$

for each $k = 1, \ldots, n$. Thus, again by (3.7), for each $1 \leq m \leq n$ we conclude that

$$\underline{\chi}(\Gamma_m^{\mathbf{v}}) = \overline{\chi}(\Gamma_m^{\mathbf{v}}) = \sum_{k=1}^{m} \chi^+(v_k).$$

This completes the proof of the theorem. □

The following example illustrates that Statement 2 of Theorem 3.1 cannot be weakened. Consider the system of differential equations

$$\dot{v}_1 = -p(t)v_1, \quad \dot{v}_2 = p(t)v_2$$

for $t > 0$, where $p(t) = \cos\log t - \sin\log t - 1$. The general solution of the system can be written in the form

$$v_1(t) = C_1 q(t)^{-1}, \quad v_2(t) = C_2 q(t),$$

for some constants C_1 and C_2, where

$$q(t) = \exp\left(\int_1^t p(\tau)\, d\tau\right) = \exp(t(\cos\log t - 1)).$$

We have

$$\liminf_{t\to+\infty} \frac{1}{t} \int_1^t p(\tau)\, d\tau = -2 \quad \text{and} \quad \limsup_{t\to+\infty} \frac{1}{t} \int_1^t p(\tau)\, d\tau = 0. \tag{3.11}$$

Observe that $\det \mathcal{A}(t) = 1$ for every t, and hence, the limit in (3.3) exists. On the other hand, it follows from (3.11) that the the limit of 1-volumes may not exist, and hence the pair of Lyapunov exponents $(\chi^+, \widetilde{\chi}^+)$ is not regular. In this case the Lyapunov exponent is equal to $\chi^+(v) = 2$ for every $v = (v_1, v_2) \neq 0$ with $C_1 \neq 0$, and $\chi^+(v) = 0$ otherwise. Hence,

$$0 = \lim_{t\to+\infty} \frac{1}{t} \log|\det \mathcal{A}(t)| < \chi_1^+ + \chi_2^+ = 2.$$

We also illustrate that Statement 2 of Theorem 3.1 cannot be weakened by replacing the limit in (3.3) by the upper limit. Consider a system of differential equations

$$\dot{v}_1 = v_2, \quad \dot{v}_2 = p(t)v_2$$

for $t > 0$. Observe that the general solution of the system can be written in the form

$$v_1(t) = C_1 + C_2 \int_1^t q(\tau)\, d\tau, \quad v_2(t) = C_2 q(t),$$

for some constants C_1 and C_2. One can easily show that for every vector $v = (v_1, v_2) \neq 0$, we have

$$\chi^+(v) = \lim_{t\to+\infty} \frac{1}{t} \log\|v(t)\| = 0.$$

On the other hand, it follows from (3.11) that the limit in (3.3) does not exist, and hence, the pair of Lyapunov exponents $(\chi^+, \widetilde{\chi}^+)$ is not regular.

We say that the Lyapunov exponent χ^+ is *forward regular* if the pair of Lyapunov exponents $(\chi^+, \widetilde{\chi}^+)$ is regular. We also say that the Lyapunov exponent χ^+ is *exact* if $\overline{\chi}(\Gamma_m^{\mathbf{v}}) = \underline{\chi}(\Gamma_m^{\mathbf{v}})$ for any $1 \leq m \leq n$ and any basis $\mathbf{v}$ of $\mathbb{C}^n$. We emphasize that exactness does not require the identity in (3.4), but only the existence of the limit in (3.4). By Theorem 3.1, if the function $t \mapsto \|A(t)\|$ is bounded (i.e., if (1.2) holds) then forward regularity is equivalent to exactness. In the case of unbounded matrix functions exactness may turn out to be weaker then forward regularity as the following example illustrates.

Consider the system of differential equations

$$\dot{v}_1 = av_1 + e^{bt}v_2, \quad \dot{v}_2 = -av_2$$

for $t \in \mathbb{R}$, where a and b are positive constants such that $2a < b$. The general solution of the system can be written in the form

$$v_1(t) = C_1 e^{at} + \frac{C_2}{b-2a} e^{(b-a)t}, \quad v_2(t) = C_2 e^{-at},$$

for some constants C_1 and C_2. One can easily show that the Lyapunov exponent is exact. In particular, for every vector $v = (v_1, v_2) \neq 0$ we have $\chi^+(v) = a$ whenever $C_1 \neq 0$, and $\chi^+(v) = b - a$ otherwise. By (3.2) we obtain $\det \mathcal{A}(t) = 1$ and

$$0 = \lim_{t\to+\infty} \frac{1}{t} \log|\det \mathcal{A}(t)| < \chi_1^+ + \chi_2^+ = b.$$

Therefore, the identity in (3.4) does not hold, and the Lyapunov exponent is not forward regular.

One can also show that forward regularity is equivalent to exactness provided the function $t \mapsto \|A(t)\|$ is integrable.

4. Lyapunov Stability Theory

We now proceed with the proof of the Lyapunov Stability Theorem 1.2. Consider the perturbed differential equation (1.6) where the function $f(t,u)$ satisfies (1.7). Let $v_1(t), \ldots, v_n(t)$ be n linearly independent solutions of (1.1) and $V(t)$ the corresponding monodromy matrix. We assume that $V(0) = \mathrm{Id}$. Denote by $\mathcal{V}(t,s)$ the Cauchy matrix of (1.1) defined by $\mathcal{V}(t,s) = V(t)V(s)^{-1}$.

THEOREM 4.1. *Assume that there are two continuous functions R and r on $[0,\infty)$ such that for every $t \geq 0$,*

$$\int_0^t R(\tau)\, d\tau \leq D_1 < +\infty, \tag{4.1}$$

$$\int_0^\infty e^{(q-1)\int_0^s (R+r)(\tau)\, d\tau}\, ds = D_2 < +\infty, \tag{4.2}$$

and for every $0 \leq s \leq t$,

$$\|\mathcal{V}(t,s)\| \leq D_3 e^{\int_s^t R(\tau)\, d\tau + (q-1)\int_0^s r(\tau)\, d\tau}. \tag{4.3}$$

Then there exist $D > 0$ and $\delta > 0$ such that for every $u_0 \in \mathbb{C}^n$ with $\|u_0\| < \delta$, there is a unique solution u of Equation (1.6) which satisfies:

1. *u is well-defined on the interval $[0,\infty)$, and $u(0) = u_0$;*
2. *for every t,*

$$\|u(t)\| \leq \|u_0\| D e^{\int_0^t R(\tau)\, d\tau}. \tag{4.4}$$

PROOF. Equation (1.6) is equivalent to the integral equation

$$u(t) = \mathcal{V}(t,0)u_0 + \int_0^t \mathcal{V}(t,s) f(s,u(s))\, ds. \tag{4.5}$$

Moreover, by (1.7) for every $u_0 \in \mathbb{C}^n$ with $u_0 \in H$ there exists a unique solution $u(t)$ satisfying the initial condition $u(0) = u_0$.

We consider the linear space $\mathcal{B}_C$ of $\mathbb{C}^n$ valued continuous functions x on the interval $[0,\infty)$ such that

$$\|x(t)\| \leq C e^{\int_0^t R(\tau)\, d\tau}$$

for every $t \geq 0$, where $C > 0$ is a constant. We endow this space with the norm

$$\|x\|_R = \sup\{\|x(t)\| e^{-\int_0^t R(\tau)\, d\tau} : t \geq 0\}.$$

It follows from (4.1) that $\|x(t)\| \leq \|x\|_R e^{D_1}$. Therefore, if $\|x\|_R$ is sufficiently small, then $x(t) \in H$ for every $t \geq 0$. This means that the operator

$$(Jx)(t) = \int_0^t \mathcal{V}(t,s) f(s,x(s))\, ds$$

is well-defined on $\mathcal{B}_C$. By (1.7), (4.2), and (4.3) we have

$$\begin{aligned}
&\|(Jx_1)(t) - (Jx_2)(t)\| \\
&\leq \int_0^t D_3 e^{\int_s^t R(\tau)\, d\tau + (q-1)\int_0^s r(\tau)\, d\tau} K(\|x_1 - x_2\|_R)^q e^{q\int_0^s R(\tau)\, d\tau}\, ds \\
&= D_3 K(\|x_1 - x_2\|_R)^q e^{\int_0^t R(\tau)\, d\tau} \int_0^t e^{(q-1)\int_0^s (R+r)(\tau)\, d\tau}\, ds \\
&\leq D_3 K D_2 (\|x_1 - x_2\|_R)^q e^{\int_0^t R(\tau)\, d\tau}.
\end{aligned}$$

It follows that

$$\begin{aligned}\|Jx_1 - Jx_2\|_R &\le D_3KD_2(\|x_1 - x_2\|_R)^q \\ &= D_3KD_2(\|x_1 - x_2\|_R)^{q-1}\|x_1 - x_2\|_R.\end{aligned}$$

Choose $\varepsilon > 0$ such that

$$\theta \overset{\text{def}}{=} D_3KD_2(2\varepsilon)^{q-1} < 1. \tag{4.6}$$

Whenever x_1, $x_2 \in \mathcal{B}_\varepsilon$ we have

$$\|Jx_1 - Jx_2\|_R \le \theta\|x_1 - x_2\|_R. \tag{4.7}$$

Choose a point $u_0 \in \mathbb{C}^n$ and set $\xi(t) = \mathcal{V}(t,0)u_0$. It follows from (4.3) that

$$\|\xi(t)\| \le \|u_0\|D_3e^{\int_0^t R(\tau)\,d\tau}$$

and hence, $\|\xi\|_R \le \|u_0\|D_3$. Consider the operator $\widetilde{J}$ defined by

$$(\widetilde{J}x)(t) = \xi(t) + (Jx)(t)$$

on the space $\mathcal{B}_\varepsilon$. Note that $\mathcal{B}_\varepsilon$ is a complete metric space with the distance $d(x,y) = \|x-y\|_R$. We have

$$\|\widetilde{J}x\|_R \le D_3\|u_0\| + \theta\|x\|_R < D_3\delta + \theta\varepsilon < \varepsilon$$

provided that δ is sufficiently small. Therefore $\widetilde{J}(\mathcal{B}_\varepsilon) \subset \mathcal{B}_\varepsilon$. By (4.6) and (4.7), the operator $\widetilde{J}$ is a contraction, and hence, there exists a unique function $u \in \mathcal{B}_\varepsilon$ which is a solution of (4.5). Furthermore, u can be obtained by

$$u(t) = \lim_{n\to\infty}(\widetilde{J}^n 0)(t) = \sum_{n=0}^{\infty}(J^n\xi)(t)$$

and hence, by (4.7),

$$\|u\|_R \le \sum_{n=0}^{\infty}\|J^n\xi\|_R \le \sum_{n=0}^{\infty}\theta^n\|\xi\|_R = \frac{\|\xi\|_R}{1-\theta} \le \frac{D_3\|u_0\|}{1-\theta}.$$

Therefore, the function $u(t)$ satisfies (4.4). □

As an immediate consequence of this theorem we obtain the following statement due to Malkin (see [**20**]).

THEOREM 4.2. *Assume that the Cauchy matrix of* (1.1) *admits the estimate*

$$\|\mathcal{V}(t,s)\| \le De^{\alpha(t-s)+\beta s}, \tag{4.8}$$

for any $0 \le s \le t$, *and some constants* α, β, *and* D, *such that*

$$(q-1)\alpha + \beta < 0. \tag{4.9}$$

Then the trivial solution of Equation (1.6) *is asymptotically and exponentially stable.*

PROOF. Set

$$R(t) = \alpha \quad\text{and}\quad r(t) = \frac{\beta}{q-1}.$$

It follows from (4.8) and (4.9) that Conditions (4.2) and (4.3) hold. In order to verify Condition (4.1) note that by setting $t = s$ in (4.3) we obtain

$$1 = \|\mathcal{V}(t,t)\| \le D_3e^{(q-1)\int_0^t r(\tau)\,d\tau}.$$

It follows that for every t,

$$\frac{\beta t}{q-1} = \int_0^t r(\tau)\,d\tau \geq -C, \tag{4.10}$$

where $C \geq 0$ is a constant. Therefore, by (4.9),

$$\alpha t = \int_0^t R(\tau)\,d\tau \leq C + \int_0^t (R(\tau) + r(\tau))\,d\tau \leq C \tag{4.11}$$

for all t. This implies Condition (4.1). Therefore, by (4.4), any solution u of (1.6) satisfies

$$\|u(t)\| \leq \|u_0\| D e^{\alpha t} \leq \|u_0\| D e^{C}. \tag{4.12}$$

This shows that the trivial solution of (1.6) is asymptotically stable.

If every solution of (1.6) is defined for all t then by (4.10) and (4.11), $\beta \geq 0$ and $\alpha \leq 0$. In view of (4.9), $\alpha < 0$, and in view of (4.12), the solution $u(t) = 0$ is exponentially stable. The desired result follows. □

Another important consequence of Theorem 4.1 is the following statement due to Lyapunov (see [**19**]).

THEOREM 4.3. *Assume that the maximal value $\chi_{\max}$ of the Lyapunov exponent of* (1.1) *is strictly negative (see* (1.5)*), and that*

$$(q-1)\chi_{\max} + \gamma < 0,$$

where γ is the regularity coefficient. Then the trivial solution of Equation (1.6) *is asymptotically and exponentially stable.*

PROOF. Consider a normal basis $(v_1(t), \ldots, v_n(t))$ for the space of solutions of (1.1), such that the numbers

$$\chi'_1 \leq \cdots \leq \chi'_n = \chi_{\max}$$

are the values of the Lyapunov exponent χ^+ for the vectors $v_1(t), \ldots, v_n(t)$. We may assume that the matrix $V(t)$ whose columns are $v_1(t), \ldots, v_n(t)$ is such that the columns $w_1(t), \ldots, w_n(t)$ of the matrix $W(t) = [V(t)^*]^{-1}$ form a normal basis for the space of solutions of the dual equation $\dot{w} = -A(t)^* w$.

Let $\mu_1, \ldots, \mu_n$ be the values of the Lyapunov exponent $\widetilde{\chi}^+$ for the vectors $w_1(t), \ldots, w_n(t)$. For every $\varepsilon > 0$ there exists a constant $D_\varepsilon > 0$ such that

$$\|v_j(t)\| \leq D_\varepsilon e^{(\chi'_j+\varepsilon)t} \quad \text{and} \quad \|w_j(t)\| \leq D_\varepsilon e^{(\mu_j+\varepsilon)t}$$

for every t. We also have

$$\gamma = \max\{\chi'_j + \mu_j : j = 1, \ldots, n\}.$$

It follows that $\chi'_j + \mu_j \leq \gamma$ for every $j = 1, \ldots, n$. Consider the Cauchy matrix

$$\mathcal{V}(t,s) = V(t)V(s)^{-1} = V(t)W(s)^*.$$

Its entries are

$$v_{ik}(t,s) = \sum_{j=1}^{n} v_{ij}(t)\overline{w_{kj}(s)},$$

where $v_{ij}(t)$ is the i-th coordinate of the vector $v_j(t)$ and $w_{kj}(s)$ is the k-th coordinate of the vector $w_j(s)$. It follows that

$$\begin{aligned}|v_{ik}(t,s)| &\le \sum_{j=1}^{n} |v_{ij}(t)| \cdot |\overline{w_{kj}(s)}| \le \sum_{j=1}^{n} \|v_j(t)\| \cdot \|w_j(s)\| \\ &\le n{D_\varepsilon}^2 e^{(\chi'_j+\varepsilon)t+(\mu_j+\varepsilon)s} = n{D_\varepsilon}^2 e^{(\chi'_j+\varepsilon)(t-s)+(\chi'_j+\mu_j+2\varepsilon)s}.\end{aligned}$$

Therefore, there exists a constant $D > 0$ such that

$$\|\mathcal{V}(t,s)\| \le De^{(\chi_{\max}+\varepsilon)(t-s)+(\gamma+2\varepsilon)s}. \tag{4.13}$$

Since ε can be chosen arbitrarily small the desired result follows from Theorem 4.2 if we set $\alpha = \chi_{\max} + \varepsilon$ and $\beta = \gamma + 2\varepsilon$. □

We conclude by observing that the Lyapunov Stability Theorem 1.2 is an immediate corollary of Theorem 4.3. Moreover, if χ^+ is forward regular, setting $\gamma = 0$ in (4.13) we obtain that for every solution $u(t)$ of (1.6) and every $s \in \mathbb{R}$,

$$\|u(t)\| \le De^{(\chi_{\max}+\varepsilon)(t-s)+2\varepsilon s}\|u(s)\| = De^{2\varepsilon s}e^{(\chi_{\max}+\varepsilon)(t-s)}\|u(s)\|. \tag{4.14}$$

In other words the contraction constant may get worse along the orbit of a solution of (1.6). This implies that the "size" of the neighborhood at time s where the asymptotic (and exponential) stability of the trivial solution is guaranteed may decay with subexponential rate. In Section 6 we will establish an analog of (4.14) for dynamical systems with discrete time (see Statement (L7) of Theorem 6.3) and illustrate the crucial role that it plays in the nonuniform hyperbolicity theory.

We now describe a generalization of the Lyapunov Stability Theorem in the case when *not all* but only *some* Lyapunov exponents are negative. Namely, let $\chi_1^+ < \cdots < \chi_s^+$ be the distinct values of the Lyapunov exponent χ^+ for Equation (1.1). Assume that

$$\chi_k^+ < 0, \tag{4.15}$$

where $1 \le k < s$ and the number χ_{k+1}^+ can be negative, positive, or equal to 0 (compare to (1.5)).

THEOREM 4.4 (see [**7**]). *Assume that the Lyapunov exponent χ^+ for Equation* (1.1) *with the matrix function $A(t)$ satisfying* (1.2), *is forward regular and satisfies Condition* (4.15). *Then there exists a local smooth submanifold $V^s \subset \mathbb{C}^n$ which passes through* 0, *is tangent at* 0 *to the linear subspace V_k (defined by* (2.1)*), and is such that the trivial solution of* (1.6) *is asymptotically and exponentially stable along V^s.*

This statement is a particular case of a more general and strong result that we discuss in Section 9.

5. The Oseledets Decomposition

Using results of Section 2, we conclude that the Lyapunov exponent χ^+ associated with Equation (1.1), as defined by (1.3), takes on only finitely many values on $\mathbb{C}^n \setminus \{0\}$. To stress that we deal with positive time we shall now denote its values by

$$\chi_1^+ < \cdots < \chi_{s^+}^+,$$

where $s^+ \le n$. We also denote by $\mathcal{V}^+$ the filtration of $\mathbb{C}^n$ associated to χ^+. We have

$$\{0\} = V_0^+ \subsetneqq V_1^+ \subsetneqq \cdots \subsetneqq V_{s^+}^+ = \mathbb{C}^n,$$

where $V_i^+ = \{v \in \mathbb{C}^n : \chi^+(v) \le \chi_i^+\}$. Finally we denote by $k_i^+ = \dim V_i^+ - \dim V_{i-1}^+$ the *multiplicity* of the value χ_i^+. Note that $\sum_{i=1}^{s^+} k_i^+ = n$.

An important advantage of Theorem 3.1 is that one can verify the regularity property of the pair $(\chi^+, \widetilde{\chi}^+)$ dealing only with the Lyapunov exponent χ^+. In this case we say that the Lyapunov exponent χ^+ is *forward regular* to stress that we allow only positive time.

Reversing the time we introduce the Lyapunov exponent $\chi^- : \mathbb{C}^n \to \mathbb{R} \cup \{-\infty\}$ by

$$\chi^-(v) = \limsup_{t \to -\infty} \frac{1}{|t|} \log \|v(t)\|,$$

where $v(t)$ is the solution of (1.1) satisfying the initial condition $v(0) = v$. The function χ^- takes on only finitely many values on $\mathbb{C}^n \setminus \{0\}$, which we denote by

$$\chi_1^- > \cdots > \chi_{s^-}^-,$$

where $s^- \le n$. We denote by $\mathcal{V}^-$ the filtration of $\mathbb{C}^n$ associated to χ^-. We have

$$\mathbb{C}^n = V_1^- \supsetneqq V_1^- \supsetneqq \cdots \supsetneqq V_{s^-}^- \supsetneqq V_{s^-+1}^- = \{0\},$$

where $V_i^- = \{v \in \mathbb{C}^n : \chi^-(v) \le \chi_i^-\}$. We also denote by $k_i^- = \dim V_i^- - \dim V_{i+1}^-$ the *multiplicity* of the value χ_i^-, and we have $\sum_{i=1}^{s^-} k_i^- = n$.

We also consider the Lyapunov exponent $\widetilde{\chi}^- : \mathbb{C}^n \to \mathbb{R} \cup \{-\infty\}$ given by

$$\widetilde{\chi}^-(w) = \limsup_{t \to -\infty} \frac{1}{|t|} \log \|w(t)\|,$$

where $w(t)$ is the solution of (3.1) satisfying the initial condition $w(0) = w$.

We say that the Lyapunov exponent χ^- is *backward regular* if the pair $(\chi^-, \widetilde{\chi}^-)$ is regular. We say that the filtrations $\mathcal{V}^+$ and $\mathcal{V}^-$ *comply* if the following properties hold:

1. $s^+ = s^- \stackrel{\text{def}}{=} s$;
2. there exists a decomposition

$$\mathbb{C}^n = \bigoplus_{i=1}^{s} E_i \tag{5.1}$$

   into subspaces E_i such that if $i = 1, \ldots, s$ then

$$V_i^+ = \bigoplus_{j=1}^{i} E_j \quad \text{and} \quad V_i^- = \bigoplus_{j=i}^{s} E_j;$$

3. $\chi_i^+ = -\chi_i^- \stackrel{\text{def}}{=} \chi_i$ for $i = 1, \ldots, s$;
4. if $i = 1, \ldots, s$ and $v \in E_i \setminus \{0\}$ then

$$\lim_{t \to \pm\infty} \frac{1}{t} \log \|v(t)\| = \chi_i$$

   with uniform convergence on $\{v \in E_i : \|v\| = 1\}$ (recall that $v(t)$ is the solution of Equation (1.1) with initial condition $v(0) = v$).

The decomposition (5.1) is called the *Oseledets decomposition* associated with the Lyapunov exponent χ^+ (or with the pair of Lyapunov exponents (χ^+, χ^-)).

We say that the Lyapunov exponent χ^+ (or the pair of Lyapunov exponents (χ^+, χ^-)) is *Lyapunov regular* if the exponent χ^+ is forward regular, the exponent χ^- is backward regular, and the filtrations $\mathcal{V}^+$ and $\mathcal{V}^-$ comply.

REMARK 5.1. We stress that simultaneous forward and backward regularity of the Lyapunov exponents does not imply the Lyapunov regularity. Roughly speaking, whether the Lyapunov exponent is forward (respectively, backward) regular does not depend only on the backward (respectively, forward) behavior of the system. On the other hand Lyapunov regularity requires some compatibility between the forward and backward behavior which is expressed in terms of the filtrations $\mathcal{V}^+$ and $\mathcal{V}^-$.

The regularity property is quite strong and requires a quite rigid behavior of the system.

THEOREM 5.2. *If the Lyapunov exponent χ^+ is Lyapunov regular then the following statements hold:*

1. *the exponents χ^+ and χ^- are exact;*
2. $\dim E_i = k_i^+ = k_i^- \stackrel{\text{def}}{=} k_i$*;*
3. *if $\mathbf{v} = (v_1, \ldots, v_{k_i})$ is a basis of E_i, then*
$$\lim_{t\to\pm\infty} \frac{1}{t} \log \Gamma_{k_i}^{\mathbf{v}}(t) = \chi_i k_i;$$
4. *if $v_i(t)$ and $v_j(t)$ are solutions of Equation (1.1) such that $v_i(0) \in E_i \setminus \{0\}$ and $v_j(0) \in E_j \setminus \{0\}$ with $i \neq j$ then*
$$\lim_{t\to\pm\infty} \frac{1}{t} \log \angle(v_i(t), v_j(t)) = 0.$$

PROOF. The first three statements follow from Theorem 3.1. For the last statement, let $\Gamma(t)$ denote the area of the rectangle formed by the vectors $v_i(t)$ and $v_j(t)$. We have
$$\Gamma(t) = \|v_i(t)\| \cdot \|v_j(t)\| \sin \angle(v_i(t), v_j(t)).$$
Since $\sin \angle(v_i(t), v_j(t)) \leq 1$, and the exponent χ^+ is exact we obtain
$$\begin{aligned}\chi_i^+ + \chi_j^+ =& \lim_{t\to+\infty} \frac{1}{t} \log \Gamma(t)\\ \leq& \chi^+(v_i) + \chi^+(v_j) + \liminf_{t\to+\infty} \frac{1}{t} \log \sin \angle(v_i(t), v_j(t))\\ \leq& \chi_i^+ + \chi_j^+ + \limsup_{t\to+\infty} \frac{1}{t} \log \sin \angle(v_i(t), v_j(t)) \leq \chi_i^+ + \chi_j^+.\end{aligned}$$
A similar argument applies to the exponent χ^- and we obtain the desired statement. □

One can construct examples of nonstationary linear differential equations whose Lyapunov exponent is forward and backward regular but is *not* Lyapunov regular. In dimension 1, an example is given by the equation $\dot{v} = a(t)v$, where $a\colon \mathbb{R} \to \mathbb{R}$ is a bounded continuous function such that $a(t) \to a_+$ as $t \to +\infty$ and $a(t) \to a_-$ as $t \to -\infty$, for some constants $a_+ \neq a_-$. The general solution of this equation is

$$v(t) = \exp\left(\int_0^t a(s)\, ds\right) v(0),$$

and hence the forward and backward exponents are respectively equal to a_+ and $-a_-$. In dimension 2, an example is given by

$$\dot{v}_1 = a(t)v_1, \quad \dot{v}_2 = a(-t)v_2,$$

with a as above. In this case the values of the exponents χ^+ and χ^- coincide (up to the change of sign required for Lyapunov regularity), but the filtrations $\mathcal{V}^+$ and $\mathcal{V}^-$ do not comply.

6. Dynamical Systems with Nonzero Lyapunov Exponents. Multiplicative Ergodic Theorem

Let φ_t be a smooth flow on a smooth compact Riemannian p-manifold M. It is generated by the vector field $\mathcal{X}$ on M given by

$$\mathcal{X}(x) = \frac{d\varphi_t(x)}{dt}|_{t=0}.$$

For every $x_0 \in M$ the trajectory $\{x(x_0, t) = \varphi_t(x_0) : t \in \mathbb{R}\}$ represents a solution of the nonlinear differential equation

$$\dot{v} = \mathcal{X}(v) \tag{6.1}$$

on the manifold M. This solution is uniquely defined by the initial condition $x(x_0, 0) = x_0$.

Given a point $x \in M$ and the trajectory $\{\varphi_t(x) : t \in \mathbb{R}\}$ passing through x we introduce the *variational differential equation*

$$\dot{w}(t) = A(x,t)w(t), \tag{6.2}$$

where

$$A(x,t) = d_{\varphi_t(x)}\mathcal{X}.$$

This is a nonstationary linear differential equation along the trajectory $\{\varphi_t(x) : t \in \mathbb{R}\}$ known also as the *equation of the first linear approximation.*

With the flow φ_t one can associate a certain collection of single linear differential equations (6.1) "along" each trajectory of the flow. The stability of a given trajectory can be described by studying small perturbations of the variational differential equation (6.2). The perturbation term is of type (1.6) and satisfies (1.7). The results of the previous section can be applied to study the stability of trajectories via the Lyapunov exponents. Although it is still a very difficult problem to verify whether a given trajectory is Lyapunov regular it turns out that "most" trajectories (in the sense of measure theory) have this property.

A similar approach is used to establish the stability of trajectories of dynamical systems with discrete time. Since it is technically simpler we describe such an approach here.

Let $f: M \to M$ be a diffeomorphism of a compact smooth Riemannian p-manifold M. Given $x \in M$, consider the trajectory $\{f^m x\}_{m\in\mathbb{Z}}$. The family of maps

$$\{d_{f^m x} f\}_{m\in\mathbb{Z}}$$

can be viewed as an analog of the variational differential equation (6.2) in the continuous time case. Given $x \in M$ and $v \in T_x M$ the formula

$$\chi(x, v) = \limsup_{m \to +\infty} \frac{1}{m} \log \|d_x f^m v\|$$

defines the *Lyapunov exponent* specified by the diffeomorphism f at the point x. The function $\chi(x, \cdot)$ takes on only finitely many values on $T_x M \setminus \{0\}$, which we denote by

$$\chi_1(x) < \cdots < \chi_{s(x)}(x),$$

where $s(x) \le p$. We denote by $\mathcal{V}_x$ the filtration of $T_x M$ associated to $\chi(x, \cdot)$. We have

$$\{0\} = V_0(x) \subsetneqq V_1(x) \subsetneqq \cdots \subsetneqq V_{s(x)}(x) = T_x M,$$

where $V_i(x) = \{v \in T_x M : \chi(x, v) \le \chi_i(x)\}$. We also introduce $k_i(x) = \dim V_i(x) - \dim V_{i-1}(x)$, which is the *multiplicity* of the value $\chi_i(x)$. We have $\sum_{i=1}^{s(x)} k_i(x) = p$. Finally, the collection of pairs

$$\operatorname{Sp} \chi(x) = \{(\chi_i(x), k_i(x)) : 1 \le i \le s(x)\}$$

is called the *Lyapunov spectrum* of the exponent $\chi(x, \cdot)$.

We note that the functions $\chi_i(x)$, $s(x)$, and $k_i(x)$ are *invariant* under f, and (Borel) *measurable* (but not necessarily continuous).

We say that a point $x \in M$ is *forward, or backward, or Lyapunov regular* if so is the Lyapunov exponent $\chi(x, \cdot)$. Note that if x is regular then so is the point $f(x)$ and thus one can speak on the whole trajectory $\{f^m(x)\}$ as being forward, backward, or Lyapunov regular.

Summarizing our discussion in the previous section we come to the following result.

THEOREM 6.1. *If a point $x \in M$ is Lyapunov regular, then there exists the Oseledets decomposition*

$$T_x M = \bigoplus_{i=1}^{s(x)} E_i(x)$$

into subspaces $E_i(x)$ such that:

1. *the subspaces $E_i(x)$ are invariant under the differential $d_x f$, i.e.,*

$$d_x f E_i(x) = E_i(f(x)),$$

and depend (Borel) measurably on x;

2. *if $v \in E_i(x) \setminus \{0\}$ then*

$$\lim_{m \to \pm\infty} \frac{1}{m} \log \|d_x f^m v\| = \chi_i(x)$$

with uniform convergence on $\{v \in E_i(x) : \|v\| = 1\}$;

3. *if $\mathbf{v} = (v_1, \ldots, v_{k_i(x)})$ is a basis of $E_i(x)$, then*

$$\lim_{n \to \pm\infty} \frac{1}{n} \log \Gamma^{\mathbf{v}}_{k_i(x)}(n) = \chi_i(x) k_i(x);$$

in particular, the Lyapunov exponent $\chi(x, \cdot)$ is exact;

4. *there exists a decomposition of the co-tangent bundle*

$$T_x^*M = \bigoplus_{i=1}^{s(x)} E_i^*(x)$$

into subspaces $E_i^*(x)$ *associated with the Lyapunov exponent* $\widetilde{\chi}^+$*; the subspaces* $E_i^*(x)$ *are invariant under the co-differential* $d_x^* f$*, that is* $d_x^* f E_i^*(x) = E_i^*(f(x))$*, and depend (Borel) measurably on* x*; moreover, if* $\{v_i(x) : i = 1, \dots, p\}$ *is a normal basis with* $v_i(x) \in E_j(x)$ *for* $n_{j-1}(x) < i \le n_j(x)$*, and* $\{v_i^*(x) : i = 1, \dots, p\}$ *is a dual basis, then* $v_i^*(x) \in E_j^*(x)$ *for* $n_{j-1}(x) < i \le n_j(x)$*.*

The following theorem due to Oseledets (see [**24**]) is a key result in studying the regularity of trajectories of dynamical systems. It shows that regularity is "typical" from the measure-theoretical point of view.

THEOREM 6.2 (Multiplicative Ergodic Theorem). *If* f *is a* C^1 *diffeomorphism of a compact smooth Riemannian manifold, then the set of Lyapunov regular points has full measure with respect to any* f*-invariant Borel probability measure on* M*.*

We stress that there may exist trajectories which are both forward and backward regular but *not* Lyapunov regular. However, such trajectories form a negligible set with respect to any f-invariant Borel probability measure.

We also emphasize that the notion of Lyapunov regularity does not require any invariant measure. Consider the set $\mathfrak{L} \subset M$ of points that are Lyapunov regular. Due to Theorem 6.2 this set is nonempty and indeed has full measure with respect to any f-invariant Borel probability measure on M.

We note that only in some exceptional cases **every** point in M is Lyapunov regular. To illustrate this consider a Smale horseshoe, i.e., a diffeomorphism $f\colon \mathbb{R}^2 \to \mathbb{R}^2$ which maps the unit square S onto a horseshoe-like set (see for example [**14**]). The map f has the locally maximal invariant set $A \stackrel{\text{def}}{=} \bigcap_{n\in\mathbb{Z}} f^n S \subset S$, and $f|A$ is topologically conjugate to the full shift on two symbols. Moreover, the map f is uniformly hyperbolic on A and hence, $T_x\mathbb{R}^2 = E_1(x) \oplus E_2(x)$ at **every** point $x \in A$, where $E_1(x)$ and $E_2(x)$ are the one-dimensional stable and unstable subspaces at x. Set $\lambda^-(x) = \log\|d_x f|E_1(x)\| < 0$ and $\lambda^+(x) = \log\|d_x f|E_2(x)\| > 0$. For every $x \in A$ and $v \in E_1(x)$ we have

$$\chi(x,v) = \limsup_{n\to\infty} \frac{1}{n} \sum_{k=0}^{n-1} \lambda^-(f^k x),$$

and for every $x \in A$ and $v \in E_2(x)$ we have

$$\chi(x,v) = \limsup_{n\to\infty} \frac{1}{n} \sum_{k=0}^{n-1} \lambda^+(f^k x).$$

For a Smale horseshoe the stable and unstable subspaces are uniformly transverse, and hence a point x is regular if and only if the Lyapunov exponent at x is exact (see Statement 3 of Theorem 6.1). Furthermore, Theorem 6.2 is a consequence of the Birkhoff ergodic theorem applied to the Hölder continuous functions λ^- and λ^+. Therefore, in the case of a linear horseshoe (when the functions λ^- and λ^+ are constant on A), or more generally when the functions λ^- and λ^+ are cohomologous to constants (i.e., $\lambda^- = a^- + \varphi^- \circ f - \varphi^-$ and $\lambda^+ = a^+ + \varphi^+ \circ f - \varphi^+$ for some constants $a^-, a^+ \in \mathbb{R}$ and some continuous functions φ^- and φ^+ on A), **every** point $x \in A$ is regular. On the other hand, when at least one of the functions λ^- and λ^+ is noncohomologous to a constant (and thus for a generic C^1 surface diffeomorphism with a Smale horseshoe), it follows from work of Barreira and Schmeling [**3**] that the set of nonregular points is nonempty and has Hausdorff dimension equal to the Hausdorff dimension of the set A.

Recall that an f-invariant measure ν is *ergodic* if for each f-invariant measurable set $A \subset M$ either A or $M \setminus A$ has zero measure. One can show that ν is ergodic if and only if every f-invariant (mod 0) measurable function (i.e., a measurable function φ such that $\varphi \circ f = \varphi$ almost everywhere) is constant almost everywhere.

Let ν be a Borel f-invariant ergodic measure. There exist numbers $\chi_i = \chi_i^\nu$ and $k_i = k_i^\nu$ for $i = 1, \ldots, s$ (where $s = s^\nu$) such that

$$\chi_i(x) = \chi_i, \quad k_i(x) = k_i, \quad s(x) = s \tag{6.3}$$

for ν-almost every x. The collection of pairs

$$\operatorname{Sp}\chi(\nu) = \{(\chi_i, k_i) : 1 \le i \le s\}$$

is called the *Lyapunov spectrum* of the measure ν.

We now consider dynamical systems whose spectrum of the Lyapunov exponent does not contain zero on some subset of M. More precisely, let

$$\begin{aligned}\Lambda = \{x \in \mathfrak{L} : \text{there exists } 1 \le k(x) < s(x) \\ \text{with } \chi_{k(x)}(x) < 0 \text{ and } \chi_{k(x)+1}(x) > 0\}.\end{aligned} \tag{6.4}$$

Note that the set Λ is f-invariant. We say that f is a *dynamical system with nonzero Lyapunov exponents* if there exists a Borel ergodic f-invariant probability measure ν on M such that $\nu(\Lambda) = 1$. The measure ν is called a *hyperbolic measure* for f. Note that if ν is hyperbolic then there is a number $1 \le k = k^\nu < s$ such that $\chi_k^\nu < 0$ and $\chi_{k+1}^\nu > 0$.

Consider the set $\widetilde{\Lambda} = \widetilde{\Lambda}^\nu$ of those points in Λ which are Lyapunov regular and satisfy (6.3). By the Multiplicative Ergodic Theorem we have $\nu(\widetilde{\Lambda}) = 1$. For every $x \in \widetilde{\Lambda}$, we set

$$E^s(x) = \bigoplus_{i=1}^{k} E_i(x) \quad \text{and} \quad E^u(x) = \bigoplus_{i=k+1}^{s} E_i(x).$$

THEOREM 6.3 (see [**27**]). *Let ν be an ergodic hyperbolic measure. The subspaces $E^s(x)$ and $E^u(x)$, $x \in \widetilde{\Lambda}$ have the following properties:*

L1. they depend measurably on x;

L2. they form a splitting of the tangent space, i.e., $T_xM = E^s(x) \oplus E^u(x)$;

L3. they are invariant:

$$d_xfE^s(x) = E^s(f(x)) \quad \text{and} \quad d_xfE^u(x) = E^u(f(x));$$

Furthermore, there exist $\varepsilon_0 > 0$ and measurable functions $C(x, \varepsilon) > 0$ and $K(x, \varepsilon) > 0$, $x \in \widetilde{\Lambda}$ and $0 < \varepsilon \le \varepsilon_0$ such that:

L4. the subspace $E^s(x)$ is stable*: if $v \in E^s(x)$ and $n > 0$, then*

$$\|d_xf^nv\| \le C(x, \varepsilon)e^{(\chi_k+\varepsilon)n}\|v\|;$$

L5. the subspace $E^u(x)$ *is* unstable*: if* $v \in E^u(x)$ *and* $n < 0$*, then*

$$\|d_x f^n v\| \le C(x,\varepsilon)e^{(\chi_{k+1}-\varepsilon)n}\|v\|;$$

L6.

$$\angle(E^s(x), E^u(x)) \ge K(x,\varepsilon);$$

L7. for every $m \in \mathbb{Z}$,

$$C(f^m(x),\varepsilon) \le C(x,\varepsilon)e^{\varepsilon|m|} \quad \text{and} \quad K(f^m(x),\varepsilon) \ge K(x,\varepsilon)e^{-\varepsilon|m|}.$$

REMARK 6.4. Condition (L7) is crucial and is a manifestation of the regularity property (it is an analog in the discrete time case of Condition (4.14)). Roughly speaking it means that the estimates (L4), (L5), and (L6) may get worse as $|m| \to \infty$ but only with subexponential rate. We stress that the rate of the contraction along stable subspaces and the rate of the expansion along unstable subspaces are exponential and hence, are substantially stronger.

PROOF OF THEOREM 6.3. We begin with the following general statement.

LEMMA 6.5. *Let* $X \subset M$ *be a Borel* f*-invariant set and* $A(x,\varepsilon)$ *a Borel positive function on* $X \times [0,\varepsilon_0)$, $0 < \varepsilon_0 < 1$ *such that for every* $\varepsilon_0 \ge \varepsilon > 0$, $x \in X$, *and* $m \in \mathbb{Z}$,

$$M_1(x,\varepsilon)e^{-\varepsilon|m|} \le A(f^m(x),\varepsilon) \le M_2(x,\varepsilon)e^{\varepsilon|m|},$$

where $M_1(x,\varepsilon)$ *and* $M_2(x,\varepsilon)$ *are Borel functions. Then one can find Borel positive functions* $B_1(x,\varepsilon)$ *and* $B_2(x,\varepsilon)$ *such that*

$$B_1(x,\varepsilon) \le A(x,\varepsilon) \le B_2(x,\varepsilon), \tag{6.5}$$

and for $m \in \mathbb{Z}$,

$$B_1(x,\varepsilon)e^{-2\varepsilon|m|} \le B_1(f^m(x),\varepsilon), \quad B_2(x,\varepsilon)e^{2\varepsilon|m|} \ge B_2(f^m(x),\varepsilon). \tag{6.6}$$

PROOF OF THE LEMMA. It follows from the conditions of the lemma that there exists $m(x,\varepsilon) > 0$ such that if $m \in \mathbb{Z}$ is such that $|m| > m(x,\varepsilon)$ then

$$-2\varepsilon \le \frac{1}{|m|}\log A(f^m(x),\varepsilon) \le 2\varepsilon.$$

Set

$$B_1(x,\varepsilon) = \min_{-m(x,\varepsilon)\le i\le m(x,\varepsilon)} \left\{1, A(f^i(x),\varepsilon)e^{2\varepsilon|i|}\right\},$$

$$B_2(x,\varepsilon) = \max_{-m(x,\varepsilon)\le i\le m(x,\varepsilon)} \left\{1, A(f^i(x),\varepsilon)e^{-2\varepsilon|i|}\right\}.$$

The functions $B_1(x,\varepsilon)$ and $B_2(x,\varepsilon)$ are Borel. Moreover, if $n \in \mathbb{Z}$ then

$$B_1(x,\varepsilon)e^{-2\varepsilon|n|} \le A(f^n(x),\varepsilon) \le B_2(x,\varepsilon)e^{2\varepsilon|n|}. \tag{6.7}$$

Furthermore, if $b_1 \le 1 \le b_2$ are such that

$$b_1 e^{-2\varepsilon|n|} \le A(f^n(x),\varepsilon) \tag{6.8}$$

and

$$b_2 e^{2\varepsilon|n|} \ge A(f^n(x),\varepsilon) \tag{6.9}$$

for all $n \in \mathbb{Z}$ then $b_1 \le B_1(x,\varepsilon)$ and $b_2 \ge B_1(x,\varepsilon)$. In other words,

$$\begin{aligned} B_1(x,\varepsilon) &= \sup\{b \le 1 : \text{inequality (6.8) holds for all } n \in \mathbb{Z}\},\\ B_2(x,\varepsilon) &= \inf\{b \ge 1 : \text{inequality (6.9) holds for all } n \in \mathbb{Z}\}. \end{aligned} \tag{6.10}$$

Inequalities (6.5) follow from (6.7) (with $n = 0$). We also have

$$A(f^{n+m}(x), \varepsilon) \leq B_2(x, \varepsilon)e^{2\varepsilon|n+m|} \leq B_2(x, \varepsilon)e^{2\varepsilon|n|+2\varepsilon|m|},$$
$$A(f^{n+m}(x), \varepsilon) \geq B_1(x, \varepsilon)e^{-2\varepsilon|n+m|} \geq B_1(x, \varepsilon)e^{-2\varepsilon|n|-2\varepsilon|m|},$$

Comparing these inequalities with (6.7) written at the point $f^m(x)$ and taking (6.10) into account we obtain (6.6). The proof of the lemma is complete. □

We now apply Lemma 6.5 to construct the function $K(x, \varepsilon)$. Let $K: \widetilde{\Lambda} \to \mathbb{R}$ be a Borel function. It is called *tempered* at the point x if

$$\lim_{m \to \pm\infty} \frac{1}{m} \log K(f^m(x)) = 0.$$

In other words the Lyapunov exponent of the function $K(x)$ is exact and is 0. It follows that the function $A(x, \varepsilon) = K(x)$ satisfies the conditions of Lemma 6.5 for all $\varepsilon > 0$.

Fix $\varepsilon > 0$ and consider the function $\gamma(x) = \angle(E^s(x), E^u(x))$ for $x \in \widetilde{\Lambda}$. Since this function is tempered (see Theorem 6.1) applying Lemma 6.5 we conclude that the function $K(x, \varepsilon) = B_1(x, \frac{1}{2}\varepsilon)$ satisfies Conditions (L6) and (L7) of the theorem.

We will now show how to construct the function $C(x, \varepsilon)$. The proof is an elaboration for the discrete time case of arguments in the proof of Theorem 4.3 (see (4.14)).

LEMMA 6.6. *There exists a Borel positive function $D(x, \varepsilon)$ ($x \in \widetilde{\Lambda}$ and $\varepsilon > 0$ is sufficiently small), such that if $m \in \mathbb{Z}$ and $1 \leq i \leq s$, then*

$$D(f^m(x), \varepsilon) \leq D(x, \varepsilon)^2 e^{2\varepsilon|m|} \tag{6.11}$$

and

$$\|df^n_{ix}\| \leq D(x, \varepsilon)e^{(\chi_i+\varepsilon)n}, \quad \|df^{-n}_{ix}\| \geq D(x, \varepsilon)^{-1}e^{-(\chi_i+\varepsilon)n}$$

for any $n \geq 0$, where $df^n_{ix} = d_x f^n | E_i(x)$.

PROOF OF THE LEMMA. Let $x \in \widetilde{\Lambda}$. By Theorem 6.1 (we use the notation of that theorem) there exists a number $n(x, \varepsilon) \in \mathbb{N}$ such that if $n \geq n(x, \varepsilon)$, then

$$\chi_i - \varepsilon \leq \frac{1}{n} \log \|df^n_{ix}\| \leq \chi_i + \varepsilon, \quad -\chi_i - \varepsilon \leq \frac{1}{n} \log \|df^{-n}_{ix}\| \leq -\chi_i + \varepsilon,$$

and

$$-\chi_i - \varepsilon \leq \frac{1}{n} \log \|d^* f^n_{ix}\| \leq -\chi_i + \varepsilon, \quad \chi_i - \varepsilon \leq \frac{1}{n} \log \|d^* f^{-n}_{ix}\| \leq \chi_i + \varepsilon,$$

where $d^* f^n_{ix} = d^*_x f^n | E^*_i(x)$ (recall that $E^*_i(x) \subset T^*_x M$ is the dual space to $E_i(x)$ and $d^* f$ is the co-differential). Set

$$D^+_1(x, \varepsilon) = \min_{1 \leq i \leq s} \min_{0 \leq j \leq n(x,\varepsilon)} \left\{1, \|df^j_{ix}\|e^{(-\chi_i+\varepsilon)j}, \|d^* f^j_{ix}\|e^{(\chi_i+\varepsilon)j}\right\},$$

$$D^-_1(x, \varepsilon) = \min_{1 \leq i \leq s} \min_{-n(x,\varepsilon) \leq j \leq 0} \left\{1, \|df^j_{ix}\|e^{(-\chi_i-\varepsilon)j}, \|d^* f^j_{ix}\|e^{(\chi_i-\varepsilon)j}\right\},$$

$$D^+_2(x, \varepsilon) = \max_{1 \leq i \leq s} \max_{0 \leq j \leq n(x,\varepsilon)} \left\{1, \|df^j_{ix}\|e^{(-\chi_i-\varepsilon)j}, \|d^* f^j_{ix}\|e^{(\chi_i-\varepsilon)j}\right\},$$

$$D^-_2(x, \varepsilon) = \max_{1 \leq i \leq s} \max_{-n(x,\varepsilon) \leq j \leq 0} \left\{1, \|df^j_{ix}\|e^{(-\chi_i+\varepsilon)j}, \|d^* f^j_{ix}\|e^{(\chi_i+\varepsilon)j}\right\},$$

and

$$D_1(x, \varepsilon) = \min\{D^+_1(x, \varepsilon), D^-_1(x, \varepsilon)\}, \quad D_2(x, \varepsilon) = \max\{D^+_2(x, \varepsilon), D^-_2(x, \varepsilon)\},$$

$$D(x,\varepsilon)=\max\{D_1(x,\varepsilon)^{-1},D_2(x,\varepsilon)\}.$$

The function $D(x,\varepsilon)$ is measurable, and if $n\geq 0$ and $1\leq i\leq s$ then

$$\begin{aligned}
D(x,\varepsilon)^{-1}e^{(\chi_i-\varepsilon)n}&\leq\|df^n_{ix}\|\leq D(x,\varepsilon)e^{(\chi_i+\varepsilon)n},\\
D(x,\varepsilon)^{-1}e^{(-\chi_i-\varepsilon)n}&\leq\|df^{-n}_{ix}\|\leq D(x,\varepsilon)e^{(-\chi_i+\varepsilon)n},\\
D(x,\varepsilon)^{-1}e^{(-\chi_i-\varepsilon)n}&\leq\|d^*f^n_{ix}\|\leq D(x,\varepsilon)e^{(-\chi_i+\varepsilon)n},\\
D(x,\varepsilon)^{-1}e^{(\chi_i-\varepsilon)n}&\leq\|d^*f^{-n}_{ix}\|\leq D(x,\varepsilon)e^{(\chi_i+\varepsilon)n}.
\end{aligned}\tag{6.12}$$

Moreover, if $d\geq 1$ is a number for which Inequalities (6.12) hold for all $n\geq 0$ and $1\leq i\leq s$ with $D(x,\varepsilon)$ replaced by d then $d\geq D(x,\varepsilon)$. Therefore,

$$\begin{aligned}
D(x,\varepsilon)=\inf\{d\geq 1:\ &\text{the inequalities (6.12) hold for all } n\geq 0\\
&\text{and } 1\leq i\leq s \text{ with } D(x,\varepsilon) \text{ replaced by } d\}.
\end{aligned}\tag{6.13}$$

We wish to compare the values of the function $D(x,\varepsilon)$ at the points x and f^mx for $m\in\mathbb{Z}$. First, let us notice that for every $x\in M$, $v\in T_xM$, and $\varphi\in T_x^*M$ with $\varphi(v)=1$ we have

$$(d_x^*f\varphi)(d_xfv)=\varphi((d_x^*f)^{-1}d_xfv)=\varphi(v)=1.\tag{6.14}$$

Second, using the Riemannian metric on the manifold M we introduce the identification map $\tau_x\colon T_x^*M\to T_xM$ such that $\langle\tau_x(\varphi),v\rangle=\varphi(v)$ where $v\in T_xM$ and $\varphi\in T_x^*M$.

Let $\{v_k^n : k=1,\dots,p\}$ be a basis of $E_i(f^n(x))$ and $\{w_k^n : k=1,\dots,p\}$ be the dual basis of $E_i^*(f^n(x))$. We have $\tau_{f^n(x)}(w_k^n)=v_k^n$. Let $A^i_{n,m}$ and $B^i_{n,m}$ be matrices corresponding to the linear maps $df^{n+1}_{if^m(x)}$ and $d^*f^{n+1}_{if^m(x)}$ with respect to those bases. It follows from (6.14) that

$$A^i_{0,m}(B^i_{0,m})^*=\mathrm{Id}$$

and hence, for every $n>0$ the matrix corresponding to the map $df^n_{if^m(x)}$ is

$$A^i_{n,m}=A^i_{0,m+n}(A^i_{0,m})^{-1}=A^i_{0,m+n}(B^i_{0,m})^*.$$

It follows from here and (6.12) that:

1. if $n>0$ then

$$\|df^n_{if^m(x)}\|\leq D(x,\varepsilon)^2e^{(\chi_i+\varepsilon)(n+m)+(-\chi_i+\varepsilon)m}=D(x,\varepsilon)^2e^{2\varepsilon m}e^{(\chi_i+\varepsilon)n},$$

$$\|df^n_{if^m(x)}\|\geq D(x,\varepsilon)^{-2}e^{(\chi_i-\varepsilon)(n+m)+(-\chi_i-\varepsilon)m}=D(x,\varepsilon)^{-2}e^{-2\varepsilon m}e^{(\chi_i-\varepsilon)n},$$

2. if $n>0$ and $m-n\geq 0$ then

$$\|df^{-n}_{if^m(x)}\|\leq D(x,\varepsilon)^2e^{(\chi_i+\varepsilon)(m-n)+(-\chi_i+\varepsilon)m}=D(x,\varepsilon)^2e^{2\varepsilon m}e^{(-\chi_i+\varepsilon)n},$$

$$\|df^{-n}_{if^m(x)}\|\geq D(x,\varepsilon)^{-2}e^{(\chi_i-\varepsilon)(m-n)+(-\chi_i-\varepsilon)m}=D(x,\varepsilon)^{-2}e^{-2\varepsilon m}e^{(-\chi_i-\varepsilon)n},$$

3. if $n>0$ and $n-m\geq 0$ then

$$\|df^{-n}_{if^m(x)}\|\leq D(x,\varepsilon)^2e^{(\chi_i+\varepsilon)(n-m)+(-\chi_i+\varepsilon)m}=D(x,\varepsilon)^2e^{2\varepsilon m}e^{(-\chi_i+\varepsilon)n},$$

$$\|df^{-n}_{if^m(x)}\|\geq D(x,\varepsilon)^{-2}e^{(\chi_i-\varepsilon)(n-m)+(-\chi_i-\varepsilon)m}=D(x,\varepsilon)^{-2}e^{-2\varepsilon m}e^{(-\chi_i-\varepsilon)n},$$

Similar inequalities hold for the maps $d^*f^n_{if^m(x)}$ for each n, $m\in\mathbb{Z}$. Comparing this with the inequalities (6.12) applied to the point $f^m(x)$ and using (6.13) we conclude that if $m\geq 0$, then

$$D(f^m(x),\varepsilon)\leq D(x,\varepsilon)^2e^{2\varepsilon m}.\tag{6.15}$$

Similar arguments show that if $m \le 0$, then

$$D(f^{-m}(x),\varepsilon) \le D(x,\varepsilon)^2 e^{-2\varepsilon m}. \tag{6.16}$$

It follows from (6.15) and (6.16) that if $m \in \mathbb{Z}$, then

$$D(f^m(x),\varepsilon) \le D(x,\varepsilon)^2 e^{2\varepsilon|m|}.$$

This completes the proof of the lemma. □

We now proceed with the proof of the theorem. Replacing in (6.11) m by $-m$ and x by $f^m(x)$ we obtain

$$D(f^m(x),\varepsilon) \ge \sqrt{D(x,\varepsilon)}e^{-\varepsilon|m|}. \tag{6.17}$$

Consider two disjoint subsets $\sigma_1, \sigma_2 \subset [1,s]\cap\mathbb{N}$ and set

$$L_1(x) = \bigoplus_{i\in\sigma_1} E_i(x), \quad L_2(x) = \bigoplus_{i\in\sigma_2} E_i(x)$$

and $\gamma_{\sigma_1\sigma_2}(x) = \angle(L_1(x), L_2(x))$. By Theorem 6.1 the function $\gamma_{\sigma_1\sigma_2}$ is tempered and hence, in view of Lemma 6.5 one can find a function $K_{\sigma_1\sigma_2}(x)$ satisfying Condition (L7) such that

$$\gamma_{\sigma_1\sigma_2}(x) \ge K_{\sigma_1\sigma_2}(x).$$

Set

$$T(x,\varepsilon) = \min K_{\sigma_1\sigma_2}(x),$$

where the minimum is taken over all pairs of disjoint subsets $\sigma_1, \sigma_2 \subset [1,s]\cap\mathbb{N}$. The function $T(x,\varepsilon)$ satisfies Condition (L7).

Let $v \in E^s(x)$. Write $v = \sum_{i=1}^k v_i$ where $v_i \in E_i(x)$. We have

$$\|v\| \le \sum_{i=1}^k \|v_i\| \le LT^{-1}(x,\varepsilon)\|v\|,$$

where $L > 1$ is a constant. Let us set $C'(x,\varepsilon) = LD(x,\varepsilon)T(x,\varepsilon)^{-1}$. It follows from (6.11) and (6.17) that the function $C'(x,\varepsilon)$ satisfies the condition of Lemma 6.5 with

$$M_1(x,\varepsilon) = \frac{2}{\pi}L\sqrt{D(x,\varepsilon)} \quad \text{and} \quad M_2(x,\varepsilon) = LD(x,\varepsilon)^2 T(x,\varepsilon)^{-1}.$$

Therefore, there exists a function $C_1(x,\varepsilon) \ge C'(x,\varepsilon)$ for which the statements of Lemma 6.5 hold.

Applying the above arguments to the inverse map f^{-1} and the subspace $E^u(x)$ one can construct a function $C_2(x,\varepsilon)$ for which the statements of Lemma 6.5 hold. The desired function $C(x,\varepsilon)$ can now be defined by

$$C(x,\varepsilon) = \max\{C_1(x,\varepsilon/2), C_2(x,\varepsilon/2)\}.$$

This completes the proof of the theorem. □

7. Nonuniform Hyperbolicity. Regular Sets

Let $f\colon M \to M$ be a diffeomorphism of a compact smooth Riemannian manifold M. A measurable f-invariant subset $\mathcal{R} \subset M$ is called *nonuniformly hyperbolic* if there exist: (a) numbers λ and μ such that $0 < \lambda < 1 < \mu$; (b) a number ε and real functions C, $K\colon \mathcal{R} \to (0,\infty)$; (c) subspaces $E^s(x)$ and $E^u(x)$ for each $x \in \mathcal{R}$, which satisfy the following conditions:

H1. the subspaces $E^s(x)$ and $E^u(x)$ depend measurably on x and form an invariant splitting of the tangent space, i.e.,

$$T_xM = E^s(x) \oplus E^u(x), \quad d_xfE^s(x) = E^s(f(x)), \quad d_xfE^u(x) = E^u(f(x));$$

H2. the subspace $E^s(x)$ is *stable*: if $v \in E^s(x)$, $m \in \mathbb{Z}$, and $n > 0$, then

$$\|d_{f^m(x)}f^n v\| \le C(f^m(x))\lambda^n e^{\varepsilon n}\|v\|;$$

H3. the subspace $E^u(x)$ is *unstable*: if $v \in E^u(x)$, $m \in \mathbb{Z}$, and $n < 0$, then

$$\|d_{f^m(x)}f^n v\| \le C(f^m(x))\mu^n e^{\varepsilon |n|}\|v\|;$$

H4. for $n \in \mathbb{Z}$,

$$\angle(E^s(f^n(x)), E^u(f^n(x))) \ge K(f^n(x));$$

H5. for m, $n \in \mathbb{Z}$,

$$C(f^{m+n}(x)) \le C(f^m(x))e^{\varepsilon|n|}, \quad K(f^{m+n}(x)) \ge K(f^m(x))e^{-\varepsilon|n|}.$$

We can summarize the discussion in the previous section by saying that for any hyperbolic measure ν the set of Lyapunov regular points with nonzero Lyapunov exponents contains a nonuniformly hyperbolic set of full ν-measure with

$$\lambda = e^{\chi_k^\nu}, \quad \mu = e^{\chi_{k+1}^\nu}, \quad C(x) = C(x,\varepsilon), \quad K(x) = K(x,\varepsilon)$$

for *any* fixed $0 < \varepsilon \le \varepsilon_0$ (see Conditions (L1)-(L7) in Section 6). In fact, finding trajectories with nonzero Lyapunov exponents seems to be a universal approach in establishing nonuniform hyperbolicity.

We emphasize that the set of points (trajectories) with nonzero Lyapunov exponents whose regularity coefficient is sufficiently small (but may not necessarily be zero) is nonuniformly hyperbolic for some $\varepsilon > 0$.

We now provide a more detailed description of a nonuniformly hyperbolic set $\mathcal{R}$. Fix $\varepsilon > 0$. Given $\ell > 0$, we introduce the *regular set* (of level ℓ) by

$$\mathcal{R}^\ell = \left\{ x \in \mathcal{R} : C(x,\varepsilon) \le \ell,\ K(x,\varepsilon) \ge \frac{1}{\ell} \right\}. \tag{7.1}$$

Regular sets can be viewed as noninvariant uniformly hyperbolic sets for the diffeomorphism f. They have the following basic properties:

R1. $\mathcal{R}^\ell \subset \mathcal{R}^{\ell+1}$;

R2. if $m \in \mathbb{Z}$, then $f^m(\mathcal{R}^\ell) \subset \mathcal{R}^{\ell'}$, where $\ell' = \ell \exp(|m|\varepsilon)$;

R3. the subspaces $E^s(x)$ and $E^u(x)$ depend continuously on $x \in \mathcal{R}^\ell$; moreover, they depend *Hölder continuously* on $x \in \mathcal{R}^\ell$ (see Appendix A by Brin). This means that

$$d(E^s(x), E^s(y)) \le C\rho(x,y)^\alpha \quad \text{and} \quad d(E^u(x), E^u(y)) \le C\rho(x,y)^\alpha,$$

where $C > 0$ and $\alpha \in (0,1]$ are constants, and d is the distance in the Grassmannian bundle of TM generated by the Riemannian metric.

We consider the sets $\mathcal{Q}^\ell = \overline{\mathcal{R}^\ell}$ that are the closures of the sets $\mathcal{R}^\ell$.

Set $\mathcal{Q} = \bigcup_{\ell\geq 1}\mathcal{Q}^\ell$. It is easy to see that $\mathcal{Q}^\ell \subset \mathcal{Q}^{\ell+1}$ and that the set $\mathcal{Q}$ is f-invariant.

Given a point $x \in \mathcal{Q}^\ell$ choose a sequence of points $x_n \in \mathcal{R}^\ell$ which converges to x. Passing to a subsequence, we may assume that the sequences of subspaces $E^s(x_n)$ and $E^u(x_n)$ converge to some subspaces at x which we denote by $E^s(x)$ and $E^u(x)$ respectively. It is easy to see that they satisfy the following properties:

R4. $T_xM = E^s(x) \oplus E^u(x)$;

R5. if $v \in E^s(x)$ and $n > 0$, then

$$\|d_xf^n v\| \leq \ell\lambda^n e^{\varepsilon n}\|v\|;$$

R6. if $v \in E^u(x)$ and $n < 0$, then

$$\|d_xf^n v\| \leq \ell\mu^n e^{\varepsilon|n|}\|v\|;$$

R7. $\angle(E^s(x), E^u(x)) \geq \frac{1}{\ell}$.

This implies that the subspaces $E^s(x)$ and $E^u(x)$ are uniquely defined (in particular, they do not depend on the choice of the sequence $x_n \to x$). They depend continuously on $x \in \mathcal{Q}^\ell$ (and indeed, Hölder continuously). Furthermore, the subspaces $E^s(x)$ and $E^u(x)$, for $x \in \mathcal{Q}$, determine a nonuniformly hyperbolic structure on the set $\mathcal{Q}$ with $C(x,\varepsilon) = \ell$ and $K(x,\varepsilon) = \frac{1}{\ell}$ if $x \in \mathcal{Q}^\ell \setminus \mathcal{Q}^{\ell-1}$.

We now consider a smooth flow φ_t on a compact smooth Riemannian manifold M which is generated by a vector field $X(x)$. A measurable φ_t-invariant subset $\mathcal{R} \subset M$ is called *nonuniformly hyperbolic* if there exist: (a) numbers λ and μ such that $0 < \lambda < 1 < \mu$; (b) real functions C, $K\colon \mathcal{R}\times(0,1) \to (0,\infty)$; (c) subspaces $E^s(x)$ and $E^u(x)$ for each $x \in \mathcal{R}$, which satisfy Conditions (H2)-(H5) and the following condition:

H1'. the subspaces $E^s(x)$ and $E^u(x)$ depend measurably on x and together with the subspace $E^0(x) = \{\alpha X(x) : \alpha \in \mathbb{R}\}$ form an invariant splitting of the tangent space, i.e.,

$$T_xM = E^s(x) \oplus E^u(x) \oplus E^0(x),$$

with

$$d_xfE^s(x) = E^s(f(x)) \quad \text{and} \quad d_xfE^u(x) = E^u(f(x)).$$

One can extend the notion of regular set to flows on nonuniformly hyperbolic subsets.

We say that a dynamical system (with discrete or continuous time) is *nonuniformly hyperbolic* if it possesses an invariant nonuniformly hyperbolic subset.

REMARK 7.1. One can generalize the notion of nonuniformly hyperbolicity from complete to partial. More precisely, we say that a set $\mathcal{R}$ is *nonuniformly partially hyperbolic* if there exist: (a) numbers λ and μ such that $0 < \lambda < \mu$ and $\min\{\lambda, \mu^{-1}\} < 1$; (b) real functions C, $K\colon \mathcal{R}\times(0,1) \to (0,\infty)$; (c) subspaces $E^s(x)$ and $E^u(x)$ for each $x \in \mathcal{R}$, which satisfy Conditions (H1)–(H5). Let us emphasize that in the case of partial hyperbolicity the vectors in the unstable subspaces $E^u(x)$ may indeed contract and have negative Lyapunov exponents since the number μ is not necessarily greater than one. In these lectures we consider only the case of complete hyperbolicity although many results can be generalized (sometimes literally) to nonuniformly partially hyperbolic sets.

One can further generalize nonuniform (complete or partial) hyperbolicity by requiring that λ and μ be f-invariant measurable functions $\lambda\colon \mathcal{R} \to \mathbb{R}$ and $\mu\colon \mathcal{R} \to \mathbb{R}$ such that $0 < \lambda(x) < 1 < \mu(x)$ for each $x \in \mathcal{R}$ (in the case of partial hyperbolicity one should assume that instead $0 < \lambda(x) < \mu(x))$ and $\min\{\lambda(x), \mu(x)^{-1}\} < 1$ for each $x \in \mathcal{R}$). In other words, the rates $\lambda(x)$ and $\mu(x)$ may vary from trajectory to trajectory. In fact, one can easily reduce this more general case to the previous one by considering the (invariant) sets of points x where $\lambda(x) \le \lambda$ and $\mu(x) \le \mu$ for fixed constants λ and μ.

REMARK 7.2. Let ν be a hyperbolic measure for a diffeomorphism f and $\mathcal{R}$ the set of Lyapunov regular points with nonzero Lyapunov exponents. The regular sets $\mathcal{R}^\ell$ consist of Lyapunov regular points which satisfy (7.1) while the sets $\mathcal{Q}^\ell$ may contain some nonregular points with sufficiently small regularity coefficient (compare to (4.13) and (4.14)).

8. Examples of Nonuniformly Hyperbolic Systems

We present some examples of dynamical systems with continuous time which are nonuniformly hyperbolic. The first such example was constructed in [**26**] by a "surgery" of an Anosov flow.

8.1. Let φ_t be an Anosov flow on a compact 3-dimensional manifold M which is defined by a vector field X and preserves a smooth ergodic measure μ. Fix a point $p_0 \in M$. One can introduce a coordinate system x, y, z in a ball $B_d(p_0)$ (for some $d > 0$) such that p_0 is the origin (i.e., $p_0 = 0$) and $X = \partial/\partial z$.

For each $\varepsilon > 0$, let $T_\varepsilon = \mathbb{S}^1 \times D_\varepsilon \subset B_d(0)$ be the solid torus obtained by rotating the disk

$$D_\varepsilon = \{(x, y, z) \in B_d(0) : x = 0 \text{ and } (y - d/2)^2 + z^2 \le (\varepsilon d)^2\}$$

around the z-axis. Every point on the solid torus can be represented as a pair (θ, y, z) with $\theta \in \mathbb{S}^1$ and $(y, z) \in D_\varepsilon$.

For every $0 \le \alpha \le 2\pi$, we consider the cross-section of the solid torus $\Pi_\alpha = \{(\theta, y, z) : \theta = \alpha\}$.

Let $\widetilde{X}$ be a smooth vector field on $M \setminus T_\varepsilon$ and $\widetilde{\varphi}_t$ the flow generated by $\widetilde{X}$. One can construct a vector field $\widetilde{X}$ such that:

1. $\widetilde{X}|(M \setminus T_{2\varepsilon}) = X|(M \setminus T_{2\varepsilon})$;
2. for any $0 \le \alpha, \beta \le 2\pi$, the vector field $\widetilde{X}|\Pi_\beta$ is the image of the vector field $\widetilde{X}|\Pi_\alpha$ under the rotation around the z-axis that moves Π_α onto Π_β;
3. for every $0 \le \alpha \le 2\pi$, the unique two fixed points of the flow $\widetilde{\varphi}_t|\Pi_\alpha$ are those in the intersection of Π_α with the hyperplanes $z = \pm\varepsilon d$;
4. for every $0 \le \alpha \le 2\pi$ and $(y, z) \in D_{2\varepsilon} \setminus \operatorname{int} D_\varepsilon$, the trajectory of the flow $\widetilde{\varphi}|\Pi_\alpha$ passing through the point (y, z) is invariant under the symmetry $(\alpha, y, z) \mapsto (\alpha, y, -z)$;
5. the flow $\widetilde{\varphi}_t|\Pi_\alpha$ preserves the measure μ_α that is the conditional measure generated by the measure μ on the set Π_α (see [**5**] for details).

See Figure 1. One can see that the orbits of the flows φ_t and $\widetilde{\varphi}_t$ coincide on $M \setminus T_{2\varepsilon}$, that the flow $\widetilde{\varphi}_t$ preserves the measure μ, and that the only fixed points of this flow are those on the circles $\{(\theta, y, z) : z = -\varepsilon d\}$ and $\{(\theta, y, z) : z = \varepsilon d\}$.

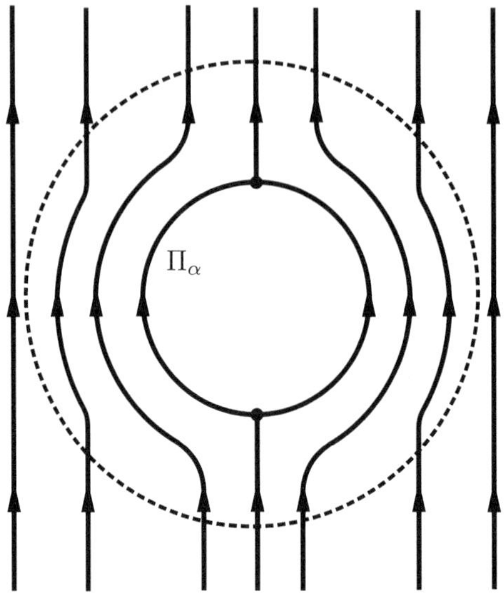

FIGURE 1. A cross-section Π_α of the solid torus and the flow $\widetilde{\varphi}_t$

On the set $T_{2\varepsilon} \setminus \operatorname{int} T_\varepsilon$ we introduce natural coordinates θ_1, θ_2, r with $0 \le \theta_1, \theta_2 < 2\pi$ and $\varepsilon d \le r \le 2\varepsilon d$ such that the set of fixed points of $\widetilde{\varphi}_t$ is composed of those for which $r = \varepsilon d$, and $\theta_1 = 0$ or $\theta_1 = \pi$.

Consider the flow on $T_{2\varepsilon} \setminus \operatorname{int} T_\varepsilon$ defined by

$$(\theta_1, \theta_2, r, t) \mapsto (\theta_1, \theta_2 + [2 - r/(\varepsilon d)]^4 t \cos\theta_1, r),$$

and let $\hat{X}$ be the corresponding vector field. Consider now the flow ψ_t on $M \setminus \operatorname{int} T_\varepsilon$ generated by the vector field Y on $M \setminus \operatorname{int} T_\varepsilon$ defined by

$$Y(x) = \begin{cases} X(x), & x \in M \setminus \operatorname{int} T_{2\varepsilon} \\ \widetilde{X}(x) + \hat{X}(x), & x \in \operatorname{int} T_{2\varepsilon} \setminus \operatorname{int} T_\varepsilon \end{cases}.$$

PROPOSITION 8.1. *The following properties hold:*

1. *The flow ψ_t preserves the measure μ and is ergodic.*
2. *The flow ψ_t has no fixed points.*
3. *The flow ψ_t is nonuniformly hyperbolic (but not uniformly hyperbolic). Moreover, for μ-almost every $x \in M \setminus T_{2\varepsilon}$,*

$$\chi(x, v) < 0 \text{ if } v \in E^s(x), \text{ and } \chi(x, v) > 0 \text{ if } v \in E^u(x),$$

where $E^u(x)$ and $E^s(x)$ are respectively stable and unstable subspaces of the flow φ_t at the point x.

PROOF. By the construction of ψ_t, it preserves μ. Since the vector fields $\widetilde{X}$ and $\hat{X}$ commute the flow is ergodic. Therefore, the first statement holds. The second statement follows from the construction of the flow ψ_t. In order to prove the third statement consider the function

$$T(x, t) = \int_0^t I_{T_{2\varepsilon}}(\varphi_\tau x)\, d\tau,$$

where $I_{T_{2\varepsilon}}$ denotes the indicator (the characteristic function) of the set $T_{2\varepsilon}$. By the Birkhoff Ergodic Theorem,

$$\lim_{t\to+\infty}\frac{T(x,t)}{t}=\mu(T_{2\varepsilon})$$

for μ-almost every $x\in M$.

Fix a point $x\in M\setminus T_{2\varepsilon}$. Consider a moment of time t_1 at which the trajectory $\psi_t x$ enters the set $T_{2\varepsilon}$ and the next moment of time t_2 at which this trajectory exits the set $T_{2\varepsilon}$. Given a vector $v\in E^u(x)$ denote by $\widetilde{v}_i$ the orthogonal projection of the vector $d_x\psi_{t_i}v$ onto the (x,y) plane for $i=1$, 2. It follows from the construction of the flows $\widetilde{\varphi}_t$ and ψ_t (see Property 4) that $\|\widetilde{v}_1\|\ge\|\widetilde{v}_2\|$. Since the unstable subspaces $E^u(x)$ depend continuously on x there exists $K\ge 1$ (independent of x, t_1, and t_2) such that

$$\|d_x\psi_t v\|\ge K\|d_x\varphi_{t-T(x,t)}v\|.$$

It follows that for μ-almost every $x\in M\setminus T_{2\varepsilon}$ and $v\in E^u(x)$,

$$\chi(x,v)=\limsup_{t\to+\infty}\frac{1}{t}\log\|d_x\psi_t v\|\ge(1-\mu(T_{2\varepsilon}))\limsup_{t\to+\infty}\frac{1}{t}\log\|d_x\varphi_t v\|>0.$$

provided ε is sufficiently small. Repeating the above argument with respect to the inverse flow ψ_{-t} one can show that $\chi(x,v)<0$ for μ-almost every $x\in M\setminus T_{2\varepsilon}$ and $v\in E^s(x)$. □

Set $M_1=M\setminus T_\varepsilon$ and consider a copy $(\widetilde{M_1},\widetilde{\psi}_t)$ of the flow (M_1,ψ_t). One can glue the manifolds M_1 and $\widetilde{M_1}$ along their boundaries ∂T_ε and obtain a three-dimensional smooth Riemannian manifold D without boundary. We define a flow F_t on D by

$$F_t x=\begin{cases}\psi_t x, & x\in M_1\\ \widetilde{\psi}_t x, & x\in\widetilde{M_1}\end{cases}.$$

It is clear that the flow F_t is nonuniformly hyperbolic and preserves the measure μ.

8.2. Our next example is geodesic flows on compact smooth Riemannian manifolds of nonpositive curvature. Let M be a compact smooth p-dimensional Riemannian manifold. We assume that for any $x\in M$ and any two vectors v_1, $v_2\in T_xM$ the sectional curvature $K_x(v_1,v_2)$ satisfies

$$K_x(v_1,v_2)\le 0. \tag{8.1}$$

We then say that M has *nonpositive curvature.*

The *geodesic flow* g_t acts on the tangent bundle TM by the formula

$$g_t v=\dot\gamma_v(t),$$

where $\dot\gamma_v(t)$ is the unit tangent vector along the geodesic $\gamma_v(t)$ defined by the vector v (i.e., such that $\dot\gamma_v(0)=v$; this geodesic is uniquely defined). The geodesic flow generates a vector field V on TM given by

$$V(v)=\frac{d(g_t v)}{dt}\bigg|_{t=0}.$$

Since M is compact the flow g_t is well-defined for all $t\in\mathbb{R}$.

We endow the second tangent space $T(TM)$ with a special Riemannian metric. Let $\pi\colon TM\to M$ be the natural projection (i.e., $\pi(x,v)=x$ for each $x\in M$ and each $v\in T_xM$) and $K\colon T(TM)\to TM$ the linear (connection) operator defined by $K\xi=(\nabla Z)(t)|_{t=0}$, where $Z(t)$ is any curve in TM such that $Z(0)=d\pi\xi$,

$\frac{d}{dt}Z(t)|_{t=0} = \xi$ and ∇ is the covariant derivative. The *canonical metric* on $T(TM)$ is given by

$$\langle \xi, \eta \rangle_v = \langle d_v\pi\xi, d_v\pi\eta \rangle_{\pi v} + \langle K\xi, K\eta \rangle_{\pi v}.$$

The set $SM \subset TM$ of the unit vectors is invariant with respect to the geodesic flow, and is a compact manifold of dimension $2p-1$. In what follows we consider the geodesic flow restricted to the subset SM.

The study of hyperbolic properties of the geodesic flow is based upon the description of solutions of the variational equation (6.2) for the flow. One can show that this equation along a given trajectory g_tv of the flow is the *Jacobi equation* along the geodesic $\gamma_v(t)$:

$$Y''(t) + R_{XY}X(t) = 0. \tag{8.2}$$

Here $Y(t)$ is a vector field along $\gamma_v(t), X(t) = \dot\gamma(t)$, and R_{XY} is the curvature tensor. More precisely, the relation between the variational equations (6.2) and the Jacobi equation (8.2) can be described as follows. Fix a vector $v \in SM$ and an element $\xi \in T_vSM$. Let $Y_\xi(t)$ be the unique solution of Equation (8.2) satisfying the initial conditions $Y_\xi(0) = d_v\pi\xi$ and $Y'_\xi(0) = K\xi$. One can show that the map $\xi \mapsto Y_\xi(t)$ is an isomorphism for which $d_{g_tv}\pi d_vg_t\xi = Y_\xi(t)$ and $Kd_vg_t\xi = Y'_\xi(t)$ (see [**9**]). This map establishes the identification between solutions of the variational equation (6.2) and solutions of the Jacobi equation (8.2).

Recall that the Fermi coordinates $\{e_i(t)\}$, for $i = 1, \ldots, p$ along the geodesic $\gamma_v(t)$ are obtained by the time t parallel translation along $\gamma_v(t)$ of an orthonormal basis $\{e_i(0)\}$ in $T_{\gamma_v(0)}M$ where $e_1(t) = \dot\gamma(t)$. Using these coordinates we can rewrite Equation (8.2) in the matrix form

$$\frac{d^2}{dt^2}A(t) + K(t)A(t) = 0, \tag{8.3}$$

where $A(t) = (a_{ij}(t))$ and $K(t) = (k_{ij}(t))$ are matrix functions, with entries $k_{ij}(t) = K_{\gamma_v(t)}(e_i(t), e_j(t))$. The boundary value problem for Equation (8.3) has a unique solution, i.e., for any numbers s_1, s_2 and any matrices A_1, A_2 there exists a unique solution $A(t)$ of Equation (8.3) satisfying $A(s_1) = A_1$ and $A(s_2) = A_2$.

PROPOSITION 8.2 (see [**9**]). *Given $s \in \mathbb{R}$, let $A_s(t)$ be the unique solution of Equation* (8.3) *satisfying the boundary conditions: $A_s(0) = \mathrm{Id}$ (where Id is the identity matrix) and $A_s(s) = 0$. Then there exists a limit*

$$\lim_{t\to\infty} \frac{d}{dt}A_s(t)\Big|_{t=0} = A^+.$$

We define the *positive limit solution* $A^+(t)$ of Equation (8.3) as the solution that satisfies the initial conditions:

$$A^+(0) = \mathrm{Id} \quad \text{and} \quad \frac{d}{dt}A^+(t)\Big|_{t=0} = A^+.$$

One can show (see [**9**]) that this solution is nondegenerate (i.e., that $\det A^+(t) \neq 0$ for every $t \in \mathbb{R}$) and that $A^+(t) = \lim_{s\to+\infty} A_s(t)$. Moreover, it can be written in the form

$$A^+(t) = C(t)\int_t^\infty C(s)^{-1}(C(s)^{-1})^* ds,$$

where $C(t)$ is the solution of Equation (8.3) satisfying the initial conditions

$$C(0) = 0 \quad \text{and} \quad \frac{d}{dt}C(t)\Big|_{t=0} = \mathrm{Id}\,.$$

Similarly, letting $s \to -\infty$ we can define the *negative limit solution* $A^-(t)$ of Equation (8.3).

For every $v \in SM$ let us set

$$E^+(v) = \{\xi \in T_vSM : \langle \xi, V(v) \rangle = 0 \text{ and } Y_\xi(t) = A^+(t) d_v\pi\xi\},$$

$$E^-(v) = \{\xi \in T_vSM : \langle \xi, V(v) \rangle = 0 \text{ and } Y_\xi(t) = A^-(t) d_v\pi\xi\},$$

where V is the vector field generated by the geodesic flow.

PROPOSITION 8.3 (see [**9**]). *The sets $E^-(v)$ and $E^+(v)$ are subspaces of T_vSM and satisfy the following properties:*

1. $\dim E^-(v) = \dim E^+(v) = p - 1$;
2. $d_v\pi E^-(v) = d_v\pi E^+(v) = \{w \in T_{\pi v}M : w$ *is orthogonal to* $v\}$;
3. *the subspaces $E^-(v)$ and $E^+(v)$ are invariant under the differential d_vg_t, i.e., $d_vg_tE^-(v) = E^-(g_tv)$ and $d_vg_tE^+(v) = E^+(g_tv)$;*
4. *if $\tau\colon SM \to SM$ is the involution defined by $\tau v = -v$, then*
$$E^+(-v) = d_v\tau E^-(v) \quad \text{and} \quad E^-(-v) = d_v\tau E^+(v);$$
5. *if $K_x(v_1, v_2) \ge -a^2$ for some $a > 0$ and all $x \in M$, then $\|K\xi\| \le a\|d_v\pi\xi\|$ for every $\xi \in E^+(v)$ and every $\xi \in E^-(v)$;*
6. *if $\xi \in E^+(v)$ or $\xi \in E^-(v)$, then $Y_\xi(t) \ne 0$ for every $t \in \mathbb{R}$;*
7. *$\xi \in E^+(v)$ (respectively, $\xi \in E^-(v)$) if and only if*
$$\langle \xi, V(v) \rangle = 0 \quad \text{and} \quad \|d_{g_tv}\pi d_vg_t\xi\| \le c$$
for every $t > 0$ (respectively, $t < 0$), for some constant $c > 0$;
8. *if $\xi \in E^+(v)$ (respectively, $\xi \in E^-(v)$) then the function $t \mapsto \|Y_\xi(t)\|$ is nonincreasing (respectively, nondecreasing).*

In view of Properties 5 and 7 we have $\xi \in E^+(v)$ (respectively, $\xi \in E^-(v)$) if and only if $\langle \xi, V(v) \rangle = 0$ and $\|d_vg_t\xi\| \le c$ for $t > 0$ (respectively, $t < 0$), for some constant $c > 0$. This observation and Property 3 justify to call $E^+(v)$ and $E^-(v)$ the *stable* and *unstable subspaces*.

In general, the subspaces $E^-(v)$ and $E^+(v)$ do not span the whole second tangent space T_vSM. If they do span T_vSM for every $v \in SM$, then the geodesic flow is Anosov (see [**9**]). This is the case when the curvature is strictly negative. However, for a general manifold of nonpositive curvature one can only expect that the geodesic flow is nonuniformly hyperbolic. To see this consider the set

$$\Delta = \left\{ v \in SM : \limsup_{t\to\infty} \frac{1}{t} \int_0^t K_{\pi g_s v}(v, w)\, ds < 0 \right. \\ \left. \text{for every } w \in SM \text{ orthogonal to } v \right\}. \tag{8.4}$$

It is easy to see that the set Δ is measurable and invariant under the flow g_t.

THEOREM 8.4 (see [**28, 29**]). *If the Riemannian manifold M has nonpositive curvature then for every $v \in \Delta$, we have $\chi(v,\xi) < 0$ if $\xi \in E^+(v)$ and $\chi(v, \xi) > 0$ if $\xi \in E^-(v)$.*

PROOF. Let $\psi\colon \mathbb{R}^+ \to \mathbb{R}$ be a continuous function. Set

$$\overline{\psi} = \limsup_{t\to\infty} \frac{1}{t} \int_0^t \psi(s)\, ds, \quad \underline{\psi} = \liminf_{t\to\infty} \frac{1}{t} \int_0^t \psi(s)\, ds,$$

$$\widetilde{\psi} = \liminf_{t\to\infty} \frac{1}{t}\int_0^t \psi(s)^2\,ds.$$

We need the following lemma.

LEMMA 8.5. *Assume that* $c = \sup_{t\geq 0}|\psi(t)| < \infty$. *Then:*

1. *if* $\psi(t) \leq 0$ *for all* $t \geq 0$ *and* $\widetilde{\psi} > 0$ *then* $\overline{\psi} < 0$;
2. *if* $\psi(t) \geq 0$ *for all* $t \geq 0$ *and* $\widetilde{\psi} > 0$ *then* $\underline{\psi} > 0$.

PROOF OF THE LEMMA. Assume that $\psi(t) \leq 0$. Then $\overline{\psi} \leq 0$. On the other hand, if $c > 0$ then

$$-\frac{\overline{\psi}}{c} = \underline{\left|\frac{\psi}{c}\right|} \geq \widetilde{\left(\frac{\psi}{c}\right)} = \frac{\widetilde{\psi}}{c^2} > 0.$$

This implies that $\overline{\psi} < 0$ and completes the proof of the first statement. The proof of the second statement is similar. □

We proceed with the proof of the theorem. Fix $v \in \Delta$, $\xi \in E^+(v)$, and consider the function $\varphi(t) = \frac{1}{2}\|Y_\xi(t)\|^2$. Using (8.3) we obtain

$$\frac{d^2}{dt^2}\varphi(t) = \frac{1}{2}\frac{d^2}{dt^2}\langle Y_\xi, Y_\xi\rangle = -K(t)\varphi(t) + \|Y_\xi'(t)\|^2,$$

where $K(t) = K_{\gamma_v(t)}(Y_\xi(t), w(t))$ and $w(t) = \dot{\gamma}_v(t)$. It follows from Proposition 8.3 that $\varphi(t) \neq 0$, and $\frac{d}{dt}\varphi(t) \leq 0$ for all $t \geq 0$. Set

$$z(t) = (\varphi(t))^{-1}\frac{d}{dt}\varphi(t).$$

It is easy to check that the function $z(t)$ satisfies the Ricatti equation

$$\frac{d}{dt}z(t) + z(t)^2 - (\varphi(t))^{-1}\|Y_\xi'(t)\|^2 + K(t) = 0. \tag{8.5}$$

By Proposition 8.3,

$$\begin{aligned}\left|\frac{d}{dt}\varphi(t)\right| &= \frac{1}{2}\left|\frac{d}{dt}\langle Y_\xi, Y_\xi\rangle\right| = |\langle Y_\xi, Y_\xi'\rangle| \\ &= |\langle d_{g_t v}\pi d_v g_t\xi, K d_v g_t \xi\rangle| \leq a\|d_{g_t v}\pi d_v g_t \xi\|^2 = 2a\varphi(t).\end{aligned}$$

It follows that $\sup_{t\geq 0}|z(t)| \leq 2a$. Integrating the Ricatti equation (8.5) on the interval $[0, t]$ we obtain that

$$z(t) - z(0) + \int_0^t z(s)^2\,ds = \int_0^t (\varphi(s))^{-1}\|Y_\xi'(s)\|^2\,ds - \int_0^t K(s)\,ds.$$

It follows that for $v \in \Delta$ (see (8.4)) we have

$$\begin{aligned}\liminf_{t\to\infty}\frac{1}{t}\int_0^t z(s)^2\,ds &\geq \liminf_{t\to\infty}\frac{1}{t}\int_0^t (\varphi(s))^{-1}\|Y_\xi'(s)\|^2\,ds \\ &\quad - \limsup_{t\to\infty}\frac{1}{t}\int_0^t K(s)\,ds > 0.\end{aligned}$$

Therefore, in view of Lemma 8.5 we conclude that

$$\limsup_{t\to\infty}\frac{1}{t}\int_0^t z(s)\,ds < 0.$$

On the other hand, using Proposition 8.3 we find that

$$\begin{aligned}\chi(v,\xi) &= \limsup_{t\to\infty}\frac{1}{t}\log\|d_v g_t\xi\| = \limsup_{t\to\infty}\frac{1}{t}\log\|d_{g_tv}\pi d_v g_t\xi\|\\ &= \limsup_{t\to\infty}\frac{1}{t}\log\|Y_\xi(t)\| = \frac{1}{2}\limsup_{t\to\infty}\frac{1}{t}\int_0^t z(s)\,ds.\end{aligned}$$

This completes the proof of the first statement of the theorem. The second statement can be proved in a similar way. □

The geodesic flow preserves the Liouville measure μ on the tangent bundle which is induced by the Riemannian metric. We denote by ν the Lebesgue measure on M. It follows from Theorem 8.4 that if the set Δ has positive Liouville measure then the geodesic flow $g_t|\Delta$ is nonuniformly hyperbolic. It is, therefore, crucial to find conditions which would guarantee that Δ has positive Liouville measure.

We first consider the two-dimensional case.

THEOREM 8.6. *Let M be a smooth compact surface of nonpositive curvature $K(x)$ and genus greater than* 1. *Then $\mu(\Delta) > 0$.*

PROOF. By the Gauss–Bonnet formula the Euler characteristic of M is

$$\frac{1}{2\pi}\int_M K(x)\,d\nu(x)\,.$$

It follows from the condition of the theorem that

$$\int_M K(x)\,d\nu(x) < 0. \tag{8.6}$$

Choose two orthogonal vectors $v,\ w \in SM$. It is easy to see that

$$\limsup_{t\to\infty}\frac{1}{t}\int_0^t K_{\pi(g_sv)}(v,w)\,ds = \limsup_{t\to\infty}\frac{1}{t}\int_0^t K(\pi(g_sv))\,ds.$$

By the Birkhoff ergodic theorem we obtain that for μ-almost every $v \in SM$ there exists the limit

$$\lim_{t\to\infty}\frac{1}{t}\int_0^t K(\pi(g_sv))\,ds = \Phi(v)$$

and that

$$\int_{SM}\Phi(v)\,d\mu(v) = \int_M K(x)\,d\nu(x).$$

The desired result follows from (8.6). □

In the multidimensional case one can establish the following criterion for positivity of the Liouville measure of the set Δ.

THEOREM 8.7. *Let M be a smooth compact Riemannian manifold of nonpositive curvature. Assume that there exist $x \in M$ and a vector $v \in S_xM$ such that*

$$K_x(v,w) < 0 \tag{8.7}$$

for any vector $w \in S_xM$ which is orthogonal to v. Then $\mu(\Delta) > 0$.

PROOF. Condition (8.7) holds in a small open neighborhood of (x, v). The desired result now follows from the Birkhoff ergodic theorem. □

We remark that Condition (8.7) is a multidimensional version of (8.6). It is easy to see that the set Δ is everywhere dense. One can also show that it is open (see Theorem 16.4 below).

9. Existence of Local Stable Manifolds

Consider a nonuniformly (partially) hyperbolic f-invariant set $\mathcal{R}$, and the associated sets $\mathcal{R}^\ell$ defined by (7.1). According to the results in Section 7 we may replace everywhere in the sequel the sets $\mathcal{R}^\ell$ by the compact sets $\overline{\mathcal{R}^\ell}$.

The hyperbolicity conditions allow one to describe the asymptotic behavior of trajectories which start in a small neighborhood of a hyperbolic trajectory. More precisely, applying Theorem 4.4 one can construct at every point $x \in \mathcal{R}$ local stable and unstable manifolds $V^s(x)$ and $V^u(x)$ such that every trajectory $f^n(y)$ with $y \in V^s(x)$ approaches $f^n(x)$ with an exponential rate, and every trajectory $f^{-n}(y)$ with $y \in V^u(x)$ approaches $f^{-n}(x)$ with an exponential rate. Roughly speaking, the behavior of trajectories that start in a neighborhood of x resembles that of the trajectories in a neighborhood of a fixed (or periodic) hyperbolic point. However, the rate of exponential convergence may vary from orbit to orbit and may also get worse along the orbit but with subexponential rate (due to Condition (H5) in Section 7).

A more precise description is given by the following theorem which is one of the key results in the hyperbolic theory.

THEOREM 9.1 (Stable Manifold Theorem). *Let $\mathcal{R}$ be a nonuniformly hyperbolic set for a $C^{1+\alpha}$ diffeomorphism f. Then for every $x \in \mathcal{R}$ there exists a local stable manifold $V^s(x)$ such that $x \in V^s(x)$, $T_xV^s(x) = E^s(x)$, and if $y \in V^s(x)$ and $n \geq 0$ then*

$$\rho(f^n(x), f^n(y)) \leq T(x)\lambda^n e^{\varepsilon n}\rho(x, y), \tag{9.1}$$

where ρ is the distance induced in M by the Riemannian metric and $T\colon \mathcal{R} \to (0, \infty)$ is a Borel function such that if $m \in \mathbb{Z}$ then

$$T(f^m(x)) \leq T(x)e^{10\varepsilon|m|}. \tag{9.2}$$

Inequality (9.2) should be compared to Condition (H5) in Section 7.

The Stable Manifold Theorem was first established by Pesin in [**27**]. His approach was an elaboration of the classical work of Perron. This approach was extended by Katok and Strelcyn in [**15**] to smooth maps with singularities (they include billiard systems and other physical models). Ruelle [**34**] provided another proof of Theorem 9.1, based on studying the perturbations of the matrix products in the Multiplicative Ergodic Theorem 6.2. Fathi, Herman, and Yoccoz [**10**] provided a detailed exposition of Theorem 9.1 which essentially follows the approaches of Pesin and Ruelle. Another proof of the Stable Manifold Theorem was provided by Pugh and Shub in [**32**] using graph transform techniques.

In [**30**], Pugh constructed an explicit example of a C^1 diffeomorphism (which is not $C^{1+\alpha}$ for any $\alpha > 0$) on a 4-dimensional manifold for which the statement of Theorem 9.1 does not hold. More precisely, there exists no manifold tangent to $E^s(x)$ such that (9.1) holds on some open neighborhood of x. This example illustrates that the hypothesis $\alpha > 0$ is crucial in Theorem 9.1.

On another direction, Liu and Qian [**18**] established a version of Theorem 9.1 for random maps. One can extend the Stable Manifold Theorem 9.1 to infinite-dimensional spaces. Ruelle [**35**] proved this theorem for Hilbert spaces, closely following his approach in [**34**], and Mañé [**22**] considered Banach spaces (under certain compactness assumptions on the dynamics).

The local stable manifold $V^s(x)$ in Theorem 9.1 is constructed via a smooth map

$$\psi^s\colon B^s(r) \to E^u(x)$$

which satisfies

$$\psi^s(0) = 0 \quad \text{and} \quad d\psi^s(0) = 0. \tag{9.3}$$

Here $B^s(r)$ is the ball of radius r in $E^s(x)$ centered at the origin; $r = r(x)$ is called the *size* of the local stable manifold. One now obtains the local stable manifold by projecting the graph of ψ^s into M by the exponential map

$$V^s(x) = \exp_x\{(x, \psi^s(x)) : x \in B^s(r)\}.$$

It follows from (9.3) that

$$x \in V^s(x) \quad \text{and} \quad T_xV^s(x) = E^s(x).$$

We now describe a construction of the function ψ^s. Fix $x \in M$ and consider the map

$$\widetilde{f}_x = \exp_{f(x)}^{-1} \circ f \circ \exp_x\colon B^s(r) \times B^u(r) \to T_{f(x)}M, \tag{9.4}$$

which is well-defined if r is sufficiently small (here $B^u(r)$ is the ball of radius r in $E^u(x)$ centered at the origin). By Condition (H1) in Section 7, the map $\widetilde{f}$ can be written in the following form:

$$\widetilde{f}_x(v, w) = (A_xv + g_x(v, w), B_xw + h_x(v, w)), \tag{9.5}$$

where $v \in E^s(x)$ and $w \in E^u(x)$. Furthermore,

$$A_x\colon E^s(x) \to E^s(f(x)) \quad \text{and} \quad B_x\colon E^u(x) \to E^u(f(x))$$

are linear maps. In view of Conditions (H2) and (H3) in Section 7 the map A_x is a contraction and the map B_x is an expansion. Since f is of class $C^{1+\alpha}$ we also have

$$\|g_x(v, w)\| \le C_1(\|v\| + \|w\|)^{1+\alpha}, \quad \|h_x(v, w)\| \le C_2(\|v\| + \|w\|)^{1+\alpha}, \tag{9.6}$$

and

$$\begin{aligned} \|dg_x(v_1, w_1) - dg_x(v_2, w_2)\| &\le C_1(\|v_1 - v_2\| + \|w_1 - w_2\|)^\alpha, \\ \|dh_x(v_1, w_1) - dh_x(v_2, w_2)\| &\le C_2(\|v_1 - v_2\| + \|w_1 - w_2\|)^\alpha \end{aligned} \tag{9.7}$$

where $C_1 > 0$ and $C_2 > 0$ are constants (which may depend on x).

In other words the map $\widetilde{f}_x$ can be viewed as a small perturbation of the linear map $(v, w) \mapsto (A_xv, B_xw)$ by the perturbation $(g_x(v, w), h_x(v, w))$ satisfying Conditions (9.6) and (9.7), which are analogous to Condition (1.7). Therefore, if the point x is Lyapunov regular then an appropriate discrete time version of Theorem 4.4 applies and gives the existence of a local stable manifold at x.

Note that if the point x is nonuniformly hyperbolic (i.e., $x \in \mathcal{R}$) but is not Lyapunov regular then Theorem 4.4 cannot be used. Furthermore, Theorem 4.4 does not provide any information on the size of the local manifold and how it varies with x.

Therefore, we describe another approach for constructing local stable manifolds. First we construct a special inner product in the tangent bundle $T\mathcal{R}$ which is known as the *Lyapunov inner product*. It provides an important technical tool in studying nonuniform hyperbolicity.

Choose numbers $0 < \lambda' < \mu' < \infty$ such that

$$\lambda e^\varepsilon < \lambda', \quad \mu' < \mu e^{-\varepsilon}. \tag{9.8}$$

We define a new inner product $\langle\cdot,\cdot\rangle'_x$, called *Lyapunov inner product*, as follows. Set

$$\langle v,w\rangle'_x=\sum_{k=0}^{\infty}\langle df^kv,df^kw\rangle_{f^k(x)}{\lambda'}^{-2k} \tag{9.9}$$

if v, $w\in E^s(x)$, and

$$\langle v,w\rangle'_x=\sum_{k=0}^{\infty}\langle df^{-k}v,df^{-k}w\rangle_{f^{-k}(x)}{\mu'}^{2k} \tag{9.10}$$

if v, $w\in E^u(x)$.

Notice that each series converges. Indeed, by the Cauchy–Schwarz inequality, Condition (H5) in Section 7, and (9.8), if v, $w\in E^s(x)$ then

$$\begin{aligned}\langle v,w\rangle'_x&\le\sum_{k=0}^{\infty}C(f^k(x))^2\lambda^{2k}{\lambda'}^{-2k}\|v\|_x\|w\|_x\\&\le C(x)^2(1-\lambda e^{\varepsilon}/\lambda')^{-1}\|v\|_x\|w\|_x<\infty,\end{aligned} \tag{9.11}$$

and if v, $w\in E^u(x)$ then

$$\begin{aligned}\langle v,w\rangle'_x&\le\sum_{k=0}^{\infty}C(f^{-k}(x))^2\mu^{-2k}{\mu'}^{2k}\|v\|_x\|w\|_x\\&\le C(x)^2(1-\mu'/(\mu e^{-\varepsilon}))^{-1}\|v\|_x\|w\|_x<\infty.\end{aligned} \tag{9.12}$$

We extend $\langle\cdot,\cdot\rangle'_x$ to all vectors in T_xM by declaring the subspaces $E^s(x)$ and $E^u(x)$ to be mutually orthogonal with respect to $\langle\cdot,\cdot\rangle'_x$, i.e., we set

$$\langle v,w\rangle'_x=\langle v^s,w^s\rangle'_x+\langle v^u,w^u\rangle'_x,$$

where $v=v^s+v^u$ and $w=w^s+w^u$ with v^s, $w^s\in E^s(x)$ and v^u, $w^u\in E^u(x)$.

We emphasize that the Lyapunov inner product, and hence, the associated norm $\|\cdot\|'$, called *Lyapunov norm*, depend on the choice of numbers λ' and μ'. The Lyapunov inner product has several important properties which determine its use in the study of nonuniform hyperbolicity.

N1. the angle between the subspaces $E^s(x)$ and $E^u(x)$ in the inner product $\langle\cdot,\cdot\rangle'_x$ is $\pi/2$ for each $x\in\mathcal{R}$;

N2. $\|A_x\|'\le\lambda'$ and $\|{B_x}^{-1}\|'\le(\mu')^{-1}$;

N3. the relation between the Lyapunov inner product and the Riemannian inner product is given by

$$\frac{1}{\sqrt{2}}\|w\|_x\le\|w\|'_x\le D(x)\|w\|_x,$$

where $w\in T_xM$, and

$$D(x)=C(x)K(x)^{-1}[(1-\lambda e^{\varepsilon}/\lambda')^{-1}+(1-\mu'/(\mu e^{-\varepsilon}))^{-1}]^{1/2}$$

is a measurable function satisfying for $m\in\mathbb{Z}$,

$$D(f^m(x))\le D(x)e^{2\varepsilon|m|}. \tag{9.13}$$

Property (N1) holds true due to the construction of the Lyapunov inner product. To show Property (N2) we use (9.9) and write for $v \in E^s(x)$,

$$\begin{aligned}\|A_x v\|'^2 &= \langle dfv, dfv\rangle' = \sum_{k=0}^{\infty}\langle df^k dfv, df^k dfv\rangle \lambda'^{-2k}\\ &= \sum_{k=1}^{\infty}\langle df^k v, df^k v\rangle \lambda'^{-2(k-1)} = \lambda'^2 \sum_{k=1}^{\infty}\langle df^k v, df^k v\rangle \lambda'^{-2k}\\ &= \lambda'^2(\|v\|'^2 - \|v\|^2) \le \lambda'^2\|v\|'^2.\end{aligned}$$

The proof of the second inequality is similar and uses (9.10). Property (N3) follows from Condition (H4) in Section 7, (9.11), and (9.12).

Properties (N1) and (N2) show that the action of the differential df on $T\mathcal{R}$ is *uniformly* hyperbolic with respect to the Lyapunov inner product. One can now use either the Hadamard method or the Perron method together with the Lyapunov inner product to construct local stable manifolds, well-known in the uniform hyperbolic theory. Note that the "perturbation map" $\widetilde{f}_x$ (see (9.4) and (9.5)) satisfies Condition (9.6) in a neighborhood U_x of the point x whose size depends on x. Moreover, the size of U_x decays along the trajectory of x with subexponential rate (see (9.13)). This requires a substantial modification of the classical Hadamard–Perron approach and forces the size of the local stable manifolds to decay along the trajectory of x with subexponential rate (see (9.2)).

We now briefly describe a modification of the Perron method.

Fix $x \in \mathcal{R}$. Consider the trajectory $f^m(x)$ and the family of maps $F_m = \widetilde{f}_{f^m(x)}$. We identify the tangent spaces $T_{f^m(x)}M$ with $\mathbb{R}^n = \mathbb{R}^k \times \mathbb{R}^{p-k}$ (recall that $p = \dim M$ and $1 \le k < p$) via an isomorphism τ_m such that $\tau_m(E^s(x)) = \mathbb{R}^k$ and $\tau_m(E^u(x)) = \mathbb{R}^{p-k}$. We can write $\widetilde{F}_m = \tau_{m+1} \circ F_m \circ {\tau_m}^{-1}$ in the form

$$\widetilde{F}_m(v, w) = (A_m v + g_m(v, w), B_m w + h_m(v, w)), \tag{9.14}$$

where $A_m \colon \mathbb{R}^k \to \mathbb{R}^k$ and $B_m \colon \mathbb{R}^{p-k} \to \mathbb{R}^{p-k}$ are linear maps, and $g\colon \mathbb{R}^n \to \mathbb{R}^k$ and $h\colon \mathbb{R}^n \to \mathbb{R}^{n-k}$ are nonlinear maps defined for each $v \in B^s(r_0) \subset \mathbb{R}^k$ and $w \in B^u(r_0) \subset \mathbb{R}^{p-k}$ (recall that $B^s(r_0)$ and $B^u(r_0)$ are balls centered at 0 of radius r_0). With respect to the Lyapunov metric these maps satisfy:

$$\|A_m\|' \le \lambda' \quad\text{and}\quad (\|{B_m}^{-1}\|')^{-1} \ge \mu', \quad\text{where}\quad 0 < \lambda' < \min\{1, \mu'\} \tag{9.15}$$

and

$$g_m(0,0) = 0, \quad dg_m(0,0) = 0, \quad h_m(0,0) = 0, \quad dh_m(0,0) = 0, \tag{9.16}$$

$$\begin{aligned}\|dg_m(v_1, w_1) - dg_m(v_2, w_2)\|' &\le C\gamma^{-m}(\|v_1 - v_2\|' + \|w_1 - w_2\|')^\alpha,\\ \|dh_m(v_1, w_1) - dh_m(v_2, w_2)\|' &\le C\gamma^{-m}(\|v_1 - v_2\|' + \|w_1 - w_2\|')^\alpha,\end{aligned} \tag{9.17}$$

where

$$\lambda'^\alpha < \gamma < 1, \quad 0 < \alpha \le 1, \quad C > 0 \tag{9.18}$$

(see (9.7) and Conditions (N2) and (N3)).

THEOREM 9.2 (see [**27**]). *Let κ be any number satisfying*

$$\lambda' < \kappa < \min\{\mu', \gamma^{\frac{1}{\alpha}}\}. \tag{9.19}$$

Then there exist constants $D > 0$ and $r_0 > r > 0$, and a map $\psi^s \colon B^s(r) \to \mathbb{R}^{p-k}$ such that:

1. *ψ^s is of class $C^{1+\alpha}$, and $\psi^s(0)=0$ and $d\psi^s(0)=0$;*
2. *$\|d\psi^s(v_1)-d\psi^s(v_2)\|'\le D(\|v_1-v_2\|')^\alpha$ for any v_1, $v_2\in B^s(r)$;*
3. *if $m\ge 0$ and $v\in B^s(r)$ then*
$$\left(\prod_{i=0}^{m-1}\widetilde{F}_i\right)(v,\psi^s(v))\in B^s(r)\times B^u(r),$$
$$\left\|\left(\prod_{i=0}^{m-1}\widetilde{F}_i\right)(v,\psi^s(v))\right\|'\le D\kappa^m\|(v,\psi^s(v))\|',$$
where $\prod_{i=0}^{m-1}\widetilde{F}_i$ denotes the composition $\widetilde{F}_{m-1}\circ\cdots\circ\widetilde{F}_0$;
4. *given $v\in B^s(r)$ and $w\in B^u(r)$, if there is a number $K>0$ such that*
$$\left(\prod_{i=0}^{m-1}\widetilde{F}_i\right)(v,w)\in B^s(r)\times B^u(r)\quad\text{and}\quad\left\|\left(\prod_{i=0}^{m-1}\widetilde{F}_i\right)(v,w)\right\|'\le K\kappa^m$$
for every $m\ge 0$, then $w=\psi^s(v)$;
5. *the numbers D and r depend only on the numbers λ', μ', γ, α, κ, and C.*

PROOF. Consider the linear space Γ_κ of sequences of vectors
$$z=\{z(m)\in\mathbb{R}^p\}_{m\in\mathbb{N}},$$
satisfying the following condition:
$$\|z\|_\kappa=\sup_{m\ge 0}(\kappa^{-m}\|z(m)\|')<\infty.$$
It is easy to verify that Γ_κ is a Banach space with the norm $\|z\|_\kappa$. Consider the open set
$$W=\{z\in\Gamma_\kappa\colon z(m)\in B^s(r)\times B^u(r)\text{ for every }m\in\mathbb{N}\}$$
and the map $\Phi_\kappa\colon B^s(r_0)\times W\to\Gamma_\kappa$ defined by
$$\Phi_\kappa(y,z)(0)=\left(y,-\sum_{k=0}^{\infty}\left(\prod_{i=0}^{k}B_i\right)^{-1}h_k(z(k))\right),$$
and if $m>0$ then
$$\Phi_\kappa(y,z)(m)=-z(m)+\left(\left(\prod_{i=0}^{m-1}A_i\right)y,0\right)$$
$$+\left(\sum_{n=0}^{m-1}\left(\prod_{i=n+1}^{m-1}A_i\right)g_n(z(n)),-\sum_{n=0}^{\infty}\left(\prod_{i=0}^{n}B_{i+m}\right)^{-1}h_{n+m}(z(n+m))\right).$$
Using Conditions (9.15)–(9.19) we will show that the map Φ_κ is well-defined and is continuously differentiable over y and z. Indeed, by (9.16) and (9.17), if $z\in B^s(r_0)\times B^u(r_0)$ and $n\ge 0$, then
$$\begin{aligned}\|g_n(z)\|'&=\|g_n(z)-g_n(0)\|'\le\|dg_n(\xi)\|'\cdot\|z\|'\\&=\|dg_n(\xi)-dg_n(0)\|'\cdot\|z\|'\\&\le C\gamma^{-n}(\|\xi\|')^\alpha\|z\|'\le C\gamma^{-n}(\|z\|')^{1+\alpha},\end{aligned}\tag{9.20}$$
where ξ lies on the interval that joins the point 0 and z. Similarly, we have that
$$\|h_n(z)\|'\le C\gamma^{-n}(\|z\|')^{1+\alpha}.\tag{9.21}$$

Using (9.15), (9.20), and (9.21) we obtain

$$
\begin{aligned}
\|\Phi_\kappa(y,z)\|_\kappa &= \sup_{m\geq 0}(\kappa^{-m}\|\Phi_\kappa(y,z)(m)\|')\\
&\leq \sup_{m\geq 0}(\kappa^{-m}\|z(m)\|')\\
&+\sup_{m\geq 0}\Bigg\{\kappa^{-m}\Bigg[\prod_{i=0}^{m-1}\|A_i\|'\cdot\|y\|'+\sum_{n=0}^{m-1}\Bigg(\prod_{i=n+1}^{m-1}\|A_i\|'\Bigg)\\
&\times C\gamma^{-n}(\|z(n)\|')^{1+\alpha}\\
&+\sum_{n=0}^{\infty}\Bigg(\prod_{i=0}^{n}\|B_{i+m}^{-1}\|'\Bigg)C\gamma^{-(m+n)}(\|z(n+m)\|')^{1+\alpha}\Bigg]\Bigg\}\\
&\leq \|z\|_\kappa + \sup_{m\geq 0}(\kappa^{-m}\lambda'^m)\|y\|'\\
&+\sup_{m\geq 0}\Bigg(\kappa^{-m}C\|z\|_\kappa{}^{1+\alpha}\Bigg[\sum_{n=0}^{m-1}\lambda'^{m-n-1}\gamma^{-n}\kappa^{(1+\alpha)n}\\
&+\sum_{n=0}^{\infty}\mu'^{-(n+1)}\gamma^{-(m+n)}\kappa^{(1+\alpha)(m+n)}\Bigg]\Bigg).
\end{aligned}
\tag{9.22}
$$

In view of (9.19) we have

$$\sup_{m\geq 0}(\kappa^{-m}\lambda'^m)=1.$$

Since the function $x\mapsto xa^x$ reaches its maximum $-1/e\log a$ at $x=-1/\log a$ we obtain

$$
\begin{aligned}
&\sup_{m\geq 0}\Bigg[\kappa^{-m}\lambda'^{m-1}\sum_{n=0}^{m-1}\Big(\lambda'^{-1}\gamma^{-1}\kappa^{(1+\alpha)}\Big)^n\Bigg]\\
&\leq\begin{cases}\lambda'^{-1}(\kappa^{-1}\lambda')^m m & \text{if } \lambda'^{-1}\gamma^{-1}\kappa^{(1+\alpha)}\leq 1\\ \lambda'^{-1}(\gamma^{-1}\kappa^{\alpha})^m m & \text{if } \lambda'^{-1}\gamma^{-1}\kappa^{(1+\alpha)}\geq 1\end{cases}\\
&\leq \lambda'^{-1}e^{-1}\left(\log\max\left\{\frac{\kappa^\alpha}{\gamma},\frac{\lambda'}{\kappa}\right\}\right)^{-1}\stackrel{\text{def}}{=}M_1.
\end{aligned}
$$

Furthermore, since $\mu'^{-1}\gamma^{-1}\kappa^{1+\alpha}=(\mu'^{-1}\kappa)(\kappa^\alpha\gamma^{-1})<1$, we have

$$
\begin{aligned}
&\sup_{m\geq 0}\left(\kappa^{-m}\gamma^{-m}\kappa^{(1+\alpha)m}\mu'^{-1}\sum_{n=0}^{m-1}\Big(\mu'^{-1}\gamma^{-1}\kappa^{(1+\alpha)}\Big)^n\right)\\
&=\frac{1}{\mu'-\gamma^{-1}\kappa^{1+\alpha}}\stackrel{\text{def}}{=}M_2.
\end{aligned}
\tag{9.23}
$$

Setting $M=M_1+M_2$ we conclude that

$$\|\Phi_\kappa(y,z)\|_\kappa\leq\|z\|_\kappa+\|y\|'+CM\|z\|_\kappa{}^{1+\alpha}.$$

This implies that the map Φ_κ is well-defined. Moreover,

$$\Phi_\kappa(0,0)=(0,0).$$

We now show that the map Φ_κ is of class C^1. Indeed, for any $y \in B^s(r_0)$ and $t \in E^s$ such that $y+t \in B^s(r_0)$, given $z \in W$ and $m \geq 0$ we have

$$\Phi_\kappa(y+t,z)(m) - \Phi_\kappa(y,z)(m) = \left(\left(\prod_{i=0}^{m-1} A_i\right) t, 0\right).$$

It follows that

$$\partial_y \Phi_\kappa(y,z)(m) = \left(\prod_{i=0}^{m-1} A_i, 0\right).$$

Now for any $y \in B^s(r_0)$, $z \in W$, $t \in \Gamma_\kappa$ such that $z+t \in W$ we can write

$$\Phi_\kappa(y,z+t) - \Phi_\kappa(y,z) = (\mathcal{A}_\kappa(z) - \mathrm{Id})t + o(z,t),$$

where Id is the identity map,

$$\begin{aligned}(\mathcal{A}_\kappa(z))t(m) = \Bigg(&\sum_{n=0}^{m-1}\left(\prod_{i=n+1}^{m-1} A_i\right) dg_n(z(n))t(n),\\ &-\sum_{n=0}^{\infty}\left(\prod_{i=0}^{n} B_{i+m}\right)^{-1} dh_{n+m}(z(n+m))t(m+n)\Bigg),\end{aligned}$$

and

$$o(z,t)(m) =$$
$$\left(\sum_{n=0}^{m-1}\left(\prod_{i=n+1}^{m-1} A_i\right) o_1(z,t)(n), -\sum_{n=0}^{\infty}\left(\prod_{i=0}^{n} B_{i+m}\right)^{-1} o_2(z,t)(m+n)\right).$$

Here $o_i(z,t)(m)$ for $i=1$, 2 are defined by

$$\begin{aligned}o_1(z,t)(m) &= g_m((z+t)(m)) - g_m(z(m)) - dg_m(z(m))t(m),\\ o_2(z,t)(m) &= h_m((z+t)(m)) - h_m(z(m)) - dh_m(z(m))t(m).\end{aligned}$$

If z_1, $z_2 \in W$ and $t \in \Gamma_\kappa$ then

$$\begin{aligned}&\|(\mathcal{A}_\kappa(z_1) - \mathcal{A}_\kappa(z_2))t\|_\kappa =\\ &\leq \sup_{m\geq 0}\Bigg\{\kappa^{-m}\Bigg[\sum_{n=0}^{m-1}\left(\prod_{i=n+1}^{m-1}\|A_i\|'\right) C\gamma^{-n}(\|z_1(n)-z_2(n)\|')^\alpha \|t(n)\|'\\ &\quad+\sum_{n=0}^{\infty}\left(\prod_{i=0}^{n}\|B_{i+m}^{-1}\|'\right) C\gamma^{-(n+m)}(\|z_1(n+m)-z_2(n+m)\|')^\alpha\\ &\quad\times\|t(n+m)\|'\Bigg]\Bigg\}\\ &\leq \sup_{m\geq 0}\Bigg\{\kappa^{-m} C\Bigg[\sum_{n=0}^{m-1}\lambda'^{m-n-1}\gamma^{-n}\kappa^{(1+\alpha)n}\\ &\quad+\sum_{n=0}^{\infty}\mu'^{-(n+1)}\gamma^{-(n+m)}\kappa^{(1+\alpha)(n+m)}\Bigg]\Bigg\}\times\|z_1-z_2\|_\kappa{}^\alpha\|t\|_\kappa.\end{aligned}$$

It follows that (see (9.22)–(9.23))

$$\|(\mathcal{A}_\kappa(z_1) - \mathcal{A}_\kappa(z_2))t\|_\kappa \leq CM\|z_1-z_2\|_\kappa{}^\alpha\|t\|_\kappa. \tag{9.24}$$

It follows from (9.17) and the mean value theorem that

$$\|o_i(z,t)(m)\|' \le C\gamma^{-m}(\|t(m)\|')^{1+\alpha}.$$

This implies that

$$\|o(z,t)\|_\kappa \le CM\|t\|_\kappa{}^{1+\alpha}.$$

We conclude that $\partial_z\Phi_\kappa(y,z) = \mathcal{A}_\kappa(z) - \mathrm{Id}$. In particular, $\partial_z\Phi_\kappa(y,0) = -\mathrm{Id}$. By (9.24), the map $\partial_z\Phi_\kappa(y,z)$ is continuous. Therefore, the map Φ_κ satisfies the conditions of the Implicit Function Theorem, and hence, there exist a number $r \le r_0$ and a map $\varphi\colon B^s(r) \to W$ of class C^1 with

$$\varphi(0) = 0 \quad\text{and}\quad \Phi_\kappa(y,\varphi(y)) = 0. \tag{9.25}$$

Let us notice that by applying an appropriate version of the Implicit Function Theorem one can obtain an explicit estimate of the number r and thus show that it depends only on the numbers λ', μ', γ, α, κ, and C. More precisely, the following statement holds.

LEMMA 9.3 (Implicit Function Theorem). *Let E_1, E_2, and G be Banach spaces and $g\colon A_1 \times A_2 \to G$ a C^1 map, where $A_i \subset E_i$ is a ball centered at 0 of radius R_i for $i = 1, 2$. Assume that $g(0,0) = 0$, and that the partial derivative (over the second coordinate) $D_2g(0,0)\colon E_2 \to G$ is a linear homeomorphism. Assume also that Dg is Hölder continuous in $A_1 \times A_2$ with Hölder constant a and Hölder exponent α. Let B be a ball in E_1 centered at 0 of radius r_0 where*

$$r_0 = \min\left\{R_1, R_2, \frac{R_2}{2bc}, \frac{1}{(1+2bc)(2ac)^{1/\alpha}}\right\}.$$

Set

$$b = \max_{x\in A_1}\|D_1g(x,0)\|, \quad c = \|(D_2g(0,0))^{-1}\|.$$

Then there exists a unique map $u\colon B \to A_2$ satisfying the following properties:

1. *u is of class $C^{1+\alpha}$, $g(x,u(x)) = 0$ for every $x \in B$, and $u(0) = 0$;*
2. *if $x_1, x_2 \in B$ then*
$$\left\|\frac{du}{dx}(x_1) - \frac{du}{dx}(x_2)\right\| \le 8ac(1+2bc)^2\|x_1 - x_2\|^\alpha;$$
3. *if $x \in B$ then*
$$\left\|\frac{du}{dx}(x)\right\| \le 1 + 2bc.$$

PROOF OF THE LEMMA. See [**27**]. □

Let us notice that the map Φ_κ satisfies the conditions of Lemma 9.3 with

$$c = 1, \quad b = 1, \quad a = CM. \tag{9.26}$$

To show this we observe that the map $\partial_z\Phi_\kappa$ is Hölder continuous. Indeed, by (9.24),

$$\begin{aligned}\|\partial_z\Phi_\kappa(y_1,z_1) - \partial_z\Phi_\kappa(y_2,z_2)\| &\le \|\partial_z\Phi_\kappa(y_1,z_1) - \partial_z\Phi_\kappa(y_1,z_2)\| \\ &\quad + \|\partial_z\Phi_\kappa(y_1,z_2) - \partial_z\Phi_\kappa(y_2,z_2)\| \\ &= 2\|\mathcal{A}_\kappa(z_1) - \mathcal{A}_\kappa(z_2)\| \le CM\|z_1 - z_2\|_\kappa{}^\alpha.\end{aligned}$$

We now describe some properties of the map φ. Differentiating the second equality in (9.25) with respect to y we obtain

$$d\varphi(y) = -[\partial_z\Phi_\kappa(y,\varphi(y))]^{-1}\partial_y\Phi_\kappa(y,\varphi(y)).$$

Setting $y=0$ in this equality yields

$$d\varphi(0)(m)=\left(\prod_{i=0}^{m-1}A_i,0\right). \tag{9.27}$$

One can write the vector $\varphi(y)(m)$ in the form

$$\varphi(y)(m)=(\varphi_1(y)(m),\varphi_2(y)(m)),$$

where $\varphi_1(y)(m)\in\mathbb{R}^k$ and $\varphi_2(y)(m)\in\mathbb{R}^{p-k}$. It follows from (9.25) that if $m\geq 0$ then

$$\varphi_1(y)(m)=\left(\prod_{i=0}^{m-1}A_i\right)y+\sum_{n=0}^{m-1}\left(\prod_{i=n+1}^{m-1}A_i\right)g_n(\varphi(y)(n)), \tag{9.28}$$

and

$$\varphi_2(y)(m)=-\sum_{n=0}^{\infty}\left(\prod_{i=0}^{n}B_{i+m}\right)^{-1}h_{n+m}(\varphi(y)(n+m)). \tag{9.29}$$

These equalities imply that

$$\varphi_1(y)(m+1)=A_m\varphi_1(y)(m)+g_m(\varphi_1(y)(m),\varphi_2(y)(m)),$$
$$\varphi_2(y)(m+1)=B_m\varphi_2(y)(m)+h_m(\varphi_1(y)(m),\varphi_2(y)(m)).$$

Indeed, iterating the first equality "forward" one easily obtains (9.28). Rewriting the second equality in the form

$$\varphi_2(y)(m)=B_m^{-1}\varphi_2(y)(m+1)-B_m^{-1}h_m(\varphi_1(y)(m),\varphi_2(y)(m)$$

and iterating it "backward" yields (9.29).

This, we obtain that the function $\varphi(y)$ is invariant under the family of maps $\widetilde{F}_m$, i.e.,

$$\widetilde{F}_m(\varphi(y)(m))=\varphi(y)(m+1). \tag{9.30}$$

The desired map ψ^s is now defined by

$$\psi^s(v)=\varphi_2(v)(0)$$

for each $v\in B^s(r)$. Note that $\varphi_1(v)(0)=v$. It follows from (9.25), (9.27), and Lemma 9.3 that the map ψ^s satisfies Statement 1 of the theorem. It follows from (9.30) that

$$\prod_{i=0}^{m-1}\widetilde{F}_i(v,\psi^s(v))=\prod_{i=0}^{m-1}\widetilde{F}_i(\varphi_1(v)(0),\varphi_2(v)(0))$$
$$=\prod_{i=0}^{m-1}\widetilde{F}_i(\varphi(v)(0)=\varphi(v)(m).$$

Applying Lemma 9.3 and (9.26) we find that

$$\left\|\prod_{i=0}^{m-1}\widetilde{F}_i(v,\psi^s(v))\right\|'\leq\kappa^m\|\varphi(v)\|_\kappa=\kappa^m\|\varphi(v)-\varphi(0)\|_\kappa$$
$$\leq\kappa^m\sup_{\xi\in B^s(r)}\|d\varphi(\xi)\|_\kappa\|v\|'\leq 3\kappa^m\|v\|'\leq 3\kappa^m\|(v,\psi^s(v))\|'$$

(we use here the fact that for every $v = (v_1, v_2) \in \mathbb{R}$ we have $\|v_1\|, \|v_2\| \le \|v\|$). This proves Statement 3. Furthermore, for any $u_1, u_2 \in B^s(r)$ we have in view of Lemma 9.3 and (9.26) that D depends only on the numbers λ', μ', γ, α, κ, and C, and that

$$\begin{aligned}\|d\psi^s(v_1) - d\psi^s(v_2)\|' &= \|d\varphi_2(v_1)(0) - d\varphi_2(v_2)(0)\|' \\ &\le \|d\varphi(v_1)(0) - d\varphi(v_2)(0)\|' \le 72CM(\|v_1 - v_2\|')^\alpha.\end{aligned}$$

This establishes Statements 2 and 5. Let $(v, w) \in B^s(r) \times B^u(r_0)$ satisfies the assumptions of Statement 4 in the theorem. Set

$$z(m) = \left(\prod_{i=0}^{m-1} \widetilde{F}_i(v, w)\right).$$

It follows that $z \in \Gamma_\kappa$ (with $\|z\|_\kappa \le K$) and that $\Phi_k(v, z) = 0$. The uniqueness of the the map φ implies that $z = \varphi(v)$ and hence,

$$w = \varphi_2(v)(0) = \psi^s(v).$$

This establishes Statement 4. The desired result now follows. □

REMARK 9.4. Theorem 9.1 holds for nonuniformly partially hyperbolic sets. The proof does not require any changes.

10. Basic Properties of Local Stable and Unstable Manifolds

In the following series of remarks we describe some basic properties of local manifolds.

REMARK 10.1. In the hyperbolic theory there is a symmetry between the objects marked by the index "s" and those marked by the index "u". Namely, when the time direction is reversed the statements concerning objects with index "s" become the statements about the corresponding objects with index "u". In particular, this allows one to define a *local unstable manifold* $V^u(x)$ at a point x as a local stable manifold for f^{-1}. Its properties are similar to those of $V^s(x)$.

REMARK 10.2. It follows from Statement 5 of Theorem 9.2 and Condition (N3) that the sizes of local stable and unstable manifolds are bounded from below on any regular set $\mathcal{R}^\ell$, i.e.,

$$r(x) \ge r_\ell > 0 \text{ for } x \in \mathcal{R}^\ell,$$

where r_ℓ depends only on ℓ. Moreover, local stable and unstable manifolds depend continuously on $x \in \overline{\mathcal{R}^\ell}$ in the C^1 topology, i.e., if $x_n \in \overline{\mathcal{R}^\ell}$ is a sequence of points converging to x then the sequence of local stable manifolds $V^s(x_n)$ converges to $V^s(x)$ and the sequence of local unstable manifolds $V^u(x_n)$ converges to $V^u(x)$ in the C^1 topology. The existence of the manifolds $V^s(x)$ and $V^u(x)$ for every point $x \in \overline{\mathcal{R}^\ell}$ follows from Theorem 9.1 and the existence of a nonuniformly hyperbolic structure on the set $\overline{\mathcal{R}^\ell}$ (see Section 7).

In addition, there exists a number $\delta_\ell > 0$ such that for every $x \in \mathcal{R}^\ell$ and $y \in \mathcal{R}^\ell \cap B(x, \delta_\ell)$ the intersection $V^s(x) \cap V^u(y)$ is not empty and consists of a single point which depends continuously on x and y.

REMARK 10.3. It also follows from Statement 5 of Theorem 9.2 and Condition (N3) in Section 9 that the sizes of the local stable and unstable manifolds at a point $x \in \mathcal{R}$ and any point $y = f^m(x)$ for $m \in \mathbb{Z}$ along the trajectory of x are related by

$$r(f^m(x)) \geq Ke^{-\varepsilon|m|}r(x), \tag{10.1}$$

where $K > 0$ is a constant.

If μ is an invariant Borel ergodic measure for f, then for all sufficiently large ℓ the regular set $\mathcal{R}^\ell$ has positive measure. Therefore, the trajectory of almost every point visits the set $\mathcal{R}^\ell$ infinitely many times. It follows that for typical points x the function $r(f^m(x))$ is an oscillating function of m which is of the same order as $r(x)$ for many values of m. Nevertheless, for some integers m the value $r(f^m(x))$ may become as small as it is allowed by (10.1). Let us emphasize that the rate of the decrease of the size of the local stable manifolds $V^s(f^m(x))$ as $m \to \infty$ is smaller than the rate of approach of the trajectories $f^m(x)$ and $f^m(y)$ for $y \in V^s(x)$.

REMARK 10.4. Local stable and unstable manifolds depend *Hölder continuously* on $x \in \mathcal{R}^\ell$; see Property (R3) in Section 7. For every $\ell \geq 1$, $x \in \mathcal{R}^\ell$, and points z_1, $z_2 \in V^s(x)$ or z_1, $z_2 \in V^u(x)$ we have

$$d(T_{z_1}V^s(x), T_{z_2}V^s(x)) \leq C\rho(z_1, z_2)^\alpha,$$

$$d(T_{z_1}V^u(x), T_{z_2}V^u(x)) \leq C\rho(z_1, z_2)^\alpha,$$

where $C > 0$ is a constant and d is the distance in the Grassmannian bundle of TM generated by the Riemannian metric.

REMARK 10.5. One can obtain a more refined information about the smoothness of the local stable manifold. More precisely, if f is of class $C^{p+\alpha}$, with $p \geq 1$ and $0 < \alpha \leq 1$ (i.e., $d^p f$ is Hölder continuous with Hölder exponent α), then $V^s(x)$ is of class C^p; in particular, if f is of class C^p for some $p \geq 2$, then $V^s(x)$ is of class C^{p-1} (and even of class $C^{p-1+\alpha}$ for any $0 < \alpha < 1$). These results are immediate consequences of the following version of Theorem 9.2.

THEOREM 10.6. *Assume that the conditions of Theorem 9.2 hold. In addition, assume that:*

1. *g_m and h_m are of class C^p for some $p \geq 2$;*
2. *there exists a constant K such that for $\ell = 1, \dots, p$,*

$$\sup_{z \in B} \|d^\ell g_m(z)\|' \leq K\gamma^{-m}, \quad \sup_{z \in B} \|d^\ell h_m(z)\|' \leq K\gamma^{-m},$$

   where $B = B^s(r_0) \times B^u(r_0)$ (see (9.14));
3. *if z_1, $z_2 \in B$, then*

$$\|d^p g_m(z_1) - d^p g_m(z_2)\|' \leq K\gamma^{-m}(\|z_1 - z_2\|')^\alpha,$$

$$\|d^p h_m(z_1) - d^p h_m(z_2)\|' \leq K\gamma^{-m}(\|z_1 - z_2\|')^\alpha$$

   for some $\alpha \in (0, 1)$.

If $\psi^s(u)$ is the map constructed in Theorem 9.2, then there exists a number $N > 0$, depending only on the numbers λ', μ', γ, α, κ, and K, such that:

1. *ψ^s is of class $C^{p+\alpha}$;*
2. $\sup_{u \in B^s(r)} \|d^\ell \psi^s(u)\|' \leq N_\ell$ *for $\ell = 1, \dots, p$.*

PROOF. It is sufficient tc show that Φ_κ is of class C^p. Indeed, a simple modification of arguments in the proof of Theorem 9.2 allows one to show that

$$d_y^\ell \Phi_\kappa(y,z) = (0,0), \quad 2 \le \ell \le p$$

and

$$d_z^\ell \Phi_\kappa(y,z)(m) =$$

$$\left(\sum_{n=0}^{m-1}\left(\prod_{i=n+1}^{m-1} A_i\right) d^\ell g_n(z(n)), -\sum_{n=0}^{\infty}\left(\prod_{i=0}^{n} B_{i+m}^{-1}\right) d^\ell h_{m+n}(z(m+n))\right)$$

(see [**27**] for more details). □

In [**32**], Pugh and Shub strengthened the above result and showed that in fact, if f is of class C^p for some $p \ge 2$, then $V^s(x)$ is also of class C^p.

REMARK 10.7. Consider a nonuniformly hyperbolic set $\mathcal{R}$ which consists of a *single* nonuniformly hyperbolic trajectory $\{f^n x\}_{n\in\mathbb{Z}}$. The Stable Manifold Theorem 9.1 applies. It characterizes the behavior of trajectories nearby an *individual* nonuniformly hyperbolic trajectory, and does not need the presence of any other nonuniformly hyperbolic trajectories. Note also that one can replace the assumption that the manifold M is compact by the assumption that the diffeomorphism f satisfies (9.4)–(9.7) along a given nonuniformly hyperbolic trajectory.

REMARK 10.8. There is another proof of the Stable Manifold Theorem which is based on a version of the Graph Transform Property — a statement that is well-known in the uniform hyperbolic theory. This approach is essentially an elaboration of the Hadamard method.

Let $x \in \mathcal{R}$. Choose numbers r_0, b_0, c_0, and d_0 and for every $m \ge 0$, set

$$r_m = r_0 e^{-\varepsilon m}, \quad b_m = b_0 e^{-\varepsilon m}, \quad c_m = c_0 e^{-\varepsilon m}, \quad d_m = d_0 \lambda'^m e^{\varepsilon m}.$$

Consider the set Ψ of $C^{1+\alpha}$ functions on $\{(m,v) : m \in \mathbb{N}, v \in B^s(r_m)\}$ such that

$$\psi(m,v) \in E^u(f^{-m}(x)) \text{ for every } m \ge 0 \text{ and } v \in B^s(r_m)$$

(where $B^s(r_m)$ is the ball in $E^s(f^{-m}(x))$ centered at 0 of radius r_m), and satisfying the following conditions:

$$\|\psi(m,0)\| \le b_m, \qquad \max_{v\in B^s(r_m)} \|d\psi(m,v)\| \le c_m,$$

and if v_1, $v_2 \in B^s(r_m)$, then

$$\|d\psi(m,v_1) - d\psi(m,v_2)\| \le d_m \|v_1 - v_2\|^\alpha. \tag{10.2}$$

THEOREM 10.9 (Graph Transform Property). *For every $\ell \ge 1$ there are positive constants r_0, b_0, c_0, and d_0, which depend only on ℓ, such that for every $x \in \mathcal{R}^\ell$ and every function $\psi \in \Psi$ there exists a function $\widetilde{\psi} \in \Psi$ for which*

$$F_m^{-1}(\{(v,\psi(m,v)) : v \in B^s(r_m)\}) \supset \{(v,\widetilde{\psi}(m+1,v)) : v \in B^s(r_{m+1})\} \tag{10.3}$$

for all $m \in \mathbb{N}$.

SKETCH OF THE PROOF. Let $(v^{(m+1)}, w^{(m+1)}) = F_m^{-1}(v,\psi(m,v))$. One can write

$$(v^{(m+1)}, w^{(m+1)}) = (A_m^{-1}v + \widetilde{g}_m(v,\psi(m,v)), B_m^{-1}\psi(m,v) + \widetilde{h}_m(v,\psi(m,v))),$$

where (compare with Section 9)

$$\widetilde{g}_m(0,0)=0,\quad d\widetilde{g}_m(0,0)=0,\quad \widetilde{h}_m(0,0)=0,\quad d\widetilde{h}_m(0,0)=0,$$
$$\|d\widetilde{g}_m(v_1,w_1)-d\widetilde{g}_m(v_2,w_2)\|'\le C\gamma^{-m}(\|v_1-v_2\|'+\|w_1-w_2\|')^\alpha,$$
$$\|d\widetilde{h}_m(v_1,w_1)-d\widetilde{h}_m(v_2,w_2)\|'\le C\gamma^{-m}(\|v_1-v_2\|'+\|w_1-w_2\|')^\alpha$$

for some constants satisfying (9.18). We have

$$\begin{aligned}\|v_1^{(m+1)}-v_2^{(m+1)}\| &\ge (\lambda')^{-1}\|v_1-v_2\|\\ &\quad -\|\widetilde{g}_m(v_1,\psi(m,v_1))-\widetilde{g}_m(v_2,\psi(m,v_2))\|\\ &=[(\lambda')^{-1}-c(1+c_m)]\|v_1-v_2\|\end{aligned}$$

for some constant $c>0$. Therefore, by choosing r_0, b_0, $c_0>0$ sufficiently small, one can define a function $v\mapsto\widetilde{\psi}(m+1,v)$ on $B^s(r_{m+1})$ by

$$\widetilde{\psi}(m+1,v^{(m+1)})=w^{(m+1)}, \tag{10.4}$$

which satisfies (10.3). Furthermore, we have

$$\begin{aligned}\|w_1^{(m+1)}-w_2^{(m+1)}\| &\le (\mu')^{-1}c_m\|v_1-v_2\|+c(1+c_m)\|v_1-v_2\|\\ &\le \frac{(\mu')^{-1}c_m+c(1+c_m)}{(\lambda')^{-1}-c(1+c_m)}\|v_1^{(m+1)}-v_2^{(m+1)}\|.\end{aligned}$$

By eventually choosing a smaller $c_0>0$, this implies that

$$\|\widetilde{\psi}(m+1,0)\|\le b_{m+1},\qquad \max_{v\in B^s(r_{m+1})}\|d\widetilde{\psi}(m+1,v)\|\le c_{m+1},$$

Taking derivatives in (10.4) and using (10.2) one can show that if v_1, $v_2\in B^s(r_{m+1})$, then

$$\|d\widetilde{\psi}(m+1,v_1)-d\widetilde{\psi}(m+1,v_2)\|\le d_{m+1}\|v_1-v_2\|^\alpha.$$

This completes the proof of the theorem. □

REMARK 10.10. Let $E=E_1\times E_2$ be a Banach space which is the product of two Banach spaces E_1 and E_2. Let also $F_m\colon E\to E$ be a family of $C^{1+\alpha}$ maps of the form (9.14) satisfying Conditions (9.15)–(9.17). Then the statement of Theorem 9.2 holds (with the obvious modifications).

REMARK 10.11. For every $x\in\mathcal{R}$ we define the *global stable and unstable manifolds* by

$$W^s(x)=\bigcup_{n=0}^\infty f^{-n}(V^s(f^n(x))),\quad W^u(x)=\bigcup_{n=0}^\infty f^n(V^u(f^{-n}(x))). \tag{10.5}$$

They are finite-dimensional immersed smooth submanifolds (of class $C^{r+\alpha}$ if f is of class $C^{r+\alpha}$) invariant under f. They have the following properties which are immediate consequences of the Stable Manifold Theorem 9.2.

THEOREM 10.12. *If x, $y\in\mathcal{R}$, then:*

1. *$W^s(x)\cap W^s(y)=\varnothing$ if $y\notin W^s(x)$; $W^u(x)\cap W^u(y)=\varnothing$ if $y\notin W^u(x)$;*
2. *$W^s(x)=W^s(y)$ if $y\in W^s(x)$; $W^u(x)=W^u(y)$ if $y\in W^u(x)$;*
3. *for every $y\in W^s(x)$ (or $y\in W^u(x)$) we have that $\rho(f^n(x),f^n(y))\to 0$ as $n\to+\infty$ (respectively, $n\to-\infty$) with an exponential rate.*

We emphasize that for dynamical systems satisfying uniform hyperbolicity conditions (Anosov diffeomorphisms or Axiom A diffeomorphisms) the global stable and unstable manifolds have the following characterization:

$$W^s(x) = \{y \in M : \rho(f^n(x), f^n(y)) \to 0 \text{ as } n \to +\infty\},$$
$$W^u(x) = \{y \in M : \rho(f^n(x), f^n(y)) \to 0 \text{ as } n \to -\infty\}.$$

This characterization may not hold for dynamical systems satisfying nonuniform hyperbolicity conditions

REMARK 10.13. Let φ_t be a smooth flow on a compact smooth Riemannian manifold M. The following is an analog of Theorem 9.1 for flows on nonuniformly hyperbolic sets (see Section 7).

THEOREM 10.14 (Stable Manifold Theorem for Flows). *Let $\mathcal{R}$ be a nonuniformly hyperbolic set for a $C^{1+\alpha}$ flow φ_t. Then for every $x \in \mathcal{R}$ there exists a local stable manifold $V^s(x)$ such that $x \in V^s(x)$, $T_xV^s(x) = E^s(x)$, and if $y \in V^s(x)$ and $t > 0$ then*

$$\rho(\varphi_t(x), \varphi_t(y)) \leq T(x)\lambda^t e^{\varepsilon t}\rho(x, y), \tag{10.6}$$

where $T\colon \mathcal{R} \to (0, \infty)$ is a Borel function such that if $s \in \mathbb{R}$ then

$$T(\varphi_s(x)) \leq T(x)e^{10\varepsilon|s|}.$$

The proof of Theorem 10.14 can be obtained by applying Theorem 9.1 (see also Remark 9.4) to the diffeomorphism $f = \varphi_1$ and verifying that the local stable manifold obtained in this way satisfies (10.6). We call $V^s(x)$ a *local stable manifold* at x. In a similar fashion, by reversing the time one can show that there exists a *local unstable manifold* $V^u(x)$ at x such that $T_xV^u(x) = E^u(x)$.

For every $x \in \mathcal{R}$ we define the *global stable* and *unstable manifolds* at x by

$$W^s(x) = \bigcup_{t>0} \varphi_{-t}(V^s(\varphi_t(x))), \quad W^u(x) = \bigcup_{t>0} \varphi_t(V^u(\varphi_{-t}(x))). \tag{10.7}$$

These are finite-dimensional immersed smooth submanifolds (of class $C^{r+\alpha}$ if φ_t is of class $C^{r+\alpha}$). They satisfy Properties 1 and 2 in Theorem 10.12. Furthermore, for every $y \in W^s(x)$ (or $y \in W^u(x)$) we have $\rho(\varphi_t(x), \varphi_t(y)) \to 0$ as $t \to +\infty$ (respectively, $t \to -\infty$) with an exponential rate.

We also define the *global weakly stable* and *unstable manifolds* at x by

$$W^{s0}(x) = \bigcup_{t\in\mathbb{R}} W^s(\varphi_t(x)), \quad W^{u0}(x) = \bigcup_{t\in\mathbb{R}} W^u(\varphi_t(x)).$$

It follows from (10.7) that

$$W^{s0}(x) = \bigcup_{t\in\mathbb{R}} \varphi_t(W^s(x)), \quad W^{u0}(x) = \bigcup_{t\in\mathbb{R}} \varphi_t(W^u(x)).$$

We remark that each of these two families of invariant immersed manifolds forms a partition of the set $\mathcal{R}$.

11. Absolute Continuity. Holonomy Map

Let f be a $C^{1+\alpha}$ diffeomorphism of a compact smooth Riemannian manifold M. We describe one of the most crucial properties of local stable and unstable manifolds which is known as *absolute continuity.*

Let $\mathcal{R}$ be the set of nonuniformly hyperbolic points for f and $\{\mathcal{R}^\ell : \ell \geq 1\}$ the associated collection of regular sets (see Section 7). We assume that $\mathcal{R}$ is not empty. Without loss of generality we may assume that each set $\mathcal{R}^\ell$ is compact (otherwise we can replace them with sets $\mathcal{Q}^\ell = \overline{\mathcal{R}^\ell}$). We have $\mathcal{R}^\ell \subset \mathcal{R}^{\ell+1}$ for every ℓ. Furthermore, the stable and unstable subspaces $E^s(x)$ and $E^u(x)$ as well as the local stable manifolds $V^s(x)$ and local unstable manifolds $V^u(x)$ depend continuously on $x \in \mathcal{R}^\ell$ and their sizes are bounded away from zero by a number r_ℓ (see Remark 10.2).

Fix $x \in \mathcal{R}^\ell$, a number r, $0 < r \leq r_\ell$ and set

$$Q^\ell(x) = \bigcup_{w \in \mathcal{R}^\ell \cap B(x,r)} V^s(w), \tag{11.1}$$

where $B(x,r)$ is the ball at x of radius r.

Consider a local open submanifold W which is *uniformly* transverse to the family of local stable manifolds $\mathcal{L}(x) = \{V^s(w) : w \in \mathcal{R}^\ell \cap B(x,r)\}$. If r is sufficiently small (in accordance with Remarks 10.3 and 10.2) the latter can be assured provided that W is chosen so that the set $\exp_x^{-1} W$ is the graph of a smooth map $\psi\colon B^u(q) \subset E^u(x) \to E^s(x)$, for some q, with a sufficiently small C^1 norm. In this case W intersects each local stable manifold $V^s(w) \in \mathcal{L}(x)$ and this intersection is transverse. We will consider only local open submanifolds constructed in this way and call them *transversals to the family* $\mathcal{L}(x)$. We also say that the map ψ *represents* W.

Let W^1 and W^2 be two transversals to the family $\mathcal{L}(x)$. We define the *holonomy map*

$$\pi\colon Q^\ell(x) \cap W^1 \to Q^\ell(x) \cap W^2$$

using the relation

$$\pi(y) = W^2 \cap V^s(w), \text{ where } y = W^1 \cap V^s(w) \text{ and } w \in Q^\ell(x) \cap B(x,r).$$

See Figure 2. Note that the holonomy map π is a homeomorphism onto its image.

We set

$$\Delta = \Delta(W^1, W^2) = \|\psi^1\|_{C^1} + \|\psi^2\|_{C^1},$$

where the maps ψ^1 and ψ^2 represent W^1 and W^2 respectively.

We recall that if ν and μ are two measures on a measurable space X than ν is said to be *absolutely continuous* with respect to μ if for every $\varepsilon > 0$ there exists $\delta > 0$ such that $\nu(E) < \varepsilon$ for every measurable set E for which $\mu(E) < \delta$. A measurable invertible transformation $T\colon X \to Y$ of measurable spaces (X, ν) and (Y, μ) is said to be *absolutely continuous* if the measure μ is absolutely continuous with respect to the measure $T_*\nu$. In this case one defines the *Jacobian* $J(T)(x)$ of T at a point $x \in X$ (specified by the measures ν and μ) to be the Radon–Nikodym derivative $\frac{d\mu}{d(T_*\nu)}$. If X is a metric space with metric ρ then for μ-almost every $x \in X$ one has

$$J(T)(x) = \lim_{r \to 0} \frac{\mu(T(B(x,r)))}{\nu(B(x,r))}. \tag{11.2}$$

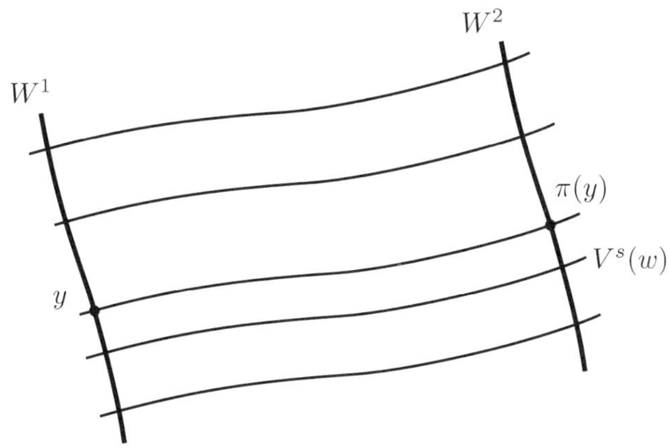

FIGURE 2. Family $\mathcal{L}(x)$ of local stable leaves $V^s(w)$ for each $w \in \mathcal{R}^\ell \cap B(x,r)$ and transversals W^1 and W^2

Given a submanifold W in M we denote by ν_W the Riemannian volume on W induced by the restriction of the Riemannian metric to W. We denote by $J^s(\pi)(y)$ the Jacobian of the holonomy map π at the point $y \in Q^\ell(x) \cap W^1$ specified by the measures ν_{W^1} and ν_{W^2}.

THEOREM 11.1 (Absolute Continuity; see [**27**]). *Given $\ell \geq 1$, $x \in \mathcal{R}^\ell$, and transversals W^1 and W^2 to the family $\mathcal{L}(x)$, the holonomy map π is absolutely continuous (with respect to the measures ν_{W^1} and ν_{W^2}) and the Jacobian $J^s(\pi)(y)$ is bounded from above and bounded away from zero.*

The first proof of the absolute continuity for Anosov diffeomorphism was obtained by Anosov in [**1**] (see also [**2**]). For nonuniformly hyperbolic systems the absolute continuity property was established by Pesin in [**27**]. We shall present a proof of Theorem 11.1 which is an elaboration of the proof of Pesin. In [**15**], Katok and Strelcyn provided an extension of Pesin's approach to smooth maps with singularities. Pugh and Shub [**32**] presented a proof of the absolute continuity of the holonomy map, modifying their approach for the horocycle foliations of Anosov actions [**31**]. Liu and Qian, [**18**] established a version of Theorem 11.1 for random diffeomorphisms following essentially the approach of Pesin.

We remark that the absolute continuity of the holonomy map does not imply directly that the Jacobian of the holonomy map is bounded (to prove this one needs the Hölder property of the family of local stable manifolds or an equivalent statement).

In the appendix to [**4**], Brin provided a detailed presentation of the proof of absolute continuity for geodesic flows on manifolds of strictly negative sectional curvature. The presentation can be considered as a distillation of the original approach by Anosov and Sinai in [**1, 2**]. In [**21**], Mañé presented a proof of the absolute continuity for Anosov diffeomorphisms, although his arguments contain a serious gap.

REMARK 11.2. It is easy to see that the holonomy map π transfers the measure $\nu_{W^1}|Q^\ell(x) \cap W^1$ into a measure which is indeed, *equivalent* to the measure $\nu_{W^2}|Q^\ell(x) \cap W^2$.

REMARK 11.3. The family of local unstable manifolds also satisfies the absolute continuity property.

REMARK 11.4. It follows from Theorem 11.1 that the stable and unstable foliations of Anosov diffeomorphisms have the absolute continuity property. The same holds true for the stable and unstable foliations of partially hyperbolic diffeomorphisms. Note that these foliations are C^r-continuous but in general, not smooth (in the latter case Theorem 11.1 will be a trivial corollary of the Fubini theorem).

REMARK 11.5. A. Katok constructed an example of a partially hyperbolic volume-preserving diffeomorphism f of a compact manifold whose central distribution is integrable. The corresponding central foliation is invariant and continuous but not absolutely continuous. The leaves of the foliation are smooth compact submanifolds (see the description of this example in [**23**]). The map f is not ergodic and the Lyapunov exponent in the central direction is zero. Shub and Wilkinson (see [**36**]) constructed an open set of partially hyperbolic (but nonuniformly hyperbolic) volume-preserving ergodic diffeomorphisms of the three-dimensional torus whose central foliation is continuous but not absolutely continuous. The leaves of this foliation are diffeomorphic to the unit circle and the Lyapunov exponent in the central direction is positive almost everywhere.

PROOF OF THEOREM 11.1. We split the proof into several steps.

Step 1. Fix a point $w \in \mathcal{R}^\ell \cap B(x,r)$ and let $y_i = V^s(w) \cap W^i$ for $i = 1, 2$. Choose $m \geq 0$, $q > 0$ and set for $i = 1, 2$

$$W^i_m = f^m(W^i), \quad w_m = f^m(w), \quad y^i_m = f^m(y_i), \quad q_m = qe^{\varepsilon m}. \tag{11.3}$$

Note that $W^i_0 = W^i$, $w_0 = w$, $y^i_0 = y^i$, and $q_0 = q$. Consider the open smooth submanifolds W^1_m and W^2_m and the point w_m. In view of Remark 10.5 for $i = 1, 2$, there exists an open neighborhood $W^i_m(w,q) \subset W^i_m$ of the point y^i_m such that

$$W^i_m(w,q) = \exp_{w_m}\{(\psi^i_m(v), v) \colon v \in B^u(q_m)\}. \tag{11.4}$$

If $q = q(m)$ is sufficiently small than for any $w \in \mathcal{R}^\ell \cap B(x,r)$ and $k = 0, \ldots, m$, we have that

$$f^{-1}(W^i_k(w,q)) \subset W^i_{k-1}(w,q),\ i = 1,\ 2. \tag{11.5}$$

We wish to compare the measures $\nu_{W^1_m}|W^1_m(w,q)$ and $\nu_{W^2_m}|W^2_m(w,q)$ for sufficiently large m.

LEMMA 11.6. *There exists a constant $K_1 > 0$ such that the following holds: for any $m > 0$ there exists $q_0 = q_0(m) > 0$ such that for any $0 < q \leq q_0$, we have*

$$\left|\frac{\nu_{W^1_m}(W^1_m(w,q))}{\nu_{W^2_m}(W^2_m(w,q))} - 1\right| \leq K_1\Delta.$$

PROOF OF THE LEMMA. The result follows from Theorem 10.9. □

Step 2. We continue with the following Covering Lemma.

LEMMA 11.7. *For any $m > 0$ there are points $w_j \in \mathcal{R}^\ell \cap B(x,r)$, $j = 1, \ldots,$ $p = p(m)$ and $q = q(m) > 0$ such that the sets $W^i_m(w_j, q)$ form an open cover of the set $f^m(Q^\ell(x) \cap W^i)$ (see (11.1)) of finite multiplicity which depends only on the dimension of W^i for $i = 1, 2$.*

PROOF OF THE LEMMA. We recall the Besicovich Covering Lemma. It states that for each $Z \subset \mathbb{R}^k$, if $r\colon Z \to \mathbb{R}^+$ is a bounded function, then the cover $\{B(x, r(x)) : x \in Z\}$ of Z contains a countable (Besicovich) subcover of finite multiplicity depending only on k. This statements readily extends to smooth manifolds via Whitneys's Embedding Theorem.

It follows from the definition of sets $W_m^i(w, q)$ (see (11.4)) that for each sufficiently small $q = q(m) > 0$ there is a number $R = R(\ell, m) > 0$ such that for every $w \in \mathcal{R}^\ell \cap B(x, r)$,

$$B_m^i(w, \frac{1}{2}R) \subset W_m^i(w, q) \subset B_m^i(w, R),$$

where $B_m^i(w, R)$ is the ball in W_m^i centered at w of radius R. Consider the cover of the set $Z^i = f^m(Q^\ell(x) \cap W^i)$ by balls $\{B_m^i(w, \frac{1}{2}R),\ w \in \mathcal{R}^\ell \cap B(x, r)\}$. Applying the Besicovich Covering Lemma we obtain a Besicovich subcover $\{B_m^i(w_j, \frac{1}{2}R),\ j = 1, \dots, p\}$, $p = p(m)$ of the set Z^i of finite multiplicity M (which does not depend on m). The sets $W_m^i(w_j, q)$ also cover Z^i and the multiplicity M_1 of this cover does not depend on m. Indeed, M_1 does not exceed the multiplicity M_2 of the cover of Z^i by balls $B_m^i(w_j, R)$. Note that every ball $B_m^i(w_j, R)$ can be covered by not more than CM balls $B_m^i(w, \frac{1}{2}R)$ where C depends only on the dimension of W^i. Furthermore, given w we have $w_j \in B_m^i(w, \frac{1}{2}R)$ for at most a number M of points w_j (otherwise at least $M+1$ balls $B_m^i(w_j, \frac{1}{2}R)$ would contain w). Therefore $w_j \in B_m^i(w, R)$ for at most CM points w_j. This implies that $M_2 \le CM$. □

Step 3. We now compute the measures that are the pullbacks under f^{-m} of the measures $\nu_{W_m^i} | W_m^i(w, q)$ for $i = 1, 2$. Note that for every $w \in \mathcal{R}^\ell \cap B(x, r)$ the points y_m^1 and y_m^2 (see (11.3)) lie on the stable manifold $V^s(w_m)$. Choose $z \in f^{-m}(W_m^i(w, q))$ (with $i = 1$ or $i = 2$). Set $z_m = f^m(z)$ and

$$T^i(z, m) = J(d_{z_m} f^{-m} | T_{z_m} W_m^i(w, q)).$$

We need the following lemma.

LEMMA 11.8. *There exist $K_2 > 0$ and $m_1(\ell) > 0$ such that for every $w \in \mathcal{R}^\ell \cap B(x, r)$ and $m \ge m_1(\ell)$ one can find $q = q(m)$ such that*

$$\left| \frac{T^2(y_m^2, m)}{T^1(y_m^1, m)} - 1 \right| \le K_2 \Delta,$$

and if $z \in W_0^1(w, q)$ then

$$\left| \frac{T^1(z_m, m)}{T^1(y_m^1, m)} - 1 \right| \le K_2 \Delta.$$

PROOF OF THE LEMMA. For any $0 < k \le m$ and $z \in W_0^1(w, q)$ we set

$$V(k, z) = T_{z_k} W_k^1(w, q).$$

We transport parallelly the space $V(k, z) \subset T_{z_k} M$ along the geodesic that connects the points z_k and y_k^1 (this geodesic is uniquely defined since these points are sufficiently close to each other). We obtain a new subspace $\widetilde{V}(k, z) \subset T_{y_k^1} M$ and we have

$$\begin{aligned}&\left|J(d_{z_k}f^{-1}|V(k,z))-J(d_{y_k^1}f^{-1}|V(k,y_1))\right|\\&\leq\left|J(d_{z_k}f^{-1}|V(k,z))-J(d_{y_k^1}f^{-1}|\widetilde{V}(k,z))\right|\\&\quad+\left|J(d_{y_k^1}f^{-1}|\widetilde{V}(k,z))-J(d_{y_k^1}f^{-1}|V(k,y_1))\right|.\end{aligned}$$

Since $f\in C^{1+\alpha}$ in view of (11.5), we obtain

$$\begin{aligned}&\left|J(d_{z_k}f^{-1}|V(k,z))-J(d_{y_k^1}f^{-1}|V(k,y_1))\right|\\&\leq C_1\rho(z_k,y_k^1)^\alpha+C_2d(V(k,z),V(k,y_1)),\end{aligned}$$

where $C_1>0$ and $C_2>0$ are constants (recall that ρ is the distance in M and d is the distance in the Grassmannian bundle of TM generated by the Riemannian metric). It follows from Remark 10.4 (see also Property (R3) in Section 7) that

$$d(V(k,z),V(k,y_1))\leq C_3\rho(z_k,y_k^1)^\alpha,$$

where $C_3>0$ is a constant. This and (9.1) imply that

$$\left|J(d_{z_k}f^{-1}|V(k,z))-J(d_{y_k^1}f^{-1}|V(k,y_1))\right|\leq C_4(\lambda^ke^{\varepsilon k}\rho(z,y_1))^\alpha,$$

where $C_4>0$ is a constant. Note that for any $0<k\leq m$ and $z\in W_0^1(w,q)$, we have

$$C_5{}^{-1}\leq|J(d_{z_k}f^{-1}|V(k,z))|\leq C_5,$$

where $C_5>0$ is a constant. We have

$$\begin{aligned}\frac{T^1(z_m,m)}{T^1(y_m^1,m)}&=\prod_{k=1}^m\frac{J(d_{z_k}f^{-1}|V(k,z))}{J(d_{y_k^1}f^{-1}|V(k,y_1))}\\&=\exp\sum_{k=1}^m\log\frac{J(d_{z_k}f^{-1}|V(k,z))}{J(d_{y_k^1}f^{-1}|V(k,y_1))}\\&\leq\exp\sum_{k=1}^m\left(\frac{J(d_{z_k}f^{-1}|V(k,z))}{J(d_{y_k^1}f^{-1}|V(k,y_1))}-1\right)\\&\leq\exp\left(\sum_{k=1}^mC_4C_5(\lambda^ke^{\varepsilon k}\rho(z,y_1))^\alpha\right)\\&\leq\exp\left(\frac{C_4C_5(\lambda e^\varepsilon\rho(z,y_1))^\alpha}{1-(\lambda e^\varepsilon)^\alpha}\right).\end{aligned}$$

The last expression can be made arbitrarily close to 1 by choosing q sufficiently small. This proves the second inequality. The first inequality can be proven in a similar fashion. □

Lemma 11.8 allows one to compare the measures of the preimages under f^{-m} of $W_m^1(w,q)$ and $W_m^2(w,q)$. More precisely, the following statement holds.

LEMMA 11.9. *There exist $K_3>0$ and $m_2(\ell)>0$ such that if $w\in\mathcal{R}^\ell\cap B(x,r)$ and $m\geq m_2(\ell)$, then one can find $q=q(m)$ such that*

$$\left|\frac{\nu_{W^1}(f^{-m}(W_m^1(w,q)))}{\nu_{W^2}(f^{-m}(W_m^2(w,q)))}-1\right|\leq K_3\Delta.$$

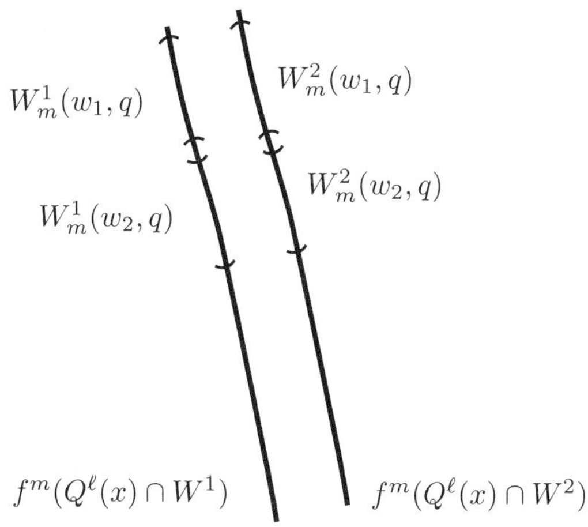

FIGURE 3. Sets $f^m(Q^\ell(x)\cap W^i)$ and their covers by sets $W^i_m(w_j,q)$ for $i=1,2$

PROOF OF THE LEMMA. For $i=1,2$ we have

$$\begin{aligned}\nu_{W^i}(f^{-m}(W^i_m(w,q))) &= \int_{W^i_m(w,q)} T^i(z,m)\,d\nu_{W^i_m}(z)\\ &= T^i(z^i_m,m)\nu_{W^i_m}(W^i_m(w,q)),\end{aligned}$$

where $z^i_m\in W^i_m(w,q)$ are some points. It follows from the assumptions of the lemma, (11.5), and Lemmas 11.6 and 11.8 that for sufficiently large m and small q (assuming without loss of generality that $\Delta\le 1$),

$$\begin{aligned}&\left|\frac{\nu_{W^1}(f^{-m}(W^1_m(w,q)))}{\nu_{W^2}(f^{-m}(W^2_m(w,q)))}-1\right|\\ &\le\left|\frac{T^1(z^1_m,m)}{T^2(z^2_m,m)}-1\right|\times\frac{\nu_{W^1_m}(W^1_m(w,q))}{\nu_{W^2_m}(W^1_m(w,q))}+\left|\frac{\nu_{W^1_m}(W^1_m(w,q))}{\nu_{W^2_m}(W^2_m(w,q))}-1\right|\\ &\le\left|\frac{T^1(z^1_m,m)}{T^1(y^1_m,m)}\frac{T^2(y^2_m,m)}{T^2(z^2_m,m)}\frac{T^1(y^1_m,m)}{T^2(y^2_m,m)}-1\right|\times(1+K_1\Delta)+K_1\Delta\\ &\le((1+K_2\Delta)^3-1)(1+K_1\Delta)+K_1\Delta\le K_3\Delta.\end{aligned}$$

The lemma follows. □

Step 4. Given $m>0$, choose points $w_j\in\mathcal{R}^\ell\cap B(x,r)$ and number $q=q(m)$ as in Lemma 11.7. Consider two sets

$$\hat W^1_m=\bigcup_{j=1}^p W^1_m(w_j,q),\quad \hat W^2_m=\bigcup_{j=1}^p W^2_m(w_j,q).$$

Note that $\hat W^i_m\supset f^m(Q^\ell(x)\cap W^i)$. See Figure 3.

We wish to compare the measures $\nu_{W^1}|f^{-m}(\hat W^1_m)$ and $\nu_{W^2}|f^{-m}(\hat W^2_m)$. Let L_i be the multiplicities of the covers $\{W^i_m(w_j,q)\}$, $i=1,2$ constructed in Lemma 11.7. Observe that the cover of the set $Q^\ell(x)\cap W^i$ by the sets $\{f^{-m}(W^i_m(w_j,q))\}$ is also of multiplicity L_i. Set $L = \max\{L_1, L_2\}$. We have

$$\frac{1}{L}\sum_{j=1}^{p} \nu_{W^i}(f^{-m}(W^i_m(w_j,q))) \le \nu_{W^i}(\hat{f}^{-m}(W^i_m))$$
$$\le \sum_{j=1}^{p} \nu_{W^i}(f^{-m}(W^i_m(w_j,q))).$$

It follows from Lemma 11.9 that

$$\begin{aligned}\frac{\nu_{W^1}(f^{-m}(\hat{W}^1_m))}{\nu_{W^2}(f^{-m}(\hat{W}^2_m))} &\le L\frac{\sum_{j=1}^{p}\nu_{W^1}(f^{-m}(W^1_m(w_j,q)))}{\sum_{j=1}^{p}\nu_{W^2}(f^{-m}(W^2_m(w_j,q)))}\\ &\le L\max\left\{\frac{\nu_{W^1}(f^{-m}(W^1_m(w_j,q)))}{\nu_{W^2}(f^{-m}(W^2_m(w_j,q)))} : j=1,\dots,p\right\}\\ &\le L\,(1+K_3\Delta),\end{aligned} \tag{11.6}$$

with a similar bound for the inverse ratio. We conclude that

$$K_4^{-1} \le \frac{\nu_{W^1}(f^{-m}(\hat{W}^1_m))}{\nu_{W^2}(f^{-m}(\hat{W}^2_m))} \le K_4, \tag{11.7}$$

where $K_4 > 0$ is a constant independent of m.

Step 5. Without loss of generality we may assume that $\nu_{W^i}(Q^\ell(x)\cap W^i) > 0$. Given $\beta > 0$, denote by U^i_β the β-neighborhood of the set $Q^\ell(x)\cap W^i$ for $i = 1, 2$. We need the following lemma from measure theory.

LEMMA 11.10. *There exists $\beta_0 > 0$ such that for every $0 < \beta \le \beta_0$ and $i = 1$, 2, we have*

$$1-\Delta \le \frac{\nu_{W^i}(Q^\ell(x)\cap W^i)}{\nu_{W^i}(U^i_\beta)}.$$

For any $\beta > 0$ and any sufficiently large $m > 0$ we have $Q^\ell(x)\cap W^i \subset f^{-m}(\hat{W}^i_m) \subset U^i_\beta$ for $i = 1, 2$. It follows from Lemma 11.10 and (11.7) that

$$\frac{1-\Delta}{K_4} \le \frac{\nu_{W^1}(Q^\ell(x)\cap W^1)}{\nu_{W^2}(Q^\ell(x)\cap W^2)} \le \frac{K_4}{1-\Delta}.$$

We emphasize that the constant K_4 does not depend on the size of the transversals W^1 and W^2 but only on their dimension. In particular, for each $y \in Q^\ell(x)\cap W^1$ we have

$$\frac{1-\Delta}{K_4} \le \frac{\nu_{W^2}(\pi(Q^\ell(x)\cap B^1(y,r)))}{\nu_{W^1}(Q^\ell(x)\cap B^1(y,r))} \le \frac{K_4}{1-\Delta},$$

where $B^1(y,r) \subset W^1$ is the ball of radius r centered at y. Letting $r \to 0$ we conclude from (11.2) that

$$\frac{1-\Delta}{K_4} \le J^s(\pi)(y) \le \frac{K_4}{1-\Delta}$$

for ν_{W^1}-almost every $y \in W^1$. Therefore the Jacobian is bounded and is bounded away from zero. This completes the proof of the theorem. ☐

REMARK 11.11. One can strengthen the statement of Lemma 11.7 and for each $i = 1, 2$, construct a cover of the set $f^m(Q^\ell(x)\cap W^i)$ by the sets $W^i_m(w_j,q)$ for $j = 1, \dots, p$ such that this cover is of multiplicity 1 on a subset D_i whose measure is arbitrarily close to the measure of the set $f^m(Q^\ell(x)\cap W^i)$ (see [**27**] for details).

One can now strengthen the inequality (11.7) (setting $L = 1$ in (11.6)) and prove that

$$\left|\frac{\nu_{W^1}(f^{-m}(\hat{W}^1_m))}{\nu_{W^2}(f^{-m}(\hat{W}^2_m))} - 1\right| \leq K_5\Delta(W^1, W^2) \tag{11.8}$$

for all sufficiently large m. Here $K_5 > 0$ is a constant which depends only on ℓ. It follows from (11.8) that the Jacobian $J^s(\pi)(y)$ for $y \in Q^\ell(x) \cap W^1$ satisfies

$$|J^s(\pi)(y) - 1| \leq K_5\Delta(W^1, W^2). \tag{11.9}$$

Moreover, one can obtain the following formula for the Jacobian of the holonomy map:

$$J^s(\pi)(y) = \prod_{k=0}^{\infty} \frac{J(d_{f^k(y)}f|E^s(f^k(y))}{J(d_{f^k(x)}f|E^s(f^k(x))}.$$

12. Absolute Continuity and Smooth Invariant Measures

One of the main manifestations of absolute continuity is the description of the ergodic properties of diffeomorphisms preserving smooth measures. In order to achieve this goal we provide in this section a more detailed description of absolute continuity property of local stable and unstable manifolds for diffeomorphisms preserving smooth measures.

Let f be a $C^{1+\alpha}$ diffeomorphism of a smooth compact Riemannian manifold M without boundary. We assume that f preserves a smooth measure ν, i.e., a measure which is equivalent to the Riemannian volume, and that this measure is hyperbolic, and thus the set $\mathcal{R}$ of nonuniformly hyperbolic points has full measure.

Let $\mathcal{R}^\ell$ be a regular set of positive measure. Fix a Lebesgue point $x \in \mathcal{R}^\ell$ and consider the family of local stable manifolds

$$\mathcal{L}(x) = \{V^s(w) : w \in \mathcal{R}^\ell \cap B(x, r)\}$$

where $r = r(\ell)$ is sufficiently small. We denote by $\nu^s(w) = \nu_{V^s(w)}$ for $w \in \mathcal{R}^\ell \cap B(x, r)$, the Riemannian volume on $V^s(w)$ induced by the Riemannian metric on M.

There exists a family of smooth submanifolds $T^s(w)$ for $w \in B(x, r)$ with the following properties:

1. $T^s(w)$ is transverse to the family of local stable manifolds $\mathcal{L}(x)$ for every $w \in B(x, r)$;
2. $T^s(w_1) \cap T^s(w_2) = \varnothing$ if $w_2 \notin T^s(w_1)$ and $T^s(w_1) = T^s(w_2)$ if $w_2 \in T^s(w_1)$;
3. $\bigcup_{w \in B(x,r)} T^s(w) \supset B(x, r)$.
4. $T^s(w)$ depends continuously on $w \in B(x, r)$.

In other words, the collection of smooth submanifolds $T^s(w)$ generates a partition of the ball $B(x, r)$. We denote this partition by T^s. We also denote by $\mu^s(w) = \mu_{T^s(w)}$ the Riemannian volume on $T^s(w)$ induced by the Riemannian metric on M.

Consider now the partition ξ^s of the set $Q^\ell(x)$ (defined by (11.1)) into local stable manifolds $V^s(w)$ for $w \in \mathcal{R}^\ell \cap B(x, r)$. It is clear that this partition is measurable. Denote by $\widetilde{\nu}^s(w)$ the conditional measure on $V^s(w)$ generated by the partition ξ^s and the measure ν. It is easy to see that the factor space $Q^\ell(x)/\xi^s$ can be identified with the subset

$$A^s(w) = \{y \in T^s(w) : \text{there exists } z \in \mathcal{R}^\ell \cap B(x,r) \\ \text{such that } y = T^s(w) \cap V^s(z)\}$$

for every $w \in B(x,r)$. We denote by $\hat{\nu}^s$ the factor measure generated by the partition ξ^s and the measure ν.

THEOREM 12.1. *The following statements hold:*

1. *the measures $\widetilde{\nu}^s(w)$ and $\nu^s(w)$ are equivalent for ν-almost every $w \in \mathcal{R}^\ell \cap B(x,r)$;*
2. *the factor measure $\hat{\nu}^s$ is equivalent to the measure $\mu^s(w)|A^s(w)$ for ν-almost every $w \in \mathcal{R}^\ell \cap B(x,r)$.*

PROOF. Consider the partition T^s. Denote by $\widetilde{\mu}^s(w)$ for $w \in B(x,r)$ the system of conditional measures and by $\hat{\mu}^s$ the factor measure generated by the partition T^s and the measure ν. One can see that the factor space $B(x,r)/T^s$ can be identified with a local stable manifold $V^s(w)$ for some $w \in \mathcal{R}^\ell \cap B(x,r)$. Since the partition T^s is smooth there exist measurable bounded functions $g(w,z)$ and $h(w)$ (where $w \in \mathcal{R}^\ell \cap B(x,r)$ and $z \in T^s(w)$) such that

$$d\widetilde{\mu}^s(w)(z) = g(w,z)\, d\mu^s(w)(z), \quad d\hat{\mu}^s(w) = h(w)\, d\nu^s(w).$$

Let $B \subset Q^\ell(x)$ be a Borel subset of positive ν-measure. We have

$$\begin{aligned}
\nu(B) &= \int_{B(x,r)/T^s} \int_{T^s(w)} \chi_B(w,z)\, d\widetilde{\mu}^s(w)(z)\, d\hat{\mu}^s(w) \\
&= \int_{B(x,r)/T^s} \int_{T^s(w)} \chi_B(w,z) g(w,z)\, d\mu^s(w)(z)\, d\hat{\mu}^s(w) \\
&= \int_{B(x,r)/T^s} \int_{T^s(w_0)} \chi_B(w,z) g(w,z) J^s(\pi_{w_0 w})(y)\, d\mu^s(w_0)(y)\, d\hat{\mu}^s(w),
\end{aligned}$$

where $\chi_B(w,z)$ is the indicator function of the set B at the point $z \in T^s(w)$, $\pi_{w_0 w}$ is the holonomy map between the transversals $T^s(w_0)$ and $T^s(w)$, $z = \pi_{w_0 w}(y)$, and $J^s(\pi_{w_0 w})(y)$ is the Jacobian of the map $\pi_{w_0 w}$ at y. Applying Fubini's theorem we obtain

$$\begin{aligned}
\nu(B) &= \int_{T^s(w_0)} \int_{B(x,r)/T^s} \chi_B(w,z) g(w,z) J^s(\pi_{w_0 w})(y)\, d\hat{\mu}^s(w)\, d\mu^s(w_0)(y) \\
&= \int_{T^s(w_0)} \int_{V^s(w)} \chi_B(w,z) g(w,z) J^s(\pi_{w_0 w})(y) h(w)\, d\nu^s(w)\, d\mu^s(w_0)(y).
\end{aligned}$$

This implies the desired result. □

As an immediate consequence of Theorem 12.1 we obtain the following results.

THEOREM 12.2. *For ν-almost every $x \in \mathcal{R}$ we have*

$$\nu^s(V^s(x) \setminus \mathcal{R}) = 0.$$

Let $x \in \mathcal{R}^\ell$ and W be a smooth submanifold which is transverse to the family of local smooth manifolds $\mathcal{L}(x)$. Let N be a set of zero Lebesgue measure in W.

THEOREM 12.3. *We have $\nu(\bigcup V^s(w)) = 0$ where the union is taken over all points $w \in \mathcal{R}^\ell \cap B(x,r)$ for which $V^s(w) \cap W \in N$.*

13. Ergodicity of Nonuniformly Hyperbolic Systems Preserving Smooth Measures

In this section we move to the core of smooth ergodic theory and consider smooth dynamical systems preserving smooth hyperbolic invariant measures. A sufficiently complete description of their ergodic properties is one of the main manifestations of the above results on local instability (see Sections 9 and 10) and absolute continuity (see Sections 11 and 12). It turns out that smooth hyperbolic invariant measures have an abundance of ergodic properties which makes smooth ergodic theory a deep and well-developed part of the general theory of smooth dynamical systems. Nevertheless, many interesting problems in this theory still remain open.

Let f be a $C^{1+\alpha}$ diffeomorphism of a smooth compact Riemannian manifold M, and ν a Borel f-invariant measure on M. We assume that ν is equivalent to the Riemannian volume on M and that the set Λ of points with nonzero Lyapunov exponents (defined by (6.4)) has positive ν-measure.

The following statement is one of the main results of smooth ergodic theory. It describes the decomposition of the measure ν into ergodic components.

THEOREM 13.1. *There exist invariant sets* Λ_0, Λ_1, ... *such that:*

1. $\bigcup_{i\geq 0} \Lambda_i = \Lambda$, *and* $\Lambda_i \cap \Lambda_j = \varnothing$ *whenever* $i \neq j$;
2. $\nu(\Lambda_0) = 0$, *and* $\nu(\Lambda_i) > 0$ *for each* $i \geq 1$;
3. $f|\Lambda_i$ *is ergodic for each* $i \geq 1$.

PROOF. We begin with the following statement.

LEMMA 13.2. *Given an f-invariant Borel function φ, there exists a set $N \subset M$ of zero measure such that if $y \in \Lambda \cap B(x, r)$ and $z, w \in V^s(y) \setminus N$ or $z, w \in V^u(y) \setminus N$ then $\varphi(z) = \varphi(w)$.*

PROOF OF THE LEMMA. Let z, $w \in V^s(y)$ and ψ be a continuous function. By Birkhoff's ergodic theorem the functions

$$\overline{\psi}(x) = \lim_{n\to\infty} \frac{1}{2n+1} \sum_{k=-n}^{n} \psi(f^k x),$$

$$\psi^+(x) = \lim_{n\to\infty} \frac{1}{n} \sum_{k=1}^{n} \psi(f^k x), \quad \text{and} \quad \psi^-(x) = \lim_{n\to\infty} \frac{1}{n} \sum_{k=1}^{n} \psi(f^{-k} x),$$

are defined for ν-almost every point x. We also have that $\overline{\psi}(x) = \psi^+(x) = \psi^-(x)$ outside a subset $N \subset M$ of zero measure.

Since $\rho(f^n z, f^n w) \to 0$ as $n \to \infty$ (see Theorem 9.2) and ψ is continuous, we obtain

$$\overline{\psi}(z) = \psi^+(z) = \psi^+(w) = \overline{\psi}(w).$$

Notice that the continuous functions are dense in $L^1(M, \nu)$ and hence, the functions of the form $\overline{\psi}$ are dense in the set of f-invariant Borel functions. The lemma follows. □

Consider the regular sets

$$\Lambda^\ell = \left\{ x \in \Lambda : C(x, \varepsilon) \leq \ell, K(x, \varepsilon) \geq \frac{1}{\ell} \right\},$$

where the functions $C(x,\varepsilon)$ and $K(x,\varepsilon)$ are given by Theorem 6.3. Let $x \in \Lambda^\ell$ be a Lebesgue point of Λ^ℓ, and for each $r > 0$ set

$$P^\ell(x,r) = \bigcup_{y\in\Lambda^\ell\cap B(x,r)} (V^s(y) \cup V^u(y)).$$

LEMMA 13.3. *There exists $r = r(\ell) > 0$ such that the map f is ergodic on the set*

$$Q(x) = \bigcup_{n\in\mathbb{Z}} f^n(P^\ell(x,r)). \tag{13.1}$$

PROOF OF THE LEMMA. Let φ be an f-invariant function, and N the set of zero measure constructed in Lemma 13.2. Choose $0 < r < \min\{r_\ell, \delta_\ell\}$ (see Remark 10.2). We also assume that the number r satisfies the requirements in Section 12. By Theorem 12.1 there exists a point $y \in (\Lambda^\ell \cap B(x,r)) \setminus N$ such that

$$\nu_y^s(V^s(y) \cap N) = 0 \quad \text{and} \quad \nu_y^u(V^u(y) \cap N) = 0,$$

where ν_y^s and ν_y^u are, respectively, the measures induced on $V^s(y)$ and $V^u(y)$ by the Riemannian volume. Let $R^s = \bigcup V^s(z)$ and $R^u = \bigcup V^u(z)$, where the unions are taken over all points $z \in \Lambda^\ell \cap B(x, r_\ell)$ for which, respectively, $V^s(z) \cap V^u(y) \in N$ and $V^s(y) \cap V^u(z) \in N$. By Theorem 12.3, we have $\nu(R^s) = \nu(R^u) = 0$.

Let z_1, $z_2 \in P^\ell(x,r) \setminus (R^s \cup R^u \cup N)$, and define a point $w_i \in \Lambda^\ell \cap B(x,r)$ so that $z_i \in V^s(w_i)$ or $z_i \in V^u(w_i)$ for $i = 1$, 2. Depending on the location of the points z_1 and z_2, there are four possible cases:

1) $z_1 \in V^u(w_1)$ and $z_2 \in V^u(w_2)$;
2) $z_1 \in V^s(w_1)$ and $z_2 \in V^u(w_2)$;
3) $z_1 \in V^u(w_1)$ and $z_2 \in V^s(w_2)$;
4) $z_1 \in V^s(w_1)$ and $z_2 \in V^s(w_2)$.

We will consider only the first two cases; the other two can be treated in a similar fashion. Notice that in the first case the intersection $V^s(y) \cap V^u(w_i)$ is nonempty and consists of a single point y_i for $i = 1$, 2. It follows from the definition of the set R^u that y_1, $y_2 \notin N$. Therefore, by Lemma 13.2,

$$\varphi(z_1) = \varphi(y_1) = \varphi(y_2) = \varphi(z_2).$$

In the second case, the intersection $V^s(w_i) \cap V^u(y)$ is nonempty and consists of a single point y_i for $i = 1$, 2. It is easy to see that y_1, $y_2 \notin N$. Since $y \notin N$, Lemma 13.2 implies that

$$\varphi(z_1) = \varphi(y_1) = \varphi(y) = \varphi(y_2) = \varphi(z_2).$$

This completes the proof of the lemma. □

LEMMA 13.4. *We have $\nu(Q(x)) > 0$ and $Q(x) \subset \Lambda$ (mod 0).*

PROOF OF THE LEMMA. Since $Q(x) \supset P^\ell(x,r) \supset \Lambda^\ell \cap B(x,r)$, we have

$$\nu(Q(x)) \geq \nu(\Lambda^\ell \cap B(x,r)) > 0.$$

By Lemma 13.3 the map $f|Q(x)$ is ergodic. Therefore

$$Q(x) = Q(x) \cap \Lambda \subset \Lambda \pmod 0$$

and the lemma follows. □

Since almost every point $x \in \Lambda$ is a Lebesgue point of Λ^ℓ for some ℓ, the invariant sets $Q(x)$ (defined by (13.1) for different values of ℓ) cover the set Λ (mod 0).

By Lemma 13.4, there is at most countable number of such sets. We denote them by $Q_1, Q_2, \ldots$. We have $\nu(Q_i) > 0$ for each $i \geq 1$, and the set $\Lambda_0 = \Lambda \setminus \bigcup_{i \geq 1} Q_i$ has zero measure.

By Lemma 13.3, the map $f|Q_i$ is ergodic for each $i \geq 1$. This yields

$$Q_i \cap Q_j = \varnothing \quad (\text{mod } 0), \tag{13.2}$$

whenever $i \neq j$. For each $n \geq 1$, let us set $\Lambda_n = Q_n \setminus \bigcup_{i=1}^{n-1} Q_i$. We have $\Lambda_i \cap \Lambda_j = \varnothing$ whenever $i \neq j$. It follows from (13.2) that $\nu(Q_i) = \nu(\Lambda_i) > 0$ for each $i \geq 1$. This completes the proof of the theorem. □

As an immediate consequence of Theorem 13.1 we obtain the following result. It is a generalization of Theorem 12.2 to the case when the set Λ has positive but not necessarily full measure.

THEOREM 13.5. *For ν-almost every $x \in \Lambda$ we have*

$$\nu^s(V^s(x) \setminus \Lambda) = 0, \quad \nu^u(V^u(x) \setminus \Lambda) = 0.$$

Theorem 13.1 is the first step in describing the ergodic properties of smooth hyperbolic invariant measures. The following *Spectral Decomposition Theorem* provides a substantially deeper information.

We remind the reader the notion of Bernoulli automorphism. Let (X, μ) be a Lebesgue space, that is, a probability measure space with at most a countable number of atoms whose union $Y \subset X$ is such that $\mu|(X \setminus Y)$ is metrically isomorphic to the Lebesgue measure on some interval $[0, a]$. One can naturally associate to (X, μ) a two-sided Bernoulli shift $\sigma\colon X^{\mathbb{Z}} \to X^{\mathbb{Z}}$ defined by $(\sigma x)_n = x_{n+1}$ for each $n \in \mathbb{N}$, which preserves the convolution measure $\otimes_{\mathbb{Z}} \mu$. A *Bernoulli automorphism* (T, ν) is an invertible (mod 0) measure preserving transformation which is metrically isomorphic to the Bernoulli shift associated to some Lebesgue space (X, μ).

We now state the Spectral Decomposition Theorem.

THEOREM 13.6 (see [**28**]). *For each $i \geq 1$ the following properties hold:*

1. *Λ_i is a disjoint union of sets Λ_i^j, for $j = 1, \ldots, n_i$, which are cyclically permuted by f, i.e., $f(\Lambda_i^j) = \Lambda_i^{j+1}$ for $j = 1, \ldots, n_i - 1$, and $f(\Lambda_i^{n_i}) = \Lambda_i^1$;*
2. *$f^{n_i}|\Lambda_i^j$ is a Bernoulli automorphism for each j.*

We remark that the proof of Statement 2 requires the more sharp estimate (11.9) of the Jacobian of the holonomy map.

We shall now consider the case of a $C^{1+\alpha}$ flow φ_t on a compact manifold M. We assume that ν is a smooth measure which is φ_t-invariant. This means that $\nu(\varphi_t A) = \nu(A)$ for any Borel set $A \subset M$ and $t \in \mathbb{R}$. In a similar fashion to that in Sections 9 and 12 one can establish the existence of families of local stable and unstable manifolds (see Remark 10.13), and show that these manifolds possess the absolute continuity property (the precise formulation of the results is entirely analogous).

Following [**29**], we shall now formulate results concerning the ergodic decomposition and the Bernoulli property for flows. Note that the Lyapunov exponent along the flow direction is zero. We assume that all other values of the Lyapunov exponent are nonzero for ν-almost every point. We also assume that ν vanishes on the set of fixed points of φ_t.

THEOREM 13.7. *There exist invariant sets Λ_0, Λ_1, ... such that:*

1. $\bigcup_{i\geq 1}\Lambda_i = \Lambda$, *and* $\Lambda_i \cap \Lambda_j = \varnothing$ *whenever* $i \neq j$;
2. $\nu(\Lambda_0) = 0$, *and* $\nu(\Lambda_i) > 0$ *for each* $i \geq 1$;
3. $\varphi_t|\Lambda_i$ *is ergodic for each* $i \geq 1$.

Recall that a *Bernoulli flow* is a flow φ_t such that every transformation φ_t is a Bernoulli automorphism.

THEOREM 13.8. *For each $i \geq 1$, one of the following exclusive alternatives holds:*

1. $\varphi_t|\Lambda_i$ *is a Bernoulli flow;*
2. $\varphi_t|\Lambda_i$ *is isomorphic to a constant-time suspension over a Bernoulli automorphism.*

14. Local Ergodicity

In the case of an Anosov diffeomorphism f preserving a smooth invariant measure on a *connected* smooth Riemannian manifold one can strengthen the Spectral Decomposition Theorem 13.6 by showing that f is indeed, ergodic.

For a general $C^{1+\alpha}$ diffeomorphism preserving a smooth hyperbolic measure one should expect that the number of open ergodic component is countable (**not** finite). Dolgopyat, Hu, and Pesin constructed such an example in Appendix B. It is therefore, an interesting problem in smooth ergodic theorem to find additional conditions which would guarantee that the ergodic components of positive measure are open (mod 0). This is known as the *local ergodicity problem.*

The main obstacles for local ergodicity are the following:

1. the stable and unstable foliations are measurable but not necessarily continuous;
2. the unstable leaves may not expand under the action of f^n (we remind the reader that they were defined as being exponentially *contracting* under f^{-n}); the same is true for stable leaves with respect to the action of f^{-n};
3. the stable and unstable distributions are measurable but not necessarily continuous.

There are two different ways to obtain sufficient conditions for local ergodicity. One was suggested by Pesin in [**28**] and the other one by Katok and Burns in [**13**]. Each of them is based on requirements which eliminate one or more of the above mentioned obstacles.

We first describe the approach developed in [**28**]. Roughly speaking it requires that the stable (or unstable) foliations are *locally continuous.*

Given a subset $X \subset M$, we call a partition ξ of X a C^1 *continuous lamination of* X if there exist continuous functions $\delta\colon X \to (0,\infty)$ and $q\colon X \to (0,\infty)$ and an integer $k > 0$ such that for each $x \in X$

1. there exists a smooth immersed k-dimensional manifold $W(x)$ containing x for which $\xi(x) = W(x) \cap X$ where $\xi(x)$ is the element of the partition ξ containing x; the manifold $W(x)$ is called the *(global) leaf* of the lamination

at x; the connected component of the intersection $W(x) \cap B(x, \delta(x))$ that contains x is called the *local leaf* at x and is denoted by $V(x)$;
2. there exists a continuous map $\varphi \colon B(x, q(x)) \to C^1(D, M)$ (where $D \subset \mathbb{R}^k$ is the unit ball) such that for every $y \in X \cap B(x, q(x))$ the manifold $V(y)$ is the image of the map $\varphi(y) \colon D \to M$.

For every $x \in X$ and $y \in B(x, q(x))$ we set $U(y) = \varphi(y)(D)$ and we call it the *local leaf* of the lamination at y. Note that $U(y) = V(y)$ for $y \in X$.

A C^1 continuous lamination of the whole manifold M is called a C^1 continuous foliation on M. The stable (as well as unstable) foliation of an Anosov diffeomorphism of a compact smooth connected Riemannian manifold M is a C^1 continuous foliation of M (see [**12**]).

THEOREM 14.1. *Let f be a $C^{1+\alpha}$ diffeomorphism of a compact smooth Riemannian manifold preserving a smooth measure ν and Λ the nonuniformly hyperbolic set for f (see Section 7). Assume that $\nu(\Lambda) > 0$ and that there exists a C^1 continuous lamination W of Λ such that $W(x) = W^s(x)$ for every $x \in \Lambda$ (where $W^s(x)$ is the global stable manifold at x; see Section 10). Then every ergodic component of f of positive measure is open* (mod 0) *in Λ (with respect to the induced topology).*

PROOF. We need the following statement which is an immediate consequence of Theorem 12.2.

LEMMA 14.2. *There exists a set $N \subset M$ of zero measure such that*
$$\nu_x^s(V^s(x) \setminus \Lambda) = \nu_x^u(V^u(x) \setminus \Lambda) = 0$$
for every $x \in \Lambda \setminus N$.

Let $Q \subset \Lambda$ be an f-invariant set with $\nu(Q) > 0$. We assume that $f|Q$ is ergodic, and we will show that Q is open (mod 0). By Lemma 13.3, there exists a density point x of the set Λ^ℓ such that $Q = Q(x)$ (mod 0) (see (13.1)). By Lemma 14.2, ν_x^u-almost every point $y \in V^u(x)$ belongs to Λ. Let $B_U(y, r)$ be the ball in $U(y)$ (with respect to the intrinsic metric) centered at y of radius r. For a ν_x^u-measurable set $Y \subset V^u(x)$ we denote
$$R(x, r, Y) = \bigcup_{y \in Y} B_U(y, r).$$
We also set
$$R(r) = R(x, r, V^u(x)), \quad \widetilde{R}(r) = R(x, r, V^u(x) \cap \Lambda)$$
and given $m \in \mathbb{N}$,
$$R^m(r) = R(x, r, V^u(x) \cap \Lambda^m).$$
Clearly,
$$R^m(r) \subset \widetilde{R}(r) \subset R(r)$$
for every $m \in \mathbb{N}$.

Since W is a C^1 continuous lamination one can find $\delta_0 > 0$ such that $B_U(y, \delta_0) \subset U(y)$ for any $y \in V^u(x)$. By Theorem 9.2 and Remark 10.2, there exists $r_m > 0$ such that $B_U(y, r_m) \subset V^s(y)$ for any $y \in V^u(x) \cap \Lambda^m$.

Fix $r \in (0, r_m]$. Given $y \in R^m(r/2)$, we denote by $n_i(y)$ the successive return times of the positive semi-trajectory of y to the set $R^m(r/2)$. We also denote by $z_i \in V^u(x) \cap \Lambda^m$ a point for which $f^{n_i(y)}(y) \in B_U(z_i, r/2)$ for $i \in \mathbb{N}$. In view of the

Poincaré recurrence theorem, one can find a subset $N \subset M$ of zero ν-measure for which the sequence $n_i(y)$ is well-defined provided that $y \in R^m(r/2) \setminus N$.

LEMMA 14.3. *For ν-almost every $y \in R^m(r/2)$ we have*

$$W^s(y) = \bigcup_{i=1}^{\infty} f^{-n_i(y)}(B_U(z_i, r)). \tag{14.1}$$

PROOF OF THE LEMMA. Let $y \in R^m(r/2) \setminus N$ and $z \in W^s(y)$. By Theorem 10.12 we have $\rho(f^{n_i(y)}y, f^{n_i(y)}z) \leq r/2$ for all sufficiently large i. Therefore,

$$f^{n_i(y)}(z) \in B_U(f^{n_i(y)}(y), r/2) \subset B_U(z_i, r)$$

and the lemma follows. □

Denote by $\xi^m(\delta_0)$ the partition of the set $R^m(\delta_0)$ into the sets $B_U(y, \delta_0)$.

LEMMA 14.4. *The partition $\xi^m(\delta_0/2)$ is measurable and has the following properties:*

1. *the conditional measure on the element $B_U(y, \delta_0/2)$ of this partition is absolutely continuous with respect to the measure ν_y^s;*
2. *the factor measure on the factor space $R^m(\delta_0/2)/\xi^m(\delta_0/2)$ is absolutely continuous with respect to the measure $\nu_x^u|V^u(x) \cap \Lambda^m$.*

PROOF OF THE LEMMA. Choose $r = \min\{\delta_0/100, r_m\}$. By Lemma 14.3, for almost every point $w \in V^u(x) \cap \Lambda^m$ one can find a point $y(w) \in B_U(w, r/2) \subset R^m(r/2)$ for which (14.1) holds. Moreover, the point $y(w)$ can be chosen in such a way that the map

$$w \in V^u(x) \cap \Lambda^m \mapsto y(w) \in R^m(r/2)$$

is measurable. For each $n \in \mathbb{N}$, set

$$R_n = \bigcup_{w \in V^u(x) \cap \Lambda^m} \bigcup_{n_i(y) \leq n} \left(f^{-n_i(y(w))}(B_U(z_i, r)) \cap R^m(3\delta_0/4) \right).$$

Observe that

$$R^m\left(\frac{3}{4}\delta_0\right) = \bigcup_{n \in \mathbb{N}} R_n.$$

Given $\varepsilon > 0$, there exists $p > 0$ and a set $Y \subset V^u(x) \cap \Lambda^m$ such that

$$\nu_x^u((V^u(x) \cap \Lambda^m) \setminus Y) \leq \varepsilon \quad \text{and} \quad R(\delta_0/2, Y) \subset \bigcup_{n \leq p} R_n. \tag{14.2}$$

It follows from Theorem 12.1 that the partition $\xi^m(\delta_0/2)|R_n$ satisfies Statements 1 and 2 of the lemma for each $n > 0$. Since ε is arbitrary the desired result follows. □

We proceed with the proof of the theorem. Denote by $\xi(\delta_0)$ the partition of the set $R(\delta_0)$ into the sets $B_U(y, \delta_0)$ and by $\widetilde{\xi}(\delta_0)$ the partition of the set $\widetilde{R}(\delta_0)$ into the sets $B_U(y, \delta_0)$. Since W is a C^1 continuous lamination, the factor space $R(\delta_0)/\xi(\delta_0)$ can be identified with $V^u(x)$ and the factor space $\widetilde{R}(\delta_0)/\xi(\delta_0)$ with $V^u(x) \cap \Lambda$.

Letting $m \to \infty$ in Lemma 14.4, we conclude that the partition $\widetilde{\xi}(\delta_0/2)$ also satisfies Statements 1 and 2 of the lemma. By Lemma 13.3, we have $Q \supset \widetilde{R}(\delta_0/2)$. This implies that $Q = \bigcup_{n \in \mathbb{Z}} f^n(\widetilde{R}(\delta_0/2))$.

Note that the set $R(\delta_0/2)$ is open. We will show that the set

$$A = \Lambda \cap (R(\delta_0/2) \setminus \widetilde{R}(\delta_0/2))) \tag{14.3}$$

has zero measure. Assuming the contrary we have that the set

$$A^m = \Lambda^m \cap (R(\delta_0/2) \setminus \widetilde{R}(\delta_0/2))) \tag{14.4}$$

has positive measure for all sufficiently large m. Therefore, for ν-almost every $z \in A^m$ we obtain that $\nu_z^u(V^u(z) \cap A^m) > 0$. Consider the set $R(z, \delta_0/2, A^m)$. Clearly,

$$R(z, \delta_0/2, A^m) \subset R(\delta_0/2) \setminus \widetilde{R}(\delta_0/2).$$

Let $B^m = R(z, \delta_0/2, A^m) \cap V^u(x)$. This set is not empty. It follows from (14.4) that $B^\ell \subset V^u(x) \setminus \Lambda$ and thus $\nu_x^u(B^\ell) = 0$ (see Lemma 14.2). Moreover, repeating arguments in the proof of Lemma 14.4 (see (14.2)) one can show that the holonomy map π that moves $V^u(z) \cap A^m$ onto B^m is absolutely continuous. Hence, $\nu_x^u(B^m) > 0$. This contradiction implies (14.3) and completes the proof of the theorem. □

Theorem 14.1 provides a way to establish ergodicity of the map $f|\Lambda$. Recall that a map f is called *topologically transitive* if given two nonempty open sets there exists a positive integer n such that $f^n A \cap B \neq \varnothing$.

THEOREM 14.5. *Assume that the conditions of Theorem 14.1 hold and that Λ is an open* (mod 0) *set (in particular is a set of full measure). Then:*

1. *every ergodic component of f of positive measure lying in Λ is open* (mod 0)*;*
2. *if $f|\Lambda$ is topologically transitive then $f|\Lambda$ is ergodic.*

PROOF. The first statement follows from Theorem 14.1. Assume that the map $f|\Lambda$ is topologically transitive. Let C, D be two distinct ergodic components of positive measure. Since $\nu(\Lambda) = 1$ we have that $C, D \subset \Lambda$ (mod 0). Moreover, $\nu(f^n C \cap D) = 0$ for any integer n. By Theorem 14.1, the sets C and D are open (mod 0) and since the map f is topologically transitive, we have $f^m C \cap D \neq \varnothing$ for some m. Furthermore, the set $f^m C \cap D$ is also open (mod 0) and since ν is equivalent to the Riemannian volume, we conclude that $\nu(f^m C \cap D) > 0$. This contradiction implies that f is ergodic. □

For a general diffeomorphism preserving a smooth hyperbolic measure, one should not expect the unstable (and stable) foliation to be locally continuous. In order to explain why this can happen consider a local unstable manifold $V^u(x)$ passing through a point $x \in \Lambda$. For a sufficiently large ℓ, the set $V^u(x) \cap \Lambda^\ell$ has positive Riemannian volume (as a subset of the smooth manifold $V^u(x)$) but is, in general, a Cantor-like set. When the local manifold is moved forward a given time n one should expect a sufficiently small neighborhood of the set $V^u(x) \cap \Lambda^\ell$ to expand. Other pieces of the local manifold (corresponding to bigger values of ℓ) will also expand but with smaller rates. This implies that the global leaf $W^u(x)$ (defined by (10.5), see Remark 10.11) may bend "uncontrollably" — this phenomenon has not yet been observed in any example but is thought to be "real" (and even "typical" in some sense). As a result the map $x \mapsto \varphi_x$ in the definition of local continuity may not be indeed continuous. Furthermore, the global manifold $W^u(x)$ may turn out to be "bounded", i.e., it may not admit an embedding of an arbitrarily large ball in $\mathbb{R}^k$ (where $k = \dim W^u(x)$). This phenomenon, again, has not yet been observed.

The local continuity of the global foliation often comes up in the following setting. Using some additional information on the system one can build a locally continuous invariant foliation whose leaves contain leaves of the unstable foliation. This along may not yet guarantee the local continuity of the unstable foliation.

However, often one may observe that local unstable leaves expand in a "controllable" way when they are moved forward. This situation occurs, for example, for geodesic flows on compact Riemannian manifolds of nonpositive curvature (see Section 16).

We now state a formal criterion for local continuity.

THEOREM 14.6 (see [**28**]). *Let f be a $C^{1+\alpha}$ diffeomorphism of a compact smooth Riemannian manifold preserving a smooth hyperbolic measure ν and Λ the nonuniformly hyperbolic set for f of full measure. Let also W be a C^1 continuous lamination of Λ with the following properties:*

1. *$W(x) \supset V^s(x)$ for every $x \in \Lambda$;*
2. *there exists a number $\delta_0 > 0$ and a measurable function $n(x)$ on Λ such that for almost every $x \in \Lambda$ and any $n \geq n(x)$,*
$$f^{-n}(V^s(x)) \supset B_U(f^{-n}(x), \delta_0).$$

Then every ergodic component of f of positive measure is open (mod 0).

PROOF. Let x be a Lebesgue point of the set Λ^ℓ for some sufficiently large $\ell > 0$. Set $A(r) = \Lambda^\ell \cap B(x,r)$ where $B(x,r)$ is the ball in M centered at x of radius r. Applying Lemma 14.3 to the set $A(r)$ for sufficiently small r and using the conditions of the theorem we find that $W^s(y) \supset B_U(y, \delta_0)$ for almost every $y \in A(r)$. One can now obtain the desired result by repeating arguments in the proof of Theorem 14.1. □

For one-dimensional foliations the second condition of Theorem 14.6 holds automatically and hence, can be omitted.

THEOREM 14.7 (see [**28**]). *Let W be a C^1 continuous one-dimensional lamination of Λ, satisfying the following property: $W(x) \supset V^s(x)$ for every $x \in \Lambda$. Then every ergodic component of f of positive measure is open* (mod 0). *Moreover, $W^s(x) = W(x)$ for almost every $x \in \Lambda$.*

PROOF. Fix $\ell > 1$ sufficiently large. For almost every point $x \in \Lambda^\ell$, the intersection $A(x) = V^s(x) \cap \Lambda^\ell$ has positive Lebesgue measure in $V^s(x)$. For every $y \in A(x)$ let $s(y)$ be the distance between x and y measured along $V^s(x)$. Then there exists a differentiable curve $\gamma\colon [0, s(y)] \to V^s(x)$ with $\gamma(0) = x$ and $\gamma(s(y)) = y$, satisfying

$$\begin{aligned}\rho_{W(f^{-n}x)}(f^{-n}x, f^{-n}y) &= \int_0^{s(y)} \|d_{\gamma(t)} f^{-n} \gamma'(t)\| \, dt \\ &\geq \int_0^{s(y)} \|d_{f^n\gamma(t)} f^n \gamma'(t)\|^{-1} \, dt \geq \ell^{-1} e^{-\varepsilon n} \lambda^{-n} \geq \delta_0\end{aligned}$$

for sufficiently large n (see Section 7), where δ_0 is a positive constant. Therefore, the second condition of Theorem 14.6 holds and the desired result follows. □

REMARK 14.8. Theorems 14.6 and 14.7 can be extended to the case when the set Λ is open (mod 0) and has positive (not necessarily full) measure. They can also be extended (with trivial modifications) to dynamical systems with continuous time.

We now describe the approach in [**13**] to study the local ergodicity. A continuous function $Q\colon TM \to \mathbb{R}$ is called an *infinitesimal eventually strict Lyapunov function* for f over a set $U \subset M$ if:

1. for each $x \in U$ the function $Q_x = Q|T_xM$ is homogeneous of degree one, and takes both positive and negative values;
2. there exist continuous distributions $D^s_x \subset C^s(x)$ and $D^u_x \subset C^u(x)$ such that $T_xM = D^s_x \oplus D^u_x$ for all $x \in U$, where
$$C^s(x) = Q^{-1}((-\infty, 0)) \cup \{0\} \quad \text{and} \quad C^u(x) = Q^{-1}((0, \infty)) \cup \{0\}$$
are called the *stable* and *unstable cones* of Q_x;
3. if $x \in U$, $n \in \mathbb{N}$, $f^n(x) \in U$, and $v \in T_xM$ then
$$Q_{f^n(x)}(d_xf^nv) \geq Q_x(v);$$
4. for ν-almost every $x \in U$ there exists $k = k(x)$, $\ell = \ell(x) \in \mathbb{N}$ such that $f^k(x) \in U$, $f^{-\ell}(x) \in U$, and if $v \in T_xM \setminus \{0\}$ then
$$Q_{f^k(x)}(d_xf^kv) > Q_x(v) \quad \text{and} \quad Q_{f^{-\ell}(x)}(d_xf^{-\ell}v) < Q_x(v).$$

A function Q is called an *infinitesimal eventually uniform Lyapunov function* for f over a set $U \subset M$ if it satisfies Conditions 1–3 and the following condition: there exists $\varepsilon > 0$ such that for ν-almost every $x \in M$ one can find $k = k(x)$, $\ell = \ell(x) \in \mathbb{N}$ for which $f^k(x) \in U$, $f^{-\ell}(x) \in U$, and if $v \in T_xM \setminus \{0\}$ then
$$Q_{f^k(x)}(d_xf^kv) > Q_x(v) + \varepsilon\|v\| \quad \text{and} \quad Q_{f^{-\ell}(x)}(d_xf^{-\ell}v) < Q_x(v) - \varepsilon\|v\|.$$
The following result of Katok and Burns gives a criterion of local ergodicity in terms of infinitesimal Lyapunov functions.

THEOREM 14.9. *The following properties hold:*

1. *If f possesses an infinitesimal eventually strict Lyapunov function Q over an open set $U \subset M$, then almost every ergodic component of f on the set $\bigcup_{n\in\mathbb{Z}} f^nU$ is open* (mod 0).
2. *If f possesses an infinitesimal eventually uniform Lyapunov function Q over an open set $U \subset M$, then every connected component of the set $\bigcup_{n\in\mathbb{Z}} f^nU$ belongs to one ergodic component of f. Moreover, if U is connected then $f|U$ is a Bernoulli transformation.*

SKETCH OF THE PROOF. When Q is an infinitesimal eventually strict Lyapunov function, given a compact set $K \subset U$ one can use the uniform continuity of $x \mapsto Q_x$ on the set K, and Condition 3 to show that the size of the stable and unstable manifolds on K is uniformly bounded away from zero. Furthermore, using Condition 4 one can show that for ν-almost every point $z \in M$ there exist $\theta = \theta(z) > 0$ and a neighborhood N of z such that for ν-almost every $x \in N$ and $y \in V^u(x) \cap N$ the tangent space $T_yV^u(x)$ is in the θ-interior of $C^u(y)$. A similar statement holds for stable manifolds.

Together with Condition 2 this implies that the stable and unstable manifolds have almost everywhere a "uniform" product structure; namely, for almost every $x \in U$ there exist a neighborhood $N(x)$ of x and $\delta > 0$ such that:

1. $V^s(y)$ and $V^u(y)$ have size at least δ for almost every $y \in N(x)$;
2. $V^s(y) \cap V^u(z) \neq \varnothing$ for $\nu \times \nu$-almost every $(y, z) \in N(x) \times N(x)$.

The proof of Statement 1 follows now from a similar argument to that in the proof of Lemma 13.3.

When Q is an infinitesimal eventually uniform Lyapunov function, the function $\theta(z)$ is uniformly bounded away from zero. This can be used to establish that for every x (and not only almost every x) there exists a neighborhood $N(x)$ of x and

$\delta > 0$ with Properties 1 and 2. A similar argument now yields the first claim in Statement 2. The last claim is an immediate consequence of Theorem 13.6. □

In conclusion we will state a few interesting open problems relevant to the above discussion. In what follows M is a compact smooth Riemannian manifold, f is a C^r diffeomorphism of M ($r \geq 2$) preserving the Riemannian volume, and Λ is a nonuniformly hyperbolic set for f.

PROBLEM 1. Construct a diffeomorphism f for which:

1. the set Λ has positive but **not** full measure;
2. the set Λ is **not** open (mod 0);
3. the set Λ has full measure but some (or all) ergodic components of positive measure are **not** open (mod 0).

PROBLEM 2. Assume that $\dim M = 2$.

1. Is it true that every ergodic component of positive measure is open (mod 0)?
2. Is it true that there can be only finitely many ergodic components of positive measure?

15. The Entropy Formula

One of the main ideas of Smooth Ergodic Theory is that sufficient instability of trajectories yields rich ergodic properties of the system. The entropy formula is in a sense a "quantitative manifestation" of this idea and is yet another pearl of Smooth Ergodic Theory. It expresses the Kolmogorov–Sinai entropy $h_\nu(f)$ of a diffeomorphism, preserving a smooth hyperbolic measure, in terms of the values of the Lyapunov exponent.

In [**33**], Ruelle established an upper bound for the metric entropy of a diffeomorphism, preserving an arbitrary Borel probability measure, in terms of the Lyapunov exponents. Independently, the same upper bound was found by Margulis (unpublished) in the case of volume-preserving diffeomorphisms.

We now briefly recall the relevant notions from the entropy theory. Let $(X, \mathcal{B}, \mu)$ be a Lebesgue measure space. This means that μ is a Lebesgue measure on X with σ-algebra $\mathcal{B}$. A finite or countable family $\xi \subset \mathcal{B}$ is called a *measurable partition* of X if $\mu(\bigcup_{C\in\xi} C) = \mu(X)$, and $\mu(C \cap D) = 0$ for every C, $D \in \xi$ such that $C \neq D$. The *entropy* of the measurable partition ξ (with respect to μ) is given by

$$H_\mu(\xi) = -\sum_{C\in\xi} \mu(C) \log \mu(C),$$

with the convention that $0 \log 0 = 0$. Given two measurable partitions ξ and ζ we also define the *conditional entropy of ξ with respect to ζ* by

$$H_\mu(\xi|\zeta) = -\sum_{C\in\xi}\sum_{D\in\zeta} \mu(C \cap D) \log \frac{\mu(C \cap D)}{\mu(D)}.$$

Given two measurable partitions ξ and η of X we shall write $\xi \subset \eta$ if for every $D \in \eta$ there exists $C \in \xi$ such that $D \subset C$ (mod 0). If ξ, η, and ζ are measurable partitions with $\xi \subset \eta$ we have $H_\mu(\xi) \leq H_\mu(\eta)$ and $H_\mu(\zeta|\xi) \geq H_\mu(\zeta|\eta)$.

Consider now a measurable function $T\colon X \to X$. We define the *entropy* of T with respect to partition ξ by the limit

$$h_\mu(T,\xi) = \lim_{n\to\infty} \frac{1}{n} H_\mu\left(\bigvee_{k=0}^{n-1} T^{-k}\xi\right), \tag{15.1}$$

which always exists. It can be shown that

$$h_\mu(T,\xi) = \inf_n \frac{1}{n} H_\mu\left(\bigvee_{k=0}^{n-1} T^{-k}\xi\right) = \lim_{n\to\infty} H_\mu\left(\xi | \bigvee_{k=1}^{n} T^{-k}\xi\right). \tag{15.2}$$

When $\xi^- = \bigvee_{k=0}^{\infty} T^{-k}\xi$ is a partition we have $h_\mu(T,\xi) = H_\mu(\xi|\xi^-)$. We finally define the *entropy* of T with respect to μ by

$$h_\mu(T) - \sup_\xi h_\mu(T,\xi),$$

where the supremum is taken over all measurable partitions ξ with finite entropy (which indeed, coincides with the supremum over all finite measurable partitions). We have $h_\mu(T,\xi) \le H_\mu(\xi)$ (by (15.2)) and $h_\mu(T^m) = mh_\mu(T)$ for every $m \in \mathbb{N}$.

When T is invertible with measurable inverse one can replace $T^{-k}\xi$ by $T^k\xi$ in the formulas (15.1) and (15.2). In particular, $h_\mu(T^{-1},\xi) = h_\mu(T,\xi)$ and $h_\mu(T^{-1}) = h_\mu(T)$. Therefore, $h_\mu(T^m) = |m|h_\mu(T)$ for every $m \in \mathbb{Z}$.

Let f be a C^1 diffeomorphism of a smooth compact Riemannian manifold and ν an invariant Borel probability measure on M. Recall that the Lyapunov spectrum of the exponent $\chi(x,\cdot)$ is (see Section 6)

$$\operatorname{Sp}\chi(x) = \{(\chi_i(x), k_i(x)) : 1 \le i \le s(x)\}.$$

THEOREM 15.1 (Margulis–Ruelle Inequality). *We have*

$$h_\nu(f) \le \int_M \sum_{i:\chi_i(x)>0} k_i(x)\chi_i(x)\,d\nu(x).$$

PROOF. By decomposing ν into its ergodic components we may assume without loss of generality that ν is ergodic. Then $k_i(x) = k_i$ and $\chi_i(x) = \chi_i$ are constant ν-almost everywhere for each i. Fix $m > 0$. Since M is compact, there exists $t_m > 0$ such that for every $0 < t \le t_m$, $y \in M$, and $x \in B(y,t)$, we have

$$\frac{1}{2} d_x f^m\left(\exp_x^{-1} B(y,t)\right) \subset \exp_{f^m x}^{-1} f^m B(y,t) \subset 2 d_x f^m\left(\exp_x^{-1} B(y,t)\right), \tag{15.3}$$

where for a set $A \subset T_zM$ and $z \in M$, we write $\alpha A = \{\alpha v : v \in A\}$.

LEMMA 15.2. *Given $\varepsilon > 0$, there exists a partition ξ of M satisfying the following conditions:*

1. $\operatorname{diam}\xi \le t_m/10$ *and* $h_\nu(f^m,\xi) \ge h_\nu(f^m) - \varepsilon$;
2. *for every element $C \in \xi$ there exist balls $B(x,r)$ and $B(x,r')$, such that $r < 2r' \le t_m/20$ and $B(x,r') \subset C \subset B(x,r)$;*
3. *there exists $0 < r < t_m/20$ such that if $C \in \xi$ then $C \subset B(y,r)$ for some $y \in M$, and if $x \in C$ then*

$$\frac{1}{2} d_x f^m\left(\exp_x^{-1} B(y,r)\right) \subset \exp_{f^m x}^{-1} f^m C \subset 2 d_x f^m\left(\exp_x^{-1} B(y,r)\right).$$

PROOF OF THE LEMMA. Given $\alpha > 0$, consider a maximal α-separated set Γ, i.e., a finite set of points for which $d(x,y) > \alpha$ whenever $x, y \in \Gamma$. For each $x \in \Gamma$ define

$$\mathcal{D}_\Gamma(x) = \{y \in M : d(y,x) \leq d(y,z) \text{ for all } z \in \Gamma \setminus \{x\}\}.$$

Obviously, $B(x,\alpha/2) \subset \mathcal{D}_\Gamma(x) \subset B(x,\alpha)$. Note that the sets $\mathcal{D}_\Gamma(x)$ corresponding to different points $x \in \Gamma$ intersect only along their boundaries, i.e., at a finite number of submanifolds of codimension greater than zero. Since ν is a Borel measure, if necessary we can move the boundaries slightly so that they have zero measure. Thus, we obtain a partition ξ with $\operatorname{diam}\xi \leq \alpha$. Moreover, we can choose a partition ξ such that

$$h_\nu(f^m,\xi) > h_\nu(f^m) - \varepsilon \quad \text{and} \quad \operatorname{diam}\xi < t_m/10.$$

This implies Statements 1 and 2. Statement 3 follows from (15.3). □

We proceed with the proof of the theorem. We have

$$\begin{aligned} h_\nu(f^m,\xi) &= \lim_{k\to\infty} H_\nu(\xi|f^m\xi \vee \cdots \vee f^{km}\xi) \\ &\leq H_\nu(\xi|f^m\xi) = \sum_{D\in f^m\xi} \nu(D)H(\xi|D) \\ &\leq \sum_{D\in f^m\xi} \nu(D)\log\operatorname{card}\{C \in \xi : C\cap D \neq \varnothing\}, \end{aligned} \tag{15.4}$$

where $H(\xi|D)$ is the entropy of ξ with respect to the conditional measure on D induced by ν. Our goal is to obtain a uniform exponential estimate for the number of those elements $C \in \xi$ which have nonempty intersection with a given element $D \in f^m\xi$, and then an exponential bound for the number of those sets $D \in f^m\xi$ which contain regular points.

LEMMA 15.3. *There exists a constant $K_1 > 0$ such that if $D \in f^m\xi$, then*

$$\operatorname{card}\{C \in \xi : D\cap C = \varnothing\} \leq K_1 \sup\{\|d_xf\|^{mn} : x \in M\}$$

where $n = \dim M$.

PROOF. By the Mean Value Theorem,

$$\operatorname{diam}(f^mC) \leq \sup\{\|d_xf\|^m : x \in M\}\operatorname{diam} C.$$

Thus, if $C\cap D \neq \varnothing$ then C is contained in the $4r'$-neighborhood of D. Therefore,

$$\operatorname{diam} C \leq (\sup\{\|d_xf\|^m : x \in M\} + 2)4r'$$

and hence, the volume of C is bounded from above by

$$Kr^n \sup\{\|d_xf\|^{mn} : x \in M\},$$

where $K > 0$ is a constant. On the other hand, by Property 2 of the partition ξ, the set C contains a ball $B(x,r')$, and hence, the volume of C is at least $K'(r')^n$, where K' is a positive constant. This implies the desired result. □

Fix $\varepsilon > 0$ and let R_m be the set of forward regular points $x \in M$ which satisfy the following condition: if $k > m$ and $v \in T_xM$, then

$$e^{k(\chi(x,v)-\varepsilon)}\|v\| \leq \|d_xf^kv\| \leq e^{k(\chi(x,v)+\varepsilon)}\|v\|.$$

LEMMA 15.4. *If $D \in f^m\xi$ has nonempty intersection with R_m then there exists a constant $K_2 > 0$ such that*

$$\operatorname{card}\{C \in \xi : D \cap C \neq \varnothing\} \leq K_2 e^{\varepsilon m} \prod_{i:\chi_i>0} e^{m(\chi_i+\varepsilon)k_i}.$$

PROOF. Let $C' \in \xi$ be such that $C' \cap R_m \neq \varnothing$ and $f^m C' = D$. Pick a point $x \in C' \cap f^{-m}(R_m)$ and let $B = B(x, 2\operatorname{diam} C')$. The set $\widetilde{B}_0 = d_x f^m(\exp_x^{-1} B) \subset T_{f^m x}M$ is an ellipsoid and $D \subset B_0 = \exp_{f^m(x)}(\widetilde{B}_0)$. If a set $C \in \xi$ has nonempty intersection with B_0 then it lies in the set

$$B_1 = \{y \in M : d(y, B_0) < \operatorname{diam} \xi\}.$$

Therefore,

$$\operatorname{card}\{C \in \xi : D \cap C \neq \varnothing\} \leq \operatorname{vol}(B_1)(\operatorname{diam} \xi)^{-n},$$

where $\operatorname{vol}(B_1)$ denotes the volume of B_1. Up to a bounded factor, $\operatorname{vol}(B_1)$ is bounded by the product of the lengths of the axes of the ellipsoid $\widetilde{B}_0$. Those of them that correspond to nonpositive exponents are at most subexponentially large. The remaining ones are of size at most $e^{m(\chi_i+\varepsilon)}$, up to a bounded factor, for all sufficiently large m. Thus,

$$\begin{aligned}\operatorname{vol}(B_1) &\leq K e^{m\varepsilon} (\operatorname{diam} B)^n \prod_{i:\chi_i>0} e^{m(\chi_i+\varepsilon)k_i} \\ &\leq K e^{m\varepsilon} (2 \operatorname{diam} \xi)^n \prod_{i:\chi_i>0} e^{m(\chi_i+\varepsilon)k_i},\end{aligned}$$

for some constant $K > 0$. The lemma follows. □

By Lemmas 15.3 and 15.4 and (15.4), we obtain

$$\begin{aligned} m h_\nu(f) - \varepsilon &= h_\nu(f^m) - \varepsilon \leq h_\nu(f^m, \xi) \\ &\leq \sum_{D \cap R_m \neq \varnothing} \nu(D) \left(\log K_2 + \varepsilon m + m \sum_{i:\chi_i>0} (\chi_i + \varepsilon) k_i \right) \\ &\quad + \sum_{D \cap R_m = \varnothing} \nu(D)(\log K_1 + nm \log \sup\{\|d_x f\| : x \in M\}) \\ &\leq \log K_2 + \varepsilon m + m \sum_{i:\chi_i>0} (\chi_i + \varepsilon) k_i \\ &\quad + (\log K_1 + nm \log \sup\{\|d_x f\| : x \in M\}) \nu(M \setminus R_m). \end{aligned}$$

By the Multiplicative Ergodic Theorem 6.2, we have

$$\bigcup_{m \geq 0} R_m = M \pmod 0$$

for all sufficiently small ε. It follows that

$$h_\nu(f) \leq \varepsilon + \sum_{i:\chi_i>0} (\chi_i + \varepsilon) k_i.$$

Letting $\varepsilon \to 0$ we obtain the desired upper bound. □

An important consequence of the Margulis–Ruelle Inequality is that any C^1 diffeomorphism with positive topological entropy has an invariant measure with at least one positive and one negative Lyapunov exponent. In particular, a surface diffeomorphism with positive topological entropy always possesses a hyperbolic invariant measure.

Notice that for an arbitrary invariant measure the Margulis–Ruelle Inequality may be strict. The reader can examine a diffeomorphism with a hyperbolic fixed point and an atomic measure concentrated at this point.

We shall now prove the lower bound and hence, the entropy formula. It was established by Pesin in [**28**].

THEOREM 15.5 (Entropy Formula). *If f is of class $C^{1+\alpha}$ and ν is a hyperbolic smooth invariant measure on M, then*

$$h_\nu(f) = \int_M \sum_{i:\chi_i(x)>0} k_i(x)\chi_i(x)\, d\nu(x). \tag{15.5}$$

PROOF. We only need to show that

$$h_\nu(f) \geq \int_M \sum_{i:\chi_i(x)>0} k_i(x)\chi_i(x)\, d\nu(x),$$

or equivalently (by replacing f by f^{-1} and using Theorem 6.2) that

$$h_\nu(f) \geq -\int_M \sum_{i:\chi_i(x)<0} k_i(x)\chi_i(x)\, d\nu(x). \tag{15.6}$$

Consider the set

$$\Gamma = \{x \in M : \chi^+(x,v) < 0 \text{ for some } v \in T_xM\}.$$

If $\nu(\Gamma) = 0$ the desired result follows immediately from (15.6). We therefore assume that $\nu(\Gamma) > 0$. Consider the subset $\widetilde{\Gamma} \subset \Gamma$ of regular points. For each $x \in \widetilde{\Gamma}$ set

$$T_n(x) = J(d_xf^n|E^s(x)) \quad \text{and} \quad g(x) = \prod_{i:\chi_i(x)<0} e^{k_i(x)\chi_i(x)}.$$

Fix $\varepsilon > 0$ and for each $n \in \mathbb{N}$ consider the invariant sets

$$\widetilde{\Gamma}_n = \{x \in \widetilde{\Gamma} : (1+\varepsilon)^{-n} < g(x) \leq (1+\varepsilon)^{-n+1}\}.$$

We shall evaluate the entropy of the restriction $f|\widetilde{\Gamma}_n$ with respect to the measure $\nu_n = \nu|\widetilde{\Gamma}_n$.

LEMMA 15.6. *Given $\varepsilon > 0$, there exists a Borel positive function $L(x)$ such that for any $x \in \widetilde{\Gamma}$ and $n \in \mathbb{N}$,*

$$T_n(x) \leq L(x)g(x)^n e^{\varepsilon n}.$$

PROOF OF THE LEMMA. This is an immediate consequence of Theorems 6.1 and 5.2. □

Fix $n \in \mathbb{N}$. For each $\ell \in \mathbb{N}$ we define the measurable sets

$$\widetilde{\Gamma}_{n,\ell} = \{x \in \widetilde{\Gamma}_n \cap \mathcal{R}^\ell : L(x) \leq \ell\},$$

and also

$$\Gamma^\ell = \bigcup_{x\in\widetilde{\Gamma}_{n,\ell}} V^s(x) \quad \text{and} \quad \hat{\Gamma} = \bigcup_{n\in\mathbb{Z}} f^n(\Gamma^\ell).$$

Note that $\hat{\Gamma}$ is f-invariant. Fix $\beta > 0$. By choosing ℓ sufficiently large, we obtain $\nu_n(\widetilde{\Gamma}_{n,\ell}) > 1 - \beta$. Since $\widetilde{\Gamma}_{n,\ell} \subset \Gamma^\ell \subset \Gamma \subset \widetilde{\Gamma}_n$ (mod 0) we also have

$$\nu_n(\hat{\Gamma} \setminus \Gamma^\ell) \leq \beta \quad \text{and} \quad \nu_n(\widetilde{\Gamma}_n \setminus \hat{\Gamma}) \leq \beta. \tag{15.7}$$

Consider now any finite measurable partition ξ of M such that each element $\xi(x)$ of ξ is homeomorphic to a ball, has piecewise smooth boundary, and has diameter at most r_ℓ (see Lemma 15.2).

We define a partition η of $\hat{\Gamma}$ composed of $\hat{\Gamma}\setminus\Gamma^\ell$, and of the elements $V^s(y)\cap\xi(x)$ for each $y \in \xi(x) \cap \widetilde{\Gamma}_{n,\ell}$.

Given $x \in \Gamma^\ell$ and a sufficiently small $r \leq r_\ell$ (see Section 11), we set

$$B_\eta(x,r) = \{y \in \eta(x) : \rho(y,x) < r\}.$$

LEMMA 15.7. *There exists $q_\ell > 0$ and a set $A^\ell \subset \Gamma^\ell$ with $\nu_n(\Gamma^\ell \setminus A^\ell) \leq \beta$ such that $\eta^-(x) \supset B_\eta(x, q_\ell)$ for every $x \in A^\ell$.*

PROOF OF THE LEMMA. For each $\delta > 0$ set

$$\partial\xi = \bigcup_{y\in M} \partial(\xi(y)) \quad \text{and} \quad \partial\xi_\delta = \{y : \rho(y, \partial\xi) \leq \delta\}.$$

One can easily show that there exists a constant $C_1 > 0$ such that

$$\nu_n(\partial\xi_\delta) \leq C_1\delta. \tag{15.8}$$

Let

$$D_q = \{x \in \Gamma^\ell : B_\eta(x,q) \setminus \eta^-(x) \neq \varnothing\}.$$

If $x \in D_q$ then there exist $m \in \mathbb{N}$ and $y \in B_\eta(x,q)$ such that $y \notin (f^{-m}\eta)(x)$. Hence, $\partial\xi \cap f^m(B_\eta(x,q)) \neq \varnothing$. Therefore, by Theorem 9.1 if $x \in D_q$ then $f^m(x) \in \partial\xi_{(C_2\lambda^m e^{\varepsilon m} q)}$ for some constant $C_2 = C_2(\ell) > 0$. Thus, in view of (15.8),

$$\nu_n(D_q) \leq C_1 \sum_{m=0}^{\infty} C_2\lambda^m e^{\varepsilon m} q \leq C_3 q$$

for some constant $C_3 = C_3(\ell) > 0$. The lemma follows by setting $q_\ell = \beta {C_3}^{-1}$ and $A^\ell = \Gamma^\ell \setminus D_{q_\ell}$. $\square$

For each $x \in \mathcal{R}^\ell$ let ν_x be the measure induced by the Riemannian volume on $V^s(x)$ as a smooth submanifold in M (see Section 12).

The following statement is a crucial step in the proof of the lower bound and exploits the absolute continuity property of the measure ν in an essential way.

LEMMA 15.8. *There exists a constant $C_5 = C_5(\ell) \geq 1$ such that if $x \in A^\ell \cap V^s(y)$ and $y \in \widetilde{\Gamma}_{\ell,n}$ then*

$${C_5}^{-1} \leq \frac{d\nu_x^-}{d\nu_y} \leq C_5.$$

PROOF OF THE LEMMA. The statement follows immediately from Theorem 12.1 and Lemma 15.7. $\square$

We also need the following statement.

LEMMA 15.9. *There exists a constant $C_4 = C_4(\ell) > 0$ such that for every $x \in \Gamma^\ell \cap V^s(y)$ with $y \in \widetilde{\Gamma}_{\ell,n}$ and for every $m \in \mathbb{N}$ we have*

$$\nu_{f^m y}(f^m(\eta(x))) \leq C_4 T_m(y).$$

PROOF OF THE LEMMA. We have

$$\nu_{f^m y}(f^m(\eta(x))) = \int_{\eta(x)} T_m(z)\, d\nu_y(z).$$

A similar argument to that in the proof of Lemma 11.8 shows that there exists a constant $C' > 0$ depending only on ℓ such that

$$\left|\frac{T_m(z)}{T_m(y)} - 1\right| \le C'$$

for every $z \in \eta(x)$ and all sufficiently large $m \in \mathbb{N}$. The desired result follows. □

We now complete the proof of the theorem. Write $\hat{f} = f|\hat{\Gamma}$. For every $m \in \mathbb{N}$ we have

$$h_{\nu_n}(f|\widetilde{\Gamma}_n) \ge h_{\nu_n}(\hat{f}) = \frac{1}{m} h_{\nu_n}(\hat{f}^m) \ge \frac{1}{m} H_{\nu_n}(\hat{f}^m \eta|\eta^-), \tag{15.9}$$

where $\eta^- = \bigvee_{k=0}^{\infty} \hat{f}^{-k}\eta$. To estimate the last expression we find a lower bound for

$$H_{\nu_n}(\hat{f}^m\eta|\eta^-(x)) = \int_{\eta^-(x)} -\log \nu_x^-(\eta^-(x) \cap (f^m\eta)(y))\, d\nu_x^-(y),$$

where ν_x^- is the conditional measure on the element $\eta^-(x)$ of the partition η^-.

Note that if $x \in A^\ell$ then

$$\nu_x(B_\eta(x, q_\ell)) \ge C_6 {q_\ell}^{\dim M}$$

for some constant $C_6 = C_6(\ell) > 0$. It follows from Lemmas 15.6, 15.9 and 15.8 that for $x \in A^\ell$ and $n \in \mathbb{N}$

$$\begin{aligned} H_{\nu_n}(\hat{f}^m\eta|\eta^-(x)) &\ge -C_5 \log\left(C_5\nu_x(\eta^-(x) \cap (f^m\eta)(x))\right) \nu_x(B_\eta(x, q_\ell)) \\ &\ge -C_5 \log(C_5 C_4 T_m(x)) \nu_x(B_\eta(x, q_\ell)) \\ &\ge -C_5 \log(C_5 C_4 L(x) g(x)^m e^{\varepsilon m}) C_6 {q_\ell}^{\dim M} \\ &\ge -C_5 \log(C_5 C_4 \ell (1+\varepsilon)^{(-n+1)m} e^{\varepsilon m}) C_6 {q_\ell}^{\dim M}. \end{aligned}$$

Choosing q_ℓ sufficiently small we may assume that $C_5 C_6 q_\ell^{\dim M} < 1$. Therefore,

$$\begin{aligned} H_{\nu_n}(\hat{f}^m\eta|\eta^-(x)) &\ge -m(-n+1)\log(1+\varepsilon) - m\varepsilon - C_7 \\ &\ge -m \log g(x) - m(\log(1+\varepsilon) + \varepsilon) - C_7 \end{aligned}$$

for some constant $C_7 > 0$ depending only on ℓ. Integrating this inequality over the elements of η^- we obtain

$$\frac{1}{m} H_{\nu_n}(\hat{f}^m\eta|\eta^-) \ge \int_{A^\ell} \left(-\log g(x) - (\log(1+\varepsilon) + \varepsilon) - \frac{C_7}{m}\right) d\nu_n(x).$$

Given $\delta > 0$, we can choose numbers m and ℓ sufficiently large and ε sufficiently small such that (in view of (15.7), (15.9), and Lemma 15.7)

$$h_{\nu_n}(f|\widetilde{\Gamma}_n) \ge -\frac{1}{\nu(\widetilde{\Gamma}_n)} \int_{\widetilde{\Gamma}_n} \sum_{i:\chi_i(x)<0} k_i(x)\chi_i(x)\, d\nu(x) - \delta.$$

Summing up all these inequalities over n (note that the sets $\widetilde{\Gamma}_n$ are disjoint and invariant) we obtain

$$
\begin{aligned}
h_\nu(f) &\geq \sum_{n=1}^{\infty} \nu(\widetilde{\Gamma}_n) h_{\nu_n}(f|\widetilde{\Gamma}_n) \\
&\geq \sum_{n=1}^{\infty}\left(-\int_{\widetilde{\Gamma}_n} \sum_{i:\chi_i(x)<0} k_i(x)\chi_i(x)\,d\nu(x) - \delta\nu(\widetilde{\Gamma}_n)\right) \\
&= \int_M \sum_{i:\chi_i(x)<0} k_i(x)\chi_i(x)\,d\nu(x) - \delta.
\end{aligned}
$$

Since δ is arbitrary the desired result follows. □

REMARK 15.10. We note that the above proof also establishes the entropy formula (15.5) when ν is not a smooth measure, i.e., a measure which is equivalent to the Riemannian volume on M but is just absolutely continuous.

In the case of Anosov diffeomorphisms, Sinai established in [**37**] a statement equivalent to the entropy formula.

We now describe a more general class of measures for which the entropy formula holds.

Consider the measurable partition ξ^s of the set $Q^\ell(x)$ (see (11.1)) into local stable manifolds for $w \in \mathcal{R}^\ell \cap B(x,r)$. We denote by $\widetilde{\nu}^s(w)$ the associated conditional measure on $V^s(w)$ generated by the partition ξ^s and the measure ν (see Section 12). In a similar fashion the measurable partition ξ^u into local unstable manifolds on $\mathcal{R}^\ell \cap B(x,r)$ generates conditional measures $\widetilde{\nu}^u(w)$ on $V^u(w)$, $w \in \mathcal{R}^\ell \cap B(x,r)$. The invariant measure ν is called an *SRB-measure* (after Sinai, Ruelle, and Bowen) if f has at least one nonzero exponent ν-almost everywhere, and has absolutely continuous conditional measures on stable manifolds (or unstable manifolds), that is, the measure $\widetilde{\nu}^s(w)$ (or the measure $\widetilde{\nu}^u(w)$) is absolutely continuous with respect to the Riemannian volume on $V^s(w)$ (or on $V^u(w)$) for ν-almost every point w. By Theorem 12.1, a smooth measure ν (or more generally a measure ν absolutely continuous with respect to the Riemannian volume) invariant under a $C^{1+\alpha}$ diffeomorphism is an SRB-measure.

Ledrappier and Strelcyn extended the entropy formula to SRB-measures invariant under a $C^{1+\alpha}$ diffeomorphism [**16**].

The following result is due to Ledrappier and Young and provides a characterization of SRB-measures in terms of the entropy formula.

THEOREM 15.11 ([**17**]). *For a Borel measure ν invariant under a C^2 diffeomorphism, the entropy formula* (15.5) *holds if and only if ν is an SRB-measure.*

16. Ergodic Properties of Geodesic Flows on Compact Surfaces of Nonpositive Curvature

As we saw in Section 8 the geodesic flow g_t on a compact surface M of nonpositive curvature and of genus greater than 1 is nonuniformly hyperbolic on the set Δ of positive Liouville measure which is defined by (8.4). Since the Liouville measure is invariant under the geodesic flow the results of Section 13 apply and show that the ergodic components are of positive Liouville measure (see Theorem 13.1). In this section we show that, indeed, the geodesic flow on Δ is ergodic.

First, we state a remarkable result by Eberlein [**8**].

PROPOSITION 16.1. *The geodesic flow g_t on M is topologically transitive.*

We now wish to show that every ergodic component of $g_t|\Delta$ is open (mod 0). To achieve this we construct C^1 continuous laminations of SM, W^- and W^+, such that $W^s(x) = W^-(x)$ and $W^u(x) = W^+(x)$ for almost every $x \in \Delta$ (W^- and W^+ are known as the stable and unstable horocycle foliations; see below). We then apply Theorems 14.5 and 14.7 (more precisely, their analog for flows; see Remark 14.8) to derive that the flow $g_t|\Delta$ is ergodic. In order to proceed in this direction we need some basic information on manifolds of nonpositive curvature.

We denote by H the universal Riemannian cover of M, i.e., a simply connected complete Riemannian manifold for which $M = H/\Gamma$ where Γ is a discrete subgroup of the group of isometries of H, isomorphic to $\pi_1(M)$. According to the Hadamard–Cartan theorem, any two points $x, y \in H$ are joined by a single geodesic which we denote by γ_{xy}. For any $x \in H$, the exponential map $\exp_x : \mathbb{R}^p \to H$ is a diffeomorphism. Hence, the map

$$\varphi_x(y) = \exp_x\left(\frac{y}{1 - \|y\|}\right)$$

is a homeomorphism of the open unit ball B onto H.

Two geodesics $\gamma_1(t)$ and $\gamma_2(t)$ in H are called *asymptotic* for $t > 0$ if

$$\sup_{t>0} \rho(\gamma_1(t), \gamma_2(t)) < \infty,$$

where ρ is the distance in H induced by the Riemannian metric. Given a point $x \in H$ there is a unique geodesic starting at x which is asymptotic for $t > 0$ to a given geodesic. The asymptoticity is an equivalence relation, and the equivalence class $\gamma(\infty)$ corresponding to a geodesic γ is called a *point at infinity*. The set of these classes is denoted by $H(\infty)$ and is called the *ideal boundary* of H. One can extend the topology of the space H to $\overline{H} = H \cup H(\infty)$ so that $\overline{H}$ becomes a compact metric space.

The map φ_x can be extended to a homeomorphism (still denoted by φ_x) of the closed ball $\overline{B} = B \cup S^{p-1}$ (where S^{p-1} is the $(p-1)$-dimensional sphere in $\mathbb{R}^p$) onto $\overline{H}$ by the equality

$$\varphi_x(y) = \gamma_y(+\infty), \qquad y \in S^{p-1}.$$

In particular, φ_x maps S^{p-1} homeomorphically onto $H(\infty)$.

For any two distinct points x and y on the ideal boundary there is a geodesic which joins them. This geodesic is uniquely defined if the Riemannian metric is of strictly negative curvature (i.e., if the inequality (8.1) is strict). Otherwise, there may exist a pair of distinct points $x, y \in H(\infty)$ which can be joined by more than one geodesic. More precisely, there exists a geodesically complete embedding into H of an infinite strip of zero curvature which consists of geodesics joining x and y. Moreover, any two geodesics on the universal cover which are asymptotic both for $t > 0$ and for $t < 0$ (i.e., they can be joined to distinct points on the ideal boundary) bound a flat strip. This statement is known as the *flat strip theorem* (see [**29, 6**]).

The fundamental group $\pi_1(M)$ of the manifold M acts on the universal cover H by isometries. This action can be extended to the ideal boundary $H(\infty)$. Namely, if $p = \gamma_v(+\infty) \in H(\infty)$ and $\zeta \in \pi_1(M)$, then $\zeta(p)$ is the equivalent class of geodesics which are asymptotic to the geodesic $\zeta(\gamma_v(t))$.

We now describe the *horocycle foliations* for the geodesic flow.

PROPOSITION 16.2 (see [**8, 29**]). *The distributions E^- and E^+ are integrable. Their integral manifolds form C^1 continuous foliations of SM which are invariant under the flow g_t.*

We denote these foliations by W^- and W^+. The foliations W^- and W^+ can be lifted from the manifold SM to the manifold SH. We denote these lifts by $\widetilde{W}^-$ and $\widetilde{W}^+$ respectively. The set $L(x,p) = \pi\widetilde{W}^-(v)$ is called the *limit cycle (horocycle)* centered at $p = \gamma_v(+\infty) \in H(\infty)$ passing through the point $x = \pi v$.

The proof of Proposition 16.2 is based upon constructing the limit cycle $L(x,p)$ ($p = \gamma_v(+\infty)$, $x = \pi v$) as a limit of circles $\mathbb{S}^1(\gamma_v(t),t)$ as $t \to +\infty$ (notice that these circles pass through x and their centers approach p). The convergence means that for any open ball $B(x,R)$ in H centered at x of radius R the family of smooth curves $B(x,R) \cap \mathbb{S}^1(\gamma_v(t),t)$ converges uniformly to the smooth curve $B(x,R) \cap L(x,p)$ as $t \to +\infty$ in the C^1 topology (see [**29**] for details). More precisely, one can prove the following result.

PROPOSITION 16.3. *The following properties hold:*

1. *For any $x \in H$ and $p \in H(\infty)$ there exists a unique limit cycle $L(x,p)$ centered at p and passing through x; this limit cycle is a limit in the C^1 topology of circles $\mathbb{S}^1(\gamma(t),t)$ as $t \to +\infty$ where γ is the unique geodesic joining x and p.*
2. *The leaf $W^+(v)$ is the framing of the limit circle $L(x,p)$ ($x = \pi v$ and $p = \gamma_v(+\infty)$) by orthonormal vectors which have the same direction as the vector v (i.e., they are "inside" the limit sphere). The leaf $W^-(v)$ is the framing of the limit circle $L(x,p)$ ($x = \pi v$ and $p = \gamma_v(-\infty) = \gamma_{-v}(+\infty)$) by orthonormal vectors which have the same direction as the vector v (i.e., they are "outside" the limit sphere).*
3. *For every $\zeta \in \pi_1(M)$, we have*
$$\zeta(L(x,p)) = L(\zeta(x),\zeta(p)),$$
$$d_v\zeta\widetilde{W}^-(v) = \widetilde{W}^-(d_v\zeta v), \quad d_v\zeta\widetilde{W}^+(v) = \widetilde{W}^+(d_v\zeta v).$$
4. *For every v, $w \in SH$, for which $\gamma_v(+\infty) = \gamma_w(+\infty) = p$, the geodesic $\gamma_w(t)$ intersects the limit circle $L(\pi(v),p)$ at some point.*

We now state our main result.

THEOREM 16.4. *Let M be a compact surface of nonpositive curvature and of genus greater than 1. Then the following properties hold:*

1. *the set Δ defined by* (8.4) *has positive Liouville measure, is open* (mod 0)*, and is everywhere dense;*
2. *the geodesic flow $g_t|\Delta$ is ergodic.*

PROOF. By Theorem 8.6, $\mu(\Delta) > 0$ where μ is the Liouville measure. We shall show that the set Δ is open (mod 0). Note that given $v \in \Delta$ there is a number t such that the curvature K_x of M at the point $x = \gamma_v(t)$ is strictly negative. Therefore there is a ball $B(x,r)$ in M centered at x of radius r such that $K_y < 0$ for every $y \in B(x,r)$. It follows that there is a ball $D(v,q)$ in SM centered at v of radius q such that $\gamma_w(t) \in B(x,r)$ for every $w \in D(v,q)$. Moreover, for almost every $w \in D(v,q)$ (with respect to the Liouville measure) there exists a sequence of numbers $t_n \to \infty$ such that $\gamma_w(t_n) \in B(x,r)$. It follows that $w \in \Delta$.

Note that since the geodesic flow is topologically transitive and the set Δ is open (mod 0) this set is everywhere dense. This proves the first statement.

For μ-almost every $v \in \Delta$, consider one-dimensional local stable and unstable manifolds $V^s(v)$ and $V^u(v)$. We denote by $\widetilde{V}^s(\widetilde{v})$ and $\widetilde{V}^u(\widetilde{v})$ their lifts to SH (with $\widetilde{v}$ being a lift of v). Given $\widetilde{w} \in \widetilde{V}^s(\widetilde{v})$, we have

$$\rho(\pi g_t\widetilde{v}, \pi g_t\widetilde{w}) \to 0 \text{ as } t \to \infty. \tag{16.1}$$

It follows that the geodesics $\gamma_{\widetilde{v}}(t)$ and $\gamma_{\widetilde{w}}(t)$ are asymptotic and hence, $\gamma_{\widetilde{v}}(+\infty) = \gamma_{\widetilde{w}}(+\infty)$. We wish to show that $\widetilde{w} \in \widetilde{W^+}(\widetilde{v})$. Assuming the contrary consider the limit circle $L(\pi\widetilde{v}, \gamma_{\widetilde{v}}(+\infty))$. Let $\widetilde{z}$ be the point of intersection of the geodesic $\gamma_{\widetilde{w}}(t)$ and this limit circle (such a point exists by Proposition 16.3). We have

$$\rho(\pi g_t\widetilde{v}, \pi g_t\widetilde{w}) \geq \rho(\pi\widetilde{w}, \widetilde{z}) > 0$$

which contradicts to (16.1). It follows that $\widetilde{V}^s(\widetilde{v}) \subset \widetilde{W^-}(\widetilde{v})$ for every $v \in \Delta$ and every lift $\widetilde{v}$ of v. Arguing similarly, one can show that $\widetilde{V}^u(\widetilde{v}) \subset \widetilde{W^-}(\widetilde{v})$ for every $v \in \Delta$ and every lift $\widetilde{v}$ of v. By Proposition 16.2 and Theorems 14.5 and 14.7 (more precisely, by their analog for flows; see Remark 14.8), we conclude that $g_t|\Delta$ is ergodic. □

References

[1] D. Anosov, *Geodesic flows on closed Riemann manifolds with negative curvature*, Proc. Steklov Inst. Math. **90** (1969), 1–235.

[2] D. Anosov and Ya. Sinai, *Certain smooth ergodic systems*, Russian Math. Surveys **22** (1967), no. 5, 103–167.

[3] L. Barreira and J. Schmeling, *Sets of "non-typical" points have full topological entropy and full Hausdorff dimension*, Israel J. Math. **116** (2000), 29–70.

[4] W. Ballmann, *Lectures on spaces of nonpositive curvature*, with an appendix by M. Brin, DMV Seminar 25, Birkhäuser Verlag, Basel, 1995.

[5] A. Blohin, *Smooth ergodic flows on surfaces*, Trans. Moscow Math. Soc. **27** (1972), 117–134.

[6] K. Burns and A. Katok, *Manifolds with nonpositive curvature*, Ergodic Theory Dynam. Systems **5** (1985), no. 2, 307–317.

[7] D. Bylov, R. Vinograd, D. Grobman and V. Nemyckii, *Theory of Lyapunov exponents and its application to problems of stability*, Izdat. "Nauka", Moscow, 1966, in Russian.

[8] P. Eberlein, *Geodesic flows on negatively curved manifolds. I*, Ann. of Math. (2) **95** (1972), 492–510.

[9] P. Eberlein, *When is a geodesic flow of Anosov type? I*, J. Differential Geom. **8** (1973), 437–463; *II*, J. Differential Geom. **8** (1973), 565–577.

[10] A. Fathi, M. Herman and J.-C. Yoccoz, *A proof of Pesin's stable manifold theorem*, Geometric Dynamics (Rio de Janeiro, 1981), Jacob Palis, ed., Lecture Notes in Math. 1007, Springer Verlag, 1983, pp. 177–215.

[11] H. Federer, *Geometric measure theory*, Springer Verlag, 1969.

[12] M. Hirsch, C. Pugh and M. Shub, *Invariant manifolds*, Lecture Notes in Math. 583, 1977.

[13] A. Katok and K. Burns, *Infinitesimal Lyapunov functions, invariant cone families and stochastic properties of smooth dynamical systems*, Ergodic Theory Dynam. Systems **14** (1994), 757–785.

[14] A. Katok and B. Hasselblatt, *Introduction to the modern theory of dynamical systems*, Cambridge University Press, 1995.

[15] A. Katok and J.-M. Strelcyn, *Invariant manifolds, entropy and billiards; smooth maps with singularities*, with the collaboration of François Ledrappier and Feliks Przytycki, Lecture Notes in Math. 1222, Springer Verlag, 1986.

[16] F. Ledrappier and J.-M. Strelcyn, *A proof of the estimate from below in Pesin's entropy formula*, Ergodic Theory Dynam. Systems **2** (1982), 203–219.

[17] F. Ledrappier and L.-S. Young, *The metric entropy of diffeomorphisms. I. Characterization of measures satisfying Pesin's entropy formula*, Ann. of Math. (2) **122** (1985), 509–539.

[18] P.-D. Liu and M. Qian, *Smooth ergodic theory of random dynamical systems*, Lecture Notes in Math. 1606, Springer Verlag, 1995.

[19] A. Lyapunov, *The general problem of the stability of motion*, Taylor & Francis, 1992.

[20] I. Malkin, *A theorem on stability via the first approximation*, Dokladi, Akademii Nauk USSR **76** (1951), no. 6, 783–784.

[21] R. Mañé, *Ergodic theory and differentiable dynamics*, Ergebnisse der Mathematik und ihrer Grenzgebiete 3. Folge·Band 8, Springer Verlag, 1987.

[22] R. Mañé, *Lyapunov exponents and stable manifolds for compact transformations*, Geometric Dynamics (Rio de Janeiro, 1981), Jacob Palis, ed., Lecture Notes in Math. 1007, Springer Verlag, 1983, pp. 522–577.

[23] J. Milnor, *Fubini foiled: Katok's paradoxical example in measure theory*, Math. Intelligencer **19** (1997), 30–32.

[24] V. Oseledets, *A multiplicative ergodic theorem. Liapunov characteristic numbers for dynamical systems* Trans. Moscow Math. Soc. **19** (1968), 197–221.

[25] O. Perron, *Die Ordnungszahlen linearer Differentialgleichungssyteme*, Math. Zs. **31** (1930), 748–766.

[26] Ya. Pesin, *An example of a nonergodic flow with nonzero characteristic exponents*, Func. Anal. and its Appl. **8** (1974), no. 3, 263–264.

[27] Ya. Pesin, *Families of invariant manifolds corresponding to nonzero characteristic exponents*, Math. USSR-Izv.**40** (1976), no. 6, 1261–1305.

[28] Ya. Pesin, *Characteristic Ljapunov exponents, and smooth ergodic theory*, Russian Math. Surveys **32** (1977), no. 4, 55–114.

[29] Ya. Pesin, *Geodesic flows on closed Riemannian manifolds without focal points*, Math. USSR-Izv. **11** (1977), no. 6, 1195–1228.

[30] C. Pugh, *The $C^{1+\alpha}$ hypothesis in Pesin theory*, Inst. Hautes Études Sci. Publ. Math. **59** (1984), 143–161.

[31] C. Pugh and M. Shub, *Ergodicity of Anosov actions*, Invent. Math. **15** (1972), 1–23.

[32] C. Pugh and M. Shub, *Ergodic attractors*, Trans. Amer. Math. Soc. **312** (1989), no. 1, 1–54.

[33] D. Ruelle, *An inequality for the entropy of differentiable maps*, Bol. Soc. Brasil. Mat. **9** (1978), no. 1, 83–87.

[34] D. Ruelle, *Ergodic theory of differentiable dynamical systems*, Inst. Hautes Études Sci. Publ. Math. **50** (1979), 27–58.

[35] D. Ruelle, *Characteristic exponents and invariant manifolds in Hilbert space*, Ann. of Math. (2) **115** (1982), no. 2, 243–290.

[36] M. Shub and A. Wilkinson, *Pathological foliations and removable zero exponents*, Invent. Math. **139** (2000), 495–508.

[37] Ya. Sinai, *Dynamical systems with countably-multiple Lebesgue spectrum II*, Amer. Math. Soc. Trans. (2) **68** (1966), 34–88.

Departamento de Matemática, Instituto Superior Técnico, 1049-001 Lisboa, Portugal

E-mail address: luis.barreira@math.ist.utl.pt

URL: http://www.math.ist.utl.pt/~barreira/

Department of Mathematics, The Pennsylvania State University, University Park, PA 16802, U.S.A.

E-mail address: pesin@math.psu.edu

URL: http://www.math.psu.edu/pesin/

Appendix A. Hölder Continuity of Invariant Distributions

M. Brin

Although the stable and unstable subspaces of a nonuniformly hyperbolic dynamical system f depend continuously on the point inside an appropriate set (see Section 7 above), in general, they are not differentiable functions of the point even if f is real analytic. This happens because the stable and unstable subspaces depend on the infinite future and past, respectively. We prove here that the stable and unstable subspaces depend Hölder continuously on the point. We consider only the stable subspaces, the Hölder continuity of the unstable subspaces follows by reversing the time.

For the uniformly hyperbolic case, the Hölder continuity was established by Anosov [**A**]. The main idea of the argument below is the same as in [**BK**].

A k-dimensional distribution E on a subset Λ of a differentiable manifold M is a family of k-dimensional subspaces $E(x) \subset T_xM$, $x \in \Lambda$. A Riemannian metric on M naturally induces distances in TM and in the space of k-dimensional subspaces in TM. The Hölder continuity of a distribution E can be defined using these distances. However, by the Whitney Embedding Theorem [**H**], every manifold M can be embedded in $\mathbb{R}^N$ with a sufficiently large N. If M is compact, the Riemannian metric on M is equivalent to the distance $\|x - y\|$ induced by the embedding. The Hölder exponent does not change if the Riemannian metric is changed for an equivalent smooth metric, the Hölder constant does change. Therefore we assume in Theorem A.3, without loss of generality, that the manifold is embedded in $\mathbb{R}^N$.

For a subspace $A \subset \mathbb{R}^N$ and a vector $v \in \mathbb{R}^N$, set

$$\operatorname{dist}(v, A) = \min_{w \in A} \|v - w\|.$$

i.e., $\operatorname{dist}(v, A)$ is the length of the difference between v and its orthogonal projection to A. For subspaces A, B in $\mathbb{R}^N$, define

$$\operatorname{dist}(A, B) = \max \left\{ \max_{v \in A, \|v\|=1} \operatorname{dist}(v, B), \max_{w \in B, \|w\|=1} \operatorname{dist}(w, A) \right\}.$$

A k-dimensional *distribution* E defined on a subset $\Lambda \subset \mathbb{R}^N$ is *Hölder continuous* with *Hölder exponent* $\alpha \in (0, 1]$ and *Hölder constant* $L > 0$ if there is $\varepsilon_0 > 0$ such that

$$\operatorname{dist}(E(x), E(y)) \le L \cdot \|x - y\|^\alpha$$

for all x, $y \in \Lambda$ with $\|x - y\| \le \varepsilon_0$.

The following two lemmas can be used to prove the Hölder continuity of invariant distributions for a variety of dynamical systems.

LEMMA A.1. *Let A_k and B_k, $k = 0, 1, \ldots$, be two sequences of real $N \times N$ matrices such that for some $\Delta \in (0, 1)$ and $a > 1$,*

$$\|A_k - B_k\| \le \Delta a^k \qquad \text{for } k = 0, 1, 2, \ldots.$$

Suppose that there are subspaces E_A, $E_B \subset \mathbb{R}^N$ and positive $\lambda < \mu$ and $C' > 1$ so that $\lambda < a$ and for each positive integer k,

$$\|A_k v\| \le C' \lambda^k \|v\| \text{ if } v \in E_A; \qquad \|A_k w\| \ge {C'}^{-1} \mu^k \|w\| \text{ if } w \perp E_A;$$

$$\|B_k v\| \le C' \lambda^k \|v\| \text{ if } v \in E_B; \qquad \|B_k w\| \ge {C'}^{-1} \mu^k \|w\| \text{ if } w \perp E_B.$$

Then $\mathrm{dist}(E_A, E_B) \le 3C'^2 \frac{\mu}{\lambda} \Delta^{\frac{\log \mu - \log \lambda}{\log a - \log \lambda}}$.

PROOF. Set

$$Q_A^k = \{v \in \mathbb{R}^N : \|A_k v\| \le 2C'\lambda^k \|v\|\}$$

and

$$Q_B^k = \{v \in \mathbb{R}^N : \|B_k v\| \le 2C'\lambda^k \|v\|\}.$$

For $v \in \mathbb{R}^N$, write $v = v^\lambda + v^\perp$, where $v^\lambda \in E_A$ and $v^\perp \perp E_A$. If $v \in Q_A^k$, then

$$\|A_k v\| = \|A_k(v^\lambda + v^\perp)\| \ge \|A_k v^\perp\| - \|A_k v^\lambda\| \ge C'^{-1}\mu^k \|v^\perp\| - C'\lambda^k \|v^\lambda\|,$$

and hence $\|v^\perp\| \le C'\mu^{-k}(\|A_k v\| + C'\lambda^k \|v^\lambda\|) \le 3C'^2(\lambda/\mu)^k \|v\|$. Therefore

$$\mathrm{dist}(v, E_A) \le 3C'^2 \left(\frac{\lambda}{\mu}\right)^k \|v\|. \tag{A.1}$$

Set $\gamma = \lambda/a < 1$. There is a unique nonnegative integer k such that $\gamma^{k+1} < \Delta \le \gamma^k$. Let $w \in E_B$. Then

$$\begin{aligned} \|A_k w\| &\le \|B_k w\| + \|A_k - B_k\| \cdot \|w\| \le C'\lambda^k \|w\| + \Delta a^k \|w\| \\ &\le (C'\lambda^k + (\gamma a)^k)\|w\| \le 2C'\lambda^k \|w\|. \end{aligned}$$

It follows that $w \in Q_A^k$ and hence $E_B \subset Q_A^k$. By symmetry, $E_A \subset Q_B^k$. By (A.1) and by the choice of k,

$$\mathrm{dist}(E_A, E_B) \le 3C'^2 \left(\frac{\lambda}{\mu}\right)^k \le 3C'^2 \frac{\mu}{\lambda} \Delta^{\frac{\log \mu - \log \lambda}{\log a - \log \lambda}}$$

□

LEMMA A.2. *Let* $f\colon M \to M$ *be a* $C^{1+\beta}$ *map of a compact, m-dimensional,* C^2 *submanifold* $M \subset \mathbb{R}^N$. *Then for every* $a > (\max_{z\in M} \|d_z f\|)^{1+\beta}$ *there is* $C_1 > 1$ *such that for each* $n \in \mathbb{N}$ *and all* $x, y \in M$

$$\|d_x f^n - d_y f^n\| \le C_1 a^n \cdot \|x - y\|^\beta.$$

PROOF. Let C_2 be such that $\|d_x f - d_y f\| \le C_2 \|x - y\|^\beta$ and set

$$a_1 = \max_{z\in M} \|d_z f\| \ge 1\,.$$

Observe that $\|f^n(x) - f^n(y)\| \le a_1^n \|x - y\|$ for all $x, y \in M$. Fix $a > a_1$. Then the lemma holds true for $n = 1$ and any $C_1 \ge C_2$. For the inductive step we have

$$\begin{aligned} \|d_x f^{n+1} - d_y f^{n+1}\| \le& \|d_{f^n(x)} f\| \cdot \|d_x f^n - d_y f^n\| \\ &+ \|d_{f^n(x)} f - d_{f^n(y)} f\| \cdot \|d_y f^n\| \\ \le& a_1 C_1 a^n \|x - y\|^\beta + C_2 \big(a_1^n \|x - y\|\big)^\beta a_1^n \\ \le& C_1 a^{n+1} \|x - y\|^\beta \left(\frac{a_1}{a} + \frac{C_2}{C_1} \frac{(a_1^{1+\beta})^n}{a^{n+1}}\right). \end{aligned}$$

If $a > a_1^{1+\beta}$, then there is $C_1 \ge C_2$ for which the last factor in parentheses is less than 1. □

The subspaces E_1, $E_2 \subset \mathbb{R}^N$ are θ-transverse if $\|v_1 - v_2\| \ge \theta$ for all unit vectors $v_1 \in E_1$ and $v_2 \in E_2$.

THEOREM A.3. *Let M be a compact, m-dimensional, C^2 submanifold of $\mathbb{R}^N$, $m < N$, and $f\colon M \to M$ a $C^{1+\beta}$ map, $\beta \in (0,1)$. Suppose that there exist a subset $\Lambda \subset M$ and real numbers $0 < \lambda < \mu$, $C > 0$, $\theta > 0$ such that for each $x \in \Lambda$ there are θ-transverse subspaces $E^s(x)$, $E^u(x) \subset T_xM$ with the following properties: $T_xM = E^s(x) \oplus E^u(x)$, $\|d_xf^kv^s\| \le C\lambda^k\|v^s\|$ and $\|d_xf^kv^u\| \ge C^{-1}\mu^k\|v^u\|$ for all $v^s \in E^s(x)$, $v^u \in E^u(x)$, and all positive integers k.*

Then for every $a > \left(\max_{z\in M}\|d_zf\|\right)^{1+\beta}$, the distribution E^s is Hölder continuous with exponent $\alpha = \frac{\log\mu - \log\lambda}{\log a - \log\lambda}\beta$.

PROOF. For $x \in M$, let $E^\perp(x)$ denote the orthogonal complement to the tangent plane T_xM in $\mathbb{R}^N$. Since $E^\perp$ is a smooth distribution, it is sufficient to prove the Hölder continuity of $\widetilde{E}^s = E^s \oplus E^\perp$.

Since $E^s(x)$ and $E^u(x)$ are θ-transverse and of complementary dimensions in T_xM, there is $\widetilde{C} > 1$ such that $\|d_xf^kw\| \ge \widetilde{C}^{-1}\mu^k\|w\|$ for all $x \in \Lambda$ and $w \perp E^o(x)$.

For x, $y \in \Lambda$ and a positive integer k, let A_k and B_k be $N \times N$ matrices such that $A_kv = d_xf^kv$ if $v \in T_xM$, $A_kw = 0$ if $w \perp T_xM$ and $B_kv = d_yf^kv$ if $v \in T_yM$, $B_kw = 0$ if $w \perp T_yM$. By Lemma A.2, $\|A_k - B_k\| \le C_1a^k\|x-y\|^\beta$. Now the theorem follows from Lemma A.1 with $\Delta = C_1\|x-y\|^\beta$, $E_A = \widetilde{E}^s(x)$, $E_B = \widetilde{E}^s(y)$, and $C' = \max(C, \widetilde{C})$. □

References

[A] D. Anosov, *Tangential fields of transversal foliations in $\mathcal{Y}$-systems*, Math. Notes **2** (1967), no. 5, 818–823.

[BK] M. Brin and Yu. Kifer, *Dynamics of Markov chains and stable manifolds for random diffeomorphisms*, Ergodic Theory Dynam. Systems **7** (1987), 351–374.

[H] M. Hirsch, *Differential topology*, Springer Verlag, 1994.

DEPARTMENT OF MATHEMATICS, UNIVERSITY OF MARYLAND, COLLEGE PARK, MD 20742-4015, U.S.A.

E-mail address: mbrin@math.umd.edu

URL: http://www.math.umd.edu/~mib/

Appendix B. An Example of a Smooth Hyperbolic Measure with Countably Many Ergodic Components

D. Dolgopyat, H. Hu and Ya. Pesin[1]

B.1. Introduction. We construct an example of a diffeomorphism with nonzero Lyapunov exponents with respect to a smooth invariant measure which has countably many ergodic components. More precisely we will prove the following result.

THEOREM B.1. *There exists a C^∞ diffeomorphism f of the three dimensional torus $\mathbb{T}^3$ such that*

1. *f preserves the Riemannian volume μ on $\mathbb{T}^3$;*
2. *μ is a hyperbolic measure;*
3. *f has countably many ergodic components which are open* (mod 0).

B.2. Construction of the Diffeomorphism f. Let $A\colon \mathbb{T}^2 \to \mathbb{T}^2$ be a linear hyperbolic automorphism. Passing if necessary to a power of A we may assume that A has at least two fixed points p and p'. Consider the map $F = A \times \mathrm{Id}$ of the three dimensional torus $\mathbb{T}^3 = \mathbb{T}^2 \times \mathbb{S}^1$. We will perturb F to obtain the desired map f.

Consider a countable collection of intervals $\{I_n\}_{n=1}^\infty$ on the circle $\mathbb{S}^1$, where

$$I_{2n} = [(n+2)^{-1}, (n+1)^{-1}], \quad I_{2n-1} = [1-(n+1)^{-1}, 1-(n+2)^{-1}].$$

Clearly, $\bigcup_{n=1}^\infty I_n = (0,1)$ and $\operatorname{int} I_n$ are pairwise disjoint.

By Proposition B.2 below, for each n one can construct a C^∞ volume preserving ergodic diffeomorphism $f_n\colon \mathbb{T}^2 \times [0,1] \to \mathbb{T}^2 \times [0,1]$ which satisfies:

1. $\|F - f_n\|_{C^n} < e^{-n^2}$;
2. for all $0 \le m < \infty$, $D^m f_n | \mathbb{T}^2 \times \{z\} = D^m F | \mathbb{T}^2 \times \{z\}$ for $z = 0$ or 1;
3. f_n has nonzero Lyapunov exponents μ-almost everywhere.

Let $L_n\colon I_n \to [0,1]$ be the affine map and $\pi_n = (\mathrm{Id}, L_n)\colon \mathbb{T}^2 \times I_n \to \mathbb{T}^2 \times [0,1]$. We define the map f by setting $f|\mathbb{T}^2 \times I_n = \pi_n^{-1} \circ f_n \circ \pi_n$ for all n and $f|\mathbb{T}^2 \times \{0\} = F|\mathbb{T}^2 \times \{0\}$. Note that for every $n > 0$ and $0 \le m \le n$ we have

$$\begin{aligned}\|D^m F|\mathbb{T}^2 \times I_n - \pi_n^{-1} \circ D^m f_n \circ \pi_n\|_{C^n} &\le \|\pi_n^{-1} \circ (D^m F - D^m f_n) \circ \pi_n\|_{C^n} \\ &\le e^{-n^2} \cdot (n+1)^n \to 0\end{aligned}$$

as $n \to \infty$. It follows that f is C^∞ on M and has the required properties.

B.3. Main Proposition. The goal of this section is to prove the following statement. Set $I = [0,1]$.

---

[1]*Key words and phrases.* Hyperbolic measure, Lyapunov exponents, stable ergodicity, accessibility.

D. Dolgopyat was partially supported by the Sloan Foundation. H. Hu was partially supported by the National Science Foundation grant #DMS-9970646. Ya. Pesin was partially supported by the National Science Foundation grant #DMS-9704564 and by the NATO grant CRG970161. D. Dolgopyat and Ya. Pesin wish to thank the Isaac Newton Institute for Mathematical Sciences (Cambridge, UK) for hospitality and support during their stay there in June-July 2000. H. Hu wishes to thank the Centre de Physique Théorique (Marseille-Luminy, France) for hospitality and support during his stay there in May-July 2000.

PROPOSITION B.2. *For any $k \geq 2$ and $\delta > 0$, there exists a map g of the three dimensional manifold $M = \mathbb{T}^2 \times I$ such that:*

1. *g is a C^∞ volume preserving diffeomorphism of M;*
2. *$\|F - g\|_{C^k} \leq \delta$;*
3. *for all $0 \leq m < \infty$, $D^m g|\mathbb{T}^2 \times \{z\} = D^m F|\mathbb{T}^2 \times \{z\}$ for $z = 0$ and 1;*
4. *g is ergodic with respect to the Riemannian volume and has nonzero Lyapunov exponents almost everywhere.*

Before giving the formal proof let us outline the main idea. The result will be achieved in two steps. First applying an argument of [**SW**] we construct a perturbation map which has nonzero *average* central exponent $\int_M \chi^c(x)\, d\mu(x) \neq 0$, where $\chi^c(x)$ denotes the Lyapunov exponent of x along the neutral subspace $E^c(x)$. We then further perturb this diffeomorphism modifying an approach in [**NT**] to ensure that it has the accessibility property and therefore, is ergodic (see Section B.4 for details).

We believe that this approach works in a more general setting. Namely, we conjecture that the following statement holds.

CONJECTURE. *Consider a one parameter family g_ε with $g_0 = F$. Then for sufficiently small ε, g_ε satisfies the conditions of Proposition B.2 except for a positive codimension submanifold in the space of one parameter families.*

PROOF OF PROPOSITION B.2. Consider the linear hyperbolic map A of the torus $\mathbb{T}^2$. We may assume that its eigenvalues are η and η^{-1}, where $\eta > 1$. Let p and p' be fixed points of A. Choose a number $\varepsilon_0 > 0$ such that $d(p, p') \geq 3\varepsilon_0$. Consider the local stable and unstable one-dimensional manifolds for A at points p and p' of "size" ε_0 and denote them respectively by $V^s(p)$, $V^u(p)$, $V^s(p')$, and $V^u(p')$.

Let us choose the smallest positive number n_1 such that the intersection

$$A^{-n_1}(V^s(p')) \cap V^u(p) \cap B(p, \varepsilon_0)$$

consists of a single point which we denote by q_1 (here $B(p, \varepsilon_0)$ is the ball in $\mathbb{T}^2$ of radius ε_0 centered at p). Similarly, we choose the smallest positive number n_2 such that the intersection $A^{n_2}(V^u(p')) \cap V^s(p) \cap B(p, \varepsilon_0)$ consists of a single point which we denote by q_2.

Given a sufficiently small number $\varepsilon \in (0, \varepsilon_0)$,

$$\varepsilon \leq \frac{1}{2} \min\{d(p, q_1), d(p, q_2)\},$$

there is $\ell \geq 2$ such that (see Figure B.1)

$$A^{-\ell}(q_1) \notin B(p, \varepsilon), \quad A^{-\ell-1}(q_1) \in B(p, \varepsilon). \tag{B.1}$$

We now choose $\varepsilon' \in (0, \varepsilon)$ such that $A^{-\ell-1}(q_1) \in B(p, \varepsilon')$.

Finally, we assume ε to be so small that for some $q \in \mathbb{T}^2$ we have

$$B(p, \varepsilon) \cap (A^{-n_1}(V^s(p')) \cup A^{n_2}(V^u(p'))) = \varnothing,$$

$$A^i(B(q, \varepsilon)) \cap B(q, \varepsilon) = \varnothing, \quad A^i(B(q, \varepsilon)) \cap B(p, \varepsilon) = \varnothing$$

for $i = 1, \ldots, N$, where $N > 0$ will be determined later, and $\varepsilon = \varepsilon(N)$.

Set $\Omega_1 = B(p, \varepsilon_0) \times I$ and $\Omega_2 = B^{uc}(\bar{q}, \varepsilon_0) \times B^s(\bar{q}, \varepsilon_0)$, where $\bar{q} = (q, 1/2)$ and $B^{uc}(\bar{q}, \varepsilon_0) \subset V^u(q) \times I$ and $B^s(\bar{q}, \varepsilon_0) \subset V^s(q)$ are balls of radius ε_0 about $\bar{q}$.

After this preliminary consideration we describe the construction of the map g.

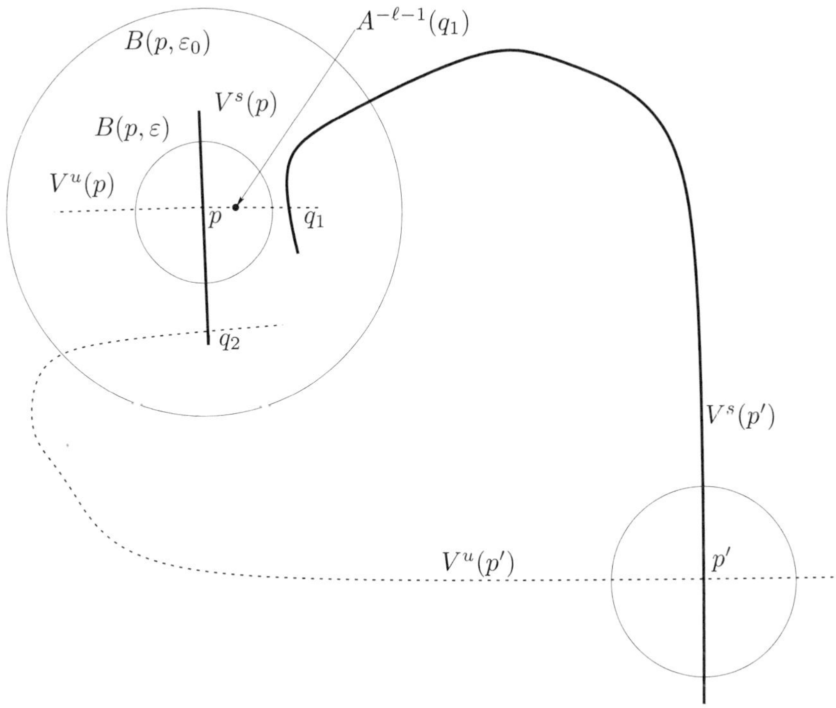

FIGURE B.1

Consider the coordinate system in Ω_1 originated at $(p,0) \in M$ with x, y, and z-axes to be unstable, stable, and neutral directions respectively for the map F. If a point $w = (x,y,z) \in \Omega_1$ and $F(w) \in \Omega_1$ then $F(w) = (\eta x, \eta^{-1}y, z)$.

Choose a C^∞ function $\xi\colon I \to \mathbb{R}^+$ satisfying:

1. $\xi(z) > 0$ on $(0,1)$;
2. $\xi^{(i)}(0) = \xi^{(i)}(1) = 0$ for $i = 0, 1, \ldots, k$;
3. $\|\xi\|_{C^k} \leq \delta$.

We also choose two C^∞ functions $\varphi = \varphi(x)$ and $\psi = \psi(y)$ which are defined on the interval $(-\varepsilon_0, \varepsilon_0)$ and satisfy

4. $\varphi(x) = \varphi_0$ if $x \in (-\varepsilon', \varepsilon')$ and $\psi(y) = \psi_0$ if $y \in (-\varepsilon', \varepsilon')$, where φ_0 and ψ_0 are positive constants;
5. $\varphi(x) = 0$ if $|x| \geq \varepsilon$; $\psi(y) \geq 0$ for any y and $\psi(y) = 0$ if $|y| \geq \varepsilon$;
6. $\|\varphi\|_{C^k} \leq \delta$, $\|\psi\|_{C^k} \leq \delta$;
7. $\int_0^{\pm\varepsilon} \varphi(s)\,ds = 0$.

We now define the vector field X on Ω_1 by

$$X(x,y,z) = \left(-\psi(y)\xi'(z)\int_0^x \varphi(s)ds, 0, \psi(y)\xi(z)\varphi(x)\right).$$

It is easy to check that X is a divergence free vector field supported on $(-\varepsilon,\varepsilon) \times (-\varepsilon,\varepsilon) \times I$.

We define the map h_t on Ω_1 to be the time t map of the flow generated by X and we set $h_t = \mathrm{Id}$ on the complement of Ω_1. It is easy to see that h_t is a

C^∞ volume preserving diffeomorphism of M which preserves the y coordinate (the stable direction for the map F).

Consider now the coordinate system in Ω_2 originated at $(q, 1/2)$ with x, y, and z-axes to be unstable, stable, and neutral directions respectively. We then switch to the cylindrical coordinate system (r, θ, y), where $x = r\cos\theta$, $y = y$, and $z = r\sin\theta$.

Consider a C^∞ function $\rho\colon (-\varepsilon_0, \varepsilon_0) \to \mathbb{R}^+$ satisfying:

8. $\rho(r) > 0$ if $0.2\varepsilon' \le r \le 0.9\varepsilon$ and $\rho(r) = 0$ if $r \le 0.1\varepsilon'$ or $r \ge \varepsilon$;
9. $\|\rho\|_{C^k} \le \delta$.

We define now the map $\widetilde{h}_\tau$ on Ω_2 by

$$\widetilde{h}_\tau(r, \theta, y) = (r, \theta + \tau\psi(y)\rho(r), y). \tag{B.2}$$

and we set $\widetilde{h}_\tau = \mathrm{Id}$ on $M \setminus \Omega_2$. It is easy to see that for every τ the map $\widetilde{h}_\tau$ is a C^∞ volume preserving diffeomorphism of M.

Let us set $g = g_{t\tau} = h_t \circ F \circ \widetilde{h}_\tau$. For all sufficiently small $t > 0$ and τ, the map $g_{t\tau}$ is C^k close to F and hence, is a partially hyperbolic (in the narrow sense) C^∞ diffeomorphism of M. It preserves the Riemannian volume in M and is ergodic by Proposition B.3. It remains to show that $g_{t\tau}$ has nonzero Lyapunov exponents almost everywhere.

Denote by $E^s_{t\tau}(w)$, $E^u_{t\tau}(w)$, and $E^c_{t\tau}(w)$ the stable, unstable, and neutral subspaces at a point $w \in M$ for the map $g_{t\tau}$. It suffices to show that for almost everywhere point $w \in M$ and every vector $v \in E^c_\tau(w)$, the Lyapunov exponent $\chi(w, v) \ne 0$.

Set $\kappa_{t\tau}(w) = Dg_{t\tau}|E^u_{t\tau}(w)$, $w \in M$. By Proposition B.6, for all sufficiently small $\tau > 0$,

$$\int_M \log \kappa_{0\tau}(w)\, dw < \log\eta.$$

The subspace $E^u_{t\tau}(w)$ depends continuously on t and τ (for a fixed w; for details see the paper by Burns, Pugh, Shub, and Wilkinson in this volume) and hence, so does $\kappa_{t\tau}$. It follows that for all sufficiently small $\tau > 0$, there is $t > 0$ such that

$$\int_M \log \kappa_{t\tau}(w)\, dw < \log\eta.$$

Denote by $\chi^s_{t\tau}(w)$, $\chi^u_{t\tau}(w)$, and $\chi^c_{t\tau}(w)$ the Lyapunov exponents of $g_{t\tau}$ at the point $w \in M$ in the stable, unstable, and neutral directions respectively (since these directions are one-dimensional the Lyapunov exponents do not depend on the vector). By the ergodicity of $g_{t\tau}$, we have that for almost every $w \in M$,

$$\chi^u_{t\tau}(w) = \lim_{n\to\infty} \frac{1}{n} \log \prod_{i=0}^{n-1} \kappa_{t\tau}(g^i_{t\tau}(w)).$$

By the Birkhoff ergodic theorem, we get

$$\chi^u_{t\tau}(w) = \int_M \log \kappa_{t\tau}(w)\, dw < \log\eta.$$

Since $E^s_{t\tau}(w) = E^s_{00}(w) = E^s_F(w)$ for every t and τ, we conclude that $\chi^s_{t\tau}(w) = -\log\eta$ for almost every $w \in M$. Since $g_{t\tau}$ is volume preserving,

$$\chi^s_{t\tau}(w) + \chi^u_{t\tau}(w) + \chi^c_{t\tau}(w) = 0$$

for almost every $w \in M$. It follows that $\chi^c_{t\tau}(w) \neq 0$ for almost every $w \in M$ and hence, $g_{t\tau}$ has nonzero Lyapunov exponents almost everywhere. This completes the proof of the proposition. □

B.4. Ergodicity of the Map $g_{t\tau}$.

PROPOSITION B.3. *For every sufficiently small $t > 0$ and $\tau \geq 0$ the map $g_{t\tau}$ is ergodic.*

PROOF. Consider a partially hyperbolic (in the narrow sense) diffeomorphism f of a compact Riemannian manifold M preserving the Riemannian volume. Two points x, $y \in M$ are called *accessible (with respect to f)* if they can be joined by a piecewise differentiable piecewise nonsingular path which consists of segments tangent to either E^u or E^s. The diffeomorphism f satisfies the *essential accessibility property* if almost any two points in M (with respect to the Riemannian volume) are accessible. We will show that the map $g_{t\tau}$ has the essential accessibility property. The ergodicity of the map will then follow from the result by Pugh and Shub (see [**PS**]; see also the paper by Burns, Pugh, Shub, and Wilkinson in this volume).

Given a point $w \in M$, denote by $\mathcal{A}(w)$ the set of points $q \in M$ such that w and q are accessible. Set $I_p = \{p\} \times (0,1)$.

LEMMA B.4. *For every $z \in (0,1)$,*

$$\mathcal{A}(p,z) \supset I_p. \tag{B.3}$$

PROOF OF LEMMA B.4. We use the coordinate system (x,y,z) in Ω_1 described above. Since the map h_t preserves the center leaf I_p, we have that

$$h_t(0,0,z) = (h_t^{(1)}(0,0,z), h_t^{(2)}(0,0,z), h_t^{(3)}(0,0,z)) = (0,0,h_t^{(3)}(0,0,z))$$

for $z \in (0,1)$. It suffices to show that for every $z \in (0,1)$,

$$\mathcal{A}(p,z) \supset \{(p,a) : a \in [(h_t^{-\ell})^{(3)}(p,z), z]\}, \tag{B.4}$$

where ℓ is chosen by (B.1). In fact, since accessibility is a transitive relation and $h_t^{-n}(p,z) \to (p,0)$ for any $z \in (0,1)$, (B.4) implies that $\mathcal{A}(p,z) \supset \{(p,a) : a \in (0,z]\}$. Since this holds true for all $z \in (0,1)$ and accessibility is a reflexive relation, we obtain (B.3).

Now we proceed with the proof of (B.4).

Let $q_1 \in V^u_{t\tau}(p)$ and $q_2 \in V^s_{t\tau}(p)$ be two points constructed in Section B.3. The intersection $V^s_{t\tau}(q_1) \cap V^u_{t\tau}(q_2)$ is not empty and consists of a single point q_3. We will prove that for any $z_0 \in (0,1)$, there exist $z_i \in (0,1)$, $i = 1,2,3,4$ such that

$$(q_1, z_1) \in V^u_{t\tau}((p,z_0)), \quad (q_3, z_3) \in V^s_{t\tau}((q_1,z_1)),$$
$$(q_2, z_2) \in V^u_{t\tau}((q_3,z_3)), \quad (p, z_4) \in V^s_{t\tau}((q_2,z_2))$$

and

$$z_4 \leq (h_t^{-\ell})^{(3)}(p,z_0). \tag{B.5}$$

See Figure B.2. This means that $(p,z_4) \in \mathcal{A}(p,z_0)$. By continuity, we conclude that

$$\{(p,a) : a \in [z_4, z_0]\} \subset \mathcal{A}(p,z_0)$$

and (B.4) follows.

Since $g_{t\tau}$ preserves the xz-plane, we have that $V^{uc}_{t\tau}((p,z_0)) = V^{uc}_F((p,z_0))$. Hence, there is a unique $z_1 \in (0,1)$ such that $(q_1,z_1) \in V^u_{t\tau}((p,z_0))$. Notice that

$$g_{t\tau}^{-n}(p,z_0) = (p, h_t^{-n}((p,z_0)), \quad g_{t\tau}^{-n}(q_1,z_1) = (A^{-n}q_1, z_1)$$

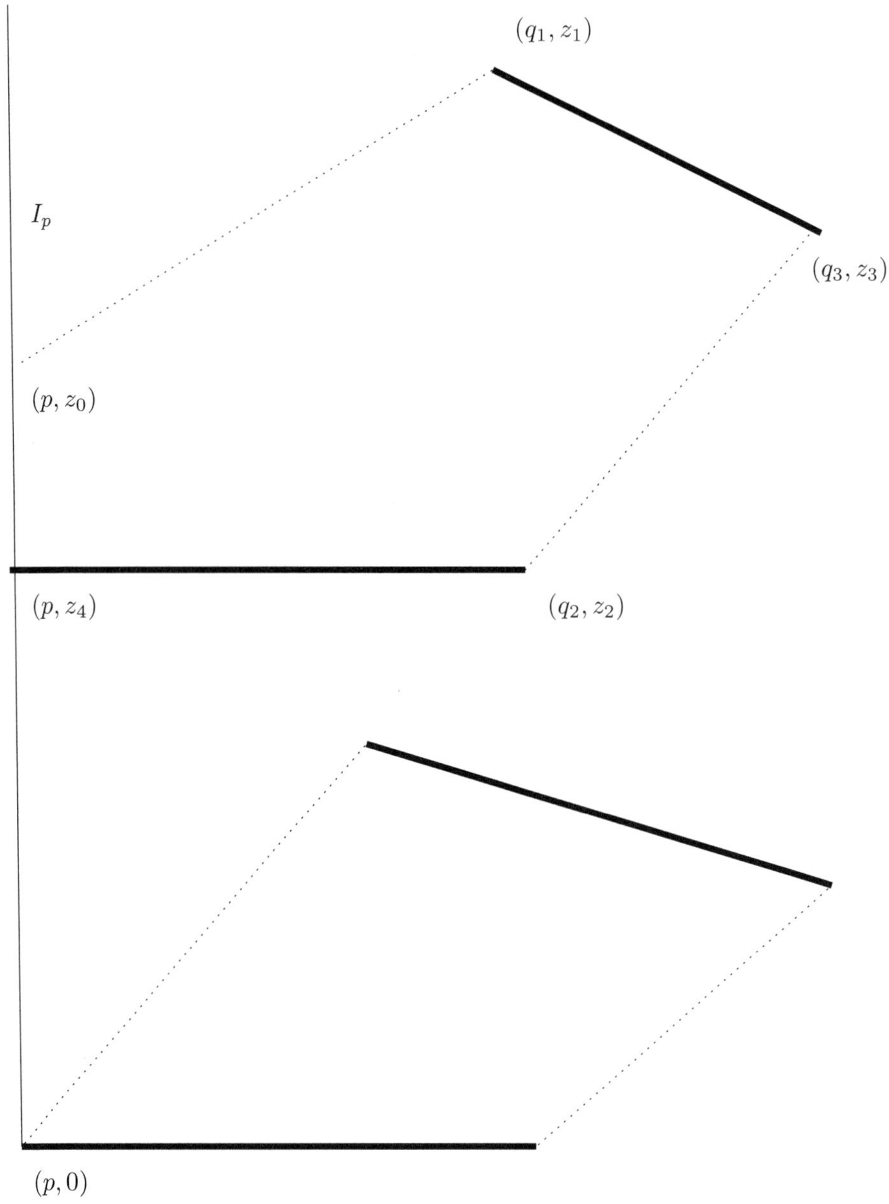

FIGURE B.2. Bold lines are stable manifolds, dotted — unstable ones

for $n \leq \ell$. This is true because the points $A^{-n}q_1$, $n = 0, 1, \ldots, \ell$ lie outside the ε-neighborhood of I_p, where the perturbation map $h_t = \mathrm{Id}$. Similarly, since the points $A^{-n}q_1$, $n > \ell$ lie inside the ε'-neighborhood of I_p, and the third component of h_t depends only on the z-coordinate, we have

$$g_{t\tau}^{-n}(q_1, z_1) = (A^{-n}q_1, h_t^{-n+\ell}z_1).$$

Since $d(g_{t\tau}^{-n}((p,z_0)), g_{t\tau}^{-n}((q_1,z_1))) \to 0$ as $n \to \infty$, we have

$$d(h_t^{-n}((p,z_0)), h_t^{-n+\ell}((p,z_1))) \to 0$$

as $n \to \infty$. It follows that $z_1 = (h_t^{-\ell})^{(3)}((p,z_0))$.

By the construction of the map h_t (that is $h_t = \text{Id}$ outside Ω_1) the sets $A^{-n_1}V_{t\tau}^s(p')$ and $A^{n_2}V_{t\tau}^u(p')$ are pieces of horizontal lines. This means that $z_2 = z_3 = z_1$.

Since the third component of h_t is nondecreasing from (q_2, z_2) to (p, z_4) along $V_{t\tau}^s(p)$, we conclude that $z_4 \le z_3 = z_1 = (h_t^{-\ell})^3(p, z_0)$ and thus (B.5) holds. □

The essential accessibility property follows from Lemma B.4 and the following statement.

LEMMA B.5 (see [**NT**]). *Assume that any two points in I_p are accessible. Then the map $g_{t\tau}$ satisfies the essential accessibility property.*

PROOF OF LEMMA B.5. It is easy to see that for any two points $x, y \in M$ which do not lie on the boundary of M one can find points $x', y' \in I_p$ such that the pairs (x, x') and (y, y') are accessible. By Lemma B.4 the points x', y' are accessible. Since accessibility is a transitive relation the result follows. □

This completes the proof of the proposition. □

B.5. Hyperbolicity of the Map $g_{0\tau}$. In this section we show that for all sufficiently small τ, the map $g_{0\tau}$ has nonzero average Lyapunov exponent in the central direction. Since this map is ergodic this implies that $g_{0\tau}$ has nonzero Lyapunov exponents almost everywhere.

PROPOSITION B.6. *For any sufficiently small $\tau > 0$,*

$$\int_M \log \kappa_{0\tau}(w)dw < \log \eta.$$

PROOF. Our approach is an elaboration of an argument in [**SW**].

For any $w \in M$, we introduce the coordinate system in T_wM associated with the splitting $E_F^u(w) \oplus E_F^s(w) \oplus E_F^c(w)$. Given $\tau \ge 0$ and $w \in M$, there exists a unique number $\alpha_\tau(w)$ such that the vector $v_\tau(w) = (1, 0, \alpha_\tau(w))^t$ lies in $E_{0\tau}^u(w)$ (where t denotes the transpose). Since the map $\widetilde{h}_\tau$ preserves the y coordinate, by the definition of the function $\alpha_\tau(w)$, one can write the vector $Dg_{0\tau}(w)v_\tau(w)$ in the form

$$Dg_{0\tau}(w)v_\tau(w) = (\bar{\kappa}_\tau(w), 0, \bar{\kappa}_\tau(w)\alpha_\tau(g_{t0}(w)))^t \tag{B.6}$$

for some $\bar{\kappa}_\tau(w) > 1$. Since the expanding rate of $Dg_{0\tau}(w)$ along its unstable direction is $\kappa_{0\tau}(w)$ we obtain that

$$\kappa_{0\tau}(w) = \bar{\kappa}_\tau(w)\frac{\sqrt{1+\alpha_\tau(g_{0\tau}(w))^2}}{\sqrt{1+\alpha_\tau(w)^2}}.$$

Since $E_{0\tau}^u(w)$ is close to $E_{00}^u(w)$ the function $\alpha_\tau(w)$ is uniformly bounded. Using the fact that the map $g_{0\tau}$ preserves the Riemannian volume we find that

$$L_\tau = \int_M \log \kappa_{0\tau}(w)\, dw = \int_M \log \bar{\kappa}_\tau(w)\, dw. \tag{B.7}$$

Consider the map $\widetilde{h}_\tau$. Since it preserves the y-coordinate using (B.2), we can write that

$$\widetilde{h}_\tau(x,y,z) = (r\cos\sigma, y, r\sin\sigma),$$

where $\sigma = \sigma(\tau, r, \theta, y) = \theta + \tau\psi(y)\rho(r)$. Therefore, the differential

$$D\widetilde{h}_\tau \colon E^u_F(w) \oplus E^c_F(w) \to E^u_F(g_{0\tau}(w)) \oplus E^c_F(g_{0\tau}(w))$$

can be written in the matrix form

$$\begin{aligned} D\widetilde{h}_\tau(w) &= \begin{pmatrix} A(\tau,w) & B(\tau,w) \\ C(\tau,w) & D(\tau,w) \end{pmatrix} \\ &= \begin{pmatrix} r_x\cos\sigma - r\sigma_x\sin\sigma & r_y\cos\sigma - r\sigma_y\sin\sigma \\ r_x\sin\sigma + r\sigma_x\cos\sigma & r_y\sin\sigma + r\sigma_y\cos\sigma \end{pmatrix}, \end{aligned}$$

where

$$\begin{aligned} r_x &= \frac{\partial r}{\partial x} = \frac{x}{r} = \cos\theta, \quad r_z = \frac{\partial r}{\partial z} = \frac{y}{r} = \sin\theta, \\ \sigma_x &= \frac{\partial \sigma}{\partial x} = \frac{-z}{r^2} + \frac{z}{r}\tau\widetilde{\rho}_r(y,r) = \frac{\sin\theta}{r} + \tau\widetilde{\rho}_r(y,r)\cos\theta, \\ \sigma_z &= \frac{\partial \sigma}{\partial z} = \frac{x}{r^2} + \frac{x}{r}\tau\widetilde{\rho}_r(y,r) = \frac{\cos\theta}{r} + \tau\widetilde{\rho}_r(y,r)\sin\theta, \end{aligned}$$

and $\widetilde{\rho}(y,r) = \psi(y)\rho(r)$. It is easy to check that

$$\begin{aligned} A &= A(\tau,w) = 1 - \tau r\widetilde{\rho}_r\sin\theta\cos\theta - \frac{\tau^2\widetilde{\rho}^2}{2} - \tau^2 r\widetilde{\rho}\widetilde{\rho}_r\cos^2\theta + O(\tau^3), \\ B &= B(\tau,w) = -\tau\widetilde{\rho} - \tau r\widetilde{\rho}_r\sin^2\theta - \tau^2 r\widetilde{\rho}\widetilde{\rho}_r\sin\theta\cos\theta + O(\tau^3), \\ C &= C(\tau,w) = \tau\widetilde{\rho} + \tau r\widetilde{\rho}_r\cos^2\theta - \tau^2 r\widetilde{\rho}\widetilde{\rho}_r\sin\theta\cos\theta + O(\tau^3), \\ D &= D(\tau,w) = 1 + \tau r\widetilde{\rho}_r\sin\theta\cos\theta - \frac{\tau^2\widetilde{\rho}^2}{2} - \tau^2 r\widetilde{\rho}\widetilde{\rho}_r\sin^2\theta + O(\tau^3). \end{aligned} \tag{B.8}$$

By Lemma B.7 below, we have

$$L_\tau = \log\eta - \int_M \log(D(\tau,w) - \eta B(\tau,w)\alpha_\tau(g_{0\tau}(w)))dw.$$

By Lemma B.8, we have

$$\left.\frac{dL_\tau}{d\tau}\right|_{\tau=0} = 0, \qquad \left.\frac{d^2L_\tau}{d\tau^2}\right|_{\tau=0} < 0.$$

So we can choose τ so small that $L_\tau \neq \log\eta$. □

LEMMA B.7.

$$L_\tau = \log\eta - \int_M \log(D(\tau,w) - \eta B(\tau,w)\alpha_\tau(g_{0\tau}(w)))dw.$$

PROOF OF LEMMA B.7. Since $g_{0\tau} = h_0 \circ F \circ \widetilde{h}_\tau = F \circ \widetilde{h}_\tau$, we have that

$$D_\tau(w) = Dg_{0\tau}(w)|E^u_{0\tau}(w) \oplus E^c_{0\tau}(w) = \begin{pmatrix} \eta A(\tau,w) & \eta B(\tau,w) \\ C(\tau,w) & D(\tau,w) \end{pmatrix}.$$

By (B.6),

$$\begin{aligned} D_\tau(w)\begin{pmatrix} 1 \\ \alpha_\tau(w) \end{pmatrix} &= \begin{pmatrix} \eta A(\tau,w) + \eta B(\tau,w)\,\alpha_\tau(w) \\ C(\tau,w) + D(\tau,w)\,\alpha_\tau(w) \end{pmatrix} \\ &= \begin{pmatrix} \kappa_\tau(w) \\ \kappa_\tau(w)\,\alpha_\tau(g_{0\tau}(w)) \end{pmatrix}. \end{aligned} \tag{B.9}$$

Since $\widetilde{h}_\tau$ is volume preserving, $AD - BC = 1$ and therefore,

$$A + B\alpha = \frac{1}{D} + \frac{B}{D}(C + D\alpha).$$

Comparing the components in (B.9), we obtain

$$\begin{aligned}\kappa_\tau(w) =& \eta(A(\tau,w) + B(\tau,w)\,\alpha_\tau(w))\\ =& \eta\left(\frac{1}{D(\tau,w)} + \frac{B(\tau,w)}{D(\tau,w)}(C(\tau,w) + D(\tau,w)\,\alpha_\tau(w))\right)\\ =& \eta\left(\frac{1}{D(\tau,w)} + \frac{B(\tau,w)}{D(\tau,w)}(\kappa_\tau(w)\,\alpha_\tau(g_{0\tau}(w)))\right).\end{aligned}$$

Solving for $\kappa_\tau(w)$, we get

$$\kappa_\tau(w) = \frac{\eta}{D(\tau,w) - \eta B(\tau,w)\,\alpha_\tau(g_{0\tau}(w))}.$$

The desired result follows from (B.7). □

LEMMA B.8.

$$\left.\frac{dL_\tau}{d\tau}\right|_{\tau=0} = 0, \qquad \left.\frac{d^2L_\tau}{d\tau^2}\right|_{\tau=0} < 0. \tag{B.10}$$

PROOF OF LEMMA B.8. In order to simplify notations we set $D'_\tau = \frac{\partial D}{\partial\tau}$, $B'_\tau = \frac{\partial B}{\partial\tau}$, $C'_\tau = \frac{\partial C}{\partial\tau}$, $D''_{\tau\tau} = \frac{\partial^2 D}{\partial\tau^2}$, and $B''_{\tau\tau} = \frac{\partial^2 B}{\partial\tau^2}$. Since the function $\alpha_\tau(w)$ is differentiable over τ (see the paper by Burns, Pugh, Shub, and Wilkinson in this volume) by Lemma B.7, we find

$$\frac{dL_\tau}{d\tau} = -\int_M \frac{D'_\tau - \eta B'_\tau\alpha(g_{0\tau}(w)) - \eta B\frac{\partial\alpha_\tau(w)}{\partial\tau}(g_{0\tau}(w))}{D(\tau,w) - \eta B(\tau,w)\alpha_\tau(w)(g_{0\tau}(w))}\,dw$$

and therefore,

$$\begin{aligned}\frac{d^2L_\tau}{d\tau^2} =& \int_M \left(\frac{D'_\tau - \eta B'_\tau\alpha(g_{0\tau}(w)) - \eta B(\tau,w)\frac{\partial\alpha_\tau(w)}{\partial\tau}(g_{0\tau}(w))}{D(\tau,w) - \eta B(\tau,w)\alpha_s(g_{0\tau}(w))}\right)^2 dw\\ &- \int_M \frac{E(\tau,w)}{D(\tau,w) - \eta B(\tau,w)\,\alpha_\tau(g_{0\tau}(w))}dw,\end{aligned}$$

where

$$\begin{aligned}E(\tau,w) =& D''_{\tau\tau} - \eta B''_{\tau\tau}\alpha(g_{0\tau}(w))\\ &- \eta B(\tau,w)\frac{\partial^2\alpha_\tau(w)}{\partial\tau^2}(g_{0\tau}(w)) - 2\eta B'_\tau\frac{\partial\alpha_\tau(w)}{\partial\tau}(g_{0\tau}(w)).\end{aligned}$$

Note that for all $w \notin \Omega_2$,

$$A(\tau,w) = D(\tau,w) = 1, \quad C(\tau,w) = B(\tau,w) = 0$$

and for all $w \in M$,

$$A(0,w) = D(0,w) = 1, \quad C(0,w) = B(0,w) = 0, \quad \alpha_0(w) = 0.$$

It follows that

$$\left.\frac{dL_\tau}{d\tau}\right|_{\tau=0} = \int_{\Omega_2} D'_\tau\,dw, \tag{B.11}$$

and also that

$$\text{(B.12)}\qquad \frac{d^2L_\tau}{d\tau^2}\Big|_{\tau=0}=\int_{\Omega_2}\left[(D'_\tau)^2-D''_{\tau\tau}+2\eta B'_\tau\frac{\partial\alpha_\tau(w)}{\partial\tau}(g_{0\tau}(w))\right]_{\tau=0}dw.$$

By (B.8), we obtain that

$$D'_\tau(0,w)=r\widetilde{\rho}_r(r)\sin\theta\cos\theta$$

and hence,

$$\int_{\Omega_2}D'_\tau dw=0.$$

Therefore, (B.11) implies the equality in (B.10).

We now proceed with the inequality in (B.10). Applying Lemma B.9 below we obtain that

$$\frac{\partial\alpha}{\partial\tau}(g_{0\tau}(w))\Big|_{\tau=0}=\frac{C'_\tau(0,w)}{\eta}+\sum_{n=1}^{\infty}\frac{C'_\tau(0,g_{00}^{-n}(w))}{\eta^{n+1}}.$$

It follows that

$$\begin{aligned}2\eta B'_\tau(0,w)\frac{\partial\alpha}{\partial\tau}(g_{0\tau}(w))\Big|_{\tau=0}=&2B'_\tau(0,w)C'_\tau(0,w)\\&+2B'_\tau(0,w)\sum_{n=1}^{\infty}\frac{C'_\tau(0,g_{00}^{-n}(w))}{\eta^n}.\end{aligned}$$

First, we evaluate the term

$$\mathcal{F}(w)=D'_\tau(0,w)^2-D''_{\tau\tau}(0,w)+2B'_\tau(0,w)C'_\tau(0,w).$$

Using (B.8), we find that

$$\text{(B.13)}\qquad\begin{aligned}\mathcal{F}(w)=&(r\widetilde{\rho}_r\sin\theta\cos\theta)^2+(\widetilde{\rho}^2+2r\widetilde{\rho}\widetilde{\rho}_r\sin^2\theta)\\&-2(\widetilde{\rho}+r\widetilde{\rho}_r\sin^2\theta)(\widetilde{\rho}+r\widetilde{\rho}_r\cos^2\theta)\\=&-\widetilde{\rho}^2-(r\widetilde{\rho}_r\sin\theta\cos\theta)^2-2r\widetilde{\rho}\widetilde{\rho}_r\cos^2\theta.\end{aligned}$$

Recall that $\Omega_2=B^{uc}(\bar{q},\varepsilon_0)\times B^s(\bar{q},\varepsilon_0)$ and $\widetilde{\rho}(r)=0$ if $r\geq\varepsilon$. We have

$$\text{(B.14)}\qquad\int_{\Omega_2}2r\widetilde{\rho}\widetilde{\rho}_r\cos^2\theta\,dw=\int_{-\varepsilon_0}^{\varepsilon_0}dy\int_0^{2\pi}2\cos^2\theta\,d\theta\int_0^{\varepsilon}r^2\widetilde{\rho}\widetilde{\rho}_r\,dr.$$

Since $0=\widetilde{\rho}(0)=\widetilde{\rho}(\varepsilon)$ (by the definition of the function ρ), we find that

$$\int_0^{\varepsilon}r^2\widetilde{\rho}\widetilde{\rho}_r\,dr=\frac{1}{2}r^2\widetilde{\rho}^2\Big|_0^{\varepsilon}-\int_0^{\varepsilon}r\widetilde{\rho}^2\,dr=-\int_0^{\varepsilon}r\widetilde{\rho}^2\,dr.$$

We also have that

$$\text{(B.15)}\qquad\int_0^{2\pi}2\cos^2\theta\,d\theta=\int_0^{2\pi}d\theta.$$

It follows from (B.14)–(B.15) that

$$\text{(B.16)}\qquad-\int_{\Omega_2}2r\widetilde{\rho}\widetilde{\rho}_r\cos^2\theta\,dw=\int_{\Omega_2}r\widetilde{\rho}^2\,dw\leq\varepsilon\int_{\Omega_2}\widetilde{\rho}^2\,dw.$$

Arguing similarly one can show that

$$\text{(B.17)}\qquad-\int_{\Omega_2}(r\widetilde{\rho}_r\sin\theta\cos\theta)^2\,dw=-\frac{1}{8}\int_{\Omega_2}(r\widetilde{\rho}_r)^2\,dw$$

Thus we conclude using (B.13), (B.16), and (B.17) that

$$\int_{\Omega_2} \mathcal{F}(w)\,dw \le -(1-\varepsilon)\int_{\Omega_2} \widetilde{\rho}_r^2\,dw - \frac{1}{8}\int_{\Omega_2} (r\widetilde{\rho}_r)^2\,dw < 0. \tag{B.18}$$

We now evaluate the remaining term

$$\mathcal{G}(w) = \sum_{n=1}^{\infty} \frac{1}{\eta^i} \int_{\Omega_2} 2B'_\tau(0,w)C'_\tau(0, g_{00}^{-n}(w))\,dw.$$

Since the map $g_{00} = F$ preserves the Riemannian volume we obtain that

$$\begin{aligned}\int_{\Omega_2} 2B'_\tau(0,w)C'_\tau(0,g_{00}^{-n}(w))\,dw &\le \int_{\Omega_2} B'_\tau(0,w)^2\,dw + \int_{\Omega_2} C'_\tau(0,g_{00}^{-n}(w))^2\,dw \\ &= \int_{\Omega_2} B'_\tau(0,w)^2\,dw + \int_{\Omega_2} C'_\tau(0,w)^2\,dw\end{aligned}$$

Applying (B.8), we find that

$$\begin{aligned}&\int_{\Omega_2} B'_\tau(0,w)^2\,dw + \int_{\Omega_2} C'_\tau(0,w)^2\,dw \\ &= \int_{\Omega_2} (\widetilde{\rho} + r\widetilde{\rho}_r \sin^2\theta)^2\,dw + \int_{\Omega_2} (\widetilde{\rho} + r\widetilde{\rho}_r\cos^2\theta)^2\,dw \\ &\le 4\left(\int_{\Omega_2} \widetilde{\rho}^2\,dw + \int_{\Omega_2} r^2\widetilde{\rho}_r^2\,dw\right).\end{aligned}$$

It follows that for sufficiently large $N > 0$ (which does not depend on ε)

$$\sum_{i=N}^{\infty} \frac{1}{\eta^i}\int_{\Omega_2} 2B'_\tau(0,w)C'_\tau(0,g_{00}^{-i}(w))\,dw \le \frac{1}{10}\left(\int_{\Omega_2} \widetilde{\rho}^2\,dw + \int_{\Omega_2} r^2\widetilde{\rho}_r^2\,dw\right). \tag{B.19}$$

Note that if $g_{00}^{-n}\Omega_2 \cap \Omega_2 = \varnothing$, then $B'_\tau(0,w)C'_\tau(0,g_{00}^{-n}(w)) = 0$ for all w. Hence,

$$\int_{\Omega_2} 2B'_\tau(0,w)C'_\tau(0,g_{00}^{-n}(w))\,dw = 0.$$

We may choose the point q and a small ε such that $g_{00}^{-n}\Omega_2 \cap \Omega_2 = F^{-n}\Omega_2 \cap \Omega_2 = \varnothing$ for all $n = 1, 2, \ldots, N$. It follows from (B.12), (B.18), and (B.19) that

$$\begin{aligned}\left.\frac{d^2L_\tau}{d\tau^2}\right|_{\tau=0} &= \int_{\Omega_2} \mathcal{F}(w)\,dw + \int_{\Omega_2} \mathcal{G}(w)\,dw \\ &\le -\left(\frac{9}{10} - \varepsilon\right)\int_{\Omega_2} \widetilde{\rho}^2\,dw - \frac{1}{40}\int_{\Omega_2} r^2\widetilde{\rho}_r^2\,dw < 0.\end{aligned}$$

The desired result follows. □

LEMMA B.9.

$$\left.\frac{\partial\alpha}{\partial\tau}(g_{0\tau}(w))\right|_{\tau=0} = \sum_{n=0}^{\infty} \frac{C'_\tau(0, g_{00}^{-n}(w))}{\eta^{n+1}}.$$

PROOF OF LEMMA B.9. Define

$$R(\tau, w, \alpha) = \frac{C(\tau,w) + D(\tau,w)\alpha}{\eta(A(\tau,w) + B(\tau,w)\alpha)}.$$

It follows from (B.6) that

$$\alpha_\tau(g_{0\tau}(w)) = R(\tau, w, \alpha_\tau(w)). \tag{B.20}$$

By (B.6) and (B.8), we have

$$\frac{\partial R}{\partial \tau}\Big|_{\tau=0} = \frac{(C'_\tau + D'_\tau\alpha)(A+B\alpha) + (C+D\alpha)(A'_\tau + B'_\tau\alpha)}{\eta(A+B\alpha)^2}\Big|_{\tau=0} = \frac{C'_\tau(0,w)}{\eta}.$$

Since $AD - BC = 1$,

$$\frac{\partial R}{\partial \alpha}\Big|_{\tau=0} = \frac{AD-BC}{\eta(A+B\alpha)^2}\Big|_{\tau=0} = \frac{1}{\eta}.$$

It follows from (B.20) that

$$\frac{\partial \alpha}{\partial \tau}(g_{0\tau}(w))\Big|_{\tau=0} = \frac{C'_\tau(0,w)}{\eta} + \frac{1}{\eta}\cdot\frac{\partial \alpha}{\partial \tau}(w)\Big|_{\tau=0}.$$

Since this inequality holds for any w, replacing w with $g_{0\tau}^{-1}(w)$ we obtain

$$\frac{\partial \alpha}{\partial \tau}(w)\Big|_{\tau=0} = \frac{C'_\tau(0,g_{0\tau}^{-1}(w))}{\eta} + \frac{1}{\eta}\cdot\frac{\partial \alpha}{\partial \tau}(g_{0\tau}^{-1}(w))\Big|_{\tau=0}.$$

The result follows by induction. □

References

[NT] V. Niţică and A. Török, *An open and dense set of stably ergodic diffeomorphisms in a neighborhood of a nonergodic one*, Topology, to appear.

[PS] C. Pugh and M. Shub, *Stable ergodicity and julienne quasi-conformality*, J. Eur. Math. Soc. **2** (2000), no. 1, 1–52.

[SW] M. Shub and A. Wilkinson, *Pathological foliations and removable zero exponents*, Invent. Math. **139** (2000), 495–508.

Department of Mathematics, The Pennsylvania State University, University Park, PA 16802, U.S.A.

E-mail address: `dolgop@math.psu.edu`

URL: `http://www.math.psu.edu/dolgop/`

Department of Mathematics, The Pennsylvania State University, University Park, PA 16802, U.S.A.

E-mail address: `hu@math.psu.edu`

URL: `http://www.math.psu.edu/hu/`

Department of Mathematics, The Pennsylvania State University, University Park, PA 16802, U.S.A.

E-mail address: `pesin@math.psu.edu`

URL: `http://www.math.psu.edu/pesin/`

Proceedings of Symposia in Pure Mathematics
Volume **69**, 2001

Cocycles, Cohomology and Combinatorial Constructions in Ergodic Theory

Anatole Katok in collaboration with E. A. Robinson, Jr.

Dedicated to the memory of Michel Herman (1942–2000)

Introduction

Cocycles and cohomological equations play a central role in ergodic theory as well as in its applications to other areas of dynamics. Among the questions which are reduced to cohomological considerations are existence of invariant measures and invariant geometric structures, time change for flows and, more generally, actions of continuous groups, orbit equivalence and its restricted versions, existence of eigenfunctions, classification of various kinds of product actions and many others. The subject is very diverse and includes many measure–theoretic, algebraic, analytic and geometric aspects. Very broadly, cohomological considerations produce two types of conclusions:

(i) if the set (it is very often not a group) of cohomology classes under consideration allows a reasonable structure then the corresponding objects allow a nice classification. In the extreme case when the set of cohomology classes is very small (e.g. if there is only one class) one speaks about *rigidity*;

(ii) if there is no good structure in the set of cohomology classes then individual classes are usually very "chaotic" and this often leads to construction of objects within a given class with various, often exotic properties. Conclusions of this type thus lead to genericity statements, as well as of counterexamples to certain naturally sounding hypotheses.

In the classical ergodic theory, which deals with measure preserving or nonsingular actions of $\mathbb{Z}$ and $\mathbb{R}$, conclusions of the first type never appear; the same is true in the more general context of ergodic theory for actions of amenable groups. In the topological context the situation changes only slightly. On the other hand, there is a variety of interesting situations in finer categories such as Hölder, C^r, $1 \le r \le \infty$, or real analytic, where a description of cohomology classes is possible, and produces crucial insights into various classification problems.

It is interesting to notice a difference between the classical cases of a $\mathbb{Z}$ or an $\mathbb{R}$ action, where such a classification only very rarely amounts to rigidity, and other cases, such as actions of higher rank abelian groups (i.e. $\mathbb{Z}^k$ or $\mathbb{R}^k$ for $k \le 2$), where rigidity is quite widespread and is often crucial for understanding of structural questions.

Furthermore, for actions of groups which are both "large" and "rigid", rigidity appears and becomes, in many respects, prevalent already at the ergodic theory level, e.g. in the measurable category. In fact, one of the central issues in dynamics of actions of such groups is a translation of the available measurable rigidity results into topological and especially differentiable context. The prime examples of such groups are noncompact semisimple Lie groups of real rank greater than one and lattices (discrete subgroups of cofinite volume) in such groups. Kazhdan property (T) is often a key ingredient in producing various rigidity properties, although by itself it leads only to a limited array of results. Still some rigidity results are being extended to specific property (T) groups beyond semisimple Lie groups of higher rank and their lattices.

In the present work we almost completely concentrate on the classical cases. A more general setting appears only in basic definitions and in some comments meant to underline the contrast with those. The first two chapters deal with the measurable setting and hence all the results we prove or discuss are of type (ii). In the third chapter we introduce a general framework for the cohomological phenomena of type (i) (stability and effectiveness) with rigidity as a special case. We illustrate how these phenomena appear in several topological, symbolic and smooth settings. In the last two chapters we return to the type (ii) phenomena but in more specialized settings. Our aim there is to demonstrate how a "controlled chaos" can be produced in a variety of simple smooth or other special situations. The crucial concept here is very fast periodic approximation in the number–theoretic (Liouvillean numbers), differentiable and measurable contexts.

The purpose of this work is to introduce the main concepts and principal techniques and illustrate those by a variety of interesting example including both general statements and the treatment of particular classes of systems. We do not aspire to present a comprehensive survey of the subject. Accordingly, we keep references to a minimum.

The present work is an updated, revised and expanded version of the second of the four parts of our work "Constructions in Ergodic Theory" originally intended to appear as a book form, which was mostly written on 1982-83, appended during the eighties and which has been circulated in the manuscript form.

"Constructions in Ergodic Theory" is dedicated to a systematic although by no means exhaustive development of several principal classes of combinatorial constructions of measure-preserving transformations which allow to obtain some nontrivial properties and which are well adapted to the realization of abstract measure–preserving transformations as smooth or real-analytic systems on compact manifolds preserving an absolutely continuous or smooth measure, and as other special classes of dynamical systems. An updated version of the first part which contains a definitive account of the general concept of periodic approximation as well as its applications to establishing genericity of various ergodic properties in a variety of categories is about to appear as [**K1**]. The third and fourth parts were left unfinished and their fate at the time of writing remains uncertain.

The developments of the last decade, especially those dealing with actions of groups other than $\mathbb{Z}$ and $\mathbb{R}$, changed the appearance, and, to a certain extent, even the basic perception of the area which is the subject of the present work. Still the program outlined and illustrated in Part II of "Constructions in Ergodic

Theory" has proved to be fundamentally sound. In fact, this program influenced some of the later developments such as the systematic use of, and the search for, invariant distributions for various classes of dynamical systems. We would like to repeat though that, revisions and additions notwithstanding, the present text to a large extent reflects the perspective of the early eighties and hence primarily describes and refers to work published by that time. We apologize for probably not sufficiently emphasizing certain more recent results.

We would like to thank Alistair Windsor who carefully read the text and made some valuable suggestions.

Michel Herman made fundamental contributions to differentiable dynamics, smooth ergodic theory, and, specifically the area covered in the present work. Aside from numerous published papers his thinking made a huge influence on the way we view these subjects. His untimely death is a tremendous loss for the world dynamics community.

1. Definitions and principal constructions

We are going to discuss a group of constructions which appear frequently both in the general theory of measure-preserving transformations and in various concrete situations. The central concept for this circle of ideas is the notion of an (untwisted) cocycle over a measure-preserving transformation (and more generally over a group action), and the corresponding notions of coboundary and cohomology. In order to explain this concept in the most natural way, we will leave, for the moment, the confines of classical ergodic theory, which deals with measure-preserving transformations and flows, and consider the actions of more general groups. We will be able to touch only a few aspects of this subject. For a systematic review of basic notions in the area see [**HaK**]. A useful introductory discussion can be found [**KH**], section 2.9.

Cocycles play a particularly important role in the ergodic theory and dynamics of actions of groups other than $\mathbb{Z}$ and $\mathbb{R}$. Various aspects of this subject are treated in e.g. [**Sch1**], [**Z1**], [**HK1**], [**KSp1**].

Another aspect of the subject which we will not be able to discuss has been developed in the remarkable papers of Herman [**H4, H5**]. It involves the use of cocycles in smooth ergodic theory, in particular to obtain below estimates of the Lyapunov characteristic exponents in nonuniformly hyperbolic situations.

1.1 Cocycles, coboundaries and Mackey range. Let us consider a Lebesgue space (X,μ) and let Γ and G be two locally compact second countable groups. Let us suppose that $S = \{S_\gamma\}_{\gamma\in\Gamma}$ is a measurable right action of Γ on (X,μ) by measure-preserving transformations. The action S is *ergodic* if any S–invariant set has measure zero or full measure or, equivalently, if any S–invariant measurable function is almost everywhere constant.

DEFINITION 1.1. A G *cocycle* over the action S is a measurable function $\alpha : X\times\Gamma \to G$ such that

$$\alpha(x,\gamma_1\gamma_2) = \alpha(x,\gamma_1)\alpha(S_{\gamma_1}x,\gamma_2) \tag{1.1}$$

It follows from (1.1) that $\alpha(x, id_\gamma) = id_G$ and $\alpha(x,\gamma^{-1}) = (\alpha(S_{\gamma^{-1}}x,\gamma))^{-1}$. From a purely formal point of view, this is a special case of a concept familiar in algebra, topology and other branches of mathematics. The same is true for the next two definitions.

DEFINITION 1.2. Two G cocycles α and β over a Γ action S are called *cohomologous* if there exists a measurable map $\psi : X \to G$ such that

$$\beta(x,\gamma) = \psi^{-1}(x)\alpha(x,\gamma)\psi(S_\gamma x). \tag{1.2}$$

Let us note that for any cocycle α and for any ψ the function β defined by (1.2) is also a cocycle.

DEFINITION 1.3. A cocycle α is called a *coboundary* if there exists a measurable map $\psi : X \to G$ such that

$$\alpha(x,\gamma) = \psi^{-1}(x)\psi(S_\gamma x). \tag{1.3}$$

Curiously, in various cohomology theories there seems to be no established name for a cochain which provides the equivalence between two cocycles, i.e. in our situation for the function ψ in (1.2). Following a prevalent, but by no means universal, usage in ergodic theory we will call those functions *transfer functions.*

Notice that if the action S is ergodic then a transfer function ψ is uniquely defined up to left multiplication by a constant function.

DEFINITION 1.4. We will call a cocycle α an *almost coboundary* if it is cohomologous to a cocycle with constant coefficients, i.e. if there exists a homomorphism $\phi : \Gamma \to G$ such that for some measurable $\psi : X \to G$

$$\alpha(x,y) = \psi^{-1}(x)\phi(\gamma)\psi(S_\gamma x) \tag{1.4}$$

For a given cocycle, it is natural to ask whether it is a coboundary or at least an almost coboundary. The answer to this question depends on the solvability of equations (1.3) and (1.4) for ψ. We will refer to such equations as *cohomological equations.*

Since the function defined by (1.3) is always a cocycle, it is easy to construct a lot of cocycles which are coboundaries. As algebraic intuition would suggest, and as is confirmed by the discussion below, coboundaries should be regarded as trivial cocycles. As long as homomorphisms of Γ into G are known one finds many almost coboundaries too. The existence of cocycles other than almost coboundaries in many cases represents a formidable problem. This may not sound surprising to those who are familiar with cohomology theories in algebra and topology. However, the structure (or rather the absence of any reasonable structure) of the set of all cohomology classes of cocycles in many cases is strikingly different from what might be expected from familiar analogies.

If G is an abelian group then the product of two cocycles (coboundaries) is again a cocycle (corr. coboundary); thus the cocycles form a group and the coboundaries a subgroup. Hence one can define the corresponding (first) cohomology group.

If $\Gamma = \mathbb{Z}$, all cocycles can be described rather easily. Namely, the cocycles are in a one-to-one correspondence with measurable functions $h : X \to G$. The formula

$$\alpha(x,n) = \begin{cases} h(x)h(Sx)\dots h(S^{n-1}x) & n \geq 0 \\ h^{-1}(S^{-1}x)\dots h^{-1}(S^n x) & n < 0 \end{cases} \tag{1.5}$$

determines a cocycle and every cocycle α can be represented in this way by making $h(x) = \alpha(x,1)$. A similar description is possible for the continuous time case ($\Gamma = \mathbb{R}$), subject to certain inessential restrictions (cf. Section 1.3).

The question of whether a cocycle over a $\mathbb{Z}$ action is a coboundary will be one of our central topics. The classification, up to cohomology, of cocycles over a $\mathbb{Z}$ action in a purely measurable setting does not make any sense as results of the next chapter will demonstrate. However, as will be seen in Chapter 3, in some finer categories it does become feasible for particular cases.

As we have noted, the construction of cocycles other than almost coboundaries for larger groups may be difficult. This is related to various rigidity phenomena which assert that within various classes any cocycle is an almost coboundary. For actions of some sufficiently large groups, such as semisimple Lie groups or lattices in such groups the situation is strikingly different from the classical cases and rigidity phenomena appear already in the measurable category, the best known example being Zimmer cocycle superrigidity theorem [**Z1**] [**FK**]. For actions of amenable groups the orbit equivalence theory [**CFW**] implies that the situation in measurable category is essentially as chaotic as for the classical cases. However for abelian groups of higher rank, e.g. $\Gamma = \mathbb{R}^h$ on $\mathbb{R}^h$ $h \geq 2$ nontrivial rigidity phenomena appear in Hölder and smooth categories [**KSp1**] [**KSch**] [**Sch1**].

The following construction, which is sometimes called the Mackey range [**Z1**], [**FK**], [**HaK**] allows us to associate with a G cocycle over a right Γ action, a left action of G. It generalizes the notion of induced action well known in the theory of group representations as well as constructions of the special flow (flow under a function), and the induced and special (integral) automorphisms, familiar in ergodic theory.

Any right Γ action $S = \{S_\gamma\}_{\gamma\in\Gamma}$ and G cocycle α over S determine a G extension $S^\alpha = \{S^\alpha_\gamma\}_{\gamma\in\Gamma}$ of S which acts on $X \times G$ by the following formula

$$S^\alpha_\gamma(x,g) = (S_\gamma x, g\alpha(x,\gamma)) \tag{1.6}$$

The cocycle equation (1.1) is equivalent to the group property for the extension

$$S^\alpha_{\gamma_1} S^\alpha_{\gamma_2} = S^\alpha_{\gamma_2\gamma_1}$$

since

$$S^\alpha_{\gamma_1} S^\alpha_{\gamma_2} = (S_{\gamma_1} S_{\gamma_2} x, g\alpha(x,\gamma_2)\alpha(S_{\gamma_2}x,\gamma_1)) = (S_{\gamma_2\gamma_1}x, g\alpha(x,\gamma_2\gamma_1)) = S^\alpha_{\gamma_2\gamma_1}(x,g).$$

There is a natural notion of isomorphism between two G extensions S^α and S^β of a Γ action S, namely an isomorphism which preserves every fiber $\{x\} \times G$, shifting it by an element of G

$$\psi(x,g) = (x, g\psi(x)). \tag{1.7}$$

Clearly

$$\psi \circ S^\beta_\gamma(x,g) = (S_\gamma x, g\beta(x,y)\psi(S_\gamma x))$$

and

$$S^\alpha_\gamma \circ \psi(x,g) = (S_\gamma x, g\psi(x)\alpha(x,\gamma)).$$

Thus two cocycles α and β are cohomologous if and only if S^α and S^β are isomorphic extensions and in particular α is a coboundary if and only if S^α is isomorphic to the trivial extension S^{id}:

$$S^{id}_\gamma(x,g) = (S_\gamma x, g) \tag{1.8}$$

and α is an almost coboundary if and only if S^α is isomorphic to a product extension S^ϕ

$$S^\phi_\gamma(x,g) = (S_\gamma x, g\phi(\gamma)) \tag{1.9}$$

where $\phi : \Gamma \to G$ is a group homomorphism.

The group G acts on $X \times G$ in a natural way by the left shifts $L_{g_0}(x,g) = (x, g_0 g)$, and this action obviously commutes with any extension S^α of the form (1.6). In particular it preserves the decomposition of $X \times G$ into orbits of the action S^α and thus, at least formally, we can consider the factor action of G on the space of these orbits. We will denote this factor action by L^α. In general, the space of its orbits may not have a good measurable structure. Even if it does, the natural invariant measure may be infinite. For example, for the trivial extension (1.8) the factor is naturally isomorphic to the group G itself, and if G is not compact there is no finite translation invariant measure. In general one takes *the measurable hull* of the partition into the orbits of S^α, i.e. the measurable partition corresponding to the σ–algebra of measurable sets consisting of whole orbits of S^α. The factor action L^α restricted to this measurable hull is called the *Mackey range* of the cocycle α.

In the case of the constant coefficient cocycle ϕ (cf. (1.9)) the action L^ϕ is known as the action induced by the homomorphism ϕ. This action has a natural finite invariant measure if the subgroup $\phi(\Gamma) \subseteq G$ is closed, unimodular, and has cofinite volume in G. In particular, this is true if $\phi(\Gamma)$ is discrete and the factor $G/\phi(\Gamma)$ is compact. We proceed to discuss a natural generalization of this last condition to arbitrary cocycles.

1.2 Lipschitz cocycles, Pseudo-isometries and the Ambrose–Kakutani theorem. Let us consider two metrics d and d' on the same topological space X. we will call these metric *uniformly equivalent* if there are positive constants A, B and C such that for $x, x' \in X$,

$$Ad(x,x') - C < d'(x,x') < Bd(x,x') + C \tag{1.10}$$

A map f between metric spaces (X, d_X) and (Y, d_Y) is called a *pseudo isometry* if for some positive constants A, B and C, and for every $x, x' \in X$, $x \neq x'$,

$$Ad_X(x,x') - C < d_Y(f(x), f(x')) < Bd_X(x,x') + C.$$

It is clear that if we replace the metrics d_X and d_Y with uniformly equivalent metrics d'_X and d'_Y then the pseudo isometry f remains a pseudo isometry with respect to these new metrics.

Let us now return to group actions and cocycles. We will assume for the remainder of this section that Γ is a finitely generated discrete group and that G is a locally compact Lie group. Under these assumptions, both G and Γ possess natural classes of uniformly equivalent left-invariant metrics. Namely for Γ we take the word-length metric determined by any finite system of generators and for G, any left invariant Riemannian metric.

DEFINITION 1.5. A G cocycle α over a Γ action S is called a *Lipschitz cocycle* if for almost any $x \in X$, the map $\alpha_x : \alpha_x(\gamma) = \alpha(\gamma, x)$ is a pseudo isometry from Γ to G with constants A, B, C independent of x.

The special cases $\Gamma = \mathbb{Z}^n$ and $G = \mathbb{R}^n$ or $\mathbb{Z}^n$ are discussed in [**K2**]. In these cases the Lipschitz condition can be replaced by weaker "integrability" conditions.

Some important properties of Lipschitz cocycles in a more general setting can be derived from the results in the unpublished paper [**BKM**] which concern the behavior of nets in metric spaces under pseudo isometries. A *net* is a discrete subset of a metric space with the property that every point is a bounded distance away from it.

THEOREM 1.6. [**BKM**], *Let Γ be a discrete cocompact (and hence finitely generated) subgroup of a connected Lie group G. Then for every Lipschitz G cocycle α over a Γ action S on a Lebesgue space (X, μ), the extension S^α has a fundamental domain $D = \bigcup_{x \in X} \{x\} \times D_x$ where D_x is a bounded set in G whose boundary has Haar measure zero.*

By taking the restriction of $\mu \times \lambda_G$ to D , where λ_G is Haar measure in G, we obtain a natural finite invariant measure for the G action L^α in the factor of $X \times G$ into orbits of S^α. Obviously this conclusion holds for any cocycle β cohomologous to a Lipschitz cocycle α since the factor actions L^α and L^β for cohomologous cocycles α and β are in a natural correspondence.

Let us illustrate this situation by the classical case $\Gamma = \mathbb{Z}$, $G = \mathbb{R}$. According to (1.5) any $\mathbb{R}$ cocycle α over a $\mathbb{Z}$ action $S = \{S^n\}_{n \in \mathbb{Z}}$ is determined by a function $h : X \to \mathbb{R}$. If we assume that

$$0 < A < h(x) < B \tag{1.11}$$

then the cocycle determined by h is Lipschitz. In this case the extension of S to $X \times \mathbb{R}$ is generated by the automorphism

$$S^h(x, t) = (Sx, t + h(x)).$$

The set

$$D = \{(x, t) : 0 \le t \le h(S^{-1}x)\}$$

is a fundamental domain for S^h and the factor action $L^h = \{L^h_t\}_{t \in \mathbb{R}}$ acts on D as the vertical flow, where pairs of points of the form $(x, 0)$ and $(S^{-1}x, h(S^{-1}x))$ are identified. In this way, the factor action can be naturally identified with the special flow over S^{-1} built under the function $h(S^{-1}x)$ and the invariant measure is induced from $X \times \mathbb{R}$. As we will show in the next chapter, for every L^1 function g such that $\int_X g d\mu > 0$, one can find a function h satisfying (1.11) such that

$$g(x) = h(x) + \psi(Sx) - \psi(x) \tag{1.12}$$

for a measurable function ψ. In other words, the real-valued cocycle generated by any integrable function with non-zero average is cohomologous to a Lipschitz cocycle. For a generalization of this fact to other groups see [**HK1**].

The concept of a Lipschitz cocycle suggests a natural notion of equivalence for Γ actions.

DEFINITION 1.7. Let Γ be a finitely generated discrete group. Two ergodic right Γ actions $S = \{S_\gamma\}_{\gamma \in \Gamma}$ and $\mathcal{T} = \{T_\gamma\}_{\gamma \in \Gamma}$ are called *Kakutani equivalent* if there exists a Lipschitz Γ cocycle α over S such that the corresponding action L^α is isomorphic to the left action $T^{-1} = \{T_{\gamma^{-1}}\}_{\gamma \in \Gamma}$.

THEOREM 1.8. *(i) Kakutani equivalence is an equivalence relation.*
(ii) Let Γ be a cocompact discrete subgroup of a Lie group G. Then two ergodic right Γ actions S and $\mathcal{T}$ are Kakutani equivalent if and only if there exist Lipschitz G cocycles α and β over S and $\mathcal{T}$ correspondingly such that the corresponding left G actions L^α and L^β are isomorphic.

This theorem, which is a relatively easy corollary of the results in [**BKM**], provides a generalization of the classical result of Kakutani [**Ka**] (cf. also [**ORW**], [**K3**]) concerning the case $\Gamma = \mathbb{Z}$, $G = \mathbb{R}$.

Another interesting question is the extent to which the Ambrose–Kakutani theorem can be generalized. This theorem asserts in particular that every ergodic flow is isomorphic to a special flow over an ergodic automorphism. It is easy to see that in addition (1.11) can be satisfied, so that the Ambrose–Kakutani theorem essentially says that every ergodic flow ($\mathbb{R}$ action) is isomorphic to a flow obtained from a Lipschitz cocycle over an ergodic automorphism via the construction of Mackey range. The generalization of this theorem to the case $\Gamma = \mathbb{Z}^n$ and $G = \mathbb{R}^n$ is obtained in [**K2**]. For further discussion of that case see [**JR**].

On the other hand, for many sufficiently large groups the Ambrose–Kakutani theorem is not true and orbit equivalence classes tend to contain lots of information about the acting group (cf. [**Z1**], [**Z2**], [**Fu**]).

1.3 Cohomological equations for measure-preserving transformations and flows. From now on, we will consider only cocycles over $\mathbb{Z}$ and $\mathbb{R}$ actions, i.e. measure-preserving transformations and flows. Any G cocycle over a $\mathbb{Z}$ action is determined by a measurable function $h : X \to G$ via (1.5). For this reason, we will sometimes call the function h itself a cocycle.

1.3.1 Cocycles over flows. Let us give a similar characterization for a special class of cocycles over a flow. We will consider the case where G is a Lie group and we will assume the existence for a.e. x of the derivative

$$a(x) \stackrel{\text{def}}{=} \left.\frac{d\alpha(x,t)}{dt}\right|_{t=0}. \tag{1.13}$$

Here a is a measurable map from X to the Lie algebra $\mathfrak{g} = T_{id}G$ of G, and the cocycle α can be recovered from a by solving the differential equation

$$\frac{d\alpha(x,t)}{dt} = \mathcal{L}_{\alpha(x,t)}a(S_t x)$$

where $\mathcal{L}_g$ is the differential of the left shift by g, which carries the Lie algebra $\mathfrak{g}$ to T_gG.

This condition, differentiability along the orbits of an action, is not very restrictive. For example any real–valued cocycle α (and hence any vector–valued cocycle too) is cohomologous to a cocycle satisfying (1.13), namely the cocycle

$$\alpha_s(x,t) = \int_0^s \alpha(S_\tau x, t)d\tau.$$

for any positive s.

We are now going to review some of the main cases where cocycles and cohomology appear in ergodic theory. See [**HaK**] for an additional discussion.

1.3.2 Ergodicity and eigenvalues. Let us first observe that T is not ergodic if and only if the $\frac{\psi(Tx)}{\psi(x)} = 1$ has a non-constant solution. Furthermore, the eigenvalue problem for the ergodic transformation T is equivalent to the question of whether the constant S^1 cocycle λ, $\lambda(x,n) = \lambda^n$, is a coboundary. A similar characterization holds for flows.

1.3.3 Compact group extensions. Next consider the compact group extension S^h of a measure-preserving transformation S

$$S^h(x,g) = (Sx, gh(x)),$$

where $g \in G$, a compact group, and $h : X \to G$ is measurable. The extension S^h can be viewed as a measure-preserving transformation on $(X \times G, \mu \times \lambda_G)$, where λ_G is the normalized Haar measure on G. If the G cocycles defined by h_1 and h_2 are cohomologous then the corresponding extensions are isomorphic as measure-preserving transformations.

If G is abelian then

$$L^2(X \times G, \mu \times \lambda_G) = \bigoplus_{\chi \in G^*} H_\chi.$$

Here G^* is the group of characters of G and $H_\chi = \{f(x)\chi(g) : f \in L^2(X,\mu)\}$. This decomposition is orthogonal and is invariant under the unitary operator U_{S^h} corresponding to S^h. This means that any eigenfunction for S^h lies in one of the subspaces H_χ. This leads to the equation

$$\chi(h(x))f(Sx) = f(x) \tag{1.14}$$

for an invariant function $f(x)\chi(g) \in H_\chi$ and

$$\chi(h(x))f(Sx) = \lambda f(x) \tag{1.15}$$

for an eigenfunction $f(x)\chi(g)$. In other words, the questions reduces to determining whether the S^1 cocycle $\chi \circ h$ over S is correspondingly a coboundary or an almost coboundary.

1.3.4 Induced and special transformations. The induced map T_A is defined on a subset A of positive measure by

$$T_A x = T^{\min\{n>0: T^n x \in A\}}.$$

The special transformation $T_{n(\cdot)}$ (or integral transformation) is defined by a measure-preserving transformation T and a positive integrable integer valued "roof" function n on X. $T_{n(\cdot)}$ is a transformation of the set $X_{n(\cdot)} = \{(x,j) : x \in X,\ 1 \le j \le n(x)\}$ defined by

$$T_{n(\cdot)}(x,j) = \begin{cases} (x, j+1) & \text{if} \quad j < n(x) \\ (Tx, 1) & \text{if} \quad j = n(x) \end{cases} \tag{1.16}$$

It preserves the measure induced on $X_{n(\cdot)}$ by the measure $\mu \times \lambda$ on $X \times \mathbb{Z}$, where λ is the uniform measure on $\mathbb{Z}$.

The induced and special transformations correspond to particular cases of the of the Mackey range construction, discussed in 1.1, for the case $\Gamma = G = \mathbb{Z}$. To obtain them in this way, we take for the induced map a $\mathbb{Z}$ cocycle over a transformation T determined by a function with values 0 and 1, (i.e. the characteristic function of a set), and for the special transformation, a $\mathbb{Z}$ cocycle with positive values. Let us note that a natural generalization of both constructions is provided by non-negative integer valued cocycles. Cocycles of this type play a central role in the theory of Kakutani equivalence (cf. [**K3**], [**ORW**]).

The isomorphism of induced maps T_A and T_B of the form $x \to T^{h(x)}x$ leads to the following cohomological equation for a $\mathbb{Z}$ valued function h:

$$\chi_A(x) = \chi_B(x) + h(Tx) - h(x). \tag{1.17}$$

Similarly, for special automorphisms with roof functions $n(x)$ and $m(x)$ this kind of isomorphism is equivalent to the existence of an integer valued solution ψ for the cohomological equation

$$n(x) = m(x) + \psi(Tx) - \psi(x). \tag{1.18}$$

The existence of an eigenfunction with eigenvalue λ for an induced or a special transformation also has a simple cohomological formulation. If the special transformation $T_{n(\cdot)}$ has an eigenvalue λ then

$$\lambda^{n(x)} = f(x)(f(Tx))^{-1} \tag{1.19}$$

where f is the restriction of an eigenfunction with eigenvalue λ to the base. In other words, the cocycle $\lambda^{n(x)}$ is a coboundary. Likewise for an induced map T_A, the existence of an eigenfunction with eigenvalue λ implies

$$\lambda^{\chi_A(x)} = f(x)(f(Tx))^{-1} \tag{1.20}$$

where f is the following extension to X of an eigenfunction g on $A : f(x) = \lambda g(T^{-i(x)}x)$, where $i(x)$ is the smallest positive number such that $T^{-i(x)}x \in A$.

1.3.5 Special flows. Two special flows built over T, $\{T_t^{h_1(\cdot)}\}_{t\in\mathbb{R}}$ and $\{T_t^{h_2(\cdot)}\}_{t\in\mathbb{R}}$ are isomorphic if h_1 and h_2, considered as $\mathbb{R}$ cocycles, are cohomologous, i.e. if

$$h_1(x) = h_2(x) + \psi(Tx) - \psi(x) \tag{1.21}$$

for a real valued measurable function ψ. In particular, if h is an almost coboundary, meaning the following cohomological equation has a measurable solution ψ

$$h(x) = h_0 + \psi(Tx) - \psi(x), \tag{1.22}$$

then the special flow $\{T_t^{h(\cdot)}\}_{t\in\mathbb{R}}$ is isomorphic to the suspension flow with a constant function. It is easy to see that $h_0 = \int_X h d\mu$. (see Proposition 2.1). The existence of an eigenfunction for the special flow implies that for some real r,

$$\exp\, irh(x) = f(x)f^{-1}(Tx) \tag{1.23}$$

for some measurable $f : X \to S^1$, so that the S^1 cocycle $\exp\, irh(x)$ is a coboundary. Let us note that by exponentiating (1.22) we obtain (1.23) for any $r = \frac{2\pi k}{h_0}$, $k \in \mathbb{Z}$; however, (1.23) may be true without (1.22).

Isomorphism conditions (1.17), (1.18), and (1.21) are only necessary but not sufficient. They correspond to special types of isomorphisms which, in a sense, preserve orbits and preserve order on these orbits (cf. [**K3**], §3). It follows from the theory of Kakutani equivalence that there are many other cases of isomorphism.

2. Structure of equivalence classes

Most of the material of this chapter is an amalgamation of results by Kočergin [**Kc**] and Ornstein and Smorodinsky [**OS**]. In [**HK1**] a completely different method is developed which yields most of these results. In the first section we arrive at the conclusion that in the measurable category each equivalence class of cocycles over an ergodic measure-preserving transformation is "uniformly distributed" within the subspace of all cocycles with a given average. Moreover, by modifying a cocycle on an arbitrarily small set, even subject to extra (sufficiently flexible) conditions, one can bring it within an arbitrary cohomology class. In the second section we push the same line of argument even further and show that every cocycle is cohomologous to a "regular" one, e.g. continuous or even continuously differentiable except for one point. Later in sections 3.3 and 3.4 we will show that the results of that kind can not be extended much further. Even a mild uniform condition stronger than continuity (e.g. a Hölder condition) in many cases changes the picture completely.

2.1 Majorization and density in L^1. We begin with two preliminary results.

PROPOSITION 2.1. *Let $h_1, h_2 : X \to \mathbb{R}^n$ be two measurable L^1 cocycles over an ergodic measure-preserving transformation T. If h_1 is cohomologous to h_2 then*

$$\int_X h_1 d\mu = \int_X h_2 d\mu.$$

PROOF. Let us assume that there is a measurable transfer function $\psi : X \to \mathbb{R}^n$

$$h_1(x) - h_2(x) = \psi(Tx) - \psi(x) \tag{2.1}$$

The statement of the proposition is obvious if ψ is integrable. In the general situation we apply the ergodic theorem for vector-valued functions to the function $h_1 - h_2$. Since by (2.1)

$$\frac{1}{n}\sum_{k=0}^{n-1} h_1(T^k x) - h_2(T^k x) = \frac{1}{n}(\psi(T^n x) - \psi(x))$$

so that by the ergodic theorem the left-hand side is close to $\int_X (h_1 - h_2) d\mu$ on a set of large measure. On the other hand, the right-hand side must be close to 0 except for a set of small measure. This implies that $\int_X h_1 d\mu = \int_X h_2 d\mu$. □

The next lemma is useful in handling cocycles with values in compact abelian groups.

Let $G = \mathbb{T}^k \oplus \bigoplus_{i=1}^{\ell} \mathbb{Z}/k_i$. There is a natural epimorphism $\exp : H = \mathbb{R}^k \oplus \mathbb{Z}^\ell \to G$ so that if the H cocycles h_1 and h_2 over a measure-preserving transformation T are cohomologous via the transfer function ψ then G cocycles $\exp \circ h_1$ and $\exp \circ h_2$ are cohomologous via $\exp \psi$.

LEMMA 2.2. *Given a G cocycle over T, $g : X \to G$ and $h_0 \in \mathbb{R}^{k+\ell}$ there exists an L^1 H cocycle h such that $\exp h = g$ and $\int_X h(x) d\mu = h_0$.*

REMARK. Although the cocycle h has discrete components we assume that the lattice $\mathbb{Z}^\ell$ is embedded into $\mathbb{R}^\ell$ so that integration makes sense.

PROOF. Since exp is a surjection on a compact subset of H we can find a bounded measurable branch for $\exp^{-1}$. By pulling back g along that branch we obtain a bounded and consequently integrable h such that $\exp h = g$. In order to correct the value of the integral we can add to h any integrable cocycle h' whose values lie in the kernel of exp. This kernel is a lattice in the continuous part of H and is a sublattice of finite rank in the discrete part. Obviously, one can find a lattice valued function with a given value of the integral so that the correction is possible. □

Now we proceed to the central result of this section. It is very close to Theorem 2 from [**Kc**]. However, the proof of that theorem indicated in [**Kc**] looks incomplete. We use instead a slight modification of the proof of a similar statement – Lemma 1 from [**OS**].

THEOREM 2.3. *Let f, g be two $\mathbb{R}$ or $\mathbb{Z}$ cocycles over an ergodic measure-preserving transformation T such that $\|f\|_{L^1} < \|g\|_{L^1}$. There exists a cocycle h cohomologous to f and such that $|h(x)| \le |g(x)|$ almost everywhere. If f is non-negative, h also can be chosen non-negative. Furthermore, if the function g is fixed then for every $\epsilon > 0$ one can find $\delta > 0$ such that if $\|f\|_{L^1} < \delta\|g\|_{L^1}$ then the transfer function ψ connecting f and h vanishes on a set of measure greater than $1 - \epsilon$.*

PROOF. We will treat the cases of $\mathbb{R}$ and $\mathbb{Z}$ cocycles simultaneously.

Let $x \in X$. Consider the set of all pairs of integers (k, ℓ) satisfying the following properties

(i) $k \le 0 \le \ell$

(ii) For every $m = k,\ k+1, \dots, \ell - 1$

$$\sum_{i=k}^{m} |f(T^i x)| \ge \sum_{i=k}^{m} |g(T^i x)|$$

(iii) $\sum_{i=k}^{\ell} |f(T^i x)| < \sum_{i=k}^{m} |g(T^i x)|$

It follows from the Birkhoff ergodic theorem and from the inequality $\|f\|_{L^1} < \|g\|_{L^1}$ that for almost every $x \in X$ at least one such pair exists and the values of k for all such pairs are bounded from below. Let $(k(x), \ell(x))$ be the pair satisfying (i)–(iii) with the minimal value of k. Such a pair is obviously unique. For $k = 0, -1, -2, \dots$, $\ell = 0, 1, 2, \dots$ let

$$A_{k,\ell} = \{x \in X,\ k(x) = k,\ \ell(x) = \ell\}$$

The sets $A_{k,\ell}$ form a partition of X up to a set of measure 0 and $T^m A_{0,\ell} = A_{-m,\ell-m}$ for $m = 1, \dots, \ell$. We are going to construct a function h, such that $|h| \le |g|$ and

$$\sum_{i=k(x)}^{\ell(x)} h(T^i x) = \sum_{i=k(x)}^{\ell(x)} f(T^i x) \tag{2.2}$$

The last condition implies that h is cohomologous to f via the transfer function

$$\psi(x) = \sum_{i=k(x)}^{0} h(T^i x) - f(T^i x). \tag{2.3}$$

In particular, $\psi = 0$ on the set $\bigcup_{k=0}^{\infty} A_{k,0}$. Let us show how to construct h satisfying (2.2). The solutions are slightly different for the real-valued and integer-valued cocycles. In the real case the easiest way is to put

$$h(x) = |g(x)| \frac{\sum_{i=k(x)}^{\ell(x)} f(T^i x)}{\sum_{i=k(x)}^{\ell(x)} |g(T^i x)|}.$$

Condition (iii) implies that $|h| < |g|$. Condition (2.2) follows from this definition automatically.

In the case of integer valued cocycles let us consider the following measurable sets

$$B_{\ell, f_0, \ldots, f_\ell, g_0, \ldots, g_\ell} = \{x \in A_{0,\ell}, f(T^i x) = f_i, \ |g(T^i x)| = g_i, \ \ i = 0, \ldots, \ell\}$$

Since by (iii) $\sum_{i=0}^{\ell} |f_i| < \sum_{i=0}^{\ell} g_i$ one can find integers $h_0, \ldots, h_\ell$ of the same sign such that $|h_i| \leq g_i \; i = 0, \ldots, \ell$ and $\sum_{i=0}^{\ell} h_i = \sum_{i=0}^{\ell} f_i$. We put for $x \in B_\ell, f_0, \ldots, f_\ell, g_0, \ldots, g_\ell$, $i = 0, \ldots, \ell$, $h(T^i x) = h_i$. Condition (2.2) is obviously satisfied.

It remains to estimate the measure of the support of the transfer function ψ defined by (2.3). Since $\psi = 0$ on the set $\bigcup_{k=0}^{\infty} A_{k,0}$ it is enough to estimate the measure of that set. We have from (ii)

$$\begin{aligned} \|f\|_{L^1} &\geq \int_{X \setminus \cup A_{k,0}} |f(x)| d\mu = \sum_{\ell=0}^{\ell} |f(T^i x)| d\mu \geq \\ &\geq \sum_{\ell=0}^{\ell} \int_{A_{0,\ell}} \sum_{i=0}^{\ell-1} |g(T^i x)| d\mu = \int_{X \setminus \cup A_{k,0}} |g(x)| d\mu \end{aligned} \tag{2.4}$$

Since $|g|$ is an integrable function there is a function $\omega(\epsilon)$ decreasing to zero at $\epsilon \to 0$ such that if $\mu(B) < \epsilon \; \int_B g d\mu < \omega(\epsilon)$. Thus (2.4) implies that $\|f\|_{L^1} > \omega(\mu(\text{supp}\, \psi))$. □

Next two corollaries are also valid for both $\mathbb{R}$ and $\mathbb{Z}$ cocycles over an ergodic measure-preserving transformation T.

COROLLARY 2.4. *Every L^1 cocycle f is cohomologous to a bounded cocycle.*

COROLLARY 2.5. *Suppose f and g are L^1 cocycles such that $\int_X f d\mu = 0$ and $\|g\|_{L^1} > 0$. Then there exists an L^1 cocycle h cohomologous to f such $|h(x)| \leq |g(x)|$ almost everywhere.*

There is one more corollary for $\mathbb{R}$-cocycles.

COROLLARY 2.6. *Given any two L^1 cocycles f and g with $\int_X f d\mu = \int_X g d\mu$ and $\epsilon > 0$, there exists an L^1 cocycle h cohomologous to f such that $|h(x) - g(x)| < \epsilon$.*

Furthermore, h can be required to coincide with g on a given set of measure less than 1.

A counterpart of that statement for $\mathbb{Z}$-cocycles is the following.

COROLLARY 2.7. *Given any two integrable $\mathbb{Z}$ cocycles f and g with $\int_X f d\mu = \int_X g d\mu$, there exists an integrable $\mathbb{Z}$ cocycle h cohomologous to f and such that*

$$|h - g| \leq 1.$$

Furthermore, h can be required to coincide with g on a given set A of measure less than 1. *If g is not constant outside of A the range of h may be required to be contained in the range of g.*

PROOF OF COROLLARIES 2.4–2.7. Corollary 2.4 follows directly from Theorem 2.3 if we take $g = (1+\epsilon)\int_X f d\mu$ for a positive ϵ.

To obtain Corollary 2.5 it is enough to show that f is cohomologous to a function with arbitrary small L^1-norm, because then Theorem 2.3 directly applies. To do that we represent $f = f_+ - f_-$ where $f_+ = \max(f, 0)$. Since $\int_X f d\mu = 0$, $\|f_+\|_{L^1} = \|f_-\|_{L^1}$. Let us fix an $\epsilon > 0$ and apply Theorem 2.3 to the pair f_-, $(1+\epsilon)f_+$. Thus, the function f_- is cohomologous to a nonnegative function h such that $h(x) \leq (1+\epsilon)f_+(x)$ and consequently $f = f_+ - f_-$ is cohomologous to $f_+ - h > -\epsilon f_+$. The last inequality implies that $\|f_+ - h\|_{L^1} = \int |f_+ - h| d\mu = 2 \int \max(0, h - f_+) d\mu < 2\epsilon \|f_+\|_{L^1}$. Since ϵ can be chosen arbitrarily small this finishes the proof of Corollary 2.5.

Corollary 2.6 follows from the previous one applied to the pair of functions $f - g$ and s where s is nonnegative, less than ϵ everywhere and is equal to zero on the given set. Thus, $f = g + (f - g)$ is cohomologous to $g + h$ where $|h(x)| \leq s(x)$.

A very similar argument applies to Corollary 2.7. Here we take as s a characteristic function of a set $B \subset X \backslash A$ such that the values of g on B are not equal to $\sup_{X\backslash A} g$. □

Applying Theorem 2.3 and its four corollaries to each coordinate of a cocycle with values in $H = \mathbb{R}^k \oplus \mathbb{Z}^\ell$ we obtain similar results for cocycles with values in H. In particular, the following theorem follows immediately.

THEOREM 2.8. *Given a set $U \subseteq X$, $\mu(U) < 1$, a measurable function $f : U \to H = \mathbb{R}^k \oplus \mathbb{Z}^\ell$, a vector $h_0 \in \mathbb{R}^{k+\ell}$ and an L^1 H cocycle g over an ergodic measure-preserving transformation T, the set of all cocycles cohomologous to g is dense in the set*

$$A(f, h_0) = \{h \in L^1(X, H),\ h = f \text{ on } U,\ \int_X h d\mu = h_0\}.$$

In particular, for $U = \emptyset$ we obtain that every cohomology class is dense in the subspace of all cocycles with a fixed average.

This theorem together with Lemma 2.2 implies a similar but even more universal density result for cocycles with values in a compact abelian group of the form

$$G = \mathbb{T}^k \times \bigoplus_{i=1}^{\ell} \mathbb{Z}/k_i.$$

COROLLARY 2.9. *Given a set $U \subseteq X$, $\mu(U) < 1$, a measurable function $f: U \to G$ and a measurable G cocycle g over T, the set of all cocycle cohomologous to g is dense in the L^1 topology in the set*

$$A(f) = \{h : h = f \ \text{ on } U\}.$$

In particular for $U = \emptyset$ we obtain that every cohomology class of G cocycles is dense in the space of all measurable cocycles.

2.2 Continuous and almost differentiable representations. In this section we consider only real-valued cocycles.

THEOREM 2.10. *Let $\mathcal{L} \subset L^1(X, \mu)$ be a linear subspace of L^1 dense in the L^1 topology and closed in the L^∞ topology (uniform convergence almost everywhere). Then for every $f \in L^1(X, \mu)$ the set $\mathcal{L}_f = \{h \in \mathcal{L},\ h$ is cohomologous to $f\}$ is dense in the L^∞ topology in the set $\{h \in \mathcal{L}, \int h d\mu = \int f d\mu\}$.*

PROOF. Let us fix a function $h \in \mathcal{L}$ such that $\int f d\mu = \int h d\mu$ and $\epsilon > 0$. By Corollary 2.6 one can find a cocycle f_1 cohomologous to f and such that $|f_1 - h| < \frac{\epsilon}{2}$. Then we began to apply Theorem 2.3 inductively. First we approximate f_1 by a function $h_1 \in \mathcal{L}$ such that

$$\|h_1 - f_1\|_{L^1} < \frac{\delta_1 \epsilon}{4}$$

where δ_1 is chosen sufficiently small to ensure that, by Theorem 2.3, $h_1 - f_1$ is cohomologous to a function r_1 such that $|f_1| < \frac{\epsilon}{4}$ via a transfer function ψ_1 whose support has measure less than $\frac{1}{2}$. The function $h_1 + r_1 \stackrel{\text{def}}{=} f_2$ is thus cohomologous to f_1 and consequently to f. Then we approximate f_2 by $h_2 \in \mathcal{L}$ such that $\|f_2 - h_2\|_{L^1} < \delta_2 \frac{\epsilon}{8}$ with an appropriately chosen δ_2 so that $f_2 - h_2$ is cohomologous to a function r_2, $|r_2| < \frac{\epsilon}{8}$ via a transfer function supported by a set of measure less than $\frac{1}{4}$, denote $h_2 + r_2 = f_3$, etc. In the limit we obtain

$$f' = \lim f_n = \lim h_n$$

and in both cases the convergence is uniform. Since $h_n \in \mathcal{L}$ and $\mathcal{L}$ is L_∞ closed, $f' \in \mathcal{L}$. On the other hand, since the transfer function connecting f_n and f_{n+1} has support of measure less 2^{-n}, by the Borel–Cantelli Lemma the sequence of transfer functions between f and f_n converges in probability to a transfer function between f and f'. □

COROLLARY 2.11. [**OS**] *Let X be a compact metric space, μ be a Borel probability nonatomic measure on $X, T : X \to X$ be a measure-preserving transformation (not necessarily continuous). Then every real-valued cocycle $f \in L^1(X, \mu)$ is cohomologous to a continuous cocycle. Moreover the set of continuous cocycles cohomologous to f is dense in uniform topology in the space of all continuous functions with the same integral as f.*

This statement follows immediately from Theorem 2.10 if we put $\mathcal{L} = C(X)$, the space of all continuous functions.

Corollary 2.11 can be strengthened by specifying the values of a continuous function cohomologous to f on any closed set F so that $\mu(X \backslash F) > 0$. The proof repeats that of Theorem 2.10 with an extra observation that f_1 may be already made to coincide with the given function on F and all the successive approximations may be chosen in order not to change that.

Pushing the method described above a bit further one obtains the following result which looks quite striking at first glance.

THEOREM 2.12. *Let M be a compact differentiable manifold, μ be a Borel probability measure on M, $T : M \to M$ be a measure-preserving transformation. Then every real-valued cocycle $f \in L^1(M, \mu)$ is cohomologous to a continuous cocycle $\bar{f}$ which is continuously differentiable except at a single point.*

SKETCH OF PROOF. First one finds a continuous cocycle f_1 cohomologous to f which is continuously differentiable outside a ball B_1 of radius, say, $1/2$ and can be extended to a continuously differentiable function. This is possible by a stronger version of Corollary 2.11 mentioned above. Then one approximates f_1 in uniform topology by a continuously differentiable cocycle g_1 which coincides with f outside B_1. If the L^1 norm of $f_1 - g_1$ is small enough one can find a cocycle f_2 cohomologous to f_1 (and hence to f) which coincides with f_1 outside a smaller ball $B_2 \subset B_1$ of radius $1/4$ and extends to a continuously differentiable function and such that the support of the transfer function ψ_1 has measure less than $1/2$. Continuing by induction one constructs on the nth step the cocycle f_n continuously differentiable outside of ball $B_n \subset B_{n-1}$ of radius 2^{n+1} which coincides with f_{n-1} outside of the ball B_{n-1} and extends to a continuously differentiable function and such that a transfer function ψ_n connecting f_n with f_{n-1} is supported on a set of measure less than 2^n. In the limit the function $\bar{f} = \lim_{n\to\infty} f_n$ is continuous everywhere and continuously differentiable outside of the single point $\bigcap\limits_{n=1}^{\infty} B_n$. By the Borel–Cantelli lemma the series $\sum\limits_{n=1}^{\infty} \psi_n$ converges and hence gives a transfer function between f_1 and $\bar{f}$. Since f_1 is cohomologous to f this finished the proof. □

REMARK. As many instances discussed in the next chapter will show, the single point of nondifferentiablity cannot be removed.

The transfer functions involved in the equivalence between continuous cocycles will very often be discontinuous, even for a homeomorphism.

For a homeomorphism f we will denote the set of f invariant Borel probability measures by $\mathcal{M}(f)$.

PROPOSITION 2.13. *Let f be a homeomorphism of a compact metric space X. The following three subspaces of $C(X)$ coincide:*

(i) $E_1 = \{\phi : \int \phi d\nu = 0 \text{ for all } \nu \in \mathcal{M}(f)\}$

(ii) $E_2 = \bar{\mathcal{C}}$, where $\mathcal{C} = \{\phi : \phi = h \circ f - h$ for a continuous function $h\}$

(iii) $E_3 = \bar{\mathcal{B}}$, where $\mathcal{B} = \{\phi : \phi = h \circ f - h$ for a bounded Borel function $h\}$.

PROOF. Obviously $E_3 \supseteq E_2$; since the space E_1 is closed we can apply Proposition 2.1 and conclude that $E_1 \supseteq E_3$. Thus it is enough to show that $E_2 \supseteq E_1$. This is a fairly straightforward argument using duality and the Hahn-Banach Theorem. We will present it here, because the same argument will appear several times in subsequent discussions. Consider the continuous linear operator $A : C(X) \to C(X)$ such that $(A\phi)(x) = \phi(f(x)) - \phi(x)$. E_2 is the closure of the image of A. The dual operator A^* acts on the dual space $C^*(X)$. Since every element of $C^*(X)$ is the difference of two measures, $\mathcal{M}(f)$ spans KerA^*. Thus $\chi \in$ Ker A^* if and only if for every $\phi \in C(X)$,

$$(A^*\chi)\phi = \chi(A\phi) = 0.$$

and $\phi \in E_1$ if and only if for every $\chi \in \mathrm{Ker}A^*$, $\chi(\phi) = 0$. In other words $\chi(\phi) = 0$ as soon as χ annihilates $\mathrm{Im}A$. By the Hahn-Banach Theorem, if $\phi \notin E_2$ one can construct χ such that $\chi(\phi) = 1$ and $\chi(\mathrm{Im}A) = 0$. □

REMARK. If the transformation f is minimal, i.e. all of its orbits are dense, then $\mathcal{C} = \mathcal{B}$. (Gottshalk, Hedlund, see e.g. [**KH**], Theorem 2.9.4.)

In view of Theorem 2.10 and Proposition 2.13, we can see that unless f is uniquely ergodic there are many continuous cocycles which are coboundaries via discontinuous and even unbounded transfer functions. Theorem 4.2 below shows that this is the case in the uniquely ergodic situation as well. At the end of section 3.3 we will give an example of a topologically transitive but not minimal transformation and a naturally defined continuous function which is a coboundary with a bounded but discontinuous transfer function.

3. Rigidity and stability

We proceed to explore various situations where equivalence classes of cocycles have a reasonable structure. As the discussion in the previous chapter shows, one can hope to find such situations only if the topology in the corresponding space of cocycles is considerably stronger than the topology of uniform convergence. The most natural situations to look upon are analytic, smooth, Lipschitz or Hölder cocycles over a transformation or a flow preserving the corresponding structure. Two basic concepts for the subsequent discussion are those of rigidity and stability. We will define these notions for cocycles over actions of fairly general groups but our discussion will be almost completely restricted to the case of a single transformation ($\mathbb{Z}$-action), with several glimpses of the case of flows.

3.1 Definitions. Let H be a topological space of G cocycles over a measurable measure-preserving action S of a locally compact second countable group Γ. We will assume that the topology in H is stronger than the topology of convergence in probability uniform on compact subsets of Γ. In other words, if a sequence of cocycles $\alpha_n \in H$ converges in H to a cocycle α, then for any compact subset $K \subseteq \Gamma$ and any neighborhood U of the identity in G, and for any sufficiently large n $\alpha_n^{-1}(x,\gamma)\alpha(x,\gamma) \in U$ for all $\gamma \in K$ and for a set of x close to full measure.

DEFINITION 3.1. The space H is *rigid with respect to the action* S if every cocycle $\alpha \in H$ is an almost coboundary. If in addition, for every $\alpha \in H$, the transfer function ψ which establishes the equivalence between α and a constant cocycle, can be chosen from a given class Ψ of maps from X to G, we will call H *Ψ-rigid with respect to S* .

It is natural to allow a certain freedom of terminology and to speak of continuous, Hölder, smooth, (C^r or C^∞), or analytic rigidity when the transfer functions possess the corresponding properties.

For $\mathbb{R}^n$-cocycles over a measure-preserving transformation rigidity means that all of the functions from H with the fixed average are cohomologous to each other.

For actions of certain groups, e.g. semisimple Lie groups of rank greater than one, or lattices in such groups, rigidity occurs for large classes of measurable cocycles

[**Z1**], but for actions of $\mathbb{Z}$, and more generally, of amenable groups, it can appear only if H has a sufficiently fine topology. There is however a big difference between the classical case ($\mathbb{Z}$ and $\mathbb{R}$ actions) and the situation for actions of higher rank abelian groups.

In the latter case cocycle rigidity in Hölder, smooth and analytic categories is quite widespread, appears for many natural examples of actions, both smooth [**KSp1**] and symbolic [**Sch2**], and is related with hyperbolic behavior in the smooth situation and with expansiveness in the symbolic one. Cocycle rigidity is a key link in the whole panoply of rigidity results for such actions including local and global differentiable rigidity [**KSp2**].

For the classical cases Hölder cocycle rigidity very likely never happens and smooth cocycle rigidity appears for Diophantine translation of the torus and probably only there. We will discuss this issue in the next section.

A property which is much more common and definitely more useful in the classical cases is stability:

DEFINITION 3.2. Under the same assumptions as in Definition 3.1, we will call the space H *stable with respect to* S if every class of cohomologous cocycles in H is closed.

In this definition we assume that arbitrary measurable G-valued functions are allowed as transfer functions between cocycles from the space H. However, equivalence classes within H may shrink considerably if we put certain restrictions on the transfer functions. In particular, it is conceivable that unrestricted equivalence classes are not closed while the restricted ones are and vice versa. This possibility suggests certain ramifications of the concept of stability. As before, let Ψ be a certain class of maps from X to G.

DEFINITION 3.3. The space H is called Ψ*-stable with respect to the action* S if every class of cocycles from H, cohomologous via transfer functions from Ψ, is closed.

Let us note that in general, Ψ-stability may not imply stability.

DEFINITION 3.4. The space H is called Ψ*-effective with respect to* H if for any two cohomologous cocycles $\alpha, \beta \in H$ the transfer function ψ belongs to Ψ.

Obviously, if H is Ψ-effective then Ψ-stability implies stability.

Terms such as smooth, analytic or Hölder stability and effectiveness have the obvious meaning. We observe that if equivalent cocycles are assumed to be defined everywhere and the transfer function is only assumed to be measurable, then the corresponding cohomological equation needs only be satisfied almost everywhere.

If both spaces H and Ψ consist of continuous maps, then there are certain obvious invariants which are preserved under cohomology via a transfer function from Ψ. For example, if G is a subgroup of $\mathbb{R}^n$ and ν is any measure on X invariant with respect to the action S, then for any $\gamma \in \Gamma$ the average

$$\int_X \alpha(x, \gamma) d\nu$$

is a cohomology invariant (Proposition 2.13). This remains true even if ψ is a bounded Borel function. These remarks together with Corollary 2.11 show that for continuous cocycles over transformations which are not uniquely ergodic equivalence classes with continuous or bounded Borel transfer functions must be much smaller than the unrestricted equivalence classes.

For $\Gamma = \mathbb{Z}$ and arbitrary G, the conjugacy class in G of the product of the values of the cocycle along any periodic orbit is invariant. In sections 3.3 and 3.4 we will discuss situations when these invariants determine the equivalence classes, thus producing stability.

While no general theory exists for stability and effectiveness of smooth or other natural classes of cocycles, there is a good understanding of these phenomena for smooth dynamical systems with hyperbolic behavior and their counterparts in topological and symbolic dynamics. To a lesser extent stability and effectiveness are understood for partially hyperbolic systems and for parabolic systems with sufficiently regular features. For a general discussion of hyperbolic, partially hyperbolic, parabolic and elliptic behavior in dynamics, see [**HaK**]. A general outline of stability and effectiveness properties is also discussed there. Presently we will glance at some characteristic phenomena which appear for each type of behavior.

3.2 Translations of the torus and smooth rigidity.

3.2.1 Diophantine and Liouvillean numbers. The example which we are going to describe plays a basic role in the theory of perturbations of completely integrable Hamiltonian systems. The rigidity result was known to Kolmogorov in the early fifties [**Ko**].

Let $\lambda = (\lambda_1, \dots, \lambda_m) \in \mathbb{T}^m$ so that each λ_i is a complex number of modulus 1. We will call the vector λ *Diophantine* if for some positive k and c and for arbitrary integers $n_1, \dots, n_m$ such that $\sum_{j=1}^{m} |n_j| > 0$,

$$\left| \sum_{j=1}^{m} \lambda_j^{n_j} \quad 1 \right| > c \left(\sum_{j=1}^{m} |n_j| \right)^{-k}. \tag{3.1}$$

If $\lambda_j = \exp 2\pi i \alpha_j$, $j = 1, \dots, m$, then (3.1) is equivalent to the following condition for the vector $\alpha = (\alpha_1, \dots, \alpha_m)$ of phases: for arbitrary integers $n_1, \dots, n_m$ such that $\sum_{j=1}^{m} |n_j| > 0$ and for every integer n,

$$\left| \sum_{j=1}^{m} \alpha_j n_j - n \right| > c \left(\sum_{j=1}^{m} |n_j| \right)^{-k}.$$

The vector of phases which is of course defined up to an integer vector is also called Diophantine. Often the vectors α (and corresponding λ's) which are not Diophantine are called *Liouvillean.*

Let $R_\lambda : \mathbb{T}^m \to \mathbb{T}^m$ be the rotation* of the torus by λ,

$$R_\lambda(z_1, \dots, z_m) = (\lambda_1 z_1, \dots, \lambda_m z_m).$$

This is the prototype example of elliptic behavior in dynamics (see [**HaK**], Chapter 7).

*This terminology suits multiplicative notations on the torus we used so far. When additive notations are used as later in this section the same map R_λ is called a *translation* and is denoted by T_α

3.2.2 Cohomological equation over a rotation.

THEOREM 3.5. *If λ is Diophantine then for every C^∞ function (real or complex) h on $\mathbb{T}^n$ there exists a C^∞ function ψ (real or complex) such that*

$$h(z) - \int_{\mathbb{T}^m} h d\mu = \psi(R_\lambda z) - \psi(z) \tag{3.2}$$

where μ is Lebesgue measure on $\mathbb{T}^m$.

Furthermore, Diophantine vectors are the only ones for which this property holds.

PROOF. We will use Fourier expansion. For $n_1, \dots, n_m \in \mathbb{Z}$, let us denote $\chi_{n_1,\dots,n_m} = \prod_{i=1}^{m} z_i^{n_i}$. Then

$$h = \sum_{n_1,\dots,n_m} h_{n_1,\dots,n_m} \chi_{n_1,\dots,n_m}$$

and h is a C^∞ function if and only if the Fourier coefficients $h_{n_1,\dots,n_m}$ decrease faster than any power of $|n_1| + \cdots + |n_m|$.

Let us look for the Fourier expansion for the solution ψ of (3.2).

$$\psi = \sum_{\substack{n_1,\dots,n_m \\ \Sigma|n_i|>0}} \psi_{n_1}, \dots, n_m \ \chi_{n_1,\dots,n_m}.$$

Its Fourier coefficients $\psi_{n_1,\dots,n_m}$ are found from (3.2):

$$\psi_{n_1,\dots,n_m} = \frac{h_{n_1,\dots,n_m}}{\prod_{i=1}^{m} \lambda_i^{n_i} - 1}. \tag{3.3}$$

If h is C^∞ and λ is Diophantine, then it follows from (3.1) that the coefficients $\psi_{n_1,\dots,n_m}$ decrease faster than any power of $|n_1| + \cdots + |n_m|$, so that ψ defined through (3.3) is a C^∞ function.

Conversely, if λ is Liouvillean, then there exists a sequence of vectors $n^k = (n_1^k, \dots, n_m^k)$ such that for any natural number s as $k \to \infty$,

$$\left| \sum_{j=1}^{m} \lambda_j^{n_j^k} - 1 \right| = o\left(\sum_{j=1}^{m} |n_j^k| \right)^{-s}.$$

Define h by choosing its only nonzero Fourier coefficients as

$$h_{n_1^k,\dots,n_m^k} = \left| \sum_{j=1}^{m} \lambda_j^{n_j^k} - 1 \right|^{1/2} \quad \text{for } k = 1, \dots.$$

These Fourier coefficients decrease faster than any power of $\sum_{j=1}^{m} |n_j^k|$, hence the function h is C^∞. On the other hand, for a solution ψ of the cohomological equation (3.2) the Fourier coefficients $\psi_{n_1^k,\dots,n_m^k}$ given by (3.3) go to infinity as $k \to \infty$ which contradicts existence of even an L^1 solution to (3.2). □

REMARKS. 1. In the Liouvillean case the Fourier series for h constructed above is lacunary. In the case of one variable Herman [**He1**] showed that if the cohomological equation (3.2) with an L^2 function h given by a lacunary series has a measurable solution, then the solution is actually L^2. Thus, in our case with $m = 1$ for a Liouvillean α there is no measurable solution at all. This can also be shown using the methods developed in Section 5.1 below.

2. Herman also obtained sharp results for the equation (3.2) for Diophantine α with functions h of finite regularity. Namely if h is C^r, (3.1) is satisfied and $r - k$ is not an integer, then there is a C^{r-k} solution, and if $r - k$ is an integer, there is a solution which is $C^{r-k-\epsilon}$ for any $\epsilon > 0$ ([**He2**], Proposition A.8.1). However, typically (for a dense G_δ of h's in the corresponding spaces) there is no $C^{r-k+\epsilon}$ solution for any positive ϵ and no C^{r-k} solution if $r - k$ is an integer ([**He2**], Proposition XIII.4.5).

3. Theorem 5.6 below contains a generalization of our argument for the Liouvillean case.

3.2.3 Smooth rigidity and effectiveness. Using the terminology introduced in section 3.1 we can reformulate Theorem 3.5 by saying that for a Diophantine λ, the rotation R_λ is both C^∞ rigid and C^∞ effective. We do not know any other examples of measure-preserving diffeomorphisms which possess these properties.

CONJECTURE 3.6. *Any C^∞ diffeomorphism of a compact connected manifold which is C^∞ rigid and C^∞ effective is C^∞ conjugate to a Diophantine translation of a torus.*

There are similar questions concerning rigidity in C^r category for finite r.

PROPOSITION 3.7. *A C^∞ rigid and C^∞ effective diffeomorphism f is uniquely ergodic and its only invariant measure is a volume with C^∞ density.*

PROOF. Unique ergodicity immediately follows from Proposition 2.13 and the density of C^∞ functions in the space of continuous functions. For, given two measures μ and ν one can always find a C^∞ function such that its integrals with respect to these measures are different. But if a function is cohomologous to a constant even via a continuous transfer function this constant must be equal to the integral of the function with respect to any invariant measure, a contradiction.

Now consider a measure μ given by a positive C^∞ density. We have $f_*\mu = J\mu$ where the Radon–Nikodym derivative J (called the Jacobian) is C^∞ and positive. The C^∞ function $\log J$ is by assumption cohomologous to a constant:

$$\log J(x) = H(fx) - H(x) + c.$$

But then the measure $\nu = \exp H\mu$ satisfies $f_*\nu = \exp c\nu$. Since the total measure is preserved this implies that $c = 0$, hence ν is invariant. □

COROLLARY 3.8. *A C^∞ rigid and C^∞ effective diffeomorphism f is minimal.*

A part of the conjecture is that a C^∞ rigid and C^∞ effective diffeomorphism have to act on a torus. A possible approach would be to try to prove that as a measure preserving transformation such a map has to have pure point spectrum. It is not difficult to show that every ergodic diffeomorphism with pure point spectrum and C^∞ eigenfunctions is conjugate to a translation on a torus. An alternative possibility would be a smoothly rigid diffeomorphism with a continuous part in the spectrum or one with pure point spectrum but with non-smooth eigenfunctions. Even without rigidity the last possibility constitutes an interesting question. Such "non-standard" models do exist for R_λ with some λ which are very well approximated by rational vectors [**AK**], but their existence even for a single Diophantine λ is an open question.

Now consider diffeomorphisms of the torus $\mathbb{T}^m$. We use additive notation now. Every such diffeomorphism is homotopic to an automorphism F_A, whose lift to $\mathbb{R}^m$ is the linear map given by A, an $m \times m$ matrix with determinant ± 1.

There are two extreme cases in which our conjecture can be proven. If the matrix A has no roots of unity among the eigenvalues then any diffeomorphism homotopic to F_A has infinitely many periodic points ([**KH**], Theorem 8.7.1) and hence no cocycle rigidity is possible.

PROPOSITION 3.9. [**LSa**] *A C^∞ rigid and C^∞ effective diffeomorphism f of $\mathbb{T}^m$ homotopic to identity is C^∞ conjugate to a Diophantine translation.*

PROOF. Since f is homotopic to identity its lift F to the universal cover $\mathbb{R}^m$ has the form $\mathrm{Id} + H$ where H is a periodic function, hence it projects to function h on the torus. Using rigidity and effectiveness we obtain $h = \psi \circ f - \psi + c$, where ψ is a C^∞ function and c is a constant. Lift ψ to the function Ψ on $\mathbb{R}^m$. Obviously $H = \Psi \circ F - \Psi + c$. Consider the map $S = \mathrm{Id} - \Psi : \mathbb{R}^m \to \mathbb{R}^m$. We have

$$S \circ F = \mathrm{Id} + H - \Psi \circ F = \mathrm{Id} + \Psi \circ F - \Psi + c - \Psi \circ F = \mathrm{Id} - \Psi + c = (\mathrm{Id} + c) \circ S$$

Let s be the projection of the map S to the torus. Projecting the last equality on the torus we obtain

$$s \circ f = T_c \circ s \tag{3.4}$$

where $T_c = \mathrm{Id} + c$ is the translation by c.

It remains to show that s is a diffeomorphism. Taking the derivative of (3.4) we see that the set of singular points of s is f–invariant and hence dense by Corollary 3.8. By the Sard Theorem the set of regular points is open and dense hence the set of singular points is empty. Thus s is a covering map but since it is homotopic to identity it is a diffeomorphism. By Theorem 3.5 c must be a Diophantine vector. □

Using a similar method one can treat the case where the matrix A is unipotent, i.e. all of its eigenvalues are equal to one. In this case one can show that C^∞ cocycle rigidity and effectiveness implies that the diffeomorphism is C^∞ conjugate to an affine map. Using the method of Section 3.6 below one can show that such a map with a unipotent $A \neq \mathrm{Id}$ has infinitely many invariant distributions and hence cannot be rigid.

The remaining difficult case is when A is reducible with eigenvalues of absolute value greater than one coexisting with roots of unity.

3.3 Stability of Hölder cocycles for transformations with specification. In this section we will consider a rather robust situation of cocycle stability in the context of topological dynamics. Dynamical conditions which appear in this and the next section are abstract versions of hyperbolicity. Thus in these sections we essentially describe certain aspects of cocycle stability and effectiveness for dynamical systems with hyperbolic behavior.

Let f be a homeomorphism of a compact metric space X.

DEFINITION 3.10. We will say that f satisfies the *strong specification property* if for any $\epsilon > 0$ there exists $N(\epsilon)$ such that for any collection of orbit segments

$$\{f^i(x_j)\}_{j=1,\ldots,L,i=0,\ldots,m_j-1}$$

and for any $n_1, \ldots, n_L > N(\epsilon)$, there exists a periodic orbit

$$\{f^i y\}_{i=0,\ldots,P-1}$$

of period

$$P = \sum_{j=1}^{L} m_j + n_j$$

such that for $j = 1, \ldots, L$,

$$\operatorname{dist}(f^i x_j, f^{\ell(i,j)} y) < \epsilon$$

where

$$\ell(i,j) = i + \sum_{k=1}^{j-1} m_k + n_k.$$

The two main examples with this property are:

(i) A topological Markov chain (subshift of finite type) Σ_A, where A is a zero-one matrix such that for some n, A^n has all positive entries; the latter condition is called transitivity (see [**KH**], Section 1.2.).

(ii) The restriction of a diffeomorphism of a compact manifold to a locally maximal (basic) hyperbolic set Λ satisfying one of the following equivalent conditions ([**KH**], Section 18.3):

(a) $F\big|_\Lambda$ is topologically mixing, i.e. for every two nonempty open sets U and V, there exists $N(U,V)$ such that $f^n U \cap V \neq \emptyset$ for all $n > N(U,V)$.
(b) Every power of f has a dense orbit on Λ.
(c) At least one stable or unstable manifold for f is dense in Λ.

Now assume that the homeomorphism f satisfies the strong specification property.

DEFINITION 3.11. We will say that a vector-function $h : X \to \mathbb{R}^m$ satisfies a *dynamical Hölder condition* if there exist $\epsilon > 0$ and $K > 0$ such that for any n, if $x, y \in X$,

$$d_n^f(x,y) \le \epsilon \text{ then } \sum_{i=0}^{n-1} \|h(f^i x) - h(f^i y)\| \le K \tag{3.5}$$

The reason for calling this property a Hölder condition is that in both examples (i) and (ii) above, this property is satisfied by Hölder functions with respect to a natural metric. In the symbolic case (i) the metric in the space

$$\Omega_A = \{\omega = (\ldots\omega_{-1}\omega_0\omega_1,\ldots) : \omega_n \in \{0,\ldots,N-1\},\ A_{\omega_{n-1}\omega_n} = 1,\ n \in \mathbb{Z}\}$$

can be chosen as $\text{dist}(\omega, \omega') = \sum\limits_{n=-\infty}^{\infty} \frac{|\omega_n - \omega'_n|}{\lambda^{|n|}}$ for any $\lambda > 1$. In the smooth case (ii) any metric on Λ induced by a smooth Riemannian metric on the ambient manifold can be chosen.

All vector-valued functions on X satisfying (3.5) with a fixed ϵ form a Banach space H_ϵ^f with respect to the norm

$$\|h\|_\epsilon^f = \max_{x \in X} \|h(x)\| + \sup_n \sup_{x,y : d_n^f(x,y) \leq \epsilon} \sum_{i=0}^{n-1} \|h(f^i x) - h(f^i y)\|.$$

The spaces H_ϵ^f will serve as H in our study of stability of cocycles.

PROPOSITION 3.12. *If f satisfies the strong specification property and $h : X \to \mathbb{R}^m$ satisfies the dynamical Hölder condition then the following two properties are equivalent*

(i) $\sum\limits_{i=0}^{n-1} h(f^i x) = 0$ *for every periodic point $x \in X$ such that $f^n x = x$.*

(ii) *There exists a bounded Borel function $\psi : X \to \mathbb{R}^m$ such that*

$$h(x) = \psi(fx) - \psi(x).$$

This proposition immediately implies the following stability result.

COROLLARY 3.13. *Let f be a homeomorphism of a compact metric space X satisfying the strong specification property and Ψ be the set of all bounded Borel maps from X to $\mathbb{R}^m$. Then the spaces H_ϵ^f are Ψ-stable.*

PROOF OF COROLLARY 3.13. By Proposition 3.12 two $\mathbb{R}^m$ cocycles $h_1, h_2 \in H_\epsilon^f$ are equivalent with the transfer function from Ψ if and only if the sums of their values along each periodic orbit are equal. Those conditions define a closed affine subspace of H_ϵ^f.

PROOF OF PROPOSITION 3.12. Condition (ii) implies (i) because if $f^n x = x$ then $\sum\limits_{i=0}^{n-1} h(f^i x) = \psi(f^n x) - \psi(x) = 0$. The converse follows easily from Lemma 3.14 and 3.15 which are formulated and proved below.

LEMMA 3.14. *Under the assumptions of Proposition 3.12, if (i) is satisfied then there exists a constant $L > 0$ such that for every $x \in X$ and every positive integer n*

$$\left\| \sum_{i=0}^{n-1} h(f^i x) \right\| < L.$$

PROOF. We apply the specification property to the orbit segment $\{f^i x\}$ $i = 0, \dots, n-1$, with $\epsilon > 0$ small enough to use the dynamical Hölder condition for h. This gives us a periodic point y of period, say, $n + N(\epsilon)$ such that $d_n^f(x, y) < \epsilon$. We have

$$\sum_{i=0}^{n-1} h(f^i x) = \sum_{i=0}^{n+N(\epsilon)-1} h(f^i y) +$$

$$+ \sum_{i=0}^{n-1} (h(f^i x) - h(f^i y)) - \sum_{n}^{n+N(\epsilon)-1} h(f^i y).$$

The first sum is equal to zero by (i); the second is uniformly bounded by the dynamical Hölder property; the third contains a bounded number of terms each of them bounded by the sup $\|h\|$. This proves the lemma. □

LEMMA 3.15. *(D. Rudolph). If $\{a_n\}_{n\in\mathbb{Z}}$ is a sequence of real numbers such that for any integers $n, m, n < m$,*

$$\left|\sum_{i=n}^{m-1} a_i\right| < L$$

then

$$a_n = b_{n+1} - b_n$$

where

$$b_n = -\sup_{N\in\mathbb{Z}} S_n^N$$

and

$$S_n^N = \begin{cases} \sum_{i=0}^{N-1} a_{n+i}, & N > 0 \\ 0, & N = 0 \\ -\sum_{i=1}^{-N} a_{n-i}, & N < 0 \end{cases}$$

so that b_n is uniformly bounded.

PROOF. One can easily see that

$$S_{n+1}^{N-1} = S_n^N - a_n$$

so that

$$b_{n+1} - b_n = -\sup_{N\in\mathbb{Z}} S_{n+1}^{N-1} + \sup_{N\in\mathbb{Z}} S_n^N = a_n. \square$$

In order to finish the proof of Proposition 3.12, we first use Lemma 3.14 and then apply Lemma 3.15 to every coordinate of h along every orbit of f. This gives us the solution $\psi = (\psi_1, \ldots, \psi_m)$ of the cohomological equation where

$$\psi_i(x) = -\sup_{N\in\mathbb{Z}} \begin{cases} \sum_{n=0}^{N-1} h_i(f^n x), & N > 0 \\ 0, & N = 0 \\ -\sum_{n=-1}^{N} h_i(f^n x), & N < 0. \end{cases} \tag{3.6}$$

These functions are obviously Borel, and by Lemma 3.14 they are bounded. □

The proof above demonstrates in the simplest possible form a general and fruitful method of establishing triviality of real- or vector-valued cocycles over group actions. If we denote the cocycle $h(x, n)$ defined by a given function h then formula (3.6) gives the value of the ith coordinate of the transfer function simply as $-\sup_n h_i(x, n)$. Of course, such a definition is only possible if the values of the cocycle at every point are bounded. They do not have to be uniformly bounded though; the resulting transfer function would then be unbounded.

Moreover, there are situations when the set of values of the cocycle at a particular point is unbounded but the cocycle at the that point as a function of n (i.e. a function on $\mathbb{Z}$, or, for a more general action, function of the acting group), has some equivariant characteristics. Here are examples of such characteristics which are actually useful in ergodic theory and topological dynamics. We restrict ourselves to $\mathbb{R}$ valued cocycles over $\mathbb{Z}$ actions.

(i) Suppose the set of values of the cocycle at x has a Cesaro average or, more generally, can be averaged using some algorithmic procedure. If $A(x)$ is the value of such an average, then $-A(x)$ can be used as the solution of the coboundary equation. This construction can also be used for actions of amenable groups. Aside from an obvious generalization to vector valued cocycles it extends to situations where a natural notion of center of gravity can be introduced in the range of the cocycle.

(ii) Suppose there is a set of of values on n of positive lower density, say $d(x)$, for which the values of the cocycle are bounded from above and below. In this case one can define the distribution function $F_x(t)$ of the values of cocycle at x, namely, the lower density of a set for which the values of the cocycle are $\leq t$. This function is monotone non-decreasing and has asymptotic values $0 \leq F_x(-\infty) < F_x(\infty) \leq 1$. Now let

$$t(x) \stackrel{\text{def}}{=} \sup\{t: \ F_x(t) < \frac{F_x(-\infty) + F_x(\infty)}{2}\}.$$

By assumption $t(x)$ is a finite number and $t(f(x)) = t(x) + h(x)$ thus producing a solution of the coboundary equation.

In the next section we will show that in certain cases, including topological Markov chains (subshifts of finite type) and hyperbolic sets for diffeomorphisms, one can prove that the transfer function ψ is continuous and even Hölder. In general, however, this is not true. We will now describe a counterexample due to B. Marcus.

EXAMPLE. The space X will be a shift-invariant closed subset of the set of doubly infinite $(+1, -1)$ sequences, containing all sequences $\omega = \{\omega_n\}$, $n \in \mathbb{Z}$, such that for all $m, n \in \mathbb{Z}$, $m < n$,

$$\left|\sum_{i=n}^{n-1} \omega_i\right| \leq K \tag{3.7}$$

for a fixed positive integer $K \geq 2$. Naturally, the transformation will be the shift on two symbols σ_2. We will consider the zero coordinate ω_0 as a function on X, so that $\omega_0(\sigma_2^n \omega) = \omega_n$.

PROPOSITION 3.16. *The shift σ_2 restricted to the space X satisfies the strong specification property, the function ω_0 satisfies dynamical Hölder condition, and ω_0 is not a coboundary with a continuous transfer function.*

PROOF. If $\omega \in X$ is periodic, i.e. if $\omega_{n+k} = \omega_n$ for some k and for all n then $\sum_{i=0}^{k-1} \omega_i = 0$, since otherwise (3.7) cannot be true. The function ω_0 satisfies the dynamical Hölder condition because, as we mentioned before, this is true for the full shift space. The specification property can be established directly, but it is easier to deduce it from a general argument.

The map we are considering is a *sofic system*, i.e. a symbolic system which is continuous factor of a subshift of finite type. To see this, we take the shift on $K+1$ symbols $\{0, 1, \ldots, K\}$, and define the transition matrix $A = (a_{ij})$ where

$$a_{ij} = 1 \quad \text{if} \quad |i - j| = 1.$$

and is zero otherwise. Denote by σ_A the restriction of the shift to Ω_A. Every element $\alpha \in \Omega_A$ determines a sequence $\omega(\alpha)$ of $+1$ and -1, namely

$$\omega_i(\alpha) = \alpha_{i+1} - \alpha_i. \tag{3.8}$$

Obviously any $\omega(\alpha)$ satisfies (3.7) and conversely, for any sequence ω satisfying (3.7) we can find $\alpha \in \Omega_A$ such that $\omega = \omega(\alpha)$. Finally, it follows from the definition that any factor of a system with specification satisfies the specification property. Thus our system has specification.

We will now show that the function ψ defined by (3.6) for $h(\omega) = \omega_0$ is not continuous. For, if ω_0 is cohomologous to zero with a continuous transfer function ψ, then the cohomological equation can be lifted to a cohomological equation on Ω_A, so that the lift $\widetilde{\psi}$ of ψ is constant on preimages of points and is still continuous. But from (3.8), we see that $\omega_0(\alpha) = \alpha_1 - \alpha_0 = \alpha_0(\sigma_A\alpha) - \alpha_0(\alpha)$. Since we assume that $\omega_0(\alpha) = \widetilde{\psi}(\sigma_A\alpha) - \widetilde{\psi}(\alpha)$, then by the uniqueness, up to a constant c of a continuous (but not a bounded Borel!) transfer function, we have $\widetilde{\psi}(\alpha) = \alpha_0 + c$. This implies that α_0 is the same for any α projected to a given ω.

If $\sup_{m<n} \left| \sum_{i=m}^{n} \omega_i \right| < K$ then there exist more than one such α with different 0 coordinates. This shows that $\widetilde{\psi}$ cannot be continuous. □

3.4 Livshitz* theory.

3.4.1 Closing lemma and continuous solutions. Let us now show how a condition slightly different from the specification property, enables us to insure the continuity of the transfer function, and also to generalize the results of the previous section to cocycles with values in more general groups. Proposition 3.18 below is an abstract and somewhat diluted version of an extract from Livshitz's work [**L1**].

DEFINITION 3.17. A homeomorphism f of a compact metric space X satisfies the *closing lemma* if there exists an $\epsilon_0 > 0$ such that for any ϵ, $0 < \epsilon \leq \epsilon_0$, there is a $\delta > 0$ such that for any orbit segment $\{f^i(x)\}_{i=0,\ldots,m}$ with $\operatorname{dist}(f^m x, x) < \delta$, there exists a periodic orbit $\{f^i(y)\}_{i=0,\ldots,m-1}$, $f^m(y) = y$, such that $d_m^f(x, y) < \epsilon$.

The closing lemma is satisfied for subshifts of finite type and for locally maximal hyperbolic sets.

We will consider a G-cocycle h, where G is an arbitrary locally compact abelian group, and we will need a slightly stronger assumption on h than the equivalent of the dynamical Hölder condition. Namely, we will assume that for any neighborhood U of the identity in G there exists $\epsilon > 0$ such that if $d_n^f(x, y) \leq \epsilon$ then

$$h(x) \ldots h(f^{n-1}x) h(f^{n-1}y)^{-1} \ldots h(y)^{-1} \in U. \tag{3.9}$$

---

*We use a natural English spelling of the name as in [**KH**] instead of the phonetic spelling Livšic which appears in English translations of [**L1**] and [**L2**]

PROPOSITION 3.18. *Suppose that f is topologically transitive* and satisfies the closing lemma. A G-cocycle h satisfying (3.9) can be represented as $\psi(fx)\psi^{-1}(x)$, where ψ is continuous, if and only if the product of the values of h along every periodic orbit is equal to the identity.*

PROOF. Again as in Proposition 3.12, the statement is obvious in one direction, i.e. the cohomology condition implies the statement about periodic orbits.

Let $x \in X$ be a point with a dense orbit. We define ψ along this orbit by

$$\psi(f^n x) = \begin{cases} id, & n = 0 \\ h(f^{n-1}x)\dots h(x), & n > 0 \\ h(f^n x)^{-1} \cdot h(f^{n+1}x)^{-1} \cdot \dots \cdot h(f^{-1}x)^{-1}, & n < 0. \end{cases}$$

we will show that the ψ defined in this way is uniformly continuous on the orbit. Then it can be extended to be a continuous function on X. Since $(\psi \circ f) \cdot \psi^{-1}$ coincides with h along the orbit, the continuity of h implies the desired result.

In order to show uniform continuity, we assume that the points $f^n x$ and $f^m x$, $n < m$ are sufficiently close. Then by the closing lemma, we can approximate the segment $\{f^{n+k}x\}_{k=0,\dots,m-n}$ by a periodic orbit $\{f^k y\}_{k=0,\dots,m-n-1}$.

We have

$$\begin{aligned} &\psi(f^m x)\psi(f^n x)^{-1} \\ &= h(f^n x)h(f^{n+1}x)\dots(h(f^{m-1}x) \\ &= (h(f^n x)\dots h(f^{m-1}x))(h(y)\dots h(f^{m+n-1}y))^{-1}(h(y))\dots h(f^{m+n-1}y)) \\ &= (h(f^n x)\dots h(f^{m-1}x)) \cdot (h(f^{m+n-1}y)^{-1}\dots h(y)^{-1}) \cdot (h(y))\dots h(f^{m+n-1}y)) \end{aligned}$$

The third product is the identity and if $d^f_{m-n}(f^n x, y)$ is small enough, then by (3.9) the product of the first and second products belong to a given small neighborhood of the identity. But by the closing lemma, we can guarantee that $d^f_{m-n}(f^n x, y)$ is small by making $\mathrm{dist}(f^n x, f^m x)$ small. □

3.4.2 Hölder regularity. The core results of the Livshitz theory establishing both stability and effectiveness of cocycles in the primary hyperbolic situations can be summarized as follows.

THEOREM 3.19. [**L1**] [**L2**] [**LS**] [**KH, Section 19.2**] *Suppose that the group G has equivalent left and right invariant metrics and let f be any topological Markov chain or a restriction of a C^2 diffeomorphism to a locally maximal hyperbolic set. In both cases we assume that all points are nonwandering (see* [**KH, Section 18.3**]. *Let furthermore h be an α-Hölder cocycle over f, $0 < \alpha \le 1$. Then the following conditions are equivalent*

(i) *If $f^n x = x$ then $h(x,n) = Id_G$, i.e. the product of the values of h along every periodic orbit is equal to the identity.*
(ii) *$h(x) = \psi(fx)\psi^{-1}(x)$, where ψ is an α-Hölder function*
(iii) *$h(x) = \psi(fx)\psi^{-1}(x)$, where ψ is a measurable function with respect to the Gibbs measure (or the equilibrium state, see* [**KH, Chapter 20**]*) μ_ϕ for a Hölder function ϕ.*

*Topological transitivity follows from the specification property.

REMARK. Notice that by Corollary 2.11 the continuity of h is not sufficient to make any conclusion about existence of transfer functions. However for cocycles satisfying the Dini condition which is weaker than Hölder the assumptions of Proposition 3.18 can be verified.

Obviously (ii) implies both (i) and (iii). To deduce (ii) from (i) one uses the same argument as in the proof of Proposition 3.18 but employs a stronger version of the closing lemma which gives the necessary estimates ([**KH, Section 6.4c**]). The deduction (ii) from (iii) uses the key *a priori* regularity argument which is also the basis of all higher regularity results. In order to avoid cumbersome notations we will only discuss cocycles with values in abelian groups.

Consider the stable and unstable manifolds $W^s(x)$ and $W^u(x)$ of a point x (see [**KH, Sections 6.2, 6.4**]). In the symbolic case these are simply the sets of points with the same nonnegative (corr. nonpositive) coordinates as x.

If ϕ is a solution of cohomological equation then for any $y \in W^s(x)$ one has for any $n > 0$,

$$\psi(y) = \psi(x) \cdot \psi^{-1}(f^n y) \cdot \psi(f^n x) \cdot \prod_{i=0}^{n-1} h(f^i x) \cdot h^{-1}(f^i y).$$

Since the distance between $f^n x$ and $f^n y$ exponentially decreases the product converges exponentially as $n \to \infty$. If ψ is uniformly continuous then $\psi^{-1}(f^n y) \cdot \psi(f^n x)$ converges to Id_G so that

$$\psi(y) = \psi(x) \cdot \lim_{n\to\infty} \prod_{i=0}^{n-1} h(f^i x) \cdot h^{-1}(f^i y). \tag{3.10}$$

If ψ is only measurable a justification for taking is limit is needed.

One can write a similar expression for the values of the transfer function along the unstable manifold $W^u(x)$ using the negative iterates of f, namely for $y \in W^u(x)$,

$$\psi(y) = \psi(x) \cdot \lim_{n\to\infty} \prod_{i=1}^{n} h(f^{-i} x) \cdot h^{-1}(f^{-i} y). \tag{3.11}$$

Legitimacy of these procedures in the case of a measurable transfer function ψ requires an argument which is based on the product structure for the measures μ_ϕ.

Notice also that these expressions allow to obtain Holder regularity of ψ as long as existence of a solution is known.

Theorem 3.19 has a natural extension for flows

3.4.3 Differentiable stability. In order to discuss further regularity let us assume that f is an Anosov diffeomorphism of a compact manifold. If the cocycle h is differentiable one can differentiate expressions (3.10) and (3.11) with respect to y. Since individual stable and unstable manifolds have the same degree of differentiability as the map, one can show using the chain rule that for a C^∞ diffeomorphism the solution ψ has as many derivatives along the stable and unstable directions as the map f, and that those derivatives are continuous This is sufficient to deduce that if h is C^1 then ψ is also C^1.

Higher regularity follows from the remarkable general result by Journé [**Jo**]:

JOURNÉ'S THEOREM. *Assume F and E are two transverse Hólder foliations with C^∞ leaves such all derivatives of all orders along the leaves are uniformly continuous. Let h be a function which has all derivatives of order up to r along the leaves of both foliations and theses derivatives are α Hölder. Then h is C^r and rth derivatives are β Hölder for some $\beta > 0$.*

This, in particular, implies C^r stability for C^{r+1} cocycles and hence C^∞ stability for C^∞ cocycles. The latter together with other regularity results was originally proved in [**dLMM**] using elliptic regularity theory. Another proof based on the Fourier transform is in [**HK2**]. Analytic stability was proved in [**dL**] based on some ideas from [**HK2**].

Thus in the hyperbolic case for cocycles with values in abelian groups the understanding of stability issues is virtually complete.

Without going into details we would like alert the reader to the fact that the picture becomes more complicated and less clear-cut for cocycles with values in nonabelian groups. The main issue here is that the exponential convergence of some orbits within the system, on which stability results are based, has to overcome possible exponential divergence within the group. Thus if the group G itself has a subexponential growth (e.g. if it is a nilpotent Lie group) or if the cocycle is sufficiently small (i.e. its values are close enough to the identity) this divergence can be overcome and stability results hold. The basic source here is the seminal paper by Livshitz [**L2**], where in addition to the correct results along the above lines a wrong claim for the general case is made. The paper by V. Niţică and A. Török [**NT**] reflects the present state of the subject including both optimal regularity results and examples showing that in general even the transfer functions between analytic cocycles may have only limited regularity.

3.5 Invariant distributions and stability of partially hyperbolic systems.

3.5.1 Invariant distributions and periodic orbits. In all cases of stability which we have discussed, from toral rotations to Anosov systems, the cohomology classes of cocycles were determined by the values of integrals with respect to the invariant measures. Invariant measures can be viewed as linear functionals on the space of continuous functions, but as we have seen, stability can only occur in smaller spaces. It is natural to wonder whether there are other invariant functionals defined on these spaces which cannot be extended to the space of all continuous functions, in particular whether there are any invariant distributions of positive order, i.e. invariant linear functionals defined on the spaces of smooth functions.

In fact, invariant distributions represent the most general possible source of stability of spaces of smooth vector-valued cocycles. For, let f be a C^r or C^∞ diffeomorphism of a compact manifold X and H be the space of C^r on C^∞ functions on X. Stability in H means that the operator $U_f - \mathrm{Id} : H \to H$ has closed image, say H_0. But then by the duality H_0 corresponds to the kernel of the dual operator in the space of all continuous linear functionals on H, i.e. H_0 is the common zero level for all f-invariant distributions (cf. Proposition 2.13)). For the flow case we consider instead of $U_f - \mathrm{Id}$ the differential operator generated by the flow.

Since in all cases mentioned in Sections 3.3 and 3.4 the space H_0 was determined by invariant measures the stability implies that in those cases invariant measures are dense in the weak topology in the corresponding spaces of invariant distributions. On the other hand, specification property implies that invariant measures carried by individual periodic orbits are dense in weak topology in the space of invariant measures. These remarks together with the results discussed in subsection 3.4.3 prove the following fact.

COROLLARY 3.20. *For any transitive Anosov diffeomorphism invariant measures carried by individual periodic orbits are dense in the weak topology in the space of all invariant distributions of all orders.*

REMARK. In [**V3**] W. Veech proved a similar statement for an arbitrary (not necessarily hyperbolic) automorphism F_A of the torus $\mathbb{T}^m$ which is ergodic with respect to Lebesgue measure. Equivalently, the matrix A has no roots of unity among its eigenvalues. Such an automorphism is necessarily partially hyperbolic which in this case means that there are eigenvalues both inside and outside of the unit circle. Each orbit $\mathcal{O}$ of the dual automorphism A^* of the lattice $\mathbb{Z}^m$ produces an invariant distribution $\delta_{\mathcal{O}}$ of correlation type similarly to the special case discussed in subsection 3.6.1 below. The system $\delta_{\mathcal{O}}$ generates the space of invariant distributions. Although periodic orbits for an ergodic toral automorphism are dense, specification property does not hold. This is the reason why in order to show that those distributions can be weakly approximated by linear combinations of invariant measures on periodic orbits, Veech has to use arithmetic techniques which are much more subtle and specialized than the arguments in the hyperbolic case.

In the next two subsections we will describe two different types of general constructions of invariant distributions which are not measures. These constructions produce interesting examples even in the hyperbolic case.

3.5.2 Periodic cycle distributions for partially hyperbolic systems. We discuss some of the constructions and results from [**KK**].

A diffeomorphism g of a manifold M with a Riemannian norm $||\cdot||$ is called *partially hyperbolic* if there exist real numbers $\lambda_1 > \mu_1 > 0$, $i = 1, 2$, $K, K' > 0$ and a continuous splitting of the tangent bundle

$$TM = E^+ \oplus E^0 \oplus E^-$$

such that for all $x \in M$, for all $v \in E^+(x)$ ($v \in E^+(x)$ respectively) and $n > 0$ ($n < 0$ respectively) we have for the differential $g_* : TM \to TM$

$$||g_*(v)|| \leq Ke^{-\lambda_1 n}||v||,$$

and, respectively

$$||g_*(v)|| \leq Ke^{-\lambda_2 |n|}||v||,$$

and for all $n \in \mathbb{Z}$ and $v \in E^0(x)$ we have

$$||g_*(v)|| \geq K'e^{-\mu_1 n}||v||,\ n > 0 \text{ and } ||g_*(v)|| \geq K'e^{-\mu_2 |n|}||v||,\ n < 0.$$

This property does not depend on the choice of Riemannian metric if the manifold M is compact.

We will call E^+ and E^- *the stable and unstable distributions* respectively. They are uniquely integrable to foliations with smooth leaves which we will denote W^s and W^u.

For any x, y such that $y \in W^s(x)$ the distance $d_M(f^i(x), f^i(y))$ decreases exponentially, thus $\phi(f^i(x)) - \phi(f^i(y))$ also decreases exponentially. Therefore, for any Hölder function ϕ the series

$$P^+(x,y)(\phi) = \sum_{i=0}^{\infty}(\phi(f^i(x)) - \phi(f^i(y)))$$

converges absolutely. (cf. (3.10)). Similarly for any x, y such that $y \in W^u(x)$ the series

$$P^-(x,y)(\phi) = -\sum_{i=-1}^{-\infty}(\phi(f^i(x)) - \phi(f^i(y))$$

also converges absolutely (cf. (3.11)).

We will call a set $S(x, y)$ of points $x_1 = x, x_2, \dots, x_k = y \in M$ a *broken path* from x to y, if $x_{i+1} \in W^a(x_i)$, $i = 1, \dots, k-1$, where $a = s$ or u. A closed broken path, i.e a set C of points $x_1, x_2, \dots, x_{2n}, x_{2n+1} = x_1 \in M$ if $x_{2k} \in W^s(x_{2k-1})$ and $x_{2k+1} \in W^u(x_{2k})$, for $k = 1, \dots, n$. will be called a *periodic cycle.*

For an Anosov system there are many periodic cycles of length four; however in the general partially hyperbolic case periodic cycles must be longer.

DEFINITION 3.21. For a periodic cycle C, we will denote by $F(C)$ the following continuous functional on the space L of Hölder functions:

$$F(C)(\phi) = P^+(x_1, x_2)(\phi) + P^-(x_2, x_3)(\phi) + \cdots$$

$$+P^+(x_{2n-1}, x_{2n})(\phi) + P^-(x_{2n}, x_1)(\phi).$$

We will call this functional a *periodic cycle functional.*

The periodic cycle functionals are invariant distributions. Thus any such functional must vanish on the space of coboundaries.

If ψ is a solution of the cohomological equation

$$\phi(x) = \psi(fx) - \psi(x),$$

then the values of ψ along every broken path are uniquely determined by its value at the beginning and are obtained by adding the value of P^+ or P^- depending on whether consecutive points on the path belong to the same stable or unstable manifold. In particular, the periodic cycle functionals represent obstructions to this process which must vanish to ensure uniqueness of the result.

One can reverse the argument and try to construct the solution for a given ϕ for which the periodic cycle functionals vanish. The best chances of success appear if any two points can be connected by a broken path. This is not always true; counterexamples are time t maps for suspensions of Anosov diffeomorphisms, and, more interestingly, ergodic but not hyperbolic automorphisms of a torus. However those cases are rather special. There are many partially hyperbolic systems for which not only connection is possible but the length of a minimal broken path connecting two points can be estimated from above by a positive power α of the distance. Such systems are called *locally (α) Hölder transitive* and in this case for any function for which the periodic cycle functionals vanish the construction of a solution along broken paths succeeds and produces a Hölder function albeit with a smaller Hölder exponent. To summarize

THEOREM 3.22. [**KK**] *If f is a partially hyperbolic diffeomorphism which is locally α-Hölder transitive, then, for any $\beta \in (0,1]$, the space of β-Hölder cocycles is $\alpha\beta$-Hölder stable and C^0-effective. The space of Hölder coboundaries is the common kernel of periodic cycle functionals.*

Coming back to the hyperbolic case one can use the periodic cycle functionals coming from the *quadrangles*, i.e. the cycles of period four. Moreover, due to the local product structure this construction works not only for Anosov diffeomorphisms but for the locally maximal hyperbolic sets as well. It this case the pair of foliations is locally Lipschitz transitive (i.e. locally 1-Hölder transitive) so that no loss in the Hölder exponent appears in agreement with Theorem 3.19(ii). It would be interesting to find a reasonably explicit procedure to recover the sums of a function along periodic orbits through the periodic cycle functionals.

3.5.3 Correlation functions. Let $f : (X, \mu) \to (X, \mu)$ be a measure preserving transformation, ϕ and ψ be L^2 functions. The inner products

$$\int_X \phi \cdot \overline{\psi \circ f^{-n}} d\mu = \int_X \phi \circ f^n \cdot \overline{\psi} d\mu$$

are called the *correlation coefficients* of ϕ and ψ. The map f is mixing if the correlation coefficients of any two functions with zero average go to zero as $n \to \infty$. It may happen that for a given ϕ and dense set of functions ψ with zero average the series

$$\sum_{n=-\infty}^{\infty} \int_X \phi \circ f^n \cdot \overline{\psi} d\mu \tag{3.12}$$

converge. Obviously exponential decay of correlation is sufficient for convergence. If this happens for all zero average C^∞ functions, (3.12) defines an invariant distribution.

THEOREM 3.23. *Let f be a transitive topological Markov chain or a restriction of a C^2 diffeomorphism to a topologically mixing locally maximal hyperbolic set and let μ_θ be the Gibbs measure for a Hölder function θ. For any two Hölder functions the correlation coefficients*

$$\int_X \phi \cdot \overline{\psi \circ f^{-n}} d\mu$$

decay exponentially.

Correlation decay provides yet another way to characterize cohomological behavior of Hölder functions for hyperbolic dynamical systems.

THEOREM 3.24. *Let f be a transitive topological Markov chain or a restriction of a C^2 diffeomorphism to a topologically mixing locally maximal hyperbolic set and let μ_θ be the Gibbs measure for a Hölder function θ. For an α-Hölder function ϕ the following conditions are equivalent*

(i) $\phi(x) = h(fx) - h(x)$, *where h is an α-Hölder function,*
(ii) *For any Hölder function ψ,*

$$\sum_{n=-\infty}^{\infty} \int_X \phi \circ f^n \cdot \overline{\psi} d\mu_\theta = 0$$

SKETCH OF PROOF. Assume (i). Then

$$\begin{aligned}\sum_{n=-\infty}^{\infty}\int_X \phi\circ f^n\cdot\overline{\psi}d\mu_\theta &= \sum_{n=-\infty}^{\infty}\int_X \phi\cdot\overline{\psi\circ f^{-n}}d\mu_\theta\\ &= \sum_{n=-\infty}^{\infty}\int_X (h\circ f-h)\cdot\overline{\psi\circ f^{-n}}d\mu_\theta\\ &= \lim_{n\to\infty}\sum_{k=-n}^{n-1}\int_X (h\circ f-h)\cdot\overline{\psi\circ f^{-k}}d\mu_\theta\\ &= \lim_{n\to\infty}\int_X (h\circ f^n-h\circ f^{-n})\cdot\overline{\psi}d\mu_\theta=0.\end{aligned}$$

Now assume (ii). By Theorem 3.19 it is sufficient to proof that the sum of values of ϕ along any periodic orbit is equal to zero. Assume the opposite. Then there exists a periodic orbit with nonzero, say, positive sum. One can assume then that ϕ is actually positive at all points of the orbit. Using certain care one can find a function ψ with zero average and with very large positive values at the orbit for which sufficiently many correlation coefficients will be positive and fairly large which prevents vanishing of (3.12)

3.6 Stability determined by invariant distributions in parabolic systems.

3.6.1 A simple example: an affine map of the torus. Now we are going to discuss in detail an example of a different sort, namely a uniquely ergodic transformation where invariant distributions provide a complete system of invariants for the cohomology, but the single invariant measure does not give enough information for this purpose. Invariant distributions which we will encounter will be of the correlation type (3.12).

Consider the affine A map of $\mathbb{T}^2$ given by

$$A(z,w)=(\lambda z, zw).$$

We will consider only real or complex valued cocycles.

If we assume that λ is Diophantine (3.1), i.e for some positive k and c, and for every positive integer q,

$$|\lambda^q-1|>c/q^k. \tag{3.13}$$

then by Theorem 3.5 functions depending only on z are C^∞ rigid and in the case of finite regularity there is a fixed loss of regularity (see remarks in subsection 3.2.2). Since the decomposition into functions depending only on z and their orthogonal complement is invariant it is sufficient to consider functions in that complement.

Let us define a family of first order A-invariant distributions $\delta_{m,n}$, for $0\le m<|n|$, $n\neq 0$, by their Fourier expansions:

$$\delta_{m,n}=\sum_{k=-\infty}^{\infty}\lambda^{km+\frac{k(k-1)n}{2}}z^{m+kn}w^n.$$

Equivalently, for $f \in C^1(\mathbb{T}^2)$, let $z = e^{2\pi i\phi}$ and $w = e^{2\pi i\theta}$ and let as before $\chi_{m,n} = z^m w^n$. Then

$$\begin{aligned}\delta_{m,n}(f) &= \lim_{K\to\infty}\int_{\mathbb{T}^2} f(z,w)\sum_{k=-K}^{K}\lambda^{-km-\frac{k(k-1)n}{2}}\overline{z}^{m+kn}\overline{w}^n\,d\phi d\theta \\ &= \lim_{K\to\infty}\int_{\mathbb{T}^2} f(z,w)\sum_{k=-K}^{K}\overline{\chi_{m,n}\circ A^k}d\phi d\theta \\ &= \lim_{K\to\infty}\sum_{k=-K}^{K} f(A^k(z,w))\overline{z}^m\overline{w}^n\,d\phi d\theta.\end{aligned}$$

Obviously every distribution $\delta_{m,n}$ is a derivative of an L^2-function.

Let H^r, $r \ge 0$, be the space of functions ψ on $\mathbb{T}^2$ such that ψ is r times differentiable and all of its rth derivatives have absolutely convergent Fourier series.

THEOREM 3.25. *Let $f, g \in C^{r+\epsilon}$, $r \ge 3$ and $\epsilon > 0$. Assume that both functions are orthogonal to the functions depending only on z. Then there exists $\psi \in H^{r-3}$ such that*

$$f = g + \psi\circ A - \psi \tag{3.14}$$

if and only if

$$\delta_{m,n}(f) = \delta_{m,n}(g)$$

for all $n \in \mathbb{Z}$, $0 \le m < |n|$. In particular, if f and g are C^∞, then ψ is C^∞. In addition, if f and g are real analytic then so is ψ. Furthermore, for $f, g \in C^1$ if equation (3.14) has an L^1 solution then $\delta_{m,n}(f) = \delta_{m,n}(g)$ for all $0 \le m < |n|$, so that the solution actually belongs to H^{r-3}.

Our estimate of the loss of regularity for the solution of the cohomological equation is not sharp but it is sufficient to establish C^∞ stability.

PROOF. It is enough to consider only the case $g = 0$. We look for the Fourier coefficients $\psi_{m,n}$ for the transfer function ψ. By our assumption $f_{m,0} = 0$ for all $m \in \mathbb{Z}$ hence we need only consider the case $n \ne 0$. Since $U_A\chi_{m,n} = \lambda^m\chi_{m+n,n}$ for $n \ne 0$ equation (3.14) implies the following infinite system of algebraic equations.

$$f_{m,n} = \lambda^{m-n}\psi_{m-n,n} - \psi_{m,n} \tag{3.15}$$

so that

$$\psi_{m,n} = -f_{m,n} + \lambda^{m-n}\psi_{m-n,n}.$$

Proceeding by induction,

$$\begin{aligned}\psi_{m,n} &= -f_{m,n} - \lambda^{m-n}f_{m-n,n} + \lambda^{2m-3n}\psi_{m-2n,n} \\ &= -\sum_{k=0}^{N-1}\lambda^{e_k(m,n)}f_{m-kn,n} + \lambda^{e_N(m,n)}\psi_{m-Nn,n}\end{aligned} \tag{3.16}$$

for all $N \ge 0$, where

$$e_k(m,n) = e_{k-1}(m,n) + m - kn$$

and

$$e_0(m,n) = 0$$

so that

$$e_k(m,n) = km - \frac{1}{2}k(k+1)n.$$

Similarly, using (3.15) again, and repeating the steps above

$$\begin{aligned} \psi_{m,n} &= \lambda^{-m}(f_{m+n,n} + \psi_{m+n,n}) \\ &= \sum_{k=1}^{N} \lambda^{-e'_k(m,n)} f_{m+kn,n} + \lambda^{-e'_N(m,n)} \psi_{m+Nn,n} \end{aligned} \tag{3.17}$$

for all $N \geq 1$, where

$$e'_k(m,n) = km + \frac{1}{2}k(k-1)n.$$

Assuming that $\psi_{m+Nn,n} \to 0$ as $N \to \pm\infty$, which is true even if $\psi \in L^1$, we obtain from (3.16) and (3.17) two expressions for $\psi_{m,n}$:

$$\psi_{m,n} = -\sum_{k=0}^{\infty} \lambda^{km - \frac{1}{2}k(k+1)n} f_{m-kn,n} \tag{3.18}$$

and

$$\psi_{m,n} = \sum_{k=1}^{\infty} \lambda^{-km - \frac{1}{2}k(k-1)n} f_{m+kn,n} \tag{3.19}$$

If the solution exists, these two expressions must coincide, so that

$$\sum_{k=-\infty}^{\infty} \lambda^{-km - \frac{1}{2}k(k-1)n} f_{m+kn,n} = 0. \tag{3.20}$$

For $0 \leq m < |n|$, the left hand side of (3.20) is exactly what we call $\delta_{m,n}(f)$. It is easy to see that if equation (3.20) is satisfied for some (m,n), it is also satisfied for $(m+kn,n)$ for every k. This gives the last statement in the Theorem.

Assuming (3.20), we can estimate $\psi_{m,n}$. For $mn < 0$ we will use (3.18) and for $mn \geq 0$ we will use (3.19). We have respectively

$$|\psi_{m,n}| \leq \sum_{k=0}^{\infty} |f_{m-kn,n}|$$

$$|\psi_{m,n}| \leq \sum_{k=1}^{\infty} |f_{m+kn,n}|.$$

If $f \in C^{r+\epsilon}$, then

$$|f_{m,n}| \leq \frac{c}{(|m|+|n|)^{r+\delta}}$$

for some $\delta > 0$, and so

$$|\psi_{m,n}| \leq \frac{c'}{(|m|+|n|)^{r-1+\delta/2}}.$$

This implies that $\psi_{m,n} \in H^{r-3}$.

If f is real analytic, then

$$|f_{m,n}| < c \exp(-\alpha(|m|+|n|))$$

for $\alpha, c > 0$. It follows that

$$|\psi_{m,n}| \leq \sum_{k=0}^{\infty} c \exp(-\alpha(|m| + k|n| + |n|) \leq c' \exp -\alpha(|m| + |n|) .$$

□

Thus, we have shown that various spaces of smooth cocycles over A are stable and "almost" effective with a certain loss of regularity which is natural to expect. The C^∞ cocycles are both C^∞ stable and C^∞ effective. We do not know whether the existence of a measurable solution to (3.14), for sufficiently smooth f and g implies that the solution is L^1, which is then smooth by the theorem. The difficulty here is that non-integrable measurable functions are not distributions so that Fourier analysis can not be applied in this case. Probably, in order to solve this problem one needs to develop a more geometric approach to the solution of (3.14). This comment applies also to the cocycles with values in groups which are not linear spaces.

It is interesting to notice that if functions f and g have absolutely convergent Fourier series and equal averages then equation (3.14) has solutions ψ which are distributions of order 1. Either of the two formulas (3.18) and (3.19) gives Fourier coefficients for such a solution.

The method of Theorem 3.25 applies to arbitrary ergodic automorphisms and affine maps on a torus of any dimension. The only restriction is the Diophantine condition for the rotational part of the affine map. In all those cases there is a natural one-to-one correspondence between invariant distributions which determine stability and infinite orbit of the lattice automorphism dual to the automorphism part of the considered map.

3.6.2 Horocycle flows. Another and even more interesting example of C^∞ stability and C^∞ effectiveness determined by invariant distributions is given by the horocycle flow on a compact surface of constant negative curvature. This flow can be equivalently described as the action of the one-parameter subgroup

$$H_t = \begin{pmatrix} 1 & t \\ 0 & 1 \end{pmatrix}$$

by right multiplications on the left factor space $B = PSL(2,\mathbb{R})/\Gamma$, where Γ is a discrete cocompact subgroup of $PSL(2,\mathbb{R})$.

The horocycle flow is a part of the right action of $PSL(2,\mathbb{R})$ on B. This action generates a unitary representation of $PSL(2,\mathbb{R})$ in $L^2(B,\chi)$ where χ is Haar measure. This representation decomposes into a direct sum of countably many irreducible ones. Obviously every representation space is invariant with respect to the operators corresponding to the horocycle flow. The space of each irreducible representation carries certain invariant distributions for the horocycle flow. Namely,

(i) for a representation from the principal series there are two invariant distributions of order less than one similar to $\delta_{m,n}$ for the affine map on $\mathbb{T}^2$;
(ii) for a representation from the discrete series there is one invariant distribution of order less than one;
(iii) for a representation form the complementary series there are infinitely many representations of growing order.

In every representation space there is an intrinsic notion of a C^∞ function which agrees with the geometric notion coming from B. More precisely, let $f \in C^\infty(B)$; and for any irreducible representation θ of $PSL(2,\mathbb{R})$, let P_θ be the orthogonal projection of $L^2(B)$ to the representation space. Then $P_\theta f$ is a C^∞ vector in the sense of the representation and the norms $\|P_\theta f\|$ decrease faster than any power of a parameter naturally ascribed to the irreducible representations. Furthermore, this condition is necessary and sufficient for f to be a C^∞ function. These remarks show that the problem can be treated in a way very similar to the way affine maps of $\mathbb{T}^2$ were treated; using certain (not always elementary) facts about irreducible unitary representations of $PSL(2,\mathbb{R})$ instead of the elementary Fourier analysis of the previous proof. See [**FlFo**] for detailed rigorous arguments in this case.

3.6.3 Flows on surfaces. Without trying to explain the substance of the situation we have to mention for the sake of completeness another remarkable instance of stability determined by invariant distributions.

Consider a compact orientable surface S of genus $g > 1$ with a smooth positive area form Ω. Since the Euler characteristic $\chi = 2 - 2g$ of S is negative any vector field on S has to have zeroes. The case which is the most interesting from the dynamical point of view is those of area preserving topologically transitive flows with finitely many fixed points. These fixed points have to be of saddle type, either simple, or multiple saddles. A topologically transitive area preserving flow with finitely many saddle points exhibit parabolic behavior since its derivative has triangular form outside of the fixed points. However this behavior is less uniform than that of affine maps or horocycle flows. Aside from point masses at each saddle there are no more than g ergodic probability measures [**KH, Theorem 14.7.6**].The transverse behavior of area preserving flows on surfaces is defined by a finite set of parameters [**KH, Theorem 14.7.4**] and is closely related to the interval exchange transformations discussed in Section 4.4. For the set of parameters of full measure there is only one nontrivial probability ergodic invariant measure. There are invariant distributions defined by the jets at the fixed points; they are invariants of the normal form of the vector field near these points. Forni [**Fo**] discovered additional invariant distributions which are supported outside of fixed points and give a complete description of these distributions. The picture is quite different from the cases we discussed before. There is only a finite dimensional space of invariant distributions for any given order r but the dimension grows to infinity with r. For certain conditions which can be interpreted as counterparts of Diophantine conditions Forni showed C^∞ stability of C^∞ cocycles.

4. Wild cochains with tame coboundaries

In this and the next chapter we will discuss the situations where correspondingly effectiveness and stability break down.

Thus, we will presently consider "regular" (e.g. continuous, smooth or piecewise constant) cocycles which are coboundaries in measure-theoretic sense but such that the transfer functions behave very wildly from the topological point of view. This phenomenon is often connected with abnormally good periodic approximation. It was known to Kolmogorov [**Ko**] who used it to construct an example of a time change for a linear flow of the two–torus with pure point spectrum but discontinuous eigenfunctions. Furstenberg [**F**] used it for his construction of minimal but not uniquely ergodic diffeomorphisms of $\mathbb{T}^2$. Anosov [**A**] studied the case of circular rotation thoroughly and systematically and in particular found examples of analytic coboundaries with wild transfer functions.

Cocycles of this sort are used to construct various examples of dynamical systems whose topological and ergodic properties differ wildly. They can also be used for more positive tasks, e.g. for producing realizations of measure-preserving transformations by volume preserving homeomorphisms of various manifolds.

We will illustrate this phenomenon with four examples. The first one deals with continuous cocycles over a topologically transitive homeomorphism of a compact metric space preserving a Borel measure positive on open sets. As we noticed in Section 2.2 if the homeomorphism is not uniquely ergodic then the existence of cohomologous continuous cocycles equivalent via discontinuous transfer functions follows from Corollary 2.11 and Proposition 2.13. The construction of Theorem 4.2 shows a specific mechanism for the occurrence of wild transfer functions between continuous cocycles. This mechanism is universal and in particular is independent of the existence of more than one ergodic invariant measure.

The results of Chapter 3 show that in general such mechanism cannot be extended to smooth cocycles. However, an extension of that sort is possible if the diffeomorphism allows a very good periodic approximation in C^∞ sense (Theorem 4.5). We will also indicate how a similar construction works for real analytic cocycles over a rotation of the circle with exceptionally well approximable rotation number (Theorem 4.7).

Our last example deals with even smaller, actually a finite-dimensional space of cocycles, namely some piecewise constant cocycles over interval exchange transformations (Theorem 4.9).

The method of proof in all four cases is rather uniform. The desired cocycle together with the transfer function will be constructed by a converging iterative (or inductive) procedure. On every step of the construction we will have a cocycle which is a coboundary within the given category. In other words, the transfer function will be correspondingly continuous, smooth, real-analytic or piecewise constant. The sequence of cocycles will converge within the category while the sequence of transfer functions will converge only in probability or in L^1 and will diverge in a certain prescribed fashion in the uniform topology. The control over the convergence will by provided by making the modifications on any step insignificant enough in measure-theoretic sense in comparison with the previous steps. On the other hand, the divergence of the sequence of transfer functions is controlled by making subsequent additions sufficiently uniformly distributed.

4.1 Continuous cocycles over measure-preserving homeomorphisms. Let X, Y be two topological spaces and ν be a Borel measure on X positive on open sets.

DEFINITION 4.1. We will say that a Borel map $\psi : X \to Y$ is *metrically dense with respect to* ν if for any pair of non-empty open sets $V \subset X$ and $W \subset Y$,

$$\nu(V \cap \psi^{-1}(W)) > 0.$$

Equivalently, this property means that the lift of the measure ν to graph $\psi \subset X \times Y$ is a measure positive on open sets.

THEOREM 4.2. *Let X be a compact metric space and U be an open subset of X. Suppose $f : X \to X$ is a topologically transitive homeomorphism preserving a Borel probability measure μ which is positive on open sets. Let G be a path connected separable locally compact abelian group. Then there exists a continuous map $\phi : X \to G$ such that*

(i) *Outside of U, $\phi = e$, the identity in G.*

(ii) *$\phi(x) = \psi(f(x))\psi^{-1}(x) = \lim_{n\to\infty} \psi_n(f(x))\psi_n^{-1}(x)$ where the functions ψ_n are continuous and ψ is measurable.*

(iii) *ψ is metrically dense with respect to μ.*

This theorem can be used to prove that every abstract measure-preserving transformation with finite entropy can be realized as a volume-preserving homeomorphism of any compact Riemannian manifold of dimension greater than one.

PROOF. *Step 1.* The desired function ϕ will be constructed as a uniform $\lim_{n\to\infty} \phi_n$ where inductively

$$\phi_{n+1}(x) = \phi_n(x)\eta_n(x) \tag{4.1}$$

such that outside of U, $\eta_n = e$ and

$$\eta_n(x) = \theta_n(f(x))\theta_n^{-1}(x). \tag{4.2}$$

In other words

$$\phi_n(x) = \psi_n(f(x))\psi_n^{-1}(x)$$

where

$$\psi_n(x) = \prod_{j=1}^{n} \theta_j(x). \tag{4.3}$$

We will ensure that the sequence ψ_n converges in probability so that (ii) will be satisfied. On the other hand, we will make the sequence ψ_n wildly divergent in the uniform topology so that on the n'th step, the function ψ_n will satisfy the property (iii$_n$), stated below, which can be considered as an approximate version of (iii). In order to formulate this property let us fix a translation invariant metric d_G on G and denote by $D(r, g)$ the ball of radius r around $g \in G$. Let $g_n,\ n = 1, 2, \dots$ be a dense sequence of elements of G. We will assume that there exists a sequence of positive numbers δ_n, $\delta_n < \delta_{n-1}/2$, such that for every $k - 1, \dots, n$ and for every disc D of radius 2^{-k} in X,

(iii$_n$) $\mu(D \cap \psi_n^{-1} D(2^{-k}, g_k)) > \delta_k \left(\frac{1}{2} + \frac{1}{2^{n-k+1}}\right)$.

In order to ensure that subsequent steps do not destroy this condition we will assume

(a_n) $\mu\{x \in X; \theta_n(x) \neq e\} < \delta_{n-1}/10$.

It is clear that the sequence ψ_n defined by (4.3), and satisfying (iii$_n$) and (a_n) converges in probability and that the limit function ψ satisfies (iii).

Step 2. We will show how to construct θ_n and δ_n inductively. The base of the induction is trivial. Now let us assume that θ_{n-1} and δ_{n-1} have been constructed. This allowed us to construct ψ_{n-1}, and by the uniform continuity of this function, we can find a positive number δ_n' such that if $d_X(x_1, x_2) < \delta_n'$ where d_X is the distance in X then

$$d_G(\psi_{n-1}(x_1), \psi_{n-1}(x_2)) < 2^{-n-10}. \tag{4.4}$$

Let

$$H = \{g \in G : gg_n^{-1} \in (\psi_{n-1}(X))^{-1}\}.$$

This set is compact, so that we can cover it by a finite set of balls of radius 2^{-n-5} and pick an element g in each of the balls. S will denote the set of all such g. For each $g \in S$ we are going to construct a function θ_n^g. The supports of θ_n^g will be disjoint for different g and θ_n^g will be supported by a very small neighborhood of a segment of a single orbit in such a way as to make θ_n^g reach the value g inside every ball of radius δ_n' in X. Furthermore the function

$$\eta_n^g(x) = \theta_n^g(f(x))(\theta_n^g(x))^{-1} \tag{4.5}$$

will be equal to the identity outside the set U and will be uniformly within $1/2^n$ of the identity everywhere. Then we will put

$$\theta_n = \prod_{g\in S} \theta_n^g$$

so that the function $\psi_n = \psi_{n-1}\theta_n$ will reach a value within 2^{-n-4} of g_n inside every δ_n' ball in X. This follows from the choice of δ_n'. On the other hand, $\mu(\text{supp}(\theta_N))$ will be chosen so small that the values near g_k, $k = 1, \dots, n-1$ achieved on the previous stages of the construction will persist.

Step 3. We will now show how to construct the functions θ_n^g and insure (a_n) and (iii_n). Let us connect every $g \in S$ by a path γ_g to the identity $e \in G$. This path may be divided into intervals such that each interval lies within a ball of radius 2^{-n}. Let K_n be the maximal number of such intervals for any $g \in S$.

For every $g \in S$ let us take a sufficiently long segment of a dense orbit, say $\Gamma_g = \{x_g, f(x_g) \dots f^N(x_g)\}$ such that all Γ_g are disjoint for different g and in addition each segment consists of three consecutive parts so that the first and the last parts contain at least K_n points from U, and the middle part intersects every ball in X of radius $\delta_n'/10$. Take the segment Γ_g and mark the first K_n iterates where $f^j x_g \in U$. Call these times $t_1(g), \dots, t_{K_n}(g)$. Do the same for the last K_n visits calling them $s_{K_n}(g) \dots s_1(g)$, where $s_1(g)$ is the last visit to U. The function θ_n^g will be different from the identity only in very small neighborhoods of the points $f^{t_i(g)}(x_g)$ and $f^{s_i(g)}(x_g)$, $i = 1, \dots, K_n$. The range of this function will be a neighborhood of the path γ_g. Pick points $e = h_1, \dots, h_{K_n} = g$ so that each interval $[h_j, h_{j+1}] \subseteq \gamma_g$ lies within a ball of radius 2^{-n}, and construct the function θ_n^g by induction in $j = 1, \dots, K_n$. Suppose that it has been constructed through time t_j. We multiply it by a continuous function η_n^g whose support $U(g,j) \subset U$ is a small neighborhood of $f^{t_{j+1}(g)^n}(x_g)$, whose range is $h_j^{-1}([h_j, h_{j+1}])$, and such that

$$\eta_n^g(f^{t_{j+1}(g)}(x_g)) = h_j^{-1}h_{j+1}$$

so that

$$\theta_n^g(f^{t_{j+1}(g)}(x_g)) = h_{j+1}.$$

We continue this procedure until $t_{K_n}(g)$ has been reached so that $\theta_n^g(f^{t_{K_n}(g)}(x_g)) = g$, stop for the middle section of Γ_g, and then "undo" the function θ_n^g along the last section of Γ_g by putting θ_n^g in a neighborhood of $f^{s_j(g)}(x_g)$ equal to

$$\theta_n^g(x) = (\theta_n^g(f^{t_j(g)-s_j(g)}(x)))^{-1}.$$

Thus, if the neighborhoods $U(g, j)$, $j = 1 \dots K_n$ are chosen sufficiently small then for any integer t, $t_{K_n}(g) \le t \le s_{K_n}(g)$

$$\theta_n^g(f^t(x_g)) = g.$$

By the choice of the interval Γ_g the function θ_n^g reaches value g inside any disc of radius $\delta_n'/10$ in X. By the choice of $U(g, j)$ one can also guarantee, that the total measure of the support of each θ_n^g (and consequently of θ_n) can be made arbitrary small and the supports of θ_n^g, $g \in S$, disjoint. In particular, the condition (a_n) will be satisfied. That ensures that the condition (iii_n) is satisfied for $k = 1, \dots, n-1$. Let us show that if δ_n is chosen sufficiently small then (iii_n) is also satisfied for $k = n$.

Let $x \in X$ and $g \in S$ be such that

$$d_G(g\psi_{n-1}(x), g_n) < 2^{-n-5} \tag{4.6}$$

By the choice of the orbit segments Γ_g the ball of radius $\frac{\delta_n'}{10}$ about x contains a point from the middle segment of Γ_g, say y, so that $\theta_n(y) = \theta_n^g(y) = g$. So that by (4.4) and (4.6)

$$d_G(\psi_n(y), g_n) < 2^{-n-4}.$$

Since $\delta_n < 2^{-n}$ this implies that for the ball $D(x)$ about x of radius 2^{-n}

$$\mu(x) \stackrel{\text{def}}{=} \mu(D(x) \cap \psi_n^{-1}(D_G(2^{-n}, g_n)) > 0$$

and by the compactness of X the measure $\mu(x)$ is bounded away from 0 for all x. Any positive lower bound for $\mu(x)$ may serve as δ_n. Obviously this choice of δ_n insures (iii_n) for $k = n$. □

Let us consider the following continuous G extension f^ϕ of f acting in the space $X \times G$:

$$f^\phi(x, g) = (f(x), g\phi(x))$$

and let $\nu = (id \times \psi)_*\mu$.

COROLLARY 4.3.
(i) *The measure ν is positive on open sets in $X \times G$.*
(ii) *The extension f^ϕ preserves ν.*
(iii) *(f^ϕ, ν) is metrically isomorphic to (f, μ).*

PROOF. By Theorem 4.2, (ii), the map $R = id \times \psi : X \to X \times G$ satisfies $R \circ f^\phi \circ R^{-1} = f$. This proves (ii) and (iii). By Theorem 4.2, (iii), the measure ν is positive on open sets, since for $V \subseteq X$, $W \subseteq G$, $\nu(V \times W) = \mu(V \cap \psi^{-1}(W))$. □

4.2 Fast approximation and C^∞ cocycles. A construction similar to that of Theorem 4.2 works in C^∞ category if the transformation in consideration possesses a C^∞ equivalent of periodic homogeneous approximation with a speed which is faster than any negative power of the characteristic parameter. While we do not need the above notions in our presentation a proper discussion can be found in [**K1**]. To simplify notations we will consider only real-valued cocycles. Let M be a compact connected m-dimensional C^∞ manifold, $C^r(M)$ be the Banach space of r times continuously differentiable real-valued functions on M provided with a norm $\|\cdot\|_r$. We will use the same notation for the associated operator norm.

DEFINITION 4.4. A C^∞ diffeomorphism $f : M \to M$ *admits fast* C^∞ *periodic approximation* if there exists a sequence $q_n \to \infty$ such that for every pair of positive integers k, r there exists a constant $c(k,r)$ such that

$$\|U_{f^{q_n}} - Id\|_r < \frac{c(k,r)}{q_n^k} \tag{4.7}$$

Equivalently, for every $h \in C^\infty(M)$, $h \neq 0$,

$$\|h \circ f^{-q_n} - h\|_r < \frac{\|h\|_r \cdot c(k,r)}{q_n^k}. \tag{4.8}$$

EXAMPLE. For the rotation of the circle $R_\lambda : z \to \lambda z$ fast C^∞ periodic approximation is equivalent to λ being Liouvillean (cf. Section 3.2).

THEOREM 4.5. *Suppose that a C^∞ diffeomorphism $f : M \to M$ preserves a measure μ given by a density bounded between two positive constants, is topologically transitive and admits fast C^∞ periodic approximation. Let U be any open non-empty subset of M. Then there exists a C^∞ cocycle h equal to 0 outside U and such that*

$$h = \psi \circ f - \psi = \lim_{n\to\infty} \psi_n \circ f - \psi_n$$

where the functions ψ_n are C^∞ and the function ψ is measurable, discontinuous at every point and for every $\lambda \in \mathbb{R}\backslash\{0\}$ the function $\exp i\lambda\psi : M \to S^1$ is metrically dense with respect to μ (cf. Definition 4.1). In addition, h can be chosen in such a way that ψ either belongs to L^1 or not.

PROOF. Let $x_0 \in U$ be a point whose semi-orbit $\{f^n x_0\}$ $n = 0, 1, 2, \ldots$ is dense in M. Let $t_1, \ldots, t_m$ be C^∞ local coordinates in a neighborhood $V \subset U$ of the point x_0 such that $t_i(x_0) = 0$, $i = 1, \ldots, m$. Obviously, every C^∞ function of the local coordinates equal to zero near ∂V can be extended to a C^∞ function on M by making it equal to zero outside V. Let $\theta : \mathbb{R} \to \mathbb{R}$ be a standard C^∞ bump function such that $\theta(0) = 1$, $\theta(t) = 0$ for $|t| \geq 1$ and θ is non-increasing for positive values of t and non-decreasing for the negative ones. Then for any sufficiently large positive q the function h_q defined by

$$h_q(x) = \begin{cases} \theta\left(q^6\left(\sum_{i=1}^m t_i^2(x)\right)\right) & \text{if} \quad x \in V \\ 0 \quad \text{otherwise} \end{cases} \tag{4.9}$$

is a C^∞ function on M.

For two versions of the statement (ψ is integrable or not) the function h is defined by

$$h = \sum_{k=1}^\infty q_{n_k}^2 \left(h_{q_{n_k}^2} \circ f^{-q_{n_k}} - h_{q_{n_k}^2}\right) \tag{4.10}$$

and

$$h = \sum_{k=1}^\infty q_{n_k}^{3m} \left(h_{q_{n_k}^2} \circ f^{-q_{n_k}} - h_{q_{n_k}^2}\right) \tag{4.11}$$

correspondingly, where q_n is a sequence which appears in (4.8) and the sequence $n_k \to \infty$ will be chosen later. To verify all the assertions of the theorem except for the metric density of the function exp $i\lambda\psi$ it is enough to observe that $q_{n_k} > k$.

First let us show that the series in the right hand part of (4.10) and (4.11) converge in the C^∞ topology.

By applying the chain rule in (4.9) we obtain

$$\|h_q\|_r < c_1(r)q^{6r} \tag{4.12}$$

where $c_1(r)$ depends on θ but not on q. Let us put in (4.8) $k = 6r + 3m + 10$. We obtain from (4.8) and (4.12)

$$\|h\|_r \leq c_1(r) \cdot c(k,r) \sum_{n=1}^{\infty} q_n^{-10}.$$

Let us show that the cocycle h is a measurable coboundary. Define

$$\psi_q \stackrel{\text{def}}{=} -\sum_{\ell=1}^{q} h_{q^2} \cdot f^{-\ell}.$$

It follows immediately from the definition that

$$h_{q^2} \circ f^{-q} - h_{q^2} = \psi_q \cdot f - \psi_q.$$

Since $\mu(\text{supp } h_q) < c_2 \cdot q^{-3m}$ we have

$$\mu(\text{supp } \psi_q) \leq q\mu(\text{supp } h_q) < c_2 q^{1-3m} \leq c_2 q^{-2} \tag{4.13}$$

so that $\sum\limits_q \mu(\text{supp}\psi_q) < \infty$.

Consequently for every sequence of constants a_q the series $\sum\limits_q a_q\psi_q$ converges in probability. This implies that $h = \psi \circ f - \psi$ where for h defined by (4.10)

$$\psi = \sum_{k=1}^{\infty} q_{n-k}^2 \psi_{q_{n_k}} \tag{4.14}$$

and for h defined by (4.11)

$$\psi = \sum q_{n-k}^{3m} \psi_{q_{n_k}}. \tag{4.15}$$

We have from (4.9) $c_r q^{-3m} < \|h_q\|_{L^1} < c_3 q^{-3m}$ and consequently

$$\|\psi_q\|_{L^1} \leq q\|h_q\|_{L^1} < c_3 q^{1-3m}$$

so that it follows that ψ defined by (4.14) is an L^1 function. On the other hand, since h_q is non-negative and

$$\|\psi_q\|_{L^1} > c_r q^{1-3m}$$

the function defined by (4.15) is not integrable. The discontinuity of ψ at every point follows from the fact that for every positive m $\psi_{q_{n_k}}(f^m x_0) \to -\infty$ so that since all $\psi_{q_{n_k}}$ are non-positive functions

$$\lim_{x \to f^m x_0} \psi(x) = -\infty$$

Since the semi-orbit of x_0 is dense this proves discontinuity.

It remains to show how an appropriate choice of the sequence q_{n_k} provides the metric density of the function $\exp i\lambda\psi$. This can be achieved by an inductive argument very similar to the one used in the corresponding part of the proof of Theorem 4.2. Since ψ_q is a non-positive continuous function, $\mu(\text{supp }\psi_q) \to 0$ as $q \to \infty$ and $\psi_q(f^\ell x_0) < -1$ for $\ell = 1, \dots, q$, it follows that for every $\epsilon > 0$ there exists $Q(\epsilon)$ such that for every $q > Q(\epsilon)$ the functions $\exp i\lambda^2\psi_q$ and $\exp i\lambda q^{3m}\psi_q$ map every disc of radius ϵ onto the circle. Moreover, this remains true for every function χ of the form $\exp i\lambda(q^2\psi_2 + \phi)$ or $\exp i\lambda(q^{3m}\psi_q + \phi)$ where ϕ is continuous and the oscillation of ϕ on any disc of radius ϵ does not exceed $1/\lambda$. If ϕ is fixed this implies that for any fixed $\delta > 0$, any ϵ-disc D in X and any δ-interval $\Delta \subset S^1$,

$$\mu(\chi^{-1}(\Delta) \cap D) > \alpha(\epsilon, \delta, q) > 0.$$

We can think now of ϕ as $\sum_{\ell=1}^{k-1} q_{n_\ell}^{3m}\psi_{q_\ell}$. Fixing ϵ_n such that the oscillation of ϕ in every ϵ-disc is less than $1/\lambda$, we can find k such that $q_{n-k} > Q(\epsilon_n)$ and apply the construction described above. In order to keep the approximate density achieved on the previous steps intact, we assume in addition that q_{n_k} grow so fast that

$$\sum_{\ell=k+1}^{\infty} \psi(\text{supp }\mu_{q_{n_\ell}}) < \frac{\alpha(\epsilon_k, w^{-k}, q_{n_K})}{2}.$$

This implies that for any interval Δ of length 2^{-k} and any ϵ_k-disc $D \subset X$

$$\mu(\psi^{-1}(\Delta) \cap D) > \frac{1}{2}\alpha(\epsilon_k, 2^{-k}, q_{n_k}). \square$$

COROLLARY 4.6. *If a diffeomorphism f satisfies all the assumptions of Theorem 4.5 and is ergodic with respect to the measure μ then there exists a C^∞ S^1 extension of f which is topologically transitive on $M \times S^1$ but not ergodic with respect to the product measure.*

PROOF. We define the extension via the cocycle $\exp ih$ where h is given by Theorem 4.5. Since this cocycle is a coboundary the extension is measure-theoretically isomorphic to the trivial extension and thus is not ergodic. On the other hand, the measure $(id \times \psi)_*\mu$ is invariant and ergodic with respect to the extension and since ψ is metrically dense this measure is positive on open subsets of $M \times S^1$. This implies topological transitivity. $\square$

4.3 Minimal nonergodic diffeomorphisms of $\mathbb{T}^2$. In certain cases, the construction of a wild coboundary can be made more global and uniform. Here is an example of how such a construction works for rotations of the circle allowing good periodic approximation. Since the result is not new (cf. [**F**], [**A**]), we give only a very brief sketch of the argument. See [**KH**], Sections 2.9 and 12.6 for a more detailed argument.

THEOREM 4.7. *There exists a λ and an analytic function ϕ such that*

$$\phi(x) = \psi(\lambda z) - \psi(z) \tag{4.16}$$

where ψ is measurable, metrically dense with respect to Lebesgue measure (cf. Definition 4.1), and ψ can be made either integrable or not integrable.

COROLLARY 4.8. *For every $r > 0$, the S^1 extension*

$$R_\lambda^{\exp ir\phi}(z,w) = (\lambda z, w \exp ir\phi(z))$$

is minimal but not strictly ergodic.

One can deduce the corollary directly from the statement of the theorem, however, it is easier to apply the general result of Gottshalk and Hedlund, (cf. [**Pa**], where Furstenberg's construction can also be found, or [**KH**], Section 2.9), which asserts that if the extension $R^{\exp ir\phi}$ is not minimal then for some positive integer n, the cocycle $\exp(irn\phi)$ is a coboundary with a continuous transfer function.

Let us outline one of the constructions which leads to a proof of Theorem 4.7. This construction is very close in spirit to the general constructions in sections 4.1 and 4.2. We will use Dirichlet kernels

$$D_{q,n}(z) = \sum_{\substack{j=-n+1 \\ j\neq 0}}^{n-1} z^{jq} + z^{-jq}$$

to produce an analytic counterpart of the bump functions used in those arguments. The transfer function is build as

$$\sum_{n=1}^{\infty} W_n D_{q_n,m_n}(z) \tag{4.17}$$

where by making $q_n, m_n \to \infty$ fast enough, we guarantee convergence in probability, since most of the mass of the Dirichlet kernel $D_{q,n}$ is concentrated around the qth roots of unity, and by controlling W_n, we can make the function (4.17) either integrable or not. The number $\lambda = \exp 2\pi i\alpha$ is built inductively as the limit of very good approximations

$$\alpha \stackrel{\text{def}}{=} \lim_{n\to\infty} p_n/q_n.$$

The inductive step works as follows. Having constructed p_n, q_n, m_n and W_n, we consider the function

$$\psi_n(z) = \sum_{j=1}^{n} W_j D_{q_j,m_j}(z)$$

as given, and assume that we have a good estimate in some complex neighborhood of S^1 for

$$\psi_n(\lambda_n z) = \psi_n(z),$$

where $\lambda_n = \exp 2\pi i p_n/q_n$. That estimate will persist if we replace λ_n by any number sufficiently close to it, in particular by $\exp 2\pi i(p_n/q_n + 1/Lq_n)$ for any sufficiently large L. We can choose $L = L_n$ and m_{n+1} in such a way that the Dirichlet kernel

$$D_{L_n q_n, m_{n+1}}$$

is on the one hand, sufficiently close to zero in probability, and on the other hand sufficiently close to being metrically dense. The first property guarantees convergence in probability, and also guarantees the persistence of the degree of metric density achieved in the previous step. The last property guarantees a better degree of metric density for

$$\psi_{n+1} = \psi_n + W_{n+1} D_{L_n q_n, m_{n+1}},$$

so that we put $q_{n+1} = L_n q_n$. Since the Dirichlet kernel has period $2\pi/q_{n+1}$,

$$\psi_{n+1}(\lambda_{n+1}z) - \psi_{n+1}(z) = \psi_n(\lambda_{n+1}z) - \psi_n(z)$$

and as we have seen the expression on the right hand side can be controlled in a complex neighborhood of S^1.

The arguments in section 5.4 below follow a very similar pattern with cosines instead of Dirichlet kernels. However, those arguments are generic, whereas the property we are discussing now can be carried out only by a construction where the next step depends thoroughly on the previous ones. Similar phenomena appear in the conjugation–approximation construction [**AK**], [**HaK**] which is the most powerful general method of constructing diffeomorphisms with prescribed, often exotic properties based on fast periodic approximation.

REMARK. Herman [**He1**] pointed out that for rotations of the circle, cohomology with nonintegrable transfer functions can not be effected by a lacunary Fourier series. This conforms well with the highly non-lacunary nature of the Dirichlet kernels and also underlines the difference between the above construction and that of section 5.4.

4.4 Minimal nonergodic interval exchange transformations. An *interval exchange transformation* (i.e.t.) is a map T of an interval Δ onto itself which preserves Lebesgue measure λ and has only a finite number of discontinuities. It owes its name to an equivalent description as a rearrangement in some fixed new order of subintervals $\Delta_1, \dots, \Delta_m$, which form a partition of Δ, with a possible change of orientation on some of the subintervals. If the orientation does not change we will call T on *oriented interval exchange transformation.* There is a certain ambiguity in the definition of i.e.t. at the points of discontinuity but we fix this ambiguity for an oriented i.e.t. by assuming that the interval Δ is right half-open and that T is continuous from the right. Interval exchange transformations appear as natural section maps for area preserving flows on the surfaces of higher genus.

If an i.e.t. T has a dense orbit then every other orbit, except possibly the orbits of the points of discontinuity, is also dense [**Ke**], see also [**KH**], Section 14.5. We will call such an i.e.t. *quasi-minimal.* If T is an oriented i.e.t. and all its semi-orbits are dense we will call it a *minimal* i.e.t. If we fix m and the order in which the subintervals $\Delta_1 \dots \Delta_m$ are permuted then an oriented i.e.t. is determined by a probability vector

$$\lambda(\Delta_1), \dots, \lambda(\Delta_m)$$

i.e. by a point of an $(m-1)$-dimensional simplex σ_{m-1}. Under some natural combinatorial assumptions all elements of the simplexes except for those belonging to a countably many linear submanifolds determine minimal i.e.t.

The total number of ergodic probability invariant measures for a quasi-minimal i.e.t. is always finite ([**Ve1**], [**KH**], section 14.5). If there is only one such measure we will call T strictly ergodic. Since Lebesgue measure is always invariant this unique measure must be Lebesgue. The central result in the theory of interval exchange transformations is that under the same natural combinatorial assumptions which yield minimality, almost every point of the simplex σ_{m-1} with respect to Lebesgue measure determines a strictly ergodic i.e.t. This was proved independently and simultaneously by W. Veech [**V2**] and H. Masur [**M**].

We are going to show that the minimal but not strictly ergodic i.e.t., although nongeneric, are fairly common.*

We will deal with two-point extension of a minimal oriented i.e.t. Let us call a function $n : \Delta \to \mathbb{Z}/2 = \{0,1\}$ a *k-step function* if it has exactly k points of discontinuity. We associate to any $n : \Delta \to \mathbb{Z}/2$ a $\mathbb{Z}/2$ extension $T^{n(\cdot)}$ of T.

$$T^{n(\cdot)}(x,j) = (Tx, j + n(x)). \tag{4.18}$$

If T is an exchange transformation of m intervals and n is a k-step function then $T^{n(\cdot)}$ can be easily interpreted as an i.e.t. exchanging not more than $2m + 2k$ intervals.

THEOREM 4.9. *Let T be a minimal oriented interval exchange transformation. There exists a 3-step function $n : \Delta \to \mathbb{Z}/2$ such that*

(i) $n(x) = h(Tx) - h(x)$
where $h : \Delta \to \mathbb{Z}/2$ is a measurable function and

(ii) *For each subinterval $I \subset \Delta$,*

$$\lambda(h^{-1}(\{0\}) \cap I) > 0 \quad \textit{and}$$
$$\lambda(h^{-1}(\{1\}) \cap I) > 0.$$

Before proving this theorem we will show how it allows us to construct minimal but not strictly ergodic i.e.t.'s.

COROLLARY 4.10. *The two-point extension $T^{n(\cdot)}$ (cf. (4.18)) is minimal but not uniquely ergodic.*

The proof of the corollary is essentially the same as that of Corollary 4.3. By Theorem 4.9 (i) $n(x)$ is a coboundary so that $T^{n(\cdot)}$ is metrically isomorphic to $T \times Id$ via $R(x,j) = (x, j + h(x))$. Therefore $T^{n(\cdot)}$ preserves two sets of positive product measure: graph (h) and graph $(1 - h)$. By Theorem 4.9 (ii) and by the minimality of T every orbit of $T^{n(\cdot)}$ visits every open subset of $\Delta \times \mathbb{Z}/2$.

PROOF OF THEOREM 4.9. Since T^{-1} is also an i.e.t. with the same number of points of discontinuity we can replace (i) by

(i′) $n(x) = h(T^{-1}x) - h(x)$,

then apply the result to T^{-1} and obtain the statement of the theorem.

In all the subsequent arguments we omit the minus sign because we are working in $\mathbb{Z}/2$.

Let us call an interval $I \subseteq \Delta$ *k-clear* if T^i is continuous on the interior of I for $|i| =\leq k$. Let a be the left endpoint of Δ and $a_n \stackrel{\text{def}}{=} T^n a$. By the minimality of T the sequence $\{a_n\}, n = 0, 1, \dots$ is dense in Δ. If $a_n < a_m$ we will denote by $\chi_{m,n} : \Delta \to \mathbb{Z}/2$ the characteristic function of the interval $[a_n, a_m)$.

As in the proof of Theorem 4.2 the function $n(x)$ is constructed inductively with very similar methods of control over convergence and divergence. However, the whole construction is much more explicit. Let $k_0 = 0$ and $k_1 > 0$ be chosen in such a way that the interval $[a, a_{k_1}]$ is disjoint from its image $[a_1, a_{k_1+1}]$. The inductive construction is determined by an increasing sequence of natural numbers

*The rest of this section appeared with minor modifications in [**KH**] as subsection 14.5e

k_m, $m = 0, 1, \ldots$. In order to formulate the conditions on k_m we need another sequence ℓ_m defined by

$$\begin{aligned} \ell_0 &= 1, \ell_1 = k_1 + 1 \\ \ell_{m+1} &= k_{m+1} - k_m + \ell_{m-1} \end{aligned} \tag{4.19}$$

so that

$$\ell_m = 1 + \sum_{i=0}^{m-1} k_{m-i}(-1)^i.$$

We will assume that

(I_1) $a < a_{k_1} < a_{k_2} < \cdots < a_{k_m} < a_1 < a_{\ell_2} < \cdots < q_{\ell_{2[\frac{m}{2}]}}$

$$< a_{\ell_1} < a_{\ell_3} < \cdots < a_{\ell_{2[\frac{m+1}{2}]-1}}$$

(I_2) The interval $[a_{k_m}, a_{k_{m+1}}]$ is $2k_m-$ clear

(I_3) $a_{k_{m+1}} - a_{k_m} < \frac{a_{k_m} - a_{k_{m-1}}}{3k_m}$.

Let us now show how conditions (I_1), (I_2) and (I_3) can be satisfied inductively. Let us assume that they are already satisfied up to m. Then by (I_1) we can find e,

$$a_{k_m} < e < c_m \stackrel{\text{def}}{=} \begin{cases} a_{\ell_m}, & m \text{ even} \\ a_{\ell_{m-1}}, & m \text{ odd} \end{cases}$$

such that $[a_{k_m}, e)$ is a k_m-clear interval. By the minimality of T, we can find $k_{m+1} > k_m$ such that $a_{k_{m+1}}$ belongs to the left half of the interval $[a_{k_m}, e)$. This implies (I_2) for $m+1$. From the definition of ℓ_m, c_{m+1} cannot be more to the left of c_m than $a_{k_{m+1}}$ is to the right of a_{k_m}. This implies (I_1). Finally, (I_3) can be achieved by simply choosing $a_{k_{m+1}}$ each time very close to a_{k_m}.

Now we are prepared to construct inductively the cocycle $n(x)$ and the transfer function $h(x)$. We begin with a 3-step function

$$\begin{aligned} n_1(x) &= \chi_{0,k_1}(x) + \chi_{1,k_1+1}(x) \\ &= \chi_{0,k_1}(x) - \chi_{0,k_1}(T^{-1}x) \end{aligned}$$

and proceed by induction

$$\begin{aligned} n_{m+1}(x) &= n_m(x) + \chi_{k_m,k_{m+1}}(x) + \chi_{\ell_{m-1},\ell_{m+1}}(x) \\ &= n_m(x) + \chi_{k_m,k_{m+1}}(x) - \chi_{k_m,k_{m+1}}(T^{\ell_{m+1}-k_{m+1}}x) \\ &= n_m(x) + g_{m+1}(x) - g_{m+1}(T^{-1}x) \end{aligned} \tag{4.20}$$

where

$$g_{m+1}(x) = \sum_{i=0}^{k_m-\ell_{m-1}-1} \chi_{k_m+j,k_{m+1}+j}(x).$$

We will represent $n_m(x)$ on one hand as a 3-step function and on the other hand as a coboundary.

To show that it is a 3-step function we have from (4.20) and (I_1),

$$\begin{aligned} n_m(x) = \sum_{i=1}^{m} \chi_{k_{i-1},k_i}(x) &= \chi_{\ell_0,\ell_1}(x) + \sum_{i=1}^{m-1} \chi_{\ell_{i-1},\ell_{i+1}}(x) \\ &= \chi_{0,k_m}(x) + \chi_{\ell_{2[\frac{m}{2}]},\ell_{2[\frac{m+1}{2}]-1}}(x). \end{aligned}$$

Thus $n_m(x)$ converges pointwise to

$$n(x) = \chi_{[a,b)}(x) + \chi_{[c,d)}(x)$$

where

$$b = \lim_{m\to\infty} a_{k_m}, \quad c = \lim_{m\to\infty} a_{\ell_{2m}}, \quad d = \lim_{m\to\infty} a_{\ell_{2m+1}}.$$

To represent n_m as a coboundary, we use (4.20) to obtain

$$n_m(x) = h_m(x) - h_m(T^{-1}x)$$

where

$$\begin{aligned} h_m(x) &= \chi_{0,k_1}(x) + \sum_{i=2}^{m} g_i(x) \\ &= \chi_{0,k_1}(x) + \sum_{i=2}^{m} \sum_{j=0}^{k_{i-1}-\ell_{i-2}-1} \chi_{k_{i-1}+j,k_i+j}(x). \end{aligned} \tag{4.21}$$

Since $h_{m+1}(x) = h_m(x) + g_{m+1}(x)$ we have from (4.20) and (I_3),

$$\begin{aligned} &\lambda(\{x \in \Delta : g_{m+1}(x) = 1\} \\ &\qquad \le (k_m - \ell_{m-1})(a_{k_{m+1}} - a_{k_m}) < \frac{a_{k_m} - a_{k_{m_1}}}{3} \end{aligned}$$

so that the sequence h_m converges in L^1 to a function h which obviously satisfies (i'). Thus, it remains to prove (ii). Again the argument resembles very much those from section 4.1, being only simpler and more explicit.

Let us call an interval $[a_{k_{m-1}+j}, a_{k_m+j})j = 0, 1, \ldots, k_{m-1} - \ell_{m-2} - 1$ *of rank* m. It follows from (I_3) that all intervals of a given rank are pairwise disjoint. Moreover for $\ell < m$ any interval of rank m is either disjoint with an interval of rank ℓ or is contained inside such an interval. To prove that let us assume that the opposite is true so that for some

$$j \in \{0, 1, \ldots, k_{m-1} - \ell_{m-2} - 1\}$$

and

$$s \in \{0, 1, \ldots, k_{\ell-1} - \ell_{\ell-2} - 1\}$$

$$a_{k_{m-1}+j} < a_{k_{\ell-1}+s} < a_{k_m+j}.$$

Since $k_{\ell-1} + s < 2k_{\ell-1} < 2k_{m-1}$ we can apply $T^{-(k_{\ell-1}+s)}$ and obtain from (I.2) that the point a_0 lies between

$$a_{k_{m-1}} + (j - k_{\ell-1} - s)$$

and

$$a_{k_m+(j-k_{\ell-1}-s)}.$$

But this is impossible because a is the left end of Δ.

It follows from the last statement and from (4.21) that the function h_m is constant on every interval of rank m.

Let us fix an interval $I \subset \Delta$. By the minimality of T every orbit segment of length N for sufficiently large N intersects I, in particular I contains an interval I' of rank m for a sufficiently large m . h_m is constant on I' and from (I_3) for $n > m$

$$\lambda(\{x \in I' \; ; \; h_n(x) = h_m(x)\}) \geq$$

$$\lambda(I') - \sum_{k=1}^{n-m} \lambda(g_{n+k}^{-1}(1)) \geq \tag{4.22}$$

$$\lambda(I')\left(1 - \frac{1}{3} - \frac{1}{9} - \cdots - \frac{1}{3^{m-n}}\right)$$

so that

$$\lambda\{x \in I' : h(x) = h_m(x)\} \geq \frac{\lambda(I')}{3}.$$

Let us now take the smallest $m' > m$ such that I' contains an interval I'' of rank m'. Then the value of $h_{m'}$ on I'' is different from that of h_m on I' so that applying the same argument to I'' we obtain

$$\lambda\{x \in I'' : h(x) = h_{m'}(x)\} > \frac{\lambda(I'')}{3}.$$

Inequalities (4.22) and (4.23) imply (ii). □

REMARK. A slight modification of the argument allows us to replace minimality by quasi-minimality in the assumption with the outcome being a 4-step function instead of a 3-step function.

5. Non-trivial cocycles

In this chapter we discuss methods of proving that a certain cocycle is not a coboundary or an almost coboundary, and various applications of those methods. In contrast with situations considered in Chapter 3 where the non-triviality of cocyles was derived from non-vanishing of certain invariant distributions we will concentrate now on the non-stable case, i.e. we will look for non-trivial cocycles which are the limits of coboundaries in various topologies. The subtlety of this problem is that there are very few means to disprove the existence of a transfer function which is assumed to be only measurable with no assumptions about continuity or integrability.

5.1 Two general criteria. The following propositions summarize the two main approaches to the problem. In both cases G is an arbitrary second countable topological group.

PROPOSITION 5.1. *Let ϕ be a G-valued cocycle over a measure-preserving transformation $T : (X, \mu) \to (x, \mu)$.*

(i) *If for a sequence of integers q_n, $T^{q_n} \to Id$ in the weak topology* and ϕ is a coboundary then the product $\phi(x)\phi(Tx)\dots\phi(T^{q_n-1}x) \to Id_G$ in probability.*

(ii) *If for any $q_n \to \infty$, any $\alpha > 0$, and for every measurable set A*

$$\mu(T^{q_n}A \cap A) > \alpha\mu(A),$$

*A transformations T for which such a sequence exists is called rigid [**HaK**], [**K1**].

then for every coboundary ϕ and every neighborhood U of $e \in G$,

$$\underline{\lim}_{n\to\infty} \mu\{x : \phi(x)\phi(Tx)\dots\phi(T^{q_n-1}x) \in U\} > \alpha.$$

PROOF. (i) If $\phi(x) = \psi(x)^{-1}\psi(Tx)$ then

$$\phi(x)\phi(Tx)\dots\phi(T^{q_n-1}x) = \psi^{-1}(x)\psi(T^{q_n}x)$$

and since $\psi(T^{q_n}x)$ converges in probability to $\psi(x)$ so that the product converges to identity.

ii) Let us take a countable partition η of G such that for every element $c \in \eta$ and $g_1, g_2 \in c$, $g_1g_2^{-1}$ and $g_1^{-1}g_2$, belong to U. Let $A = \psi^{-1}(c)$, for all $c \in \eta$ and apply a diagonal argument. □

PROPOSITION 5.2. *Let G be provided with a metric compatible with the uniform structure. If a G-valued cocycle over T is a coboundary then for every $\epsilon > 0$ there exists R such that for all n*

$$\mu\{x \in X; dist(\phi(x)\phi(Tx)\dots\phi(T^{n-1}x), e) \le R\} > 1 - \epsilon$$

PROOF. Again if ϕ is a coboundary

$$\phi(x)\phi(Tx)\dots\phi(T^{n-1}x) = (\psi(x))^{-1}\psi(T^n x).$$

Since ψ and ψ^{-1} have the property in question, T^n preserves the measure so that $\psi \cdot T^n$ also has that property. Finally, since the product is continuous for every R there is $R_1 = R_1(R)$ such that if $\mathrm{dist}(x_i, e) < R$ $i = 1, 2$ then $\mathrm{dist}(x_1x_2, e) < R_1$. □

Since we will be interested in finding cocycles which are not almost coboundaries we will formulate explicitly the contrapositives of Proposition 5.1(i) and 5.2 for that case.

COROLLARY 5.3. *Under the same conditions as Proposition 5.1, if there exits a neighborhood U of the identity $e \in G$, a sequence $q_n \to \infty$, sequences of closed subsets V_n and W_n of G, and $\epsilon > 0$ such that*

(i) $T^{q_n} \to Id$ *in the weak topology*

(ii) $V_nW_n^{-1} \cap U = \phi$

(iii) $\mu\{x : \phi(x)\phi(Tx)\dots\phi(T^{q_n-1}x) \in V_n\} > \epsilon$

(iv) $\mu\{x : \phi(x)\phi(Tx)\dots\phi(T^{q_n-1}x) \in W_n\} > \epsilon.$

Then ϕ is not cohomologous to any constant.

COROLLARY 5.4. *Under the same conditions as Proposition 5.2, let U_k be an increasing sequence of neighborhoods of the identity in G such that $\bigcup_{k=1}^{\infty} U_k = G$. If there exists a sequence $n_k \to \infty$, sequences of closed subsets V_k and W_k of G, and $\epsilon > 0$ such that*

(i) $V_kW_k^{-1} \cap U_k = \phi$

(ii) $\mu(\{x : \phi(x)\phi(Tx)\dots\phi(T^{n_{k-1}}x) \in V_k\}) > \epsilon$

(iii) $\mu(\{x : \phi(x)\phi(Tx)\dots\phi(T^{n_{k-1}}x) \in W_k\}) > \epsilon$

Then ϕ is not cohomologous to any constant.

The method based on Corollary 5.3 applies to cocycles with values in arbitrary groups including compact ones, but it requires the transformation to be rigid. The second method is free of that restriction but it works only for cocycles with values in some non-compact (e.g. compactly generated) groups.

In sections 5.2–5.5 we will explore various situations where a cocycle $\phi = \lim \phi_n$ fails to be an almost coboundary, although ϕ_n's are coboundaries and L^1 convergence of ϕ_n to ϕ is very fast. Our strategy will be as follows. Although ϕ_n is a coboundary, for large values of n the corresponding transfer function ψ_n will be huge so that ϕ_n will satisfy the condition (ii)–(iv) of Corollaries 5.3 or 5.4 for a finite but growing with n number of iterates.

Due to fast convergence all the alterations on the subsequent steps will be small enough to keep the properties for the limit cocycle for the given iterates and, naturally, they will produce similar properties for higher iterates.

In addition to Corollary 5.4 the following observation is useful in this situation. Let $T : (X, \mu) \to (X, \mu)$ be an ergodic measure-preserving transformation and let U and V be two disjoint closed subsets of G. Let $\phi(x) = (\psi(x))^{-1}\psi(Tx)$ and let $u = \mu(\psi^{-1}(U))$, $v = \mu(\psi^{-1}(V))$.

LEMMA 5.5. *Suppose that*

$$\epsilon_0 = u(1-u) - 1 + u + v > 0. \tag{5.1}$$

Then there exists an arbitrarily large k such that for every $\delta > 0$

$$\mu(\{x : \phi(x)\phi(Tx)\dots\phi(T^{k-1}x) \in V^{-1}U\}) \geq \epsilon_0 - \delta$$

and

$$\mu(\{x : \phi(x)\phi(Tx)\dots\phi(T^{k-1}x) \in U^{-1}V\}) \geq \epsilon_0 - \delta.$$

REMARK. Condition (5.1) is satisfied if $u + v$ is sufficiently close to one and neither u or v is very small.

PROOF. We write as usual

$$\phi(x)\phi(Tx)\dots\phi(T^{k-1}x) = \psi(x)^{-1}\psi(T^k x)$$

and take sets $A = \psi^{-1}(U)$ and $B = \psi^{-1}(V)$. From the ergodic theorem,

$$\lim_{N\to\infty} \frac{1}{N} \sum_{k=0}^{N-1} \mu(A \cap T^{-k}A) = \mu(A)^2 = u^2. \tag{5.2}$$

In particular we can find an arbitrarily large k such that $\mu(A \cap T^{-k}A) \leq u^2 + \delta$ so that $\mu((X\backslash A) \cap T^{-k}A) \geq 1 - u(1-u) - \delta$ and by (5.1), $\mu(B \cap T^{-k}A) \geq \mu((X\backslash A) \cap T^k A) - \mu(X\backslash(A \cup B)) \geq u(1-u) - \delta - 1 + u + v = \epsilon_0 - \delta$. If $x \in B$ and $T^k x \in A$ then

$$\phi(x)\phi(Tx)\dots\phi(T^{k-1}x) \in V^{-1}U$$

Similarly, we show that $\mu(A \cap T^{-k}B) \geq$

$$\begin{aligned} &\mu(A \cap T^{-k}(X \backslash A)) - \mu(X \backslash (A \cup B)) \\ &= \mu(T^k A \cap (X \backslash A)) - 1 + u + v \\ &= \mu(T^k A) - \mu(T^k A \cap A) - 1 + u + v \\ &= \mu(A) - \mu(A \cap T^{-k}A) - 1 + u + v \\ &\geq u(1-u) - 1 + u + v - \delta = \epsilon_0 - \delta. \end{aligned}$$

□

This lemma can be used to verify the conditions of Corollary 5.4 if, e.g. G is an abelian group provided with a translation invariant metric and U, V are such that

$$\operatorname{dist}(U^{-1}V,\ V^{-1}U)$$

is sufficiently large. (Cf. the proof of Theorem 5.6 below).

5.2 The case of fast C^∞ approximation. We now return to the situation discussed in section 4.2.

THEOREM 5.6. *Suppose that f is a C^∞ diffeomorphism of a compact connected m-dimensional manifold M which preserves a measure μ given by a density bounded between two positive constants, and is ergodic with respect to μ. If f admits a fast C^∞ periodic approximation then there exists a real valued C^∞ cocycle $h = \lim_{n\to\infty}(\psi_n \circ f - \psi_n)$, where ψ_n is C^∞, which is not cohomologous to any constant.*

PROOF. We begin with a "cubic triangulation" of M; in other words we form a mod 0 partition of M into diffeomorphic images of the standard m-dimensional cube $I^m = \{(t_1, \dots t_m) : 0 \leq t_i \leq 1,\ i = 1 \dots m\}$. Then for a given number n we can subdivide the standard cube by n^m equal cubes by dividing the value of each coordinate into n equal intervals. The images of this subdivision form a partition of M which we will denote by ξ_n. We associate with every element $c \in \xi_n$ a function α_c which is somewhat similar to the function defined by (4.9).

The element c is an image under a fixed (independent on n) diffeomorphism of a cube in $\mathbb{R}^m$ with a center at a point $(t_1^0 \dots t_m^0)$ and with sides parallel to the coordinate axes. We define α_c as the image under that diffeomorphism of the function $n \prod_{i=1}^{m} \theta_n(t_i - t_i^0)$ extended by zero outside of the image of I^m. Here $\theta_n(t)$ is a C^∞ bump function equal to 1 for $|t| < \frac{1}{2n}$, to 0 for $|t| > \frac{1}{2n} + \frac{1}{2n^{10}}$ and such that its r-th derivative does not exceed $C(r)n^{100}$ for $r = 1, 2, \dots$.

Due to the fast periodic approximation for $c \in \xi_{q_n}$ the function $\alpha_c \circ f^{q_n} - \alpha_c$ allows the C^r estimate

$$\|\alpha_c \circ f^{q_n} - \alpha_c\|_r < c_1(k,r) q_n^{-k} \tag{5.3}$$

(cf. (4.8)). Details of this estimate are the same as in the proof of Theorem 4.5.

We have

$$\alpha_c \circ f^{q_n} - \alpha_c = \psi_c \circ f - \psi_c$$

where

$$\psi_c = \sum_{k=0}^{q_n - 1} \alpha_c \circ f^k.$$

We are going to show that one can choose a collection of elements of ξ_{q_n}, $c_1, \ldots, c_\ell$ so that the function $\psi = \sum_{i=1}^{\ell} \psi_{c_i}$ has the following property

$$\begin{aligned} \mu(\psi^{-1}([q_n, \infty))) &> \frac{2}{5} \\ \mu(\psi^{-1}(\{0\})) &> \frac{2}{5}. \end{aligned} \tag{5.4}$$

Then we can apply Lemma 5.5 and obtain a number $k = k_n$ such that for the function

$$\phi = \sum_{i-1}^{\ell} \alpha_{c_i} \circ f - \alpha_{c_i} = \psi \circ f - \psi, \tag{5.5}$$

$$\begin{gathered} \mu(\{x : \sum_{j=0}^{k_n - 1} \phi(T^j x) \leq -q_n\}) > \frac{1}{50} \\ \text{and} \\ \mu(\{x : \sum_{j=0}^{k_n - 1} \phi(T^j x) \geq q_n\}) > \frac{1}{50}. \end{gathered} \tag{5.6}$$

To verify (5.4) we order the elements of ξ_{q_n} and consider successively the functions $\psi_{c_1}, \psi_{c_1} + \psi_{c_2}, \psi_{c_1} + \psi_{c_2} + \psi_{c_3}, \ldots$.

We denote $\psi^{(k)} = \sum_{i=1}^{k} \psi_{c_i}$.

Let us look at the distribution of the values of two successive functions $\psi^{(k)}$ and $\psi^{(k+1)}$. Both functions are non-negative and $\psi^{(k+1)} - \psi^{(k)} = \psi_{c_k}$. the support of ψ_{c_k} is the union of q_n successive iterates of c_k with its neighborhood. We will disregard all the neighborhoods and their images since the total measure of their union does not exceed a constant multiple of q_n^{-9}.

By neglecting this small set we can reduce the question to considering distributions of iterates for characteristic functions of the elements $c_k \in \xi_{q_n}$ multiplied by the constant q_n. These iterates cover increasingly large fractions of the total measure. The distribution of a sum is the sum of the distributions. Adding the sum of the iterates of $q_n \chi_{c_k}$, we change the value by at least q_n on a set of measure between $\mu(c_k)$ and $q_n \mu(c_k)$.

The last number is small since by our assumptions $\mu(c_n) < q_n^{-m} \cdot d$ for a constant d. Eventually the value on a set of measure close to 1 becomes greater than q_n. Since we start from zero, that means that for some k the measure of support is within dq_n^{-m+1} of $1/2$. By making n large enough we obtain (5.4).

Let us point out that

$$\|\phi\|_r \leq d_1 q_n^{m-1} \max_{c \in \xi_{q_n}} \|\alpha_c \circ f - \alpha_c\|_r$$

for another constant d_1 and thus $\|\phi\|_r$ decreases faster than any power of q_n as $n \to \infty$ by (4.8).

Now we proceed to build the cocycle h inductively as a sum of $\phi^{(k)}$ where each $\phi^{(k)}$ is defined by (5.5) for a certain recurrence time q_{n_k}. By choosing the numbers q_{n_k} sufficiently quickly increasing we guarantee the convergence of the series $h = \sum_{k=1}^{\infty} \phi^{(k)}$. On the other hand we want to insure that for every k

$$\mu(\{x : \sum_{j=0}^{q_{n_k}-1} h(T^j x) \leq -\frac{q_{k_n}}{2}\}) > \frac{1}{50}$$

(5.7) and

$$\mu(\{x : \sum_{j=0}^{q_{n_k}-1} h(T^j x) \geq \frac{q_{k_n}}{2}\}) > \frac{1}{50}.$$

By Corollary 5.4 these inequalities imply that h is not cohomologous to a constant. We have

$$h = \sum_{i=0}^{k-1} \phi^{(i)} + \phi^{(k)} + \sum_{i=k+1}^{\infty} \phi^{(i)}.$$

Since $\phi^{(i)} = \psi^{(i)} \circ f - \psi^{(i)}$ (cf. (5.5)) for every ℓ

$$\left| \sum_{j=0}^{\ell-1} \sum_{i=0}^{k-1} \phi^{(i)}(T^j x) \right| < 2 \sum_{i=1}^{k-1} \max |\psi^{(i)}| \tag{5.8}$$

and the expression in the right hand part of the last inequality is known by the time we make the choice of q_{n_k} so that we can choose

$$q_{n_k} > 20 \sum_{i=1}^{k-1} \max |\psi^{(i)}|.$$

On the other hand by choosing the subsequent return times far away we can insure that for $i > k$

$$|\psi^{(i)}| < 2^{-1} \cdot k_{n-K}^{-1}$$

so that

$$\left| \sum_{j=0}^{k_{n_k}-1} \sum_{i=k+1}^{\infty} \phi^{(i)}(T^j x) \right| < 1. \tag{5.9}$$

Since

$$\left| \sum_{j=0}^{k_{n_k}-1} h(t^j x) \right| > \left| \sum_{j=0}^{k_{n_k}-1} \phi^{(k)}(T^j(x)) \right| -$$
$$\left| \sum_{j=0}^{k_{n_k}-1} \sum_{i=0}^{k-1} \phi^{(i)}(T^j x) \right| - \left| \sum_{j=0}^{k_{n_k}-1} \sum_{i=k+1}^{\infty} \phi^{(i)}(T^j(x)) \right|,$$

inequalities (5.6), (5.8) and (12.9) imply (5.7). □

5.3 Weakly mixing flows on $\mathbb{T}^2$. Kolmogorov's Theorem 3.5 implies that for any Diophantine (in other words, not exceptionally well rationally approximable) number α every C^∞ special flow build over the rotation R_λ, $\lambda = \exp 2\pi i\alpha$ has discrete spectrum (cf. (1.22)). Now we are going to show that for very well approximable rotation numbers there exist not only C^∞ but even real-analytic special flows which have continuous spectrum. To that end we will find a criterion which guarantees that for a real-valued function h on the circle the function $\exp irh$ for every real r is not a coboundary. We will use the approach suggested by Corollary 5.3. Let us consider the special flow over R_λ build under a function $h_0 + h(z)$ where h_0 is a constant and $h(z)$ is a function with zero average

$$h(z) = \sum_{n\neq 0} h_n z^n.$$

We consider a sequence of iterates, say q_n, $n = 1, 2, \dots$ of the rotation R_λ corresponding to a very good rational approximation p_n/q_n, $0 \le p_n < q_n$ of α. To make the notation lighter we omit the index n in the subsequent computation.

We have

$$\begin{aligned}\sum_{k=0}^{q-1} h(\lambda^k z) = \sum_{m=-\infty}^{\infty} h_m \left(\frac{1-\lambda^{qm}}{1-\lambda^m}\right) z^m = \sum_{q\neq lm} h_m \left(\frac{1-\lambda^{qm}}{1-\lambda^m}\right) z^m +\\ \sum_{\ell\neq 0}^{\infty} h_{\ell q}\left(\frac{1-\lambda^{q^2\ell}}{1-\lambda^q}\right) z^{\ell q}.\end{aligned} \tag{5.10}$$

We approximate λ by $\lambda_0 = \exp 2\pi i\frac{p}{q}$ and write the expression for $\sum\limits_{k=0}^{q-1} h(\lambda_0^h z)$ corresponding (5.10). The sum over m not divisible by q vanishes and the second sum becomes

$$\sum_{\ell=-\infty}^{\infty} q h_{\ell q} z^{\ell q}.$$

We denote by a_q the L^2 norm of this power series, by b_q the expression $q \sum\limits_{k=-\infty}^{\infty} |h_{\ell q}|$ and by $e_q = |\alpha - \frac{p}{q}|$. Obviously $a_q \le b_1$. We will assume a certain regularity condition on the decrease of Fourier coefficients $h_{\ell q}$, namely

$$\frac{|h_q|}{\sum\limits_{\ell=1}^{\infty} |h_{\ell q}|} > c > 0 \tag{5.11}$$

where the constant c is independent of q. Since $|h_q| < a_q$ (5.11) implies that

$$b_1 \le \sqrt{2}c^{-1} a_q \tag{5.12}$$

Obviously for every integer m

$$|\lambda^m - \lambda_0^m| \le |m| \cdot |\lambda - \lambda_0| \le |m| \cdot e_q$$

so that if h' is absolutely summable

$$\begin{aligned}
&\left|\sum_{k=0}^{q-1} h(\lambda^k z) - h(\lambda_0^k z)\right| \\
&\leq \left|\sum_{m=-\infty}^{\infty}\right| h_m \left|\sum_{k=0}^{q-1}\right| \lambda^{km} - \lambda_0^{km} \,|\,| \\
&\leq e_q \frac{q(q-1)}{2} \sum_m m|h_m| < c' e_q\, q^2
\end{aligned} \tag{5.13}$$

where c' is a constant which depends on h but not on q.

Now we make an assumption on the relationship between the speed of approximation for α and the decrease of Fourier coefficients for h. Namely we assume

$$\lim_{n\to\infty} \frac{e_{q_n} q_n^2}{b_{q_n}} = 0. \tag{5.14}$$

By (5.14) and (5.12) we see that for the function

$$h^{(q)}(z) = \sum_{k=0}^{q-1} h(\lambda^k z)$$

which has zero average both the maximum and the L^2-norm are of order b_q. Thus it has to have both positive and negative values of that order on sets of measure bounded away from zero by a constant independent of q. Up to an error of order $e_q q^2$ the function $h^{(q)}$ coincides with the function

$$\widetilde{h}^{(q)} = q \sum_{\ell=-\infty}^{\infty} h_{\ell q} z^{\ell q}.$$

Let us now take $L = \epsilon_0 \left[\frac{1}{a_q}\right]$ where ϵ_0 is a sufficiently small constant and consider

$$\sum_{k=0}^{Lq-1} h(\lambda^k z) = \sum_{j=0}^{L-1} \sum_{k=0}^{q-1} h(\lambda^k(\lambda^{jq} z)) = \sum_{j=0}^{L-1} h^{(q)}(\lambda^{jq} z).$$

Replacing in the last expression $h^{(q)}$ by $\widetilde{h}^{(q)}$ we allow an error of order $e_q \cdot q^2 \cdot L = o(1)$.

On the other hand, we have

$$\sum_{j=0}^{L-1} \widetilde{h}^{(q)}(\lambda^{jq} z) = q \sum_{\ell=-\infty}^{\infty} h_{\ell q} \sum_{j=0}^{L-1} \lambda^{j\ell q^2} z^{\ell q}. \tag{5.15}$$

We want again to compare uniform and L^2 norms for a function, this time for the one given by (5.15).

The uniform norm does not exceed Lb_q. The L^2 norm is greater than

$$q|h_q| \sum_{j=0}^{L-1} \lambda^{jq^2}.$$

We have by (5.12) and (5.14) for $0 \le j \le L-1$

$$|\lambda^{jq^2} - 1| \le jq^2|\lambda - \lambda_0| \le Lq^2 e_q = o(1).$$

Thus if q is sufficiently large,

$$\left|\sum_{j=0}^{L-1} \lambda^{jq^2}\right| > \frac{L}{2}$$

so that by (5.11) and the definition of L the L^2 norm of our function is greater than a certain constant which depends on h and ϵ_0 and can be arbitrary small by the choice of ϵ_0. What is important is that the ratio of the above estimate of the uniform norm and the below estimate for the L^2 norm is a constant independent of q ϵ_0. Since the average of the function (5.15) is zero it reaches both positive and negative values of order ϵ_0 on sets of the measure separated from zero. Taking into account the remark about the error we conclude that the same is true for the function

$$\sum_{k=0}^{Lq-1} h(\lambda^k)$$

But by our choice of L and by (5.14) $\lambda^{Lq} \to 1$ as $q \to \infty$ so that the function

$$\sum_{k=0}^{Lq-1} h_0 + h(\lambda^k z)$$

is not close in probability to any constant. However the variation of that function is estimated from above by a multiple of ϵ_0. When we pass to the function $\exp ir(h_0 + h(z))$ we see that the spread of values for

$$\prod_{k=0}^{Lq-1} \exp ir(h_0 + h(\lambda^k z))$$

persists if ϵ_0 is chosen small enough. Applying Corollary 5.3 we see that the special flow over R_λ build under $h_0 + h$ is weakly mixing. Let us summarize the discussion.

THEOREM 5.7. *Let $h(z) = \sum_{n \ne 0} h_n z^n$ be a C^2 real valued function on S^1 with zero average. Let R_λ be a rotation on S^1 $\lambda = \exp 2\pi i\alpha$. Suppose for a certain sequence of rational numbers p_n/q_n,*

$$\frac{q_n|\alpha - p_n/q_n|}{\sum_{k=1}^{\infty} |h_{kq_n}|} \to 0 \tag{5.16}$$

and

$$\frac{|h_{q_n}|}{\sum_{k=1}^{\infty} |h_{kq_n}|} > c > 0. \tag{5.17}$$

Then for any h_0 and r the cocycle $\exp ir(h_0 + h(z))$ is not a coboundary and consequently the special flow over R_λ build under the function $h_0 + h(z)$ is weakly mixing.

REMARK. Since conditions (5.16) and (5.17) remain true if we multiply h by a constant, those conditions also guarantee that S^1 extension of R_λ determined by the function exp ih has no eigenfunctions except for those lifted from the base (cf. (1.15)).

It is interesting to notice that for real analytic functions with a regular decrease of Fourier coefficients the sufficient condition for weak mixing given by (5.16) is very close in terms of the speed of approximation for λ to the negation of the sufficient condition for the discrete spectrum namely

$$\sum_n \frac{|h_n|^2}{|\lambda^n - 1|^2} < \infty \tag{5.18}$$

(cf. Theorem 3.5).

Let us consider for example the function

$$h(z) = \sum_{n \neq 0} 2^{-|n|} z^n = \frac{2\cos\phi - 2}{5 - 2\cos\phi}$$

where $z = \exp 2\pi i\phi$. This function obviously satisfies (5.17) and (5.16) becomes for it

$$2^{q_n} q_n \left| \alpha - \frac{p_n}{q_n} \right| \to 0$$

which is sufficient for weak mixing for the special flow. On the other hand (5.18) converges if for some $c > 0$ and for all $n > 0$

$$\frac{2^{-n}}{|\lambda^n - 1|} < \frac{c}{n} \quad \text{or, equivalently,} \quad 2^q \left| \alpha - \frac{p}{q} \right| > c$$

for all p and q.

Let us conclude with several comments.

Weak mixing is typical in the category sense for the special flows over circle rotations and hence for time changes in linear flows on the torus [**Fa1**].

Mixed spectrum can appear even for an analytic roof function (A. Katok; unpublished) while it is not known whether for an analytic of smooth roof function an exotic pure point spectrum can appear, i.e. whether it is possible that the special flow has a pure point spectrum, but the roof function is not additively cohomologous to a constant.

It is possible that for the roof functions with a sufficiently regular decay of Fourier coefficients there is a dichotomy between the solvability of the additive cohomological equation and weak mixing.

Finally mixing is possible for a smooth time change in a linear flow on a torus of dimension greater than two [**Fa2**].

5.4 Ergodicity of analytic cylindrical cascades. A problem closely connected with that discussed in section 5.3 concerns the $\mathbb{R}$ extension of the rotation R_λ

$$R_\lambda^h(z,t) = (\lambda z, t + h(z)). \tag{5.19}$$

This transformation, which is sometimes called a cylindrical cascade, preserves infinite Lebesgue measure. We are going to discuss the possibility of T being ergodic. It is obviously necessary that h have zero average. On the other hand, if the cocycle h is a coboundary, then R_λ^h is isomorphic to the direct product $R_\lambda \times Id$, and thus not ergodic. Furthermore, if

$$\exp\, irh(z) = \psi(z)^{-1}\psi(\lambda z)$$

then the function

$$\psi(z)^{-1} \exp\, irt$$

is R_λ^h invariant. Thus the ergodicity on R_λ^h implies that the special flow over R_λ build under the function $h(z) + 2\pi\alpha k + 2\pi\ell$ is weakly mixing for any integers k and ℓ, where $\lambda = \exp\, 2\pi i\alpha$. It looks as though the ergodicity of R_λ^h is a stronger property than the weak mixing of the special flows; however, no counterexamples are known.

Krygin [**Kr**] proved the existence of an analytic cocycle over an irrational rotation such that the extension (5.19) is ergodic. This result was generalized by Herman [**He3**] to cocycles with values in more general groups. Herman's proof is essentially categorical. Another interesting feature of that proof is that it interprets the approximation of the cocycle by coboundaries with diverging transfer functions as the approximation of a diffeomorphism by elements of the actions of a compact group via diverging conjugations as in [**AK**].

We will now formulate and sketch a proof of a categorical version of Krygin's theorem. Let us fix a complex annular neighborhood U of the circle S^1 and consider the space of all functions analytic in U and continuous on the boundary, with the topology of uniform convergence. Let $\mathcal{A}$ be the product of this space with circle, with the product topology.

THEOREM 5.8. *The set of pairs $(h, \lambda) \in \mathcal{A}$ such that the extension (5.19) is ergodic is a residual set in $\mathcal{A}$.*

We follow the same course as in the categorical theorems in [**K1**]. First, the ergodicity of R_λ^h on the cylinder follows from the ergodicity of the induced map $(R_\lambda^h)_{B_M}$ on every band $B_M = [-M, M] \times S^1$. The categorical argument rests on the following observation which is an almost immediate corollary of the ergodic theorem.

PROPOSITION 5.9. *Let $T : (X, \mu) \to (X, \mu)$ be a measure-preserving transformation of a Lebesgue probability space and let $\{\phi_1, \phi_2 \dots\}$ be a countable dense subset of $L^1(X, \mu)$. Suppose for any positive integers K and N and any $\epsilon > 0$, there exists a set $A_{K,N,\epsilon} \subset X$ of measure greater than $1 - \epsilon$ such that for every $x \in A_{K,N,\epsilon}$ there exists $n(x) > N$ such that for $k = 1, \dots, K$*

$$\left| \frac{1}{n(x)} \sum_{j=0}^{n(x)-1} \phi_k(T^j x) - \int \phi_k d\mu \right| < \epsilon. \tag{5.20}$$

Then T is ergodic.

To set up the categorical argument, we fix a band B_M, a natural number N, an $\epsilon > 0$, a continuous function ϕ on B_M, and a neighborhood A in $\mathcal{A}$. We will construct below a pair $(h, \lambda) = (h^{(A)}, \lambda^{(A)}) \in A$ such that if for $T = (R_\lambda^h)_{B_M}$ and for the given function (5.20) holds, then (5.20) remains true for $T(R_{\lambda'}^{h'})_{B_M}$ for any other (h', λ') from a sufficiently small neighborhood of (h, λ). Let us denote this neighborhood by

$$P(h, \lambda, \phi, N, \epsilon).$$

Let $A_1, A_2, A_3, \dots$ be a base of neighborhoods in $\mathcal{A}$ and let $\phi_1, \phi_2, \dots$ be a dense set in $C(B_M)$.

The set

$$\mathcal{B} = \bigcap_{J=1}^{\infty} \bigcap_{N=1}^{\infty} \bigcap_{k=1}^{\infty} \bigcup_{\ell=1}^{\infty} P(h^{(A_\ell)}, \lambda^{(A_\ell)}, \phi_k, N, 2^{-J}).$$

is a dense G_δ set in $\mathcal{A}$.

According to Proposition 5.9, for every $(h, \lambda) \in \mathcal{B}$ $(R_\lambda^h)_{B_M}$ is ergodic. Taking the intersection over all integers M, we obtain the result.

It remains to prove the approximation step.

We can think of A as a neighborhood of a pair (h, λ) where h is a coboundary and λ is rational. This is because for every irrational λ, the trigonometric polynomials are all coboundaries with real analytic transfer functions, and thus coboundaries are dense for the given λ. Keeping the transfer function fixed, we approximate λ by a rational number. Thus we can think of A as a neighborhood of $(h, \exp 2\pi i\, p_0/q_0)$ where $h(z) = \psi(z \exp 2\pi i\, p_0/q_0) - \psi(z)$.

Let us note that if g is a coboundary, $g(z) = \phi(\lambda z) - \phi(z)$, then the corresponding cylindrical cascade R_λ^g satisfies

$$R_\lambda^g = \Phi \circ (R_\lambda \times Id) \circ \Phi^{-1}$$

where

$$\Phi(z, t) = (z, t + \phi(z)).$$

The map R_λ^g is embedded into the periodic flow

$$S_s = \Phi \circ (R_x \times Id) \circ \Phi^{-1},$$

$s \in \mathbb{R}$, whose orbits are graphs of the functions $\phi + const.$ In particular, for any sufficiently large q and p such that $(p, q) = 1$, the orbits of $S_{p/q}$ fill those graphs with high uniform density. Both of these remarks remain true when we pass from the flow S_s to the flow induced by S_s on any band B_M.

For any $\epsilon > 0$, any M and any given function ψ, one can find Q and W such that for any $q \geq Q$ and $w \geq W$ the function

$$\psi_{w,q}(z) = \psi(z) + w(z^q + z^{-q})$$

has the property that all but a set of measure of less than ϵ of the points in B_M belong to graphs of the functions $\psi_{w,q} + const.$ whose intersections with B_M are ϵ uniformly distributed. The last property means that for any intervals $\Delta \subseteq S^1$ and $\Sigma \subseteq [-M, M]$,

$$\left| \lambda(\{z \in \Delta : \psi_{w,q}(z) + c \in \Sigma\}) - \frac{\lambda(\Delta)\lambda(\Sigma)}{2M} \right| < \epsilon.$$

We first perturb p_0/q_0, replacing it with p_1/q_1 for q_1 sufficiently large and then replace h by

$$\begin{aligned} h_1(z) &= \psi(z \exp 2\pi i\, p_1/q_1) - \psi(z) \\ &= \psi(z \exp 2\pi i\, p_1/q_1) + w((z \exp 2\pi i\, p_1/q_1)^{q_1} \\ &\quad + (z \exp 2\pi i\, p_1/q_1)^{-q_1}) \\ &\quad - \psi(z) - w(z^{q_1} + z^{-q_1}) \\ &= \widetilde{\psi}(z \exp 2\pi i\, p_1/q_1) - \widetilde{\psi}(z) \end{aligned}$$

where

$$\widetilde{\psi}(z) = \psi(z) + w(z^{q_1} + z^{-q_1})$$

and we choose q_1 and w according to the previous remark. Then we choose a very large (actually arbitrarily large) L, and replace h_1 by

$$\widetilde{h}(z) = \widetilde{\psi}\left(z \exp 2\pi i \left(\frac{p_1}{q_1} + \frac{1}{Lq_1}\right)\right) - \widetilde{\psi}(z).$$

Now the pair $(h^{(A)}, \lambda^{(A)})$, needed for the category argument, can be taken as $(\widetilde{h}, \widetilde{\lambda})$ where $\widetilde{\lambda} = \exp 2\pi i \left(\frac{p_1}{q_1} + \frac{1}{Lq_1}\right)$. Due to the remark about graphs, the map induced by $R_{\widetilde{\lambda}}^{\widetilde{h}}$ on B_M satisfies the assumptions of Proposition 5.9 with good precision. □

Let us note that in a constructive version of this argument, the cocycle may be built as a lacunary Fourier series.

5.5 Weak mixing of special flows over interval exchange transformations. Let $T : \Delta \to \Delta$ be an interval exchange transformation (cf. 4.4). We will consider a special flow built over T under a function h which we assume to be piecewise continuously differentiable; so that at every point of nondifferentiability, $h'(x)$ has limits from the left and right. let us identify the ends of the interval Δ and fix an orientation on the circle obtained this way. We denote

$$\mathcal{J}(h) = \sum f(x+0) - f(x-0)$$

where the summation is taken over the finite set of all points of discontinuity.

Let us remark that the first examples of measure-preserving transformations and flows with continuous spectrum were constructed by J. Von Neumann in his celebrated paper [**N**] which established the foundation of modern ergodic theory. He proved that the special flow built over an irrational rotation under a piecewise C^1 function h such that $\mathcal{J}(h) \neq 0$ is weakly mixing. In his proof, von Neumann used the rigidity of the base transformation (i.e. the rotation). Although not every interval exchange transformation is rigid*, they always possess a recurrence property which is strong enough to carry through an argument similar to von Neumann's. Namely, we will use the following property [**K4**].

PROPOSITION 5.10. *There exists a positive α and a sequence of positive integers $n_k \to \infty$ such that for every measurable set $A \subseteq S^1$*

$$\mu(A \cap T^{n_k} A) > \alpha\mu(A)$$

We can now formulate the generalization of von Neumann's result.

THEOREM 5.11. *If $T : \Delta \to \Delta$ is an interval exchange transformation ergodic with respect to Lebesgue measure and h is a piecewise C^1 function such that $\mathcal{J}(h) \neq 0$, then the special flow over T build under h is weakly mixing but not mixing.*

---

*cf. Del Junco [**J**], who proved that some exchanges of 3 intervals have minimal self-joinings and consequently their centralizers consist only of powers. On the other hand, a rigid transformation always has an uncountable centralizer.

PROOF. Since h is a function of bounded variation, the absence of mixing follows from [**K4**].

To prove weak mixing, we show as usual that for every $r \neq 0$ the function $\exp irh(x)$ is not cohomologous to any constant. We will use Proposition 5.1, ii), whose assumption in this case, is satisfied due to Proposition 5.10. Let $\theta_n(x) = \exp ir \sum_{k=0}^{n-1} h(T^k x)$. We will show that any weak limit point of the sequence of measures

$$(\theta_n)_* \lambda$$

is a nonatomic measure, where λ denotes Lebesgue measure on Δ.

Let ζ be the partition of Δ into subintervals formed by all points of discontinuity of T and all points where the derivative $h'(x)$ is discontinuous and let $\zeta_n = \zeta \vee T^{-1}\zeta \vee \dots T^{-n+1}\zeta$. Let $\omega(\epsilon)$ be a common modulus of continuity for h' on all its intervals of continuity. If $c \in \zeta_n$ then all functions $h \circ T^i$; $i = 0, \dots, n-1$ are differentiable on c and since $T^i\big|_c$ has derivative ± 1 we have for $x, y \in c$

$$|h'(T^i x) - h'(T^i y)| \leq \omega(|x - y|)$$

and consequently

$$\left| \sum_{i=0}^{n-1} h'(T^i x) = \sum_{i=0}^{n-1} h'(T^i y) \right| < n\omega(|x - y|). \tag{5.21}$$

We have

$$\int_\Delta h'(x)dx = -\mathcal{J}(h). \tag{5.22}$$

Let us denote for an integer n and $\epsilon > 0$

$$A_{n,\epsilon} = \left\{ x \in \Delta : \left| \frac{1}{n} \sum_{i=0}^{n-1} h(T^i x) + \mathcal{J}(h) \right| < \epsilon \right\}. \tag{5.23}$$

By the ergodicity of T we have for any $\epsilon > 0$

$$\lambda(A_{n,\epsilon}) \to 1 \quad \text{or} \quad n \to \infty.$$

Let us now fix $\beta > 0$ and consider all elements of the partition ζ_n of length $> \frac{\beta}{n}$. Let $R = \operatorname{card} \zeta$. Since the total number of elements in ζ does not exceed Rn (every iterate of T adds as many new elements as the number of points of discontinuity of T), the total measure of these elements is greater than $1 - R\beta$. Let us divide now every element into intervals of length between $\frac{\beta}{n}$ and $\frac{w\beta}{n}$ and call those of the intervals, which intersect the set $\Delta_{n,\epsilon}$, (n, ϵ, β)-admissible ones. The total measure of (n, ϵ, β)-admissible intervals is clearly greater than $\lambda(\Delta_{n,\epsilon}) - \beta R$. For every (n, ϵ, β)-admissible interval σ we have by (5.21) and (5.22)

$$\begin{gathered}\left(-\mathcal{J}(h) - \epsilon - \omega\left(\frac{2\beta}{n}\right)\right) n < \min_\sigma \sum_{i=0}^{n-1} h'(T^i x) \leq \max_\sigma \sum_{i=0}^{n-1} f'(T^i x) < \\ \left(-\mathcal{J}(h) + \epsilon + \omega\left(\frac{2\beta}{n}\right)\right) n\end{gathered} \tag{5.24}$$

If ϵ is chosen small enough and n is big enough then in particular (5.24) implies that

$$\max_{\sigma}\sum_{i=0}^{n-1} h'(T^i x) < 2|\mathcal{J}(h)|n \tag{5.25}$$

so that by choosing $\beta < \frac{1}{4r|\mathcal{J}(h)|}$ (recall that the length of σ is less than $\frac{2\beta}{n}$) we can guarantee that the function θ_n is injective on every (n,ϵ,β)-admissible interval.

On the other hand, the length of the image $\theta_n(\sigma)$ is bigger than $\frac{r}{2}\beta\mathcal{J}(h)|$ if we guarantee by our choice of n, ϵ that

$$\min_{\sigma}\sum_{i=0}^{n-1} h'(T^i x) > \frac{1}{2}|\mathcal{J}(h)|n \tag{5.26}$$

It follows from (5.25) and (5.26) that the density of the measure $(\theta_n)_*(\lambda|_\sigma) = \mu_{n,\sigma}$ can oscillate on the interval $\theta_n(\sigma)$ ratio of at most 4, so that for every interval $\delta \in \Delta$ of length ℓ

$$\mu_{n,\sigma}(\delta) < \frac{4\ell}{\theta_n(\sigma)} < \frac{4\ell}{r\beta|\mathcal{J}(h)|}.$$

This means in particular that

$$\begin{aligned}
&\lambda(\{x \in \Delta : \theta_n(x) \in \delta\}) \leq \\
&\lambda(\{x \in \Delta : \theta_n(x) \in \delta\ ,\ x \text{ belongs to an } (n,\epsilon,\beta)\text{-admissible} \\
&(\text{interval})\}) + \mu(\{x \in \Delta\ ,\ x \text{ does not belong to an} \\
&(n,\epsilon,\beta)\text{-admissible interval}\}) < \frac{4\ell}{r\beta|\mathcal{J}(h)|} + R\beta + 1 - \lambda(A_{n,\epsilon}).
\end{aligned} \tag{5.27}$$

Fixing first a sufficiently small ϵ and a sufficiently big n, and then sufficiently small β we can always find ℓ such that the right hand part of (5.27) is less than any given positive number. This shows that any weak limit power of the sequence of measures $(\theta_n)_*\lambda$ cannot contain an atom. □

References

[A] D. V. Anosov, *Functional homology equation connected with an ergodic rotation of the circle*, Math. USSR Izvestija **7** (1973), 1257–1271.

[AK] D. V. Anosov and A. B. Katok, *New examples in smooth ergodic theory. Ergodic diffeomorphisms*, Transactions of the Moscow Mathematical Society **123** (1970), 1–35.

[BKM] D. Bernstein, A. Katok, and G. D. Mostow, *Nets and Quasiisometries*, unpublished.

[CFW] A. Connes, J. Feldman and B. Weiss, *An amenable equivalence relation is generated by a single transformation*, Erg. Th. & Dynam. Syst. **1** (1981), 431–450.

[Fa1] B. Fayad, *Weak mixing in reparametrized linear flows on the torus*, to appear, Erg. Th. & Dynam. Syst..

[Fa2] B. Fayad, *Analytic mixing reparametrizations of the irrational flows on the torus* $\mathbb{T}^n$, $n \geq 3$, to appear, Erg. Th. & Dynam. Syst..

[FK] R. Feres and A. Katok, *Ergodic theory and dynamics of G-spaces*, Handbook in Dynamical Systems, vol. 1, Elsevier, 2001.

[FlFo] G. Forni, *Invariant distributions for horocycles flows*, in preparation.

[Fo] G. Forni, *Solutions of the cohomological equation for area-preserving flows on compact surfaces of higher genus* **146** (1997), 295–344.

[Fu] A. Furman, *Orbit equivalence rigidity*, Ann. of Math **150** (1999), 1083–1108..

[F] H. Furstenberg, *Strict ergodicity and transformations of the torus*, Amer. J. Math. **83** (1961), 573–601.

[HaK] B. Hasselblatt and A. Katok, *Principal structures*, Handbook in Dynamical Systems, vol. 1, Elsevier, 2001.

[He1] M. R. Herman, *L^2 regularity of measurable solutions of a finite difference equation of the circle*, preprint, University of Warwick (1976).

[He2] M. R. Herman, *Sur la conjugaison diffrentiable des diffomorphismes du cercle des rotations*, Publ. Math. IHES **49** (1979), 5–233.

[He3] M. R. Herman, *Constructions de diffeomorphismes ergodiques*, Unpublished manuscript.

[He4] M. R. Herman, *Construction d'un diffeomorphisme minimal d'entropie topologique non nulle*, Erg. Th. & Dynam. Syst. **1** (1981), 65–76.

[He5] M. R. Herman, *Une mthode pour minorer les exposants de Lyapounov et quelques exemples montrant le caractre local d'un thorme d'Arnol'd et de Moser sur le tore de dimension* 2, Comm. Math. Helv. **58** (1983), 453–502..

[HK1] S. Hurder and A. Katok, *Ergodic theory and Weil measures for foliations*, Ann. Math. **126** (1987), 221–275.

[HK2] S. Hurder and A. Katok, *Differentiability, rigidity and Godbillon-Vey classes for Anosov flows*, Publ. Math. IHES **72** (1990)), 5–61.

[Jo] J.-L. Journé, *On a regularity problem occurring in connection with Anosov diffeomorphisms*, Comm. Math. Phys. **106** (1986), 345–351.

[J] A. del Junco, *A family of counterexamples in ergodic theory*, Israel J. Math **44** (1983), 160–188.

[JR] A. Del Junco and D. Rudolph, *Kakutani equivalence of ergodic $\mathbb{Z}^n$ actions*, Erg. Th. & Dynam. Syst. **4** (1984), 89–104.

[Ka] S. Kakutani, *Induced measure preserving transformations*, Proc. Japan Acad. **19** (1943), 635–641.

[K1] A. Katok, *Combinatorial constructions in ergodic theory. I. Approximation and genericity*, Proc. Steklov Inst. Math (2001).

[K2] A. B. Katok, *The special representation theorem for multi-dimensional group actions*, Asterisque **49** (1977), 117-140.

[K3] A. B. Katok, *Monotone equivalence in ergodic theory*, Math. USSR, Izvestija **11** (1977), 99–146.

[K4] A. Katok, *Interval exchange transformations and some special flows are not mixing,*, Isr. J. Math **35** (1980), 301-310.

[KH] A. Katok and B. Hasselblatt, *Introduction to the modern theory of dynamical systems*, Cambridge university press, 1995.

[KK] A.Katok and A. Kononenko, *Cocycles' stability for partially hyperbolic systems.*, Math. Res. Lett **3** (1996), 191–210.

[KSch] A. Katok and K. Schmidt, *The cohomology of expansive $\mathbb{Z}^d$-actions by automorphisms of compact abelian groups*, Pacific J. Math. **170** (1995), 105-142.

[KSp1] A.Katok and R. Spatzier, *First cohomology of Anosov actions of higher rank abelian groups and applications to rigidity*, Publ. Math. IHES **79** (1994), 131-156.

[KSp2] A.Katok and R. Spatzier, *Differential rigidity of Anosov actions of higher rank abelian groups and algebraic lattice actions*, Proc. Steklov ath. Inst. **216** (1997), 287-314.

[Ke] M. Keane, *Interval exchange transformations*, Math. Zeitsch. **141** (1975), 25–31.

[Kc] A. V. Kočergin, *The homology of functions over dynamical systems (in Russian)*, Dokl. Akad. Nauk SSSR **231** (1976), 795–798.

[Ko] A. N. Kolmogorov, *On dynamical systems with integral invariant on the torus*, in Russian, Doklady Akad. Nauk SSSR **93** (1953), 763–766.

[Kr] A. Krygin, *Examples of ergodic cylindrical cascades*, Math. Notes USSR Acad. Sci **16** (1974), 1180–1186.

[dl] R. de la Llave, *Analytic regularity of solutions of Livsic's cohomology equation and some applications to analytic conjugacy of hyperbolic dynamical systems*, Erg. Th. & Dynam. Syst **17** (1997), 649–662.

[dLMM] R. de la Llave, J. M. Marco, and R.Moriyn, *Canonical perturbation theory of Anosov systems and regularity results for the Livšic cohomology equation*, Ann. of Math. **123** (1986), 537–611.

[L1] A. Livšic, *Some homology properties of U–systems*, Math. Notes USSR Acad. Sci **10** (1971), 758–763.

[L2] A. Livšic, *Cohomology of dynamical systems*, Izv. Akad. Nauk SSSR Ser. Mat. **36** (1972), 1296–1320.

[LS] A. Livšic and Ja. G. Sinaĭ, *On invariant measures compatible with smooth structure for transitive U-systems. Soviet Math. Doklady* **13** (1972), 1656–1659.

[LSa] R. U. Luz, and N. M. dos Santos,, *Cohomology-free diffeomorphisms of low-dimension tori*, Erg. Th. & Dynam. Syst. **18** (1998), 985–1006.

[M] H. Masur, *Interval exchange transformations and measured foliations*, Ann. of Math **115** (1982), 169–200.

[N] J. von Neumann, *Zur Operatorenmethode in der klassischen Mechanik*, Ann. Math. **33** (1932), 587–642.

[NT] V. Niţică and A. Török, *Regularity of the transfer map for cohomologous cocycles*, Erg. Th. & Dynam. Syst. **18** (1998), 1187–1209.

[ORW] D. Ornstein, D. Rudolph, and B. Weiss, *Equivalence of measure preserving transformations*, Memoirs of the AMS **37** (1982), no. 262.

[OS] D. Ornstein and M. Smorodinsky, *Continuous speed changes for flows*, Israel J. Math. **31** (1978), 161–168.

[Pa] W. Parry, *Topics in ergodic theory*, Cambridge University Press, 1881.

[Sch1] K. Schmidt, *Dynamical systems of algebraic origin*, Birkhäuser, 1995.

[Sch2] K. Schmidt, *The cohomology of higher-dimensional shifts of finite type*, Pacific J. Math **170** (1995), 237–269.

[V1] W. Veech, *Interval exchange transformations*, J. Analyse Math. **33** (1978), 222–272.

[V2] W. Veech, *Gauss measures for transformations on the space of interval exchange maps*, Ann. of Math. **115** (1982), 201–242.

[V3] W. Veech, *Periodic points and invariant pseudomeasures for toral endomorphisms*, Erg. Th. & Dynam. Syst. **6** (1986), 449–473.

[Z1] R. Zimmer, *Ergodic theory and semisimple groups*, Birkhäuser, 1984.

[Z2] R. Zimmer, *Actions of semisimple groups and discrete subgroups*, Proceedings of the International Congress of Mathematicians, Amer. Math. Soc., Providence, RI, 1987, pp. 1247–1258.

Department of Mathematics, Pennsylvania State University, University Park, PA 16802

Department of Mathematics, George Washington University, Washington, D. C. 20052

Proceedings of Symposia in Pure Mathematics
Volume **69**, 2001

A Tutorial on KAM Theory

Rafael de la Llave

ABSTRACT. This is a tutorial on some of the main ideas in KAM theory. The goal is to present the background and to explain and compare somewhat informally some of the main methods of proof.

It is an expanded version of the lectures given by the author in the Summer Research Institute on *Smooth Ergodic Theory* Seattle, 1999. The style is pedagogical and expository and it only aims to be an introduction to the primary literature. It does not aim to be a systematic survey nor to present full proofs.

CONTENTS

1. Introduction

The goal of these lectures is to present an introduction to some of the main ideas involved in KAM theory on the persistence of quasiperiodic motions under perturbations. The name comes from the initials of A. N. Kolmogorov, V. I. Arnol'd and J. Moser who initiated the theory. See [**Kol54**], [**Arn63a**], [**Arn63b**], [**Mos62**], [**Mos66b**], [**Mos66a**] for the original papers.

2000 *Mathematics Subject Classification.* Primary 37J40.

Key words and phrases. KAM theory, stability, Perturbation theory, quasiperiodic orbits, Hamiltonian systems.

By now, it is a full fledged theory and it provides a systematic tool for the analysis of many dynamical systems and it also has relations with other areas of analysis.

The conclusions of the theory are, roughly, that in C^k – k rather high depending on the dimension – open sets of of dynamical systems satisfying some geometric properties – e.g. Hamiltonian, volume preserving, reversible, etc. – there are sets of positive measure covered by invariant tori. In particular, since sets with a positive measure of invariant tori is incompatible with ergodicity, we conclude that for the systems mentioned above, ergodicity cannot be a C^k generic property [**MM74**].

Of course, the existence of the quasiperiodic orbits, has many other consequences besides preventing ergodicity. The invariant tori are important landmarks that guide the motion.

Besides its applications to mechanics, dynamical systems and ergodic theory, KAM theory has grown enormously and has very interesting ramifications in dynamical systems and in Analysis. For example, averaging theory gives rise to estimates for very long times valid for all initial conditions, one can use partially hyperbolic tori to show existence of orbits that escape. On the analytical side, the theory leads to functional analysis methods that can be used to solve a variety of functional equations, many of which have interest in ergodic theory and in related disciplines such as differential geometry.

There already exist excellent surveys, systematic expositions and tutorials of KAM theory.

We quote in chronological order: [**Arn63b**], [**Mos66b, Mos66a, Mos73**], [**AA68**], [**Rüs70**], [**Rüs**], [**Zeh75, Zeh76a**], [**Dou82a**], [**Bos86**], [**Sal86**], [**Pös82**], [**Pös92**] [**Yoc92**], [**AKN93**], [**dlL93**]. [**BHS96b**], [**Gal83a**], [**Gal86**], [**Way96**], [**Rüs98**], and [**Mar00**].

Hence, one has to justify the effort in writing and reading yet another exposition.

I decided that each of the surveys above has picked up a particular point of view and tried to either present a large part of KAM theory from this point of view or to provide a particularly enlightening example.

Given the high quality of all (but one) of the above surveys and tutorials, there seems to be little point in trying to achieve the same goals. Therefore, rather than presenting a point of view with full proofs, this tutorial will have only the more modest goal of summarizing some of the main ideas entering into KAM theory and describing and comparing the main points of view. Therefore, it is not a substitute for the full papers we reference.

One of the disadvantages of covering such wide ground is that the presentation will have to be sketchy at some points. Hopefully, we have flagged a good fraction of these sketchy points and referred to the relevant literature. I would be happy if these lectures provide a road map (necessarily omitting important details) of a fraction of the literature that encourages somebody to enter into the field. Needless to say, this is not a survey and we have not made any attempt to be systematic nor to reach the forefront of research.

It should be kept in mind that KAM theory has experienced spectacular progress in recent years and that it is a very active area of research.

1.1. Some recent developments. Let me mention some of these new developments (in no particular order and, with no claim of completeness of the list and omitting classical results – i.e. more that 15 years old –). They will not be covered in the lectures, which will be concerned only with the most classical results.

The novice that is reading the paper to get initiated to KAM theory is encouraged to skip it for the moment and only come back to it as suggestions for future reading. Of course, the experts will notice many omissions. The only point we are trying to make is that the theory is still finding exciting results and that there is work to do.

- The "lack of parameters" which was considered inaccessible has been solved very elegantly [**JS92**], [**Eli88**]. (See [**BHS96b**] for a recent survey, and also [**BHS96a**] [**Sev99**].) This has lead to remarkable progress in the existence of lower dimensional tori, specially elliptic tori – a theory of hyperbolic tori has been known for a long time – (see e.g. [**JV97b**], [**JV97a**].)
- As a corollary of this, one can get a reasonable KAM theory for volume preserving systems getting tori of codimension one. Hence blocking diffusion in many problems in hydrodynamics, etc. (See [**CS90**], [**BHS96b**], [**DdlL90**], [**Xia92**], [**Yoc92**].)
- The KAM theory for infinite dimensional systems has made remarkable progress.

  Note that in infinite dimensional systems, the most interesting tori are of lower dimension than the number of degrees of freedom.

  The subject of infinite dimensional KAM by itself would require a review of its own longer than these notes. We just refer to [**CW93**], [**CW94**], [**Pös96**], [**Bou99a**], [**Bou99b**], and [**Kuk93**] as representative references, where the interested reader can find further references.
- Many systems in applications – e.g in statistical mechanics – have the structure that they consists of arrays of systems connected by local couplings.[1] For these systems one can take advantage of this structure and develop a more efficient KAM theory than the simple application of the general results. [**Way84**], [**Pös90**] [**FSW86**]. Other KAM methods for these systems are developed in [**AFS88**] and [**AF88**], [**AF91**] which consider the existence of periodic solutions. See also [**Way86**] for an Nekhoroshev theorem for these systems. Conjectures and preliminary estimates (a challenge for rigorous proofs) on these systems can be found in [**BGG85b**], [**BGG85a**], [**HdlL00**],[**CCSPC97**].
- The non-degeneracy conditions needed for the KAM theorems have been greatly weakened in [**Rüs90**], [**CS94**], [**Rüs98**]. See also [**BHS96b**], [**BHS96a**], [**Sev95**], [**Sev96**].
- Modern techniques of PDE's such as viscosity solutions have been used to study the Hamilton-Jacobi equation [**Lio82**], [**CL83**], [**CEL84**], leading to a weak version of KAM theory that has deep relations with Aubry-Mather theory [**Fat97b**], [**Fat97a**].
- There has been quite spectacular progress in the problem of *reducibility* of linear equations with quasiperiodic coefficients (That is, the study whether an equation with the form $\dot{x} = A(\phi + \omega t)x$ where $A \to \mathbf{T}^d \mathcal{M}_{n\times n}$ and $\omega \in$

[1]These systems, under the name of *coupled lattice maps* have also been the subject of very intense research when they have hyperbolicity properties, in some sense opposite to the situation considered in KAM theory

$\mathbf{R}^d$ is an irrational vector. can be transformed into constant coefficients. After the original work of [**DS75**], two important recent developments were [**MP84**] which introduced the deep idea of using transformations which are not close to the identity to eliminate small terms and [**Ryc92**] which introduced a renormalization mechanism. After that, many more new important refinements were introduced in several works (one needs to find ways to combine perturbative steps with non-perturbative ones). This is still a very active area and progress is being made constantly. We refer to the lectures of prof. Eliasson in this volume for up to date references. See also [**Eli**],[**Kri99**].

- The problem of reducibility is related to the problem of existence of pure point spectrum of one-dimensional Schrödinger operators with quasiperiodic coefficients. This area has experienced quite significant progress. Besides some of the papers mentioned in the previous paragraphs, let us mention [**CS91**], [**FSW90**], [**Eli97**].
- For Schrödinger operators in higher dimensions with random or quasi-periodic potential the theory of localization also has advanced greatly thanks to a multi-scale analysis which is quite reminiscent of KAM theory [**FS83**],[**FS84**]. Indeed, this analogy has been pursued quite fruitfully. [**Alb93**].
- Even if the symplectic forms that appear in mechanical systems admit a primitive (see later in Section 3.5), there are other symplectic forms without this feature. For such forms without a primitive, one has the possibility of finding persistent tori of more dimension than the degrees of freedom. This has important consequences and leads to very interesting examples in ergodic theory. See [**Yoc92**], [**Her91**] (See also [**Par84**], [**Par89**], [**FY98**].)
- KAM methods have been extended to elliptic PDE's – they are not evolution equations. The role of time in KAM has been taken by spatial variables. (See [**Koz83**], [**Mos88**], [**Mos95**].) This has also been related to a variational structure of the equations [**Mos86**], [**Ban89**].
- There are some proofs of KAM type theorems based on different principles, notably renormalization group, [**BGK99**], [**Koc99**], [**Kos91**]. This is perhaps related to some recent proofs that do not even use Fourier analysis [**KS86**], [**KS87**], [**SK89**], [**KO93**], [**KO89b**], [**KO89a**], [**Sta88**], [**Hay90**] [**Sti94**].
- More interestingly, renormalization group has been used to describe the breakdown of invariant circles, starting with [**McK82**] – which includes a beautiful picture in terms of fixed points and manifolds of operators and makes very detailed predictions about scalings at breakdown – or [**ED81**], which contains a simpler approach that gives less detailed predictions. Much of what is known at this level remains at the level of numerical well founded conjectures. Indeed, there are still quite important issues that are not even known at this level. Among the rigorous work in this area, we mention [**Sti93**], [**Sti97**].
- KAM theory has started to become a tool of applied mathematics with the advent of constructive methods to asses the reliability of numerical computations [**CC95**], [**dlLR91**], [**Sch95**], [**Jor99**].

- For some special cases of KAM theory, there has also been very important progress examining the limits of validity; the role of the arithmetic conditions has been clarified for complex mappings – specially quadratic – [**Yoc95**]. See also [**PM00**]. The study of the radius of convergence of the linearization in the same mappings [**MMY97**] has also been quite well understood.

  In some twist mappings, there has been a very significant advances in the study of non-existence of tori [**Mat88**], [**MP85**], [**Jun91**]. The domains of convergence of the perturbative expansions have been analyzed using tools similar to those used for analytic complex mappings starting in [**Dav94**] – a map which has features between those of a complex analytic map and those of a twist map – and then in [**MS92**], [**BM95**], [**BG99**].
- Two different techniques to study quasiperiodic orbits on twist maps are the variational methods of Mather [**MF94**] and the renormalization group [**Koc99**];

  In many cases, these theories have ranges of validity much greater than those covered by KAM theory and, therefore provide some glimpse into what happens at the breakdown of KAM theory.
- There has been great progress in using "direct methods", which are based on writing a perturbative expansion and showing it converges by studying more deeply the structure of small denominators.

  In the study of iterations of analytic functions, these methods led to the original proof of Siegel [**Sie42**], which was the first problem in which small denominators were understood. They were also used in the first proof of the optimal arithmetic conditions [**Brj71**].

  In the study of Lindstedt series (see Section 2.1), the proof of convergence by exhibiting explicitly cancellations of the series was accomplished in [**Eli96**] (the preprint circulated much earlier). The proof of the convergence of the Lindstedt series in [**Eli96**] is much more subtle than that of [**Sie42**]. Contrary to the terms in the expansions considered in [**Sie42**], the terms in the Lindstedt series do grow very fast and one cannot establish convergence by just bounding sizes but one needs to exhibit cancellations in the terms.

  Expositions and simplifications of this work relating it also to techniques of perturbative Quantum Field Theory can be found in [**Gal94b**], [**GG95**], [**CF94**] and extensions to some PDE's in [**CF96**].

  Direct methods not only provide alternative proofs of known facts, but also have been used to prove several results, which at the moment do not seem to have proofs using rapidly convergent methods. To my knowledge, the following results established using direct methods do not have rapidly convergent proofs: The existence of some invariant manifolds contained in center manifolds in [**Pös86**] was proved using cancellations similar to those of Siegel. It seems that there are no rapidly convergent proofs of these results (however, see [**Sto94a**], [**Sto94b**] which solve a very related problem.)

  The deeper cancellations of [**Eli96**] have been used to give a proof of the Gallavotti conjectures (which imply, among other consequences, the amusing result that an analytic Hamiltonian near an elliptic fixed point

is the sum of two integrable systems – of course integrated in different coordinates.) [**Eli89**] and to prove the existence of quasi-flat intersections in [**Gal94a**]. A problem that remains open is the fact that the Lindstedt series for lower dimensional KAM tori involve less small divisors conditions than the KAM proof. (See [**JdlLZ99**] for a discussion of this problem.)
- Subjects closely related to KAM theory such as averaging and Nekhoroshev theory have also experienced a great deal of development.
- Even if this is somewhat out of the line of topics to be discussed here, we note that related fields such as averaging theory and Nekhoroshev estimates has also experienced very important developments. Let us just mention very quickly: An elegant proof of the theorem based on approximation by periodic orbits [**Loc92**], the proof of what are conjectured to be the optimal exponents [**LN92**], [**Pös93**], [**DG96**] – the later paper contains a unified point of view for KAM and Nekhoroshev theorems – and the proof of Nekhoroshev estimates in a neighborhood of an elliptic fixed point [**GFB98**], [**FGB98**], [**Nie98**], [**Pös**]. In a more innovative direction, Nekhoroshev type theorems for PDE's have been established [**BN98**], [**Nek99**], [**Bam99b**], [**Bam99a**].
- The list could (perhaps should) be continued, with other topics that are related to KAM theory and connecting it to other theories of mechanics, such as averaging theory, Aubry-Mather theory, quantum versions of KAM theory, rigidity theory, exponential asymptotics or Arnol'd diffusion and many others which are not even mentioned mainly because of the ignorance of the lecturer, which he is the first to regret.

Needless to say in this tutorial, we cannot hope to do justice to all the topics above. (Indeed, I have little hope that the above list of topics and references is complete.) The only goal is to provide an entry point to the main ideas that will need to be read from the literature and, possibly, to convey some of the excitement and the beauty of this area of research.

Clearly, I cannot (and I do not) make any claim of originality or completeness. This is not a systematic survey of topics of current research. The modest goal I set for these notes is to help some readers to get started in the beautiful and active subject of KAM theory by giving a crude road map. I just hope that the many deficiencies of this tutorial will incense somebody into writing a proper review or a better tutorial. In the mean time, I will be happy to receive comments, corrections and suggestions for improvement of this tutorial which I will make available electronically.

2. Some motivating examples

2.1. Lindstedt series for twist maps. One of the original motivations of KAM theory was the study of quasi-periodic solutions of Hamiltonian systems. In this Section we will cover some elementary and well-known examples.

One particularly motivating example is the so-called *standard map*. This is a map from $\mathbf{R} \times \mathbf{T}$ to itself. We denote the real coordinate by p and the angle one by q. Denoting by p_n, q_n the values of these coordinates at the discrete time n, the map can be written as:

$$\begin{aligned} p_{n+1} &= p_n - \varepsilon V'(q_n) \\ q_{n+1} &= (q_n + p_{n+1}) \mod 1, \end{aligned} \tag{2.1}$$

where $V(x) = V(x+1)$ is a smooth (for our purposes in this Section, analytic) function. We will also use a more explicit expression for the map.

$$T_\varepsilon(p,q) = (p - \varepsilon V'(q),\, q + p - \varepsilon V'(q)) \ . \tag{2.2}$$

Substituting the expression for p_{n+1} given in the second equation of (2.1) into the first, we see that the system (2.1) is equivalent to the second order equation.

$$q_{n+1} + q_{n-1} - 2q_n = -\varepsilon V'(q_n) \ , \tag{2.3}$$

The first, "Hamiltonian", formulation (2.1) appears naturally in some mechanical systems (e.g., the kicked pendulum). The second, "Lagrangian", one (2.3) appears naturally from a variational principle, namely, it is equivalent to the equations

$$\partial \mathcal{L} / \partial q_n = 0 \tag{2.4}$$

with

$$\mathcal{L}(q) = -\sum_n \left[\frac{1}{2}(q_{n+1} - q_n - a)^2 + \varepsilon V(q_n) \right] \ . \tag{2.5}$$

The equations (2.4) – often called Euler-Lagrange equations – express that $\{q_n\}$ is a critical point for the action (2.5).

The model (2.5) has appeared in solid state physics under the name *Frenkel-Kontorova model.* (See e.g. [**Aub83**].) One physical interpretation (not the only possible one) that has lead to many heuristic insights is that q_n is the position of the nth atom in a chain. These atoms interact with their nearest neighbors by the quadratic potential energy $\frac{1}{2}(q_{n+1} - q_n - a)^2$ (corresponding to springs connecting the nearest neighbors) and with a substratum by the potential energy $\varepsilon V(q_n)$. The parameter a is the equilibrium length of each spring. Note that a drops from the equilibrium equations (2.3) but affects which among all the equilibria corresponds to a minimum of the energy.

Another interpretation, of more interest for the theme of these lectures, is that q_n are the positions at consecutive times of a one-degree of freedom twist map. The general term in the sum $S(q_{n+1}, q_n)$ are the generating functions of the map. (See Section 3.7.) Then, the Euler-Lagrange equations for critical points of the functional are equivalent to the sequence $\{q_n\}$ being the projection of an orbit.

The first formulation (2.1) is area preserving whenever V' is a periodic function of the cylinder – not necessarily the derivative of a periodic function (i.e., the Jacobian of the transformation $(p_n, q_n) \mapsto (p_{n+1}, q_{n+1})$ is equal to 1). When, as we have indicated, V' is indeed the derivative of a periodic function, then the map is *exact*, a concept that we will discuss in greater detail in Section 3.5 and that has great importance for KAM theory.

If we look at the map (2.1) for $\varepsilon = 0$, we note that it becomes

$$\begin{aligned} p_{n+1} &= p_n \\ q_{n+1} &= q_n + p_n \ , \end{aligned} \tag{2.6}$$

so that the "horizontal" circles $\{p_n = \text{const}, \, n \in \mathbf{Z}\}$ in the cylinder are preserved and the motion of each q_n in each circle is a rigid rotation that is faster in the circles with larger p_n. Note that when p_0 is an irrational number, a classical elementary theorem in number theory shows that the orbit is dense on the circle. (A deeper theorem due to Weyl shows that it is actually equidistributed in the circle.)

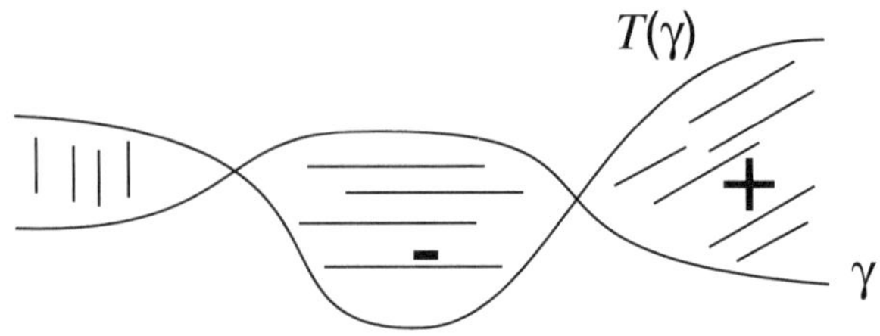

FIGURE 1. The flux is the oriented area between a circle and its image.

We are interested in finding whether, when we turn on the perturbation ε, some of this behavior persists. More concretely, we are interested in knowing whether there are quasi-periodic orbits that persist and that fill a circle densely.

Problems that are qualitatively similar to (2.1) appear in celestial mechanics [**SM95**] and the role of these quasi-periodic orbits have been appreciated for many years. One can already find a rather systematic study in [**Poi93**] and the treatment there refers to many older works.

We note that the existence of quasi-periodic orbits is hopeless if one allows general perturbations of (2.6). For example, if we take a map of the form

$$\begin{aligned} p_{n+1} &= p_n - \varepsilon p_n \\ q_{n+1} &= q_n + p_{n+1} \ , \end{aligned} \tag{2.7}$$

we see that applying repeatedly (2.7), we have

$$p_n = (1 - \varepsilon)^n p_0$$

so that, when $0 < \varepsilon < 2$, all orbits concentrate on the very small set $p = 0$ and that we get at most only one frequency. When $\varepsilon < 0$ or $\varepsilon > 2$, all the orbits except those in $p = 0$, blow up to infinity. Hence, we can have maps with radically different dynamical behavior by making arbitrarily small perturbations.

More subtly, the orbits of

$$\begin{aligned} p_{n+1} &= p_n + \varepsilon \\ q_{n+1} &= q_n + p_{n+1} \end{aligned} \tag{2.8}$$

escape towards infinity and never come back to themselves (in particular, can never be quasi-periodic).

The first example is not area preserving and the motion is concentrated in a smaller area (in particular, it does not come back to itself). The second example is area preserving but has non-zero "flux".

DEFINITION 2.1. The "flux" of an area preserving map T of the cylinder is defined as follows: given a continuous circle γ on the cylinder, the flux of T is the oriented area between $T(\gamma)$, the image of the circle, and γ — see Figure 1.

The fact that the map is area preserving implies easily that this flux is independent of the circle (hence it is an invariant of the map). Clearly, if the map T had a continuous invariant circle, the flux should be zero, so we cannot find an invariant circle in (2.8) for $\varepsilon \neq 0$ since the flux is ε.

REMARK 2.2. Note that if a map has a homotopically nontrivial invariant curve, then the flux is zero (compute it for the curve). Conversely, if the flux is zero, any homotopically non-trivial curve has to have an intersection with its image. (If it did not have any intersection, by Rolle's theorem, then the image would always be in above or below the curve.) The property that every curve intersects its image plays an important role in KAM theory and is sometimes called *intersection property*. Besides area preserving and zero flux, there are other geometric assumptions that imply the intersection property, notably, reversibility of the map (see [**AS86**])

As a simple calculation shows, that perturbation in (2.1) is of the form $V'(q_n)$, with V 1-periodic — therefore $\int_0^1 V'(q_n)\,dq_n = V(1) - V(0) = 0$ — the flux of (2.1) is zero.

We see that even the possibility that there exist these quasi-periodic orbits filling an invariant circle depends on geometric invariants.

Indeed, when we consider higher dimensional mechanical systems, the analogue of area preservation is the preservation of a symplectic form, the analogue of the flux is the Calabi invariant [**Cal70**] and the systems with zero Calabi invariant are called exact.

We point out, however, that the relation of the geometry to KAM theory is somewhat subtle. Even if the above considerations show that some amount of geometry is necessary, they by no means show what the geometric structure is, and much less hint on how it is to be incorporated in the proof.

The first widely used and generally applicable method to study numerically quasi-periodic orbits seems to have been the method of Lindstedt. (We follow in this exposition [**FdlL92**].)

The basic idea of Lindstedt's method is to consider a family of quasiperiodic functions depending on the parameter ε and to impose that it becomes a solution of our equations of motion. The resulting equation is solved – in the sense of power series in ε – by equating terms with same powers of ε on both sides of the equation. We will see how to apply this procedure to (2.1) or (2.3).

In the Hamiltonian formulation (2.1), (2.2) we seek $K_\varepsilon : \mathbf{T}^1 \to \mathbf{R} \times \mathbf{T}^1$ in such a way that

$$T_\varepsilon \circ K_\varepsilon(\theta) = K_\varepsilon(\theta + \omega)\ . \tag{2.9}$$

We set:

$$K_\varepsilon(\theta) = \sum_{n=0}^{\infty} \varepsilon^n K_n(\theta) \tag{2.10}$$

and try to solve by matching powers of ε on both sides of (2.9), (after expanding $T_\varepsilon \circ K_\varepsilon(\theta)$ as much as possible in ε using the Taylor's theorem).[2] That is,

$$\begin{aligned} T_\varepsilon \circ K_\varepsilon(\theta) \ &= \ T_0 \circ K_0 + \varepsilon[T_1 \circ K_0 + (DT_0 \circ K_0)K_1] \\ &\quad + \varepsilon^2[T_2 \circ K_0 + (DT_0 \circ K_0)K_2 \\ &\qquad + (DT_1 \circ K_0)K_1 + \frac{1}{2}(D^2T_0 \circ K_0)K_1^{\otimes 2}] + \dots\ . \end{aligned}$$

[2]The notation is somewhat unfortunate since K_n could mean both the n term in the Taylor expansion and K_ε evaluated for $\varepsilon = n$. In the discussion that follows, K_1,K_2, etc. will always refer to the Taylor expansion. Note that K_0 is the same in both meanings.

In the Lagrangian formulation (2.3) we seek $g_\varepsilon : \mathbf{R} \to \mathbf{R}$ satisfying $g_\varepsilon(\theta + 1) = g_\varepsilon(\theta) + 1$ — or, equivalently, $g_\varepsilon(\theta) = \theta + \ell_\varepsilon(\theta)$ with $\ell_\varepsilon(\theta + 1) = \ell_\varepsilon(\theta)$, i.e., $\ell_\varepsilon : \mathbf{T}^1 \to \mathbf{T}^1$ — in such a way that

$$\ell_\varepsilon(\theta + \omega) + \ell_\varepsilon(\theta - \omega) - 2\ell_\varepsilon(\theta) = -\varepsilon V'(\theta + \ell_\varepsilon(\theta)) \ . \tag{2.11}$$

If we find solutions of (2.11), we can ensure that some orbits q_n solving (2.3) can be written as

$$q_n = n\omega + \ell_\varepsilon(n\omega) \ .$$

Note that the fact that, when we choose coordinates on the circle, we can put the origin at any place, implies that $K_\varepsilon(\cdot + \sigma)$ is a solution of (2.9) if K_ε is, and that $\ell_\varepsilon(\cdot + \sigma) + \sigma$ is a solution of (2.11) if ℓ_ε is. Hence, we can – and will – always assume that

$$\int_0^1 \ell_\varepsilon(\theta)\, d\theta = 0 \ . \tag{2.12}$$

This assumption, will not interfere with existence questions, since it can always be adjusted, but will ensure uniqueness.

If we now write[3]

$$\ell_\varepsilon(\theta) = \sum_{n=0}^{\infty} \ell_n(\theta)\varepsilon^n$$

and start matching powers, we see that matching the zero order terms yields

$$\begin{aligned} L_\omega \ell_0(\theta) &\equiv \ell_0(\theta + \omega) + \ell_0(\theta - \omega) - 2\ell_0(\theta) = 0 \ , \\ &\int_0^1 \ell_0(\theta)\, d\theta = 0 \ . \end{aligned} \tag{2.13}$$

The operator L_ω n (2.13), which will appear repeatedly in KAM theory, can be conveniently analyzed by using Fourier coefficients. Note that

$$L_\omega e^{2\pi i k\theta} = 2(\cos 2\pi k\omega - 1)\, e^{2\pi i k\theta} \ .$$

Hence, if $\eta(\theta) = \sum_k \hat\eta_k e^{2\pi i k\theta}$, then the equation

$$L_\omega \varphi(\theta) = \eta(\theta)$$

reduces formally to

$$2(\cos 2\pi k\omega - 1)\, \hat\varphi_k = \hat\eta_k \ .$$

We see that if $\omega \notin \mathbf{Q}$, the equation (2.13) can be solved formally in Fourier coefficients and $\ell_0 = 0$. (Later we will develop an analytic theory and describe precisely conditions under which these solutions can indeed be interpreted as functions.)

When $\omega \notin \mathbf{Q}$, we see that $\cos 2\pi k\omega \neq 1$ except when $k = 0$. Hence, even to write a solution we need $\hat\eta_0 = 0$, and then we can write the formal solutions as

$$\hat\varphi_k = \frac{\hat\eta_k}{2(\cos 2\pi k\omega - 1)} \ , \qquad k \neq 0 \tag{2.14}$$

Note, however, that the status of the solution (2.14) is somewhat complicated since $2\pi k\omega$ is dense on the circle and, hence, the denominator in (2.14) becomes arbitrarily small. Nevertheless, provided that η is a trigonometric polynomial, (See Exercise 2.6 , where this is established under certain circumstances) and ω is irrational, we can solve the equation (2.13). In case that the R.H.S. is analytic and

[3] The same remark about the unfortunate notation we made in (2.10) also applies here.

that the number ω satisfies certain number theoretic properties, in Exercise 2.16, we can show that the solution is analytic.

The equation obtained by matching ε^1 is:

$$L_\omega \ell_1(\theta) = -V'(\theta) \; ; \qquad \int_0^1 \ell_1(\theta)\, d\theta = 0 \; . \tag{2.15}$$

Since $\int_0^1 V'(\theta)\, d\theta = 0$, we see that (2.15) admits a formal solution. (Again, we note that the fact that $\int_0^1 V'(\theta)\, d\theta = 0$ has a geometric interpretation as zero flux.)

Matching the ε^2 terms, we obtain

$$L_\omega \ell_2(\theta) = -V''(\theta)\ell_1(\theta) \; ; \qquad \int_0^1 \ell_2(\theta)\, d\theta = 0 \; , \tag{2.16}$$

and, more generally,

$$L_\omega \ell_n(\theta) = S_n(\theta) \; ; \qquad \int_0^1 \ell_n(\theta)\, d\theta = 0 \; , \tag{2.17}$$

where S_n is an expression which involves derivatives of V and terms previously computed. It is true (but by no means obvious) that

$$\int_0^1 S_n(\theta)\, d\theta = 0, \tag{2.18}$$

so that we can solve (2.17) and proceed to compute the series to all orders (when ω is irrational and S is a trigonometric polynomial or when ω is Diophantine (see later) and S is analytic). The fact that (2.18) holds was already pointed out in Vol II of [**Poi93**].

We will establish (2.18) directly by a seemingly miraculous calculation, whose meaning will become clear when we study the geometry of the problem. (We hope that going through the messy calculation first will give an appreciation for the geometric methods. Similar calculations will appear in Section 6.3.)

The desired result (2.18) follows if we realize that denoting

$$\ell_\varepsilon^{[\le n]}(\theta) = \sum_{i \le n} \varepsilon^i \ell_i(\theta) \; ,$$

we have:

$$L_\omega \ell_\varepsilon^{[\le n]} = \varepsilon^n S_n \tag{2.19}$$

Hence, multiplying (2.19) by $\left[1 + \ell_\varepsilon^{[\le n]\prime}(\theta)\right]$ and integrating, we obtain

$$\begin{aligned} 0 = & \int_0^1 L_\omega \ell_\varepsilon^{[\le n]}(\theta)\, d\theta + \int_0^1 L_\omega \ell_\varepsilon^{[\le n]}(\theta)\, \ell_\varepsilon^{[\le n]\prime}(\theta)\, d\theta \\ & + \int_0^1 V'(\theta + \ell_\varepsilon^{[\le n]}(\theta)) \left[1 + \ell_\varepsilon^{[\le n]\prime}(\theta)\right] d\theta \\ & - \varepsilon^n \int_0^1 S_n(\theta)\, \ell_\varepsilon^{[\le n]\prime}(\theta)\, d\theta \\ & - \varepsilon^n \int_0^1 S_n(\theta)\, d\theta \\ & + O(\varepsilon^{n+1}) \; . \end{aligned} \tag{2.20}$$

Now, we are going to use different arguments to show that all the terms in (2.20) except $\int S_n(\theta)\,d\theta$ vanish. This will establish the desired result.

By changing variables in the integral we have:

$$\int_0^1 V'(\theta + \ell_\varepsilon^{[\leq n]}(\theta)) \left[1 + \ell_\varepsilon^{[\leq n]\prime}(\theta)\right] d\theta = 0. \tag{2.21}$$

Furthermore, it is clear that $\int_0^1 L_\omega \ell_\varepsilon^{[\leq n]}(\theta)\,d\theta = 0$ because for any periodic function f $\int_0^1 f(\theta)\,d\theta = \int_0^1 f(\theta+\omega)\,d\theta = \int_0^1 f(\theta-\omega)\,d\theta$

Noting that

$$\int_0^1 \ell_\varepsilon^{[\leq n]}(\theta)\,\ell_\varepsilon^{[\leq n]\prime}(\theta) = \int_0^1 \frac{1}{2}\left(\left[\ell_\varepsilon^{[\leq n]}(\theta)\right]^2\right)' d\theta = 0$$

and that

$$\begin{aligned}\int_0^1 \ell_\varepsilon^{[\leq n]}(\theta+\omega)\,\ell_\varepsilon^{[\leq n]\prime}(\theta)\,d\theta &= -\int_0^1 \ell_\varepsilon^{[\leq n]\prime}(\theta+\omega)\,\ell_\varepsilon^{[\leq n]}(\theta)\,d\theta \\ &= -\int_0^1 \ell_\varepsilon^{[\leq n]\prime}(\theta)\,\ell_\varepsilon^{[\leq n]}(\theta-\omega)\,d\theta\,,\end{aligned}$$

we obtain that

$$\int_0^1 L_\omega \ell_\varepsilon^{[\leq n]}(\theta)\,\ell_\varepsilon^{[\leq n]\prime}(\theta)\,d\theta = 0.$$

It is also clear that, because ℓ_0 is a constant,

$$\varepsilon^n S_n(\theta)\ell_\varepsilon^{[\leq n]\prime}(\theta) = O(\varepsilon^{n+1}) \tag{2.22}$$

Hence, putting together (2.20) and the subsequent identities, we obtain the desired conclusion that $\int_0^1 S_n(\theta)\,d\theta$ vanishes.

□

REMARK 2.3. There is a geometric interpretation for the vanishing of this integral. One can compute the flux over the curve in the Hamiltonian formalism predicted by $\ell_\varepsilon^{[\leq n]}(\theta)$. The fact that the flux vanishes is equivalent to the fact that the integral vanishes.

REMARK 2.4. Note that it is rather remarkable that for every irrational frequency we can find formal solutions (when the perturbation is a polynomial), or for Diophantine frequencies for analytic perturbations. Heuristically, this can be explained by the fact that, in area preserving systems, we do not have small parts of the system controlling the long term behavior (as it is the case in dissipative systems) and, hence, perturbations still have to leave open many possibilities for motion of the system.

When one applies the Lindstedt method to dissipative systems, [**RA87**], typically one sees that, except for a few frequencies, the perturbation equations do not have a solution.

REMARK 2.5. The Lindstedt method can be used for dissipative systems. (Code for easy to use, general purpose implementations is available from [**RA92**].) Then, one considers

$$T_\varepsilon \circ K_\varepsilon(\theta) = K_\varepsilon(\theta + \omega_\varepsilon)\,.$$

with $\omega_\varepsilon = \sum \varepsilon^n \omega_n$. One has to choose the terms $\omega_0, \ldots, \omega_n$, so that the equations (2.17) have solutions. It is a practical and easily implementable method to compute limit cycles.

EXERCISE 2.6. Show that if V is a trigonometric polynomial, then l_n is also a trigonometric polynomial. Moreover, $\deg(l_n) \leq An + B$ where A and B are constants that depend only on the degree of V. (For a trigonometric polynomial, $V(\theta) = \sum_{|k| \leq M} \hat{V}_k \exp(2\pi i k\theta)$, the degree is M when $\hat{V}_M \neq 0$ or $\hat{V}_{-M} \neq 0$.)

As a consequence, if V is a trigonometric polynomial and ω is irrational, then the Lindstedt procedure can be carried out to all orders.

REMARK 2.7. The above procedure can be carried out even in the case that the function $V(x)$ is $e^{2\pi i x}$.

In this case, we obtain the so-called semi-standard map. It can be easily shown that the trigonometric polynomials that appear in the series only contain terms with positive frequencies. This makes the terms in the Lindstedt series easier to analyze than those of the case $V(x) = e^{2\pi i x} + e^{-2\pi i x}$. Indeed, the analytical properties of the term of the series for $V(x) = e^{2\pi i x}$ very similar to those of the normalization problem for a polynomial.

We refer to [**GP81**] for numerical explorations, to [**Dav94**] for rigorous upper bounds of the radius of convergence and to [**BM95**], [**BG99**] for a method to transfer results from this complex case to the real one.

The convergence of the expansions obtained remains at this stage of the argument we have presented highly problematic. Note that, at every stage, (2.17) involves small divisors. Worse still, the S_n's are formed by multiplying terms obtained through solving small divisor equations. Hence, the S_n could be much bigger than the individual terms.

Poincaré undertook in [**Poi93**], Paragraph 148 a study of the convergence of these series. He obtained negative results for uniform convergence in a parameter that also forced the frequency to change. His conclusions read (I transcribe the French as an example of the extremely nuanced way in which Poincaré formulated the result.) Roughly, he says that one can conclude that the series does not converge, then points out that this has not been proved rigorously and that there are cases that could be left open, including quadratic irrationals. The conclusion is that, even if the divergence has not been proved, it is quite improbable.

> Il semble donc permis de conclure que les series (2) ne convergent pas.
>
> Toutefois le raisonement qui précède ne suffit pas pour établir ce point avec une rigueur complète.
>
> En effect, ce que nous avons démontré au n° 42 c'est qu'il ne peut pas arriver que, pour toutes les valeurs de μ inferieurs a une certaine limite, il y ait une double infinité de solutions périodiques, et il nous suffirait ici que cette double infinité existâit pour une valeur de $\hat{\mu}$ determinée, different de 0 et généralment trés petite.
>
> [....]
>
> Ne peut-il pas arriver que les series (2) convergént quand on donne aux x_i^0 certaines valeurs convenablement choisies?

> Supposons, pour simplifier, qu'il y ait deux degrees de liberté; les series ne pourraient-elles pas, par example, converger quand x_1^0 et x_2^0 ont été choisis de telle sorte que le rapport $\frac{n_1}{n_2}$ soit incommensurable, et que son carré soit au contraire commensurable. (ou quand le rapport $\frac{n_1}{n_2}$ est assujetti á une autre condition analogue à celle que je viens d'ennoncer un peu au hassard)?
>
> Les raisonnements de ce Chapitre ne me permettent pas d'affirmer que ce fait ne se présentera pas. Tout ce qu'il m'est permis de dire, cest qu'il es fort inversemblable.

This was remarkably prescient since indeed the series do converge for Diophantine numbers. In particular, for algebraic irrationals (see Section 3.3, Theorem 3.6).

It is not difficult to show that, for Diophantine frequencies, these series satisfy estimates that fall short of showing analyticity

$$\|\ell_n\|_\sigma \le (n!)^\nu \tag{2.23}$$

where ν is a positive number. These estimates are sometimes called Gevrey estimates and they appear very frequently in asymptotic analysis.

It is not difficult to construct examples (indeed we present one in Exercise 4.23) which have a similar structure and that the linearized equation that we have to solve at each step satisfy similar estimates. Nevertheless they saturate (2.23). Indeed, in many apparently similar problems with a very similar structure (e.g. Birkhoff normal forms near a fixed point, normal forms near a torus, jets of center manifolds) the bounds (2.23) are saturated. We will not have time to discuss these problems in these notes.

The proof of convergence of Lindstedt series was obtained in [**Mos67**] in a somewhat indirect way. Using the KAM theory, it is shown that the solutions produced by the KAM theory are analytic on the perturbation parameter. It follows that the coefficients of the expansion have to be the terms of the Lindstedt series and, therefore, that the Lindstedt series are convergent.

The example in Exercise 4.23 shows that the convergence that one finds in KAM theory has to depend on the existence of massive cancellations.

The direct study of the Lindstedt series was tackled successfully in [**Eli96**]. One needs to exhibit remarkable cancellations. The papers [**Gal94b**] and [**CF94**] contain another version of the cancellations above relating it to methods of quantum field theory.

We note that the transformations that reduce a map to its normal Birkhoff normal form either near a fixed point or near a torus were known to diverge for a long time. (See [**Sie54**], [**Mos60**].)

Examples of divergence of asymptotic series were constructed in [**Poi93**]. To justify their empirically observed usefulness, the same reference developed a theory of asymptotic series, which has a great importance even today.

It should be remarked that, at the moment of this writing, the convergence of Lindstedt series in slightly different situations (lower dimensional tori [**JdlLZ99**] or the jets for center manifolds of positive definite systems [**Mie91**], p. 39) are still open problems.

2.2. Siegel disks. The following example is interesting because the geometry is reduced to a minimum and only the analytical difficulties remain. Not surprisingly, it was the first small divisors problem to be solved [**Sie42**], albeit with a technique very different from KAM. (Even if we will not discuss the original Siegel technique in these notes, we point out that, besides the original paper, there are more modern expositions and extensions. [**Brj71**],[**Pös86**].)

This problem is quite paradigmatic both for KAM theory and for the theory of holomorphic dynamics. In these lectures, we will discuss only the KAM aspects and not the holomorphic dynamics. A very good introduction to the problems connected with Siegel theorem is including both the KAM aspects and the holomorphic dynamics aspect is [**Her87**]. More up to date references are [**PM92**], [**Yoc95**]. The lectures [**Mar00**] contain a great deal of material on the Siegel problem.

We consider analytic maps $f : \mathbf{C} \to \mathbf{C}$, $f(z) = az + N(z)$ with $N(0) = 0$, $N'(0) = 0$, and we are interested in studying their dynamics near the origin.

When $|a| \neq 1$, it is easy to show that the dynamics, up to an analytic change of variables is that of az. More precisely, there exists an $h : U \subset \mathbf{C} \to \mathbf{C}$, $h(0) = 0$, $h'(0) = 1$ and

$$f \circ h = h(az) \tag{2.24}$$

in a neighborhood of the origin.

The proof for $|a| > 1$ can be easily obtained as follows (the case $0 < |a| < 1$ case follows by considering f^{-1} in place of f).

We seek a fixed point of $h \mapsto f \circ h \circ a^{-1}$ on a space of functions $h(z) = z + \Delta(z)$ with $\Delta(z) = O(z^2)$.

That is, we seek fixed points of the operator

$$\mathcal{T}(\Delta) = a\,\Delta \circ a^{-1} + N \circ (\mathrm{Id} + \Delta) \circ a^{-1}\ .$$

We note that, on a space of functions with $\|\Delta\|_r = \sup_{|z| \leq r} |\Delta(z)/z^2|$, the operator $\mathcal{T}$ is a contraction if r is sufficiently small. Note that then $\mathcal{T}^r(0)$ converges uniformly on a ball and the limit is analytic.

REMARK 2.8. Note that the previous argument works without any significant change when $f : \mathbf{C}^d \to \mathbf{C}^d$ and a is a matrix all eigenvalues of which have modulus less than 1. Indeed, a very similar result for flows already appears in Poincaré's thesis [**Poi78**], where it was established using the majorant method. (Remember that the concept of Banach spaces had not been yet formalized, so that fixed point proofs were unthinkable). The method in [**Poi78**] can be adapted without too much difficulty to cover the theorem started above. Hence, the situation when all the eigenvalues are smaller than one is sometimes called *the Poincaré domain.*

The situation that remains to be settled is that when $|a| = 1$.

REMARK 2.9. Building up on case for $|a| < 1$, there is a lovely proof by Yoccoz [**Her87**] using complex function theory that one can extend the conjugacies for $|a| < 1$ to a positive measure set with $|a| = 1$. Several elements of this proof can be used to obtain a very fast algorithm to compute the so called *Siegel radius.* (See the definition in Proposition 2.13.)

Another cute proof of a particular case of Siegel's theorem is in [**dlL83**] adapting a method of [**Her86**]. This method can be applied to a variety of one-dimensional problems.

The method of [**Sie42**] has been quite refined and extended in [**Brj71**], [**Brj72**]. We will not discuss the above proofs here, because, in contrast with KAM ideas that have a wide range of applications, they seem to be rather restricted.

It is typical of complex dynamics that there are very few possibilities for the dynamics. Either it is very unstable or it is a rigid rotation (up to a change of variables).

We will prove something more general.

LEMMA 2.10. *Let $f : \mathbf{C}^d \to \mathbf{C}^d$ be analytic in a neighborhood of the origin and*

$$f(0) = 0 \ , \qquad Df(0) = A \ ,$$

where A is a diagonal matrix with all the diagonal elements of unit modulus (hence $\|A^{-n}\| = 1 \ \forall n \in \mathbf{Z}$).

Assume that there is a domain U, $0 \in U$ and a constant $K > 0$ such that for all $n \in \mathbf{N}$

$$\sup_{z \in U} |f^n(z)| \le K \ . \tag{2.25}$$

Then there exists an analytic function $h : U \to \mathbf{C}^d$ such that $h(0) = 0$,

$$h'(0) = \mathrm{Id} \qquad , h \circ f = A \circ h. \tag{2.26}$$

Of course, by the implicit function theorem a solution of (2.26) implies that there is a solution of (4.4) (h in (4.4) is the inverse of h in (2.26)).

Note also that the assumption (2.25) implies, by Cauchy estimates that

$$|Df^n(0)| \le K' \ ,$$

hence, that all the eigenvalues are inside the closed unit circle and that the eigenvalues on the unit circle have trivial Jordan blocks. If rather than assuming (2.25) for $n \in \mathbf{N}$, we assumed it for $n \in \mathbf{Z}$, this would imply the assumption that A is diagonal and has the eigenvalues on the unit circle.

PROOF. Consider

$$h^{(n)}(z) = \frac{1}{n} \sum_{i=0}^{n} A^{-n} f^n(z) \ .$$

Note that, using the definition of A and (2.25) and we have:

$$h^{(n)}(0) = 0 \ , \quad h^{(n)\prime}(0) = 1 \ , \tag{2.27}$$

$$\sup_{z \in U} |h^{(n)}(z)| \le K \ , \tag{2.28}$$

$$h^{(n)} \circ f(z) = Ah^{(n)}(z) + (1/n)[A^{-(n+1)} f^{n+1}(z) - z] \ . \tag{2.29}$$

By (2.28), $h^{(n)}$ restricted to U is a normal family and we can find a subsequence converging uniformly on compact sets to a function h. Using (2.27), we obtain that $\tilde{h}(0) = 0$, $\tilde{h}'(0) = 1$.

Note also that, since $|f^n(z)|$ is bounded independently of n, by (2.25) and so is z for $z \in U$, we have that

$$\frac{1}{n}[A^{-(n+1)} f^{n+1}(z) - z]$$

converges to zero uniformly on any compact set contained in U as $n \to \infty$. Therefore, taking the limit $n \to \infty$ of (2.29), we obtain $h \circ f = A \circ h$. □

EXERCISE 2.11. Show that one can always assume that U is to be simply connected. (Somewhat imprecisely, but pictorially, if we are given are given a U with holes, we can always consider $\tilde{U}$ obtained by filling the holes of U. The maximum modulus principle shows that f^n is uniformly bounded in $\tilde{U}$.)

In one dimension, show that the Riemann mapping that sends U into the unit disk and 0 to itself should satisfy (2.26) except the normalization of the derivative.

PROPOSITION 2.12. *If the product of eigenvalues of A is not another eigenvalue, then the function $\tilde{h}$ satisfying* (2.26) *is unique even in the sense of formal power series.*

Note that, when $d = 1$ the condition of Proposition 2.12 reduces to the fact that A is not a root of unity. In particular, it is satisfied when the modulus of A is not equal to one. When the modulus equals to 1, the hypothesis of Proposition 2.12 reduces to a not being a root of unity, which is the same as $a = \exp(2\pi i\theta)$ with $\theta \in \mathbf{R} - \mathbf{Q}$.

PROOF. If we expand using the standard Taylor formula for multi-variable functions,

$$f(z) = \sum_{n=0}^{\infty} f_n z^{\otimes n}$$

(where f_n is a symmetric n-linear form taking values in $\mathbf{C}^d$) and seek a similar expansion for $\tilde{h}$, we notice that

$$A\tilde{h}_n - \tilde{h}_n A^{\otimes n} = S_n \ ,$$

where S_n is a polynomial expression involving only the coefficients of f and $\tilde{h}_1 = \mathrm{Id}, \dots, \tilde{h}_{n-1}$.

As it turns out, the spectrum of the operator $\mathcal{L}_A$ acting on n-multilinear forms by

$$\tilde{h}_n \mapsto A\tilde{h}_n - \tilde{h}_n A^{\otimes n} \tag{2.30}$$

is:

$$\mathrm{Spec}(\mathcal{L}_A) = a_i - a_{\sigma_1} \dots a_{\sigma_n}, \quad i \in \{1, \dots, d\}, \quad \sigma_1, \dots, \sigma_n \in \{1, \dots, d\}, \tag{2.31}$$

where a_i denotes the eigenvalues of A.

See, e.g., [**Nel69**] for a detailed computation which also leads to interesting algorithms. We just indicate that the result can be obtained very easily when the matrix is diagonalizable since one can construct a complete set of eigenvalues of (2.30) by taking products of eigenvalues of A. The set of diagonalizable matrices is dense on the space of matrices. Hence the desired identity between the spectrum of (2.30) and the set described in (2.31) holds in a dense set of matrices. We also note that the spectrum is continuous with respect to the linear operator.

□

When $d = 1$ and $|a| = 1$, as we mentioned before, the condition for Proposition 2.12 (usually referred to as *non-resonance condition*) reduces to:

$$a = e^{2\pi i\omega} \ , \qquad \omega \in \mathbf{R} - \mathbf{Q} \ .$$

We note that, even if $a(a^{n-1}-1) \neq 0$, it can be arbitrarily close to zero, because $e^{2\pi i\omega(n-1)}$ is dense in the unit circle. Hence, we also have small divisors in the computation of the $\tilde{h}_n$'s.

We note that when $d > 1$, we can have small divisors if there is some $|a_i| > 1$, $|a_j| < 1$ even if they are real. When all $|a_j| = 1$, $a_j = e^{2\pi i\omega_j}$, the non-resonance condition amounts to

$$\sum_j k_j\omega_j \neq \omega_i \ , \ \forall k_j \in \mathbf{N}, \sum_j k_j \geq 2 \tag{2.32}$$

We now investigate a few of the analyticity properties of h. Of course, the power series expansion converges in a disk (perhaps of zero radius) but we could worry about whether it is possible to perform analytic continuation and obtain h defined on a larger domain.

PROPOSITION 2.13. *If f is entire, the maximal domain of definition of h is invariant under A.*

In particular, when $d = 1$, $|a| = 1$, $a^n \neq 1$, the domain of convergence is a disk. (The radius of the disk of convergence of the function h such that $h'(0) = 1$ is called the Siegel radius.)

Moreover, when $d = 1$, $|a| \leq 1$, $a^n \neq 1$, the function h is univalent in the domain of convergence.

PROOF. To prove the first point, we just observe that if f is entire and h is analytic in the neighborhood of a point z_0, we can use the functional equation (2.24) to define the function h in a neighborhood of $A\,z_0$.

Hence, if h was defined in domain D and $z_0 \in D$ was connected to the origin by a path $\gamma \subset D$, we see that Az_0 is connected to the origin by $a\gamma \subset aD$. We conclude that it is defined in $AD \cup D$ and that the analytical continuation is unique. If we consider the maximal domain of definition $AD \cup D \subset D$. Hence $AD = D$.

The second statement follows by observing that the only domains invariant under an irrational rotation are disks.

To prove univalence, we assume that if $h(z_1) = h(z_2)$ and one of them – say z_2 – different from 0. We want to conclude that $z_1 = z_2$.

Using (2.24), we obtain $h(az_1) = h(az_2)$. Repeating the process, $h(a^n z_1) = h(a^n z_2)$.

Hence, when $z \in \{a^n z_2\}$, we have

$$h(z) = h(z\alpha) \tag{2.33}$$

with $\alpha = z_1/z_2$. Since the set where (2.33) holds has an accumulation point: when $|a| < 1$, it accumulates to 0, when $|a| = 1$ since it is an irrational rotation, the orbits are dense on circles), we conclude that it holds all over the unit disk. Taking derivatives at $z = 0$, using $h'(0) = 1$, we obtain $\alpha = 1$. □

EXERCISE 2.14. Show that the conclusions of Proposition 2.13 remain true if we consider $d > 1$ and A a diagonalizable matrix with all eigenvalues in the unit disc and satisfying (2.32). Namely

- i') The domain of definition is a polydisk.
- ii') The function is univalent in its domain of definition.

EXERCISE 2.15. Once we know that the domain of the function h in (2.24) is a disk, the question is to obtain estimates of the radius.

Lower bounds are obtained from KAM theory.

Obtain upper bounds also using the fact that by the Bieberbach-De Branges theorem, the Taylor coefficients of a univalent function satisfy upper bounds that depend on the radius of the disk. On the other hand, we know the coefficients explicitly.

Also obtain upper bounds when $f(z) = az + z^2$ using the area formula for univalent functions $\text{Area}\, h(B_r(0)) = \pi \sum_{i=1}^{\infty} |h_i|^2 r^{2i-2}$ knowing that the range of h – orbits that are bounded – cannot include any point outside of the disk of radius 2 and that we know the coefficients h_k.

This exercise is carried out in great detail in [**Ran87**], which established upper and lower bounds of the radius for rotation by the golden mean.

It turns out to be very easy to produce examples where the series diverges. We will discuss what we think is oldest one [**Cre28**] (reproduced in [**Bla84**]). Other examples of [**Cre38**] can be found in [**SM95**] Chapter 25 in a more modern form. A different line of argument appears in [**Ily79**], using more complex analysis. This argument has been recently extended considerably [**PM00**].

Consider $f(z) = az + z^2$ with $a = e^{2\pi i\omega}$, then its nth iteration is

$$f^n(z) = a^n z + \cdots + z^{2^n} .$$

If we seek fixed points of f^n, different from zero, they satisfy $(a^n - 1) + \cdots + z^{2^n - 1} = 0$. The product of the $2^n - 1$ roots of this equation is $a^n - 1$. Hence, there is at least one root with modulus smaller or equal to $|a^n - 1|^{1/(2^n-1)}$. It is possible to find numbers $\omega \in \mathbf{R} - \mathbf{Q}$ such that

$$\liminf_{n\to\infty} [\text{dist}(n\omega, \mathbf{N})]^{1/(2^n-1)} = 0 .$$

Hence, the f above has periodic orbits different from zero in any neighborhood of the origin. This is a contradiction with f being conjugate to an irrational rotation in any neighborhood of the origin. This shows that the perturbation expansions may diverge if the rotations are very well approximated by rational numbers.

For complex polynomials in one variable it has been shown [**Yoc95**], (see also [**PM92**]) that if ω does not satisfy the Brjuno conditions (2.34) below, the series for the quadratic polynomial diverges. The Theorem 4.1 which we will prove later will establish that if the condition is met, then the series for all the non-linearities converges.

We say that ω satisfies a Brjuno condition when there exists an Ω increasing and log convex (the later properties are just for convenience and can always be adjusted) such that

$$\Omega(n) \geq \sup_{k\leq n} |a^k - 1|^{-1}$$

$$\sum_n \frac{\log \Omega(2^n)}{2^n} < \infty \iff \sum_n \frac{\log \Omega(n)}{n^2} < \infty \tag{2.34}$$

The equivalence of the two forms of the condition is very easy from Cauchy test for the convergence of series. An example of functions $\Omega(n)$ satisfying (2.34) is:

$$\Omega(n) = \exp(An/(\log(n) \log\log(n) \cdots [\log^k(n)]^{1+\epsilon}))$$

for large enough n, where by $\log^k$ we denote the function log applied k times.

Indeed, [**Yoc95**] shows that if ω fails to satisfy the condition (2.34) then $f(z) = e^{2\pi i\omega} z + z^2$ is not linearizable in any neighborhood of the origin.

REMARK 2.16. In [**Yoc95**] one can find the result that if, a function $f(z)$ with $f(0) = 0$, $f'(0) = a$, with $|a| = 1$ is not linearizable, near 0, then, the quadratic function $az + z^2$ is not linearizable.

See also [**PM00**].

In the case of one dimensional variables, one can use the powerful theory of continued fractions to express the Brjuno condition in an equivalent form.

If $\omega \in \mathbf{R} - \mathbf{Q}$ can be written $\omega = [a_0, a_1, a_2, \cdots, a_n, \cdots]$ with $a_i \in \mathbf{N}^+$, we call $[a_0, a_1, \cdots, a_n] = p_n/q_n$ the convergents.

Brjuno condition is equivalent to

$$\mathcal{B}(\omega) \equiv \sum_n (\log q_{n+1})/q_n \leq \infty \tag{2.35}$$

A very similar condition

$$\sum_n (\log \log q_{n+1})/q_n \leq \infty \tag{2.36}$$

has been found in [**PM91**] [**PM93**] to be necessary and sufficient for the existence of the Cremer's phenomenon of accumulation of periodic orbits near the origin in the sense that if condition (2.36) is satisfied, then, all non-linearizable functions have a sequence of periodic orbits accumulating at the origin. If condition (2.36) is not satisfied, there exists a non-linearizable germ with no periodic orbits other than zero in a neighborhood of zero.

REMARK 2.17. We note that the formula (2.35) has very interesting covariance properties under modular transformations. They have been used quite successfully in [**MMY97**].

Without entering in many details, we point out that another function very closely related to the one we have defined satisfies (setting $\tilde{\mathcal{B}}(x) = +\infty$ when $x \in \mathbf{Q}$)

$$\begin{aligned} \tilde{\mathcal{B}}(\omega) &= -\log(x) + x\tilde{\mathcal{B}}(1/x) \quad x \in (0, 1/2) \\ \tilde{\mathcal{B}}(\omega)(-x) &= \tilde{\mathcal{B}}(x) \quad x \in (-1/2, 0) \\ \tilde{\mathcal{B}}(\omega)(x+1) &= \tilde{\mathcal{B}}(x) \end{aligned}$$

Similar invariance properties are true for the sum appearing in (2.36). Nevertheless, it does not seem to have been investigated as extensively.

Unfortunately, this one dimensional theory does not have analogues in higher dimensions. Some preliminary numerical explorations for the higher dimensional case were done in [**Tom96**].

REMARK 2.18. There is a very similar theory of changes of variables that reduce the problem to linear – or some canonical – form for differential equations.

Of course, these normalizations resemble the normalizations of singularity theory and are basic for many applied questions such as *bifurcation theory*.

Similarly, there is a theory of these questions in the C^∞ or C^r categories under assumptions, which typically include that there are no eigenvalues of unit length. This theory usually goes under the name of Sternberg theory.

The reduction of maps and differential equations to *normal form* by means of changes of variables can also be done when the map is required to preserve a symplectic – or another geometric – structure and one requires that the change of variables preserve the same structure.

We will not discuss much of these interesting theories. For more information on many of these topics we refer to [**Bru89**], [**Bib79**].

3. Preliminaries

In this Section, we will collect some background in analysis, number theory and (symplectic and volume preserving) geometry. Experts will presumably be familiar with most of the material and will only need to read this as it is referenced in the following text. Of course, this chapter is not a substitute for manuals in geometry or on analysis. I have found [**Thi97**], [**AM78**], [**GP74**] useful for geometry and [**Ste70**] [**Kat76**] useful for analysis. Many of the techniques are discussed in other papers in KAM theory which we will mention as we proceed. Specially the papers [**Mos66b**], [**Mos66a**] contain an excellent background in many of the analytical techniques.

In the previous discussion of Lindstedt series we saw that we had to consider repeatedly equations of the form

$$L_\omega \varphi = \eta.$$

(The formal solution was given in (2.14).)

In this Section, we will also study equations

$$D_\omega \varphi = \eta, \quad \text{where} \quad D_\omega = \sum_i \omega_i \frac{\partial}{\partial \theta_i} ,$$

which also appears in KAM theory.

A first step towards obtaining proofs of the KAM theorem is to devise a theory of these equations. That is, find conditions in ω and η so that the function defined by (2.14) has a precise meaning.

The guiding heuristic principles are very simple:

1) The smoother the function η, the faster its Fourier coefficients $\hat{\eta}_k$ decay.
2) Some numbers ω are such that the denominators appearing in the solution (2.14) do not grow very fast with k.
3) Hence, for the numbers alluded to in 2), we will be able to make sense of the formal solutions (2.14) when the function considered is smooth.

We devote Sections 3.1, 3.3, 3.4 to making precise the points above. We will need to discuss number theoretic properties (usually called Diophantine properties) that quantify how small the denominators can be as a function of k. We will also need to study characterizations of regularity in terms of Fourier coefficients.

Since the result in KAM theory depends on the geometric properties of the map – as illustrated in (2.7) and (2.8) – it is clear that we will need to understand which geometric properties enter in the conclusions. Moreover, many of the traditional proofs indeed use a geometric formalism. Hence, we have devoted a Section 3.5 to collect the facts we will need from differential geometry.

3.1. Preliminaries in analysis. In modern analysis, it is customary to measure the regularity of a function by saying that it belongs to some space in a certain scale of spaces. Some scales that are widely used on compact manifolds are:

$$
\begin{aligned}
C^r &\equiv \left\{\eta \,\middle|\, D^r\eta \text{ is continuous, } \|\eta\|_{C^r} \equiv \max\left(\sup_x |\eta(x)|, \dots, \right.\right.\\
&\qquad \left.\left. \sup_x |D^r\eta(x)|\right)\right\}, \\
C^{r+\alpha} &\equiv \left\{\eta \,\middle|\, \|\eta\|_{C^r} \equiv \max\left(\sup_x |\eta(x)|, \dots, \sup_x |D^r\eta(x)|, \right.\right.\\
&\qquad \left.\left. \sup_{x\neq y} \frac{|D^r\eta(x) - D^r\eta(y)|}{|x-y|^\alpha}\right)\right\}, \\
A_\delta &\equiv \left\{\eta \,\middle|\, \eta \text{ analytic on } |\operatorname{Im}\theta| < \delta, \text{ continuous on } |\operatorname{Im}\theta| \le \delta, \right.\\
&\qquad \left. \|\eta\|_\delta \equiv \sup_{|\operatorname{Im}\theta|\le\delta} |\eta(\theta)|\right\}, \\
H^s &\equiv \left\{\eta \,\middle|\, \eta \in L^2,\ (-\Delta+1)^{s/2}\eta \in L^2,\ \|\eta\|_{H^s} \equiv \|(-\Delta+1)^{s/2}\eta\|_{L^2}\right\},
\end{aligned}
$$

where $r = 0, 1, 2, \dots$, $\alpha \in [0,1]$, $\delta \in \mathbf{R}^+$, $s \in \mathbf{R}$ and we have used Δ to denote the Laplacian

Note that this notation (even if in wide usage) has certain ugly points. C^{r+0} and C^{r+1} are ambiguous and can be considered according to the first or the second definition. Indeed, C^{r+0} consider according to the two definitions agrees as a space (that is, the functions in one are functions in the other and the topologies are the same), but the norms differ (they are equivalent). On the other hand, C^{r+1} differs when the we interpret it in the first or in the second sense. To avoid that, we will use $C^{r+\mathrm{Lip}}$ to identify the second definition.

All the above scales of spaces have advantages and disadvantages. Against $C^{r+\alpha}$ we note that, even if for $r = 0$, these are the Hölder spaces which can be defined in great generality (e.g. metric spaces), when $r \ge 1$, the definition needs to be done in a differentiable system of coordinates. This is because, for $r \ge 1$, $D^r\eta(x)$ and $D^r\eta(y)$ are multilinear operators in T_xM and T_yM, so that the differences in the definition are comparing operators in different spaces. Even though the different choices of coordinates lead to equivalent norms, some of the geometric considerations are somehow cumbersome. Also the composition operator — ubiquitous in KAM theory — has properties which are cumbersome to trace in $C^{r+\alpha}$. For example, the mapping $x \to f(x+\cdot)$ can be discontinuous in $C^{r+\alpha}$ when f is $C^{r+\alpha}$.

It is somewhat unfortunate that the notations C^r ($r \in \mathbf{N}$) and $C^{r+\alpha}$, ($r \in \mathbf{N}, \alpha \in [0,1) \cup \{\mathrm{Lip}\}$ suggest that one can consider perhaps C^s ($s \in \mathbf{R}$) which includes both. If one proceeds in this way, one obtains very bad properties for the scale of spaces. In colorful words, "the limit of the space $C^{k+\alpha}$ as $\alpha \to 0$ is not C^k". More precisely, several important inequalities such as interpolation inequalities which relate the different norms in a scale fail to hold. Many characterizations – e.g. in terms of approximations by analytic functions – break down for the case that r is an integer.

A possible way of breaking up the unfortunate C^r vs. $C^{r+\alpha}$ notation is to introduce the spaces called Λ_r in [**Ste70**], or $\hat{C}^r$ in [**Zeh75**], [**Mos66b**], [**Mos66a**].

In general we define:

$$
\begin{aligned}
&\Lambda_0 = C^0 \\
&\Lambda_1 = \Big\{ f \Bigm| \sup_{\substack{1>|h|>0 \\ x\in\mathbf{R}}} \frac{|f(x+h)+f(x-h)-2f(x)|}{h} \equiv \|f\|_{\Lambda_1} < \infty \Big\} \\
&\Lambda_r = \{ f \mid Df \in \Lambda_{r-1},\ \|f\|_{\Lambda_r} = \max(\|f\|_{C^0}, \|Df\|_{\Lambda_{r-1}}) \Big\} \quad r \in \mathbf{N} \\
&\Lambda_{r+\alpha} = C^{r+\alpha} \qquad r+\alpha \notin \mathbf{N}
\end{aligned} \tag{3.1}
$$

Here $[r]$ means the integer part of r and $\{r\}$ means the fractional part of r.

There are many reasons why the Λ_α spaces are the natural scale of spaces to consider when one is considering a space that includes the usual $C^{r+\alpha}$. For example, one can obtain very nice approximation theory, interpolation inequalities, and generalize naturally to several variables. Note that

$$C^1 \subsetneq C^{0+\mathrm{Lip}} \subsetneq \Lambda_1 \ .$$

Again, we point out that it is not easy to define these spaces on manifolds except through patches. Choosing different patches leads to different norms. Fortunately, all of them are equivalent and, hence define the same topology in the spaces.

Note that C^r norms can be defined naturally on any smooth Riemannian manifold. (The norm of derivatives can be defined since it is the norm of multilinear operators in the tangent bundle.)

The main inconvenience of C^r (r is, by assumption, an integer) is that the characterization by Fourier series is rather cumbersome. It is easy to show integrating by parts that

$$
\begin{aligned}
\hat{\eta}_k &= \int_0^1 \eta(\theta) e^{-2\pi i k\theta}\, d\theta \\
&= (-2\pi i)^{-r} k^{-r} \int_0^1 D^r \eta(\theta) e^{-2\pi i k\theta}\, d\theta \ .
\end{aligned}
$$

Hence, if $\eta \in C^r$, we have

$$\sup_k \big(|\hat{\eta}_k|\, |k|^r \big) \le C_r \|\eta\|_{C^r} \ . \tag{3.2}$$

where C_r is a constant that depends only on r.

In the other direction, we have for any $\delta > 0$

$$
\begin{aligned}
\|\eta\|_{C^r} &\le \tilde{C}_r \sum_k |\hat{\eta}_k|\, |k|^r = \tilde{C}_r \sum_k \frac{1}{|k|^{1+\delta}} |\hat{\eta}_k|\, |k|^{r+1+\delta} \\
&\le \tilde{C}_r \left(\sum_k \frac{1}{|k|^{1+\delta}} \right) \sup_k \big(|\hat{\eta}_k|\, |k|^{r+1+\delta} \big) \\
&\le \tilde{\tilde{C}}_{r,\delta} \sup_k \big(|\hat{\eta}_k|\, |k|^{r+1+\delta} \big) \ .
\end{aligned} \tag{3.3}
$$

Both inequalities (3.2), (3.3) are essentially optimal in the following sense. Inequality (3.2) is saturated by trigonometric polynomials, while the usual square wave — or iterated integrals of it — shows that it is impossible to reduce the exponent on the right hand side of (3.3) to $r+1$. This discrepancy is worse when we consider functions on $\mathbf{T}^d$, $d > 1$. In that case, to obtain convergence of the series, in (3.3) one needs to take $\delta > d$. This shows that studying regularity in terms of just the size of the coefficients will lead to less than optimal results.

EXERCISE 3.1. Show that given any sequence a_n of positive numbers converging to zero, the set of continuous functions f with $\limsup_k |\hat{f}_k|/a_k = \infty$ is residual in C^0.

The spaces of analytic functions A_δ are better behaved in respect of characterizations of norms of the function in terms of its Fourier coefficients. Integrating over an appropriate contour, we have Cauchy inequality

$$|\hat{\eta}_k| \le e^{-2\pi\delta|k|} \|\eta\|_{A_\delta} \,. \tag{3.4}$$

On the other hand,

$$\begin{aligned} \|\eta\|_{A_{\delta-\sigma}} &\le \sum_{k\in\mathbf{Z}} e^{2\pi(\delta-\sigma)|k|} |\hat{\eta}_k| \\ &\le \left(\sum_{k\in\mathbf{Z}} e^{-2\pi\sigma|k|}\right) \sup_k e^{2\pi\delta|k|}|\hat{\eta}_k| \\ &\le C\sigma^{-1} \sup_k e^{2\pi\delta|k|}|\hat{\eta}_k| \end{aligned} \tag{3.5}$$

Of course, for Sobolev spaces, the characterization in terms of Fourier coefficients is extremely clean:

$$\|\eta\|_{H^s} = \left(\sum_{k\in\mathbf{Z}} (|k|^2+1)^s |\hat{\eta}_k|^2\right)^{1/2} .$$

Sobolev spaces have other advantages. For example, they are very well suited for numerical work and they also work nicely with partial differential operators. Many of the tools that we used in Λ_α spaces also carry through to Sobolev spaces.

For example, we have the interpolation inequalities:

$$\|u\|_{H^j} \le K \|u\|_{H^m}^{j/m} \|u\|_{H^0}^{1-j/m} \,. \tag{3.6}$$

This inequality is a particular case of the following Nirenberg inequality

$$\|D^i u\|_{L^r(\mathbf{R}^n)} \le C \|u\|_{L^p(\mathbf{R}^n)}^{1-i/m} \cdot \|D^m u\|_{L^q(\mathbf{R}^n)}^{i/m} \,, \tag{3.7}$$

where $1/r = (1-i/m)(1/p) + (i/m)(1/q)$. We refer to [**Ada75**], p. 79.

These interpolation inequalities both for Λ_α and for Sobolev spaces are part of the more general "complex interpolation method" and the scales of spaces are "interpolation spaces". Even if this is quite important for certain problems of analysis in these spaces, we will not go into these matters here.

As we will see later, some of the abstract versions of KAM as an implicit function theorem work perfectly well for Sobolev spaces. I think it is mainly a historical anomaly that these spaces are not used more frequently in the KAM theory of dynamical systems. (Notable exceptions are [**Her86**], [**KO89b**].) Of course, for the applications of Nash-Moser theory to PDE's or geometric problems, Sobolev spaces are used quite often.

One of the most useful tools in the study of $C^{r+\alpha}$ spaces is that they can be characterized by their approximation properties by analytic functions.

The following characterization of Λ_r spaces (remember that they agree with the Hölder spaces $C^{[r]+\{r\}}$ when $\{r\} \neq 0$) comes from [**Mos66b, Mos66a**] (see also [**Zeh75**], Lemma 2.2).

LEMMA 3.2. *Let $h \in C^0(\mathbf{T}^1)$. Then $h \in \Lambda_r$ if and only if for some $\sigma > 0$ we can find a sequence $h_i \in A_{\sigma 2^{-i}}$ such that*

i) $\|h_i - h_0\|_{C^0} \to 0$

ii) $\sup_{i \geq 1}(2^{ir}\|h_i - h_{i-1}\|_{A_{\sigma 2^{i-1}}}) < \infty$

Moreover, it is possible to arrange that the sup in ii) is equivalent to $\|h\|_{\Lambda_r}$ if one chooses the h_i appropriately.

If we denote the sup in ii) by M, we have that for $h \in \Lambda_r$ it is possible to find a sequence h_i in such a way that $M \leq C_{\sigma,r}\|h\|_{\Lambda_r}$. Conversely, for any sequence h_i as above, we have $\|h\|_{\Lambda_r} \leq \tilde{C}_{\sigma,r}M$. Given a function $h \in \Lambda_r$ there are canonical ways of producing the desired h_j. For example, in [**Ste70**] and [**Kra83**] is shown that one can use convolution with the Poisson kernel to produce the h_j. In that case, the sup in ii) can be taken to define a norm equivalent to $\| \ \|_{\Lambda_r}$.

Another important feature of the Λ_α spaces is that they admit a very efficient approximation theory.

The first naive idea that occurs to one when trying to approximate a function by a smoother one is just to expand in Fourier series and to keep only a finite number of terms corresponding to the harmonics of small degree. Indeed, for some methods of proof of the KAM theorem that emphasize geometry this is the method of choice. (See Section 6.2.) Unfortunately, keeping only a finite number of the low order Fourier terms is a much less efficient method of approximation (from the point of view of the number of derivatives required) than convolving with a smooth kernel. Recall that summing a Fourier series is just convolution with the Dirichlet kernel,

$$\begin{aligned} \sum_{k=-N}^{N} \hat{\eta}_k e^{2\pi i k\theta} &= \int_0^1 \eta(\varphi)\mathcal{D}_N(\theta - \varphi)\, d\varphi = (\eta * D_N)(\theta) \\ \mathcal{D}_N(\theta) &= \frac{\sin(2N+1)\pi\theta}{\sin \pi\theta}, \end{aligned}$$

which is large and oscillatory and hence generates more oscillations upon convolution than smooth kernels.

Hence the method of choice of approximating functions by smoother ones is to choose an positive analytic function $K : \mathbf{R}^d \to \mathbf{R}$ decaying at infinity rather fast and with integral 1 and define $K_t(x) \equiv \frac{1}{t^d}K(x/t)$.

We define smoothing operators S_t by convoluting with the kernels K_t. That is:

$$S_t\phi = K_t * \phi.$$

The properties of these smoothing operators that are useful in KAM theory are (we express them in terms of the Λ_r spaces introduced in (3.1)):

$$\begin{aligned} &\text{i)} && \lim\nolimits_{t\to\infty} \|S_t u - u\|_{\Lambda_0} = 0\,, \quad u \in \Lambda_0 \\ (3.8) \qquad &\text{ii)} && \|S_t u\|_{\Lambda_\mu} \leq t^{\mu-\lambda}C_{\lambda\mu}\|u\|_{\Lambda_\lambda}\,, \quad u \in \Lambda_\lambda,\ 0 \leq \lambda \leq \mu \\ &\text{iii)} && \|(S_t - 1)u\|_{\Lambda_\lambda} \leq t^{-(\mu-\lambda)}C_{\lambda\mu}\|u\|_{\Lambda_\mu}\,, \quad x \in \Lambda_\mu\,, 0 \leq \lambda \leq \mu \end{aligned}$$

We note that a slightly weaker version of these properties is:

$$\begin{aligned} (3.9) \qquad &\text{ii}') && \|S_t u\|_{\Lambda_{t^{-1}}} \leq k(\ell)\|u\|_{\Lambda_\ell} \qquad t \geq 0 \\ &\text{iii}') && \|(S_\tau - S_t)u\|_{\Lambda_{\tau^{-1}}} \leq t^{-\ell}k(\ell)\|u\|_{\Lambda_\ell} \qquad u \in \Lambda_\ell \quad \tau \geq t \geq 1 \end{aligned}$$

Note that it is easy to show that ii) $\Rightarrow$ ii′), iii) $\Rightarrow$ iii′). In [**Zeh75**] operators S_t satisfying (3.8) are said to constitute a C^∞ smoothing and those satisfying i), ii′), iii′) a C^ω smoothing.

There are other smoothing operators and other scales of spaces that satisfies the same inequalities. Indeed, in the most abstract version of KAM theory, which we discuss in Section 5, one can even abstract these properties and obtain a general proof which also applies to many other situations.

One important consequence of the existence of smoothing operators is the existence of interpolation inequalities (see [**Zeh75**]). Even if this inequality were proved directly long time ago, and can be obtained by different methods, it is interesting to note that they are a consequence of the existence of smoothing operators. As we mentioned, this happens in other situations and for other spaces than Λ_r. In the following, we denote $\|u\|_r \equiv \|u\|_{\Lambda_r}$.

LEMMA 3.3. *For any* $0 \le \lambda \le \mu$, $0 \le \alpha \le 1$, *denoting*

$$\nu = (1-\alpha)\lambda + \alpha\mu$$

we have for any $u \in \Lambda_\mu$:

$$\|u\|_\nu \le C_{\alpha,\lambda,\mu} \|u\|_\lambda^{1-\alpha}\, \|u\|_\mu^\alpha \tag{3.10}$$

PROOF. We clearly have:

$$\|u\|_\nu \le \|S_t u\|_\nu + \|(\mathrm{Id} - S_t)u\|_\nu \ .$$

Applying *ii*) of (3.8) to the first term and *iii*) to the second, we obtain:

$$\|u\|_\nu \le t^{\nu-\lambda} C_{\alpha\lambda,\mu} \|u\|_\lambda + t^{-(\mu-\nu)} C_{\alpha\mu,\nu} \|u\|_\mu$$

and we obtain (3.10) by optimizing the right hand side in t.

□

These inequalities are descendents of inequalities for derivatives of functions which were proved, in different versions, by Hadamard and Kolmogorov and others. For Λ_r, $r \notin \mathbf{N}$ and for C^r, $r \in \mathbf{N}$, the proofs can be done by elementary methods and extend even to functions defined in Banach spaces [**dlLO99**]. For analytic spaces, these interpolation inequalities are classical in complex analysis and are a consequence of the fact that the $\log |f(z)|$ is sub-harmonic when $f(z)$ is analytic [**Rud87**].

In KAM theory the interpolation inequalities (3.10) are useful because if we have a smooth norm ($\| \ \|_\mu$) blowing up and a not so smooth one ($\| \ \|_\lambda$) going to zero, we can still get that other norms smoother than λ still converge.

All the above results about Λ_α spaces of functions on the real line can be generalized to spaces of functions on $\mathbf{R}^n$. Indeed, one of the nicest things of these spaces is that the theory for them can be reduced to the study of one dimensional restrictions of the function. We refer to [**Ste70, Kra83**] for more details.

For analytic spaces, the theory can be also extended with minor modifications. In KAM theory we often have to consider functions defined in $\mathbf{T}^m \times \mathbf{R}^n$ (often $n = m = d$). In such a case, it is very convenient to use expansions which are Taylor expansions in the real variables and Fourier expansions in the angle:

$$f(\theta, I) = \sum_{j \in \mathbf{N}^n, k \in \mathbf{Z}^m} f_{j,k} I^j \exp(2\pi i k \cdot \theta) \tag{3.11}$$

For these functions, it is convenient to define norms

$$\|f\|_\sigma = \sup_{|I| \le e^{2\pi\sigma}, |\operatorname{Im}(\theta)| \le \sigma} |f(\theta, I)| \tag{3.12}$$

With this definition, we have the Cauchy bounds

$$\begin{aligned} |f_{j,k}| &\le \exp(-2\pi\delta(|j| + |k|))\|f\|_\sigma \\ \left\| \frac{\partial^{|r|+|s|}}{\partial I^r \partial \theta^s} f \right\|_{\sigma-\delta} &\le C_{r,s,n,m} \delta^{-|r|-|s|} \|f\|_\sigma \end{aligned} \tag{3.13}$$

The proof of these inequalities is quite standard in complex analysis and will not be given in detail here. It suffices to express the derivatives as integrals over an $n+m$ dimensional torus which is close to the boundary of the domain in which $f(\theta, I)$ is controlled by $\|f\|_\sigma$. The only subtlety is that for some $l \in \{1, \ldots, m\}$, $k_l > 0$ one needs to choose the torus $\operatorname{Im}(\theta_l) = \ \sigma$. (Similarly for the case when $k_l < 0$ one needs to choose the torus $\operatorname{Im}(\theta_l) = \sigma$.)

It is also obvious that, under these supremum norm the spaces constitute a Banach algebra, that is:

$$\|fg\|_\sigma \le \|f\|_\sigma \|g\|_\sigma \ . \tag{3.14}$$

Therefore, if $\|f\|_\sigma < 1$, then $\|(1+f)^{-1}\|_\sigma \le (1 - \|f\|_\sigma)^{-1}$.

3.2. Regularity of functions defined in closed sets. The Whitney extension theorem. In KAM theory, we often have to study functions defined in Cantor sets. In particular, sets with empty interior. In this situation, the concept of Whitney differentiability plays an important role.

A reasonable notion of smooth functions in closed sets is that they are the restriction of smooth functions in open sets that contain them. This definition is somewhat unsatisfactory since the extension is not unique.

In the paper [**Whi34a**], one can find an intrinsic characterization of smooth functions in a closed set.

DEFINITION 3.4. We say that a function f is C^k in the sense of Whitney in a compact set $F \subset \mathbf{R}^d$ when for every point $x \in F$ we can find polynomials P_x of degree less that k such that

$$\begin{aligned} &f(x) = P_x(x) \quad x \in F \\ &|D^i P_x(y) - D^i P_x(x)| \le |x-y|^{r-i} \sigma(|x-y|) \quad x, y \in F \end{aligned} \tag{3.15}$$

where σ is a function that tends to zero.

It is clear that if a function is the restriction of a C^k function the Taylor polynomials will do.

The deep theorem of [**Whi34a**] is that the converse is true. That is,

THEOREM 3.5. *Let $F \subset \mathbf{R}^d$ be a compact set.*

If for a function f we can find polynomials satisfying (3.15) *and such that $f(x) = P_x(x)$ then the function f can be extended to an a C^r function in $\mathbf{R}^d$.*

Note that if a function is C^r in $\mathbf{R}^d$, then one can find polynomials satisfying (3.15) by taking just the Taylor expansions of f.

Contrary with what happened with the ordinary derivatives, the polynomials satisfying (3.15) may not be unique. (For example, if we take F to be the x-axis in $\mathbf{R}^2$, we can take polynomials with a a very different behavior in the y direction.)

There are other variants of the definitions in which rather than using $D^i P_x$ one introduces another polynomial P_x^i which is then, required to satisfy compatibility conditions with the other polynomials.

Another variant useful for KAM theory appears in [**dlLV00**]. It roughly states that, for Cantor sets with a certain geometric structure, one just needs to verify (3.15) for $i = 0$.

The assumption that F is compact can be removed. It suffices to require (3.15) in each compact subset of F, allowing σ to depend on the compact subset.

In [**Ste70**] one can find a version of this theorem in which the extensions can be implemented via a linear extension operator. (There is a different extension operator $\mathcal{E}_k$ for each k.) In [**Ste70**], one can also find versions for $C^{k+\alpha}$. The C^∞ version can be found in [**Whi34b**].

Even if adapting Whitney's theorem from real valued function to functions taking values in a Banach space is well known, (e.g [**Fed69**] p. 225 ff.) I do not know how to prove a similar result when F lies on an infinite dimensional space.

3.3. Diophantine properties. In this Section, we want to study the existence of vectors $\omega \in \mathbf{R}^n$ so that we can obtain upper bounds of $[\mathrm{dist}(\omega \cdot k, \mathbf{N})]^{-1}$ and of $|\omega \cdot k|^{-1}$ when $k \in \mathbf{Z}^n - \{0\}$. These are the small divisors that appear respectively in the solution of the equations (3.27), (3.26), which appear often in KAM theory.

When we are studying problems such as those in Section 2.2, we need only to consider $k \in \mathbf{N}^n$.

When $n = 1$ for (3.27) (and for $n = 2$ for (3.26)) one can get quite good results using classical tools of number theory, notably continued fractions, which we will not review here, in spite of their importance in 1-dimensional dynamics.

For example, the classical result of Liouville states

THEOREM 3.6. *Let $\omega \in \mathbf{R} - \mathbf{Q}$ satisfy $P(\omega) = 0$ with P a polynomial of degree ℓ with integer coefficients. Assume that $P'(\omega) = 0, \dots, P^{(j}(\omega) = 0$, $P^{(j+1}(\omega) \neq 0$. Then for some $C > 0$*

$$\left|\omega - \frac{m}{n}\right| \geq C n^{-\ell/(j+1)} \qquad \forall\, m, n \in \mathbf{Z}\,. \tag{3.16}$$

PROOF. The zeroes of polynomials are isolated, hence $P(\frac{m}{n}) \neq 0$ when $\frac{m}{n}$ is close enough to ω. This together with the fact that $n^\ell P(\frac{m}{n}) \in \mathbf{Z}$ implies that $|n^\ell P(\frac{m}{n})| \geq 1$ and, therefore, $|P(\frac{m}{n}) - P(\omega)| \geq n^{-\ell}$. On the other hand, by the Taylor's theorem,

$$\left|P\Big(\frac{m}{n}\Big) - P(\omega)\right| \leq C\left|\omega - \frac{m}{n}\right|^{j+1}$$

for some $C > 0$. (The R.H.S. is the remainder of Taylor's theorem.) This yields the desired result for $\frac{m}{n}$ close to ω. For $\frac{m}{n}$ far from ω, the result is obvious. □

Theorem 3.6 was significantly improved by Roth, who showed that, if ω is an algebraic irrational, $|\omega - \frac{m}{n}| \geq C_\varepsilon n^{-2-\varepsilon}$ for every $\varepsilon > 0$.

The numbers that satisfy the equation (3.16) in the conclusions of Theorem 3.6 are quite important in number theory and in KAM theory and are called Diophantine. As we will see in Lemma 3.9, Diophantine numbers occupy positive measure, hence, there are some of them which do not satisfy the hypothesis of Theorem 3.6.

DEFINITION 3.7. A number ω is called *Diophantine* of type (K, ν) for $K > 0$ and $\nu \geq 1$, if

$$\left|\omega - \frac{p}{q}\right| > K\, |q|^{-1-\nu} \tag{3.17}$$

for all $\frac{p}{q} \in \mathbf{Q}$. We will denote by $\mathcal{D}_{K,\nu}$ the set of numbers that satisfy (3.17). We denote by $\mathcal{D}_\nu = \cup_{K>0}\mathcal{D}_{K,\nu}$.

A number which is not Diophantine is called a *Liouville number.*

The numbers ω for which $|\omega - \frac{m}{n}| \geq Cn^{-2}$ are called "constant type" and the previous result shows that quadratic irrationals are constant type. It is an open problem to decide whether $\sqrt[3]{2}$ is constant type or not. Indeed, it would be quite interesting to produce any non-quadratic algebraic number which is of constant type.

In higher dimensions, there are two types of Diophantine conditions that appear in KAM theory, namely:

$$|\omega \cdot k|^{-1} \leq C|k|^\nu \quad \forall k \in \mathbf{Z}^n - \{0\} \tag{3.18}$$

$$|\omega \cdot k - \ell|^{-1} \leq C|k|^\nu \quad \forall (k, \ell) \in \mathbf{Z}^n \times \mathbf{Z} - \{(0,0)\} \tag{3.19}$$

The first condition (3.18) appears when we consider the KAM theory for flows, the second one (3.19) when we consider KAM theory for maps. As we will see the arguments are very similar in both cases.

REMARK 3.8. One important difference between these Diophantine conditions is that the first condition (3.18) is maintained – with only different constants – if the vector ω is multiplied by a constant. Nevertheless, the second one is not. Indeed, if we take advantage of this to set one of the coordinates to 1, then, we see that (3.18) becomes (3.19) for the vector in one dimension less obtained by keeping the coordinates not set to 1.

The arguments that study geometry of these Diophantine conditions are identical. Nevertheless, we point out that the scale invariance of (3.18) will have some consequences later, namely that KAM tori for flows often appear in smooth one-dimensional families, whereas those for maps are isolated.

For us, the most important result is

LEMMA 3.9. *Let $\Omega : \mathbf{R} \to \mathbf{R}$ be an increasing function satisfying*

$$\sum_{r=1}^{\infty} \Omega(r)^{-1} r^{n-1} < a(n) \tag{3.20}$$

where $a(n)$ is an explicit function of the dimension n. Then the set $\mathcal{D}_\Omega$ of $\omega \in \mathbf{R}^n$ such that

$$\left(\inf_{\ell \in \mathbf{N}} |\omega \cdot k - \ell|\right)^{-1} \leq \Omega(|k|) \qquad \forall\, k \in \mathbf{Z}^n \setminus \{0\} \tag{3.21}$$

has the property that, given any unit cube $\mathcal{C}$

$$|\mathcal{C} \cap \mathcal{D}_\Omega| \geq 1 - a(n)^{-1} \sum_{r=1}^{\infty} \Omega(r)^{-1} r^{n-1} \tag{3.22}$$

where $|\ |$ denotes the Lebesgue measure.

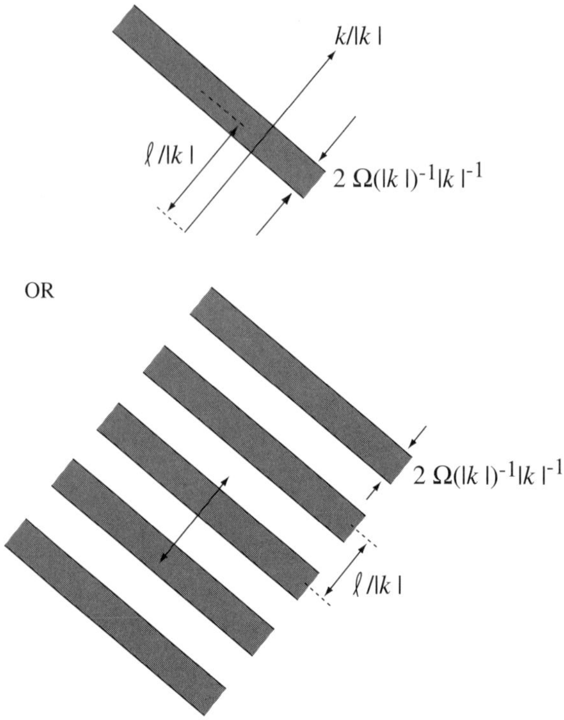

FIGURE 2

Note that when we take $\Omega(|k|) = K^{-1}|k|^{\nu}$, (3.21) reduces to (3.17). The condition (3.20) is satisfied for $\nu > n$ and for K sufficiently big. This shows that the set of Diophantine numbers $\mathcal{D}_\nu$ has full measure for $\nu > n$. Indeed

$$|\mathcal{C} \cap \mathcal{D}_{K,\nu}| \geq 1 - Kb(\nu, n) \tag{3.23}$$

PROOF. We will denote by σ_n constants that depend only on the dimension n. The same symbol can be used for different constants.

For $k \in \mathbf{Z}^n \setminus \{0\}$, $\ell \in \mathbf{Z}$ we consider the set

$$\mathcal{B}_{k,\ell} = \{\omega \in \mathbf{R}^n \mid |\omega \cdot k - \ell| \leq \Omega(|k|)^{-1}\}$$

consisting of the ω's for which the desired inequality (3.21) fails precisely for k, ℓ. The desired set will be the intersection of the complements of these sets.

Geometrically $\mathcal{B}_{k,\ell}$ is a strip bounded by parallel planes which are at a distance $2\Omega(|k|)^{-1}|k|^{-1}$ apart (see Figure 3.3). Thus, given a unit cube $\mathcal{C} \in \mathbf{R}^n$, the measure of $\mathcal{C} \cap \mathcal{B}_{k,\ell}$ cannot exceed $\sigma_n \Omega(|k|)^{-1}|k|^{-1}$.

We also observe that given $k \in \mathbf{Z}^n - \{0\}$, there is only a finite number of ℓ such that $\mathcal{C} \cap \mathcal{B}_{k,\ell} \neq \emptyset$. Indeed, this number can be bounded by $\sigma_n|k|$.

Therefore, for any $k \in \mathbf{Z}^n \setminus \{0\}$

$$\sum_{\ell \in \mathbf{Z}} |\mathcal{B}_{k,\ell} \cap \mathcal{C}| \leq \sigma_n \Omega(|k|)^{-1} ,$$

hence,

$$
\begin{aligned}
1-|\mathcal{C}\cap\mathcal{D}_\Omega| &= \sum_{k\in\mathbf{Z}^n}\sum_{\ell\in\mathbf{Z}\setminus\{0\}} |\mathcal{B}_{k,\ell}\cap\mathcal{C}| \\
&\le \sigma_n \sum_{k\in\mathbf{Z}^n\setminus\{0\}} \Omega(|k|)^{-1} \\
&\le \sigma_n \sum_{r=1}^{\infty} \Omega(r)^{-1} r^{n-1}\,.
\end{aligned}
\tag{3.24}
$$

Under the hypothesis that the R.H.S. of the above equation is smaller than 1, the conclusions hold. □

An important generalization of the above argument [**Pja69**] leads to the conclusion that a submanifold of Euclidean space that has curvature (or torsion or any other higher order condition) in such a way that planes cannot have a high order tangency to it (see below or see the references) then the submanifold has to contain Diophantine numbers. Even if the proof is relatively simple, the abundance of Diophantine numbers in lower dimensional curves has very deep consequences since it allows one to reduce the number of free parameters needed in KAM proofs.

LEMMA 3.10. *Let Σ be a compact C^{l+1} submanifold of $\mathbf{R}^n$.*

Assume that at every point $x\in\Sigma$ of the manifold

$$
T_x\Sigma + T_x^2\Sigma + \cdots + T_x^l\Sigma = T_x\mathbf{R}^n \tag{3.25}
$$

where by $T_x^j\Sigma$ we denote the j tangent plane to Σ.

Then, we can find a constant C_Σ that depends only on the manifold such that:

$$
|\Sigma - \mathcal{D}_\Omega| \le C_\Sigma \sum_{r=1}^{\infty} \Omega(r)^{1/l} r^{n-1}
$$

where by $|\cdot|$ we denote the Riemannian volume of the manifold.

The geometric meaning of the hypothesis (3.25) is that the manifold is not too flat and that it has curvature and torsion (or torsion of high order) so that every neighborhood of a point has to explore all the directions in space. In particular, we will have a lower bound on the area of the portion of the manifold that can be trapped in a resonant region, which in the space of ω is a flat plane.

The remaining details of the proof is left as an exercise for the interested reader. See also the lectures on number theory in this volume. The proof follows by noting that because of (3.25) we can bound the measure of the regions $\sum_{l\in\mathbf{Z}}\Sigma\cup\mathcal{B}_{k,l}\le C_\Sigma\Omega(k)^{-1/l}$. The worst case happens when the manifold is tangent to a very high order to one of the resonant regions. Since the order of tangency – as well as the constants involved – are uniformly bounded, we obtain the desired result.

REMARK 3.11. Notice that the formulation of the Diophantine properties (3.18) and (3.19) also makes sense if we allow ω to take complex values. This sometimes appears when we study complex maps and it is a useful tool. Notice that the argument we have presented works very similarly for the case of ω taking complex values. Indeed, the norm of the inverse can be bounded by the norm of inverse of the real part (or the norm of the inverse of the imaginary part) so, when the real or imaginary parts of an ω vector are Diophantine, the vector is Diophantine.

Sometimes, when studying problems with polynomials we will also need the inequalities only for $k \in \mathbf{N}^n$. Needless to say, these are much easier to satisfy since the signs have less possibilities to compensate and lead to small numbers.

EXERCISE 3.12. Construct a complex vector which is Diophantine, but whose imaginary and real parts are not Diophantine.

REMARK 3.13. The same simple minded argument used in the proof of Lemma 3.9 can be used to obtain estimates not only on the Lebesgue measure of the set of Diophantine numbers but also other geometric properties (for example Hausdorff measure), of sets satisfying Diophantine properties, and that are forced to belong to a manifold, have a resonance, etc.

3.4. Estimates for the linearized equation. In this subsection, we will consider estimates for the following equations (3.26), (3.27) that occur very frequently in KAM theory. We have encounter them already in the study of Lindstedt series and we will encounter them again as linearized equations.

We will consider equations of the form:

$$D_\omega \varphi = \eta \qquad \left(D\omega \equiv \omega_1 \frac{\partial}{\partial \theta_1} + \cdots + \omega_n \frac{\partial}{\partial \theta_n} \right) \tag{3.26}$$

$$\begin{aligned} L_\omega \varphi = \eta \qquad & \Big(L_\omega \varphi(\theta_1, \ldots, \theta_n) \\ & \equiv \varphi(\theta_1 + \omega_1, \ldots, \theta_n + \omega_n) - \varphi(\theta_1, \ldots, \theta_n) \Big) , \end{aligned} \tag{3.27}$$

where $\eta : \mathbf{T}^n \to \mathbf{R}$ is given and the unknown function to be found is φ.

For the sake of simplicity we will only discuss in detail (3.26). The same considerations apply for (3.27) and we will indicate the minor differences – in fact simplifications – that enter in the discussion of (3.27).

Recall that these equations have a formal solution in terms of Fourier series. Namely, if

$$\eta(\theta) = \sum_{k \in \mathbf{Z}^n} \hat{\eta}_k e^{2\pi i k \cdot \theta} , \qquad \hat{\eta}_0 = 0 ,$$

then any reasonable solution of (3.26) for which one can define unique Fourier coefficients (e.g. any distribution) has to satisfy:

$$\hat{\varphi}_k 2\pi i k \cdot \omega = \hat{\eta}_k .$$

Hence, if $k \cdot \omega \neq 0$, then

$$\hat{\varphi}_k = \frac{\hat{\eta}_k}{2\pi i k \cdot \omega} . \tag{3.28}$$

We restrict our attention to cases when $k \cdot \omega \neq 0$ for any $k \in \mathbf{Z}^n - \{0\}$. In that case φ is determined by (3.28) up to an additive constant since we can take any $\hat{\varphi}_0$. To avoid unnecessary complications, we will set $\hat{\varphi}_0 = 0$.

It is not difficult to see that, unless we impose some quantitative restriction on how fast $|k \cdot \omega|^{-1}$ can grow, the solutions given by (3.28) may fail to be even distributions. E.g., take $\hat{\eta}_k = e^{-|k|}$ and arrange that there are infinitely many k for which $|k \cdot \omega|^{-1} \geq e^{e^{|k|}}$.

EXERCISE 3.14. Given any sequence a_n of positive terms tending to infinity construct an $\omega \in \mathbf{R}^n - \mathbf{Q}^n$ such that, for infinitely many $k \in \mathbf{Z}^n$

$$|\omega \cdot k|^{-1} \geq a_{|k|}. \tag{3.29}$$

Show that the ω constructed above are dense (even if, as we have shown, they will be of measure zero for sequences a_n which grow fast enough).

We will consider ω which satisfy

$$|k \cdot \omega|^{-1} \leq \gamma |k|^{\nu} \ . \tag{3.30}$$

These numbers were studied in Section 3.3.

It is not difficult to obtain some crude bounds for analytic or finite differentiable functions (we will do better later). Recall that for $\eta \in A_\delta$

$$|\hat{\eta}_k| \leq e^{-2\pi\delta|k|} \|\eta\|_{A_\delta} \ ,$$

while for $\eta \in C^r$

$$|\hat{\eta}_k| \leq (2\pi)^{-r} |k|^{-r} \|\eta\|_{C^r} \ .$$

Hence, if ω satisfies (3.30), we have for $\eta \in A_\delta$

$$|\hat{\varphi}_k| \leq (2\pi)^{-1} \gamma |k|^{\nu} e^{-2\pi\delta|k|} \|\eta\|_{A_\delta} \ ,$$

and for $\eta \in C^r$

$$|\hat{\varphi}_k| \leq (2\pi)^{-r-1} \gamma |k|^{\nu - r} \|\eta\|_{C^r} \ .$$

These estimates do not allow us to conclude that φ belongs to the same space as η, but allow us to conclude that it belongs to a slightly weaker space.

As mentioned before, the characterization of the analytic spaces in terms of their Fourier series is very clean, so that we can obtain estimates of the solutions in these spaces. Then, we will use Lemma 3.2 to obtain the results for Λ_r spaces.

Since for $0 < \sigma < \delta$ we have:

$$\|e^{2\pi i k \cdot \theta}\|_{\delta - \sigma} \leq e^{2\pi(\delta - \sigma)|k|} \ ,$$

we have

$$\begin{aligned} \|\varphi\|_{A_{\delta-\sigma}} &\leq \sum_{k \in \mathbf{Z}^n \setminus \{0\}} |\hat{\varphi}_k| e^{2\pi|k|(\delta-\sigma)} \\ &\leq \sum_{k \in \mathbf{Z}^n \setminus \{0\}} \frac{1}{2\pi |k \cdot \omega|} \|\eta\|_{A_\delta} e^{-2\pi\sigma|k|} \\ &\leq \frac{1}{2\pi} \gamma \|\eta\|_{A_\delta} \sum_{k \in \mathbf{Z}^n \setminus \{0\}} |k|^{\nu} e^{-2\pi\sigma|k|} \qquad (3.31) \\ &\leq C\gamma \|\eta\|_{A_\delta} \sum_{\ell \in \mathbf{N}} \ell^{\nu+n-1} e^{-2\pi\sigma\ell} \\ &\leq C\gamma\sigma^{-(\nu+n)} \|\eta\|_{A_\delta} \ , \end{aligned}$$

where in the fourth inequality we have just used that we do first the sum in the k with $|k| = \ell$ (the number of terms in this sum can be bounded by $C\ell^{n-1}$). We denote by C constants that depend only on ν and the dimension n and are independent of γ, k, etc.

Similarly, using that

$$\|e^{2\pi i k \cdot \theta}\|_{C^s} \leq C|k|^s \ ,$$

we have[4]

$$\begin{aligned}\|\varphi\|_{C^s} &\leq C\gamma\|\eta\|_{C^r}\sum_{k\in\mathbf{Z}^n}|k|^{\nu-r+s}\\ &\leq C\gamma\|\eta\|_{C^r}\sum_{\ell\in\mathbf{N}}\ell^{\nu-r+s+n-1}\ .\end{aligned}$$

The sum in the R.H.S. converges provided that

$$r>\nu+s+n\ .$$

Actually, one can do significantly better that these crude bounds if one notices that the small divisors have to appear rather infrequently (see [**Rüs75, Rüs76b**]).

Note that $\omega\cdot(k+\ell)=\omega\cdot k+\omega\cdot\ell$. Hence, if $\omega\cdot k$ happens to be very small, $\omega\cdot(k+\ell)\approx\omega\cdot\ell$, so that if $|\ell|<<|k|$, $\omega\cdot(k+\ell)\approx\omega\cdot\ell$.

In other words, the really bad small divisors appear surrounded by a ball on which the divisors are not that small. Hence, if instead of estimating the size as in (3.31) using the estimates (3.18) in the third step we use a Cauchy-Schwartz inequality, which takes into account the sum of terms, not just the the sup and that can profit from the fact that (3.18) cannot be saturated very often, we obtain the result of [**Rüs75, Rüs76b**], which reads:

LEMMA 3.15. *Assume that ω satisfies* (3.18), *with $\nu\geq n-1$ and that $\tilde\omega$ satisfies* (3.19). *Let $\eta,\tilde\eta$ be analytic functions with zero average.*

Then, we can find $\varphi,\tilde\varphi$ solving (3.26), (3.27). *Namely*

$$\begin{aligned}D_\omega\varphi&=\eta\\ L_{\tilde\omega}\tilde\varphi&=\tilde\eta.\end{aligned}\tag{3.32}$$

and φ, $\tilde\varphi$ have zero average.

Moreover, we have for all $\delta>0$:

$$\begin{aligned}\|\varphi\|_{\sigma-\delta}&\leq C\delta^{-\nu}K_{\nu,n}\|\eta\|_\sigma\\ \|\tilde\varphi\|_{\sigma-\delta}&\leq C\delta^{-\nu}K_{\nu,n}\|\tilde\eta\|_\sigma\end{aligned}\tag{3.33}$$

Where the C are the same constants that appear in (3.18), (3.19) *and K are constants that depend (in a very explicit formula) only on the exponent in* (3.18), (3.19) *and the dimension of the space.*

If we assume that η, $\tilde\eta$ are in Λ_r, $r>\nu$, we obtain:

$$\begin{aligned}\|\varphi\|_{\Lambda_{r-\nu}}&\leq CK_{\nu,n}\|\eta\|_{\Lambda_r}\\ \|\tilde\varphi\|_{\Lambda_{r-\nu}}&\leq CK_{\nu,n}\|\tilde\eta\|_{\Lambda_r}\end{aligned}\tag{3.34}$$

We just note that the part (3.34) is a consequence of (3.33) using the the characterization of differentiable functions by properties of the approximation by analytic functions in Lemma 3.2.

When studying analytic problems, one can be sloppy with the exponents obtained and still arrive at the same result. However, as (3.34) shows, taking care of the exponents is crucial if we are studying finitely differentiable problems and want to obtain regularity which is close to optimal.

EXERCISE 3.16. Read the argument in [**Rüs76b**]. Do you obtain some improvement using the Hölder inequality in place of Cauchy-Schwartz?

[4]Here, C depends on s even if it is independent of k. We, however do not include the s dependence in the notation to avoid clutter.

EXERCISE 3.17. In the study of Lindstedt series (e.g. (2.17)) we encountered second order equations for φ given η of the form:

$$\varphi(x+\omega) + \varphi(x-\omega) - 2\varphi(x) = \eta(x) \tag{3.35}$$

where φ and η are periodic and ω is a Diophantine number.

Develop a theory of the equation (3.35) along the theory of the theory developed in Lemma 3.15.

Do it either by treating it directly in Fourier series or by factoring it as two equations:

$$\begin{aligned} w(x) - w(x-\omega) &= \eta(x) \\ \varphi(x+\omega) - \varphi(x) &= w(x) \end{aligned} \tag{3.36}$$

Are there any differences between the estimates or the solvability conditions you get by the two methods?

What happens if instead of using the naive estimates presented in the text you use the estimates of [**Rüs76a**]?

3.5. Geometric structures. There are several structures that play an important role in KAM theory. In this Section, we will discuss symplectic and, more briefly, volume preserving and reversible systems (there are other geometric structures that have come to play a role in KAM theory, but we will not discuss them here).

In this Section, the emphasis will be on the geometric structures and not on the differentiability properties, so we will assume that vector fields generate flows, for which variational equations are valid, etc. (i.e., that they have some mild differentiability properties).

Here we will use Cartan calculus of differential forms rather than the old-fashioned notation. Since Cartan calculus uses only geometrically natural operations, it is conceptually simpler. This is a great advantage in mechanics, where one frequently uses changes of variables, restriction to submanifolds given by regular values of the integrals of motion, etc..

The traditional notation — in which one writes functions as functions of the coordinates, e.g., $H(p,q)$ — is perfectly adequate when the coordinates are fixed. On the other hand, when one changes coordinates, one has to decide whether $H(p',q')$ denotes the same function of new arguments or whether $H(p',q')$ is a different function of p' and q' which produces the same numerical value as the old function H produced with the old variables p and q. The ambiguity increases enormously when one needs to compute partial derivatives — a great deal of the complications in traditional books and papers on mechanics and thermodynamics arises from this.

For KAM theory these considerations are not so crucial because many of the operations one has to perform require using Fourier coefficients and the like, which forces the fixing of a certain system of coordinates. Nevertheless, we think the conceptual simplification provided by the geometric notation is worth the effort required in introducing it.

Now let us start with some important definitions.

DEFINITION 3.18. A *symplectic structure* in a manifold is given by a 2-form ω_2 satisfying the conditions

i) ω_2 is non degenerate

ii) ω_2 is closed, i.e., $d\omega_2 = 0$.

A *volume form* in a manifold of dimension n is an n-form ω_n that satisfies

i$'$) ω_n is non degenerate.

Naturally, an n-form ω_n in an n-dimensional manifold automatically satisfies

ii$'$) $d\omega_n = 0$.

Much of the geometric theory goes through just under the conditions i) and ii) — or i$'$) and ii$'$). When we do not need to distinguish between the symplectic and the volume preserving cases, we will use ω to denote either ω_2 or ω_n.

Properties i) and i$'$) allow us to identify a vector field v with a 1- and $(n-1)$-form, respectively, by

$$\mathrm{i}_v\omega_2 := \omega_2(v,\cdot) = \gamma_1\ , \qquad \mathrm{i}_v\omega_n = \gamma_{n-1}\ . \tag{3.37}$$

We will denote the identifications (3.37) by $\mathcal{I}_{\omega_2}$ and $\mathcal{I}_{\omega_n}$, respectively.

Fundamental examples of a symplectic form ω_2 on $\mathbf{R}^k \times \mathbf{R}^k$ and a volume form ω_n on $\mathbf{R}^n$ are

$$\begin{aligned} \omega_2 &= \sum_{i=1}^{k} dp_i \wedge dq_i\ , \\ \omega_n &= dx_1 \wedge \ldots \wedge dx_n\ . \end{aligned} \tag{3.38}$$

REMARK 3.19. The name symplectic seems to have been originated as a pun on the name complex. Indeed, there is a sense in which symplectic geometry is a complexification of Riemannian geometry. This is actually quite deep and there is a wonderful new area of research using methods of complex analysis in symplectic topology.

Since these notes are focused on KAM theory, it suffices to note that in mechanics one often finds the matrix $J \equiv= \begin{pmatrix} 0 & \mathrm{Id}_d \\ -\mathrm{Id}_d & 0 \end{pmatrix}$ which satisfies $J^2 = -1$ and which, therefore, is quite analogous to multiplication by i in complex analysis.

The identification of vector fields with forms plays a very important role because it allows us to describe the vector fields whose flow preserves the structure. Denote by Φ_t a family of diffeomorphisms of the manifold generated by the time-dependent vector field v_t, i.e.,

$$\frac{d}{dt}\Phi_t = v_t \circ \Phi_t\ , \qquad \Phi_0 = \mathrm{Id}\ .$$

In particular, if v_t is independent of t, Φ_t is a flow: $\Phi_{t+s} = \Phi_t \circ \Phi_s$. (Again we recall that in this Section we are assuming the objects to be differentiable enough, in this case v_t to be C^1.)

Using the definition of Lie derivative, Cartan's so called "magic formula" to express the Lie derivative

$$L_X\gamma = d(\mathrm{i}_X\gamma) + \mathrm{i}_X(d\gamma), \tag{3.39}$$

the closedness of ω and the definition of $\mathcal{I}_\omega$ we obtain:

$$\frac{d}{ds}|_{s=0}\Phi^*_{t+s}\omega = \Phi^*_t L_{v_t}\omega = \Phi^*_t(d\,\mathrm{i}_{v_t}\omega + \mathrm{i}_{v_t}\,d\omega) = \Phi^*_t\,d\,\mathcal{I}_\omega v_t\ .$$

Thus, if ω is invariant under the flow Φ_t (i.e., $\Phi^*_t\omega = \omega$), we conclude that $\mathcal{I}_\omega v_t$ is closed.

The above result is quite interesting because the Φ_t-invariance of ω seems at first sight to be a non-linear and non-local constraint for the flow Φ_t. The vector field v_t is perfectly linear and local.

Of particular importance for KAM theory are the vector fields (called *exact symplectic*, resp. *exact volume preserving*) for which $\mathcal{I}_\omega v_t$ is exact, i.e.,

$$\mathcal{I}_\omega v_t = \mathrm{d}\gamma_t$$

with γ_t a function (symplectic case) or an $(n-2)$-form (volume preserving case). Sometimes these are called *globally Hamiltonian* vector fields to indicate that one can find a Hamiltonian that generates them globally and not only locally. All the flows that preserve the symplectic or volume structure can be expressed locally as a Hamiltonian flow, but perhaps not globally. We will come back to this in more detail when we consider some extra structure of the space.

Of course, when we are considering local problems, by Poincaré's lemma, we do not need to distinguish between symplectic and exact symplectic vector fields.

In the symplectic case, for (3.38), we have that $\mathcal{I}_{\omega_2} v \equiv \mathrm{i}_v\omega_2 = -\mathrm{d}H$ reduces to the standard Hamilton's equations

$$v_{p_i} = -\frac{\partial H}{\partial q_i}\,, \qquad v_{q_i} = \frac{\partial H}{\partial p_i}\,.$$

The function H is called the *Hamiltonian* of the vector field v. Vector fields satisfying locally $\mathrm{i}_v\omega_2 = -\mathrm{d}H$ for some function H are called *locally Hamiltonian vector fields.* If the function H can be defined globally, the vector field v is called *globally Hamiltonian.*

An important consequence of the preservation of symplectic or volume form is that if a diffeomorphism f preserves the form ω and $\mathrm{i}_X\omega = -\mathrm{d}H$, we have

$$\mathrm{i}_{f_*X}\omega = \mathrm{i}_{f_*X} f_*\omega = f_*(\mathrm{i}_X\omega) = -f_*\mathrm{d}H = -\mathrm{d}f_*H = -\mathrm{d}(H\circ f^{-1})\,, \tag{3.40}$$

so that f_*X is also a Hamiltonian flow for $H\circ f^{-1}$.

In old fashioned language, this was described as saying that "canonical transformations preserve the form of Hamilton's equations" or some similar sentence. (In old fashioned books the name canonical transformations referred to diffeomorphisms preserving the symplectic form, or sometimes to what we have referred to as exact symplectic.)

The importance of the formula (3.40) is that to make canonical changes of variables to a Hamiltonian vector field, it suffices to make changes of variables in the Hamiltonian functions. This is conceptually much simpler and computationally more efficient. As we will see, canonical perturbation theory owes its success to this remark. Note that this calculation goes through both for symplectic and volume forms. Using Cartan calculus, it is possible to develop perturbation theories for symplectic and volume preserving flows which are completely analogous.

Notice that in 2 dimensions the volume form and the symplectic structure are the same and that, when $n = 2k$,

$$\omega_n^{\wedge k} := \omega_2\wedge\omega_2\wedge\cdots\wedge\omega_2 \qquad (k \text{ times})$$

is a volume form.

Clearly, a flow that preserves ω_2, also preserves $\omega_n^{\wedge k}$. This fact is usually referred to in mechanics as *Liouville's theorem* and is of fundamental importance since it makes a connection of mechanics with ergodic theory. Indeed, ergodic theory was introduced in the study of the relations of this observation with statistical mechanics.

In the study of Hamiltonian flows, it is also of interest to study the form μ_E defined in the regular energy surfaces $\Sigma_E = \{H = E\}$ – assumed that dH is not degenerate so that it is an smooth manifold – by $\omega^{\wedge k} = \mu_E \wedge dH$. Since H is invariant under the flow, so is dH and μ_E is invariant.

The intermediate forms, $\omega_2 \wedge \cdots \wedge \omega_2$ (ℓ times, $\ell < k$) are also invariant. It seems that not much use has been made of them ([**Poi93**], Chapters XXII–XXVII is devoted to this question).

One of the first consequences of the identification between vector fields and forms (3.37) is a simple proof of the Darboux theorem. (See [**MS95**]. The original proof along this lines was done for the volume case in [**Mos65**].)

THEOREM 3.20. *Given a symplectic or volume preserving form ω and a point x_0, there exists a local diffeomorphism f on a neighborhood of x_0 to $\mathbf{R}^n$ such that $f^*\omega$ is of the form in* (3.38).

Note that the Darboux theorem implies that there are no local symplectic or volume invariants (so that the recent but already very rich theory of symplectic invariants and obstructions is eminently global). Moreover, an argument of [**Mos65**] shows that for volume preserving geometry in a compact manifold the only invariant is the total volume. This is in great contrast with Riemannian geometry where the "theorema egregium" of Gauss shows that there are local invariants for isometry[5].

Of particular importance for KAM theory will be the study of exact transformations. They can only be defined on manifolds where ω is exact, i.e., manifolds for which

$$\omega = \mathrm{d}\theta \ .$$

One important example is

$$\omega_2 = \sum_{i=1}^{k} \mathrm{d}p_i \wedge \mathrm{d}q_i \ , \qquad \theta = \sum_{i=1}^{k} p_i \, \mathrm{d}q_i$$

with $q \in \mathbf{T}^k$, $p \in \mathbf{R}^k$, so that $M = \mathbf{T}^k \times \mathbf{R}^k$.

More generally, if $M = T^*N$ is the cotangent bundle of the k-dimensional manifold N, and $\pi : T^*N \to N$ is the projection, one can define θ intrinsically as the only 1-form in T^*N with the property

$$\gamma^*\theta = \gamma$$

for all 1-forms γ on N [**AM78**], Pro. 3.2.11. (Here, γ is considered as a map from N to T^*N, satisfying $\pi \circ \gamma = \mathrm{Id}$, so that γ^* maps the 1-forms in T^*N into 1-forms in N.) One can easily check that this is equivalent to the standard prescription of taking a local trivialization of T^*N with coordinates (p, q) and then setting $\theta = \sum_{i=1}^{k} p_i \, \mathrm{d}q_i$. One needs to check that the definition is independent of the system of coordinates chosen.

For volume preserving maps, our main example will be

$$M = \mathbf{T}^{n-1} \times \mathbf{R} \ , \qquad \theta = p \, \mathrm{d}q_1 \wedge \cdots \wedge \mathrm{d}q_{n-1} \ ,$$

[5]Note that the condition $d\omega = 0$ is some sort of curvature condition, so that perhaps it is fairer to compare symplectic geometry to a sort of Riemannian geometry of flat manifolds.

where $(q_1, \dots, q_{n-1}) \in \mathbf{T}^{n-1}$, $p \in \mathbf{R}$. We note that given ω, θ is determined up to a closed form.

When ω is exact (i.e., $\omega = \mathrm{d}\theta$) we use that $f^*\mathrm{d} = \mathrm{d}f^*$ to obtain that for any diffeomorphism f preserving ω

$$\mathrm{d}(f^*\theta - \theta) = 0 \ .$$

We say that the ω-preserving diffeomorphism f is *exact* when

$$f^*\theta - \theta = \mathrm{d}S \ . \tag{3.41}$$

Once we fix θ, S is defined up to a form of zero exterior derivative; in the symplectic case, this means up to a constant.

Conversely, it turns out that the function S determines to a large extent the diffeomorphism. If we know S in the whole manifold and and the diffeomorphism restricted to a *Lagrangian* submanifold, it is possible to reconstruct the diffeomorphism in the whole manifold. (See [**Har99**].)

In exact symplectic (or volume) manifolds ($\omega = d\theta$), there is a very close relationship between exact families and hamiltonian flows. Families of exact diffeomorphisms are generated by a Hamiltonian flow and vice versa.

To show the first statement, note that if f_t is an smooth family, we have that ${f_t}^*\theta - \theta = \mathrm{d}S_t$ and we can choose S_t smooth in t. If we take derivatives of this relation with respect to t and introduce the vector field $\mathcal{F}_t$ generating f_t by $\frac{d}{dt} f_t = \mathcal{F}_t \circ f_t$, we have

$${f_t}^* L_{\mathcal{F}_t}\theta = \mathrm{d}\dot{S}_t \ , \tag{3.42}$$

where L denotes the Lie derivative and $\dot{S}_t = \frac{d}{dt} S_t$. Using Cartan's formula for the Lie derivative, we have

$${f_t}^*[\mathrm{d}\mathrm{i}_{\mathcal{F}_t}\theta + \mathrm{i}_{\mathcal{F}_t} d\theta] = \mathrm{d}\dot{S}_t \ . \tag{3.43}$$

Therefore,

$$\mathrm{i}_{\mathcal{F}_t}\omega = \mathrm{d}\left[({f_t}^*)^{-1}\dot{S}_t - \mathrm{i}_{\mathcal{F}_t}\theta\right] \ . \tag{3.44}$$

Hence, we conclude that a family of exact maps is generated by a Hamiltonian flow of Hamiltonian given by the formula:

$$H_t = \mathrm{i}_{\mathcal{F}_t}\theta - ({f_t}^*)^{-1}\dot{S}_t \ . \tag{3.45}$$

In the exact symplectic case, the last formula reads

$$H_t = \mathrm{i}_{\mathcal{F}_t}\theta - \dot{S}_t \circ f_t^{-1} \ .$$

The converse is proved by a very similar calculation. Note that if we are given H_t and an exact ω-preserving diffeomorphism f_0 with an initial primitive S_0, and f_t is generated by the Hamiltonian flow of H_t, then the deformation f_t is also exact, and the primitive S_t satisfies the differential equation

$$\dot{S}_t = f_t^*(\mathrm{i}_{\mathcal{F}_t}\theta - H_t) \ . \tag{3.46}$$

Notice that the obstruction for a symplectic or volume preserving diffeomorphism f to be exact is just the cohomology class with real coefficients of $f_*\theta - \theta$. For example, in the map we considered before,

$$f(q,p) = (q, p + a)$$

with a a constant n-vector, we have

$$\begin{aligned} f_*\theta_1 - \theta_1 &= \sum_{i=1}^{k} a_i \, dq_i \ , \\ f_*\theta_{n-1} - \theta_{n-1} &= a \, dq_1 \wedge \cdots \wedge dq_{n-1} \ . \end{aligned}$$

In this case, one can see that the cohomology obstruction vanishes if and only if the flux that we considered in Definition 2.1 vanishes.

If f is a diffeomorphism close to the identity in the C^r $(r = 1, 2, \ldots, \infty)$ topology, it is not hard to show for M as in the example that there is an exact family of vector fields interpolating with the identity.

This can also be proved for analytic functions. however, it is far from trivial. See [**KP94**].

The reason why exactness plays an important role in KAM theory can be understood from the simple example (already mentioned) in $\mathbf{R} \times \mathbf{T}$,

$$f(p,q) = (p + \varepsilon, q + p) \ , \qquad f^n(p,q) = (p + n\varepsilon, q + np) \ ,$$

which does not admit any quasiperiodic orbits for $\varepsilon \neq 0$ (all the orbits escape to infinity).

A consequence of great importance for us later is that if we choose a function, resp. an $(n-1)$-form, γ and form a vector field by

$$v = \mathcal{I}_\omega \, d\gamma \ ,$$

then the time one map of the vector field, Φ_1, is exact. This gives a convenient way to generate transformations close to the identity.

Since commutators of vector fields are an ingredient of the variational equations, it is quite interesting to study how commutators interact with the geometric structures (volume forms and symplectic).

Recall that the commutator of two vector fields can be considered as the commutator of the vector fields considered as differential operators. That is, the commutator of C^1 vector fields is defined as

$$[X, Y] = XY - YX, \tag{3.47}$$

when we consider the vector fields as first order differential operators in a manifold i without boundary. (It is somewhat surprising, but of course true, that the commutator is first order operator, the R.H.S. of (3.47) looks like a second order operator!) The commutator can also be defined as

$$[X, Y] = \lim_{t \to 0} t^{-2} (Y_{-t} \circ X_{-t} \circ Y_t \circ X_t - \mathrm{Id})$$

where X_t denotes the flow generated by X and similarly for Y. We have also taken the usual liberty of employing additive notation rather than a more geometric one to denote comparisons.

The following well known result relates the commutators to geometry. We have followed the presentation of [**BdlLW96**].

LEMMA 3.21. *Let ω be a non-degenerate closed form as before.*

(i) *If X, Y are locally Hamiltonian vector fields, then $[X, Y]$ is a globally Hamiltonian vector field with Hamiltonian $\mathrm{i}_Y(\mathrm{i}_X\omega) = \omega(X, Y)$.*

(ii) *If X has H as a Hamiltonian and Y is locally Hamiltonian, then $-L_Y H$ is a Hamiltonian for $[X, Y]$.*

(iii) *If Y has F as a Hamiltonian and X is locally Hamiltonian, then $L_X F$ is a Hamiltonian of $[X, Y]$.*

PROOF. First recall the identities $L_X \mathrm{d}\alpha = \mathrm{d}L_X\alpha$ and

$$\mathrm{i}_{[X,Y]}\alpha = L_X \mathrm{i}_Y \alpha - \mathrm{i}_Y L_X \alpha$$

which are valid for each m-form α and vector fields X and Y. Also, observe that a locally Hamiltonian vector field X satisfies

$$L_X \omega = 0$$

which follows easily from Cartan's "magic formula" (3.39).

To prove (i), compute:

$$\begin{aligned} \mathrm{i}_{[X,Y]}\omega &= L_X \mathrm{i}_Y \omega - \mathrm{i}_Y L_X \omega = L_X \mathrm{i}_Y \omega \\ &= \mathrm{i}_X \mathrm{d}\,\mathrm{i}_Y \omega + \mathrm{d}\,\mathrm{i}_X \mathrm{i}_Y \omega = \mathrm{d}\,\mathrm{i}_X \mathrm{i}_Y \omega \\ &= \mathrm{d}(\omega(Y, X)) = -\mathrm{d}(\omega(X, Y)) \ . \end{aligned}$$

For (ii), we know that $\mathrm{i}_X\omega = -\mathrm{d}H$ and use (i):

$$\begin{aligned} -\mathrm{d}(-L_Y H) &= L_Y(\mathrm{d}H) = -L_Y \mathrm{i}_X \omega = -\mathrm{i}_Y \mathrm{d}\,\mathrm{i}_X \omega - \mathrm{d}\,\mathrm{i}_Y \mathrm{i}_X \omega \\ &= \mathrm{i}_Y \mathrm{d}(\mathrm{d}H) - \mathrm{d}(\omega(X, Y)) = \mathrm{i}_{[X,Y]}\omega \ . \end{aligned}$$

The proof of (iii) is analogous to that of (ii): from $\mathrm{i}_Y\omega = -\mathrm{d}F$ we obtain

$$-\mathrm{d}(L_X F) = -L_X(\mathrm{d}F) = L_X \mathrm{i}_Y \omega = \mathrm{i}_{[X,Y]}\omega \ .$$

□

Let X and Y be Hamiltonian vector fields with Hamiltonians H and F, respectively (i.e., $\mathrm{i}_X\omega = -\mathrm{d}H$, $\mathrm{i}_Y\omega = -\mathrm{d}F$). The *Poisson bracket* of H and F is defined as

(3.48) $$\{H, F\} := -L_X F \ ,$$

or, equivalently, as

$$\{H, F\} = -\mathrm{d}H(X) = (\mathrm{i}_Y\omega)(X) = \omega(Y, X) \ .$$

The antisymmetry of ω yields

(3.49) $$\{H, F\} = -\{F, H\}$$

as well as the formula $\{H, F\} = L_Y H$.

In coordinates,

(3.50) $$\{H, F\} = \sum_{i=1}^{k} \left(\frac{\partial H}{\partial q_i} \frac{\partial F}{\partial p_i} - \frac{\partial H}{\partial p_i} \frac{\partial F}{\partial q_i} \right) \ .$$

Using (3.50), one can easily check that the Poisson bracket satisfies the Jacobi identity,

$$\{H, \{F, G\}\} + \{F, \{G, H\}\} + \{G, \{H, F\}\} = 0 \ ,$$

which, together with the linearity and the antisymmetry of the Poisson bracket, implies that the functions on the phase space of a dynamical systems with the Poisson bracket are a Lie algebra. Moreover, the Poisson bracket is a derivation of this Lie algebra.

This means:

(3.51) $$\{H, \{F, G\}\} = \{H, F\}G + F\{H, G\}.$$

The property (ii) (or (iii)) of Lemma 3.21 implies that the Hamiltonian vector field corresponding to $\{H, F\}$ is equal to $-[X, Y]$:

$$\mathrm{i}_{[X,Y]} = \mathrm{d}\{H, F\} \; .$$

This means that the map from the functions on the phase space to their Hamiltonian vector fields (i.e., $H \mapsto X$ such that $\mathrm{i}_X\omega = -\mathrm{d}H$) is a morphism of Lie algebras (the Lie-algebraic operations being respectively the Poisson bracket and the commutator of vector fields).

Note that $L_X F$ means the Lie derivative of F along the flow of X and, similarly, $L_Y H$ is the Lie derivative of H along the flow of Y.

By the identities above,

$$L_X F = -L_Y H,$$

which indicates that the derivative of a Hamiltonian form along the Hamiltonian flow of another Hamiltonian form is related by a sign change to the situations when the roles are reversed. This is a somewhat surprising property of Hamiltonian systems.

One way to look at the above calculations is to realize that the exact transformations are a group and that the vector fields $\mathcal{I}_\omega \, \mathrm{d}\gamma$ are a Lie algebra. (In the old fashioned language, the vector fields of the form $\mathcal{I}_\omega \, \mathrm{d}\gamma$ were called "infinitesimal transformations" or, given that "infinitesimal" is a somewhat dirty word in some circles, "transformations close to the identity.")

Unfortunately, even if this point of view is heuristically correct, it is not without problems. First of all, composition of transformations is not a differentiable operation in almost any precise sense. Indeed, note that $f \circ (g + \Delta) - f \circ g \approx f' \circ g \Delta$, so that the derivative of composition should be a multiplication by $f' \circ g$. Hence, if we consider composition in a space as C^r, Λ_δ, etc., then $f' \circ g$ may not belong to this space; if we consider $C^{r+\alpha}$ spaces, then the composition is not even continuous!

More importantly, the exponential of the Lie algebra does not cover an open neighborhood of the identity. That is, in any arbitrarily small neighborhood of the identity in Λ_δ, there exist exact maps that cannot be written as time one maps of a differentiable vector field.

Another important geometric structure that plays a role in KAM theory is the so called *reversible* systems. They appear in any applied problem in which time can be "run backwards" (i.e., if $\gamma(t)$ is a trajectory, then $\gamma(T - t)$ also is). This happens in mechanical problems without friction or in electric circuits without resistors and in other problems. Examples of reversible systems also appear in finite dimensional truncations of fluid mechanics problems when there is no viscosity. In general, physical problems in which there is no dissipation are often reversible. When the systems are not mechanical, there is no reason why we should have also a symplectic structure. In particular, in the example of circuits, it is possible to find interesting examples with odd dimensions.

A map is said to be *reversible* when there exists an involution R (that is, $R^2 = \mathrm{Id}$) for which $A^{-1} = R^{-1}AR$, i.e., A is conjugate by R to its own inverse.

Since $R^{-1} = R$, the above condition can be expressed as $A^{-1} = RAR = RAR^{-1}$. Note also that reversibility implies that $S = AR$ is an involution. Hence, A is a product of two involutions, $A = SR$. One can also check that the product of two involutions is reversible with respect to either of them, so that one can just as well define a reversible map as the product of two involutions, even if this obscures the physical interpretation and the origin of the name.

Sometimes one does not require that R is an involution. These systems are sometimes called *weakly reversible*. The KAM theory only needs weak reversibility. (Actually, in many occasions that KAM theory applies, we can use KAM theory to show that the systems are actually reversible.)

For flows, the definition is similar: the flow f_t is *reversible* if there exists an involution R such that $R^{-1} f_t R = f_t^{-1} = f_{-t}$. Taking derivatives, we obtain the reversibility condition in terms of the vector field $\mathcal{F}_t$ generating the flow: $R_* \mathcal{F}_t = -\mathcal{F}_t$.

One very important example of a reversible system is a mechanical system without friction whose forces depend only on the position of the particles. If we reverse the velocities and keep the positions the same, the system runs backwards. Hence we can take $R(x, v) = (x, -v)$. Clearly, R is an involution. Reversible mappings have recently received a great deal of interest in the context of statistical mechanics since many slightly dissipative models are reversible. This reversibility leads to very amusing consequences such as pairing rules for Lyapunov exponents. See [**BCP98**] for some applications to Statistical Mechanics and references.

Good surveys of reversible systems in general are [**Sev86**] and [**AS86**] and recent developments in the KAM theory for reversible systems are covered in [**Sev98**].

3.6. Canonical perturbation theory. The goal of perturbation theory is to understand the dynamics of a "perturbed" system which is close to another well understood system. Usually these well understood systems are chosen among "integrable" systems but this is not necessarily the case. As we will see in later proofs of KAM theorem, sometimes we want to take as unperturbed systems systems of a particular kind that have an interesting feature. In the case of integrable systems, the feature of interest for the study is quasi-periodic orbits.

The most naive approach to perturbation theory is to develop the solutions in powers of the perturbation parameter. That is, if we have a vector field

$$X_\varepsilon = X_0 + \varepsilon X_1 + \varepsilon^2 X_2 + \cdots ,$$

we try to find solutions of

$$\dot{x}_\varepsilon = X_\varepsilon(x_\varepsilon) , \qquad x_\varepsilon(0) = a(\varepsilon)$$

by setting

$$\begin{aligned} x_\varepsilon(t) &= x_0(t) + \varepsilon x_1(t) + \cdots , \\ a(\varepsilon) &= a_0 + \varepsilon a_1 + \varepsilon^2 a_2 + \cdots , \end{aligned} \tag{3.52}$$

substituting in the equation and solving.

That is,

$$\begin{aligned} \dot{x}_0 &= X_0(x_0) &, \quad & x_0(0) = a_0 \\ \dot{x}_1 &= X_1(x_0) + DX_0(x_0)x_1 &, \quad & x_1(0) = a_1 \\ \dot{x}_2 &= X_2(x_0) + DX_1(x_0)x_1 &, \quad & x_2(0) = a_2 \\ &\quad + DX_0(x_0)x_2 + \tfrac{1}{2} D^2 X_0(x_0) x_1^{\otimes 2} \\ &\quad \cdots \quad . \end{aligned} \tag{3.53}$$

Provided that X_ε is analytic in ε and its argument, this series was shown to converge by Cauchy (but before that, it was used regularly by Newton).

Note that all the equations in the hierarchy have the form

$$\dot{x}_n - DX_0(x_0)x_n = R_n \ , \qquad x_n(0) = a_n \ ,$$

where R_n is a polynomial expression involving only terms $x_0 \dots x_{n-1}$ and (known!) derivatives of X_i.

In spite of its ancient pedigree and the theorems of convergence this method has shortcomings.

It is an easy exercise that taking the second order problem

$$\ddot{x}_\varepsilon = -(1 + 2\varepsilon + \varepsilon^2)x_\varepsilon \ , \qquad x_\varepsilon(0) = 1 \ , \ \dot{x}_\varepsilon(0) = 0 \ ,$$

the solution is

$$\begin{aligned} x_0(t) &= \cos t \\ x_1(t) &= -t \sin t \\ x_2(t) &= -t^2 \cos t \\ &\cdots \ . \end{aligned}$$

This series indeed converges to the right solution $x_\varepsilon(t) = \cos((1+\varepsilon)t)$ as well as one can expect (it is entire in ε and in t) but, if one truncates, one can see that the approximate solution thus obtained blows up. Indeed, the more terms one takes, the more severe the blow up is. On the other hand, the true solution remains bounded for all times.

Hence, these series are unable to predict long term behavior, even in those extremely favorable examples where the function is linear and the solutions are entire. Of course, this phenomenon only becomes worse if one considers other more complicated non-linear problems.

This phenomenon caused consternation when the phenomena above appeared in the study of the solar system and the instability of the solar system was confirmed to all orders in perturbation theory. The terms with powers in t became dominant for t of order of centuries, which gave then the name "secular" terms (in Latin "saeculum" means – among other things – century).

A more careful examination of the convergence proof and the quantitative estimates that lead to it, shows that one cannot trust this perturbation theory to order n except when $\varepsilon^n t \ll 1$. For non-entire perturbations, one should not use this naive perturbation method except when $\varepsilon t \ll 1$. (Fortunately, we will be using some more effective methods that can give information on perturbations over longer time scales.)

The Lindstedt series we have seen in Section 2.1 originated with the goal of obtaining a perturbation series that produced series which were always periodic or quasi-periodic (that is, free of secular terms).

A much more effective method to ascertain the long term behavior of systems is the following:

We try to find transformations g_ε in such a way that

$$g_{\varepsilon *} X_\varepsilon = X_0 \ . \tag{3.54}$$

This method is not restricted to Hamiltonian systems. Indeed, the very influential book [**BM61**] develops many applications to non-Hamiltonian systems (one can also find there Lindstedt series for dissipative systems).

This method is, however, very well suited for Hamiltonian systems because it is very easy to keep track of families of transformations of Hamiltonian systems and vector fields.

In the case where g_ε are canonical transformations and X_ε are Hamiltonian vector fields (i.e., $\mathrm{i}_{X_\varepsilon}\omega = \mathrm{d}H_\varepsilon$), as we saw in (3.40), the equation (3.54) reduces to

$$H_\varepsilon \circ g_\varepsilon = H_0 \ . \tag{3.55}$$

One should emphasize that in contrast with the more elementary "secular method", the validity of this method is not limited by the length of the orbit but rather by whether the orbit leaves the region where the transformation g_ε is defined.

In some cases, especially when there is some contraction (of course, this never happens for Hamiltonian systems), one can use the perturbation theory itself to show that this region is never left by the trajectories.

Note that if (3.55) is solved, then we have

$$g_\varepsilon^{-1} \circ \Phi_t^\varepsilon \circ g_\varepsilon = \Phi_t^0 \ ,$$

where Φ_t^ε and Φ_t^0 denote the flows of H^ε and H^0, respectively.

To solve (3.55), it is of paramount importance to parameterize the families g_ε in such a way that (3.55) can be solved order by order.

One possibility followed in old fashioned books (but not very practical in many applications) is to parameterize g_ε by their generating functions (see Section 3.7). One shortcoming of generating functions is that one needs to assume existence of a system of global coordinates which are mixed variables (or work in patches). Another shortcoming of this method is that the rule of composition is awkward and it involves solving implicit equations (see Section 3.7).

Another alternative, which is more geometrical is that of the Lie series. The basic idea is that we try to consider transformations as time one maps of Hamiltonian vector fields. Some more detailed tutorials on Lie series are [**MH92**], [**Mey91**], [**DF76**]. Some reviews of canonical perturbation theory from the point of view of Physicists including a variety of applications are [**Car81**], [**Omo86**].

It is customary to write the time one maps of a vector field $\mathcal{L}$ as $\exp(\mathcal{L})$. This notation is motivated by the remark that the space vector fields can be considered as the Lie algebra of the space of diffeomorphisms. Also, if we identify g with the operator

$$U_g : L^2(M) \to L^2(M) \ : g \mapsto U_g\varphi = \varphi \circ g \ ,$$

then $\exp(\mathcal{L}) = U_g$ in the usual sense of operator theory when $\mathcal{L}$ is a complete flow preserving volume.

This notation is very suggestive and one would also like to use tools of Lie group theory such as Baker-Cambell-Hausdorff formula

$$\begin{aligned}\exp(\varepsilon\mathcal{L})\exp(\varepsilon\mathcal{L}^1) = \exp\Big(&\varepsilon(\mathcal{L}+\mathcal{L}^1) + \varepsilon^2\frac{1}{2}[\mathcal{L},\mathcal{L}^1] + \varepsilon^3\mathcal{T}_3 \\ &+\cdots+\varepsilon^n\mathcal{T}_n+\cdots\Big) \ ,\end{aligned} \tag{3.56}$$

where [,] denotes the commutator and $\mathcal{T}_n$ is a sum of iterated commutators of $\mathcal{L}, \mathcal{L}^1$.

Even if the sums in (3.56) cannot be considered as convergent, the formula can be justified in an appropriate weak sense ([**dlLMM86**]), and it is true that when applying the formula up to order n, we have, for sufficiently differentiable vector fields,

$$\left\| \left[\exp(\varepsilon\mathcal{L}) \exp(\varepsilon\mathcal{L}^1) - \exp\left(\sum_{n=1}^{N} \varepsilon^N \mathcal{T}_n \right) \right] \varphi \right\|_{C^r} \leq \tag{3.57}$$
$$\leq \varepsilon^{N+1} C_{\mathcal{L},\mathcal{L}^1,N} \|\varphi\|_{C^{N+r+2}} \ .$$

In spite of (3.57), it is not true that $\exp\left(\sum_{n=1}^{N} \varepsilon^n \mathcal{T}_n\right) \varphi$ converges as $N \to \infty$ even for an analytic φ (a sketch of a proof will be given later). It is, however, not difficult to obtain bounds $C_{\mathcal{L},\mathcal{L}^1,N} \leq (N!)^k$ for some $k > 0$, so that (3.57) can be used quite quantitatively.

In connection with (3.56) it is interesting to note that the commutator of two locally Hamiltonian vector fields is globally Hamiltonian (see Proposition 3.21). Hence, even if $\mathcal{L}$, $\mathcal{L}^1$ are only locally Hamiltonian, all the $\mathcal{T}_n$'s are globally Hamiltonian and can, therefore, be described by the Hamiltonian function.

There are several variants of the method of Lie transforms that have been considered in the literature depending on how we write our candidate map in terms of exponentials (time-one maps) of Hamiltonian vector fields. In order of historical appearance some of the methods proposed in the literature are:

$$g_\varepsilon = \exp(\varepsilon\mathcal{L}_1 + \varepsilon^2\mathcal{L}_2 + \cdots + \varepsilon^n\mathcal{L}_n + \cdots) \ , \tag{3.58}$$

$$g_\varepsilon = \cdots \exp(\varepsilon^n\mathcal{L}_n) \cdots \exp(\varepsilon^2\mathcal{L}_2) \exp(\varepsilon\mathcal{L}_1) \ , \tag{3.59}$$

$$g_\varepsilon = \cdots \exp\left(\sum_{i=2^n}^{2^{n+1}-1} \varepsilon^i \mathcal{L}_i \right) \cdots \exp(\varepsilon^3\mathcal{L}_3 + \varepsilon^2\mathcal{L}_2) \exp(\mathcal{L}_1) \ . \tag{3.60}$$

(See [**Dep70**],[**DF76**],[**dlLMM86**] respectively.)

The recursive equation for the perturbation expansions can be computed rather straightforwardly if we use with abandon — we can if we interpret the formulas in the asymptotic sense – the formulas $\exp\mathcal{L} = \sum_{i=0}^{\infty} \frac{1}{n!}\mathcal{L}^n$, think of the $\mathcal{L}^n$ as differential operators and rearrange the expressions according to the rules of of non-commutative algebra.

For example, in (3.58) we obtain:

$$\exp(\varepsilon\mathcal{L}_1 \cdots + \varepsilon^n\mathcal{L}_n) H_\varepsilon = H_0$$

$$\mathcal{L}_1 H_0 + H_1 = 0$$

$$\left(\tfrac{1}{2}\mathcal{L}_1^2 + \mathcal{L}_2\right) H_0 + \mathcal{L}_1 H_1 + H_2 = 0$$

$$\left[\tfrac{1}{6}\mathcal{L}_1^3 + \tfrac{1}{2}(\mathcal{L}_1\mathcal{L}_2 + \mathcal{L}_2\mathcal{L}_1) + \mathcal{L}_3\right] H_0$$
$$+ \left(\tfrac{1}{2}\mathcal{L}_1^2 + \mathcal{L}_2\right) H_1 + \mathcal{L}_1 H_2 + H_3 = 0 \ .$$

A point that we would like to emphasize is that the equation that we obtain in the three schemes (3.58), (3.59), (3.60) for Lie series is always

$$\mathcal{L}_n H_0 + H_n = R_n \ , \tag{3.61}$$

where R_n is an expression that depends only on previously computed terms.

Using (3.49), we can transform (3.61) into

$$-\mathcal{H}_0 L_n + H_n = R_n \ . \tag{3.62}$$

Note that, if we have a theory for the solutions of equations of the form (3.62), we can proceed along the perturbation schemes above.

Note that if we take

$$H_0(p,q) = \omega \cdot p \ ,$$

then (3.62) reduces to the equation (3.27) that we have studied (under Diophantine assumptions on ω) in Section 3.4. Both the data and the unknown in (3.62) have an extra variable, but since it enters as a parameter, we can discuss the regularity of the equation in terms of the theory that we have developed.

Perhaps more importantly, we note that if we have a good theory of approximate solutions of (3.62) we can solve the hierarchies of equations approximately. This is important in practice as well as in some proofs on KAM theorem.

We also note that an integrable system $H_0(p)$ can be written using the Taylor expansion

$$H_0(p) = H_0(0) + \omega \cdot p + O(p^2)$$

Hence, we can solve very approximately (3.62) in a sufficiently small neighborhood of $\{p = 0\}$. This is what is actually used in KAM theory.

These algorithms are also practical tools that can and have been implemented numerically. The next two remarks are concerned with some issues about numerical implementations.

In [**dlLMM86**] one can find an appendix where it is shown that the theories based in the three schemes above (and in others) are equivalent in the sense that they give results which are equivalent in the sense of asymptotic series.

REMARK 3.22. We emphasize that although all the schemes (3.58), (3.59), (3.60) are formally equivalent in the sense that they require solving the same equations, they are not at all equivalent from the point of view of efficiency and stability of the numerical implementation or from the point of view of detailed estimates or even convergence.

As we pointed out, the exponential of vector fields does not cover any neighborhood of the origin in the group of diffeomorphisms so that (3.58) does not provide with a good parameterization of a neighborhood of the identity and, perhaps relatedly, it is known to be outperformed in stability etc. by (3.59) [**DF76**].

The method (3.60) [**dlLMM86**] is actually convergent in many cases. Indeed, the KAM theorem asserts it does converge in certain cases as we will see. For example, it is convergent for the perturbation series that are based in Kolmogorov's method' that will be discussed in Section 6.1.

The only numerical implementations of (3.60) that I know of are some tentative ones carried out by A. Delshams and the author, but it seems that the scheme (3.60) has a very good chance to be very efficient and stable. Indeed, it seems to be the only method for which it is possible to establish convergence.

REMARK 3.23. Sometimes in the numerical solution of the equations (3.58), (3.59), (3.60) it is sometimes advantageous – both from the point of view of speed and of reliability – not to proceed order by order but rather to take groups of orders $[2^n, 2^{n+1} - 1]$.

This is tantamount to solving the equations by a Newton method in the space of families. It has the disadvantage over the order by order algorithm that at every stage one has to solve a different equation. This inconvenience is sometimes offset by the advantage that one linear equation allows one to study many orders and because the equations that need to be solved may be more stable than those of other methods.

These quadratic algorithms can be used for all the three methods described above. Nevertheless, they are somewhat easier to implement in (3.60) which has some quadratic convergence already in place.

We emphasize that all the methods can be studied either order by order or quadratically.

I think that it would be quite important to have a better theory of these algorithms.

One lemma that we will be using later is that it is possible to approximate the action of the Lie transform on functions by just the first term in the series of the exponential.

LEMMA 3.24. *Let H, G be functions on $\mathbf{T}^n \times \mathbf{R}^n$ endowed with the canonical symplectic structure. We use the notation of* (3.12) *for the analytic norms of functions.*

Assume that:

i) $\|H\|_\sigma$ *is finite.*

ii) *For a constant C which depends only on the dimension, we have for $\delta > 0$*

$$\delta^2 > C\|G\|_\sigma \; . \tag{3.63}$$

Then, for another constant $\tilde{C}$ depending only on the dimension, we have:

$$\|H \circ \exp \mathcal{L}_G - H - \{H, G\}\|_{\sigma-\delta} \leq \tilde{C}\delta^{-4}\|G\|_\sigma^2 \|H\|_\sigma \tag{3.64}$$

PROOF. By Cauchy estimates, (3.13), we have:

$$\|\nabla G\|_{\sigma-\delta/2} \leq \hat{C}\delta^{-1}\|G\|_\sigma . \tag{3.65}$$

with $\hat{C}$ a constant that depends only on the dimension.

The constant in (3.63) is chosen so that the R.H.S. of (3.65) is smaller than $\delta/2$.

Therefore, all the trajectories of the Hamiltonian flow generated by G which start in the region

$$\mathcal{D}_{\sigma-\delta} \equiv \{|I| \leq e^{2\pi(\sigma-\delta)}, |\operatorname{Im}(\phi)| \leq \sigma - \delta\}$$

do not leave the region $\mathcal{D}_{\sigma-\delta/2}$ for a time smaller than one (note that they are moving at an speed that does not allow them to transverse the region separating the domains in a unit of time). Hence,

$$\exp(\mathcal{L}_G)(\mathcal{D}_{\sigma-\delta}) \subset \mathcal{D}_{\sigma-\delta/2} \; .$$

In particular, we can define the composition $H \circ \exp(\mathcal{L}_G)$ in $\mathcal{D}_{\sigma-\delta}$.

For any point (I, ϕ), we can estimate the difference along a trajectory by using the Taylor theorem with remainder along a trajectory. It suffices to estimate the second derivative of H and the square of the displacement. The second derivative of H can be estimated by Cauchy estimates (3.13) $\|\nabla^2 H\|_{\sigma-\delta/2} \leq \tilde{C}\delta^{-2}\|H\|_\sigma$.

The displacement can be estimated by $\|\nabla G\|_{\sigma-\delta/2}$, which by Cauchy estimates (3.13) can be estimated by $\tilde{C}\delta^{-1}\|G\|_\sigma$.

Putting these two estimates together, obtains the desired result. □

REMARK 3.25. Analogues of Lemma 3.24 are true in any analytic symplectic manifold. One just needs to define appropriately norms of analytic functions, Cauchy inequalities, etc. In the versions of KAM theory that we will cover in this tutorial, the version we have stated is enough, but the reader is encouraged to formulate and prove the more general versions.

It is also possible to develop a canonical perturbation theory for maps. Again, the main idea is to change variables so that the system becomes close to the system which is "well understood".

The perturbative equation in this case becomes

$$g_\varepsilon^{-1} \circ f_\varepsilon \circ g_\varepsilon = f_0 \ . \tag{3.66}$$

We should think of those equations as equations for g_ε given f_ε.

These equations have been dealt with traditionally by parameterizing f_ε using the generating functions method, and similarly for the g_ε.

A more geometric method to use in perturbation theory is the method of deformations which was introduced in singularity theory. (In the book [**MH92**], one can also find this method introduced in the Lie transform method.) It seems particularly well suited to discuss conjugacy equations of a geometric nature. (See [**dlLMM86, BdlLW96**] for some global geometric applications.) We write

$$\frac{d}{d\varepsilon} f_\varepsilon = \mathcal{F}_\varepsilon \circ f_\varepsilon \ , \qquad \mathcal{F}_\varepsilon = \mathcal{I}_\omega(\mathrm{d}F_\varepsilon) \ .$$

We refer to f_ε as a family, $\mathcal{F}_\varepsilon$ as the generator and to F_ε as the Hamiltonian and adopt the typographical convention of using the same letter to denote the objects associated with the same family but using lowercase to denote the family, calligraphic font to denote the generator and capital to denote the Hamiltonian.

We note that, under the assumption that $\mathcal{F}_\varepsilon$ is C^1, given the generator and the initial point f_0 of the family, we can reconstruct f_ε in a unique way. Hence, given $F_\varepsilon \subset C^2$, and f_0 we can reconstruct f_ε.

If we express equation (3.66) in terms of the generators, it becomes

$$-\mathcal{G}_\varepsilon + \mathcal{F}_\varepsilon + f_{\varepsilon *}\mathcal{G}_\varepsilon = 0 \ . \tag{3.67}$$

Expressed in terms of Hamiltonians, it reads

$$-G_\varepsilon + F_\varepsilon + f_{\varepsilon *}G_\varepsilon = 0 \ . \tag{3.68}$$

(In the Hamiltonian case, we recall $f_{\varepsilon *}G_\varepsilon = G_\varepsilon \circ f_\varepsilon$.)

There are several advantages in expressing equation (3.66) in terms of the generators and the Hamiltonians:

- The equations in terms of the generators are linear. This is natural if we think that the vector fields are infinitesimal quantities which can, therefore, enter only linearly.
- The geometric structure — not only symplectic, but also volume preserving and contact (which we have not and will not discuss in these lectures) are taken care without any extra constraint.
- These equations are geometrically natural and can be formulated globally.

The proof that (3.67) and (3.68) are equivalent to (3.66) follows easily from the observation that

$$
\begin{aligned}
k_\varepsilon &= f_\varepsilon \circ g_\varepsilon \\
\Longleftrightarrow \mathcal{K}_\varepsilon &= \mathcal{F}_\varepsilon + f_{\varepsilon *}\mathcal{G}_\varepsilon \quad ; \quad k_0 = f_0 \circ g_0 \\
\Longleftrightarrow K_\varepsilon &= F_\varepsilon + f_{\varepsilon *}G_\varepsilon \quad ; \quad k_0 = f_0 \circ g_0 \ .
\end{aligned}
\tag{3.69}
$$

Even if the equations (3.68) is linear in the Hamiltonian F_ε, we should keep in mind that f_ε depends on F_ε through the very non-linear process of solving the corresponding ODE.

Nevertheless, one can approximate (3.68) by

$$
F_\varepsilon - G_\varepsilon + f_{0*}G_\varepsilon = 0 \ . \tag{3.70}
$$

When $f_0(I,\phi) = (I, \phi+\omega)$, this equation – for a fixed I – has the form of (3.27) the difference equations which were studied in Section 3.4. Since I can be considered as just a parameter in the data for the equation, we can use the regularity theory derived for (3.27).

If G_ε is a solution of (3.70), we note that

$$
F_\varepsilon - G_\varepsilon + f_{\varepsilon *}G_\varepsilon = (f_{\varepsilon *} - f_{0*})G_\varepsilon \ . \tag{3.71}
$$

The intuition is that if F_ε is small, we can think that G_ε (obtained by solving a linear equation with F_ε as R.H.S.) is small and that $f_{\varepsilon *} - f_0$ (obtained by solving a differential equation which involves derivatives of F_ε) is also small. Hence, the term in the R.H.S. of (3.71) is "quadratically" small.

Using the estimates in Lemma 3.15 and mean value theorem etc., we can prove the estimate in the analytic spaces

$$
\|(f_{\varepsilon *} - f_{0*})G_\varepsilon\|_{\sigma-\delta} \le C\delta^{-2\nu-4}\|F_\varepsilon\|_\sigma \ .
$$

Similarly, for the finitely differentiable case,

$$
\|(f_{\varepsilon *} - f_{0*})G_\varepsilon\|_{\Lambda^r} \le C\|F_\varepsilon\|^2_{\Lambda^{r+\nu+4}} \ .
$$

Note also that if we write

$$
F_\varepsilon = \varepsilon F_1 + \varepsilon^2 F_2 + \cdots
$$

and try to find

$$
G_\varepsilon = \varepsilon G_1 + \varepsilon^2 G_2 + \cdots \ ,
$$

then (3.68) can be turned into a hierarchy of equations for the G_n's. All the equations are of the form

$$
G_n - f_{0*}G_n + F_n = R_n \ ,
$$

where R_n is an expression involving previously computed terms.

REMARK 3.26. For later developments, it is important to note that both (3.66) and (3.68) (and (3.67), (3.68)) have a "group structure".

This means that if we can find an approximate solution g_ε (e.g., by solving the first order equations), we can perform the (3.70), (3.62) change of variables and set

$$
\begin{aligned}
\tilde{f}_\varepsilon &= g_\varepsilon^{-1} \circ f_\varepsilon \circ g_\varepsilon \\
\tilde{H}_\varepsilon &= H_\varepsilon \circ g_\varepsilon \ .
\end{aligned}
\tag{3.72}
$$

If we solve the problem for $\tilde{f}_\varepsilon$, $\tilde{H}_\varepsilon$, i.e.,

$$\begin{aligned} \tilde{g}_\varepsilon^{-1} \circ \tilde{f}_\varepsilon \circ \tilde{g}_\varepsilon &= f_0 \\ \tilde{H}_\varepsilon \circ \tilde{g}_\varepsilon &= H_0 \ , \end{aligned} \tag{3.73}$$

then, we have solved the original problem since joining (3.72) and (3.73), we obtain

$$\begin{aligned} (g_\varepsilon \circ \tilde{g}_\varepsilon)^{-1} \circ f_\varepsilon \circ g_\varepsilon \circ \tilde{g}_\varepsilon &= f_0 \\ H_\varepsilon \circ g_\varepsilon \circ \tilde{g}_\varepsilon &= H_0 \ . \end{aligned}$$

The importance of the above observation, which will be appreciated later, is that, by making successive changes of variables, we can eliminate all the linear terms of the error by solving an equation which is just the linearized equation *at the integrable system.*

This is an important difference with the standard Newton method since the standard Newton method requires that we solve the linearized equation in a neighborhood.

The fact that we can obtain a method that, for all purposes is like a Newton method but which nevertheless only requires that we know how to solve one linearized equation depends crucially on the fact that the equations that we are studying have a particular structure which is called *group structure* and that will be discussed much more in Section 5, in particular, Remark 5.7 and Exercise4.23.

3.7. Generating functions. One of the reasons why Hamiltonian mechanics is so practical is because of the ease with which one can generate enough canonical transformations.

In old fashioned books ([**Whi88**], [**Gol80**], [**LL76**]) one can find that canonical transformations are described in terms of generating functions. We will describe those briefly and only for purposes of comparing with older books. It should be remarked however, that generating functions, even if not so useful from the point of view of transformation theory (there are better tools such as Lie transforms) are still quite useful tools in the variational formulation of Hamiltonian mechanics, providing thus a valuable link to Lagrangian mechanics. Moreover, some of the constructions that appear in generating functions are quite natural in optics. See [**BW65**].

The equation

$$f^*\theta - \theta = \mathrm{d}S$$

is written in old fashioned notations as

$$p'\,\mathrm{d}q' - p\,\mathrm{d}q = \mathrm{d}S \ , \tag{3.74}$$

where $p\,\mathrm{d}q := \sum_{i=1}^k p_i\,\mathrm{d}q_i$, etc. This should be interpreted as saying that we consider the coordinate functions p_i, q_i and the transformed functions $p_i' = p_i \circ f$, $q_i' = q_i \circ f$. Then, $\theta = p\,\mathrm{d}q$ and $f^*\theta = p'\,\mathrm{d}q'$; S is a function on the manifold.

When q, q' are a good coordinate system (i.e. p can be expressed as a function of q and q', $p = p(q, q')$), we can define a function $\mathcal{S} : \mathbf{R}^n \times \mathbf{R}^n \to \mathbf{R}$ by setting $\mathcal{S}(q, q') := S(q, p(q, q'))$. Usually, in old fashioned notations, this is described as "expressing S in terms of q and q'" or simply by writing "$S = S(q, q')$" or something to that effect. Very often the same letter is used for S, $\mathcal{S}$.

REMARK 3.27. In old fashioned notation in mechanics, the same letter is used for the functions that give the same result irrespective of the arguments. Of course, even if this is almost manageable and one understand what is meant by $S(q,p)$, $S(q,q')$, by paying attention to the arguments this notation wrecks havock when one tries to evaluate at concrete points. For example, what is meant by $S(2,\pi)$ when one is considering at the same time $S(q,p)$, $S(q,q')$?

Note however, that the assumption that q, q' is a system of coordinates is far from trivial. To begin with, it is not obvious that the manifold on which we are working admits a system of coordinates. Even if it does, or if we work just on a neighborhood so that we have local coordinates, there are other conditions to be imposed. For example, it is false for the identity and for transformations close to identity, it may be a system of coordinates with undesirable properties. It is, however, true for $(p,q) \mapsto (p, q+p)$ and small perturbations.

In that case, when we compute the differential in (3.74), we have

$$\mathrm{d}S = \partial_1 \mathcal{S}(q,q')\,\mathrm{d}q + \partial_2 \mathcal{S}(q,q')\,\mathrm{d}q' ,$$

hence

$$p = -\partial_1 \mathcal{S}(q,q') , \qquad p' = \partial_2 \mathcal{S}(q,q') . \tag{3.75}$$

We think of (3.75) as of an equation for p', q' in terms of p, q. If the implicit function theorem applies (for which it suffices that q, q' provide a good system of coordinates on the manifold) and indeed the equations (3.75) can be solved differentiably, $\mathcal{S}$ determines the transformation. Note that the implicit function theorem will apply in a C^2 open set of functions $\mathcal{S}$, so that we can think of this procedure as giving a chart of some subset of the space of symplectic mappings. Also note that we parameterize the transformation by one scalar function. Moreover, the changes of variables given by (3.75) are automatically symplectic. Keeping track of transformations – in an open set – which satisfy some non-linear and non-local constraints (preserving the symplectic structure) by just keeping track of a function is a great simplification.

However, one important shortcoming of these generating functions is that for the identity transformation, q, q' is not a good system of coordinates on the manifold and we cannot use (3.75) to represent the identity or near identity transformations. As we have seen, near identity transformations play an important role in canonical perturbation theory, so, it is necessary to devise variants of the method to incorporate them.

In the case that the coordinate functions p, q are global (or that we just work on a neighborhood), we can write

$$p\,\mathrm{d}q = -q\,\mathrm{d}p + \mathrm{d}(pq) .$$

Hence (3.74) reads

$$p'\,\mathrm{d}q' + q\,\mathrm{d}p = \mathrm{d}(S + pq) . \tag{3.76}$$

In the case that p, q' is a good system of coordinates (as happens in a neighborhood of the identity), we can write

$$S + pq = \tilde{\mathcal{S}}(p,q')$$

and from (3.76) we see that

$$q = \partial_1 \tilde{\mathcal{S}}(p,q') , \qquad p' = \partial_2 \tilde{\mathcal{S}}(p,q') .$$

Again, we can consider this as a system of implicit equations defining p', q' in terms of p, q.

Note that if q is an angle, then $\mathcal{S}(q+k, q'+\ell) = \mathcal{S}(q, q')$ for all $k, \ell \in \mathbf{Z}^n$. On the other hand, $\tilde{\mathcal{S}}(p, q'+\ell) = \mathcal{S}(p, q') + p\,\ell$. Even if this generating function works in neighborhoods of the identity, it does not work at all for the map $(p, q) \mapsto (-q, p)$.

One can use similar procedures to obtain many other generating functions.

For example, one can use for a partition of $\{1, \ldots, d\}$ into two sets $\mathcal{A}$ and $\mathcal{B}$ the formula:

$$-\sum_{i\in\mathcal{B}} p'_i \,\mathrm{d}q'_i + \sum_{i\in\mathcal{A}} p_i \,\mathrm{d}q_i = \sum_{i\in\mathcal{B}} q'_i \,\mathrm{d}p'_i - \sum_{i\in\mathcal{A}} q_i \,\mathrm{d}p_i + \mathrm{d}\left(-\sum_{i\in\mathcal{B}} p'_i q'_i + \sum_{i\in\mathcal{A}} p_i q_i\right)$$

to change some of the p_i's for q_i's in the push-forward.

Even if these procedures are quite customary in old fashioned mechanics treatises, they will not be very useful for us. Again, we emphasize that even if the q, q' generating function can be defined in any exact manifold, the others seem to require some extra structure, which can be arranged in small neighborhoods.

We note however, that the function S has a well defined intrinsic meaning as evidenced in [**BW65**] — this is sometimes described as Hamilton-Jacobi equation or "the action as a function of coordinates" depending on what interpretation one gives. We refer to [**Har99**] for much more information on this primitive function.

In Hamiltonian optics [**BW65**], S represents the phase of the wave. Indeed, Hamiltonian mechanics was developed as a byproduct of Hamiltonian optics. This explains why so much of Hamiltonian mechanics, especially in earlier treatises is based on studying S and its relatives.

More modern treatments ([**Arn89**], [**AM78**]) prefer to start from the symplectic geometry and postulate it without any other motivation that it eventually works. This is certainly expeditious.

4. Two KAM proofs in a model problem

In this section we will discuss one of the technically simplest applications of the KAM methodology: the Siegel center theorem.

The main goal of this application is to show in action perhaps the most basic heuristic principle of the KAM method:

Quadratic convergence can overcome small divisors.

Roughly speaking this means that if we have a method of improvement that reduces the error to something that is quadratic in the original error, even if the solution requires solving an equation which involves small denominators, we can still obtain convergence.

The fact that the convergence does indeed take place is rather subtle. In our opinion, the only way to appreciate the subtlety of the convergence achieved by KAM theory is to give a serious try to several other seemingly reasonable schemes and see them fail. At the end of the proof, we have suggested several of these schemes as exercises.

Besides those exercises, we have also included some exercises which admit easy solutions and provide extensions to the material in the text.

We also emphasize that the fact that one can get a quadratically convergent method solving only one small denominator equation is far from trivial and it requires that the equations we consider have some special structure. This will be elaborated in more detail in Section 5 and in particular in Remark 5.7.

In this section, we will present two versions of the Siegel theorem – one using just Diophantine conditions in one dimension, and another using approximation functions and decomposition in scales in higher dimensions.

The second proof will be formulated as a set of exercises. The main ideas of this section follow [**Mos66a**], [**Zeh77**] and [**Arn88**]. Indeed, we follow these references rather closely.

These two proofs will illustrate the main features of KAM proofs and contain the essential analytic and number theoretic difficulties even if they do not involve any geometry.

We will start with a one dimensional problem. See [**Mos66a**] for more details on the proof we present and [**Zeh77**] for a higher dimensional version.

THEOREM 4.1. *Let $f : U \subset \mathbf{C} \to \mathbf{C}$ be and analytic function of the form*

$$f(z) = az + \hat{f}(z) \tag{4.1}$$

with $\hat{f}(z) = O(z^2)$.

Assume that

$$|(a^n - 1)^{-1}| \leq n^\nu K \tag{4.2}$$

and that

$$\|\hat{f}\|_1 \leq \rho(\nu) K^2 \ , \tag{4.3}$$

where $\rho(\nu)$ is an explicit function.

Then there exists a unique function

$$h(z) = z + \hat{h}(z)$$

with $\hat{h}(z)$ analytic in a disc of radius

$$\sigma = 1 - 2\rho(\nu)$$

such that

$$f \circ h(z) = h(az) \ . \tag{4.4}$$

Moreover, we have

$$\|\hat{h}\|_\sigma \leq \|\hat{f}\|_1 C \ . \tag{4.5}$$

REMARK 4.2. The uniqueness for h claimed in Theorem 4.1 means that if there are two functions satisfying this they have to agree in an open set of the origin. As we have seen already, the condition (4.2) and (4.4) determine the jet of h uniquely.

REMARK 4.3. Condition (4.2) is automatic when $|a| \neq 1$. In that case, we have presented a simple proof already. So we will restrict ourselves to the case when $|a| = 1$.

REMARK 4.4. It is a standard observation that, assuming that f is defined in a ball of radius 1 and small is the same as considering a small neighborhood.

Heuristically, in a small neighborhood, the linear part is the dominant term and it is natural to try to describe the behavior of the whole system in terms of the behavior of the linear one.

More precisely, given f, consider for λ small

$$f_\lambda = \lambda^{-1} f(\lambda z) \ .$$

Notice that f_λ has the same linear part and is defined in $\lambda^{-1}B_r$ if f is defined on B_r. Since $|\hat{f}(z)| = O(|z|^2)$, we have $\|\hat{f}_\lambda\|_{B_1} \le C\lambda$.

If we apply Theorem 4.1 to f_λ, we obtain a h_λ. Then h will satisfy (4.4).

REMARK 4.5. Condition (4.2) is not optimal. Later we will discuss how to obtain the same result when the arithmetic condition (4.2) is replaced by the Brjuno condition, which is indeed optimal as shown in [**Yoc95**],[**PM92**].

The fact that if Brjuno condition fails one can construct counterexamples is considerably deeper and out of the scope of these notes. See the references above.

Before embarking in the proof, we note that all the methods are based in estimates for the equation

$$\varphi(az) - a\varphi(z) = \eta; \quad \varphi(0) = 0 \tag{4.6}$$

in which we consider η and a as given and we are to determine φ.

The analysis of this equation is very similar to the analysis of (3.27) in Section 3.4. Since it is not completely identical, we need to start by revising slightly the definitions of norms and the setup.

We define the norm of an analytic function by[6]

$$\|f\|_r = \sup_{|z|\le r} |f(z)|.$$

LEMMA 4.6. *Assume that a satisfies* (4.2). *Then, if $\eta(0) \equiv \eta_0 = 0$ we can find a solution of* (4.6). *Moreover*

$$\|\varphi\|_{re^{-\delta}} \le CK|\delta|^{-\tau}\|\eta\|_r \tag{4.7}$$

PROOF. This follows from the results in Section 3.4.

It suffices to write $z = \exp(2\pi i\theta)$. Then, the result stated is a particular case of Lemma 3.15 applied to a Fourier series which only has positive terms. ☐

EXERCISE 4.7. Give a direct proof of the Lemma 4.6

One can follow the sketch in the beginning of Section 3.4.

Start by observing that the solution of (4.6) is $\varphi_k = \eta_k(a^k - a)^{-1}$. Estimate $|\varphi_k|$ using the above formula, (4.2), and the estimates for $|\eta_k|$ in terms of $\|\eta\|_r$ obtained using Cauchy estimates.

Estimate $\|\varphi\|_{re^{-\delta}}$ by the sup of the coefficients. Then, one ends up with the desired result with $\tau = \nu + 1$.

Since we are dealing with analytic estimates, this is enough to get through the proof. The ambitious reader is invited to carry out an analysis similar to that in [**Rüs76b**] and obtain the optimal exponent.

Now, we proceed to the proof of Theorem 4.1. The proof we present here follows [**Zeh77**] – it is a particular case of the results of that paper.

PROOF. Proceeding heuristically for the moment, we can think of (4.4) as an implicit equation in a space of functions

$$0 = \mathcal{T}(f,h) \equiv f \circ h - h \circ a$$

[6]These norms are slightly inconsistent with those in Section 3.1 in which we took $\|f\|_\sigma = \sup_{|z|\le e^\sigma} |f(z)|$. The convention of Section 3.1 is more natural when one is using at the same time Fourier series and Taylor series. For the present section, the convention we now take is more natural.

(by a we denote either the constant or the function $a(z) = az$). Note that $\mathcal{T}(a, \mathrm{Id}) = 0$.

We consider f fixed (but close to a) and we are given an approximate solution h

$$\mathcal{T}(f, h) \equiv f \circ h - h \circ a = R \ , \tag{4.8}$$

where R is the remainder which we would like to think of as small (the precise sense in which it is small will not be made explicit in this heuristic discussion).

We would like to obtain a Δ that eliminates most of R so that $\mathcal{T}(f, h+\Delta) \ll R$. This amounts to a Newton's method. Since

$$\mathcal{T}(f, h + \Delta) \approx \mathcal{T}(f, h) + D_2\mathcal{T}(f, h)\Delta \ ,$$

we are lead to consider the equation for Δ

$$R + D_2\mathcal{T}(f, h)\Delta = 0 \ . \tag{4.9}$$

In our case – remember that we are, for the moment, just proceeding heuristically, but this step is not difficult to justify – we have that the derivatives will be:

$$D_2\mathcal{T}(f, h)\Delta = (f' \circ h)\Delta - \Delta \circ a \ .$$

Hence, in our case (4.9) becomes:

$$(f' \circ h)\Delta - \Delta \circ a = -R \ . \tag{4.10}$$

If the factor $f' \circ h = a + \hat{f}' \circ h$ were just a, the equation (4.10) would reduce to those considered in Lemma 4.6.

One way that succeeds in reducing the annoying $\hat{f}' \circ h$ to a constant is the following: (in the exercises we examine several seemingly natural methods which do not work).

Take derivatives with respect to z of (4.10) and obtain the identity

$$f' \circ h\, h' - a h' \circ a = R' \ . \tag{4.11}$$

If rather than looking for Δ, we look for w defined by $\Delta = h' w$ (remember h is close to the identity, so that indeed $1/h'$ is an analytic function so that looking for Δ and for w is equivalent), equation (4.10) becomes

$$f' \circ h\, h' w - h' \circ a\, w \circ a = -R \ . \tag{4.12}$$

Substituting (4.11) in (4.12), we are lead to

$$a\, h' \circ a\, w - h' \circ a\, w \circ a = -R - R' w \tag{4.13}$$

or

$$a\, w - w \circ a = -(h' \circ a)^{-1} R - (h' \circ a)^{-1} R' w \ . \tag{4.14}$$

If we ignore the term $(h' \circ a)^{-1} R' w$ (the intuition, which we will later turn into rigorous estimates, says that $h' \circ a$ is of order one, R and R' are small, hence w is small and $R'w$ is much smaller), we simplify the problem to studying

$$aw - w \circ a = -(h' \circ a)^{-1} R \ , \tag{4.15}$$

which indeed is an equation of the type we considered in Lemma 4.6.

Hence, the prescription that we have derived heuristically to obtain a more approximate solution is:

(1) Take w solving (4.15);
(2) Form $\Delta = h'w$;

(3) Then, $h + \Delta$ should be a better solution to the problem.

Now, we turn to making all the previous ideas rigorous. We will need to show that the procedure improves (that is, show estimates for the remainder after one step given estimates on the remainder before starting). We will also need to show that the procedure can be repeated infinitely often and that it leads to a convergent procedure.

If we are given a system with an remainder and run the procedure outlined above, the following lemma will establish bounds for the new remainder in terms of the original one.

We will follow standard practice in KAM theory and denote by C throughout the proof constants that depend only on the dimension and other parameter which are fixed in our proof. In our case, since we are paying special attention to the dependence of the domain loss parameter on the size of the Diophantine constants and the smallness assumptions, C will not depend on them. Other KAM proofs which emphasize other features may allow C to stand for constants that could also depend on the Diophantine constants.

LEMMA 4.8. *Let f be as in Theorem 4.1, $h(z) = z + \hat{h}(z)$, $(\hat{h}(z) = O(|z|^2))$ defined in a ball of radius $\frac{1}{2} < \sigma < 1$ satisfy*

$$\|\hat{h}'\|_\sigma \le M \le 1/2 \; . \tag{4.16}$$

with

$$\sigma + M < 1 \; , \tag{4.17}$$

$$\|f \circ h - h \circ a\|_\sigma \le \varepsilon \; . \tag{4.18}$$

Assume furthermore that $\delta > 0$ is such that

$$KC\delta^{-\nu-1}\varepsilon + \sigma e^{-\delta} < \sigma \; . \tag{4.19}$$

Then, the prescription above can be carried out and we have:

$$\|f \circ (h + \Delta) - (h + \Delta) \circ a\|_{\sigma e^{-\delta}} \le KC\delta^{-\nu-1}\varepsilon^2 + 2\|f\|_1 (KC\delta^{-\nu-1}\varepsilon M)^2 \; . \tag{4.20}$$

REMARK 4.9. Notice that since for $\delta \ge 0$, $\sigma(1 - e^{-\delta}) \le \sigma\delta$, condition (4.19) is implied by

$$\sigma\delta \ge CK\delta^{-\nu-1}\varepsilon \; . \tag{4.21}$$

which, once we have σ, just tells us that δ cannot be smaller than a power of ε.

REMARK 4.10. Note that if we assume without loss of generality that $\|\hat{f}\|_1 \le 2$, $K \le K^2$, $\delta < 1$, the R.H.S. of (4.20) is less or equal to

$$CK^2\varepsilon^2\delta^{-2(\nu+1)} \; . \tag{4.22}$$

PROOF. To check that the prescription can indeed be carried out, we just need to check that the function $f \circ (h + \Delta)$ can be defined. Hence, our first goal will be to obtain estimates on Δ and show that the image of the ball of radius $re^{-\delta}$ under $h + \Delta$ is contained in the domain of f. Indeed, the estimates for the range will allow us also to obtain estimates for the derivative of f via the Cauchy theorem which will later prove to be useful.

Then, we will obtain the estimates in (4.20) and (4.22) provided that we have suitable estimates on $\|\Delta\|_{\sigma e^{-\delta}}$.

To obtain the estimates on $\|\Delta\|_{\sigma e^{-\delta}}$, we note that using the Banach algebra property of the norms and the inductive assumption (4.16), we can bound the R.H.S. of (4.15) by

$$\|(h' \circ a)^{-1} R\|_\sigma \leq (1 - 1/2)^{-1} \|R\|_\sigma \ .$$

By Lemma 4.6 we have that

$$\|w\|_{\sigma e^{-\delta}} \leq K C \delta^{-\nu} \|R\|_\sigma \ .$$

By Cauchy estimates, (see Lemma 3.13, but take into account that now we are in an slightly different situation), we have:

$$\begin{aligned} \|h' \circ a\|_{\sigma e^{-\delta}} &\leq K\delta^{-1}\|h\|_\sigma \ , && (4.23)\\ \|R'\|_{\sigma e^{-\delta}} &\leq K\delta^{-1}\varepsilon \ . && (4.24) \end{aligned}$$

Hence, taking into account (3.14), and that we had called $\|R\|_\sigma = \varepsilon$, we obtain from the previous results:

$$\begin{aligned} \|\Delta\|_{\sigma e^{-\delta}} &\leq KC\delta^{-\nu-1}\varepsilon M \ , && (4.25)\\ \|R'w\|_{\sigma e^{-\delta}} &\leq KC\delta^{-\nu-1}\varepsilon^2 \ . && (4.26) \end{aligned}$$

Note that the assumption (4.16),

$$\|h + \Delta\|_{\sigma e^{-\delta}} < 1 \ ,$$

so that, as claimed, the composition in (4.20) indeed makes sense.

To obtain the estimates in (4.20), we consider the term to be estimated in (4.20) and the obvious identity obtained just by adding and subtracting terms to it and grouping the result conveniently.

$$\begin{aligned} f \circ (h + \Delta) &- (h + \Delta) \circ a \\ &= f \circ h - h \circ a + f' \circ h\, \Delta - \Delta \circ a \\ &+ [f \circ (h + \Delta) - f \circ h - f' \circ h\, \Delta] \ . \end{aligned} \tag{4.27}$$

The first four terms in the R.H.S. of (4.27), using (4.11) and (4.12) amount to:

$$R + h' \circ aw + R'w - ah' \circ aw = R'\, w \ .$$

The term in braces in (4.27) can be estimated because, by a calculus identity (Taylor theorem with the Lagrange form of the remainder)

$$\begin{aligned} f(h(z) + \Delta(z)) - f(h(z)) - f'(h(z))\Delta(z) &= \\ &= -\int_0^1 (s-1)\, f''(h(z) + s\Delta(z))\, \Delta^2(z)\, ds \ . \end{aligned} \tag{4.28}$$

Since, again by Cauchy bounds and (4.16) we have

$$\|f''(h(z) + s\Delta(z))\|_{\sigma e^{-\delta}} \leq C\delta^{-2}\|f\|_1 \ ,$$

we can bound the $\| \ \|_{\sigma e^{-\delta}}$ of (4.28) by

$$1/2\|f\|_1(KC\delta^{-\nu-1}\varepsilon M)^2 \tag{4.29}$$

If we estimate (4.27) putting together (4.26) and (4.29), and remembering the standing assumptions on M, $\|f\|_1$, we obtain (4.20).

□

To finish the proof of Theorem 4.1, we just need to show that if $\|\hat{f}\|_1$ is sufficiently small, we can repeat the iterative procedure arbitrarily often and that we converge to a limit which satisfies (4.5).

We will denote by subindices n the objects after n steps of the iterative process (assuming that it can be carried out this far). For example, σ_n will be the domain of definition of h_n and we have $\sigma_{n+1} = \sigma_n e^{-\delta_n}$. To simplify the discussion, we will use the condition (4.21) which implies (4.16) and the bounds (4.22).

The main thing that we have to do is to choose the δ_n's. Notice that if we choose δ_n going to zero slowly, we lose more domain than needed and end up with a weaker theorem –of course, if we lose too fast, we end up with an empty domain. On the other hand, the smaller that we choose δ_n, the worse (4.22) becomes.

A reasonable compromise that is neither too fast so that we end up with no domain nor too slow so that we can still converge is to choose an exponential rate of decay. In the exercises, we will explore other choices.

We will choose

$$\delta_n = \delta_0 2^{-n} , \tag{4.30}$$

and then, will show how to choose δ_0.

With this choice of δ_n, (4.22) implies easily

$$\varepsilon_{n+1} \le CK^2 \varepsilon_n^2 \delta_0^{-2\mu} A^{2n} \tag{4.31}$$

where $\mu = \nu + 1$, $A = 2^\mu$.

We assume by induction that the iterative step can be carried out n times (i.e., that hypothesis (4.21) is verified for the first n steps). We will show that, under certain assumptions on the size of δ_0, ε_0, which will be independent of n, hypothesis (4.21) will be verified for $n+1$. Moreover, we will show that ε_{n+1} decreases very fast. Then, by repeated application of (4.31) we have:

$$\begin{aligned} \varepsilon_{n+1} &\le CK^2 \delta_0^{-2\mu} \varepsilon_n^2 A^n \\ &\le (CK^2 \delta_0^{-2\mu})^{1+2} A^{n+2(n-1)} \varepsilon_{n-1}^{2\cdot 2} \\ &\le \cdots \\ &\le (CK^2 \delta_0^{-2\mu})^{1+2+2^2+\cdots+2^{n-1}\cdot 1} A^{n+2(n-1)+\cdots+2^{n-1}} \varepsilon_0^{2^{n+1}} . \end{aligned} \tag{4.32}$$

Note that $1 + 2 + 2^2 + \cdots + 2^n \le 2^{n+1}$ and without loss of generality, we can assume that $CK^2\delta_0^{-2\mu} > 1$. Similarly,

$$\begin{aligned} n + 2(n-1) + \cdots + 2^{n-1} \cdot 1 \\ &= 2^n [n 2^{-n} + (n-1)2^{-(n-1)} + \cdots + 2^{-1} \cdot 1] \\ &\le 2^n \sum_{k=1}^{\infty} k 2^{-k} = 2^n \cdot 2 = 2^{n+1} , \end{aligned}$$

hence

$$\varepsilon_{n+1} \le \left(CK^2 \delta_0^{-2\mu} \varepsilon_0 A \right)^{2^{n+1}} . \tag{4.33}$$

Notice that if $\rho \equiv CK^2 \delta_0^{-2\mu} A \varepsilon_0 < 1$, then (4.33) converges to zero extremely fast (faster than any exponential).

The equation that we need to satisfy to be able to perform the next step is

$$\delta_{n+1} \equiv \delta_0 2^{-(n+1)} \ge CK\delta_0^{-\mu} 2^{-n\mu} \varepsilon_{n+1} = CK\delta_0^{-\mu} 2^{-n\mu} \rho^{2^{n+1}}$$

or

$$CK\delta_0^{-\mu-1} \le 2^{n\mu-(n+1)}\rho^{-2^{n+1}} . \tag{4.34}$$

By now, it should be clear that if we take $\delta_0 = \frac{1}{2}$ (so that $\sigma_n \ge e^{-1}$), if we assume that ε_0 is sufficiently small, we can satisfy (4.34).

Moreover, since by (4.25),

$$\|\Delta_n\|_{e^{-1}} \le KC\delta_0 2^{\mu n}\rho^{2^n} ,$$

we see that $\sum \Delta_n < \infty$ Hence

$$\Delta \equiv \sum \Delta_n$$

converges uniformly in the space of functions in the disk of radius e^{-1} and we can easily bound $\|\Delta\|_{e^{-1}}$.

□

At the end of this subsection, we have collected some exercises that explore alternatives for the present proof and for another that will be presented.

Let us highlight some of the remarkable points of the proof.

REMARK 4.11. We call attention to the remarkable fact that the derivatives of (4.10) could be used to transform the equation (4.11) into a much simpler equation (with an error which is small if the remainder is small and of *quadratic order*).

This is what allowed us to solve the step with quadratic error. In turn, this quadratic error was crucial in being able to deal with the small divisors (see the following remark). See exercise 4.23 for an example of a problem with very similar analytical properties but without group structure for which the result is false.

The possibility of performing this remarkable simplification comes from the group structure of the equations, as was emphasized in [**Zeh76a**].

This remarkable cancellation has other justifications, for example, in the context of Lagrangian principles. Indeed, one can see that it is related to the symmetry that we used in (2.18). With a bit of hindsight we can see that the factor $(1+\ell^{[<n]'})$ used there is really an *infinitesimal translation* on the right for the data of the problem and that the cancellation is just a reflection of the fact that the original problem is invariant under translations (see the classical Noether theorem about variational principles with continuous symmetries). In Quantum Field Theory the identities that come from changes of variables are called Ward identities. The relation between Ward identities and the identities used in Lindstedt series has been emphasized in [**Gal94a**], [**BGK99**], which are papers designed to bridge the gap between the language of Quantum Field Theory and KAM theory. Of course, in QFT one often does not consider the objects as defined per se, but rather as formal power series.

REMARK 4.12. Once we have the iterative step and the estimates that give quadratic convergence, the rest is (even if miraculous and quite remarkable) by now well understood.

Indeed, there are several abstract formulations, some of which we will discuss later. See Section 5.

In what follows, we will emphasize the steps required to reach the quadratic convergence and leave to the reader the checking that the convergence indeed takes place.

Experience shows that, once one has worked out a few quadratically converging arguments it becomes faster and more reliable to work out a proof by oneself than to read the proofs by others. It is certainly more instructive for the reader and more comfortable for the writer. In this case, the reader should be assured by the existence of properly written papers that we reference where he/she is encouraged to look for extra details.

Obtaining the quadratically convergent algorithm in classical KAM theory is not obvious since it depends on cancellations given by the geometry or the structure of the problem which eliminate some terms which would result in a linearly convergent method.

Note that in the classical KAM theory, we are constrained by the fact that we know only how to solve one linearized equation (in contrast with the usual Newton method, where we can solve the linearized equation in a whole neighborhood.

In the remainder of this subsection, we will present a proof of the multidimensional case of Theorem 4.1 following [**Arn88**] chapter 28. The one-dimensional version of this proof is covered in [**SM95**] chapter 25.

THEOREM 4.13. *Let $f : U \subset \mathbf{C}^d \to \mathbf{C}^d$, be analytic in a polydisk. $f(0) = 0$. Denote $Df(0) = A$ and assume that A is diagonal and $\sigma = (\sigma_1, \sigma_2, \dots, \sigma_d)$, the spectrum of A satisfies:*

$$|\sigma^k - \sigma_i|^{-1} \le C|k|^\nu \quad k \in \mathbf{N}^d \quad |k| \ge 2, \quad i \in \{1, \cdots n\} \tag{4.35}$$

(where we use the customary multi-index notation $\sigma^k = \sigma_1^{k_1} \cdot \sigma_2^{k_2} \cdot \dots \cdot \sigma_d^{k_d}$, $|k| = k_1 + \cdots k_d$).

Then, we can find an $h : V \subset \mathbf{C}^d \to \mathbf{C}^d$, $h(0) = 0$, $Dh(0) = \mathrm{Id}$ such that in a neighborhood we have:

$$h^{-1} \circ f \circ h = A \tag{4.36}$$

The conclusion of (4.36) is again that f is just the linear map in other coordinates.

REMARK 4.14. Notice that we are not assuming that the σ_i have modulus 1. The multidimensional case can have several interesting examples in which some of the σ_i are smaller than 1 and others are greater than 1. In the case that there are no eigenvalues equal to 1 and that no product of eigenvalues is another eigenvalue, Sternberg theorem will guarantee us that there exists a C^∞ change of variables that reduces the system to a linear one. To obtain that the change of variables is analytic, we need not only that the products are not eigenvalues but also some quantitative estimates on how far they are such as (4.35). Note also that the C^∞ changes of variables produced by Sternberg theorem are not unique, whereas, as pointed out before, the analytic ones are unique.

REMARK 4.15. Note that, implicitly, condition (4.35) requires that there is no eigenvalue 0, hence A is invertible.

It is very easy to show that if one eigenvalue is 0 one should not expect the conclusion to be true.

REMARK 4.16. If we write $\sigma_j = \exp(2\pi i \omega_j)$ – with ω_j possibly complex numbers – we see that the (4.35) is equivalent to the fact that ω satisfies (3.19) but we only need it for $k \in \mathbf{N}^d$ rather than $k \in \mathbf{Z}^d$.

We will discuss the different stages of the proof but leave many details to the reader since this will provide some training and, moreover, it can be found in the references indicated.

Proceeding heuristically, we will note that if $h(z) = z + \hat{h}(z)$, we have $h^{-1}(z) = z - \hat{h}(z) + O([\hat{h}]^2)$. (Here in the O notation we allow to include derivatives. For example, $\hat{h}'\hat{h}''$ will be a term allowed in $O([\hat{h}]^2)$.)

If we assume that $f(z) = Az + \hat{f}(z)$ and that $\hat{f}$ is small, if we want to make the changes of variables that reduce f to linear with an smaller error, we have

$$h^{-1} \circ f \circ h(z) = A(z) + \hat{f}(z) - \hat{h} \circ A(z) - A\hat{h}(z) + O([\hat{h}]^2, [\hat{f}][\hat{h}]) \tag{4.37}$$

This suggests the following iterative step

1) Solve the following equation for $\hat{h}$

$$\hat{f}(z) = \hat{h} \circ A(z) - A\hat{h}(z) \tag{4.38}$$

2) Consider now the the map

$$\tilde{f} = h^{-1} \circ f \circ h(z)$$

If all works according to plan, $\tilde{f}$ will be much closer to the linear map A.

The approximations we have taken can be readily estimated by adding and subtracting as follows. (Ignore for the moment questions of domains of definition. Suffice it to say that the simple minded identities we obtain are supposed to hold in a domain near the origin. Later we will need to worry about how big we can choose the domain.)

$$f \circ h(z) = Az + A\hat{h}(z) + \hat{f}(z) + R_1(z) \tag{4.39}$$

where $R_1(z) = \hat{f} \circ h(z) - \hat{f}(z)$.

$$(\mathrm{Id} - \hat{h}) \circ f \circ h(z) = Az + \hat{h}(z) + \hat{f}(z) + R_1(z) - \hat{h}(Az) + R_2(z) \tag{4.40}$$

where $R_2(z) = \hat{h}(Az) - \hat{h} \circ f \circ h(z)$.

$$h^{-1} \circ f \circ h(z) = (\mathrm{Id} - \hat{h}) \circ f \circ h(z) + R_3(z) \tag{4.41}$$

where $R_3(z) = \Big(h^{-1} - (\mathrm{Id} - \hat{h})\Big) \circ f \circ h(z)$.

Hence, from (4.39), (4.40), (4.41), we obtain:

$$h^{-1} \circ f \circ h(z) = Az - \hat{h} \circ (Az) + \hat{f}(z) + \hat{h}(z) + R_1(z) + R_2(z) + R_3(z)\ . \tag{4.42}$$

Now we turn to the task of obtaining estimates that quantify how this step indeed improves the situation and how we can use it repeatedly to converge to a solution. We highlight the main arguments.

i) Estimates on $\hat{h}$ obtained solving (4.38).

By carrying out exactly the same procedure indicated before estimating the sizes of the Taylor coefficients by the size of the function, solving the small divisor equation for the coefficients and then, estimating the size in an slightly smaller domain we obtain:

$$\|\hat{h}\|_{re^{-\delta}} \le C\delta^{-\tau}\|\hat{f}\|_r \tag{4.43}$$

Note that using (4.38) we immediately obtain estimates

$$\|\hat{h} \circ A\|_{re^{-\delta}} \leq C\delta^{-\tau}\|\hat{f}\|_r \tag{4.44}$$

Having control of $\hat{h}$ both in the polydisk and in its image under A will be quite important to be able to check that compositions, etc. make sense.

ii) Estimates obtained using (4.43) and the implicit function theorem.

Note that this requires that we assume some smallness condition in $\hat{h}$ that ensures that we can indeed define the compositions.

$$\|h^{-1} - (\mathrm{Id} - \hat{h})\|_{re^{-2\delta}} \leq C\delta^{-2\tau-1}\|\hat{h}\|^2_{re^{-\delta}} \tag{4.45}$$

and, similarly using (4.44) and the implicit function theorem (again, we need some conditions that ensure that we can define the compositions needed to apply the implicit function theorem).

$$\|h^{-1} - (\mathrm{Id} - \hat{h})\|_{re^{-2\delta}} \leq C\delta^{-2\tau-1}\|\hat{h} \circ A\|^2_{re^{-\delta}} \tag{4.46}$$

For the conditions that allow us to use the implicit function theorem, it is enough to assume that we have:

$$\begin{aligned} \|\hat{h}\|_{re^{-\delta}} &\leq K\delta \\ \|\hat{h} \circ A\|_{re^{-\delta}} &\leq K\delta \end{aligned} \tag{4.47}$$

which, in view of (4.43) (4.44) are implied by:

$$\begin{aligned} \|\hat{f}\|_{re^{-\delta}} &\leq K\delta^{\tau+1} \\ \|\hat{f} \circ A\|_{re^{-\delta}} &\leq K\delta^{\tau+1} \end{aligned} \tag{4.48}$$

iii) Easy estimates using the mean value theorem.

Note that we can estimate

$$\|f \circ h_1(z) - f \circ h_2(z)\|_r \leq \sup_{z \in \Sigma} |f'(z)| \|h_1 - h_2\|_r \ , \tag{4.49}$$

where Σ is a convex domain that includes the image of the polydisk of radius r under h_1 and h_2. In particular, we can take Σ to be the polydisk of radius $r + \max(\|h_1\|_r, \|h_2\|_r)$.

If we use Cauchy estimates, can obtain

$$\|f \circ h_1(z) - f \circ h_2(z)\|_r \leq \delta^{-1}\|f\|_{r'}\|h_1 - h_2\|_r \tag{4.50}$$

where $r' = [r + \max(\|h_1\|_r, \|h_2\|_r)]\, e^{\delta}$.

As it turns out, to be able to make sure that the domains match we need a condition of the same form as (4.48).

Hence, one can prove a lemma that ensures that, provided (4.48) holds, we can perform the step and obtain estimates

$$\|\tilde{f} - A\|_{re^{-3\delta}} \leq CK^2\delta^{-\tau'}\|f - A\|_r^2 \ , \tag{4.51}$$

where, as usual, we have denoted by C a constant that depends only on the dimension, K is the constant in the Diophantine inequality (4.2) and τ' is an exponent related to the Diophantine exponent (roughly twice, since we are squaring the result of Lemma 4.6 and we are applying Cauchy bounds twice).

This statement is usually called the *iterative lemma*.

Once we have an iterative lemma, we need to show

iv) Choose $\delta_n = \delta_0 2^{-n}$. If you assume that $r_0 \geq 3\delta_0$ and that $\|f - A\|_{r_0}$is sufficiently small, then the iterative lemma can be applied repeatedly to obtain a sequence $\{f_n\}$ defined on $r_n = r_{n-1}e^{-\delta_{n-1}}$. This sequence satisfies

$$\|f_n - A\|_{r_n} \leq C\alpha^{2^n} \tag{4.52}$$

for some $0 < \alpha < 1$, which can be made arbitrarily small by assuming $\|f_0 - A\|_{r_0}$ is sufficiently small.

v) We need to show that the compositions $h^{(n)} \equiv h_n \circ h_{n-1} \circ \cdots h_0$ converge on a non-trivial domain.

This follows because $h_n - h_{n-1} = \hat{h}_n \circ h_{n-1} \circ \cdots h_0$ and we can estimate

$$\|\hat{h}_n\|_{r_n e^{-2\delta_n}} \leq \|\hat{f}\|_{r_n} 2^{n\tau'}$$

and using (4.48) show that

$$\|\hat{h}_n \circ h_{n-1} \circ \cdots h_0\|_{r_n e^{-3\delta_n}} \leq \|\hat{h}_n\|_{r_n e^{-2\delta_n}}$$

Besides the fact that the quadratic convergence is allowing us to dominate the small divisors, we want to highlight some features of the algorithm.

Note again that we can only solve the linearized equation at precisely the identity. Nevertheless, the progress that we are making, allows us to reduce the problem closer to the identity so that we are starting at a problem which is even more favorable. Again, this is the group structure of the problem. The successive changes of variables has been applied very often in the proofs involving Hamiltonian systems with preference to the proofs that involve just solving functional equations. This is due, in part to the fact that Hamiltonian systems have a very nice transformation theory. It is also true that reducing to normal forms, even if only approximately has very interesting byproducts. For example, the Nekhoroshev theorem.

Note that the analytic part of convergence was extremely similar. We obtained the estimates which are *quadratic* but which contain the bad term which has to grow unbounded. All these estimates were proved under some inductive assumptions that allow one to perform the algorithm. The quadratic nature of the estimates can be used to show that if we start with small enough error, the growing terms due to the solution of the linearized equation do not spoil the convergence and that indeed we recover the inductive assumption that allows us to keep on improving our linearization. Once we obtain that the remainder goes to zero extremely fast, it is possible to show that the composition of the transformations converges.

EXERCISE 4.17. When A has a non-trivial Jordan block and the spectrum satisfies (4.35), show that the cohomology equation $A\hat{h} - \hat{h} \circ A = \hat{f}$ is solvable as a formal power series.

What type of estimates do you obtain?

Are the estimates you obtain enough to prove Theorem 4.13 without the assumption that A is diagonalizable?

If not, can you construct a counterexample?

Note: Answers to this exercise are known in the literature. A recent paper that the reader can consult is [**DG00**].

EXERCISE 4.18. Obtain optimal estimates in the Rüssman style for the linear equation for analytic functions in the several variables case. The case when A is a diagonalizable matrix with all eigenvalues equal to 1 is very similar to the one we have discussed so far. Much more interesting are the cases when the matrix has eigenvalues of modulus 1 and non-trivial Jordan blocks.

In preparation of arguments to come, note that, when A has eigenvalues of modulus different from 1, if the domain of φ is a polydisk, the domain of $\varphi \circ A$ is a different set.

EXERCISE 4.19. Formulate the improved estimates of Exercise 4.18 in the language of approximation functions Ω. Do they lead to some improvement in Brjuno conditions?

EXERCISE 4.20. Some of the estimates in the proof of Theorem 4.1 we have presented are rather wasteful.

Notice in particular that we estimated in (4.24)

$$\|R'\|_{\sigma e^{-\delta}} \leq K\delta^{-1}\varepsilon \, .$$

We can observe that, as we iterate, the remainder vanishes at higher and higher orders. This will allow us to use sharper Cauchy estimates, which we detail below.

Note that if a function $f(z) = z^N g(z)$, we have $f'(z) = Nz^{N-1}g(z) + z^N g'(z)$. Also, we have $\|f\|_1 = \|g\|_1$. Hence

$$\begin{aligned} \|f'\|_r &\leq Nr^{N-1}\|g\|_r + Cr^N r^{-1}\|g\|_1 \\ &\leq CNr^{N-1}\|f\|_1 \end{aligned} \tag{4.53}$$

Carry on the proof using these improved estimates and see if one obtains something better.

EXERCISE 4.21. There is a certain arbitrariness in the speed at which domain is lost in the proofs.

What happens is you take $\delta_n = \delta_0 n^{-\alpha^n}$ with $\alpha > 0$?

What happens with $\delta_n = \delta_0 n^{-\alpha}$, $\alpha > 1$, or $\delta_n = \delta_0 n^{-1}(\log n)^{\alpha}$, $\alpha > 1$?

EXERCISE 4.22. Fix $a = \exp\left\{2\pi i \frac{\sqrt{5}-1}{2}\right\}$ and consider

$$f_N(z) = az + z^N \, .$$

What are the asymptotic of the Siegel radius as $N \to \infty$?

EXERCISE 4.23. In the classical Newton method, we use the fact that if the derivative $D_2\mathcal{T}(a, \mathrm{Id})$ is invertible, then $D_2\mathcal{T}(f, h)$ is invertible when (f, h) are in a neighborhood of $(a\,\mathrm{Id})$ and, moreover, the norm of the inverse is bounded.

We can try to apply the same ideas involved in the proof that in the classical case the invertibility of the derivative is an open condition $(A + B)^{-1} = A^{-1}\sum_{i=0}(-BA^{-1})^i$. (sometimes called the Neumann series) to solve the equation

$$f' \circ h\Delta - \Delta \circ a = -R$$

by iterating the solution of

$$a\Delta - \Delta \circ a = -R - \hat{f}' \circ h\Delta$$

Try to carry out the procedure and decide whether it can be applied as an ingredient in a KAM proof. (e.g. one can try to take more stages in the proof as one progresses etc.

To the best of the knowledge of the author it cannot be made to work (unless one uses cancellations similar to those used in the quadratically convergent methods or those of the direct methods) but attempting this will give an appreciation of the cleverness of the use of rapidly convergence methods.

Of course, if there is a proof that succeeds in accomplishing this, the result will be quite interesting.

5. Hard implicit function theorems.

Before proceeding to more geometric considerations, it will be convenient to abstract some of the properties that made the previous argument work and isolate them in an abstract implicit function theorem. This will streamline a good deal of the arguments and illustrate quite strikingly the principle that the quadratic convergence can dominate the small divisors.

Even if implicit function theorems take care very nicely of the analysis of the convergence, they ignore the geometric considerations and particularities of the problem at hand. This particularities are crucial to obtain the general framework of the implicit function theorem. Nevertheless, it is useful to introduce the difficulties one at a time.

Later we will have to spend time making sure that we can fit a problem or an algorithm to solve a problem into the functional framework of a theorem.

We emphasize however that the usefulness of these implicit function theorems is not restricted to KAM theory and they have been used in a variety of problems in geometry, PDE, etc. and that in any case, they are a very useful strategic guide on how to organize the proofs of the problem at hand.

There are different versions of implicit function theorems adapted to work in KAM theory. We just mention [**Zeh75**], [**Ham82**], [**Hör90**].(See also [**Hör85**]). The main variation we have included is that we have used the approximation functions (introduced seemingly in [**Rüs80**]) in the implicit function theorem. Some parts of the exposition are based on [**dlLV00**]. A very good recent exposition – regretfully, not easy to obtain – of Nash-Moser theorems including detailed comparisons and examples of applications, specially to PDE's is [**HM94**]. Also very important for the relation with PDE's are [**AG91**], [**Hör90**]. (Of course, one should also consider the work of [**CW93**], [**Bou99b**], even if it has not been formulated as an abstract implicit function theorem and I am not sure it fits easily into the existing ones.)

The theorem that most closely models the problem we have discussed so far (and those that we will discuss later) is that of [**Zeh75**], which he calls *analytic smoothing* which we now, reproduce, with an small improvement to deal with the Rüsmann conditions rather than just the Diophantine conditions.

REMARK 5.1. In the case of one-dimensional dynamics, one can use the theory of continued fractions to show that the Rüsmmann conditions that we use in the implicit function theorem (expressed in terms of solvability of equations) are equivalent to the conditions used in [**Brj71**] (expressed in terms of number theoretic properties). I do not know if this equivalence is true in higher dimensions.

Note that to abstract the spaces of analytic functions defined on balls of different radius, we will consider not just a single Banach space, but rather a family of Banach spaces. In the following, it will be good to keep in mind the proof of Theorem 4.1 as motivation for the definitions and the assumptions.

THEOREM 5.2. *We will consider scales of Banach spaces $\{X_\sigma\}_{\sigma\in[0,1]}$ such that for $0 \le \sigma' \le \sigma \le 1$ we have:*

$$X_0 \supseteq X_{\sigma'} \supseteq X_\sigma \supseteq X_1 \tag{5.1}$$

$$\|x\|_{X_{\sigma'}} \le \|x\|_{X_\sigma} \ , \tag{5.2}$$

and analogously for $\{Y_\sigma\}_{\sigma\in[0,1]}$, $\{Z_\sigma\}_{\sigma\in[0,1]}$.

Assume that we have $F : X_0 \times Y_0 \to Z_0$

1) *$F(f_0, u_0) = 0$ for some $f_0 \in X_1, u_0 \in Y_1$.*
2) *The domain of F contains the sets*

$$\mathcal{B}_\sigma = \{(f,u) \in X_\sigma \times Y_\sigma \mid \|f - f_0\|_{X_\sigma} \le A, \ \|u - u_0\|_{X_\sigma} \le B\} \ .$$

3) *$F(\mathcal{B}_\sigma) \subset Z_\sigma$ and it is continuous when the range and the domain are given the natural topologies.*

In what follows, $M \ge 1, \gamma > 0, \alpha \ge 0$ will denote fixed constants.

Assume furthermore:

H1) *F satisfies a so called* "Taylor estimate"*. More precisely:*

H1.1) *The mapping*

$$F(f,\cdot) : Y_\sigma \cap B_\sigma \to Z_{\sigma'}$$

is Frechet differentiable for every $\sigma' < \sigma$.

Denote by $D_2F(f,u)$ the Frechet derivative and

$$Q(f;u,v) \equiv F(f,u) - F(f,v) - D_2F(f,v)(u-v) \tag{5.3}$$

H1.2) *We have the bounds:*

$$\|Q(f;u,v)\|_{\sigma'} \le \Upsilon(\sigma - \sigma')\|u - v\|_\sigma^2 \tag{5.4}$$

where Υ is a decreasing function. (We will assume without loss of generality and to avoid complications in algebraic expressions, that $\Upsilon > 1$.) The function Υ is called an approximation function*. It will also enter in subsequent hypothesis and in (H4) it will be required to satisfy certain conditions.*

H3) Approximate right inverse *We can find an approximate right inverse for the derivative.*

That is we can find a linear operator η that maps Z_σ into $X_{\sigma'}$ for all $\sigma' < \sigma$ and that satisfies:

$$\begin{aligned} &\|\eta(f,u)z\|_{\sigma'} \le \Upsilon(\sigma - \sigma')\|z\|_\sigma \\ &\|D_2F(f,u)\eta(f,u)z - z\|_{\sigma'} \le \Upsilon(\sigma - \sigma')\|F(f,u)\|_\sigma\|z\|_\sigma \end{aligned} \tag{5.5}$$

H4) *The approximation function satisfies the* Brjuno-Rüssmann conditions*:*

The function Υ in (5.5) satisfies that there is a sequence $\delta_n > 0$ such that $\sum_n \delta_n = 1/2$, $\sum 2^{-n}|\log(\delta_n/2)| < \infty$ and such

$$\sum_n 2^{-n}\log(\Upsilon(\delta_n)) < \infty \tag{5.6}$$

Then, there exists a constant C, depending only on M, α and Υ such that if u_0 is an approximate solution. That is:

$$\|F(f,u_0)\|_1 \equiv \varepsilon \tag{5.7}$$

is sufficiently small, then, we can find $u^* \in X_{1/2}$ *solving exactly the equation*

$$F(f, u^*) = 0$$

Moreover,

$$\|u - u^*\|_{1/2} \le C\|F(f, u_0)\|_1 \tag{5.8}$$

REMARK 5.3. The theorem in [**Zeh75**] included also a hypothesis H2 that allowed one to obtain information on the dependence of the solutions u in terms of f.

We have eliminated the dependence of u on f from the conclusions of the main theorem and relegated it to remarks (see Remark 5.13). Hence, we suppressed H2 from the main theorem, but kept the numbering to allow easy comparisons. On the other hand, the hypothesis H4 here is different from that of [**Zeh75**], but it plays the same role.

REMARK 5.4. There are several equivalent formulations of hypothesis H4). For all practical purposes, it suffices to take δ a fixed exponential sequence. See the exercises.

The proof of this theorem is very simple since we have abstracted away many of the complications of the previous theorem. We will present it and then, we will highlight some of the subtle points and indicate some of the applications.

REMARK 5.5. One of the important features of the proof in [**Zeh75**], which we have eliminated for this pedagogical presentation, is that the final result is expressed in a form which is independent of the space considered. This requires one to assume that $\Upsilon(t) = Ct^{-\alpha}$ for some positive C, α. This case has very important consequences such as the finitely differentiable case. We will develop these improvements in the exercises.

PROOF. We use a quasi-Newton method defined by the iterative procedure

$$u_{n+1} = u_n - \eta(f, u_n)F(f, u_n) \tag{5.9}$$

in which η takes the place of the inverse of the derivative in the regular Newton method.

We set $\sigma_{n+1} = \sigma_n - \delta_n$ and $\sigma_0 = 1$. We will obtain recursively estimates of $\|F(f, u_n)\|_{\sigma_n}$ $\|u_n\|_{\sigma_n}$, and of $\|u_n - u_{n+1}\|_{\sigma_{n+1}}$.

Since $\sigma_n \ge 1/2$, the later estimates will imply that u_n converges in $X_{1/2}$.

Adding and subtracting, we have:

$$\begin{aligned} F(f, u_{n+1}) = F(f, u_{n+1}) - F(f, u_n) - D_2F(f, u_n)\eta(f, u_n)F(f, u_n) \\ + F(f, u_n) + D_2F(f, u_n)\eta(f, u_n)F(f, u_n) \ . \end{aligned} \tag{5.10}$$

We can estimate the terms in the second line in (5.10) using the second part of (5.5)

$$\|F(f, u_n) - D_2F(f, u_n)\eta(f, u_n)F(f, u_n)\|_{\sigma_{n+1}} \le \Upsilon(\delta_n)\|F(f, u)\|^2_{\sigma_n} \ .$$

Using the first part of (5.5), we obtain: for $\tau_n = (\sigma_n + \sigma_{n+1})/2$

$$\|\eta(f, u_n)F(f, u_n)\|_{\tau_n} \le \Upsilon(\delta_n/2)\|F(f, u_n)\|_{\sigma_n} \ . \tag{5.11}$$

This estimate allows us to apply (5.4) to the terms in the first line of (5.10).

Hence, we obtain (bounding $\Upsilon(\delta_n/2) > \Upsilon(\delta_n)$)

$$\|F(f, u_{n+1})\|_{\sigma_{n+1}} \leq 2\Upsilon(\delta_n/2)^2 \|F(f, u_n)\|^2_{\sigma_n} \ . \tag{5.12}$$

If we iterate (5.12), we obtain

$$\begin{aligned} \|F(f, u_{n+1})\|_{\sigma_{n+1}} \leq & 2\Upsilon(\delta_n/2)^2 \times (2\Upsilon(\delta_{n-1}/2)^2)^2 \times \cdots \times \\ & \times (2\Upsilon(\delta_0/2)^2)^{2^n} \|F(f, u_0)\|^{2^{n+1}}_{\sigma_0} \\ = & 2^{1+2+\cdots+2^n} \Upsilon(\delta_n/2)^2 \times \Upsilon((1/2)\delta_{n-1})^{2^2} \times \cdots \times \\ & \times \Upsilon((1/2)\delta_0)^{2^{n+1}} \|F(f, u_0)\|^{2^{n+1}}_{\sigma_0} \ . \end{aligned} \tag{5.13}$$

We can estimate the logarithm of the factor of $\|F(f, u_0)\|^{2^{n+1}}_{\sigma_0}$ in the R.H.S. of (5.13) by:

$$\begin{aligned} 2^{n+1} \big[\log(2) + \log \Upsilon((1/2)\delta_n) 2^{-n} + \cdots + \log \Upsilon((1/2)\delta_{n-1}) 2^{-(n-1)} \\ + \cdots + \log \Upsilon((1/2)\delta_0) 2^0 \big] \ . \end{aligned}$$

We see that under our assumption H4) (see (5.6)), the term in braces can be bounded by a constant (the sum of the series). Hence (5.13) yields

$$\|F(f, u_{n+1})\|_{\sigma_{n+1}} \leq (A \|F(f, u_0)\|_{\sigma_0})^{2^{n+1}} \tag{5.14}$$

where A is a constant depending on the properties of the approximation function and the other constants involved in the set up of the problem.

We see that, if we $\|F(f, u_0)\|_{\sigma_0}$ is sufficiently small, the right hand side of (5.14) converges to zero extremely fast.

Using (5.11), we have:

$$\|u_n - u_{n+1}\|_{\sigma_{n+1}} \leq \Upsilon(\delta_n/2)(A\|F(f, u_0)\|_{\sigma_0})^{2^n} \tag{5.15}$$

where A is also a constant depending only on the properties of the approximation function and the other constants involved in the set up of the problem. (It will be different from the A in (5.14), but we follow the standard practice of denoting all such constants by the same letter.)

The R.H.S. of (5.15) is a convergent series because by our assumption (5.6) the general term of the series is bounded. Therefore, $\log \Upsilon(\delta_n/2) 2^{-n} \leq B$ (where, again, B is another constant depending only on the constants of the problem and the approximation function).

When $A\|F(f, u_0)\|_{\sigma_0} < 1$, the second factor converges to zero faster than any exponential.

Note also that, if $\|F(f, u_0)\|_{\sigma_0}$ small enough, the series obtained summing (5.15) has a sum as small as desired. In particular, we can verify that the limit is close to u_0 in $X_{1/2}$.

Hence:

$$\Upsilon(1/2\delta_n)(A\|F(f, u_0)\|_{\sigma_0})^{2^n} \leq (Ae^B \|F(f, u_0)\|_{\sigma_0})^{2^n}$$

This establishes the claim. □

EXERCISE 5.6. Many classical proofs of the classical implicit function theorem are based not in the Newton method, which is quadratically convergent, but rather in a contraction mapping principle (which is called linearly convergence since the remainder after one step is only a fixed factor smaller than the remainder before the step.)

Can one base a method that beats small denominators on a linearly convergent procedure?

Similarly, one can get algorithms whose convergence is faster than quadratic. (For example, solving the equation given by the second order Taylor expansion or interpolating several of the previous steps of the algorithm.) Can one base a hard implicit function theorem on these algorithms?

It is interesting to check how the previous result compares with the proof we have presented of Theorem 4.1. The scales of spaces are just spaces of analytic functions on balls of different radii. The approximate inverse corresponds to the solving of the linearized equation by comparing it with the equation obtained by taking derivatives of the remainder. Checking that the scales map into each other is roughly the same as our inductive hypothesis.

In the presentation of Theorem 4.1, we have, of course taken

$$\Upsilon(\delta) = M\delta^{-\tau} \tag{5.16}$$

This is a very important particular case of the whole theorem since it not only appears in interesting situations but also leads to further consequences which we will discuss in the following remarks. (Of course, the reader should also consult [**Zeh75**] and the other references.)

The choice of a general Υ satisfying (5.6) corresponds to the small divisors satisfying (2.34), whereas (5.16) corresponds to Diophantine conditions. (For more details see [**Rüs90**], [**DeL97**], [**dlLV00**].)

REMARK 5.7. The existence of approximate inverses is a general feature of conjugacy problems or of problems having a group structure.

As pointed out in [**Zeh75**] p. 133 ff. existence of approximate inverses assuming only the existence of an inverse in the trivial case is a general feature of conjugacy problems, at least at the heuristic level. This indeed gives a guiding principle for the cancellations that we found e.g. in the proof of Theorem 4.1 in which we used that comparing the prescription suggested by the heuristic Newton method with the derivative of the remainder the linearized equation suggested by the heuristic Newton method can be reduced to constant coefficients up to quadratically small errors.

Notice that the functionals we are solving are conjugacy equations. Hence they satisfy the identity

$$F(f, u \circ v) = F(F(f, u), v) \tag{5.17}$$

If we take $v = \mathrm{Id} + \hat{v}$ and we think of $\hat{v}$ as infinitesimal, we obtain

$$D_2F(f, u)u'\hat{v} = D_2F(F(f, u), \mathrm{Id})\hat{v} \tag{5.18}$$

If we assume that $\eta D_2(f_0, \mathrm{Id}) = \mathrm{Id}$, we obtain that:

$$\eta D_2F(f, u)u'\hat{v} = \mathrm{Id} + \eta[D_2F(F(f, u), \mathrm{Id}) - D_2F(f_0, \mathrm{Id})] \tag{5.19}$$

Notice that we can expect that, if D_2F satisfies some Lipschitz conditions on the first argument, the term in braces in the R.H.S. of (5.19) satisfies the bounds we wanted for an approximate inverse provided that η satisfies the desired bounds.

The importance of this remark is that by knowing the existence of η, which is just an inverse of $D_2F(f_0, \mathrm{Id})$ we can deduce, for functionals with a group structure, the existence of approximate inverse in a whole neighborhood, which the hypothesis needed by Theorem 5.2.

Of course, F satisfies assumption (5.17) when $F(f,u) = u^{-1} \circ f \circ u$ but it could also be the action by u on vector fields or more complicated objects and indeed it happens quite frequently when one is considering geometrical problems.

REMARK 5.8. strategy for KAM (Discussed in more detail in Section 6.1) can be formulated as reducing the Hamiltonian to a Hamiltonian of a particular kind.

Hence, we are not interested in just solving the equation $F(f,u) = 0$ but rather $F(f,u) = N$, where N is a submanifold of infinite codimension. Indeed, this is the problem that is considered in [**Zeh75**] and especially in [**Zeh76a**].

REMARK 5.9. Even if most of the classical KAM problems (certainly all that will be discussed in this notes) are conjugacy problems and, therefore, have the group structure, this is not completely necessary to have a quadratic algorithm not completely necessary to have a quadratic scheme.

A review of problems in geometry which are not conjugacy problems can be found in [**Ham82**].

A very interesting recent development is the observation that variational problems with symmetry also present another general structure that allows to obtain quadratic convergence. See, for example [**Koz83**], [**Mos88**] for PDE's or, in the context of KAM [**SZ89**]. (We will present an account of that work in Section 6.3.)

Much more interesting is the fact that in [**CW93**], [**CW94**], another mechanism to obtain quadratic convergence was introduced. At the moment, I do not know of a functional analytic framework that encompasses these remarkable results.

REMARK 5.10. In the applications of the implicit function theorems to problems of persistence of tori – and to some geometric problems – we are not interested in the equation $F(f,u) = 0$ but rather in the equation $F(f,u) \in N$ where N is an appropriate submanifold.

See exercises 5.25, 5.26

REMARK 5.11. Note that the structure of Theorem 5.2 is that the input is just an approximate solution (with some extra mild requirements) and that the output is an exact solution not too far from the original approximate solution.

In the most commonly quoted applications, the input is the exact solution for an integrable system, which is an approximate solution for a quasi-integrable system. Nevertheless, other applications are possible. Among them, we mention:

1) Numerical algorithms:

If carefully implemented and successfully, numerical algorithms produce approximate solutions (i.e. ssomething that, when plugged into the equations satisfies them approximately).

Hence, using a theorem with the structure of Theorem 5.2, one can justify that the approximate solutions produced by a computer algorithm indeed correspond to a true solution nearby. In numerical analysis, this is sometimes called a-posteriori bounds. (See [**BZ82**], [**dlLR91**], [**CC95**].) We discuss some numerical issues involved in Section 7.

2) Justification of asymptotic expansions (e.g. LLindstedt series)

These expansions produce objects that satisfy the equations approximately. Hence, a theorem similar to Theorem 5.2, can be used to justify asymptotic expansions. That is, show that one can indeed find tori which are not far away from the truncations of the Lindstedt series. For the KAM tori, one can find this type of arguments in [**Mos67**]. In these case, it is also shown that the Lindstedt series converge (since the torus should be analytic as a function of the parameter). We emphasize that to justify the asymptotic nature of the series one just needs that the series produce objects that satisfy the equation with smaller errors and that are not too complicated. The Lindstedt series of lower dimensional tori are studied by this method in [**JdlLZ99**]. In that case, we do not know whether the series converges or not, but following the argument sketched here, it is possible to show that they are asymptotic in a certain complex domain.

3) Establishing continuity or Whitney regularity of the solutions with respect to parameters – assuming that F is more regular in both its arguments –.

This application is worked out explicitly in [**dlLV00**]. The latter arguments require some certain amount of uniqueness, which is not provided by the theorem in the way we have stated and proved it, but which we obtain in Exercise 5.24.

4) Obtaining a result for finitely differentiable problems out of the analytic ones.

An application that can already be found in [**Mos66b**], [**Mos66a**] is that, as we saw in Lemma 3.2, we can characterize finitely differentiable functions by their approximation properties by analytic functions. We just sketch the argument.

Given a smooth f, we study the problem $F(f, u)$ by considering a sequence of problems $F(f_n, u_n) = 0$ where f_n are constructed approximating the smooth function f by analytic functions.

Using that $||f_n - f_{n+1}||_{2^{-(n+1)}} \leq C2^{-(n+1)r}$ it is often possible (using the structure of F) to show that

$$||F(f_n, u_n) - F(f_{n+1}, u_n)||_{2^{-(n+1)}} = ||F(f_{n+1}, u_n)||_{2^{-(n+1)}} \leq C2^{-(n+1)r'}$$

We consider u_n as an approximate solution for the problem with f_{n+1}. In the case that Υ is a power, it follows that

$$||u_n - u_{n+1}||_{2^{-(n+2)}} \leq C2^{-(n+1)r''}$$

from which, appealing again to Lemma 3.2 we obtain that there $u = \lim u_n$ which solves the desired equation and which is analytic.

This method has the advantage that one always works with analytic functions for which estimates are often easier and, as we have seen sharper if one needs to use Fourier coefficients.

We refer to [**Mos66b**], [**Mos66a**], [**Zeh75**], [**Zeh76a**] for more details (such as how to get the induction started), somewhat different versions of the argument, and applications to concrete problems.

The quantitative estimates needed to carry out this strategy are explained in Exercise5.19.

5) Bootstrapping the regularity.

A solution which is moderately smooth, if approximated by an analytic one is an analytic approximate solution.

Applying a theorem of this sort, one can conclude that given an analytic problem, if there is a sufficiently smooth solution (so that the smoothings are indeed very good approximations), then there is an analytic one. Of course, if one does have uniqueness of the problem, one obtains that any solution that has a certain regularity, is analytic.

Of course, if we start with a problem that is very regular, we can also show that given a solution which is beyond a certain critical regularity, there will be another one which is as as the problem allows, and if there is uniqueness, we conclude that all the solutions beyond a certain regularity are as smooth as the problem allows.

Arguments of this type are worked out explicitly in [**SZ89**]. Again, we refer to Exercise5.19 for some of the quantitative estimates needed.

REMARK 5.12. Notice that the Theorem 5.2 only assumes the existence of an approximate right inverse.

One should not expect that the solution one produces in the theorem to be unique. Indeed, in some problems such as the Nash embedding theorem which motivated a good deal of the original research one only has an approximate right inverse and, indeed the solution is not unique. In many geometric problems, the results we seek are in any case invariant under diffeomorphisms, so that it is to be expected that the solution is not unique.

Under moderate assumptions – e.g. under the existence of an approximate left inverse – one gets uniqueness. See the remarks in [**Zeh75**] and see Exercise 5.24. These assumptions are often satisfied in KAM theory or uniqueness of the objects we are interested in can be obtained by other means. (Often one seeks geometrical objects in coordinate systems, so that the geometric objects may be unique even if their coordinate representation is not.)

One situation when these considerations play a role is the proof of the KAM theorem following Kolmogorov's strategy. (See Section 6.1.)

In this method, we seek a change of variables in which the resulting system manifestly has an invariant torus. That is, we try to reduce the system to the the Kolmogorov normal form (6.6). Such change of variables is manifestly not unique since the normal form does not specify what are the higher order terms and one can make changes of variables that only depend to higher order in the actions. Therefore, one cannot expect uniqueness in the change of variables nor in the term of the normal form and a formulation of the theorem based on this formalism cannot aspire to obtain uniqueness. Nevertheless, it is true that, under moderate non-degeneracy assumptions, the torus that has a prescribed frequency is unique.

REMARK 5.13. The theorem of [**Zeh75**] has an extra hypothesis H2 that requires that F is Lipschitz in the first argument Then, one obtains Lipschitz dependence on the solution on the function f (in some appropriate spaces).

We note that, in the case that there is no uniqueness, the only claim made is that the algorithm (5.9) leads to a solution that depends in a Lipschitz manner on f. Clearly, when there is no uniqueness one could make different choices of solutions for different f and end with a u that depends discontinuously on f.

A detailed treatment of these ideas can be found in [**Zeh76b**].

We point out that there are other methods to obtain smooth dependence with respect to parameters that do not involve following the proof of Theorem 5.2 and checking the differentiability with respect to parameters of all the steps.

1) One can also obtain quickly higher regularity with respect to parameters by applying Theorem 5.2 in spaces that consists of smooth families of functions. Of course, one needs that the approximate inverse also maps smooth families into smooth families. This is somewhat tricky since approximate inverses are not uniquely defined, so one could make different choices for different values of the parameter and spoil even continuity.

Nevertheless, for problems with group structure, the prescription given by (5.18) gives a way of accomplishing the solution in spaces of smooth families of functions. Arguments of this sort are carried out in detail in [**dlLO00**] to solve a problem in differential geometry.

2) When there is uniqueness, one can follow other sort of arguments such as finding formal derivatives for the solution and then, showing that these formal derivatives satisfy the hypothesis of Whitney theorem [**dlLV00**].

REMARK 5.14. When one has some regularity – at least Lipschitz – with respect to the parameters, one can start discussing issues – important in the applications – such as the measures in the space of parameter covered.

EXERCISE 5.15. Write precisely the reduction of Theorem 4.1 to Theorem 5.2 by making explicit choices of spaces, etc.

EXERCISE 5.16. A challenging variant of the previous exercise is to show that, if the number ω satisfies the conditions (2.34), the approximate inverse we constructed in the proof of Theorem 4.1 satisfies (5.6).

If independent study fails, see [**Rüs90**], [**DeL97**] for estimates that go from arithmetic conditions to approximation functions.

REMARK 5.17. In practical applications, e.g. when one is computing numerically solutions to a problem defined implicitly one of course, does not compute the inverse of the matrix, but rather solves numerically the system.

In numerical practice, this usually entails a factorization of the matrix. Traditionally, one uses the LU factorization (Gaussian elimination), even if in KAM theorems that tend to be ill conditioned one should, perhaps, prefer the SVD decomposition.

In any case, it is convenient not to have to recompute these factorizations – which may much more costly than the application to a function –. Of course, we would not like to lose the quadratic convergence which, e.g. in continuation methods that require great precision is much more practical that a method that converges more slowly.

The following two schemes, which avoid having to recompute factorizations but which get convergence faster than linear are studied in [**Mos73**] p. 151. The second one comes from [**Hal75**]. A geometric interpretation of these methods as a Newton method in the space of jets is discussed in [**McG90**].

$$\begin{aligned} u_{n+1} &= u_n - \eta_n F(f, u_n) \\ \eta_{n+1} &= \eta_n - \eta_n(\mathrm{Id} - D_2 F(f, u_n))\eta_n \end{aligned} \tag{5.20}$$

$$\begin{aligned} u_{n+1} &= u_n - \eta_n F(f, u_n) \\ \eta_{n+1} &= \eta_n - \eta_n(\mathrm{Id} - D_2 F(f, u_{n+1}))\eta_n \end{aligned} \tag{5.21}$$

(In numerical applications, one does not compute the product of matrices in (5.20), (5.21). Note that it suffices to apply the matrices to vectors.)

EXERCISE 5.18. Show in finite dimensions that, under smoothness assumptions and smallness assumptions: (5.20) leads to

$$\|F(f, u_{n+1})\| \le C|F(f, u_n)|^{(\sqrt{5}+1)/2}$$

and (5.21) leads to

$$\|F(f, u_{n+1})\| \leq C|F(f, u_n)|^2.$$

Applications of these schemes to hard implicit function theorems and other modifications of the basic algorithm will be developed in the following exercises.

The following exercises are designed to show that the quadratic convergence is rather forgiving and that there are many variants that also work. We have also included some variants in which the results fail so that the reader can start to develop a feeling for the range of applicability of the techniques.

EXERCISE 5.19. Consider the following improvements to Theorem 5.2 (either separately or several at the same time, for the most ambitious reader).

As we will note in the exercises some of them have important consequences, beyond serving as training.

- Modify the hypothesis and the conclusions so that the approximate solution is assumed to satisfy

$$\|F(f, u_0)\|_{\delta_0} \equiv \varepsilon$$

  instead of (5.7) and the conclusion about u^* reads

(5.22) $$\|u - u^*\|_{\delta_0/2} \leq C(\delta_0)\|F(f, u_0)\|_{\delta_0}$$

  instead of (5.8).

  Hint: This result can be deduced from the statement of the theorem just by a relabeling of the spaces.

- Show that in case that we take $\Upsilon(t) = Ct^{-\alpha}$ for some $C, \alpha > 0$, we have $C(\delta_0) = C' t^{-\alpha'}$ for some $C', \alpha' > 0$.

  REMARK 5.20. The previous two items are quite important since they allow to obtain finite differentiability out of the analytic result. The strategy to obtain that is explained in Remark5.11.

  They are worked out in [**Zeh75**]. It can also be worked out from the statement that we have given by a rescaling argument.

- Show that in case that we take $\Upsilon(t) = s\Upsilon_0(t)$ where we consider Υ_0 as a fixed function and s as a variable, the smallness conditions required in (5.7) are

$$\epsilon s^2 < K$$

  and that the conclusions (5.8) read

$$\|u - u^*\|_{1/2} \leq Ks\|F(f, u_0)\|_1$$

  where now K is a constant depending on all the other properties of the hypothesis, but independent of s.

  REMARK 5.21. In applications to the KAM theorem, the meaning of the parameter s is teh allowes size of the Diophantine constant.

  This improvement is worked in [**Zeh76b**].

  It leads rather directly to estimates on the measure of the set of tori covered by KAM theorem. (See [**dlLV00**].)

- Consider that in (5.4), (5.5) we have three different Υ functions. For example, three different powers.

(This appears in practice. Some of the powers come from the Diophantine approximations whereas others come from the differentiation of composition and the like.)

- Modify the second equation of (5.5) to read

$$\|dF(f,u)\eta(f,u)z - z\|_{\sigma'} \le \Upsilon(\sigma-\sigma')\|F(f,u)\|_\sigma^{\kappa'}\|z\|_\sigma$$

for some $\kappa' > 0$.

- Modify (5.4) to read

(5.23) $$\|Q(f;u,v)\|_{\sigma'} \le \Upsilon(\sigma-\sigma')\|u-v\|_\sigma^{1+\kappa}$$

for some $\kappa > 0$.

- One can also have a different approximate inverse during the iteration.

$$\begin{aligned}\|dF(f,u_n)\eta_n(f,u_n)z - z\|_{\sigma'} \le &\Upsilon(\sigma-\sigma')\|F(f,u_n)\|_\sigma\|z\|_\sigma \\ &+ \exp(-a(1+\kappa'')^n)\end{aligned}$$

for some $\kappa'' > 0$.

A variant is to choose

(5.24) $$\begin{aligned}\|dF(f,u_n)\eta_n(f,u_n)z - z\|_{\sigma'} \le &\Upsilon(\sigma-\sigma')\|F(f,u_n)\|_\sigma\|z\|_\sigma \\ &+ \exp(-4^n(\sigma-\sigma'))\end{aligned}$$

This appears in some proofs (e.g. in Arnol'd type proofs) when one tries to do some truncation of the problem. This improvement is not too tricky to do by itself, but it is not so easy to understand how it does work with the others. It is quite enlightening to understand how it works with the method of obtaining finite differentiability.

with some of the others.

Under these modifications, one has to modify slightly the conditions (5.6).

EXERCISE 5.22. Formulate precisely the assumptions of domain loss etc. to obtain a proof of the implicit function theorem using an iteration as in (5.21).

EXERCISE 5.23. Taking into account the improvement suggested in (5.23) give a proof of the theorem using the scheme of (5.20).

EXERCISE 5.24. Show that if one supplements the assumption H3 of Theorem 5.2 with the existence of a left approximate inverse satisfying the same estimates, one obtains that the solution is unique in an appropriate sense.

Formulate a precise theorem in which the domains in which uniqueness holds are explicitly specified.

(Some version of this is done in [**Zeh75**].)

EXERCISE 5.25. State and prove an implicit function theorem in which we do not attempt to solve $F(f,u) = 0$ but rather $F(f,u) \in N$ as explained in Remark 5.10.

In that generality, one should not expect uniqueness, hence, continuity and differentiability with respect to parameters is presumably not very clean.

EXERCISE 5.26. When considering the normal form problem one should also modify the assumption H3 of Theorem to be: 5.2

$$\|dF(f,u)\eta(f,u)z - z\|_{\sigma'} \le \Upsilon(\sigma-\sigma')d_\sigma(N, F(f,u))\|z\|_\sigma$$

where d_σ denotes the distance between sets measured with the norm $\|\cdot\|_\sigma$.

This observation appears in [**Mor82**].

EXERCISE 5.27. A classical theorem in KAM theory is the theorem of [**Arn61**] which states that given a diffeomorphism of the circle with a rotation number ρ, which is Diophantine and sufficiently close to the rotation by ρ in an analytic topology, then, there is an analytic change of variables that transforms it in the rotation by ρ.

Formulate it in terms of an abstract implicit function theorem.

The main difficulty is that, when we start proving this theorem, we do not know that the set diffeomorphisms with rotation number ρ is a manifold. (We know it after we prove the theorem!.)

Note also that the conjugacy is not unique since all rotations conjugate a rotation to itself.

I know several ways to do it, but all of them require some dirty tricks. (A good source for those – and for almost anything having to do with circle maps – is [**Her79**] and [**Her83a**]).

Note also that for this problem there are proofs that do not use KAM. Besides the renormalization proofs and [**KO89a**], already mentioned in Section 1.1, we mention [**Her85**] and, for constant type numbers (those that have a bounded continued fraction expansion) [**Her83b**].

REMARK 5.28. Note that, the estimates we have made to prove Theorem 5.2 do not use that $\|\cdot\|_\sigma$ is a norm. They would have worked just as well if $\|\cdot\|_\sigma$ had been a semi-norm.

Of course, in order that the result is meaningful, we would need that the family of seminorms $\{\|\cdot\|_\sigma\}_{\sigma\in(0,1/2)}$ defines a useful space, i.e., they define a Fréchet space. See [**Ham82**] for more details about such improvement and also for applications.

The original proof of KAM theorems for finite differentiability were based on different schemes than the proof we have presented.

Note that, for example, the proof of Theorem 4.13 follows a different scheme. At every step, the linear operator we have to solve does have an inverse (not just an approximate inverse). The problem is that the operator is unbounded and, hence, simple-minded iterations such as those of the classical Newton method do not work.

This situation happens also in PDE's. A notable example was the celebrated Nash embedding theorem [**Nas56**].

The method used in [**Mos66b**] and [**Mos66a**] was to combine steps of the linearized operator with smoothings. The method allows a norm – in a space of somewhat smooth functions – to blow up, whereas a norm – in a space of rougher functions – decreases. By using interpolation inequalities, one can recover good behavior of some intermediate norms. (The decrease may not be exactly quadratic, but it is still is faster than exponential.) This technique has been highly formalized in [**Ham82**], which also includes a wealth of applications, mainly to geometry. See [**Ham82**], Section 3, for a comparison with the methods of Zehnder [**Zeh75**].

In the following, we present a proof along these lines, which follows rather closely [**Sch69**]. This book also contains a very nice discussion of the Nash embedding theorem and on other problems of nonlinear functional analysis.

In the sequel, we shall refer to a certain range $m-\alpha \le r \le m+10\alpha$ of spaces Λ_r (defined in Section 3.1), and to a certain constant $M > 1$. We suppose that M

is sufficiently large so that the smoothing operators S_t satisfy

$$\begin{aligned}\|S_t u\|_\rho &\le M t^{\rho-r}\|u\|_r \quad u\in\Lambda_r\\ \|(\mathrm{Id}-S_t)u\|_r &\le M t^{r-\rho}\|u\|_\rho \quad u\in\Lambda_\rho\end{aligned} \tag{5.25}$$

for $m-\alpha\le r\le\rho\le m+10\alpha$ ($\|\cdot\|_r$ stands for the norm in Λ_r).

THEOREM 5.29. *Let B_m be the unit ball in Λ_m and $f : B_m \to \Lambda_{m-\alpha}$ be a map that satisfies:*

(i) *$f(B_m\cap\Lambda_r)\subset\Lambda_{r-\alpha}$, for $m\le r\le m+10\alpha$;*
(ii) *$f_{|B_m\cap\Lambda_r} : B_m\cap\Lambda_r\to\Lambda_{r-\alpha}$ has two continuous Fréchet derivatives, both bounded by M, for $m\le r\le m+10\alpha$;*
(ii) *There exists a map $L : B_m \to \mathcal{B}(\Lambda_m,\Lambda_{m-\alpha})$, where $\mathcal{B}(\Lambda_m,\Lambda_{m-\alpha})$ is the space of bounded linear operators on Λ_m to $\Lambda_{m-\alpha}$ such that:*
   (ii.a) *$\|L(u)h\|_{m-\alpha}\le M\|h\|_m$, $\quad u\in B_m, h\in\Lambda_m$;*
   (ii.b) *$df(u)L(u)h = h$, $\quad u\in B_m, h\in\Lambda_{m+\alpha}$;*
   (ii.c) *$\|L(u)f(u)\|_{m+9\alpha}\le M(1+\|u\|_{m+10\alpha})$, $\quad u\in B_m\cap\Lambda_{m+10\alpha}$.*

Then, if $E := \|f(0)\|_{m+9\alpha}$ is sufficiently small, there exists $u\in\Lambda_m$ such that $f(u)=0$.

PROOF. Let $\kappa>1$, $\beta,\mu,\nu>0$ be real numbers to be specified later. We will need that they satisfy a finite set of inequalities relating them and the constants appearing in the assumptions of the problem.

We construct a sequence $\{u_n\}_{n\ge1}\subset\Lambda_m$ by taking $u_0=0$ and

$$u_{n+1}=u_n-S_nL(u_n)f(u_n)$$

where $S_n=S_{t_n}$ and $t_n=e^{\beta\kappa^n}$. Later on, we will prove that this sequence satisfies, for $n\ge1$:

(p1;n) $u_{n-1}\in B_m$,
(p2;n) $\|u_n-u_{n-1}\|_m\le e^{-\mu\alpha\beta\kappa^n}$,
(p3;n) $u_n\in\Lambda_{m+10\alpha}$ and $1+\|u_n\|_{m+10\alpha}\le e^{\nu\alpha\beta\kappa^n}$.

Notice that, then, $\{u_n\}_{n\ge1}\subset\Lambda_m$ converges to some $u\in\Lambda_m$ and, moreover:

$$\begin{aligned}\|f(u_n)\|_{m-\alpha} &=\|df(u_n)(u_{n+1}-u_n)-df(u_n)(\mathrm{Id}-S_n)L(u_n)f(u_n)\|_{m-\alpha}\\ &\le M\|(u_{n+1}-u_n)\|_m+M^2t_n^{-9\alpha}\|L(u_n)f(u_n)\|_{m+9\alpha}\\ &\le Me^{-\mu\alpha\beta\kappa^{n+1}}+M^2e^{(\nu-9)\alpha\beta\kappa^n}.\end{aligned} \tag{5.26}$$

Hence, the R.H.S. of the previous inequality (5.26) converges to zero when n goes to infinity, provided that

$$\nu<9. \tag{5.27}$$

We are going two prove by induction the three properties satisfied by the sequence $\{u_n\}_{n\ge1}$. For $n=1$, condition(p1;1) is trivial.

Condition (p2;1) reads

$$\begin{aligned}\|u_1-u_0\|_m &=\|S_0L(0)f(0)\|_m\le Mt_0^\alpha\|L(0)f(0)\|_{m-\alpha}\\ &\le M^2e^{\alpha\beta}E\\ &\le e^{-\mu\alpha\beta\kappa},\end{aligned} \tag{5.28}$$

where the last inequality holds if

$$E\le M^{-2}e^{-(1+\mu\kappa)\alpha\beta}. \tag{5.29}$$

Condition (p3;1) reads

$$
\begin{aligned}
1+\|u_1\|_{m+10\alpha} =&1+\|S_0L(0)f(0)\|_{m+10\alpha} \le 1+Mt_0^\alpha\|L(0)f(0)\|_{m+9\alpha}\\
\le&1+M^2e^{\alpha\beta}\le 2M^2e^{\alpha\beta}\\
\le&e^{\nu\alpha\beta\kappa},
\end{aligned}
$$

where the last inequality holds if

$$1\le \frac{1}{2}e^{\alpha\beta(\nu\kappa-1}M^{-2}, \tag{5.30}$$

that is, if $\nu\kappa>1$ and β is sufficiently large.

Suppose now that conditions (p1;j),(p2;j) and (p3;j) are true for $j\le n$. Then,

$$\|u_n\|_m\le \sum_{j=1}^{\infty} e^{-\mu\alpha\beta\kappa^j}\le \sum_{j=1}^{\infty} e^{-\mu\alpha\beta(\kappa-1)j} = \frac{e^{-\mu\alpha\beta(\kappa-1)}}{1-e^{\mu\alpha\beta(\kappa-1)j}}<1, \tag{5.31}$$

If we require that

$$\frac{e^{-\mu\alpha\beta(\kappa-1)}}{1-e^{\mu\alpha\beta(\kappa-1)j}}<1 \tag{5.32}$$

which holds when

$$\mu\beta>>1$$

we obtain that the R.H.S. of (5.31) is bounded from above by 1 and, therefore, we recover (p1:n+1).

To prove (p3;n+1) note that

$$
\begin{aligned}
1+\|u_{n+1}\|_{m+10\alpha}&\le 1+\sum_{j=0}^{n}\|S_jL(u_j)f(u_j)\|_{m+10\alpha}\\
&\le 1+M^2\sum_{j=0}^{n} e^{(1+\nu)\alpha\beta\kappa^j}\ .
\end{aligned}
$$

Hence,

$$
\begin{aligned}
(1+\|u_{n+1}\|_{m+10\alpha})e^{-nu\alpha\beta\kappa^{n+1}}&\le e^{-nu\alpha\beta\kappa^{n+1}}+M^2\sum_{j=0}^{n} e^{(1+\nu-\nu\kappa)\alpha\beta\kappa^j}\\
&\le 1\ ,
\end{aligned}
$$

where the last inequality holds (and so (p3;n+1)) if $\nu>\frac{1}{\kappa-1}$ and β is sufficiently large.

Finally, we come to the proof of (p2;n+1). We have:

$$
\begin{aligned}
\|(u_{n+1}-u_n)\|_m =&\|S_nL(u_n)f(u_n)\|_m\le M^2e^{\alpha\beta\kappa^n}\|f(u_n)\|_m\\
\le&M^2e^{\alpha\beta\kappa^n}(\|f(u_{n-1})-df(u_{n-1})S_{n-1}L(u_{n-1})f(u_{n-1})\|_m\\
&\quad+M\|(u_n-u_{n-1})\|_m^2)\\
\le&M^5(e^{(\nu-9+\kappa)\alpha\beta\kappa^{n-1}}+e^{(1-2\mu)\alpha\beta\kappa^n})
\end{aligned}
$$

Therefore, if

$$M^5(e^{(\nu-9+\kappa)\alpha\beta\kappa^{n-1}}+e^{(1-2\mu)\alpha\beta\kappa^n})e^{-\mu\alpha\beta\kappa^{n+1}} \tag{5.33}$$

we recover (p2;n+1)).

The condition (5.33) is true when $\kappa < 2$, $\mu > \frac{1}{2-\kappa}$, $\nu > 9 - \kappa - \mu\kappa^2$ and β is sufficiently large.

Therefore, we have established that, when the parameters μ, ν, κ satisfy (5.27), (5.29), (5.30), (5.32), (5.33) then we can carry out the induction and establish the theorem.

This is satisfied if we take $1 < \kappa < 2$, $\mu > \frac{1}{2-\kappa}$ and $\frac{1}{\kappa} < \frac{1}{\kappa-1} < \nu < 9-\kappa-\mu\kappa^2 < 9$. For instance, $\kappa = \frac{3}{2}$, $\mu = \frac{20}{9}$ and $\nu = \frac{9}{4}$) then, choose β sufficiently large). □

REMARK 5.30. The above methods of proof can also produce results for C^∞ functions. This is significantly more complicated than the ideas used so far and we will not discuss them.

REMARK 5.31. In many applications the embeddings of scales of spaces considered are not just continuous but also compact. This allows one to improve several of the steps. See [**Hör90**] which also includes very nice ideas on how to use paradifferential calculus and several interesting new ideas to obtain very sharp results on the differentiability.

6. Persistence of invariant tori for quasi-integrable systems.

In this section, we will present several proofs of the theorem that made KAM theory famous. This theorem is very useful in mechanics and in ergodic theory.

Basically the theorem says that an integrable system which is not degenerate (See below for a precise definitions) and sufficiently differentiable has the property that many of the quasi-periodic orbits persist under small perturbations.

The theorem has versions for Hamiltonian flows and for exact symplectic maps.

The simple minded versions that we will discuss can be stated as follows:

THEOREM 6.1. *Consider the symplectic manifold $M = \mathbf{R}^n \times \mathbf{T}^n$ endowed with the canonical symplectic form.*

Let $H : M \to \mathbf{R}$ be an analytic function such that:

$$H(I, \phi) = h(I) + R(I, \phi) \tag{6.1}$$

Let $\omega \in \mathbf{R}^n$ satisfy (3.18), *and $\omega = \nabla h(I_0)$ for some I_0.*

Assume that for I in a neighborhood of I_0 we have:

$$\left| \det \frac{\partial^2}{\partial I_i \partial I_j} h(I) \right| \geq \kappa > 0. \tag{6.2}$$

Then, if $\|R\|_\sigma$ is sufficiently small, the Hamilton equations for (6.1) *admit a quasiperiodic solution of frequency ω.*

This solution lies on an analytic torus $\mathcal{T}$, which it fills densely. Moreover, if $\|R\|_\sigma$ is sufficiently small $\mathcal{T}$ is arbitrarily close to the torus $\{I_0\} \times \mathbf{T}^n$

The version for exact symplectic maps reads as follows:

THEOREM 6.2. *Consider the symplectic manifold $M = \mathbf{R}^n \times \mathbf{T}^n$ endowed with the canonical symplectic form.*

Consider the map $F_0 : M \to M$ given by:

$$F_0(I, \phi) = (I, \phi + \Delta(I)) \tag{6.3}$$

where $\Delta : \mathbf{R}^n \to \mathbf{R}^n$ is an analytic function,

$$\Delta_j(I) = \frac{\partial}{\partial I_j} \Phi(I) \tag{6.4}$$

Assume that $\omega \in \mathbf{R}^n$ *satisfies* (3.19) *and*

$$\omega = \Delta(I_0)$$

for some I_0 *and that*

$$\left|\det \frac{\partial}{\partial I_j}\Delta_i(I)\right| \geq \kappa > 0 \tag{6.5}$$

in a neighborhood of I_0.

Let $F : M \to M$ *be an analytic, exact symplectic map.*

If $\|F - F_0\|_\sigma$ *is sufficiently small, then the map* F *admits a quasiperiodic orbit of frequency* ω. *This orbit is dense in an analytic torus which (if* $\|F - F_0\|_\sigma$ *is sufficiently small) is arbitrarily close in the analytic topology to the torus* $\{I_0\} \times \mathbf{T}^n$ *which is filled densely by the orbit of frequency* ω *of* F_0.

REMARK 6.3. The condition (6.4) is imposed so that the unperturbed map is exact symplectic.

An obvious consequence of (6.4) is that the matrix in (6.5) is symmetric.

In what follows we will indicate several proofs of the above theorems. The ideas and techniques of the proofs in both cases are roughly the same. Moreover, one can pass from one to the other by an ingenious construction [**Dou82b**], so that they are indeed equivalent in a precise sense.

Since proofs of these theorems have been in the literature for several decades, and many of the estimates have been covered in the previous sections, we will leave many of the details to the reader, indicating the most interesting ones as exercises.

Of course, the theorems, as stated above are quite far from the state of the art, but we hope that they still contain enough difficulty to illustrate the techniques of the theory and to fulfill the pedagogical goal of these notes. We will also present references to places in the literature where more elaborate arguments, which we will try to sketch, lead to sharper results.

6.1. Kolmogorov's method. The original paper has been translated in [**Kol54**]. A translation of a much less detailed account can be found in an appendix of [**AM78**]. Very good modern implementations of the method can be found in [**BGGS84**], [**Bar70**]. A generalized discussion of the ideas, putting them in a much wider context can be found in [**Mos67**], [**Zeh76a**]. (See Remark 6.6.)

We observe that a Hamiltonian system of the form[7]

$$H(I, \phi) = \omega I + O(I^2) \tag{6.6}$$

has Hamiltonian equations of motion

$$\begin{aligned} \dot{\phi} &= \omega + O(I) \\ \dot{I} &= O(I^2) \end{aligned}$$

Hence $\phi = \phi_0 + \omega t$, $I = 0$ is a solution.

A quasi-integrable system has the form

$$H(I, \phi) = h(I) + R(I, \phi)$$

with $R(I, \phi)$ "small" in some sense that will be made precise later.

Clearly we can consider Hamiltonians defined up to constants.

[7]We use the notation $O(I^2)$ to denote functions $A(I, \phi)$ such that $A(0, \phi) = 0$, $\frac{\partial}{\partial \phi} A(0, \phi) = 0$ and similarly for other orders

We will write

$$h(I) = \omega I + h_2 I^2 + \cdots + h_n I^n + h_{[>n]}(I)$$

where $h_i I^i$ stands for the homogeneous polynomial of degree i in the Taylor expansion of h (think of I^i as standing for all the monomials of degree i), and $h_{[>n]}(I) = o(I^n)$.

Similarly, we will write, performing a Taylor expansion in the variable I,

$$R(I, \phi) = R_0(\phi) + R_1(\phi) I + \cdots + R_n(\phi) I^n + R_{[>n]}(I, \phi) \ .$$

Then, we can write the quasi-integrable Hamiltonian as

$$H(I, \phi) = R_0(\phi) + \omega I + R_1(\phi) I + h_2 I^2 + R_2(\phi) I^2 + H_{[>2]}(I, \phi) \tag{6.7}$$

We observe that, if $R_0(\phi)$ and $R_1(\phi)$ were zero, we would be in the situation described in (6.6). To add a bit of color to the description of the proof, we will refer to these terms as the "*bad*" terms since their presence spoils the easy argument for existence of quasi-periodic orbits.

The idea of the proof of Theorem 6.1 by this method is to find a canonical transformation C – which will be close to the identity – in such a way that $H \circ C^{-1}$ will not have the bad terms.

The canonical transformation will be constructed as the limit of a sequence of canonical transformations $C^{(n}$ defined recursively by:

$$C^{(n+1} = \exp\left(-\mathcal{L}_{G^{(n}}\right) \circ (T^{(n)})^{-1} \circ C^{(n} \ ,$$

where $T^{(n}$ is a canonical transformation of the form

$$T^{(n}(I, \phi) = (I + k_n, \phi) \ , \qquad (k_n \text{ constant}) \ . \tag{6.8}$$

and $\exp\left(-\mathcal{L}_{G^{(n}}\right)$ is the time one map of the Hamiltonian flow corresponding to the Hamiltonian $-G^{(n}$. The theory of these canonical transformations was developed in Section 3.6.[8]

We will denote by $H^{(n}$ the Hamiltonian expressed in the coordinates given by $C^{(n}$. That is, $H^{(n} \circ C^{(n} = H$. Hence, $H^{(n+1} = H^{(n} \circ T^{(n} \circ \exp(\mathcal{L}_{G^{(n}})$.

We will choose the $G^{(n}$ and the $T^{(n}$ in such a way that they reduce as much as possible the bad terms of the Hamiltonian $H^{(n}$.

We will try to find the Hamiltonians of these transformations among linear functions in I

$$G^{(n}(I, \phi) = G_0^{(n}(\phi) + G_1^{(n}(\phi) I \ . \tag{6.9}$$

This is a reasonable choice to try first since this is the form of the terms that we want to eliminate. As we will see rather quickly, it works. (If not, we would have gone back and chosen a more complicated $G^{(n}$.)

Even if for the proof that we have discussed here it is enough to verify that the above form works, the reader that plans to study new problems may be interested in the fact that there is a theory to predict what terms will work and we have sketched it in Remark 6.6. See also [**Mos73**] p. 138.

---

[8]Notice that the canonical transformation in (6.8) cannot be generated by a time one map of a vector field generated by a Hamiltonian function since it is not an exact symplectic transformation. One can develop the proof considering the exponential of a locally Hamiltonian vector field that combines $\exp\left(-\mathcal{L}_{G^{(n}}\right) \circ (T^{(n)})^{-1}$. (See [**BGGS84**].) We prefer to keep the translations separate with a view to proving translated curve theorems later.

We first describe semi-formally the step to construct the transformation $G^{(n}$. Since

$$H^{(n} \exp G^{(n} = H^{(n} + \{H^{(n}, G^{(n}\} + \text{"second order in } G^{(n}\text{"} , \tag{6.10}$$

we try to eliminate the bad terms in the main part of (6.10). Expanding (6.10) more explicitly, taking into account (6.9) and (6.7), we have:

$$\begin{aligned} H^{(n} + \{H^{(n}, G^{(n}\} = {} & \omega I \\ & + \{\omega I, G_0^{(n}\} + R_0^{(n}(\phi) \\ & + \{\omega I, G_1^{(n}(\phi) I\} + \{h_2^{(n} I^2, G_0^{(n}(\phi)\} + R_1^{(n}(\phi) I \\ & + \{H_{[>1]}^{(n}(I,\phi), G_1^{(n}(\phi) I\} \\ & + \{H_{[>2]}^{(n}(I,\phi), G_0^{(n}(\phi)\} + H_{[>1]}^{(n}(I,\phi) \\ & + \text{"second order in } G^{(n}\text{"} . \end{aligned} \tag{6.11}$$

Notice that the "bad" terms of (6.11) (i.e. those that do not include I or include it only to the first power) are precisely those on the second and third lines (up to "second order in $G^{(n}$" terms).

The goal will be to choose $G^{(n}$ in such a way that the bad terms in the resulting Hamiltonian are much smaller than those in the original system. If we manage to eliminate the bad terms in the main part of (6.11), the Hamiltonian in (6.10) will only have bad terms which are "second order in $G^{(n}$".

We claim that it is always possible to find a $G_0^{(n}$ in such a way that we eliminate the bad term with no powers of I. Equating the second line of (6.11) to zero, we obtain the following equation for $G_0^{(n}$:

$$\{\omega I, G_0^{(n}\} + R_0^{(n}(\phi) = 0 \tag{6.12}$$

Equation (6.12) is, of the form (3.26) for which we have developed a theory in Lemma 3.15. Note that $\{\omega I, G_0^{(n}\} = L_\omega G_0^{(n}(\phi)$ where L_ω is defined in (3.26).)

The main conclusion of the theory of Lemma 3.15 is that the equation (6.12) can always be solved (with an slightly regular function), if the R.H.S. has average zero. Notice that, since Hamiltonians are defined only up to the addition of a constant, we can always ensure that $R_0^{(n}$ has average zero and, hence, that equation (6.12) can be solved for $G_0^{(n}$.

Eliminating the second bad term in (6.11) is more subtle. The equation to eliminate this term is

$$\{\omega I, G_1^{(n}(\phi) I\} + \{h_2^{(n} I^2, G_0^{(n}(\phi)\} + R_1^{(n}(\phi) I = 0 \tag{6.13}$$

The $G_0^{(n}$ appearing in (6.13) is known since we found it by solving (6.12) so that the equation (6.13) is only an equation for $G_1^{(n}$ and all the other terms in it are known.

Noting that all the terms have the structure of the dot product of a vector (depending on ϕ) with I and eliminating this vector, we can write the equation (6.13) as

$$L_\omega G_1^{(n}(\phi) = -2h_2^{(n} \nabla G_0^{(n}(\phi) - R_1^{(n}(\phi) . \tag{6.14}$$

The equation for each of the components of (6.14) is just one equation of the form (3.26).

We see that (6.14) will have a solution when and only when the average of the right-hand side is equal to zero. The average of the term $h_2^{(n}\nabla G_0^{(n}(\phi)$ is automatically zero. Hence, we conclude that if

$$\int_0^1 R_1^{(n}(\phi)\,d\phi = 0 \ , \tag{6.15}$$

we can indeed solve (6.14) and, hence, eliminate the second part of bad terms.

Of course, (6.15) is very restrictive. It is very easy to construct perturbations that do not satisfy the condition. Here is when the translations $T^{(n}$ come into play. Given any Hamiltonian of the form (6.7), provided that the $h_2^{(n}$ satisfies the non-degeneracy assumptions, it is possible to choose a translation $T^{(n}$ of the form (6.8) in such a way that the average of $R_1^{(n}$ vanishes. This is an application of the implicit function theorem provided that $h_2^{(n}$ is an invertible matrix and that, of course, all the R terms are small. Notice that the "vertical" translation by k_n is roughly given by $k_n \approx -\frac{1}{2}\left(h_2^{(n)}\right)^{-1}\bar{R}_1^{(n}$. (We call attention to the fact that the conditions that need to be adjusted in (6.15) is exactly the number of parameters at our disposal when we apply a translation.)

The magnitude of the translation required to adjust the average of $R_1^{(n}$ can be bounded by a constant times the size of $R_1^{(n}$ (provided that $h_2^{(n}$ is invertible and that the other terms are small, so that we can apply the implicit function theorem).

Hence, the algorithm for the iterative proof is:

(1) To determine the translation so that $H^{(n}\circ T^{(n}$ satisfies the normalization

$$\int_0^1 \frac{\partial}{\partial I} H^{(n}\circ T^{(n}\Big|_{I=0}\,d\phi = \omega \ .$$

(2) For the "new" $H^{(n}$ (i.e. for $H^{(n}\circ T^{(n)}$) find $G_0^{(n}$ and $G_1^{(n}$ in such a way that we eliminate the two "bad terms" in (3.52) up to quadratic error.

We have already seen that step 2 involves small divisors and unbounded operators. Nevertheless, we have also seen several times that the quadratic convergence can overcome the effect of small denominators (for Diophantine numbers). Compared with the previous cases we have dealt with, the only new complication of the present algorithm is that we have to deal with the extra complication of having to adjust the translation so that (6.14) becomes solvable.

The main complication of the translation is that terms that were high order generate lower order terms. For example, a "good term" $H(I,\phi) = f(\phi)I^2$, with $f(\phi)$ a ϕ-dependent quadratic form becomes upon translation

$$H\circ T = f(\phi)I^2 + 2f(\phi)Ik + f(\phi)k^2 \ . \tag{6.16}$$

The last two terms of (6.16) are "bad".

The fact that find a translation to eliminate the average in (6.14) depends on the fact that the quadratic term $h_2^{(n}$ is invertible. We need to keep track of the fact that this remain so under the successive changes of variables. This is not so difficult since the condition is an open condition.

From the analytic point of view, we note that the procedure involves solving (twice) equations of the form (3.26) and applying the implicit function theorem. As we did in Theorem 4.1, the *second order* terms can be estimated in analyticity domains using Cauchy estimates.

In summary, we have sketched a procedure that given a perturbation that satisfies certain non-degeneracy conditions, makes a change of variable that reduces the bad terms and whose resulting error is smaller. More precisely, given estimates of the bad terms in a domain, we can obtain estimates of the resulting bad terms in an slightly smaller domain. The estimates will be of the form $||New||_{\sigma e^{-\delta}} \leq C\delta^{-\tau}||Original||_{\sigma}$ Note also that, in order to match domains etc. we need that δ and the size of the remainder are suitably related.

The proof consists in showing that if the original error is sufficiently small, then we can carry out indefinitely the iterative procedure sketched above and it converges in a non-trivial domain.

Here we sketch the main considerations that need to be taken into account converting the above remarks into a proof. The reader is urged to either work them out alone or to use this as a reading guide for excellent expositions in the literature (some of them are discussed below).

A) We start by deciding that we consider domains loses of the form $\delta_0 2^{-n}$, and that we will do estimates on domains parametrized by a r_n defined by $r_{n+1} = r_n - \delta_0 2^{-n}$.

B) We will need to assume inductively that

B.1) We have bounds:

$$\|(h_2^{(n})^{-1}\| \leq C_1 \ ,$$

and that the derivatives of R are sufficiently small so that they do not affect the application of the implicit function theorem (to ensure the existence of the translation $T^{(n)}$).

We take C_1 to be twice the initial constant: $C_1 := 2\,\|(h_2^{(0})^{-1}\|$. We will need to check that, if the initial error is small enough, the iterative procedure keeps the assumption being valid.

B.2) Assume inductively that:

$$\|R^{(n}\|_{r_n} \leq C_2 \ ,$$

with C_2 being twice the initial value: $C_2 := 2\,\|R^{(0}\|_{r_0}$.

B.3) We will also assume that we have bounds similar to those in the study of the Siegel problem

$$\|\nabla G^{(n}\|_{r_n} \leq \delta_0 2^{-n\tau} \tag{6.17}$$

The goal of the latter bounds (6.17) is to ensure that when we perform the composition of $H^{(n} \circ T^{(n} \circ \exp(\mathcal{L}_{G^{(n}})$, the composition is still defined in the smaller domain.

C) Using assumption B.1, we are able to control the size of the translation by $\|R^{n)}\|_{r_n}$ times an universal constant.

Given B.2, we see that the size of the remainder of $H^{(n} \circ T^{(n}$ is still of the same order of magnitude as $\|R^{(n}\|_{r_n}$. (The new lower order terms generated are bounded by the size of the translation.)

D) Solving the small divisors equation, we obtain $G_1^{n)}$, $G_0^{n)}$. We can bound

$$\|G_1^{(n}\|_{r_{n+1}} + \|G_0^{(n}\|_{r_{n+1}} \le CK^2 \frac{2^{n\tau'}}{\delta_0^{\tau'}} \left(\|R_0^{(n}\|_{r_n} + \|R_1^{(n}\|_{r_n} \right)^2 ,$$

The factor $2^{n\tau'}\delta_0^{\tau'}$ is the usual small divisor factor when we take domain losses as in A).

E) The heuristics can be justified by adding and subtracting and applying the mean value theorem pretty much in the same way that we did in the proof of Siegel theorem but using the estimates we developed in Section 3.6.

We obtain:

$$\|R_0^{(n+1}\|_{r_{n+1}} + \|R_1^{(n+1}\|_{r_{n+1}} \le CK^2 \frac{2^{n\tau'}}{\delta_0^{\tau'}} \left(\|R_0^{(n}\|_{r_n} + \|R_1^{(n}\|_{r_n} \right)^2 . \tag{6.18}$$

F) The rest is essentially mopping up:

F.1) We need to show that the quadratic convergence implied by (6.18) implies that the inductive assumptions in B) remain valid (if we start with a small enough error). This is accomplished in a similar manner as that in the Siegel theorem (the only delicate one is (6.17) and this is exactly the same as in the Siegel domain).

F.2) We need to show that the accumulated transformation converge. Again, this is not very delicate since the quadratic convergence implies that $C^{(n}$ are converging to the identity extremely rapidly.

We urge the reader to compare the above sketch with the papers [**BGGS84**] and with [**Bar70**] which contain very readable full proofs.

The main difference in the strategies of those papers with the presentation here is that [**Bar70**] uses generating functions to deal with canonical transformations. Both of [**BGGS84**] [**Bar70**] do not make a distinction between the translations and the exact exact transformations and they use just one locally hamiltonian transformation that accomplishes the effect of the two steps that we discussed. This is, of course, perfectly fine for the problem at hand. we have, however, preferred to keep the two types of transformations separate with a view in translated curve theorems.

A very pedagogical proof of a particular case of the result (that nevertheless contains the most essential difficulties) is [**Thi97**]. The paper [**Zeh76a**] contains a detailed reduction of the proof based in the Kolmogorov method to an abstract implicit function theorem very similar to Theorem 5.2.

REMARK 6.4. The Kolmogorov method of proof has the advantage that it is quite direct and very well suited to functional analysis. We always deal with the same linearized equation with the same frequency. In particular, it leads to very good regularity results.

The main disadvantages arise from the fact that every different frequencies require different transformations. Moreover, the form (6.6) is not unique.

Natural question, which are important for applications, but that do not follow directly from the results are what is the measure covered by the tori and determining whether tori of similar frequencies are close together. (Indeed, so far, we have not shown that there is only one torus with the a given frequency. Note that there are many hamiltonians with the same form (6.6).)

The question about the measure occupied by tori can be answered by showing that the mapping which associates to a frequency ω satisfying (3.18) the torus with frequency ω produced in the Theorem 6.1 is Lipschitz.

Moreover, the tori can be expressed as the graph of a function of ϕ,

$$I = W_\omega(\phi). \tag{6.19}$$

Clearly, given one torus, there is only one function W, whereas, given one torus, there will be several hamiltonians of the form (6.6) and several transformations reducing the original flow to them.

It is true that $W_\omega(\phi)$ turns out to be Lipschitz with Lipschitz constant close to $\|h_2^{-1}\|$.

The proof of these Lipschitz properties can be obtained rather easily if we note that the system of the form (6.6) is also an approximate solution to the equation for $\tilde{\omega}$ in the plane of ω. Hence, if ω and $\tilde{\omega}$ are close enough, we can consider the torus for frequency ω as an approximate solution for the equation that would produce a torus of frequency $\tilde{\omega}$. The error of the approximation is controlled by $\omega - \tilde{\omega}$. Hence, applying the procedure, we see that we produce a solution which differs not more than something that can be controlled by $\omega - \tilde{\omega}$. This type of argument also leads to uniqueness results of the torus with a given frequency. More details on this type of argument can be found in [**Dou88**], [**Sev95**].

As we will see later, it is true that the map $(\omega, \phi) \mapsto W_\omega(\phi)$ introduced above is differentiable in the sense of Whitney.

REMARK 6.5. Another aspect in which the method of proof we have discussed is not optimal is that it requires very strong non-degeneracy conditions.

Notice that we want to ensure that the size of the translation required to adjust the error to zero average is commensurate with the error. In a degenerate situation, the size of the translation would be a root of the size of the error and then, the method as we have presented it, would collapse.

As a matter of fact, one can get a better non degeneracy condition if one does not fix the frequency, but fixes it up to a multiple. Hence, the only thing that we require is that $\operatorname{Span}(\omega) + \operatorname{Range}(h_2) = \mathbf{R}^n$.

One can also use clever tricks to reduce degenerate situations to non-degenerate ones. For example, in [**BH91**].

As we will see later, one can do significantly better than that by using other methods. For example, [**Sev95**].

REMARK 6.6. There is an interesting interpretation of the method of proof we have presented above in terms of geometry in infinite dimensional spaces.

This interpretation can certainly serve as a heuristic guide and many KAM theorems can be fit into this form. It was proposed in [**Mos67**] and developed quite forcefully in [**Zeh76a**], which developed in this language the main KAM theorems. In [**Ham82**], a similar philosophy is applied to many geometric problems.

The idea is to think of (6.6) as defining a manifold $\mathcal{N}$ in the space of Hamiltonians $\mathcal{H}$. All the elements of this manifold have a feature that we are interesting in studying. In this case, having an invariant torus of frequency ω.

We also have an action Ψ of a group. In this case, the action by canonical transformations. The proof we have sketched shows that given a neighborhood $\mathcal{U}$ of $\mathcal{N}$ in $\mathcal{H}$ all the elements of $\mathcal{U}$ have an orbit under Ψ that intersects $\mathcal{N}$.

Even if this is not completely trivial to make precise, (one has to define the topologies of the spaces of hamiltonians and mappings, check that they are manifolds, check the properties of the action of the group of transformations on it, etc.) it can serve as a heuristic principle to decide which theorems are possible. (Note that, if we were considering a finite dimensional problem, we could just decide what was true by deciding whether the tangent spaces of $\mathcal{N}$ and of the orbits of the action span the tangent space of $\mathcal{H}$.)

We note that this line of reasoning and these heuristic principles apply to other problems outside mechanics. Indeed, a good part of singularity theory can be formulated in this way. Similarly, many problems in geometry and PDE can be reduced to implicit function theorems by applying this heuristic picture. (See [**Ham82**].)[9]

The idea of deciding which theorems in KAM theory could be true by just looking at when the tangent spaces span leads very quickly to the problem of counting parameters. (See the discussion in [**Mos67**].) Roughly one needs that the normal form $\mathcal{N}$ and the group acting contain enough free parameters to overcome all the obstructions imposed by the geometry.

One of the important developments of later years is that in this counting of parameters, one should include the frequency [**Eli88**] or the perturbation parameter [**JS92**]. One reason why this is not obvious is that these extra parameters have a Cantor structure, hence at first sight, notions based on the geometry of tangent spaces etc. do not seem workable. Nevertheless, it turns out to be true that one can use these Cantor parameters very much in the same way as continuous families supplementing the standard geometric arguments based on implicit function theorems with measure theoretic estimates. Indeed, the next method of proof which we discuss can be used to cope with this type of problems. We refer to [**Sev99**] for an account of recent developments in the lack of parameters problem and, relatedly on the problem of study of degenerate systems.

EXERCISE 6.7. Try to carry out the proof choosing the translation $T^{(n}$ given by $k_n = -\frac{1}{2}\left(h_2^{(n)}\right)^{-1}\bar{R}_1^{(n}$. Notice that in such a choice we kill the average of $R_1^{(n}$ up to second order terms in $R^{(n}$.

6.2. Arnol'd method. In [**Arn63a**], V. I. Arnol'd introduced a method prove the persistence of quasi-periodic solution quite different from the method of proof by Kolmogorov that we have discussed in the previous Section 6.1.

Rather than trying to perform a change of variables that produces one torus, the method of [**Arn63a**] produces changes of variables that reduce the system to approximately integrable in a region of space. Hence, the method of [**Arn63a**] produces all the tori at the same time.

The main complication that arises with respect to the method of Kolmogorov is that the intermediate steps require to study transformations that are defined in rather complicate regions. The fist transformation is defined in a domain that excludes the low order resonances. (The places where $\partial_I H(I)\cdot k \ll 1$ for $|k|$ not big.)

[9] Incidentally, in singularity theory one has a very powerful implicit function theorem [**Mat69**], which allows to deal in some cases with operators that loose fraction of the derivatives. The method of [**Mat69**] paper has, to my knowledge not been used in KAM theory, even if [**dlLMM86**], which considers perturbations theories for Hamiltonian systems that were, previously done using KAM theory, was very inspired by it.

In successive steps of the iterative procedure, one performs another transformation that reduces the system much more closely to integrable, but in a more complicated region since we need to take into account more resonances. At the end of the process one ends up with a transformation defined on a Cantor family of invariant tori. (A set which is locally diffeomorphic to the Cartesian product of a Cantor set and a torus. Each of the torus in a connected component of the set is invariant.)

An alternative way to describe the whole process is to say that we have a smooth canonical transformation defined in the whole space which reduces the perturbed differential equation to integrable in a smaller set. At intermediate steps of the iteration we just keep estimates of how the system differs from integrable in a smaller and smaller set with increasingly complicated geometry. In the limit, we obtain control on just a Cantor family of tori, on which the system can be considered as integrable.

The basic strategy, which we will detail later, has several advantages with respect to the Kolmogorov one:

One of them is that one obtains more information on the way that the tori are organized. For example, it follow rather naturally that the tori constitute a family that is differentiable in the sense of Whitney. (This was observed in [**CG82**] [**Gal83a**]. Similar results can were obtained by other methods in [**Pös82**].)

Another advantage is that if we stop the process after a finite number of steps, we may still have quite good information about the system. For example, under the assumption that $\frac{\partial^2}{\partial I_i \partial I_j} H(I)$ is a positive definite matrix, in [**Neh77**] (as a matter of fact, the assumption in [**Neh77**] is sharper but more complicated to state than the positive definiteness, which is enough in many applications and which is the assumption used in many more modern proofs.) one can find the result that, denoting by $\epsilon = \|R\|_\sigma$ in (6.1) we have, for times $t \leq \exp(A\epsilon^{-a})$, **all** the orbits of the perturbed system (6.1) remain at a distance not more than ϵ^b of those of the perturbed system.

The method of exclusion of parameters near the resonances and continuing the transformation in the rest of the space, has had many applications in other KAM problems. For example, in the problem of changing a system with quasi-periodic coefficients into constant coefficients, usually called the reducibility problem, most of the papers (see specially the early ones [**DS75**]) are quite influenced by the method. We refer to the lectures of Prof. Eliasson in this meeting for an up to date review of this problem. The strategy of [**Arn63a**] was also employed in the first proofs that started to study the problem of lack of parameters and the related problem of studying systems which are rather degenerate.

From the point of view of the regularity assumptions needed the main shortcoming of the method is that the analytic part of the proof is based on truncating the Fourier series of the perturbation which produces bad results in finitely differentiable systems. Even it it is not too difficult, I know of no place in the literature where the Arnol'd strategy is implemented for finitely differentiable systems. (I wrote some very preliminary notes on that for a graduate course.)

Another shortcoming arises from the fact that one of the elements of the iterative step is the domain of the definition on which the changes of variables are defined. Keeping track of this domain is much more complicated than keeping track of the sizes of the functions. Hence, the proofs are more complicated and often one obtains worse estimates on the sizes of perturbations allowed and other quantitative results. (Nevertheless the method was used in the first proofs of several sharp estimates such as [**Ne˘i81**] [**Way84**].)

I do not think that the method of [**Arn63a**] has been formalized in such a way that it leads to an abstract implicit function theorem in the style of Theorem 5.2 which takes care of the detailed estimates in applications or, at least provided with a detailed strategy to carry them out.

Besides [**Arn63a**], a very pleasant and instructive modern exposition of this method of proof is [**Gal83a**] (see also [**CG82**].) The Nekhoroshev theorem proved by this method is nicely explained in [**BGG85c**] and a unified exposition of Nekhoroshev and KAM theorems is in [**DG96**]. Other proofs of Nekhoroshev theorems are covered in [**Pös93**], [**Loc92**].

In somewhat more, but still insufficient, detail: At the nth step of Arnold's method, we keep track of:

1) An excluded set, on which we do not expect to define the transformation.
2) In the complement of the excluded set we have defined a transformation $C^{(n}$ in such a way that
$$H \circ C^{(n} = \bar{H}^{(n}(I) + R^{(n} \ ,$$
3) We keep track of $\|\nabla H^{(0} - \nabla \bar{H}^{(n}\|_{\sigma_n}$ and $\|R^{(n}\|_{\sigma_n}$. We assume by induction that $\|\nabla H^{(0} - \nabla \bar{H}^{(n}\|_{\sigma_n}$ remains bounded and that $\|R^{(n}\|_{\sigma_n}$ is bounded by a superexponentially decreasing function. (The $\|\cdot\|_{\sigma_n}$ norms will refer to complex extensions of the excluded set, not a fixed set.)

Filling in more details about 1): The excluded set consists of bands given by

$$\left|\frac{\partial \bar{H}^{(i}}{\partial I}(I) \cdot k\right| \geq C_{n,i}|k|^{-\nu} \ , \quad 2^{i-1} < |k| \leq 2^i \ . \tag{6.20}$$

In particular, it is a set with piecewise smooth boundary and the angles of the corners are bounded from below by $C4^{-n}$ (where, C again is a constant that depends on the inductive assumptions).

This lower bound on the angles comes from the fact that a bound of this sort is what one would get for planes whose normals are integer vectors of total length 2^n and the fact that $\bar{H}^{(i}$ are uniform diffeomorphisms and, therefore only change the angles by a factor which remains uniformly bounded through all the iteration.

We denote the excluded set by $\mathcal{E}_n$ and

$$\begin{aligned} \mathcal{D}_{n,\sigma} &:= \{\, z \in \mathbf{C}^d \times \mathbf{C}^d/\mathbf{Z}^d \,|\, d(z, \mathcal{E}_n) \leq \sigma \,\} \\ \|f\|_{n,\sigma} &:= \sup_{z \in \mathcal{D}_{n,\sigma}} |f(z)| \ . \end{aligned}$$

Once we fix a sequence $\{\sigma_n\}$ (we will take $\sigma_n = \sigma_0(1 - \frac{1}{2}\sum_{n=0}^{\infty}(\frac{1}{3})^n)$), we denote the norm $\|\cdot\|_{n,\sigma_n}$ by $\|\cdot\|_n$.

The main difference between these norms and the regular ones is that, due to to the small angles, the Cauchy estimates are worse. Nevertheless, given the lower bound on the angles, they do not get too much worse:

$$\|\nabla f\|_{\sigma_n e^{-\delta_n}} \leq C\left(\delta_n e^{-4^n}\right)^{-1} \|f\|_{\sigma_n} \ .$$

To go from one step to the next, we exclude an slightly larger region and define a new transformation $C^{(n+1} = C^{(n} \circ \exp(\mathcal{L}_{G^{(n}})$ so that the new remainder is much smaller (here, we will need to make a small modification to our usual notion of smaller, meaning quadratic times powers of the domain loss).

We see that

$$H^{(n} \circ \exp(\mathcal{L}_{G^{(n}}) = \bar{H}^{(n} + R^{(n} + \{\bar{H}^{(n}(I), G^{(n}\} \\ + \{R^{(n}(I), G^{(n}\} + O((G^{(n})^2)$$

(a precise estimate for $O((G^{(n})^2)$ appears in Lemma 3.24).

A new idea of the method is to modify the prescription of Newton method by restricting only to a finite number of frequencies and include a truncation of the Fourier series so that, at every stage, we only have to deal with a finite (but growing) number of denominators. The error incurred by the truncation can be estimated if we increase the order of truncation at the right speed.

We write:

$$R^{(n\,[\le 2^n]}(I,\phi) = \sum_{|k|\le 2^n} \hat{R}_k^{(n} e^{2\pi i k\phi}$$

$$R^{(n\,[> 2^n]}(I,\phi) = \sum_{|k|> 2^n} \hat{R}_k^{(n} e^{2\pi i k\phi}$$

Hence, we solve:

$$\{\bar{H}^{(n}(I), G^{(n}\} + R^{(n\,[\le 2^n]}(I,\phi) = \Delta^{(n}(I)\ . \tag{6.21}$$

The equation (6.21) can be solved by setting

$$\hat{G}_k^{(n}(I) = \hat{R}_k^{(n}(I)/(\frac{\partial \bar{H}^{(n}}{\partial I}\cdot k)\ , \quad |k|\le 2^n\ . \tag{6.22}$$

By the definition of the excluded set, we can bound the denominators (6.22) over the complement of the excluded set.

Notice also that we can bound

$$\|R^{(n\,[>2^n]}\|_{\sigma_n e^{-\delta_n}} \le \|R^{(n}\|_{\sigma_n} e^{-\delta_n 2^n}\ .$$

This allows us to define the generator of the transformation that eliminates $R^{(n\,[\le 2^n]}$ (up to quadratic orders).

We have estimates

$$\|G^{(n}\|_{\sigma_n e^{-\delta}} \le C\delta^{-\tau}\|R^{(n}\|_{\sigma_n}\ ,$$

where, as usual, τ is roughly ν plus something depending on the dimension. We use the letter τ to denote similarly constants that depend only on the Diophantine exponent and the dimension.

To study the domain of $\exp(\mathcal{L}_{G^{(n}})$, we note that if we set

$$C_{n+1,\,i} = \delta^{-\tau} 2^n \|R^{(n}\|_{\sigma_n} + C_{n,\,i}\ ,$$

we can define the transformation from the set

$$\left|\frac{\partial \bar{H}^{(i}}{\partial I}\cdot k\right| \ge C_{n+1,\,i}|k|^{-\nu}\ , \quad 2^{i-1} < |k| \le 2^i \quad i = 1,2,\dots,n \tag{6.23}$$

to the set

$$\left|\frac{\partial \bar{H}^{(i}}{\partial I}\cdot k\right| \ge C_{n,\,i}|k|^{-\nu}\ , \quad 2^{i-1} < |k| \le 2^i \quad i = 1,2,\dots,n\ .$$

In that case, we have

$$\bar{H}^{(n+1}(I) = \bar{H}^{(n}(I) + \bar{\Delta}^{(n+1}(I)\ ,$$

from which it is clear that

$$\|\bar{H}^{(n+1}\|_{\sigma_{n+1}, \mathcal{D}_n} \le \|\bar{H}^{(n}\|_{\sigma_n} + \|R^{(n}\|_{\sigma_n}$$

and

$$\|\nabla \bar{H}^{(n+1} - \nabla \bar{H}^{(n}\|_{\sigma_{n+1}, \mathcal{D}_n} \le C 2^\tau \|R^{(n}\|_{\sigma_n} .$$

Most importantly, we have:

$$\|R^{(n+1}\|_{\sigma_{n+1}, \mathcal{D}_{n+1}} \le C 2^{n\tau} \|R^{(n}\|_{\sigma_n, \mathcal{D}_n} + 2^{-\delta_n 2^n} . \tag{6.24}$$

To define the next excluded set, the only thing we have to do is to add to the excluded regions corresponding to

$$\left| \frac{\partial \bar{H}^{(n+1}}{\partial I} \cdot k \right| \ge C |k|^{-\nu} , \quad 2^n < |k| \le 2^{n+1} .$$

Of course, excluding more regions makes the suprema in the left-hand side of (6.24) and all the other estimates even smaller.

The recursion (6.24) leads still to superexponential convergence choosing $\delta_n = \delta_0 (2/3)^n$. Establishing this was proposed in Exercise 5.19, see (5.24).

Once we have the superexponential convergence of the reminders, we obtain that the $C_{n,i}$'s remain bounded and so does $\|\nabla \bar{H}^{(n}\|_{\sigma_n}$ Indeed, $\|\nabla \bar{H}^{(0} - \nabla \bar{H}^{(n}\|_{\sigma_n}$ is small (arbitrarily small if we assume that $\|R^{(0}\|_{\sigma_0}$ is sufficiently small. Similarly, it is easy to check that $\|(\nabla^2 \bar{H}^{(n)})^{-1}\|_{\sigma_n}$ remains bounded and that the bound is close to the one for $\|(\nabla^2 \bar{H}^{(0)})^{-1}\|_{\sigma_n}$ if $\|R^{(0}\|_{\sigma_0}$ is sufficiently small. Hence, under the assumption that $\|R^{(0}\|_{\sigma_0}$ is sufficiently small, we can verify the inductive assumption on $\|(\nabla^2 \bar{H}^{(n)})^{-1}\|_{\sigma_n}$.

The passage to the limit in this procedure is somewhat subtle.

In the original coordinates, we have to study the sets $(C^{(n)})^{-1}\mathcal{E}^n$. These sets will be dense. By increasing slightly the excluded sets at each stage so that we exclude also the mismatches of the domain, we can arrange that $(C^{(n)})^{-1}\mathcal{E}^n$ are increasing. (note that this extra exclusion will be decreasing superexponentially since the transformations that we need to carry out in each step are decreasing superexponentially) Hence, $(C^{(n)})^{-1}(\mathbf{T}^d \times \mathbf{R}^d - \mathcal{E}^n)$ is a decreasing sequence of compact sets. On the other hand, their measure remains bounded away from zero as follows from the fact that $\|(\nabla^2 \bar{H}^{(n)})^{-1}\|_{\sigma_n}$ remains uniformly bounded so that we can use the same arguments as in Section 3.9).

It is slightly more subtle, but we can also estimate the derivatives of the transformations $C^{(n}$ to show that the derivatives remain bounded (it follows by an argument very similar to that used in the proof of Theorem 4.13 part v)) This shows that the sets $(C^{(n)})^{-1}(\mathbf{T}^d \times \mathbf{R}^d - \mathcal{E}^n)$ get closer and closer to being invariant. The limiting set will be invariant.

If one keeps track of all the derivatives in the closed sets, one can show that the limiting transformation $C^{(\infty}$ is differentiable in the sense of Whitney (see [**CG82**] or [**Gal83b**]). An interesting remark [**Val98**] is that one can use the fact that the gaps between the sets are much larger than the corrections to show directly the Whitney extension theorem. This remark could be important when studying infinite dimensional systems (e.g. PDE's). In infinite dimensions, the Whitney extension theorem is not a available, but the method of [**Val98**] could still work to produce tori that lie in a smooth family.

For more details of this method of proof we refer to the original paper [**Arn63a**], and the more expository paper [**Arn63b**], which also contains applications to celestial mechanics.

An early development of the method with several improvements is [**Sva80**]. More modern expositions (including the Whitney differentiability) of Arnol'd method are [**CG82**] and [**Gal83b**]. An exposition of the Arnol'd method that, at the same time proves Nekhoroshev's theorem and clarifies the geometry of the domains, is [**DG96**].

The method also lies at the heart of several other papers. One paper that incorporates the exclusion of parameters but is free of many geometric complications is [**DS75**]. This paper also shows that the method can allow some frequencies that are not Diophantine (they allow $|\omega \cdot k|^{-1} \geq \exp \frac{A|k|}{\log |k|^{1+\varepsilon}}$).

The method of transformations and exclusion of parameters is the basis of many modern developments in KAM theory related to lower dimensional tori, e.g. [**Eli89**], [**JS92**], [**JV97a**].

6.3. Lagrangian proof. In this section, we study a proof of of the Theorem 6.2 which has a different flavor from the proofs already presented. We will present the proof only in the case $d = 1$ and only for the particular case of the map given in (2.2). Similar proofs in any dimension and for more general maps are in the literature and we refer to that.

The proof differs substantially from the previous proofs of Theorem 6.1 in that it does not use compositions. Of course, the proof we presented of Theorem 4.1 does not require compositions either, even if the proofs we have presented so far for Theorems 6.1 do rely on transformation theory/ More interesting is that it is based on Lagrangian formalism. (That is, on second order equations rather than in systems of first order equations. The structure that is used is the fact that the equations solve a Lagrange variational principle, not that they come from a Hamiltonian formalism.)

The proof we present is based on unpublished notes of J. Moser for a course he gave in Zürich. A generalization of these results is included in the paper [**SZ89**]. We follow very closely the presentation in one of the chapters of [**Ran87**] (which in turn followed the presentation of the Moser's course.)

In [**Ran87**], one can also find the implementation of computer assisted proofs based on this method. In particular the result that the map given in (2.2) for $V(x) = \varepsilon\frac{1}{2\pi}\sin(2\pi x)$ has an invariant circle with golden mean rotation for $\varepsilon = .93$ (this was later improved to $\varepsilon = 0.935$). This is very close to the values for which [**Jun91**] showed that there can be no invariant circle. We discuss some of these issues in Section 7.[10]

The Lagrangian formalism for KAM theory has several other applications. For example, many elliptic PDE's have a very natural Lagrangian formalism but not a simple Hamiltonian one. (Note that in this case, the independent variable is multidimensional, while in Mechanics, the independent variable is the time, which

[10]The conjecture in [**Gre79**], given a theoretical – but not yet rigorous – basis in [**McK82**] is that there are smooth invariant circles with rotation golden mean when $\varepsilon < \varepsilon^*$ and not when $\varepsilon > \varepsilon*$. For $\varepsilon = \varepsilon^*$ there is an invariant circle which is not very differentiable. It is believed that $\varepsilon^* \approx .971635$. Of course, in other families it could – probably does – happen that the set of parameters for which one can find an smooth invariant circle is a more complicated set, perhaps with infinitely many components.

is one-dimensional) There is no easy canonical transformation theory for elliptic PDE's.

We will try to find solutions to (2.11) which read

$$\ell_\varepsilon(\theta+\omega)+\ell_\varepsilon(\theta-\omega)-2\ell_\varepsilon(\theta)=-\varepsilon V'(\theta+\ell_\varepsilon(\theta))\ . \tag{6.25}$$

We refer to Section 2.1 for the interpretation of this equation as a parameterization of a set in which the motion is quasiperiodic of frequency ω.

Somewhat informally, what we will do is to show that there is a procedure that, given an an approximate solution of (6.25) (which is not too badly behaved) we can produce another function that solves the equation even more approximately. Then, we will have to show that the whole process can be iterated indefinitely and that it converges to a solution.

Of course, making precise the notion of close, will involve introducing analytic norms. The statement that the result of the algorithm is closer to being a solution will mean to prove that in an slightly smaller domain, we will have the usual bounds which are quadratic in the previous error and have powers of the loss of analyticity. The fact that the iterative step can be performed and that it will lead to the desired improvement will require that certain expressions are not too large. (This is what we alluded to when mentioning that the solution is well behaved.) Of course, we will need to check that, if the initial error is small enough, the quadratic convergence that ensues, allows us to recover the inductive hypothesis indefinitely.

The theorem whose proof we will sketch is:

THEOREM 6.8. *Let $\ell_0 : \mathbf{T}^1 \to \mathbf{T}^1 \times \mathbf{R}^1$ be such that $\|\ell_0\|_\sigma < \infty$. Assume that*

$$\begin{aligned}&\|\ell_0'+1\|_\sigma \le M_+,\\&\|(\ell_0'+1)^{-1}\|_\sigma \le M_-\end{aligned} \tag{6.26}$$

$$\frac{1}{2}\|V''\|_{\sigma e^{(M_+ + \frac{3}{M_-})}} \le D \tag{6.27}$$

and

$$\|\ell_0(x+\omega)-2\ell_0(x)+\ell_0(x-\omega)-V'(\ell_0(x)+x)\|_\sigma \le \epsilon \tag{6.28}$$

where M_+, M_-, D, and ϵ are finite positive constants.

Let $\Gamma(M_+, M_-, D, K, \nu)$ be a function which will be made rather explicit during the proof, where K and ν are the constant and the exponent appearing in the Diophantine properties of ω.

If

$$\epsilon \le \Gamma(M_+, M_-, D, K, \nu)$$

then, there is a periodic function ℓ, $\ell(x+1)=\ell(x)$, solving (6.25).

Moreover

$$\|\ell-\ell_0\|_{\sigma_0/2} \le C\epsilon$$

where C is a constant that depends on M_+, M_-, D, K, ν.

The proof will be done using a quasi-Newton method. The method will be rather similar to the proof of Theorem 4.1. We try to solve the infinitesimal equation suggested by the Newton method. This will lead to an equation which is not immediately solvable with the method of Section 3.4. Nevertheless, by manipulating the equation with the remainder, we will arrive at a factorization of the equation that will be solvable by applying repeatedly the theory for the equation (3.27).

Denote

$$T(\ell)(x) \equiv \ell(x+\omega) - 2\ell(x) + \ell(x-\omega) - V'(\ell(x)+x) \tag{6.29}$$

We assume that we are given an approximate solution ℓ such that

$$T(\ell) = R \tag{6.30}$$

where R is small.

The prescription of the Newton method would be to find a Δ periodic solving

$$\Delta(x+\omega) + \Delta(x-\omega) - (2 + V''(\ell(x)+x))\Delta(x) = -R(x) \tag{6.31}$$

This equation (6.31) is not readily solvable in terms of Fourier coefficients (as indicated in Exercise 3.17) since the term $(2+V''(\ell(x)+x))\Delta(x)$ is not diagonal in Fourier coefficients. Our next task is to manipulate the equation so that it becomes solvable using the Fourier methods. The manipulations that will follow, even if rather straightforward and indeed convenient for numerical work will perhaps look mysterious, but in later sections we will argue that the success is due to natural geometric reasons.

If we take derivatives with respect to x of (6.30), we obtain, denoting by $T_\omega(x) = x + \omega$ and by $g'(x) = \ell'(x) + 1$

$$g' \circ T_\omega + g' \circ T_{-\omega} - (2 + V'' \circ g)g' = R' \tag{6.32}$$

Substituting the expression for $2 + V'' \circ g$ from (6.32) into (6.31), we obtain

$$g'\Delta \circ T_\omega + g'\Delta \circ T_{-\omega} - [-R' + g' \circ T_\omega + g' \circ T_{-\omega}]\Delta = -g'R \tag{6.33}$$

Ignoring the term $R'\Delta$, which is quadratic in the error, yields the system of equations:

$$\left(\frac{\Delta}{\ell'+1}\right) \circ T_{-\omega} - \left(\frac{\Delta}{\ell'+1}\right) = \frac{W}{(\ell'+1)(\ell'_0+1)\circ T_{-\omega}} \tag{6.34}$$

with:

$$W \circ T_\omega \quad W - (\ell'+1)R \tag{6.35}$$

The above system of equations consists of equations of the form (3.27) which can be studied using Lemma 3.15. We first solve (6.35) for W and we take the solution and substitute it in the R.H.S. of (6.34). We then solve (6.34) for $\frac{\Delta}{\ell'+1}$, out of which Δ is obtained just multiplying by $\ell' + 1$.

Of course, in order to carry out the above plan, we need to check that the equations we plan to solve are indeed solvable (i.e. that their R.H.S. has average zero). Later, we will have to worry about obtaining estimates of the solution thus obtained.

The fact that (6.35) can be solved is a calculation which we have done in Section 2.1 when we wanted to show the solvability of equation (2.17). Once we have that the equation (6.35) is solvable up to an additive constant, we can determine the additive constant in W in such a way that the R.H.S. of (6.34) has average zero. (Note that adding a constant $\bar{W}$ to W changes the average of the R.H.S. of (6.34) by $\bar{W} \cdot \int((\ell'+1)(\ell'+1)\circ T_{-\omega})^{-1}$. The integral is not zero, since we assumed by induction that the denominator in the integrand is bounded away from zero and hence, it is positive. Indeed, we can have a bound for it under the assumptions given in our inductive hypothesis.

The fact that this procedure works can be shown using the familiar method.

Adding and subtracting, we show that given some inductive assumptions on bounds on $||\ell' + 1||_\sigma$, $||(\ell' + 1)^{-1}||_\sigma$, the above procedure leads to a new quadratic remainder. That is, as usual, we have

$$||R^{n+1}||_{\sigma_n - \delta_n} \leq ||R^n||^2_{\sigma_n} CK^2 \delta_n^{-\tau}$$

The bounds we assumed on the derivatives deteriorate slightly, but again the quadratic convergence ensures that they remain bounded during the iteration.

REMARK 6.9. The remarkable cancellations (see (6.32), (6.31)) between the derivative of the remainder and the linearized equation which allowed us to obtain a quadratically convergent method solving only the linear equation (and performing easy multiplication and divisions by known functions) are not a coincidence. In [**SZ89**] one can find how they work for twist maps if we uses the equations given by the generating functions, linking them to a Lagrangian formalism.

Indeed there are deeper reasons. For example, the cancellations apply to partial differential equations. See [**Koz83**], [**Mos88**].

REMARK 6.10. If one is interested in obtaining the existence of invariant circles for numerical values that are as close to the optimal value as possible, one should be prepared to cope with the difficulty of having the quality of the solution get worse and worse. Indeed, the domains of analyticity shrink and function becomes more and more close to having zeros. Indeed, there are precise predictions – not rigorous but supported by numerical evidence that at the breakdown of the invariant circle for the map given in (2.2), all the difficulties happen at the same time and indeed all the quantities that need to be estimated blow up as powers of the distance of the parameter to the critical one.

See [**BCCF92**] for numerical results and [**dlL92**] for a non-rigorous explanation and precise conjectures.

6.4. Proof without changes of variables. In this section, we will present another proof of theorem 6.2 This proof is based on [**GJdlLV00**]. A version of the method for lower dimensional tori was presented in [**JdlLZ99**].

The proof actually proves something more general since the main result does not require that the map is exact. Of course, without assuming exactness, we cannot expect to have invariant tori as was shown in the examples. The conclusion of the main theorem is that for symplectic maps that satisfy all the other assumptions of Theorem 6.2 there is a torus that gets translated rigidly in the direction of the actions. The points of the torus are, roughly, rotated.

This is a generalization to higher dimensions of the translated curve theorem of [**Rüs76b**].

If we assume that the map is exact, it will be very easy to show that the translation has to vanish and that the torus is indeed invariant and that the motion on it is conjugated to a rigid rotation.

We will consider the symplectic manifold $\mathbf{T}^d \times \mathbf{R}^d$ endowed with the standard symplectic structure.

We will consider a map $F : \mathbf{T}^d \times \mathbf{R}^d \to \mathbf{T}^d \times \mathbf{R}^d$ which is symplectic (not necessarily exact) analytic (and other conditions somewhat weaker than those of Theorem 6.2 which we will formulate when the heuristic discussion motivates them). We fix ω Diophantine and seek a mapping $K : \mathbf{T}^d \to \mathbf{T}^d \times \mathbf{R}^d$ and a vector $a \in \mathbf{R}^d$

in such a way that

$$F \circ K(\theta) = K(\theta + \omega) + (0, a) \tag{6.36}$$

Note that, of course, the equation (6.36) expresses that the image of the torus is translated in the direction of the action by a rigid displacement a.

In the case that F is an integrable map, all the tori given by a parameterization

$$K_0(\theta) = (\theta, I_0) \tag{6.37}$$

are "vertically" translated. So, we expect that the functions we will have to consider will be close to that.

Later, we will show that if F is exact (and there are other conditions), then, a should vanish. This is very similar to the line of argument in [**Rüs76b**]). When $d = 1$, it can be seen that the zero flux condition indeed implies that $a = 0$. We note that, even if the proof does not use the exactness of the symplectic structure, it uses the symplectic structure. Under appropriate redefinition of translation, one can have similar theorems in other symplectic manifolds.

We will sketch the proof of the following theorem:

THEOREM 6.11. *Assume that F is an analytic symplectic map of $\mathbf{T}^d \times \mathbf{R}^d$ endowed with the canonical symplectic structure and that ω is a Diophantine number.*

Assume that F is close to an integrable map and that it satisfies the hypothesis of non-degeneracy of Theorem 6.2. Assume that we can find an approximate (K, a) solution of (6.36).

If the residual of (6.36) *is small enough (depending on properties of F and of $\|K - K_0\|$) where K_0 is the solution in* (6.37)*), then we can find an exact solution of* (6.36).

In particular, if we take as approximate solution K_0 as in (6.37) *the hypothesis are satisfied when F is sufficiently close to an integrable map.*

The fact that F is close to integrable is not really necessary as it will transpire from the proof. At this stage it is only introduced to avoid using a more complicated notion of non-degeneracy than that used in Theorem 6.2. As the proof in Section 6.3 it can apply to all the maps of the form (2.2). Even that can be generalized by formulating appropriately the degeneracy. See [**GJdlLV00**].

A very simple calculation (a more general version appears in [**JdlLZ99**]) shows that if we have an exact system, then the translation a for an true solution has to be zero.

PROPOSITION 6.12. *If the K solving* (6.36) *is close to K_0 in an analytic norm and F is exact, then $a = 0$.*

Before starting the discussion of the Theorem 6.11, we discuss the proof of Proposition 6.12.

PROOF. Let α be the symplectic potential form $\alpha = \sum_i I_i \, d\phi_i$. Assume also that $F^*\alpha = \alpha + dS$.

We consider the loops in the ith angle coordinate given by

$$L_{\theta_1, \cdots, \theta_{i-1}, \theta_{i+1}, \cdots, \theta_d}(\theta) := K(\theta_1, \cdots, \theta_{i-1}, \theta, \theta_{i+1}, \cdots, \theta_d),$$

where $\theta_1, \cdots, \theta_{i-1}, \theta_{i+1}, \cdots, \theta_d \in \mathbf{T}$.

Because of (6.36) and the exactness of the map, we have

$$\int_{L_{\theta_1,\cdots,\theta_{i-1},\theta_{i+1},\cdots,\theta_d}} \alpha = a_i + \int_{L_{\theta_1+\omega_1,\cdots,\theta_{i-1}+\omega_{i-1},\theta_{i+1}+\omega_{i+1},\cdots,\theta_d+\omega_d}} \alpha$$

And, integrating over the variables $\theta_1, \cdots, \theta_{i-1}, \theta_{i+1}, \cdots \theta_d$, we obtain

$$a_i = 0\,.$$

□

REMARK 6.13. This Theorem (and the Proposition) are much weaker than what can be proved by the method.

For example, the hypothesis that the system is close to integrable can be replaced by several quantitative statements about the approximate solution. This is quite important for several applications. We note that approximate embeddings can be obtained with the computer. One can try to solve a discretized Fourier series or, a big advantage for the present method, just compute orbits and compute the Fourier transform. This improvement is discussed in more detail in Remark 6.19.

Once this improvement is in place, it should be apparent that the way that the torus is embedded does not play any role. We do not need that the system is close to integrable or that the tori are close to integrable. (This is the main difference with the exposition of [**Bos86**].)

In particular, we can justify the existence of tori which have different topology than the tori of the unperturbed system. (Recently there has been some interest in these secondary tori since there are numerical experiments that suggest that secondary tori are very important for the statistical properties of coupled systems [**HdlL00**].)

Also, an important advantage of the method is that it allows one to deal will more degenerate situations than the twist mapping. Indeed, one can use it to deal with non-twist maps and with even more degenerate situations.

We refer the reader to [**GJdlLV00**] for these precisions as well as for more details about the proof. We also refer to [**JdlLZ99**] for another application of similar techniques to discuss lower dimensional tori.

Now, we start describing the main ideas of the proof. Again, we refer to [**GJdlLV00**] for more details.

The method of proof will be an iterative procedure in which we start from (6.36) being satisfied with a certain error and return a solution that satisfies the equation with an smaller error. Of course, as usual in this theory, what we mean by smaller error is that the size of the new error will be bounded (in a smaller domain than the original one) by the square of the size of the original error times a factor that is the domain loss parameter to a negative power. Of course, by now the convergence of the procedure should be well understood. Actually, since we do not need to make changes of variables and we do not need to keep track of much the geometric structures, the inductive hypothesis will be very mild.

We will begin with a heuristic discussion.

If we start with an approximate solution of (6.36) that is:

$$F \circ K(\theta) - K(\theta + \omega) - (0, a) = R(\theta) \tag{6.38}$$

where R is small in some appropriate norm that we will make precise later.

The Newton method prescription, would be to change K into $K + \Delta$, a into $a + \alpha$ in such a way that

$$DF \circ K(\theta)\ \Delta(\theta) - \Delta(\theta + \omega) - (0, \alpha) = -R(\theta)\ . \tag{6.39}$$

Unfortunately, this equation is not readily solvable by easy methods such as comparing Fourier coefficients since it involves the non-constant coefficient factor $DF \circ K(\theta)$.

Hence, we try to compare it with the equation obtained taking derivatives of (6.38):

$$DF \circ K(\theta)\partial_\theta K(\theta) - \partial_\theta K(\theta + \omega) = \partial_\theta R(\theta). \tag{6.40}$$

At this point, we are going to introduce some notation (which is not completely necessary but which will make the geometry more concrete). We define $D(\theta) := DF \circ K(\theta)$ and let $K_1(\theta)$ an orthogonal basis for $\partial_\theta K(\theta)$. The previous equation (6.40) reads $D(\theta)K_1(\theta) - K_1(\theta + \omega) = R_1(\theta)$. As usual, we define the matrix

$$J = \begin{pmatrix} 0 & \mathrm{Id}_d \\ -\mathrm{Id}_d & 0 \end{pmatrix},$$

which is the representation in coordinates of the symplectic form.

We define then the symplectic matrices.

$$M(\theta) = [K_1(\theta), JK_1(\theta)] \tag{6.41}$$

Notice that from the fact that K_1 is almost invariant under $DF(\theta)$ we obtain that:

$$DF(\theta)M(\theta) = M(\theta + \omega)\begin{pmatrix} \mathrm{Id} & A(\theta) \\ 0 & B(\theta) \end{pmatrix} + O(R) \tag{6.42}$$

We will introduce the assumption that $M(\theta)$ is invertible for all θ.

This is reasonable assumption in view of the fact that, for integrable systems[11] using (6.37) we have:

$$M(\theta) = \begin{pmatrix} \mathrm{Id}_d & 0 \\ 0 & \mathrm{Id}_d \end{pmatrix}.$$

Recall also that the assumption that the map F preserves the symplectic form is equivalent to

$$JDF(x) = \left[DF(x)^t\right]^{-1} J\ . \tag{6.43}$$

This gives that $B(\theta) = \mathrm{Id}_d$.

We note that in the integrable case, the matrix $A(\theta)$ will be a constant $d \times d$ matrix A and, the twist condition implies that A is invertible. Hence, in the proof of the theorem, we will assume that $A(\theta)$ is not very far from a constant, invertible matrix in the sense that $\bar{A}(\theta)$ is an invertible matrix. Indeed, this is the only non-degeneracy condition that we will assume.

We call attention to the fact that the non-degeneracy assumption only amounts to the invertibility of M and the fact that $\bar{A}(\theta)$ is invertible. These assumptions could be checked a posteriori on a numerically computed solution or on an approximate solution produced by any other means. Other than that, we do not need any

[11] Indeed, this is the only reason why we assumed that F was close to integrable. If we formulate the theorem assuming that M is invertible, we could have eliminated the assumption of close to integrability. Later, we will need to formulate the non-degeneracy assumption using this matrix M. This is explained in more detail in Remark 6.19 and in the references quoted there.

property of the map F. See Remark 6.19 and the references quoted there for an explanation of this alternative approach.

REMARK 6.14. It is an easy exercise to show that, under Diophantine conditions we can reduce the block $A(\theta)$ to a constant, so that the matrix is is indeed reducible, Nevertheless, for the applications that we have in mind, this does not help. Indeed, by doing it, we incur in extra small denominator estimates, which can worsen the result.

REMARK 6.15. A more geometric interpretation of the previous calculations is to say that $DF \circ K(\theta)$ is a reducible matrix whenever K is a parameterization of an invariant torus by a rigid rotation.

We want to give a geometric argument that shows that the linearization of the equations around an invariant torus is reducible. The argument will show that for an approximate solution, the equation will be approximately reducible and, hence that one can start an iterative procedure in which in the iterative step we improve the solution of the main equation and its reducibility.

That is, our goal find a system of coordinates on the tangent of the torus so that the matrix representing $DF \circ K(\theta)$ has constant coefficients.

Since the vectors along the direction of θ are moved just by a rotation in the torus, this is an invariant field that can be lifted to the space by the embedding. By the preservation of the symplectic structure, we also have that the plane spanned by the the vector and its symplectic conjugate is also preserved. We can see that in the plane spanned by a vector and its symplectic the matrix has to be upper diagonal (one vector is preserved.) The dilation along the symplectic conjugate has to be the inverse of the dilation along the preserved direction due to the requirement that the two-area in the plane is preserved. This gives us the diagonal blocks of the matrix. The upper diagonal does not bother.

This system of coordinates provides with a system in which the derivative is upper triangular.

Once that we have that the diagonal blocks are constant, then it is easy to see that the linearized equation can be solved by using equations of the form (3.27).

The above geometric interpretation makes it clear that we do not need the symplectic form to be constant. Moreover, it is clear that it does not require that the symplectic form has action-angle variables and that it can accommodate certain singularities.

The algorithm is now very easy. If we write $\Delta(\theta) = M(\theta)w(\theta)$ and substitute in (6.39) we obtain

$$D(\theta)M(\theta)w(\theta) - M(\theta+\omega)w(\theta+\omega) - (0,\alpha) = -R(\theta) \tag{6.44}$$

which using (6.41) becomes:

$$M(\theta+\omega)\left[\begin{pmatrix} \mathrm{Id}_d & A(\theta) \\ 0 & \mathrm{Id}_d \end{pmatrix} w(\theta) - w(\theta+\omega)\right] - (0,\alpha) = -R(\theta) - N(\theta)w(\theta)\,. \tag{6.45}$$

Therefore, ignoring the last term of the R.H.S. of (6.45), which is quadratic, we are lead to the study of the equation:

$$\begin{pmatrix} \mathrm{Id}_d & A(\theta) \\ 0 & \mathrm{Id}_d \end{pmatrix} [w(\theta) - w(\theta+\omega)] = -M(\theta+\omega)^{-1}\left[R(\theta) - (0,\alpha)\right] \tag{6.46}$$

We claim that this equation for w, α can be studied using the methods that we have developed in Section 3.4. This will constitute our iterative step. Of course, after this heuristic derivation, we will need to go back and justify the estimates of the step and show that it can be iterated. This, even if being the essential part of the proof, we hope will bring no surprises anymore for the reader.

If we write (6.46) in components, denoting the components of

$$w(\theta) = (w_\phi(\theta), w_I(\theta))$$

and by Π_ϕ,Π_I the projections over the components, we have:

$$\begin{aligned} w_\phi(\theta) + A(\theta)w_I(\theta) - w_\phi(\theta+\omega) &= -\Pi_\phi M(\theta+\omega)^{-1}\left[R(\theta) - (0,\alpha)\right]) \\ w_I(\theta) - w_I(\theta+\omega) &= -\Pi_I M\left(\theta+\omega\right)^{-1}\left[R(\theta) - (0,\alpha)\right]) \end{aligned} \tag{6.47}$$

If we look at the second equation in (6.47) (recall that it is an equation for w_ϕ and α) we see that it is an equation of the form (3.27) which we have already studied. We chose α in such a way that the R.H.S. has average 0. (This can be done if M is close to the identity, but otherwise it can be made into an assumption to be checked a posteriori on the approximate solution.)

Note that we have bounds

$$|\alpha| \le C\|R\|_\sigma \|M^{-1}\|_\sigma \|(M^{-1} - \mathrm{Id})^{-1}\|_\sigma \tag{6.48}$$

If we assume for convenience (somewhat sharper assumptions could also work, see Remark 6.19) that the factors in the R.H.S. of (6.48) satisfy:

$$\|M^{-1}\|_\sigma \|(M^{-1} - \mathrm{Id})^{-1}\|_\sigma \le 333\ . \tag{6.49}$$

Then we can apply Lemma 3.15 to obtain w_I up to a constant, which we will determine in the next equation.

We have, denoting by $\tilde{w}_I$ the solution with zero average:

$$\|\tilde{w}_I\|_{\sigma-\delta} \le C\delta^{-\nu}\|R\|_\sigma\ . \tag{6.50}$$

If we look at first equation of (6.47) we see that, at this stage of the argument is an equation only for w_ϕ and the average of of w_I. Hence, we write it as

$$w_\phi(\theta) - w_\phi(\theta+\omega) = -\Pi_\phi M(\theta+\omega)^{-1}\left[R(\theta) - (0,\alpha)\right]) - A(\theta)w_I(\theta)\ . \tag{6.51}$$

If we assume that

$$\|(\bar{A})^{-1}\| \le 333\ , \tag{6.52}$$

we can determine $\bar{w}_I$ so that the terms in the R.H.S. of (6.51) have average 0. We have:

$$|\bar{w}_I| \le C\|R\|_\sigma$$

We will furthermore assume that

$$\|A\|_\sigma \le C \tag{6.53}$$

Hence, we can apply Lemma 3.15 and obtain a w_ϕ with zero average which satisfies:

$$\|w_\phi\|_{\sigma-2\delta} \le C\delta^{-2\nu}\|R\|_\sigma\ . \tag{6.54}$$

Note that the power of δ in this case is twice as high as that in the previous one since the R.H.S. of (6.51) involves the solution of the previous one.

From (6.50)(6.54), using the inductive assumptions on the size of M, we obtain

$$\|\Delta\|_{\sigma-2\delta} \le C\delta^{-\tau}\|R\|_\sigma .$$

From this, the rest of the proof of the translated tori theorem is very similar to the previous proofs, in particular to the proof of Theorem 4.1.

Under the assumption that

$$\|K\|_\sigma + \|\Delta\|_{\sigma-4\delta} \le \Sigma - \delta \tag{6.55}$$

where Σ denotes the size of the domain of analyticity of F, we can define the composition $F \circ (K + \Delta)$ and indeed the range of $K + \Delta$ is at least a distance δ from the boundary of the domain of definition of F.

Note that adding and subtracting and using Taylor's theorem to control the terms neglected to derive (6.39), (and Cauchy bounds to control the size or the derivatives involved) we get:

$$\|\tilde{R}\|_{\sigma-4\delta} \le C\delta^{-\tau'}\|R\|_\sigma^2 . \tag{6.56}$$

From this, we can conclude as in the previous cases that if the original remainder is small, then the iteration can be carried out an arbitrarily large number of times, moreover, the final remainder in its domain of definition keeps decreasing.

Note also that this proof – in contrast with those based on composition – does not require any subtle inductive hypothesis to ensure that the domains of the composition match. These assumptions, that we had to consider in the proofs based on composition are subtle because they require that the errors decrease faster than the analyticity losses.

In this case, the only assumptions that we have to check are (6.55), (6.49), (6.52), (6.53).

We can see that if we start with an small enough residual, the iterative procedure does not change A or M much, so that using in the step bounds which are twice the ones at the start, the estimates of the step remain valid if the original error is small enough.

REMARK 6.16. We emphasize that the only thing that we need to get the proof started is an approximate solution of the functional equation.

This can be obtained in a variety of ways. For example if the system was close to integrable, one could take as an initial guess the parameterization of the integrable system.

Other choices are possible. One could use a few steps of the Lindstedt series. In such a case, the proof will establish that the Lindstedt series is asymptotic.

More audaciously, one could use the results of a non-rigorous, numerical algorithm. Provided that one can verify rigorously that one has an approximate solution, one then obtains a rigorous proof of the existence of these circles. These issues will be explored in more detail in Section 7.

We also note that the present proof does not require much from the function except that it gives a parameterization of an invariant torus. In particular, it can apply to tori of topological types not present in the original system.

REMARK 6.17. Another proof without changes of variables can be found in [**Bos86**] which is based in unpublished work of M. Herman. This proof contains also a translated curve theorem.

The main difference with the proof presented here is that that method parameterizes the curves by the graph of a function. When studying tori that are not graphs, it requires that one performs a preliminary change of variables.

REMARK 6.18. The twist hypothesis in this method of proof can be bargained away considerably.

REMARK 6.19. A variant of the method that is useful in the study of lower dimensional tori or for some degenerate situations, is to take as a starting point of the procedure not just the K but the K and the M, which, respectively, almost solve the equation and almost reduce the equation to constant coefficients.

The iterative step, uses the M to solve the equation and then updates the M so that it reduces the new linearized equation to an even higher approximation.

By intertwining the improvement in K and in M it is possible to achieve quadratic convergence.

One advantage of this improvement is that, if one studies this for lower dimensional tori, both the K and the M can be computed perturbatively. The approximate K is a polynomial in the perturbation parameter, nevertheless, the M is a polynomial in the square root. Hence, the iterative method based on both approximations can capture the singularity structure much better than the approximation we have discussed here.

We refer to [**JdlLZ99**] for more details about the method for lower dimensional tori.

REMARK 6.20. One feels that these methods of reducing the equation to constant coefficients is a bit of overkill. When one tries to invert an operator, diagonalizing it is rather more than what is needed.

Indeed, the great advances in KAM for PDE's started when the emphasis went from diagonalizing the linear operator as was done usually in KAM theory to just using estimates from the inverse. (See [**CW93**],[**Bou99a**]) Even if we will not discuss it in these notes, when one considers elliptic lower dimensional tori some of the resonances that appear in some proofs are obstructions to the diagonalization of the operator, not to the invertibility. Therefore, they can be eliminated from the proofs of the existence of the torus if one relies on inverting the operator rather than just diagonalizing it. (See [**Bou97**].) Related to this issue we call attention to the lectures of prof. Eliasson in this meeting. He shows that even if it could happen that the tori are not reducible, using his non-perturbative results, they are arbitrarily close to reducible. This is enough to continue the iterative procedure.

REMARK 6.21. One issue that still is quite puzzling to me is that if one performs the Lindstedt series for lower dimensional tori, one encounters only small denominators related to the frequencies of the motion on the torus. This is significantly less small denominators than those appearing in the proofs mentioned above in which one needs to take into account denominators which happen when harmonics of the intrinsic frequencies of the torus are close to a normal frequency. The proofs in which one also proves reducibility of the lower dimensional torus, require even more small denominators conditions. In them, one has to take into account the cases when differences of two normal frequencies become a combination of the frequencies of motion in the torus.

In [**JdlLZ99**], one can find a proof of the fact that the Lindstedt series is asymptotic and defines an analytic function in a large sector. (One has to exclude an exponentially thin wedge.) Nevertheless, the convergence or not of these series has not been settled.

Note that at the same time that one develops the series for the torus, one develops also a series for the reducing matrix which also does not present other small divisors than those of the intrinsic frequencies. The convergence of this series has not been settled either.

7. Some remarks on computer assisted proofs

The existence or non-existence of invariant tori in a system appearing in a concrete application could have enormous practical importance.

For example, there are many systems such as accelerators or plasma devices that are modeled rather well by Hamiltonian systems. The existence of tori in these systems has very drastic effects in their long term behavior. For example, if the system is a two dimensional map, the existence of invariant circles, will imply that one region of phase space will remain trapped for ever. This is of great interest for plasma devices whose goal is to confine a plasma or for accelerators that try to keep a beam of particles in place. Indeed, many of these devices are designed in such a way that they maximize the *abundance* and robustness of invariant tori. One hopes that, even if the Hamiltonian approximation is not completely accurate, the KAM tori will survive somehow.

In celestial mechanics, one is interested also in finding regions with invariant tori since they are suitable for parking orbits.

The judicious numerical experimentation with dynamical system has been a great source of insight and inspiration, even if, of course, much of the work is non-rigorous and, hence, does not fit well with this tutorial. We refer the reader to [**Hén83**], [**Sim98**] for some study of the issues involved in numerical computations and to [**Mei92**] for a point of view closer to the physical applications

Of course, in these applications, one also wants to get, besides the existence, information about the shape of the torus and more details about its properties. What we want to discuss in this section is how some of these non-rigorous calculations can be turned into theorems.

The basic observation is that some of the KAM proofs we have presented here have the structure that they formulate a functional equation and show that, given an approximate solution which is not too bad from the analytic point of view, then there is a true solution, which, moreover is not too far from the approximate solutions. These constructive methods do not require that the system is close to integrable.

Note that these proofs do not care about how we have produced the approximate solutions. The only thing that we need to verify rigorously is that these approximate solutions indeed solve the functional equation to up to an small error and that their analyticity properties are adequate. Hence the problem of justifying that these computed solution correspond to a true one reduces to showing rigorously that these numerically specified functions indeed solve the desired functional equation with a good accuracy and verifying rigorously their analyticity properties.

Of course, given one polynomial with a few coefficients one could imagine studying its properties with respect to an easy equation such as (6.30) with a pencil and a notepad. (See [**Her86**] for an example of these verifications with pencil and paper.) Nevertheless, if the number of coefficients approaches those needed for what is considered good accuracy in numerical calculations (this is often a few hundred or a few thousand coefficients), using a notepad becomes impossible.

One would like to use a computer. The problem with using a computer is that, as they are used most commonly computers do not deal with real numbers and they do not perform on them the mathematical arithmetic operations. In their normal mode of working, computers deal only with a finite set of numbers, the *representable* numbers[12]. On these representable numbers, we perform arithmetic operations which are approximations of the arithmetic operations among real numbers.

These operations produce an approximation to the true answer if at all possible[13] or if it is impossible to give a reasonable answer in terms of representable numbers (e.g. if you ask to multiply by 10 the largest representative number or to divide by zero) they do not return an answer, but they *raise an exception* which typically does something drastic such as causing the program to terminate abruptly, perhaps copying the state of the memory to a file (*dumping a core*) that can be examined to trace the problem. (A good discussion of the subtleties involved in the implementation of floating point arithmetic is [**Knu97**].)

One problem with this approximate way of proceeding is that approximate of approximate may not be approximate enough. Much less if one repeats the process of approximation a large number of times. Of course, given that a computer nowadays produces over one hundred million operations in a second, we have to worry about the effect that performing millions of approximations may lead us away from a good approximation.

As every good numerical analyst knows, producing numbers is not too difficult. Unless the computer catches fire, you will get numbers. The real difficult issue is to produce numbers that can be trusted. More difficult even is to device methods that ensure the numbers produced can be trusted. One should keep in mind that most of the technology and research happens at the borderline regions when the algorithms are about to break. (If the problems we are studying were safely solvable, we would fix the situation going to a more challenging problem.)

The problem of reliability of arithmetic calculations is significantly more pressing for the problems involving small divisors. We have seen already that the Lindstedt series manage to converge only through massive amounts of cancellations. Cancellations are one of the worst enemies of accuracy in floating point calculations. Since computers keep a fixed number of digits, adding numbers that cancel almost exactly, will lead to a catastrophic lack of precision (e.g. if we have 1.00001 and 1.00000 exact up to six digits, their difference will only have one exact digit.)

---

[12]In modern computers, there are almost universally around 2^{64} representable numbers, those which can be written in 8 bytes – there are a few delicate and complicated issues such as denormalized numbers. Most computers also use for certain calculations numbers with 80 bits, which are, 2^{80} numbers.

There is a rather detailed standard by IEEE [**IEE85**] on how to perform arithmetic in numbers. It does not only specify the precision to be used, but also rounding and how to report troubles such as attempted division by zero or overflow. This standard is now almost universally implemented in the chips and the languages (rater inexplicably Java did not include it) and there are good tests of compliance so that one can asses one's arithmetic. See [**Kah96**].

[13]The process of taking the true result and producing a representable number is called *rounding*. Returning an representable number that is larger than the true result is called rounding up, similarly rounding down, rounding to nearest, rounding towards zero etc. The IEEE standard mentioned above specifies that the user can control the properties of the rounding and of the exceptions by setting a control word.

Many of the problems with small divisors are such that the numerics deteriorates in a complicated way until the algorithms blow up or start behaving erratically.

EXERCISE 7.1. One of the standard programs to asses the characteristics of a computer is

```
epsilon = 1.0;
oneplus = 1.0 + epsilon;
count = 0;
while (oneplus > 1.0){
epsilon /= 2.0;
oneplus = one + epsilon;
count++;
}
printf("%d", count);
```

Run it in your computer.
Run also

```
epsilon = 1.0;
count = 0;
while ( 1.0 + epsilon > 1.0){
epsilon /= 2.0;
count++;
}
printf("%d", count);
```

Chances are that the results will be quite different. Explain why.

EXERCISE 7.2. The computer program Mathematica uses a numerical scheme in which high precision numbers drop precision if the last figures cannot be kept.
This leads to some unexpected effects.
Run

```
a = N[Pi,40]
Do[ a = 2*a -a , {100}]
```

and discuss the results.

One way of obtaining reliable results from a computer without sacrificing too much performance is to use *interval arithmetic* (See [**Moo79**], [**KM84**].) The idea is that a real variable is represented by two representable numbers which are supposed to mean an upper and a lower bound for the value of the variable we are interested in.

Once one has bounds for the values of a variable, one can operate on these bounds in such a way that one always keeps obtaining bounds. The only subtlety is that when adding upper bounds, one has to round up, adding lower bounds, one has to round down, etc. This can be done by reprogramming pieces of the arithmetic, or, in systems that conform to the IEEE standard by setting appropriately the control word. This quickly leads to an arithmetic among intervals that can produce bounds of arithmetic expressions given bounds on the variables.

One can pass from bounds on arithmetic expressions to bounds on sets in functional spaces. For example, one can specify a set in function space. For example, if we specify a set of analytic functions by

$$U_{v_1,\dots,v_n;\varepsilon} = \{f(z) \quad | \quad f(z) = \sum_{i=0}^{N} f_i z^i + f_e(z), f_i \in v_i, ||f_e||_1 \leq \varepsilon\} \tag{7.1}$$

where v_i are intervals (i.e. pairs of representable numbers) and ε is a representable number. (There are, of course, many variants. One can for example, take into account that some errors are high order, use other norm for the error or even several norms at the same time.)

It is reasonably easy to imagine how can one define operations on sets of the type in (7.1) such that the numerical operations bound the real operations on sets. With a bit more of imagination, one can do compositions, integrals, and other operations. In particular, one can implement the operations involved in the evaluation of the terms in (6.25).

If starting with the numerically produced non-rigorous guess one can use the rigorous interval arithmetic to verify the hypothesis of Theorem 6.8 – or some other theorem enjoying a similar structure – then, one can guarantee that there is a true solution near the computed one. This strategy has been implemented in [**Ran87**], [**dlLR91**]. Similar ideas have been implemented in [**CC95**].

Indeed, by now, starting with the inspiring proof of [**Lan82**] (it relied on the usual contraction mapping theorem rather than in the hard implicit function theorems) there has been a number of significant theorems proved with similar techniques. A survey of these developments is [**KSW96**].

One of the main difficulties of the method is that it requires to spend a great deal of time in coding carefully the problems. One can hope that some of the tasks could be automated but there are difficulties. Even if automatic translation of arithmetic expressions produces a valid answer, arithmetic expressions that are equivalent under the ordinary rules of arithmetic are not equivalent under interval arithmetic. For example in intervals

$$(a + b) \times c \subset a \times c + b \times c \tag{7.2}$$

and the inclusion can be strict. A classic problem in interval arithmetic is to find fast algorithms to compute accurately the image of the unit disk under a polynomial.

EXERCISE 7.3. Give a proof of (7.2) and find examples when it is strict.

I personally think that computer assisted proofs and is a very interesting area in which it is possible to find a meaningful collaboration between Mathematicians (proving theorems of the right kind), Computer Scientists (developing good software tools that relieve the tedium of programming the variants required) and applied scientists that have challenging real life problems.

8. Acknowledgements

The work of the author was supported in part by NSF grants. I received substantial assistance in the preparation of these notes from A. Haro, N. Petrov, J. Vano. Parts of this work are based on unpublished joint work with other people that we intend to publish in fuller versions. Comments from H. Eliasson, T. Gramchev, and many other participants in the SRI and by A. Jorba, M. Sevryuk, R. Perez-Marco removed many mistakes and typos. Needless to say, they are not to be blamed for those that escaped their eye or for missing those that were introduced in ulterior revisions which they did not see. I also want to express my appreciation of the great amount of work put by the AMS staff, specially W. Drady and by A. Katok, Y. Pesin and, specially H. Weiss to organize this Summer Institute. It was a privilege and a humbling experience to witness their dedication. The enthusiasm of the participants in the SRI was contagious.

References

[AA68] V. I. Arnol'd and A. Avez. *Ergodic Problems of Classical Mechanics.* W. A. Benjamin, New York-Amsterdam, 1968.

[Ada75] R. A. Adams. *Sobolev Spaces.* Academic Press, New York-London, 1975.

[AF88] C. Albanese and J. Fröhlich. Periodic solutions of some infinite-dimensional Hamiltonian systems associated with nonlinear partial differential equations. I. *Comm. Math. Phys.*, 116(3):475–502, 1988.

[AF91] Claudio Albanese and Jürg Fröhlich. Perturbation theory for periodic orbits in a class of infinite-dimensional Hamiltonian systems. *Comm. Math. Phys.*, 138(1):193–205, 1991.

[AFS88] Claudio Albanese, Jürg Fröhlich, and Thomas Spencer. Periodic solutions of some infinite-dimensional Hamiltonian systems associated with nonlinear partial difference equations. II. *Comm. Math. Phys.*, 119(4):677–699, 1988.

[AG91] Serge Alinhac and Patrick Gérard. *Opérateurs pseudo-différentiels et théorème de Nash-Moser.* InterEditions, Paris, 1991.

[AKN93] V. I. Arnol'd, V. V. Kozlov, and A. I. Neishtadt. Mathematical aspects of classical and celestial mechanics. In *Dynamical Systems, III*, pages vii–xiv, 1–291. Springer, Berlin, 1993.

[Alb93] Claudio Albanese. KAM theory in momentum space and quasiperiodic Schrödinger operators. *Ann. Inst. H. Poincaré Anal. Non Linéaire*, 10(1):1–97, 1993.

[AM78] R. Abraham and J. E. Marsden. *Foundations of Mechanics.* Benjamin/Cummings, Reading, Mass., 1978.

[Arn61] V. I. Arnol'd. Small denominators. I. Mapping the circle onto itself. *Izv. Akad. Nauk SSSR Ser. Mat.*, 25:21–86, 1961. English translation: *Amer. Math. Sos. Transl. (2)*, 46:213–284, 1965.

[Arn63a] V. I. Arnol'd. Proof of a theorem of A. N. Kolmogorov on the invariance of quasi-periodic motions under small perturbations. *Russian Math. Surveys*, 18(5):9–36, 1963.

[Arn63b] V. I. Arnol'd. Small denominators and problems of stability of motion in classical and celestial mechanics. *Russian Math. Surveys*, 18(6):85–191, 1963.

[Arn88] V. I. Arnol'd. *Geometrical Methods in the Theory of Ordinary Differential Equations.* Springer-Verlag, New York, second edition, 1988.

[Arn89] V. I. Arnol'd. *Mathematical Methods of Classical Mechanics.* Springer-Verlag, New York, second edition, 1989.

[AS86] V.I. Arnol'd and M. Sevryuk. Oscillations and bifurcations in reversible systems. In R. Z. Sagdeeev, editor, *Nonlinear Phenomena in Plasma Physics and Hydrodynamics*, pages 31–64. Mir, Moscow, 1986.

[Aub83] S. Aubry. The twist map, the extended Frenkel-Kontorova model and the devil's staircase. *Phys. D*, 7(3):240–258, 1983.

[Bam99a] D. Bambusi. Nekhoroshev theorem for small amplitude solutions in nonlinear Schrödinger equations. *Math. Z.*, 230(2):345–387, 1999.

[Bam99b] D. Bambusi. On long time stability in Hamiltonian perturbations of non-resonant linear PDEs. *Nonlinearity*, 12(4):823–850, 1999.

[Ban89] V. Bangert. On minimal laminations of the torus. *Ann. Inst. H. Poincaré Anal. Non Linéaire*, 6(2):95–138, 1989.

[Bar70] R. B. Barrar. Convergence of the von Zeipel procedure. *Celestial Mech.*, 2(4):494–504, 1970.

[BCCF92] A. Berretti, A. Celletti, L. Chierchia, and C. Falcolini. Natural boundaries for area-preserving twist maps. *J. Statist. Phys.*, 66(5-6):1613–1630, 1992.

[BCP98] F. Bonetto, E. G. D. Cohen, and C. Pugh. On the validity of the conjugate pairing rule for Lyapunov exponents. *J. Statist. Phys.*, 92(3-4):587–627, 1998.

[BdlLW96] A. Banyaga, R. de la Llave, and C. E. Wayne. Cohomology equations near hyperbolic points and geometric versions of Sternberg linearization theorem. *J. Geom. Anal.*, 6(4):613–649 (1997), 1996.

[BG99] A. Berretti and G. Gentile. Scaling properties for the radius of convergence of Lindstedt series: Generalized standard maps. `http://www.ma.utexas.edu/mp_arc` , 99–377, 1999.

[BGG85a] G. Benettin, L. Galgani, and A. Giorgilli. Classical perturbation theory for systems of weakly coupled rotators. *Nuovo Cimento B (11)*, 89(2):89–102, 1985.

[BGG85b] G. Benettin, L. Galgani, and A. Giorgilli. Numerical investigations on a chain of weakly coupled rotators in the light of classical perturbation theory. *Nuovo Cimento B (11)*, 89(2):103–119, 1985.

[BGG85c] G. Benettin, L. Galgani, and A. Giorgilli. A proof of Nekhoroshev's theorem for the stability times in nearly integrable Hamiltonian systems. *Celestial Mech.*, 37(1):1–25, 1985.

[BGGS84] G. Benettin, L. Galgani, A. Giorgili, and J.-M. Strelcyn. A proof of Kolmogorov's theorem on invariant tori using canonical transformations defined by the Lie method. *Nuovo Cimento B (11)*, 79(2):201–223, 1984.

[BGK99] J. Bricmont, K. Gawędzki, and A. Kupiainen. KAM theorem and quantum field theory. *Comm. Math. Phys.*, 201(3):699–727, 1999.

[BH91] H. W. Broer and G. B. Huitema. A proof of the isoenergetic KAM-theorem from the "ordinary" one. *J. Differential Equations*, 90(1):52–60, 1991.

[BHS96a] H. W. Broer, G. B. Huitema, and M. B. Sevryuk. Families of quasi-periodic motions in dynamical systems depending on parameters. In *Nonlinear dynamical systems and chaos (Groningen, 1995)*, pages 171–211. Birkhäuser, Basel, 1996.

[BHS96b] H. W. Broer, G. B. Huitema, and M. B. Sevryuk. *Quasi-Periodic Motions in Families of Dynamical Systems.* Order Amidst Chaos. Springer-Verlag, Berlin, 1996.

[Bib79] Yu. N. Bibikov. *Local Theory of Nonlinear Analytic Ordinary Differential Equations.* Springer-Verlag, Berlin, 1979.

[Bla84] P. Blanchard. Complex analytic dynamics on the Riemann sphere. *Bull. Amer. Math. Soc. (N.S.)*, 11(1):85–141, 1984.

[BM61] N. N. Bogoliubov and Y. A. Mitropolsky. *Asymptotic Methods in the Theory of Non-Linear Oscillations.* Hindustan Publ. Corp., Delhi, Gordon and Breach, New York, 1961.

[BM95] A. Berretti and S. Marmi. Scaling, perturbative renormalization and analyticity for the standard map and some generalizations. *Chaos Solitons Fractals*, 5(2):257–269, 1995.

[BN98] D. Bambusi and N. N. Nekhoroshev. A property of exponential stability in nonlinear wave equations near the fundamental linear mode. *Phys. D*, 122(1-4):73–104, 1998.

[Bos86] J.-B. Bost. Tores invariants des systèmes dynamiques hamiltoniens (d'après Kolmogorov, Arnold, Moser, Rüssmann, Zehnder, Herman, Pöschel, ...). *Astérisque*, No. 133–134:113–157, 1986. Seminar Bourbaki, Vol. 1984/85.

[Bou97] J. Bourgain. On Melnikov's persistency problem. *Math. Res. Lett.*, 4(4):445–458, 1997.

[Bou99a] J. Bourgain. Nonlinear Schrödinger equations. In *Hyperbolic Equations and Frequency Interactions (Park City, UT, 1995)*, pages 3–157. Amer. Math. Soc., Providence, RI, 1999.

[Bou99b] Jean Bourgain. Periodic solutions of nonlinear wave equations. In *Harmonic analysis and partial differential equations (Chicago, IL, 1996)*, pages 69–97. Univ. Chicago Press, Chicago, IL, 1999.

[Brj71] A. D. Brjuno. Analytic form of differential equations. I. *Trudy Moskov. Mat. Obšč.*, 25:119–262, 1971. English translation: *Trans. Moscow Math. Soc.*, 25(1971):131–288, 1973.

[Brj72] A. D. Brjuno. Analytic form of differential equations. II. *Trudy Moskov. Mat. Obšč.*, 26:199–239, 1972. English translation: *Trans. Moscow Math. Soc.*, 26(1972):199–239, 1974.

[Bru89] A. D. Bruno. *Local Methods in Nonlinear Differential Equations.* Springer-Verlag, Berlin, 1989.

[BW65] M. Born and E. Wolf. *Principles of Optics: Electromagnetic Theory of Propagation, Interference and Diffraction of Light.* Pergamon Press, Oxford, revised edition, 1965.

[BZ82] D. Braess and E. Zehnder. On the numerical treatment of a small divisor problem. *Numer. Math.*, 39(2):269–292, 1982.

[Cal70] E. Calabi. On the group of automorphisms of a symplectic manifold. In *Problems in Analysis (Lectures at the Symposium in Honor of Salomon Bochner, Princeton Univ., Princeton, N.J., 1969)*, pages 1–26. Princeton Univ. Press, Princeton, N.J., 1970.

[Car81] John R. Cary. Lie transform perturbation theory for Hamiltonian systems. *Phys. Rep.*, 79(2):129–159, 1981.

[CC95] A. Celletti and L. Chierchia. A constructive theory of Lagrangian tori and computer-assisted applications. In *Dynamics Reported*, pages 60–129. Springer, Berlin, 1995.

[CCSPC97] L. Casetti, M. Cerruti-Sola, M. Pettini, and E. G. D. Cohen. The Fermi-Pasta-Ulam problem revisited: stochasticity thresholds in nonlinear Hamiltonian systems. *Phys. Rev. E (3)*, 55(6, part A):6566–6574, 1997.

[CEL84] M. G. Crandall, L. C. Evans, and P.-L. Lions. Some properties of viscosity solutions of Hamilton– Jacobi equations. *Trans. Amer. Math. Soc.*, 282(2):487–502, 1984.

[CF94] L. Chierchia and C. Falcolini. A direct proof of a theorem by Kolmogorov in Hamiltonian systems. *Ann. Scuola Norm. Sup. Pisa Cl. Sci. (4)*, 21(4):541–593, 1994.

[CF96] L. Chierchia and C. Falcolini. A note on quasi-periodic solutions of some elliptic systems. *Z. Angew. Math. Phys.*, 47(2):210–220, 1996.

[CG82] L. Chierchia and G. Gallavotti. Smooth prime integrals for quasi-integrable Hamiltonian systems. *Nuovo Cimento B (11)*, 67(2):277–295, 1982.

[CL83] M. G. Crandall and P.-L. Lions. Viscosity solutions of Hamilton-Jacobi equations. *Trans. Amer. Math. Soc.*, 277(1):1–42, 1983.

[Cre28] H. Cremer. Zum zentrumproblem. *Math. Ann.*, 98:151–163, 1928.

[Cre38] H. Cremer. Über die häufigkeit der nichtzentren. *Math. Ann.*, 115:573–580, 1938.

[CS91] V. A. Chulaevsky and Ya. G. Sina˘i. The exponential localization and structure of the spectrum for 1D quasi-periodic discrete Schrödinger operators. *Rev. Math. Phys.*, 3(3):241–284, 1991.

[CS94] Chong Qing Cheng and Yi Sui Sun. Existence of KAM tori in degenerate Hamiltonian systems. *J. Differential Equations*, 114(1):288–335, 1994.

[CS90] Chong Qing Cheng and Yi Sui Sun. Existence of invariant tori in three-dimensional measure-preserving mappings. *Celestial Mech. Dynam. Astronom.*, 47(3):275–292, 1989/90.

[CW93] W. Craig and C. E. Wayne. Newton's method and periodic solutions of nonlinear wave equations. *Comm. Pure Appl. Math.*, 46(11):1409–1498, 1993.

[CW94] W. Craig and C. E. Wayne. Periodic solutions of nonlinear Schrödinger equations and the Nash-Moser method. In *Hamiltonian Mechanics (Toruń, 1993)*, pages 103–122. Plenum, New York, 1994.

[Dav94] A. M. Davie. The critical function for the semistandard map. *Nonlinearity*, 7(1):219–229, 1994.

[DdlL90] A. Delshams and R. de la Llave. Existence of quasi-periodic orbits and absence of transport for volume preserving transformations and flows. *Preprint*, 1990.

[DeL97] D. DeLatte. Diophantine conditions for the linearization of commuting holomorphic functions. *Discrete Contin. Dynam. Systems*, 3(3):317–332, 1997.

[Dep70] A. Deprit. Canonical transformations depending on a small parameter. *Celestial Mech.*, 1:12–30, 1969/1970.

[DF76] A. J. Dragt and J. M. Finn. Lie series and invariant functions for analytic symplectic maps. *J. Mathematical Phys.*, 17(12):2215–2127, 1976.

[DG96] A. Delshams and P. Gutiérrez. Effective stability and KAM theory. *J. Differential Equations*, 128(2):415–490, 1996.

[DG00] D. Delatte and T. Gramchev. Normal forms of maps with linear parts having jordan blocks. *Preprint*, 2000.

[dlL83] R. de la Llave. A simple proof of a particular case of C. Siegel's center theorem. *J. Math. Phys.*, 24(8):2118–2121, 1983.

[dlL92] R. de la Llave. A renormalization group explanation of numerical observations of analyticity domains. *J. Statist. Phys.*, 66(5-6):1631–1634, 1992.

[dlL93] R. de la Llave. Introduction to K.A.M. theory. In *Computational Physics (Almuñécar, 1992)*, pages 73–105. World Sci. Publishing, River Edge, NJ, 1993.

[dlLMM86] R. de la Llave, J. M. Marco, and R. Moriyón. Canonical perturbation theory of Anosov systems and regularity results for the Livšic cohomology equation. *Ann. of Math. (2)*, 123(3):537–611, 1986.

[dlLO99] R. de la Llave and R. Obaya. Regularity of the composition operator in spaces of Hölder functions. *Discrete Contin. Dynam. Systems*, 5(1):157–184, 1999.

[dlLO00] R. de la Llave and R. Obaya. Decomposition theorems for groups of diffeomorphisms in the sphere. *Trans. Amer. Math. Soc.*, 352(3):1005–1020, 2000.

[dlLR91] R. de la Llave and D. Rana. Accurate strategies for K.A.M. bounds and their implementation. In *Computer Aided Proofs in Analysis (Cincinnati, OH, 1989)*, pages 127–146. Springer, New York, 1991.

[dlLV00] R. de la Llave and J. Vano. A Whitney-Zehnder implicit function theorem. *Manuscript*, 2000.

[Dou82a] R. Douady. Applications du théorème des tores invariants. *Univ. Paris VII, These 3 cycle*, 1982.

[Dou82b] R. Douady. Une démonstration directe de l'équivalence des théorèmes de tores invariants pour difféomorphismes et champs de vecteurs. *C. R. Acad. Sci. Paris Sér. I Math.*, 295(2):201–204, 1982.

[Dou88] R. Douady. Regular dependence of invariant curves and Aubry-Mather sets of twist maps of an annulus. *Ergodic Theory Dynamical Systems*, 8(4):555–584, 1988.

[DS75] E. I. Dinaburg and Ja. G. Sinai. The one-dimensional Schrödinger equation with quasiperiodic potential. *Funkcional. Anal. i Priložen.*, 9(4):8–21, 1975. English translation: *Functional Anal. Appl.*, 9(1975):279–289, 1976.

[ED81] D. F. Escande and F. Doveil. Renormalization method for computing the threshold of the large-scale stochastic instability in two degrees of freedom Hamiltonian systems. *J. Statist. Phys.*, 26(2):257–284, 1981.

[Eli] L. H. Eliasson. Reducibility and point spectrum for linear quasi-periodic skew-products. In *Proceedings of the International Congress of Mathematicians, Vol. II (Berlin, 1998)*, pages 779–787 (electronic).

[Eli88] L. H. Eliasson. Perturbations of stable invariant tori for Hamiltonian systems. *Ann. Scuola Norm. Sup. Pisa Cl. Sci. (4)*, 15(1):115–147 (1989), 1988.

[Eli89] L. H. Eliasson. Hamiltonian systems with linear normal form near an invariant torus. In *Nonlinear Dynamics (Bologna, 1988)*, pages 11–29. World Sci. Publishing, Teaneck, NJ, 1989.

[Eli96] L. H. Eliasson. Absolutely convergent series expansions for quasi periodic motions. *Math. Phys. Electron. J.*, 2:Paper 4, 33 pp. (electronic), 1996.

[Eli97] L. H. Eliasson. Discrete one-dimensional quasi-periodic Schrödinger operators with pure point spectrum. *Acta Math.*, 179(2):153–196, 1997.

[Fat97a] A. Fathi. Solutions KAM faibles conjuguées et barrières de Peierls. *C. R. Acad. Sci. Paris Sér. I Math.*, 325(6):649–652, 1997.

[Fat97b] A. Fathi. Théorème KAM faible et théorie de Mather sur les systèmes lagrangiens. *C. R. Acad. Sci. Paris Sér. I Math.*, 324(9):1043–1046, 1997.

[FdlL92] C. Falcolini and R. de la Llave. Numerical calculation of domains of analyticity for perturbation theories in the presence of small divisors. *J. Statist. Phys.*, 67(3-4):645–666, 1992.

[Fed69] H. Federer. *Geometric Measure Theory.* Springer-Verlag New York Inc., New York, 1969.

[FGB98] F. Fassò, M. Guzzo, and G. Benettin. Nekhoroshev-stability of elliptic equilibria of hamiltonian systems. *Comm. Math. Phys.*, 197(2):347–360, 1998.

[FS83] J. Fröhlich and T. Spencer. Absence of diffusion in the Anderson tight binding model for large disorder or low energy. *Comm. Math. Phys.*, 88(2):151–184, 1983.

[FS84] J. Fröhlich and T. Spencer. A rigorous approach to Anderson localization. *Phys. Rep.*, 103(1-4):9–25, 1984.

[FSW86] J. Fröhlich, T. Spencer, and C. E. Wayne. Localization in disordered, nonlinear dynamical systems. *J. Statist. Phys.*, 42(3-4):247–274, 1986.

[FSW90] J. Fröhlich, T. Spencer, and P. Wittwer. Localization for a class of one-dimensional quasi-periodic Schrödinger operators. *Comm. Math. Phys.*, 132(1):5–25, 1990.

[FY98] Cong Fuzhong and Li Yong. Existence of higher-dimensional invariant tori for Hamiltonian systems. *J. Math. Anal. Appl.*, 222(1):255–267, 1998.

[Gal83a] G. Gallavotti. Perturbation theory for classical Hamiltonian systems. In *Scaling and Self-Similarity in Physics (Bures-sur-Yvette, 1981/1982)*, pages 359–426. Birkhäuser, Boston, Mass., 1983.

[Gal83b] G. Gallavotti. Perturbation theory for classical Hamiltonian systems. In *Scaling and Self-Similarity in Physics (Bures-sur-Yvette, 1981/1982)*, pages 359–426. Birkhäuser Boston, Boston, Mass., 1983.

[Gal86] G. Gallavotti. Quasi-integrable mechanical systems. In *Phénomènes critiques, systèmes aléatoires, théories de jauge, Part I, II (Les Houches, 1984)*, pages 539–624. North-Holland, Amsterdam, 1986.

[Gal94a] G. Gallavotti. Twistless KAM tori. *Comm. Math. Phys.*, 164(1):145–156, 1994.

[Gal94b] G. Gallavotti. Twistless KAM tori, quasi flat homoclinic intersections, and other cancellations in the perturbation series of certain completely integrable Hamiltonian systems. A review. *Rev. Math. Phys.*, 6(3):343–411, 1994.

[GFB98] M. Guzzo, F. Fassò, and G. Benettin. On the stability of elliptic equilibria. *Math. Phys. Electron. J.*, 4:Paper 1, 16 pp. (electronic), 1998.

[GG95] G. Gallavotti and G. Gentile. Majorant series convergence for twistless KAM tori. *Ergodic Theory Dynam. Systems*, 15(5):857–869, 1995.

[GJdlLV00] A. Gonzalez, A. Jorba, R. de la Llave, and J. Villanueva. KAM theory for non action-angle Hamiltonian systems. *Manuscript*, 2000.

[Gol80] H. Goldstein. *Classical Mechanics.* Addison-Wesley, Reading, Mass., second edition, 1980.

[GP74] Victor Guillemin and Alan Pollack. *Differential topology.* Prentice-Hall Inc., Englewood Cliffs, N.J., 1974.

[GP81] J. M. Greene and I. C. Percival. Hamiltonian maps in the complex plane. *Phys. D*, 3(3):530–548, 1981.

[Gre79] J. M. Greene. A method for determining a stochastic transition. *Jour. Math. Phys.*, 20:1183–1201, 1979.

[Hal75] O. H. Hald. On a Newton-Moser type method. *Numer. Math.*, 23:411–426, 1975.

[Ham82] R. S. Hamilton. The inverse function theorem of Nash and Moser. *Bull. Amer. Math. Soc. (N.S.)*, 7(1):65–222, 1982.

[Har99] A. Haro. Interpolation of an exact symplectomorphism by a Hamiltonian flow. `http://www.ma.utexas.edu/mp_arc` , 99–100, 1999.

[Hay90] N. T. A. Haydn. On invariant curves under renormalisation. *Nonlinearity*, 3(3):887–912, 1990.

[HdlL00] A. Haro and R. de la Llave. New mechanisms for lack of equipartion of energy. *Phys. Rev. Lett.*, 89(7):1859–1862, 2000.

[Hén83] Michel Hénon. Numerical exploration of Hamiltonian systems. In *Chaotic Behavior of Deterministic Systems (Les Houches, 1981)*, pages 53–170. North-Holland, Amsterdam, 1983.

[Her79] M.-R. Herman. Sur la conjugaison différentiable des difféomorphismes du cercle à des rotations. *Inst. Hautes Études Sci. Publ. Math.*, (49):5–233, 1979.

[Her83a] M.-R. Herman. *Sur les courbes invariantes par les difféomorphismes de l'anneau. Vol. 1.* Société Mathématique de France, Paris, 1983.

[Her83b] M.-R. Herman. Sur les difféomorphismes du cercle de nombre de rotation de type constant. In *Conference on Harmonic Analysis in Honor of Antoni Zygmund, Vol. I, II (Chicago, Ill., 1981)*, pages 708–725. Wadsworth, Belmont, CA, 1983.

[Her85] M.-R. Herman. Simple proofs of local conjugacy theorems for diffeomorphisms of the circle with almost every rotation number. *Bol. Soc. Brasil. Mat.*, 16(1):45–83, 1985.

[Her86] M.-R. Herman. Sur les courbes invariantes par les difféomorphismes de l'anneau. Vol. 2. *Astérisque*, (144):248, 1986. With a correction to: *On the curves invariant under diffeomorphisms of the annulus, Vol. 1* (French) [Astérisque No. 103-104, Soc. Math. France, Paris, 1983; MR 85m:58062].

[Her87] M.-R. Herman. Recent results and some open questions on Siegel's linearization theorem of germs of complex analytic diffeomorphisms of $\mathbf{c}^n$ near a fixed point. In *VIIIth International Congress on Mathematical Physics (Marseille, 1986)*, pages 138–184. World Sci. Publishing, Singapore, 1987.

[Her91] M.-R. Herman. Exemples de flots hamiltoniens dont aucune perturbation en topologie C^∞ n'a d'orbites périodiques sur un ouvert de surfaces d'énergies. *C. R. Acad. Sci. Paris Sér. I Math.*, 312(13):989–994, 1991.

[HM94] J. Hounie and P. Malagutti. O teorema de Nash-Moser e aplicacoes. *Coloquio Matematico Brasileiro*, 19, 1994.

[Hör85] L. Hörmander. On the Nash–Moser implicit function theorem. *Ann. Acad. Sci. Fenn. Ser. A I Math.*, 10:255–259, 1985.

[Hör90] L. Hörmander. The Nash-Moser theorem and paradifferential operators. In *Analysis, et cetera*, pages 429–449. Academic Press, Boston, MA, 1990.

[IEE85] IEEE. Ieee standard no.:754-1985 pdf, standard for binary floating point arithmetic, 1985.

[Ily79] Ju. S. Ilyashenko. Divergence of series that reduce an analytic differential equation to linear normal form at a singular point. *Funktsional. Anal. i Prilozhen.*, 13(3):87–88, 1979.

[JdlLZ99] À. Jorba, R. de la Llave, and M. Zou. Lindstedt series for lower-dimensional tori. In *Hamiltonian Systems with Three or More Degrees of Freedom (S'Agaró, 1995)*, pages 151–167. Kluwer Acad. Publ., Dordrecht, 1999.

[Jor99] À. Jorba. A methodology for the numerical computation of normal forms, centre manifolds and first integrals of Hamiltonian systems. *Experiment. Math.*, 8(2):155–195, 1999.

[JS92] À. Jorba and C. Simó. On the reducibility of linear differential equations with quasiperiodic coefficients. *J. Differential Equations*, 98(1):111–124, 1992.

[Jun91] I. Jungreis. A method for proving that monotone twist maps have no invariant circles. *Ergodic Theory Dynamical Systems*, 11(1):79–84, 1991.

[JV97a] À. Jorba and J. Villanueva. On the normal behaviour of partially elliptic lower-dimensional tori of Hamiltonian systems. *Nonlinearity*, 10(4):783–822, 1997.

[JV97b] À. Jorba and J. Villanueva. On the persistence of lower-dimensional invariant tori under quasi-periodic perturbations. *J. Nonlinear Sci.*, 7(5):427–473, 1997.

[Kah96] W. Kahan. Lecture notes on the status of ieee standard for binary floating point arithmetic, 1996. `http://www.cs.berkeley.edu/~kahan/`.

[Kat76] Yitzhak Katznelson. *An introduction to harmonic analysis*. Dover Publications Inc., New York, corrected edition, 1976.

[KM84] E. W. Kaucher and W. L. Miranker. *Self-validating Numerics for Function Space Problems*. Academic Press Inc., Orlando, Fla., 1984.

[Knu97] Donald E. Knuth. *The art of computer programming. Vol. 2: Seminumerical algorithms*. Addison-Wesley Publishing Co., Reading, Mass.-London-Don Mills, Ont, third revised edition, 1997.

[KO89a] Y. Katznelson and D. Ornstein. The absolute continuity of the conjugation of certain diffeomorphisms of the circle. *Ergodic Theory Dynamical Systems*, 9(4):681–690, 1989.

[KO89b] Y. Katznelson and D. Ornstein. The differentiability of the conjugation of certain diffeomorphisms of the circle. *Ergodic Theory Dynamical Systems*, 9(4):643–680, 1989.

[KO93] Y. Katznelson and D. S. Ornstein. A new method for twist theorems. *J. Anal. Math.*, 60:157–208, 1993.

[Koc99] H. Koch. A renormalization group for Hamiltonians, with applications to KAM tori. *Ergodic Theory Dynam. Systems*, 19:1–47, 1999.

[Kol54] A. N. Kolmogorov. On conservation of conditionally periodic motions for a small change in Hamilton's function. *Dokl. Akad. Nauk SSSR (N.S.)*, 98:527–530, 1954. English translation in *Stochastic Behavior in Classical and Quantum Hamiltonian Systems (Volta Memorial Conf., Como, 1977)*, Lecture Notes in Phys., 93, pages 51–56. Springer, Berlin, 1979.

[Kos91] D. V. Kosygin. Multidimensional KAM theory from the renormalization group viewpoint. In *Dynamical Systems and Statistical Mechanics (Moscow, 1991)*, pages 99–129. Amer. Math. Soc., Providence, RI, 1991.

[Koz83] S. M. Kozlov. Reducibility of quasiperiodic differential operators and averaging. *Trudy Moskov. Mat. Obshch.*, 46:99–123, 1983. English translation: *Trans. Moscow Math. Soc.*, Issue 2:101–126, 1984.

[KP94] Sergei Kuksin and Jürgen Pöschel. On the inclusion of analytic symplectic maps in analytic Hamiltonian flows and its applications. In *Seminar on Dynamical Systems (St. Petersburg, 1991)*, pages 96–116. Birkhäuser, Basel, 1994.

[Kra83] S. G. Krantz. Lipschitz spaces, smoothness of functions, and approximation theory. *Exposition. Math.*, 1(3):193–260, 1983.

[Kri99] Raphaël Krikorian. Réductibilité des systèmes produits-croisés à valeurs dans des groupes compacts. *Astérisque*, (259):vi+216, 1999.

[KS86] K. Khanin and Ya. G. Sinai. Renormalization group method and Kolmogorov-Arnold-Moser theory. In R. Z. Sagdeev, editor, *Nonlinear Phenomena in Plasma Physics and Hydrodynamics*, pages 31–64. Mir, Moscow, 1986.

[KS87] K. M. Khanin and Ya. G. Sinai. A new proof of M. Herman's theorem. *Comm. Math. Phys.*, 112(1):89–101, 1987.

[KSW96] H. Koch, A. Schenkel, and P. Wittwer. Computer-assisted proofs in analysis and programming in logic: a case study. *SIAM Rev.*, 38(4):565–604, 1996.

[Kuk93] S. B. Kuksin. *Nearly Integrable Infinite-Dimensional Hamiltonian Systems.* Springer-Verlag, Berlin, 1993.

[Lan82] O. E. Lanford, III. A computer-assisted proof of the Feigenbaum conjectures. *Bull. Amer. Math. Soc. (N.S.)*, 6(3):427–434, 1982.

[Lio82] P.-L. Lions. *Generalized Solutions of Hamilton–Jacobi Equations.* Pitman (Advanced Publishing Program), Boston, Mass., 1982.

[LL76] L. D. Landau and E. M. Lifshitz. *Course of Theoretical Physics. Vol. 1. Mechanics.* Pergamon Press, Oxford, third edition, 1976.

[LN92] P. Lochak and A. I. Neishtadt. Estimates of stability time for nearly integrable systems with a quasiconvex Hamiltonian. *Chaos*, 2(4):495–499, 1992.

[Loc92] P. Lochak. Canonical perturbation theory: an approach based on joint approximations. *Uspekhi Mat. Nauk*, 47(6(288)):59–140, 1992. English translation: *Russian Math. Surveys*, 47(6):57–133, 1992.

[Mar00] Stefano Marmi. *An introduction to small divisors problems.* Istituti Editoriali e Poligrafici Internazionali Pisa-Roma, 2000. `http://www.ma.utexas.edu/mp_arc`.

[Mat69] J. N. Mather. Stability of C^∞ mappings. II. Infinitesimal stability implies stability. *Ann. of Math. (2)*, 89:254–291, 1969.

[Mat88] J. N. Mather. Destruction of invariant circles. *Ergodic Theory Dynamical Systems*, 8 (Charles Conley Memorial Issue):199–214, 1988.

[McG90] R. McGehee. A note on the Moser-Hald variation of Newton's method. In *Analysis, et cetera*, pages 495–499. Academic Press, Boston, MA, 1990.

[McK82] R. S. McKay. *Renormalisation in Area Preserving Maps.* PhD thesis, Princeton University, 1982.

[Mei92] J. D. Meiss. Symplectic maps, variational principles, and transport. *Rev. Modern Phys.*, 64(3):795–848, 1992.

[Mey91] K. R. Meyer. Lie transform tutorial. II. In *Computer Aided Proofs in Analysis (Cincinnati, OH, 1989)*, pages 190–210. Springer, New York, 1991.

[MF94] J. N. Mather and G. Forni. Action minimizing orbits in Hamiltonian systems. In *Transition to Chaos in Classical and Quantum Mechanics (Montecatini Terme, 1991)*, pages 92–186. Springer, Berlin, 1994.

[MH92] K. R. Meyer and G. R. Hall. *Introduction to Hamiltonian Dynamical Systems and the N-body Problem.* Springer-Verlag, New York, 1992.

[Mie91] A. Mielke. *Hamiltonian and Lagrangian Flows on Center Manifolds.* With Applications to Elliptic Variational Problems. Springer-Verlag, Berlin, 1991.

[MM74] L. Markus and K. R. Meyer. *Generic Hamiltonian dynamical systems are neither integrable nor ergodic.* American Mathematical Society, Providence, R.I., 1974. Memoirs of the American Mathematical Society, No. 144.

[MMY97] S. Marmi, P. Moussa, and J.-C. Yoccoz. The Brjuno functions and their regularity properties. *Comm. Math. Phys.*, 186(2):265–293, 1997.

[Moo79] R. E. Moore. *Methods and Applications of Interval Analysis*. Society for Industrial and Applied Mathematics (SIAM), Philadelphia, Pa., 1979.

[Mor82] R. Moriyón. Regularity of the Dirichlet problem for the degenerate complex Monge-Ampère equation. *Comm. Pure Appl. Math.*, 35(1):1–27, 1982.

[Mos60] J. Moser. On the integrability of area preserving Cremona mappings near an elliptic fixed point. *Bol. Soc. Mat. Mexicana (2)*, 5:176–180, 1960.

[Mos62] J. Moser. On invariant curves of area-preserving mappings of an annulus. *Nachr. Akad. Wiss. Göttingen Math.-Phys. Kl. II*, 1962:1–20, 1962.

[Mos65] J. Moser. On the volume elements on a manifold. *Trans. Amer. Math. Soc.*, 120:286–294, 1965.

[Mos66a] J. Moser. A rapidly convergent iteration method and non-linear differential equations. II. *Ann. Scuola Norm. Sup. Pisa (3)*, 20:499–535, 1966.

[Mos66b] J. Moser. A rapidly convergent iteration method and non-linear partial differential equations. I. *Ann. Scuola Norm. Sup. Pisa (3)*, 20:265–315, 1966.

[Mos67] J. Moser. Convergent series expansions for quasi-periodic motions. *Math. Ann.*, 169:136–176, 1967.

[Mos73] J. Moser. *Stable and Random Motions in Dynamical Systems*. Princeton University Press, Princeton, N. J., 1973.

[Mos86] J. Moser. Minimal solutions of variational problems on a torus. *Ann. Inst. H. Poincaré Anal. Non Linéaire*, 3(3):229–272, 1986.

[Mos88] J. Moser. A stability theorem for minimal foliations on a torus. *Ergodic Theory Dynamical Systems*, 8*(Charles Conley Memorial Issue):251–281, 1988.

[Mos95] J. Moser. On the persistence of pseudo-holomorphic curves on an almost complex torus (with an appendix by Jürgen Pöschel). *Invent. Math.*, 119(3):401–442, 1995.

[MP84] Jürgen Moser and Jürgen Pöschel. An extension of a result by Dinaburg and Sinai on quasiperiodic potentials. *Comment. Math. Helv.*, 59(1):39–85, 1984.

[MP85] R. S. MacKay and I. C. Percival. Converse KAM: theory and practice. *Comm. Math. Phys.*, 98(4):469–512, 1985.

[MS92] Stefano Marmi and Jaroslav Stark. On the standard map critical function. *Nonlinearity*, 5(3):743–761, 1992.

[MS95] D. McDuff and D. Salamon. *Introduction to Symplectic Topology*. The Clarendon Press Oxford University Press, New York, 1995.

[Nas56] J. Nash. The imbedding problem for Riemannian manifolds. *Ann. of Math. (2)*, 63:20–63, 1956.

[Neh77] N. N. Nehorošev. An exponential estimate of the time of stability of nearly integrable Hamiltonian systems. *Uspehi Mat. Nauk*, 32(6(198)):5–66, 287, 1977. English translation: *Russian Math. Surveys*, 32(6):1–65, 1977.

[Ne˘i81] A. I. Ne˘ishtadt. Estimates in the Kolmogorov theorem on conservation of conditionally periodic motions. *Prikl. Mat. Mekh.*, 45(6):1016–1025, 1981.

[Nek99] N. N. Nekhoroshev. Exponential stability of the approximate fundamental mode of a nonlinear wave equation. *Funktsional. Anal. i Prilozhen.*, 33(1):80–83, 1999. English translation will be published in *Funct. Anal. Appl.*

[Nel69] E. Nelson. *Topics in Dynamics. I: Flows*. Princeton University Press, Princeton, N.J., 1969.

[Nie98] Laurent Niederman. Nonlinear stability around an elliptic equilibrium point in a Hamiltonian system. *Nonlinearity*, 11(6):1465–1479, 1998.

[Omo86] Stephen M. Omohundro. *Geometric perturbation theory in physics*. World Scientific Publishing Co., Singapore, 1986.

[Par84] I. O. Parasyuk. Preservation of multidimensional invariant tori of Hamiltonian systems. *Ukrain. Mat. Zh.*, 36(4):467–473, 1984.

[Par89] I. O. Parasyuk. Deformations of a symplectic structure, and co-isotropic invariant tori of Hamiltonian systems. *Mat. Fiz. Nelinein. Mekh.*, (12(46)):35–37, 94, 1989.

[Pja69] A. S. Pjartli. Diophantine approximations of submanifolds of a Euclidean space. *Funkcional. Anal. i Priložen.*, 3(4):59–62, 1969.

[PM91] R. Pérez Marco. Sur la structure des germes holomorphes non linéarisables. *C. R. Acad. Sci. Paris Sér. I Math.*, 312(7):533–536, 1991.

[PM92] R. Pérez Marco. Solution complète au problème de Siegel de linéarisation d'une application holomorphe au voisinage d'un point fixe (d'après J.-C. Yoccoz). *Astérisque*, 206:Exp. No. 753, 4, 273–310, 1992. Séminaire Bourbaki, Vol. 1991/92.

[PM93] R. Pérez Marco. Sur les dynamiques holomorphes non linéarisables et une conjecture de V. I. Arnol'd. *Ann. Sci. École Norm. Sup. (4)*, 26(5):565–644, 1993.

[PM00] R. Pérez Marco. Total convergence or general divergence in small divisors. *Preprint*, 2000.

[Poi78] H. Poincaré. Note sur les propriétés de fonctions définies par les equations différentielles. *Jour. de la Ec. Polyth.*, pages 13–26, 1878.

[Poi93] H. Poincaré. *New Methods of Celestial Mechanics. Vols. 1–3.* American Institute of Physics, New York, 1993.

[Pös] Jürgen Pöschel. On Nekhoroshev's estimate at an elliptic equilibrium. *Internat. Math. Res. Notices*, 1999(4):203–215.

[Pös82] J. Pöschel. Integrability of Hamiltonian systems on Cantor sets. *Comm. Pure Appl. Math.*, 35(5):653–696, 1982.

[Pös86] J. Pöschel. On invariant manifolds of complex analytic mappings near fixed points. *Exposition. Math.*, 4(2):97–109, 1986.

[Pös90] J. Pöschel. Small divisors with spatial structure in infinite-dimensional Hamiltonian systems. *Comm. Math. Phys.*, 127(2):351–393, 1990.

[Pös92] Jürgen Pöschel. A lecture on the classical kam theorem. 1992.

[Pös93] J. Pöschel. Nekhoroshev estimates for quasi-convex Hamiltonian systems. *Math. Z.*, 213(2):187–216, 1993.

[Pös96] J. Pöschel. A KAM-theorem for some nonlinear partial differential equations. *Ann. Scuola Norm. Sup. Pisa Cl. Sci. (4)*, 23(1):119–148, 1996.

[RA87] R. H. Rand and D. Armbruster. *Perturbation Methods, Bifurcation Theory and Computer Algebra.* Springer-Verlag, New York, 1987.

[RA92] R. Rand and D. Armbruster. Programs from the book "Perturbation Methods, Bifurcation Theory and Computer Algebra". `http://www.ma.utexas.edu/mp_arc` , 92–113, 1992.

[Ran87] D. Rana. *Proof of Accurate Upper and Lower Bounds to Stability Domains in Small Denominator Problems.* PhD thesis, Princeton University, 1987.

[Rud87] W. Rudin. *Real and Complex Analysis.* McGraw-Hill, New York, third edition, 1987.

[Rüs] H. Rüssmann. Kleine Nenner. II. Bemerkungen zur Newtonschen Methode. *Nachr. Akad. Wiss. Göttingen Math.-Phys. Kl. II*, 1972:1–10.

[Rüs70] H. Rüssmann. Kleine Nenner. I. Über invariante Kurven differenzierbarer Abbildungen eines Kreisringes. *Nachr. Akad. Wiss. Göttingen Math.-Phys. Kl. II*, 1970:67–105, 1970.

[Rüs75] H. Rüssmann. On optimal estimates for the solutions of linear partial differential equations of first order with constant coefficients on the torus. In *Dynamical Systems, Theory and Applications (Battelle Rencontres, Seattle, Wash., 1974)*, pages 598–624. Lecture Notes in Phys., Vol. 38, Berlin, 1975. Springer.

[Rüs76a] H. Rüssmann. On a new proof of Moser's twist mapping theorem. In *Proceedings of the Fifth Conference on Mathematical Methods in Celestial Mechanics (Oberwolfach, 1975), Part I*, volume 14, pages 19–31, 1976.

[Rüs76b] H. Rüssmann. On optimal estimates for the solutions of linear difference equations on the circle. *Celestial Mech.*, 14(1):33–37, 1976.

[Rüs80] H. Rüssmann. On the one-dimensional Schrödinger equation with a quasiperiodic potential. In *Nonlinear Dynamics (Internat. Conf., New York, 1979)*, pages 90–107. New York Acad. Sci., New York, 1980.

[Rüs90] H. Rüssmann. Nondegeneracy in the perturbation theory of integrable dynamical systems. In *Stochastics, Algebra and Analysis in Classical and Quantum Dynamics (Marseille, 1988)*, pages 211–223. Kluwer Acad. Publ., Dordrecht, 1990.

[Rüs98] H. Rüssmann. Invariant tori in the perturbation theory of weakly non-degenerate integrable Hamiltonian systems. *Mainz preprint*, 14, 1998.

[Ryc92] Marek Rychlik. Renormalization of cocycles and linear ODE with almost-periodic coefficients. *Invent. Math.*, 110(1):173–206, 1992.

[Sal86] D. Salamon. The Kolmogorov-Arnold-Moser theorem. *Zürich preprint*, 1986.

[Sch69] J. T. Schwartz. *Nonlinear Functional Analysis.* Gordon and Breach, New York, 1969.

[Sch95] D. S. Schmidt. Computing the motion of the moon accurately. In *Hamiltonian Dynamical Systems (Cincinnati, OH, 1992)*, pages 341–361. Springer, New York, 1995.

[Sev86] M. B. Sevryuk. *Reversible Systems*. Springer-Verlag, Berlin, 1986.

[Sev95] M. B. Sevryuk. KAM-stable Hamiltonians. *J. Dynam. Control Systems*, 1(3):351–366, 1995.

[Sev96] M. B. Sevryuk. Invariant tori of Hamiltonian systems that are nondegenerate in the sense of Rüssmann. *Dokl. Akad. Nauk*, 346(5):590–593, 1996.

[Sev98] M. B. Sevryuk. The finite-dimensional reversible KAM theory. *Phys. D*, 112(1-2):132–147, 1998.

[Sev99] M. B. Sevryuk. The lack-of-parameters problem in the KAM theory revisited. In *Hamiltonian Systems with Three or More Degrees of Freedom (S'Agaró, 1995)*, pages 568–572. Kluwer Acad. Publ., Dordrecht, 1999.

[Sie42] C. L. Siegel. Iteration of analytic functions. *Ann. of Math. (2)*, 43:607–612, 1942.

[Sie54] Carl Ludwig Siegel. Über die Existenz einer Normalform analytischer Hamiltonscher Differentialgleichungen in der Nähe einer Gleichgewichtslösung. *Math. Ann.*, 128:144–170, 1954.

[Sim98] C. Simó. Effective computations in celestial mechanics and astrodynamics. In *Modern Methods of Analytical Mechanics and Their Applications (Udine, 1997)*, pages 55–102. Springer, Vienna, 1998.

[SK89] Ya. G. Sinai and K. M. Khanin. Smoothness of conjugacies of diffeomorphisms of the circle with rotations. *Uspekhi Mat. Nauk*, 44(1(265)):57–82, 247, 1989. English translation: *Russian Math. Surveys*, 44(1):69–99, 1989.

[SM95] C. L. Siegel and J. K. Moser. *Lectures on Celestial Mechanics*. Springer-Verlag, Berlin, 1995. Reprint of the 1971 translation.

[Sta88] J. Stark. Smooth conjugacy and renormalisation for diffeomorphisms of the circle. *Nonlinearity*, 1(4):541–575, 1988.

[Ste70] E. M. Stein. *Singular Integrals and Differentiability Properties of Functions*. Princeton University Press, Princeton, N.J., 1970.

[Sti93] Andreas Stirnemann. Renormalization for golden circles. *Comm. Math. Phys.*, 152(2):369–431, 1993.

[Sti94] Andreas Stirnemann. A renormalization proof of Siegel's theorem. *Nonlinearity*, 7(3):943–958, 1994.

[Sti97] Andreas Stirnemann. Towards an existence proof of MacKay's fixed point. *Comm. Math. Phys.*, 188(3):723–735, 1997.

[Sto94a] Laurent Stolovitch. Sur un théorème de Dulac. *Ann. Inst. Fourier (Grenoble)*, 44(5):1397–1433, 1994.

[Sto94b] Laurent Stolovitch. Sur un théorème de Dulac. *C. R. Acad. Sci. Paris Sér. I Math.*, 319(3):253–256, 1994.

[Sva80] N. V. Svanidze. Small perturbations of an integrable dynamical system with an integral invariant. *Trudy Mat. Inst. Steklov.*, 147:124–146, 204, 1980. English translation: *Proc. Steklov Inst. Math.*, 1981, no. 2.

[SZ89] D. Salamon and E. Zehnder. KAM theory in configuration space. *Comment. Math. Helv.*, 64(1):84–132, 1989.

[Thi97] W. Thirring. *Classical Mathematical Physics*. Springer-Verlag, New York, third edition, 1997.

[Tom96] S. Tompaidis. Numerical study of invariant sets of a quasiperiodic perturbation of a symplectic map. *Experiment. Math.*, 5(3):211–230, 1996.

[Val98] E. Valdinoci. Poster presented in Aussois. 1998.

[Way84] C. E. Wayne. The KAM theory of systems with short range interactions. I, II. *Comm. Math. Phys.*, 96(3):311–329, 331–344, 1984.

[Way86] C. E. Wayne. Bounds on the trajectories of a system of weakly coupled rotators. *Comm. Math. Phys.*, 104(1):21–36, 1986.

[Way96] C. E. Wayne. An introduction to KAM theory. In *Dynamical Systems and Probabilistic Methods in Partial Differential Equations (Berkeley, CA, 1994)*, pages 3–29. Amer. Math. Soc., Providence, RI, 1996.

[Whi34a] H. Whitney. Analytic extensions of differentiable functions defined in closed sets. *Trans. Amer. Math. Soc.*, 36(1):63–89, 1934.

[Whi34b] H. Whitney. Differentiable functions defined in closed sets. I. *Trans. Amer. Math. Soc.*, 36(2):369–387, 1934.

[Whi88] E. T. Whittaker. *A Treatise on the Analytical Dynamics of Particles and Rigid Bodies.* With an Introduction to the Problem of Three Bodies. Cambridge University Press, Cambridge, 1988.

[Xia92] Zhihong Xia. Existence of invariant tori in volume-preserving diffeomorphisms. *Ergodic Theory Dynamical Systems*, 12(3):621–631, 1992.

[Yoc92] J.-C. Yoccoz. Travaux de Herman sur les tores invariants. *Astérisque*, 206:Exp. No. 754, 4, 311–344, 1992. Séminaire Bourbaki, Vol. 1991/92.

[Yoc95] J.-C. Yoccoz. Théorème de Siegel, nombres de Bruno et polynômes quadratiques. *Astérisque*, 231:3–88, 1995.

[Zeh75] E. Zehnder. Generalized implicit function theorems with applications to some small divisor problems. I. *Comm. Pure Appl. Math.*, 28:91–140, 1975.

[Zeh76a] E. Zehnder. Generalized implicit function theorems with applications to some small divisor problems. II. *Comm. Pure Appl. Math.*, 29(1):49–111, 1976.

[Zeh76b] Eduard Zehnder. Moser's implicit function theorem in the framework of analytic smoothing. *Math. Ann.*, 219(2):105–121, 1976.

[Zeh77] E. Zehnder. A simple proof of a generalization of a theorem by C. L. Siegel. In *Geometry and Topology (Proc. III Latin Amer. School of Math., Inst. Mat. Pura Aplicada CNPq, Rio de Janeiro, 1976)*, pages 855–866. Lecture Notes in Math., Vol. 597, Berlin, 1977. Springer.

Department of Mathematics, The University of Texas at Austin, Austin, TX 78712-1082

E-mail address: llave@math.utexas.edu

Part II

Survey–Expository Articles

Part IIa

Systems with Hyperbolic Behavior

Proceedings of Symposia in Pure Mathematics
Volume **69**, 2001

Decay of Correlations

Viviane Baladi

ABSTRACT. This introduction to rates of mixing of discrete time dynamical systems contains an overview of known results (mostly for uniformly or nonuniformly hyperbolic maps and their SRB measures) and techniques (symbolic dynamics, towers with recurrence times, spectrum of Perron-Frobenius-Ruelle operator and dynamical zeta functions, projective metrics on Birkhoff cones, probabilistic coupling).

1. What are these notes about? Invariant measures, physical measures, and rates of mixing

We are interested in discrete-time dynamical systems represented by the iterates $f^n = f \circ f^{n-1}$ (where $n \in \mathbb{Z}_+$ represents time) of a transformation $f : M \to M$ which is "chaotic," in the sense that arbitrarily close distinct initial points become "separated" (assuming a metric structure or at least a topology on M) if one waits long enough. This separation (sometimes referred to as "sensitive dependence on initial conditions") often takes place at exponential speed, as in the elementary paradigm of the "angle-doubling" map $z \mapsto z^2$ on the circle $\{z \in \mathbb{C} \mid |z| = 1\}$, where the distance between $f^n(x)$ and $f^n(y)$ is 2^n times the distance between x and y if they are close enough.

Our aim is to describe the long-time behaviour of generic initial conditions. Generic is understood in a measure-theoretical sense, so that this task of *statistically describing* the asymptotics of "most" initial data is not rendered completely hopeless by the sensitive dependence property. We shall mostly discuss the case when M is a compact Riemann manifold, so that we have a natural a priori probability measure on M: the Lebesgue measure. We therefore seek to understand invariant probability measures μ (i.e., $\mu(f^{-1}(E)) = \mu(E)$ for every Borel set) which are somehow related to the a priori measure.

The first key concept is, of course, ergodicity (see [Wal] for this, and other, ergodic-theoretical notions). Recall that the celebrated *Birkhoff theorem* says that

2000 *Mathematics Subject Classification.* Primary 37A25; Secondary 37C05, 37C30, 37D20, 37D25.

Key words and phrases. Chaotic dynamics, physical measures, SRB measures, rates of mixing, correlation functions, transfer operators.

The author was supported in part by the Fonds National de la Recherche Scientifique Suisse #20-50493.97.

if μ is an f invariant Borel probability measure such that (f,μ) is *ergodic* (i.e., "indecomposable" in the sense that $E = f^{-1}E$ only if $\mu(E) = 0$ or 1), then for each continuous "test" function (also called "observable") $\varphi : M \to \mathbb{C}$ and μ-almost all $x \in M$ the "time average converges to the space average:"

$$\lim_{n\to\infty} \frac{1}{n}\sum_{i=0}^{n-1} \varphi(f^i x) = \int \varphi \, d\mu \, . \tag{1.1}$$

(In fact, convergence does not hold only for continuous observables, $L^1(d\mu)$ would suffice.) Of course, μ can be supported on a set of zero Lebesgue measure (the simplest example being a periodic orbit). In this case "μ-almost everywhere" is not very meaningful! Suppressing the observable, one may consider the set of points x (called the *ergodic basin* of μ) such that the convergence

$$\lim_{n\to\infty} \frac{1}{n}\sum_{i=0}^{n-1} \delta_{f^i x} = \mu \, . \tag{1.2}$$

holds (weakly, with δ_y denoting the Dirac mass at y). If the ergodic basin of x has positive Lebesgue measure, μ is called a *physical measure.* In recent literature, physical measures are sometimes called *SRB measures,* for Sinai-Ruelle-Bowen. We prefer to restrict the SRB terminology to its historical technical definition (absolute continuity with respect to Lebesgue of the measures induced along the unstable directions, see, e.g., [Yo2], see also [PeS, BoV, ABV] for generalisations) although it is often equivalent to the above concept of physical measure.

We shall discuss below classes of dynamical systems possessing a single, or finitely many, physical measure(s) the (union of the) ergodic basin(s) of which have total Lebesgue measure in M. (For saddles, or systems with "holes," the notion of physical measure is not appropriate and one should condition by the property of remaining in a neighbourhood. This gives rise to the notion of *quasi-SRB* or *generalised SRB* measure, see, e.g., Ruelle [Ru10] for the concept, Collet–Martínez–Schmitt [CMS] and Chernov–Markarian–Troubetzkoy [CMT] for examples.)

For an ergodic physical measure, one would like to quantify the spatial nature of the convergence (1.1–1.2). For a real-valued observable, assuming that $\int \varphi \, d\mu = 0$ (replacing if necessary φ by $\varphi - \int \varphi \, d\mu$), one says that the *central limit theorem* (CLT) holds for (f,μ) and φ if the "random variable"

$$x \mapsto \frac{1}{\sqrt{n}} \sum_{i=0}^{n-1} \varphi \circ f^n(x)$$

(where x is distributed with law μ) converges in law to a normal distribution $\mathcal{N}(0,\sigma)$ for some $0 < \sigma < \infty$. Recall that in probability theory the central limit theorem holds, e.g., for independent identically distributed random variables [Dur], but also under weaker "independence" assumptions. Of course, a deterministic dynamical system does not behave like an i.i.d. random variable! Nevertheless we shall encounter below "weak independence" assumptions involving the *correlation functions* or *rates of mixing* which are the main topic of these notes, and which are sufficient for the CLT.

A statistical property which implies ergodicity is *mixing*, a kind of asymptotic independence: (f, μ) is mixing if for every pair of observables $\varphi, \psi \in L^2(d\mu)$

$$\lim_{n\to\infty} \int (\varphi \circ f^n)\, \psi \, d\mu = \int \varphi \, d\mu \int \psi \, d\mu \,. \tag{1.3}$$

For the special case when the observables are characteristic functions χ_A and χ_B of Borel sets $A, B \subset M$, this means

$$\lim_{n\to\infty} \mu(f^{-n} A \cap B) = \mu(A)\mu(B) \,, \tag{1.4}$$

which is indeed the classical notion of (asymptotic) independence. Often, (1.3) is equivalent to a more natural (and often more convenient) formulation of mixing (using dx for Lebesgue measure again):

$$\lim_{n\to\infty} \int (\varphi \circ f^n)\, \psi \, dx = \int \varphi \, d\mu \int \psi \, dx \,. \tag{1.5}$$

Indeed, the left hand side is now a well-defined expression even if μ is unknown. In particular, taking $\psi \equiv 1$ we see that (1.5) corresponds to *convergence to equilibrium:*

$$\lim_{n\to\infty} \int (\varphi \circ f^n)\, dx = \int \varphi \, d\mu \,. \tag{1.6}$$

The expression

$$\rho_{\varphi\psi}(n) = \int (\varphi \circ f^n)\, \psi \, d\mu - \int \varphi \, d\mu \int \psi \, d\mu \,, \tag{1.7}$$

viewed as a function of $n \in \mathbb{Z}_+$ is called the *correlation function* of (f, μ) for the observables φ, ψ. By a slight abuse of language the difference

$$\rho'_{\varphi,\psi}(n) = \int (\varphi \circ f^n)\, \psi \, dx - \int \varphi \, d\mu \int \psi \, dx \,, \tag{1.8}$$

may also be called a correlation function. We shall see that sometimes the correlation function $\rho_{\varphi\psi}(n)$ decays exponentially with n for suitable φ, ψ. It is thus natural to consider its (a priori formal) Fourier transform

$$\widehat{\rho}_{\varphi\psi}(\omega) = \sum_{n\in\mathbb{Z}} e^{in\omega} \rho_{\varphi\psi}(n) \tag{1.9}$$

(setting $\rho_{\varphi\psi}(n) = \rho_{\psi\varphi}(-n)$ for negative n if f is not invertible). Exponential decay (with a uniform rate in φ, ψ) would imply that $\widehat{\rho}_{\varphi\psi}(\omega)$ is analytic in a strip $|\Im\omega| < \eta$, with η independent of φ, ψ in some function class. In some cases $\widehat{\rho}_{\varphi\psi}(\omega)$ admits an extension meromorphic in a strip $\Im(\omega) < \eta'$ where the possible location of its poles (the *Ruelle resonances,* or the *correlation spectrum*) only depend on the dynamical system (f, μ) (and not on the observables). They are often described by the set of poles of a dynamical zeta function or the set of zeroes of a dynamical Fredholm determinant, as we shall explain in these notes.

Mixing alone is in general *not* enough to imply the CLT: we need to quantify this notion by studying the *speed* at which the convergence (1.3) (or (1.5)) takes place. This is one of the motivations to study the rate of decay of correlations functions. For this, it is in general necessary to restrict the test functions to a functional subspace $\mathcal{B}$ of $L^2(d\mu)$. It is often convenient to take for $\mathcal{B}$ a Banach space (suitable Hilbert spaces are usually not readily available). In many cases, $\rho'_{\varphi,\psi}(n) \sim \rho_{\varphi\psi}(n)$

when $n \to \infty$ if the observables belong to an appropriate function space. In Section 4 we shall give a catalogue of results for speeds of mixing. The rule of thumb is that if it is possible to find a space of observables $\mathcal{B}$ such that $|\rho_{\varphi\psi}(n)| \le C(n)$ for all $n \in \mathbb{Z}_+$ and all $\varphi, \psi \in \mathcal{B}$, then the central limit theorem holds for observables in $\mathcal{B}$ *if the speed of mixing* $C(n)$ is summable, i.e., if $\sum_n C(n) < \infty$. We refer to [Li2], and also [Yo2, Yo3], [V2] for precise statements, their proofs, and references (in particular to previous work of Gordin, and to a probabilistic Lindeberg-Levy central limit theorem for martingales [Bil], as well as an important martingale approximation theorem by Kipnis and Varadhan).

2. A pedestrian's view: Three elementary examples

Before quoting (mostly without proof) a series of known results, let us discuss three examples which can be treated "by hand." We hope this will ease the beginners' entrance into this topic, but we would like to point out that this "hands on" approach is *not* characteristic of the field. Most results are obtained by rather abstract and sometimes long proofs which require, besides the usual geometric and topological toolbox of dynamics, a not completely trivial amount of functional analysis (spectral theory of bounded operators on Banach spaces, projective metrics on convex cones). Probability theory also enters the picture in the more advanced results, e.g., via the implementation of large deviation ideas.

Example # 1: *Rotations on the circle.* (Ergodic isometry with no mixing.)

For $\alpha \in [0,1)$, consider the rotation $R_\alpha(z) = ze^{2i\pi\alpha}$ on the circle $\{z \in \mathbb{C} \mid |z| = 1\}$. Lebesgue measure μ is clearly invariant under R_α. We shall see that (R_α, μ) is ergodic if and only if α is irrational (it is then an ergodic physical measure), but that it is *not* mixing.

If α is irrational then for any $\varphi \in L^2(d\mu)$ such that $\varphi = \varphi \circ R_\alpha$ (Lebesgue almost everywhere), we have, using the Fourier expansion $\varphi(z) = \sum_{\mathbb{Z}} b_n z^n$ of φ:

$$\varphi(e^{2i\pi\alpha} z) = \sum_{n\in\mathbb{Z}} b_n e^{2i\pi n\alpha} z^n , \tag{2.1}$$

so that $\varphi \circ R_\alpha = \varphi$ if and only if for all $n \in \mathbb{Z}$

$$b_n(e^{2i\pi n\alpha} - 1) = 0 . \tag{2.2}$$

This implies that $b_n = 0$ except perhaps for $n = 0$. Thus, φ is constant. (And if it is the characteristic function of a set, this set has either zero or full measure.) Now if $\alpha = p/q$ is rational then R_α^q is the identity map, which is clearly not ergodic.

We now check that (R_α, μ) is not mixing, assuming $0 < \alpha < 1/2$ (the case $\alpha \ge 1/2$ is similar). For this, take an arc of circle A of length $\alpha/10$ and let B be the arc of circle which is the complement of an arc of length $\alpha/2$ centered at A. Thus, the length of B is $> 2\alpha$. Clearly, there are arbitrarily large iterates n so that $R_\alpha^n(A)$ is contained in B but also arbitrarily large iterates m such that R_α^m is contained in the complement of B. (Draw a picture!) Thus, $\mu(R_\alpha^n A \cap B) = \mu(A) > 0$ but $\mu(R_\alpha^m A \cap B) = 0$, which is incompatible with mixing.

Example # 2: *Angle doubling on the circle.* (Linear expanding map with super-exponential mixing.)

We now move to an example of dynamics which is locally expanding but confined to a compact set: the angle-doubling map $f(z) = z^2$ on the circle $\{z \in \mathbb{C} \mid$

$|z| = 1\}$. Again, Lebesgue measure μ on the circle is preserved. (Draw the graph of the map $x \mapsto 2x$ modulo 1 on the interval.) Let us show that (f, μ) is mixing (and thus ergodic), again by a Fourier series argument. Write two elements φ, ψ of $L^2(d\mu)$ as

$$\varphi = \sum_{m \in \mathbb{Z}} a_m z^m, \quad \psi = \sum_{\ell \in \mathbb{Z}} b_\ell z^\ell, \tag{2.3}$$

and assume that $\int \psi \, d\mu = 0$, that is, $b_0 = 0$. Then, since $\varphi \circ f^n = \sum_m a_m (z^{2^n})^m$, we have

$$\int \varphi \circ f^n \, \psi \, d\mu = \sum_m \sum_\ell \int a_m (z^{2^n})^m b_\ell z^\ell \, d\mu. \tag{2.4}$$

The only nonvanishing terms in the double sum are those for which $\ell = -2^n m$, $\ell \neq 0$, and we are left with

$$\int \varphi \circ f^n \, \psi \, d\mu = \sum_{m \neq 0} a_m b_{-2^n m}, \tag{2.5}$$

which clearly goes to zero as $n \to \infty$. The reader is invited to prove that the convergence to zero is faster than any exponential if φ, ψ are analytic. (If they are only Lipschitz, exponential mixing with rate $1/2$ holds in general. The statement for C^r observables, with $0 < r \leq \infty$, is that mixing takes place at least with exponential rate $1/2^r$.)

The same property of superexponential mixing for analytic observables and exponential mixing for finitely differentiable observables occurs for linear uniformly hyperbolic maps, e.g., the Arnold "cat map" $\begin{pmatrix} 2 & 1 \\ 1 & 1 \end{pmatrix}$ on the two-dimensional torus, which also preserves Lebesgue measure. For nonlinear but smooth (C^2, or at least $C^{1+\epsilon}$) uniformly expanding or uniformly hyperbolic dynamics, the physical measure first has to be constructed, and one can then prove that it mixes exponentially. (See Sections 3 and 4.)

Example # 3: *Full shift, finitely many symbols.* (Symbolic hyperbolic/expanding model.)

Subshifts of finite type will be discussed further later on. Let us restrict for the moment to the simplest case of *full shifts*. Let then $S = \{1, 2, \ldots, N\}$ and $\Sigma = S^{\mathbb{Z}}$ be the set of biinfinite sequences with coefficients in S. Define the (two-sided) shift σ to the left on Σ by $\sigma(x)_n = x_{n+1}$. We also consider the (nonivertible) one-sided shift σ_+ on $\Sigma_+ = S^{\mathbb{Z}_+}$. For each $0 < \theta < 1$ we may use the metric

$$d_\theta(x, y) = \sum_k \theta^{|k|}(1 - \delta(x_k, y_k)), \tag{2.6}$$

where $\delta(\cdot, \cdot)$ is the Kronecker δ. (The shifts σ and σ_+ are continuous. The shift σ_+ is locally expanding, while σ is "hyperbolic" for this metric.)

If $N = 2$ the one-sided full shift is a symbolic model for the angle-doubling map f of Example 2. The reader who has never seen this is invited to construct a Lipschitz surjective map $\pi : \Sigma_+ \to S^1$ such that $f\pi = \pi\sigma_+$, i.e., π is a *semi-conjugacy*, by considering the partition of the circle into the two arcs $\{e^{2i\pi\alpha} \mid 0 < \alpha < 1/2\}$ and $\{e^{2i\pi\alpha} \mid 1/2 < \alpha < 1\}$.

The *Bernoulli measure* μ on Σ, respectively Σ_+ is an invariant probability measure for σ, respectively σ_+, constructed by assigning mass $(1/N)^k$ to each k-*cylinder* Z_k ($a_i \in S$, and ℓ in $\mathbb{Z}$ or $\mathbb{Z}_+$ respectively):

$$Z_k = Z_k(a_1, \ldots, a_k, \ell) = \{x \in \Sigma_{(+)} \mid x_{\ell+j-1} = a_j \, , j = 1, \ldots, k\} \, . \tag{2.7}$$

Standard results from probability theory guarantee that this defines a Borel probability measure on Σ or Σ_+, and it is not difficult to verify that this measure is shift-invariant. To see that the one-sided shift is mixing (and thus ergodic), express observables φ, $\psi \in L^2(d\mu)$ as weighted sums of characteristic functions of cylinders (at $\ell = 0$) of increasing length k, and use that

$$\int \chi_Z \circ (\sigma_+)^n \chi_{Z'} \, d\mu = \mu(Z) \, \mu(Z') \, ,$$

for two cylinders Z, Z' centered (at $\ell = 0$), if n is larger than the sum of the lengths of Z and Z'. The details are left to the reader.

The connection with physical measures is given by the semi-conjugacy π which lifts Lebesgue measure to the ergodic invariant measure μ.

By showing that the "transfer operator" $\mathcal{L}$ associated to this dynamical system has a "spectral gap" when acting on Lipschitz observables $\varphi : \Sigma_+ \to \mathbb{C}$ for d_θ, one proves that correlation functions associated to (μ, σ_+) and Lipschitz observables on Σ_+ (and similarly for Σ) decay exponentially fast with rate at least θ. (For Hölder observables of exponents $0 < \alpha < 1$ the rate is θ^α.) See Sections 3 and 4.

3. A road atlas: results for deterministic maps

Here are the main examples of possible speeds of mixing (we concentrate on the rate of decay of correlations (1.7)) for discrete-time dynamical systems $f : M \to M$ posessing a mixing *physical measure* μ:

Speed # 1: *Exponential.*

If there is a Banach space $\mathcal{B}$ of functions or distributions on M and a "rate" $0 < \tau < 1$ so that for each φ, $\psi \in \mathcal{B}$ there is $K(\varphi, \psi) > 0$ so that for all $n \in \mathbb{Z}_+$

$$\rho_{\varphi\psi}(n) \le K(\varphi, \psi) \, \tau^n \, , \tag{3.1}$$

the correlation function is said to enjoy *exponential decay.* (It is then summable so that our "rule of thumb" condition for the CLT holds.) The constant K usually depends on φ, ψ through their norms, a typical example is $K(\varphi, \psi) \le C \int |\varphi| \, dx \, \|\psi\|_{\mathcal{B}}$, where $C > 1$ only depends on f.

Exponential decay of correlations holds for the physical measure(s) of:

- *Smooth uniformly expanding dynamics.*

More precisely: assume that M is compact and connected, $f : M \to M$ is $C^{1+\epsilon}$ for some $\epsilon > 0$ and that there are $C > 0$ and $\lambda > 1$ with $\|Df^n v\| \ge C\lambda^n$ for all $v \in TM$. Then f possesses a unique (ergodic) absolutely continuous invariant measure μ, with a positive density with respect to Lebesgue. It is thus the unique physical measure, and also an SRB measure. This measure is mixing [KS]. The rate of mixing is exponential if the observables are at least Hölder (see, e.g., [Ru5] for a relatively recent reference). The result also holds for random compositions of smooth expanding maps [BKS].

• *One-dimensional piecewise smooth and piecewise (non)uniformly expanding dynamics.*

More precisely, $M = I$ is a compact interval, $f : I \to I$ is continuous and there is a finite partition of I into subintervals such that f restricted to each subinterval J is C^2 (or at least $C^{1+\epsilon}$), and $|f'|_J| \geq \lambda > 1$. In this context, Lasota and Yorke [LaY] showed existence of an ergodic absolutely continuous invariant measure μ, which is thus a physical measure. Later Li and Yorke [LiY] studied the number of ergodic absolutely continuous invariant measures for such systems. Rychlik [Ry] studied the case when there are countably many (maximal) intervals of monotonicity. Using a transfer operator approach, Hofbauer and Keller [HoK] obtained exponential decay of correlations for (f, μ), and for observables φ, ψ in the Banach space BV of functions of bounded variation, if f is topologically mixing. Keller [Ke] later refined the spectral analysis. One can reduce to the topological mixing case by taking a suitable iterate of f, using important work of Hofbauer (see [Ho] and references therein), in particular a "tower extension" of the dynamics which inspired much later work. See Buzzi [Bu1] for a proof of exponential decay of correlations for an absolutely continuous quasi-invariant measure for random compositions of piecewise monotone transformations of the interval. We refer to Section 5 for other recent results and connections with dynamical zeta functions.

The uniform expansion assumption has been weakened to include, e.g., the logistic map $f_a(x) = a - x^2$ for a positive measure set of parameters a. For the full parabola $a = 2$ existence and mixing of an absolutely continuous measure are not difficult to obtain: $2 - x^2$ may be explicitly conjugated to a "tent map" of slopes ± 2 (Ruelle [Ru2]), Misiurewicz extended this to logistic maps for which the postcritical orbit is bounded away from the critical point. Jakobson and Benedicks-Carleson showed that f_a admits an absolutely continuous invariant (physical) ergodic measure μ for a positive measure set of as (see, e.g., [dMvS]). Then, using a change of metric in a tower extension, Young [Yo1] and, independently, Keller and Nowicki [KeN] showed that if f is topologically mixing the correlation function of (f, μ) decays exponentially for observables of bounded variation. (Their results apply in fact to a larger class of nonuniformly expanding unimodal maps. See also Section 9 of [Yo2].) Exponential decay of correlations is stochastically stable [BV] for such nonuniformly hyperbolic unimodal maps.

Piecewise expanding maps in higher dimensions are discussed under the heading "expanding systems with singularities" below.

• *Smooth uniformly hyperbolic dynamics.*

More precisely, M is compact and f is a $C^{1+\epsilon}$ Anosov or Axiom A diffeomorphism. Existence of (finitely many) ergodic physical measures was obtained by Sinai (Anosov diffeomorphisms) and Ruelle (Axiom A attractors). If f is topologically mixing (one may reduce to this case via the Smale spectral decomposition) then (f, μ) is mixing and correlations decay exponentially fast for Hölder observables (see, e.g., [Bow], [Ru3]). Pollicott [Po1], Ruelle [Ru4] and Haydn [Ha1, Ha2] obtained important results relating the correlation function and the zeta function. The first proofs used Markov partitions and symbolic dynamics to reduce to a transfer operator defined from a subshift of finite type and a Hölder weight. Different proofs using Birkhoff cones [Li1], [V2], or the usual spectral gap approach but on towers with recurrence estimates [Yo2], were recently carried out.

• *Some smooth nonuniformly hyperbolic dynamical systems.*

The Hénon family is a two-parameter family $H_{a,b}$ of diffeomorphisms which is in some sense a two-dimensional version of the unimodal family f_a. Benedicks and Young [BeY3] recently showed that, for a positive measure set of parameters a, b, the physical measure μ that they had previously exhibited [BeY2] for $H_{a,b}$ enjoyed exponential decay of correlations for Hölder observables. (Using Young's tower construction [Yo2].) A recent work of Wang and Young [WY] extends this analysis to an axiomatic class of nonuniformly hyperbolic surface attractors.

• *Some partially hyperbolic maps.*

Some partially hyperbolic maps (such as time-one maps of hyperbolic geodesic flows or certain skew product constructions) naturally preserve volume, so that the physical measure is given a priori. (See under polynomial mixing for Dolgopyat's [Do5] rapid mixing results for some compact group extensions.) In more abstract, non necessarily volume-preserving situations, Bonatti-Viana [BoV] (assuming that the central direction is "mostly contracting") and more recently Alves-Bonatti-Viana [ABV] (if the central direction is "mostly expanding") recently showed that there are finitely many physical measures the union of the ergodic basins of which has full measure (building on previous work by Pesin and Sinai [PeS]). In the mostly contracting (topologically mixing) case, using methods in [Yo2], Castro [Cas] (under additional technical assumptions) and Dolgopyat [Do4] (under natural additional assumptions) showed that such physical measures enjoy exponential decay of correlations for Hölder observables. (Castro introduced a combinatorical variant of Young's tower construction.)

• *Some expanding/hyperbolic systems with "singularities."*

This class extends piecewise monotone interval maps to higher dimensions. Such dynamical systems appear naturally, e.g., when one considers billiards flows (which preserve volume by construction). In dimension two, Buzzi [Bu2] and Tsujii [Tsu1] independently proved that piecewise expanding and analytic maps admit an absolutely continuous invariant probability measure. An example of Tsujii ([Tsu2], see also [Bu3]) demonstrates that assuming only piecewise expansion and arbitrary high (but finite) differentiability is not sufficient to garantee existence of (a finite number of) such invariant measures. Exponential decay of correlations for a large class of piecewise smooth hyperbolic area-preserving systems was shown by Liverani [Li1]. Existence of a physical measure and exponential decay of its correlations was obtained for certain piecewise expanding or hyperbolic maps in higher dimensions by Young [Yo2] and then Chernov [Ch3], and for the Poincaré maps of certain billiards by Young [Yo2], Chernov [Ch4]. For recent work regarding weak sufficient conditions to obtain an absolutely continuous invariant measure for higher-dimensional piecewise expanding dynamical systems we mention Saussol [Sau]. See also Cowieson's generic transversality condition [Co]. Recently, Buzzi and Maume [BM] obtained weak sufficient conditions for exponential decay of the corresponding correlations.

Sometimes lower bounds $\rho_{\varphi\psi}(n) \geq K'\tau^n$ for the decay of correlations can be obtained by exhibiting eigenvalues of modulus τ with eigenfunctions in $\mathcal{B}$ for a suitable transfer operator. Then, mixing is "truly exponential."

Speed # 2: *Stretched exponential.*

If there is a Banach space $\mathcal{B}$ of functions or distributions on M, a "rate" $0 < \tau < 1$ and an "exponent" $0 < \xi < 1$ so that for each φ, $\psi \in \mathcal{B}$ there is $K(\varphi, \psi) > 0$ so that for all $n \in \mathbb{Z}_+$

$$\rho_{\varphi\psi}(n) \le K(\varphi, \psi)\, \tau^{n^\xi}\,, \tag{3.2}$$

the correlation function is said to enjoy *stretched exponential decay.* (It is then always summable so that the "rule of thumb" for the TCL is satisfied.) To the author's knowledge there are few concrete examples of systems with "true" stretched exponential decay of correlations (with lower bounds). Early estimates (by Bunimovich–Sinai–Chernov) for the decay of correlation of Poincaré maps of billiards gave stretched exponential upper bounds (Liouville volume is ergodic and invariant and thus a physical measure), but the more recent results [Yo2], [Ch4] mentioned above yield exponential asymptotics. Another example is given by the decay of correlation of certain hyperbolic flows: the stretched exponential bounds for correlations of Anosov flows obtained by Chernov [Ch2] and Liverani [Li3] were subsumed by the exponential estimates of Dolgopyat [Do1] *when the dynamical foliations are C^1.* (Note, however, that in the more generic case of Lipschitz foliations the stretched exponential bounds of Chernov [Ch2] are stronger than the rapid decay obtained by Dolgopyat [Do2, Do3]. Also, Liverani's results [Li3] apply to Hölder foliation, giving polynomial bounds.)

Stretched exponential upper bounds have recently been obtained for some random nonuniformly hyperbolic systems (Baladi-Benedicks-Maume [BBM], lower bounds are not available), and may be conjectured for the systems studied by Viana [V2] and Alves [A].

In Young's symbolic tower model, for systems with stretched exponential return asymptotics [Yo3] the rates of decay of correlations and the speed of approach to equilibrium are truly stretched exponential (see Theorem 4.2(3) below).

Speed # 3: *Polynomial.*

If there is a Banach space $\mathcal{B}$ of functions or distributions on M and an "exponent" $\gamma > 0$ so that for each φ, $\psi \in \mathcal{B}$ there is $K(\varphi, \psi) > 0$ so that for all $n \in \mathbb{Z}_+$

$$\rho_{\varphi\psi}(n) \le K(\varphi, \psi)\, \frac{1}{n^\gamma}\,, \tag{3.3}$$

the correlation function is said to enjoy *polynomial decay* of correlations. The correlation function is summable if $\gamma > 1$.

Polynomial decay of correlations has been proved for some "parabolic" or "intermittent" systems having an absolutely continuous invariant probability measure. They include (mixing) one-dimensional maps expanding everywhere except at a fixed point at the origin where the map has local expansion $f(x) \sim x + Cx^{1+\delta}$, and $0 < \delta < 1$. Then, $\gamma = 1 - 1/\delta$, so that the CLT holds if $0 < \delta < 1/2$ (we refer to Hu [Hu], who also obtained lower bounds, and Young [Yo3], as well as Isola [Is1, Is2] for strong results on a piecewise linear model). See also Liverani–Saussol–Vaienti [LSV2] for earlier estimates using a different approach. Hu has recently suggested the terminology *almost hyperbolic* for a class of dynamical systems including this model. Higher dimensional dynamical systems with neutral periodic orbits appear in connection with continued fraction algorithms from number theory. (There is in particular work of Yuri [Yu1, Yu2] regarding the existence of ergodic absolutely continuous measures and speed of approach to equilibrium.) We also mention a paper by Aaronson-Denker-Urbański [ADU]. Sometimes the presence of a nonhyperbolic periodic orbit is an obstruction to the existence of a good finite physical measure (see the two-dimensional example of Hu and Young [HuY]). Fayad [F] has very recently obtained polynomial rates of mixing for the a.c.i.m. of a class of smooth flows of the two-torus.

Dolgopyat [Do5] obtained rapid mixing (i.e., polynomial of arbitrary exponent) for C^∞ test functions and a class of Diophantine compact group extensions (here SRB is just volume), using asociated Brin groups.

In a recent paper, using the coupling technique on towers from [Yo3], Bruin, Luzzatto and van Strien [BLS] exhibit various possible rates of mixing (exponential to polynomial) within the quadratic family.

Other equilibrium states:

The physical measure is not the only invariant measure of natural interest. Other Gibbs states and equilibrium states (for Hölder and non Hölder weights) have been studied, together with their rates of mixing. We mention a few references into the vast literature dealing with thermodynamic formalism: Bowen [Bow], Ruelle [Ru5], Hofbauer-Keller [HK], Keller-Nowicki [KN], Maume [Mau1, Mau2], Mayer [Ma1], Liverani-Saussol-Vaienti [LSV1]. See also [Bal]. We shall not discuss this (interesting) problem further in these notes.

4. Choosing a vehicle and a highway: Two models and three approaches to study speed of mixing

Three different approaches have been applied successfully to estimate the rate of decay of correlation and the speed of approach to equilibrium. We shall describe them briefly, pointing out relations with dynamical zeta functions when relevant and giving a few references. Some of these methods (in particular cone methods) may be implemented on the dynamical system under study directly. Others require to construct first a "model" for the dynamics $f : M \to M$, usually in the form of an extension map $F : \Delta \to \Delta$ with a semi-conjugacy $\pi : \Delta \to M'$ with $M' \subset M$, such that $f \circ \pi = \pi \circ F$. Such a model should embody the main features of the system, and the projection (although neither necessarily surjective nor injective) should be such that not too much information is lost. Before picking a highway, we need to discuss cars types: Let us thus describe two such models: a very classical one (symbolic Markov dynamics) and a more recent and perhaps more flexible one (Markov towers with asymptotics of recurrence times).

4.A Two models: Symbolic dynamics and Towers with return times.

Model # 1: *Symbolic dynamics (SFT).*

Uniformly expanding or uniformly hyperbolic dynamical systems (such as Axiom A diffeomorphisms $f : M \to M$ restricted to an attracting transitive basic set) admit finite Markov partitions. This enables construction of a semi-conjugacy $\pi : \Sigma_A \to M$ with a subshift of finite type (SFT): the restriction of the full shift σ on N symbols to a compact invariant subset $\Sigma_A \subset \Sigma$ defined by a transition matrix A. It is known [Bow] that the unique "equilibrium state" of $\log |\det Df|_{E^u}|$ (where E^u denotes as usual the unstable bundle) is the physical measure of a transitive Axiom A diffeomorphisms.

Lifting the function $\log|\det Df|_{E^u}|$ to Σ_A, one gets a Lipschitz function (if θ is chosen in an appropriate way). Therefore, it suffices to understand equilibrium states of Lipschitz functions for a subshift of finite type. We refer, e.g., to the books of Bowen [Bow], Ruelle [Ru3], or Parry and Pollicott [PaPo]. Generalisations of this model consist in considering countably many symbols ([Br], [Mau2], [Sar]) or weakening ([Mau1], [Po2]) the Hölder/Lipschitz assumption.

Model # 2: *Markov towers with recurrence time asymptotics.*

The use of towers with return times to the basis is not new in dynamics (especially continuous-time dynamics): The grandfather of such constructions in ergodic theory is probably due to Rokhlin. More recently, Hofbauer [Ho] constructed Markov tower extensions of discrete-time interval maps. They were used, e.g., by Hofbauer-Keller [HoKe] and Keller–Nowicki [KN] to study the statistical properties of interval maps. Other tower constructions are due to Young [Yo1] and Baladi-Viana [BV]. We describe here Young's [Yo2, Yo3] recent axiomatised version :

Let $(\Delta_0, \mathcal{B}, m)$ be a probability space, which will be the "basis" of our tower. (The reader can imagine that Δ_0 is a topological space and $\mathcal{B}$ the Borel σ-algebra, but this is not necessary.) We view m as our reference measure, in applications it is often Lebesgue measure or a measure induced by Lebesgue. Assume we are given a countable partition $\Delta_0 = \cup_{i\in\mathbb{Z}}\Delta_{0,i}$, modulo m, together with a positive integer-valued function R (the return-time), defined m-almost everywhere on Δ_0 and constant on each $\Delta_{0,i}$. The *tower* is obtained by layering $R|_{\Delta_{0,i}}$ copies of $\Delta_{0,i}$ above itself:

$$\Delta = \{(x,\ell) \in \Delta_0 \times \mathbb{Z}_+\,,\ \ell+1 \le R(x)\}\,. \tag{4.1}$$

Set $\Delta_{\ell,i}$ for the subsets of $\Delta_\ell = \{x \in \Delta_0 \mid R(x) \ge \ell\}$ induced by the partition $\Delta_{0,i}$, and call $\mathcal{Z}$ the resulting partition of Δ. (We sometimes abuse slightly notation, writing $\Delta = \cup_{\ell\ge 0}\Delta_\ell$.)

The *tower* dynamics $F : \Delta \to \Delta$ takes the following form:

$$F(x,\ell) = \begin{cases} (x,\ell+1) & \ell+1 < R(x), \\ (y(x),0) & \ell+1 = R(x), \end{cases} \tag{4.2}$$

where the map $x \mapsto y(x)$ is such that the restriction of $(x,0) \to F^{R(x)}(x,0)$ to $\Delta_{0,i}$ is a bijection $\Delta_{0,i} \to \Delta_0$, for all i.

In view of defining a "dynamical distance," introduce a separation time $s : \Delta \times \Delta \to \mathbb{Z}_+ \cup \{\infty\}$ by setting

$$s(x,y) = \min\{n \ge 0 \mid F^n(x) \text{ and } F^n(y) \text{ lie in distinct elements of } \mathcal{Z}\}\,. \tag{4.3}$$

The definition implies in particular that if (x,ℓ) and (y,ℓ) are both in some $\Delta_{\ell,j}$, then $s((x,0),(y,0)) \ge \ell$ and for any $(x,0),(y,0) \in \Delta_{0,j} \times \{0\}$

$$s((x,0),(y,0)) = R(x) + s(F^{R(x)}(x,0), F^{R(x)}(y,0))\,. \tag{4.4}$$

The two main assumptions on this abstract model are:

- *Bounded distortion:* For each $\Delta_{0,i}$, the map $F^R|_{\Delta_{0,i}}$ and its inverse are nonsingular with respect to the measure m so that the Jacobian JF^R exists (and almost never vanishes). There are $C > 0$ and $0 < \beta < 1$ so that for all i and each $x, y \in \Delta_{0,i}$

$$\left|\frac{JF^R(x)}{JF^R(y)} - 1\right| \le C\beta^{s(F^R(x),F^R(y))} \tag{4.5}$$

• *Integrability of return times:* $\int_{\Delta_0} R(x)dm(x) < \infty$.

Consider the "dynamical Lipschitz" function space on Δ:

$$\mathcal{F}_\beta(\Delta) = \{\varphi : \Delta \to \mathbb{C} \mid \exists C_\varphi , |\varphi(x) - \varphi(y)| \le C_\varphi \beta^{s(x,y)} , \forall x, y \in \Delta\} . \tag{4.6}$$

Young [Yo2], [Yo3] showed that these assumptions are enough to guarantee the existence of an F-invariant measure ν absolutely continuous with respect to m on Δ. The density of this measure belongs to $\mathcal{F}_\beta$ (see Theorem 4.2 (1) below). The interest of this method lies in its rather broad applicability to actual dynamical systems $f : M \to M$ possessing "some" hyperbolicity. The difficulty there is to find a subset $\Lambda \subset M$ and a return-time function R so that f^R has good properties on Λ, in order to construct a semi-conjugacy $\pi : \Delta \to \cup_{\ell \ge 0} f^R \Lambda$ between f and F, showing that F satisfies the desired axioms. (The semi-conjugacy map is neither surjective nor injective in general.) Then, if the return-times satisfy exponential asymptotic bounds (see Theorem 4.2 (4)) the transfer operator associated to F has a spectral gap on a suitable space [Yo2], and one may show exponential decay of correlations for (F, ν) and observables in (a modified version of) $\mathcal{F}_\beta$. The physical measure μ may be defined by projecting ν, and the result on correlations carries over to Lipschitz observables (Lipschitz functions are lifted to $\mathcal{F}_\beta$). See Theorem 4.2(5) (6) for the nonexponential case. Note that there are slight technical differences between the definitions of the tower models in [Yo2] and [Yo3].

4.B Three methods to estimate speeds of mixing.

Method # 1: *Spectrum of transfer operator.*

Spectral gap (spectral approach).

The spectrum of the Ruelle transfer operator $\mathcal{L}_g$, associated to a subshift of finite type and a Hölder or Lipschitz weight g, and acting on the Banach space of Lipschitz functions for a metric d_θ are discussed in the notes of M. Pollicott on *Zeta functions,* where its relationship with the zeta function

$$\zeta(g, z) = \exp \sum_{n \ge 1} \frac{z^n}{n} \sum_{x \in \mathrm{Fix} f^n} \prod_{k=0}^{n-1} g(f^k(x))$$

was presented. Let us sketch here how a spectral gap for a transfer operator is related to exponential decay of correlations, in the simplest possible case: smooth expanding dynamics on the circle S^1. Let then $f : S^1 \to S^1$ be C^r for some $r \ge 2$, and assume that f is uniformly expanding in the sense that there is $C > 0$ and $\lambda > 1$ so that $|(f^n)'| \ge C\lambda^n$ for all $n \in \mathbb{Z}_+$ (an example is given by considering a small C^r perturbation of the angle-doubling map from Section 2). The transfer operator $\mathcal{L}$, associated to f and the weight $g = 1/|f'|$, acting on the Banach space $\mathcal{B}$ of C^s functions on M (with $0 \le s \le r-1$, with fractal exponents denoting Hölder continuity) φ by

$$\mathcal{L}\varphi(x) = \sum_{fy=x} \varphi(y)g(y) = \sum_{fy=x} \frac{\varphi(y)}{|f'(y)|} , \tag{4.7}$$

is appropriate to study absolutely continuous invariant measures for f. Indeed, its dual $\mathcal{L}^*$ acting on $\mathcal{B}^*$ by $\mathcal{L}\nu(\varphi) = \nu(\mathcal{L}\varphi)$ for $\varphi \in \mathcal{B}$ preserves Lebesgue measure dx:

$$\int \mathcal{L}\varphi \, dx = \int \varphi \, dx .$$

(By the change of variable formula, the above equality holds in fact for all $\varphi \in L^1(dx)$, since $|f'|$ is the Jacobian of f with respect to Lebesgue measure.)

It can be shown (by arguments similar to those used to prove the Ruelle operator theorem in the notes of M. Pollicott on *Zeta functions*) that the spectral radius of $\mathcal{L}$ is equal to 1 and that its *essential spectral radius* is not larger than λ^{-s}. (This fact also holds for $s = 0$. In this case each point in the open disc of radius 1 is an eigenvalue of infinite multiplicity! Hence, the Banach space of continuous functions and a fortiori $L^1(dx)$ or $L^2(dx)$ are "too big"). See Section 5 for a proof of an analogous result. We just mention that the basic ingredient for the bound for the essential spectral radius in the case $s = 1$ is the "Lasota-Yorke" [LaY] inequality: there is $C > 0$ so that for all C^1 functions φ and all $n \in \mathbb{Z}^+$

$$\sup_M \left| \frac{d}{dx}(\mathcal{L}^n \varphi)(x) \right| \le C\lambda^{-n} \sup_M \left| \frac{d}{dx}\varphi(x) \right| + C \sup_M |\varphi(x)| . \tag{4.8}$$

The above estimate may be proved by using the Leibniz formula (derivation of a product), chain rule (derivation of a composition). The fact that the inverse branches of f^n contract by a factor at least λ^{-n} obviously enters the computation. (In other words: composing a C^1 function φ with a C^1 contraction *improves* the smoothness of φ in the sense that it decreases the C^1 norm.)

Assume from now on that $s > 0$. Since the spectrum of a bounded linear operator and its dual coincide outside of the disc of essential spectral radius, $\mathcal{L}$ has a fixed point $\varphi_0 \in \mathcal{B}$. In fact, using positivity of $\mathcal{L}$ and generalising the finite-dimensional *Perron-Frobenius* theorem (see, e.g., [Wal]), one shows that φ_0 is a positive function, which can be normalised to ensure $\int \varphi_0 \, dx = 1$. Easy algebra then shows that the absolutely continuous probability measure $\mu = \varphi_0 \, dx$ is f-invariant:

$$\int \varphi \circ f \varphi_0 \, dx = \int \mathcal{L}(\varphi \circ f \varphi_0) \, dx = \int \varphi(\mathcal{L}\varphi_0) \, dx = \int \varphi \, d\mu .$$

The eigenvalues of $\mathcal{L}$ on the unit circle consist in a finite set of roots of unity, they are simple because smooth expanding circle maps are topologically transitive. Using that smooth expanding circle maps are in fact topologically mixing, one shows that the eigenvalue 1 is simple and is the only eigenvalue of modulus 1 of $\mathcal{L}$ acting on $C^s(S^1)$ if $s > 0$. In other words, we have the following *spectral decomposition:*

$$\mathcal{L}\varphi = \varphi_0 \int \varphi \, dx + \mathcal{R}\varphi , \tag{4.9}$$

where $\mathcal{R}$ is a bounded operator of spectral radius $\tau < 1$ on $C^s(S^1)$. (Either τ corresponds to the modulus of an eigenfunction of $\mathcal{L}$ or it is just the essential spectral radius.) Thus, since $\mathcal{L}^n(\varphi \circ f^n \psi) = \varphi \mathcal{L}^n \psi$, for $\varphi, \psi \in C^s$,

$$\begin{aligned} |\int \varphi \circ f^n \psi \, dx - \int \varphi \, d\mu \int \psi \, dx| &= |\int \varphi(\mathcal{L}^n \psi - \varphi_0 \int \psi dx) dx| \\ &\le \sup |\mathcal{R}^n \psi| \int |\varphi| \, dx \end{aligned}$$

which proves exponential decay of correlations (for any rate $> \tau$) since $\sup |\varphi| \le \|\varphi\|$ for the Banach space $\mathcal{B} = C^s(S^1)$.

Beyond first gap: zeta function (spectral approach).

A more careful spectral decomposition than (4.9), writing out explicitly a sum over eigenvalues of modulus larger than $\lambda^{-s} + \epsilon$ for small ϵ, is the starting point to see that the Fourier transform

$$\widehat{\rho}(\omega) = \sum_{n \in \mathbb{Z}} e^{in\omega} \rho_{\varphi\psi}(n)$$

of the correlation function associated to φ, $\psi \in \mathcal{B}$ (setting $\rho_{\varphi\psi}(n) = \rho_{\psi\varphi}(-n)$ for negative n) is meromorphic in the strip $\Im(\omega) < \log \lambda^s$, where its poles (the *Ruelle resonances*) may only be located at ω with $\exp(i\omega)$ or $\exp(-i\omega)$ an eigenvalue of $\mathcal{L}$. Also, these eigenvalues of $\mathcal{L}$ on $C^s(S^1)$ are described (including multiplicity) by the poles of the zeta function $\zeta_g(z)$ for $g = 1/|f'|$ (which is known to be meromorphic in the disc of radius λ [Ru5])

Further spectrum: Fredholm determinants (spectral approach).

Let us see what more can be said if f and g are of higher smoothness. In the case of analytic expanding maps on S^1, the transfer operator $\mathcal{L}$ acting on the Banach space of holomorphic bounded functions in a small annulus obtained by "thickening" the unit circle is a *compact* operator (in particular its essential spectral radius vanishes). This can be shown by using Montel's theorem and is due to the fact that composing a holomorphic function with a holomorphic contraction extends the domain of holomorphy ("improves analyticity"). In fact, as observed by Ruelle [Ru1] in the seventies, $\mathcal{L}$ is in this case a *nuclear operator in the sense of Grothendieck* [Gr1, Gr2]. This also holds for analytic Anosov diffeomorphisms with analytic invariant foliations (unfortunately a very strong assumption): see Ruelle [Ru1], Fried [Fr1], and Mayer's extensive study of the thermodynamic formalism of the Gauss map $x \mapsto \{1/x\}$ [Ma1] (and references therein). We especially recommend Mayer's very clear survey [Ma2]. Nuclear operators generalise classical Fredholm operators such as

$$\mathcal{K} : \varphi \to \int_0^1 K(\cdot, y)\varphi(y)\, dy\,, \qquad K(x,y) \text{ continuous on } [0,1] \times [0,1]\,, \tag{4.10}$$

acting on $C^0([0,1])$. In particular, nuclear operators possess a *trace* $\mathrm{tr}\,\mathcal{K}$ which has many of the features of the trace (i.e., sum of eigenvalues) of a finite matrix. The most important of these features is perhaps the fact that the power series

$$\exp - \sum_{n=1}^{\infty} \frac{z^n}{n} \mathrm{tr}\,\mathcal{K}^n \tag{4.11}$$

is an entire function, called the Fredholm determinant of the operator $\mathcal{K}$, the zeroes of which are exactly the inverse of the eigenvalues of $\mathcal{K}$. (Clearly, this generalises the formula $\det(I - zA) = \exp - \sum_{n=1}^{\infty} \frac{z^n}{n} \mathrm{tr}\, A^n$ from linear algebra, for a finite matrix A.) In the case of the Fredholm operator $\mathcal{K}$ from (4.10), the trace is just given by the, intuitively obvious, "sum along the diagonal" $\int K(x,x)\, dx$. In the case of dynamically defined transfer operators (4.7) associated to analytic expanding dynamics f and analytic weights g (e.g., $1/|f'|$, the modulus being dealt with by a suitable argument), the trace can be computed by using Cauchy integrals:

$$\sum_{x \in \mathrm{Fix}\, f} \frac{g(x)}{1 - (f_x^{-n})'(x)} = \sum_{x \in \mathrm{Fix}\, f} \frac{1}{\det(Df_x^n(x) - \mathrm{Id})}\,,$$

where the inverse branch f_x^{-n} is uniquely defined by the periodic orbit of x.

The expression in the denominator *morally* comes from the Jacobian of the change of variable associated to the map $\mathrm{Id} - f^{-1}$ arising from the operator with singular kernel:

$$\mathcal{L}\varphi(x) = \int \varphi(y)\delta(f(y)-x)g(y)|f'(y)|\,dy = \int \varphi(y)\delta(f(y)-x)g(y)|\mathrm{det}Df(y)|\,dy\,,$$

where δ is the Dirac delta.

Therefore, the Fredholm determinant whose zeroes describe the spectrum of $\mathcal{L}_g$ acting on holomorphic functions for $g = 1/|\mathrm{det}Df|$ (i.e., the correlation spectrum of the physical measure of the expanding analytic map f) is just

$$\begin{aligned}(4.12)\qquad &\exp - \sum_{n=1}^{\infty} \frac{z^n}{n} \sum_{x \in \text{ Fix } f^n} \frac{1}{|\mathrm{det}Df^n(x)|} \frac{1}{\det(\mathrm{Id} - Df_x^{-n}(x))} \\ &= \exp - \sum_{n=1}^{\infty} \frac{z^n}{n} \sum_{x \in \text{ Fix } f^n} \frac{1}{\det(Df^n(x) - \mathrm{Id})}\,.\end{aligned}$$

The denominator $\det(\mathrm{Id} - Df_x^{-n}(x))$, which is present here but not in the zeta function $\zeta_g(x)$, improves the convergence of the series and extends the domain in which its zeroes describe the inverse eigenvalues of $\mathcal{L}$. (The reader is invited to compute both the zeta function and the dynamical determinant of the angle doubling map for the weight $g = 1/2$.) See [Ru1], [Fr1].

Although the transfer operators (4.7) associated to finitely differentiable expanding maps and/or weights (say, C^r for $r \geq 1$, and assuming that the spectral radius is 1 for simplicity) are not compact, expressions of the type (4.12) have been shown to decribe their inverse spectrum in the disc of radius λ^r (Ruelle [Ru5, Ru6]), a larger radius than the value λ corresponding to the domain of "visibility" (meromorphy) of the zeta function.

Two-sided or hyperbolic dynamics (spectral approach).

The situation of two-sided shifts (or hyperbolic maps as opposed to expanding maps) is more delicate. Indeed, coexistence of contracting and expanding directions means that the transfer operator does not improve regularity of smooth functions. We first discuss the *"historical" approach.* It has been used in statistical mechanics, see Bowen's book [Bow], but also very recently in the axiomatic framework of Young [Yo2], e.g., for good Hénon maps [BeY3]. It consists in reducing to the one-sided dynamics by expressing the measure $\hat{\mu}$ under study as the natural extension of a one-sided measure μ (see below). The idea is thus to "quotient out" the stable directions and some information ("half of the smoothness") is lost in the process. On must also approximate a "two-sided" function (depending on both past and future) by sums of "one-sided" functions. Let us describe this in detail for an equilibrium measure $\hat{\mu}$ of a two-sided subshift of finite type $\sigma : \Sigma_A \to \Sigma_A$ and a positive Lipschitz weight $\hat{g}$. Let $\mathcal{C}_\theta$ and $\mathcal{C}_\theta^+$ be the Banach spaces of Lipschitz functions for d_θ on the two-sided and one-sided shift spaces Σ_A, Σ_A^+, writing $|\cdot|_\theta$ for the Lipschitz constant and $\|\cdot\|_\theta = \max(|\cdot|_\theta, \sup|\cdot|)$ for the norm. For $\varphi, \psi \in \mathcal{C}_\theta$, assuming that $\int_{\Sigma_A} \psi\, d\hat{\mu} = 0$, we wish to analyse

$$(4.13)\qquad \rho_{\varphi\psi}(n) = \int \varphi \circ \sigma^n\, \psi\, d\hat{\mu}\,.$$

and the formal Fourier transform $\widehat{\rho}_{\varphi\psi}(\omega)$.

Our starting point is the fact that for each $\phi \in \mathcal{C}_\theta$, there is a sequence of functions $\phi_n \in \mathcal{C}_\theta^+$, $n \ge 0$ such that ϕ_n is constant on n-cylinders at 0, $\sup_{\Sigma_A^+} |\phi_n| \le \theta^n \|\phi\|_\theta$ (hence, $|\phi_n|_\theta \le 2\|\phi\|_\theta$) for all $n \ge 0$, and with

$$\phi = \sum_{n=0}^{\infty} \phi_n \circ \Pi \circ \sigma^{-n} , \tag{4.14}$$

where $\Pi : \Sigma_A \to \Sigma_A^+$ is the map which simply "forgets" x_i for $i < 0$. We may in fact construct the sequence ϕ_n explicitly: For each $1 \le j \le N$, fix $a^{(j)} \in \Sigma_A$ with $(a^{(j)})_0 = j$, and define a function $r : \Sigma_A \to \Sigma_A$ by setting

$$r(x)_k = \begin{cases} x_k & \text{if } k \ge 0 \\ (a^{(x_0)})_k & \text{if } k \le 0 . \end{cases} \tag{4.15}$$

Abusing notation, the function r can be viewed as $r : \Sigma_A^+ \to \Sigma_A$ with $\Pi \circ r = \mathrm{Id}$. Then, we may take $\phi_0 = \phi \circ r$, and for each $n \ge 1$, inductively define,

$$\phi_n = (\phi - \sum_{k=0}^{n-1} \phi_k \circ \Pi \circ \sigma^{-k}) \circ \sigma^n \circ r . \tag{4.16}$$

We finally associate a formal power series to $\phi \in \mathcal{C}_\theta$:

$$\tilde{\phi}(z) = \sum_{n=0}^{\infty} z^n \phi_n . \tag{4.17}$$

It is not difficult to see that $\tilde{\phi}(z)$ is an analytic $\mathcal{C}_\theta^+$-valued functions for $|z| < 1$, and is an analytic function with values in the space of bounded functions for $|z| < \theta^{-1}$. (See [DS, Definition VI.10.5] for the concept of analytic functions from $\mathbb{C}$ to an arbitrary Banach space.)

Now, it is known that there is $h \in \mathcal{C}_{\sqrt{\theta}}$ with $\hat{g} = g + h - h \circ \sigma$ where g only depends on the future and can thus be viewed as an element of $\mathcal{C}^+_{\sqrt{\theta}}$. (Note that one loses "one-half" of the differentiability: a Lipschitz function for $d_{\sqrt{\theta}}$ is 1/2-Hölder for d_θ.) Also, the one-sided shift invariant equilibrium state μ for g admits $\hat{\mu}$ as natural extension (see, e.g., [Bow]). The transfer operator $\mathcal{L}_g$ associated to σ^+ and g acts on $\mathcal{C}^+_{\sqrt{\theta}}$, with maximal eigenvalue $e^P > 0$ and essential spectral radius $< \sqrt{\theta} e^P$, and $\mu = \varphi_0 \, \nu_0$ with $\mathcal{L}_g \varphi_0 = e^P \varphi_0$ and $\mathcal{L}_g^* \nu_0 = e^P \nu_0$. There is a normalised operator $\tilde{\mathcal{L}}$ which satisfies $\tilde{\mathcal{L}}^* \mu = \mu$, $\tilde{\mathcal{L}} 1 = 1$, and which has the same spectrum as $e^{-P} \mathcal{L}_g$. (See, e.g., [PaPo].)

Finally, we get the following sequence of equalities between power series:

$$\begin{aligned} \hat{\rho}_{\varphi\psi}(\omega) &= \sum_{k=-\infty}^{\infty} e^{ki\omega} \int_{\Sigma_A} (\varphi \circ \sigma^{-k}) \, \psi \, d\hat{\mu} \\ &= \sum_{k=-\infty}^{\infty} \sum_{m,n=0}^{\infty} e^{ki\omega} \int_{\Sigma_A} (\varphi_m \circ \Pi \circ \sigma^{-k-m})(\psi_n \circ \Pi \circ \sigma^{-n}) \, d\hat{\mu} \end{aligned}$$

$$= \sum_{m,n=0}^{\infty} e^{(n-m)i\omega} \Big[\sum_{k\le n-m} e^{(k+m-n)i\omega} \int_{\Sigma_A^+} (\varphi_m \circ (\sigma^+)^{n-m-k})\, \psi_n \, d\mu$$

$$+ \sum_{k\ge n-m} e^{(k+m-n)i\omega} \int_{\Sigma_A^+} \varphi_m (\psi_n \circ (\sigma^+)^{k+m-n})\, d\mu - \int_{\Sigma_A^+} \varphi_m \, \psi_n \, d\mu \Big]$$

$$= \sum_{\ell=0}^{\infty} e^{-\ell i\omega} \int (\tilde\varphi(e^{-i\omega}) \circ (\sigma^+)^{\ell}) \tilde\psi(e^{i\omega})\, d\mu$$

$$+ \sum_{\ell=0}^{\infty} e^{\ell i\omega} \int \tilde\varphi(e^{-i\omega})(\tilde\psi(e^{i\omega}) \circ (\sigma^+)^{\ell})\, d\mu - \int \tilde\varphi(e^{-i\omega})\tilde\psi(e^{i\omega})\, d\mu$$

$$= \sum_{\ell=0}^{\infty} e^{-\ell i\omega} \int \tilde\varphi(e^{-i\omega}) \tilde{\mathcal{L}}^{\ell}(\tilde\psi(e^{i\omega}))\, d\mu$$

$$+ \sum_{\ell=0}^{\infty} e^{\ell i\omega} \int \tilde\psi(e^{i\omega}) \tilde{\mathcal{L}}^{\ell} \tilde\varphi(e^{-i\omega})\, d\mu - \int \tilde\varphi(e^{-i\omega})\, \tilde\psi(e^{i\omega})\, d\mu$$

$$= \int \tilde\varphi(e^{-i\omega})(1 - e^{-i\omega}\tilde{\mathcal{L}})^{-1} \tilde\psi(e^{i\omega})\, d\mu$$

$$+ \int \tilde\psi(e^{i\omega})(1 - e^{i\omega}\tilde{\mathcal{L}})^{-1} \tilde\varphi(e^{-i\omega})\, d\mu - \int \tilde\varphi(e^{-i\omega})\, \tilde\psi(e^{i\omega})\, d\mu\,. \tag{4.18}$$

The spectral properties of $\mathcal{L}_g : \mathcal{C}^+_{\sqrt{\theta}} \to \mathcal{C}^+_{\sqrt{\theta}}$, together with the mentioned analyticity properties of the maps $z \mapsto \tilde\varphi(z)$, $\tilde\psi(z)$, imply that the right-hand side of (4.18) is an analytic function of ω in the strip $|\Im\omega| < \log(\tau^{-1})$ (τ as after (4.9)) which admits a meromorphic extension to the strip $|\Im\omega| < \log(\sqrt{\theta}^{-1})$. (For explicit formulas of the corresponding residues as well as an analysis of the order of the pole, see [Ru4], and also [Ha1].)

The above computations can be applied directly via symbolic dynamics and Markov partitions to show exponential decay for the physical measure (or other equilibrium measures) of Axiom A attractors ([Bow], [Po1], [Ru4]). We already proved that the corresponding ideas (quotienting out stable directions and approaching two sided functions by sums of one-sided functions) have been adapted by Young [Yo2] to towers with asymptotics of return times, in particular to prove exponential mixing for good Hénon maps [BeY3].

Another more *modern approach* to attack the *two-sided or hyperbolic* situation was initiated in the early nineties by Rugh [Rug1, Rug2] for analytic Axiom A diffeomorphisms (without assuming smoothness of the invariant foliations) and analytic weights. One of the main ideas was to introduce a Banach space of functions analytic in the unstable directions and "dual of analytic" in the stable directions. (In an analytic framework one can morally think of Laurent series.) The transfer operator associated to the full hyperbolic dynamics (*without* quotienting out stable directions) then has good spectral properties on this Banach space, it is even nuclear in the sense of Grothendieck. One thus obtains Fredholm determinants which are entire functions in this framework. The method was then extended further by Fried [Fr3].

Assuming only finite differentiability (here, partitions of unity are available), Kitaev [Kit] obtained a large radius of convergence (of the order of $\lambda^{r/2}$) for the dynamical determinant of C^r hyperbolic diffeomorphisms with C^r weights g. Fried has very recently (June 1999, IHES talk) announced a spectral interpretation of the zeroes of the corresponding Fredholm determinant.

Beyond uniform expansion (spectral approach).

In the list in Section 3 we mentioned that exponential decay holds for some nonuniformly expanding dynamical systems. The spectral approach has also been implemented there, by first constructing a *tower* model for the map. (Young [Yo1, Yo2], Benedicks-Young [BeY3], Keller-Nowicki [KN]). We refer to Section 5 for a decription of a form of regularisation, the kneading approach (Baladi-Ruelle [BR1, BR2]) inspired by Milnor-Thurston [MT].

Method # 2: *Projective (Hilbert) metrics on Birkhoff cones.*

One way to prove the finite-dimensional Perron-Frobenius theorem is by using the fact that any $n \times n$ matrix A with strictly positive coefficients maps the positive "quadrant" in $\mathbb{R}^n$ strictly into itself. Representing half-lines in the quadrant by their intersection with the simplex $\sum x_i = 1$, it is convenient to use the projective metric on the quadrant given by the cross ratio on the line joining x and y. One then proves that this projective metric is strictly decreased by the action of A (see, e.g., [Fur, § 15]). There exists an infinite dimensional version of this theory, where the positive quadrant is replaced by a convex cone Λ in a topological vector space (for example a space of functions or distributions), and A is replaced by a linear operator mapping Λ into itself. The cones are called Birkhoff cones in honor of Garrett Birkhoff (the main original reference by Birkhoff is [Bir], see also [Li1] for a reader-friendly exposition). These techniques were introduced into dynamical systems in particular by Ferrero-Schmitt in their pioneering paper [FS]. They were more recently used by Liverani [Li1] to prove exponential rates of mixing of some expanding and hyperbolic dynamical systems. This approach was generalised by Viana and is very well presented in his book [V2]. Castro [Cas] defined cones (in a tower setting) for some partially hyperbolic systems. In all these cases, a function cone is exhibited which is mapped strictly into itself with "finite diameter," allowing to prove simultaneously existence (and uniqueness) of the physical measure *and* exponential approach to equilibrium. Much more recently, Maume was able to use cone techniques to obtain other speeds of mixing (slower than exponential): first in a symbolic framework [Mau2] (with countably many symbols) and then in the axiomatic setting of towers with return times described in Section 4.A [Mau3]. In this recent work, the function cone is *not* mapped strictly into itself after a single iteration.

As there is no (apparent) direct relation between the invariant cone approach and zeta functions, we shall only give a very brief introduction to the main notions here, referring the interested reader to [Li1] or [V2].

DEFINITION (CONES AND PARTIAL ORDER). Let $\mathcal{B}$ be a Banach space with the topology induced by its norm. A subset $\Lambda \subset \mathcal{B} \setminus \{0\}$ is called a *cone* if $\lambda\varphi \in \Lambda$ for all $\varphi \in \Lambda$ and all $\lambda \in \mathbb{R}_+^*$. The cone Λ is said to be *convex* if $\varphi + \psi \in \Lambda$ whenever both φ and ψ are in Λ. Any convex cone Λ defines a *partial ordering* compatible with the vector space structure $\preceq$ on $\mathcal{B}$ by setting

$$\varphi \preceq \psi \Leftrightarrow \psi - \varphi \in \Lambda \cup \{0\}\,.$$

DEFINITION(HILBERT PSEUDO-METRIC Θ_Λ). For φ, ψ in a cone Λ, set

$$\alpha(\varphi,\psi) = \sup\{\lambda \in \mathbb{R}_+ \mid \lambda\varphi \preceq \psi\}\,,\ \beta(\varphi,\psi) = \inf\{\lambda \in \mathbb{R}_+ \mid \psi \preceq \lambda\varphi\}\,,$$

(putting $\alpha = 0$ and $\beta = \infty$ when the corresponding sets are empty). Then let

$$\Theta_\Lambda(\varphi,\psi) = \log\frac{\beta(\varphi,\psi)}{\alpha(\varphi,\psi)}\,.$$

The Hilbert pseudo-metric Θ_Λ induces a distance on the quotient $\Lambda/\sim$ where $\varphi\sim\psi$ if $\psi = \lambda\varphi$ for some $\lambda\in\mathbb{R}_+^*$.

The key result is the fact that any linear operator $\mathcal{L}$ mapping a convex cone Λ inside itself does not increase the Hilbert pseudo-distance, with a strict contraction property whenever the Θ_Λ-diameter of $\mathcal{L}\Lambda$ is finite (write $\mathrm{diam}_{\Theta_\Lambda}(A) = \sup_{\varphi,\psi\in A}\Theta_\Lambda(\varphi,\psi)$):

THEOREM 4.1 (BIRKHOFF'S INEQUALITY). *Let Λ be a convex cone in a Banach space $\mathcal{B}$. If $\mathcal{L}:\mathcal{B}\to\mathcal{B}$ is a linear operator such that $\mathcal{L}(\Lambda)\subset\Lambda$, then for each $\varphi,\psi\in\Lambda$ we have*

$$\Theta_\Lambda(\mathcal{L}\varphi,\mathcal{L}\psi) \le \tanh\left(\frac{\mathrm{diam}_{\Theta_\Lambda}(\mathcal{L}\Lambda)}{4}\right)\cdot\Theta_\Lambda(\varphi,\psi)\,.$$

For example, to prove exponential mixing for C^2 expanding maps (see Speed #1 in Section 3), one may use a one-parameter family of Lipschitz cones Λ_L, indexed by $L>0$:

$$\Lambda_L = \{\varphi\in C^0(M) \mid \varphi(x)>0\,,\ \frac{\varphi(x)}{\varphi(y)} \le e^{Ld(x,y)}\,,\ \forall x,y\in M\}\,.$$

As mentioned above, in the hyperbolic case, more interesting cones were introduced by Liverani [Li1] and generalised by Viana [V2]. We refer to these sources for more, in particular discussions on how the pseudo-metric may be compared to the L^1 distance or some other metric on the function space.

Method # 3: *Probabilistic coupling.*

"Coupling" is a technique from probability theory [Lind] which is useful to compare sequences of probability measures in particular to estimate at which speed the distance between them decreases. This approach has been introduced into dynamical systems by Young [Yo3]. More precisely, she was able to apply it to the tower model with recurrence asymptotics described in Section 4.A. (See Bressaud-Fernandez-Galves [BFG], Bruin-Luzzatto-van Strien [BLS], and Baladi-Benedicks-Maume [BBM] for recent applications.) As there is no (apparent) direct relation between the invariant cone approach and zeta functions, we shall only state the main result in [Yo3] here.

Consider the tower model $F:\Delta\to\Delta$ with return time R and the function space $\mathcal{F}_\beta$ from (4.6). Let $\mathcal{F}_\beta^+$ denote those elements of $\mathcal{F}_\beta$ whose restriction to each $\Delta_{\ell,j}$ is either strictly positive or identically zero (see [Yo3] for a formal definition). Denote by $|\rho|$ the total mass of a signed measure ρ on Δ. (If ρ is absolutely continuous with respect to m then $|\rho| = \int |\frac{d\rho}{dm}|\, dm$.) Finally, define ($m$ almost everywhere) $\widehat{R} : \Delta \to \mathbb{Z}_+$ by setting $\widehat{R}(x, \ell) = R(x) - \ell$. Note that

$$m(\{x \in \Delta \mid \widehat{R}(x) > n\}) = \sum_{\ell > n} m(\Delta_\ell) = \sum_{\ell > n} m(\{x \in \Delta_0 \mid R(x) > \ell\}).$$

The main result is:

THEOREM 4.2 (RECURRENCE TIMES AND RATES OF MIXING [Yo3]). *(1) $F : \Delta \to \Delta$ admits an invariant probability measure ν which is absolutely continuous with respect to m. The density $d\nu/dm$ is an element of $\mathcal{F}^+_\beta(\Delta)$ and* $\inf(d\nu/dm) > 0$.

(2) The pair (F, ν) is exact (and thus mixing) if and only if the greatest common divisor of the return times $\{R(x), x \in \Delta_0\}$ is equal to one.

(3) Assume that (F, ν) is exact. Then there is a probability measure λ on Δ, absolutely continuous with respect to m, with $\frac{d\lambda}{dm} \in \mathcal{F}^+_\beta$ and $C > 0$ with

$$|F^n_* \lambda - \nu| \geq C\, m\{\widehat{R} > n\}, \forall n \in \mathbb{Z}_+ .$$

(4) Assume that (F, ν) is exact and that there is $0 < \tilde{\tau} < 1$ with $m(\{\widehat{R} > n\}) = O(\tilde{\tau}^n)$. Then there is $\tilde{\tau} \leq \tau < 1$ such that for each absolutely continuous probability measure λ on Δ with $\frac{d\lambda}{dm} \in \mathcal{F}^+_\beta$

$$|F^n_* \lambda - \nu| \leq C\, \tau^n, \forall n \in \mathbb{Z}_+ .$$

(5) Assume that (F, ν) is exact and that there are $0 < \tilde{\tau} < 1$ and $0 < \tilde{\xi} \leq 1$ with $m(\{\widehat{R} > n\}) = O(\tilde{\tau}^{n^{\tilde{\xi}}})$. Then there are $\tilde{\tau} \leq \tau < 1$ and $\xi \leq \tilde{\xi}$ such that for each probability measure λ on Δ with $\frac{d\lambda}{dm} \in \mathcal{F}^+_\beta$

$$|F^n_* \lambda - \nu| \leq C\, \tau^{n^\xi}, \forall n \in \mathbb{Z}_+ .$$

(6) Assume that (F, ν) is exact, and that for some $\gamma > 0$ and all n, $m(\{\widehat{R} > n\}) = O(n^{-\gamma})$. Then for each probability measure λ on Δ with $\frac{d\lambda}{dm} \in \mathcal{F}^+_\beta$

$$|F^n_* \lambda - \nu| \leq \frac{C}{n^\gamma}, \forall n \in \mathbb{Z}_+ .$$

In the above theorem, the statements on rates of mixing are formulated in terms of bounds for the approach to equilibrium $|F^n_* \lambda - \nu|$. The corresponding upper bounds for decay of correlations are easily deduced.

Other methods:

We do not claim that the three approaches described are the only ones in existence! In particular "Markov approximation" techniques have been much investigated (see Chernov [Ch2, Ch1], Lecture 12 in Sinai's book [Sin] and also some work of Yuri [Yu1, Yu2]). See also the results of Jakobson and Newhouse [JN] for two-dimensional maps.

Also, there exist a few "tricks" to improve the spectral properties of the transfer operator. One of them involves "regularising" the operator (an instance of this method is the kneading operator approach presented in Section 5, see also the work of Rugh [Rug3] on analytic maps with a neutral fixed point). Finally, some of the results discussed involve constructing a transfer operator which is not directly associated to the dynamics to be studied ([Li3], [LSV2], [Po2]) but to a combination of a suitable iterate n of the dynamics and some "noise," letting the noise level ϵ go to zero and taking a suitable iterate $N(\epsilon)$. In some sense the method used by Chernov [Ch1] and Hu [Hu] also involves this idea.

5. Walking around a picturesque neighbourhood: One-dimensional dynamics and kneading operators

We discuss here a relatively recent strategy, inspired by the Milnor-Thurston kneading matrix, to implement the spectral approach (control of the essential spectral radius and link with dynamical zeta function) which has been studied in one-dimension. Let $I \subset \mathbb{R}$ be a compact interval, and Ω be a finite set of indices. For each $\omega \in \Omega$, let $I_\omega \subset I$ be an open interval, $\psi_\omega : I_\omega \to \psi_\omega(I_\omega)$ a homeomorphism (setting also $\epsilon_\omega = 1$ if ψ_ω preserves orientation, $\epsilon_\omega = -1$ otherwise). Finally, let $g_\omega : I \to \mathbb{C}$ be of bounded variation (in particular, dg_ω is a complex measure), continuous, and supported in I_ω. The assumption that each g_ω is both continuous and supported in I_ω is technical, to avoid boundary terms when integrating by parts. It is in general not satisfied if $g_\omega = |\psi'_\omega|$ with ψ_ω the inverse branches of a single map. Fortunately, it can basically be suppressed for Theorem 5.1 (see [Bal]) and greatly weakened for Theorem 5.2 [Rey].

We define a transfer operator acting on the space $\mathcal{B}$ of functions $\varphi : I \to \mathbb{C}$ of bounded variation modulo countable sets (or just on bounded functions) by

$$\mathcal{L}\varphi = \sum_{\omega\in\Omega} (\varphi \circ \psi_\omega)\, g_\omega\,. \tag{5.1}$$

We also introduce a formal dual of $\mathcal{L}$

$$\widehat{\mathcal{L}}\varphi = \sum_{\omega\subset\Omega} \epsilon_\omega\, (\varphi \circ \psi_\omega^{-1})\, (g_\omega \circ \psi_\omega^{-1})\,. \tag{5.2}$$

The main example is when the ψ_ω are the local inverse branches of a piecewise monotone interval map f, i.e., there is a partition of a compact interval $I = [a_0, a_n]$, given by $a_0 < a_1 < \ldots < a_N$, and $f : I \to I$ so that $f|_{[a_i,a_{i+1}]}$ is continuous and monotone, $\Omega = \{1, \ldots, N\}$, $\psi_\omega = (f|_{[a_{\omega-1},a_\omega]})^{-1}$, and $g_\omega = g \circ \psi_\omega$. Then the formal dual is just a weighted composition operator $\widehat{\mathcal{L}}\varphi = (\varphi \circ f) g \epsilon_f$, where ϵ_f is the "sign of the slope of f."

Finally, writing $\|\cdot\|_\infty$ for the operator norm of $\mathcal{L}$ or $\widehat{\mathcal{L}}$ acting on bounded functions, we define

$$R := \lim_{n\to\infty} (\|\mathcal{L}^n\|_\infty)^{1/n}\,, \quad \widehat{R} := \lim_{n\to\infty} (\|\widehat{\mathcal{L}}^n\|_\infty)^{1/n}\,. \tag{5.3}$$

Up to exchanging $\mathcal{L}$ and $\widehat{\mathcal{L}}$, we can assume to fix ideas that $\widehat{R} \le R$. The inequality $\widehat{R} \le R$ is automatic in the setting of piecewise monotone and piecewise continuous interval maps: In this case $\widehat{R} = \lim_{n\to\infty} (\sup_x |g^{(n)}(x)|)^{1/n}$. (We write $g^{(n)}(x) = \prod_{k=0}^{n-1} g(f^k x)$.)

5.A Bounding the essential spectral radius.

THEOREM 5.1 (SPECTRUM OF $\mathcal{L}$ ON BV, RUELLE). *Assume $\widehat{R} \le R$.*
(1) The essential spectral radius of $\mathcal{L}$ on BV is not larger than $\widehat{R}$.
(2) The spectral radius $\rho(\mathcal{L})$ of $\mathcal{L}$ on BV satisfies $\rho(\mathcal{L}) \le R$.

(3) If all functions g_ω are real and nonnegative, then $\rho(\mathcal{L}) = R$. If, additionally, $\widehat{R} < R$ then R is an eigenvalue with nonnegative eigenfunction φ_g, and there is an eigenfunctional ν_g of $\mathcal{L}^*$ for R, with ν_g a Borel probability measure and $\nu_g(\varphi_g) > 0$. The measure $\varphi_g \nu_g$ is f-invariant if the ψ_ω are the inverse branches of a given f.

There is no "hyperbolicity" or "expansion" assumption in Theorem 5.1. Note however that if the ψ_ω are the inverse branches of f a piecewise differentiable map, and $g = 1/|f'|$ then $R = 1$, and the strict inequality $\widehat{R} < R$ is satisfied if $|f'| \geq \lambda > 1$. Under such assumptions, the result above had been proved previously by Hofbauer-Keller [HK] (see also Baladi-Keller [BK] and Ruelle [Ru7]).

PROOF OF THEOREM 5.1. (1) We start by sketching Ruelle's elegant conceptual proof [Ru8] of the upper bound for the essential spectal radius, which emphasizes the role played by the Leibniz formula (also essential in the proof of (4.8)). See [Bal] for more details.

The key remark is that addition of a compact operator does not change the essential spectral radius of a bounded operator. Let $d : \mathcal{B} \to C(I)^*$ be the Banach space isomorphism between functions of bounded variation and signed (or complex) finite measures. We shall prove that the essential spectral radius of the conjugated operator $\mathcal{M} = d\mathcal{L}d^{-1}$ acting on $C(I)^*$ is at most $\widehat{R}$ by writing $\mathcal{M} = \mathcal{M}_1 + \mathcal{M}_2$, with $\mathcal{M}_2$ compact and $\mathcal{M}_1$ having spectral radius at most $\widehat{R}$.

By definition,

$$\begin{aligned} \mathcal{M}(\mu) &= d\mathcal{L}(d^{-1}\mu) \\ &= d \sum_\omega g_\omega \, (d^{-1}\mu) \circ \psi_\omega \, . \end{aligned} \tag{5.4}$$

We may assume that the functions appearing in the sum have at most regular discontinuities. Therefore, the Leibniz rule for functions of bounded variation, together with a change of variable formula, produce a decomposition $\mathcal{M} = \mathcal{M}_1 + \mathcal{M}_2$, where $\mathcal{M}_1$ has spectral radius on $C(I)^*$ bounded by $\widehat{R}$ and where

$$\mathcal{M}_2 \mu = \sum_\omega (d^{-1}\mu) \circ \psi_\omega \, dg_\omega \tag{5.5}$$

is compact. Compactness of $\mathcal{M}_2$ follows from finiteness of the sum and an intuitively obvious property of "compactness of integration." In the case when the ψ_ω are inverse branches of f, e.g., the bound on the spectral radius of $\mathcal{M}_1$ is an easy exercise left to the reader, since

$$\mathcal{M}_1 \mu = f_*(\epsilon_f \cdot g)\mu \, . \tag{5.6}$$

(2) Since we have just shown that $\rho_{\mathrm{ess}}(\mathcal{L}) \leq \widehat{R} \leq R$, to check that $\rho(\mathcal{L}) \leq R$ it suffices to see that whenever λ is an eigenvalue of $\mathcal{L}$ on $\mathcal{B}$ then $|\lambda| \leq R$. This is evident, since λ is then an eigenvalue for $\mathcal{L}$ acting on bounded functions modulo countable sets, and R is the spectral radius on that space.

(3) To construct $\varphi_g \in \mathcal{B}$ and a Borel probability measure ν_g with $\mathcal{L}(\varphi_g) = R\varphi_g$ and $\mathcal{L}^*(\nu_g) = R\nu_g$, one can use general remarks on positive operators, following Ruelle, Keller, and others (see, e.g., the proof of the Ruelle-Perron-Frobenius theorem in [Bal1]). □

5.B Kneading operators and sharp traces.

Up to this point, the dynamical zeta functions and generalised Fredholm determinants whose analytical properties were related to the spectral properties of transfer operators involved sums over all periodic points, requiring in particular the set of periodic points to be at most countable. In their pioneering paper on one-dimensional dynamics, Milnor and Thurston [MT] associated to any piecewise monotone interval map $f : I \to I$ (with finitely many, say N, monotonicity intervals) a *negative zeta function*

$$\zeta^-(z) = \exp \sum_{n=1}^{\infty} \frac{z^n}{n} 2 \, \# \operatorname{Fix}^- f^n \,, \tag{5.7}$$

where the set of *negative fixed points* of f^n is

$$\operatorname{Fix}^- f^n = \{x \in I \mid f^n x = x \,, f^n \text{ monotone decreasing in a neighbourhood of } x\} \,.$$

(See also the notes on *Zeta functions* by M. Pollicott.) The important fact, of course, is that $\operatorname{Fix}^- f^n$ is a finite set. The principle "what goes down must go up" explains why it is natural to double the negative periodic points. If f is piecewise *expanding,* one can show that $\zeta^-(z)$ is just the usual unweighted dynamical zeta function (1.1) of f, up to a simple polynomial factor.

Milnor-Thurston [MT] proved (using a homotopy argument) the surprising equation

$$\zeta^-(z) \det(1 - D(z)) = 1 \,, \tag{5.8}$$

where $D(z)$ is the *kneading matrix,* a finite $(N+1) \times (N+1)$ matrix, with coefficients power series associated to the itineraries of the turning points. These power series (the *kneading invariants*) embody a rather complete description of the map f, and the one-dimensional kneading theory is by now well developed, see, e.g., de Melo-van Strien [MS]. (Milnor and Thurston's version of (5.8) involves a polynomial correction, due to the fact that they worked with an $N-1 \times N-1$ matrix.)

One can rewrite $\zeta^-(z)$ as a *Lefschetz zeta function,* $\zeta^L(z)$ where all periodic points x are counted, but with a weight $L(x) \in \{0, -1, 1\}$ (cancellations may occur, in particular in homtervals). This second formulation (Baladi-Ruelle [BR1]) makes it easier to define a weighted negative (Lefschetz) dynamical zeta function especially if the weight is locally constant. Formula (5.8) was extended to the weighted and "nonfunctional" case (ψ_ω, g_ω) situations in a series of papers (in particular Baladi-Ruelle [BR1], Ruelle [Ru8], Baladi-Ruelle [BR2]). The "usual" relationship between the poles of the zeta function and the inverse eigenvalues of a transfer operator acting on functions of bounded variation (or C^r with r-th derivative of bounded variation, Ruelle [Ru8]) was established in this framework. Again we limit ourselves to simplest statements. We shall use the sign function

$$\operatorname{sgn}(x) = -1 \text{ if } x < 0 \,, \operatorname{sgn}(x) = 0 \text{ if } x = 0 \,, \operatorname{sgn}(x) = 1 \text{ if } x > 0 \,, \tag{5.9}$$

which has the property that $\frac{1}{2} d(\operatorname{sgn})$ is δ_0 the Dirac mass at 0. Define also the *sharp trace* of the data $(\Omega, I_\omega, \psi_\omega, g_\omega)$ by

$$\operatorname{tr}^{\#} \mathcal{L} := \sum_{\omega \in \Omega} \int_{I_\omega} dg_\omega(x) \, \frac{1}{2} \operatorname{sgn}(\psi_\omega x - x) \,, \tag{5.10}$$

Definition (5.10) clearly extends to the iterates $\mathcal{L}^n$. The sharp trace has the trace property

$$\text{tr}^{\#}(\mathcal{L}_1\mathcal{L}_2) = \text{tr}^{\#}(\mathcal{L}_2\mathcal{L}_1) \tag{5.11}$$

for any transfer operators $\mathcal{L}_1$, $\mathcal{L}_2$ of the form (5.1).

THEOREM 5.2 (SHARP TRACES AND SHARP DETERMINANTS [BR2]). *Assume that $\widehat{R} \le R$. The sharp determinant*

$$det^{\#}(1 - z\mathcal{L}) = \exp - \sum_{m=1}^{\infty} \frac{z^m}{m} tr^{\#}\mathcal{L}^m . \tag{5.12}$$

defines a holomorphic function in the disc of radius $\widehat{R}^{-1}$ where its zeroes are exactly the inverses of the eigenvalues of $\mathcal{L}$ acting on $\mathcal{B}$ of modulus at least $\widehat{R}$. The order of λ^{-1} as a zero of $det^{\#}(1 - z\mathcal{L})$ coincides with the algebraic multiplicity of λ as an eigenvalue of $\mathcal{L}$.

An application of integration by parts and change of variables rule in an integral yields $\text{tr}^{\#}\widehat{\mathcal{L}} = -\text{tr}^{\#}\mathcal{L}$. We thus have the *functional equation*

$$\det{}^{\#}(1 - z\widehat{\mathcal{L}}) = \frac{1}{\det^{\#}(1 - z\mathcal{L})} . \tag{5.13}$$

Note also that one can rewrite $\det^{\#}(1 - z\mathcal{L})$ as a (weighted) Lefschetz zeta function whenever the set of "periodic orbits" is countable (using $d\,(\text{sgn}) = \delta_0$). More precisely, if $\text{Fix}\psi^{(n)}$ is finite for each relevant composition $\psi^{(n)} = \psi_{\omega_n} \circ \cdots \circ \psi_{\omega_1}$ then

$$\text{tr}^{\#}\mathcal{L}^n = \sum_{\omega_1, \dots, \omega_n} \sum_{x \in \text{Fix}\psi^n} L(x, \psi^{(n)}) g^{(n)}(x) ,$$

where the Lefschetz number $L(x, \psi)$, for x an isolated fixed point of a monotone function $\psi : (a, b) \to \mathbb{R}$ is defined by:

$$L(x, \psi) = \begin{cases} 0 & \text{if graph of } \psi \text{ doesn't cross the diagonal at } x, \\ \epsilon_\psi \lim_{y \to x} \frac{\text{sgn}(\psi y - y)}{\text{sgn}(x - y)} & \text{otherwise,} \end{cases}$$

with $\epsilon_\psi = 1$ if ψ preserves orientation and -1 otherwise.

SKETCH OF PROOF OF THEOREM 5.2. The proof is by regularisation. Specifically, it uses a family of almost trace class operators, the *kneading operators* $\mathcal{D}(z)$, which are analogues of Milnor and Thurston's kneading matrix. These operators act on $L^2(d\mu)$ for an auxiliary measure $\mu = \sum_\omega |dg_\omega| + \sum_\omega |d(g_\omega \circ \psi_\omega^{-1})|$, guaranteeing the existence of the Radon-Nikodym derivatives $dg_\omega/d\mu$. For any z not in the spectrum of $\mathcal{L}$, the operator $\mathcal{D}(z)$ is defined by

$$\mathcal{D}(z) = z\mathcal{N}(1 - z\mathcal{L})^{-1}\mathcal{S} , \tag{5.14}$$

where the (smoothness improving, compact) integration operator S is defined by

$$\mathcal{S}\varphi(x) = \int_I \frac{1}{2}\text{sgn}\,(x - y)\varphi(y)\, d\mu(y) , \tag{5.15}$$

and where the auxiliary operator $\mathcal{N}$ is given by

$$\mathcal{N}\varphi = \sum_{\omega\in\Omega}(\varphi\circ\psi_\omega)\frac{dg_\omega}{d\mu}\,. \tag{5.16}$$

We define $\widehat{\mathcal{D}}(z)$ analogously (replacing g_ω, ψ_ω and $\mathcal{M}$ by $\epsilon_\omega g_\omega\circ\psi_\omega^{-1}$, ψ_ω^{-1} and $\widehat{\mathcal{M}}$).

Since $\mathcal{D}(z)\varphi(x) = \int \mathcal{D}_{xy}(z)\varphi(y)\,d\mu(y)$ has a bounded kernel, whenever $1/z$ is not in the spectrum of $\mathcal{L}$, it is a Hilbert-Schmidt operator in $L^2(d\mu)$ so that the regularised determinant $\det_2(1+\mathcal{D}(z))$ of order 2 is well-defined (Simon [Sim]). We may thus set

$$\det{}^*(1+\mathcal{D}(z)) := e^{\int \mathcal{D}_{xx}(z)\,d\mu(x)}\det{}_2(1+\mathcal{D}(z))\,.$$

The key identity in the proof of Theorem 5.2 is the following analogue of (5.8):

$$\det{}^{\#}(1-z\mathcal{L})\det{}^*(1+\mathcal{D}(z)) = 1\,. \tag{5.17}$$

The first proof of (5.17) in Baladi-Ruelle [BR2] was by a series resummation argument (involving repeated use of integration by parts and change of variables). A more conceptual proof can be extracted from the method used in a one-dimensional complex partial analogue of Theorem 5.2 (Baladi et al. [BKRS]). We do not state the complex result for lack of space, mentioning only that $\mathrm{sgn}(x)/2$ is replaced by the function $\sigma(z) = 1/(\pi z)$, and derivation is replaced by $\bar\partial$ (using in particular $\bar\partial\sigma = \delta_0$), and that in the simplest cases we have the formula $\mathrm{tr}^{\#}\mathcal{L} = \sum_{\omega\in\Omega}\sum_{x\in\mathrm{Fix}\psi_\omega(x)} g_\omega(x)/(1-\partial\psi_\omega(x))$ (note the absence of absolute value).

We sketch the conceptual proof of (5.17) now: after unifying the sharp trace (5.10) of transfer operators with the "trace" of kernel operators defined by

$$\mathrm{tr}^*\mathcal{D}(z) := \int \mathcal{D}_{xx}(z)\,d\mu(x)\,,$$

we check that this unified trace, noted tr^*, satisfies (5.11). Hence, the usual determinant formulas hold for $\det^*(1-\mathcal{M}) = \exp - \sum_n \frac{1}{n}\mathrm{tr}^*\mathcal{M}^n$. The argument is then quite literally a three-line proof:

$$\begin{aligned}\det{}^*(1+\mathcal{D}(z))\det{}^{\#}(1-z\mathcal{L}) &= \det{}^*(1+z\mathcal{N}(1-z\mathcal{L})^{-1}S)\det{}^*(1-z\mathcal{L})\\ &= \det{}^*(1+zS\mathcal{N}(1-z\mathcal{L})^{-1})\det{}^*(1-z\mathcal{L})\\ &= \det{}^*(1+z(S\mathcal{N}-\mathcal{L})) = 1\,,\end{aligned}$$

where we used $\mathrm{tr}^{\#}\mathcal{L} = \mathrm{tr}^*(S\mathcal{N})$, and more generally the less trivial identities $\mathrm{tr}^*(S\mathcal{N}-\mathcal{L})^n = 0$ for $n\geq 1$, to get the last equality. □

Notational warning: We use g instead of $\exp(g)$ for the weight function in these notes, because we want to admit the possibility that $g(x) = 0$. We have reserved the letter $\mathcal{F}$ for the function spaces in [Yo2, Yo3] and use $\mathcal{C}$ for the Lipschitz functions in the subshift of finite type model.

Final Comment: These notes and in particular the bibliography do not have any pretension to being exhaustive. We are grateful to the referee for pointing out some glaring omissions and apologize to authors for any remaining ones. The book [Ba1] contains a more detailed account of some topics.

References

[ADU] J. Aaronson, M. Denker, and M. Urbański, *Ergodic theory for Markov fibered systems and parabolic rational maps*, Trans. Amer. Math. Soc. **337** (1993), 495–548.

[A] J.F. Alves, *SRB measures for nonhyperbolic systems with multidimensional expansion*, Annales scient. Ecole normale sup. (4) **33** (2000), 1–32.

[ABV] J.F. Alves, C. Bonatti, and M. Viana, *SRB measures for partially hyperbolic systems whose central direction is mostly expanding*, Invent. Math. **140** (2000), 351–398.

[Bal] V. Baladi, *Positive Transfer Operators and Decay of Correlations*, To appear (Vol. 16, Advanced Series in Nonlinear Dynamics), World Scientific, Singapore.

[BBM] V. Baladi, M. Benedicks, and V. Maume, *Almost sure rates of mixing for i.i.d. unimodal maps*, Preprint (1999).

[BK] V. Baladi and G. Keller, *Zeta functions and transfer operators for piecewise monotone transformations*, Comm. Math. Phys. **127** (1990), 459–479.

[BKRS] V. Baladi, A. Kitaev, D. Ruelle, and S. Semmes, *Sharp determinants and kneading operators for holomorphic maps*, Proc. Steklov Inst. Math. **216** (1997), 186–228.

[BKS] V. Baladi, A. Kondah, and B. Schmitt, *Random correlations for small perturbations of expanding maps*, Random & Computational Dynamics **4** (1996), 179–204.

[BR1] V. Baladi and D. Ruelle, *An extension of the theorem of Milnor and Thurston on the zeta functions of interval maps*, Ergodic Theory Dynam. Systems **14** (1994), 621–632.

[BR2] V. Baladi and D. Ruelle, *Sharp determinants*, Invent. Math. **123** (1996), 553–574.

[BV] V. Baladi and M. Viana, *Strong stochastic stability and rate of mixing for unimodal maps*, Annales scient. Ecole normale sup. (4) **29** (1996), 483–517.

[BaY] V. Baladi and L.-S. Young, *On the spectra of randomly perturbed expanding maps*, (see also Erratum, Comm. Math. Phys., **166**, 219–220 (1994)), Comm. Math. Phys. **156** (1993), 355–385.

[BeY1] M. Benedicks and L.-S. Young, *Absolutely continuous invariant measures and random perturbations for certain one-dimensional maps*, Ergodic Theory Dynam. Systems **12** (1992), 13–37.

[BeY2] M. Benedicks M. and L.-S. Young, *Sinai-Bowen-Ruelle measures for certain Hénon maps*, Invent. Math. **112** (1993), 541–576.

[BeY3] M. Benedicks and L.-S. Young, *Markov extensions and decay of correlations for certain Hénon maps*, Astérisque **261** (2000), 13–56.

[Bil] P. Billingsley, *The Lindeberg-Lévy theorem for martingales*, Proc. Amer. Math. Soc. **12** (1961), 788–792.

[Bir] G. Birkhoff, *Lattice theory (3rd ed.)*, Amer. Math. Soc. Colloquium Publications, Amer. Math. Soc., Providence, RI, 1967.

[Bov] C. Bonatti and M. Viana, *SRB measures for partially hyperbolic systems whose central direction is mostly contracting*, Israel J. Math. **115** (2000), 157–193.

[Bow] R. Bowen, *Equilibrium States and the Ergodic Theory of Anosov Diffeomorphisms*, Springer Lect. Notes in Math. **470**, Berlin-Heidelberg-New York, 1975.

[Br] X. Bressaud, *Subshifts on an infinite alphabet*, Ergodic Theory Dynam. Systems **19** (1999), 1175–1200.

[BFG] X. Bressaud, R. Fernández, and A. Galves, *Decay of correlations for non Hölderian dynamics: a coupling approach*, Electron. J. Probab. **4** (1999), 19 pp..

[BLS] H. Bruin, S. Luzzatto, and S. van Strien, *Decay of correlations in one-dimensional dynamics*, Preprint (1999).

[Bu1] J. Buzzi, *Exponential decay of correlations for random Lasota-Yorke maps*, Comm. Math. Phys **208** (1999), 25–54.

[Bu2] J. Buzzi, *Piecewise expanding piecewise and $\mathbb{R}$-analytic mappings of the plane*, Preprint IML (1998), `http://iml.univ-mrs.fr/~buzzi`, to appear Ergodic Theory Dynam. Systems.

[Bu3] J. Buzzi, *No or infinitely many a.c.i.p. for piecewise expanding C^r maps in higher dimensions*, Preprint (2000).

[BM] J. Buzzi and V. Maume-Deschamps, *Decay of correlations for piecewise expanding maps in higher dimensions*, Preprint `mp_arc` (1999).

[Cas] A. Castro, *Backward inducing and exponential decay of correlations for partially hyperbolic attractors with mostly contracting direction*, Ph.D. thesis, IMPA (1998).

[Ch1] N. Chernov, *Limit theorems and Markov approximations for chaotic dynamics*, Probab. Theory Related Fields **101** (1995), 321–362.

[Ch2] N. Chernov, *Markov approximations and decay of correlations*, Ann. of Math. (2) **147** (1998), 269–324.

[Ch3] N. Chernov, *Statistical properties of piecewise smooth hyperbolic systems in high dimensions*, Discrete Contin. Dynam. Systems **5** (1999), 425–448.

[Ch4] N. Chernov, *Decay of correlations and dispersing billiards*, J. Statist. Phys. **94** (1999), 513–556.

[CMT] N. Chernov, R. Markarian, and S. Troubetzkoy, *Conditionally invariant measures for Anosov maps with small holes*, Ergodic Theory Dynam. Systems **18** (1998), 1049–1074.

[CMS] P. Collet, S. Martínez, and B. Schmitt, *The Pianigiani-Yorke measure for topological Markov chains*, Israel J. Math **97** (1997), 61–70.

[Co] W. Cowieson, *Piecewise smooth expanding maps in* $\mathbb{R}^d$, Ph.D. thesis UCLA (1999).

[Do1] D. Dolgopyat, *On decay of correlations in Anosov flows*, Annals of Math. **147** (1998), 325–355.

[Do2] D. Dolgopyat, *Prevalence of rapid mixing in hyperbolic flows*, Ergodic Theory Dynam. Systems **18** (1998), 1097–1114.

[Do3] D. Dolgopyat, *Prevalence of rapid mixing-II: topological prevalence*, Preprint (1998), to appear Ergodic Theory Dynam. Systems.

[Do4] D. Dolgopyat, *On dynamics of mostly contracting diffeomorphisms*, Preprint (1998), to appear Comm. Math. Phys.

[Do5] D. Dolgopyat, *On mixing properties of compact group extensions of hyperbolic systems*, Preprint (2000).

[DS] N. Dunford and J.T. Schwartz, *Linear Operators, Part I: General theory*, Wiley Classics Library Edition, Wiley, New York, 1988.

[Dur] R. Durrett, *Probability: theory and examples*, Second edition, Duxbury Press, Belmont, Ca, 1996.

[F] B. Fayad, *Polynomial correlations for a class of smooth flows on the two torus*, Preprint (2000).

[FS] P. Ferrero and B. Schmitt, *Produits aléatoires d'opérateurs matrices de transfert*, Probab. Theory Related Fields, **79** (1988), 227–248.

[Fr1] D. Fried, *The zeta functions of Ruelle and Selberg I*, Ann. Sci. École Norm. Sup. (4) **19** (1986), 491-517.

[Fr2] D. Fried, *The flat-trace asymptotics of a uniform system of contractions*, Ergodic Theory Dynam. Systems **15** (1995), 1061–1073.

[Fr3] D. Fried, *Meromorphic zeta functions for analytic flows*, Comm. Math. Phys. **174** (1995), 161–190.

[Fur] H. Furstenberg, *Stationary processes and prediction theory*, Princeton U. Press, Princeton, NJ, 1960.

[Gr1] A. Grothendieck, *Produits tensoriels topologiques et espaces nucléaires*, Mem. Amer. Math. Soc. **16**), Amer. Math. Soc., 1955.

[Gr2] A. Grothendieck, *La théorie de Fredholm*, Bull. Soc. Math. France **84** (1956), 319–384.

[Ha1] N.T.A. Haydn, *Gibbs' Functionals on Subshifts*, Comm. Math. Phys. **134** (1990), 217–236.

[Ha2] N.T.A. Haydn, *Meromorphic extension of the zeta function for Axiom A flows*, Ergodic Theory Dynam. Systems **10** (1990), 347–360.

[Ho] F. Hofbauer, *Piecewise invertible dynamical systems*, Probab. Theory Related Fields **72** (1986), 359–386.

[HoK] F. Hofbauer and G. Keller, *Ergodic properties of invariant measures for piecewise monotonic transformations*, Math. Z. **180** (1982), 119–140.

[Hu] H. Hu, *Decay of correlations for piecewise smooth maps with indifferent fixed points*, Preprint (1998).

[HuY] H. Hu and L.-S. Young, *Nonexistence of SBR measures for some diffeomorphisms that are "almost Anosov"*, Ergodic Theory Dynam. Systems **15** (1995), 67–76.

[Is1] S. Isola, *Renewal sequences and intermittency*, J. Stat. Phys. **97** (1999), 263–280.

[Is2] S. Isola, *On the rate of convergence to equilibrium for countable ergodic Markov chains*, Preprint (1997).

[JN] M. Jakobson and S. Newhouse, *Asymptotic measures for hyperbolic piecewise smooth mappings of a rectangle*, Astérisque **261** (2000), 103–159.

[Ke] G. Keller, *On the rate of convergence to equilibrium in one-dimensional systems*, Commun. Math. Phys. **96** (1984), 181–193.

[KN] G. Keller and T. Nowicki, *Spectral theory, zeta functions and the distribution of periodic points for Collet-Eckmann maps*, Comm. Math. Phys. **149** (1992), 31-69.

[Kit] A. Kitaev, *Fredholm determinants for hyperbolic diffeomorphisms of finite smoothness*, (See also Corrigendum, 1717–1719), Nonlinearity **12** (1999), 141–179.

[KS] K. Krzyżewski and W. Szlenk, *On invariant measures for expanding differentiable mappings*, Studia Math. **23** (1969), 83–92.

[LaY] A. Lasota and J. A. Yorke, *On the existence of invariant measures for piecewise monotonic transformations*, Trans. Amer. Math. Soc. **186** (1973), 481–488.

[LiY] T.Y. Li and J. A. Yorke, *Ergodic transformations from an interval to itself*, Trans. Amer. Math. Soc. **235** (1978), 183–192.

[Lin] T. Lindvall, *Lectures on the coupling method*, Wiley Series in Probability and Mathematical Statistics, Wiley, New York, 1992.

[Li1] C. Liverani, *Decay of correlations*, Ann. of Math. **142** (1995), 239–301.

[Li2] C. Liverani, *Central limit theorem for deterministic systems*, Proceedings of the International Conference on Dynamical Systems (Montevideo, 1995), Longman, Harlow, 1996.

[Li3] C. Liverani, *Flows, random perturbations and rate of mixing*, Ergodic Theory Dynam. Systems **18** (1998), 1421–1446.

[LSV1] C. Liverani, B. Saussol, and S. Vaienti, *Conformal measure and decay of correlation for covering weighted systems*, Ergodic Theory Dynam. Systems **18** (1998), 1399–1420.

[LSV2] C. Liverani, B. Saussol, and S. Vaienti, *A probabilistic approach to intermittency*, Ergodic Theory Dynam. Systems **19** (1999), 671–686.

[Mau1] V. Maume-Deschamps, *Equilibrium states for non Hölderian random dynamical systems*, Random Comput. Dynam. **5** (1997), 319–335.

[Mau2] V. Maume-Deschamps, *Correlations decay for Markov maps on a countable states space*, Preprint `http://www.unige.ch/math/biblio/preprints/pp98.html` (1998), to appear Ergodic Theory Dynam. Systems.

[Mau3] V. Maume-Deschamps, *Projective metrics and mixing properties on towers*, Preprint (1999), to appear Ergodic Theory Dynam. Systems.

[Ma1] D. Mayer, *On the thermodynamic formalism for the Gauss map*, Comm. Math. Phys. **130** (1990), 311–333.

[Ma2] D. Mayer, *Continued fractions and related transformations*, Ergodic Theory, Symbolic Dynamics and Hyperbolic Spaces (T. Bedford, M. Keane, and C. Series, ed.), Oxford University Press, Oxford, 1991.

[MS] W. de Melo and S. van Strien, *One-dimensional Dynamics*, Springer-Verlag, Berlin, 1993.

[MT] J. Milnor and W. Thurston, *On iterated maps of the interval*, Lecture Notes in Mathematics **1342**, Springer, 1988.

[PaPo] W. Parry and M. Pollicott, *Zeta Functions and the Periodic Orbit Structure of Hyperbolic Dynamics*, Astérisque **187–188**, Soc. Math. de France, Paris, 1990.

[PeS] Ya.B. Pesin and Ya.G. Sinai, *Gibbs measures for partially hyperbolic attractors*, Ergodic Theory Dynam. Systems **2** (1982), 417–438.

[Po1] M. Pollicott, *On the rate of mixing of Axiom A flows*, Invent. Math. **81** (1985), 413–426.

[Po2] M. Pollicott, *Rates of mixing for potentials of summable variation*, Preprint (1997).

[Rey] L. Rey-Bellet, *Unpublished notes on kneading operators (1995).*

[Ru1] D. Ruelle, *Zeta functions for expanding maps and Anosov flows*, Invent. Math. **34** (1976), 231–242.

[Ru2] D. Ruelle, *Applications conservant une mesure absolument continue*, Comm. Math. Phys. **55** (1977), 47–51.

[Ru3] D. Ruelle, *Thermodynamic Formalism*, Addison Wesley, Reading MA, 1978.

[Ru4] D. Ruelle, *One-dimensional Gibbs states and Axiom A diffeomorphisms*, J. Diff. Geom. **25** (1987), 117–137.

[Ru5] D. Ruelle, *The thermodynamic formalism for expanding maps*, Comm. Math. Phys. **125** (1989), 239–262.

[Ru6] D. Ruelle, *An extension of the theory of Fredholm determinants*, Inst. Hautes Etudes Sci. Publ. Math. **72** (1990), 175–193.

[Ru7] D. Ruelle, *Dynamical zeta functions for piecewise monotone maps of the interval*, Amer. Math. Soc., Providence, RI, 1994.

[Ru8] D. Ruelle, *Sharp determinants for smooth interval maps*, Proceedings International Congress on Dynamical Systems (Montevideo, 1995), F. Ledrappier, J. Lewowicz, and S. Newhouse eds., Pitman Research Notes in Mathematics Series **362**, Addison Wesley Longman, Harlow, 1996.

[Ru10] D. Ruelle, *Differentiation of SRB states*, Comm. Math. Phys. **189** (1997), 365–371.

[Rug1] H.H. Rugh, *The correlation spectrum for hyperbolic analytic maps*, Nonlinearity **5** (1992), 1237–1263.

[Rug2] H.H. Rugh, *Generalized Fredholm determinants and Selberg zeta functions for Axiom A dynamical systems*, Ergodic Theory Dynam. Systems **16** (1996), 805–819.

[Rug3] H.H. Rugh, *Intermittency and regularized Fredholm determinants*, Invent. Math. **135** (1999), 1–25.

[Ry] M. Rychlik, *Bounded variation and invariant measures*, Studia Math. **LXXVI** (1983), 69–80.

[Sar] O. Sarig, *Thermodynamic formalism for countable Markov shifts*, Ergodic Theory Dynam. Systems **19** (1999), 1565–1593.

[Sau] B. Saussol, *Absolutely continuous invariant measures for multidimensional expanding maps*, Israel J. Math **116** (2000), 223–248.

[Sim] B. Simon, *Trace Ideals and their Applications*, Cambridge University Press, Cambridge, 1979.

[Sin] Ya.G. Sinai, *Topics in Ergodic Theory*, Princeton University Press, Princeton, 1994.

[Tsu1] M. Tsujii, *Absolutely continuous invariant measures for piecewise real-analytic expanding maps on the plane*, Comm. Math. Phys. **208** (2000), 605–622.

[Tsu2] M. Tsujii, *Piecewise expanding maps on the plane with singular ergodic properties*, Preprint (1998), to appear Ergodic Theory Dynam. Systems.

[V1] M. Viana, *Multidimensional nonhyperbolic attractors*, Inst. Hautes Etudes Sci. Publ. Math. **85** (1997), 63–96.

[V2] M. Viana, *Stochastic dynamics of deterministic systems*, Col. Bras. de Matemática, 1997.

[Wal] P. Walters, *An Introduction to Ergodic Theory*, Springer-Verlag, New York, 1982.

[WY] Q. Wang and L.-S. Young, *Analysis of a class of strange attractors*, Preprint (1999).

[Yo1] L.-S. Young, *Decay of correlations for certain quadratic maps*, Comm. Math. Phys. **146** (1992), 123–138.

[Yo2] L.-S. Young, *Statistical properties of systems with some hyperbolicity including certain billiards*, Annals. of Math. **147** (1998), 585–650.

[Yo3] L.-S. Young, *Recurrence times and rates of mixing*, Israel. J. Math. **110** (1999), 153–188.

[Yu1] M. Yuri, *On the convergence to equilibrium states for certain nonhyperbolic systems*, Ergodic Theory Dynam. Systems **17** (1997), 977–1000.

[Yu2] M. Yuri, *Statistical properties for nonhyperbolic maps with finite range structure*, Trans. Amer. Math. Soc. **352** (2000), 2369–2388.

Mathématiques, Université de Genève, CH-1211 Geneva 24, SWITZERLAND

Current address: Laboratoire de Mathématiques, CNRS, UMR 8628, Université de Paris-Sud, F-91405 Orsay, FRANCE

E-mail address: viviane.baladi@math.u-psud.fr

Proceedings of Symposia in Pure Mathematics
Volume **69**, 2001

Recent Results about Stable Ergodicity

Keith Burns, Charles Pugh, Michael Shub, and Amie Wilkinson

Introduction

The ergodic theory of uniformly hyperbolic, or Axiom A, diffeomorphisms has been studied extensively, beginning with the pioneering work of Anosov, Sinai, Ruelle and Bowen [**An, Si, Rue, Bow**]. Much work in dynamical systems since then, including this paper, has been directed toward extending their results beyond Axiom A.

Three basic ways to relax the condition of uniform hyperbolicity on a diffeomorphism $f : M \to M$ are:

(a) *Nonuniform hyperbolicity.* Consider Lyapunov exponents and assume that none of them are zero. Most orbits are hyperbolic, but the hyperbolicity is nonuniform.

(b) *Partial hyperbolicity.* Permit some tangent directions on which Tf acts neutrally, but require that Tf be uniformly hyperbolic in some other tangent directions. No orbits need be completely hyperbolic, but all have uniformly hyperbolic parts.

(c) *Dominated splitting.* Keep a Tf-invariant splitting into two subbundles, but relax hyperbolicity to a domination condition: maximal expansion in one bundle is uniformly bounded by least expansion in the other.

Approach (a) is due to Pesin. For an introduction to the theory of nonuniform hyperbolicity (often called Pesin theory) we refer the reader to the survey by Barreira and Pesin in this volume. Approach (b) is due to Brin and Pesin, and is the one we follow here. Of course, it is possible to combine (a) and (b) to produce the concept of partial non-uniform hyperbolicity. Approach (c) is due to Mañé [**Mañ2**]. There is now a large and important body of work in this area. It is closely related to the results that we discuss, but lies a little outside the scope of this survey; we refer the reader to [**AlBonVi**], [**BonDí**], [**BonDíPuj**], [**BonVi**], [**Cas**], [**DíPujUr**], [**DíRo**], [**Do1**], [**He1**], [**He2**], [**PujSam**], [**Vi**].

Our central theme is, to quote [**PugSh5**], "that a little hyperbolicity goes a long way in guaranteeing stably ergodic behavior." By "a little hyperbolicity" we mean

2000 *Mathematics Subject Classification.* Primary 37D30.

The first author was supported in part by NSF Grant #DMS 9803346.
The third author was supported in part by NSF Grant #DMS-9616120.
The fourth author was supported in part by an NSF Postdoctoral Fellowship.

partial hyperbolicity, i.e. (b) above. More precisely there is a Tf-invariant splitting $TM = E^u \oplus E^c \oplus E^s$, in which at least two of the subbundles are nontrivial, such that Tf uniformly expands all vectors in E^u and uniformly contracts all vectors in E^s, while vectors in E^c are neither contracted as strongly as any nonzero vector in E^s nor expanded as strongly as any nonzero vector in E^u. We believe that the following should be true:

CONJECTURE 0.1. [**PugSh5**, Conjecture 1] and [**PugSh6**, Conjecture 1] On any compact manifold, ergodicity holds for an open and dense set of C^2 volume preserving partially hyperbolic diffeomorphisms (provided E^u and E^s are both nontrivial).

This conjecture can be split into two parts using the concept of accessibility. If we have a partially hyperbolic diffeomorphism $f : M \to M$ with associated splitting $TM = E^u \oplus E^c \oplus E^s$, let us say that a point y is accessible from a point x if there is a C^1 path from x to y whose tangent vector always lies in $E^u \cup E^s$ and vanishes at most finitely many times. Accessibility is an equivalence relation. We say that f has the *accessibility property* if there is only one equivalence class. If every measurable set which is a union of equivalence classes has either full measure or measure zero, then f has the *essential accessibility property.* It is clear that accessibility implies essential accessibility.

CONJECTURE 0.2. [**PugSh5**, Conjecture 4] and [**PugSh6**, Conjecture 2] Accessibility holds for an open and dense subset of C^2 partially hyperbolic diffeomorphisms, volume preserving or not (provided E^u and E^s are both nontrivial).

CONJECTURE 0.3. [**PugSh6**, Conjecture 3] A partially hyperbolic C^2 volume preserving diffeomorphism with the essential accessibility property is ergodic.

The paper presents the current evidence in favor of these conjectures. In Section 1, we introduce the concept of partial hyperbolicity and give examples of partially hyperbolic diffeomorphisms. In Section 2 we discuss the result of Pugh and Shub from [**PugSh6**], which is the closest approach so far to a general proof of Conjecture 0.3. Section 3 is devoted to a discussion of topological ideas which can be used to approach Conjecture 0.2. Section 4 surveys the applications that have been made of the Pugh-Shub theorem and these ideas. In particular, we will see that Conjecture 0.1 holds within several interesting families of examples. In Section 5, we discuss the pathological center foliation of the partially hyperbolic diffeomorphisms in [**ShWi2**]. In the last section, we raise a few questions, make some conjectures, and propose a vague idea of how one might proceed to prove them.

1. Partially hyperbolic diffeomorphisms

Throughout this paper, M is a compact, connected and boundaryless Riemannian manifold, and $f : M \to M$ is a C^2 diffeomorphism. We will focus here on the ergodic properties of f as it is perturbed inside $\mathrm{Diff}^2(M)$, the space of all C^2 diffeomorphisms of M. For a smooth measure μ on M, let $\mathrm{Diff}^2_\mu(M)$ be those f in $\mathrm{Diff}^2(M)$ that preserve μ. We say that $f \in \mathrm{Diff}^2_\mu(M)$ is *stably ergodic* if there is a neighborhood $\mathcal{U}$ of f in $\mathrm{Diff}^2_\mu(M)$ such that every $g \in \mathcal{U}$ is ergodic with respect to

μ.[1] More generally f is stably ... if there is a neighborhood $\mathcal{U}$ of f in $\mathrm{Diff}^2_\mu(M)$ such that every $g \in \mathcal{U}$ is ....

The first diffeomorphisms proven to be stably ergodic were Anosov diffeomorphisms [**An**]. Partial hyperbolicity is a weakening of the Anosov condition that allows for much more dynamical complexity, and yet, as we shall see, often provides enough structure for stable ergodicity. A diffeomorphism f is *partially hyperbolic* if the tangent bundle to M splits as a Tf-invariant sum:

$$TM = E^u \oplus E^c \oplus E^s,$$

with at least 2 of the subbundles in the sum nontrivial, and there exist constants $a < b < 1 < c < d$, and a Finsler structure $\|\cdot\|$ on M such that, for all $p \in M$ and all $v \in T_pM$,

$$\begin{aligned} v \in E^u(p) &\Rightarrow d\|v\| < \|T_pfv\| \\ v \in E^c(p) &\Rightarrow b\|v\| \le \|T_pfv\| \le c\|v\| \\ v \in E^s(p) &\Rightarrow \|T_pfv\| \le a\|v\|. \end{aligned}$$

The bundles E^u and E^s are called the *stable* and *unstable* bundles, respectively, and E^c the *center* bundle for f. It is not necessary to assume these bundles are continuous in the definition; continuity follows from invariance and the growth conditions given above. A similar argument shows that the splitting changes continuously when f varies in the C^1 topology; see e.g. [**HiPugSh**]. Equivalent definitions of partial hyperbolicity are given by Brin and Pesin in [**BriPe1**] and by Hirsch, Pugh, and Shub in [**HiPugSh**].

Several classically-studied dynamical systems are partially hyperbolic systems, including time-t maps of Anosov flows, frame flows for negatively curved manifolds, and certain algebraic systems. We give here a list of examples. They are grouped into catogories, which have some overlap.

Anosov diffeomorphisms. An Anosov diffeomorphism is by definition a partially hyperbolic diffeomorphism with a trivial center bundle. Every Anosov diffeomorphism is partially hyperbolic in at least three ways: first, the obvious way, with E^c trivial, E^u = the expanding bundle from the Anosov splitting, and E^s = the contracting bundle from the Anosov splitting; second, with E^s trivial, E^c = contracting bundle, and E^u = expanding bundle; third, with E^u trivial, E^c = expanding bundle, and E^s = contracting bundle. Some Anosov diffeomorphisms are even partially hyperbolic with *three* nontrivial invariant bundles; the center bundle can include invariant weakly stable or unstable directions. Unless otherwise specified, when we discuss Anosov diffeomorphisms, we will assume that it is the center bundle that is trivial. Anosov showed in 1967 that C^2 Anosov diffeomorphisms that preserve a smooth invariant measure μ are ergodic [**An**].

Time-t maps of Anosov flows. Closely related to the Anosov diffeomorphisms are the *time-t maps of Anosov flows.* Recall that a flow $\varphi_t : M \to M$ is *Anosov* if for every $t \neq 0$, the derivative $T\varphi_t$ leaves invariant a splitting $TM = E^u \oplus \langle\dot{\varphi}\rangle \oplus E^s$ with E^u and E^s uniformly expanded and contracted, respectively, by $T\varphi_t$. For every $t \neq 0$, the diffeomorphism φ_t is partially hyperbolic, with the

[1]We use the C^2 topology on $\mathrm{Diff}^2_\mu(M)$ to define stable ergodicity. We remark that, in most of the results discussed in this paper, properties that are open in $\mathrm{Diff}^r(M)$ are in fact C^1-open, and dense properties are C^r-dense.

center bundle tangent to the flow lines. Note that, because of the neutral direction $\langle\dot{\varphi}\rangle$, the time-$t$ map of an Anosov flow is not an Anosov diffeomorphism. Basic examples of Anosov flows are: geodesic flows on unit tangent bundles of closed Riemannian manifolds of negative sectional curvature, and special flows built over Anosov diffeomorphisms. Geodesic flows preserve Liouville measure, and a special flow built over an Anosov diffeomorphism f_0 preserves a smooth volume if f_0 does.

Anosov proved that for certain Anosov flows, for example the geodesic flows, the time-t map, for $t \neq 0$ is ergodic [**An**]. The precise condition that implies ergodicity of φ_t is essential accessibility, which we discuss in the next section.

Algebraic systems. Let G be a connected Lie group and let Γ be a discrete cocompact subgroup of G, i.e. a discrete subgroup such that G/Γ is compact. For $g \in G$ denote by $L_g : G/\Gamma \to G/\Gamma$ the left translation $L_g(a\Gamma) = ga\Gamma$. Let $A : G \to G$ be an automorphism such that $A(\Gamma) = \Gamma$; then A induces a diffeomorphism $A : G/\Gamma \to G/\Gamma$. An *affine diffeomorphism* of G/Γ is a map of the form

$$L_g \circ A : G/\Gamma \to G/\Gamma.$$

Since Γ is discrete and cocompact, Haar measure on G projects to a finite measure μ on G/Γ, invariant under left translations and under the action of A. Then the affine diffeomorphism $L_g \circ A$ preserves μ. Let $\mathfrak{g}$ be the Lie algebra of G and let $f = L_g \circ A : G/\Gamma \to G/\Gamma$, be an affine diffeomorphism. Then f induces the Lie algebra automorphism $l(f) : \mathfrak{g} \to \mathfrak{g}$:

$$l(f) = ad(g) \circ T_e(A),$$

where $ad(g) : \mathfrak{g} \to \mathfrak{g}$ is the adjoint action of g on $\mathfrak{g}$. The Lie algebra $\mathfrak{g}$ splits into generalized eigenspaces for $l(f)$:

$$\mathfrak{g} = \mathfrak{g}^u \oplus \mathfrak{g}^c \oplus \mathfrak{g}^s,$$

where the eigenvalues of $l(f)$ are respectively outside, on, or inside the unit circle.

The Lie subalgebra $\mathfrak{h}$ of $\mathfrak{g}$ generated by $\mathfrak{g}^u \oplus \mathfrak{g}^s$ is an ideal in $\mathfrak{g}$. The following is proved[2] in [**PugSh6**].

PROPOSITION 1.1. *The affine diffeomorphism $f = L_g \circ A$ of G/Γ is partially hyperbolic if and only if $\mathfrak{h} \neq 0$.*

The ergodicity of L_g on a homogeneous space G/Γ was studied extensively in [**Mo**] and [**BreMo**]. It follows easily from their work that the condition $\mathfrak{h} = \mathfrak{g}$ implies the ergodicity of L_g on G/Γ.

Systems that fiber over partially hyperbolic diffeomorphisms. Let $f_0 : M_0 \to M_0$ be a partially hyperbolic diffeomorphism with splitting $TM_0 = E_0^u \oplus E_0^c \oplus E_0^s$, and let $\pi : M \to M_0$ be a fiber bundle over M_0. If $f : M \to M$ is a lift of f_0, then f is partially hyperbolic, provided that:

$$\|Tf|_{\ker(T\pi)}\| \cdot \|Tf_0^{-1}|_{E_0^u}\| < 1 \quad \text{and} \quad \|Tf^{-1}|_{\ker(T\pi)}\| \cdot \|Tf_0|_{E_0^s}\| < 1.$$

Indeed, the splitting $TM = E^u \oplus E^c \oplus E^s$ for f projects under $T\pi$ to the splitting $TM_0 = E_0^u \oplus E_0^c \oplus E_0^s$ for f_0. The stable and unstable bundles are found by a standard graph transform argument, and the center bundle is $T\pi^{-1}(E_0^c)$.

---

[2] A more general proposition with Γ replaced by a closed cocompact subgroup appears in [**PugSh6**]. The arguments given there are incomplete in that generality, but are valid when the subgroup is discrete.

Falling into this category are direct products, compact group extensions of Anosov diffeomorphisms, and time-t maps of frame flows for negatively-curved manifolds.

Direct products. Let $f_0 : M_0 \to M_0$ be an Anosov diffeomorphism with associated splitting $TM_0 = E_0^u \oplus E_0^s$ and let $f_1 : M_1 \to M_1$ be a diffeomorphism of a Riemannian manifold. Suppose that no vector in TM_1 is contracted as strongly as any vector in E_0^s or expanded as strongly as any vector in E_0^u, i.e.

$$\|Tf_0^{-1}|_{E_0^s}\|^{-1} < \|Tf_1^{-1}\|^{-1} \leq \|Tf_1\| < \|Tf_0|_{E_0^u}\|.$$

Then $f_0 \times f_1$ is partially hyperbolic with $E^u = E_0^u \times \{0\}$, $E^c = \{0\} \times TM_1$, and $E^s = E_0^s \times \{0\}$. In this simple situation, the unstable and stable bundles through $(p_0, p_1) \in M_0 \times M_1$ are tangent to the submanifold $M_0 \times \{p_1\}$, and the center bundle is the tangent space to $\{p_0\} \times M_1$.

Compact group extensions of Anosov diffeomorphisms. Let G be a compact Lie group, and let $f_0 : M_0 \to M_0$ be an Anosov diffeomorphism. Each smooth function $\varphi : M_0 \to G$ defines a *skew product transformation* $f_\varphi : M_0 \times G \to M_0 \times G$ by the formula:

$$f_\varphi(p_0, g) = (f_0(p_0), \varphi(p_0)g).$$

Left translations are isometries of G in the bi-invariant metric, and so f_φ is partially hyperbolic. If f_0 preserves a smooth probability measure μ_0, then f_φ preserves the smooth probability measure $\mu = \mu_0 \times \nu_G$, where ν_G is (normalized) Haar measure on G. Skew products over f_0 are also called *G-extensions of f_0*. The set of all C^2 skew products over f_0,

$$\mathrm{Skew}(f_0, G) = \{f_\varphi \mid \varphi \in C^2(M_0, G)\},$$

is a closed subset of $\mathrm{Diff}^2_\mu(M_0 \times G)$; its connected components correspond to homotopy classes of φ.

The ergodic properties of such skew products were initially studied by Brin [**Bri2**]. He proved that ergodic diffeomorphisms form an open and dense subset of $\mathrm{Skew}(f_0, G)$ for any volume preserving Anosov diffeomorphism f_0 and any compact group G.

Frame flows. Let V be a closed oriented manifold of negative sectional curvatures, and let $M_0 = SV$ be the unit tangent bundle of V. Let M be the space of positively oriented orthonormal n-frames in TV. This gives a fiber bundle where the natural projection $\pi : M \to M_0$ takes a frame to its first vector. The associated structure group $SO(n-1)$ acts on fibers by rotating the frames, keeping the first vector fixed. In particular, we can identify each fiber M_x with $SO(n-1)$. Let $\hat{\varphi}_t : M \to M$ denote the frame flow, which acts on frames by moving their first vectors according to the geodesic flow and moving the other vectors by parallel transport along the geodesic defined by the first vector. The projection π is a semi-conjugacy from $\hat{\varphi}_t$ to the geodesic flow φ_t; i.e. $\pi \circ \hat{\varphi}_t = \varphi_t \circ \pi$ for each $t \in \mathbf{R}$. The frame flow $\hat{\varphi}_t$ preserves the measure that is (locally) a product of Liouville measure with (normalized) Haar measure on $SO(n-1)$. The time one map of the frame flow is a partially hyperbolic diffeomorphism. The center bundle has dimension $1 + \dim SO(n-1)$ and is spanned by the flow direction and the fiber direction. The map $\hat{\varphi}_1$ is an example of a principal $SO(n-1)$-bundle extension of φ_1.

Brin together with Gromov and Karcher [**Bri1, BriGr, BriK**] studied ergodicity of the frame flow. Ergodicity holds for an open and dense set of metrics that includes all metrics with constant negative curvature. The only known examples with nonergodic frame flow are very special, e.g. quotients of complex hyperbolic spaces. These results will be described in more detail in Section 4.

Anosov-like diffeomorphisms. Let $f_0 : M \to M$ be a partially hyperbolic diffeomorphism and let $f_t : M \to M$, for $t \in [0,1]$, be an isotopy through diffeomorphisms. Suppose that during the isotopy the expansion of the unstable bundle and the contraction of the stable bundle remain stronger than any expansion or contraction in the center bundle. Then f_1 will also be partially hyperbolic. These conditions can be verified by the construction of suitable cone fields. This construction first appears in Shub's example of non-Anosov stably topologically transitive diffeomorphisms [**Sh**]. It has also been exploited in [**Mañ1, Carva, BonVi**]. In all of these cases f_0 is an Anosov diffeomorphism and the center bundle is a sum of invariant stable and unstable directions.

Perturbations of partially hyperbolic diffeomorphisms. If f is partially hyperbolic and the C^1 distance from f to g is small enough, then g is also partially hyperbolic. Thus, perturbations of the partially hyperbolic examples are partially hyperbolic. Since perturbations don't always preserve the categories we considered (e.g. a C^1 perturbation of an algebraic system is generally not algebraic), the ergodicity results mentioned above don't apply to all volume-preserving perturbations. What is known about the counterparts of these results for stable ergodicity will be described in Section 4.

A few remarks about these examples.

- Partially hyperbolic diffeomorphisms are generally not structurally stable (take the product of an Anosov diffeomorphism with any non-structurally stable map).
- On some manifolds, there are partially hyperbolic maps arbitrarily close to the identity map (consider the time-t maps of an Anosov flow as $t \to 0$).
- The decomposition of TM into stable, unstable, and center bundles is not necessarily unique (see the discussion of Anosov diffeomorphisms).
- The center bundle E^c is not always integrable — see [**Wi**] (while, as we explain in the next section, the unstable and stable bundles are always integrable). An open question is whether a bunching condition on the central part of the spectrum implies integrability.
- Volume-preserving partially hyperbolic maps are not always ergodic (take the product of a volume-preserving Anosov diffeomorphism with the identity map).
- There are partially hyperbolic diffeomorphisms with trivial stable bundles. Examples based on the derived from Anosov construction are described in [**BonVi**]. This paper also contains examples of stably ergodic diffeomorphisms that are not partially hyperbolic (although they do possess a dominated splitting).

2. Pugh and Shub's theorem

This section will give conditions under which a partially hyperbolic diffeomorphism is ergodic. The stability of these conditions will be considered in the next section.

The most important condition is essential accessibility, which was already defined in the introduction. The definition that we give here is equivalent to the earlier one.

If f is C^k partially hyperbolic, then its stable and unstable bundles are uniquely integrable and are tangent to foliations $\mathcal{W}^u$ and $\mathcal{W}^s$, whose leaves are C^k. These foliations are sometimes referred to as the *strong unstable* and *strong stable* foliations, especially if f is the time-t map of an Anosov flow.

A set is *us-saturated* if it is a union of unstable leaves and is also the union of stable leaves. The smallest *us*-saturated set that contains a given point is its *us-accessible set.* These accessible sets partition M. If the partition has just one element, M, we say that f has the *accessibility property.* If each measurable set that is the union of accessible sets has measure zero or one, we say that f has the *essential accessibility property.* Accessibility is widely used in control theory. As far as we know it was first used in the dynamical systems world by Brin and Pesin [**BriPe1**]. Their work was motivated in part by a paper of Sacksteder [**Sa**] where an infinitesimal version of the idea appears (in Theorem 6.2).

A simple description of accessibility can be given using *us*-paths. A *us-path* is a path $\psi : [0,1] \to M$ consisting of a finite number of consecutive arcs — called *legs* — each of which is a curve that lies in a single leaf of $\mathcal{W}^u$ or $\mathcal{W}^s$. The endpoints of the legs of the *us*-path will be called *corner points.* It is obvious that the set of points that can be reached from a given point by travelling along *us*-paths is exactly the *us*-accessible set of the point, as defined above. This means that f has the accessibility property if and only if any two points can be joined by a *us*-path. It is also clear that two points can be joined by a *us*-path if and only if they can be joined by a a C^1 path whose tangent vector always lies in $E^u \cup E^s$ and vanishes at most finitely many times. Thus the definition of accessibility given in the introduction coincides with the definition given here.

We now recall Conjecture 0.2 from the introduction.

CONJECTURE 2.1. [**PugSh6**, Conjecture 3] Let $f \in \mathrm{Diff}^2_\mu(M)$ where M is compact. If f is partially hyperbolic and essentially accessible, then f is ergodic.

Pugh and Shub have shown that this conjecture holds under some relatively mild technical hypotheses, which we describe below.

THEOREM 2.2. [**PugSh6**, Theorem A] *Let $f \in \mathit{Diff}^2_\mu(M)$ where M is compact. If f is center bunched, partially hyperbolic, dynamically coherent, and essentially accessible, then f is ergodic.*

Theorem 2.2 also holds when f is a C^2 diffeomorphism that preserves any measure equivalent to a smooth measure on M.

All of the ergodicity and stable ergodicity results in this paper are special cases or applications of this theorem. The special case when the central bundle is trivial is the classical result that a C^2 volume preserving Anosov diffeomorphism is ergodic [**An**]; the hypotheses of center bunching, dynamical coherence and essential accessibility are automatically satisfied in this case. Brin and Pesin [**BriPe1**] proved a version of Theorem 2.2 with the additional, very powerful, hypothesis that the central foliation is Lipschitz. The first case of Theorem 2.2 without such an assumption was proved by Grayson, Pugh and Shub in [**GrPugSh**]. This result was improved by Wilkinson [**Wi**] and Pugh and Shub [**PugSh5**], and reached its current form in [**PugSh6**].

Before proceeding, we discuss the technical conditions "center bunched" and "dynamically coherent".

A partially hyperbolic diffeomorphism f is *center bunched* if the action of Tf on the center bundle E^c is close enough to isometric, i.e., $\|T_pf|_{E^c}\|$ and $m(T_pf|_{E^c})$ are close enough to 1. (Here $m(A) = \|A^{-1}\|^{-1}$ denotes the conorm of the operator A.) The details can be found in section 4 of [**PugSh6**]. This property is immediately satisfied if Tf does act isometrically on E^c.[3]

A partially hyperbolic diffeomorphism f is *dynamically coherent* if:

(i) the distributions E^c, $E^c \oplus E^u$, and $E^c \oplus E^s$ are integrable, and everywhere tangent to foliations $\mathcal{W}^c$, $\mathcal{W}^{cu}$, and $\mathcal{W}^{cs}$, called the *center*, *center-unstable*, and *center-stable* foliations, respectively; and

(ii) $\mathcal{W}^c$ and $\mathcal{W}^u$ subfoliate $\mathcal{W}^{cu}$ while $\mathcal{W}^c$ and $\mathcal{W}^s$ subfoliate $\mathcal{W}^{cs}$.

This is the definition given in [**PugSh3**, §2]. It actually follows automatically from results in [**HiPugSh**] that the strong unstable and strong stable foliations sub-foliate $\mathcal{W}^{cu}$ and $\mathcal{W}^{cs}$ respectively. But since the center foliation may not be uniquely integrable, we cannot avoid explicitly assuming that $\mathcal{W}^c$ sub-foliates $\mathcal{W}^{cu}$ and $\mathcal{W}^{cs}$. It follows from [**HiPugSh**, Theorem 7.6] that if the center foliation $\mathcal{W}^c$ exists and is C^1, then f is dynamically coherent, and in fact stably dynamically coherent as we shall see in Proposition 3.1.

2.1. E. Hopf's argument for ergodicity. Underlying many proofs of ergodicity for systems with hyperbolic behavior is an argument originally due to E. Hopf. This argument is simple yet powerful. Let X be a compact metric space, $f : X \to X$ be a homeomorphism, and μ be an f-invariant Borel probability measure.

We denote by $\mathcal{B}$ the Borel sigma algebra on X and by $\mathcal{T}$ the trivial subalgebra consisting of Borel sets which have measure zero or one. Given a subalgebra $\mathcal{A} \subset \mathcal{B}$, its *saturation with respect to sets of measure* 0 is

$$\mathrm{Sat}_0(\mathcal{A}) = \{B \in \mathcal{B} : \exists\, A \in \mathcal{A} \text{ with } \mu(\mathrm{A} \,\Delta\, \mathrm{B}) = 0\}.$$

Thus $\mathcal{T}$ is the saturation of the two element subalgebra consisting of X and the empty set.

Using the homeomorphism f we define two equivalence relations $\sim_u, \sim_s$ on X by

$$x \sim_u y \quad \Leftrightarrow \quad d(f^n(x), f^n(y)) \to 0 \text{ as } n \to -\infty$$

and

$$x \sim_s y \quad \Leftrightarrow \quad d(f^n(x), f^n(y)) \to 0 \text{ as } n \to +\infty.$$

Correspondingly, we have Borel subalgebras $\mathcal{U}, \mathcal{S}$ that consist of whole $\sim_u$ and whole $\sim_s$ equivalence classes. Note that the definitions are purely topological; it is not assumed that f is differentiable let alone partially hyperbolic. Also the convergence in the definitions of $\sim_u$ and $\sim_s$ is not required to be exponential. Of course, if f is partially hyperbolic, and x and y are in the same unstable (resp. stable) leaf, then $x \sim_u y$ (resp. $x \sim_s y$). But there are many partially hyperbolic maps for which the $\sim_u$ and $\sim_s$ equivalence classes are larger than the unstable and stable leaves respectively. A trivial example is the case where E^c is the sum of a weakly expanding and a weakly contracting subbundle.

THEOREM 2.3. *If* $\mathrm{Sat}_0(\mathcal{U}) \cap \mathrm{Sat}_0(\mathcal{S}) = \mathcal{T}$ *then* f *is ergodic.*

[3]An incorrect description of center bunching is given in [**BuPugWi, BuWi, ShWi1**]. The statements made there that center bunching holds when $\mu_c = 1$ or E^c is one dimensional are false.

Hopf's proof of Theorem 2.3 uses the Birkhoff Ergodic Theorem. We summarize various versions of it in the next theorem.

THEOREM 2.4. *Let $\phi : X \to \mathbf{R}$ be in $L^1(X, \mu)$.*

a) $B_+(\phi)(x) = \lim_{n\to\infty} \frac{1}{n} \sum_{j=0}^{n-1} \phi(f^j(x))$ *exists a.e.*
b) $B_-(\phi)(x) = \lim_{n\to-\infty} \frac{1}{|n|} \sum_{j=n+1}^{0} \phi(f^j(x))$ *exists a.e.*
c) $B_+(\phi) = B_-(\phi)$ *a.e.*
d) B_+ *and* B_- *define continuous operators on* $L^1(X, \mu)$.

In order to prove Theorem 2.3, it suffices to show that the image of B_+, or equivalently of B_-, consists of functions which are a.e. constant. Since $C^0(X)$, the space of continuous, real-valued functions, is dense in $L^1(X, \mu)$, it is enough to show that $B_+(C^0(X))$ and $B_-(C^0(X))$ are contained in the constant functions. For this we have a lemma.

LEMMA 2.5. *Let ϕ be continuous.*

a) *If $x \sim_s y$ then $B_+(\phi)(x) = B_+(\phi)(y)$.*
b) *If $x \sim_u y$ then $B_-(\phi)(x) = B_-(\phi)(y)$.*

PROOF. We prove a) only. We claim that if $\lim_{n\to\infty} \frac{1}{n} \sum_{j=0}^{n-1} \phi(f^j(x))$ exists, then so does $\lim_{n\to\infty} \frac{1}{n} \sum_{j=0}^{n-1} \phi(f^j(y))$, and they are equal.

Given $\epsilon > 0$, there exists $\delta > 0$ such that if $d(w, z) < \delta$, then $|\phi(x) - \phi(y)| < \epsilon$. Since $d(f^n(x), f^n(y)) \to 0$ as $n \to \infty$, it follows that there is an N such that for $M \geq N$,

$$\frac{1}{M}\Big|\sum_{j=0}^{M-1}(\phi(f^j(x)) - \phi(f^j(y)))\Big| < 2\epsilon.$$

We are done. □

Now we prove Theorem 2.3.

PROOF. Suppose that there is a continuous function ϕ such that $B_+(\phi)$ is not a.e. constant. For a real number r, let

$$\begin{aligned} G_+(r) &= \{x \in X : \text{ the limit } B_+(\phi)(x) \text{ exists and } B_+(\phi)(x) < r\} \\ G_-(r) &= \{x \in X : \text{ the limit } B_-(\phi)(x) \text{ exists and } B_-(\phi)(x) < r\}. \end{aligned}$$

They are f-invariant Borel sets that belong to $\mathcal{S}$ and $\mathcal{U}$ respectively, and which by Birkhoff's Theorem differ by a zero set. Thus

$$G(r) = G_-(r) \cap G_+(r) \in \mathrm{Sat}_0(\mathcal{U}) \cap \mathrm{Sat}_0(\mathcal{S}).$$

Since $B_+(\phi)$ is not constant a.e. there is a real number r such that

$$0 < \mu\{x \in X \mid B_+(\phi)(x) < r\} < 1,$$

and so $G(r) \notin \mathcal{T}$, which is a contradiction. It follows that after all $B_+(\phi)$ is a.e. constant. This is true for all continuous ϕ and hence for all integrable ϕ, which gives ergodicity of f. □

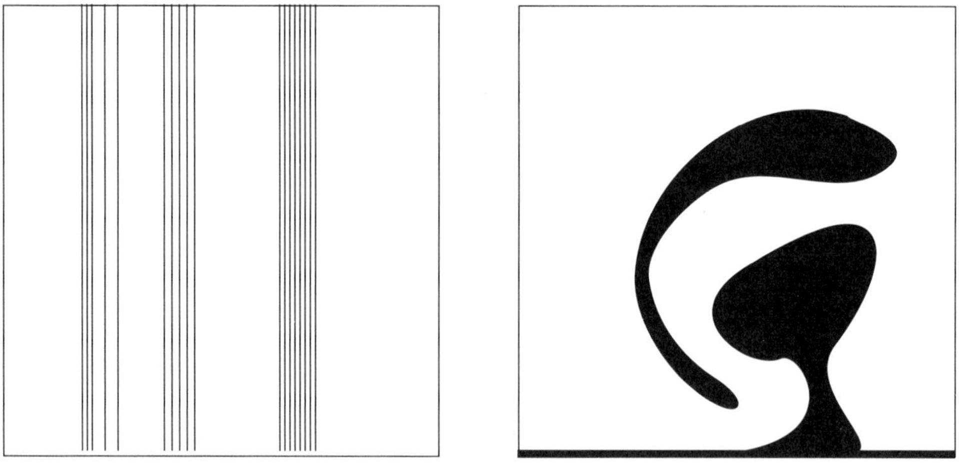

FIGURE 1. Sets in the the Borel subalgebras $\mathcal{A}$ and $\mathcal{C}$.

2.2. Ergodicity of partially hyperbolic systems. To prove the ergodicity of a volume preserving partially hyperbolic diffeomorphism f with the essential accessibility property would now seem to be a simple matter. By Hopf's argument in Theorem 2.3, it suffices to show that each f-invariant Borel set which, up to a zero set, is the union of whole strong unstable manifolds and which is also, up to a possibly different zero set, the union of whole strong stable manifolds, has measure zero or one. But there is a problem. Essential accessibility gives a different, weaker, and hence more easily satisfied condition. It requires the two zero sets to be *the same*.

In terms of subalgebras this can be expressed as follows. Let $\mathcal{M}^u$ and $\mathcal{M}^s$ be the subalgebras of Borel sets that consist of whole strong unstable and whole strong stable leaves respectively. Essential accessibility gives

$$\mathrm{Sat}_0(\mathcal{M}^{\mathrm{u}} \cap \mathcal{M}^{\mathrm{s}}) = \mathcal{T}. \tag{2.1}$$

Since $\mathcal{U} \subset \mathcal{M}^u$ and $\mathcal{S} \subset \mathcal{M}^s$, Hopf's hypothesis, $\mathrm{Sat}_0(\mathcal{U}) \cap \mathrm{Sat}_0(\mathcal{S}) = \mathcal{T}$, is implied by

$$\mathrm{Sat}_0(\mathcal{M}^{\mathrm{u}}) \cap \mathrm{Sat}_0(\mathcal{M}^{\mathrm{s}}) = \mathcal{T}. \tag{2.2}$$

Clearly the saturate of an intersection is contained in the intersection of the saturates; hence $\mathrm{Sat}_0(\mathcal{M}^{\mathrm{u}} \cap \mathcal{M}^{\mathrm{s}}) \subset \mathrm{Sat}_0(\mathcal{M}^{\mathrm{u}}) \cap \mathrm{Sat}_0(\mathcal{M}^{\mathrm{s}})$ and

$$(2.2) \Rightarrow (2.1).$$

The fact that the converse is false is shown by the following example.

EXAMPLE 2.6. Let X be the unit square I^2 equipped with Lebesgue measure, and let $\mathcal{A}$ be the Borel subalgebra of sets of whole vertical segments $x \times I$, while $\mathcal{C}$ is the Borel subalgebra of sets that either contain the horizontal $I \times 0$ or are disjoint from it.

One easily checks that $\mathrm{Sat}_0(\mathcal{C}) = \mathcal{B}$, while $\mathcal{A} \cap \mathcal{C} = \{\emptyset, X\}$. Thus

$$\mathrm{Sat}_0(\mathcal{A} \cap \mathcal{C}) = \mathcal{T} \neq \mathcal{A} \subset \mathrm{Sat}_0(\mathcal{A}) = \mathrm{Sat}_0(\mathcal{A}) \cap \mathrm{Sat}_0(\mathcal{C}).$$

In essence, what Pugh and Shub show in [**PugSh6**] is that under the additional hypotheses on f of dynamical coherence and center bunching,

$$(2.1) \Rightarrow (2.2),$$

so Hopf's idea works after all. Namely, they prove

THEOREM 2.7. *If f is C^2, partially hyperbolic, dynamically coherent, and center bunched then*

$$\mathrm{Sat}_0(\mathcal{M}^{\mathrm{u}}) \cap \mathrm{Sat}_0(\mathcal{M}^{\mathrm{s}}) \subset \mathrm{Sat}_0(\mathcal{M}^{\mathrm{u}} \cap \mathcal{M}^{\mathrm{s}}).$$

Equality of the two sets under the hypotheses of the theorem is an immediate corollary, and we obtain

COROLLARY 2.8. *If f is C^2, volume-preserving, partially hyperbolic, dynamically coherent, center bunched, and essentially accessible then f is ergodic.*

Proof. Since $\mathcal{U} \subset \mathcal{M}^u$ and $\mathcal{S} \subset \mathcal{M}^s$, we have

$$\mathrm{Sat}_0(\mathcal{U}) \cap \mathrm{Sat}_0(\mathcal{S}) \subset \mathrm{Sat}_0(\mathcal{M}^{\mathrm{u}}) \cap \mathrm{Sat}_0(\mathcal{M}^{\mathrm{s}}) = \mathrm{Sat}_0(\mathcal{M}^{\mathrm{u}} \cap \mathcal{M}^{\mathrm{s}}),$$

which, by essential accessibility, is $\mathcal{T}$. Then Theorem 2.3 implies that f is ergodic. □

REMARK 2.9. This argument applies whenever f preserves a measure equivalent to a smooth measure.

2.3. The Kolmogorov Property. The conclusion of Theorem 2.3 can be strengthened when f is partially hyperbolic. An invertible μ-preserving map $f : X \to X$ has the *Kolmogorov property* if there is a sub-σ-algebra $\mathcal{A}$ of the Borel σ-algebra $\mathcal{B}$ such that $f^{-1}\mathcal{A} \subseteq \mathcal{A}$, $\bigcup_{n=-\infty}^{\infty} f^{-n}\mathcal{A}$ generates $\mathcal{B}$, and $\bigcap_{n=0}^{\infty} f^{-n}\mathcal{A}$ is the trivial σ-algebra $\mathcal{T}$. An equivalent definition is that f has the Kolmogorov property if and only if f has no nontrivial factors of zero entropy. A diffeomorphism f that has the Kolmogorov property is also called a *K-system*. K-systems are mixing of all orders. See [**Par1**].

THEOREM 2.10. *Let f be partially hyperbolic. If* $\mathrm{Sat}_0(\mathcal{M}^{\mathrm{u}}) \cap \mathrm{Sat}_0(\mathcal{M}^{\mathrm{s}}) = \mathcal{T}$ *then f has the Kolmogorov property.*

Proof: Let $\mathcal{P}$ be the Pinsker subalgbra, the largest subalgebra for which $f|_{\mathcal{P}}$ has zero entropy. According to Proposition 5.1 of [**BriPe1**], if f is partially hyperbolic, then $\mathcal{P}$ is contained mod 0 in $\mathrm{Sat}_0(\mathcal{M}^{\mathrm{u}}) \cap \mathrm{Sat}_0(\mathcal{M}^{\mathrm{s}})$. If $\mathrm{Sat}_0(\mathcal{M}^{\mathrm{u}}) \cap \mathrm{Sat}_0(\mathcal{M}^{\mathrm{s}}) = \mathcal{T}$, then $\mathcal{P}$ is trivial, and f is a K-system. □

Theorem 2.10 implies the following strengthening of Corollary 2.8.

COROLLARY 2.11. *If f is C^2, volume-preserving, partially hyperbolic, dynamically coherent, center bunched, and essentially accessible, then f has the Kolmogorov property.*

Notice that the hypothesis of Theorem 2.10 is $\mathrm{Sat}_0(\mathcal{M}^{\mathrm{u}}) \cap \mathrm{Sat}_0(\mathcal{M}^{\mathrm{s}}) = \mathcal{T}$, rather than the (possibly weaker) condition of Hopf that $\mathrm{Sat}_0(\mathcal{U}) \cap \mathrm{Sat}_0(\mathcal{S}) = \mathcal{T}$. This is because the results in [**BriPe1**] which we use in the proof apply only to partially hyperbolic maps. We do not know if there is a general proof along the lines of the Hopf argument that $\mathrm{Sat}_0(\mathcal{U}) \cap \mathrm{Sat}_0(\mathcal{S}) = \mathcal{T}$ implies the Kolmogorov property. The closest approach is provided by Theorem B of [**LeYou**]. This theorem applies to any C^2 diffeomorphism and any invariant Borel probability measure. It says that $\mathcal{P} = \mathrm{Sat}_0(\mathcal{B}^{\mathrm{u}}) = \mathrm{Sat}_0(\mathcal{B}^{\mathrm{s}})$, where $\mathcal{P}$ is the Pinsker subalgebra, and $\mathcal{B}^s$ and $\mathcal{B}^u$ are the subalgebras of Borel sets that are unions of entire Pesin stable and unstable manifolds respectively. Note that $\mathcal{S} \subset \mathcal{B}^s$ and $\mathcal{U} \subset \mathcal{B}^u$, but the containments may be proper. Theorem 2.10 is a special case of this result because $\mathcal{B}^s \subset \mathcal{M}^s$ and $\mathcal{B}^u \subset \mathcal{M}^u$ when f is partially hyperbolic.

It is natural to ask whether the Kolmogorov property can be replaced by the Bernoulli property in Corollary 2.11. We suspect that the answer is "no", but have not been able to adapt any of the established methods for constructing examples that are K but not Bernoulli in order to prove this. Other stochastic properties have been studied by Dolgopyat in [**Do3**].

2.4. Absolute continuity. Even in the case when the center bundle is trivial and f is an Anosov diffeomorphism, the proof of Theorem 2.2 uses a major idea: absolute continuity of foliations.

Let $\mathcal{W}$ be a topological foliation with smooth d dimensional leaves in an n dimensional Riemannian manifold. A transversal to $\mathcal{W}$ is a smooth submanifold with dimension $n - d$ that intersects all leaves of $\mathcal{W}$ transversely. If T_1 and T_2 are two nearby transversals, there is a natural map, known as the holonomy or Poincaré map, defined by moving from T_1 to T_2 along short paths in the leaves of $\mathcal{W}$. Let μ_1 and μ_2 be the measures on T_1 and T_2 by the Riemannian structure. The foliation $\mathcal{W}$ is *absolutely continuous* if the measure μ_1^* obtained by using the holonomy map to transport μ_1 from T_1 to T_2 is absolutely continuous with respect to μ_2 for any choice of T_1 and T_2. We say that $\mathcal{W}$ is absolutely continuous *with bounded jacobians* if in addition the Radon-Nikodym derivative $d\mu_1^*/d\mu_2$ is always essentially bounded on compact subsets of T_2.

It is easily seen that a Lipschitz foliation is absolutely continuous with bounded jacobians. But continuous foliations are usually not absolutely continuous, no matter how smooth their leaves are. One of the fundamental properties of partially hyperbolic diffeomorphisms, first obtained by Anosov [**An**] in the uniformly hyperbolic context and independently by Brin and Pesin [**BriPe1**] and Pugh and Shub [**PugSh2**] in the partially hyperbolic setting, is:

THEOREM 2.12. *The stable and unstable foliations of a C^2 partially hyperbolic diffeomorphism are absolutely continuous with bounded jacobians.*

Ergodicity of C^2 volume preserving Anosov diffeomorphisms is an easy consequence of this theorem and the Hopf argument. The general proof of Theorem 2.2, however, uses a second major idea, namely juliennes and julienne density points, which we describe in the next subsection.

The main difficulty is that the center foliation $\mathcal{W}^c$, in contrast to $\mathcal{W}^u$ and $\mathcal{W}^s$, is not always absolutely continuous. Explicit examples of the this phenomenon have been constructed. They are described in Section 5. In the appendix to this paper we give a proof of Theorem 2.2 under the additional assumption that $\mathcal{W}^c$ is absolutely continuous.[4]

[4]The argument in the appendix is a corrected version of the "first false proof" in [**GrPugSh**]. The julienne argument in the next subsection is a corrected version of the "second false proof" in [**GrPugSh**].

2.5. Juliennes and julienne quasi-conformality. As we saw in Section 2.2, proving Theorem 2.2 reduces to proving Theorem 2.7. The proof of Theorem 2.7 proceeds via julienne density points. It is carried out in [**PugSh6**]. We give here a brief indication of what is involved.

Given a smooth Riemannian metric on M and a measurable subset X of M, a point $y \in M$ is a *density point* of X if

$$\frac{\mu(D(y,r)\cap X)}{\mu(D(y,r))} \to 1 \qquad \text{as } r \to 0.$$

Here $D(y,r)$ is the ball of radius r around y in M. Let $D(X)$ be the set of density points of X.

The Lebesgue-Vitali theorem says that $D(X) = X \bmod 0$. We would like to assert that if X is mod 0 a union of strong stable manifolds, then $D(X)$ is a union of whole strong stable manifolds and similarly that if X is mod 0 a union of strong unstable manifolds, then $D(X)$ is a union of whole strong unstable manifolds. This would prove Theorem 2.7.

The trouble is that strong stable and strong unstable holonomy maps don't quasi preserve the family of discs; the distortion in shape produced by the holonomy can become worse and worse as the size of the disc shrinks. This makes it difficult to conclude that y is a density point of X if z is a density point of X for some $z \in W^s(y)$ or $W^u(y)$.

Instead of using discs to define density points, we define for each $p \in M$ a family of neighborhoods of p, $J_n(p)$, for $n \in \mathbf{N}$. They are called juliennes because their shapes are reminiscent of julienned vegetables. The juliennes are dynamically defined so that they are not too badly distorted by the holonomy maps; the long thin shape is essential for this purpose. For each p we have $J_1(p) \supset J_2(p) \supset \cdots$ and $\operatorname{diam} J_n \to 0$ exponentially as $n \to \infty$. As $n \to \infty$ the shape of $J_n(p)$ changes: it becomes relatively longer in the center direction and relatively shorter in the stable and unstable directions.

The volume of $J_n(p)$ shrinks as n increases in a uniform way. For each $j \in \mathbf{N}$

$$\mu(J_n(p))/\mu(J_{n+j}(p)) \text{ is uniformly bounded}$$

for all $p \in M$ and, more significantly, for all $n \in \mathbf{N}$.

A point $y \in M$ is a *julienne density point* of a measurable set $X \subset M$ if

$$\frac{\mu(J_n(y)\cap X)}{\mu(J_n(y))} \to 1 \qquad \text{as } n \to 0.$$

We let $D_J(X)$ be the set of julienne density points of X. The family of juliennes forms what is called a differentiation basis. An analogous differentiation basis that defines density points in the ordinary sense is the family of discs $D_n(p) = D(p, \delta^n)$, for $p \in M$ and $n \in \mathbf{N}$, where $\delta < 1$ is a constant.

The analogue of the Lebesgue-Vitali theorem holds for julienne density points: $D_J(X) = X \bmod 0$ for any measurable X. This fact is non-trivial because in general there are counter-examples to the prevalence of density points when one strays too far from the use of round or quasi-round sets like balls and cubes as differentiation bases. The main step in proving Lebesgue-Vitali for juliennes is to show that there is a constant $l \in \mathbf{N}$ such that (for all $p, q \in M$ and $n \in \mathbf{N}$)

$$J_{n+l}(p) \cap J_{n+l}(q) \neq \emptyset \quad \Rightarrow \quad J_{n+l}(q) \subset J_n(p).$$

The corresponding statement for the discs $D_n(p)$ is easy to prove and implies the usual Lebesgue-Vitali theorem in the same way that this statement implies the prevalence of julienne density points.

Unlike ordinary density points, julienne density points of a set X are preserved under strong stable (resp. strong unstable) holonomy provided that X is mod 0 a union of complete strong stable (resp. strong unstable) manifolds. The basis for this is the following quasi-conformality property, which we formulate in the case of strong unstable holonomy (the stable case is completely analogous). Let $J_n^{cs}(p)$ be the component of p in $J_n(p) \cap \mathcal{W}^{cs}(p)$. Suppose $q \in \mathcal{W}^u(p)$ and there is a curve from p to q with length at most 1 that lies in $\mathcal{W}^u(p)$. Let $\theta : \mathcal{W}^{cs}(p) \to \mathcal{W}^{cs}(p)$ be the holonomy along leaves of $\mathcal{W}^u$. Then there is a $k \in \mathbf{N}$ such that for all large enough n we have

$$J_{n+k}^{cs}(q) \subset \theta(J_n^{cs}(p)) \subset J_{n-k}^{cs}(q).$$

The constant k can be chosen so that this holds for all p and q with the above properties. The corresponding property does not hold for the discs $D_n(p)$.

In summary, the following is proved in [**PugSh6**]:

THEOREM 2.13. *(a)* $D_J(X) = X$ mod 0 *for any measurable set* X.

(b) If X *is* mod 0 *a union of strong stable manifolds, then* $D_J(X)$ *is a union of whole strong stable manifolds.*

(c) If X *is* mod 0 *a union of strong unstable manifolds, then* $D_J(X)$ *is a union of whole strong unstable manifolds.*

Theorem 2.7 is an immediate corollary of this theorem and, as we saw earlier, Theorem 2.2 follows from Theorem 2.7.

Finally a few words about the juliennes. Note that a homeomorphism within ϵ of the identity may violently distort the shape of small discs, but does not distort large discs. The image of a disc of radius 1 is contained in a disc of radius $1+\epsilon$ and contains a disc of radius $1-\epsilon$. Suppose we are interested in holonomy at distance 1 along a strong unstable leaf. Here is a very simplified version of what is done.

Let p, q be points of a common strong unstable leaf such that their distance apart along that leaf is 1. See Figure 2.

Choose n large so that the distance $d(f^{-n}(p), f^{-n}(q))$ is small. Take sets in $W^{cs}(f^{-n}(p))$, $W^{cs}(f^{-n}(q))$ which are small but large compared to this distance and such that the unstable holonomy quasi preserves their shape. These sets are essentially small product rectangles in the center stable leaves at $f^{-n}(p)$ and $f^{-n}(q)$ with respect to the sub-foliations of the center stable leaves by the center and strong stable foliations. See Figure 3.

Then we define the n'th center stable juliennes at p and q, $J_n^{cs}(p)$ and $J_n^{cs}(q)$, as the image of these sets. See Figure 4.

By construction their "shapes" are quasi preserved. Do the same for $W^{cu}(p)$ to define $J_n^{cu}(p)$. Then define $J_n(p)$ to be the "product" of $J_n^{cu}(p)$ and $J_n^{cs}(p)$.

3. Stable accessibility and stable ergodicity

Theorem 2.2 gives stable ergodicity in any situation where its hypotheses are stably satisfied, i.e. they hold in an open subset of $\mathrm{Diff}^2_\mu(M)$. We now consider what happens to the hypotheses when f is perturbed.

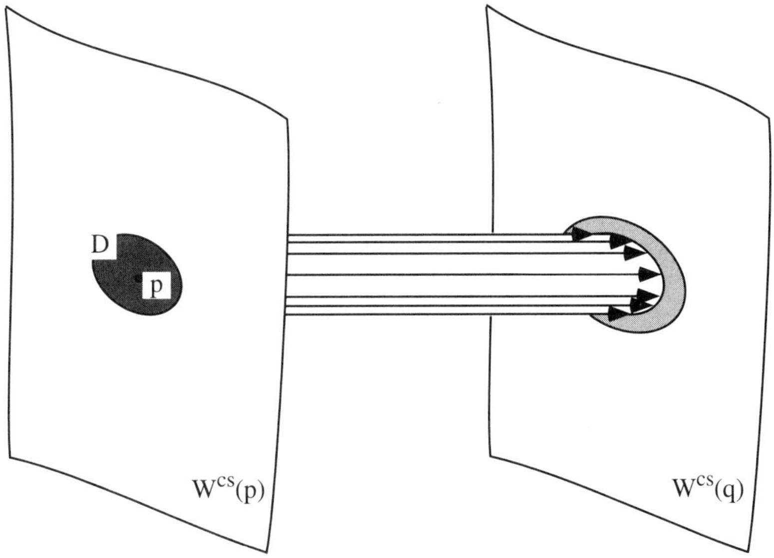

FIGURE 2. Unit holonomy along the strong unstable foliation.

Partial hyperbolicity is an open property in the C^1 topology on diffeomorphisms of M, and so any diffeomorphism g of M that is sufficiently C^1-close to the partially hyperbolic diffeomorphism f has stable and unstable foliations $\mathcal{W}_g^u$ and $\mathcal{W}_g^s$.

The property of center bunching is obviously C^1 open.

We suspect that dynamical coherence is also always C^1 open. But the state of the art is:

PROPOSITION 3.1. [**PugSh5**, Proposition 2.3] *If the center foliation $\mathcal{W}^c$ exists and is of class C^1, then f is dynamically coherent, as is any g close enough to f in the C^1 topology.*

Combining this proposition with Theorem 2.2 and the above remarks gives:

THEOREM 3.2. *Let $f \in \mathit{Diff}_\mu^2(M)$ where M is compact. Suppose f is*

(i) *partially hyperbolic,*

(ii) *center bunched, and*

(iii) *f has a central foliation and this foliation is C^1.*

If f is stably essentially accessible, then f is stably ergodic.

Another form of this result is useful in situations where one wishes to show that f is approximated by stably ergodic diffeomorphisms.

THEOREM 3.3. *If f satisfies (i)–(iii) and there are stably essentially accessible diffeomorphisms arbitrarily close to f, then f is in the closure of the stably ergodic diffeomorphisms.*

Applying these theorems reduces to verifying stable essential accessibility. We now discuss this point. Essential accessibility is not always stable. A skew product example that demonstrates this was constructed by Brin in the 1970's [**Bri1**, end of §2]. It will be described in more detail at the end of this section.

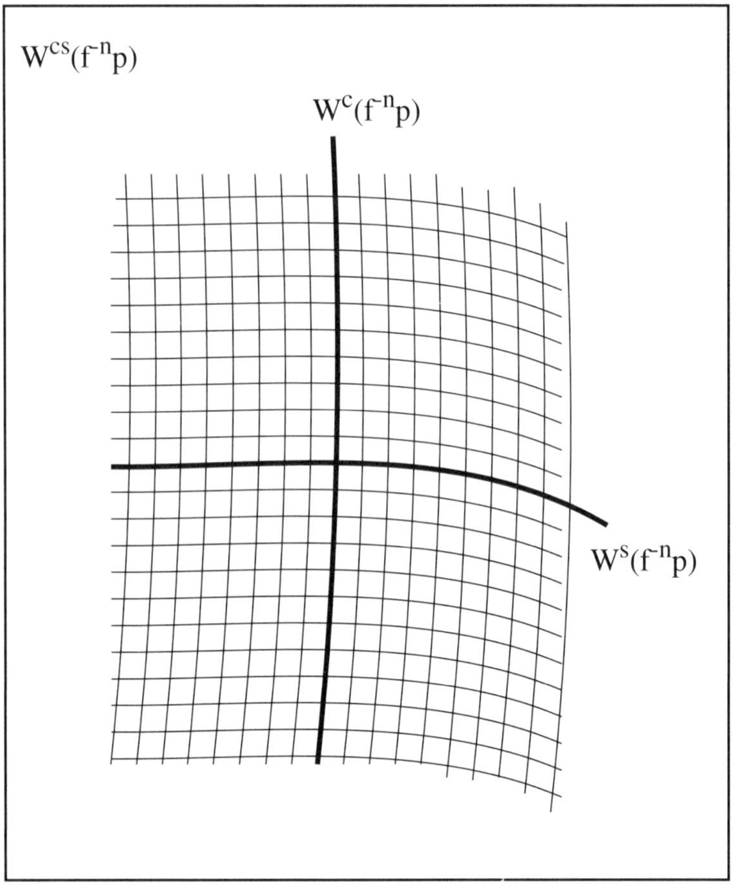

FIGURE 3. The center-stable julienne at the point $f^{-n}(p)$.

Essential accessibility is implied by accessibility, a topological property, whose stability can be verified with the use of topological tools. It is possible that accessibility is always stable, but this is an open question. There is one important but very exceptional case in which stability is automatic:

PROPOSITION 3.4. [**PugSh5**, Corollary 4.2] *If $\mathcal{W}^u$ and $\mathcal{W}^s$ are C^1 foliations and have the accessibility property, then f is stably accessible.*

Even if accessibility is not always stable, it seems quite likely that it should be a dense property:

CONJECTURE 3.5. [**PugSh5**, Conjecture 4] Stable accessibility should hold for a dense set of C^2 partially hyperbolic diffeomorphisms, volume preserving or not.

This is just a reformulation of Conjecture 0.3. In the case when the leaves of the center foliation $\mathcal{W}^c$ are one dimensional, the conjecture has been proved by Niţică and Török [**NiTö**] under some mild additional assumptions; see Section 5. The conjecture has been proven in full generality by Dolgopyat [**Do2**] in the case when E^u, E^c and E^s are all one dimensional; in particular he does not assume

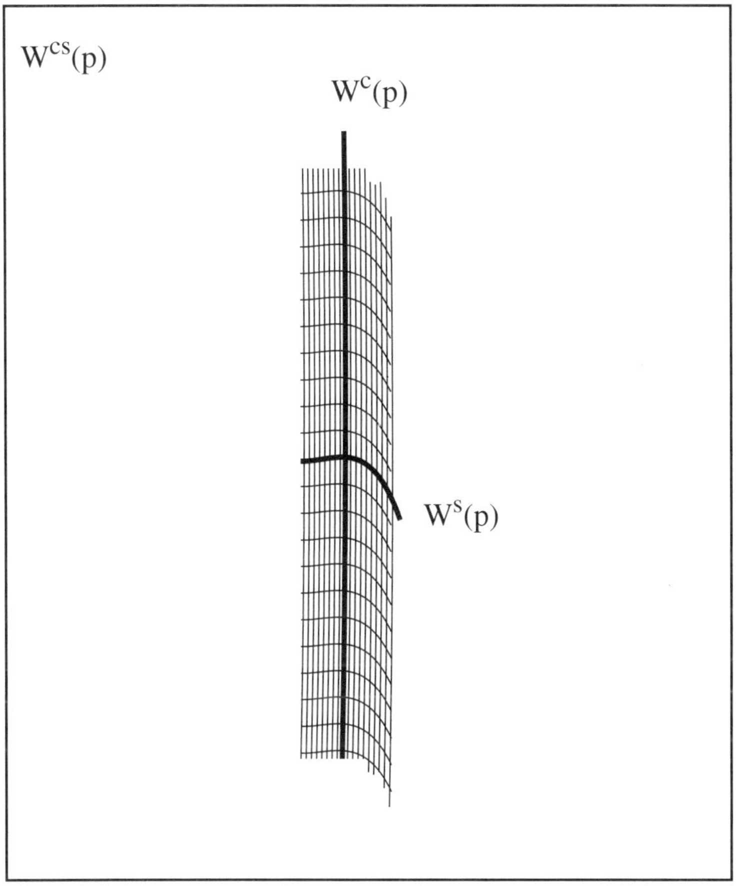

FIGURE 4. The center-stable julienne at the point p.

dynamical coherence. As we shall see in Section 4, the conjecture holds within several interesting special classes of maps.

We now describe some topological ideas which have been useful in verifying stable accessibility. This discussion is based on [**BuPugWi**] and [**BuWi**, §5].

Recall that the accessibility property means that any two points can be joined by a us-path. This leads us to consider conditions under which the fact that there is a us-path from a point p_0 to a point q_0 will be preserved under C^1 small changes of the map, in other words under small perturbations of the foliations $\mathcal{W}^u$ and $\mathcal{W}^s$. It is enough to have a continuous, dim M-parameter family of us-paths emanating from p_0 whose endpoints surround q_0. One way to make this idea precise is the following:

DEFINITION 3.6. A point q_0 can be *engulfed* from a point p_0 if there is a continuous map $\Psi : Z \times [0,1] \to M$ such that:

(1) Z is a compact, connected, orientable, manifold with boundary (e.g. a disc) whose dimension is $n = \dim M$.
(2) For each $z \in Z$, the curve $\psi_z(\cdot) = \Psi(z, \cdot)$ is a us-path with $\psi_z(0) = p_0$.
(3) There is a constant C such that every path ψ_z has at most C legs.

(4) $\psi_z(1) \neq q_0$ for all $z \in \partial Z$.
(5) The map $(Z, \partial Z) \to (M, M \setminus \{q_0\})$ defined by $z \mapsto \Psi(z, 1)$ has nonzero degree. The degree is the unique integer $l \geq 0$ such that a generator for $H_n(Z, \partial Z)$ is mapped to $\pm l$ times a generator for $H_n(M, M \setminus \{q_0\})$.

Recall two properties of degree: it does not change under small perturbations of the map; and nonzero degree implies that the map is surjective. Thus q_0 can be reached from p_0 along a us-path if q_0 can be engulfed from p_0. Moreover engulfing is stable under small perturbations of p_0, q_0 and the foliations $\mathcal{W}^u$ and $\mathcal{W}^s$. Another useful property, which is minor extension of [**BuWi**, Proposition 5.1][5], is:

PROPOSITION 3.7. *Suppose that q_0 can be engulfed from p_0 and there are us-paths from p_1 to p_0 and from q_0 to q_1. Then q_1 can be engulfed from p_1.*

Engulfing can be reduced to a simpler condition.

DEFINITION 3.8. A point q_0 can be *centrally engulfed* from a point p_0 if there is a continuous map $\Psi : Z \times [0, 1] \to M$ such that:
(1) Z is a compact, connected, orientable manifold with boundary whose dimension c is the dimension of the fibers of $\mathcal{W}^c$.
(2) For each $z \in Z$, the curve $\psi_z(\cdot) = \Psi(z, \cdot)$ is a us-path with $\psi_z(0) = p_0$ and $\psi_z(1) \in \mathcal{W}^c(q_0)$.
(3) There is a constant C such that every path ψ_z has at most C legs.
(4) $\psi_z(1) \neq q_0$ for all $z \in \partial Z$.
(5) The map $(Z, \partial Z) \to (\mathcal{W}^c(q_0), \mathcal{W}^c(q_0) \setminus \{q_0\})$ defined by $z \mapsto \Psi(z, 1)$ has nonzero degree.

It is easy to extend a central engulfing to a full engulfing by adding one unstable and one stable leg to each of the paths from the central engulfing. Thus we have

LEMMA 3.9. [**BuWi**, Lemma 5.2][6] *Suppose q_0 can be centrally engulfed from p_0. Then q_0 can be engulfed from p_0.*

An immediate consequence of the above results is

COROLLARY 3.10. *Let $f : M \to M$ be a dynamically coherent, partially hyperbolic diffeomorphism. Suppose that there is a point p_0 such that any point of M can be reached from p_0 along a us-path and p_0 can be centrally engulfed from p_0. Then f is stably accessible.*

The construction of central engulfings is based on the *quadrilateral argument* first used by Brin in [**Bri1**, Proposition 2.3]. This argument produces a strong conclusion, namely central engulfing, from the surprisingly mild hypotheses that $E^u \oplus E^s$ is not integrable and E^c is one dimensional. Thus we suppose that the central foliation has one dimensional leaves and there is a short us-path with 4 legs that begins at some point p_0, ends in $\mathcal{W}^c(p_0)$ and is *not* closed. Then it is easy to create a central engulfing of p_0 from itself. First choose a second us-quadrilateral that passes through the same central fibers as the first path but does so in the opposite order; its endpoint will be in $\mathcal{W}^c(p_0)$, on the other side of p_0 from the endpoint of the first path. Then contract each of these us-quadrilaterals down to

[5] We take the opportunity to point out that the proof given in [**BuWi**] used but did not state the fact that foliations in question have differentiable leaves.

[6] We have corrected the typographical error in [**BuWi**].

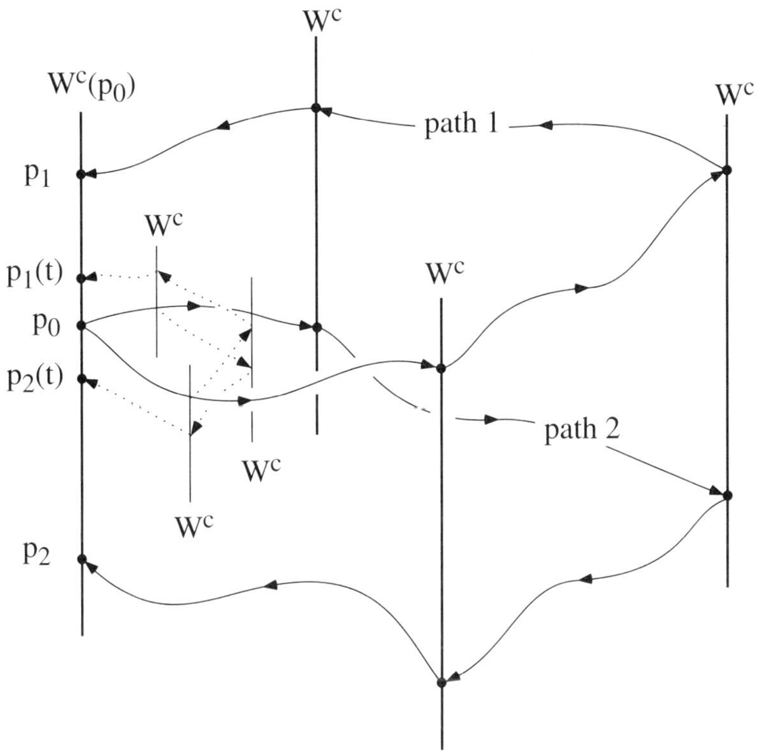

FIGURE 5. A Brin quadrilateral in which the arrows indicate the orientations of the two paths.

the trivial quadrilateral, as shown in Figure 5. See [**BuPugWi, KaKo**] for a more detailed description.

The argument is well suited to showing that stable accessibility is a dense property, because it is easy to introduce small perturbations that break up closed us-quadrilaterals. When the leaves of the central foliation have dimension greater than one, it is necessary to consider several different families of loops. This introduces significant technical difficulties, which have been overcome in several cases; see [**ShWi1**], [**BuWi**, §9] and the discussion in this paper about Theorems 4.8, 4.9, and 4.11.

Brin's skew product example. We now describe the example of Brin [**Bri1**, end of §2] mentioned earlier in this section. Let $f_0 : M_0 \to M_0$ be a C^2 volume preserving Anosov diffeomorphism of a compact manifold. We can choose a C^2 map $\varphi : M_0 \to \mathbf{R}$ such that the quadrilateral argument can be applied to the skew product $f_\varphi : M_0 \times \mathbf{R} \to M_0 \times \mathbf{R}$. Indeed the set of such φ is open and dense; see Proposition 2.3 of [**Bri1**] or the discussion at the end of §9 in [**BuWi**]. Since the base map f_0 obviously has the accessibility property, it is easy to see that f_φ also has the accessibility property (in fact any point in $M_0 \times \mathbf{R}$ can be centrally engulfed from any other).

Now let $i : \mathbf{R} \to \mathbf{T}^2$ be a group homomorphism whose image is a line with irrational slope. Consider the skew product $f_\psi : M_0 \times \mathbf{T}^2 \to M_0 \times \mathbf{T}^2$, where

$\psi = i \circ \varphi$. Then f_ψ does not have the accessibility property because the set of points that can reached from a given point (p, g) intersects each fiber in an irrational line. But f_ψ does have essential accessibility because the Lebesgue measurable subsets of the torus that are unions of translates of a given irrational line have either full measure or measure zero. It is immediate from the results in Section 2 that f_ψ is ergodic and K.

The essential accessibility of f_ψ is not stable, because one can rotate the immersion i through an arbitrarily small angle and make its image become a line of rational slope.

It is possible to replace $\mathbf{R}$ by $\mathbf{R}^k$ and $\mathbf{T}^2$ by $\mathbf{T}^l$ with $l > k$ in the construction.

4. Applications

For most of the examples discussed in Section 1, the Pugh-Shub Theorem, more specifically Theorems 3.2 and 3.3, can be applied to obtain stable ergodicity results. We survey these applications in this section. In all of these applications, the verification of center bunching and dynamical coherence is a simple matter. Stable accessibility is the central issue.

Time-t maps of Anosov flows. In 1994, Grayson, Pugh and Shub introduced the idea of juliennes in [**GrPugSh**] and used it to show that if S is a surface of constant negative curvature and φ_t is the geodesic flow on the unit tangent bundle of S, then the time-t map φ_t is stably ergodic, for $t \neq 0$. In 1995, Wilkinson [**Wi**] generalized this result to surfaces of variable negative curvature.

It is not difficult to see that the time-t map of an Anosov flow, for any $t \neq 0$, is stably dynamically coherent and center bunched; dynamical coherence follows from the fact that the center foliation by orbits is C^1, and center bunching is immediate, since the center foliation is 1-dimensional. Thus, the Pugh-Shub Theorem reduces stable ergodicity of φ_t to stable accessibility.

Stable accessibility was proven for contact Anosov flows, a class which includes geodesic flows for metrics with negative curvature, by Katok and Kononenko [**KaKo**, Proposition 5.2] . In the contact case, the distribution $E^u \oplus E^s$ is smooth and nowhere integrable. This means that *any* point of M centrally engulfs itself, and Brin's quadrilateral argument described in the previous section can be applied to prove stable accessibility. It turns out that for Anosov flows, non-integrability of $E^u \oplus E^s$ *at a single point* is sufficient for stable accessibility; this was proven by Burns, Pugh and Wilkinson.

THEOREM 4.1. [**BuPugWi**] *Let φ_t be a C^1 Anosov flow. Suppose that the strong stable and unstable foliations for the flow are not jointly integrable. Then the time-t map φ_t, for $t \neq 0$, is stably accessible.*

For codimension-1 Anosov flows, the non-integrability condition in Theorem 4.1 is equivalent to topological mixing. Since Plante [**Pl**] showed that the only non-mixing codimension-1 Anosov flows are special flows under constant height functions, there is the corollary:

COROLLARY 4.2. [**BuPugWi**] *Let $\varphi_t : M \to M$ be a C^2, volume-preserving Anosov flow on a compact 3-manifold. The time-t map φ_t is stably ergodic ($t \neq 0$) if and only if the flow φ_t is not a special flow under a constant height function.*

Thus:

Virtually all volume-preserving 3-dimensional Anosov flows have stably ergodic time-1 map.

Algebraic systems. Let $f : G/\Gamma \to G/\Gamma$ be a volume-preserving affine diffeomorphism, and let $\mathfrak{h}$ be the hyperbolic ideal in $\mathfrak{g}$ defined in Section 1. By Proposition 1.1, f is partially hyperbolic if and only if $\mathfrak{h} \neq 0$. A simple argument shows that such systems are always dynamically coherent and center bunched. Because the stable and unstable bundles are smooth, accessibility implies stable accessibility, by Proposition 3.4. Accessibility then can be reduced to an algebraic condition:

PROPOSITION 4.3. [**PugSh6**] *The affine diffeomorphism $f = L_g \circ A$ of G/Γ is partially hyperbolic, dynamically coherent, center bunched and has the stable accessibility property if and only if $\mathfrak{h} = \mathfrak{g}$.*

In view of the Pugh-Shub Theorem, this has as a corollary:

COROLLARY 4.4. *The affine diffeomorphism $f = L_g \circ A$ of G/Γ is stably ergodic if $\mathfrak{h} = \mathfrak{g}$.*

Here are some special cases of Corollary 4.4:

PROPOSITION 4.5. *Suppose that G is a simple Lie group (its Lie algebra has no non-trivial ideals), and that Γ is a discrete cocompact subgroup of G. Let $g \in G$ be given. Then L_g is stably ergodic if $Ad(g)$ has at least one eigenvalue of modulus different from 1.*

COROLLARY 4.6. *Let Γ be a discrete cocompact subgroup of $SL(n, \mathbf{R})$ and let $g \in SL(n, \mathbf{R})$ be given. Left multiplication by g on $SL(n, \mathbf{R})/\Gamma$ is stably ergodic if g has at least one eigenvalue of modulus different than 1.*

In [**BreS**] it was shown that some hyperbolicity is necessary for left multiplication by g to be stably ergodic even among the left translations. The results in [**BreS**], combined with Corollary 4.5 completely classify the stably ergodic affine transformations of compact homogeneous spaces of simple Lie groups. For these spaces,

stable ergodicity is equivalent to stable ergodicity among affine transformations.

Nonetheless, there are very basic examples which are not covered by this corollary, and for which stable ergodicity is not known to hold. Let

$$A = \begin{pmatrix} 0 & 0 & 0 & -1 \\ 1 & 0 & 0 & 8 \\ 0 & 1 & 0 & -6 \\ 0 & 0 & 1 & 8 \end{pmatrix}.$$

The matrix A induces a volume-preserving diffeomorphism of the 4-torus $f_A : \mathbf{T}^4 \to \mathbf{T}^4$. The map $l(f_A) = A$ has eigenvalues $\{\exp \pm 2\pi\alpha, \lambda^{\pm 1}\}$, where $\alpha \approx .206827$ is irrational and $\lambda \approx 7.32763 > 1$; since none of these are roots of 1, it follows from a classical theorem that f_A is ergodic. Alternatively, one can observe that f_A is center bunched, is dynamically coherent because it has a C^1 center foliation, and has the essential accessibility property because the stable and unstable eigenspaces of A span an irrational plane. Ergodicity now follows from Theorem 2.2.

For this example, $G = \mathbf{R}^4$, $\Gamma = \mathbf{Z}^4$, and the hyperbolic subalgebra $\mathfrak{h}$ is two dimensional, and so f_A does not have the accessibility property. We cannot apply Corollary 4.4 to obtain stable ergodicity. Stable ergodicity would follow from Theorem 3.2 if the essential accessibility of f_A were stable.

QUESTION 4.7. Is either the ergodicity or the essential accessibility of f_A stable?

We find this a fascinating but difficult problem. In fact f_A was the first example examined by Pugh and Shub with respect to stable ergodicity, beginning in 1969.

Although we do not know that f_A is stably ergodic, it does lie in the closure of the stably ergodic diffeomorphisms.

THEOREM 4.8. [**ShWi1**] *f_A can be approximated arbitrarily well (in the C^∞ topology) by a stably accessible, stably ergodic diffeomorphism.*

We sketch the proof of Theorem 4.8 here. Since f_A is center bunched and stably dynamically coherent, stable ergodicity in Theorem 4.8 follows if we can approximate f_A by a stably accessible diffeomorphism. Fix x, and let $h_0, k_0 : \mathcal{W}^c_{loc}(x) \to \mathcal{W}^c(x)$ be the holonomy maps defined by two different us-quadrilaterals γ_1 and γ_2 that start at x and end in $\mathcal{W}^c_{loc}(x)$. Integrability of E^u and E^s implies that the holonomy around any sufficiently small us-path beginning and ending in the same local center leaf is the identity map. Thus $h_0 = id = k_0$. The main idea in the proof of Theorem 4.8 is to perturb f_A in a careful way so that these holonomy maps become nontrivial.

More precisely, there is a one parameter family f_ϵ of perturbations of f_A for which the corresponding holonomy maps $h_\epsilon, k_\epsilon : \mathcal{W}^c_{loc}(x) \to \mathcal{W}^c(x)$ have the following strong engulfing property: there is a $\delta_0 > 0$ such that, for each $\epsilon > 0$,

$$\{h_\epsilon^m \circ h_\epsilon^n(x) \mid |n|, |m| < \delta_0/\epsilon\}$$

is an ϵ-dense net in $\mathcal{W}^c_{\delta_0}(x)$ (the δ_0-ball around x in $\mathcal{W}^c_{loc}(x)$). This is possible because h_0 and k_0 are the identity map; one perturbs f_A in neighborhoods of γ_1 and γ_2 so that the diffeomorphisms h_ϵ and k_ϵ are approximately ϵ-translations of $\mathcal{W}^c_{\delta_0}(x)$ in orthogonal directions inside $\mathcal{W}^c(x)$.

Once this is arranged, it is not difficult to construct a center engulfing from x to any point in $\mathcal{W}^c_{\delta_0}(x)$. Since $f_\epsilon \to f_A$ as $\epsilon \to 0$ and stable and unstable leaves for f_A are dense, the stable and unstable leaves are δ_0-dense for ϵ sufficiently small. This proves stable accessibility.

Systems that fiber over partially hyperbolic diffeomorphisms. Suppose that f fibers over the partially hyperbolic diffeomorphism f_0. If the center foliation for f_0 is C^1, then so is the center foliation for f, and f is stably dynamically coherent. In the examples discussed here, this is always the case. Center bunching is also an easy condition to verify for these examples.

Direct products. As was remarked in Section 1, partially hyperbolic diffeomorphisms are not always ergodic. A dramatic illustration of this fact is obtained by taking the product $f \times g_\lambda$, where f is an area-preserving Anosov diffeomorphism of the 2-torus $\mathbf{T}^2 = \mathbf{R}^2/\mathbf{Z}^2$ and $g_\lambda : \mathbf{T}^2 \to \mathbf{T}^2$ is the *standard map:*

$$g_\lambda(z, w) = (z + w, w + (\lambda \sin(2\pi(z + w)))).$$

By KAM theory, for all values of λ near 0, g_λ has a positive-measure set of invariant circles. For such parameter values, this map is persistently not ergodic; any sufficiently nearby C^∞ symplectic map will fail to be ergodic.

Since g_λ is not ergodic, the product $f \times g_\lambda$ is not ergodic; it has a positive measure set of invariant, codimension-1 tori. But while g_λ is stably non-ergodic, $f \times g_\lambda$ may be approximated by a stably accessible, stably ergodic diffeomorphism, and all of these invariant tori disappear. The hyperbolicity of the base map f dominates the KAM center behavior of g_λ.

THEOREM 4.9. [**ShWi1**] *Let $f : \mathbf{T}^{2n} \to \mathbf{T}^{2n}$ be a C^r symplectic Anosov diffeomorphism, $r \geq 2$, and let $G : \mathbf{T}^{2m} \to \mathbf{T}^{2m}$ be a symplectic linear map whose eigenvalues lie on the unit circle in $\mathbf{C}$.*

Then there is a neighborhood $\mathcal{U}$ of G in the space of symplectic C^r diffeomorphisms $Diff^r_\omega(\mathbf{T}^{2m})$ such that for every $g \in \mathcal{U}$, the diffeomorphism $f \times g : \mathbf{T}^{2(m+n)} \to \mathbf{T}^{2(m+n)}$ is partially hyperbolic, dynamically coherent and center bunched. Further, for every neighborhood $\mathcal{V}$ of $f \times g$ in $Diff^r_\omega(\mathbf{T}^{2m})$, there exists an $h \in \mathcal{V}$ such that such that h is stably accessible and stably ergodic.

COROLLARY 4.10. [**ShWi1**] *For any C^∞ symplectic Anosov diffeomorphism f, the map $f \times g_\lambda$ can be C^∞ approximated arbitrarily well by a symplectic, stably ergodic diffeomorphism if λ is sufficiently close to 0.*

The word "symplectic" may be changed to "volume-preserving" in Theorem 4.9.

Since they have C^1 central foliations, partially hyperbolic direct products are stably dynamically coherent. To approximate a center bunched direct product by a stably ergodic diffeomorphism, it suffices, by the Pugh-Shub Theorem, to approximate it by a stably accessible diffeomorphism. The apparent difficulty in doing this is that direct products are never accessible, or even essentially accessible (unless the second factor is trivial). Indeed, it is not hard to see that the accessibility classes for $f \times g$ are the compact leaves of the "horizontal foliation" formed by products of the first factor with points of the second factor — far from the whole manifold! But it is precisely the triviality of the unperturbed system that gives enough control over the perturbation to obtain stable accessibility. The perturbation proceeds much as in the proof of Theorem 4.8 outlined above. The argument in the proof of Theorem 4.9 uses nothing about the dynamics of the second factor g and works for all partially hyperbolic direct products. Thus:

Partially hyperbolic direct products are in the closure of the stably accessible diffeomorphisms. Center bunched direct products are in the closure of the stably ergodic diffeomorphisms.

Skew products. We have seen in the sections on time-t maps of Anosov flows and affine diffeomorphisms that stable ergodicity is often dense within a family of partially hyperbolic diffeomorphisms. In particular, stable ergodicity is dense among time-1 maps of Anosov flows of three manifolds, and among left-translations of compact homogeneous spaces of simple lie groups. These results are a baby version of Conjecture 0.1: they show density of stable ergodicity within a (nowhere-dense) submanifold of the volume preserving diffeomorphisms. Another natural family to study in this context is the space $\text{Skew}(f_0, G)$ of C^2 extensions of an Anosov diffeomorphism f_0 by a compact Lie group G.

In 1975, Brin showed that ergodicity is open and dense in $\text{Skew}(f_0, G)$. Brin studied the holonomy subgroup H of G obtained by following us-paths beginning at a given point (x_0, g_0) and ending in $\mathcal{W}^c(x_0)$. More precisely

$$H = \{g \in G \mid \text{there is a } us\text{-path from } (x_0, g_0) \text{ to } (x_0, gg_0)\}.$$

Let H^0 be the connected component of the identity in H and let $\overline{H}$ be the closure of H in G. Then H^0 is the subgroup of H corresponding to us-paths whose projections to M_0 are contractible. It follows that H^0 is path connected and hence is a Lie subgroup of G by the theorem of Kuranishi and Yamabe. It is not difficult to see that $\overline{H} = G$ if and only if the skew product is essentially accessible.

Brin showed that $\overline{H} = G$ is a necessary and sufficient condition for a skew product to be a K-system. He also showed by a direct construction that $H^0 = G$ for an open and dense set in Skew(f_0, G). Similar results, using different techniques, have subsequently been obtained by Adler, Kitchens and Shub [**AdKiSh**], Parry and Pollicott [**ParPo**], Field and Parry [**FiPa**], Field and Nicol [**FiNi1, FiNi2**], and Walkden [**Wa**]; in this work the base map f_0 is often a basic set for an Axiom A diffeomorphism.

The ergodicity results for partially hyperbolic systems at the time of Brin's work did not apply to perturbations of skew products, and Brin did not investigate whether $H^0 = G$ (or equivalently $H = G$) implies stable accessibility. Burns and Wilkinson showed that $H^0 = G$ does imply stable accessibility, and thus obtained the result:

THEOREM 4.11. [**BuWi**] *Stable ergodicity is open and dense in Skew(f_0, G).*

Another question is whether the notions of "stable ergodicity" and "stable ergodicity among skew products" coincide (certainly the first implies the second). Burns and Wilkinson show that under an additional assumption on f_0, they are equivalent:

THEOREM 4.12. [**BuWi**] *Suppose $f_0 : N_0 \to N_0$ is an Anosov diffeomorphism of a compact infranilmanifold. Then $f_\varphi \in$ Skew(f_0, G) is stably ergodic (in Diff$^2_\mu(N_0 \times G)$) if and only if f_φ is stably ergodic among skew products (that is, inside of Skew(f_0, G)).*

This theorem is based on a classification of skew products that are not stably ergodic.

THEOREM 4.13. [**BuWi**] *Suppose $f_0 : N_0 \to N_0$ is an Anosov diffeomorphism of a compact infranilmanifold and $f_\varphi \in$ Skew(f_0, G) is not stably ergodic (in Diff$^2_\mu(N_0 \times G)$). Then f_φ has a factor that is either:*

(1) *$f_0 \times id_{S^1}$;*
(2) *$f_0 \times R$, where R is a rational rotation of S^1; or*
(3) *a skew product of the type described at the end of Section 3.*

Frame flows. The frame flow for a manifold with negative sectional curvatures is not always ergodic. Kähler manifolds with negative curvature and real dimension at least 4 — in particular quotients of the complex and quaternionic hyperbolic spaces — have nonergodic frame flows because the complex structure is invariant under parallel translation [**BriGr**]. The curvature tensor in these examples is invariant under parallel translation and, if the metric is suitably scaled, every vector lies in both a plane with curvature -1 and a plane with curvature $-1/4$. These seem to be the only known examples with negative curvature and nonergodic frame flow. It has been conjectured that if the sectional curvatures are strictly pinched between -1 and $-1/4$, then the frame flow is ergodic [**BriK, Bri4**]. Brin also conjectures in [**Bri4**] that the frame flow should be ergodic unless the holonomy of the manifold is a proper subgroup of $SO(n)$.

Here are the cases in which the frame flow is known to be ergodic:

THEOREM 4.14. *The frame flow of a compact n-dimensional Riemannian manifold V with negative sectional curvatures pinched between* $-\Lambda^2$ *and* $-\lambda^2$ *is ergodic:*

(1) *if V has constant negative curvature* (Brin [**Bri1**])*;*
(2) *for a set of metrics with negative curvature that is open and dense in the* C^3 *topology* (Brin, [**Bri1**])*;*
(3) *if n is odd and* $n \neq 7$ (Brin and Gromov [**BriGr**])*;*
(4) *if n is even,* $n \neq 8$*, and* $\lambda/\Lambda > 0.93$ (Brin and Karcher [**BriK**]).
(5) *if* $n = 7$ *and* $\lambda/\Lambda > 0.999714\ldots$ (Burns and Pollicott [**BuPo**])*, or*
(6) *if* $n = 8$ *and V and* $\lambda/\Lambda > 0.999785\ldots$ (Burns and Pollicott [**BuPo**]).

Ergodicity of the frame flow and stable ergodicity of the time one map of the frame flow are very closely related. It is not difficult to see that both properties hold if there is one point in the frame bundle M which can be centrally engulfed from itself. The central direction is spanned by the flow direction and the fiber direction. The fact that the geodesic flow in negative curvature is a contact Anosov flow makes engulfing in the flow direction automatic. Central engulfing therefore reduces to engulfing in the fiber direction. The study of fiber engulfing is completely analogous to the study of central engulfing for skew products. The original proofs by Brin, Gromov and Karcher in [**Bri1, BriGr, BriK**] do not explicitly construct fiber engulfings but they can all be adapted to do so; these adaptations are described in [**BuPo**]. The proofs of 5. and 6. in [**BuPo**] are also based on the construction of a fiber engulfing.

Thus in all known examples where the frame flow is ergodic, its time one map is stably ergodic. In particular we have another partial verification of Conjecture 0.1: the time one map of the frame flow is stably ergodic for an open and dense set of metrics with negative curvature. We conjecture that the frame flow is ergodic if and only if its time one map is stably ergodic.

Perturbations of partially hyperbolic systems. There are a few stable ergodicity results that apply to a dense set of diffeomorphisms in a neighborhood of a *non*ergodic diffeomorphism. By combining the Brin quadrilateral argument with the dynamics on the center manifolds, Niţică and Török were able to show:

THEOREM 4.15. [**NiTö**] *Let* $f_0 : N_0 \to N_0$ *be an Anosov diffeomorphism, and let* S^1 *be the circle. Then there is a neighborhood* $\mathcal{U}$ *of* $f_0 \times id_{S^1} : N_0 \times S^1 \to N_0 \times S^1$ *in* $Diff^2(N_0 \times S^1)$ *such that stable accessibility is open and dense in* $\mathcal{U}$.

They also prove, under some additional assumptions, that stable accessibility is dense among partially hyperbolic diffeomorphisms with one-dimensional center. As a corollary, they obtain:

COROLLARY 4.16. [**NiTö**] *Assume that* f_0 *preserves the volume* μ_0*, and let* μ *be the product of* μ_0 *with any smooth measure on* S^1*. Then there is a neighborhood* $\mathcal{U}$ *of* $f_0 \times id_{S^1} : N_0 \times S^1 \to N_0 \times S^1$ *in* $Diff^2_\mu(N_0 \times S^1)$ *such that stable ergodicity is open and dense in* $\mathcal{U}$.

Dolgopyat [**Do2**] has shown that the techniques of Niţică and Török can be generalized to show that among *all* partially hyperbolic diffeomorphisms of 3-manifolds (with both stable and unstable bundles nontrivial), stable accessibility is dense. Then, using other techniques, he shows that stable ergodicity (in fact, stable Bernoullicity) is dense. He does not assume dynamical coherence.

5. Pathological foliations

Shub and Wilkinson [**ShWi2**] have found an open set of partially hyperbolic diffeomorphisms of $\mathcal{T}^3$ that are ergodic and Bernoulli with respect to Lebesgue measure and have non absolutely continuous center foliation. Their examples are small perturbations of the map $f_0 = A \times id_{S^1}$, where A is the Anosov diffeomorphism of $\mathcal{T}^2$ induced by the matrix $\begin{pmatrix} 2 & 1 \\ 1 & 1 \end{pmatrix}$.

Any diffeomorphism f that is close enough to f_0 in the C^1 topology is partially hyperbolic. The three invariant subbundles E^s, E^c and E^u into which $T\mathcal{T}^3$ splits all have one dimensional fibers. The leaves of the corresponding foliations are all C^1 and the leaves of the center foliation $\mathcal{W}^c$ are circles, which are small perturbations of the fibers of the product foliation. If f preserves Lebesgue measure, there will be Lyapunov exponents λ^s, λ^c and λ^u associated to these subbundles and defined Lebesgue almost everywhere. It is clear that $\lambda^s < 0 < \lambda^u$ almost everywhere. Since the leaves of the foliation $\mathcal{W}^c$ are all C^1 embedded circles, whose lengths are uniformly bounded from above and from below, it is clear that that the length of a finite arc of a central leaf cannot grow exponentially under iteration by f. Nevertheless, *it is possible for f to be ergodic with respect to μ and have $\lambda^c > 0$ almost everywhere.*

Such paradoxical behavior is possible only if the set $P = \{p \in \mathcal{T}^3 : \lambda^u(p) > 0\}$, which has full Lebesgue measure, intersects each leaf of $\mathcal{W}^c$ in a set that has measure 0 with respect to arclength along the leaf. Thus we cannot use Fubini's theorem to compute the Lebesgue measure of P by integrating the length measures of its intersections with the leaves of $\mathcal{W}^c$; this means that the foliation $\mathcal{W}^c$ is not absolutely continuous. (Of course the foliation $\mathcal{W}^c$ does not satisfy the hypotheses of Fubini's theorem — even though its leaves are C^1, they vary in a way that is only Hölder continuous.) The existence of such non absolutely continuous foliations has been called "Fubini's nightmare" by Flaminio. Katok had previously constructed a dynamically invariant foliation that is not absolutely continuous; it was presented by Milnor [**Mi**] under the title *Fubini foiled.* The important features of Shub and Wilkinson's example persist under small perturbations. Thus Fubini's nightmare can arise for an open set of diffeomorphisms, and one must face the fact that it is a natural, perhaps even typical, phenomenon.

QUESTION 5.1. How typical is it?

A partial answer is given by the work of Dolgopyat [**Do2**] described at the end of Section 4. His results show, in particular, that the time one map of the geodesic flow of a surface of negative curvature has a neighborhood within which all maps in an open dense subset have nonzero Lyapunov exponent in the center direction. It then follows from an argument of Mañé [**Mañ4**], which exploits the fact that the center is one dimensional, that these maps have non absolutely continuous center foliations. The idea is that if the center foliation is absolutely continuous, then iteration of the map will exponentially expand a typical leaf. But this is impossible since the map is close to the time one map of the geodesic flow, which uniformly translates all center leaves.

Katok's example in [**Mi**] has the property that there is a set which has full measure in the ambient space but meets each leaf of the foliation in at most one point. Ruelle and Wilkinson [**RueWi**] and Katok have shown that the Shub-Wilkinson examples have a similar property: for each such example there is a finite number k such that the set P described above meets almost every leaf of the center foliation exactly k points. Katok has pointed out that it is possible to have $k > 1$; this is achieved by versions of the Shub-Wilkinson example that commute with a finite group of symmetries.

Ruelle and Wilkinson obtain their result by applying the following theorem to the inverses of the Shub-Wilkinson examples.

THEOREM 5.2. [**RueWi**] *Let $F : X \times M \to X \times M$ be a skew product whose base X is a compact metric space and whose fiber M is a closed Riemannian manifold. Assume that F is invertible, ergodic with respect to a Borel probability measure ν and has all Lyapunov exponents in the M direction negative. Then there exist an integer k and a set S such that $\nu(S) = 1$ and $S \cap (\{x\} \times M)$ has precisely k points for every x such that S intersects $\{x\} \times M$.*

It is essential that F be invertible; an example of Kifer [**Ki1**], which is described in [**RueWi**], shows that Theorem 5.2 is false if F is not invertible.

We now outline some of the ideas from Shub and Wilkinson's construction. They give an explicit example of a one parameter variation f_a of f_0 with the property that the average value of the center exponent λ_a^c for f_a is positive unless $a = 0$. The average value of λ_a^c is the integral of the function $r_a^c(p)$ that is defined by $\|D_p f_a(v)\| = r_a^c(p)\|v\|$ for all $v \in E_{f_a}^c(p)$; the exponent λ_a^c is the Birkhoff average of r_a^c. Estimating the integral of r_a^c is a nontrivial but feasible calculation, whereas pointwise estimates on λ_a^c would be hopelessly difficult.

By the result of Niţică and Török described in Section 4, these f_a lie in the closure of the stably ergodic diffeomorphisms for any small enough a. Thus if a is small enough, f_a is approximated by a stably ergodic diffeomorphism f with the property that the exponent for the central direction has positive average[7].

The diffeomorphism f is stably ergodic and has the average value of the exponent λ^c positive. The average values of the exponents change continuously when we perturb f, so all diffeomorphisms close enough to f are ergodic and have λ^c positive on average. Since exponents are invariant functions, all diffeomorphisms close enough to f have λ^c positive almost everywhere and thus are nonuniformly hyperbolic on a set of full measure (since we certainly have λ^u and λ^s a.e. non zero). It follows from Pesin theory that they are Bernoulli. Thus f is stably Bernoulli. As explained above the center foliation $\mathcal{W}^c$ must be non absolutely continuous when $\lambda^c > 0$ almost everywhere.

6. Further questions and conjectures

In this section we try to put our work in a broader perspective. A very optimistic goal for dynamical systems theory would be to understand most dynamical systems using a few basic properties which hold quite generally. Various unsuccessful attempts were made starting around 1960 to achieve this goal via uniform hyperbolicity and the concepts of structural and Ω stability [**Sm2**].

One of the major achievements in the theory of uniformly hyperbolic dynamical systems was the theorems of Sinai, Ruelle and Bowen about invariant measures on

[7] Shub and Wilkinson actually proceed in a different way and extend f_a to a 2-parameter family.

the attractors of a dynamical system. These measures and attractors are now called SRB measures and SRB attractors.

Much of the work in dynamical systems since the mid-seventies has been concerned with extending the domain of existence of SRB measures beyond the uniformly hyperbolic case. By now we have quite a bit of evidence coming from low dimensions such as the quadratic family and Hénon maps, and scattered examples in higher dimension where SRB measures do in fact generally exist [**BonVi, AlBonVi, Carva, Cas, Do1, FiNi2**]. The evidence is good enough so that by now various conjectures have been made. The earliest precise version of one is in [**Pal**]. This is preceded by [**PugSh4**] and followed by [**ShWi2**] which have variants. We attempt to explain the latter two.

We would like to be able to say that the typical orbit of the typical dynamical system has such and such dynamical behavior, even though it may be difficult to verify that a given system or point is typical. Stable ergodicity is a good property from this point of view. Almost every orbit is equi-distributed. Moreover, this fact is true stably or robustly. That is to say nearby systems have the same property.

SRB measures are an analog of ergodic measures in the general or non-volume preserving case. So their existence begins to tell us about the behavior of typical orbits. The proofs in low dimensions of the existence of these measures are a major achievement. The proofs give credence to the existence of SRB measures in general, but they are so particular to low dimensions and rely on so detailed an understanding of the local geometric properties of the iterates of the mappings, that they are discouraging of ever finding a proof in higher dimensions of the general existence of these SRB measures. In particular, it appears hopeless to attempt to find a proof by exhaustion of cases.

Instead we look for a few dynamical properties

- which might be proven to hold for the typical system, without first undertaking an extensive understanding of the dynamics of the system,
- which yet might allow us to conclude useful and even robust information about the dynamics, and
- which might eventually imply the existence of SRB measures.

The simple idea is to relax hyperbolicity in various ways and see what conclusions we may still make. We began in the volume preserving case with partially hyperbolic systems. An example of the type of property we have in mind is stable accessibility of partially hyperbolic systems. As we have conjectured in the introduction, this property should be open and dense among partially hyperbolic systems. We have not proven the conjecture in general and have only done some cases, described in Section 4, but we imagine that the conjecture might be proven by some transversality argument or hands on engulfing argument not too terribly tied to dynamics.

The general plan is to relax uniform partial hyperbolicity to non-uniform partial hyperbolicity, then in the discussion below to consider a variant of accessible sets. There are two frameworks: the volume preserving and the general or dissipative case. Several questions are naturally raised in each framework. Quite a bit of the following discussion is taken from [**ShWi2**].

We discuss the volume preserving case first. While uniformly hyperbolic systems enjoy strong mixing properties, they are not dense among C^1 diffeomorphisms [**Sm1**], [**AbSm**]. Using the concept of Lyapunov exponents, Pesin introduced a weaker form of hyperbolicity, which he termed *nonuniform hyperbolicity.* At almost all points, all the Lyapunov exponents are non-zero. Nonuniformly hyperbolic diffeomorphisms share several mixing properties with uniformly hyperbolic ones.

Pesin [**Pe**] proved that if $f : M \to M$ is a volume-preserving C^2 nonuniformly hyperbolic diffeomorphism, then M may be written as the disjoint union of countably many invariant sets of positive measure on which f is ergodic. He asked if nonuniform hyperbolicity is generic in $\mathrm{Diff}^r_\mu(M)$, the space of C^r, volume-preserving diffeomorphisms of M.

Pesin's question is answered in the negative for large r by Cheng-Sun [**ChSu**], Herman and Xia ([**He1**], [**Xia**]; see also [**Yoc**]). In particular, on any manifold M of dimension at least 2, and for sufficiently large r, there are open sets of volume preserving C^r diffeomorphisms of M all of which possess positive measure sets of codimension one invariant tori; on each such torus, the diffeomorphism is C^1 conjugate to a diophantine translation. In these examples *all* of the Lyapunov exponents are 0 on the invariant tori.

We use the following terminology to discuss Lyapunov exponents. An invariant measure ν is *neutral* if all its Lyapunov exponents are ν-a.e. zero, *partially non-uniformly hyperbolic* if some of them are ν-a.e. non-zero, and *(fully) non-uniformly hyperbolic* if all of them are ν-a.e. non-zero.

The example of Shub and Wilkinson described in Theorem 4.9 gives some hope that a variant of Pesin's original vision holds true for volume-preserving diffeomorphisms: either all exponents are zero (Lebesgue almost everywhere) or, as with our examples, the system may be perturbed to become stably nonuniformly hyperbolic. We express this in the following two questions, the second being a special case of the first.

QUESTION 6.1. Is it a generic property of f in $\mathrm{Diff}^r_\mu(M)$, $r \geq 1$, that almost every ergodic component is either neutral (all Lyapunov exponents zero) or fully non-uniformly hyperbolic (all Lyapunov exponents non-zero)?

QUESTION 6.2. Is the answer to Question 6.1 affirmative for the generic *ergodic f*?

Mañé [**Mañ2, Mañ5**] has outlined an affirmative answer to Question 6.1 when M is a surface and $r = 1$: the generic diffeomorphism in $\mathrm{Diff}^1_\mu(M^2)$ either has all of its Lyapunov exponents zero or is an Anosov diffeomorphism. But Mañé only sketches the proofs and as far we know nobody has filled in all the details.[8]

In another direction, Theorem 4.9 recalls [**BriFeKa**] in which Brin, Feldman and Katok construct diffeomorphisms on every manifold that are Bernoulli. Brin [**Bri3**] went on to show that such examples can be made to have all but one Lyapunov exponent not equal to zero.

QUESTION 6.3. [Brin] Does every manifold of dimension greater than or equal to two admit a *fully* nonuniformly hyperbolic Bernoulli diffeomorphism?

[8]A proof of Mañé's result was recently announced by Jairo Bochi.

If we consider general dynamical systems, i.e. $\mathrm{Diff}^r(M)$ with no invariant measure assumed, then we replace ergodicity by Sinai-Ruelle-Bowen (SRB) measures and attractors. They are also called ergodic attractors.[9]

Given $f \in \mathrm{Diff}^r(M)$ (not necessarily preserving μ), a closed, f-invariant set $A \subset M$ and an f-invariant ergodic measure ν on A, we define $B(A, \nu)$, the *basin* of A, to be the set of points $x \in M$ such that $f^n(x) \to A$ as $n \to \infty$, and for every continuous function $\phi : M \to \mathbf{R}$

$$\lim_{n\to\infty} \frac{1}{n}[\phi(x) + \cdots + \phi(f^n(x))] = \int_A \phi(x)\, d\nu.$$

Definition: ν is an *SRB measure* and A is an *SRB attractor* (we also refer to it as an *ergodic attractor*) if the Lebesgue measure of $B(A, \nu)$ is positive.

It follows from the definition that a diffeomorphism has at most countably many SRB measures. Sinai, Ruelle and Bowen proved that for $r \geq 2$ and f an Axiom A, no-cycle diffeomorphism (see [**Si**], [**Rue**], [**Bow**]), the number of SRB attractors is finite and the union of their basins has full Lebesgue measure in M.

QUESTION 6.4. For the generic f in $\mathrm{Diff}^r(M)$, $r \geq 2$, does the union of its SRB basins have full Lebesgue measure in M?

It is also conjectured in [**Pal**] that for almost every element in a generic finite dimensional family of diffeomorphisms there are only finitely many such SRB attractors. Conjecture 3 of [**PugSh4**] was intended to yield the same.

As alluded to above, this natural question is on the minds of quite a few people. See [**PugSh4**], [**Pal**], [**BonVi**] for discussions and examples. As we have suggested it might be approached along the lines of [**Pe**], [**PugSh3**], [**PugSh6**] by considering weak limits of averages of push-forwards

$$\nu = \lim_{n\to\infty} \frac{1}{n} \sum_0^n f_*^j \mu.$$

QUESTION 6.5. For the generic f in $\mathrm{Diff}^r(M)$, $r \geq 1$, is almost every ergodic component of ν partially non-uniformly hyperbolic, or even fully non-uniformly hyperbolic?

If Question 6.5 has a positive answer then one could proceed to use Lyapunov exponents, amalgamating points as in [**PugSh4**] via a variant of the unstable-stable accessibility relation. For $x, y \in M$ we say that $x \geq_1 y$ if $\mathcal{W}^u(x) \cap W^s(y) \neq \emptyset$. Let $\geq$ be the order obtained by transitivizing $x \geq_1 y$. We say that $x \sim y$ if $x \geq y$ and $y \geq x$. Now the equivalence classes themselves are ordered and minimal elements are candidates to be ergodic attractors.

Note that in the last question we have not included the possibility that all exponents be zero. There are no persistent examples we know of in $\mathrm{Diff}^r(M)$ analogous to the examples of Cheng-Sun and Herman, i.e. persistent neutral invariant measures with positive Lebesgue measure basins.

Here is a vague idea of how some of these genericity questions on Lyapunov exponents might be attacked. Instead of proving theorems about the properties of

[9] We take some of the conclusions of the theorems of Sinai, Ruelle, and Bowen as a definition and warn the reader that the use of SRB measure or attractor is not uniform in the literature. For a survey of SRB measures (using a different definition) see [**You**].

the iterates of a single diffeomorphism we would like to deduce them from properties of a family of diffeomorphisms.

Let $\mathcal{F}$ be a family of C^r diffeomorphisms of a compact manifold M. We would like to deduce that most elements of $\mathcal{F}$ have nonzero exponents. We formulate our strategy in the following principle: *if the family $\mathcal{F}$ is sufficiently rich, then good properties of $\mathcal{F}$ are inherited by many of the elements of $\mathcal{F}$.* The good property we have in mind is positive Lyapunov exponents. To make this strategy more precise, we need to formulate what it means for the family $\mathcal{F}$ to have the "good property," and then find an appropriate notion of richness of $\mathcal{F}$.

A theorem that exploits this strategy is the Abraham transversality theorem (see [**AbRo**]), where the good property is transversality. Earlier proofs of the transversality theorems proceed by perturbing individual elements. Instead, Abraham deduces transversality of most individual mappings in a family $\mathcal{F}$ whenever the joint evaluation map $(x, f) \mapsto f(x)$ has a transversality property. The family $\mathcal{F}$ must be rich in the sense that it forms a smooth (possibly infinite dimensional) manifold along which the evaluation map has the transversality property.

Further illustrations of the principle are provided by ergodic theory. The Mautner phenomenon [**Mo**] quite generally concludes the ergodicity of individual elements of a transitive semi-simple Lie group action. If $\mathbf{R}^n$ acts ergodically then almost all individual elements are ergodic [**PughSh1**]. In these examples, the good property is ergodicity (or transitivity) of the G-action, and the richness of the family comes from the Lie group structure.

We propose a notion of richness that takes into account not only the elements of $\mathcal{F}$ but their distribution, given by a probability measure ν supported on $\mathcal{F}$. Using this measure, there is a natural way to define the exponents of the family $\mathcal{F}$; namely, we consider the exponents of random products of elements of $\mathcal{F}$.

In some situations the behavior of random products is easier to understand than the behavior of the powers of individual elements. One example of this phenomenon is the relative simplicity of Franks' theorem [**Fr**] characterizing structural stability (defined with respect to random products) in comparison to Mañé's [**Mañ3**] (where structural stability is defined the standard way, using powers). Other contexts in which the behavior of random products is understood are the exponents of matrices [**GoMar**], SRB measures for diffeomorphisms [**Ar**], and the exponents of volume preserving diffeomorphisms [**Carve**]. We now describe the latter result in more detail.

Let ν be a probability measure supported on $\mathcal{F}$ and let $f_1, f_2, \ldots$ be a sequence of diffeomorphisms in $\mathcal{F}$, independent and identically distributed with respect to ν. Let $f^{(1)} = f_1$ and let $f^{(n)} = f_n \circ f^{(n-1)}$. Then the *maximal random Lyapunov exponent for $\mathcal{F}$ at x*:

$$R(\nu, x) = \lim_{n\to\infty} \frac{1}{n} \ln \|T_x f^{(n)}\|$$

exists almost surely and is positive unless ν is fairly degenerate [**Carve**]. This degeneracy condition is easily violated. Hence we *expect* random Lyapunov exponents to be nonzero.

While it is fairly easy to construct families with nonzero random Lyapunov exponents, it is more difficult to have individual iterates inherit the property. Part of the problem may be the need for quantitative estimates. Part of the problem might have to do with the richness of the family.

Here is a simple example to illustrate what happens when the family is not sufficiently rich. In this example, random exponents are nonzero but individual elements have only zero exponents. Let $M = \mathbf{T}^2$, let $\mathcal{F}$ consist of two elements, the automorphisms given by $A = \begin{pmatrix} 1 & 1 \\ 0 & 1 \end{pmatrix}$ and $B = \begin{pmatrix} 1 & 0 \\ 1 & 1 \end{pmatrix}$. Let ν assign positive probability to each of the elements of $\mathcal{F}$. These matrices preserve the positive vectors in $\mathbf{R}^2$. A random product of the elements of $\mathcal{F}$ will almost surely contain the matrix AB with positive frequency, and this matrix expands the first coordinate of any positive vector by a factor ≥ 2. Hence, by the ergodic theorem, the random exponents are nonzero. On the other hand, the exponents of both elements of $\mathcal{F}$ are zero. The family is not very rich.

We introduce now a notion of richness of $\mathcal{F}$ which might, in some situations, be sufficient to deduce properties of the exponents of elements of $\mathcal{F}$ from those of the random exponents. This notion was suggested to us by some preliminary numerical experiments and by actual results in the setting of random matrix products [**DeSh**].

We focus on the problem for $M = S^n$, the n-sphere. Let μ be Lebesgue measure on S^n normalized to be a probability measure, and let m be Liouville measure on $T_1(S^n)$, the unit tangent bundle of S^n, similarly normalized to be a probability measure. The orthogonal group $O(n+1)$ acts by isometries on the n-sphere and so induces an action on the space of μ-preserving diffeomorphisms by

$$f \mapsto O \circ f, \qquad \text{for } O \in O(n+1).$$

Let ν be a probability measure supported on $\mathcal{F} \subset \mathrm{Diff}^r_\mu(S^n)$. We say that ν is *orthogonally invariant* if ν is preserved by every element of $O(n+1)$ under the action described above.

For example, let

$$\mathcal{F} = O(n+1)f = \{O \circ f \mid O \in O(n+1)\},$$

for a fixed $f \in \mathrm{Diff}^r_\mu(S^n)$. Defining ν by transporting Haar measure on $O(n+1)$ to $\mathcal{F}$, we obtain an orthogonally-invariant measure. Because $O(n+1)$ acts transitively on $T_1(S^n)$, a random product of elements of $\mathcal{F}$ will pick up the behavior of f in almost all tangent directions — the family is reasonably rich in that sense.

Let ν be an orthogonally invariant measure on $\mathcal{F}$. From orthogonal invariance and the ergodic theorem, it is easy to see that the mean maximal random Lyapunov exponent for $\mathcal{F}$, which we will denote by $R(\nu)$, can be expressed as an integral:

$$R(\nu) = \int R(\nu, x)\, d\mu = \int_{\mathrm{Diff}^r_\mu(S^n)} \int_{T_1(S^n)} \ln \|Df(x)v\|\, dm\, d\nu.$$

We define the *mean Lyapunov exponent* to be

$$\Lambda(\nu) = \int_{\mathrm{Diff}^r_\mu(S^n)} \int_{S^n} \lambda(f, x)\, d\mu\, d\nu$$

where $\lambda(f, x)$ is the maximal Lyapunov exponent of f at x.

QUESTION 6.6. Is there a positive constant $C(n)$ — perhaps 1 — depending on n alone such that $\Lambda(\nu) \geq C(n)R(\nu)$?

If the answer to Question 6.6 were affirmative, then a positive measure set of elements of $\mathcal{F}$ would have areas of positive exponents, assuming a nondegeneracy condition on ν. We add here that this type of question has been asked before and has been the subject of a lot of research. What is new is the notion of richness which allows us to express the relation between exponents as an inequality in integrals.

The question is already interesting for S^2. Express S^2 as the sphere of radius 1/2 centered at $(1/2, 0)$ in $\mathbf{R} \times \mathbf{C}$, so that the coordinates $(r, z) \in S^2$ satisfy the equation

$$|r - 1/2|^2 + |z|^2 = 1/4.$$

In these coordinates define a twist map $f_\epsilon : S^2 \to S^2$, for $\epsilon > 0$, by

$$f_\epsilon(r, z) = (r, \exp(2\pi i r \epsilon) z).$$

Let $\mathcal{F}$ be the orbit $O(3)f$ and let ν be the push forward of Haar measure on $O(3)$. A very small and inconclusive numerical experiment seemed to indicate that for ϵ close to 0 the inequality may hold with $C(n) = 1$. It seemed the constant may decrease as the twist increases speed.

Michel Herman has pointed out the references [**Neĭ, Neh, Lo**] to us. In light of these references, he suggests that it is extremely likely that Question 6.6 has a negative answer, precisely for the twist map example f_ϵ, for ϵ very small. If he is correct, it would be interesting to know if some similar lower bound estimate is available with an appropriate concept of richness of the family.

There is a somewhat less speculative version of this question in linear algebra. Now we let ν be an orthogonally invariant measure on the space of $n \times n$ real matrices, $M(n, n)$. Then we let

$$R(\nu) = \int_{M(n,n)} \int_{S^{n-1}} \ln \|M(x)\| \, d\mu \, d\nu$$

and

$$\Lambda(\nu) = \int_{M(n,n)} \ln \rho(M) \, d\nu,$$

where $\rho(M)$ is the spectral radius of M. Now we repeat the question, with some more confidence that the constant is 1.

QUESTION 6.7. Is there a positive constant $C(n)$ — perhaps 1 — depending on n alone such that $\Lambda(\nu) \geq C(n) R(\nu)$?[10]

It is a pleasant exercise to prove the inequality in Question 6.7 for $n = 2$ with $C(2) = 1$. See [**Ki1, Ki2**] and [**BouLa**] for background on random diffeomorphisms and random matrix products.

We end with one last question of a very different nature. We have used both the strong unstable and strong stable foliations in our proof of ergodicity. We don't know an example where this is strictly necessary.

QUESTION 6.8. For a partially hyperbolic C^2 ergodic diffeomorphism f with the essential accessibility property, is the strong unstable foliation already uniquely ergodic?

Unique ergodicity of $\mathcal{W}^u$ was proved by Bowen and Marcus [**BowMar**] in the case where f is the time one map of a hyperbolic flow. For homogeneous space examples the question is related to the Mautner phenomenon discussed in [**Mo**] and [**Par2**].

---

[10]The answer to this question is "yes" if $M(n, n)$ is the space of *complex* matrices and ν is invariant under the unitary group $U(n)$ [**DeSh**].

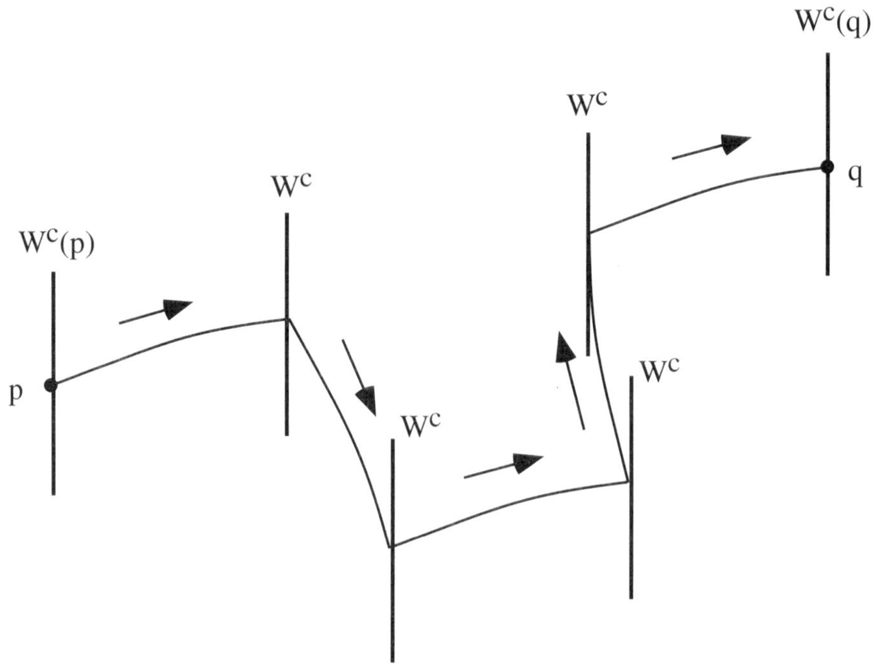

FIGURE 6. It is implausible that a is a density point of A and b is a density point of $B = M \setminus A$.

7. Appendix: Turning the corner

An intentionally over-simplified proof of ergodicity is outlined in [**GrPugSh**]. It is based on Figure 6 which purports to show that if p is a center density point (see below for the definition) of an essentially us-saturated set A then so is any point q which can be connected to p by a us-path. It is assumed that the center foliation is absolutely continuous, since otherwise the set of center density points can be empty, as is the case in the examples of Katok and Shub-Wilkinson described in Section 5. The obstacle to making Figure 6 into a proof of ergodicity is that the property of being a center density point of A may be lost at the corners of the path. We get around this obstacle, constructing us-paths with "good corners", and proving the following result.

THEOREM 7.1. *If $f \in \mathit{Diff}^2_\mu(M)$ is partially hyperbolic, dynamically coherent, center bunched, and essentially accessible then absolute continuity of the center foliation implies ergodicity of f.*

Proof. M is compact and Riemannian. Its Riemann structure induces smooth measures $\mu_u, \mu_{cu}, \mu_c, \mu_{cs}, \mu_s$ on the leaves of the invariant foliations, and induces the smooth measure μ on M. The key property of an absolutely continuous foliation is Cavalieri's Principle: if $Z \subset M$ is a μ-zero set then almost every leaf meets Z in a set of leaf measure zero. See [**PugSh2**]. The unstable and stable foliations are always absolutely continuous, and now by assumption the center foliation is absolutely continuous. This makes the center unstable and center stable foliations absolutely continuous too.

As we saw in Section 2.2, the assertion of Theorem 7.1 amounts to the inclusion

$$\text{Sat}_0(\mathcal{M}^u) \cap \text{Sat}_0(\mathcal{M}^s) \subset \text{Sat}_0(\mathcal{M}^u \cap \mathcal{M}^s). \tag{7.1}$$

Let $A \in \text{Sat}_0(\mathcal{M}^u) \cap \text{Sat}_0(\mathcal{M}^s)$ be given. This means that A is measurable and

(a) Modulo a μ-zero set A is the union of almost whole unstable leaves.
(b) Modulo a μ-zero set A is the union of almost whole stable leaves.

We must show that A is sandwiched as $A_0 \subset A \subset A_1$ such that $A_1 \setminus A_0$ is a zero set and us-paths starting in A_0 stay in A_1.

The union of the unstable leaves that meet A in sets of full leaf measure is denoted A^u. The unstable holonomy maps preserve A^u, and absolute continuity of the unstable foliation implies that A^u differs from A by a zero set. Corresponding definitions and facts hold for A^s. The complement $B = M \setminus A$ also belongs to $\text{Sat}_0(\mathcal{M}^u) \cap \text{Sat}_0(\mathcal{M}^s)$, and we have the corresponding sets B^u, B^s.

Absolute continuity of the center foliation implies that almost every point of $p \in A$ is a *center density point* of A. That is,

$$\lim_{r \to 0} \frac{\mu_c(W^c(p,r) \cap A)}{\mu_c(W^c(p,r))} = 1.$$

The corresponding fact is true for B.

Definition: A center leaf W^c is *good* if

(a) $A^u \cap W^c = A \cap W^c = A^s \cap W^c$ modulo center leaf zero sets.
(b) $B^u \cap W^c = B \cap W^c = B^s \cap W^c$ modulo center leaf zero sets.

A us-path can safely turn the corner at a good leaf: center density of A is preserved.

Let G_1 be the union of the good center leaves. Since the symmetric differences $A \Delta A^u$, $A \Delta A^s$, $B \Delta B^u$, $B \Delta B^s$ are zero sets, absolute continuity of the center foliation implies that $\mu(M \setminus G_1) = 0$. Discard from G_1 those center leaves $W^c p$ for which $G_1 \cap W^{cu} p$ fails to have full leaf measure in $W^{cu} p$. (For example, if there are only countably many good center leaves in a center unstable leaf, we discard all of them.) Discard also those center leaves $W^c p$ for which $G_1 \cap W^{cs} p$ fails to have full leaf measure in $W^{cs} p$. The union of the remaining good center leaves, G_2, has full measure. Repeat the process, and discard from G_2 those center leaves $W^c p$ for which $G_2 \cap W^{cu} p$ or $G_2 \cap W^{cs} p$ fails to have full leaf measure. Induction gives $G_1 \supset G_2 \supset \cdots$. Then $E = \bigcap G_k$ is a set of full measure that consists of good center leaves $W^c p$ for which $G \cap W^{cu} p$ and $G \cap W^{cs} p$ have full leaf measure. We refer to the points and center leaves in E as *excellent*. Excellent center leaves are good, and moreover if $W^{cu} p$ contains one excellent center leaf then it consists almost entirely of excellent center leaves. The same holds for center stable leaves.

To arrive at a contradiction, suppose that A and $B = M \setminus A$ both have positive measure. Then

$$A^* = \{a \in A : W^c a \text{ is excellent and } a \text{ is a center density point of } A\}$$

$$B^* = \{b \in B : W^c b \text{ is excellent and } b \text{ is a center density point of } B\}$$

differ from A and B by zero sets, and so have positive measure. Essential accessibility implies that there is a us-path γ from $p \in A^*$ to $q \in B^*$. Let n be the number of legs of γ. The path γ can be approximated by a us-path γ' that starts at p, ends at a point q' arbitrarily near q, and whose corners occur at excellent center leaves. (This is what makes excellent center leaves excellent: turning a corner at an excellent center leaf permits turning the next corner also at an excellent center leaf.) The holonomy maps $\theta : W^c p \to W^c q$ and $\theta' : W^c p \to W^c q'$ are composites of n diffeomorphisms of class C^1. For the unstable and stable foliations restricted to the center unstable and center stable leaves respectively are of class C^1. This follows from Theorem B in [**PugShWi**] and the center bunching assumption. In fact, θ' C^1-approximates θ. Thus there is an $r > 0$ that depends on n but is independent of q' as $q' \to q$, such that the concentration of A in $W^c(q', r)$ is $\geq .9$ and the concentration of B in $W^c(q, r)$ is also $\geq .9$.

The stable holonomy map $h : W^{cu}q' \to W^{cu}q$ preserves center leaves and is absolutely continuous. Thus $W^{cu}q'$ consists almost entirely of excellent center leaves, the same is true of $W^{cu}q$, and h sends almost every excellent center leaf in $W^{cu}q'$ to an excellent center leaf in $W^{cu}q$. In particular, the path γ' extends to a us-path $\gamma' \cup \epsilon$ where ϵ is an arbitrarily short three legged path with excellent corners that starts at q' and ends at a point $q'' \in W^c(q, r)$. As $q' \to q$, the path ϵ shrinks to q, the point q'' converges to q, and the us-holonomy map $W^c q' \to W^c q$ around ϵ converges in the C^1 sense to the identity. Thus A has concentration $\geq .8$ in $W^c(q'', r)$. Since r is fixed while $q'' \to q$, this contradicts the fact that B has concentration $\geq .9$ in $W^c(q, r)$.

Here is the upshot. We have

$$A^* = A_0 \subset A \subset A_1 = M \setminus B^*,$$

where $A_1 \setminus A_0$ is a zero set and us-paths starting in A_0 stay in A_1. Thus $A \in \mathrm{Sat}_0(\mathcal{M}^u \cap \mathcal{M}^s)$, which completes the proof of (7.1) and hence of the theorem. □

We now describe very briefly a second approach to Theorem 7.1, which is in the spirit of the julienne density point argument used to prove Theorem 2.2. The strategy is to prove that a partial analogue for Lebesgue density points of Theorem 2.13 holds when the center foliation is absolutely continuous. Theorem 7.1 then follows in the same way that Theorem 2.2 follows from Theorem 2.13. Recall that $D(X)$ is the set of Lebesgue density points for a measurable set X.

THEOREM 7.2. *Let $f \in \mathit{Diff}^2_\mu(M)$ where M is compact. Suppose f is partially hyperbolic, dynamically coherent, and center bunched. Assume that the center foliation $\mathcal{W}^c$ is absolutely continuous with bounded jacobians. Let $A \in \mathrm{Sat}_0(\mathcal{M}^{\mathrm{u}}) \cap \mathrm{Sat}_0(\mathcal{M}^{\mathrm{s}})$. Then $D(X) \in \mathcal{M}^u \cap \mathcal{M}^s$.*

In this theorem, we assume the hypotheses of parts (b) and (c) of Theorem 2.13 and obtain the conclusions of both (b) and (c). The proof uses arguments similar to those in the proof of Theorem 7.1.

We conclude with a technical remark. The bunching assumption that is needed in Theorems 7.1 and 7.2 is weaker than the bunching assumption in Theorem 2.2. All that is used in the arguments in this appendix is that the stable and unstable holonomies between center leaves are C^1. With the notation of Section 4 of [**PugSh6**], it is sufficient to have $\nu < \gamma^2$, whereas Theorem 2.2 assumes that $\nu < \gamma^{2+2\theta}$.

References

[AbRo] Abraham, R. and J. Robbin, *Transversal Mappings and Flows*, W. A. Benjamin, Inc., New York-Amsterdam 1967.

[AbSm] Abraham, R. and S. Smale, *Nongenericity of Ω-stability*, 1970 Global Analysis (Proc. Sympos. Pure Math., Vol. XIV, Berkeley, Calif., 1968) pp. 5–8 Amer. Math. Soc., Providence, R.I.

[AdKiSh] Adler, R., B. Kitchens and M. Shub, *Stably ergodic skew products*, Discr. and Contin. Dynam. Syst. **2** (1996), 349–350.

[AlBonVi] Alves, J.F., C. Bonatti and M. Viana, *SRB measures for partially hyperbolic systems whose central direction is mostly expanding*, Invent. Math. **140** (2000), 351–398.

[An] Anosov, D. V., *Geodesic flows on closed Riemannian manifolds of negative curvature*, Proc. Steklov. Inst. Math. **90** (1967).

[Ar] de Araujo, V.D.M., *Attractors and Time Averages for Random Maps*, Tese de Doutorado, IMPA (1998).

[BonDí] Bonatti, C. and L. Díaz, *Persistent nonhyperbolic transitive diffeomorphisms*, Ann. of Math. **143** (1996), 357–396.

[BonDíPuj] Bonatti, C., Díaz, L., and E. Pujals, *A C^1-generic dichotomy for diffeomorphisms: weak forms of hyperbolicity or infinitely many sinks or sources*, Preprint (1999).

[BonVi] Bonatti, C. and M. Viana, *SRB measures for partially hyperbolic systems whose central direction is mostly contracting*, Israel J. Math. **115** (2000), 157–194.

[BouLa] Bougerol, P. and J. Lacroix, *Products of random matrices with applications to Schrdinger operators.* Progress in Probability and Statistics **8**, Birkhäuser Boston, Inc., Boston, MA, 1985.

[Bow] Bowen, R., *Equilibrium states and the ergodic theory of Anosov diffeomorphisms*, Lecture Notes in Mathematics **470**, Springer-Verlag, Berlin-New York, 1975.

[BowMar] Bowen, R. and B. Marcus, *Unique ergodicity for horocycle foliations*, Israel J. Math. **26** (1977), 43–67.

[BreMo] Brezin, J. and C. Moore, *Flows on homogeneous spaces: a new look.* Amer. J. Math. **103** (1981), 571–613.

[BreS] Brezin, J. and M. Shub, *Stable ergodicity in homogeneous spaces.* Bol. Soc. Brasil. Mat. (N.S.) **28** (1997), no. 2, 197–210.

[Bri1] Brin, M., *Topological transitivity of one class of dynamical systems and flows of frames on manifolds of negative curvature*, Func. Anal. Appl. **9** (1975), 9–19.

[Bri2] Brin, M., *The topology of group extensions of C systems*, Mat. Zametki **18** (1975), 453–465.

[Bri3] Brin, M., *Bernoulli diffeomorphisms with $n-1$ nonzero exponents*, Ergod. Th. and Dynam. Syst. **1** (1981), 1–7.

[Bri4] Brin, M., *Ergodic theory of frame flows*, In Ergodic theory and dynamical systems, II (College Park, Md., 1979/1980), pp. 163–183, Progr. Math., 21, Birkhäuser, Boston, Mass., 1982.

[BriFeKa] Brin, M., J. Feldman and A. Katok, *Bernoulli diffeomorphisms and group extensions of dynamical systems with nonzero characteristic exponents*, Ann. of Math. **113** (1981), 159–179.

[BriGr] Brin, M. and M. Gromov, *On the ergodicity of frame flows*, Invent. Math. **60** (1980), 1–7.

[BriK] Brin, M. and H. Karcher, *Frame flows on manifolds with pinched negative curvature*, Compos. Math. **52** (1984), 275–297.

[BriPe1] Brin, M. and Ja. Pesin, *Partially hyperbolic dynamical systems*, Math. USSR Izvestija **8** (1974), 177–218.

[BriPe2] Brin, M. and Ja. Pesin, *Flows of frames on manifolds of negative curvature*, Uspehi Mat. Nauk. **28** (1973), 169–170.

[BuPo] Burns, K. and M. Pollicott, *Stable ergodicity and frame flows*, Preprint (2000).

[BuPugWi] Burns, K., C. Pugh and A. Wilkinson, *Stable ergodicity and Anosov flows*, Topology **39** (2000), 149–159.

[BuWi] Burns, K. and A. Wilkinson, *Stable ergodicity of skew products*, Ann. Sci. École Norm. Sup. **32** (1999), 859–889.

[Carva] Carvalho, M., *Sinai-Ruelle-Bowen measures for N-dimensional derived from Anosov diffeomorphisms*, Ergod. Th. and Dynam. Syst. 13 (1993), 21–44.

[Carve] Carverhill, A., *Furstenberg's theorem for non-linear stochastic systems*, Probability Theory and Related Fields, 74 (1987), 529-534.

[Cas] Castro, A., *Backwards inducing and decay of correlations for certain partially hyperbolic maps whose central direction is mostly contracting*, Tese de Doutorado, IMPA (1998).

[ChSu] Cheng, C.-Q., and Y.-S. Sun, *Existence of invariant tori in three dimensional measure-preserving mappings*, Celestial Mech. Dynam. Astronom. **47** (1989/90), 275–292.

[DeSh] Dedieu, J.P. and M. Shub, *On random and mean exponents for unitarily invariant probability measures on $GL(n, \mathbf{C})$*, preprint.

[DíPujUr] Díaz, L., E. Pujals and R. Ures, *Partial hyperbolicity and robust transitivity*, Acta Math. **183** (1999), 1–43.

[DíRo] Díaz and J. Rocha, *Partially hyperbolic and transitive dynamics generated by heteroclinic cycles*, To appear: Ergod. Th. and Dynam. Syst.

[Do1] Dolgopyat, D., *On dynamics of mostly contracting diffeomorphisms*, To appear: Comm. Math. Phys.

[Do2] Dolgopyat, D., *On dynamics of partially hyperbolic systems on three manifolds*, Preprint (1999).

[Do3] Dolgopyat, D., *Limit theorems for partially hyperbolic systems*, Preprint (1999).

[FiNi1] Field, M. and M. Nicol, *Ergodic theory of equivariant diffeomorphisms I: Markov partitions*, Preprint (1999).

[FiNi2] Field, M. and M. Nicol, *Ergodic theory of equivariant diffeomorphisms II: stable ergodicity*, Preprint (1999).

[FiPa] Field, M. and W. Parry, *Stable ergodicity of skew extensions by compact Lie groups*, Topology **38** (1999), 167–187.

[Fr] Franks, J., *Time dependent structural stability*, Invent. Math. **24** (1974), 163–172.

[FuKi] Furstenberg, H. and T. Kifer,*Random matrix products and measures on projective spaces*, Israel J. Math. **46** (1983), 12–32.

[GrPugSh] Grayson, M., C. Pugh and M. Shub, *Stably ergodic diffeomorphisms*, Ann. Math. **140** (1994), 295–329.

[He1] Herman, M., *Stabilité topologique des systèmes dynamiques conservatifs*, Preprint (1990).

[He2] Herman, M., *Some open problems in dynamical systems*, Proc. of Int. Congress of Math., Vol. II (Berlin,1998). Doc. Math 1998, Extra Vol.II, 797-808 (electronic).

[HiPugSh] Hirsch, M., C. Pugh and M. Shub, *Invariant manifolds*, Lecture Notes in Mathematics, **583**, Springer-Verlag, 1977.

[KaKo] Katok, A. and A. Kononenko, *Cocycles' stability for partially hyperbolic systems*, Math. Res. Lett. **3** (1996), 191–210.

[Ki1] Kifer, Y., *Ergodic theory of random transformations*, Progress in Probability and Statistics, 10. Birkhäuser Boston, Inc., Boston, MA, 1986.

[Ki2] Kifer, Y., *Random dynamics and its applications*, Proc. of Int. Congress of Math., Vol. II (Berlin,1998). Doc. Math 1998, Extra Vol. II, 809-818 (electronic).

[LeYou] Ledrappier, F. and L.-S. Young *The metric entropy of diffeomorphisms. I. Characterization of measures satisfying Pesin's entropy formula*, Ann. of Math. **122** (1985), 509–539.

[Lo] Lochak, P. *Canonical perturbation theory via simultaneous approximations*, Russian Math. Surveys **47** (1992), no. 6, 57–133.

[Mañ1] Mañé, R., *Contributions to the stability conjecture*, Topology **17** (1978), 383–396.

[Mañ2] Mañé, R., *Oseledec's theorem from the generic viewpoint*, Proc. of Int. Congress of Math. (1983) Warszawa 1269–1276.

[Mañ3] Mañé, R., *A proof of the C^1 stability conjecture*. Inst. Hautes Etudes Sci. Publ. Math. No. 66, (1988), 161–210.

[Mañ4] Mañé, R., private communication.

[Mañ5] Mañé, R., *The Lyapunov exponents of generic area preserving diffeomorphisms*, in International Conference on Dynamical Systems: Montevideo 1995, a Tribute to Ricardo Mañé, F. Ledrappier et al, eds. Pitman Research Notes in Math. Series **362** (1996), Longman, 110–119.

[GoMar] Golć dsheĭd, I. Ya. and G. Margulis, *Lyapunov exponents of a product of random matrices, Russian Math. Surveys* **44** (1989), 11–71.

[Mi] Milnor, J. *Fubini foiled: Katok's paradoxical example in measure theory*, Math. Intelligencer, **19** (1997), 30–32.

[Mo] Moore, C.C. *Ergodicity of flows on homogeneous spaces*, Amer. J. Math. **88** (1966), 154–178.

[Neh] Nehorošev, N. N. *An exponential estimate of the time of stability of nearly integrable Hamiltonian systems*, Russian Math Surveys, **32-6** (1977), 1–65.

[Neĭ] Neĭshtadt, A. I. *The separation of motions in systems with rapidly rotating phase*, J. Appl. Math. Mech. **48** (1984), no. 2, 133–139.

[NiTö] Niţică, V. and A. Török, *An open dense set of stably ergodic diffeomorphisms in a neighborhood of a non-ergodic one*, Preprint (1998).

[Pal] Palis, J., *A global view of dynamics and a conjecture on the denseness of finitude of attractors*, Astérisque (1998).

[Par1] Parry, W., *Entropy and Generators in Ergodic Theory*, W. A. Benjamin, Inc. 1969.

[Par2] Parry W., *Dynamical systems on nilmanifolds*. Bull. London Math. Soc. **2** (1970), 37–40.

[ParPo] Parry W., and M. Pollicott, *Stability of mixing for toral extensions of hyperbolic systems*, Tr. Mat. Inst. Steklova **216** (1997), Din. Sist. i Smezhnye Vopr., 354–363.

[Pe] Pesin, Ya., *Characteristic Lyapunov exponents and smooth ergodic theory*, Russian Math. Surveys **32** (1977), no. 4(196), 55–112, 287.

[Pl] Plante, J., *Anosov flows*, Amer. J. Math. **94** (1972), 729–754.

[PughSh1] Pugh, C. and M. Shub, *Ergodic elements of ergodic actions*, Compositio Math. **23** (1971), 115–122.

[PugSh2] Pugh, C. and M. Shub, *Ergodicity of Anosov actions*, Invent. Math. **15** (1972), 1–23.

[PugSh3] Pugh, C. and M. Shub, *Ergodic attractors*, Trans. Amer. Math. Soc. **312** (1989), 1–54.

[PugSh4] Pugh, C. and M. Shub, *Stable ergodicity and partial hyperbolicity*, in International Conference on Dynamical Systems: Montevideo 1995, a Tribute to Ricardo Mañé Pitman Res. Notes in Math. **362** (F.Ledrappier et al, eds.) 182-187.

[PugSh5] Pugh, C. and M. Shub, *Stably ergodic dynamical systems and partial hyperbolicity*, J. of Complexity **13** (1997), 125–179.

[PugSh6] Pugh, C. and M. Shub, *Stable ergodicity and julienne quasiconformality*, J. Eur. Math. Soc. **2** (2000), 1–52.

[PugShWi] Pugh, C., M. Shub and A. Wilkinson, *Hölder Foliations*, Duke Math. J. **86** (1997), 517–546.

[PujSam] Pujals, E. and M. Sambarino, *Homoclinic tangencies and hyperbolicity for surface diffeomorphisms: a conjecture of Palis*, IMPA preprint (1998).

[Rud] Rudolph, D., *Classifying the isometric extensions of a Bernoulli shift*, J. Analyse Math. **34** (1978), 36–60.

[Rue] Ruelle, D., *A measure associated with Axiom-A attractors*, Amer. J. Math., **98** (1976), 619–654.

[RueWi] Ruelle, D. and A. Wilkinson, *Absolutely singular dynamical foliations*, Preprint (2000).

[Sa] Sacksteder, R. *Strongly mixing transformations*, Global Analysis (Proc. Sympos. Pure Math., Vol. XIV, Berkeley, Calif., 1968), pp. 245–252. Amer. Math. Soc., Providence, R. I., 1970.

[Sh] Shub, M., *Topologically transitive diffeomorphisms of* T^4, Symposium on Differential Equations and Dynamical Systems, (D. Chillingworth ed.) Springer Lecture Notes in Mathematics **206** (1971), 39–40.

[ShWi1] Shub, M. and A. Wilkinson, *Stably ergodic approximation: two examples*, Ergod. Th. and Dyam. Syst. **20**, (2000), 875–894.

[ShWi2] Shub, M. and A. Wilkinson, *A stably Bernoullian diffeomorphism that is not Anosov*, Preprint (1998).

[ShWi3] Shub, M. and A. Wilkinson, *Pathological foliations and removable zero exponents*, Invent. Math. **139**, (2000), 495–508.

[Si] Sinai, Ya., *Gibbs measures in ergodic theory*, Russian Math. Surveys **27** (1972), 21–69.

[Sm1] Smale, S., *Structurally stable systems are not dense*, Amer. J. Math. **88** (1966), 491–496.

[Sm2] Smale, S., *Differentiable dynamical systems*, Bull. A.M.S. **73** (1967), 747-817.

[Vi] Viana, M., *Multidimensional nonhyperbolic attractors*, Inst. Hautes Études Sci. Publ. Math. No. 85 (1997), 63–96.

[Wa] Walkden, C., *Stable ergodic properties of cocycles over hyperbolic attractors*, Comm. Math. Phys. **205** (1999), 263–281.

[Wi] Wilkinson, A. *Stable ergodicity of the time one map of a geodesic flow*, Ergod. Th. and Dynam. Syst. **18** (1998), 1545–1588.

[Xia] Xia, Z., *Existence of invariant tori in volume-preserving diffeomorphisms*, Ergod. Th. and Dynam. Syst. **12** (1992), 621–631.

[Yoc] Yoccoz, J.-C., *Travaux de Herman sur les tores invariants*, Séminaire Bourbaki, Vol. 1991/92. Astérisque No. 206 (1992), Exp. No. 754, **4**, 311–344.

[You] Young, L.-S., *Ergodic theory of chaotic dynamical systems*, Int. Congress of Math. Phys., 1997.

Department of Mathematics, Northwestern University, Evanston, IL 60208
E-mail address: burns@math.northwestern.edu

Department of Mathematics, University of California, Berkeley, CA 94720
E-mail address: pugh@math.berkeley.edu

IBM T. J. Watson Research Center, P. O. Box 218, Yorktown Heights, NY 10598
E-mail address: mshub@us.ibm.com

Department of Mathematics, Northwestern University, Evanston, IL 60208
E-mail address: wilkinso@math.northwestern.edu

Proceedings of Symposia in Pure Mathematics
Volume **69**, 2001

Statistical Properties of Some Almost Hyperbolic Systems

Huyi Hu

ABSTRACT. Almost hyperbolic systems are smooth dynamical systems that are hyperbolic everywhere except for a finite set of points. Evidences from low-dimensional cases show that the topological properties of such systems are similar to that of uniformly hyperbolic systems, and the ergodic properties may be quite different. It admit SRB measures or infinite SRB measures. In the latter case, the systems are statistically deterministic in the sense that almost every orbit spends one hundred percent of its time arbitrarily close to the indifferent fixed points, though they are still topologically mixing. Even in the former case, correlation decay may change from exponential to power law.

Introduction

A smooth dynamical system is almost hyperbolic if it is hyperbolic everywhere except at a finite set of points. Ergodic properties of an almost hyperbolic system may be quite different from that of a uniformly hyperbolic system, even though we can make them topologically conjugate.

Let us imagine that we slowly deform a hyperbolic toral automorphism near the origin until the differential becomes identity. For some deformation the system does not admit any Sinai-Ruelle-Bowen measure. Instead, it admits a σ-finite measure that has absolutely continuous conditional measures on unstable manifolds. We call it an *infinite SRB measure.* For some other deformation, the resulting system still admits an SRB measure. However, the rate of decay of correlations of the system changes from exponential to polynomial. Similar phenomena also exist in some expanding maps with indifferent fixed points.

These properties imply different statistical behavior for the system. If the system has an infinite SRB measure, then for Lebesgue almost every initial condition, the orbit spends nearly one hundred percent of its time arbitrarily close to the indifferent fixed point, though it is also true that almost every orbit is dense on the torus. Hence, the system is deterministic from statistical point of view, and chaotic from topological point of view. The origin is a saddle, but statistically it

2000 *Mathematics Subject Classification.* Primary 37A25; Secondary 37D25, 37C40.

Key words and phrases. Almost hyperbolic system, almost Anosov system, SRB measure, infinite SRB measure, decay of correlations.

The work is partially supported by the National Science Foundation grant #DMS-9970646.

looks like a sink. For the case that the system admits an SRB measure, the orbits visit every open set with positive frequency. However, they spend longer time near the indifferent fixed point, since the density of the measure is unbounded there. Therefore, the rate of mixing of the system, if measured by the rate of decay of correlations, is slower. So we can think that the system is "less chaotic" than the undeformed system.

The behavior of such systems mainly depends on the distribution of the physically observable measure μ in the neighborhoods of the indifferent fixed point. More precisely, it depends on how fast the measure escape from the neighborhoods. Let P_n be the sets of points that escape from a neighborhood at time n. If $\sum_{n=1}^{\infty} \mu P_n$ diverges, then the system has an infinite SRB measure, and it is statistically deterministic. If $\sum_{n=1}^{\infty} \mu P_n$ converges, but μP_n decrease in a polynomial rate, then the system has an SRB measure and has polynomial decay of correlations. This is the "less chaotic" case. If μP_n decrease exponentially fast, then the system has an SRB measure and has exponential decay of correlations. This happens in uniformly hyperbolic systems and some other nonuniformly hyperbolic systems people have studied.

Sometimes the information can be observed directly from the dynamics. Let us consider a piecewise smooth almost expanding map with an indifferent fixed point p at which $|\det Df_p| = 1$. The potential functions ϕ we are interested in (for example, $\phi = -\log|\det Df|$,) is negative and generically is of order n^{-1} on P_n. If on P_n, $\phi \sim -(\beta+1)/n$, then the absolutely continuous invariant measure μ is infinite for $0 < \beta \le 1$, and is finite with $O(n^{-(\beta-1)})$ as the rate of decay of correlations for $\beta > 1$. The systems we will discuss in Theorem 3.11, Fact 3.14 and Theorem 3.17 belong to the case. However, β may vary with x on P_n for systems in higher dimensional spaces. Therefore more detailed analysis is required.

In this paper we mainly consider almost Anosov systems on the surfaces and almost expanding maps on the intervals. We will describe the local behavior of the systems near the indifferent fixed points, and introduce the main idea for existence of SRB measures and rates of decay of correlations. Similar results should hold in general almost hyperbolic systems, however, we don't know very much beyond these maps.

1. Definitions

We start with hyperbolic case. We define the rank of a cone in $\mathbb{R}^n$ as the maximal dimension of a linear subspace contained in the cone. A pair of cones is called complete if the sum of their ranks is n.

DEFINITION 1.1. Let f be a C^r, $r > 1$, diffeomorphism from a manifold M to itself. A closed invariant subset $\Lambda \subset M$ is called an *almost hyperbolic set*, if there exist two continuous families of complete closed cones $\mathcal{C}^u_x$ and $\mathcal{C}^s_x$ on the tangent bundle $T_\Lambda M$ such that except at a finite set $S \subset \Lambda$,

i) (invariance) $Df_x\mathcal{C}^u_x \subset \operatorname{int}\mathcal{C}^u_{fx}$, and $Df_x\mathcal{C}^s_x \supset \operatorname{int}\mathcal{C}^s_{fx}$;

ii) (hyperbolicity) $|Df_x(v)| > |v|$, $|Df^n_x(v)| \to \infty$ $\forall v \in \mathcal{C}^u_x$, and $|Df_x(v)| < |v|$, $|Df^n_x(v)| \to 0$ $\forall v \in \mathcal{C}^s_x$.

If $\Lambda = M$, then we call the system an *almost Anosov diffeomorphism*.

It is possible that $|Df_x(v)| = |v|$ for $|v| \in \mathcal{C}_x^u$ or $\mathcal{C}_x^s$ if $x \in S$. So the systems are not uniformly hyperbolic.

The reason that we use cones rather than decomposition $TM = E^u \oplus E^s$ is that the unstable and stable bundles may not be continuous. In uniformly hyperbolic systems, continuity of $x \to \mathcal{C}_x^u$ and $\mathcal{C}_x^s$ follows from invariance and hyperbolicity. If we allow $Df_p = \text{id}$ for some $p \in S$, then the cone families are not necessary continuous. In fact, the curvature of unstable and stable leaves depends on hyperbolicity. Near these points, the leaves can have a u-turn. This happens in an example constructed by A. Katok [K] of a C^∞ Bernoulli diffeomorphism on the 2-dimensional sphere that is hyperbolic everywhere except at four points. In [GK] M. Gerber and A. Katok constructed a C^∞ smooth model for pseudo-Anosov homeomorphisms in which there are more than one unstable and stable directions near the exceptional points. (See also [G] for analytic models.) These are impossible if the cone families continuous at every point in the manifolds.

One may also consider the case that S is a submanifold of M. (See [N] for a generalization of Theorem 3.8 to the case.)

For noninvertible systems, we consider piecewise almost expanding maps.

DEFINITION 1.2. Let f be a C^r, $r > 1$, piecewise smooth map from a manifold M to itself, and $\Lambda \subset M$ be a closed invariant subset. We say that f is *almost expanding* on Λ if f is uniformly expanding away from a finite set S. That is, for any $\epsilon > 0$, there is $\kappa > 1$ such that if $x \in \Lambda \backslash B(S, \epsilon)$, then

$$|Df_x(v)| > \kappa |v| \qquad \forall v.$$

Without loss generality, we assume that in both cases S is an f-invariant set. By considering f^n, we assume that S consists of fixed points. We are interested in the case that the fixed points are indifferent, because otherwise the systems become uniformly hyperbolic or uniformly expanding. Further, we assume that S consists of only one indifferent fixed point p except when otherwise stated.

DEFINITION 1.3. A fixed point p is called *indifferent* if Df_p has an eigenvalue λ with $|\lambda| = 1$.

In this paper we only consider the systems that are topologically transitive, and allow Markov partitions $\mathcal{R}$ into finite number of elements $\{R_j\}$. We denote $\mathcal{R}_n = \bigvee_{i=0}^{n-1} f^{-i}\mathcal{R}$, and then let $\mathcal{R}_n(x) \in \mathcal{R}_n$ be the element that contains x. Restricted to each element, f is C^r with $r > 1$. So the systems are smooth or piecewise smooth. We also assume that R_0 is the element containing p. We denote

$$P_n = \{x \in R_0 : f^k x \in R_0, 0 \le k \le n-1, f^n x \notin R_0\}, \tag{1.1}$$

$Q_n = \bigcup_{i=0}^{n} P_i$ and $\mathcal{O}_n = \bigcup_{i=n}^{\infty} P_i$. We will regard $P_0 = \Lambda \backslash R_0$

We will use the notation "$\sim$" very often, where $a(n) \sim b(n)$ means that $a(n) = b(n) + O(n^{-(1+t)})$ for some $t > 0$.

2. Dynamics near Indifferent Fixed Points

Consider an almost expanding map f on the unit interval I. Assume that 0 is the fixed point and I_0 is the element of the Markov partition that contains 0.

Take $x \in I_0$. Consider the backwards orbit $\{x_n\}_{n=0}^{\infty}$ of x, where $x_n = f^{-n}x \cap I_0$. We know that if $f'(0) > 1$, then x_n converges to 0 exponentially fast. In our case, the rate is polynomial.

FACT 2.1. *Suppose*

$$fx = x\big(1 + x^{\gamma} + o(x^{\gamma})\big), \qquad x \to 0. \tag{2.1}$$

Then for any $x \in I_0$, the backwards orbit $\{x_n\}_0^{\infty}$ satisfies $x_n^{\gamma} \sim (\gamma n)^{-1}$.

REMARK 2.2. By the fact we can see that on P_n,

$$f'x \sim 1 + \frac{1+\beta}{n}, \qquad \beta = 1/\gamma. \tag{2.2}$$

We cannot expect bounded distortion on any interval that contains 0, since $(f^n)'(0) = 1$ and $(f^n)'(x_n)$ is unbounded by Fact 2.1 and (2.2). However, the following fact says that restricted to P_k, distortion is bounded. This is enough for us to get absolute continuity of invariant measures and to estimate the rates of decay of correlations.

PROPOSITION 2.3. *There exists constant $J > 0$ such that for any $x, y \in P_k$, for any $n > 0$, if $x_n \in f^{-n}x$ and $y_n \in f^{-n}y \cap \mathcal{R}_n(x_n)$, then*

$$\log \frac{\big|(f^n)'(y_n)\big|}{\big|(f^n)'(x_n)\big|} \le \frac{J}{x} d(x, y).$$

Now we consider almost hyperbolic systems on surfaces. Let p be an indifferent fixed point. If one of eigenvalues of Df_p is not equal to 1, then there is a gap between the spectrum of Df. So we can apply the stable manifold theorem ([HPS]) to obtain unstable and stable manifolds, which form C^1 foliations. The latter means that the holonomy maps by sliding along the unstable or stable leaves are C^1. So near p, by choosing a suitable coordinate system, the map can be written as

$$f(x, y) = \big(x + ax^{2k+1} + \text{h.o.t.},\ \lambda y + \text{h.o.t.}\ \big), \tag{2.3}$$

for some $\lambda > 1$, $a < 0$, or $\lambda < 1$, $a > 0$. Hence the orbits change polynomially in one direction and exponentially in the other direction. (Note that we need odd power for the second term in x-component to keep the map hyperbolic.)

For the case that 1 is the only eigenvalues of Df_p, there are two possibilities: the eigenspace of Df_p is one or two dimensional. In the first case the unstable and stable manifolds are tangent at p. These maps were studied by J. Lewowicz ([Lo]). Recently, it is proved by E. Catsigeras and H. Enrich [CE] that with some conditions on the coefficient of the Taylor expression, f have an SRB measure.

In the second case, which is simpler and has been understood better, $Df_p = \text{id}$. We assume the systems satisfy nondegenerate conditions below. It seems that similar results are also true if we use higher order jet instead of the third order one. However, there is no reference about it.

DEFINITION 2.4. An almost Anosov diffeomorphism $f : M \to M$ is said to be *nondegenerate*, if there are constants $\kappa^+, \kappa^- > 0$ such that for all x near p,

$$|Df_x v| \ge \big(1 + \kappa^+ d(x, p)^2\big)|v| \qquad \forall v \in \mathcal{C}_x^u,$$
$$|Df_x v| \le \big(1 - \kappa^- d(x, p)^2\big)|v| \qquad \forall v \in \mathcal{C}_x^s\ .$$

Note that $Df_p = \mathrm{id}$ by the assumption, $D^2 f_p = 0$ by hyperbolicity, and $D^3 f_p \neq 0$ by the nondegenerate conditions. So near p the motion is dominated by $D^3 f_p$, which makes the dynamics quite different (see [H1] for more details).

Since $D^3 f_p$ is a 3-form, the immediate consequence is that near p, f^{n^2} acts on x/n just like f acting on x up to a scale of n times. That is, $f^{n^2}(tx/n) \sim f(tx)/n$, or more precisely, $\lim_{t\to 0} \dfrac{f^{n^2}(tx/n)}{f(tx)/n} = 1$. Df^{n^2} has the same property. We should recall that near a hyperbolic fixed point, the corresponding relation is $f(tx/n) \sim f(tx)/n$, since f is close to a hyperbolic linear map. The facts are very helpful for us to understand the behavior of orbits when they are near p.

The system allows a decomposition of the tangent bundle into $TM = E^u \oplus E^s$. For $x \neq p$, existence and continuity of E^u_x and E^s_x can be obtained in the same way as for uniformly hyperbolic systems. Further, E^u_x and E^s_x satisfy the Hölder condition outside any neighborhood of p. (The Hölder coefficient increases as the size of the neighborhood shrinks.) We can also define E^u_p and E^s_p: they are the only lines such that $D^3 f_p(v, v, v)$ and v collinear for any v in the line. From the similarity between Df at x and Df^{n^2} at x/n, we can see that for any x, E^u_{tx} and E^s_{tx} has a limit as $t \to 0$, and the limits are different if the directions of x are different. This is because near p, $Df_x = \mathrm{id} + (1/2)D^3 f(x, x, \cdot)$, and $D^3 f(x, x, \cdot)$ change as the direction of x change, while near a hyperbolic fixed point, all Df_x are close to Df_p.

Note that between unstable and stable parts, there is no gap in the spectrum of Df. The stable manifold theorem doesn't apply. However, (weak) unstable and stable manifolds can be obtained by graph transformation for $x \neq p$. $W^u_{\mathrm{loc}}(p)$ and $W^s_{\mathrm{loc}}(p)$ can be defined as the set of points whose backward or forward orbits stay inside a small neighborhood of p. They are smooth curves and tangent to E^u_p and E^s_p respectively.

We have similar distortion estimates as for one dimensional case. The following result is proved in [H1] and refined in [H4].

Let P be a rectangle with $p \in \mathrm{int}\, P$. Recall the definition of P_n in (1.1).

PROPOSITION 2.5. *There exists constant $I > 0$, depending on P, $\theta \in (0, 1)$ such that for any $x, y \in P_k$, $k \geq 0$, with $y \in W^u_{\mathrm{loc}}(x)$,*

$$\log \frac{|Df_y^{-n}|_{E^u_y}|}{|Df_x^{-n}|_{E^u_x}|} \leq I d^u(x, y)^\theta \qquad \forall n > 0.$$

Away from p, the unstable and stable foliations are Lipschitz. That is, the holonomy maps $\pi_{xy} : W^{u(s)}_{\mathrm{loc}}(x) \to W^{u(s)}_{\mathrm{loc}}(y)$ are Lipschitz, if they stay away from $W^u_{\mathrm{loc}}(p)$ and $W^s_{\mathrm{loc}}(p)$. (Of cause, the Lipschitz constants increase as they get closer to p.) On the other hand, for any x close to p, the holonomy maps from $W^u_{\mathrm{loc}}(x)$ to $W^u_{\mathrm{loc}}(p)$ is only Hölder. Let us assume that in a neighborhood of p the map f can be written as the following in a local coordinate system,

$$f\begin{pmatrix} x \\ y \end{pmatrix} = \begin{pmatrix} x + x\big(ax^2 + by^2 + O((x^2 + y^2)^{3/2})\big) \\ y - y\big(cx^2 + dy^2 + O((x^2 + y^2)^{3/2})\big) \end{pmatrix}. \tag{2.4}$$

Note that for any $z \in W^u_{\mathrm{loc}}(p)$, $d(p, f^{-n}z)$ have the order $n^{-1/2}$ by Fact 2.1.

PROPOSITION 2.6. *Let $q \in W^s_{\mathrm{loc}}(p)$ with $q \neq p$. Then for any $z \in W^u_{\mathrm{loc}}(p)$, $d\big(q, \pi_{pq}(f^{-n}z)\big) = O(n^{-b/2d})$.*

The non-Lipschitzness of the stable foliation is very important for the statistical behavior of the systems. It is $d(q, \pi_{pq}(f^{-n}z))$, not $d(p, f^{-n}z)$, that plays the role, as we will see in Proposition 3.4 and Theorem 4.5. (If we construct a tower as in [Y1] or [Y2], then $d(q, \pi_{pq}(f^{-n}z))$ is the size of the n'th level.)

3. Invariant Measures

3.1. SRB Measures and Infinite SRB Measures. For an Anosov system $f : M \to M$, a result of Sinai (see e.g. [S]) says that f admits a unique invariant Borel probability measure μ with the property that μ has absolutely continuous conditional measures on unstable manifolds. This is the invariant measure that is observed physically, for if $\phi : M \to \mathbb{R}$ is a continuous function, then for Lebesgue almost every point $x \in M$,

$$\lim_{n\to\infty} \frac{1}{n} \sum_{i=0}^{n-1} \phi(f^i x) = \int \phi d\mu.$$

These results have been extended to Axiom A attractors by Ruelle, Bowen, etc. (See e.g. [B].) This measure is called an Sinai-Ruelle-Bowen measure or an SRB measure. Now there are some results about existence of SRB measures for nonuniformly hyperbolic systems such as Hénon maps and hyperbolic systems with singularities. (See [BY], [P], [JN], [BV], etc.)

DEFINITION 3.1. An f-invariant probability measure μ is called an *SRB measure* for $f : \Lambda \to \Lambda$ if

i) (f, μ) has positive Lyapunov exponents almost everywhere;
ii) μ has absolutely continuous conditional measures on unstable manifolds.

For the precise meaning of the second condition above, we refer [LS].

Some almost Anosov systems do not admit SRB measures. Instead, they admit σ-finite measures that have similar properties.

DEFINITION 3.2. An f-invariant σ-finite measure μ is called an *infinite SRB measure* for $f : \Lambda \to \Lambda$ if $\mu\Lambda = \infty$, and for any $R \subset \Lambda$ with $\mu(R) < \infty$ and $R = \overline{\operatorname{int} R}$,

i) (f_R, μ_R) has positive Lyapunov exponents almost everywhere, and
ii) μ_R has absolutely continuous conditional measures on unstable manifolds of f.

where f_R is the first return map of f on R, and μ_R is the conditional measure of μ on R.

In this paper, the term "SBR measure" without any qualifications will be reserved for probability measures.

We will see in Corollary 3.5 that orbit distributions are quite different between systems with SRB measures and those with infinite SRB measures.

For almost Anosov systems on compact surfaces, we have the following.

THEOREM 3.3. *Every nondegenerate almost Anosov diffeomorphism on a surface has either an SRB measure or an infinite SRB measure.*

PROOF. Take a suitable rectangle P that contains p as its interior point. Let $g : M\backslash P \to M\backslash P$ be the first return map.

Take a piece of unstable curve γ. Let m_γ be the Lebesgue measure on γ and let $g_*^n m_\gamma$ be the push-forward of m_γ, i.e., $(g_*^n m_\gamma)(E) = m_\gamma(g^{-n}E)$. Then any accumulation point $\bar{\mu}$ of $n^{-1}\sum_{i=0}^{n-1} g_*^i m_\gamma$ in the weak* topology is a g-invariant measure.

Let ρ_i denote the density of $g_*^i m_\gamma$ with respect to the Lebesgue measure on $g^i\gamma$. Then by distortion estimates in Proposition 2.5 for any small rectangle $R \subset M\backslash P$, for all x, y in a same component of $g^i\gamma \cap R$ that crosses R,

$$e^{-I} \le \frac{\rho_i(x)}{\rho_i(y)} \le e^I. \tag{3.1}$$

This bound on ρ_i is passed on to the limit measure $\bar{\mu}$. Then we get absolute continuity.

We can extend $\bar{\mu}$ to an f-invariant measure μ in an obvious way. If μ is finite, then we get an SRB measure (after normalization), otherwise we get an infinite SRB measure. □

To determine if μ is finite, we only need check if the return time on $f^{-1}P\backslash P$ is integrable, or equivalently, to see if μP is finite. The following proposition gives a direct way to estimate it, in which the notations are the same as in Proposition 2.6.

PROPOSITION 3.4. *Let $q \in W^s_{\text{loc}}(p)$ with $q \in f^{-1}P\backslash P$. Then μ is finite if and only if $\sum_{n=1}^{\infty} d(q, \pi_{pq}(f^n z))$ is convergent.*

PROOF. Let $P'_n = \{x \notin P : f^i x \in P,\ 1 \le i \le n\}$. We have $\mu P = \sum_{n=1}^{\infty} \bar{\mu} P'_n$ since $P = \bigcup_{n=1}^{\infty} f^n P'_n$. $\mu P'_n$ is proportional to the length $l(\gamma_n^u)$ of any unstable curve $\gamma_n^u \subset P'_n$ that crosses P'_n, because (3.1) also holds for the density function of the conditional measures of μ. Now we use the fact that $l(\gamma_n^u)$ is proportional to $d(q, \pi_{pq}(f^n z))$ □

COROLLARY 3.5. *Suppose $f : M \to M$ is an almost Anosov diffeomorphism with only one indifferent fixed point p. Let $\phi : M \to \mathbb{R}$ be a continuous function.*

If f admits an SRB measure μ, then for Leb-a.e. $x \in M$,

$$\lim_{n\to\infty} \frac{1}{n} \sum_{i=0}^{n-1} \phi(f^i x) = \int \phi \, d\mu.$$

If f admits an infinite SRB measure μ, then for Leb-a.e. $x \in M$,

$$\lim_{n\to\infty} \frac{1}{n} \sum_{i=0}^{n-1} \phi(f^i x) = \phi(p).$$

The second part of the corollary is true because the measure of any neighborhood of p is infinite, while the measure of its complement is finite.

As we mentioned before that if f admits an infinite SRB measure, then it is statistically deterministic and topologically chaotic. Here is another feature.

COROLLARY 3.6. *If f has an infinite SRB measure, then the forward Lyapunov exponents are 0 for Leb-a.e. $x \in M$.*

PROOF. Take $\phi(x) = \log |Df_x|_{E^u_x}|$ and $\log |Df_x|_{E^s_x}|$ in the above corollary. Continuity of ϕ follows from continuity of E^u_x and E^s_x for $x \neq p$ and the fact $Df_x \to Df_p = \text{id}$ as $x \to p$. □

It seems that whether a system has an SRB measure depends on whether expansion is stronger than contraction near the indifferent fixed point. Here we discuss some special cases. Assume that Df_p has eigenvalues $\overline{\lambda} \geq \underline{\lambda} > 0$.

FACT 3.7. *If $\overline{\lambda} > 1$, then f has an SRB measure.*

This is because the system is uniformly expanding along unstable directions. So distortion for $Df|_{E^u}$ is bounded. Therefore the same "push-forward" method gives an probability invariant measure.

If $\underline{\lambda} < 1$ besides $\overline{\lambda} > 1$, then the system is an Anosov system.

THEOREM 3.8 ([HY]). *If $\overline{\lambda} = 1$, $\underline{\lambda} < 1$, then f has an infinite SRB measure.*

PROOF. By (2.3) and Fact 2.1, $d(p, f^{-n}z)$ is of order $n^{-1/2k}$ $\forall z \in W^u_{\text{loc}}(p)$. Then the result follows from Proposition 3.4 and the fact that the stable foliation is C^1. □

If $Df_p = \text{id}$, then we need look at the third order terms of the Taylor expression of the map at p. We only have some sufficient conditions to guarantee that both cases in Theorem 3.3 do occur. Assume that we can choose a suitable coordinate system such that f can be written as

$$f\begin{pmatrix} x \\ y \end{pmatrix} = \begin{pmatrix} x + x\big(ax^2 + rxy + by^2 + O((x^2+y^2)^{3/2})\big) \\ y - y\big(cx^2 + sxy + dy^2 + O((x^2+y^2)^{3/2})\big) \end{pmatrix}.$$

THEOREM 3.9 ([H1]). *With the above notations,*

i) *f has an SRB measure if $b > 2d$, $r = 0 = s$, and $ad > bc$;*

ii) *f has an infinite SRB measure if $2b < d$ and $rs \neq 0$.*

PROOF. Part i) follows from Proposition 2.6 and 3.4. For Part ii), we can proof that $d\big(q, \pi_{pq}(f^{-n}z)\big)$ is bounded below by $O(n^{-1/2})$. □

The conditions in the first part say $a/c > b/d > 2$. So expansion is stronger. For the second part, we can see that at least there is a cone on the plane in which contraction is stronger.

If the only eigenvalue of Df_p is 1 and $Df_p \neq \text{id}$, then near p, f has the form

$$f(x, y) = (x + ay + r(x, y), y + s(x, y)),$$

where $a \neq 0$, and r and s consist of higher order terms of x and y.

THEOREM 3.10 ([CE]). *If the coefficient of x^2 in r is zero and coefficient of x^3 in s is nonzero, then f has an SRB measure.*

3.2. Absolutely Continuous Invariant Measures. It is well-known that piecewise smooth uniformly expanding systems on the unit interval have absolutely continuous invariant measures. (A folklore theorem, also see e.g. [KS], [W] for higher dimensional results.) For almost expanding maps on the unit interval, it is proved by G. Pianigiani [Pi] that absolutely continuous invariant measures exist, which may be finite or σ-finite.

The proof in [Pi] uses the first return maps. Here we give an alternative one. This method can be easily adopted to a more general setting.

THEOREM 3.11. *Let $f : I \to I$ be a piecewise smooth expanding map with an indifferent fixed point 0. Assume that near 0, f satisfies (2.1). Then f has an absolutely continuous invariant measure μ, and μ is finite if $0 < \gamma < 1$, σ-finite if $\gamma \geq 1$.*

PROOF. Take $\phi(x) = -\log f'(x)$. Define the Perron-Frobenius operator by

$$\mathcal{L}_\phi g(x) = \sum_{fx_1 = x} e^{\phi(x_1)} g(x_1).$$

Since the fixed point of $\mathcal{L}_\phi$ is unbounded near 0, we take a Banach space $\mathcal{B} = C^0(0,1]$ but with the norm $\|g\| = \sup\{x^{2\gamma J} g(x) : x \in I\}$. Then we construct a convex set

$$\begin{aligned} \mathcal{G} = \{g \in \mathcal{B} :\ & g > 0,\ g(1) = 1,\ \exists H_0 \text{ s.t. } x^\gamma g(x) \leq H_0\ \forall x \in I_0, \\ & g(y) \leq g(x)(1 + Jx^{-1}d(x,y)\ \forall x, y \in P_n\}. \end{aligned} \tag{3.2}$$

Note that as $x \to 0$, $x^{2\gamma J} g(x) \to 0$ uniformly for all $g \in \mathcal{G}$. So $\mathcal{G}$ is compact in $\mathcal{B}$.

Let $\tilde{\mathcal{L}}_\phi g = \mathcal{L}_\phi g / \mathcal{L}_\phi g(1)$. Clearly, $\tilde{\mathcal{L}}_\phi$ is continuous. It can be proved that $\tilde{\mathcal{L}}_\phi \mathcal{G} \subset \mathcal{G}$. So by Schauder-Tychonoff fixed point theorem, $\tilde{\mathcal{L}}_\phi$ has a fixed point $h \in \mathcal{G}$. Hence, $\mathcal{L}_\phi h = h$. Using h as the density function, we get an invariant measure μ. By the corollary below, h is integrable if and only if $\gamma < 1$. □

The following result tells that on P_n, $h(x)$ increases like $\gamma n \tilde{h}(0)$, where $\tilde{h}(x) = \sum_{\hat{x}_1 \in f^{-1}x \setminus I_0} \frac{h(\hat{x}_1)}{f'(\hat{x}_1)}$. (Recall Fact 2.1.)

COROLLARY 3.12. $\lim_{x \to 0} x^\gamma h(x) = \tilde{h}(0)$.

PROOF. Observe that $\mathcal{L}_\phi h = h$ implies that for $x \in I_0$, if $x_1 = f^{-1}x \cap I_0$, then

$$x_1^\gamma h(x_1) \sim x^\gamma h(x) \frac{x_1^\gamma}{x^\gamma} \cdot f'(x_1) - \tilde{h}(0) x_1^\gamma f'(x_1),$$

where $\sim$ means that the difference between the two sides is of order higher than x^γ. Since $f'(x) \sim 1 + (1+\gamma)x^\gamma$ and $x^\gamma / x_1^\gamma \sim 1 + \gamma x_1^\gamma$, we have

$$x_1^\gamma h(x_1) \sim x^\gamma h(x)(1 + x_1^\gamma) - \tilde{h}(0) x_1^\gamma.$$

So if there is $c > 0$ such that $x^\gamma h(x) \geq \tilde{h}(0)(1 + c)$ for a small x, then

$$x_1^\gamma h(x_1) \geq \tilde{h}(0)(1 + c + 0.5 c x_1^\gamma)$$

because $(1 + c)(1 + x_1^\gamma) - x_1^\gamma = 1 + c + c x_1^\gamma$. Inductively, $x_n^\gamma h(x_n) \geq h(0)(1 + c + 0.5c \sum_{i=1}^n x_i^\gamma)$. Since $\sum_{i=1}^n x_i^\gamma$ diverges, it contradicts the fact that $x^\gamma h(x) \leq H_0$ for all $x \in I_0$. So the upper limit is bounded by $\tilde{h}(0)$. The lower limit can be proved similarly. □

By checking the proof carefully, we can actually get the following.

THEOREM 3.13. *Let $f : \Lambda \to \Lambda$ be a piecewise smooth almost expanding map with an indifferent fixed point. Suppose $\phi = -\log|\det Df_x|$ satisfies the following.*

i) *There is $J_\phi > 0$ such that $S_n\phi(y_n) - S_n\phi(x_n) \leq J_\phi x^{-1} d(x,y)\ \ \forall x, y \in P_n, x_n \in f^{-n}x, y_n \in f^{-n}y \cap \mathcal{R}_n(x_n)$.*

ii) *On P_n, $\phi(x) \leq -\dfrac{\beta + 1}{n}$ for some $\beta > 0$.*

Then f has an absolutely continuous invariant measure μ. Further, μ is finite if $\beta > 1$, and σ-finite if moreover $\phi(x) \geq -\dfrac{\beta'+1}{n}$ on P_n for some $0 < \beta' \leq 1$.

Theorem 3.11 says that we can use γ to determine finiteness of absolutely continuous invariant measures of almost expanding systems on intervals. It is unclear what the higher dimensional versions are. However, in some simple cases, we can get similar results, which can be regarded as an application of Theorem 3.13.

Assume that an almost expanding map $f : I^m \to I^m$ has the form

$$f(x) = x(1 + |x|^\gamma) + o(x^{1+\gamma})$$

as x near 0. That is, f is a map that expands at the same rate in all directions at 0, plus a higher order perturbation. Then with the same methods, we have:

FACT 3.14. *f has an absolutely continuous invariant measure μ, which is finite if $0 < \gamma < m$, and σ-finite if $\gamma \geq m$.*

PROOF. Check that on P_n, $\phi \sim -\dfrac{1+m/\gamma}{n}$. This also implies the first condition in the above theorem. □

However, in higher dimensional spaces, ϕ may varies, say, from $-(1+\beta_1)/n$ to $-(1+\beta_2)/n$, $\beta_1 \neq \beta_2$, on P_n. If so, the first condition in Theorem 3.13 usually fails. In this case we need modify the proof. Here is an example.

FACT 3.15. *Suppose $f : I^2 \to I^2$ is an almost expanding map and near 0,*

$$f\begin{pmatrix} x_1 \\ x_2 \end{pmatrix} = \begin{pmatrix} x_1 + x_1^{1+\gamma_1} + o(x_1^{1+\gamma_1}) \\ x_2 + x_2^{1+\gamma_2} + o(x_2^{1+\gamma_2}) \end{pmatrix}.$$

Then f has an absolutely continuous invariant measure μ, which is finite if $\gamma_1^{-1} + \gamma_2^{-1} > 1$, and σ-finite if $\gamma_1^{-1} + \gamma_2^{-1} \leq 1$.

PROOF. We can partition each P_n into countable many smaller pieces, such that on each of them the map has bounded distortion. To make the set $\mathcal{G}$ compact, we use $x_1^{2\gamma_1 J} x_2^{2\gamma_2 J} g(x_1, x_2)$ to define the norm in the space $\mathcal{B}$. To compute integral of return time, we compute the sum of the area of $\mathcal{O}_n$, which are rectangles of size approximately $(\gamma_1 n)^{-1/\gamma_1} \times (\gamma_2 n)^{-1/\gamma_2}$ by Fact 2.1. □

If an almost expanding system has an absolutely continuous invariant measure μ, and it contains only one indifferent fixed point p, then a statement similar in Corollary 3.5 can be made. That is, for Lebesgue almost every point, the time average is equal to $\int \phi d\mu$ if μ is finite, and is equal to $\phi(p)$ if μ is σ-finite.

If f has two indifferent fixed points that cause σ-finite measure, then from the case of one dimensional systems, the limits for time average may not exist.

THEOREM 3.16 ([I]). *Suppose f is a piecewise smooth expanding map on the interval with indifferent fixed points p and q such that*

$$\begin{aligned} |f(x) - x| &= a|x-p|^{1+\alpha} + o(|x-p|^{1+\alpha}), \qquad && x \to p; \\ |f(x) - x| &= b|x-q|^{1+\beta} + o(|x-q|^{1+\beta}), \qquad && x \to q. \end{aligned}$$

Then for any small neighborhoods U and V of p and q respectively, Leb-a.e. x,

$$\lim_{n\to\infty} \frac{\sum_{i=0}^n 1_U(f^i(x))}{\sum_{i=0}^n 1_V(f^i(x))} = \begin{cases} 0 & \text{if } \beta > \alpha \geq 1, \\ c > 0 & \text{if } \beta = \alpha = 1, \\ \text{doesn't exist} & \text{otherwise.} \end{cases}$$

3.3. Measures of Full Hausdorff Dimensions. A measure ν on a bounded subset Λ in $\mathbb{R}^n$ is called an measure of full Hausdorff dimension if $HD(\nu) = HD(\Lambda)$, where $HD(\Lambda)$ is the Hausdorff dimension of Λ, and

$$HD(\nu) = \inf\{HD(\Lambda_0) : \Lambda_0 \subset \Lambda,\ \nu(\Lambda\backslash\Lambda_0) = 0\}.$$

If Λ is an invariant set of a conformal expanding map, then f has an invariant probability measure μ such that $HD(\mu) = HD(\Lambda)$. However, for an invariant set of an almost expanding map, the measure of full Hausdorff dimension may be σ-finite.

THEOREM 3.17 ([U]). *Let $0 \in \Lambda \subset I$ be an invariant set of a piecewise smooth almost expanding map f with $\delta = HD(\Lambda)$. Assume that near 0, f satisfies (2.1). Then f has an invariant measure μ of full Hausdorff dimension, and μ is finite if $\delta > 2\gamma/(1+\gamma)$; μ is σ-finite if $\delta \le 2\gamma/(1+\gamma)$.*

PROOF. Take $\phi(x) = -\delta \log f'(x)$. On P_n, we have $\phi(x) \sim \delta(1+\gamma^{-1})/n$. Then we use a similar method for Theorem 3.13. The role of the Lebesgue measure is replaced by a conformal measure ν which is a fixed point of the dual operator of $\mathcal{L}_\phi$ on the space $C^0(\Lambda)$. □

By works of M. Urbański *et al*, Λ differ from hyperbolic Cantor sets in many properties. (See [U] for details and related references.) HD(Λ) is always bounded below by $\gamma/(1+\gamma)$, while the Hausdorff dimensions of hyperbolic Cantor sets can be arbitrary small. The topological pressure $P(f|_\Lambda, -t\log f')$ is nonnegative. (This is because near 0, $e^{S_n\phi(x_n)}$ decreases in a polynomial rate.) The pressure is positive if $t <$ HD(Λ), and 0 if $t \ge$ HD(Λ). That is, HD(Λ) is the smallest solution of $P(f|_\Lambda, -t\log f') = 0$. For hyperbolic Cantor sets, $P(f|_\Lambda, -t\log f')$, as a function of t, is strictly decreasing, and HD(Λ) is the unique solution of the equation.

3.4. Weak Gibbs States and Equilibrium States. We cannot expect the measures obtained in Section 3.1- 3.3 to be Gibbs States for the potential functions. This is because at an indifferent fixed point p, we may have $e^{S_n\phi(p)} = 1$ for any n, and $\mu\mathcal{O}_n \to 0$ in a polynomial rate. Therefore there is no uniform bound for the ratios between $\mu\mathcal{R}_n(p)$ and $\exp\{-nP + S_n\phi(p)\}$. Instead, we consider week Gibbs states.

The notion of weak Gibbs state was introduced by R. Dobrushin, and then studied by C. Maes *et al* (see e.g. [MRM] for related references). The definition we use below is from [Yu2]. Recall that $\mathcal{R}_n(x)$ was defined in Section 1.

DEFINITION 3.18. A Borel probability measure μ is called a *weak Gibbs state* for a function ϕ with constant P if there is a sequence $\{C_n\}$ of positive number with $\lim_{n\to\infty} (1/n)\log C_n = 0$ such that for μ-a.e. x,

$$C_n^{-1} \le \frac{\mu\mathcal{R}_n(x)}{\exp\{-nP + S_n\phi(x)\}} \le C_n. \tag{3.3}$$

It is proved in [Yu2] and [MRTMV] that absolute continuous invariant probability measures for systems discussed in Theorem 3.11 are week Gibbs states. Actually, all probability measures we discussed in previous sections are weak Gibbs states.

A weak Gibbs state of ϕ is an equilibrium state of the function, and the constant P is the topological pressure. In fact, since μ is a probability measure, (3.3) gives

$$C_n^{-1} \le e^{-nP}\sum e^{S_n\phi(x_B)} \le C_n,$$

where the sum is taken over all elements $B \in \mathcal{R}_n$ and x_B is any point in B. By the definition, we know that P is the topological pressure. Also, by (3.3),

$$\frac{1}{n}S_n\phi(x) - \frac{1}{n}\log C_n \le P + \frac{1}{n}\log \mu\mathcal{R}_n(x) \le \frac{1}{n}S_n\phi(x) + \frac{1}{n}\log C_n.$$

Since $\lim_{n\to\infty}(1/n)S_n\phi(x) = \int \phi d\mu$ and $\lim_{n\to\infty} -(1/n)\log \mathcal{R}_n(x) = h_\mu(\sigma)$ for μ-a.e.x by the Birkhoff ergodic theorem and the Shannon-McMillan-Breiman theorem respectively, the inequality implies $P = \int \phi d\mu + h_\mu(\sigma)$.

It is clear that equilibrium states are not unique for almost hyperbolic systems since obviously the point mass δ_p is another one.

4. Convergence Rate

We discuss decay rates of correlations and convergence rates of functions under the Perron-Frobenius operators.

4.1. Decay of Correlations. The correlation of two functions $g, \hat{g} : M \to \mathbb{R}$ is

$$\rho_{g,\hat{g}}(n) = \int g \cdot (\hat{g}\circ f^n) d\mu - \int g d\mu \int \hat{g} d\mu.$$

If a system (f, μ) is mixing, then for any $g, \hat{g} \in L^2(\mu)$, $\lim_{n\to 0} \rho_{g,\hat{g}}(n) = 0$. So we can think that the speed of the convergence reflects the speed of mixing. (We refer the paper by V. Baladi in this volume for general information and related references about decay of correlations.)

One of important characters for almost hyperbolic systems is that they can have polynomial decay of correlations.

DEFINITION 4.1. We say that (f, μ) has *polynomial decay of correlations* with degree α for functions in $\mathcal{F}$, if for all $g, \hat{g} \in \mathcal{F}$, $\exists C = C(g, \hat{g}) > 0$ s.t.

$$|\rho_{g,\hat{g}}(n)| \le Cn^{-\alpha} \qquad \forall n \ge 1.$$

Polynomial decay of correlations was first proved in 1993 by Mori ([M]) and by Lambert-Siboni-Vaienti ([LaSV]) for piecewise linear maps on the unit interval with an indifferent fixed point (Takahashi model). For piecewise smooth almost expanding maps, the result was proved by Young [Y2], Liverani-Saussol-Vaienti ([LiSV]), Pollicott-Yuri ([PY]), and the author [H2], all with different approaches. The lower bounds of the decay rates for the systems are also polynomial. (See Theorem 4.3.) For invertible systems, polynomial decay is proved for some almost Anosov systems on surfaces (Theorem 4.5).

Applying the results in ([Y2]) to an almost hyperbolic system, we get that decay rates of correlations are determined by $m\mathcal{O}_n$, where m is a reference measure. Here we introduce slightly different conditions, which is given in terms of invariant measures ([H3]).

Let f be an almost expanding map preserving an invariant measure μ. Suppose $\mu \circ f << \mu$. Put

$$\psi(x) = -\log \frac{d\mu\circ f}{d\mu}(x).$$

If μ is an absolutely continuous measure with density function h, then it is easy to check that $\psi(x) = -\log\big(|\det Df_x| \cdot h(fx)/h(x)\big)$.

Let

$$\mathcal{L}_\psi g = \sum_{x_1 \in f^{-1}x} e^{\psi(x_1)} g(x_1)$$

be the Perron-Frobenius operator. Since μ is an f-invariant measure, $\mathcal{L}_\psi c = c$ for any constant c and $\mu(\mathcal{L}_\psi g) = \mu(g)$ for any integrable function g.

THEOREM 4.2 ([H3]). *Suppose ψ satisfies the following:*

i) *there is $J_\psi > 0$, $\theta \in (0,1)$ such that $S_n\psi(y_n) - S_n\psi(x_n) \le J_\psi d(x,y)^\theta$ $\forall x, y \in P_n$, where $x_n \in f^{-n}x$, $y_n \in f^{-n}y \cap \mathcal{R}_n(x_n)$;*

ii) *on P_n, $\psi(x) \le -\frac{\beta}{n} + O(\frac{1}{n^{1+r}})$, $r > 0$, for some $\beta > 1$, or*

ii') *$\forall k > 0$, $\exists C_\psi > 0$ such that $e^{S_n\psi(x_n)} \le C_\psi(n+k)^{-\beta}$, $\forall x \in P_k$.*

Then (f, μ) has polynomial decay of correlations with degree $\beta - 1$.

Clearly, condition ii) implies ii'). The latter implies $\mu\mathcal{O}_n = O(n^{-\beta})$.

Proof of Theorem 4.2: It follows directly from Theorem 4.6 in the next subsection, because $\rho_{g,\hat{g}}(n) = \int (\hat{g} \circ f^n) \cdot (g - \mu(g)) d\mu = \int \hat{g} \cdot (\mathcal{L}_\psi^n g - \mu(g)) d\mu$, where we use the fact that $\mathcal{L}_\psi$ is the dual of the operator $F : L^2(\mu) \to L^2(\mu)$ defined by $Fg(x) = g(fx)$. □

For one dimensional systems, we have the following, where $\beta = 1/\alpha$.

THEOREM 4.3. *Let f be a piecewise smooth expanding map on the interval with an indifferent fixed point 0. Assume that f satisfies (2.1) with $0 < \gamma < 1$. Then*

i) ([Y2], [H2]) *for all Lipschitz functions g, $\hat{g}$, there exists $C = C(g, \hat{g})$ s.t.*

$$|\rho_{g,\hat{g}}(n)| \le Cn^{-(\beta-1)} \quad \forall n \ge 1;$$

ii) ([H2]) *there exist C^∞ functions g, $\hat{g}$ and a constant $C' > 0$ s.t.*

$$|\rho_{g,\hat{g}}(n)| \ge C'n^{-(\beta-1)} \quad \forall n \ge 1.$$

Proof of Part i): Distortion estimates for ψ follow from that for ϕ and the fact that $h \in \mathcal{G}$, defined in (3.2). For the second condition, we consider $x \in P_n$. We know $\phi(x) \sim -\frac{\beta+1}{n}$ by Fact 2.1. By Corollary 3.12, $\frac{h(fx)}{h(x)} \sim \frac{n-1}{n}$. So $\psi(x) \sim -\frac{\beta+1}{n} + \frac{1}{n} = -\frac{\beta}{n}$. □

The proof for Part ii) needs different method. For lower bounds in general case, we have:

CONJECTURE 4.4. *Under the circumstances of Theorem 4.2, if moreover $\psi(x) \ge -\frac{\beta'}{n} + O(\frac{1}{n^{1+r}})$, $r > 0$, for some $\beta' \ge \beta$, then the lower bounds of decay rates of the system are of order $O(n^{-(\beta'-1)})$.*

The systems discussed in Fact 3.14 and Theorem 3.17 are also have polynomial decay of correlations with the degrees $m\gamma^{-1} - 1$ and $\delta(1 + \gamma^{-1}) - 1$ respectively, whenever the measures we discussed are finite.

Pollicott and Yuri ([PY]) studied decay rate of correlations of the inhomogeneous Diophantine approximation transformation, defined by

$$f(x,y) = \Big(\frac{1}{x} - [\frac{1-y}{x}] + [-\frac{y}{x}], -[-\frac{y}{x}] - \frac{y}{x}\Big),$$

where $0 \le y \le 1$, $-1 \le x \le -y+1$, and $[x]$ is the integer part of x. This map has two indifferent periodic points $(1,0)$ and $(-1,1)$, and admits an absolutely continuous measure with density $1/(2\log 2)(1-x^2)$. The rate of decay of correlations is proved to have bounds $O(n^{-1+\epsilon})$ $\forall\epsilon > 0$. If we look at the behavior of the map near the periodic orbit, the bounds should be of order $O(n^{-1})$.

For invertible systems, we have:

THEOREM 4.5 ([H5]). *Let $f : M \to M$ be an almost Anosov diffeomorphism on the surface that satisfies conditions in Theorem 3.9. i). Then f has polynomial decay of correlations with degree $b/2d - 1$.*

PROOF. (f, μ) can be reduced to an almost expanding map $(\tilde{f}, \tilde{\mu})$ by identify points on each stable curve in every rectangle of a Markov partition. Then we can check that the resulting measure $\tilde{\mu}$ satisfies the conditions in Theorem 4.2. In Particular, if $x \in P_n$, then $e^{\psi(x)}$ is proportional to $\mu P_{n-1}/\mu P_n = \mu P'_{n-1}/\mu P'_n$ by the definition of ψ. By measuring the length of unstable leaves of P'_n (Proposition 2.6), we know that it is bounded by $\left(\frac{n-1}{n}\right)^\beta$, $\beta = b/2d$. So we get $\psi(x) \le -\beta/n$, and therefore $(\tilde{f}, \tilde{\mu})$ has decay rate bounded by $O(n^{-(\beta-1)})$. Then we pass the results from $(\tilde{f}, \tilde{\mu})$ to (f, μ). □

4.2. Convergence to Density Functions. In this section we study the rate of convergence of $\mathcal{L}_\psi^n g \to \mu(g)$. This is the rate that a system converges to its equilibrium, and the rate that determines the rate of decay of correlations.

THEOREM 4.6. *Under the circumstance of Theorem 4.2, for any Lipschitz function g, there is a constant $C > 0$ such that for all $n > 0$,*

$$\int \left|\mathcal{L}_\psi^n g - \mu(g)\right| d\mu \le \frac{C}{n^{\beta-1}};$$

and for any closed subset E with $p \notin E$, there is $C_1 > 0$ such that for all $n > 0$,

$$\left|\mathcal{L}_\psi^n g(x) - \mu(g)\right| \le \frac{C_1}{n^{\beta-1}} \qquad \forall x \in E.$$

Recall the definition of $\mathcal{L}_\psi$. Denote $g_n = \mathcal{L}_\psi^n g$. For $x \in P_n$, $n > 0$, write

$$\mathcal{L}_\psi g(x) = e^{\psi(x_1)} g(x_1) + (1 - e^{\psi(x_1)})\bar{g}(x_1),$$

where $\bar{g}(x) = \sum_{\tilde{x}\in f^{-1}(fx)\cap P_0} e^{\psi(\tilde{x})} g(\tilde{x}) / \sum_{\tilde{x}\in f^{-1}(fx)\cap P_0} e^{\psi(\tilde{x})}$, the average of g with the weight e^ψ at all preimages of fx except for x itself. It is easy to see that

$$\mathcal{L}_\psi^n g(x) = g(x_n) e^{S_n\psi(x_n)} + \sum_{j=1}^n \bar{g}_{n-j}(x_j)\left(1 - e^{\psi(x_j)}\right) e^{S_{j-1}\psi(x_{j-1})}. \tag{4.1}$$

Proof of Theorem 4.6: Note that $\mathcal{O}_{k+1}$ is the complement of Q_k. We proof the following for some large k by induction on n:

$(\mathcal{A}_n)$: $\left|g_n(x) - \mu(g)\right| \le \dfrac{C}{(n+1)^{\beta-1}}$ for any $x \in Q_k$;

$(\mathcal{B}_n)$: $\displaystyle\int_{\mathcal{O}_{k+1}} \left|g_n(x) - \mu(g)\right| d\mu \le \frac{C}{(n+1)^{\beta-1}}$.

"$(\mathcal{A}_j)$, $j = 0, \cdots, n-1$, $\Rightarrow$ $(\mathcal{B}_n)$": Apply (4.1) with the function $g(x) - \mu(g)$. Since $\int_{\mathcal{O}_{k+1}} e^{S_n\psi(x_n)} d\mu = \mu\mathcal{O}_{k+n+1}$ has the order $O(n^{-(\beta-1)})$ by the condition ii') in Theorem 4.2, so is $\int_{\mathcal{O}_{k+1}} e^{S_n\psi(x_n)} \big|g_n(x) - \mu(g)\big| d\mu$.

For the second term, we note that

$$\int_{\mathcal{O}_{k+1}} (1 - e^{\psi(x_j)}) e^{S_{j-1}\psi(x_{j-1})} = \mu\mathcal{O}_{k+j} - \mu\mathcal{O}_{k+j+1} = \mu\mathcal{P}_{k+j}$$

has the order $O((k+j)^{-\beta})$, and $g_{n-j}(x) - \mu(g)$ has the order $O((n-j)^{-(\beta-1)})$ by the induction hypotheses $(\mathcal{A}_{n-j})$. So the sum is bounded by $C'n^{-(\beta-1)}$ for some small C' if k is large enough.

"$(\mathcal{B}_n) \Rightarrow (\mathcal{A}_n)$": By $(\mathcal{B}_n)$,

$$\int_{Q_k} (g_n(x) - \mu(g)) d\mu \le C(n+1)^{-(\beta-1)}.$$

By Proposition 4.7 below, $g_n(y) - g_n(x)$ is of order $O(n^{-\beta})$ for any $x, y \in Q_k$. So $(\mathcal{A}_n)$ must be true. □

By the proof we can see that the first term in (4.1), and therefore $\mu\mathcal{O}_n$, determines the rate of convergence.

PROPOSITION 4.7. *Let $g > 0$ be a Lipschitz function, and let $k > 0$. Then there is $A > 0$ such that for any $x, y \in Q_k$,*

$$\frac{g_n(y)}{g_n(x)} \le 1 + \frac{A}{(n+1)^\beta} \qquad \forall n > 0.$$

This is the main step towards estimates of the speed of convergence $\mathcal{L}_\psi^n g \to \mu(g)$. We use projective metric (Hilbert metric) to prove the proposition, since under this metric $\mathcal{L}_\psi$ is a contracting operator. Let

$$\mathcal{C}_J(S) = \Big\{ g \in C^0(S) : g(x) > 0 \ \forall x \in S, \ g(y) \le g(x) e^{Jd(x,y)} \ \forall x, y \in S \Big\}$$

be a cone in $C^0(S)$, where $C^0(S)$ is the space of continuous functions on a metric space S. The projective metric of two functions $g, \tilde{g} \in \mathcal{C}_J(S)$ is given by

$$\Delta_J(g, \tilde{g}) = \log \frac{a(g, \tilde{g})}{b(g, \tilde{g})},$$

where

$$a(g, \tilde{g}) = \inf\{a : \ a\tilde{g} - g \in \mathcal{C}_J(S)\}, \quad b(g, \tilde{g}) = \sup\{b : \ g - b\tilde{g} \in \mathcal{C}_J(S)\}.$$

We will always take $\tilde{g} \equiv 1$. So we denote $\Theta_J(g) = \Delta_J(g, 1) = \log(a_g/b_g)$, where

$$a_g = \inf\Big\{a : \ \frac{a - g(y)}{a - g(x)} \le e^{Jd(x,y)}\Big\}, \quad b_g = \sup\Big\{b : \ \frac{g(y) - b}{g(x) - b} \le e^{Jd(x,y)}\Big\}.$$

It is easy to see that to measure how much a function g differs from a constant function, we can use $\Theta_J(g)$, or $g(y) - g(x)$, or $1 - g(y)/g(x)$.

Proof of Proposition 4.7: Suppose k is large. Take l large enough. Recall $Q_k = \bigcup_{i=0}^{n} P_i$. Define $Q_{k,l} = \bigcup_{i=k}^{l} P_i$. We prove the following for all $n \ge 0$:

$(\mathcal{C}_n)$: $\Theta_{2J}(g_n|_{Q_{k+l}}) \le \dfrac{A}{(n+1)^\beta}$;

$(\mathcal{D}_n)$: $\Theta_J(g_n|_{Q_{k,l}}) \le \frac{c}{k^{1/\gamma}} \cdot \frac{A}{(n+1)^\beta}$;

$(\mathcal{E}_n)$: $\Theta_J(g_n|_{Q_k}) \le \left(1 - \frac{c}{k^{1/\gamma}}\right) \cdot \frac{A}{(n+1)^\beta}$.

Clearly, $(\mathcal{D}_n)$ and $(\mathcal{E}_n)$ imply $(\mathcal{C}_n)$. Since $\mathcal{L}_\psi$ is contracting in the projective metric, we can get that $(\mathcal{C}_n)$ implies $(\mathcal{E}_{n+l})$, if l is large enough. To complete the induction process, we only need the fact that $(\mathcal{C}_j)$, $j = 0, \cdots n-1$, imply $(\mathcal{D}_n)$. It is enough to have $g_n(y) - g_n(x) \le cA(n+1)^{-\beta}$ for some small $c > 0$.

Recall (4.1). For the first term, since $e^{S_n\psi(x_n)}$ has the order $O(n^\beta)$ by the assumption in Theorem4.2 ii'), $e^{S_n\psi(y_n)}g(y_n) - e^{S_n\psi(x_n)}g(x_n)$ has the same order.

For the second term, we note that $d(x_j, y_j)$ has the order $j^{-\alpha}$ for some $\alpha > 1$ since f is expanding in a polynomial rate near the indifferent fixed point. Then we use $(\mathcal{C}_{n-j})$ to obtain that $\bar{g}_{n-j}(y_j) - \bar{g}_{n-j}(x_j) \le C(n-j+1)^{-\beta}d(x_j, y_j) \le C_1(n-j+1)^{-\beta}j^{-\alpha} \le C_2(n+1)^{-\beta}$. Since $\left(1 - e^{\psi(x_j)}\right)e^{S_{j-1}\psi(x_{j-1})}$, $j = 1, \cdots, n$, has the sum $1 - e^{S_n\psi(x_n)}$, it can be treated by the same way as the first term. □

If μ is a σ-finite measure, we cannot expect $\mathcal{L}_\psi^n g \to \mu(g)$ since $\mathcal{L}_\psi^n g(x) \to 0$ for Lebesgue almost every x whenever $|\mu(g)| < \infty$. However, if we multiply the left side by suitable factors, we may still have convergence to the density function.

THEOREM 4.8 ([CF]). *Suppose f is an almost expanding map on $[0, 1]$ that maps both $(0, 1/2)$ and $(1/2, 1)$ to $(0, 1)$. Assume that near 0, $fx = x + ax^2 + O(x^3)$. Then for any real function g bounded away from 0 and ∞,*

$$\lim_{n\to\infty} \frac{A_n}{n} \sum_{i=0}^{n-1} \mathcal{L}_\phi^i g = h$$

uniformly on any compact subset $E \subset (0, 1]$ for some h with $\mathcal{L}_\phi h = h$, where A_n is of order $\log n$.

By Proposition 4.7, if $0 < \gamma < 1$, then restricted to Q_k, the difference between $\mathcal{L}_\psi^n g$ and a constant is of order $O(n^{-\beta})$. Based on this observation, we make the following conjecture for one dimensional case.

CONJECTURE 4.9. *Under the circumstance of Theorem 4.3 with $\gamma \ge 1$, for any Lipschitz function g, there exist $a_n = O(n^{-\beta})$, $\beta = 1/\gamma$, such that*

$$\lim_{n\to\infty} A_n \mathcal{L}_\phi^n g = h$$

uniformly on any compact subset $E \subset (0, 1]$ for some h with $\mathcal{L}_\phi h = h$, where $A_n = \prod_{i=0}^{n-1}(1 + a_i)$. Further, the convergence rate is of order $n^{-\beta}$.

Since $\{A_n\}$ is convergent if $\gamma \in (0, 1)$, the conjecture is consistent with Theorem 4.6.

References

[B] R. Bowen, *Equilibrium States and the Ergodic Theory of Anosov Diffeomorphisms Lecture Notes in Math. Vol. 470* (Springer-Verlag, Berlin-New York, 1975).

[BY] M. Benedicks and L.-S. Young, Sinai-Bowen-Ruelle measure for certain Hénon maps, *Invent. Math.* **112** (1993), 541–576.

[BV] C. Bonatti and M. Viana, SRB measures for partially hyperbolic systems whose central direction is mostly contracting, *Israel J. Math.* **115** (2000), 157–193.

[CE] E. Catsigeras and H. Enrich, SRB measures of certain almost hyperbolic diffeomorphisms with a tangency, *Preprint.*

[CF] P. Collet and P. Ferrero, Some limit ratio theorem related to a real endomorphism in case of neutral fixed point, *Ann. Inst. Henri Poicaré* **52** (1990), 283–301.

[G] M. Gerber, Conditional stability and real analytic pseudo-Anosov maps, *Mem. Amer. Math. Soc.* **54** (1985), no. 321.

[GK] M. Gerber and A. Katok Smooth models of Thurston's pseudo-Anosov maps, *Ann. Sci. École Norm. Sup. (4)*, **15** (1982), 173–204.

[H1] H. Hu, Conditions for the Existence of SBR Measures for "Almost Anosov" Diffeomorphisms, *Trans. Amer. Math. Soc.* **352** (2000) 2331–2367.

[H2] H. Hu, Decay of Correlations for Piecewise Smooth Maps with Indifferent Fixed Points, *Preprint.*

[H3] H. Hu, Projective metrics and polynomial decay of correlations, *Preprint.*

[H4] H. Hu, Equilibrium states with polynomial decay of correlations, *Preprint.*

[H5] H. Hu, Decay of correlations for some almost Anosov diffeomorphisms, *Preprint.*

[HPK] M. Hirsch, C. Pugh and M. Shub, *Invariant manifolds, Lecture Notes in Math. Vol. 583* (Springer-Verlag, Berlin-New York, 1977).

[HY] H. Hu and L.-S. Young, Nonexistence of SBR measures for some diffeomorphisms that are "almost Anosov", *Ergod. Th. & Dynam. Sys.* **15** (1995), 67–76.

[I] T. Inoue, Ratio ergodic theorems for maps with indifferent fixed points, *Ergod. Th. & Dynam. Sys.* **17** (1997), 625–642.

[JN] M. Jakobson and S. Newhouse, On the structure of non-hyperbolic attractors, in *Dynamical systems and chaos, Vol. 1*, ed. N. Aoki, etc. (World Sci. Publishing, River Edge, 1972), pp 103–111.

[K] A. Katok, Bernoulli diffeomorphisms on surfaces, *Ann. of Math. (2)* **110** (1979), 529–547.

[KS] K. Krzyzewski and W. Szlenk, On invariant measure for expanding differentiable mappings, *Studia Math.* **33** (1969), 83–92.

[LS] F. Ledrappier and J.-M. Strelcyn, Estimation from below in Pesin's entropy formula, *Ergod. Th. & Dynam. Sys.* **2** (1982), 203–219.

[LaSV] A. Lambert, S. Siboni and S. Vaienti, Statistical properties of a nonuniformly hyperbolic map of the interval, *J. Statist. Phys.* **72** (1993), 1305–1330.

[LiSV] C. Liverani, B.Saussol and S. Vaienti, A probabilistic approach to intermittency, *Ergod. Th. & Dynam. Sys.* **19** (1999), 671–68.

[Le] J. Lewowicz, Lyapunov functions and topological stability, *J. Differential Equations* **38** (1980), 192–209.

[M] M. Mori, On the intermittency of a piecewise linear map, *Tokyo J. Math.* **16** (1986), 411–428.

[MRM] C. Maes, F. Redig and A. Moffaert, Almost Gibbsian versus weakly Gibbsian measures, *Stoch. Proc. Appl.* **79** (1999), 1-15.

[MRTMV] C. Maes, F. Redig, F.Takens, A. Moffaert and E. Verbitski , Intermittency and weak Gibbs states, *Preprint.*

[N] S. Naoya, Diffeomorphisms approximated by Anosov on the 2-torus and their SBR measures, *Trans. Amer. Math. Soc.* **351** (1999), 3373–3385.

[P] Ya. B. Pesin, Dynamical systems with generalized hyperbolic attractors: hyperbolic, ergodic and topological properties, *Ergod. Th. & Dynam. Sys.* **12** (1992), 123-151.

[Pi] G. Pianigiani, First return map and invariant measures in case of neutral fixed point, *Israel J. Math.* **35** (1980), 32–48.

[PY] M. Pollicott and M. Yuri, Statistical Properties of maps with indifferent periodic points, *Preprint.*

[S] Ya. G. Sinai, Gibbs measure in ergodic theory, *Russian Math. Surveys* **27** (1972), 21–69.

[U] M. Urbański, Parabolic Cantor sets , *Fund. Math.* **151** (1996), 241-277.

[W] P. Walters, Invariant measures and equilibrium states for some mappings which expand distances, *Trans. Amer. Math. Soc.* **236** (1978), 121–153.

[Y1] L.-S. Young, Statistical properties of dynamical systems with some hyperbolicity, *Ann. of Math.* **147** (1998), 585–650.

[Y2] L.-S. Young, Recurrence time and rate of mixing, *Israel J. Math.* **110** (1999), 153–188.

[Yu1] M. Yuri, Invariant measures for certain multi-dimensional maps, *Nonlinearity* **7** (1994), 1093–1124.

[Yu2] M. Yuri, Weak Gibbs measures for certain nonhyperbolic systems, *Ergod. Th. & Dynam. Sys.* (to appear).

Department of Mathematics, Pennsylvania State University, University Park, PA 16802

E-mail address: hu@math.psu.edu

Current address: Department of Mathematics, University of Southern California, Los Angeles, CA 90089

E-mail address: huyihu@math.usc.edu

Proceedings of Symposia in Pure Mathematics
Volume **69**, 2001

Random f-Expansions

Yuri Kifer

ABSTRACT. I define and study expansions of real numbers from $[0, 1]$ corresponding to random functions and derive the thermodynamic formalism for corresponding random piece-wise expanding in average transformations. I consider also random $\beta-$expansions. A particular case: random base expansions, were considered in [Ki3]. The results have also applications to continued fraction expansions with random digits considered in [Ki6]. In Appendix I discuss some motivations for random dynamics and suggest few open problems in it.

Introduction

More than 55 years ago Bissinger [Bi] noticed that a continued fraction expansion of a number $x \in (0, 1)$ can be written in the form $x = f(a_1 + f(a_2 + \cdots) \dots)$, where $f(t) = 1/t$, which enabled him to consider expansions corresponding to other decreasing functions. A bit later Everett [Ev] showed that the same can be done for increasing functions which generalizes base m expansions corresponding to the function $f(t) = t/m$. Since then numerous papers were written on this subject, in particular, ergodic properties of such expansions were first studied by Rényi [Re], the thermodynamic formalism for the corresponding piecewise expanding transformations was developped by Walters in [Wa1,2], and decay of correlations and various limit theorems for functions of digits were derived in Hofbauer and Keller [HK] and in Heinrich [He] though for more special cases (in particular, for fixed base and continued fraction expansions) they were known earlier. The latter paper studies even more general expansions of the form $x = f_1(a_1 + f_2(a_2 + \cdots) \dots)$ with different functions f_i satisfying certain conditions.

In this paper I consider expansions of the form

$$x = f_\omega(a_1 + f_{\theta\omega}(a_2 + f_{\theta^2\omega}(a_3 + \cdots) \dots)$$

where θ is an ergodic invertible transformation of a probability space $(\Omega, \mathcal{F}, P)$ and $f = f_\omega(t)$ is a measurable function on $\Omega \times (0, \infty]$ satisfying certain conditions. On the first sight this set up is a particular case of the one from [He] mentioned

2000 *Mathematics Subject Classification.* Primary 37H99; Secondary 37E05, 37D35.
Key words and phrases. Random transformations, f-expansions, thermodynamic formalism.

above but the lack of any connection between different functions f_i in [He] does not allow to study any ergodic properties of such expansions and other results require uniform assumptions on these functions which in the random case can be replaced by much more natural average conditions. Actually, the results are obtained for random functions f_ω which give rise to piecewise expanding in average random transformations $T_\omega x = f_\omega^{-1}x - [f_\omega^{-1}x]$. In the deterministic case ergodic properties of such transformations were studied, first, in Rényi [Re] and, later, in many other papers (see, for instance, [Wa1,2], [HK], [He]). In the random setup the role of invariant measures is played by families of measures μ^ω, $\omega \in \Omega$ satisfying $T_\omega \mu^\omega = \mu^{\theta\omega}$ $P-$almost surely (a.s) and it is especially interesting to find a family $\{\mu^\omega\}$ such that $P-$almost all μ^ω are equivalent to the Lebesgue measure. Recently, such measures together with their decay of correlations properties were studied by Buzzi [Bu1,2] for random Lasota–Yorke type transformations which correspond to random $f-$expansions with finite number of digits.

A particular case of random $f-$expansions, random base expansions, which correspond to the increasing functions $f_\omega(x) = x/m(\omega)$, where m is a positive integer valued random variable, were considered in [Ki3]. The decreasing functions $f_\omega(x) = m(\omega)/x$ provide another example which in the case $m \equiv 1$ leads to the continued fraction expansion. In the next section I give a precise definition of random f-expansions and provide conditions which ensure their convergence. In Section 2 I derive the relative thermodynamic formalism for the corresponding piecewise expanding (locally homeomorphic) random transformations and extend to this case Rényi's result on the existence of an invariant measure equivalent to the Lebesgue one. In Section 3 I exhibit applications to limit theorems and to Hausdorff dimensions of measures and sets defined via digits of random $f-$expansions. In particular, I show that the distribution of a continued fraction with independent digits whose distributions form a stationary sequence has the Hausdorff dimension strictly less than 1. In Section 4 I consider random $\beta-$expansions which lead to special problems since the corresponding random transformations T_ω are not locally homeomorphic. In the deterministic case $\beta-$expansions were studied in many papers (see, for instance, [Re], [Pa], [Wa2]). For the random setup the existence of invariant measures with absolutely continuous disintegrations follows from [Bu1,2] since random Lasota–Yorke maps considered there include random $\beta-$transformations. In Appendix I discuss few open problems in smooth random dynamical systems with some hyperbolicity properties.

1. Convergence of random f-expansions

Let $A = A_\omega$ and $B = B_\omega$ be random variables such that $1 \leq A_\omega + 1 \leq B_\omega \leq \infty$ where A_ω takes on only integer values. Let $f_\omega(x)$ be a measurable family of strictly decreasing (or strictly increasing) in x functions on $[A_\omega, B_\omega]$, where $A_\omega \geq 1$ (or $A_\omega \geq 0$), such that $f_\omega(A_\omega) = 1$ and $f_\omega(B_\omega) = 0$, which means $\lim_{x\to\infty} f_\omega(x) = 0$ if $B_\omega = \infty$, (or $f_\omega(A_\omega) = 0$ and $f_\omega(B_\omega) = 1$) so that f_ω^{-1} is defined on $[0,1]$. In the strictly decreasing case I assume that there exist positive random variables $\lambda = \lambda_\omega$ and $\eta = \eta_\omega$ such that

$$\sup_{A_\omega \leq s < t \leq B_\omega} |f_\omega(t) - f_\omega(s)| \leq \eta_\omega |t - s|, \tag{1.1}$$

$$\sup_{A_\omega + f_{\theta\omega}(A_{\theta\omega}+1) \leq s < t \leq B_\omega} |f_\omega(t) - f_\omega(s)| \leq \lambda_\omega |t - s|, \tag{1.2}$$

and if $\Lambda_\omega = \max(\lambda_\omega \eta_{\theta\omega}, \eta_\omega \lambda_{\theta\omega})$ then

$$\int \log \Lambda_\omega dP(\omega) < 0, \quad \int |\log \eta_\omega| dP(\omega) < \infty. \tag{1.3}$$

In the strictly increasing case I assume only (1.1) but with η_ω satisfying

$$\int \log \eta_\omega dP(\omega) < 0. \tag{1.4}$$

For any real x I consider its random $f-$expansion defined by

$$x = a_0(x) + f_\omega(a_1(\omega, x) + f_{\theta\omega}(a_2(\omega, x) + \cdots)\dots) \tag{1.5}$$

where $a_0(\omega, x) = a_0(x) = [x]$, $r_0(\omega, x) = r_0(x) = \{x\}$, $a_{n+1}(\omega, x) = [f^{-1}_{\theta^n\omega}(r_n(\omega, x))]$, $r_{n+1}(\omega, x) = \{f^{-1}_{\theta^n\omega}(r_n(\omega, x))\}$, where $[z]$ and $\{z\}$ are the integer and the fractional part of the real number z. If $n_0 = \min\{n \geq 0 : r_n(\omega, x) = 0\} < \infty$ then I say that x has the finite $f-$expansion

$$x = a_0(x) + f_\omega(a_1(\omega, x) + f_{\theta\omega}(a_2(\omega, x) + \cdots + f_{\theta^{n_0-1}\omega}(a_{n_0}(\omega, x))\dots).$$

In particular, if x is an integer then it has the $f-$expansion $x = a_0(x)$.

Observe that $A_{\theta^n\omega} \leq a_{n+1}(\omega, x) < B_{\theta^n\omega}$ for all $n \geq 0$ and $\omega \in \Omega$. Define $f_{1,\omega}(z_1) = f_\omega(z_1)$ and

$$f_{n,\omega}(z_1, ..., z_n) = f_{n-1,\omega}(z_1, ..., z_{n-1} + f_{\theta^{n-1}\omega}(z_n)). \tag{1.6}$$

Set $C_{n,\omega}(x) = a_0(x) + f_{n,\omega}(a_1(\omega, x), ..., a_n(\omega, x))$ and call it the $n-$th convergent of x. The proof of the following result is a slight modification of [Re].

1.1 PROPOSITION. *Under the above conditions every real number x has either finite or infinite random $f-$expansion (unique by the definition) which in the latter case means that*

$$\lim_{n \to \infty} C_{n,\omega}(x) = x \quad P - a.s. \tag{1.7}$$

PROOF. Clearly, it suffices to consider $x \in (0, 1)$ and such that $r_n(\omega, x) \neq 0$ for all n. It is easy to see by induction that

$$x = f_{n,\omega}(a_1(\omega, x), ..., a_{n-1}(\omega, x), a_n(\omega, x) + r_n(\omega, x)).$$

Hence

$$\begin{aligned} x - C_{n,\omega}(x) = &f_{n,\omega}(a_1(\omega, x), ..., a_{n-1}(\omega, x), a_n(\omega, x) + r_n(\omega, x)) \\ &- f_{n,\omega}(a_1(\omega, x), ..., a_n(\omega, x)) \end{aligned}$$

and setting

$$u_k = a_{k+1}(\omega, x) + f_{n-k-1,\theta^{k+1}\omega}(a_{k+2}(\omega, x), ..., a_n(\omega, x) + r_n(\omega, x))$$

and

$$v_k = a_{k+1}(\omega, x) + f_{n-k-1,\theta^{k+1}\omega}(a_{k+2}(\omega, x), ..., a_n(\omega, x))$$

for $k = 0, 1, ..., n-1$ we have

$$x - C_{n,\omega}(x) = r_n(\omega, x) \prod_{k=0}^{n-1} \left(\frac{f_{\theta^k\omega}(u_k) - f_{\theta^k\omega}(v_k)}{u_k - v_k} \right). \tag{1.8}$$

Next, I consider the case of strictly decreasing f_ω's. Observe, that

$$u_k = a_{k+1}(\omega, x) + f_{\theta^{k+1}\omega}(a_{k+2}(\omega, x) + f_{\theta^{k+2}\omega}(u_{k+2}));$$

$$u_{k+1} = a_{k+2}(\omega, x) + f_{\theta^{k+2}\omega}(u_{k+2})$$

and

$$v_k = a_{k+1}(\omega, x) + f_{\theta^{k+1}\omega}(a_{k+2}(\omega, x) + f_{\theta^{k+2}\omega}(v_{k+2})),$$

$$v_{k+1} = a_{k+2}(\omega, x) + f_{\theta^{k+2}\omega}(v_{k+2}).$$

If $a_{k+1}(\omega, x) \geq A_{\theta^k\omega} + 1$, then

$$u_k \geq A_{\theta^k\omega} + 1 > A_{\theta^k\omega} + f_{\theta^{k+1}\omega}(A_{\theta^{k+1}\omega} + 1)$$

and

$$v_k \geq A_{\theta^k\omega} + 1 > A_{\theta^k\omega} + f_{\theta^{k+1}\omega}(A_{\theta^{k+1}\omega} + 1),$$

and so

$$(f_{\theta^k\omega}(u_k) - f_{\theta^k\omega}(v_k))(u_k - v_k)^{-1} \leq \lambda(\theta^k\omega).$$

If $a_{k+1}(\omega, x) = A_{\theta^k\omega}$ and $a_{k+2}(\omega, x) \geq A_{\theta^{k+1}\omega} + 1$ then it follows similarly that

$$(f_{\theta^{k+1}\omega}(u_{k+1}) - f_{\theta^{k+1}\omega}(v_{k+1}))(u_{k+1} - v_{k+1})^{-1} \leq \lambda(\theta^{k+1}\omega).$$

Finally, if $a_{k+1}(\omega, x) = A_{\theta^k\omega}$ and $a_{k+2}(\omega, x) = A_{\theta^{k+1}\omega}$ then

$$u_k = A_{\theta^k\omega} + f_{\theta^{k+1}\omega}(A_{\theta^{k+1}\omega} + f_{\theta^{k+2}\omega}(u_{k+2})) \geq A_{\theta^k\omega} + f_{\theta^{k+1}\omega}(A_{\theta^{k+1}\omega} + 1)$$

and

$$v_k = A_{\theta^k\omega} + f_{\theta^{k+1}\omega}(A_{\theta^{k+1}\omega} + f_{\theta^{k+2}\omega}(v_{k+2})) \geq A_{\theta^k\omega} + f_{\theta^{k+1}\omega}(A_{\theta^{k+1}\omega} + 1).$$

It follows that

$$|x - C_{n,\omega}(x)|^2 \leq \eta_\omega \eta_{\theta^n\omega} \prod_{i=0}^{n-1} \Lambda_{\theta^i\omega}.$$

Hence, by (1.3) and the ergodic theorem P–a.s.,

$$\lim_{n\to\infty} \frac{1}{n} \log |x - C_{n,\omega}(x)| \leq \frac{1}{2} \int \log \Lambda_\omega dP(\omega) < 0,$$

and so (1.7) holds true.

Next, consider the case of strictly increasing f_ω's. Then from (1.8),

$$|x - C_{n,\omega}(x)| \leq \prod_{k=0}^{n-1} \eta_{\theta^k\omega},$$

and so by (1.3), (1.4) and the ergodic theorem P–a.s.,

$$\lim_{n\to\infty} \frac{1}{n} \log |x - C_{n,\omega}(x)| \leq \int \log \eta_\omega dP(\omega) < 0$$

which yields (1.7). □

2. Relative thermodynamic formalism (Markov case)

In this section I assume that B_ω takes on only integer values including ∞. In the modern terminology of dynamical systems this case is naturally to call Markov rather than to use the name "f−expansions with independent digits" suggested in [Re] which is, in addition, confusing since one can consider also random variables whose f−expansions have independent digits (see the next section). I impose also additional conditions on random functions f_ω which will enable me to derive the thermodynamic formalism for the corresponding random transformation.

Set $I_k^\omega = f_\omega(k, k+1)$ for $A_\omega \le k < B_\omega$. Then the transformation $T_\omega x = f_\omega^{-1}x - [f_\omega^{-1}x]$ maps homeomorphically each I_k^ω onto $(0,1)$. Furthermore, by (1.5), $a_i(\theta\omega, T_\omega x) = a_{i+1}(\omega, x)$. I need the following conditions which are more convenient to formulate in terms of the transformations T_ω rather than in terms of the function f_ω. Namely, I assume that P−a.s.

(i) The restriction of T_ω to each I_k^ω is C^2, and so $T_\omega|_{I_k^\omega}$ is a diffeomorphism of I_k^ω onto $(0,1)$;

(ii) If $t_n(\omega) = \inf_{x\in I_k^\omega, k\in\mathbb{N}} \log|(T_\omega^n)'(x)|$ then $\int t_1 dP < \infty$ and there exists $\ell \in \mathbb{N}$ such that $\kappa_\ell = E_P(t_\ell|\mathcal{F}_\ell) > 0$ P−a.s., where $E_P(\cdot|\cdot)$ is the conditional expectation with respect to the probability P and $\mathcal{F}_\ell$ is the σ−algebra of θ^ℓ−invariant sets.

Set $J^\omega = \cup_{A_\omega \le k < B_\omega} I_k^\omega$ and $J = \{(\omega, x) : x \in J^\omega, \omega \in \Omega\}$. Denote by $\mathcal{V}$ the set of measurable functions $\varphi = \varphi_\omega(x)$ on J such that for P−a.a.ω the function φ_ω is continuous on J^ω and there exist random variables $K = K(\omega) = K_\varphi(\omega)$, $W = W(\omega) = W_\varphi(\omega)$, $M_k = M_k(\omega) = M_k^\varphi(\omega)$, $A_\omega \le k < B_\omega$ and a number $\alpha > 0$ such that

(iii) $\int |\log K| dP < \infty$ and $\sum_{y\in T_\omega^{-1}x} e^{\varphi_\omega(y)} \le K(\omega)\ \forall c \in (0,1)$;

(iv)

$$|\varphi_\omega(y) - \varphi_\omega(z)| \le M_k(\omega)|y - z|^\alpha,\ \forall\, y, z \in I_k^\omega,$$

$$\sup_{v\in I_k^\omega, A_\omega \le k < B_\omega} M_k(\omega)|T_\omega'(v)|^{-\alpha} \le W(\omega) < \infty, \text{ and } \int |\log W| dP < \infty.$$

For each bounded continuous function q on $(0,1)$ and $\varphi \in \mathcal{V}$ define the random transfer operator $\mathcal{L}_\varphi^\omega$ by

$$\mathcal{L}_\varphi^\omega q(x) = \sum_{y\in T_\omega^{-1}x} e^{\varphi_\omega(y)} q(y) \tag{2.1}$$

which P−a.s. is again a bounded continuous function and $\|\mathcal{L}_\varphi^\omega q\| \le K(\omega)\|q\|$ where $\|\cdot\|$ is the supremum norm. Furthermore,

$$|\mathcal{L}_\varphi^\omega q(x) - \mathcal{L}_\varphi^\omega q(x')| \le \|q\| K(\omega) e^{W(\omega)|x-x'|}$$
$$+ K(\omega)\big(\inf_{z\in I_k^\omega, k\in\mathbb{N}} |T_\omega'(z)|\big)^{-1} |q(x) - q(x')|,$$

and so if q is uniformly continuous on $(0,1)$ then P−a.s. $\mathcal{L}_\varphi^\omega q$ is also a uniformly continuous function on $(0,1)$ and thus $\mathcal{L}_\varphi^\omega$ can be extended to a positive continuous linear operator $\bar{\mathcal{L}}_\varphi^\omega$ mapping the space $C[0,1]$ of continuous functions on $[0,1]$ into itself. Together with the operator $\bar{\mathcal{L}}_\varphi^\omega$ one has also the dual operator $(\bar{\mathcal{L}}_\varphi^\omega)^*$ acting on measures on $[0,1]$.

Denote by $\xi_\omega = \{I_k^\omega\}$ the partition of J^ω into intervals I_k^ω, $A_\omega \le k < B_\omega$ and by ξ the partition of J into the sets $\Gamma_k = \{(\omega, x) : x \in I_k^\omega\}$. It follows from (ii) that ξ is a relative generator, i.e. $P-$a.s. the partition $\xi_\omega^- = \vee_{i=0}^\infty (T_\omega^i)^{-1}\xi_{\theta^i\omega}$ is the partition ε_ω of J^ω into the points. Let Θ be the skew product transformation acting by $\Theta(\omega, x) = (\theta\omega, T_\omega x)$ which maps J to $\Omega \times (0,1)$. If μ belongs to the space $\mathcal{M}_P(\Theta)$ of $\Theta-$invariant probability measures on $\Omega \times (0,1)$ with the marginal P on Ω and has the disintegrations μ^ω, $\omega \in \Omega$ then since ξ is a relative generator, the relative entropy of Θ with respect to μ can be written in the form

$$\mathbf{h}_\mu^{(r)}(\Theta) = \int \mathbf{H}_{\mu^\omega}(\xi_\omega | T_\omega^{-1}\varepsilon_{\theta\omega})dP(\omega) \tag{2.2}$$

where $\mathbf{H}_{\mu^\omega}(\cdot|T_\omega^{-1}\varepsilon_{\theta\omega})$ is the conditional entropy with respect to the $\sigma-$algebra $T_\omega^{-1}\mathcal{B}_{\theta\omega}$ and $\mathcal{B}_\omega$ is the restriction to J^ω of the Borel $\sigma-$algebra on $(0,1)$. Indeed, since ξ is a relative generator then

$$\begin{aligned}\mathbf{h}_\mu^{(r)}(\Theta) = \mathbf{h}_\mu^{(r)}(\Theta, \xi) &= \lim_{n\to\infty} \frac{1}{n} \int \mathbf{H}_{\mu^\omega}(\xi_\omega^{(n)})dP(\omega) \\ &= \lim_{n\to\infty} \frac{1}{n} \sum_{k=0}^{n-1} \int \mathbf{H}_{\mu^\omega}(\xi_\omega | T_\omega^{-1}\xi_{\theta\omega}^{(k)})dP(\omega),\end{aligned} \tag{2.3}$$

where $\xi_\omega^{(k)} = \vee_{i=0}^{k-1}(T^i)^{-1}\xi_{\theta^i\omega}$, and so (2.2) follows since the conditional entropies in the last integral converge $P-$a.s. as $k \to \infty$ to $\mathbf{H}_{\mu^\omega}(\xi_\omega | T_\omega^{-1}\xi_{\theta\omega}^-)$.

Next, define the conditional information

$$\mathbf{I}_{\mu^\omega}(\xi_\omega | T_\omega^{-1}\varepsilon_{\theta\omega}) = -\sum_{A_\omega \le k < B_\omega} \mathbb{I}_{I_k^\omega} \log \mu^\omega(I_k^\omega | T_\omega^{-1}\mathcal{B}_{\theta\omega})$$

where $\mathbb{I}_U(x) = 1$ if $x \in U$ and $= 0$, otherwise, and $\mu^\omega(\cdot|\cdot)$ is the conditional probability. Then

$$\mathbf{H}_{\mu^\omega}(\xi_\omega | T_\omega^{-1}\varepsilon_{\theta\omega}) = \int \mathbf{I}_{\mu^\omega}(\xi_\omega | T_\omega^{-1}\varepsilon_{\theta\omega})d\mu^\omega. \tag{2.4}$$

Since entropies here may be infinite I say, following [Wa1], that $\mu \in \mathcal{M}_P(\Theta)$ with disintegrations μ^ω is a (generalized) equilibrium state for a function $\varphi = \varphi_\omega(x)$ on J with φ_ω continuous on J^ω if

$$\int (\mathbf{I}_{\mu^\omega}(\xi_\omega | T_\omega^{-1}\varepsilon_{\theta\omega}) + \varphi)d\mu \ge \int (\mathbf{I}_{\eta^\omega}(\xi_\omega | T_\omega^{-1}\varepsilon_{\theta\omega}) + \varphi)d\eta \tag{2.5}$$

for any $\eta \in \mathcal{M}_P(\Theta)$ with disintegrations η^ω. If $\mathbf{h}_\mu^{(r)}(\Theta)$ and $\int |\varphi| d\mu$ are finite then by (2.2)–(2.5),

$$\mathbf{h}_\mu^{(r)}(\Theta) + \int \varphi d\mu \ge \mathbf{h}_\eta^{(r)}(\Theta) + \int \varphi d\eta \tag{2.6}$$

for any $\eta \in \mathcal{M}_P(\Theta)$ with finite $\mathbf{h}_\eta^{(r)}(\Theta)$ and $\int |\varphi| d\eta$.

2.1. THEOREM. *Let $\varphi \in \mathcal{V}$ and $T_\omega, \omega \in \Omega$ satisfy (i) and (ii). Then there exist a probability measure $\nu = \nu_\varphi$ on $\Omega \times [0,1]$ with atomless disintegrations ν^ω on $[0,1]$, a random variable $\lambda^\omega = \lambda^\omega_\varphi > 0$, and a measurable function $h_\omega(x) = h^\varphi_\omega(x) > 0$ on $\Omega \times [0,1]$ continuous in x and such that $P-$a.s.,*

$$(2.7) \qquad \bar{\mathcal{L}}^\omega_\varphi h_\omega = \lambda^\omega h_{\theta\omega}, \ (\bar{\mathcal{L}}^\omega_\varphi)^* \nu^{\theta\omega} = \lambda^\omega \nu^\omega, \ \text{and} \int h_\omega d\nu^\omega = 1$$

and these relations determine the triple ν, λ, h uniquely. The probability measure $\mu = \mu_\varphi$ with atomless disintegrations $\mu^\omega = h_\omega \nu^\omega$ satisfies $P-$a.s.,

$$(2.8) \qquad T_\omega \mu^\omega = \mu^{\theta\omega} \ \text{and} \ (\bar{\mathcal{L}}^\omega_{\log g})^* \mu^{\theta\omega} - \mu^\omega$$

where $g_\omega = e^{\varphi_\omega} h_\omega (\lambda^\omega h_{\theta\omega} \circ T_\omega)^{-1}$. In particular, μ is invariant under the skew product transformation Θ acting on $\Omega \times J$ by $\Theta(\omega, x) = (\theta\omega, T_\omega x)$. Furthermore, for any $q \in C[0,1]$ $P-$a.s.,

$$(2.9) \qquad \lim_{n\to\infty} (\lambda^{\theta^{n-1}\omega} \cdots \lambda^{\theta\omega}\lambda^\omega)^{-1} (h_{\theta^n\omega})^{-1} \bar{\mathcal{L}}^{\omega,n}_\varphi q = \int q d\nu,$$

where $\bar{\mathcal{L}}^{\omega,n}_\varphi = \bar{\mathcal{L}}^{\theta^{n-1}\omega}_\varphi \circ \cdots \circ \bar{\mathcal{L}}^{\theta\omega}_\varphi \circ \bar{\mathcal{L}}^\omega_\varphi$, and weakly

$$(2.10) \qquad (T^n_\omega)^{-1} \nu^{\theta^n\omega} \to \mu^\omega \ \text{as} \ n \to \infty.$$

Finally, μ is ergodic, uniquely determined by (2.8) and it is the unique equilibrium state of Θ for the function φ.

PROOF. The proof proceeds similarly to [Ki2] and [KK] which extend to random expanding in average transformations most of arguments from [Wa1]. Still, the set up in [Ki2] and [KK] concerns random transformations defined on manifolds and here T_ω's are defined on unions J^ω of random intervals. Nevertheless, the arguments can be adjusted to this case similarly to the main ingradient of the proof discussed below. Observe, that the general set up for the deterministic case treated in [Wa1] contains piecewise expanding transformations, as well. This could be further generalized to the random situation but since the one dimensional case related to random $f-$expansions is more transparent and represents the main model I restrict to it the discussion here.

For any $x, x' \in (0,1)$ define the bijection between $y \in (T^n_\omega)^{-1}x$ and $y' \in (T^n_\omega)^{-1}x'$ assuming that $T^i_\omega y$ an $T^i_\omega y'$ belong to the same interval $I^{\theta^i\omega}_{k_i}$ for $i = 0, 1, ..., n-1$. For such y, y' one has

$$(2.11) \qquad \begin{aligned} |\varphi_{\theta^i\omega}(T^i_\omega y) - \varphi_{\theta^i\omega}(T^i_\omega y')| &\le M_{k_i}(\theta^i\omega)|T^i_\omega y - T^i_\omega y'|^\alpha \\ &\le \frac{M_{k_i}(\theta^i\omega)}{|(T^{n-i}_{\theta^i\omega})'(T^i_\omega y_i)|^\alpha}|x - x'|^\alpha \end{aligned}$$

for some y_i between y and y'. Hence

(2.12)
$$\begin{aligned}\sum_{i=0}^{n-1}|\varphi_{\theta^i\omega}(T^i_\omega y)-\varphi_{\theta^i\omega}(T^i_\omega y')| &\le |x-x'|^\alpha \sum_{i=0}^{n-1}\frac{W(\theta^i\omega)}{|(T^{n-i-1}_{\theta^{i+1}\omega})'(T^{i+1}_\omega y_i)|}\\ &\le |x-x'|^\alpha\sum_{i=0}^{n-1}\exp\big(\log W(\theta^i\omega)+s(\theta^n\omega)\\ &\quad -\sum_{0\le j<[\ell^{-1}(n-\ell-i-1)]} t_\ell(\theta^{i+1+j\ell}\omega)\big)\\ &\le |x-x'|^\alpha e^{s(\theta^n\omega)}A(\theta^n\omega)\end{aligned}$$

where

$$A(\omega)=\sum_{i=-\infty}^{-1}\exp\big(\log W(\theta^i\omega)-\sum_{0\le j<[-\ell^{-1}(i+1)]-1} t_\ell(\theta^{i+1+j\ell}\omega)\big),$$

$s(\omega)=\sum_{j=1}^{\ell}|t_1(\theta^{-j}\omega)|$ and $[a]$ denotes the integral part of a number a.

Next, similarly to Proposition 2.1 in [KK] I show that $A<\infty$ $P-$a.s. Indeed, by (ii) together with the ergodic theorem it follows that there exists a random variable $N=N(\omega)$ such that for all $j\ge N$,

$$\sum_{0\le j<[-\ell^{-1}(i+1)]-1} t_\ell(\theta^{i+1+j\ell}\omega)\ge\frac{|i|}{2\ell}\tilde\kappa_\ell(\omega) \tag{2.13}$$

where $\tilde\kappa_\ell=\min_{0\le r<\ell}\kappa_\ell(\theta^r\omega)>0$ $P-$a.s. Since $\log W\in\mathbb{L}^1(\Omega,P)$ then

$$\lim_{i\to-\infty}|i|^{-1}\log W(\theta^i\omega)=0\ \ P-\text{a.s.} \tag{2.14}$$

which togeter with (2.13) yield that $A<\infty$ $P-$a.s.

Set

$$B(\omega)=\sum_{i=1}^{\infty}W(\theta^{-i}\omega)e^{-t_{i-1}(\theta^{-i+1}\omega)}. \tag{2.15}$$

Since

$$t_{i-1}(\theta^{-i+1}\omega)\ge\sum_{j=0}^{[\ell^{-1}(i-1)]-1}t_\ell(\theta^{-j\ell}\omega)-s(\theta^{-i+\ell+1}\omega)$$

then by (ii) and the ergodic theorem

$$\liminf_{i\to\infty} i^{-1}t_{i-1}(\theta^{-i+1}\omega)\ge\ell^{-1}\kappa_\ell(\omega). \tag{2.16}$$

It follows that $B<\infty$ $P-$a.s. Now, in the same way as in Proposition 2.5 from [Ki2] and Theorem 3.1 from [KK] I derive that there exists a triple ν,λ,h satisfying (2.7) and, in addition,

$$|\log h_\omega(x)-\log h_\omega(x')|\le B(\omega)|x-x'|^\alpha. \tag{2.17}$$

Set $\Psi_C=\{\omega:\max(A(\omega),B(\omega))<C\}$ and choose $C>0$ large enough so that $P(\Psi_C)>0$. Let $n_0(\omega)=0$ and, inductively,

$$n_{i+1}(\omega)=\min\{n>n_i(\omega):\theta^n\omega\in\Psi_C\}.$$

Then, in the same way as in Theorem 3.1 of [KK], it follows that for any $q \in C[0,1]$ the sequence $\mathcal{L}^{\omega,n_i}_{\log g_\omega} q$, $i = 1, 2, \ldots$ is equicontinuous, where g is the same as in (2.8), which yields the construction of measures μ_ω satisfying (2.8), so that $\int q d\mu_\omega$ is the limit of the above sequence (see Proposition 2.3 in [Ki2]), and provides the uniqueness of the triple ν, λ, h.

The following argument yields that P–a.s. the measures μ^ω have no atoms (which was not stated explicitly in [Ki2] and [KK]). Namely, by (2.8) for any $z \in J^\omega$,

$$\mu^\omega(\{z\}) = (\mathcal{L}^\omega_{\log g})^* \mu^{\theta\omega}(\{z\}) = \int \sum_{y \in T_\omega^{-1} x} g_\omega(y) \mathbb{I}_{\{z\}}(y) d\mu^{\theta\omega}(x) = g_\omega(z) \mu^{\theta\omega}(\{T_\omega z\})$$

where $\mathbb{I}_U(v) = 1$ if $v \in U$ and $=0$, otherwise. Observe, that $g_\omega > 0$ and

$$\sum_{y \in T_\omega^{-1} x} g_\omega(x) = (\lambda^\omega h_{\theta\omega})^{-1} \mathcal{L}^\omega_\varphi h_\omega = 1$$

which gives $g_\omega < 1$. Set $a_\omega = \max_{z \in J^\omega} \mu^\omega(\{z\})$. It follows that $a_\omega \le a_{\theta\omega}$ and $a_\omega < a_{\theta\omega}$ if $a_\omega > 0$. By ergodicity of θ I conclude that a is constant P–a.s. and so $a \equiv 0$ P–a.s.

Other assertions of Theorem 2.1 above follow similarly to the corresponding arguments in [Wa1] modified according to [Ki2], [KK] and the discussion above, in particular, ergodicity of μ is proved in the same way as in Theorem 3.2 from [KK]. □

Set now $\varphi^{(1)}_\omega(x) = \varphi_\omega(x) = -\log |T'_\omega(x)|$ for $x \in J^\omega$ and assume

(v) $\sup_{k \in \mathbb{N}} \sup_{x,y,z \subset I^\omega_k} \left| \frac{T''_\omega(x)}{T'_\omega(y) T'_\omega(z)} \right| = Q(\omega)$ with $\int |\log Q(\omega)| dP(\omega) < \infty$.

Observe that in my circumstances T'_ω is either positive or negative on each I^ω_k. Then for any $x, x' \in (0,1)$,

(2.18)

$$\sum_{y \in T_\omega^{-1} x} e^{\varphi_\omega(y)} = \sum_{y \in T_\omega^{-1} x} |T'_\omega(y)|^{-1} \ge \sum_{y' \in T_\omega^{-1} x'} |T'_\omega(y')|^{-1} - \sum_{y \in T_\omega^{-1} x} \left| \frac{1}{T'_\omega(y)} - \frac{1}{T'_\omega(y')} \right|$$
$$= \sum_{y' \in T_\omega^{-1} x'} e^{\varphi_\omega(y')} - \sum_{y \in T_\omega^{-1} x} \left| \frac{T''_\omega(\tilde{y})(y' - y)}{T'_\omega(y) T'_\omega(y')} \right| \ge \sum_{y' \in T_\omega^{-1} x'} e^{\varphi_\omega(y')} - Q(\omega)$$

where $\tilde{y}$ is a point between y and y' and I employed (v) and the fact that $\sum_{y \in T_\omega^{-1} x} |y' - y| \le 1$. Since (2.18) holds true for any pair $x, x' \in (0,1)$ and

$$\int_0^1 \sum_{y \in T_\omega^{-1} x} |T'_\omega(y)|^{-1} dm(x) = \int_0^1 d(T_\omega m) = 1, \tag{2.19}$$

where m is the Lebesgue measure on $[0,1]$, I conclude that

$$\sum_{y \in T_\omega^{-1} x} e^{\varphi_\omega(y)} \le 1 + Q(\omega). \tag{2.20}$$

Hence (iii) is satisfied for such φ.

Now, for any $y, y' \in I_k^\omega$, $k \in \mathbb{N}$ I have

$$|\varphi_\omega(y) - \varphi_\omega(y')| = |\varphi'_\omega(\tilde{y})(y - y')| = \frac{|T''_\omega(\tilde{y})|}{|T'_\omega(\tilde{y})|}|y - y'|$$

for some $\tilde{y}$ between y and y' which together with (v) yields (iv).

2.2. Corollary. *Let T_ω, $\omega \in \Omega$ satisfy (i),(ii), and (v). Then there exists a unique $\mu_1 = \mu_{\varphi^{(1)}} \in \mathcal{M}_P(\Theta)$ whose disintegrations μ_1^ω satisfy (2.8) with $\varphi_\omega(x) = \varphi_\omega^{(1)}(x) = -\log|T'_\omega(x)|$, μ_1 is the unique equilibrium state of Θ for the function $\varphi^{(1)}$, it is ergodic, and the following relative entropy formula*

$$\mathbf{h}_{\mu_1}^{(r)}(\Theta) = \int\int \log|T'_\omega(x)| d\mu_1^\omega(x) dP(\omega) \tag{2.21}$$

holds true. Furthermore, $P-$almost all μ^ω are equivalent to the Lebesgue measure m and if $h_\omega = \frac{d\mu_1^\omega}{dm}$ then $\nu^\omega \equiv m$, $\lambda^\omega \equiv 1$, h_ω is the corresponding triple given by Theorem 2.1 so that

$$\bar{\mathcal{L}}_{\varphi^{(1)}}^\omega h_\omega = h_{\theta\omega} \text{ and } (\bar{\mathcal{L}}_{\varphi^{(1)}}^\omega)^* m = m. \tag{2.22}$$

In particular, with these notations (2.9) and (2.10) hold true, as well.

Proof. It suffices to check only (2.21) and (2.22) since then Corollary 2.2 will follow from Theorem 2.1 by uniqueness. Now, for any $q \in C[0,1]$,

$$\begin{aligned} \int_0^1 qd(\bar{\mathcal{L}}_{\varphi^{(1)}}^\omega)^* m &= \int_0^1 \bar{\mathcal{L}}_{\varphi^{(1)}}^\omega q dm \\ &= \int_0^1 \sum_{y \in T_\omega^{-1}x} |T'_\omega(y)|^{-1} q(y) dm(x) \\ &= \int_{A_\omega \le k < B_\omega} \int_{I_k^\omega} q dm = \int_0^1 q dm \end{aligned} \tag{2.23}$$

and

$$\begin{aligned} \int_0^1 q\bar{\mathcal{L}}_{\varphi(1)}^\omega h_\omega dm &= \int_0^1 \bar{\mathcal{L}}_{\varphi(1)}^\omega ((q \circ T_\omega) h_\omega) dm \\ &= \int_0^1 (q \circ T_\omega) h_\omega dm = \int_0^1 q \circ T_\omega d\mu^\omega \\ &= \int q dT_\omega \mu^\omega = \int q d\mu^{\theta\omega} = \int q h_{\theta\omega} dm, \end{aligned} \tag{2.24}$$

and so (2.22) follows. It follows from (ii), (v), and the ergodic theorem that $P-$a.s. $\lim_{n\to\infty} \mathbf{H}_{\mu_1^\omega}(\xi_\omega^{(n)})$ equals the right hand side of (2.21) which concludes the proof in view of (2.3). □

2.3. Remark. Theorem 2.1 and Corollary 2.2 yield also new type of results for deterministic transformations, for instance, for the Gauss map $Tx = \frac{1}{x}$ (mod 1) corresponding to the continued fraction expansions. Namely, for each family of random functions $\varphi = \{\varphi_\omega\} \in \mathcal{V}$ Theorem 2.1 supplies a unique measure μ_φ whose disintegrations μ_φ^ω satisfy (2.7)–(2.10) and $T\mu_\varphi^\omega = \mu_\varphi^{\theta\omega}$. Random measures with the latter property include $T-$invariant measures and they form a larger but still interesting for study class of measures. Corollary 2.2 yields, in particular, that if T is the Gauss map and $G(\Gamma) = \frac{1}{\ln 2}\int_\Gamma \frac{dx}{x+1}$, $\Gamma \subset [0,1]$ is the Gauss measure then $P \times G$ is the unique $\theta \times T-$invariant measure having absolutely continuous disintegrations. Similar results hold true for random expanding endomorphisms and random hyperbolic diffeomorphisms (see [KK] and [GK]).

3. Applications to fractal dimensions and limit theorems

Let $p^\omega = (p_i^\omega,\ A_\omega \le i < B_\omega)$ be a measurable in ω family of probability vectors, i.e. $p_i^\omega \ge 0\,\forall i$ and $\sum_{A_\omega \le i < B_\omega} p_i^\omega = 1$,

$$\Xi^\omega = \{\xi = (\xi_i,\, i = 1,2,...) :\ A_{\theta^i\omega} \le \xi_i < B_{\theta^i\omega}\}$$

be the space of sequences, and $\Pi^\omega = \prod_{i=1}^\infty \{p_j^{\theta^i\omega},\ A_\omega \le j < B_\omega\}$ be the product measure on Ξ^ω. Set $\alpha_i^\omega(\xi) = \xi_i$ for each $\xi \in \Xi^\omega$ then $\alpha_1^\omega, \alpha_2^\omega, ...$ are independent random variables on (Ξ^ω, Π^ω) with distributions $p^\omega, p^{\theta\omega}, ...$ For each $\omega \in \Omega$ let $X^\omega \in [0,1]$ be a random point given by the random $f-$expansion

$$X^\omega = f_\omega(\alpha_1^\omega + f_{\theta\omega}(\alpha_2^\omega + \cdots)...) \tag{3.1}$$

and denote the distribution of X^ω by μ^ω, i.e. $\mu^\omega(\Gamma) = \Pi^\omega\{X^\omega \in \Gamma\}$. The following result gives a formula for the Hausdorff dimensions $\mathrm{HD}(\mu^\omega)$ of measures μ^ω, which is, recall, the smallest Hausdorff dimension of sets of full measure, and it generalizes Proposition 7.4 from [Ki7] obtained for the special case of continued fraction expansions employing a quite different approach of products of random matrices.

3.1. THEOREM. *Assume that*

$$P\{p_k^\omega = 1\} < 1\ \forall k. \tag{3.2}$$

Then the measure μ on $\Omega \times [0,1]$ with disintegrations μ^ω defined above is the unique equilibrium state of Θ for the function $\psi_\omega(x) = \log p^\omega_{a_1(\omega,x)}$ (see (1.5)) and

$$\mathbf{h}_\mu^{(r)}(\Theta) = -\int \Big(\sum_{A_\omega \le k < B_\omega} p_k^\omega \log p_k^\omega\Big) dP(\omega). \tag{3.3}$$

If $\mathbf{h}_\mu^{(r)}(\Theta) < \infty$ then $P-$a.s.,

$$HD(\mu^\omega) = \mathbf{h}_\mu^{(r)}(\Theta)/\gamma \tag{3.4}$$

where $\gamma = \int \log|T'_\omega(x)| d\mu^\omega(x) dP(\omega)$ is the Lyapunov exponent corresponding to μ. Furthermore,

$$HD(\mu^\omega) < 1 \quad P - a.s. \tag{3.5}$$

unless μ coincides with the measure μ_1 constructed in Corollary 2.2. The latter holds true if and only if the map $S_\omega = F_{\theta\omega} \circ T_\omega \circ F_\omega^{-1}$, where $F_\omega(t) = \mu_1^\omega((0,t))$, is linear on each of its monotonicity intervals and $p_k^\omega = \mu_1^\omega(I_k^\omega)$. In particular, if $T = T_\omega$ with $Tx = \frac{1}{x}$ (mod 1) then $S = S_\omega$ is not piecewise linear, and so $\mu_1 = P \times G \ne \mu$, where G is the Gauss measure, implying that (3.5) holds true, i.e. the Hausdorff dimension of the distribution of X^ω for the random continued fraction expansion above is strictly less than one (cf. [Ki6]).

PROOF. The function ψ is defined on J and ψ_ω is constant on each I_k^ω. Taking into account that

$$\sum_{y\in T_\omega^{-1}x} e^{\psi_\omega(y)} = \sum_{A_\omega\le k<B_\omega} p_k^\omega = 1 \tag{3.6}$$

I conclude that ψ satisfies (iii) and (iv). Introduce cylinder sets

$$C^\omega_{a_1,...,a_n} = \{x\in(0,1):\, a_1(\omega,x)=a_1,...,a_n(\omega,x)=a_n\}$$

where $A_{\theta^i\omega}\le a_i < B_{\theta^i\omega}$. Then

$$\mathcal{L}_\psi^*\mu^{\theta\omega}(C^\omega_{a_1,...,a_n}) = \int \sum_{y\in T_\omega^{-1}x} e^{\psi_\omega(y)}\mathbb{I}_{C^\omega_{a_1,...,a_n}}(y)d\mu^{\theta\omega}(x)$$
$$= p^\omega_{a_1}\mu^{\theta\omega}(C^{\theta\omega}_{a_2,...,a_n}) = \mu^\omega(C^\omega_{a_1,...,a_n}).$$

where, again, $\mathbb{I}_U$ is the indicator of a set U. Since this holds true for any cylinder set then

$$\mathcal{L}_\psi^*\mu^{\theta\omega} = \mu^\omega. \tag{3.7}$$

Now, (3.6) and (3.7) together with the uniqueness part of Theorem 2.1 yield that $\lambda_\psi^\omega = 1$, $h_\omega^\psi\equiv 1$, $\mu_\psi^\omega=\nu_\psi^\omega=\mu^\omega$ $P-$a.s. and, in particular, μ is ergodic and μ^ω is atomless $P-$a.s.

Recall, that $\xi_\omega = \{I_k^\omega,\, A_\omega\le k<B_\omega\}$ is a fiber of the relative generating partition ξ of J (see [KW]) and $\xi_\omega^{(n)} = \vee_{j=0}^{n-1}(T_\omega^j)^{-1}\xi_{\theta^j\omega}$ is the partition of J^ω into cylinders $C^\omega_{a_1,...,a_n}$. The entropy of the latter partition is given by the formula

$$\mathbf{H}_{\mu^\omega}(\xi_\omega^{(n)}) = -\sum_{i=0}^{n-1}\sum_{A_{\theta^i\omega}\le k<B_{\theta^i\omega}} p_k^{\theta^i\omega}\log p_k^{\theta^i\omega}$$

and by the ergodic theorem I arrive at (3.3).

Set $C_n^\omega(x) = C^\omega_{a_1,...,a_n}$ if $x\in C^\omega_{a_1,...,a_n}$. By the ergodic theorem together with (3.2) $\mu-$a.s.,

$$\lim_{n\to\infty}\frac{1}{n}\log\mu^\omega(C_n^\omega(x)) = -\mathbf{h}_\mu^{(r)}(\Theta) \tag{3.8}$$

and

$$\lim_{n\to\infty}\frac{1}{n}\log|C_n^\omega(x)| = -\gamma \tag{3.9}$$

where γ is the same as in (3.4) and it follows easily from (ii) that $\gamma>0$. Now (3.4) follows in the usual way (see [Ki3] and [Ki6]). If $\mu\ne\mu_1$ then (3.5) holds true since by Corollary 2.2 μ_1 is the only equilibrium state for the function $\varphi_\omega^{(1)}(x) = -\log|T'_\omega(x)|$, and so by (2.16),

$$0 = \mathbf{h}_{\mu_1}^{(r)}(\Theta) + \int\varphi^{(1)}d\mu_1 > \mathbf{h}_\mu^{(r)}(\Theta) + \int\varphi^{(1)}d\mu$$

implying (3.5).

Next, I employ the following argument whose deterministic version appears in [KPW]. Observe, that F_ω is a diffeomorphism $P-$a.s. since $F'_\omega = h_\omega$ is positive and continuous by Theorem 2.1 and Corollary 2.2. It is clear that $S_\omega = F_{\theta\omega}\circ T_\omega\circ F_\omega^{-1}$

is monotone on each $F_\omega I_k^\omega$ with the same type of monotonicity as T_ω has on I_k^ω and $S_\omega(F_\omega I_k^\omega) = (0,1)$. Observe also that $F_\omega \mu_1^\omega = m$, where m is the Lebesgue measure, and so $S_\omega m = m$ for all ω. It follows from Corollary 2.2 that for $m-$a.a.x,

$$S'_\omega(x) = h_{\theta\omega}(T_\omega F_\omega^{-1}x)T'_\omega(F_\omega^{-1}x)(h_\omega(F_\omega^{-1}x))^{-1} \tag{3.10}$$

and the right hand side of (3.10) is continuous in x on each $F_\omega I_k^\omega$. Set $b_1(\omega,x) = k$ if $x \in F_\omega I_k^\omega$ and, recursively, $b_{i+1}(\omega,x) = b_i(\theta\omega, S_\omega x)$ for $i = 1,2,\dots$. Then it is easy to see that for $m-$a.a. $z \in F_\omega I_k^\omega$,

$$m\{b_1(\omega,\cdot) = k \mid b_2(\omega,\cdot), b_3(\omega,\cdot),\dots\}(z) = S'_\omega(z) \tag{3.11}$$

where the left hand side is the conditional probability of the event $\{b_1(\omega,\cdot) = k\}$ with respect to the $\sigma-$algebra generated by the random variables $b_2(\omega,\cdot), b_3(\omega,\cdot),\dots$ on the probability space $((0,1),m)$. Observe that $b_i(\omega,x) = a_i(\omega, F_\omega^{-1}x)$. Assume now that $\mu = \mu_1$. Then $a_1(\omega,\cdot), a_2(\omega,\cdot),\dots$ is the sequence of independent random variables on the probability space $((0,1),\mu_1^\omega)$ and it follows that $b_1(\omega,\cdot), b_2(\omega,\cdot),\dots$ is the sequence of independent random variables on the probability space $((0,1),m)$. This together with (3.11) imply that $S'_\omega(z) = m\{x : b_1(\omega,x) = k\}$ for $m-$a.a. $z \in F_\omega I_k^\omega$, i.e. $S'_\omega(z)$ is $m-$a.s. constant on each $F_\omega I_k^\omega$ and since S_ω is continuous there it must be linear on each $F_\omega I_k^\omega$.

Conversely, assume that S_ω is linear on each $F_\omega I_k^\omega$. Since $S_\omega(F_\omega I_k^\omega) = (0,1)$ then $S'_\omega(z) = (p_k^\omega)^{-1}$ for each $z \in F_\omega I_k^\omega$ where $p_k^\omega = m(F_\omega I_k^\omega) = \mu_1^\omega(I_k^\omega)$. It is clear that

$$\begin{aligned} &m\{z : b_1(\omega,z) = k_1, b_2(\omega,z) = k_2, \dots, b_n(\omega,z) = k_n\} \\ &= m\Big(\bigcap_{i=0}^{n-1}(S_\omega^i)^{-1}F_{\theta^i\omega}I_{k_{i+1}}^{\theta^i\omega}\Big) = \prod_{i=0}^{n-1} p_{k_{i+1}}^{\theta^i\omega}, \end{aligned}$$

and so $b_1(\omega,\cdot), b_2(\omega,\cdot),\dots$ is the sequence of independent random variables on the probability space $((0,1),m)$. Hence, $a_i(\omega,x) = b_i(\omega, F_\omega x)$, $i = 1,2,\dots$ form a sequence of independent random variables on the probability space $((0,1),\mu_1^\omega)$ and

$$\mu_1^\omega\{x : a_i(\omega,x) = k\} = m\{x : b_i(\omega,x) = k\} = p_k^{\theta^{i-1}\omega}.$$

It follows that $\mu_1 = \mu$.

Next, let $f_\omega(x) = x^{-1}$ for all ω so that $T_\omega = T$ with $Tx = x^{-1}$ (mod 1). Now $\mu_1^\omega = G$ for $P-$a.a.ω and $F_\omega = F$ with $Fx = \frac{1}{\ln 2}\ln(1+x)$. It follows that $S_\omega = S$ with $Sx = \frac{1}{\ln 2}\ln(1 + (2^x-1)^{-1} - k)$ for all $x \in FI_k$ and $I_k = (\frac{1}{k+1}, \frac{1}{k})$. It is easy to see that S is not linear on FI_k (the derivative, clearly, is not constant there), and so $\mu_1 \neq \mu$ and (3.5) holds true. □

3.2. Remark. Let the measure μ in Theorem 3.1 be constructed for the continued fraction expansion, i.e. when $f(x) = \frac{1}{x}$ and $Tx = \frac{1}{x}$ (mod 1). Since μ is $\theta \times T-$invariant and ergodic then for any nonrandom cylinder set $C_{a_1,\dots,a_k} \subset [0,1]$ corresponding to the continued fraction expansions for $P-$a.a.ω and $\mu^\omega-$a.a.x,

$$\lim_{n\to\infty} n^{-1}\sum_{i=0}^{n-1}\mathbb{I}_{C_{a_1,\dots,a_k}}(T^i x) = \bar\mu(C_{a_1,\dots,a_k})$$

where $\bar\mu = \int \mu^\omega dP(\omega)$. Similarly to [KPW] it would be possible to give better than (3.5) estimates for $HD(\mu^\omega)$ if $\bar\mu(C_{a_1,\dots,a_k}) \neq G(C_{a_1,\dots,a_k})$ but I do not know whether $\bar{\mu} \neq G$ for all measures μ with product type disintegrations as above. In the deterministic case one has to check only that G itself is not a product type measure which is easy by a direct computation considering already 2-cylinders.

3.3. REMARK. Many results on dimensions from [MU] can be extended to random infinite iterated function systems (see definitions in [Ki4]) and it would be worthwhile to write detailed proofs for the corresponding random setup.

Next, I give an application of the formalism of Section 2 to ω−wise limit theorems. Under uniform in ω conditions on transformations T_ω such theorems follow from [He] but here I have only integrability and expanding in average conditions.

3.4. THEOREM. *Let $\varphi \in \mathcal{V}$ and μ_φ be the corresponding (relative) equilibrium state given by Theorem 2.1 with disintegrations μ_φ^ω. There exists a random variable $L = L(\omega)$, which can be specified explicitly in terms of T_ω's and φ, such that if $Q = Q_K = \{\omega : L(\omega) < K\}$, $P(Q) > 0$, $k_1(\omega) = \min\{n : \theta^n\omega \in Q\}$ and $\psi \in \mathcal{V}$ satisfy $\int \psi_\omega d\mu_\varphi^\omega = 0$ $P-a.s.$ and*

$$E_P \mathbb{I}_Q \Big(\sum_{j=0}^{k_1(\omega)-1} (\|\psi_{\theta^j\omega}\| + W_\psi(\theta^j\omega)) \Big)^2 < \infty$$

then for $P-a.a.\omega$ the $\mu^\omega-$distribution of $n^{-1/2}\sum_{i=0}^{n-1}(\psi\circ\Theta^i)_\omega$ converges to the normal distribution with zero mean and the variance $\sigma^2 = \lim_{n\to\infty} \frac{1}{n} E_{\mu^\omega}(\sum_{j=0}^{n-1} \psi_{\theta^j\omega} \circ T_\omega^j)^2$, which in the case $\sigma = 0$ is understood as the unit mass at 0. Furthermore, $\sigma = 0$ if and only if there exists $\eta \in \mathbb{L}^2(J,\mu)$ such that $\Psi = \sum_{j=0}^{k_1(\omega)-1} \psi \circ \Theta^j$ satisfies $\Psi \circ \Theta^{k_1} = \eta \circ \Theta^{k_1} - \eta$. Moreover, the invariance principles in the central limit theorem and in the law iterated logarithm hold true, as well (see [Ki5).

This result follows checking conditions of Theorem 2.1 in [Ki5] similarly to Section 6 there by deriving exponential decay of correlations via corresponding contracting properties of the random operator $\mathcal{L}_\varphi^\omega$ (cf. [HK] for the deterministic case).

4. Random β−expansions

In this section I consider $f_\omega(x) = \frac{x}{\beta(\omega)}$ where β is a random variable such that

$$\beta > 0, \ \log\beta \in L^1(\Omega, P), \text{ and } \int \log\beta dP > 0. \tag{4.1}$$

If β takes on only integer values I arrive at random base expansions considered in [Ki3]. By Proposition 1.1 every number $x \in [0, 1)$ has a (unique) random β−expansion of the form

$$x = \sum_{i=1}^{\infty} \frac{a_i(\omega, x)}{\beta(\omega)\beta(\theta\omega)\cdots\beta(\theta^{i-1}\omega)} \tag{4.2}$$

with $a_i(\omega, x) \in \{0, 1, ..., [\beta(\theta^{i-1}\omega)]\}$ having the representations $a_1(\omega, x) = [\beta(\omega)x]$, ..., $a_n(\omega, x) = [\beta(\omega)T_\omega^{n-1}x]$, ... where $T_\omega x = \beta(\omega)x - [\beta(\omega)x]$. The latter is a random expanding in average map of [0,1] but, in general, it is not a local homeomorphism, and so this case cannot be treated directly in the framework of Section

2. The following result is a particular case of [Bu1,2] where $\Theta-$invariant measures with absolutely continuous disintegrations have been constructed for random Lasota–Yorke maps and their exponential decay of correlations together with some multidimentional extensions have been studied.

4.1. THEOREM. *There exists a unique $\Theta-$invariant probability measure μ on $\Omega \times [0,1]$ with the marginal P whose disintegrations μ^ω for $P-a.a.\omega$ are absolutely continuous with respect to the Lebesgue measure m on $[0,1]$. The measure μ is ergodic and the Radon–Nikodim derivative $h_\omega(x) = \frac{d\mu^\omega(x)}{dm(x)}$, which can be chosen so that $\sup_x h_\omega(x) < \infty$ $P-a.s.$, satisfies*

$$\beta(\omega)h_{\theta\omega}(x) = \sum_{T_\omega y=x} h_\omega(y) = \sum_{m=0}^{[\beta(\omega)-x]} h_\omega(x_\omega(m)) \tag{4.3}$$

where $x_\omega(m) = (x+m)(\beta(\omega))^{-1}$.

It seems that, in general, there are no explicit formulas for the density h similar to the one derived in [Pa] for the deterministic case. Still, some bounds for h can be obtained. Set $L = \sum_{n=0}^{\infty} \beta^{(-n)}$ where $\beta^{(-n)} = (\beta \circ \theta^{-n}) \cdots (\beta \circ \theta^{-2})(\beta \circ \theta^{-1})$

4.2. PROPOSITION. *$P \times m-a.s.$*

$$1 - \int \beta^{-1} dP \leq h_\omega(x) \leq \int L dP. \tag{4.4}$$

PROOF. Of course, the left hand side of (4.4) is not trivial only when $\int \beta^{-1} dP < 1$ and the right hand side of (4.4) is not trivial only when $\int L dP < \infty$. Now, call a sequence $(a_1, a_2, ...)$ $\omega-$canonical if there exists a number x $(0 \leq x < 1)$ such that the digits of its random $\beta-$expansion are $a_1, a_2, ...$, i.e. $a_i(\omega, x) = a_i$ $\forall i \geq 1$. Call also $(a_1, ..., a_n)$ an $\omega-$canonical sequence of order n if $(a_1, ..., a_n, 0, 0, ...)$ is $\omega-$canonical. Let $S_\omega(n)$ denotes the number of $\omega-$canonical sequences of order $n \geq 0$ and put $S_\omega(0) \equiv 1$. Similarly to §4 in [Re] I derive easily that

$$S_\omega(n) - S_\omega(n-1) \leq \beta^{(n)}(\omega) \leq S_\omega(n) \ \forall n \geq 1$$

where $\beta^{(n)}(\omega) = \beta(\omega)\beta(\theta\omega) \cdots \beta(\theta^{n-1}\omega)$ and $\beta^{(0)}(\omega) = 1$. Hence

$$\beta^{(n)}(\omega) \leq S_\omega(n) \leq \sum_{k=0}^{n} \beta^{(k)}(\omega) \ \forall n \geq 0. \tag{4.5}$$

Next, in the same way as in [Re] I conclude that for any measurable subset $E \subset (0,1)$,

$$\frac{S_\omega(n) - S_\omega(n-1)}{\beta^{(n)}(\omega)} m(E) \leq m((T_\omega^n)^{-1}E) \leq \frac{S_\omega(n)}{\beta^{(n)}(\omega)} m(E). \tag{4.6}$$

It follows from (4.5) and (4.6) that

$$(1 - \frac{1}{n}\sum_{k=0}^{n-1} \frac{1}{\beta(\theta^k\omega)}) m(E) \leq \frac{1}{n}\sum_{k=0}^{n-1} m((T_\omega^k)^{-1}E) \leq \frac{1}{n}\sum_{k=0}^{n-1} L(\theta^k\omega) \tag{4.7}$$

with L defined above. Since all weak limits as $n \to \infty$ of the sequence of measures $\frac{1}{n}\sum_{k=0}^{n-1} \Theta^k(P \times m)$ (evaluated on functions from $\mathbb{L}^1(\Omega, C[0,1])$) are $\Theta-$invariant (see [Ar], Section 1.5) then (4.7) together with Theorem 4.1 and the ergodic theorem yield (4.4). □

The measure μ from Theorem 4.1 is the equilibrium state of Θ corresponding to the function $\varphi_\omega(x) = -\log \beta(\omega)$ since it follows easily that

$$\mathbf{h}_\mu^{(r)}(\Theta) = \int \log \beta dP. \tag{4.8}$$

As a next step it is natural to construct equilibrium states for a large class of functions. If $\varphi = \varphi_\omega(x)$ is Lipschitz continuous in x then employing the Schauder–Tichonoff fixed point theorem similarly to Lemma 3 in [Wa2] (with modifications in the spirit of Lemma 2.2 from [Ki2]) it follows that there exists a positive random variable λ^ω and a measure $\nu \in \mathcal{P}(\Omega \times [0,1))$ with disintegrations ν^ω such that for any continuous function on $[0,1]$ (and, in fact, for a larger class of functions) $P-$a.s.

$$\int \mathcal{L}_\varphi^\omega q d\nu^{\theta\omega} = \lambda^\omega \int q d\nu^\omega \tag{4.9}$$

where the definition (2.1) of the transfer operator can be written here in the form

$$\mathcal{L}_\varphi^\omega q(x) = \sum_{m=0}^{[\beta(\omega)-x]} e^{\varphi_\omega(x_\omega(m))} q_\omega(x_\omega(m)) \tag{4.10}$$

with $x_\omega(m)$ the same as in (4.3).

Next, applying again the Schauder–Tychonoff fixed point theorem similarly to Proposition 2.5 in [Ki2] one obtains a positive measurable function $h = h_\omega(x)$ such that $P-$a.s.

$$\mathcal{L}_\varphi^\omega h_\omega = \lambda^\omega h_{\theta\omega}. \tag{4.11}$$

Now, (4.9) together with (4.11) already yield that the measure μ having disintegrations $\mu^\omega = h_\omega \nu^\omega$ is $\Theta-$invariant, i.e. $T_\omega \mu^\omega = \mu^{\theta\omega}$ $P-$a.s. Next, similarly to Proposition 3.1 in [Ki2] one can show that μ is an equilibrium state of Θ corresponding to φ.

The final crucial step consists in showing that $P-$a.s. the triple ν^ω, λ^ω, h_ω is unique which would imply the uniqueness of μ, as well, together with the fact that μ is an equilibrium state of Θ corresponding to φ. The corresponding deterministic proof in [Wa2] proceeds in different ways for the cases when 1 is preperiodic and when it is not. The former can be reduced to local homeomorphisms case of [Wa1] (whose random counterpart is considered in Section 2 above) and can also be represented as a subshift of finite type. The treatment of the other case in [Wa2] relies on the assumption that all points in the orbit of β are different which is used in the key Lemma 3 there in order to show that the corresponding deterministic measure ν has no atoms. In the random setup, as a rule, arguments relying on periodic orbits do not work. Since, in fact, this "no atoms" property holds true in the deterministic situation always no matter whether the orbit of 1 is infinite or finite, it is worthwhile to look for a proof which would deal with both cases simultaneously. Such a proof, most likely, could be extended to the random setup above, as well. Other parts of the deterministic proof in [Wa2] yielding uniqueness and nice properties of equilibrium states can be adapted to the random setup similarly to [Ki2], [KK] and Section 2 above.

Observe, that if there exist nonrandom $n, k \in \mathbb{N}$ and $x_0 \in I$ such that $P-$a.s. $T^n_\omega 1 = T^{n-1}_{\theta\omega}(\beta(\omega) - [\beta(\omega)]) = x_0$ and $T^k_\omega x_0 = x_0$, in particular if $T^n_\omega 1 = 0$, then the problem reduces to the random subshifts of finite type case for which the required results (as formulated in Theorem 2.1 above) are already known (see [KK]). Assume, for instance, that $n = 2$, $x_0 = 0$, and so $k = 1$. Then $\beta(\omega) - [\beta(\omega)] = \frac{l}{\beta(\theta\omega)}$ for some $l = 0, 1, ..., [\beta(\theta\omega)]$ and the dynamics can be described by the left shift on the sequence spaces $\Xi_\omega = \{\xi = (\xi_m, m \geq 0) : \pi_{\xi_m \xi_{m+1}}(\theta^m\omega) = 1 \ \forall m \geq 0\}$ where $\Pi(\omega) = (\pi_{ij}(\omega))$ are $r(\omega) \times r(\theta\omega)$ matrices, where $r(\omega) = [\beta(\omega)] + 1$ if $\beta(\omega) > [\beta(\omega)]$ and $r(\omega) = \beta(\omega)$ if $\beta(\omega) = [\beta(\omega)]$, with entries taking on only 0 and 1 values and such that $\pi_{ij}(\omega) = 1$ if and only if either $0 \leq i \leq [\beta(\omega)]$ or $i = [\beta(\omega)] + 1$, $\beta(\omega) > [\beta(\omega)]$ and $j \leq \beta(\theta\omega)(\beta(\omega) - [\beta(\omega)]) \in \mathbb{N}$.

Appendix: Some open problems in random dynamics

Evolution by its nature is time dependent so physical systems can be better described by compositions of different maps rather than by repeated application of exactly the same transformation. It is natural to study such problems for typical, in some sense, sequences of maps which are picked at random in a stationary fashion. This leads directly to the framework of random transformations. Random diffeomorphisms emerge from stochastic flows which solve stochastic differential equations (see [Ar]) and they also appear naturally as a model for random perturbations of dynamical systems. Results in dynamical systems which have natural extensions to the random dynamical systems case can be viewed as more robust and their study may shed additional light on the stability questions, as well, as produce more powerful methods of proof. A quite natural extension of methods of hyperbolic dynamics to skew product systems with some hyperbolic actions in fibers leads to random hyperbolic systems (see [GK]) but this theory is not completed yet. A definition of a random hyperbolic set can be found in [GK] but the global definition of random Axiom A diffeomorphisms and their properties are not quite clear yet except for the case when they are picked at random from a small C^2 neighborhood of a deterministic Axion A diffeomorphism. In the latter case some results can be proved for all and not just for typical sequences of diffeomorphisms (see [Bak] and [AF]). I note that some important general notions of dynamical systems such as nonwandering points and, in particular, periodic points may not play a natural role in the random setup. It seems that some deterministic results do not have natural extensions for the random setup and, on the other hand, some results on random transformations may have no counterparts for the deterministic case assuming that transformations are truly random, for instance, that θ acts on Ω aperiodically or has positive entropy etc. As an example of such results I can mention the existence of finite relative topological generators (a relative topological version of Krieger's theorem) obtained in [KW]. The theory of random transformations is a relatively new subject and it is still on the stage of building its base, and so I expect that its further development will produce new interesting sometimes may be surprising results. A statistical physics model involving random diffeomorphisms was considered in [Ru2]. The ideology of random transformations extends beyond dynamical systems and it leads to the study of Markov chains and more general processes in random (dynamical) environments (see [Ki5] and references there), as well, as of products of independent random matrices with stationary changing distributions and, more generally, of random walks on groups with stationary changing increment distributions (see [Ki7]). For a survey on random dynamics I refer the reader to [Ki6] (see also [Ki8]).

I start the discussion of open problems with random Anosov diffeomorphisms T_ω of the d–dimensional torus $\mathbb{T}^d$. A general definition can be found in [GK] and it requires the existence of splittings $\mathbb{R}^d = \Gamma_\omega^- \oplus \Gamma_\omega^+$ such that $DT_\omega\Gamma_\omega^\pm = \Gamma_{\theta\omega}^\pm$, where DS denotes the differential of S, $\|D(T_\omega^n)^{\mp 1}\xi\| \le \lambda^{(n)}(\omega)\|\xi\|$ for $\xi \in \Gamma_\omega^\pm$ where $\lambda^{(n)}(\omega) = \lambda(\omega)\lambda(\theta\omega)\cdots\lambda(\theta^{n-1}\omega)$ and $-\infty < \int \log \lambda dP < 0$. In addition, the minimal angle $\alpha(\omega)$ between Γ_ω^+ and Γ_ω^- should satisfy $\int \log \alpha dP > -\infty$. Random Anosov diffeomorphisms can be viewed as mildly nonuniform in time but uniform in space hyperbolic systems. A class of such random Anosov diffeomorphisms which are naturally to call random Anosov automorphisms is given by random matrices A_ω with integer entries such that $\det A_\omega = 1$ P–a.s. and that the random matrix product $A_{\theta^{n-1}\omega}\cdots A_{\theta\omega}A_\omega$ has nonzero Lyapunov exponents. A more specific example from [AF] yields random Anosov automorphisms of $\mathbb{T}^2$ with matrices A_ω having the form

$$A_\omega = \begin{pmatrix} 1 + n(\omega)n(\theta\omega) & n(\theta\omega) \\ n(\omega) & 1 \end{pmatrix}.$$

Here n is a positive integer valued random variable with $\log n \in \mathbb{L}^1(\Omega, P)$ and θ^2 is also supposed to be ergodic. Then $\lambda(\omega)$ and the stable and unstable directions $\Gamma_\omega^\mp$ can be expressed via continued fractions (see [GK], Example 4). The deterministic version of the following is a well known theorem (see, for instance, Section 18.6 in [KH]).

A1. CONJECTURE. *For any random Anosov diffeomorphism T_ω, $\omega \in \Omega$ of $\mathbb{T}^d$ there exists a random Anosov automorphism A_ω, $\omega \in \Omega$ and a random homeomorphism h_ω, $\omega \in \Omega$ of $\mathbb{T}^d$ such that $A_\omega = h_{\theta\omega}^{-1}T_\omega h_\omega$ and A_ω is in the same homotopy class as T_ω, i.e. they induce the same action in the fundamental group $\mathbb{Z}^d$ of $\mathbb{T}^d$.*

In order to prove this result, one shows, first, that the linear actions $\tilde{T}_\omega$ induced by T_ω in the fundamental group produce a random Anosov automorphism. Next, it is necessary to construct a random semiconjugacy which should be similar to arguments in the random Hartman–Grobman theorem (see Section 7.4 in [Ar]). The final part of the proof should yield that, in fact, this semiconjugacy is the required random conjugacy. The deterministic proof employs, usually, periodic points which are rare in the random setup and, it seems, some form of shadowing should be used instead. Observe, that if all realizations of a random Anosov (more generally, hyperbolic) diffeomorphism are close to the same deterministic one then the former is randomly conjugate to the latter which follows via the shadowing (see [Li]). The existence of P–a.s. smooth conjugations h_ω is less clear (cf. [KH], p.p. 640–641). It is easy to see that if h_ω is a random conjugation and probability measures m^ω satisfy $A_\omega m^\omega = m^{\theta\omega}$ P–a.s. then the measures $\mu^\omega = h_\omega m^\omega$ satisfy $T_\omega \mu^\omega = \mu^{\theta\omega}$ P–a.s. and if h_ω is smooth P–a.s. then Lyapunov exponents of A with respect to m and of T with respect to μ must be the same. The following would be a random extension of [Ll].

A2. CONJECTURE. *Let $T_\omega, S_\omega, \omega \in \Omega$ be random C^k Anosov diffeomorphisms of a two dimensional compact manifold M and there exists a random homeomorphism $h_\omega, \omega \in \Omega$ of M such that $S_\omega = h_{\theta\omega}^{-1}T_\omega h_\omega$. Suppose that the Lyapunov exponents of S_ω (i.e. of the cocycle $DS_\omega^{n+m} = DS_{\theta^m\omega}^n \circ DS_\omega^m$) with respect to any probability measure m such that $dm(\omega, x) = dm^\omega(x)dP(\omega)$ and $S_\omega m^\omega = m^{\theta\omega}$ are the* *same as of* T_ω *with respect to the measure* μ *given by* $d\mu(\omega,x)=d(h_\omega m^\omega)(x)dP(\omega)$. *Then* h_ω *is* $C^{k-\varepsilon},\varepsilon>0$ $P-a.s.$

Another rigidity statement which would be interesting to verify in the random situation concerns atypically smooth stable and unstable foliations.

A3. CONJECTURE. *Any* C^∞ *random Anosov diffeomorphism of* $\mathbb{T}^d$ *with* P*-a.s.* C^∞ *random stable and unstable foliations is* C^∞ *random conjugate to a random Anosov automorphism of* $\mathbb{T}^d$.

In the deterministic case this conjecture is known to be true assuming that some C^∞ affine connection (for instance, a simplectic 2 form) is left invariant (see [BL]). It would be interesting to obtain a characterisation of random Anosov diffeomorphism via an appropriate gap in a random version of the Mather's spectrum (see [Ma]) whose role may play then the infinite dimensional Lyapunov spectrum of the cocycle DS^n_ω acting on the space of vector fields. As the next step, one may apply pinching conditions on this spectrum to the rigidity problem of Conjecture A3 similarly to the deterministic case (see [Fe]). Another problem related to the nonexistent yet spectral theory of random operators concerns random Perron–Frobenius operators which are the main tool in the thermodynamic formalism type constructions of equilibrium states (see [Ki2], [KK], [Ki5], [KG] and Section 2 above). In the deterministic case proofs of analitic dependence on parameters of different quantities emerging there are based on the well known perturbation theory of linear operators which is not yet worked out properly in the random case.

The topological pressure of a Hölder continuous potential g, the corresponding equilibrium state and its entropy for an Anosov diffeomorphism T are known to depend smoothly both on g and on T in an appropriate sense (see [Ru1] and [Co] for a more difficult flow case). A random extension of this result has not been dealt with properly as yet.

A4. PROBLEM. *Show that the relative topological pressure of a random Hölder potential for a random Anosov diffeomorphism depend on them smoothly and that the same is true for the corresponding equilibrium state and its relative entropy.*

Periodic orbits play an extremely important role in the deterministic dynamics, especially, for hyperbolic dynamical systems where already definition of some important objects, such as the $\zeta-$function, relies on periodic orbits. In the random setup it is not quite clear what should play their role, in general. As a certain replacement for periodic orbits here A. Katok suggested to study invariant measures μ of the skew product transformation $\Theta(\omega,x)=(\theta\omega,T_\omega x)$ whose disintegrations μ^ω have supports consisting of finite number of points. More generally, one may be interested in invariant measures with zero relative entropy.

Such measures can be constructed for certain systems which have random symbolic representations. Consider, for instance, the (Markov) setup of random $f-$expansions from Section 2 above where I need only that B_ω takes on values in $\mathbb{N}\cup\{\infty\}$ and that the assumptions (i) and (ii) there hold true. Let $\Pi(\omega)=(\pi_{kl}(\omega),\ A_\omega\le k<B_\omega,\ A_{\theta\omega}\le l<B_{\theta\omega})$ be a random (measurable in ω) matrix with 0 and 1 entries having $P-$a.s. no zero columns and $b_i(\omega)$, $i=1,2,...,m$ (where $m\in\mathbb{N}$ is a constant) be integer valued random variables satisfying $A_\omega\le b_i(\omega)<B_\omega$ and $\pi_{b_i(\omega)b_{i+1}(\theta\omega)}(\omega)=1$ $P-$a.s. for all $i=1,2,...,m$ where $b_{m+1}(\omega)=b_1(\omega)$. Assume also that with positive probability either $\max_i b_i(\omega)>A_\omega$ or $\min_i b_i(\omega)<B_\omega-1$. Let $\Gamma_\omega=\{x=f_\omega(a_1+f_{\theta\omega}(a_2+f_{\theta^2\omega}(a_3+\cdots)...):\ A_{\theta^i\omega}\le a_{i+1}<$

$B_{\theta^i\omega}, \pi_{a_{i+1}a_{i+2}}(\theta^i\omega) = 1 \forall i = 0,1,2,...\}$ then, clearly, $T_\omega\Gamma_\omega = \Gamma_{\theta\omega}$ and so the set $\Gamma = \{(\omega,x) : x \in \Gamma_\omega\}$ satisfies $\Theta\Gamma = \Gamma$. Now set

$$x_0(\omega) = f_\omega(a_1(\omega) + f_{\theta\omega}(a_2(\omega) + f_{\theta^2\omega}(a_3(\omega) + \cdots)...)$$

where $a_{jm+i}(\omega) = b_i(\theta^{jm+i-1}\omega)$ for all $i = 1,2,...,m$ and $j = 0,1,....$ Put also $x_i(\omega) = T^i_{\theta^{-i}\omega}x_0(\theta^{-i}\omega)$ for $i = 1,2,...,m-1$ where, again, $T_\omega x = f_\omega^{-1}x - [f_\omega^{-1}x]$. It is clear that $x_i(\omega) \in \Gamma_\omega$ and $T^j_\omega x_i(\omega) = x_{(j+i) \bmod m}(\theta^j\omega)$. Set

$$R_\omega = \{x_0(\omega), x_1(\omega), ..., x_{m-1}(\omega)\}$$

then, clearly, $T_\omega R_\omega = R_{\theta\omega}$, and so the set $R = \{(\omega,x) : x \in R_\omega\}$ satisfies $\Theta R = R$. Let μ^ω be the uniform measure on R_ω, i.e. $\mu^\omega(x_i(\omega)) = \frac{1}{m}$ for $i = 0,1,...,m-1$, then $T_\omega\mu^\omega = \mu^{\theta\omega}$, and so the probability measure μ on $\Omega \times [0,1]$ with the marginal P and the disintegrations μ^ω is $\Theta-$invariant.

This construction provides a replacement for periodic orbits for some (but not all) systems whose symbolic representations are random subshifts of finite type. In fact, such "periodic" orbits are dense whenever the random subshift of finite type is uniformly topologically mixing which means that there exists a constant $N \in \mathbb{N}$ such that $\Pi^N(\omega) = \Pi(\omega)\Pi(\theta\omega)\cdots\Pi(\theta^{N-1}\omega)$ is a positive matrix $P-$a.s. Indeed, let $b_i(\omega)$, $i = 1,2,...,k$ be integer valued random variables satisfying $A_\omega \le b_i(\omega) < B_\omega$ and $\pi_{b_i(\omega)b_{i+1}(\theta\omega)} = 1$ $P-$a.s. for $i = 1,2,...,k-1$ but not necessarily $\pi_{b_k(\omega)b_1(\theta\omega)}(\omega) = 1$. Since $\Pi^N(\omega)$ is a positive matrix $P-$a.s. I can choose integer valued random variables $b_i(\omega)$ with $A_\omega \le b_i(\omega) < B_\omega$, $i = k+1,...,k+N-1$ such that $\pi_{b_i(\omega)b_{i+1}(\theta\omega)} = 1$ $P-$a.s. for $i = k,k+1,...,k+N$ where $b_{k+N}(\omega) = b_1(\omega)$. This provides the above construction with $m = k+N-1$ and any permissible initial string $b_1(\omega), b_2(\theta\omega), ..., b_k(\theta^{k-1}\omega)$.

In general, symbolic representations of expanding in average random transformations and of random hyperbolic diffeomorphisms lead to random subshifts of finite type where $\Pi^N(\omega)$ is a positive matrix for some (nonconstant) integer valued random variable N (and, again, P-a.s. $\Pi(\omega)$ does not have zero columns) and the full implementation of the above construction becomes unclear. Still, the following induced transformation construction may do the job. Choose $n \in \mathbb{N}$ large enough so that $Q = Q_n = \{\omega : N(\omega) \le n\}$ satisfies $P(Q) > 0$ and let $k(\omega) = \min\{k \ge 0 : \theta^k\omega \in Q\}$. Since $\Pi^{k(\omega)+n}(\omega)$ is a positive matrix $P-$a.s. then replacing θ, T_ω, Θ by $\tilde\theta = \theta^{k(\omega)+n}$, $\tilde T_\omega = T_\omega^{k(\omega)+n}$, $\tilde\Theta = \Theta^{k(\omega)+n}$ with θ^nQ in place of Ω I arrive at the above setup for $\tilde\theta$, $\tilde T_\omega$, $\tilde\Theta$. This construction yields in the natural way invariant measures with zero relative entropy for the original skew product transformation Θ, since the relative topological entropy of the latter is $(n + \int kdP)-$times the relative topological entropy of $\tilde\Theta$. Nevertheless, it is not clear whether these constructions suffice for applications to the problems related to random hyperbolic and expanding transformations discussed in this section.

One of problems whose solution is given in the deterministic case via periodic orbits is Livschitz's theorem (see [KH], Section 19.2). Its random counterpart can be described in the following way. Let T_ω, $\omega \in \Omega$ be a random Anosov diffeomorphism of $\mathbb{T}^d$ (or, more generally, T_ω has a random hyperbolic set Λ_ω, $T_\omega : \Lambda_\omega \to \Lambda_{\theta\omega}$ defined in [GK]).

A5. CONJECTURE. *Let $\varphi_\omega = \varphi_\omega(x)$ be a measurable function satisfying the random Hölder condition $|\varphi_\omega(x) - \varphi_\omega(y)| \le K_\varphi(\omega)(d(x,y))^\alpha$ where $\alpha > 0$ is a constant and $\int \log K_\varphi dP < \infty$. Assume that $\int \varphi d\mu = 0$ for any $\Theta-$invariant measure* *having the marginal P on Ω. Then there exist a random variable $h = h(\omega)$ with $\int h dP = 0$ and a measurable function $\psi_\omega = \psi_\omega(x)$ satisfying the random Hölder condition as above with the same $\alpha > 0$ and some* log *integrable random variable $K_\psi(\omega)$ such that $\varphi = \psi \circ \Theta - \psi + h$.*

Discrete time random hyperbolic dynamical systems are still in good shape since random Markov partitions there enable one to develop the relative thermodynamic formalism and produce equilibrium states, in particular, random Sinai-Ruelle-Bowen (SRB) measures μ^{SRB} (see [GK]) which can be defined via the corresponding Pesin entropy formula: the relative entropy of μ^{SRB} equals the sum of positive Lyapunov exponents. In the continuous time case very little has been done. A general definition of random hyperbolic flows was suggested in [GK] but in view of noninvariance of the flow direction no substantial results were obtained for this model as yet. Let, for instance, B be a vector field on a compact manifold M which generates a transitive Anosov flow on M and $B_1, B_2, ..., B_n$ be vector fields sufficiently close to B in the C^2 topology. If ν_t is a stationary continuous time Markov chain on the state space $\{1, 2, ..., n\}$ then the differential equation $\frac{dT^t_\omega x}{dt} = B_{\nu_t}(T^t_\omega x)$ yields a random flow T^t_ω which has expanding and contracting random subbundles close to the corresponding subbundles for the original vector field B since there are cones preserved by differentials of all flows generated by $B_i, i = 1, ..., n$. In fact, usually the flow direction (also random) is expanding, as well, and this case resembles a partially hyperbolic situation. Recently, quite a few results were obtained on deterministic partially hyperbolic dynamical systems and it would be interesting to extend them to the random setup.

A6. PROBLEM. *How to construct SRB measures for random hyperbolic flows as above and for random partially hyperbolic systems? When they are unique? What can be said about other equilibrium states? Which limit theorems can be proved for these models?*

Among other interesting problems of random dynamics there are some questions related to dimensions of invariant measures and sets and many dimension constructions for dynamical systems appearing in [Pe] can be extended to the random setup, as well (cf. [Ki3] and [Ki4]). Still, the extension of some results such as the existence of pointwise dimensions for hyperbolic measures known as Eckmann–Ruelle conjecture which in the deterministic case was settled recently in [BPS] may require substantial work.

A7. CONJECTURE. *Let T_ω, $\omega \in \Omega$ be a random diffeomorphism of a compact manifold M and μ be an ergodic invariant with respect to the skew product transformation Θ probability measure on $\Omega \times M$ having the marginal P on Ω and disintegrations μ^ω. Assume that all Lyapunov exponents corresponding to μ (see [Ar]) are not zero. Then for $\mu-$almost all (ω, x) the limit*

$$d_\mu = \lim_{r \to 0} \frac{\log \mu^\omega(B_r(x))}{\log r}$$

exists, where $B_r(x)$ is the $r-$ball in M centered at x.

The limit d_μ is constant in view of ergodicity, it is called the pointwise dimension of μ^ω and its existence should imply (similarly to the deterministic case) the coincidence of main dimensions related to the measures μ^ω. It would be also useful, especially for dimension related applications, and seems possible to extend the non-additive thermodynamic formalism from [Ba] to the random setup. Probably, some dimension results such as those relying on delicate number theoretic properties of Lyapunov exponents and other parameters (see, for instance, PV numbers related results in [PW] and references there) do not have natural counterparts in the random setup in the sense that typically exceptional cases disappear (cf. [KS]).

A rich and natural source of random diffeomorphisms is provided by stochastic flows generated by stochastic differential equations (see [Ar]). It is still not quite clear which hyperbolicity properties can nontrivially emerge in this situation and what can be said here in a generic (in some sense) case. It follows from [LY1,2] that random SRB measures exist rather often for stochastic flows.

The last problem deals with one dimensional transformations.

A8. PROBLEM. *Describe random maps $T_\omega x = 4\lambda_\omega x(1-x)$ of $[0,1]$ possessing measurable families of absolutely continuous probability measures μ^ω on $[0,1]$ satisfying $T_\omega \mu^\omega = \mu^{\theta\omega}$ in the form generalizing well known results for the deterministic logistic family $Tx = 4\lambda x(1-x)$.*

A somewhat different and easier to deal with type of random one dimensional maps was considered in a number of papers where in the typical situation $T_\omega = 4\lambda x(1-x) + \omega_0$ with $\omega = (\omega_i, i \in \mathbb{Z}) \in (-\varepsilon, \varepsilon)^{\mathbb{Z}}$ (see [BBM] and references there). Here for a positive measure set of parameters λ the corresponding skew product transformations Θ have invariant measures with absolutely continuous disintegrations which possess strong decay of correlations properties (in an appropriate form). It may be interesting to check whether this suffice to derive here an $\omega-$wise central limit theorem via the machinery from [Ki5] or some of its modifications.

References

[Ar] L. Arnold, *Random Dynamical Systems*, Springer–Verlag, New York, 1998.

[AF] P. Arnoux and A. Fisher, *Anosov families, renormalization and the scenery flow*, Preprint.

[Bak] V. I. Bakhtin, *Random processes generated by a hyperbolic sequence of mappings. I,II*, Russian Acad. Sci. Izv. Math. **44** (1995), 247–279, 617–627.

[Ba] L. M. Barreira, *A non-additive thermodynamic formalism and applications to dimension theory of hyperbolic dynamical systems*, Ergod. Th.& Dynam. Sys. **16** (1996), 871–927.

[Bi] B. H. Bissinger, *A generalization of continued fractions*, Bull. Amer. Math. Soc. **50** (1944), 868–876.

[Bu1] J. Buzzi, *Absolutely continuous S.R.B. measures for random Lasota–Yorke maps*, Trans. Amer. Math. Soc. **352** (2000), 3289–3303.

[Bu2] J. Buzzi, *Exponential decay of correlations for random Lasota–Yorke maps*, Comm. Math. Phys. **208** (1999), 25–54.

[BL] Y. Benoist and F. Labourie, *Sur les difféomorphismes d'Anosov affines à feuilletages stable et instable différentiables*, Invent. Math. **111** (1993), 285–308.

[BPS] L. Barreira, Ya. Pesin, J. Schmeling, *Dimension and product structure of hyperbolic measures*, Ann. of Math. **149** (1999), 755–783.

[Co] G. Contreras, *Regularity of topological and metric entropy of hyperbolic flows*, Math. Z. **210** (1992), 97–111.

[Ev] C. I. Everett, *Representations for real numbers*, Bull. Amer. Math. Soc. **52** (1946), 861–869.

[Fe] R. Feres, *The invariant connection of a $\frac{1}{2}-$pinched Anosov diffeomorphism and rigidity*, Pacific J. Math. **171** (1995), 139–155.

[GK] V. M. Gundlach and Y. Kifer, *Random hyperbolic systems*, Stochastic Dynamics (H.Crauel and M.Gundlach eds.), Springer–Verlag, New York, 1999, pp. 117–145.

[He] L. Heinrich, *Mixing properties and central limit theorem for a class of non-identical piecewise monitonic C^2-transformations*, Math. Nachr. **181** (1996), 185–214.

[HK] F. Hofbauer and G. Keller, *Ergodic properties of invariant measures for piecewise monotonic transformations*, Math. Z. **180** (1982), 119–140.

[KH] A. Katok and B. Hasselblatt, *Introduction to the Modern Theory of Dynamical Systems*, Cambridge Univ. Press, Cambridge, 1995.

[KP] J. R. Kenney and T.S. Pitcher, *The dimension of some sets defined in terms of f-expansions*, Z. Wahrsch. verw. Geb. **4** (1966), 293–315.

[Ki1] Y. Kifer, *Ergodic Theory of Random Transformations*, Birkhäuser, Boston, 1986.

[Ki2] Y. Kifer, *Equilibrium states for random expanding transformations*, Random&Comp. Dyn. **1** (1992), 1-31.

[Ki3] Y. Kifer, *Fractal dimensions and random transformations*, Trans. Amer. Math. Soc. **348** (1996), 2003–2038.

[Ki4] Y. Kifer, *Fractals via random iterated function systems and random geometric constructions*, in: Fractal Geometry and Stochastics (C. Bandt, S. Graf, and M. Zähle, eds.), Birkhäuser, Basel, 1995, pp. 145–164.

[Ki5] Y. Kifer, *Limit theorems for random transformations and processes in random environments*, Trans. Amer. Math. Soc. **350** (1998), 1481–1518.

[Ki6] Y. Kifer, *Random dynamics and its applications*, Documenta Mathematica, Extra Volume ICM 1998 **II**, 809–818.

[Ki7] Y. Kifer, *"Random" random matrix products*, J. D'Analyse Math. **83** (2001).

[Ki8] Y. Kifer, *Book review on [Ar]*, Ergod. Th.& Dynam. Sys. **20** (2000), 947–950.

[KK] K. Khanin and Y. Kifer, *Thermodynamic formalism for random transformations and statistical mechanics*, in: Sinai's Moscow Seminar on Dynamical Systems (L. A. Bunimovich, B. M. Gurevich, Ya. B. Pesin, eds.), American Mathematical Society Translations–Series 2, 1995, pp. 107–140.

[KPW] Y. Kifer, Y. Peres, and B. Weiss, *A dimension gap for continued fractions with independent digits*, Israel J. Math. (2001).

[KW] Y. Kifer and B. Weiss, *Generating partitions for random transformations*, submitted, Ergod. Th.& Dynam. Sys..

[KS] J. P. Kahane and R. Salem, *Sur la convolution d'une infinité de distributions de Bernoulli*, Colloq. Math. **6** (1958), 193–202.

[Li] P.-D. Liu, *Random perturbations of Axiom A basic sets*, J. Stat. Phys. **90** (1998), 467-490.

[Ll] R. de la Llave, *Smooth conjugacy and S-R-B measures for uniformly and non-uniformly hyperbolic systems*, Comm. Math. Phys. **150** (1992), 289–320.

[LY1] F. Ledrappier and L.-S. Young, *Entropy formula for random transformations*, Prob. Th. and Rel. Fields **80** (1988), 217–240.

[LY2] F. Ledrappier and L.-S. Young, *Stability of Lyapunov exponents*, Ergod. Th& Dynam. Sys. **11** (1991), 469–484.

[Ma] J. Mather, *Characterization of Anosov diffeomorphisms*, Indag. Math. **30** (1968), 479–483.

[MU] D. Mauldin and M. Urbański, *Dimensions and measures in infinite iterated function systems*, Proc. London Math. Soc. **(3) 73** (1996), 105–154.

[Pe] Ya. Pesin, *Dimension Theory in Dynamical Systems*, Univ. of Chicago Press, Chicago, 1997.

[Pa] W. Parry, *On the $\beta-$expansions of real numbers*, Acta Math. Sci. Hungar. **11** (1960), 401–416.

[Re] A. Rényi, *Representations for real numbers and their ergodic properties*, Acta Math. Sci. Hungar. **8** (1957), 477–493.

[Ru1] D. Ruelle, *Differentiation of SRB states*, Comm. Math. Phys. **187** (1997), 227–241.

[Ru2] D. Ruelle, *Positivity of entropy production in the presence of a random thermostat*, J. Stat. Phys. **86** (1997), 935–951.

[Wa1] P. Walters, *Invariant measures and equilibrium states for some mappings which expand distances*, Trans. Amer. Math. Soc. **236** (1978), 121-153.

[Wa2] P. Walters, *Equilibrium states for $\beta-$transformations and related transformations*, Math. Z. **159** (1978), 65–88.

INSTITUTE OF MATHEMATICS, HEBREW UNIVERSITY, JERUSALEM 91904, ISRAEL
E-mail address: kifer@math.huji.ac.il

Proceedings of Symposia in Pure Mathematics
Volume **69**, 2001

Dynamical Zeta Functions

Mark Pollicott

Given a diffeomorphism $f : M \to M$ of a compact manifold it is a classical problem to count the number of periodic orbits $\{x, fx, \dots, f^{n-1}x\}$, with $f^n x = x$. For each $n \geq 1$, we denote by $N(n)$ the number of periodic points of period n.

GENERAL PROBLEM. What can we say about the numbers $N(n)$? Furthermore, what do they tell us about $f : M \to M$?

Of course, without additional assumptions there is no reason to believe that each $N(n)$ is finite.

In sections 1-5 we shall recall some ideas associated to this problem, focusing, in particular, on the context of hyperbolic diffeomorphisms where the results are most satisfactory. In sections 6-8, we study the corresponding problems for flows.

1. Zeta function for diffeomorphisms

1.1 Analytic extensions of zeta functions. An important approach to studying the number $N(n)$ of isolated periodic points was proposed by Artin and Mazur [**AM**] They defined a function of a single complex variable:

DEFINITION. The *Artin-Mazur zeta function* is defined by

$$\zeta_f(z) = \exp\left(\sum_{n=1}^{\infty} \frac{z^n}{n} N(n)\right).$$

Although, a priori, there is no guarantee that this series converges, Artin and Mazur established the following result [**AM**] (cf. also [**Ka**]).

THEOREM 1.1. *(Artin-Mazur) For a C^1 dense set of diffeomorphisms $f : M \to M$ the associated zeta function $\zeta_f(z)$ has a non-zero radius of convergence, and therefore converges to an analytic function for $|z|$ sufficiently small.*

An important class of C^1 diffeomorphisms for which the zeta function $\zeta_f(z)$ is better understood are *Axiom A diffeomorphisms*. A diffeomorphism $f : M \to M$ is said to satisfy *Axiom A* if the non-wandering set Ω has the following properties:

1. The periodic points are dense in Ω,
2. The restriction $f : \Omega \to \Omega$ is hyperbolic (i.e. there exists a Df-invariant splitting $T_\Omega M = E^s \oplus E^u$ and $C > 0$, $0 < \lambda < 1$ such that $||Df^n|E^s||$, $||Df^{-n}|E^u|| \leq C\lambda^n$, for $n \geq 0$)

The following result was proved by Manning [**Mn**] in 1971, following earlier partial results by Guckenheimer [**Gu**] and others.

THEOREM 1.2. *Let $f : M \to M$ be a Axiom A diffeomorphism then the zeta function $\zeta_f(z)$ is always well-defined for $|z|$ small and has an extension as a rational function, i.e., $\zeta_f(z) = P(z)/Q(z)$, for $z \in \mathbb{C}$ ($P, Q \in \mathbb{C}[z]$).*

An alternative proof, which applies in a more general setting, is due to Fried [**Fd3**]. However, at this point the reader may ask the following natural question.

QUESTION. Why care if $\zeta_f(z)$ is rational or not?

Assume that $\zeta_f(z) = \prod_{i=1}^k (1 - z/\alpha_i)^{n_i} / \prod_{i=1}^l (1 - z/\beta_i)^{m_i}$. If α_i, β_j are all distinct then taking the logarithmic derivative gives that

$$\begin{aligned}\sum_{n=0}^{\infty} z^n N(n+1) &= \sum_{i=1}^{k} \frac{n_i}{(z-\alpha_i)} - \sum_{i=1}^{l} \frac{m_i}{(z-\beta_i)} \\ &= \sum_{n=0}^{\infty} z^n \left(\sum_{i=1}^{k} \frac{n_i}{\alpha_i^{n+1}} - \sum_{i=1}^{l} \frac{m_i}{\beta_i^{n+1}} \right)\end{aligned}$$

and comparing coefficients gives an asymptotic estimates of the form:

COROLLARY.

$$N(n) = \sum_{i=1}^{k} \frac{n_i}{\alpha_i^n} - \sum_{i=1}^{l} \frac{m_i}{\beta_i^n}.$$

2. Zeta functions elsewhere in mathematics

In this section we want to address the following general question about the rôle of zeta functions.

QUESTION. What is the motivation for studying zeta functions?

Let us consider two famous examples of zeta functions which lie outside of the dynamical setting (but are, never the less, very instructive).

2.1 Zeta functions in algebraic geometry. Let V be a non-singular projective variety algebraic over a finite field k, say, with q elements (defined by homogeneous polynomial equations $x_0, \ldots, x_m$, say). The variety V is invariant under the Frobenius map

$$F : (x_0, \ldots, x_m) \mapsto (x_0^q, \ldots, x_m^q)$$

and we have the zeta function:

DEFINITION. The Hasse-Weil zeta function is defined by

$$\zeta(z, V) := \exp\left(\sum_{n=1}^{\infty} \frac{z^n}{n} \#\mathrm{Fix}(F^n) \right)$$

PROPOSITION (WEIL CONJECTURES). *The zeta function $\zeta(z, V)$ is rational.*

This was proved by Weil-Dwork-Grothendieck-Deligne. More precisely, they showed $\zeta(z, V) = \prod_{i=1}^{2m} P_l(z)^{(-1)^l}$ with $P_l(z) = \det(1 - zF^* | H^l(V))$, with zeros of modulus $q^{-l/2}$.

2.2 Algebraic Topology and the Lefschetz Theorem. Let $L(f)$ denote the Lefschetz number of f defined in terms of the induced actions $f_{i*} : H_i(M, \mathbb{Z}) \to H_i(M, \mathbb{Z})$ on homology by

$$L(f) = \sum_{i=0}^{\dim M} (-1)^i \mathrm{trace}(f_{i*}).$$

Similarly, let $L(f^n)$ denote the Lefschetz number of the n-th iterate $f^n : M \to M$. The Lefschetz fixed point theorem gives that

$$L(f^n) = \underbrace{\sum_{f^n x = x} L_x(f^n)}_{\substack{\text{Periodic orbits} \\ \text{with weights } \pm 1}}$$

where $L_x(f^n) = \pm 1$ is the Lefschetz index of the isolated periodic point $f^n x = x$ [**Br**].

WARNING. Although $L(f^n)$ counts periodic points it does so with a suitable weighting ± 1, which reflects some additional topological feature of the orbit.

DEFINITION. We define the *Lefschetz zeta function* by

$$\zeta_L(z) = \exp\left(\sum_{n=1}^{\infty} \frac{z^n}{n} L(f^n)\right)$$

Observe that we can write

$$\zeta_L(z) = \exp\left(\sum_{n=1}^{\infty} \frac{z^n}{n}\left(\sum_{i=0}^{\dim M} (-1)^i \mathrm{trace}(f^n_{i,*})\right)\right) = \frac{\prod_{i \text{ odd}} \det(1 - z f_{i*})}{\prod_{i \text{ even}} \det(1 - z f_{i*})}.$$

Since the last term is rational, we have obtained the following classical result.

PROPOSITION. *The Lefschetz zeta function $\zeta_L(z)$ is always a rational function.*

An application: Lefschetz zeta functions and twisted orbits. A particularly simple, yet interesting, dynamical variant on the Lefschetz zeta function involves the idea of twisted orbits.

Assume that $f : M \to M$ is an Anosov diffeomorphism (or more generally any diffeomorphism for which all periodic points $f^n x = x$ are isolated and there is a splitting $D_x f^n = E^u_x \oplus E^s_x$, where E^u and E^s correspond to the expanding and contracting eigenvalues).

We say that the periodic orbit $\tau = \{f^n x = x, fx, \ldots, f^{n-1}x\}$ is *twisted* if $D_x f^n : E^u \to E^u$ reverses the orientation. Correspondingly, we say that the periodic orbit τ is *untwisted* if $D_x f^n : E^u \to E^u$ preserves the orientation. If the periodic point $f^n x = x$ is twisted we write $\Delta(\tau) = -1$, and if the point is untwisted then we write $\Delta(\tau) = 1$. For an orbit $\tau = \{f^n x = x, fx, \ldots, f^{n-1}x\}$ we denote $|\tau| = n$.

In this context the Lefschetz zeta function can be very conveniently written as Euler product:

$$\zeta_L(z) = \prod_{\tau} \left(1 - \Delta(\tau) z^{|\tau|}\right)^{(-1)^{\dim E^u}}$$

(i.e. the exponent is -1 if $\dim E^u$ is odd and 1 if $\dim E^u$ is even).

The following result is due to Franks [**Fr2**].

THEOREM. *Let $f : M \to M$ be an Axiom A diffeomorphism.*

(i) *Assume there exists an odd $d \geq 1$ such that every periodic orbit τ of period $|\tau| = d$ has $\Delta(\tau) = -1$. Then if any such τ has least period $p|d$ then there are twisted periodic points of prime period $2^k d$, $k \geq 0$.*

(ii) *Assume there exists an odd $d \geq 1$ such that every periodic orbit τ of period $|\tau| = d$ has $\Delta(\tau) = 1$. Then if any such τ has least period $p|d$ then there are periodic points of prime period $2^k d$, $k \geq 0$.*

REMARK. More results of this nature can be found in [**Fr1, Fr2**]. More generally, mixtures of dynamical and topological methods to prove the existence of periodic points are important in such recent results as new proofs of the Poincaré-Birkhoff theorem and the Klingenberg Conjecture on closed geodesics on S^2.

3. Proving rationality

3.1. Zeta functions for subshifts of finite type. As a starting point, we need to consider a simple symbolic model. Consider a subshift of finite type $\sigma : X_A \to X_A$, where A is an irreducible matrix with entries either 0 or 1, $X_A = \{x = (x_n)_{n=0}^{\infty} : A(x_n, x_{n+1}) = 1, n \in \mathbb{Z}\}$ and $(\sigma x)_n = x_{n+1}$. The number of fixed points is given by the trace $N(1) = \text{trace}(A)$. Similarly the number of periodic points of period n is given by $N(n) = \text{trace}(A^n)$, where $A^n = \underbrace{A \cdots A}_{\times n}$ denotes n-fold matrix multiplication. We see

$$\zeta_\sigma(z) = \exp\left(\sum_{n=1}^{\infty} \frac{1}{n}\text{trace}(A^n)\right) = \frac{1}{\det(I - zA)}.$$

In particular, we have that the zeta function $\zeta_\sigma(z)$ associated to a subshift of finite type is the reciprocal of a polynomial [**BL**]. Moreover, the poles of $\zeta_\sigma(z)$ can be identified with the reciprocals $1/\lambda$ of the eigenvalues λ of the matrix A. The zeta function is an interesting invariant in the classification of subshifts of finite type.

3.2 Manning's theorem. Manning showed how this rationality could be carried over to Axiom A diffeomorphisms [**Mn**]. Let $f : \Lambda \to \Lambda$ be a hyperbolic diffeomorphism. Assume that we have a Markov Partition $\mathcal{T} = \{T_1, \dots, T_k\}$ for Λ, with $k \times k$ transition matrix A defined by

$$A(i,j) = \begin{cases} 1 & \text{if } f(\text{int}T_i) \cap \text{int}T_j \neq \emptyset \\ 0 & \text{otherwise.} \end{cases}$$

Using the transition matrix A we can associate a subshift of finite type $\sigma : X_A \to X_A$.

Unfortunately, there is not quite a one-to-one correspondence between the closed orbits of $f : \Lambda \to \Lambda$ and $\sigma : X_A \to X_A$. This discrepancy is caused by the complication of closed orbits for f which are contained in the boundaries of elements of the Markov Partition $\mathcal{T}$. To overcome this problem, Manning introduced auxiliary matrices to account for these points. For example, one could define a new

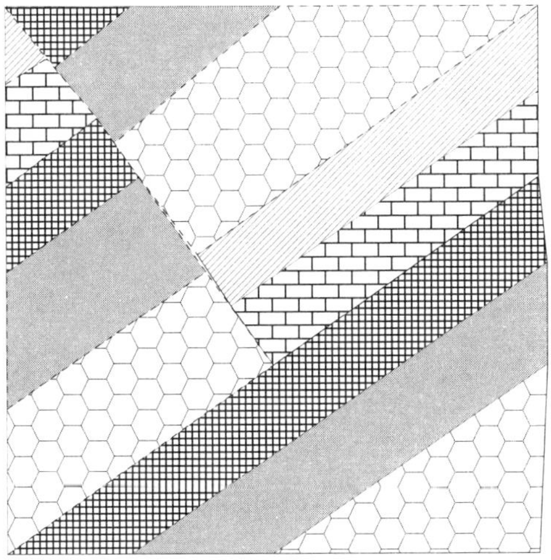

FIGURE. A Markov partition with 5 pieces for $\begin{pmatrix} 2 & 1 \\ 1 & 1 \end{pmatrix}$. The fixed point 0 is coded by two distinct rectangles.

subshift whose symbols are pairs of neighbouring rectangles $\{T_i, T_j\}$ and whose transition matrix is

$$A_{(2)}(\{T_i, T_j\}, (\{T_k, T_l\}) = \begin{cases} 1 & \text{if } A(T_i, T_k) = 1 = A(T_j, T_l) \\ & \text{or } A(T_i, T_l) = 1 = A(T_j, T_k) \\ 0 & \text{otherwise} \end{cases}$$

A periodic point in $T_i \cap T_j$ would also give a periodic point for $\sigma : X_{A_{(2)}} \to X_{A_{(2)}}$. What Manning showed was that the original zeta function $\zeta_f(z)$ can be written precisely in terms of the functions $\zeta_A(z)$, $\zeta_{A_{(2)}}(z)$, and other (necessarily rational) symbolic zeta functions constructed in a similar fashion.

EXAMPLE. Consider the map $T : \mathbb{R}^2/\mathbb{Z}^2 \to \mathbb{R}^2/\mathbb{Z}^2$ defined by $T(x, y) = (2x + y, x+y) \pmod 1$. The zeta function is $\zeta(z) = (1-z)^2/\det(1-zA)$ where $A = \begin{pmatrix} 2 & 1 \\ 1 & 1 \end{pmatrix}$.

4. Zeta functions with weightings

There are natural generalizations of the Artin-Mazur zeta function which allow for the possibility of weighting a periodic orbit $\{x, fx, \dots, f^{n-1}x\}$ by a (Hölder) continuous function $g : \Lambda \to \mathbb{R}$. Thus, in addition to the period n of the orbit we also have a useful weighting given by $g^n(x) := g(x) + g(fx) + \dots + g(f^{n-1}x)$.

DEFINITION. We define the *Ruelle zeta function* for an Axiom A diffeomorphism $f : \Lambda \to \Lambda$ and a Hölder continuous function $g : \Lambda \to \mathbb{R}$ by

$$\zeta_f(z, g) = \exp\left(\sum_{n=1}^{\infty} \frac{z^n}{n} \sum_{f^n x = x} e^{g^n(x)}\right)$$

This complex function converges to an analytic function on the disc $|z| < e^{-P(g)}$, where

$$P(g) = \limsup_{n \to +\infty} \frac{1}{n} \log \left(\sum_{f^n x = x} e^{g^n(x)} \right)$$

denotes the *pressure of the real valued function* $g : \Lambda \to \mathbb{R}$.

The main result on the domain of $\zeta_f(z, g)$ is the following.

THEOREM. *The zeta function $\zeta_f(z, g)$ has a non-zero meromorphic extension to a disk $|z| \leq \theta^{-1} e^{-P(g)}$, for some $0 < \theta < 1$, with a simple pole at $z = e^{-P(g)}$.* [**Ru3**], [**PP2**], [**Ha**]

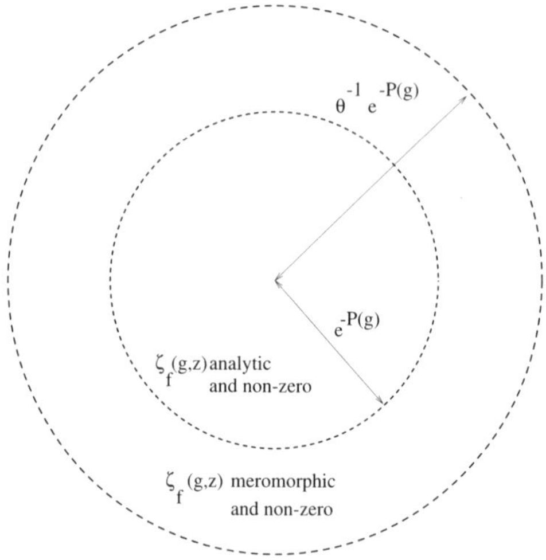

FIGURE. The analytic and meromorphic domains of $\zeta_f(z, g)$

REMARK. By meromorphic, we mean analytic except for possibly finitely many poles of finite order.

The remainder of this section is devoted to outlining the proof of the above Theorem.

4.1 Transfer operators and Hölder functions. Using Markov partitions we can model the Axiom A diffeomorphism $f : \Lambda \to \Lambda$ by a subshift of finite type $\sigma : X_A \to X_A$ (i.e., there is a semi-conjugacy $\pi : X \to \Lambda$). We can also induce $\hat{g} = g \circ \pi : X \to \mathbb{C}$.

To proceed we need to study a slightly different space. Let $X_A^+ = \{x = (x_n)_{n=0}^{\infty} : A(x_n, x_{n+1}) = 1, n \geq 0\}$ and define the *one-sided subshift* $\sigma : X_A^+ \to X_A^+$ by $(\sigma x)_n = x_{n+1}$. Without loss of generality, we can assume $\hat{g} = g \circ \pi : X_A^+ \to \mathbb{C}$ We can construct a suitable Banach space of Hölder functions on X_A^+ as follows. Given $0 < \theta < 1$, we define for $h : X_A^+ \to \mathbb{C}$ a semi-norm

$$|h|_\theta = \sup \left\{ \frac{|h(x) - h(y)|}{\theta^n} : x_i = y_i, \quad \forall i = 0, \ldots, n-1 \right\}$$

Let $\mathcal{F}_\theta = \{h : X_A^+ \to \mathbb{C} : |h|_\theta < +\infty\}$, then this is a Banach space with norm $||h|| = |h|_\theta + |h|_\infty$ (where $|h|_\infty$ denotes the usual supremum norm).

For suitable $0 < \theta < 1$ we can assume that $\hat{g} \in \mathcal{F}_\theta$ and define the *(Ruelle) transfer operators*

$$\begin{cases} L_{\hat{g}} : \mathcal{F}_\theta \to \mathcal{F}_\theta \\ L_{\hat{g}} h(x) = \sum_{Ty=x} e^{\hat{g}(y)} h(y). \end{cases}$$

(i.e. we average the function over pre-images by the one sided shift $\sigma : X_A^+ \to X_A^+$ with a weighting by e^g)

The operators L_g are not compact, but are "quasi-compact" in the following sense.

PROPOSITION. *Let $Re(\hat{g}) : X^+ \to \mathbb{R}$ be the real part of $\hat{g} : X^+ \to \mathbb{C}$.*
(i) *The spectrum of $L_{\hat{q}} : \mathcal{F}_\theta \to \mathcal{F}_\theta$ is contained in the disk $|z| \le e^{P(Re(\hat{g}))}$.*
(ii) *The essential spectrum of $L_{\hat{g}}$ is contained in the disk $|z| \le \theta e^{P(Re(\hat{g}))}$.*
(iii) *If $\hat{g}$ is actually real valued then $L_{\hat{g}}$ has $e^{P(\hat{g})}$ as a simple isolated eigenvalue. Moreover, the associated eigenfunction h satisfying $\mathcal{L}_{\hat{g}} h = e^{P(\hat{g})} h$ is strictly positive (Ruelle operator Theorem* [**Ru3**]*).*

We recall that the essential spectrum of a bounded linear operator consists of that part of the spectrum that remains once isolated eigenvalues (with eigenprojections of finite dimensional kernel and range) are removed.

A direct link between the definition of the zeta functions and the Ruelle transfer operators is the following identity

$$\sum_{\sigma^n x = x} e^{g^n(x)} = \sum_{i=1}^{k} L_g^n \chi_{[i]}(x_i) + O\left((\theta + \epsilon)^n\right),$$

for any $\epsilon > 0$. This application of transfer operators to the zeta function then shows the following result.

THEOREM. *The zeta function $\zeta_f(z, g)$ has a meromorphic extension to the region $|z| \le \theta^{-1} e^{-P(Re(g))}$. Moreover, z_0 is a singular value in this region iff $\lambda_0 = 1/z_0$ is an eigenvalue for $L_{\hat{g}}$.*

5. Zeta functions for other maps

For other types of dynamical systems we might expect different results on analyticity of the zeta function. Let us consider two different classes of interval maps.

5.1 Unimodal maps and the Milnor-Thurston zeta function.

DEFINITION. We call a continuous map $T : I \to I$ on the unit interval *unimodal* if $T(0) = 1 = T(1)$ and $\exists 0 < c < 1$ such that $T : [0, c] \to I$ is increasing and $T : [c, 1] \to I$ is decreasing. We call c the critical point.

For $x \in I$ we define the *itinerary* of this point to be the sequence $(x_n)_{n=0}^\infty$ where we write
(i) $x_n = L$ if $T^n(x) \in [0, c)$
(ii) $x_n = R$ if $T^n(x) \in (c, 1]$
(iii) $x_n = c$ if $T^n(x) = c$

The itinerary gives a natural way to code orbits as sequences. Let $\Sigma = \{(x_n)_{n=0}^{\infty} : x \in I\}$. We can then define a shift map $\sigma : \Sigma \to \Sigma$ by $\sigma(x_n)_{n=0}^{\infty} = (x_{n+1})_{n=0}^{\infty}$. The next two results show that the shift map σ reflects the dynamics of the interval map T.

LEMMA. *If x has itinerary $(x_n)_{n=0}^{\infty}$ then $T(x)$ has itinerary $\sigma(x_n)_{n=0}^{\infty}$.*

PROPOSITION. *If a point x is periodic under T (i.e. $\exists m \geq 1$ with $T^m(x) = x$) then its itinerary $(x_n)_{n=0}^{\infty}$ is periodic under σ (i.e. $\exists m \neq 1$ with $\sigma^m(x) = T^m(x)$).*

Conversely, assume that an itinerary $(x_n)_{n=0}^{\infty}$ is periodic under σ then these exists a periodic point w under T with this itinerary.

In the present context Milnor and Thurston introduced the following definition of a zeta function (which can be compared with the earlier definition of the Lefschetz zeta function).

DEFINITION. Let

$$N^-(f^k) = \#\{f^x = x : f^n \text{ is decreasing in a neighbourhood of } x\}$$

and define the *Milnor-Thurston modified zeta function* by

$$\zeta^-(z) = \exp\left(\sum_{n=1}^{\infty} \frac{z^n}{n} 2N^-(f^n)\right)$$

The main result on such zeta functions is the due to Milnor-Thurston [**MT**].

THEOREM (MILNOR-THURSTON). *The zeta function takes the form $\zeta^-(z) = 1/\det(1 - D(z))$, where $D(z)$ is a square matrix whose entries are explicit formal power series in z.*

5.2 Expanding interval maps. Aside from Axiom A diffeomorphisms, the setting in which most success has been achieved in the study of dynamical zeta functions is that of interval maps.

Assume that $T : I \to I$ is a interval map which is expanding on each of the intervals $I = I_1 \cup \ldots \cup I_k$.

EXAMPLE (β-TRANSFORMATION). Let $I = [0, 1]$ and choose $2 > \beta > 1$. We can define an interval map $f : I \to I$ by

$$fx = \begin{cases} \beta x & \text{if } 0 \leq x \leq \frac{1}{\beta} \\ \beta x - 1 & \text{if } \frac{1}{\beta} < x \leq 1 \end{cases}$$

If β is transcendental, say, then this map cannot have a Markov partition. In particular, we cannot use our ideas on zeta functions for subshifts of finite type (as we did in the case of Axiom A diffeomorphisms) and so a new approach is required. The key technical idea of the *Markov extension*, was introduced by Hofbauer [**Ho**], and developed by Keller and others.

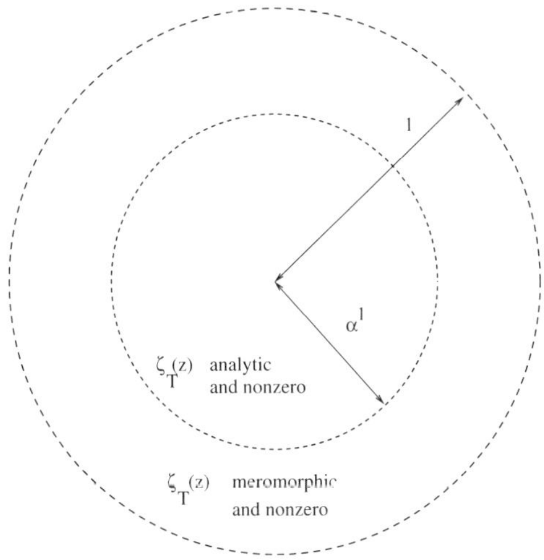

FIGURE. The domain of the zeta function for expanding interval maps

THEOREM. *Let f be a piecewise expanding interval map with $|f'| \geq \alpha > 1$. The zeta function $\zeta(z)$ is analytic for $|z| < 1/\alpha$ and has a meromorphic extension to the disk $z < 1$. (cf.* [**BK, HK**]*)*

The following corollary shows that the asymptotic formula for the number of periodic points can again be estimated quite well (although not quite as well as when the zeta function was known to be rational cf. §1.1).

COROLLARY. *For any $\delta > 0$ there exist constants*
(i) $\lambda_1, \ldots, \lambda_M$, *with $|\lambda_i| > 1 + \delta$ ($i = 1, \ldots, M$), and*
(ii) $C_1, \ldots, C_M \in \mathbb{C}$
such that

$$N(n) = \sum_{i=1}^{M} C_i \lambda_i^n + O((1+\delta)^n).$$

REMARK. This method can be adapted to Collet-Eckmann unimodal maps [**KN**], which are not necessarily expanding, but satisfy the weaker hypothesis

$$\liminf_{n \to +\infty} \frac{1}{n} \log |(f^n)'(f(0))| > 0.$$

6. Zeta functions for Axiom A flows

Let $\phi_t : M \to M$ be a C^1 flow on a compact manifold M. We call a closed ϕ-invariant set Λ a *basic set* if

1. there is a $D\phi$-invariant splitting $T_\Lambda M = E^s \oplus E^u \oplus E^u$ where E^0 is one-dimensional bundle tangent to the flow, and there are constants $C, \lambda > 0$ such that $||D\phi_t|E^s|| \leq Ce^{-\lambda t}$, $||D\phi_{-t}|E^u|| \leq Ce^{-\lambda t}$, for $t \geq 0$
2. the closed orbits in Λ are dense
3. $\phi_t : \Lambda \to \Lambda$ is dense
4. there is a neighbourhood $U \supset \Lambda$ such that $\Lambda = \cup_{t \in \mathbb{R}} \phi_t U$

The flow $\phi_t : M \to M$ satisfies *Axiom A* if the non-wandering set Ω is a finite union of basic sets and hyperbolic fixed points. We call $\phi_t : M \to M$ a transitive *Anosov flow* if $\Lambda = M$.

Given a closed orbit τ we define the *prime period* $\lambda(\tau) > 0$ to be the smallest value for which

$$\phi_{\lambda(\tau)}(x) = x, \text{ for any } x \in \tau.$$

EXAMPLE. The canonical example of an Axiom A flow is the geodesic flow on the unit tangent bundle of manifolds with negative sectional curvatures. The values $\lambda(\tau)$ are then precisely the lengths of closed geodesics.

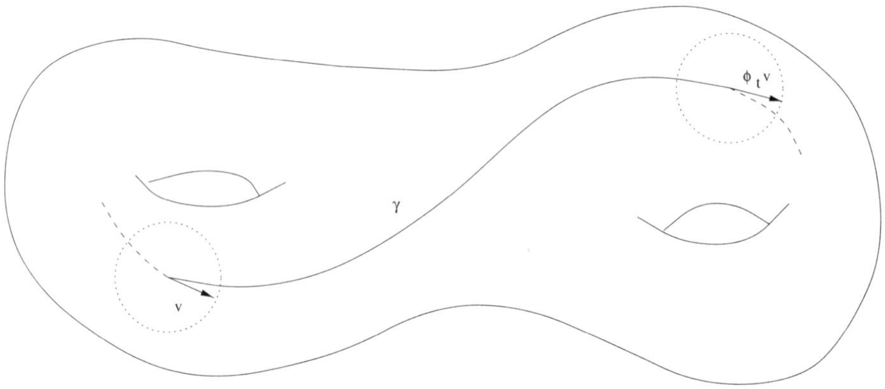

FIGURE. Geodesic flows: Tangent vectors of unit length are transported along unit speed geodesics

6.1 Zeta functions counting closed orbits. When we turn to the study of zeta functions for flows the obvious first question we must ask is:

QUESTION. How do we define a zeta function for a flow?

DEFINITION. We define the *zeta function* for the flow $\phi_t : M \to M$ by

$$\zeta_\phi(s) = \prod_\tau \left(1 - e^{-s\lambda(\tau)}\right)^{-1}, \quad s \in \mathbb{C}$$

(whenever it is well-defined and converges).

For Axiom A flows $\phi_t : \Lambda \to \Lambda$ we shall denote by $h = h(\phi)$ the *topological entropy* (of the time one flow). In this case it is easy to see that $\zeta_\phi(s)$ converges to a non-zero analytic function on the half-plane $Re(s) > h$.

A LITTLE BACKGROUND. This definition is actually due to Ruelle [**Ru1, Ru2**], following slightly different forms due to Smale and Selberg (for geodesic flows on negatively curved manifolds). However, it is strikingly reminiscent of the Euler product form of the Riemann zeta function for prime numbers:

$$\zeta_{\text{Riemann}}(s) = \prod_{p \text{ prime}} \left(1 - p^{-s}\right)^{-1}, \quad s \in \mathbb{C}.$$

This similarity will be exploited in §7.

6.2 Sections and the reduction to discrete expanding maps. To simplify slightly the explanation, let us restrict ourselves to considering (transitive) Anosov flows $\phi_t : M \to M$. For each $x \in M$, the strong stable manifolds

$$W^{ss}(x) = \{y \in M : \lim_{t\to+\infty} d(\phi_t(x), \phi_t(y)) = 0\}$$

are always C^1 immersed submanifolds. Similarly, the strong unstable manifolds

$$W^{su}(x) = \{y \in M : \lim_{t\to-\infty} d(\phi_t(x), \phi_t(y)) = 0\}$$

are C^1 immersed submanifolds.

Given a family $\mathcal{R} = \{T_1, \dots, T_k\}$ of (Markov) Poincaré sections for flow, we can assume without loss of generality that each of the sections is foliated by pieces of strong stable manifold, i.e., the sections are *aligned* in the sense that for each T_i there exists $x_i \in \operatorname{int}(T_i)$ with $T_i = \cup_{x\in W^{su}(x_i,T_i)} W^{ss}(x, T_i)$ (if this is not the case then it is easily arranged by an appropriate perturbation of the sections in the flow direction).

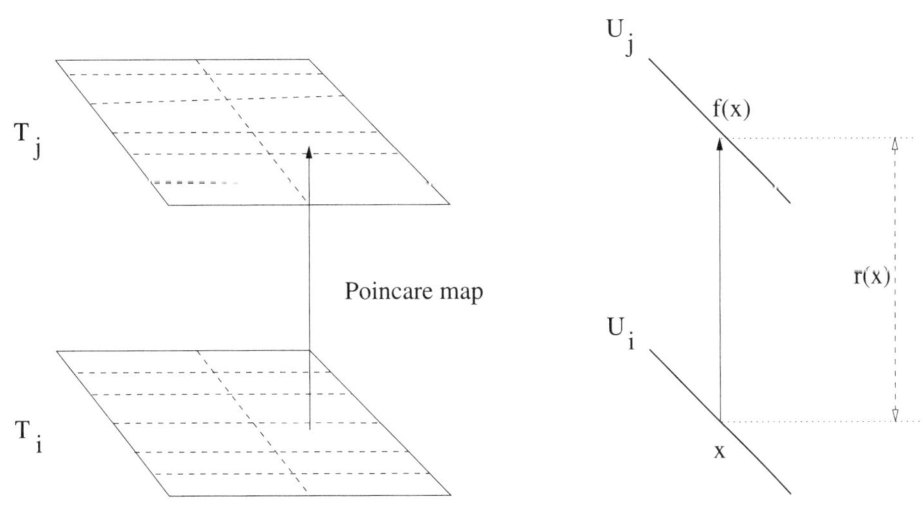

FIGURE. Poincaré sections $T_1, \dots, T_k$ give rise to an expanding map $f : U \to U$ and "transition time" function $r : U \to \mathbb{R}$.

Using Markov sections we can get a well-defined C^1 expanding interval map $f : U \to U$ by the following simple device:

1. We can identify all points on a section T_i which lie on the same stable piece. This gives an equivalence relation $\sim$ which collapses T_i to a set U_i.
2. The Poincaré map between sections defined by ϕ induces a map $f : U \to U$ (where $U = \cup_i U_i$ represents the disjoint union of the intervals U_i).
3. The transition time for a point to move from one section to the next under the flow is constant on stable pieces. This induces a map $r : U \to \mathbb{R}^+$.

A typical periodic orbit τ for $\phi_t : M \to M$ will give rise to a periodic point $f^n x = x$, say, for the map $f : U \to U$. Moreover, the period $\lambda(\tau)$ of τ can be recovered by the equality $\lambda(\tau) = r(x) + r(fx) + \ldots + r(f^{n-1}x)$ (although this correspondence may break down with those exceptional closed orbits which pass through the boundary of one of the sections from $\mathcal{R}$). We can now rewrite

$$\zeta_\phi(s) = \exp\left(\sum_{n=1}^{\infty} \frac{1}{n} \sum_{f^n x = x} e^{-sr^n(x)}\right) \eta(s) = \zeta_f(-sr, 1)\eta(s)$$

(where $\eta(s)$ is a correction for orbits passing through boundaries of sections).

Of course, there is no chance that the function $\zeta_\phi(s)$ will be rational. Therefore, let us settle for a more modest question.

QUESTION. When does $\zeta_\phi(s)$ have a meromorphic extension to $\mathbb{C}$?

6.3. C^ω horocycle foliations and C^ω sections. We shall first consider the case where $\phi : M \to M$ is a real analytic (C^ω) Anosov flow. Initially, we shall assume a stronger technical hypothesis. The strong stable manifolds $W^{ss}(x)$ are C^ω always immersed submanifolds, but we shall also assume that the foliation $\mathcal{W}^{ss} = \{W^{ss}(x)\}$ is C^ω (i.e., the associated holonomy map is C^ω). The following result is due to Ruelle [**Ru2**].

THEOREM. *For C^ω Anosov flows $\phi_t : M \to M$ for which the foliation $\mathcal{W}^{ss}$ are real analytic the zeta function $\zeta(s)$ has a meromorphic extension to the entire complex plane.*

We shall give an outline of the proof of this Theorem in §6.4 below.

REMARKS.

(a) The very stringent assumption that the foliations $\mathcal{W}^{ss}$ are real analytic holds only for very special examples (e.g. geodesic flows on manifolds with constant negative sectional curvatures). However, there have been subsequent developments which have allowed this assumption to be relaxed. In particular, work of Rugh [**Rg**] (introducing "pinning co-ordinates") and Fried [**Fd2**] shows that the conclusion of the Theorem is the same for any C^ω Anosov flow without any additional assumption on $\mathcal{W}^{ss}$.

(b) For the suggestive case of expanding C^k maps, Tangerman showed that although the associated transfer operator is not nuclear (a crucial feature in the C^ω case) the essential part of its spectrum shrinks to zero as k increases, and that this leads to corresponding improved estimates on the domain of the zeta function [**Ta**] (cf. also [**Ru4**]). Kitaev has results pertaining to Axiom A transformations in the C^∞ case [**Kitaev**].

(c) There exist examples of C^1 Axiom A flows for which the zeta function has essential singularities [**Ga**], [**PP2**].

6.4 Transfer operators and analytic functions. Recall that for a C^ω Anosov flow (with C^ω horocycle foliations) we can associate a C^ω expanding map $f : U \to U$ and a C^ω function $r : U \to \mathbb{R}$. More precisely, because of the assumption that $\mathcal{W}^{ss}$ is an analytic foliation the sets $U_i = T_i/\sim$ can be considered as a C^ω d-dimensional manifolds (where $d = \dim(E^u)$). Similarly, the C^ω Poincaré map which induces a C^ω expanding map $f : U \to U$, where $U = \cup_{i=1}^k U_i$. Finally, the piecewise C^ω return time function induces a positive C^ω function $r : X \to \mathbb{R}$.

Since U_i is real analytic this means we can choose an open neighbourhood (in the complexification) $V_i \supset U_i$ and denote $V = \cup_{i=1}^k V_i$. Moreover, we can assume V is chosen so that the function r and the local inverses to f have (complex) analytic extensions to V.

Consider the Banach space $\mathcal{A}_0$ of bounded analytic functions $h : U \to \mathbb{C}$ with the supremum norm $|h|_\infty = \sup_{z\in\mathbb{C}} |h(z)|$. By complete analogy with the definition for subshifts of finite type, we define the *transfer operator*

$$\begin{cases} L_{-sr} : \mathcal{A}_0 \to \mathcal{A}_0 \\ L_{-sr}w(x) = \sum_{Ty=x} e^{-sr(y)}h(y). \end{cases}$$

Ruelle showed the following result:

THEOREM. *The operator $L_{-sr} : \mathcal{A}_0 \to \mathcal{A}_0$ is nuclear (or trace class, in the sense of Grothendieck). In particular, the determinant* $\det(I - zL_{-sr})$ *exists, is entire in $s \in \mathbb{C}$ and satisfies*

$$\det(I - zL_g) = \exp\left(\sum_{n=1}^{\infty} \frac{z^n}{n} \sum_{f^n x = x} \frac{e^{-sr^n(x)}}{det(I - (D_x f)^{-1})}\right).$$

For the proof that the determinants are entire functions we can write $\det(I - zL_{-sr}) = 1 + \sum_{n=1}^{\infty} a_n z^n$, where $|a_n| = O(\theta^{n^{1+1/d}})$, for some $0 < \theta < 1$ (and $d = \dim E^u$).

To obtain an extension for $\zeta_f(-sr, z)$ one needs to get rid of the terms in the denominators by considering the natural generalizations $L_{g,j} : \mathcal{A}_j \to \mathcal{A}_j$, $1 \le j \le n$ of the transfer operator to C^ω j-forms. More precisely, a simple calculation gives

$$\zeta_\sigma(-sr, z) = \prod_{j=0}^{n} \operatorname{Det}(I - zL_{-sr,j})^{-1}.$$

The Bowen-Manning lemma again relates $\zeta_\phi(s)$ to $\zeta_\sigma(-sr, z)$. Altogether, this gives the Theorem in §6.3

An application: Algebraic topology and L-functions for flows. We can also define a generalization of the zeta function $\zeta_\phi(s)$ above which carries additional information about the algebraic topology of the manifold M, and the location therein of individual orbits.

Given a closed orbit τ we can denote by $[\tau] \in \pi_1(M)$ the free homotopy class in which it lies. We can consider the unstable bundle E^u restricted to the orbit τ and if this bundle is orientable we write $\Delta(\tau) = 1$, and if the bundle is not orientable then we write $\Delta(\tau) = -1$. Finally, let $R_\chi : \pi(M) \to U(n)$ be any unitary representation of fundamental group. We can define the associated *L-function* by

$$L(s, \chi) = \prod_\tau \det\left(I - \Delta(\tau)e^{-sh\lambda(\tau)}R_\chi[\tau]\right)^{(-1)^{\dim(E^u|\tau)}},$$

which converges for $Re(s) > 1$.

This definition depends only on the character $\chi = \operatorname{trace} R_\chi$, rather than on the particular choice of representation. In the special case that $R_\chi = I$ is the trivial representation we see that $L(s, \chi)$ simply reduces to the zeta function $\zeta_\phi(s)$.

The results and analysis we described in the previous section carry over to L-functions for C^ω Anosov flows (cf. [**Rg**], [**Fd2**]).

THEOREM. *For C^ω Anosov flows $\phi_t : M \to M$ for which the foliations $\mathcal{W}^{ss}$ are real analytic the L-function $L(s, R_\chi)$ is the reciprocal of an entire function.*

REMARK. As before, the assumption that the foliations $\mathcal{W}^{ss}$ are real analytic can be relaxed to $\phi_t : M \to M$ just being a C^ω Anosov flow.

In the particular case of L-functions, there is some interest in understanding the value of $L(s, R_\chi)^{-1}$ at $s = 0$. Whereas for zeta functions the value $\zeta(0)^{-1}$ is typically zero, for the case of more general L-functions $L(0, R_\chi)^{-1}$ typically takes a value which has a simple, purely topological, interpretation which is independent both of the geometry of the manifold and the particular flow $\phi_t : M \to M$.

Consider a triangularization of M and a differential complex $\mathcal{C} = \{d_i : C_i \to C_{i-1}\}$. For a given unitary representation $R_\chi : M \to U(n)$ we say that $\mathcal{C}$ is *cyclic* if $H^*(M, R_\chi) = 0$. Alternatively, we can choose a chain contraction $\delta = \{\delta_i : C_i \to C_{i+1}\}$ such that $\delta_{i-1}d_i + d_{i+1}\delta_i = I$. By a suitable choice of basis we can write $d_{\text{odd}} = \oplus_i d_{2i+1}$ and $\delta_{\text{odd}} = \oplus_i \delta_{2i+1}$ as square matrices. Finally, we can define the *Reidemeister torsion* by $\tau = \pm\det(d_{\text{odd}} + \delta_{\text{odd}})$ which, except for an ambiguity in the sign, is independent of all choices.

There are a number of interesting results by Fried relating the special value $L(0, R_\chi)$ to the torsion. The following illustrates the theme [**Fd1, Fd2**].

THEOREM. *Let $\phi_t : M \to M$ be a C^ω geodesic flow and let $R_\chi : M \to U(n)$ be an acyclic unitary representation. Then $L(0, R_\chi)^{-1} = \tau$.*

7. Extensions for Axiom A flows

Let us return to the general setting of C^1 Axiom A flows.

QUESTION. What can we say about the domain of $\zeta_\phi(s)$ for general Axiom A flows?

7.1 Domain of the ζ-function for Axiom A flows. For any Axiom A flow we have the following result.

THEOREM. *Let $\phi_t : \Lambda \to \Lambda$ be a Axiom A flow (restricted to a basic set).*

(i) *The zeta function $\zeta_\phi(s)$ is analytic for $Re(s) > h$;*
(ii) *The zeta function $\zeta_\phi(s)$ has a meromorphic extension to $Re(s) > h - \epsilon$ (for some $\epsilon > 0$);*
(iii) *$s = h$ is a simple pole, and if the flow is weak-mixing (i.e. the lengths of closed orbits are not all multiples of a constant) then there are no more zeros or poles on the line $Re(s) = h$.*

The idea of the proof is essentially the following. We can again use (Markov) sections to construct a simpler model for the flow $\phi_t : \Lambda \to \Lambda$. This time, we can associate to the flow a subshift of finite type $\sigma : X_A \to X_A$ (modelling the Poincaré map on Markov sections) and a Hölder semi-conjugacy $\pi : X_A^+ \to \cup_i T_i$ and a Hölder function $r : X_A^+ \to \mathbb{R}$ (modelling the transition time between sections) [**Bo1**].

The construction of Manning (or by cruder methods) immediately generalizes to give

$$\zeta_\phi(s) = \zeta_\sigma(-sr, 1)\eta(s)$$

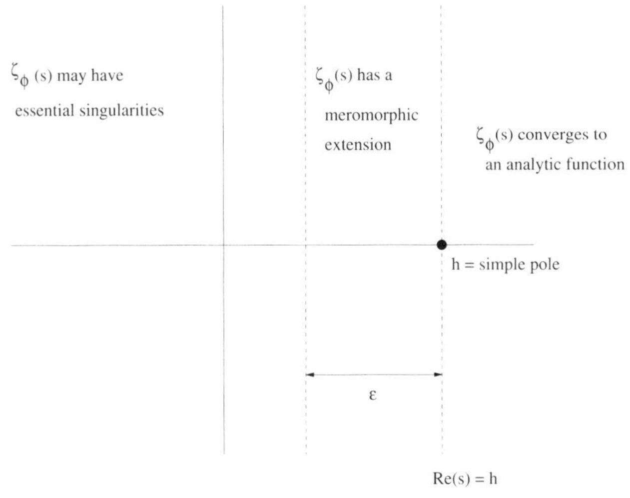

FIGURE. The extension of the domain of $\zeta_\phi(s)$.

where $\eta(s)$ is analytic for $Re(s) > h - \epsilon$, say. The results on the domain of the Ruelle zeta function $\zeta_\sigma(-sr, 1)$ (in §4) then imply the results for flows [**PP2**].

QUESTION. Why are the domains of zeta functions interesting?

An application. Let $N(T) = \#\{\tau : \lambda(\tau) \leq T\}$ be the number of closed orbits of least period at most $T > 0$. Then we have the following counting result [**Mr**], [**PP1**]:

THEOREM. *For a weak-mixing Axiom A flow*

$$N(T) \sim \frac{e^{hT}}{hT}, \quad \text{as } T \to +\infty.$$

For other variants (involving distribution of closed orbits by homology classes) see [**An**], [**KS1, KS2**], [**Ly**], [**PS2**], [**Sp**].

One proof of this Theorem follows the lines of the analytic proof of the prime number theorem (i.e., $\#\{\text{primes } p \leq x\} \sim x/\log x$, as $x \to +\infty$), which depends on the analyticity properties of probably the most famous of all zeta functions. We define the *Riemann zeta function* to be

$$\zeta_{\text{Riemann}}(s) = \sum_{n=1}^{\infty} \frac{1}{n^s} = \prod_{p \text{ prime}} \left(1 - p^{-s}\right)^{-1}.$$

This converges to an analytic function in the half-plane $Re(s) > 1$, with a meromorphic extension $\mathbb{C}$, In fact, it has an analytic extension to the entire complex plane, except for a simple pole at $s = 1$ [**El**].

Associated with the Riemann zeta function is one of the most famous conjectures in number theory:

RIEMANN HYPOTHESIS (CONJECTURE). The only zeros for $\zeta_{\text{Riemann}}(s)$ in the "critical strip" $0 \leq Re(s) \leq 1$ occur on the line $Re(s) = 1/2$.

The nearest analogue for dynamical zeta functions of this conjecture is the following positive result (due to Dolgopyat [**D2**]).

PROPOSITION. *In the special case of geodesic flows on surfaces, $\zeta_\phi(s)$ has an analytic extension to $Re(s) > h - \epsilon$ (for some $\epsilon > 0$) except for the simple pole at $s = h$.*

Consider a geodesic flow on a negatively curved surface. We can write

$$N(T) = \mathrm{Li}(e^{hT}) + O(e^{(h-\epsilon)T}), \text{ as } T \to +\infty$$

where $\mathrm{Li}(e^{hT}) = \int_2^{e^{hT}} \frac{1}{\log u} du$ [**D1**], [**PS1**].

8. Distribution of orbits and mixing

In the previous section we considered the usefulness of zeta functions in counting closed orbits. Let us now consider questions of the following nature.

QUESTION. Where to periodic points lie on the manifold?

8.1 Distribution of periodic orbits. In a similar way to that in which the Artin-Mazur zeta function influenced the asymptotic behaviour of the number $N(n)$ of periodic points for an Axiom A diffeomorphism f of period n, so the zeta function $\zeta_f(z, g)$ gives information on the distribution of orbits (relative to the Bowen-Margulis measure of maximal entropy μ). In particular, we have the following result of Bowen [**Bo1**].

PROPOSITION. *For an Axiom A diffeomorphism the probability measures*

$$\mu_n = \frac{1}{N(n)} \sum_{T^n x = x} \delta_x$$

(supported on periodic points of period n) converge in the weak-star topology to the measure of maximal entropy μ, as $n \to +\infty$.

To understand the connection with zeta functions, let us assume that $g : \Lambda \to \mathbb{R}$ is a Hölder continuous function. Then one can show

$$\left.\frac{\partial \zeta_f(z, tg)}{\partial t}\right|_{t=0} = \sum_{n=1}^{\infty} \frac{z^n}{n} \sum_{f^n x = x} g^n(x) = \frac{\int g d\mu}{z - e^{h(f)}} + \chi(z),$$

where $\chi(z)$ is analytic in a neighbourhood of $|z| \leq e^{P(g)}$ (and the residue $\int g d\mu$ is precisely the integral with respect to the measure μ). Expanding the righthand side as a power series in z and comparing coefficients we see that

$$\frac{1}{e^{nh(f)}} \sum_{f^n x = x} \frac{g^n(x)}{n} \to \int g d\mu, \text{ as } n \to +\infty.$$

We can also consider the analogous statement for flows. Let μ be the measure of maximal entropy (Bowen-Margulis measure) which is characterized by the following result of Bowen [**PP2**]:

PROPOSITION. *For an Axiom A flow consider the probability measures*

$$\mu_T = \frac{1}{N(T)} \sum_{\lambda(\tau) \leq T} \mu_\tau,$$

where $\int F d\mu_T = \int_0^{\lambda(\tau)} F(\phi_t x_\tau) dt$ *(supported on periodic orbits of period at most* T*) and* $N(T) = \sum_{\lambda(\tau) \leq T} \lambda(T)$*. Then* μ_T *converges in the weak-star topology to* μ*, as* $T \to +\infty$*.*

8.2 Rates of mixing for diffeomorphisms. To complete this section, we shall consider a problem which is very closely related to that of zeta functions. Given Hölder continuous functions $F, G : M \to \mathbb{R}$ we can define the *correlation function*

$$\rho(n) := \int F \circ f^n . G d\mu - \int F d\mu . \int G d\mu$$

QUESTION. How fast does $\rho(n)$ tend to zero?

The key to understanding this question is the complex function:

$$\hat{\rho}(w) = \sum_{n=-\infty}^{\infty} \rho(n) w^n, \quad w \in \mathbb{C}$$

(which easily converges for $|w| < 1$). There is an intimate relationship between $\hat{\rho}(w)$ and $\zeta(z)$ given by the following result [**Ru5**].

PROPOSITION.
(i) $\hat{\rho}(w)$ *has a meromorphic extension to some region* $1/\beta < |w| < \beta$ *(for some* $\beta > 1$*)* ;
(ii) $\hat{\rho}(w)$ *has poles* w_i *corresponding to those* z_i *for* $\zeta(z)$ *by* $w_i = z_i^{\pm 1}$.

Using the familiar trick of comparing the coefficients in the expansion of $\hat{\rho}(w)$, this leads to the following.

COROLLARY.
(a) *There exists* $0 < \gamma < 1$ *such that* $\rho(n) = O(\gamma^n)$,
(b) *We can write*

$$\rho(n) = \sum_i C_i w_i^n + O(1/\beta^n), \text{ as } n \to +\infty$$

(for some $C_i = C_i(F, G)$*).*

REMARK. The residues of the poles for $\hat{\rho}(w)$ give rise to what Ruelle termed "resonances" [**Ru5**].

8.3 Rates of mixing for flows. We can consider the analogous mixing problem for Axiom A flows. Given $F, G \in C^\infty(M)$ we can define the *correlation function*

$$\rho(t) := \int F \circ \phi_t . G d\mu - \int F d\mu . \int G d\mu$$

PROBLEM. *How fast does* $\rho(t)$ *tend to zero?*

Let us introduce another complex function (the Laplace transform of $\rho(t)$):

$$\hat{\rho}(w) = \int_0^\infty \rho(t) e^{-wt} dt, \quad w \in \mathbb{C}$$

(which converges for $Re(w) > 0$). There is an intimate relationship between $\hat{\rho}(w)$ and $\zeta(s)$ given by the following [**Po**].

PROPOSITION.
(i) $\hat{\rho}(w)$ *has a meromorphic extension to* $Re(s) > -\epsilon$*;*
(ii) $\hat{\rho}(w)$ *has poles* w_i *corresponding to those* s_i *for* $\zeta_\phi(s)$ *by* $w = s_i - h$.

For geodesic flows it is possible to show the following result [**D2, D3**].

THEOREM (DOLGOPYAT). *Consider a geodesic flow on a compact negatively curved surface. There exists* $\lambda > 0$ *such that* $\rho(t) = O(e^{-\lambda t})$.

References

[An] N. Anantharaman, *Precise orbit counting for closed orbits of Anosov flows*, Ann. Sci. École Norm. Sup. **33** (2000 pages 33-56).

[AM] M. Artin and B. Mazur, *On periodic points*, Annals of Math. **81** (1965), 82-99.

[BL] M. Babillot and F. Ledrappier, *Lalley's theorem on periodic orbits of hyperbolic flows*, Ergod. Th. and Dynam. Sys. **18** (1998), 17-39.

[Ba] V. Baladi, *Periodic orbits and dynamical spectra*, Ergod. Th. and Dynam. Sys. **18** (1998), 255-292.

[BK] V. Baladi and G. Keller, *Zeta functions and transfer operators for piecewise monotone transformations*, Comm. Math. Phys **127** (1990), 459-477.

[Bo1] R. Bowen, *Periodic points and measures for Axiom A diffeomorphisms*, Trans. Amer. Math. Soc. **154** (1971), 377-397.

[Bo2] R. Bowen, *Symbolic dynamics for hyperbolic flows*, Amer. J. Math. **95** (1973), 429-460.

[BL] R. Bowen and O. Lanford, *Zeta functions of restrictions of the shift transformation*, Global Analysis (Berkeley, Calif., 1968), Proc. Sympos. Pure Math. **14** (1970), 43-49.

[Br] R. Brown, *The Lefschetz fixed point theorem*, Scott, Foresman and Co., London, 1971.

[D1] D. Dolgopyat, *On the statistical properties of geodesic flows on negatively curved surfaces*, Thesis (1997).

[D2] D. Dolgopyat, *On decay of correlations in Anosov flows*, Annals of Math. **147** (1998), 357-390.

[D3] D. Dolgopyat, *Prevalence of rapid mixing: I and II*, Ergodic Th. and Dynam. Sys..

[DP] D. Dolgopyat and M. Pollicott, *Addendum to: "Periodic orbits and dynamical spectra" by V. Baladi*, Ergod. Th. and Dynam. Sys. **18** (1998), 293-301.

[El] W. Ellison and F. W. Ellison, *Prime Numbers*, Wiley, Paris, 1985.

[Fr1] J. Franks, *Homology and Dynamical systems*, Regional Conf. Series vol.49, Amer. Math. Soc., Providence, 1982.

[Fr2] J. Franks, *Period doubling and the Lefschetz formula*, Trans. Amer. Math. Soc. **287** (1985), 275-283.

[Fd1] D. Fried, *The zeta functions of Ruelle and Selberg I*, Ann. Sc. Ec. Norm. Sup. **19** (1986), 491-517.

[Fd2] D. Fried, *The flat-trace asymptotics of a uniform system of contractions*, Ergod. Th. and Dynam. Sys. **15** (1995), 1061-1071.

[Fd3] D. Fried, *Rationality for isolated expansive sets*, Advances in Math. **65** (1987), 35-38.

[Ga] G. Gallovotti, *Funzioni zeta ed insiemi basilari*, Accad. Naz. Lincei Rend. Cl. Sci. Fis. Mat. Natur **61** (1976), 309–317.

[Gu] Guckenheimer, *Axiom A + no cycles* $\Rightarrow \zeta_f(t)$ *rational*, Bull. Amer. Math. Soc. **76** (1970), 592-594.

[Ha] G. Hardy, *Ramanujan. Twelve lectures on subjects suggested by his life and work*, C.U.P., Cambridge, 1940.

[Ha] N. Haydn, *Meromorphic extension of the zeta function for Axiom A flows*, Ergod. Th. and Dynam. Sys. **10** (190), 347-360.

[Ho] F. Hofbauer, *On intrinsic ergodicity of piecewise monotonic transformations with positive entropy*, Israel J. Math. **34** (1979), 213-237.

[HK] F. Hofbauer and G. Keller, *Zeta-functions and transfer-operators for piecewise linear transformations*, J. Reine Angew. Math **352** (1984), 100–113.

[Hu] H. Huber, *Riemannsche Flchen von hyperbolischem Typus im euklidischen Raum*, Math. Ann. **138** (1959), 1–26.

[Ka] V. Kaloshin, *An extension of the Artin-Mazur theorem*, Annals. Math. **150** (1999), 729-741.

[KS1] Katsuda and Sunada, *Homology and closed geodesics in a compact Riemann surface*, Amer. J. Math. **110** (1988), 145-155.

[KS2] Katsuda and Sunada, *Closed orbits in homology classes.*, Publ. Math. **20** (1989), 5-32.

[KN] G. Keller and Nowicki, *Spectral theory, zeta functions and the distribution of periodic points for Collet-Eckmann maps*, Commun. Math. Phys. **149** (1992), 31-69.

[Ki] A. Kitaev, *Fredholm determinants for hyperbolic diffeomorphisms of finite smoothness*, Nonlinearity **12** (1999), 141-149.

[Ly] Lalley, *Closed geodesics in homology classes on surfaces of variable negative curvature.*, Duke J. Math **58** (1989), 795-821.

[Mn] A. Manning, *Axiom A diffeomorphisms have rational zeta functions*, Bull. London Math. Soc. **3** (1971), 215-220.

[Mr] G. Margulis, *Certain applications of ergodic theory to the investigation of manifolds of negative curvature*, Fun. Anal. Appl. **3** (1969), 89-90.

[MT] J. Milnor and W. Thurston, *On iterated maps of the interval*, Dynamical systems (College Park, MD, 1986–87), Lecture Notes in Math., 1342, Springer, Berlin], 1988., pp. 465–563.

[PP1] W. Parry and M. Pollicott, *An analogue of the prime number theorem for closed orbits of Axiom A flows.*, Annals of Math. **118** (1983), 573-591.

[PP2] W. Parry and M. Pollicott, *Zeta functions and the periodic orbit structure of hyperbolic dynamics*, Asterisque **187-188** (1990), 1-268.

[Po] M. Pollicott, *On the rate of mixing of Axiom A flows*, Invent. Math. **81** (1985), 413-426.

[PS1] M. Pollicott and R. Sharp, *Exponential error terms for growth functions on negatively curved surfaces*, Amer. J. Math **120** (1998), 1019-1042.

[PS2] M. Pollicott and R. Sharp, *Asymptotic expansions for closed orbits in homology classes*, Geom. Dedicata (To appear).

[Ru1] D. Ruelle, *Generalized zeta-functions for Axiom A basic sets*, Bull. Amer. Math. Soc. **82** (1976), 153-156.

[Ru2] D. Ruelle, *Zeta-functions for expanding maps and Anosov flows*, Invent. Math. **3** (1976), 231-242.

[Ru3] D. Ruelle, *Thermodynamic Formalism*, Addison Wesley, Reading, 1978.

[Ru4] D. Ruelle, *An extension of the theory of Fredholm determinants.*, Inst. Hautes Etudes Sci. Publ. Math **72** (1990), 175-193.

[Ru5] D. Ruelle, *Resonances for Axiom A flows.*, J. Differential Geom. **25** (1987), 99–116.

[Rg] H. Rugh, *Generalized Fredholm determinants and Selberg zeta functions for Axiom A dynamical systems*, Ergod. Th. and Dynam. Sys. **16** (1996), 805-819.

[Sp] R. Sharp, *Closed orbits in homology classes for Anosov flows*, Ergod. Th. and Dynam. Sys. **13** (1993), 387–408.

[Sh] M. Shub, *Global stability of dynamical systems*, Springer, 1987.

[Se] A. Selberg, *Harmonic analysis and discontinuous groups in weakly symmetric Riemannian spaces with applications to Dirichlet series*, J. Indian Math. Soc. **20** (1956), 47-87.

[Sm] S. Smale, *Differentiable dynamical systems*, Bull. Amer. Math. Soc. **73** (1967), 747–817.

[Ta] F. Tangerman, Ph. D. Thesis (1986).

Department of Mathematics, University of Manchester, Oxford Road, Manchester, Ma3 9PL, ENGLAND

Proceedings of Symposia in Pure Mathematics
Volume **69**, 2001

An Overview of the Dimension Theory of Dynamical Systems

Jörg Schmeling and Howard Weiss

CONTENTS

2000 *Mathematics Subject Classification.* Primary 37C45.

The authors want to thank the NSF and the AMS. Jörg Schmeling was supported by the Leopoldina-Forderpreis.

Introduction

The *geometry* of many invariant sets of dynamical systems (including attractors, repellers, level sets of dynamical characteristics) look very complicated at *all scales*, and their geometry is impossible to describe using standard geometric tools. Several notions of dimension have proven to be quite useful to quantify the complicated geometry of these sets. For some important classes of dynamical systems, some of these notions of dimension are intimately related to fundamental dynamical characteristics. Such characteristics include Lyapunov exponents, entropies, and pressure.

For example (see Section 1.3.3), the Hausdorff dimension of the limit set for a Fuchsian group (a discrete subgroup of $PSL(2,\mathbb{R})$ acting via Möbius transformations on the hyperbolic disk $\triangle$) essentially coincides with the measure theoretic entropy of the geodesic flow on the quotient Riemann surface with respect to the Patterson measure (the measure of maximal entropy), the topological entropy of the geodesic flow, the critical exponent for the Poincare series for the Fuchsian group, and the first eigenvalue of the Laplacian on the quotient Riemann surface. In this case the Hausdorff dimension of the limit set is not just an important dynamical characteristic, but it is also a fundamental geometric characteristic.

Tools from dynamical systems, especially from ergodic theory and thermodynamic formalism, are extremely useful to compute and estimate various dimensions. This includes dimensions of limit sets of geometric constructions (the standard Cantor set being the most famous example), which a priori, are not related to dynamical systems [**36, 80**]. One of the most useful techniques in the subject is to obtain a *Bowen formula* for the Hausdorff dimension of a set, i.e., to obtain the Hausdorff dimension as the zero of an expression involving thermodynamic pressure (see Section B). Most dimension formulas for limit sets of dynamical systems and geometric constructions in the literature are obtained, or can be viewed, as Bowen formulas.

The dimension of a set is a subtle characteristic which measures the *geometric complexity* of the set at arbitrarily fine scales. There are many notions of dimension, and most definitions involve a measurement of geometric complexity at scale ε (which ignores the irregularities of the set at size less than ε) and then considers the limiting measurement as $\varepsilon \to 0$. This will be clear when we define various notions of dimension in Section 1.1.1. For *general* sets there is no reason why any of these dimensions should coincide. However, for certain classes of hyperbolic dynamical systems, some notions of dimension do coincide. A recent important result is the solution of the higher dimensional Eckmann-Ruelle conjecture by Barreira, Pesin and Schmeling [**6**], which says that for smooth nonuniformly hyperbolic systems, the pointwise dimension is almost everywhere constant with respect to a hyperbolic measure. This result implies that many dimension characteristics for the measure coincide. In a highly influential paper, Young [**116**] proved this result in dimension two.

The deep connection between dynamical systems and dimension theory seems to have been first discovered by Billingsley [**12**] through several problems in number theory. One of the first applications was a study of the exceptional set in the Borel normal number theorem. Borel showed that almost all numbers are normal with respect to every base, and thus almost all numbers have an asymptotic frequency of $1's$ equal to $1/2$ in their diadic expansion. It is very natural to study the Hausdorff dimension of the set of numbers whose asymptotic frequency of $1's$ is equal to p, for $0 \leq p \leq 1$. Eggleston [**29**] discovered a striking dimension formula which Billingsley interpreted in terms of ergodic theory and reproved using tools from ergodic theory. We discuss several more connections with metric number theory in Section 1.3.4.

It has been known for many years that one can construct dynamical systems exhibiting various pathological behaviors by requiring a key dynamical characteristic to have special number theoretic properties (such as requiring a rotation number to admit abnormally fast rational approximations). In a similar spirit, one can sometimes construct pathological examples of dimension in dynamical systems using ideas in number theory [**87, 88**].

Another link between dynamical systems and dimension theory is through Pesin's theory of dimension-like characteristics. This general theory is a unification of many notions of dimension along with many fundamental quantities in dynamical system such as entropies, pressure, etc. In Section 1.1.4, we illustrate this general theory with a concrete example: topological entropy.

In this short survey we attempt to take the reader, who possesses a basic knowledge of hyperbolic dynamical systems and ergodic theory, on a tour of some of the important ideas, models, tools, and results in the interplay of dimension theory and dynamical systems, as well as to give a glimpse of applications in disparate areas of mathematics. There are many interesting results and applications we do not mention. Furthermore, many of the results we do mention are not stated in their most general form: we have strived for comprehension and insight over completeness. As a general reference, we strongly recommend the recent book by Y. Pesin [**73**] on Dimension Theory in Dynamical Systems. This book provides a detailed treatment of many topics of this survey and also contains a comprehensive set of references.

Part 1. The role of dimension theory in dynamical systems

1.1. Quick introduction to dimension theory

1.1.1. Basic definitions.

1.1.1.1. *Hausdorff Dimension.* Let $U \subset \mathbb{R}^N$. The **diameter** of U is defined as $|U| = \sup\{|x-y| : x, y \in U\}$. If $\{U_i\}$ is a countable collection of sets of diameter at most δ that cover Z, i.e., $Z \subset \bigcup_i U_i$ with $0 < |U_i| \leq \delta$ for each i, we say that $\{U_i\}$ is a δ-cover of Z.

Suppose that $Z \subset \mathbb{R}^N$. For any $s > 0$ we define

$$m_H(s, Z) = \lim_{\delta \to 0} m_H(s, \delta, Z),$$

where

$$m_H(s, \delta, Z) = \inf_{\{U_i\}} \left\{ \sum_i |U_i|^s : \{U_i\} \text{ is a } \delta\text{-cover of } Z \right\}.$$

The limit exists as an extended real number since $m_H(s, \delta, Z)$ is a nondecreasing function of δ. We call $m_H(s, Z)$ the **s-dimensional Hausdorff measure**

of Z. One can check that $m_H(s,\cdot)$ is an outer measure on $\mathbb{R}^N$ and satisfies the Carathéodory separation condition, and hence defines a Borel measure. It is also easy to check that for $t > s$ one has $m_H(t,\delta,Z) \le \delta^{t-s} m_H(s,\delta,Z)$, and it immediately follows that there exists a unique critical value s^* of s at which $m_H(s,Z)$ jumps from ∞ to 0. This critical value is called the **Hausdorff dimension of Z** and is written $\dim_H Z$. We call $m_H(s^*,Z)$ the Hausdorff measure of Z. The precise value of $m_H(s^*,Z)$ is usually quite hard to determine and often not very useful. However, it is sometimes useful to know whether $m_H(s^*,Z)$ is finite and strictly positive.

If $s = \dim_H Z$, then $m_H(s,Z)$ may be $0, \infty$, or finite. Hence $\dim_H Z = \sup\{s : m_H(s,Z) = \infty\} = \inf\{s : m_H(s,Z) = 0\}$.

Note that in the definition of Hausdorff dimension one must take the infimum over all $\delta-$covers of the set. This is a very difficult task. A natural and important question is how to find the optimal cover. For some conformal dynamical systems an optimal cover is the **Moran cover** (see Section 2 and [**64, 80, 81, 82, 83**]).

1.1.1.2. *Box Dimension.* Box dimension (aka Minkowski dimension, entropy dimension, Kolmogorov entropy, or capacity) is a cruder measure of geometric complexity than Hausdorff dimension, but it has the advantage that it is usually easier to compute and to estimate. We first give a nonstandard definition of box dimension, but which clearly shows the connection to Hausdorff dimension. This definition is the Carathéodory-Pesin dimension-like characteristic construction of box dimension.

To define box dimension, one slightly modifies the definition of Hausdorff dimension by replacing the covers of Z by sets of diameter at most δ by covers of Z by sets of diameter *precisely* δ. In this case $m_H(s,\delta,Z) = \delta^s N_\delta(Z)$, where $N_\delta(Z)$ is the minimal number of sets of diameter δ needed to cover the set Z. The limit defining $m_H(s,Z)$ need not exist, so we take lower and upper limits and denote the lower limit $\underline{m}_B(s,Z)$ and upper limit $\overline{m}_B(s,Z)$. There exist unique jump points $\underline{s} = \underline{\dim}_B Z$ and $\overline{s} = \overline{\dim}_B Z$, which we call the **lower and upper box dimensions**. It immediately follows from the definition that

$$\dim_H Z \le \underline{\dim}_B Z \le \overline{\dim}_B Z.$$

If $\underline{\dim}_B Z = \overline{\dim}_B Z$, we call the common value the **box dimension on Z** and denote it by $\dim_B Z$.

The more common formula for box dimension easily follows:

$$\dim_B Z = \lim_{\delta\to 0} \frac{\log N_\delta(Z)}{-\log \delta}, \tag{1}$$

provided the limit exists.

For some computations of box dimension, it is not difficult to establish the limit in (1) for a geometric subsequence. For instance, for the usual Cantor set, this limit is trivial to calculate for $\delta_k = 1/3^k$. Furstenberg observed [**35**] that the limit along any geometrically decreasing subsequence determines the limit. This immediately implies that the box dimension of the Cantor set is 1/3.

An interesting question is whether the Hausdorff dimension of an invariant set coincides with the box dimension. By exploiting special properties of Pisot numbers, in [**86, 88**] Pollicott and Weiss construct one parameter families of affine horseshoes in $\mathbb{R}^3$ with one Lyapunov exponent a Pisot number, along which the two dimensions differ. See also the recent thesis of Neunhäuserer [**66**]. On the other hand, it is known that the Hausdorff and box dimensions coincide for two dimensional hyperbolic sets [**69, 71, 111**].

As for whether or not these two dimensions coincide for a *general* closed set in $\mathbb{R}^N$, Pollicott and Weiss easily observed that with respect to the Hausdorff metric on subsets of say the unit square, there is a dense collection of subsets for which that Hausdorff dimension is strictly less than the lower box dimension which is strictly less than the upper box dimension, and that there is also a dense collection of subsets for which all three dimensions coincide. In general, Hausdorff dimension has poor semicontinuity properties, and hence it is not known whether one collection of subsets, say those for which the three dimensions are distinct, is residual.

For the graph Γ_f of a real valued continuous function f on an interval, Schmeling and Winkler [**99**] showed that the noncoincidence of dimensions is typical in the sense of Baire. More precisely, they show that for a residual set of continuous functions f one has

$$1 = \dim_H(\Gamma_f) = \underline{\dim}_B(\Gamma_f) < \overline{\dim}_H(\Gamma_f) = 2.$$

1.1.1.3. *Pointwise Dimension.* Most invariant sets which arise in dynamical systems are not self-similar or homogeneous, meaning the local geometry at arbitrarily fine scales looks different from point to point. To study the fine structure of such sets it is useful to have a notion of pointwise or local dimension which quantifies the geometric complexity of a set in an arbitrarily small neighborhood of a point.

It is useful to define a notion of pointwise dimension with respect to a Borel probability measure. This notion of pointwise dimension is a refinement of the classical geometric measure theory notion of density of a measure and is an important tool to estimate Hausdorff dimension.

Let μ be a Borel probability measure supported on $Z \subset \mathbb{R}^N$. We define the **lower and upper pointwise dimensions** by

$$\underline{d}_\mu(x) = \varliminf_{\varepsilon\to 0} \frac{\log \mu(B(x,\varepsilon))}{\log \varepsilon} \quad \text{and} \quad \overline{d}_\mu(x) = \varlimsup_{\varepsilon\to 0} \frac{\log \mu(B(x,\varepsilon))}{\log \varepsilon},$$

where $B(x,\varepsilon)$ denotes the Euclidean ball of radius ε about the point x. If $\underline{d}_\mu(x) = \overline{d}_\mu(x)$, we call the common value the **pointwise dimension** and denote it by $d_\mu(x)$. If $d_\mu(x) = d$, then for small ε the measure of small balls scales as $\mu(B(x,\varepsilon)) \asymp \varepsilon^d$.

One should not take existence of local dimension (even for *good* measures) for granted. In [**82**], Pesin and Weiss construct a Hölder continuous (with Hölder exponent arbitrarily close to one) expanding map of a compact metric space with positive topological entropy that possesses a unique measure of maximal entropy, such that pointwise dimension with respect to this measure does not exist at almost every point (see also [**23, 55**] for examples of smooth maps but with zero topological entropy).

A challenge with any such pointwise characteristic is to encode the data to obtain useful information. The dimension spectrum, one of the two main objects of study in classical multifractal analysis, provides a natural way to encode the pointwise dimension.

1.1.1.4. *Correlation Dimensions.* Another notion of dimension, which seems to be experimentally and numerically the most accessible, involves the study of correlations of the distributions of $q-$tuples along a typical orbit for $q = 2, 3, \ldots$.

This notion was introduced in [**38, 39**]. Let $g\colon X \to X$ be a map on a metric space (X,ρ) preserving a Borel probability measure μ. We set

$$C(x,q,r,n) = \frac{1}{n^q}\mathrm{card}\{(i_1 \dots i_q) \mid \rho(g^{i_j}x, g^{i_k}x) \le r \text{ for all } 0 \le i_j \le i_k < n\}.$$

We define the **correlation dimension** of order q by

$$C_q(x) = \frac{1}{1-q}\lim_{r\to 0}\lim_{n\to\infty}\frac{\log C(x,q,r,n)}{\log r},$$

provided the limits exist. If μ is ergodic, it was shown in [**74**] (see also [**79**]) that for μ almost every x

$$\lim_{n\to\infty} C(x,q,r,n) = \int_X \mu(B(y,r))^{q-1}\,d\mu(y),$$

where $B(y,r)$ denotes the ball of radius r centered at the point y. Clearly this limit depends on the metric ρ on X and the ergodic measure μ, but not explicitly on the dynamical system. Thus, for $q = 2,3,\dots$

$$C_q(x) = C_q \equiv \frac{1}{1-q}\lim_{r\to 0}\frac{\log\int_X \mu(B(y,r))^{q-1}\,d\mu(y)}{\log r},$$

provided the limit exists. In [**81**] the authors show that this limit exists for a broad class of measures including equilibrium measures on conformal repellers and for Axiom-A surface diffeomorphisms.

The natural extension of the correlation dimension of order $q = 2,3,\dots$ to all real values $q > 1$ was introduced by Hentschel and Procaccia in [**44**]. Let μ be a Borel probability measure on a metric space (X,ρ). For $q > 1$ we define the **HP-spectrum for dimensions** by

$$HP_\mu(q) = \frac{1}{1-q}\lim_{r\to 0}\frac{\log\int_X \mu(B(y,r))^{q-1}\,d\mu(y)}{\log r}.$$

For equilibrium measures on conformal repellers and for Axiom-A surface diffeomorphisms, the HP-spectrum for dimensions coincides with the Rényi spectrum of dimensions, which is one the two principle objects of study in multifractal analysis (see Section 2.4 and in particular the comments of what comprises a *complete* multifractal analysis.)

1.1.2. Basic properties of dimension. It is straight forward to verify that Hausdorff dimension and box dimension both satisfy the following properties [**32, 73**]:

(i) If $A \subset B$, then $\dim(A) \le \dim(B)$. (monotonicity)
(ii) If A is a finite set, then $\dim(A) = 0$.
(iii) If A is an open set or a $k-$dimensional smooth submanifold, then $\dim(A) = k$.
(iv) $\dim(\bigcup_{i=1}^N A_i) = \sup_{1\le i\le N}\dim(A_i)$. (finite stability)
(v) If $f\colon A \to \mathbb{R}^N$ is a Lipschitz mapping, then $\dim(f(A)) \le \dim(A)$. If $f\colon A \to f(A)$ is bi-Lipschitz, then $\dim(f(A)) = \dim(A)$.

In addition to these common properties shared by box dimension and Hausdorff dimension, the following properties are useful:

(1) Hausdorff dimension is countably stable, i.e., $\dim_H(\bigcup_i A_i) = \sup_i \dim_H(A_i)$. This property implies that the Hausdorff dimension of a countable set is zero.

However, this is not necessarily true for box dimension. For example, one can show that for the set $S = \{0, 1/2, 1/3, 1/4, \cdots\}$, the Hausdorff dimension is zero, while the box dimension is $\frac{1}{2}$.

(2) Box dimension has the property that the box dimension of a set coincides with the box dimension of its closure. Thus $\dim_H(\mathbb{Q}) = 0$, and $\dim_B(\overline{\mathbb{Q}}) = \dim_B(\mathbb{R}) = 1$, by (iii). Since many invariant sets of hyperbolic dynamical systems are dense, box dimension plays a limited role.

(3) If $f\colon X \to X$ is a C^1 mapping, then $\underline{d}_\mu$ and $\overline{d}_\mu$ are both f−invariant functions.

(4) If $A \subset \mathbb{R}$ and $0 < \dim_H(A) < 1$, then the set A is totally disconnected.

(5) Knowing the behavior of dimensions under products is quite important. It was believed, and there are published proofs with false claims, that Hausdorff dimension is always additive under products. In general, this is false. However, one can show that

$$\dim_H(A \times B) \geq \dim_H A + \dim_H B$$

and

$$\dim_H(A \times B) \leq \dim_B A + \dim_H B.$$

Hence if $\dim_H A = \dim_B A$, then the Hausdorff dimension is additive under products, i.e.,

$$\dim_H(A \times B) = \dim_H A + \dim_H B.$$

(6) The box and Hausdorff dimensions do not satisfy any semicontinuity properties for general sets in $\mathbb{R}^N$. It is usually easy to provide a counterexample to any desired semicontinuity property.

In later sections, we discuss some additional notions of dimension including the Rényi dimensions (Section 2.3), Hentschel-Procaccia dimensions (Section 2.4), and the Hausdorff dimension for a measure (Section 2.2).

1.1.3. Estimating dimension. There are three main methods to estimate Hausdorff dimension from below: the uniform mass distribution principle (Frostman), the nonuniform mass distribution principle (Young), and the potential principle (Frostman). We have seen that the Hausdorff dimension of a set can be estimated from above by the box dimension. However, if the box dimension is strictly greater than the Hausdorff dimension, a more refined upper estimate is required to compute the Hausdorff dimension. This is a difficult problem and we know just one main method to do this. We now describe these methods.

METHOD 1.1.1 (**U1**: Uniform Mass Distribution Principle [**34**]). *Let μ be a Borel probability measure supported on $Z \subset \mathbb{R}^N$ and suppose that for some $s \geq 0$ there are numbers $C > 0$, and $\delta > 0$ such that for all Borel sets $U \subset Z$ with $|U| \leq \delta$, one has*

$$\mu(U) \leq C\,|U|^s.$$

Then $m_H(s, Z) \geq 1/C$ implies that $\dim_H Z \geq s$.

PROOF. If $\{U_i\}$ is a cover of Z then

$$1 = \mu(Z) \leq \mu(\cup_i U_i) \leq \sum_i \mu(U_i) \leq C \sum_i |U_i|^s.$$

Taking the infimum over all covers one obtains that the measure $m_H(s,\delta,Z)\geq 1/C$ provided δ is small enough, so $m_H(s,Z)\geq 1/C$. □

Example: The standard 1/3-Cantor set **K**

> Let μ denote the natural probability measure on K such that each of the 2^n subintervals of length 3^{-n} at step n have mass 2^{-n}. Let U be a set with $|U|<1$ and choose $k\in\mathbb{N}$ such that $3^{-(k+1)}\leq |U|<3^{-k}$. Then U intersects at most 1 of the subintervals at step k and thus
>
> $$\mu(U)\leq 2^{-k}=(3^{-k})^{\log 2/\log 3}\leq (3|U|)^{\log 2/\log 3}.$$
>
> Hence by the uniform mass distribution principle, the Hausdorff measure $m_H(\log 2/\log 3,K)>0$ and thus $\dim_H K\geq \log 2/\log 3$.
>
> Combining this result with the upper bound for the Hausdorff dimension of the Cantor set provided by the box dimension (see Section1.1.1), we obtain that the Hausdorff dimension of the standard Cantor set is one third.

A similar argument shows that a Cantor set formed by p^n subintervals of length λ^{-n} at step n has Hausdorff dimension $\log p/\log\lambda$. It is useful to note that $\dim_H K=\log p/\log\lambda$ is a solution of the equation $p\cdot\lambda^{-s}=1$ or $P_{(\Sigma_p^+,\sigma)}(-s\log\lambda)=0$, where P denotes thermodynamic pressure (see Appendix B).

The second method is a nonuniform version of the first method.

METHOD 1.1.2 (**UII**: Non-Uniform Mass Distribution Principle [**116**]). *Suppose μ is a Borel probability measure supported on Z. If $\underline{d}_\mu(x)\geq d$ for all $x\in Z$, then $\dim_H Z\geq d$.*

The following folklore theorem [**73**] is quite useful in relating the notion of Markov pointwise dimension (see informal definition in Section 1.2) to pointwise dimension.

THEOREM 1.1.3. *Suppose μ is a Borel probability measure supported on Z and such that $d_\mu(x)=d$ for $\mu-$almost all $x\in Z$ and $\overline{d}_\mu(x)\leq d$ for $\mu-$almost all $x\in Z$. Then $\dim_H Z=d$.*

One may wonder why this method is called the nonuniform mass distribution principle. The following is a corollary of the method of proof.

COROLLARY 1.1.4. *Assume there exists $d>0$ and a Borel probability measure μ on Z such that for any $\varepsilon>0$ and $\mu-$almost every $x\in Z$ one can find a constant $C(x,\varepsilon)>0$ satisfying for any $r>0$ the measure $\mu(B(x,r))\leq C(x,\varepsilon)r^{d-\varepsilon}$. Then $\dim_H Z\geq d$.*

The third method is called the potential method.

METHOD 1.1.5 (**UIII**: Potential Principle [**34**]). *Suppose μ is a Borel probability measure supported on Z. For $s\geq 0$ we define the s-potential at a point $x\in\mathbb{R}^N$ by*

$$\varphi_s(x)=\int_{\mathbb{R}^N}\frac{d\mu(y)}{|x-y|^s}$$

and the s-energy of μ is defined by

$$I_s(\mu) = \int_{\mathbb{R}^N} \varphi_s(x)d\mu(x) = \int_{\mathbb{R}^N}\int_{\mathbb{R}^N} \frac{d\mu(x)d\mu(y)}{|x-y|^s}.$$

Let $Z \subset \mathbb{R}^N$.

- *If the measure μ satisfies $I_s(\mu) < \infty$, then $m_H(s,Z) = \infty$ and thus $dim_H(Z) \geq s$.*
- *The Hausdorff dimension satisfies*

$$dim_H(Z) = \sup\{t \mid \text{ there exists } \mu \text{ with } \mu(Z) > 0 \text{ and } I_t(\mu) < \infty\}.$$

To apply any of these methods to obtain good estimates, one needs to choose a suitable measure. In many cases of interest the optimal measure is the *measure of full dimension* (see Section 1.2).

Besides the upper estimate of Hausdorff dimension provided by box dimension (let us call this **Method LI**), the following is the only general technique to obtain upper estimates for Hausdorff dimension. We stress that the estimate must hold at every point $x \in Z$ while in Method UII the corresponding estimate needs to hold at μ-almost every point $x \in Z$.

METHOD 1.1.6 (**LII**: Non-uniform Mass Distribution Upper Estimate [**116**]). *Suppose μ is a Borel probability measure supported on Z. If $\underline{d}_\mu(x) \leq d$ for all $x \in Z$, then* $\dim_H Z \leq d$.

1.1.4. Dimension-like characteristics and topological entropy. Pesin's theory of dimension-like characteristics provides a unified treatment of dimensions and important dynamical quantities like entropies and pressure. It is particularly useful to study characteristics, such as pressure, of noncompact sets.

Fix functions ξ, η, ψ, and a subcollection $\mathcal{G} \subset \mathcal{F}$ of covers of Z, and consider the quantities

$$m_{cp}(s,Z) = \lim_{\delta\to 0} m_{cp}(s,\delta,Z),$$

where

$$m_{cp}(s,\delta,Z) = \inf_{\mathcal{G}} \sum_{U\in\mathcal{G}} \xi(U)\eta(U)^s;$$

the infimum is taken over countable subcollections $\mathcal{G} \subset \mathcal{F}$ covering Z with $\psi(\mathcal{G}) \leq \delta$. Clearly one needs strong hypotheses on the functions ψ, ξ, and η to obtain meaningful results. Rather than clutter our exposition with several pages of hypotheses, we provide two key examples, and refer the reader to [**73**] for all the details.

One obtains the Hausdorff dimension of the set Z by taking $\mathcal{F}$ = open sets, $\xi(U) = 1$, and $\eta(U) = \psi(U) = |U|$.

Under suitable hypotheses, there exists a unique critical value of s at which $m_{cp}(s,Z)$ jumps from ∞ to 0. This critical value is called the **Carathéodory-Pesin dimension** of Z and is written $\dim_{cp}Z$.

We show how to obtain the topological entropy of a continuous map f of a compact metric space (X,ρ) as a dimension-like characteristic. For each $n \in \mathbb{N}$ and $\varepsilon > 0$, define the **Bowen ball**

$$B_n(x,\varepsilon) = \{y \in X \colon \rho(f^i(x), f^i(y)) \leq \varepsilon \text{ for } 0 \leq i \leq n\}.$$

Choose $\mathcal{F} = \mathcal{F}_\varepsilon = \{B_n(x,\varepsilon) \colon x \in X, n \geq 0\}$, $\xi(B_n(x,\varepsilon)) = 1$, $\eta(B_n(x,\varepsilon)) = \exp(-n)$, and $\psi(B_n(x,\varepsilon)) = 1/n$. Define the $\varepsilon-$topological entropy of f on a set Z by

$$h_{top}(f, Z, \varepsilon) = \text{ jump point of } m_{cp}(s, Z).$$

Then the CP-topological entropy f on the set Z is defined by

$$h_{top,cp}(f, Z) = \overline{\lim}_{\varepsilon \to 0}\, h_{top}(f, Z, \varepsilon).$$

If Z is $f-$invariant and compact, this definition of CP-topological entropy coincides with the usual definition of topological entropy [**113**].

The proof of the Manning and McClusky dimension formula (see Section 2.3) involves computing dimensions and topological entropy of some noninvariant sets. The CP-topological entropy lends itself well to this task and can be used to simplify and clarify the original proof. The CP-topological entropy is also useful in types of multifractal analyses (Section 2.4).

1.2. Dimension of conformal repellers: Ruelle's pressure formula, the Moran cover, and measures of full dimension

Computing or estimating a dimension via a *pressure formula* is a fundamental technique when studying dimension for low-dimensional dynamical systems. For this reason, properties of pressure such as smoothness, convexity, and explicit formulas for derivatives play an important role in the dimension theory of dynamical systems. For example, the smoothness and convexity of the dimension spectrum and Rényi spectrum (the two main quantities in a classical multifractal analysis) are manifestations of the smoothness and convexity of pressure on the space of Hölder continuous functions on a subshift of finite type. Also, the smooth dependence of the Hausdorff dimension of basic sets for Axiom-A surface diffeomorphisms follows from smoothness of pressure.

Ruelle proved the following pressure formula for the Hausdorff dimension of a conformal repeller.

THEOREM 1.2.1 ([**90**]). *Let $f \colon J \to J$ be a conformal repeller, and consider the function $t \to P(-t \log|Df|)$, where P denotes pressure. Then*

$$P(-s \log|Df|) = 0 \qquad \Longleftrightarrow \qquad s = \dim_H(J)$$

We sketch a proof of this result and along the way introduce some of the main tools to estimate dimensions. We begin with the definition of a Markov ball. This definition requires the existence of a Markov partition $\mathcal{R} = \{R_1, \cdots, R_l\}$ for the conformal repeller (see Appendix C).

DEFINITION 1.2.2. *A* **Markov ball** *is an iterate of the rectangles in the Markov partition, i.e., a set of the form*

$$R_{i_1 \cdots i_n} = R_{i_1} \cap f^{-1}(R_{i_2}) \cap \cdots \cap f^{-n+1}(R_{i_n}).$$

We let $R_{i_1 \cdots i_n}(x)$ denote the $n-$step Markov ball that contains the point x.

For an equilibrium measure, by definition, one has uniform estimates for the measure of Markov balls. For some dynamical systems (including conformal repellers), the following equilibrium measure provides optimal estimates for Hausdorff dimension in Methods UI-UIII and LII.

DEFINITION 1.2.3. *The equilibrium* **measure of full dimension**, μ_{dim}, *is the equilibrium measure for the potential* $-s\log|Df|$, *where* s *is defined implicitly as the solution of* $P(-s\log|Df|)=0$.

Using the definition of measure of full dimension and the Jacobian estimate (see Appendix C), one can easily estimate the measure and diameter of Markov balls and relate them through a uniform asymptotic relation, i.e.,

$$\mu_{dim}(R_{i_1\cdots i_n}(x)) \asymp |Jacf^n(x)|^{-s} \asymp |R_{i_1\cdots i_n}(x)|^s, \tag{2}$$

where $\asymp$ means that the quotients are uniformly bounded above and below for all x and n.

If we know that relation (2) holds for all small Euclidean balls *(we know it only for Markov balls!)*, then Young's dimension estimates UIII and LII, using the measure of full dimension, would immediately yield Ruelle's dimension formula. We now need to carefully relate the two covers.

The Markov balls at step n have vastly different diameters, so they are not very useful for estimating box and Hausdorff dimensions. For each $\varepsilon>0$, one can construct an essentially disjoint cover $\mathcal{C}_\varepsilon$ of J using Markov balls from different generations (different n) each having diameter of order ε. This special cover is called the **Moran cover** at level n. By essentially disjoint, we mean elements of the cover can only intersect along their boundaries. The Moran cover has the following crucial property: given a point $x\in J$ and $\varepsilon>0$, the number of basic sets in the Moran cover $\mathcal{C}_\varepsilon$ that have a nonempty intersection with the ball $B(x,\varepsilon)$ is uniformly bounded from above, independent of x and ε (for further discussions see [**4**]).

The Moran cover is ideally suited for computing box dimension, where one covers the set by balls (sets) of the same diameter. It is also a convenient and sometimes optimal cover when computing Hausdorff dimension. Even when it provides an optimal cover for computing Hausdorff dimension, many other covers are also optimal. Finally, the Moran cover is an important technical tool when computing pointwise dimension.

The Moran cover is one of two new tools (the other is the doubling property of equilibrium measures (see Appendix B) which allow one to use uniform estimates for the measures of Markov balls to compute pointwise dimension and Hausdorff dimension, and thus prove Ruelle's dimension formula. This is also a major tool in the multifractal analysis (the next paragraph is written with this application in mind) and can be used to prove the special case of Young's formula in Section 2.2. For Markov maps, one can show that except for at most countably many points (the boundary points of the Markov partition), the actual pointwise dimension coincides with the auxiliary or **Markov pointwise dimension** [**87**] one obtains using Markov balls

$$\lim_{n\to\infty}\frac{\log\mu_{dim}(R_{i_1\cdots i_n}(x))}{\log|R_{i_1\cdots i_n}(x)|}.$$

Since the boundaries of Markov partitions for higher dimensional conformal repellers (and Axiom-A diffeomorphisms) typically have positive Hausdorff dimension, one needs to work harder, and the relation between pointwise dimension and Markov pointwise dimension is not as direct as in the one-dimensional case (On the other hand Schmeling proved that the Hausdorff dimension of *any* subset of the repeller can be computed by coverings with Markov balls of the symbolic preimage [**97**]). One needs to first work on the symbolic level and compute the Hausdorff dimension of the level sets for Markov pointwise dimension via pointwise dimension estimates. The folklore theorem 1.1.3 is useful here. One then uses the doubling property of equilibrium measures to show that the pushforward under the coding map of the level sets for the Markov pointwise dimension on the symbolic level coincide with the level sets for the pointwise dimension on the repeller. The result is that the Markov pointwise dimension computed using Markov balls provides the correct estimates of the Hausdorff dimension of invariant sets. The details can be found in [**73, 80, 81, 82, 83, 97**].

The reader might wonder about the origin of the term *measure of full dimension*. It is useful to define the Hausdorff dimension of a Borel probability measure μ by

$$\dim_H(\mu) = \inf_{\mu(X)=1} \dim_H(X). \tag{3}$$

It is obvious that $\dim_H(\mu) \leq \dim_H(J)$ for any measure μ supported on J. However, for the very special measure of full dimension, one has that

$$\dim_H(\mu_{dim}) = \dim_H(J).$$

For Axiom-A diffeomorphisms in dimension at least two, such an invariant measure almost never exists [**62**]. This fact is one of the major obstacles which makes studying dimension in higher dimensions much more difficult than for conformal systems (which are essentially one-dimensional).

One possible strategy is to *disintegrate* the system into the stable (contracting) and unstable (expanding) directions, and make a separate analysis for these two cases. Some people believe [**37**] that a smooth expanding map (noninvertible) admits a measure of full dimension. If this is true, the product of the full dimension measures on the stable and unstable manifolds would be a candidate for a noninvariant dynamical measure of full dimension. We consider this question to be one of the major open problems in the dimension theory of dynamical systems.

Question: Does a conformal repeller for a C^2 expanding map admit a measure of full dimension?

1.3. Brief tour through some examples

In this section we present several examples to illustrate how dimension theory naturally arises in dynamical systems. It is interesting to observe the disparate roles that dimension plays in the different examples.

1.3.1. Homoclinic bifurcations for dissipative surface diffeomorphisms. Homoclinic tangencies and their bifurcations play a fundamental role in Dynamical Systems [**69, 70**]. Palis has conjectured that the set of (nonuniformly) hyperbolic surface diffeomorphisms together with the diffeomorphisms exhibiting homoclinic tangencies are dense in the space of all surface diffeomorphisms. Systems exhibiting homoclinic tangencies can exhibit more complicated and more subtle quasi-local behavior than systems possessing transverse homoclinic points or homoclinic points with topological crossings. Gavrilov and Shilnikov showed that horseshoes (locally maximal hyperbolic sets) may exist near the homoclinic tangency. Newhouse showed that these homoclinic tangencies typically generate secondary tangencies which persist under small perturbations, and that the Gavrilov and Shilnikov horseshoes may co-exist with infinitely many sinks in a neighborhood of the homoclinic orbit.

Consider a generic family of locally dissipative surface diffeomorphisms $\{f_\mu\}$ unfolding a quadratic homoclinic tangency at $\mu = 0$ as illustrated in Figure 1. This phenomenon results in what is sometimes called an Ω–explosion of the nonwandering set. Suppose tangency is associated with a periodic point p_0 and a hyperbolic basic set $p_0 \in \Lambda_0$.

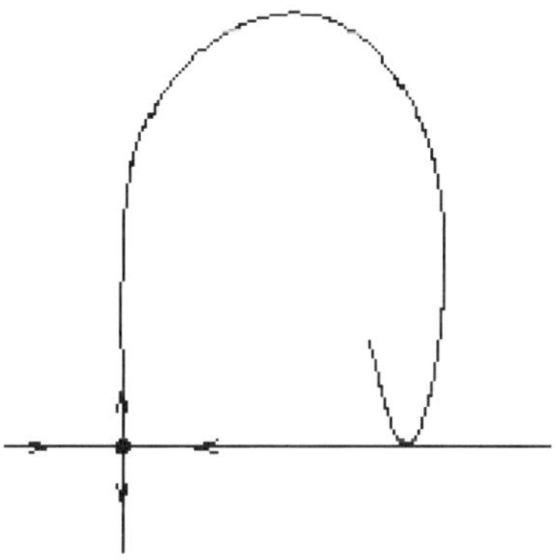

FIGURE 1

The Hausdorff dimension of the hyperbolic set for f_0 determines whether hyperbolicity is the *typical* dynamical phenomenon near f_0 or not. If $\dim_H \Lambda_0 < 1$, then hyperbolicity is the prevalent dynamical phenomenon near f_0. This is not the case if $\dim_H \Lambda_0 > 1$.

More precisely, let $\mathcal{NW}_\mu$ denote the set of nonwandering points of f_μ in an open neighborhood of $\Lambda_0 \cup \mathrm{orb}(f_0)$ after the homoclinic bifurcation. Let ℓ denote Lebesgue measure.

THEOREM 1.3.1 ([**69, 70**]). *If* $\dim_H \Lambda_0 < 1$, *then*

$$\lim_{\varepsilon \to 0} \frac{\ell\{\mu \in [0, \varepsilon] : \mathcal{NW}_\mu \text{ is hyperbolic}\}}{\varepsilon} = 1.$$

The hyperbolicity co-exists with the Newhouse phenomena (infinitely many sinks) for a residual set of parameter values.

THEOREM 1.3.2 ([**72**]). *If* $\dim_H \Lambda_0 > 1$, *then*

$$\lim_{\varepsilon \to 0} \frac{\ell\{\mu \in [0, \varepsilon] : \mathcal{NW}_\mu \text{ is hyperbolic}\}}{\varepsilon} < 1.$$

Questions about unfoldings of homoclinic tangencies associated with basic sets lead to questions on the arithmetic difference of dynamically defined Cantor sets (e.g., cookie-cutters) on $\mathbb{R}$. The proof of the above *full density* theorem uses the fact that if C_1 and C_2 are two dynamically defined Cantor sets on the line such that the sum of their Hausdorff dimensions is smaller than one, then $C_1 - C_2$ (the arithmetic difference) is a set of zero Lebesgue measure. Palis asked whether it is true that if C_1 and C_2 are dynamically defined Cantor sets, then either $C_1 - C_2$ has Lebesgue measure zero or else it contains intervals (corresponding to open sets of homoclinic tangencies). Moreira and Yoccoz [**65**] recently showed this is true in a *generic* sense.

1.3.2. Navier-Stokes equation in dimension two. Consider the two dimensional Navier-Stokes differential equation with periodic boundary conditions. It is well known that this equation can be written in the pressure-less form:

$$\frac{du}{dt} + vAu + B(u,u) = f, \qquad u(0) = u_0,$$

where u and v are functions possessing suitable regularity, vAu is the dissipative term and $B(u,u)$ is the nonlinear term. Under certain assumptions, one can estimate the Hausdorff dimension of the **universal attractor** X.

To define the universal attractor X one considers a semigroup of evolution operators $S(t)u_0 \equiv u_t$ on a certain Sobolov space $\mathcal{H}$ and defines

$$X = \cap_{t>0} S(t)\mathcal{A},$$

where $\mathcal{A}$ is an absorbing set. Any solution of the Navier-Stokes equation which is initially in $\mathcal{A}$ eventually stays in $\mathcal{A}$ for all positive time. Then [**21**] X is compact, connected, positive time invariant, contains all bounded positive time invariant sets, and for all $u_0 \in \mathcal{H}$ the distance (in the Sobolov metric)

$$\lim_{t\to\infty} d(S(t)u_0, X) = 0.$$

Seemingly standard energy and a priori estimates [**21**] allow one to prove that for some k, the k volume elements of X decay exponentially, i.e., the sum of the smallest k Lyapunov exponents is negative. One wants to relate this (in the spirit of the Kaplan-Yorke formalism [**33, 48, 53**]) to an upper bound for (some) dimension of X.

For Hausdorff dimension, this was effected by Douady and Oesterle [**26**] and also Ilyashenko, where they show that if a system exponentially contracts $k-$dimensional volumes, then the Hausdorff dimension of the maximal attractor is $\leq k$. Recently Blinchevskaya and Ilyashenko [**13**] proved this for box dimension (generalizing the finite dimensional analog of this result by Hunt [**45**]). Thus

THEOREM 1.3.3. *The box dimension* $\dim_B X$ *is finite.*

An challenging problem is to find good upper bounds for $\dim_B X$. A useful reference for this subject is [**21**], although do not take the dimension estimates seriously – the relevant dimension theory was in its infancy.

In [**27**], Eckmann-Ruelle speculated that turbulence might be due to a (finite dimensional) strange attractor in the three dimensional Navier-Stokes equation. There has been a great deal of interest in this question during the last 15 years, but we seem far from being able to determine whether this claim is indeed true. In fact, little is known about the existence of a finite dimensional universal attractor for the Navier-Stokes equation in three dimensions.

1.3.3. Geodesic flow on Riemann surfaces. About 20 years ago, Sullivan established some remarkable connections between dimension, dynamics, geometry, and Fuchsian groups. Let us first review some definitions. Excellent references for Fuchsian groups are [**49**] and [**41**].

Let Δ denote the unit disk in $\mathbb{R}^2$ with *ideal* boundary S^1. A Fuchsian group is a discrete subgroup of $PSL(2,\mathbb{R})$ which acts on $\triangle$ via Möbius transformations. The *limit set* $L(\Gamma)$ of a Fuchsian group Γ is the subset of S^1 containing all limit points of Γ-orbits $\Gamma z, z \in \triangle$. Either $L(\Gamma) = S^1$, in which case Γ is called a Fuchsian group of the first kind, or $L(\Gamma)$ is a nowhere dense subset of S^1, in which case Γ is called a Fuchsian group of the second kind.

The quotient $S = \triangle/\Gamma$ is a surface, and since the hyperbolic metric on $\triangle$ is invariant under elements of Γ, this metric descends to a hyperbolic (constant negative curvature) metric on the quotient surface $\triangle/\Gamma$, endowing S with a Riemann surface structure.

A Möbius transformation is called *parabolic* if it is conjugate to a translation in $PSL(2,\mathbb{R})$. For a Fuchsian group Γ, the quotient Riemann surface $S = \triangle/\Gamma$ will have cusps if and only if the group Γ contains parabolic elements.

Finally, Patterson gave an explicit description of the measure of maximal entropy for the geodesic flow on the unit tangent bundle to the quotient Riemann surface $S = \gneqq/\Gamma$ for a Fuchsian group of the second kind without parabolic elements. One first constructs a probability measure μ on $L(\Gamma)$ such that for all $\gamma \in \Gamma$, the measure transforms as $\gamma^*\mu = |\gamma'|^d\mu$, for a very special positive value of d (see Theorem 1.3.4). One then shows that the measure $\mu(x) \times \mu(y)/|x-y|^{2d}$ is $\Gamma-$invariant on $S^1 \times S^1$. This invariant measure is combined with arc length to yield a natural invariant measure for the geodesic flow on the unit tangent bundle T_1S.

THEOREM 1.3.4 ([**107, 108**]). *Let Γ be a Fuchsian group of the second kind without parabolic elements. The quotient Riemann surface $S = \gneqq/\Gamma$ has infinite area and the geodesic flow on the unit tangent bundle T_1S is completely nonrecurrent.*

Let $d =$ Hausdorff dimension of $L(\Gamma)$. Then:

- *$d =$ metric entropy of the geodesic flow on T_1S (with respect to Patterson measure).*
- *$d =$ the topological entropy of the geodesic flow on T_1S.*
- *$d =$ critical exponent for the Poincare series $\sum_{\gamma\in\Gamma}|\gamma'(p)|^s$ for Γ, i.e., the Poincare series converges for $s > d$ and converges for $s < d$. It is a deep result of Sullivan that the Poincare series converges for $s = d$.*
- *$d(d-1) =$ first eigenvalue of the Laplacian on S. Furthermore, one can build the first eigenfunction for the Laplacian using the $d-$dimensional Hausdorff measure of the limit set $L(\Gamma)$.*
- *$d =$ Hausdorff dimension of the set of bounded geodesics on T_1S.*

1.3.4. Some applications to number theory. A number of dimension results in number theory can be recast as dimension results for dynamical systems and attacked using tools from dynamics.

Example: Diadic expansion of numbers

Consider the diadic expansion of $x \in [0,1]$, i.e.,

$$x = \sum_{k=0}^{\infty} a_k(x) 2^{-k}\ , \ \text{where } a_k \in \{0,1\}\,.$$

Let

$$X_p = \left\{ x\colon \lim_{n\to\infty} \frac{1}{n} \sum_{k=0}^{n-1} a_k = p \right\}.$$

Borel showed that $\ell(X_{1/2}) = 1$, where ℓ denotes Lebesgue measure. Thus almost all numbers are normal, i.e., have asymptotic frequency $1/2$ for the distribution of $1's$ in their diadic expansion. These days, this result is an easy consequence of the Birkhoff ergodic theorem applied to the characteristic function of a $1-$cylinder set.

It is very natural to ask what is the Hausdorff dimension of the set of numbers whose asymptotic frequency of $1's$ is equal to p, for $0 \le p \le 1$. Eggleston [**29**] discovered the following wonderful dimension formula, which Billingsley [**12**] interpreted in terms of dynamics and reproved using tools from ergodic theory. In [**84**], Pesin and Weiss provide a new proof using a multifractal analysis for Birkhoff averages for a subshift of finite type.

THEOREM 1.3.5 ([**29**]). *The Hausdorff dimension of X_p is given by*

$$\dim_H(X_p) = (1/\log 2)[-p\log p - (1-p)\log(1-p)].$$

This formula has an obvious dynamical interpretation: $\dim_H(X_p) = h_{\mu_p}(E_2)/\lambda$, where $E_2(x) = 2x \bmod 1$, $\lambda = \log|DE_2| = \log 2$, and μ_p is the Bernoulli p measure. Thus the underlying dynamical system corresponding to dyadic expansions is the linear doubling map.

The above-mentioned multifractal analysis for Birkhoff averages is only applicable to study the diadic expansion of numbers. Barreira, Schmeling and Saussol [**10**] developed a type of multifractal analysis for multiple Birkhoff averages, which allowed them to extend this result to expansions with respect to arbitrary integer bases.

Rényi proposed a generalization of the base d expansion from integer base d to noninteger base d. In this case the underlying dynamical system is the beta shift $\beta(x) = \beta x \bmod 1$. For noninteger β, this map is expanding but is not Markov. With the help of dimension estimates, Schmeling [**94**] proves that every nonzero number is normal in almost every base.

Pollicott and Weiss extended Eggleston's result from the diadic expansion to the continued fraction expansion. The underlying dynamical system in this case is the Gauss map $T(x) = (1/x) \bmod 1$. This map is uniformly expanding, but has infinitely many branches. For this reason, the thermodynamic formalism is more subtle, although it does possess a finite absolutely continuous invariant measure. For example, the expression $P(-t\log T')$ is only defined for $t > 1/2$.

Example: Continued fraction approximation of numbers

Consider the continued fraction expansion of $x \in [0,1]$, i.e.,

$$x = \cfrac{1}{a_1 + \cfrac{1}{a_2 + \cfrac{1}{a_3 + \cfrac{1}{a_4 + \ddots}}}} = [a_1, a_2, a_3, a_4, \cdots],$$

and the approximants $r_n \equiv p_n(x)/q_n(x) = [a_1, a_2, \ldots, a_n]$. For each $n \in \mathbb{N}$, the $n-th$ continued fraction approximant is a rational number with the property that $|x - r_n| \leq 1/q_n^2$. The Gauss map acts by the shift $T([a_1, a_2, a_3, a_4, \cdots]) = [a_2, a_3, a_4, a_5, \cdots]$.

Let us now consider the set of numbers which admit a *faster approximation* by rational numbers. For $\tau \geq 2$, let $\mathcal{F}_\tau$ denote the set of $\tau-$well approximable numbers, i.e., those that satisfy

$$\mathcal{F}_\tau = \left\{ x \in I \colon \left| x - \frac{p}{q} \right| \leq \frac{C(x)}{q^\tau} \quad \text{for infinitely many } p/q \right\},$$

where p and q are relatively prime.

One can show that the rationals p/q in $\mathcal{F}_\tau$ are all continued fraction approximants. It is easy to prove that this set has zero measure for each $\tau \geq 2$. Jarnik [**46**] computed the Hausdorff dimension of $\mathcal{F}_\tau$ and showed that $\dim_H(\mathcal{F}_\tau) = 2/\tau$.

While Jarnik computes the Hausdorff dimension of the set of numbers x such the approximation error $|x - p/q|$ is bounded above by $q^{-\tau}$ for infinitely many p/q, he does not study the set of numbers for which the continued fraction algorithm converges with a precise exponential speed. We now consider the precise rate of convergence.

Let

$$\Lambda_p = \left\{ x \colon \lim_{n\to\infty} \frac{1}{n} \log \left| x - \frac{p_n(x)}{q_n(x)} \right| = p \right\}.$$

We wish to study the *precise* exponential rate of convergence of the continued fraction algorithm. Khinchin showed that almost all numbers admit the same exponential rate of convergence of the continued fraction algorithm, i.e., $\ell\left(\Lambda_{\pi^2/6\log 2}\right) = 1$. Again these days, this result is a routine consequence of the Birkhoff ergodic theorem [**12**].

It is easy to show [**87**] that the Lyapunov exponent for the Gauss map has the following number theoretic representation:

$$\lambda(x) \equiv \lim_{n\to\infty} \frac{1}{n} \log |(T^n)'(x)| = \lim_{n\to\infty} \frac{1}{n} \log \left| x - \frac{p_n(x)}{q_n(x)} \right| = \lim_{n\to\infty} \frac{1}{n} \log q_n(x).$$

Applying a multifractal analysis for the Lyapunov exponent for the Gauss map, Pollicott and Weiss show the following result.

THEOREM 1.3.6 ([**87**]). *For the Gauss map the following holds.*

- *The Lyapunov exponent $\lambda(x)$ (and precise exponential rate of approximation of the continued fraction algorithm for x) attains the interval of values*

$$[2\log\gamma, \infty) = [2\log\gamma, \lambda_0] \cup [\lambda_0, \infty),$$

where $\lambda_0 = \pi^2/6 \log 2$ *and* $\gamma = (\sqrt{5}+1)/2$ *is the Golden Mean.*

- *For* $p \in [\lambda_0, \infty)$, *the level set* Λ_p *is uncountable, dense, and has positive Hausdorff dimension. There is an explicit formula for the Hausdorff dimension of each level set in* $[\lambda_0, \infty)$: *For* $p = \int_0^1 \log |T'| d\mu_s$,

$$\dim_H \left\{ x \in [0,1] : \lim_{n\to\infty} \frac{1}{n} \log \left| x - \frac{p_n(x)}{q_n(x)} \right| = p \right\} = \frac{h_{\mu_s}(T)}{p},$$

where μ_s *is the equilibrium state for* $-s \log |T'|$.

- *The set of points for which* $\lambda(x)$ *does not exists has Hausdorff dimension* 1 *(applying the result in* [**9**]*).*

It is easy to see that the numbers for which $\lambda(x)$ does not exist include the Liouville numbers. A similar analysis works for other number theoretic algorithms including the inhomogeneous Diophantine algorithm.

A number x is called *badly approximable* if it does not belong to any $\mathcal{F}_\tau$ for $\tau \geq 2$, i.e., there exists $C > 0$ such that $|x - p/q| \geq C/q^2$ for all but finitely many relatively prime p/q. Jarnik showed that such numbers form a set of Hausdorff dimension one. It is not hard to see that the badly approximable numbers are precisely those numbers such that the coefficients in their continued fraction expansions are bounded.

Hensley considered the following refined question on the asymptotics in Jarnik's dimension one theorem. Let G_n denote the set of numbers whose continued fraction expansion uses only the numbers $\{1, 2, \cdots, n\}$. By Khintchine, $\ell(G_n) = 0$ for all n. What can one say about the Hausdorff dimension of G_n and in particular, can one find an asymptotic formula for the Hausdorff dimension of G_n as $n \to \infty$? By studying spectral theory for a variant of the transfer operator for the Gauss map, Hensley found the following asymptotic formula.

THEOREM 1.3.7 ([**43**]). *The Hausdorff dimension of* G_n *satisfies*

$$\dim_H(G_n) = 1 - \frac{6}{\pi^2 n} - \frac{72 \log n}{\pi^4 n^2} + O(\frac{1}{n^2}).$$

It has been known for many years that there is a geometric interpretation for continued fractions using geometry and dynamics (geodesic flow) on the modular surface $M = SL(2,\mathbb{R})/SL(2,\mathbb{Z})$. Artin first showed the existence of a dense geodesic on M using continued fractions [**3**]. Series [**102**] explicitly shows how the continued fraction expansion for the endpoint of a geodesic on the upper half plane is determined by the sequence of intersections of the geodesic with the Farey fundamental domain for $SL(2,\mathbb{R})$ consisting of geodesics and horocycles on the modular surface. One obtains that the geodesic flow on M is equivalent to a special flow over the Gauss map. Thus a number of results in Diophantine approximation can be proven using result about geodesics and horocycles on M, i.e., the dynamics of the geodesic flow on M.

The above theorem on the multifractal analysis of the Lyapunov exponent provides some new insights on the begavior of geodesics on the modular surface, including a new dimension formula for the set of geodesics on the modular surface which spend, on *average*, more or less time than is typical on cuspital excursions. Other authors [**63, 106, 109**] have related results quantifying the set of exception geodesics in terms of their times for *maximal* extensions up the cusp. These results are complementary, but neither seems to imply the other. Thus the results which follow from the theorem are of a new and different nature.

Jarnik's theorems follow from analogous statements about the dynamics of the geodesic flow on M [**63**]. Schmidt [**100**] proved a two dimensional version of Jarnik's theorem for badly approximable systems of linear forms. Although his proof did not use any ideas from dynamical systems, Kleinbock and Margulis [**51**] recently gave a proof which covers all dimensions using the dynamics for special one parameter diagonal subgroups of $SL(n, \mathbb{R})$ acting on $SL(n, \mathbb{R})/SL(n, \mathbb{Z})$.

There are a number of other continued fraction expansions which arise in number theory and applications [**101**]. Examples include backward continued fractions, Hirzebruch continued fractions, Japanese continued fractions, multidimensional continued fractions, etc.

For instance, each $x \in [0, 1)$ has a unique infinite backward continued fraction expansion of the form $[b_1, b_2, \cdots] = 1-(1/(b_1-1/(b_2-1/\cdots)))$, with integers $b_i > 1$. The underlying dynamical system is the map $S(x) = (1/1-x) \mod 1$, which has a neutral fixed point at 0 and is thus nonuniformly hyperbolic. The map S acts as by the shift on backward continued fractions $S([b_1, b_2, b_3, b_4, \cdots]) = [b_2, b_3, b_b, b_5, \cdots]$. Renyi showed that the *infinite* measure dx/x is invariant for S. This map does not possess a finite absolutely continuous invariant measure, and the usual *inducing* trick to study its thermodynamic formalism does not work. For this reason we consider the analog of Theorem 2.4.3 for this map to be quite difficult. The map S arrives as a factor of a suitable cross-section map of the geodesic flow on the modular surface.

Problem: Effect a multifractal analysis for the Lyapunov exponent for the map $S = (1/1-x) \mod 1$.

1.3.5. Limit sets of geometric constructions. There is a large collection of literature on the dimension theory of limit sets of deterministic and stochastic geometric constructions (see references in [**73**]). The vast majority of geometric constructions studied in the literature are generated by a finite collection of p maps which are either contractions or similarities and are modeled by the full shift on p symbols (or at most a subshift of finite type). It is important to note that limit sets of general geometric constructions are almost **never** self-similar! Figure 2 illustrates a *typical* geometric construction.

In [**80, 81**], Pesin and Weiss (see also Barreira [**5**]) consider quite general classes of geometric constructions: the basic sets may have a complicated geometry and topology; the placement of the basic sets at each step of the construction can be arbitrary and they need not be disjoint. Moreover, their constructions are modeled by arbitrary symbolic dynamical systems. They also consider nonstationary constructions, random constructions (determined by an arbitrary ergodic stationary distribution), and combinations of the above.

The prototype for their analysis is a geometric construction which defines a Cantor-like set of the form

$$F = \bigcap_{n=1}^{\infty} \bigcup_{(i_1 \cdots i_n)} \Delta_{i_1 \cdots i_n},$$

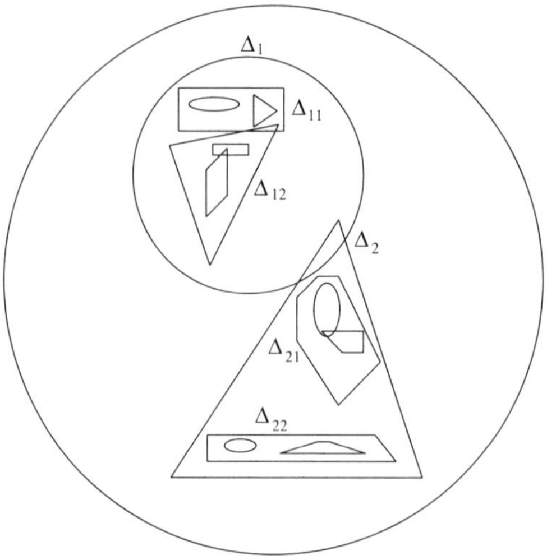

FIGURE 2. Geometric Construction

where the basic sets on the nth step of the construction, $\Delta_{i_1\cdots i_n}$, $i_k = 1, \cdots, p$, are closed, disjoint, and $\Delta_{i_1\cdots i_n j} \subset \Delta_{i_1\cdots i_n}$, for $j = 1, \cdots, p$. We assume that the $\operatorname{diam}(\Delta_{i_1\cdots i_n}) \to 0$ as $n \to \infty$, and we emphasize that the placement of the sets $\Delta_{i_1\cdots i_n}$ can be arbitrary as long as they satisfy the above conditions. See Figure 2.

About 50 years ago, Moran [**64**] computed the Hausdorff dimension of limit sets of geometric constructions in $\mathbb{R}^N$ given by p basic sets $\Delta_{i_1\ldots i_n}$ satisfying

- each basic set is the closure of its interior
- at each level the basic sets do not overlap (their interiors are disjoint)
- a basic set $\Delta_{i_1\ldots i_n j}$ is geometrically similar to the basic set $\Delta_{i_1\ldots i_n}$ for every j and n
- $\operatorname{diam}(\Delta_{i_1\ldots i_n j}) = \lambda_j \operatorname{diam}(\Delta_{i_1\ldots i_n})$ where $0 < \lambda_j < 1$ for $j = 1, \ldots, p$ are the *ratio coefficients*.

These constructions are called *Moran constructions.* Moran discovered the formula $s = \dim_H F$, where s is the unique root of the equation

$$\sum_{i=1}^{p} \lambda_i^t = 1 \tag{4}$$

and $\dim_H F$ is the Hausdorff dimension of the limit set F. He also showed that the $s-$dimensional Hausdorff measure of the limit set is finite and strictly positive.

Moran's major idea was to construct an optimal cover of the limit set (*Moran cover*) which is determined by the ratio coefficients. In [**81**], the main insight into the Moran approach is that many of the strict conditions in the definition of Moran construction are not required to build the Moran cover. For example, the geometric similarity of basic sets may be greatly weakened. Furthermore, although Moran only considered constructions modeled by the full shift, his approach can be generalized to constructions modeled by arbitrary symbolic dynamical systems.

The important techniques used in [**81**] include the Moran cover, the measure of full dimension, and *pressure formulas* for dimension (see Section 1.2). Note that the dimension formula (4) is equivalent to $P(s\varphi) = 0$ for $\varphi(x) = s \log \lambda_{x_1}$.

Clearly for totally general geometric constructions, such as the one illustrated in Figure 2, only crude dimension estimates can exist. If one desires actual dimension formulas in any generality, one generally needs to consider Moran-like constructions which are not *too far* from being *conformal.*

It is sometimes possible to construct pathological examples of dynamical systems by first constructing a suitable geometric construction and then *embedding* the limit set of the geometric construction into the limit set for a dynamical system. For instance, in [**81**] the authors provide an example of a dynamical system which shows that the smoothness hypothesis is essential in the Eckmann-Ruelle conjecture by first constructing a suitable geometric construction. See also [**23, 55, 86**].

1.3.6. Embedology and computational aspects of dimension. Tools from dynamical systems are becoming increasingly important to study the time evolution of deterministic systems in engineering and the physical and biological sciences. One of the main ideas is to assume that an underlying real world system (at least operating in certain regimes) can be accurately modeled by a smooth dynamical system $f\colon M^n \to M^n$, which possesses some type of strange attractor Λ with a natural ergodic invariant measure. When studying a complicated real world system, one can measure only a very small number of variables. The first major challenge is how to reconstruct the underlying attractor from the time measurement of a scaler quantity of the real world system. We can think of an idealized measurement as a function $h\colon M^n \to \mathbb{R}$.

The main tool researchers currently use to *reconstruct* the model system is called attractor reconstruction (see papers and references in [**68**]). This method is based on embedding with time delays (see the influential paper by [**22**], where the authors attribute the idea of delay coordinates to Ruelle), where one attempts to reconstruct the attractor for the model using a single long trajectory. The idea is that given a time series $\{x_k\}$, one considers the set points in R^{p+1} defined for each integer n by $(x_k, x_{k+\tau}, x_{k+2\tau}, \dots, x_{k+p\tau})$, and one then uses the data to try to approximate the local linear dynamics on the attractor. Most readers have seen the impressive looking pictures obtained by sampling one coordinate of a dense trajectory on the Lorenz attractor, and using that to reconstruct the attractor in $\mathbb{R}^3$ via a time delay embedding.

Assuming such a modeling scheme is valid for a particular real world system, there are many tough practical questions. Real data always has finite precision and is contaminated by noise. One also needs to find the optimal values of embedding dimension $p+1$ and the time delay τ. The usual method of estimating the optimal embedding dimension is to perform the embedding in larger and larger dimensions, estimate the correlation dimension C_2 (see Section 1.1.1.4) of the embedded set, and take the value where the correlation dimension levels off [**24**].

One usually wishes to estimate the Lyapunov exponents and various dimensions of the underlying attractor using the embedding and the approximation of the local linear dynamics on the attractor. In [**27, 19**] the authors propose and analyze a computationally viable method of estimating Lyapunov exponents, and in [**39**] the authors propose an algorithm to estimate the correlation dimension of the attractor. In [**28**] the authors show that very long data sets of high quality are necessary to obtain reasonable estimates. If the embedding dimension is too large (say more than ten), even for relatively long time series, the sparsity of data causes fatal statistical sampling problems with any possible estimation algorithm.

One foundational question is whether such an f can be reconstructed in any reasonable sense by a single smooth function h along an orbit of f? Takens [**110**] provides some rigorous mathematical underpinnings for this by showing that for a smooth $f\colon M^n \to M^n$ and for typical smooth h, the mapping $\varphi\colon M^n \to \mathbb{R}^{2n+1}$ defined by $x \to (h(x), h(f^\tau(x)), \cdots, h(f^{2n\tau}(x))$ is an embedding. Since the box dimension of the attractor may be much less than the dimension of the ambient manifold n, an interesting mathematical question with important practical applications is whether there exists $p < 2n+1$ such that the mapping on the attractor $\varphi\colon \Lambda \to \mathbb{R}^p$ defined by $x \to (h(x), h(f(x)), \cdots, h(f^p(x))$ is injective? Injectivity is important since one usually wants to predict the future behavior of the real world system (this is a model after all) using the approximation of the local linear dynamics on the attractor. It is known that for a typical smooth h the mapping φ is injective for $p > 2\dim_B(\Lambda)$ [**17**].

We briefly mention an important ingredient of the proof, which is of independent interest. Let A be a compact subset of R^n. If $p > 2\dim_B(A)$, then almost every linear mapping of R^n to R^p is injective on A, and thus it preserves all notions of dimension. This is false if the box dimension of A is replaced by the Hausdorff dimension of A. Kan (in an appendix to [**17**]) constructed a compact subset A of R^n with zero Hausdorff dimension such that no linear projection to R^p for $p < n$ is injective. However, Mãné [**60**] showed that if $p > \dim_H(A \times A)$, then almost every linear mapping of R^n to R^p is injective on A.

Some (nonmathematical) authors even speculate that the knowledge (estimates) of various dimensions and Lyapunov exponents *determines* the underlying model system. Any mathematical justification for such beliefs would be intimately connected with some type of *multifractal rigidity.*

1.3.7. Some brief remarks on other interesting examples. There are quite a few more interesting examples of dimension theory in dynamical systems.

A number of authors have studied the billiard flow on polygons in the plane whose vertex angles are rational multiples of π. Each such rational billiard table possesses a one-parameter family of billiard flows – each a flow in a fixed direction. Kerckhoff, Masur and Smillie [**47**] showed that the set of directions for which the billiard flow is nonuniquely ergodic has Lebesgue measure zero. Masur [**61**] quantified this result by showing that the Hausdorff dimension of the set of directions corresponding to nonuniquely ergodic flows is never larger than $1/2$.

A related model is the one parameter family of billiard tables consisting of a 2×1 rectangular table with a wall of length $1 - \lambda$ inserted parallel to the shorter edges and touching a longer edge at its midpoint. Each such table again possesses a one-parameter family of billiard flows – each a flow in a fixed direction. The above stated results of Masur and Kerckhoff, Masur and Smillie also apply to these billiard tables. Recently Cheung [**18**] showed that if λ is not a Liouville number, then the Hausdorff dimension of the set of directions corresponding to (minimal and) nonuniquely ergodic flows is precisely $1/2$. Cheung reduces his problem to computing the Hausdorff dimension of an affine Cantor-set arising from an interesting *new* geometric construction. What is novel about the construction is that (using the notation from Section 1.3.5) there is no upper bound for the j where $\Delta_{i_1\cdots i_n j} \subset \Delta_{i_1\cdots i_n}$.

Boshernitzan [**14**] used dimension theory in an interesting way to quantify the rate of recurrence in the Poincare recurrence theorem. Let (X, T, μ) be a measurable transformation on a compact metric space (X, d) which preserves a Borel probability measure. The Poincare recurrence theorem states that for μ-almost all $x \in X$, one has that $\underline{\lim}_{n\geq 1} d(x, T^n x) = 0$. Boshernitzan showed that if the $s-$Hausdorff measure, $m_H(s, X)$, of X is finite for some $s = s^*$, then for μ-almost all $x \in X$ one has that $\underline{\lim}_{n\geq 1} n^{1/s^*} d(x, T^n x) < \infty$. This result was recently generalized in [**8**].

Finally, several authors, including Dolgopyat [**25**] and Urbanski [**112**], studied the Hausdorff dimension of the set of nondense orbits for a hyperbolic dynamical system. For instance, Urbanski proved that for every C^2 transitive Anosov diffeomorphism, expanding mapping or Anosov flow on a compact manifold M, the set of points with nondense orbit (forward orbit in the expanding case) starting from an arbitrary open set, has full Hausdorff dimension, i.e., Hausdorff dimension on equal to $\dim(M)$.

Part 2. Dimension theory of low-dimensional dynamical systems

By low-dimensional dynamical system, we mean one or two dimensional or conformal systems (which in many ways act as a one-dimensional system). For several reasons which are explicitly listed in Section 2.5 the dimension theory for low-dimensional dynamical systems is considerably easier than for higher dimensional dynamical systems. We begin with the simplest example of a hyperbolic set in two dimensions.

2.1. Affine horseshoes in $\mathbb{R}^2$

We begin by constructing an affine horseshoe in the plane. Let $S = [0, 1] \times [0, 1]$ and define $f \colon S \to \mathbb{R}^2$ so that $f(S) \cap S$ has two vertical components V_1 and V_2 which are mapped affinely by f (see Figure 3). The inverse image $f^{-1}(S \cap f(S)) = S \cap f^{-1}(S)$ has two horizontal components H_1 and H_2, and on each component the map f has constant Jacobian

$$\begin{pmatrix} \pm\mu & 0 \\ 0 & \pm\lambda \end{pmatrix}, \text{ with } \lambda > 2 \text{ and } 0 < \mu < 1/2.$$

The set $\Lambda_n = \bigcap_{k=-n}^{n} f^k(S)$ (see Figure 4) consists of 4^n perfectly aligned rectangles of width μ^n and height λ^{-n}, and the limit set $\Lambda = \bigcap_{n=1}^{\infty} \Lambda_n = K^s \times K^u$, where K^s and K^u are two affine (standard) Cantor sets. Note that the map f restricted to Λ is not conformal if $\lambda \neq 1/\mu$. The map f on the horseshoe Λ is naturally topologically equivalent to the shift map on the full two shift $\sigma \colon \Sigma^2 \to \Sigma^2$.

It follows from the simple product structure of Λ that

$$\begin{aligned} \dim_H \Lambda = \dim_B \Lambda &= \dim_H(K^s) + \dim_H(K^u) \\ &= \frac{\log 2}{\log \lambda} + \frac{\log 2}{-\log \mu}. \end{aligned}$$

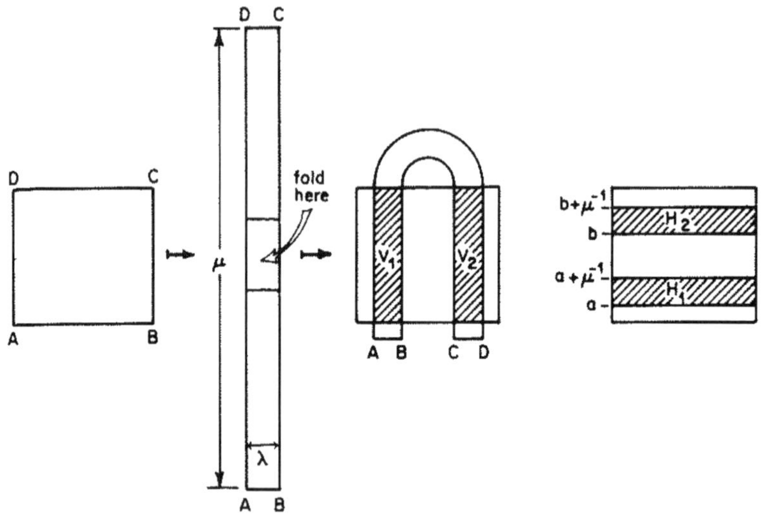

FIGURE 3. First step in construction of affine horseshoe

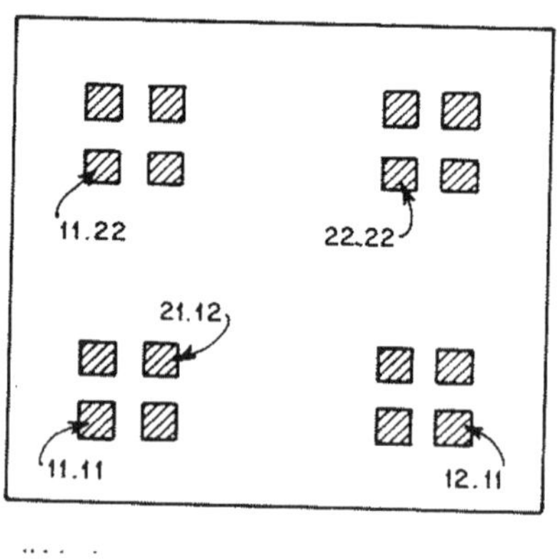

FIGURE 4. Λ_n

Let $E^u(x)$ and $E^s(x)$ denote the unstable and stable directions at the point x. Since the expansion and contraction coefficients $\log \|Df_x|E_x^u\|$ and $\log \|Df_x|E_x^s\|$ are both constant, it follows immediately from the definition of pressure (see Appendix B) that

$$\dim_H \Lambda = s_1 + s_2,$$

where s_1 and s_2 satisfy

$$P(-s \log \|Df_x|E_x^u\|) = P(-s \log \lambda) = 0$$

$$P(s \log \|Df_x|E_x^s\|) = P(s \log \mu) = 0,$$

and E_x^u, E_x^s denote the unstable and stable distributions at the point x, respectively. This formula is a special case of the Manning-McCluskey dimension formula (see Section 2.3) and holds for nonlinear horseshoes in dimension two (basic sets of

Axiom-A surface diffeomorphism). There is no known higher dimensional analog since the Hausdorff dimension may depend on the geometry of the basic set and number theoretic properties of the Lyapunov exponents [**86**].

The Hausdorff dimension of the affine horseshoe $\dim_H \Lambda$ can be expressed in terms of the topological entropy of f and the two Lyapunov exponents

$$\dim_H \Lambda = \log 2 \left(\frac{1}{\log \lambda} + \frac{1}{-\log \mu} \right) = h_{top}(f) \left(\frac{1}{\lambda^u} - \frac{1}{\lambda^s} \right).$$

A nonlinear analog of this formula (for nonaffine basic sets) does not a priori make sense since the Lyapunov exponents are not constant. It follows from a multifractal analysis for the Lyapunov exponent that Lyapunov exponents each typically attain an interval of values (see Section 2.4). Perhaps one needs to consider a suitable ergodic invariant measure, in which case the Lyapunov exponents will be constant almost everywhere, and the almost every value for the Lyapunov exponents will appear in the dimension formula. Since $h_{top}(f)$ appears in the formula, one might be tempted to consider μ_{max}, the measure of maximal entropy. This leads nowhere. We recall that the measure of full dimension μ_{dim} played a key role when studying conformal expanding maps. However, Manning and McCluskey showed that this measure does not exist for typical basic sets of Axiom-A surface diffeomorphisms.

2.2. Young's dimension formula

In this section we discuss a remarkable dimension formula of Young [**116**] for the dimension of a measure. Young proved the result for hyperbolic measures for nonuniformly hyperbolic surface diffeomorphisms, but we will only consider equilibrium measures for uniformly hyperbolic surface diffeomorphisms. The formula is for the Hausdorff dimension of a hyperbolic measure, so we need a short detour on this new notion of dimension.

DEFINITION 2.2.1. *Let $Z \subset \mathbb{R}^N$ and let μ be a probability measure supported on Z. We define the* **Hausdorff dimension of the measure μ** *by*

$$\dim_H(\mu) \equiv \inf_{K \subset Z:\, \mu(K)=1} \dim_H(K).$$

Thus $\dim_H(\mu)$ is the smallest Hausdorff dimension of any set of full measure. It immediately follows from the definition that $\dim_H(\mu) \leq \dim_H Z$. The following results of Young [**116**] show how to estimate $\dim_H(\mu)$ using the pointwise dimension of μ (compare with estimation methods UII and LII for $\dim_H Z$).

PROPOSITION 2.2.2. *Suppose μ is a probability measure supported on $Z \subset \mathbb{R}^N$.*

- *If $\underline{d}_\mu(x) \geq d$ for μ–almost all $x \in Z$, then $\dim_H(\mu) \geq d$.*
- *If $\overline{d}_\mu(x) \leq d$ for μ–almost all $x \in Z$, then $\dim_H(\mu) \leq d$.*
- *If $d_\mu(x) = d$ for μ–almost all $x \in Z$, then $\dim_H(\mu) = d$.*

We now state and outline a scheme of proof of the main result.

THEOREM 2.2.3. *Let $f\colon M^2 \to M^2$ be a C^2 Axiom-A surface diffeomorphism with basic set Λ and let μ be an equilibrium measure. Then*

$$\dim_H(\mu) \equiv \inf_{K \subset \Lambda:\, \mu(K)=1} \dim_H(K) = h_\mu(f) \left(\frac{1}{\lambda^1} - \frac{1}{\lambda^2} \right),$$

where λ^1 and λ^2 are the almost everywhere values of the two Lyapunov exponents for f and $\lambda^1 < \lambda^2$.

IDEA OF PROOF. (following [**83**])

For a person familiar with basic properties of hyperbolic sets, Markov partitions, and symbolic dynamics, the proof is quite similar to the proof of Ruelle's pressure formula discussed in Section 1.2. Since equilibrium measures possess a local product structure and the stable and unstable foliations are C^1, the pointwise dimension for μ is the sum of the pointwise dimensions for conditional *equilibrium* measures for μ on the stable and unstable manifolds. The problem essentially reduces to studying pointwise dimension for one-dimensional expanding maps. The key technical tool is (again) the construction of *stable and unstable Moran covers* which enables one to estimate pointwise dimension in terms of Markov pointwise dimension.

We note that Markov pointwise dimensions are quite easy to compute. Given a Markov partition for f, it follows from the Shannon-McMillan-Breiman theorem that for $\mu-$almost every x

$$-\lim_{n\to\infty}\frac{\log\mu(R_{i_1\cdots i_n}(x))}{n}=h_\mu(f).$$

From the Jacobian estimate (see Appendix C) it follows that for $\mu-$almost every x

$$-\lim_{n\to\infty}\frac{n}{\log|R_{i_1\cdots i_n}(x)|}=\left(\frac{1}{\lambda^1}-\frac{1}{\lambda^2}\right).$$

Thus for $\mu-$almost every x, the pointwise dimension using Markov balls is

$$\lim_{n\to\infty}\frac{\log\mu(R_{i_1\cdots i_n}(x))}{\log|R_{i_1\cdots i_n}(x)|}=h_\mu(f)\left(\frac{1}{\lambda^1}-\frac{1}{\lambda^2}\right).$$

□

The two dimensional uniformly hyperbolic version of the Eckmann Ruelle conjecture for equilibrium measures immediately follows from this formula, i.e., the pointwise dimension of an equilibrium measure exists and is constant almost everywhere. The general statement of the formula proves the full Eckmann Ruelle conjecture in dimension two.

2.3. Manning and McClusky dimension formula

In [**62**], Manning-McCluskey prove the following dimension formula for the basic set of an Axiom-A surface diffeomorphism.

THEOREM 2.3.1. *Let Λ be a basic set for a C^2 Axiom-A surface diffeomorphism $f\colon M^2\to M^2$. Then $\dim_H(\Lambda)=s_1+s_2$, where s_1 and s_2 satisfy*

$$P_{(f,\Lambda)}(-s\log\|Df_x|E^u_x\|)=0$$
$$P_{(f,\Lambda)}(s\log\|Df_x|E^s_x\|)=0.$$

IDEA OF PROOF. For a basic set of an Axiom-A surface diffeomorphism, the stable and unstable manifolds are C^1, so the local product structure provides C^1 coordinates. Hence, locally the basic set Λ is C^1 diffeomorphic to a product of its unstable and stable manifolds $W^u_{\rm loc}(x), W^s_{\rm loc}(x)$ around x

$$\Lambda\approx(W^u_{\rm loc}(x)\cap\Lambda)\times(W^s_{\rm loc}(x)\cap\Lambda).$$

We will show that

$$\dim_H(W^{u,s}_{\rm loc}(x)\cap\Lambda)=\dim_B(W^{u,s}_{\rm loc}(x)\cap\Lambda),$$

and thus by (v) in Section 1.1.2, it follows that

$$\dim_H(\Lambda) = \dim_H(W^u_{\text{loc}}(x) \cap \Lambda) + \dim_H(W^s_{\text{loc}}(x) \cap \Lambda).$$

Let μ^u denote the unique equilibrium measure for the potential $s_1 \log \|Df_x|E^u_x\|$ and let μ^u_x denote the conditional measure on $W^u_{\text{loc}}(x)$. To show that $\dim_H(W^u_{\text{loc}}(x) \cap \Lambda) \geq s_1$, one shows that the conditional measure μ^u_x satisfies a uniform mass distribution principle (Method UI in Section 1.1.4). The proof uses properties of the Moran cover and the Jacobian estimate.

To show the opposite inequality, one uses the Moran cover and the Jacobian estimate to show that

$$P\left(-(\overline{\dim}_B(W^u_{\text{loc}}(x) \cap \Lambda) - 2\varepsilon) \log \|Df_x|E^u_x\|\right) \geq 0,$$

for all $\varepsilon > 0$. It follows from well known properties of pressure (see Appendix B) that $\overline{\dim}_B(W^u_{\text{loc}}(x) \cap \Lambda) \leq s_1$.

An analogous argument proves that $\dim_H(W^s_{\text{loc}}(x) \cap \Lambda) = s_2$. □

Manning and McClusky also show that s_1 and s_2 vary continuously when the map f is smoothly perturbed. Mañé [**59**] showed that s_1 and s_2 actually vary smoothly (with one loss of derivative) when the map f is smoothly perturbed (see [**115**] for an explicit derivative formula).

2.4. Multifractal analysis

We divide the subject of multifractal analysis into three parts: the speculation, the traditional multifractal analysis, and the dynamical characteristic view of multifractal analysis.

2.4.1. The speculation. About 15 years ago, a group of physicists [**40**] suggested the idea of a multifractal analysis. Since then, there has been a great deal of speculation that a multifractal analysis will do everything from explaining turbulence to explaining the distribution of galaxies in our universe to explaining scaling laws at phase transitions. *As far as we know, nobody has been able to utilize this type of multifractal analysis to* substantially *improve the understanding of new physical or biological phenomena or to obtain new serious physical insights.* Some leading physicists now doubt whether a multifractal analysis can actually provide any new insights into *thermodynamic systems* [1].

2.4.2. The traditional notion of multifractal analysis. We have stressed that invariant sets of most dynamical systems in general are not self-similar in the strict sense. However, sometimes part of these sets can be decomposed into (perhaps uncountably many) subsets each supporting a Borel probability measure possessing a type of scaling symmetry. This means that the measure admits a group of scale symmetries which reproduces copies of the set (or its significant part of full measure) on arbitrarily small scales (up to a given precision which decays to zero with scale). Sets that admit such structure are called *multifractals*. The Hausdorff dimension of each subset can be used to encode this structure.

The two major components in the traditional notion of multifractal analysis are the $f_\mu(\alpha)$ dimension spectrum and the Rényi spectrum, $R_\mu(\alpha)$, of dimensions. The multifractal analysis unifies these two spectra via the Legendre transform. Once

[1] A turbulent fluid is not an example of a thermodynamics system

the Legendre transform relation between the two spectra is established, one can try to compute (estimate) the delicate and seemingly intractable $f_\mu(\alpha)$ dimension spectrum through the Rényi spectrum, which for ergodic dynamical systems is completely determined by the statistics of a typical trajectory.

We now define the highlights of the traditional multifractal analysis for a set $X \subset \mathbb{R}^N$ and a Borel probability measure μ supported on X. The multifractal analysis is a description of the fine-scale geometry of the set X whose constituent components are the sets

$$K_\alpha = \{x \in X \mid d_\mu(x) = \alpha\},$$

for $\alpha \in \mathbb{R}$. Let us assume that X is a conformal repeller for g and suppose μ is an equilibrium measure for a Hölder potential φ.

The $\mathbf{f_\mu(\boldsymbol{\alpha})}$**-spectrum for dimensions** is defined by

$$f_\mu(\alpha) = \dim_H K_\alpha,$$

where $\dim_H K_\alpha$ denotes the Hausdorff dimension of the set K_α. One wishes to understand the topological and dimension properties of the very complicated decomposition

$$X = \bigcup_{-\infty<\alpha<\infty} K_\alpha \cup \{x \in X \mid d_\mu(x) \text{ does not exist}\}.$$

We define the **Rényi spectrum for dimensions** as follows. For each $r > 0$ cover the set X by a *grid* [2] of mesh size r. Given a family of grids define

$$R_\mu(q) = \frac{1}{1-q} \lim_{r\to 0} \frac{\log \sum_{i=1}^{N(r)} \mu(C_r^i)^q}{\log r}, \tag{5}$$

provided the limit exists, where $N(r)$ is the number of partition elements C_r^i of the grid with $\mu(C_r^i) > 0$.

Finally we define the following useful auxiliary quantity: the one parameter family of Hölder potentials $\varphi_q(x) = -T(q)\log\|Dg\| + q(\varphi - P(\varphi))$, where $T(q)$ is chosen such that $P(\varphi_q) = 0$.

What most authors consider a multifractal analysis is to show that the functions (spectra) f_μ, R_μ, and T are smooth and convex on some intervals, that f_μ and T form a Legendre transform pair, and that $T(q) = (1-q)R_\mu(q)$.

The first rigorous multifractal-type analysis of dynamical systems was carried out in [**20**] for a special class of measures invariant under some one-dimensional Markov maps. In [**58**], Lopes effects a multifractal analysis for the measure of maximal entropy for a hyperbolic Julia set. In [**103**], Simpelaere effects a multifractal analysis for equilibrium measures of Axiom-A surface diffeomorphisms and in [**82, 83**], Pesin and Weiss, using different methods, obtain a *complete* multifractal analysis for equilibrium measures of conformal repellers and Axiom-A surface diffeomorphisms.

The equilibrium measures μ_q for the potentials φ_q play a fundamental role. Each measure μ_q has pointwise dimension $\alpha(q) = -T'(q)$ for each point on the level set $K_{\alpha(q)}$. This provides the scaling symmetries on parts of the fractal repeller.

The limit defining $R_\mu(q)$ may depend on the family of grids. Most authors make a particular choice of grid system. Also, the some authors work with the

[2] By a grid we me a partition of the set such that each partition element contains a ball of radius $\frac{1}{2}r$ and is contained in a ball of radius r, where both balls are centered at the same point.

Hentschel-Procaccia spectrum for dimensions [**44**] instead of the Rényi spectrum for dimensions. The difference is essentially working with covers instead of partitions. It is useful to know whether these two spectra coincide, what the *maximal* interval is on which these functions are defined, and how large the set of points is where the pointwise dimension does not exist. Schmeling [**96**] proved that the maximal interval coincides with the one predicted by the use of thermodynamic formalism, while Barreira and Schmeling [**9**] showed that in general the set of points where the pointwise dimension does not exist is large. By a *complete* multifractal analysis, we mean addressing these, and other issues, which very few authors address.

Despite introducing some nice mathematics and obtaining some new results in dynamical systems of independent interest (like equilibrium measures have the doubling property), these types of multifractal analyses have not yielded any *significant* new insights about repellers or attractors of hyperbolic dynamical system. However, aspects of this traditional multifractal analysis play a major role in the dynamical characteristic view of multifractal analysis.

2.4.3. The dynamical characteristic view of multifractal analysis. The new dynamical characteristic notion of multifractal analysis is an attempt to understand the fine structure of the level sets of the fundamental asymptotic quantities in ergodic theory (e.g., Birkhoff averages, local entropy, Lyapunov exponents). Important elements of a *complete* multifractal analysis entail determining the range of values these characteristics attain, a thorough analysis of the topological and dimension properties of the level sets, and an understanding of the sets where the limits do not exist.

For conformal repellers (there is an analogous formula for Axiom-A surface diffeomorphisms), the following formula provides a link for a Hölder continuous potential φ between the pointwise dimension of the equilibrium measure $d_{\mu_\varphi}(x)$, the Birkhoff average $\overline{\varphi}(x)$, and the Lyapunov exponent $\lambda(x)$.

THEOREM 2.4.1 ([**114**]). *Let $g\colon J \to J$ be a C^2 conformal expanding map and let μ_φ be the equilibrium state corresponding to the Hölder continuous potential φ. Then*

$$d_{\mu_\varphi}(x) = \frac{P(\varphi) - \overline{\varphi}(x)}{\lambda(x)}, \tag{6}$$

provided that $\overline{\varphi}(x)$ and $\lambda(x)$ exist.

There are a few special cases of this formula which allow one to obtain a multifractal analysis for some important dynamical characteristics from the traditional multifractal analysis based on $d_{\mu_\varphi}(x)$.

COROLLARY 2.4.2. *The following statements are true.*

- *For the potential $\varphi_s(x) = -s\log\|Dg_x\|$, we obtain the following formulas for the Lyapunov exponent in terms of the pointwise dimension of special equilibrium measures:*

$$\lambda(x) = \frac{P(-s\log|Dg|)}{d_{\mu_s}(x) - s}, \quad \textit{and when } s = 0 \textit{ this becomes} \quad \lambda(x) = \frac{h_{top}(g)}{d_{\mu_{max}}(x)},$$

where μ_s is the equilibrium measure for $-s\log\|Dg_x\|, h_{top}(g)$ denotes the topological entropy of g, and μ_{max} denotes the measure of maximal entropy for g.

- *If $g = \sigma\colon \Sigma_A^+ \to \Sigma_A^+$ is the shift mapping on a mixing subshift of finite type, the proof of formula 6 yields the following formula for the Birkhoff average for a Hölder continuous function φ in terms of the pointwise dimension of μ_φ :*
$$\overline{\varphi}(x) = P(\varphi) - \log\lambda \cdot d_{\mu_\varphi}(x),$$
where the constant $\lambda > 1$ depends on the choice of metric on Σ_A^+.

The question of the simultaneous existence of *both* the Lyapunov exponent and the Birkhoff average at a point (the denominator and the numerator in (6) is quite delicate and is currently not well understood for smooth hyperbolic maps.

The following theorem summarizes several results on the dynamical characteristic multifractal analysis of dynamical systems [3]. The proofs combine the corollaries above with parts of traditional multifractal analyses. We refer the reader to the papers for precise statements.

THEOREM 2.4.3. *We have:*

- [**83, 114**] *For conformal repellers and Axiom-A surface diffeomorphisms, a complete multifractal analysis exists for the Lyapunov exponent.*
- [**7, 83, 114**] *For mixing subshift of finite type, a complete multifractal analysis exists for the Birkhoff average for a Hölder continuous potential and for the local entropy for a Gibbs measure with Hölder continuous potential.*
- [**78**] *There is a multifractal analysis for three-dimensional hyperbolic flows.*
- [**87**] *For the Gauss map, a complete multifractal analysis exists for the Lyapunov exponent.*

Finally, there are other ways to encode all the pointwise data than using the Hausdorff dimension of the level sets. For instance, one can encode using the topological entropy of the map restricted to the (not necessarily invariant) level sets. For a subshift of finite type this is studied in [**7**]. We note that the definition of topological entropy as a dimension-like characteristic (Section 1.1.4) lends itself well to this type of analysis.

2.4.4. Multifractal analysis and large deviation theory. There are deep connections between large deviation theory and multifractal analysis (in both the traditional and dynamical characteristic views). The variational formula for pressure is an important tool in the analysis and can be viewed (and proven) as a large deviation result [**31**]. Some authors use large deviation theory as a tool to effect their multifractal analysis. For instance, Simpelaere [**103**] uses a large deviation theorem of Ellis [**30**] to show that $\dim_H(K_\alpha)$ is bounded above by the Legendre transform of the Rényi spectrum for dimensions. The theorem of Ellis allows one to start with the Rényi spectrum for dimensions and obtain refined information on the distribution of $\log\mu(R_n^r)/\log r$ as $r \to 0$, where R_n^r are elements of a family of Moran covers, and to identify the Legendre transform of the Rényi spectrum with the dimension spectrum.

---

[3] Since these are introductory lecture notes, we do not include some related multifractal results by Barreira, Schmeling, Soussol, Takens, Verbitsky, etc., on the entropy spectrum, where the "size" of sets is encoded by topological entropy instead of Hausdorff dimension. The entropy spectrum is an important tool in the number theoretic analysis in [**10**].

In the multifractal analysis in [**82, 83**], the authors do not use large deviation theory as a tool. However, a posteriori, they apply a version of the Ellis large deviation theorem to obtain an intriguing *counting* expression for the dimension spectrum ([**83**], Theorem IV.1). In his recent thesis, Kesseböhmer [**50**] found global and local (multiplicative) formulas for subshifts of finite type which relate large deviations of the Birkhoff average for a Hölder continuous function to the *Birkhoff spectrum* for the corresponding equilibrium measure. The following less sophisticated global formula of Kesseböhmer follows easily from the above-mentioned formula in [**83**] and the second assertion of Corollary 2.4.2.

THEOREM 2.4.4. *Let $\varphi\colon \Sigma_A^+ \to \mathbb{R}$ be a Hölder continuous function for a mixing subshift of finite type. For each $x \in \Sigma_A^+$ and $r > 0$, let $n_r(x)$ be the positive integer defined by*

$$e^{S_{n_r}\varphi(x)} \geq r \quad \textit{and} \quad e^{S_{n_r+1}\varphi(x)} < r,$$

where $S_n\varphi(x) = (1/n)\sum_{k=0}^{n-1}\varphi(\sigma^k x)$. For $\alpha(q) = -T'(q)$, where $T(q)$ is defined by $P(-T(q)\log 2 + q(\varphi - P(\varphi)) = 0$, one has that for $t > 0$

$$\dim_H \left\{x\colon \overline{\varphi}(x) = \frac{P(\varphi) - \alpha(q)}{\log 2}\right\} - \dim_H(\Sigma_A^+)$$
$$= \lim_{r \to 0} \frac{\log\left(\mu_{max}\left\{x\colon \frac{S_{n_r(x)}\varphi(x)}{-\log r} \geq -\alpha(q)\right\}\right)}{-\log r}.$$

2.5. Some remarks on dimension theory for low-dimensional versus high dimensional dynamical systems

Unlike lower dimensions (one, two, or conformal repellers), for higher dimensional dynamical systems there are no general dimension formulas (besides the Ledrappier-Young formula), and in general dimension theory is much more difficult. This is due to several factors:

1) The geometry of the hyperbolic set enters into dimension calculations in ways it does not in lower dimensions.

2) Number theoretic properties of the Lyapunov exponents [**88, 86**] enter into dimension calculations in ways they do not in lower dimensions.

3) The measure of full dimension typically does not exist for Axiom-A diffeomorphisms [**62**].

4) A useful tool when computing dimension for Axiom-A surface diffeomorphisms is to restrict the dynamics to the one-dimensional stable and unstable manifolds, compute the dimensions there, and then combine to obtain the dimension of say the basic set. This works because the local product structure provides C^1 coordinates. In other words, the stable and unstable foliations are C^1. It is known that the stable and unstable foliations for higher dimension Axiom-A diffeomorphisms are typically not C^1 [**42, 93**].

Part 3. Dimension theory of higher-dimensional dynamical systems

In this part, we investigate the dimension theory of dynamical systems on higher-dimensional manifolds. The different rates of expansion or contraction lead to tough new difficulties, and there is little general theory. We begin with some important special examples which will help to illustrate the new difficulties.

3.1. An affine horseshoe in $\mathbb{R}^3$

Consider an affine map $f\colon \mathcal{R} \subset \mathbb{R}^3 \to \mathbb{R}^3$ of a cube $\mathcal{R} = [0,1] \times [0,1]^2$ as indicated in Figure 5. For simplicity, assume that the two foliations

$$\left(\{x_1\} \times [0,1]^2\right)_{x_1 \in [0,1]} \text{ and } \left([0,1] \times \{(x_2,x_3)\}\right)_{(x_2,x_3)\in[0,1]^2}$$

are invariant under f. We use the notations $\mathcal{W}^s_{loc}(x) := \{x\} \times [0,1]^2$ and $\mathcal{W}^u_{loc}(x) := [0,1] \times \{(x_2,x_3)\}$. These sets are called the *local stable and local unstable manifolds*, respectively.

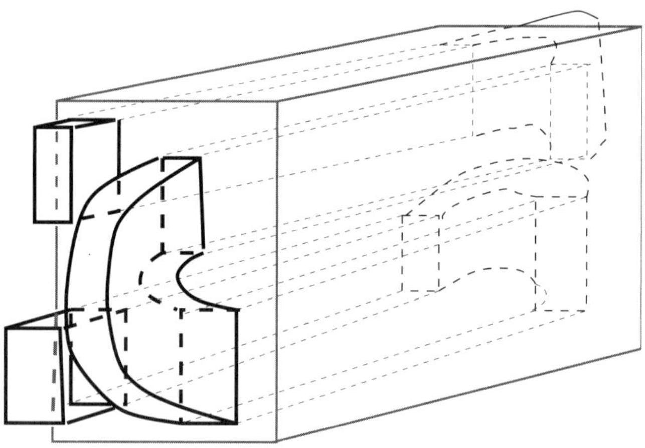

FIGURE 5. An affine horseshoe in $\mathbb{R}^3$

We define the limit set

$$\Lambda := \bigcap_{n\in\mathbb{Z}} f^n(\mathcal{R}),$$

and the unstable and stable sections of the limit set by:

$$\Lambda^u_x := \Lambda \cap \mathcal{W}^u_{loc}(x) \quad \text{and} \quad \Lambda^s_x := \Lambda \cap \mathcal{W}^s_{loc}(x).$$

REMARK 3.1.1. *The limit set Λ is diffeomorphic to the product of the two Cantor sets $\Lambda^u_x \times \Lambda^s_x$ since the horizontal lines are invariant. For general horseshoes, even those which are the locally maximal hyperbolic sets of a diffeomorphism close to f, this is not the case, and Λ is only homeomorphic to the product of two Cantor sets* [**93**].

Since the unstable manifold $\mathcal{W}^u_{loc}(x)$ is one-dimensional, we can apply Bowen's Pressure Formula to the unstable section to obtain

$$P(-\dim_H \Lambda^u_x \log \|Df|_{E^u}\|) = 0.$$

Moreover, since

$$\dim_H \Lambda_x^u = \dim_B \Lambda_x^u,$$

we have that

$$\dim_H \Lambda = \dim_H \Lambda^u + \dim_H \Lambda^s.$$

The corresponding dimension of the stable section Λ_x^s is more subtle. There are two (possibly) different contraction rates, and this case is not covered by Bowen's pressure formula.

Problem: How can one calculate $\dim_H \Lambda^s$?

In general there is no known method to compute $\dim_H \Lambda^s$! Thus the problem of computing the Hausdorff dimension of general affine horseshoes seems to be intractable at present. However, we can compute the Hausdorff dimension in special cases when Λ_x^s possesses a special structure.

3.1.1. Case 1: The conformal case. We begin with the simplest case which is essentially one-dimensional. We assume that for $x_1 \in [0,1]$, the intersection $\{x_1\} \times [0,1] \cap f(\mathcal{R})$ consists of four homothetic copies of the unit square contracted by rates λ_i; $i = 1,2,3,4$, as indicated in Figure 6. In this case, the assumption that the sections Λ_x^s are identical for all x is not required for Λ to be diffeomorphic to the product of two Cantor sets.

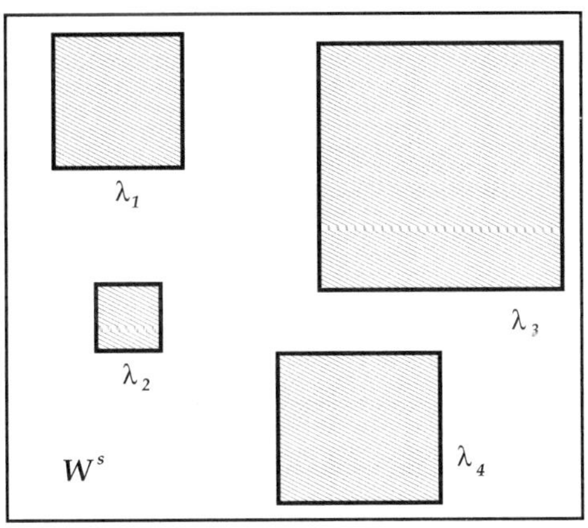

FIGURE 6. The conformal case

We can apply Bowen's Pressure Formula and obtain

$$P(\dim_H \Lambda^s \log \|Df|_{E^s}\|) = 0.$$

By (10) this is equivalent to

$$\lambda_1^{\dim_H \Lambda^s} + \lambda_2^{\dim_H \Lambda^s} + \lambda_3^{\dim_H \Lambda^s} + \lambda_4^{\dim_H \Lambda^s} = 1.$$

We observe that $h_{\text{top}}(f) = \log 4$, since Λ is modelled by the full shift on 4 symbols. If we assume that the contraction rates are identical, i.e., $\lambda_i \equiv \lambda$, we obtain the formula

$$\dim_H \Lambda^s = -\frac{\log 4}{\log \lambda} = -\frac{h_{\text{top}}(f)}{\log \lambda}. \tag{7}$$

3.1.2. Case 2: The aligned case. We now drop the assumption that the images of the square are conformal, and thus allow different contraction rates in different directions. For simplicity we still assume that the images are affine copies of the square, i.e., rectangles. We consider the two subcases when the images of the square in $\mathcal{W}^s_{loc}(x)$ are aligned along the two different contracting directions.

3.1.2.1. **Subcase A:** *Alignment along* U^1. Assume that for all $x_1 \in [0,1]$ the intersection $\{x_1\} \times [0,1]^2 \cap f(\mathcal{R})$ consists of four rectangles aligned along the axis U^1 of the stronger contraction, as indicated in Figure 7.

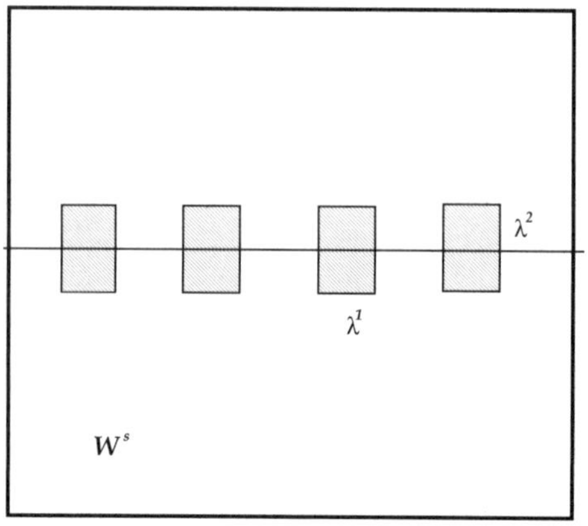

FIGURE 7. Subcase A

In this case $\Lambda^s_x = \Lambda^s_x \cap U^1$ and U^1 is invariant: $f(U^1) = U^1$. Our construction is essentially one-dimension, since only the stronger direction U^1 enters into our dimension calculation. We can apply Bowen's Pressure Formula to the one-dimensional geometric construction in U^1 and obtain

$$P(\dim_H(\Lambda^s \cap U^1) \log \|Df|_{U^1}\|) = 0.$$

This yields

$$\dim_H \Lambda^s = \dim_H(\Lambda^s \cap U^1) = -\frac{\log 4}{\log \lambda^1} = -\frac{h_{\text{top}}(f)}{\log \lambda^1}. \tag{8}$$

REMARK 3.1.2. *This situation is not stable! Small perturbations of the map f destroy the alignment along the strong stable direction and the limit set Λ^s cannot be modelled by a one-dimensional construction. The next case will be more stable and, as we will see later, is one of the examples which motivates several conjectures about dimension formulas for typical situations.*

3.1.2.2. **Subcase B:** *Alignment along* U^2. Assume that for all $x_1 \in [0,1]$ the intersection $\{x_1\} \times [0,1]^2 \cap f(\mathcal{R})$ consists of four rectangles all aligned along a line U^2 transversal to the direction of the stronger contraction, as indicated in Figure 8.

This case may be treated in a similar fashion to the previous one, where the role of the stronger contraction is replaced by the weaker contraction. In this case we have that $\Lambda^s_x = \Lambda^s_x \cap U^2$ and $f(U^2) = U^2$. This invariance is built into our special

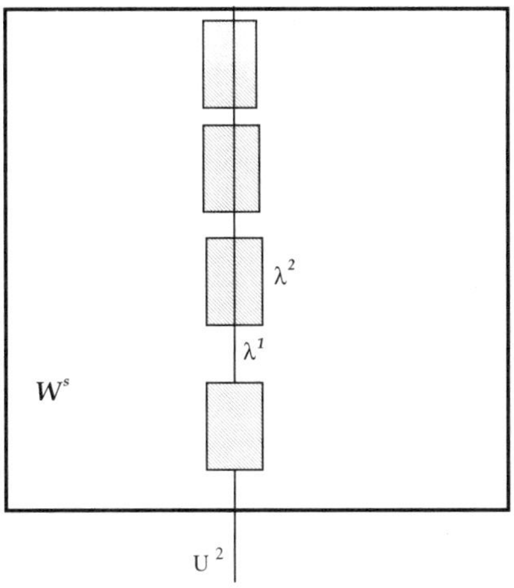

FIGURE 8. Subcase B

construction. However, one can prove that this invariance is stable (see [**93**]). Thus, our limit set is modelled by a one-dimensional construction on U^2 and

$$P(\dim_H(\Lambda^s \cap U^2) \log \|Df|_{U^2}\|) = 0.$$

It follows that

$$\dim_H \Lambda^s = \dim_H(\Lambda^s \cap U^2) = -\frac{\log 4}{\log \lambda^2} = -\frac{h_{\text{top}}(f)}{\log \lambda^2}. \tag{9}$$

This formula differs from the previous formula by substituting λ^1 with λ^2.

REMARK 3.1.3. *This situation is more stable than case A as will become clear in later discussions. The reason is that the alignment is along an arbitrary transversal, rather than a particular line. This indicates that the weak contraction rate plays a stronger role than the weaker rate.*

3.1.3. Case 3: The product case.

In this case we assume that both the horizontal and the vertical foliations are preserved: $f(U^1) = U^1$ and $f(U^2) = U^2$ (see Figure 9).

Moreover, we assume that for all x, the stable section Λ^s_x possesses a product structure

$$\Lambda^s_x = (\Lambda^s_x \cap U^1(x)) \times (\Lambda^s_x \cap U^2(x)).$$

We can apply Bowen's Pressure Formula to the one-dimensional geometric constructions for $\Lambda^s_x \cap U^1(x)$ and $\Lambda^s_x \cap U^2(x)$, respectively (both modeled by the full two-shift), to obtain

$$P(\dim_H(\Lambda^s_x \cap U^1(x)) \log \|Df|_{U^1}\|) = 0$$

and

$$P(\dim_H(\Lambda^s_x \cap U^2(x)) \log \|Df|_{U^2}\|) = 0.$$

Since

$$\dim_H(\Lambda^s_x \cap U^i(x)) = \dim_B(\Lambda^s_x \cap U^i(x)) \quad \text{for } i = 1, 2,$$

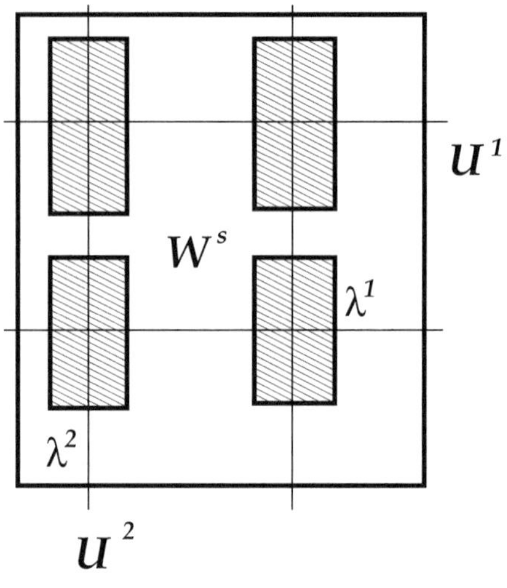

FIGURE 9. Case 3

we obtain

$$\begin{aligned}\dim_H(\Lambda^s) &= \dim_H\left((\Lambda^s \cap U^1)\times(\Lambda^s \cap U^2)\right) = \dim_H(\Lambda^s \cap U^1) + \dim_H(\Lambda^s\cap U^2)\\ &= -\frac{\log 2}{\log \lambda^1} - \frac{\log 2}{\log \lambda^2} = -\log 2\left(\frac{1}{\log\lambda^1} + \frac{1}{\log \lambda^2}\right).\end{aligned}$$

The similarity of this formula to Young's formula (Section 2.2) is not accidental. In both cases the invariant set (measure) is the product of two one-dimensional components. If we assume that the stable set Λ^s_x is *essentially one-dimensional* we recover *Young's formula*, but for the invariant set.

REMARK 3.1.4. *If in all the previous cases the exponents* $\lambda^1 = \lambda^2 = \lambda$ *and* $\|D_x f|_{E^u}\| \equiv 4 = e^{h_{top}(f)}$*, one obtains the common formula*

$$\dim_H \Lambda = \dim_H(\Lambda^u \times \Lambda^s) = h_{top}(f)\left(\frac{1}{\log\gamma} - \frac{1}{\log\lambda}\right).$$

This is Young's Formula for the measure of maximal entropy, which is not very surprising since our constructions are so special (with constant contracting and expansion rates, and in particular, the measure of maximal entropy is also a measure of maximal dimension). In general the dimension of the invariant set and of an invariant measure are different (see Section 2).

3.1.4. Case 4: The invariant axes case. In this case, as in the previous cases, we assume that there is a smooth invariant coordinate system $(U^1(x), U^2(x))$ in $\mathcal{W}^s_{loc}(x)$. However, we drop the product assumption. We assume that

$$0 < \max_{x\in\Lambda} \|D_x f|_{U^1}\| < \min_{x\in\Lambda}\|D_x f|_{U^2}\| < 1.$$

Even in this situation very little is known about $\dim_H \Lambda^s_x$.

If $\Pi\colon \Lambda^s_x \to U^2(x)$ denotes the projection along U^1, the dimension is very sensitive to the number of rectangles having the same projection (see Figure 10).

This leads to number theoretic peculiarities. In particular, if all four rectangles $\{x_1\} \times [0,1]^2 \cap f(\mathcal{R})$ have sidelength $\lambda^2 > 1/4$ and $\mu^2 < 1/4$ results of Solomyak [**105**], Przytycki and Urbanski [**88**] and Pollicott and Weiss [**86**] imply that for Lebesgue-almost every λ with $\frac{1}{2} < \lambda^2 < 1$, the Hausdorff dimension of the stable slice satisfies

$$\dim_H \Lambda_x^s = \dim_B \Lambda_x^s = 1 - \frac{\log(2\lambda)}{\log \mu}.$$

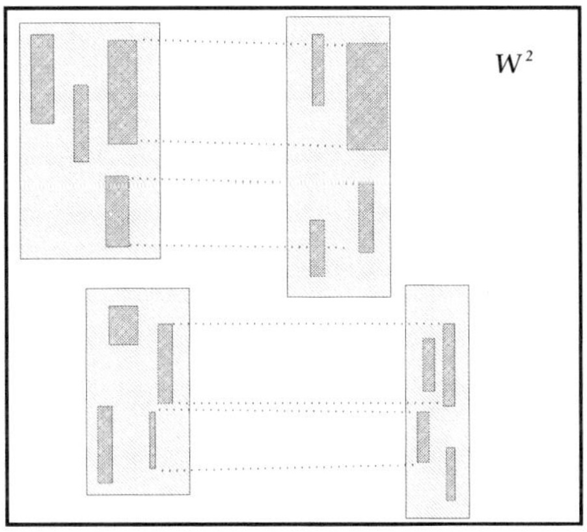

FIGURE 10. Case 4

On the other hand, if λ^{-1} is a Pisot number (a Pisot number is an algebraic integer with all of its conjugates lying strictly inside the unit circle), then

$$\dim_H \Lambda_x^s < \dim_B \Lambda_x^s = 1 - \frac{\log(2\lambda)}{\log \mu}.$$

REMARK 3.1.5. *This example shows that the Hausdorff dimension of the stable slice need not depend continuously on f.*

CONCLUSIONS 3.1.6. *Even for the simplest examples of higher-dimensional dynamical systems we are far from a general theory of the Hausdorff dimension of invariant sets. However, there is a related notion of dimension, the Hausdorff dimension of invariant measures, that seems far more tractable.*

3.2. The Bothe-Simon theorem

Here we consider an example of a hyperbolic attractor in $\mathbb{R}^3$. Let $\triangle$ denote the unit disc in $\mathbb{R}^2$. Let $f\colon S^1 \times \triangle \to S^1 \times \triangle$ be of the form (see Figure 11)

$$f(\theta, x, y) = (2\theta \bmod 1, \psi^1(\theta, x), \psi^2(\theta, y)),$$

with

$$0 < \max_{S^1\times\triangle} \frac{\partial}{\partial x}\psi^1(\theta, x) < \min_{S^1\times\triangle} \frac{\partial}{\partial y}\psi^2(\theta, y) < \lambda < 1.$$

The limit set

$$\Lambda := \bigcap_{n\in\mathbb{N}} f^n(S^1 \times \triangle)$$

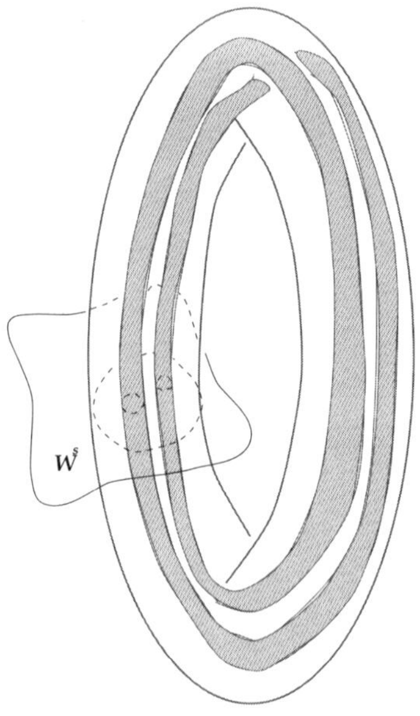

FIGURE 11. The solenoid

is called the attractor or the *solenoid.* It is an example of a structurally stable basic set and is one of the fundamental examples of a uniformly hyperbolic attractor. A major difference between solenoids and affine horseshoes is the lack of an affine structure. As for the horseshoe in Case 4, the solenoid possesses a smooth invariant coordinate system in Λ^s. It can be shown ([**15, 93**]) that this situation is structurally stable.

Bothe [**15**] and Simon [**104**] proved the following result.

THEOREM 3.2.1. *For sufficiently small λ and all θ, the thermodynamic pressure*

$$P(\dim_H \Lambda_\theta^s \log|\frac{\partial}{\partial y}\psi^2(\theta, y)|) = 0.$$

In particular, the stable dimension is independent of the stable section.

In contrast to the affine horseshoe, the Hausdorff dimension of the solenoid in $\mathbb{R}^3$ does not depend on the particular placement of the images of the torus. This is a consequence of special geometric properties of the embedding of the image of the torus into itself. The theorem implies that the weaker contraction rate plays the crucial role for calculating the dimension. Later (see section 3.4) we give a heuristic justification for this fact.

A famous family of such attractors are the Smale-Williams solenoids, where $f_\varepsilon(\theta, x, y) = (2\theta \bmod 1, \lambda x + \varepsilon\cos(2\pi\theta), \mu x + \varepsilon\sin(2\pi\theta))$. The following formulas follow from the work of Bothe and Simon.

THEOREM 3.2.2. *Consider the Smale-Williams solenoids with $\lambda, \mu < \varepsilon$ and $\lambda, \mu < 1/2$.*

- *The Hausdorff dimension of the solenoid is given by*

$$\dim_H \Lambda = 1 + \frac{\log 2}{-\log \max\{\lambda, \mu\}}.$$

- *If $\lambda, \mu < 1/8$, the Hausdorff dimension of each stable slice of the solenoid $\Lambda^s_\varphi = \{(\theta, x, y) \colon \theta = \varphi$ and $(x, y) \in \Delta\}$ is given by*

$$\dim_H \Lambda^s_\varphi = \frac{\log 2}{-\log \max\{\lambda, \mu\}}.$$

We feel the following is an important open problem.

Question: Does the formula in the second part of the previous theorem for the Hausdorff dimension of each stable slice of the solenoid hold for all $\lambda, \mu < 1/2$?

3.3. Hyperbolic Measures

The remaining part of these lectures deal with the dimension theory of invariant measures. We begin by giving some basic definitions. Let M be a smooth compact Riemannian manifold without boundary and let $f \colon M \to M$ be a diffeomorphism which preserves an invariant ergodic measure μ, i.e., $\mu(A) = \mu(f(A))$.

Given $x \in M$ and a vector $v \in T_x M$ we define the *Lyapunov exponent*

$$\chi(x, v) = \overline{\lim_{n \to \infty}} \frac{1}{n} \log \|D_x f^n v\|.$$

For fixed x, the numbers $\chi(x, \cdot)$ attain only finitely many values. By ergodicity they are μ-almost everywhere constant, and we denote the values by

$$\chi_1 \leq \cdots \leq \chi_p,$$

where $p = \dim M$.

DEFINITION 3.3.1. *The invariant ergodic measure μ is said to be* **hyperbolic** *if $\chi_i \neq 0$ for every $i = 1, \ldots, p$.*

We set

$$s := \max\{i : \chi_i < 0\} \quad \text{and} \quad u := \min\{j \colon \chi_j > 0\}.$$

Hyperbolicity implies that $u = s + 1$. For $r = r(x)$ small and $1 \leq i \leq s$, the sets

$$\mathcal{W}^i_{loc}(x) = Comp_x \left\{ y \in B(x, r) \colon \overline{\lim_{n \to \infty}} \frac{\log d(f^n x, f^n y)}{n} < \chi_i \right\},$$

where $Comp_x$ denotes the connected component containing x, are immersed local manifolds and are called the *local i^{th} stable manifolds.* Similarly one defines the *local unstable manifolds.* Figure 12 shows the strong stable foliation $\mathcal{W}^1$ inside a leaf $\mathcal{W}^2(x)$.

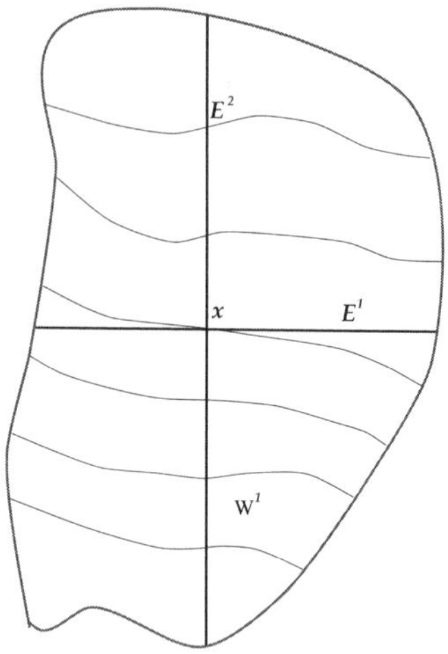

FIGURE 12. Invariant foliations

3.4. The Kaplan-Yorke conjecture

We now give a heuristic argument of how one can estimate the dimension of an invariant ergodic measure. These ideas are analogous to those developed by Kaplan and Yorke [**48, 33**] for the estimation of the dimension of an invariant set. Their estimates always provide an upper bound for the dimension. Kaplan and Yorke conjectured that in typical situations these estimates also provide a lower bound for the dimension of an attractor. Ledrappier and Young [**56, 57**] showed that if these estimates give the dimension of an invariant measure, then this measure has to be a very special measure, called the SBR-measure (after Sinai, Ruelle, and Bowen).

Consider a small ball B in the phase space. The image $f^n B$ is nearly an ellipsoid with axes of approximate length

$$e^{\chi_1 n}, \cdots, e^{\chi_p n}.$$

For $1 \le i \le s$, cover $f^n B$ by balls of radius $e^{\chi_i n}$. The main ingredient for a dimension estimate is to answer the following question: *How many balls are needed?*

Let us denote this number by N_n. Then

$$N_n := \min \# \left\{ k \colon f^n B \subset \bigcup_{l=1}^{k} B(x_l, e^{\chi_i n}) \right\}.$$

It is easy to show that

$$N_n \sim \frac{\exp[\chi_{i+1} n]}{\exp[\chi_i n]} \cdots \frac{\exp[\chi_p n]}{\exp[\chi_i n]}.$$

We can estimate the box dimension of an invariant set Λ under the additional hypothesis that $f^n B$ covers a *large* part of Λ for sufficiently large n, e.g., it would be sufficient to cover a subset which has the same dimension as Λ. This is reasonable if the dimension of open subsets of Λ is constant. We can then estimate the box dimension

$$\dim_B \Lambda \le \overline{\lim_{n\to\infty}} \frac{\log N_n}{-\log e^{\chi_i n}} = \frac{\sum_{j>i}\chi_j - (p-i)\chi_i}{-\chi_i}$$
$$= \frac{\sum_{j>i}\chi_j}{|\chi_i|} + (p-i) := \dim_L^i \Lambda,$$

where $\dim_L^i \Lambda$ denotes the Lyapunov dimension.

The main problem with this argument is in estimating the size of $f^n B$ using the Lyapunov exponents. The exponents are in general not constant everywhere. Hence, we can only obtain an estimate on the part of Λ which consists of *typical* points for the invariant ergodic measure. If the exponents are constant everywhere on Λ, the number

$$\dim_L \Lambda := \min_{1\le i\le s} \dim_L^i \Lambda$$

is called the *Lyapunov dimension* of the invariant set Λ. It is not hard to see that for $l = \min\{i\colon \sum_{j>i}\chi_j \ge 0\}$

$$\dim_L \Lambda = \dim_L^l \Lambda.$$

REMARK 3.4.1. *For an invariant ergodic measure, the above formula provides an estimate for the dimension of this measure. It is also called the Lyapunov dimension of the measure.*

We now estimate the Lyapunov dimension in our examples.

3.4.1. Examples. For the affine horseshoes from Section 3 we have that the Lyapunov exponents are constant everywhere (this is the example which lead Kaplan and Yorke to conjecture that the Lyapunov dimension and the Hausdorff dimension coincide for attractors). Thus

$$\dim_L \Lambda = \frac{\log 4}{-\log \lambda^2} + 1 > \dim_H \Lambda.$$

The noncoincidence of $\dim_L \Lambda$ and $\dim_H \Lambda$ is due to the fact that the horseshoe is not an attractor and does not carry an SBR-measure. However, we can estimate the dimension of the stable sections. For Cases 1, 2B and (for the almost everywhere situation) 4, we obtain

$$\dim_L \Lambda = \dim_H \Lambda^s + 1.$$

This approach is not valid for Cases 2A and 3.

For the solenoid (in the case of constant derivatives), the Bothe-Simon Theorem implies that

$$\dim_H \Lambda = \dim_B \Lambda = \dim_L \Lambda.$$

3.5. General Theory

In this section we investigate the dimension theory of higher dimensional dynamical systems. As we remarked earlier, it is more fruitful to study the dimension of invariant measures. We are going to establish an important connection between Lyapunov exponents and the measure theoretic entropy via dimensions of measures. Although this theory was developed for arbitrary invariant measures, to simplify the exposition, we are going to restrict ourselves to ergodic measures. One uses the ergodic decomposition to extend the results to general measures.

Let μ be an ergodic invariant measure. The Oseledec and Pesin Theory guarantees that local stable manifolds exist at μ-almost every point. This means the locally ith-stable (or unstable) manifolds partition M up to a set of measure zero. The main idea is to extract the contributions to the entropy and to the dimension in the directions of χ_i . We want to disintegrate our measure with respect to these partitions. However, these partitions are in general not measurable (in the sense of Rokhlin) because of the strong recurrence of the leaves. To overcome these difficulties, Ledrappier and Young constructed measurable partitions ξ^i, $1 \leq i \leq s$ of M with the following properties:

- $\xi^i(x) \subset (f^{-1}\xi^i)(x)$
- $\xi^i(x) \subset \xi^{i+1}$
- $\bigvee_{n=1}^{\infty} f^n \xi^i$ is a generating partition
- $\xi^i(x) \subset \mathcal{W}^i(x)$ contains an open neighborhood.

One can define conditional measures μ_x^i relative to this partition. In an important and deep paper ([**56, 57**]) Ledrappier and Young defined entropies and pointwise dimensions $d_\mu^i(x)$ with respect to conditional measures supported on the i-th unstable leaves. We first define the conditional entropies using a local approach. Fix $\varepsilon > 0$ and consider the set defined by

$$V^i(x,n) := \left\{ y \in \mathcal{W}^i(x) \colon d^i(f^{-k}x, f^{-k}y) < \varepsilon\, , \, 0 \leq k < n \right\}.$$

The ith *upper local entropy* is defined as

$$\overline{h}^i(x) := \overline{\lim_{n\to\infty}} \, -\frac{1}{n} \log \mu_x^i V^i(x,n),$$

and the ith *lower local entropy* is defined similarly.

We use a local approach to define the conditional dimensions. The ith *upper pointwise dimension* is defined as

$$\overline{d}^i(x) := \overline{\lim_{r\to 0}} \, \frac{\log \mu_x^i B^i(x,r)}{\log r},$$

and the ith *lower pointwise dimension* is defined similarly.

REMARK 3.5.1. *These definitions do not depend on ξ^i or on the choice of μ_x^i. Moreover, since these quantities are invariant, they are constant almost everywhere, and we denote by $\overline{h}^i$, $\underline{h}^i$, $\overline{d}^i$, $\underline{d}^i$ as the almost everywhere values of these quantities.*

The next theorem provides a deep connection between entropy, dimension, and Lyapunov exponents and shows that dimension is a fundamental dynamical characteristic.

THEOREM 3.5.2 (Ledrappier-Young). *Let $f \colon M \to M$ be a C^2 diffeomorphism with invariant ergodic measure μ. Then the following holds.*

1) *There exist $h^i = \overline{h}^i = \underline{h}^i$ and $d^i = \overline{d}^i = \underline{d}^i$;*
2) *$h^1 = -\chi_1 d^1$;*
3) *$h^i - h^{i-1} = -\chi_i \left(d^i - d^{i-1}\right)$;*
4) *$h^s = h_\mu(f)$.*

The next corollary is an immediate consequence of this theorem.

COROLLARY 3.5.3. *Set $d^0 = 0$. Then*

$$h_\mu(f) = -\sum_{i=1}^{s} \chi_i \left(d^i - d^{i-1}\right).$$

The proof of this theorem is difficult and uses the theory of *nonuniform hyperbolic systems.* In the next section we discuss some of the ideas and tools in the proof.

If we consider the inverse map, f^{-1}, which has the same invariant ergodic measures as f (since f is a diffeomorphism), we obtain a similar formula connecting the conditional entropies and dimensions on the ith-unstable leaves with the positive Lyapunov exponents.

The first connection between exponents and entropy was discovered independently by Margulis and Ruelle [**91**]. They proved that

$$h_\mu(f) \leq -\sum_{i=1}^{s} \chi_i.$$

Pesin [**75**] showed that this inequality is actual an equality if the measure μ is absolutely continuous with respect to the Riemannian volume of M. Ledrappier and Young ([**56**]) proved that equality is equivalent to the measure possessing absolutely continuous conditional measures on unstable manifolds, i.e., the measure is the SBR measure.

Before discussing elements of the proof of this result, we examine what the theorem provides in some special examples.

3.5.1. Examples.

3.5.1.1. *The One-dimensional case.* In this case the Ledrappier-Young theorem states that

$$\dim_H \mu = \frac{h_\mu(f)}{\chi_\mu}$$

This is equivalent to formulas in the sections on lower dimensional phenomena (Chapter 2).

We will now see what the Ledrappier-Young theorem provides for the special affine horseshoes we considered in section 3.1. In these examples the Lyapunov exponents are constant everywhere. Let $m_{\max}$ be the measure which gives equal weight to cylinders of the same length. More precisely, $m_{\max}$ is the pushforward of the maximal measure on symbolic space under the coding map.

3.5.1.2. *The conformal case.* For simplicity we assume that the contraction coefficients are everywhere constant and equal to λ. In this case the measure $m_{\max}$ is also the measure of maximal dimension. Here we have that $\mathcal{W}^1 = \mathcal{W}^2$ is a two-dimensional manifold. The Ledrappier-Young theorem implies that

$$d^s_\mu = -\frac{h_\mu(f)}{\chi_s},$$

and if $\mu = m_{\max}$, then

$$d^s_{m_{\max}} = -\frac{\log 4}{\log \lambda} = \dim_H \Lambda^s.$$

3.5.1.3. *Case 2A.* We can easily calculate the conditional entropies and obtain

$$h^1 = \log 4 \quad \text{and} \quad \chi_1 = \log \lambda^1.$$

Also,

$$h^2 = h^s = h_{m_{\max}}(f) = \log 4 \quad \text{and} \quad \chi_2 = \log \lambda^2$$

The Ledrappier-Young theorem implies that

$$d^1 = -\frac{\log 4}{\log \lambda^1} \quad \text{and} \quad d^2 = d^1 = -\frac{\log 4}{\log \lambda^1}.$$

3.5.1.4. *Case 2B.* We can also easily calculate the conditional entropies and obtain that

$$h^1 = 0 \quad \text{and} \quad \chi_1 = \log \lambda^1.$$

Also,

$$h^2 = h^s = h_{m_{\max}}(f) = \log 4 \quad \text{and} \quad \chi_2 = \log \lambda^2.$$

The LY theorem implies that

$$d^1 = 0 \quad \text{and} \quad d^2 = -\frac{\log 4}{\log \lambda^2}$$

3.5.1.5. *Case 3.* The conditional entropies can be calculated to be

$$h^1 = \log 2 \quad \text{and} \quad \chi_1 = \log \lambda^1.$$

Also,

$$h^2 = h^s = h_{m_{\max}}(f) = \log 4 \quad \text{and} \quad \chi_2 = \log \lambda^2$$

The LY theorem implies that

$$\begin{aligned} h^2 = \log 4 &= -\log \lambda^2 \left(d^2 - d^1\right) + h^1 \\ &= -\log \lambda^2 \left(d^2 + \frac{\log 2}{\log \lambda^1}\right) + \log 2 \end{aligned}$$

Hence,

$$d^2 = -\log 2 \left(\frac{1}{\log \lambda^1} + \frac{1}{\log \lambda^2}\right).$$

REMARK 3.5.4. *We have seen that the Ledrappier-Young theorem yields the Hausdorff dimension of the horseshoe in all the special cases. This is not true in general. In general, the dimension that the LY formula yields is strictly less than the dimension of the support of the measure (see Chapter 2). It is not known when a measure with full dimension, i.e, a measure for which the Hausdorff dimension of the measure coincides with the Hausdorff dimension of the underlying invariant set, exists. Even for two-dimensional horseshoes such a measure does not need to exist* [**62**].

3.5.2. Ideas of the proof of the Ledrappier-Young Theorem 3.5.2. We restrict ourselves to the main heuristic ideas in the proof. Assume we are in the idealized (and restrictive) case and have almost everywhere an *invariant coordinate system* $U^i(x)$ such that for $y \in U^i(x)$

$$\lim_{n\to\infty} \frac{1}{n} \log d(f^n x, f^n y) = \chi_i.$$

3.5.2.1. *Proof of 4 in 3.5.2.* By the Shannon-McMillan-Breiman theorem, for μ-almost every x, the measure theoretic entropy

$$h_\mu(f) = \lim_{n\to\infty} \frac{\log \mu\left\{\xi_n(x)\colon \xi_n(x) \in \bigvee_{k=0}^{n} f^n\xi\right\}}{-n}.$$

Since all points $f^{-k}y$ in the unstable direction stay close to $f^{-k}x$, they are in the same partition element provided $y \in \xi(x)$. Hence, the atom $\xi_n(x)$ is close to

$$\xi_n(x) \sim (\mathcal{W}^u(x) \cup \xi(x)) \times \xi_n^s(x).$$

Since for *typical* points x the measure of $\xi(x)$ is comparable to 1, this yields that

$$\mu(\xi_n(x)) \sim \mu_x^u(\xi(x)) \cdot \mu_x^s(\xi_n^s(x))$$

(see Figure 13). The previous step would be immediate if we knew that the measure μ is locally the product of μ^s and μ^u. This is not generally the case for an ergodic measure and essentially means the measure is an equilibrium measure. Nevertheless, the measure of the atoms ξ_n can be estimated by the above product.

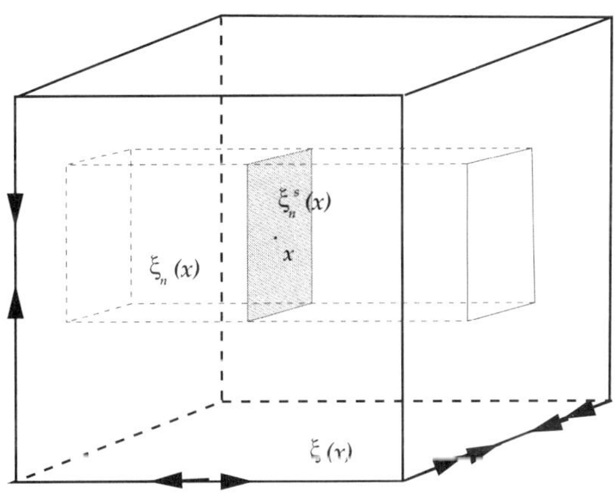

FIGURE 13. The different partition elements

Consequently, we obtain that

$$h_\mu(f) = -\lim_{n\to\infty} \frac{1}{n} \log \mu(\xi_n(x))$$

$$= \lim_{n\to\infty} \frac{\log \mu_x^u(\xi^u(x))}{-n} + \lim_{n\to\infty} \frac{\log \mu_x^s(\xi_n^s(x))}{-n} = h^s.$$

This gives 4.

3.5.2.2. *Proof of the first part of 1) in 3.5.2.* The proof that $\overline{h}^i = \underline{h}^i$ is a generalization of the *Shannon-McMillan-Breiman theorem.* The existence of the conditional pointwise dimension uses several sophisticated arguments in the Ledrappier and Young manuscript and will not be given here.

3.5.2.3. *Proof of 2) in 3.5.2.* This proof is the same as the one-dimensional case which was discussed in section 1.1.2.

3.5.2.4. *Proof of 3) in 3.5.2.* The proof is by induction on i. Let $1 \le i < s$ and define

$$B^i = B^i(x, e^{\chi_{i+1}n}) \quad B^{i+1} = B^{i+1}(x, e^{\chi_{i+1}n})$$

For typical points x and y the Shannon-McMillan-Breiman theorem implies that

$$\mu^i(\xi_n^i(y)) = \mu^i(\xi_n^{i+1}(y)) \sim e^{-nh^i} \quad \text{and} \quad \mu^{i+1}(\xi_n^{i+1}(y)) \sim e^{-nh^{i+1}}.$$

The symbol $\sim$ means that the quantities are of the same exponential order for large n. If the pointwise conditional dimensions exist we have that

$$\mu^i(B^i) \sim e^{nd^i\chi_{i+1}} \quad \text{and} \quad \mu^{i+1}(B^{i+1}) \sim e^{nd^{i+1}\chi_{i+1}}.$$

We observe that

$$B^{i+1} \sim \bigcup \{\xi^{i+1}(y) \colon \xi^{i+1}(y) \cap B^i \neq \varnothing\},$$

(see Figure 14)

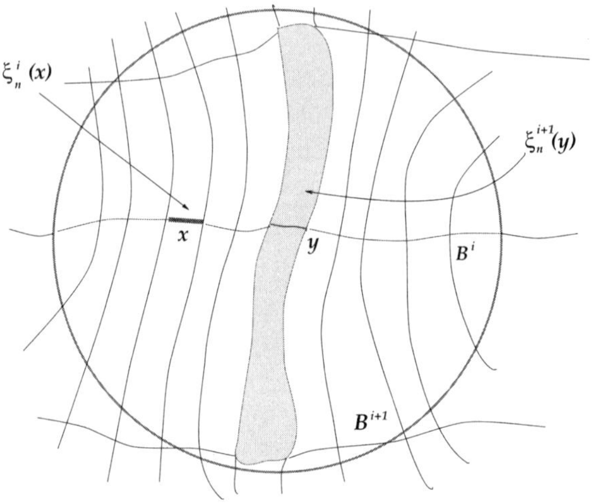

FIGURE 14.

and hence,

$$\mu(B^{i+1}) \sim \mu\left(\bigcup \{\xi^{i+1}(y) \colon \xi^{i+1}(y) \cap B^i \neq \varnothing\}\right).$$

This implies that

$$\begin{aligned}
e^{nd^{i+1}\chi_{i+1}} &\sim \mu(B^{i+1}) \\
&\sim \#\left\{\xi^{i+1}(y) \colon \xi^{i+1}(y) \cap B^i \neq \varnothing \text{ and } y \text{ is typical}\right\} \cdot \mu^{i+1}(\xi^{i+1}(y)) \\
&\sim \frac{\mu^i(B^i)}{\mu^i(\xi_n^{i+1}(y))} \cdot \mu^{i+1}(\xi^{i+1}(y)) \\
&\sim e^{nd^i\chi_{i+1}} \cdot e^{nh^i} \cdot e^{-nh^{i+1}}
\end{aligned}$$

Letting $n \to \infty$, we obtain 3).

The theorem of Ledrappier and Young tells us that the Lyapunov exponents, entropies, and the stable and unstable dimensions exist almost everywhere and provides a connection between the conditional quantities. There is also a connection between the unconditional quantities if the pointwise dimension exists. The following is a natural question.

Question: Does the μ-almost everywhere existence of d^s and d^u imply that $d_\mu(x)$ exists? If so, it would coincide with the Hausdorff dimension of the measure.

In the next section we investigate this question.

3.5.3. The existence of the pointwise dimension for hyperbolic measure. For a hyperbolic measure (i.e., no zero Lyapunov exponents), a natural guess is that the pointwise dimension exists almost everywhere, and equals the sum of the stable and unstable dimensions. A simple Fubini-type argument does not work since the foliations are typically only Hölder continuous (see [**93**]).

In [**6**], Barreira, Pesin, and Schmeling prove that every hyperbolic measure invariant under a $C^{1+\alpha}$ diffeomorphism of a smooth Riemannian manifold possesses asymptotically an *almost* local product structure, i.e., the measure of a small ball can be approximated by the product of the stable conditional measure of the stable component and the unstable conditional measure of the unstable component, up to small exponential error (see Theorem 3.5.5). This property of hyperbolic measures is the key ingredient in their proof of the long-standing Eckmann-Ruelle conjecture in dimension theory of smooth dynamical systems: the pointwise dimension of every hyperbolic measure invariant under a $C^{1+\alpha}$ diffeomorphism exists almost everywhere. This implies that most dimension-type characteristics of the measure (including the Hausdorff dimension, box dimension, and information dimension) coincide, and provides a rigorous mathematical justification of the concept of fractal dimension for hyperbolic measures.

In [**116**], Young proved this result in dimension two. Moreover, she found the following remarkable formula (see section 1.1.2): for almost every point x

$$\underline{d}_\mu(x) = \overline{d}_\mu(x) = h_\mu(f)\left(\frac{1}{\chi_1} - \frac{1}{\chi_2}\right),$$

where $h_\mu(f)$ is the measure theoretic entropy of f and $\chi_1 > 0 > \chi_2$ are the Lyapunov exponents of μ.

In [**54**], Ledrappier established the existence of the pointwise dimension for arbitrary *SRB-measures.* In [**85**], Pesin and Yue extended Ledrappier's approach and proved the existence of pointwise dimension for hyperbolic measures satisfying the so-called *semi-local product structure.* This class includes all equilibrium measures on locally maximal hyperbolic sets.

The main result in [**6**] is the following theorem.

THEOREM 3.5.5. *Let f be a $C^{1+\alpha}$ diffeomorphism on a smooth Riemannian manifold M without boundary, and μ an f-invariant compactly supported ergodic Borel probability measure. If μ is hyperbolic, then the following properties hold:*

- *for every $\delta > 0$, there exist a set $\Lambda \subset M$ with $\mu(\Lambda) > 1 - \delta$ such that for every $x \in \Lambda$ and every sufficiently small r (depending on x), one has that*

$$\begin{aligned} r^\delta \mu_x^s\left(B^s(x,r)\right)\mu_x^u\left(B^u(x,r)\right) &\le \mu\left(B(x,r)\right) \\ &\le r^{-\delta}\mu_x^s\left(B^s(x,r)\right)\mu_x^u(B^u(x,r)); \end{aligned}$$

- *For the measure μ the pointwise dimension exists almost everywhere and equals to the sum of the stable and unstable pointwise dimensions, i.e.,*
$$\underline{d} = \overline{d} = d^s + d^u.$$

Neither of the hypotheses of this theorem can be omitted. Ledrappier and Misiurewicz [**55**] constructed an example of a smooth map of a circle, preserving an ergodic measure with zero Lyapunov exponent, for which the pointwise dimension is not constant almost everywhere. In [**80**], Pesin and Weiss present an example of a Hölder homeomorphism with Hölder constant arbitrarily close to one, where the pointwise dimension for the unique measure of maximal entropy does not exist almost everywhere. There is also a one–dimensional example by Cutler [**23**]. See also [**52**].

Statement 1 of the theorem establishes an interesting new property of arbitrary hyperbolic measures. Loosely speaking, it means that every hyperbolic invariant measure splits into a product of its stable and unstable part with an exponentially small error.

3.6. Endomorphisms

After reading the previous sections, one might get the impression that the dimension theory of diffeomorphisms with a hyperbolic measure is well understood. But how close to the *real world* are these systems? Often partial differential equations (PDEs) generate only semi-flows, and the corresponding dynamical system is a priori noninvertible. Moreover, after somehow reducing the PDE to a system of ODEs, the restriction to Poincaré sections sometimes introduces singularities. For such a dynamical system with singularities, we are far from the setting of diffeomorphisms of manifolds. The first problem is to find physically meaningful measures (these measures are frequently called *SBR-measures*).

In this section we consider noninvertible maps of a manifold which may have singularities. We are not able to apply the theory of diffeomorphisms directly since unstable manifolds need not exist (each different preimage path gives a different unstable set). Nevertheless, we use an approach which allows us to apply the theory for diffeomorphisms.

REMARK 3.6.1. *The results in the section are fairly general (see* [**98**]*). For simplicity, we illustrate the theory through one example.*

3.6.1. The Belykh map. The (three-parameter) Belykh family contains both invertible and noninvertible maps. The subcollection of invertible maps in this family were introduced and studied by Belykh in [**11**] as a simple model for phase synchronization. Belykh was attempting to model the Poincaré map of a particular system of differential equations. In [**76**], Pesin studied the ergodic properties of these invertible maps, and these maps served as a guiding example in the investigations of Sataev in [**92**] and later in q[**1**].

The noninvertible maps in this family exhibit many properties of the maps in the class introduced in [**98**]. Alexander and Yorke [**2**] studied special (completely) noninvertible maps in this family and called these Fat Baker's transformations. The family of Belykh maps contains maps which have different *degrees* of noninvertibility. It appears that this family generally manifests most of the properties of generic parametrized families of hyperbolic piecewise-smooth maps.

Let $\mathbf{Q} = [-1,1] \times [-1,1]$ and $f = f_{\lambda,\gamma,k} \colon Q \to Q$ defined by

$$f(x,y) = \begin{cases} (\lambda x + 1 - \lambda, \gamma y + 1 - \gamma) & y > kx \\ (\lambda x + \lambda - 1, \gamma y + \gamma - 1) & y < kx \end{cases}$$

A representation of a noninvertible Belykh map is shown in Figure 15.

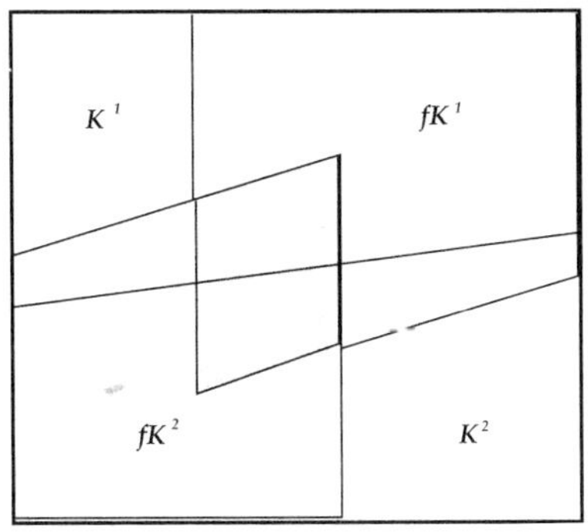

FIGURE 15.

We define a three-dimensional lift of f to $\mathbf{C} = \mathbf{Q} \times [0,1]$ by

$$F(x,y,z) := (f(x,y), \tau z + (1-\tau)(i-1)),$$

whenever $(x,y) \in K^i$ (see Figure 16).

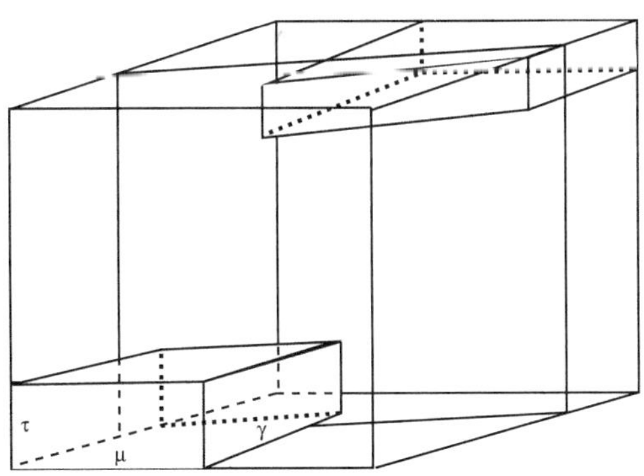

FIGURE 16.

One can easily check that the *lifted* map F is invertible, and thus may apply the dimension theory of diffeomorphisms to study F. The singularities of F are not hard to handle. Moreover, the lifted system has the invariant coordinate system $U^1(r,s,t) = \{x = r,\ y = s\}$, $U^2(r,s,t) = \{y = s,\ z = t\}$, $U^u(r,s,t) = \{x = r,\ z = t\}$ and

$$\chi_1 = \log \tau, \quad \chi_2 = \log \lambda, \quad \chi_u = \log \gamma.$$

In [**98**], the following theorem is proved.

THEOREM 3.6.2 (Schmeling-Troubetzkoy). *There exist special invariant measures (SBR measures) μ_{SBR} and $\hat{\mu}_{SBR}$ for f and F, respectively. The measure $\mu_{SBR} = \Pi_z^*(\hat{\mu}_{SBR})$ and*

$$d^u_{\hat{\mu}_{SBR}} = 1,$$

for $\hat{\mu}_{SBR}$-almost every x.

This theorem might serve as a starting point for a general dimension theory of endomorphisms.

Consider a ball $B_n := B(x, y; e^{(\chi_1 - \chi_2)n})$ contained in $\mathbf{Q}$. Then

$$\mu_{\text{SBR}}(B_n) = \hat{\mu}_{\text{SBR}}(P_n),$$

where $P_n = B_n \times [0, 1]$. The following theorem implies the existence of a type of pointwise dimension for the measure μ_{SBR}, where the usual Euclidean balls are replaced by parallelepipeds.

THEOREM 3.6.3 (Schmeling-Troubetzkoy). *The limit*

$$\lim_{n\to\infty} \frac{\log \hat{\mu}_{SBR}(P_n)}{n}$$

exists $\hat{\mu}_{SBR}$-almost everywhere.

3.6.2. Dimension and invertibility. In this section we illustrate how some dimension properties of a system reflect the fact that it is *almost* invertible. We begin with a definition.

DEFINITION 3.6.4. *A system is* almost surely invertible *if it is invertible on a set of full measure. This implies that a full measure set of points has unique forward and backward trajectories.*

We recall Young's Formula for surface diffeomorphisms

$$\dim_H \mu = h_\mu \left(\frac{1}{\chi_u} - \frac{1}{\chi_s} \right).$$

If our (two-dimensional) system is indeed invertible on a set of full measure we can apply Young's formula to the SBR-measure and obtain

$$\dim_H \mu_{\text{SBR}} = 1 - \frac{\log \gamma}{\log \lambda} = \dim_L \mu_{\text{SBR}}.$$

In [**98**], Schmeling and Troubetzkoy proved the reverse implication:

THEOREM 3.6.5 (Schmeling-Troubetzkoy). *A Belykh map is almost surely invertible (with respect to μ_{SBR}) if and only if Young's formula holds.*

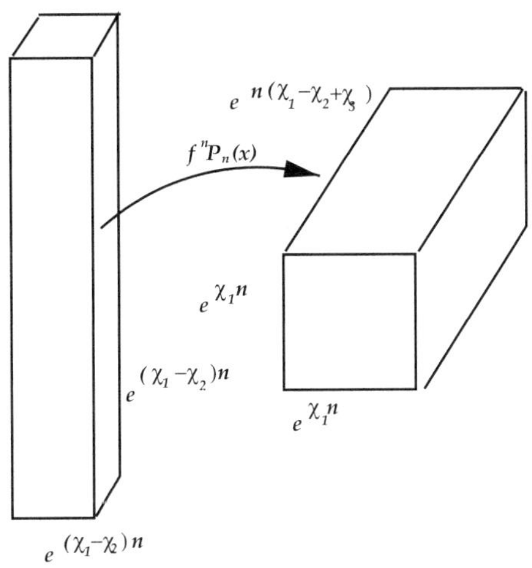

FIGURE 17.

3.6.2.1. *Sketch of proof.* Let us estimate the measure of the ball B_n. If we assume the existence of the pointwise dimension $d = \dim_{\mu_{\rm SBR}}$ at $\mu_{\rm SBR}$-almost every point we can estimate (see Figure 17):

$$\begin{aligned} e^{(\chi_1-\chi_2)nd} &\sim \mu_{\rm SBR}(B_n) \sim \hat{\mu}_{\rm SBR}(P_n) \\ &\sim \hat{\mu}_{\rm SBR}(f^n P_n) \\ &\sim \hat{\mu}^s_{\rm SBR}(f^n P_n) \cdot \hat{\mu}^u_{\rm SBR}(f^n P_n) \\ &\sim e^{\chi_1 n d^s_{\mu_{\rm SBR}}} \cdot e^{(\chi_1-\chi_2+\chi_u)nd^u_{\mu_{\rm SBR}}}, \end{aligned}$$

where d^s and d^u denote the stable and unstable dimension, respectively. We note that the *stable face* of $f^n P_n$ is nearly a ball. By the Ledrappier-Young formula we have that

$$h^u_{\mu_{\rm SBR}} = \chi_u = -\chi_1 d^1_{\mu_{\rm SBR}} - \chi_2 d^s_{\mu_{\rm SBR}} + \chi_2 d^1_{\mu_{\rm SBR}}$$

and

$$d^s_{\mu_{\rm SBR}} = \frac{\chi_2 - \chi_1}{\chi_2} d^1_{\mu_{\rm SBR}} - \frac{\chi_u}{\chi_2}.$$

We obtain that

$$\begin{aligned} (\chi - \chi_2)d &= \chi_1 d^s_{\mu_{\rm SBR}} + \chi_1 - \chi_2 + \chi_u \\ &= \frac{\chi_1}{\chi_u}(\chi_2 - \chi_1)d^1_{\mu_{\rm SBR}} - \frac{\chi_1}{\chi_u}\chi_u + \chi_1 - \chi_2 + \chi_u. \end{aligned}$$

It follows that

$$\begin{aligned}
d &= -\frac{\chi_1}{\chi_2}\left(d^1_{\mu_{\mathrm{SBR}}} - \frac{\chi_u}{\chi_1 - \chi_2}\right) + 1 + \frac{\chi_u}{\chi_1 - \chi_2} \\
&= 1 - \frac{\chi_u}{\chi_2} + \frac{h^1_{\mu_{\mathrm{SBR}}}}{\chi_2} \\
&= 1 - \frac{\log \gamma}{\log \lambda} + \frac{h^1_{\mu_{\mathrm{SBR}}}}{\log \lambda}.
\end{aligned}$$

We conclude that Young's formula holds if and only if $h^1_{\mu_{\mathrm{SBR}}} = 0$. But this means that the σ-algebra generated by the partition U^1 is completely determined by its future at almost every point. Since the partition U^1 generates the preimages, the assertion follows.

3.6.2.2. *Application to the Belykh map.* A natural question is how the Hausdorff dimension of the SBR-measure of a Belykh map depends on the parameters. By the above criterion this is connected to the almost sure invertibility of the map. In [**95**], the following theorem is proved.

THEOREM 3.6.6. *Consider the family of Belykh maps.*

- *If $\lambda \cdot \gamma^2 < 1$, then $f_{\lambda,\gamma,k}$ is fully invertible on the attractor for Lebesgue almost every (λ, γ, k). Hence,*
$$\dim_H(\mu_{SBR}) = 1 - \frac{\log \gamma}{\log \lambda}.$$
- *If $\lambda \cdot \gamma < 1$, then $f_{\lambda,\gamma,k}$ is almost surely invertible on the attractor for Lebesgue almost every (λ, γ, k). Hence,*
$$\dim_H(\mu_{SBR}) = 1 - \frac{\log \gamma}{\log \lambda}.$$
- *If $\lambda \cdot \gamma > 1$, then for Lebesgue almost every (λ, γ, k) one has that*
$$\dim_H(\mu_{SBR}) = 2 < \dim_L(\mu_{SBR}).$$

REMARK 3.6.7. *The above theorem shows that the Lyapunov dimension equals the Hausdorff dimension of the SBR-measure for almost all Belykh maps with $\lambda \cdot \gamma < 1$. This is the well-known Kaplan-Yorke conjecture for the special case of the Belykh attractor.*

The dimension theory of dynamical systems is far from being complete. Although there is now a general theory of the dimension of hyperbolic invariant measures, the corresponding theory for invariant sets is completely open. There are also many difficulties extending the dimension theory to noninvertible dynamical systems. It is also especially difficult to apply any general theory to concrete examples, for instance, to actually compute dimension. We emphasize that while the general theory may ensure the existence of the dimension, it does not provide any hint how to calculate it.

Part 4. Appendix

Appendix A. Conformal repellers

A smooth map g is **conformal** if $Dg_x = a(x) \cdot Isom_x$, where $Isom_x$ denotes an isometry of the tangent space T_xM. A smooth conformal map g is called an **expanding map** if $|a(x)| > 1$ for all points x. The repeller J for a conformal expanding map g is called a **conformal repeller**.

The following are several examples of conformal repellers.

1) *Rational Maps.* Let $R\colon \widehat{\mathbb{C}} \to \widehat{\mathbb{C}}$ be a rational map of degree ≥ 2, where $\widehat{\mathbb{C}}$ denotes the Riemann sphere. The map R, being holomorphic, is clearly conformal. The *Julia set* J of R is the closure of the set of repelling periodic points of R (recall that a periodic point p of period m is repelling if $|(R^m)'(p)| > 1$). One says that R is *hyperbolic* (or that the Julia set is *hyperbolic*) if the map R is expanding on J. It is known that the map $z \mapsto z^2 + c$ is hyperbolic provided $|c| < \frac{1}{4}$. Figure 18 illustrates the Julia set of a hyperbolic rational map. It is conjectured that there is a dense set of hyperbolic quadratic maps.

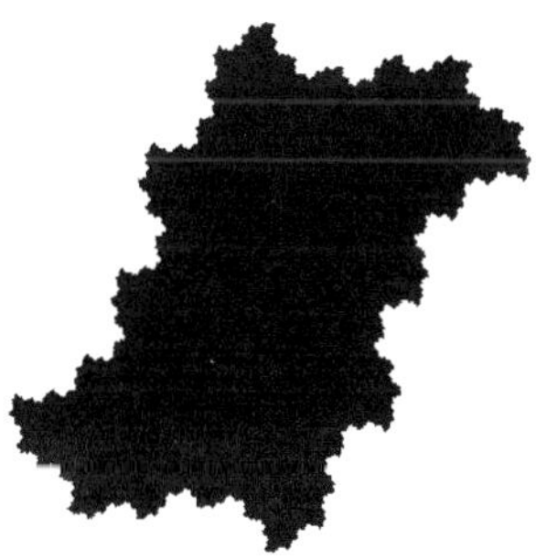

FIGURE 18. The Boundary of The *Black Spot* is the Julia Set for the Polynomial $z^2 + c$ with $c = -\frac{1}{10} + \frac{1}{5}i$

2) *One-Dimensional Markov Maps.* Let $I = [0, 1]$ and f be a **Markov map**. This means that there exists a finite family $I_1, I_2, \dots I_p \subset I$ of disjoint closed intervals such that

- for every $j \in \{1, \dots, p\}$, there is a subset $K = K(j)$ of indices with $f(I_j) = \cup_{k \in K} I_k \pmod 0$,
- for every $x \in \cup_j \overset{\circ}{I}_j$, the derivative of f exists and satisfies $|\, f'(x) \,| \geq \alpha$ for some fixed $\alpha > 0$,
- there exists $\lambda > 1$ and $n_0 > 0$ such that if $f^m(x) \in \cup_j \overset{\circ}{I}_j$, for all $0 \leq m \leq n_0 - 1$ then $|(f^{n_0})'(x)| \geq \lambda$.

Let $J = \{x \in I \mid f^n(x) \in \cup_j \overset{\circ}{I}_j \text{ for all } n \in \mathbb{N}\}$. The set J is a **repeller** for the map f. It is conformal because the domain of f is one-dimensional (see Figure 19)).

3) *Conformal Toral Endomorphism.* This is a map of a multi-dimensional torus defined by a diagonal matrix $(k, \dots, k)$, where k is an integer and $|k| > 1$.

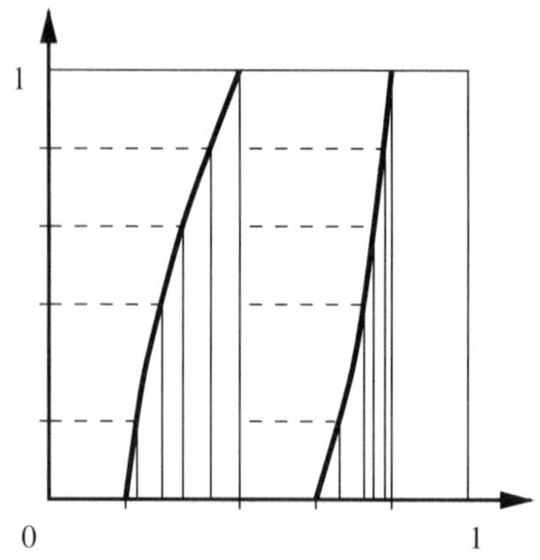

FIGURE 19. A One-dimensional Markov Map

Appendix B. Elements of thermodynamic formalism

This section contains some essential definitions and facts from symbolic dynamics and thermodynamic formalism. For details consult the excellent references [**16, 89, 77**]. Let X denote a compact metric space and let $C(X)$ denote the space of real valued continuous functions on X.

- Let $g\colon X \to X$ be a continuous map. We define the **pressure** $P\colon C(X) \to \mathbb{R}$ by
$$P_X(\varphi) = \sup_{\mu \in M(X)} \left(h_\mu(g) + \int_X \varphi \, d\mu \right),$$
where $M(X)$ denotes the set of shift-invariant probability measures on X and $h_\mu(f)$ denotes the measure theoretic entropy of the map g with respect to the measure μ. A Borel probability measure $\mu = \mu_\varphi$ on X is called an **equilibrium measure** for the potential $\varphi \in C(X)$ if
$$P_X(\varphi) = h_\mu(g) + \int_X \varphi \, d\mu.$$
For any continuous function, an equilibrium measure exists, but may not be unique. It is unique if the function is Hölder continuous. For hyperbolic maps, e.g., conformal repellers or Axiom-A diffeomorphisms, equilibrium measures are often called Gibbs measures since the pullback of an equilibrium measure under a coding map (defined using a Markov partition) is Gibbs.
- The pressure function $P\colon C^\alpha(\Sigma_A^+, \mathbb{R}) \to \mathbb{R}$ is real analytic, and there exist formulas for the first two derivatives.
- Let $\varphi, \psi \in C^\alpha(\Sigma_A^+, \mathbb{R})$. The map $\mathbb{R} \to \mathbb{R}$ defined by $t \to P(\psi + t\varphi)$ is convex. It is strictly convex unless φ is cohomologous to a constant, i.e., there exists $C > 0$ and $g \in C^\alpha(\Sigma_A^+, \mathbb{R})$ such that $\varphi(x) = g(\sigma x) - g(x) + C$.
- Let $\varphi \in C(\Sigma_A^+)$. A Borel probability measure $\mu = \mu_\varphi$ on Σ_A^+ is called a **Gibbs measure** for the potential φ if there exist constants $D_1, D_2 > 0$

such that

$$D_1 \leq \frac{\mu\{y \mid y_i = x_i, \ i = 0, \cdots, n-1\}}{\exp(-nP(\varphi) + \sum_{k=0}^{n-1} \varphi(\sigma^k x))} \leq D_2$$

for all $x = (x_1 x_2 \cdots) \in \Sigma_A^+$ and $n \geq 0$. For subshifts of finite type, Gibbs measures exist for any Hölder continuous potential φ, are unique, and coincide with the equilibrium measure for φ.

- A Gibbs measure μ is *diametrically regular* [**82**], i.e., for each $A > 1$ there exists $K > 0$ such that for any sufficiently small $r > 0$ and every x we have

$$\mu(B(x, Ar)) \leq K\mu(B(x, r)).$$

Let X be a horseshoe. Then it can be coded as a full shift Σ on n symbols (see next section). Let the potential $\varphi\colon \Sigma \to \mathbb{R}$ depend only on the first coordinate. In this case the pressure has a very simple expression

$$P(\varphi) = \log\left(\sum_{i=1}^{n} e^{\varphi([i])}\right) \tag{10}$$

A. Construction of Gibbs measures. Gibbs measures can be constructed as eigenfunctions of a certain linear operator usually called the Ruelle Transfer Operator. Let $\sigma\colon \Sigma_A^+ \to \Sigma_A^+$ be a mixing subshift of finite type, and let $\varphi \in C^0(\Sigma_A^+, \mathbb{R})$. Define the linear operator $L_\varphi\colon C^0(\Sigma_A^+, \mathbb{R}) \to C^0(\Sigma_A^+, \mathbb{R})$ by

$$L_\varphi(g)(x) \equiv \sum_{\sigma y = x} g(y) \exp \varphi(y).$$

Ruelle's Perron Frobenius theorem yields an isolated maximal eigenvalue $\exp(P(\varphi))$, $h \in C^0(\Sigma_A^+, \mathbb{R})$, and a Borel probability measure ν such that

$$L_\varphi(h) = \exp(P(\varphi))h$$

$$L_\varphi^*(\nu) = \exp(P(\varphi))\nu.$$

The measure $\mu_\varphi \equiv h\,d\nu$ is the unique Gibbs measure for the potential φ.

The smoothness of pressure follows from this construction, since standard perturbation theory implies that isolated maximal eigenvalues vary smoothly under smooth perturbations.

Appendix C. Markov partitions and the Jacobian estimate

A. Expanding maps. For simplicity we assume that the expanding map g on J is topologically mixing. Expanding maps have Markov partitions [**89, 90**] consisting of partition elements called **rectangles**, $\{R_1, \ldots, R_p\}$ of (arbitrarily small) diameter δ, such that

- each rectangle R is the closure of its interior $\overset{\circ}{R}$,
- $J = \cup_i R_i$,
- $\overset{\circ}{R}_i \cap \overset{\circ}{R}_j = \varnothing$ for $i \neq j$,
- each $g(R_i)$ is a union of rectangles R_j.

A Markov partition $\mathcal{R} = \{R_1, \ldots, R_p\}$ generates a symbolic model of the repeller by a subshift of finite type (Σ_A^+, σ), where $A = (a_{ij})$ is the transfer matrix of the Markov partition, i.e., $a_{ij} = 1$ if $\overset{\circ}{R}_i \cap g^{-1}(\overset{\circ}{R}_j) \neq \varnothing$ and $a_{ij} = 0$ otherwise.

The union over all iterates of the boundaries of the rectangles is a nowhere dense set. However, Bowen showed that this set is typically a *fractal-like* set and can have positive Hausdorff dimension.

CLAIM A.1 (Jacobian estimate for Markov expanding Map). *Let $f\colon J \to J$ be a Markov expanding map. Given $\psi \in C^\gamma(J, \mathbb{R})$ such that $\psi(x) \geq C > 0$ for all x. There exist $L_1, L_2 > 0$ (depending on ψ) such that for any $n-$tuple $(i_1 \ldots i_n)$ and any $x, y \in R_{i_1 \ldots i_n} = R_{i_1} \cap f^{-1}(R_{i_1}) \cap \cdots \cap f^{-n}(R_{i_n})$*

$$L_1 \leq \frac{\psi(f^k(x))}{\psi(f^k(y))} \leq L_2.$$

The following is the special case when $\psi(x) = Jac(x)$.

There exist $L_1, L_2 > 0$ such that for any $n-$tuple $(i_1 \ldots i_n)$ and any $x, y \in R_{i_1 \ldots i_n}$

$$L_1 \leq \frac{Jac(f^k(x))}{Jac(f^k(y))} \leq L_2.$$

B. Axiom-A diffeomorphisms. Axiom-A surface diffeomorphisms have Markov partitions [**89, 90**] consisting of partition elements usually called **rectangles**, $\{R_1, \ldots, R_p\}$, of (arbitrarily small) diameter δ, such that

- $\overset{\circ}{R}_i \cap \overset{\circ}{R}_j = \varnothing$ unless $i = j$;
- for each $x \in \overset{\circ}{R}_i \cap f^{-1}(\overset{\circ}{R}_j)$ we have

$$f(W^s_{\text{loc}}(x) \cap R_i) \subset W^s_{\text{loc}}(f(x)) \cap R_j,$$
$$f(W^u_{\text{loc}}(x) \cap R_i) \supset W^u_{\text{loc}}(f(x)) \cap R_j.$$

References

1. V. S. Afraimovich, N. I. Chernov and E. A. Sataev, *Statistical Properties o f 2D Generalized Hyperbolic Attractors*, Chaos, Vol 5 (1995), pp. 238–252
2. J. Alexander and J. Yorke, *Fat Baker's Transformations*, Erg. Th. Dyn. Syst., Vol 4 (1984), pp. 1–23
3. E. Artin, *Ein Mechanisches System mit quasi-ergodischen Bahnen*, Collected papers, Addison Wesley, 1965, pp. 499–501
4. L. Barreira, *Cantor Sets with Complicated Geometry and Modeled By General Symbolic Dynamics*, Random and Comp. Dyn., Vol. 3 (1995), pp. 213–239
5. L. Barreira, *A Non-additive Thermodynamic Formalism and Applications to Dimension Theory of Hyperbolic Dynamical Systems*, Erg. Th. Dyn. Syst., Vol. 16 (1996), pp. 871–927
6. L. Barreira, Y. Pesin, and J. Schmeling, *Dimension and Product Structure of Hyperbolic Measures*, Annals Math. Vol. 149 (1999), pp. 755–783
7. L. Barreira, Y. Pesin and J. Schmeling, *On A General Concept of Multifractal Rigidity: Multifractal Spectra for Dimensions, Entropies, and Lyapunov Exponents. Multifractal Rigidity.*, Chaos, Vol 7 (1997), pp. 27–38
8. L. Barreira and B. Saussol, *Hausdorff Dimension od Measure Via Poincaé Recurrence*, IST Preprint (2000)
9. L. Barreira and J. Schmeling, *Sets of "Non-typical" Points Have Full Topological Entropy and Full Hausdorff Dimension*, Israel Journal of Mathematics, Vol. 116 (2000), pp. 29–70
10. L. Barreira, B. Saussol, and J. Schmeling, *Higher–dimensional Multifractal Analysis*, in preparation

11. V. P. Belykh, *Models of Discrete Systems of Phase Synchronization*, in Systems of Phase Synchronization, Eds. V. V. Shakhildyan and L. N. Belynshina, Radio i Svyaz, Moscow 1982, pp. 61–176
12. P. Billingsley, *Ergodic Theory and Information*, Krieger 1978
13. M. Blinchevskaya and Yu Ilyashenko, *Estimate for the Entropy Dimension of The Maximal Attractor for $k-$Contracting Systems In An Infinite-Dimensional Space*, preprint 1998
14. M. Boshernitzan, *Quantitative Recurrence*, Inv. Math., Vol. 113 (1993), pp. 617–631
15. H.-G. Bothe, *The Hausdorff Dimension of Certain Solenoids*, Erg. Th. Dyn. Syst., Vol. 15 (1995), pp. 449–474
16. R. Bowen, *Hausdorff Dimension of Quasi-circles*, Publ. Math. IHES, Vol 50 (1979), pp. 11–25
17. M. Casdagli, T. Sauer, and J. Yorke, *Embedology*, J. Stat. Phys., Vol 65 (1991), pp. 589–616
18. Y. Cheung, *Optimal Hausdorff Dimension for the Set of Nonergodic Directions*, 2000 thesis
19. S. Ciliberto, J. P. Eckmann, S. Kamphorst, and D. Ruelle, *Liapunov Exponents from Times*, Phys. Rev. A, Vol 34 (1971)
20. P. Collet, J. L. Lebowitz, and A. Porzio, *The Dimension Spectrum of Some Dynamical Systems*, J. Stat. Physics, Vol 47 (1987), pp. 609–644
21. P. Constantin and C. Foias, *Navier-Stokes Equations*, Chicago U. Press 1988
22. J. Cruchfield, D. Farmer, N. Packard, and R. Shaw, *Geometry from a Time Series*, Phys. Rev. Lett., Vol 45 (1980), pp. 712–
23. C. Cutler, *Connecting Ergodicity and Dimension in Dynamical Systems*, Ergod. Th. and Dynam. Systems, Vol 10 (1990), pp. 451–462
24. M. Ding, C. Grebogi, E. Ott and J. Yorke, *Estimating Correlation Dimension From a Chaotic Ttmes Series: When Does the Plateau Onset Occur?*, Physica D, Vol 69 (1993), pp. 404–424
25. D. Dolgopyat, *Bounded Orbits of Anosov Flows*, Duke J. Math., Vol 87 (1997), pp. 87–114
26. A. Douady and J. Oesterle, *Dimension de Hausdorff Des Attracteurs*, CRAS, Vol 290 (1980), pp. 1135–1138
27. J. P. Eckmann and D. Ruelle, *Ergodic Theory of Chaos and Strange Attractors*, Rev. Mod. Phys, Vol 57 (1985), pp. 617–656
28. J. P. Eckmann and D. Ruelle, *Fundamental Limitations for Estimating Dimensions and Lyapunov Exponents in Dynamical Systems*, Physica D, Vol 56 (1992), pp. 185–
29. H. G. Eggleston, *Sets of Fractional Dimension Which Occur in Some Problems of Number Theory*, Proc. Lond. Math. Soc., Vol 54 (1952), pp. 42–93
30. R. Ellis, *Large Deviations for a General Class of Random Vectors*, Ann. Prob., Vol 12 (1984), pp. 1–12
31. R. Ellis, *Entropy, Large Deviations, and Statistical Mechanics*, Springer Verlag 1985
32. K. Falconer, *Fractal Geometry, Mathematical Foundations and Applications*, Cambridge U. Press 1990
33. P. Frederickson, J. Kaplan, E. Yorke, and J. Yorke, *The Liapunov Dimension of Strange Attractors*, J. Differential Equations, Vol 49 (1983), pp. 185–207
34. O. Frostman, *Potential d'équilibre Et Capacité des Ensembles Avec Quelques Applications à la Théorie des Fonctions*, Meddel. Lunds Univ. Math. Sem., Vol 3 (1935), pp. 1–118
35. H. Furstenberg, *Intersections of Cantor Sets and Transversality of Semigroups I*, in: Problems in Analysis (Sympos. Salomon Bochner, Princeton Univ), Princeton Univ. Press 1970, pp. 41–59 F
36. H. Furstenberg, *Disjointness in Ergodic Theory, Minimal Sets, and a Problem in Diophantine Approximation*, Mathematical Systems Theory, Vol 1 (1967), pp. 1–49
37. D. Gatzouras and Y. Peres, *The Variational Principle for Hausdorff Dimension: A survey*, in Ergodic theory of $\mathbb{Z}^d$ actions, Proc. of the Warwick symposium, M. Pollicott (ed.) et al, Cambridge University Press, Lond. Math. Soc. Lect. Note Ser. 228 (1996), pp. 113–125
38. P. Grassberger, Phys. Lett., Vol 97A (1983), pp. 227
39. P. Grassberger, I. Procaccia, H. Hentschel, *On the Characterization of Chaotic Motions*, Lect. Notes Physics, Vol 179 (1983), pp. 212–221
40. T. Halsey, M. Jensen, L. Kadanoff, I. Procaccia and B. Shraiman, *Fractal Measures and Their Singularities: The Characterization of Strange Sets*, Physical Review A, Vol 33, N2 (1986), pp. 1141-1151

41. W. Harvey, *Discrete Groups and Automorphic Functions*, Proceedings of NATO Advanced Study Institute, 1977, Academic Press
42. B. Hasselblatt, *Regularity of the Anosov Splitting and of Horospheric Foliations*, Erg. Th. Dyn. Syst., Vol 14 (1994), pp. 645–666
43. D. Henley , *Continued Fraction Cantor Sets, Hausdorff Dimension, and Functional Analysis*, J. Number Theory, Vol 40 (1992), pp. 336–358
44. H. G. E. Hentschel and I. Procaccia, *The Infinite Number of Generalized Dimensions of Fractals and Strange Attractors*, Physica 8D (1983), pp. 435–444
45. B. Hunt, *Maximal Local Lyapunov Dimension Bounds The Box Dimension of Chaotic Attractors*, Nonlinearity, Vol 9 (1996), pp. 845–852
46. V. Jarnik, *Über die simultanen diophantischen Approximationen* Math. Zeitschr., Vol 33 (1931), pp. 505–543
47. V. Jarnik, *Über die simultanen diophantischen Approximationen* Math. Zeitschr., Vol 33 (1931), pp. 505–543
S. Kerckhoff, H. Masur and J. Smillie, *Ergodicity of Billiard Flows and Quadratic Differentials*, Ann. of Math, Vol 124 (1986), pp. 293–311
48. J. Kaplan and J. Yorke, *Chaotic Behavior of Multidimensional Difference Equations*, in Functional Differential Equations and Approximation of Fixed Points (Proc. Summer School and Conf., Univ. Bonn, Bonn, 1978), Lecture Notes in Math., 730, 204–227, Springer, Berlin, 1979
49. S. Katok, Fuchsian Groups, U. Chicago Press, 1992
50. M. Kesseböhmer, *Multifrakale und Asymptotiken grosser Deviationen*, Thesis U. Göttingen, 1999
51. D. Kleinbock and G. Margulis, *Flows on Homogeneous Spaces and Diophantine Approximation on Manifold*, Annals of Math., Vol 148 (1998), pp. 339–360
52. B. Kra and J. Schmeling, *Diophantine numbers, dimension and Denjoy maps*, in preparation
53. F. Ledrappier, *Some Relations Between Dimension and Lyapunov Exponents*, Comm. Math. Phys., Vol 81 (1981), pp. 229–238
54. F. Ledrappier, *Dimension of invariant measures*, Proceedings of the conference on ergodic theory and related topics II (Georgenthal, 1986), Teubner–texte Math, vol 94, 137–173
55. F. Ledrappier and M. Misiurewicz, *Dimension of Invariant Measures for Maps with Exponent Zero*, Ergod. Th. and Dynam. Systems, Vol 5 (1985), pp. 595–610
56. F. Ledrappier and L. S. Young, *The Metric Entropy of Diffeomorphisms. I Characterization of Measures Satisfying Pesin's Entropy Formula*, Annals Math., Vol 122 (1985), pp. 509–539
57. F. Ledrappier and L. S. Young, *The Metric Entropy of Diffeomorphisms. II Relations Between Entropy, Exponents and Dimension*, Annals Math., Vol 122 (1985), pp. 540–574
58. A. Lopes, *The Dimension Spectrum of the Maximal Measure*, SIAM J. Math. Analysis, Vol 20 (1989), pp. 1243–1254
59. R. Mãné, *The Hausdorff Dimension of Horseshoes of Surfaces*, Bol. Soc. Bras. Math., Vol 20 (1990), pp. 1–24
60. R. Mãné, *On the Dimension of Compact Invariant Sets for Certain Nonlinear Maps*, Lecture Notes in Mathematics, Vol 898 (1981), Springer Verlag
61. H. Masur, *Hausdorff Dimension of the Set of Nonergodic Foliations of a Quadratic Differential*, Duke Math. J., Vol 66 (1992), 387–442
62. H. McCluskey and A. Manning, *Hausdorff Dimension for Horseshoes*, Erg. Th. Dyn. Syst., Vol 3 (1983), pp. 251–260
63. M. Melián and D. Pestana, *Geodesic Excursions into Cusps in Finite-Volume Hyperbolic Manifolds*, Michigan Math. J., Vol 40 (1993), pp. 77–93
64. P. Moran, *Additive Functions of Intervals and Hausdorff Dimension*, Proceedings of Cambridge Philosophical Society, Vol 42 (1946), pp. 15–23
65. C. Moreira and J. Yoccoz, *Stable Intersections of Cantor Sets with Large Hausdorff Dimension*, preprint
66. J. Neunhäuserer, *An Analysis of Dimensional Theoretical Properties of Some Affine Dynamical Systems*, Thesis, Free University Berlin, 1999
67. P. Nicholls, *The Ergodic Theory of Discrete Groups*, Cambridge U. Press 1989
68. E. Ott, T. Sauer, and J. Yorke, *Coping with Chaos*, Wiley 1994
69. J. Palis and F. Takens, *Hyperbolicity and Sensitive Chaotic Dynamics At Homoclinic Bifurcations*, Cambridge U. Press 1993

70. J. Palis and F. Takens, *Hyperbolicity and The Creation of Homoclinic Orbits*, Annals of Math, Vol 125 (1987), pp. 337–374
71. J. Palis and M. Viana, *On the Continuity of Hausdorff Dimension and Limit Capacity for Horseshoes*, Lecture Notes in Math., Vol 1331 (1988), Springer Verlag
72. J. Palis and J-C. Yoccoz, *Homoclinic Tangencies for Hyperbolic Sets of Large Hausdorff Dimension*, Acta Math., Vol 172 (1994), pp. 91–136
73. Y. Pesin, *Dimension Theory In Dynamical Systems: Rigorous Results and Applications*, Cambridge U. Press 1997
74. Y. Pesin, *On Rigorous Mathematical Definition of Correlation Dimension and Generalized Spectrum for Dimensions*, Journal of Stat. Phys., Vol 7 (1993)
75. Y. Pesin,*Characteristic Exponents and Smooth Ergodic Theory*, Russian Math Surveys, 32 no 4 (1977), pp. 55–114
76. Y. Pesin, *Dynamical Systems Eith Generalized Hyperbolic Attractors: Hyperbolic, Ergodic and Topological Properties*, Erg. Th. Dyn. Syst., Vol 12 (1992), pp. 123–151
77. Y. Pesin and B. Pitskel', *Topological Pressure and the Variational Principle for Noncompact Sets*, Funct. Anal. Appl., Vol. 18 (1984), 307–318
78. Y. Pesin and V. Sadovskaya, *Multifractal analysis of conformal axiom A flows*, preprint
79. Y. Pesin and A. Tempelman, *Correlation Dimension of Measures Invariant Under Group Actions*, Random and Computational Dynamics, Vol 3:3 (1995), pp. 137–156
80. Y. Pesin and H. Weiss, *On The Dimension of Deterministic and Random Cantor-like Sets, Symbolic Dynamics, and The Eckmann-Ruelle Conjecture*, Comm. Math. Phys., Vol 182 (1996), pp. 105–153
81. Y. Pesin and H. Weiss, *On the Dimension of Deterministic and Random Cantor-like Sets*, MathResearch Letters, Vol 1 (1994), pp. 519–529
82. Y. Pesin and H. Weiss, *A Multifractal Analysis of Equilibrium Measures for Conformal Expanding Maps and Moran-like Geometric Constructions*, J. Stat. Phys., Vol 86 (1997), pp. 233–275
83. Y. Pesin and H. Weiss, *The Multifractal Analysis of Gibbs Measures: Motivation, Mathematical Foundation and Examples*, Chaos, Vol 7 (1997), pp. 89–106
84. Y. Pesin and H. Weiss, *Some Remarks on The Fine Structure of the Exceptional Set in the Birkhoff Ergodic Theorem for Subshifts of Finite Type and Applications*, preprint 1999
85. Y. Pesin and Ch. Yue, *The Hausdorff Dimension of Measures with Non–zero Lyapunov Exponents and Local Product Structure*, preprint
86. M. Pollicott and H. Weiss, *The Dimensions of Some Self Affine Limit Sets In The Plane and Hyperbolic Sets*, J. Stat. Phys., Vol 77 (1994), pp. 841–866
87. M. Pollicott and H. Weiss, *Multifractal Analysis for the Continued Fraction and Manneville-Pomeau Transformations and Applications to Diophantine Approximation*, Comm. Math. Phys., (2000)
88. F. Przytycki and M. Urbański, *On Hausdorff Dimension of Some Fractal Sets*, Studia Mathematica, Vol 93 (1989), pp. 155–186
89. D. Ruelle, *Thermodynamic Formalism*, Addison-Wesley 1978
90. D. Ruelle, *Repellers for Real Analytic Maps*, Erg. Th. Dyn. Syst., Vol 2 (1982), pp. 99–107
91. D. Ruelle, *An Inequality for the Entropy of Differentiable Maps*, Bol. Soc. Brasil. Mat., Vol 9 (1978), pp. 83–87
92. E. A. Sataev, *Invariant Measures for Hyperbolic Systems with Singularities*, Russ. Math. Surv., Vol 47 (1992), pp. 191–251
93. J. Schmeling, *Hölder Continuity of the Holonomy Maps for Hyperbolic basic Sets. II.*, Math. Nachr., Vol 170 (1994), pp. 211–225
94. J. Schmeling, *Symbolic Dynamics for Beta-shifts and Self-Normal Numbers*, Erg. Th. Dyn. Syst., Vol 17 (1997), pp. 675–694
95. J. Schmeling, *A Dimension Formula for Endomorphisms – the Belykh Family*, Erg. Th. Dyn. Syst., Vol 18 (1998), pp. 1283–1309
96. J. Schmeling, *On the Completeness of Multifractal Spectra*, Erg. Th. Dyn. Syst., Vol 19 (1999), pp. 1–22
97. J. Schmeling, *Entropy Preservation under Markov Codings*, J. Stat. Phys., to appear
98. J. Schmeling and S. Troubetzkoy, *Dimension and Invertibility of Hyperbolic Endomorphisms with Singularities*, Erg. Th. Dyn. Syst., Vol 18 (1998), pp. 1257–1282

99. J. Schmeling and R. Winkler, *Typical dimension of the graph of certain functions*, Monatshefte für Mathematik, Vol. 119 (1995), pp. 303–320
100. W. Schmidt, *Badly Approximable Systems of Linear Forms*, J. Number Theory, Vol 1 (1969), pp. 139–154
101. F. Schweiger, *Ergodic Theory of Fibred Systems and Metric Number Theory*, Oxford Science Press, 1995
102. C. Series, *The Modular Surface and Continued Fractions*, J. London Math. Soc., Vol 31 (1985), pp. 69–80
103. K. Simon, *The Hausdorff Dimension of the General Smale–Williams Solenoid with Different Contraction Coefficients*, Proc. Am. Math. Soc., Vol. 125 (1997), pp. 1221–1228
104. D. Simpelaere, *Dimension Spectrum of Axiom-A Diffeomorphisms. II. Gibbs Measures*, Journal of Stat. Phys., Vol 76 (1994), pp. 1359–1375
105. B. Solomyak, *On the Random Series* $\sum \pm\lambda^i$ *(An Erdös Problem)*, Ann. Mathh., Vol. 142 (1995)
106. B. Stratmann, *Fractal Dimensions for Jarnik Limit Sets of Geometrically Finite Kleinian Groups; The Semi-Classical Approach*, Ark. Mat., Vol 33 (1995), pp. 385–403
107. D. Sullivan, *Discrete Conformal Groups and Measurable Dynamics*, Bull. AMS, Vol 6 (1982), pp. 57–73
108. D. Sullivan, *The Density At Infinity of A Discrete Group of Hyperbolic Motions*, Bull. AMS, Vol 50 (1979), pp. 171–202
109. D. Sullivan, *Disjoint Spheres, Approximation by Imaginary Quadratic Numbers and the Logarithmic Law for Geodesics*, Acta. Math., Vol 149 (1982), pp. 215–237
110. F. Takens, *Detecting Strange Attractors in Turbulence*, Lecture Notes in Mathematics, Vol 898 (1981), Springer Verlag
111. F. Takens, *Limit Capacity and Hausdorff Dimension of Dynamically Defined Cantor Sets*, Lecture Notes in Math., Vol 1331 (1988), Springer Verlag
112. M. Urbanski, *The Hausdorff Dimension of the Set of Points with Nondense Orbit Under a Hyperbolic Dynamical System*, Nonlinearity, Vol 4 (1991), pp. 385–397
113. P. Walters, *Introduction to Ergodic Theory*, Springer Verlag 1982
114. H. Weiss, *The Lyapunov Spectrum of Equilibrium Measures for Conformal Expanding Maps and Axiom-A Surface Diffeomorphisms*, J. Stat. Phys., (1998), to appear
115. H. Weiss, *Some Variational Formulas for Hausdorff Dimension, Topological Entropy, and SRB Entropy for Hyperbolic Dynamical System*, JStatPhys., Vol 69 (1992), pp. 879–886
116. L. S. Young, *Dimension, Entropy, and Lyapunov Exponents*, Erg. Th. Dyn. Syt., Vol 2 (1982), pp. 109–124

Fachbereich Mathematik und Informatik, Freie Universität Berlin, Arnimallee 2-6, D–14195 Berlin, Germany
E-mail address: schmeling@math.fu-berlin.de

Center for Dynamical Systems, The Pennsylvania State University, University Park, PA 16802
E-mail address: weiss@math.psu.edu
URL: http://www.math.psu.edu/~weiss/

Proceedings of Symposia in Pure Mathematics
Volume **69**, 2001

Collet-Eckmann Condition in One-Dimensional Dynamics

Grzegorz Świątek

ABSTRACT. Results on smooth maps in one real or complex dimension which satisfy the Collet-Eckmann condition are surveyed. Some questions and remarks concerning the future development of the area are made.

1. Introduction

The purpose of this survey is to show the motivation for the study of Collet-Eckmann maps, the current condition of the theory, and a few possible directions in which it might continue to develop. Not all results that are going to be mentioned will be stated, or stated in their strongest form. In such cases we provide references to original articles.

Acknowledgements. Discussions with Jacek Graczyk were of great help with this work. I also thank the referee for prompt and useful comments.

Notation.

DEFINITION 1.1. *For k real and greater than 1 and m a positive integer, let End^k_m denote the set of C^k maps from an interval I into itself with exactly m critical points, at least one of which is a local extremum and all of which belong to the interior of I. In a neighborhood U_c of each critical point c, f can be represented so that*

$$|f(x) - f(c)| = |h(x)|^\ell$$

where h is a C^k diffeomorphism of U_c onto its image and satisfies $h(c) = 0$ while $\ell > 1$ is the order *of c and not necessarily an integer.*

Finally, it is assumed that $f(\partial I) \subset \partial I$.

We will write End^k for $\bigcup_{m=1}^{\infty} \mathrm{End}^k_m$.

DEFINITION 1.2. *We say that a C^2 map f of an interval into $\mathbf{R}$ satisfies the S-condition if $\frac{1}{\sqrt{|f'(x)|}}$ is a convex, not necessarily strictly convex, function on each interval on which $f'(x) \neq 0$.*

The S-condition is equivalent to the non-positive weak Schwarzian derivative of f. It is very well known, see [**15**] for an overview. Accordingly, we consider the set S^k_m, $k \geq 2$, which is the subset of End^k_m consisting of maps which satisfy

---

Partially supported by NSF grant DMS-0072312.

the S-condition. The S_m^k is not dense in End_m^k, but one directly verifies that the monomial $|x|^\ell$ satisfies the S-condition, which causes it to hold near critical points of maps of End^k provided that $\ell \le k-1$ at such points.

1.1. Non-uniform hyperbolic theory. The non-uniformly hyperbolic theory of diffeomorphisms developed in the course of the 1970's takes the existence of Lyapunov exponents with non-zero real parts as its starting point, from which the existence of a non-uniform absolutely continuous stable foliation is derived. These concepts do not directly apply in the one-dimensional context, where there is no dimension for stable local stable manifolds to exist in, and which are rarely interesting as diffeomorphisms. Nevertheless, the underlying concept of non-uniform hyperbolicity can be applied resulting for example in the following theorem.

Ledrappier's theorem. For $f \in \mathrm{End}^k$, $k > 1$, we say that $x \in I$ is positively regular if the following three conditions are satisfied:

- the sequence $\frac{1}{n}\sum_{i=0}^{n-1}\delta_{f^i(x)}$ weakly converges towards and ergodic measure μ_x,
- the Lyapunov exponent
$$\lambda_x := \lim_{n\to\infty}\frac{\log|(f^n)'(x)|}{n}$$
exists and is positive,
- $$\lambda_x = \int \log|f'(y)|\, d\mu_x(y)\,.$$

THEOREM 1.1. *If $f \in \mathrm{End}^k$, then the set of positively regular points has positive Lebesgue measure if and only if f admits a probabilistic invariant measure of positive entropy which is absolutely continuous with respect to the Lebesgue measure.*

In fact, in all results concerning real one-dimensional dynamics "absolutely continuous" means with respect to the Lebesgue measure.

A stronger result under much stronger hypotheses is due to Keller in [**10**]:

THEOREM 1.2. *Define the upper Lyapunov exponent at x:*
$$\overline{\lambda}(x) := \limsup_{n\to\infty}\frac{\log|(f^n)'(x)|}{n}\,.$$
If $f \in S_1^3$, then $\overline{\lambda}(x)$ is constant almost everywhere with respect to the Lebesgue measure and this constant is positive if and only if f has an absolutely continuous probabilistic invariant measure of positive entropy.

Both results indicate that in one dimension positive Lyapunov exponents of the map manifest themselves in the existence of a probabilistic absolutely continuous invariant measure (pacim for short).

Expansion by default. A particular feature of a one-dimensional map is that it is expanding in the absence of certain well-defined and obvious obstacles. Consider the following theorem which is taken from the book [**15**], where it is attributed to Mañe, see [**14**], who in turn generalized a similar theorem of Misiurewicz, see [**16**], for S_1^3 maps.

THEOREM 1.3. *Let $f \in End^2$ and $K \subset I$ be compact, not necessarily forward invariant. Assume however that K is disjoint from the set of critical points, non-repelling periodic orbits and the union of immediate basins of all attracting periodic points.*

For every $x \in I$ and positive integer n whenever points

$$x, f(x), \cdots, f^{n-1}(x)$$

belong to K, then $|(f^n)'(x)| \geq C\lambda^n$ with constants $C > 0$ and $\lambda > 1$ which do not depend on x or n.

In other words, expansion along an orbit must occur provided that the orbit keeps away from non-repelling periodic orbits or critical points. It is often natural to assume that there are no non-repelling periodic orbits, or to ignore critical orbits which are asymptotic to them. So the most serious obstacle to expansion occurs when an orbit accumulates on the set of critical points. This difficulty, however, might be alleviated if one assumes expansion along orbits of critical values. It might then be true that an orbit which gets close to a critical point then follows its orbit closely for a long time and makes up for the lost expansion. Thus, positive Lyapunov exponents will be expected for almost all orbits. This may be seen as an "explanation" of the power of the Collet-Eckmann condition.

1.2. Statement of Collet-Eckmann conditions. The Collet-Eckmann conditions are relevant not only in the context of real dynamics, but for holomorphic maps as well. So in the definitions to follow f is either in End^k or a rational map of the Riemann sphere. Furthermore, I denotes either the interval in the real case, or the Julia set in the complex setting.

The forward condition.

DEFINITION 1.3. *A critical point c of f satisfies the* Collet-Eckmann *condition if and only if c is not periodic and either $(f^n)'(f(c)) = 0$ for some $n > 0$, or there exist $C > 0$ and $\lambda > 1$ with which $|(f^n)'(f(c))| \geq C\lambda^n$ holds for every $n > 0$.*

The backward condition. The forward Collet-Eckmann condition is supplemented by the following.

DEFINITION 1.4. *We say that a critical point c of f satisfies the* backward Collet-Eckmann *condition if and only if c is not periodic and for some constants $C > 0$ and $\lambda > 1$ and every $x \in I$ and $n > 0$, if $f^n(x) = c$ then either $(f^n)'(x) = 0$ or $|(f^n)'(x)| \geq C\lambda^n$.*

The conditions were originally stated for unimodal maps where the provision for points hitting other critical points of f was needed. The definitions we stated for an individual point are shared by most authors in real dynamics. What makes a *mapping* f Collet-Eckmann varies by the context, and generally it is not required that all critical points satisfy the condition of Definition 1.3. For example, in complex theory critical orbits which converge to sinks are allowed.

The original result. Using these conditions Collet and Eckmann proved, see [**3**], the following theorem:

THEOREM 1.4. *If $f \in S_1^3$ and the critical point satisfies both the backward and forward Collet-Eckmann conditions, then f has a probabilistic absolutely continuous invariant measure.*

As we have seen from Theorem 1.2, this is equivalent to positive Lyapunov exponents. As we will see from Theorem 2.1, the assuming that the backward Collet-Eckmann condition holds is superfluous.

2. Real Collet-Eckmann Maps

2.1. The S-unimodal case. The theory of Collet-Eckmann maps is best developed by far in the case when they belong to S_1^2. First, there a number of properties which are equivalent to the Collet-Eckmann condition.

DEFINITION 2.1. *A map $f \in End^k$ has a* uniform hyperbolic structure *if for some $\lambda > 1$ and every $x \in I$, $p > 0$, if $f^p(x) = x$ then $|(f^p)'(x)| \geq \lambda^p$.*

THEOREM 2.1. *If the critical point of $f \in S_1^2$ satisfies the Collet-Eckmann condition, then it also satisfies the backward Collet-Eckmann condition. On the other hand, a map in S_1^2 with the critical point which satisfies the backward Collet-Eckmann condition has uniform hyperbolic structure.*

See [**18**] for the backward C-E, and [**17**] for the statement about uniform hyperbolic structure.

Let us recall that a probabilistic measure invariant by f has an *exponential decay of correlations* provided for some constants $C > 0$ and $\lambda > 1$, every function F which is continuous of bounded variation and G which is just continuous the condition

$$|\int F \cdot (G \circ f^n)\, d\mu - \Big(\int F\, d\mu\Big) \cdot \Big(\int G\, d\mu\Big)| \leq C\lambda^{-n} \operatorname{Var}(F)\|G\|_0$$

holds for every positive integer n.

THEOREM 2.2. *If the critical point of $f \in S_1^2$ satisfies the Collet-Eckmann condition, then it has a pacim with exponential decay of correlations. On the other hand, if such a measure exists, then the uniform hyperbolic structure is also present for f.*

See [**12**] for the first part, and [**17**] for the second.

A longer list of similar statements can be found in [**17**]. A somewhat surprising development occurred in [**17**] where the following theorem was proved:

THEOREM 2.3. *If $f \in S_1^2$ has the uniform hyperbolic structure, then its critical point satisfies the Collet-Eckmann condition.*

Thus, implications in Theorems 2.1, 2.2 and 2.3 are equivalences.

Topological invariance. Theorem 2.3 also plays a crucial role in the following surprising result, see [**20**]:

THEOREM 2.4. *In the class of $f \in S_1^2$, the Collet-Eckmann condition for the critical point is a topological invariant.*

Note that a topological equivalence class in S_1^2 includes maps f with various orders ℓ of the critical point, but the value of ℓ makes no difference in the Collet-Eckmann condition being satisfied or not. The proof of Theorem 2.4 also uses the so-called topological Collet-Eckmann condition which comes from holomorphic dynamics.

Topological Collet-Eckmann condition. Let f be either in End^k, $k > 1$, or a rational map of the Riemann sphere. Correspondingly, let I be either the interval or the Julia set.

Fix $\delta > 0$, a positive integer n and $z \in I$. Define $L(n, \delta, z)$ to be the number of integers i, $0 \leq i < n$ for which some connected component of the set $f^{i-n}(D(f^n(z), \delta))$ contains both $f^i(z)$ and some critical point of f.

Loosely speaking, we pull $D(f^n(z), \delta)$ back along the orbit of z and count the number of times the preimage hits a critical point.

DEFINITION 2.2. *The* topological *Collet-Eckmann condition is that for some* L_0, *positive constants* δ *and* η, *and every* $z \in I$, *the condition*

$$\liminf \frac{\#\{1 \leq n \leq N : L(n, \delta, z) \leq L_0\}}{N} \geq \eta$$

holds.

The condition of Definition 2.2 is obviously a topological invariant. It was established in [**5**] that the Collet-Eckmann condition implies the topological condition for rational maps, and the proof goes through with modifications for S-unimodal maps. A key observation in [**20**] was that for $S \in S_1^2$ the topological Collet-Eckmann condition implies the uniform hyperbolic structure, and so is equivalent to the Collet-Eckmann condition by Theorem 2.3.

Remarks on the role of the S-condition. If f satisfies the S-condition, then its inverse branches satisfy the opposite condition: if h is an inverse branch, then $1/\sqrt{|h'|}$ is concave on any interval where it is defined. This leads to a distortion lemma analogous to Koebe's distortion theorem for univalent mappings, see [**6**]. Hence, large parts of the holomorphic theory become applicable to maps which satisfy the S-condition. On the other hand, if critical point are assumed to be of polynomial type, the S-condition is forced to hold in their neighborhoods, and since critical points often control the dynamics of interval maps, a strange phenomenon occurs in which most results that are true in S^3 are also true in End^2. The reader can consult [**15**] for many examples and extensive explanations of that fact. In this context, the following question occurs:

Question: *Fix* k *and let* $\infty \geq k \geq 2$. *If* $f \in \mathrm{End}_1^k$ *is topologically conjugate to some* $g \in S_1^2$ *and the critical point of* g *satisfies the Collet-Eckmann condition, does the critical point of* f *also satisfy the Collet-Eckmann condition?*

A negative answer would provide the first interesting instance in which the principle that the S-condition is dispensable is violated.

2.2. Existence of absolutely continuous invariant measures. A lot of effort has been made in order to extend Theorem 1.4. The following theorem appeared in [**21**]:

THEOREM 2.5. *Suppose that for some* $f \in \mathrm{End}_1^2$ *all periodic orbits are repelling and that the following* summability condition *is satisfied:*

$$\sum_{n=1}^{\infty} \frac{1}{\sqrt[\ell]{|(f^n)'(f^n(0))|}} < \infty\,.$$

The f *has an absolutely continuous invariant probability measure.*

However, no condition of this type is known to be both sufficient and necessary for the existence of a pacim. It was shown, see [**1**], that the existence of a pacim is not a topological invariant, even in S_1^3. The same work showed a sufficient topological condition which may be satisfied even if the summability hypothesis of 2.5 is not.

The multi-modal case. In the multi-modal case it is not absolutely clear what the definition of the Collet-Eckmann condition should be. The natural generalization is to assume that all critical points of the transformation satisfy the Collet-Eckmann condition from Definition 1.3. Under this condition, one can prove the existence of an absolutely continuous invariant probability measure. The following result was recently reported by Graczyk and Smirnov, see [**9**]:

THEOREM 2.6. *Suppose that $f \in S^\omega$ and every critical point of f satisfies the Collet-Eckmann condition. Then f has an absolutely continuous invariant probability measure.*

In Theorem 2.6, the Collet-Eckmann condition can be weakened to a summability condition analogous to one stated in the hypothesis of Theorem 1.4.

The existence of an absolutely continuous invariant probability measure for multi-modal maps under an expansion hypothesis analogous to the Collet-Eckmann condition was also established by Bruin, Luzzato and Van Strien, see [**2**], under finite smoothness of f, but with some additional restrictions, notably on the orders of critical points.

The issue of topological invariance. It is clear however that a condition which assumes that all critical points are Collet-Eckmann cannot be topologically invariant. This is mentioned in [**20**], but without an actual counterexample. The following example was learned from J. Graczyk.

Consider a bi-modal map from S_2^2 for which one critical point c_1 has a neighborhood U which is mapped into itself by an iterate and whose orbit is disjoint from the other critical point. The first return map g into U is then a unimodal map and can be a Collet-Eckmann unimodal map. The second critical point c_2 is assumed to eventually map to a point $x \in U$, where x is arbitrary. If two maps of this type are topologically conjugate by a map h, then the critical point corresponding to c_1 must satisfy the Collet-Eckmann condition in view of Theorem 2.4. However, if the orders of criticality at c_1 and $h(c_1)$ are different, as are the eigenvalues of the repelling points at the endpoints of U and $h(U)$, then h on U does not preserve the property of a positive lower Lyapunov exponent at every point, in particular does not preserve the Collet-Eckmann property at c_2.

Recurrent Collet-Eckmann. This condition is intended to rule out examples of the type described in the preceding paragraph.

We will say that $f \in \mathrm{End}^k$ is a *recurrent* Collet-Eckmann map provided that all periodic orbits are repelling and every critical point c whose ω-limit set contains c satisfies the Collet-Eckmann condition given by Definition 1.3.

CONJECTURE 1. *If $f \in S^\omega$ satisfies the recurrent Collet-Eckmann condition, then it has a probabilistic absolutely continuous invariant measure. Moreover, in the class S^ω the recurrent Collet-Eckmann condition is a topological invariant.*

2.3. Abundance of Collet-Eckmann maps. Finally, the importance of Collet-Eckmann maps stems from the fact that the condition is satisfied on a set of positive measure for many families. This should be contrasted with the Misiurewicz condition which generally holds on a null set, see [**26**].

The fact that for many one-parameter families of unimodal map the Collet-Eckmann condition is satisfied on a set of positive measure was proved in [**28**]. We will state a version of a somewhat later and stronger result due to Tsujii, see [**29**].

Tsujii's theorem.

DEFINITION 2.3. *We say that a mapping $f \in End^2$ satisfies the Exp condition provided that:*

- *all critical orbits are disjoint,*
- *for every critical point the forward and backward Collet-Eckmann conditions are satisfied,*
- *all periodic orbits are repelling,*
- *for every critical point c, the condition*

$$\liminf \frac{1}{n} \log |f'(f^n(a)| = 0$$

holds.

THEOREM 2.7. *Consider a one-parameter family $f_t(x)$ of maps which is C^2, parameterized by the interval $(-1, 1)$, and assume that for every $t \in (-1, 1)$ map $f_t \in End^2$. For $t \in (-1, 1)$ consider the following* non-degeneracy *condition:*

for every critical point c of $f_t(x)$

$$\sum_{j=0}^{\infty} \frac{\frac{\partial f_t^j(c)}{\partial t}}{(f_t^j)'(f_t(c))} \neq 0 \, .$$

If for a family f_t condition Exp holds for $t = 0$ and the non-degeneracy condition listed above is satisfied at $t = 0$, then both conditions are in fact satisfied for a set of parameter values for which 0 is a density point.

One notes in addition that under the Exp condition there exists an ergodic probabilistic absolutely continuous invariant measure.

3. Holomorphic Theory

In the 1990's a theory of the Collet-Eckmann condition for rational maps of the Riemann sphere was developed.

3.1. Invariant measures. When f is a rational map of the Riemann sphere, its Julia set in many cases has zero measure. Thus, it would be unreasonable to look for measures which absolutely continuous with respect to the two-dimensional Lebesgue measure.

Geometric measure.

DEFINITION 3.1. *Let f be rational map of the Riemann sphere. Define the* essential *Hausdorff dimension of its Julia set, $HDE(J_f)$ to be the supremum of Hausdorff dimensions of all Borel invariant probability measures with positive Lyapunov exponent. Recall that the Lyapunov exponent is $\chi(\mu) = \int |f'(z)| \, d\mu(z)$ with derivative $|f'|$ computed with respect to the spherical metric on $\hat{\mathbf{C}}$.*

In addition, a Borel probability measure μ is an α-*conformal* measure for f if and only if for every Borel set B on which f is 1-1, the change of variables formula

$$\mu(f(B)) = \int_B |f'(z)|^\alpha \, d\mu(z)$$

holds. Assume additionally that μ is supported on J_f.

The following theorem is due to Denker, Urbański and Przytycki, see [**4**] and [**22**]:

THEOREM 3.1. *If f is a rational map of the Riemann sphere, then there is a conformal measure with exponent equal to $HDE(J_f)$. Any other conformal measure has exponent at least equal to $HDE(J_f)$.*

Existence theorem. The following theorem is proved by Graczyk and Smirnov, [**9**]:

THEOREM 3.2. *Suppose that f is a rational map of the Riemann sphere and each critical point of f either satisfies the Collet-Eckmann condition or is asymptotic to a sink. Then there is a unique, ergodic, non-atomic, conformal measure μ with exponent $HDE(J_f) = HD(J_f) = HD(\mu)$, and an invariant probability measure ν which is absolutely continuous with respect to μ.*

An earlier weaker version of this result was proved in [**23**]. In both theorems the Collet-Eckmann condition can be replaced with a weaker summability condition.

3.2. Geometric consequences. A geometric characterization of Collet-Eckmann polynomials is found by joining the results of [**8**] and [**24**]:

THEOREM 3.3. *Let $f(z) = z^n + c$, $n \geq 2$ and $c \in \mathbf{C}$ be a uni-critical complex polynomial. Then f satisfies the Collet-Eckmann condition if and only if the complement of J_f is a* Hölder domain, *that its Riemann map has a Hölder-continuous extension to the boundary.*

In particular, if J_f is the boundary of a Hölder domain, its Hausdorff dimension is less than 2. Theorem 3.3 has a real analogue.

THEOREM 3.4. *Let $f \in S_1^2$. Then f satisfies the Collet-Eckmann condition if and only if f is bi-Hölder conjugate to a tent map, that is a map in the form*

$$g_\beta(x) = |\beta x| + 1 - \beta$$

for $x \in [-1, 1]$ and $1 < \beta \leq 2$.

The analogy holds because the boundary extension of the Riemann map semi-conjugates the action of f on J_f to the map $x \to nx \pmod 1$ on the circle. Theorem 3.4 was proved in [**19**] and [**17**].

The result about the Hausdorff dimension of the Julia set can be stated more generally.

THEOREM 3.5. *Suppose that f is a rational map of the Riemann sphere and each critical point of f either satisfies the Collet-Eckmann condition or is asymptotic to a sink. Then either J_f is the whole Riemann sphere or its Hausdorff dimension is less than 2.*

A proof can be found in [**25**]. Earlier, Theorem 3.5 was proved in [**8**] in the case when the Fatou set of f has finitely many connected components. In fact, the Hausdorff dimension being less than 2 follows from the fact that $J(f)$ is mean-porous, see [**11**].

3.3. An abundance result. There is also a result about the abundance of the Collet-Eckmann condition for one-parameter families of uni-critical maps. If $d \geq 2$ and $f_c(z) = z^d + c$, recall that $\mathcal{M}_d$ is the connectedness locus, i.e. the set of all c for which the set $f^n(c)$, $n = 1, 2, \cdots$ is bounded. The complement of $\mathcal{M}_d$ is a topological disk, and one can consider its Riemann map $R_d : D(0,1) \to \hat{\mathbf{C}}$ normalized so that $R_d(z) - 1/z$ is holomorphic.

THEOREM 3.6. *There is a set $H \subset [0,1)$ of Lebesgue measure 1 so that for every integer $d \geq 2$ and every $\theta \in H$ there are constants $C > 0$ and $\lambda > 1$ with which for every z in the closure of the set $\{R_d(re^{2\pi i\theta}) : 0 < r < 1\}$ and $n > 0$ the condition*

$$|(f^n)'(c)| \geq C\lambda^n$$

holds.

The proof of this result will be published in [**7**]. One obvious consequence is that for almost every $z \in \partial\mathcal{M}_d$ in the sense of the harmonic measure the Collet-Eckmann condition holds. This last corollary holds in an even stronger version, and namely that the set of non-Collet Eckmann maps on the boundary of $\mathcal{M}_d$ has harmonic Hausdorff dimension 0, see [**27**].

References

[1] Bruin, H.: *Invariant measures of interval maps*, PhD thesis, Technical University of Delft, 1-112 (1994)

[2] Bruin, H., Luzzato, S. & Van Strien, S.: *Decay of correlations in one-dimensional dynamics*, preprint Warwick (1999)

[3] Collet, P. & Eckmann, J.-P.: *Positive Lyapunov exponents and absolute continuity for maps of the interval*, Ergod. Th. & Dyn. Sys. **3**, 13-46 (1983)

[4] Denker, M. & Urbański, M.: *On Sullivan's conformal measures for rational maps of the Riemann sphere*, Nonlinearity **4** (1991), 365-384

[5] Denker, M., Przytycki, F. & Urbański, M.:*On the transfer operator for rational functions on the Riemann sphere*, Ergod. Th. & Dyn. Sys. **16** (1996), 255-266

[6] Graczyk, J. & Świątek, G. : *SURVEY: Smooth unimodal maps in the 1990s*, Erg. Th. Dyn. Sys. **19** (1999), 263-287

[7] Graczyk, J. & Świątek, G. : *Harmonic measure and expansion on the boundary of the connectedness locus*, to appear in Invent. Math.

[8] Graczyk, J. & Smirnov, S.: *Collet, Eckmann, & Hölder*, Invent. Math. **133** (1998), 69-96

[9] Graczyk, J. & Smirnov, S.: *Weak expansion and geometry of Julia sets*, manuscript (1998)

[10] Keller, G.: *Exponents, attractors and Hopf decompositions for interval maps*, Ergod. Th. & Dynam. Sys. **10** (1990), 717-744

[11] Koskela, P. & Rohde, S. : *Hausdorff dimension and mean porosity*, Math. Ann. **309** (1997), 593-609

[12] Keller, G. & Nowicki, T. *Spectral theory, zeta functions and the distribution of periodic points for Collet-Eckmann maps*, Commun. Math. Phys. **149** (1992), 31-69

[13] Ledrappier, F.: *Some properties of absolutely continuous invariant measures on an interval*, Erg. Th. Dyn. Sys., **1** (1981), 77-93

[14] Mañé, R.: *Hyperbolicity, sinks and measure in one dimensional dynamics*, Commun. Math. Phys. **100** (1985), 495-524; *Erratum*, Commun. Math. Phys. **112** (1987), 721-724

[15] De Melo, W. & Van Strien, S.: *One-Dimensional Dynamics*, Springer-Verlag, New York (1993)

[16] Misiurewicz, M.: *Structure of mappings of an interval with zero entropy*, Publ. Math. IHES **53** (1981), 5-17

[17] Nowicki, T. & Sands, D.: *Non-uniform hyperbolicity and universal bounds for S-unimodal maps*, Invent. Math. **132** (1998), 633-680

[18] Nowicki, T. *Symmetric S-unimodal mappings and positive Liapunov exponents*, Erg. Th. Dyn. Sys. **5** (1985), 611-616

[19] Nowicki, T. & Przytycki, F.: *The conjugacy of Collet-Eckmann's map of the interval with the tent map is Hölder continuous*, Ergod. Th. & Dyn. Sys. **9** (1989), 379-388

[20] Nowicki, T. & Przytycki, F.: *Topological invariance of the Collet-Eckmann property for S-unimodal maps*, Fund. Math. **155** (1998), 33-43

[21] Nowicki, T. & van Strien, S.: *Absolutely continuous invariant measures under the summability condition*, Invent. Math. **105**, 123-136 (1991)

[22] Przytycki, F.: *Lyapunov characteristic exponents are non-negative*, Proc. Amer. Math. Soc. **119** (1993), 309-317

[23] Przytycki, F.: *Iterations of holomorphic Collet-Eckmann Maps: conformal and invariant measures. Appendix: on non-renormalizable quadratic polynomials*, Trans. Amer. Math. Soc. **350** (1998), 717-742

[24] Przytycki, F.: *Hölder implies Collet-Eckmann*, in Geométrie complexe et systèmes dynamiques, Astérisque **261** (2000), 385-403

[25] Przytycki, F. & Rohde, S.: *Porosity of Collet-Eckmann Julia sets*, Fund. Math. **155** (1998), 189-199

[26] Sands, D.: *Misiurewicz maps are rare*, Commun. Math. Phys. **1997** (1998), 109-129

[27] Smirnov, S. *Symbolic dynamics and Collet-Eckmann conditions*, Internat. Math. Res. Notices **2000** (2000), 333-351

[28] Thieullen, P. & Tresser, C. & Young, L.-S.: *Exposant de Lyapunov positif dans des familles à un paramétre d'applications unimodales*, C.R. Acad. Sci. Paris Série I **315** (1992), 69-72

[29] Tsujii, M.: *Positive Lyapunov exponents in families of one-dimensional dynamical systems*, Invent. Math. **111** (1993), 113-137

Department of Mathematics, Pennsylvania State University, University Park, PA 16802, USA

Proceedings of Symposia in Pure Mathematics
Volume **69**, 2001

Monotonicity, $\mathcal{J}$-algebra of Potapov and Lyapunov Exponents

Maciej P. Wojtkowski

ABSTRACT. We present a new approach and a generalization of the estimates of Lyapunov exponents developed first in [**W2**] in the symplectic case. The work of Lewowicz [**L**], Markarian [**M**], and our [**W1**], [**W2**], [**W5**], are combined with the $\mathcal{J}$-algebra of Potapov, [**P1**], [**P2**], [**P3**]. We obtain a general theory which we then specify to the symplectic case. The appendix contains a simple application to the gas of hard spheres.

0. Introduction

We will discuss a general method for establishing hyperbolic properties in concrete dynamical systems.

Hyperbolicity is defined in the language of the linearized system (the derivative in the case of discrete time). It amounts to exponential growth or decay of solutions of the linearized system. To capture such a behavior we use indefinite quadratic forms. If an indefinite quadratic form (more precisely a field of such forms on the tangent bundle of the phase space) evaluated on any nonzero solution grows, then it provides us with a distinguished cone (sector) of positive tangent vectors which is taken into itself by the linearized dynamics. Once a solution enters this cone the quadratic form can be used as a measure of the magnitude of the solution and the growth can be usually strengthened to exponential growth with the help of recurrence. The decay of some solutions is obtained by reversing time and using the absolute value of our quadratic form.

This crucial idea was put forward by Lewowicz [**L**] and independently in [**W1**]. It also appeared in the study of exponential dichotomies in linear differential equations, Coppel, [**C**], Samoilenko and Kulik, [**Sa-K**].

We will consider a more general situation when solutions of the linearized system stay in the positive cone forever, once they enter it, but the form defining the cone is not necessarily increased.

To illustrate the contents of this paper let us consider the linear system of differential equations (with periodic coefficients) of the form.

$$\begin{aligned} \dot{x} &= \alpha(t)x + B(t)y, \\ \dot{y} &= C(t)x + \delta(t)y, \end{aligned} \tag{0.1}$$

where $x \in \mathbb{R}^m, y \in \mathbb{R}^m$, the functions α and δ have scalar values and B, C are $m \times m$ matrix valued functions.

We introduce the quadratic form $\mathcal{Q}(x,y) = \langle x, y\rangle$ where $\langle \cdot, \cdot \rangle$ is the arithmetic scalar product on $\mathbb{R}^m$. We have

$$\frac{d}{dt}\mathcal{Q} = (\alpha(t) + \delta(t))\langle x, y\rangle + \langle B(t)y, y\rangle + \langle C(t)x, x\rangle.$$

Under the assumption that the symmetric parts of B and C are positive definite we obtain the following property

$$\frac{d}{dt}\mathcal{Q} > 0, \quad \text{if} \quad \mathcal{Q} = 0.$$

If this property holds we will say that the linear flow defined by (0.1) is strictly $\mathcal{Q}$-separated. We will explore the consequences of this property.

We will establish that if the system (0.1) is strictly $\mathcal{Q}$-separated then the linear flow has a dominated splitting, i.e., there is an invariant splitting of $\mathbb{R}^m \times \mathbb{R}^m$ into m-dimensional subspaces E^- and E^+ such that the largest Floquet multiplier on E^- is strictly smaller (in absolute value) than the smallest Floquet multiplier on E^+. The quadratic form $\mathcal{Q}$ is negative on E^- and positive on E^+. We will also introduce quantitative description of the spectral gap.

We will use the language of matrix cocycles rather than that of differential equations and we will obtain estimates of Lyapunov exponents. The results can be readily translated into the language of differential equations.

The system (0.1) with $\alpha = \delta \equiv 0, B(t) \equiv I$ and symmetric positive definite $C(t)$ appears in the study of geodesic flows on manifolds of negative sectional curvature. In this situation we obtain the uniform expansion on E^+ and the uniform contraction on E^-, i.e., the Anosov property.

More generally in the symplectic case, if a system is $\mathcal{Q}$-separated then it is Anosov.

The symplectic case was studied earlier in [**W2**],[**W5**], where a special case of the estimates of Theorem 3.10 was obtained.

This method of estimating Lyapunov exponents for the Hamiltonian dynamics has two major applications, the Boltzmann-Sinai gas of hard spheres [**Ch-S**], [**W2**], [**W5**], [**S-Sz**], and the systems of falling balls in 1-dimension, [**W3**], [**W4**], [**S1**], [**W6**], [**W7**].

In the Appendix we present an application of the estimates of Lyapunov exponents from Theorem 3.10 to the gas of hard spheres.

The primary motivation for this paper was our study of the geodesic flows on Weyl manifolds, which we call W-flows, [**W9**], [**W8**]. The linearization of these flows has (in appropriate coordinates) the form (0.1) with $\delta \equiv 0, B(t) \equiv I$, and $C(t)$ not necessarily symmetric.

1. The $\mathcal{J}$–algebra of Potapov and contraction/expansion properties

Let us consider an n-dimensional linear space V over $\mathbb{R}$ with a chosen pseudo Euclidean structure, i.e., a nondegenerate quadratic form $\mathcal{J}$ with the positive index of inertia equal to p and the negative index of inertia equal to q, $p + q = n, p \geq 1, q \geq 1$. Depending on the sign of $\mathcal{J}$ we have positive and negative vectors. We denote by

$$\mathcal{C}_\pm = \{v \in V | \pm \mathcal{J}(v) > 0\} \cup \{0\}$$

the open cones of, respectively, positive and negative vectors (with the zero vector included), and by $\mathcal{C}_0$ their common boundary, $\mathcal{C}_0 = \{v \in V | \mathcal{J}(v) = 0\}$.

A linear subspace is called positive (negative), if it is contained in $\mathcal{C}_+$ ($\mathcal{C}_-$). The maximal dimension of a positive subspace is p and of a negative subspace q. Let P be a positive subspace of dimension p (a maximal positive subspace) and let Q be the $\mathcal{J}$–orthogonal complement of P. The dimension of Q is q and it is by necessity a negative subspace. We can introduce an auxiliary scalar product in V by using $\mathcal{J}$ in P, $-\mathcal{J}$ in Q and declaring P and Q to be orthogonal. We choose coordinates in $P = \mathbb{R}^p$ and $Q = \mathbb{R}^q$ in such a way that $\mathcal{J}$ and $-\mathcal{J}$ become the arithmetic scalar products in $\mathbb{R}^p$ and $\mathbb{R}^q$ respectively. As a result we obtain the following coordinate representation: $V = \mathbb{R}^p \times \mathbb{R}^q$ and for $v = (v_1, v_2), v_1 \in \mathbb{R}^p, v_2 \in \mathbb{R}^q$

$$\mathcal{J}(v) = \langle Jv, v \rangle = v_1^2 - v_2^2, \quad \text{where } J = \begin{bmatrix} I_p & 0 \\ 0 & -I_q \end{bmatrix}$$

I_p, I_q are identity matrices in $\mathbb{R}^p$ and $\mathbb{R}^q$ respectively. We will use freely this coordinate representation when we find it convenient. It is important though for discussing diffeomorphisms and flows on manifolds that the following theory depends on the choice of the form $\mathcal{J}$ alone.

DEFINITION 1.1. *A linear nonsingular operator $S : V \to V$ is*
1. *$\mathcal{J}$–separated, if $S\mathcal{C}_+ \subset \mathcal{C}_+$,*
2. *strictly $\mathcal{J}$–separated, if $S(\mathcal{C}_0 \cup \mathcal{C}_+) \subset \mathcal{C}_+$,*
3. *$\mathcal{J}$–monotone, if $\mathcal{J}(Sv) \geq \mathcal{J}(v)$ for every $v \in V$,*
4. *strictly $\mathcal{J}$–monotone, if $\mathcal{J}(Sv) > \mathcal{J}(v)$ for every $v \in V, v \neq 0$,*
5. *$\mathcal{J}$–isometry, if $\mathcal{J}(Sv) = \mathcal{J}(v)$ for every $v \in V$,*

Note that the five classes of operators are semigroups with respect to composition. The inclusions between them are as follows

$$\begin{array}{ccccc} (1) & \supset & (3) & \supset & (5) \\ \cup & & \cup & & \\ (2) & \supset & (4) & & \end{array}$$

We extend the definitions of properties 1 - 5 to one parameter subgroups of linear operators (elements of the Lie algebra $gl(n)$) in the natural way. An element G of $gl(n)$ has property (i) if the operator e^{tG} has property (i) for all $t > 0$.

Let us further define a $\mathcal{J}$–symmetric operator S as a linear operator on V such that $\langle JS\cdot, \cdot \rangle$ is a symmetric bilinear form on V.

Potapov, [**P1**],[**P2**],[**P3**], obtained fundamental results about the structure and properties of $\mathcal{J}$–monotone operators, which he called $\mathcal{J}$–noncontractive. Potapov developed his theory for a pseudo unitary form $\mathcal{J}$ rather than pseudo Euclidean, but we will concentrate on the real case. We will comment along the way on the complex versions of theorems.

We will formulate a series of propositions about the $\mathcal{J}$-algebra of Potapov. We will prove only the infinitesimal versions. The (harder) proofs in the general case can be found in the papers of Potapov, [**P1**], [**P2**], [**P3**].

1. If S is $\mathcal{J}$–separated then the set of positive numbers r such that $\frac{1}{r}S$ is $\mathcal{J}$–monotone is a closed interval $[r_-, r_+], r_- > 0$, possibly degenerating to a point.

2. If S is strictly $\mathcal{J}$–separated then the set of positive numbers r such that $\frac{1}{r}S$ is strictly $\mathcal{J}$–monotone is an open interval $(r_-, r_+), r_- > 0$.

These properties are the consequence of the theorem of Kühne.

THEOREM 1.2. *If a symmetric bilinear form $F(\cdot,\cdot)$ is nonnegative on $\mathcal{C}_0$ then*

$$\inf_{v\in\mathcal{C}_+} \frac{F(v,v)}{\langle Jv,v\rangle} \geq \sup_{u\in\mathcal{C}_-} \frac{F(u,u)}{\langle Ju,u\rangle},$$

and for every number r between the infimum and the supremum $F(v,v) \geq r\langle Jv,v\rangle$, for all v.

PROOF. Suppose to the contrary that there is a positive vector $v_0 \in \mathcal{C}_+$ and a negative vector $u_0 \in \mathcal{C}_-$ which violate the inequality, i.e., assuming $\langle Jv_0, v_0\rangle = 1, \langle Ju_0, u_0\rangle = -1$ we have $F(v_0,v_0) + F(u_0,u_0) < 0$. We can find the angle α such that the linear combinations $v_1 = \cos\alpha v_0 + \sin\alpha u_0$ and $v_2 = -\sin\alpha v_0 + \cos\alpha u_0$ are in $\mathcal{C}_0$. Then $F(v_1,v_1) \geq 0,\ F(v_2,v_2) \geq 0$. We obtain the contradiction

$$F(v_1,v_1) + F(v_2,v_2) = F(v_0,v_0) + F(u_0,u_0) \geq 0.$$

□

The proof of the pseudo unitary version of Theorem 1.2 can be found in [**P2**].

3. Every $\mathcal{J}$–separated operator S has a unique representation $S = RU$, where U is a $\mathcal{J}$–isometry and R is $\mathcal{J}$–symmetric and has only positive eigenvalues. The operator R is equal to $(SJS^*J)^{\frac{1}{2}}$ and it is called the modulus of S. Note that although the adjoint S^* depends on the auxiliary Euclidean structure in V the modulus R depends on the form $\mathcal{J}$ alone.

4. For a strictly $\mathcal{J}$–separated S there is a $\mathcal{J}$–isometry W such that the modulus R of S is diagonalized by W, i.e., $W^{-1}RW$ is the diagonal matrix $\begin{bmatrix} R_+ & 0 \\ 0 & R_- \end{bmatrix}$. Moreover $R_+ \geq r_+ I_p$ and $0 < R_- \leq r_- I_q$, where $r_+ > r_- > 0$ were introduced in 2.

5. For a $\mathcal{J}$–separated S the p largest eigenvalues of the modulus R are not smaller than r_+ and the q smallest eigenvalues do not exceed r_-. Let us denote these eigenvalues by

$$r_p^+ \geq \cdots \geq r_2^+ \geq r_1^+ \geq r_+ \geq r_- \geq r_1^- \geq r_2^- \geq \cdots \geq r_q^- > 0.$$

We have moreover $r_1^+ = r_+$ and $r_1^- = r_-$.

Clearly a $\mathcal{J}$–separated S is (strictly) $\mathcal{J}$–monotone if and only if $r_1^+ = r_+ \geq 1\ (> 1)$ and $r_1^- = r_- \leq 1\ (< 1)$.

The proofs depend on the celebrated theorem of Potapov, that S is $\mathcal{J}$–separated if and only if S^* is $\mathcal{J}$–separated. The infinitesimal version of this fact is easy to prove. Indeed, $G \in gl(n)$ is $\mathcal{J}$–separated if and only if

$$\langle Je^{tG}v, e^{tG}v\rangle \geq 0 \quad \text{for all} \quad t \geq 0, \quad \text{if} \quad \langle Jv,v\rangle \geq 0.$$

It follows that at $t = 0$

$$\frac{d}{dt}\langle Je^{tG}v, e^{tG}v\rangle = \langle (G^*J + JG)v, v\rangle \geq 0, \quad \text{if} \quad \langle Jv,v\rangle \geq 0.$$

Now by Theorem 1.2 there is a number a (not necessarily positive) such that

$G^*J + JG \geq aJ$. Conversely, if the last inequality holds then $G - \frac{a}{2}I$ is $\mathcal{J}$–monotone and hence G is $\mathcal{J}$–separated.

With this characterization of infinitesimal $\mathcal{J}$–separatedness we check immediately that multiplying the inequality $G^*J + JG \geq aJ$ by J on the left and on the right we obtain $JG^* + GJ \geq aJ$.

The infinitesimal versions of the claims 3 – 5 are equally simple to prove. First of all any element G of the Lie algebra $gl(n)$ can be split uniquely into the $\mathcal{J}$–symmetric $G_s = \frac{1}{2}(G + JG^*J)$ and the $\mathcal{J}$–antisymmetric $G_a = \frac{1}{2}(G - JG^*J)$.

The $\mathcal{J}$–antisymmetric elements of $gl(n)$ are the infinitesimal $\mathcal{J}$–isometries. If G is strictly $\mathcal{J}$–separated then $2JG_s = G^*J + JG > aJ$. It follows that the eigenvalues of G_s, being real, are bigger than $\frac{a}{2}$, if the eigenvector is a positive vector, and smaller than $\frac{a}{2}$, if the eigenvector is a negative vector. (To establish that the spectrum of G_s is real we need to use the pseudo unitary $\mathcal{J}$-form in the complexified space. It goes through exactly as in the case of symmetric matrices and we leave it to the reader.)

If G_s is strictly $\mathcal{J}$–separated then it is diagonalizable by a $\mathcal{J}$–isometry. Indeed it must have an invariant subspace P which is maximal positive and an invariant subspace Q which is maximal negative. (It is easier to think about $S = e^{\epsilon G_s}$. $S(S^{-1})$ takes maximal positive(negative) subspaces into maximal positive(negative) subspaces. Hence by some kind of fixed point theorem there must be invariant subspaces. One can use for example the contraction of any of the metrics discussed in Theorem 1.5 and thereafter.) Similarly to the case of symmetric matrices the subspaces P and Q are $\mathcal{J}$–orthogonal. Further any linear mapping that preserves the $\mathcal{J}$–form on the $\mathcal{J}$–orthogonal subspaces P and Q, and maps them onto two $\mathcal{J}$–orthogonal subspaces, must be a $\mathcal{J}$–isometry. Hence there is a $\mathcal{J}$–isometry U such that

$$U^{-1}G_sU = \begin{bmatrix} G_+ & 0 \\ 0 & G_- \end{bmatrix}.$$

Moreover $G_\pm$ are symmetric. So we can further diagonalize them. Also $G_+ > \frac{1}{2}aI_p$ and $G_- < \frac{1}{2}aI_q$. Indeed since the defining property for a $\mathcal{J}$-isometry U is $U^{-1} = JU^*J$, we have

$$2J(U^{-1}G_sU) = U^*2JG_sU \geq aU^*JU = aJ.$$

We proceed with the development of further properties of $\mathcal{J}$-separated operators; we will freely use properties 1 – 5. For a d–dimensional positive linear subspace $L \subset \mathcal{C}_+$ the restriction of the $\mathcal{J}$-form defines a scalar product. If S is $\mathcal{J}$–separated then $SL \subset \mathcal{C}_+$ is also a positive subspace with the scalar product furnished by $\mathcal{J}$. Hence the subspaces L and SL are equipped with (unoriented) volume elements, which we call the $\mathcal{J}$–volumes. Let $\alpha_d(L;S)$ be the coefficient of the $\mathcal{J}$–volume expansion for the linear map S restricted to L. Further let $\sigma_d(S)$ be the infimum of $\alpha_d(L;S)$ over all positive d–dimensional subspaces L. The important property of the coefficient σ_d is the supermultiplicativity: for $\mathcal{J}$–separated S_1 and S_2 we have

$$\sigma_d(S_1S_2) \geq \sigma_d(S_1)\sigma_d(S_2). \tag{1.1}$$

Using the notation introduced in 5. we have

PROPOSITION 1.3. $\sigma_d(S) = r_1^+r_2^+ \dots r_d^+$.

PROOF. We begin by establishing the inequality

$$\alpha_d(L, S) \geq r_1^+r_2^+ \dots r_d^+, \tag{1.2}$$

for all $\mathcal{J}$–separated S and positive d-dimensional subspaces L. Since the modulus of a $\mathcal{J}$–separated S is a continuous function of S then both sides of this inequality are continuous functions of S. It follows that it is sufficient to prove (1.2) for strictly $\mathcal{J}$–separated S, for which we can diagonalize the modulus as described in 4. We will be using this diagonalization.

Since S differs from its modulus R by a $\mathcal{J}$–isometry U, i.e., $S = RU$, we get that $\alpha_D(L;S)$ is equal to $\alpha_d(UL;R)$. Every d–dimensional positive subspace L can be included in a p–dimensional positive subspace T (maximal positive subspace). The subspace T is the graph of the linear mapping $T : \mathbb{R}^p \to \mathbb{R}^q$ such that $T^*T < I_p$. We will slightly abuse the notation denoting by the same letter T the linear map and its graph. In the coordinate representation the mapping T is given by a $q \times p$ matrix. Using $v_1 \in \mathbb{R}^p$ as a coordinate on the graph of T we obtain that the scalar product $\mathcal{J}$ on this subspace is equal to $\langle \Pi v_1, v_1\rangle$, where $\Pi = I_p - T^*T$. The action of R on T is given by $v_1 \mapsto R_+ v_1$. The image space RT is the graph of the linear map $\widetilde{T} = R_- T R_+^{-1}$, with the scalar product $\mathcal{J}$ given by $\langle \widetilde{\Pi} v_1, v_1\rangle$, where $\widetilde{\Pi} = I_p - \widetilde{T}^*\widetilde{T}$. We change the v_1 coordinate in T and $\widetilde{T}$ to $w = \Pi^{\frac{1}{2}} v_1$ and $\widetilde{w} = \widetilde{\Pi}^{\frac{1}{2}} v_1$ respectively, so that the scalar products in these subspaces become the standard arithmetic scalar product. In the coordinates w and $\widetilde{w}$ the action of R is given by $\widetilde{\Pi}^{\frac{1}{2}} R_+ \Pi^{-\frac{1}{2}}$. Hence the the coefficients of $\mathcal{J}$–volume expansion under the action of the restriction of R to the subspace T are given by the singular values of the matrix

$$K = \widetilde{\Pi}^{\frac{1}{2}} R_+ \Pi^{-\frac{1}{2}} = (I_p - R_+^{-1} T^* R_-^2 T R_+^{-1})^{\frac{1}{2}} R_+ (I_p - T^*T)^{-\frac{1}{2}},$$

i.e. by the eigenvalues of

$$K^*K = \Pi^{-\frac{1}{2}} R_+ \widetilde{\Pi} R_+ \Pi^{-\frac{1}{2}} = (I_p - T^*T)^{-\frac{1}{2}} (R_+^2 - T^* R_-^2 T)(I_p - T^*T)^{-\frac{1}{2}},$$

To estimate the eigenvalues of K^*K we use

$$K^*K \geq (I_p - T^*T)^{-\frac{1}{2}} R_+^2 (I_p - T^*T)^{-\frac{1}{2}}.$$

The eigenvalues of the last matrix will be estimated by the application of the following Lemma.

LEMMA 1.4. *For nonsingular symmetric matrices X, Y and a symmetric matrix Z, if*

$X^2 \leq Y^2$ then the spectrum of YZY is not smaller than the spectrum of XZX, i.e., the i-th eigenvalue (in ascending order) of XZX does not exceed the i-th eigenvalue of YZY.

Note that we do not claim that $XZX \leq YZY$, which is not guaranteed by our assumptions.

PROOF. The eigenvalues of XZX and YZY can be obtained as min-max of the Rayleigh quotients

$$\frac{\langle XZXv, v\rangle}{\langle v, v\rangle} = \frac{\langle Zu, u\rangle}{\langle X^{-2}u, u\rangle} \leq \frac{\langle Zu, u\rangle}{\langle Y^{-2}u, u\rangle} = \frac{\langle YZYw, w\rangle}{\langle w, w\rangle},$$

for $u = Xv$ and $u = Yw$. □

Applying this Lemma with $Z = R_+^2, Y = (I_p - T^*T)^{-\frac{1}{2}}$ and $X = I_p$ we obtain that the i-th, in ascending order, eigenvalue of K^*K cannot be smaller than $(r_i^+)^2$. It gives us the desired estimate (1.2) for all $\mathcal{J}$–separated S. Further for strictly

$\mathcal{J}$–separated S we obtain the equality for the infimum $\sigma_d(S)$ by considering the subspace $T = 0$.

Hence the Proposition is established for all strictly $\mathcal{J}$–separated S. To extend it to all $\mathcal{J}$–separated S let us consider the operator $C_\epsilon = \begin{bmatrix} (1+\epsilon)I_p & 0 \\ 0 & (1-\epsilon)I_q \end{bmatrix}$. C_ϵ is strictly $\mathcal{J}$–monotone for $0 < \epsilon < 1$. If S is $\mathcal{J}$–separated then $C_\epsilon S$ is strictly $\mathcal{J}$–separated. By (1.1) we have

$$\sigma_d(C_\epsilon S) \geq \sigma_d(C_\epsilon)\sigma_d(S) \geq (1+\epsilon)^d r_1^+ r_2^+ \dots r_d^+.$$

Our Proposition is already proven for $\sigma_d(C_\epsilon S)$. Hence by the continuity of the modulus we obtain it also for $\sigma_d(S)$ by taking the limit $\epsilon \to 0$.

□

Proposition 1.3 allows the interpretation of the eigenvalues of the modulus as pseudo Euclidean singular values. We will call them simply positive and negative singular values of the $\mathcal{J}$–separated operator S (note that both positive and negative singular values are positive numbers). In particular they are not changed if we compose S on the left and/or on the right with $\mathcal{J}$–isometries.

In the pseudo unitary case we consider d-dimensional complex subspaces which have $2d$-dimensional real $\mathcal{J}$-volume. Essentially the same proof will give us $\sigma_d(S) = (r_1^+ r_2^+ \dots r_d^+)^2$.

Corollary 1.5. *For $\mathcal{J}$-separated S_1 and S_2 we have*

$$r_1^+(S_1S_2) \geq r_1^+(S_1)r_1^+(S_2) \quad \text{and} \quad r_1^-(S_1S_2) \leq r_1^-(S_1)r_1^-(S_2).$$

Proof. The first part follows immediately from Proposition 1.3 for $d = 1$ and (1.1). For a $\mathcal{J}$-separated S the operator S^{-1} is $(-\mathcal{J})$-separated and the positive pseudo Euclidean singular values of S^{-1} (with respect to $-\mathcal{J}$) are equal to $\frac{1}{r_1^-(S)}, \dots, \frac{1}{r_q^-(S)}$. The second part follows now from the first applied to $S_2^{-1}S_1^{-1}$ and $-\mathcal{J}$. □

Let us consider the manifold $\mathcal{L}_+$ of all p-dimensional subspaces contained in $\mathcal{C}_+$, It can be identified with the manifold of $q \times p$ matrices T such that $T^*T < I_p$. This manifold carries the Riemannian metric of the symmetric space, [**Py**],

$$dT^2 = tr\left(\Pi^{-1} dT^* \Gamma^{-1} dT\right), \quad \text{where } \Pi = I_p - T^*T \text{ and } \Gamma = I_q - TT^*. \tag{1.3}$$

To get (1.3) we need to observe that the $\mathcal{J}$–isometries act transitively on $\mathcal{L}_+$. The isometry that takes T into the zero subspace $\mathbb{R}^p \times \{0\}$ is given by

$$\begin{bmatrix} (I_p - T^*T)^{-\frac{1}{2}} & (I_p - TT^*)^{-\frac{1}{2}} T^* \\ -(I_q - TT^*)^{-\frac{1}{2}} T & (I_q - TT^*)^{-\frac{1}{2}} \end{bmatrix} = \begin{bmatrix} (I_p - T^*T)^{-\frac{1}{2}} & -T^*(I_q - TT^*)^{-\frac{1}{2}} \\ T(I_p - T^*T)^{-\frac{1}{2}} & (I_q - TT^*)^{-\frac{1}{2}} \end{bmatrix}^{-1}.$$

Hence the metric $tr(dT^*dT)$ at 0 is transported to T as $tr(\Pi^{-\frac{1}{2}} dT^* \Gamma^{-1} dT \Pi^{-\frac{1}{2}})$.

A $\mathcal{J}$–separated operator S maps naturally $\mathcal{L}_+$ into itself. This mapping is a contraction. More precisely we have

THEOREM 1.6. *For a $\mathcal{J}$–separated operator S and $T_1, T_2 \in \mathcal{L}_+$*

$$dist(ST_1, ST_2) \leq \frac{r_1^-}{r_1^+} dist(T_1, T_2),$$

and the contraction coefficient is sharp.

PROOF. By continuity it is sufficient to prove the theorem for a strictly $\mathcal{J}$–separated operator S. We take advantage of the polar form of $S = RU$ where R is the modulus and U is the $\mathcal{J}$–isometry. Hence it is enough to show the contraction for an operator of the form described in 4. We will establish the infinitesimal contraction property. The action of this operator on $\mathcal{L}_+$ is given by $T \mapsto \widetilde{T} = R_- T R_+^{-1}$. The derivative of this action is $dT \mapsto R_- dT R_+^{-1}$. Hence we need to compare the spectra of

$$\Pi^{-\frac{1}{2}} dT^* \Gamma^{-1} dT \Pi^{-\frac{1}{2}} \quad \text{and} \quad \widetilde{\Pi}^{-\frac{1}{2}} R_+^{-1} dT^* R_- \widetilde{\Gamma}^{-1} R_- dT R_+^{-1} \widetilde{\Pi}^{-\frac{1}{2}}, \tag{1.4}$$

where Π and Γ are defined in (1.3) and $\widetilde{\Pi}, \widetilde{\Gamma}$ are defined in the same way with $\widetilde{T}$ replacing T, i.e.,

$$\widetilde{\Pi} = I_p - R_+^{-1} T^* R_-^2 T R_+^{-1}, \quad \widetilde{\Gamma} = I_q - R_- T R_+^{-2} T^* R_-.$$

The second operator in (1.4) has the same spectrum as

$$R_+^{-1} \widetilde{\Pi}^{-1} R_+^{-1} dT^* R_- \widetilde{\Gamma}^{-1} R_- dT,$$

and hence also the same spectrum as

$$(R_+^2 - T^* R_-^2 T)^{-\frac{1}{2}} dT^* (R_-^{-2} - T R_+^{-2} T^*)^{-1} dT (R_+^2 - T^* R_-^2 T)^{-\frac{1}{2}}. \tag{1.5}$$

Let us note that

$$R_+ \geq r_1^+ I_p, \; R_+^2 \geq (r_1^+)^2 I_p, \quad R_- \leq r_1^- I_q, \; R_-^2 \leq (r_1^-)^2 I_q,$$
$$R_+^2 - T^* R_-^2 T \geq (r_1^+)^2 \Pi, \quad R_-^{-2} - T R_+^{-2} T^* \geq (r_1^-)^{-2} \Gamma.$$

Applying Lemma 1.4 with $X = (R_+^2 - T^* R_-^2 T)^{-\frac{1}{2}}$ and $Y = \frac{1}{r_1^+} \Pi^{-\frac{1}{2}}$ we obtain that the spectrum of (1.5) does not exceed the spectrum of

$$\frac{1}{(r_1^+)^2} \Pi^{-\frac{1}{2}} dT^* (R_-^{-2} - T R_+^{-2} T^*)^{-1} dT \Pi^{-\frac{1}{2}} \leq \left(\frac{r_1^-}{r_1^+} \right)^2 \Pi^{-\frac{1}{2}} dT^* \Gamma^{-1} dT \Pi^{-\frac{1}{2}}.$$

We conclude that $tr \left(\widetilde{\Pi}^{-1} d\widetilde{T}^* \widetilde{\Gamma}^{-1} d\widetilde{T} \right) \leq \left(\frac{r_1^-}{r_1^+} \right)^2 tr \left(\Pi^{-1} dT^* \Gamma^{-1} dT \right)$, which gives us the infinitesimal contraction.

To establish that the coefficient of contraction is sharp, it is enough to calculate the infinitesimal coefficient of contraction at $T = 0$, which is equal to $\frac{r_1^-}{r_1^+}$. □

This theorem extends the result of Bougerol, [**B**], about the contraction of the space of positive Lagrangian subspaces under the action of monotone symplectic operators.

In the proof we have obtained the estimate of all eigenvalues and not just the trace. It leads to the contraction property for other metrics, for instance the Finsler metric defined by $\|dT\| = \sqrt{\|dT^* dT\|}$ at the zero subspace $T = 0$ and invariant under the action of $\mathcal{J}$–isometries, in parallel with [**L-W**].

Theorem 1.6 holds also in the pseudo unitary case and the proof is essentially the same. In such a case the space of complex positive subspaces of dimension p is a classical bounded domain. As shown by Suzuki, [**Su**], for this domain the Caratheodory and Kobayashi metrics coincide with the described above Finsler metric equal to $\|dT\|$ at the zero subspace $T = 0$. Hence this metric must be contracted by any holomorphic map of the domain into itself and not only by the Möbius maps as in Theorem 1.6. The Kahler metric of the symmetric space is not contracted by all holomorphic maps (cf. [**Ko**]), but it is contracted by Möbius maps defined by $\mathcal{J}$-separated operators, which is the content of Theorem 1.6 in the pseudo unitary case.

2. The symplectic case

Let us apply these general ideas to the symplectic space and symplectic operators. We have $p = q$, $V = \mathbb{R}^p \times \mathbb{R}^p$ and the symplectic form $\omega(v, w) = \langle v_1, w_2\rangle - \langle v_2, w_1\rangle$. We consider a special type of a $\mathcal{J}$–form, which we call a $\mathcal{Q}$–form. It was defined geometrically in [**W5**] . A $\mathcal{Q}$–form is defined by a pair of transversal Lagrangian subspaces L_1, L_2. Using the standard splitting $v = v_1+v_2, v_i \in L_i, i = 1, 2$, we put $\mathcal{Q}(v) = \omega(v_1, v_2)$.

The positive cone $\mathcal{C}_+$ of a $\mathcal{Q}$–form is called a symplectic sector between L_1 and L_2. The boundary of a symplectic sector contains many Lagrangian subspaces, but only two isolated ones, L_1 and L_2. These two Lagrangian subspaces act as the "sides" of the sector, which they actually are only when $p = 1$.

We can choose symplectic coordinates in such a way that $L_1 = \mathbb{R}^p \times \{0\}, L_2 = \{0\} \times \mathbb{R}^p$ and accordingly $\mathcal{Q}(v) = \langle v_1, v_2\rangle$. We will call it the standard $\mathcal{Q}$-form.

The following theorem was published by Potapov in [**P2**], and independently in [**W1**]. The proof below is essentially taken from [**W1**].

THEOREM 2.1. *If a symplectic operator is (strictly) $\mathcal{Q}$–separated then it is (strictly) $\mathcal{Q}$–monotone.*

Before providing the proof we develop a useful factorization of $\mathcal{Q}$–separated symplectic operators. We will use the standard symplectic form and the standard $\mathcal{Q}$-form in $V = \mathbb{R}^p \times \mathbb{R}^p$.

Let $S = \begin{bmatrix} A & B \\ C & D \end{bmatrix}$ be a symplectic operator on $V = \mathbb{R}^p \times \mathbb{R}^p$. Hence A, B, C, D are $p \times p$ matrices such that $C^*A = A^*C, D^*B = B^*D,$ and $D^*A - B^*C = I$.

LEMMA 2.2. *If a symplectic S is $\mathcal{Q}$–separated then A and D are invertible.*

PROOF. Suppose that there is $\widehat{v}_1 \in \mathbb{R}^p$ such that $A\widehat{v}_1 = 0$. It follows that $\widehat{v}_1 = (D^*A - B^*C)\widehat{v}_1 = -B^*C\widehat{v}_1$.

For $v = (\widehat{v}_1, v_2) \in \mathcal{C}_+$ we must have $Sv = (Bv_2, C\widehat{v}_1 + Dv_2) \in \mathcal{C}_+$, i.e., if $\langle\widehat{v}_1, v_2\rangle > 0$ then

$$0 < \langle Bv_2, C\widehat{v}_1 + Dv_2\rangle = -\langle\widehat{v}_1, v_2\rangle + \langle v_2, B^*Dv_2\rangle.$$

Choosing $v_2 = \epsilon\widehat{v}_1, \epsilon > 0$ we obtain

$$\langle\widehat{v}_1, \widehat{v}_1\rangle < \epsilon\langle\widehat{v}_1, B^*D\widehat{v}_1\rangle,$$

which leads to $\widehat{v}_1 = 0$. □

The symplectic operators of the form $\begin{bmatrix} A & 0 \\ 0 & (A^*)^{-1} \end{bmatrix}$ are also $\mathcal{Q}$–isometries.

By Lemma 2.2 given a $\mathcal{Q}$–separated symplectic S we can factor out the following $\mathcal{Q}$-isometry on the right

$$S = \begin{bmatrix} A & B \\ C & D \end{bmatrix} = \begin{bmatrix} I & K \\ H & \cdot \end{bmatrix} \begin{bmatrix} A & 0 \\ 0 & (A^*)^{-1} \end{bmatrix}$$

Symplecticity of S forces H and K to be symmetric and allows the further unique factorization

$$S = \begin{bmatrix} I & 0 \\ H & I \end{bmatrix} \begin{bmatrix} I & K \\ 0 & I \end{bmatrix} \begin{bmatrix} A & 0 \\ 0 & (A^*)^{-1} \end{bmatrix}. \tag{2.1}$$

Moreover H and K must be positive semidefinite ($H \geq 0$, $K \geq 0$). It follows from the following

PROOF OF THEOREM 2.1. Using the factorization (2.1) we get for $v = (v_1, v_2)$

$$\mathcal{Q}(Sv) = \langle v_1, v_2 \rangle + \langle K v_2, v_2 \rangle + \langle H(v_1 + K v_2), v_1 + K v_2 \rangle.$$

Taking $v_2 = 0$ we obtain $\mathcal{Q}(v) = 0$ and hence $\mathcal{Q}(Sv) = \langle H v_1, v_1 \rangle \geq 0$. Taking $v_1 = -K v_2$ we obtain $\mathcal{Q}(Sv) = 0$ which implies that $\mathcal{Q}(v) = -\langle K v_2, v_2 \rangle \leq 0$. □

As a byproduct of the proof we get the following useful observation

PROPOSITION 2.3. *A $\mathcal{Q}$–monotone symplectic operator S maps vectors from L_1 (L_2) either into $\mathcal{C}_+$ or into L_1 (L_2). Moreover if S maps L_1 and L_2 into $\mathcal{C}_+$ then it is strictly $\mathcal{Q}$–monotone.*

It turns out that the modulus of a symplectic $\mathcal{Q}$-monotone operator is also symplectic and hence the positive and negative singular values are inverses of each other. Indeed, in (2.1) we have the further factorization

$$\begin{bmatrix} I & K \\ H & I + HK \end{bmatrix} = \begin{bmatrix} (I+KH)^{\frac{1}{2}} & K(I+HK)^{-\frac{1}{2}} \\ H(I+KH)^{\frac{1}{2}} & (I+HK)^{\frac{1}{2}} \end{bmatrix} \begin{bmatrix} (I+KH)^{-\frac{1}{2}} & 0 \\ 0 & (I+HK)^{\frac{1}{2}} \end{bmatrix}.$$

Since the last matrix is a $\mathcal{Q}$–isometry and the next to last matrix is $\mathcal{Q}$–symmetric we obtain that the modulus of S is equal to the next to last matrix. We proceed to calculate its eigenvalues.

PROPOSITION 2.4. *The positive singular values of a $\mathcal{Q}$–monotone symplectic operator S are equal to $\sqrt{1+t_i} + \sqrt{t_i} = \exp(\sinh^{-1}\sqrt{t_i})$, where $t_1, \ldots, t_p$ are the eigenvalues of CB^*, which are real and nonnegative.*

PROOF. To calculate the pseudo Euclidean singular values of S we will use the fact that they are not changed if we multiply S on left and on the right by $\mathcal{Q}$–isometries. In the factorization (2.1) $CA^{-1} = H$ and $BA^* = K = AB^*$, so that $HK = CB^*$. Since $H \geq 0$ and $K \geq 0$, all the eigenvalues of HK must be real and nonnegative. Let us assume that $K > 0$, the general case will follow by continuity. We get

$$\begin{bmatrix} K^{-\frac{1}{2}} & 0 \\ 0 & K^{\frac{1}{2}} \end{bmatrix} \begin{bmatrix} I & K \\ H & I + HK \end{bmatrix} \begin{bmatrix} K^{\frac{1}{2}} & 0 \\ 0 & K^{-\frac{1}{2}} \end{bmatrix} = \begin{bmatrix} I & I \\ F & I + F \end{bmatrix}$$

where $F = K^{\frac{1}{2}}HK^{\frac{1}{2}}$ has the same eigenvalues as HK. Finally if Z is the orthogonal matrix which diagonalizes F, i.e., $Z^{-1}FZ = T$ is diagonal, then

$$\begin{bmatrix} Z^{-1} & 0 \\ 0 & Z^{-1} \end{bmatrix} \begin{bmatrix} I & I \\ F & I+F \end{bmatrix} \begin{bmatrix} Z & 0 \\ 0 & Z \end{bmatrix} = \begin{bmatrix} I & I \\ T & I+T \end{bmatrix} =$$
$$\begin{bmatrix} (I+T)^{\frac{1}{2}} & (I+T)^{-\frac{1}{2}} \\ T(I+T)^{\frac{1}{2}} & (I+T)^{\frac{1}{2}} \end{bmatrix} \begin{bmatrix} (I+T)^{-\frac{1}{2}} & 0 \\ 0 & (I+T)^{\frac{1}{2}} \end{bmatrix}.$$

It shows that the singular values of S are equal to the eigenvalues of the next to last matrix and they are easy to calculate. □

It follows from the proof of Proposition 2.4 that

PROPOSITION 2.5. *If a $\mathcal{Q}$ monotone symplectic operator S maps all vectors in L_1 and L_2 into $\mathcal{C}_+$ except for a subspace of L_1 of codimension k then at least k largest singular values of S are strictly larger than one.*

3. The Lyapunov exponents

We are going to translate the properties of $\mathcal{J}$-separated operators into the properties of Lyapunov exponents of matrix cocycles with values in the semigroup of $\mathcal{J}$-separated matrices.

Let X be a measurable space with a probabilistic measure ν and let $\Phi : X \to X$ be an ergodic map preserving the measure ν.

Let further $S : X \to Gl(V)$ be a measurable map such that $\log \|S(x)\|$ and $\log \|S(x)^{-1}\|$ are ν–integrable functions. We will refer to this requirement as *integrability condition.* We define the matrix valued cocycle

$$S^m(x) = S(\Phi^{m-1}x)\dots S(\Phi x)S(x), \quad \text{for } m \geq 0,$$
$$S^m(x) = S(\Phi^m x)^{-1}\dots S(\Phi^{-1}x)^{-1}, \quad \text{for } m \leq -1.$$

By the Oseledets Multiplicative Ergodic Theorem, [**Ra**],[**Ru**], there are numbers

$\lambda_1 < \dots < \lambda_N$, called the Lyapunov exponents of the measurable cocycle $S(x)$, $x \in X$, and for almost all x there is an invariant splitting

$$V = V_1(x) \oplus \dots \oplus V_N(x),$$

such that for all vectors $0 \neq v \in V_k(x)$

$$\lim_{m\to\pm\infty} \frac{1}{|m|} \log \|S^m(x)v\| = \pm\lambda_k.$$

The invariance of the splitting means that $S(x)V_k(x) = V_k(\Phi x)$.

To simplify the notation we will omit the dependence of the subspaces on $x \in X$,

The dimensions d_k of the subspaces V_k are called the multiplicities of the Lyapunov exponents.

Moreover, if we denote by $W_k = \bigoplus_{i=1}^k V_i$, we have that for any linear subspace $L \subset W_k$ having the trivial intersection with W_{k-t-1} and such that $W_{k-t} + L = W_k$

the exponential rate of growth of volume on S is equal to the sum of appropriate Lyapunov exponents taken with multiplicities,

$$\begin{aligned}(3.1)\qquad &\lim_{m\to+\infty}\frac{1}{m}\log|det\,(S^m(x)|_L)\,| = \widetilde{d}\lambda_{k-t} + d_{k-t+1}\lambda_{k-t+1} + \cdots + \cdots + d_k\lambda_k,\\ &\widetilde{d} = \dim W_{k-t}\cap L = \dim L - (d_{k-t+1} + \cdots + d_k).\end{aligned}$$

From now on we will be assuming that all the matrices $S(x), x\in X$, are $\mathcal{J}$–separated. Let us denote by $r_p^+(x) \ge \cdots \ge r_1^+(x) \ge r_1^-(x) \ge \cdots \ge r_q^-(x) > 0$ the pseudo Euclidean singular values of $S(x)$.

LEMMA 3.1. *All the functions* $\log r_k^{\pm}(x)$ *are* ν*-integrable.*

PROOF. It is enough to check integrability for $\log r_p^+(x)$ and $\log r_q^-(x)$. We have

$$(r_p^+)^2 = \|R^2\| = \|SJS^*J\| \le \|S\|^2,\ (r_q^-)^{-2} = \|R^{-2}\| = \|J(S^*)^{-1}JS^{-1}\| \le \|S^{-1}\|^2,$$

which proves the claim due to the integrability of $\log\|S(x)\|$ and $\log\|S(x)^{-1}\|$. □

The Lyapunov exponents $\lambda^{\pm}(v)$ (forward and backward in time) are defined for almost every initial point $x\in X$ and for all nonzero vectors $v\in V$. Let us denote by $l_{\pm} = \int \log r_1^{\pm}(x)d\nu(x)$.

PROPOSITION 3.2. *For any vector* $v\in \mathcal{C}_{\pm}, v\neq 0$ *we have* $\lambda^{\pm}(v)\ge \pm l_{\pm}$.

PROOF. Since $\frac{1}{r_1^+(x)}S$ is $\mathcal{J}$-monotone we have $\mathcal{J}(S(x)v) \ge \left(r_1^+(x)\right)^2\mathcal{J}(v)$ for all vectors v. In the positive cone the square of norm is not smaller than $\mathcal{J}$. Hence the exponential growth of $\mathcal{J}$ implies the exponential growth of the norm. More precisely we have for $v\in\mathcal{C}_+$,

$$\|S^m(x)v\|^2 \ge \mathcal{J}(S^m(x)v) \ge \prod_{i=0}^{m-1}\left(r_1^+(\Phi^i x)\right)^2\mathcal{J}(v).$$

It follows from Lemma 3.1 and the Birkhoff Ergodic Theorem that

$$\lambda^+(v) \ge \int \log r_1^+(x)d\nu(x) = l_+.$$

To obtain the inequality $\lambda^-(v)\ge -l_-$. we need to use the fact that the operators $S(x)^{-1}$ are (-$\mathcal{J}$)-separated and apply the argument above to $S(x)^{-1}$. □

PROPOSITION 3.3. *If* $l_+ > l_-$ *then for almost every* $x\in X$ *there is* $m\ge 1$ *such that* $S^m(x)$ *is strictly* $\mathcal{J}$*-separated.*

A matrix cocycle with the last property will be called *eventually strictly* $\mathcal{J}$*-separated.*

PROOF. It follows immediately from Corollary 1.4 since a $\mathcal{J}$-separated operator S is strictly $\mathcal{J}$-separated if and only if $r_1^+(S) > r_1^-(S)$. □

We introduce another labeling of the Lyapunov exponents and the subspaces of the splitting.

If $d_1 > q$ then we define $s_- = 0$. Similarly if $d_N > p$ we define $s_+ = 0$. Otherwise we let

$$s_- = \max\{k | \sum_{i=1}^{k} d_i \le q\} \quad \text{and} \quad s_+ = \max\{k | \sum_{i=N-k+1}^{N} d_i \le p\}.$$

When $s_- + s_+ = N$ we relabel the subspaces as

$$V_{-s_-} \oplus \cdots \oplus V_{-1} \oplus V_1 \oplus \cdots \oplus V_{s_+},$$

and when $s_- + s_+ = N - 1$ we use also the zero label

$$V_{-s_-} \oplus \cdots \oplus V_{-1} \oplus V_0 \oplus V_1 \oplus \cdots \oplus V_{s_+}.$$

Hence the subspace V_0 may be present or absent.

The interval of separation $[r_1^-(x), r_1^+(x)]$ varies with the operator $S(x)$. We get a gap in the Lyapunov spectrum if $l_+ = \int \log r_1^+ d\nu > \int \log r_1^- d\nu = l_-$ and in such a case V_0 is absent.

PROPOSITION 3.4. *If V_0 is present then $l_- = l_+ = \lambda_0$. If $l_+ > l_-$, then V_0 is absent, $\lambda_1 \ge l_+ > l_- \ge \lambda_{-1}$ and*

$$V_{-s_-} \oplus \cdots \oplus V_{-1} \subset \mathcal{C}_-, \quad V_1 \oplus \cdots \oplus V_{s_+} \subset \mathcal{C}_+. \tag{3.2}$$

PROOF. If V_0 is present then the dimensions of $V_{-s_-} \oplus \cdots \oplus V_{-1} \oplus V_0$ and $V_0 \oplus V_1 \oplus \cdots \oplus V_{s_+}$ exceed q and p respectively. It follows that the former subspace must contain positive vectors and the latter negative vectors. Using Proposition 3.2 we conclude that $\lambda_0 \ge l_+$ and $-\lambda_0 \ge -l_-$.

In the same way we obtain that if V_0 is absent then $\lambda_1 \ge l_+$ and $l_- \ge \lambda_{-1}$.

It remains to show the inclusions (3.2). First of all we have the weaker inclusions

$$V_{-s_-} \oplus \cdots \oplus V_{-1} \subset \mathcal{C}_- \cup \mathcal{C}_0, \quad V_1 \oplus \cdots \oplus V_{s_+} \subset \mathcal{C}_+ \cup \mathcal{C}_0. \tag{3.3}$$

Indeed, if for example $V_{-s_-} \oplus \cdots \oplus V_{-1}$ contains any positive vectors then by Proposition 3.2 $\lambda_{-1} \ge l_+$, which violates the assumption that $l_+ > l_-$.

Further by Proposition 3.3 the assumption $l_+ > l_-$ implies that the matrix cocycle is eventually strictly $\mathcal{J}$-separated. By the invariance of the splitting we get (3.2) almost everywhere. □

It seems plausible that the weaker inclusions (3.3) must hold for every $\mathcal{J}$-separated matrix cocycles without any further assumptions, but we were unable to prove it.

PROPOSITION 3.5. *If the matrix cocyle is eventually strictly $\mathcal{J}$-separated then V_0 is absent, $\lambda_1 > \lambda_{-1}$ and the inclusions* (3.2) *hold.*

PROOF. By Proposition 3.4 the claims must hold for strictly $\mathcal{J}$-separated cocycles. The general case of the eventually strictly $\mathcal{J}$-separated cocycles reduces to the special case by considering the subset

$$X_m = \{x \in X | S^m(x) \quad \text{is strictly} \quad \mathcal{J}\text{-separated}\}.$$

For large $m \ge 1$ the set X_m must have measure close to 1 (the full measure). We further consider the return map to such a set X_m and the induced matrix cocycle which is automatically strictly $\mathcal{J}$-separated. But the splitting for the induced cocycle coincides with the original splitting restricted to the subset X_m and the Lyapunov exponents are changed by the factor equal to the average return time (cf. [**K-H**], Lemma S.2.8, page 665). □

Let us introduce $d_0^- = q - d_{-1} - \cdots - d_{-s_-}$ and $d_0^+ = p - d_1 - \cdots - d_{s_+}$. Clearly $d_0^- + d_0^+ = \dim V_0$.

THEOREM 3.6. *If a measurable cocycle $S(x)$, $x \in X$, satisfies the integrability condition and it has values in the semigroup of $\mathcal{J}$–separated matrices then for $0 \leq k \leq s_+$*

$$d_0^+\lambda_0 + d_1\lambda_1 + \cdots + d_k\lambda_k \geq \sum_{i=1}^{D} \int \log r_i^+ d\nu,$$

where $D = d_0^+ + d_1 + \cdots + d_k$ and for $0 \leq k \leq s_-$

$$d_0^-\lambda_0 + d_{-1}\lambda_{-1} + \cdots + d_{-k}\lambda_{-k} \leq \sum_{i=1}^{D} \int \log r_i^- d\nu.$$

where $D = d_0^- + d_{-1} + \cdots + d_{-k}$.

PROOF OF THEOREM 3.6. We will use the formula (3.1) for the rate of volume growth on subspaces. According to the new labeling we let $W_j = \bigoplus_{i=-s_-}^{j} V_i$ for $j = -s_-, \ldots, s_+$. We proceed as in the proof of Proposition 3.2. For $0 \leq k \leq s_+$ let us consider the family of linear subspaces $L \subset W_k$ of dimension $D = \dim W_k - q = d_0^+ + d_1 + \cdots + d_k$ (or equivalently of codimension q in W_k) transversal to W_{-1} and such that $W_0 + L = W_k$. We can choose such a subspace in the positive cone $\mathcal{C}_+$. Indeed, this family of subspaces of W_k is open and dense in the respective Grassmanian. At the same time the intersection of W_k with any maximal positive subspace (of dimension p) must be of dimension at least D. Hence there are positive subspaces in W_k of dimension D. Moreover the family of such subspaces is open in the respective Grassmanian. It follows that there must be a positive subspace in our family.

If the subspace L is in the positive cone then the Euclidean volume on it is not smaller than the $\mathcal{J}$–volume. Using Proposition 1.3 we get

$$\log |det\,(S^m(x)|_L)\,| \geq \log \alpha_D(L, S^m(x)) + c = \sum_{i=0}^{m-1} \log \alpha_D(L, S(\Phi^i x)) + c$$

$$\geq \sum_{i=0}^{m-1} \log \sigma_D(S(\Phi^i x)) + c = \sum_{i=0}^{m-1} (\log r_1^+(\Phi^i x) + \cdots + \log r_D^+(\Phi^i x)) + c,$$

where the constant c is the logarithm of the ratio of the $\mathcal{J}$-volume and the Euclidean volume on the subspace L. Hence in view of (3.1) we obtain the first inequality. The other inequality is obtained by observing that $S(x)^{-1}$ are $(-\mathcal{J})$–separated and applying the same argument. □

If we introduce the n Lyapunov exponents repeated according to the multiplicities and labeled as $\mu_{-q} \leq \mu_{-q+1} \leq \cdots \leq \mu_{-1} \leq \mu_1 \leq \cdots \leq \mu_p$ we can formulate the following

COROLLARY 3.7. *For $1 \leq k_1 \leq p$ and $1 \leq k_2 \leq q$*

$$\mu_1 + \cdots + \mu_{k_1} \geq \sum_{i=1}^{k_1} \int \log r_i^+ d\nu \quad \text{and} \quad \sum_{i=1}^{k_2} \int \log r_i^- d\nu \geq \mu_{-1} + \cdots + \mu_{-k_2}.$$

PROOF. It follows formally from all the inequalities of Theorem 3.6. One can also repeat the proof of this theorem starting from a full noninvariant flag of subspaces as introduced in the proof of Theorem 3.9 below. □

Let us now formulate the symplectic version of the estimates. Hence we assume that the matrix cocycle $S(x), x \in X$, has values in the symplectic group. In such a case the Lyapunov exponents come in pairs of opposite numbers. The following proof of this fact is taken from [**W-L**].

LEMMA 3.8. *For an integrable cocycle $S(x)$, $x \in X$, with values in the symplectic group $Sp(\mathbb{R}^{2p})$, for any pair of non skew-orthogonal vectors $u, v \in \mathbb{R}^{2p}$, i.e., $\omega(u,v) \neq 0$, we have*

$$\lambda(u) + \lambda(v) \geq 0.$$

PROOF. For the standard Euclidean norm $\|\cdot\|$ we have $|\omega(u,v)| \leq \|u\|\|v\|$. By symplecticity

$$\log|\omega(S^m(x)u, S^m(x)v)| = \log|\omega(u,v)|,$$

and

$$\frac{1}{m}\log|\omega(S^m(x)u, S^m(x)v)| \leq \frac{1}{m}\log\|S^m(x)u\| + \frac{1}{m}\log\|S^m(x)v\|.$$

□

THEOREM 3.9. *If an integrable cocycle $S(x)$, $x \in X$, has values in the symplectic group $Sp(\mathbb{R}^{2p})$ then we have the following symmetry of the Lyapunov spectrum*

$$\lambda_0 = 0, \quad \text{if } V_0 \text{ is present}, \quad s_+ = s_- = s, \ \lambda_{-k} + \lambda_k = 0,$$

and the multiplicities of λ_{-k} and λ_k are equal, for $k = 1, 2, \ldots, s$. Moreover the subspace $W_k = \oplus_{i=-s}^{k} V_i$ is the skew-orthogonal complement of $W_{-(k+1)} = \oplus_{i=-s}^{-(k+1)} V_i$.

PROOF. Let $\mu_1 \leq \mu_2 \leq \cdots \leq \mu_{2p}$ be the Lyapunov exponents listed with repetitions according to their multiplicities. We have $\mu_1 + \mu_2 + \cdots + \mu_{2p} = 0$.

We can choose a flag of subspaces

$$\{0\} = Z_0 \subset Z_1(x) \subset \cdots \subset Z_{2p-1}(x) \subset Z_{2p} = \mathbb{R}^{2p},$$

such that $\dim Z_l = l$ and for all vectors $v \in Z_l(x) \setminus Z_{l-1}(x)$ the Lyapunov exponent $\lambda(v) = \mu_l$, for $l = 1, 2, \ldots, 2p$. (Note that except in the case of all multiplicities equal to 1 there is a continuum of such flags.)

Since for any $l \leq p$, $\dim Z_l + \dim Z_{2p-l+1} = 2p + 1$, there are vectors $u \in Z_l$ and $v \in Z_{2p-l+1}$ such that $\omega(u,v) \neq 0$. By continuity there must be also vectors $\tilde{u} \in Z_l \setminus Z_{l-1}$ and $\tilde{v} \in Z_{2p-l+1} \setminus Z_{2p-l}$ such that $\omega(\tilde{u}, \tilde{v}) \neq 0$. It follows from Lemma 3.8 that

$$\mu_l + \mu_{2p-l+1} \geq 0,$$

for $l = 1, 2, \ldots, p$. Adding these inequalities together, we get

$$0 = \sum_{l=1}^{p} (\mu_l + \mu_{2p-l+1}) \geq 0,$$

which shows that all the inequalities must be actually equalities. It follows immediately that for any $k = 1, \ldots, s$, the multiplicities of λ_k and λ_{-k} are equal and $\lambda_k + \lambda_{-k} = 0$.

To show that the subspace W_k is the skew-orthogonal complement of the subspace $W_{-(k+1)}$ (denoted by $W^{\angle}_{-(k+1)}$) we observe that $\omega(u, v) = 0$ for any $u \in W_k$ and $v \in W_{-(k+1)}$. Indeed, if this is not the case we could use Lemma 3.8 to claim that $\lambda_k + \lambda_{-(k+1)} \geq 0$, which leads to the contradiction

$$0 = \lambda_k + \lambda_{-k} > \lambda_k + \lambda_{-(k+1)} \geq 0.$$

We obtain $W_k \subset W^{\angle}_{-(k+1)}$. Since the dimensions of these subspaces are equal they must coincide. □

If all the symplectic matrices $S(x) = \begin{bmatrix} A(x) & B(x) \\ C(x) & D(x) \end{bmatrix}, x \in X$, are also $\mathcal{Q}$–monotone (with the standard form $\mathcal{Q}$) then using Proposition 2.4 we can reformulate Corollary 3.7 as

THEOREM 3.10. *For any* $1 \leq k \leq p$

$$\mu_1 + \cdots + \mu_k \geq \sum_{i=1}^{k} \int \sinh^{-1} \sqrt{t_i} d\nu,$$

where $0 \leq t_1(x) \leq \cdots \leq t_p(x)$ *are the eigenvalues of* CB^*.

4. The $\mathcal{J}$–separated diffeomorphisms and flows

Let us move the discussion to a manifold of dimension n and a diffeomorphism $\Phi : M^n \to M^n$. We assume that the manifold is equipped with a continuous pseudo Riemannian metric of type $(p, q), p + q = n$, which we will denote again by $\mathcal{J}$.

We obtain the fields of positive and negative cones $\mathcal{C}_{\pm}(x), x \in M$. A diffeomorphism Φ is called $\mathcal{J}$–separated if $D\Phi\mathcal{C}_+(x) \subset \mathcal{C}_+(\Phi x)$. Similarly we define diffeomorphisms which are strictly $\mathcal{J}$–separated, $\mathcal{J}$–monotone and strictly $\mathcal{J}$–monotone.

If a diffeomorphism is $\mathcal{J}$–separated then the pseudo Riemannian singular values are well defined for the derivative $D\Phi$ at any point $x \in M$. We will denote them by

$$0 < r_q^-(x) \leq \cdots \leq r_1^-(x) \leq r_1^+(x) \leq \cdots \leq r_p^+(x).$$

PROPOSITION 4.1. *If a diffeomorphism* Φ *is strictly* $\mathcal{J}$*–separated then it has a dominated splitting, i.e,, it has a continuous invariant splitting of the tangent bundle*

$$T_x M = E^-(x) \oplus E^+(x),$$

such that for some $0 < \lambda < 1, c > 0$, *and all* $m \geq 1$

$$\|D\Phi^m{}_{|E^-(x)}\| \|D\Phi^{-m}{}_{|E^+(\Phi^m x)}\| \leq c\lambda^m. \tag{4.1}$$

PROOF. At every point $x \in M$ we have the manifold of positive subspaces $\mathcal{L}_+(x)$. We will use the contraction property of the action of $D\Phi$ on $\mathcal{L}_+(x)$. The diameter of $D\Phi\mathcal{L}_+(x)$ is finite because $D\Phi$ is strictly $\mathcal{J}$–separated, and hence it is uniformly bounded by a constant c_0. It follows from Theorem 1.6 that the diameter of $D\Phi^m\mathcal{L}_+(\Phi^{-m}x)$ does not exceed $c_0\prod_{i=1}^{m-1}\frac{r_1^-(\Phi^{-i}x)}{r_1^+(\Phi^{-i}x)}$. Since the subsets

$$D\Phi^{m+1}\mathcal{L}_+(\Phi^{-m-1}x) \subset D\Phi^m\mathcal{L}_+(\Phi^{-m}x)$$

are nested and their diameter decays exponentially we get a unique point of intersection

$$E^+(x) = \bigcap_{m=1}^{\infty} D\Phi^m\mathcal{L}_+(\Phi^{-m}x).$$

It is clear that the subspace depends continuously on x (by the standard argument it is actually Hölder continuous). To obtain the other subspace of the splitting we need to observe that Φ^{-1} is strictly $(-\mathcal{J})$–separated and repeat the above argument.

The property (4.1) does not depend on the choice of a norm in the tangent bundle, a change of the norm leads to a change of the constant $c > 0$. Let us define a continuous Riemannian metric by the conditions that it is equal to $\mathcal{J}$ on E^+, $-\mathcal{J}$ on E^- and these subspaces are orthogonal. For this metric (4.1) holds with $c = 1$ and $\lambda = \sup_x \frac{r_1^-(x)}{r_1^+(x)} < 1$. Indeed, for every vector $v \in E^+(\Phi x)$ we have $D\Phi^{-1}v \in E^+(x)$ and by the definition of the pseudo Euclidean singular values (Proposition 1.3 with $d = 1$)

$$\frac{\|v\|}{\|D\Phi^{-1}v\|} \geq r_1^+(x).$$

We conclude that

$$\|D\Phi^{-1}{}_{|E^+(\Phi x)}\| \leq \frac{1}{r_1^+(x)}.$$

Similarly we can estimate $\|D\Phi_{|E^-(x)}\| \leq r_1^-(x)$. □

As a corollary we obtain the theorem of Lewowicz, [**L**],

THEOREM 4.2. *If a diffeomorphism is strictly $\mathcal{J}$–monotone then it is Anosov.*

PROOF. By Proposition 4.1 we obtain the continuous invariant splitting. On the unstable subspace the $\mathcal{J}$–form increases exponentially and it can be compared with any norm. □

Establishing that a dynamical system is $\mathcal{J}$–separated is an important step, but this property alone allows varied and complicated behavior different from an Anosov system. For example the DA attractors can be made $\mathcal{J}$–separated, [**Ro**]. Also the examples studied by Hu and Young, [**H-Y**], are $\mathcal{J}$–separated without being Anosov.

In the piecewise differentiable setup with a chosen invariant measure ν we get for a strictly $\mathcal{J}$–separated mapping Φ a measurable invariant splitting $T_xM = E^-(x) \oplus E^+(x)$ defined ν almost everywhere and $E^+(x) \subset \mathcal{C}_+$, $E^-(x) \subset \mathcal{C}_-$. Now the contraction in the manifold of positive subspaces is not uniform but by the Birkhoff Ergodic Theorem it is pointwise exponential almost everywhere.

For an ergodic invariant measure ν we can obtain the estimates of the Lyapunov exponents in terms of the pseudo-Riemannian singular values of the $\mathcal{J}$–separated $D\Phi$. We need to assume only that $D\Phi$ is $\mathcal{J}$–separated almost everywhere with respect to the measure.

THEOREM 4.3.
For a piecewise differentiable $\mathcal{J}$–separated Φ and an an ergodic invariant measure ν, if $\log \|D\Phi^{\pm 1}\|$ are integrable functions then the Lyapunov exponents

$$\mu_{-q} \leq \mu_{-q-1} \leq \cdots \leq \mu_{-1} \leq \mu_1 \leq \mu_2 \leq \cdots \leq \mu_p,$$

satisfy

$$\mu_1 + \cdots + \mu_{k_1} \geq \sum_{i=1}^{k_1} \int \log r_i^+ d\nu \quad \text{and} \quad \sum_{i=1}^{k_2} \int \log r_i^- d\nu \geq \mu_{-1} + \cdots + \mu_{-k_2},$$

for any $k_1 \leq p, k_2 \leq q$, where the coefficients

$$0 < r_q^-(x) \leq \cdots \leq r_1^-(x) \leq r_1^+(x) \leq \cdots \leq r_p^+(x)$$

are the singular values of the $\mathcal{J}$–separated operators $D_x\Phi$.

PROOF. We have an auxiliary Riemannian metric on M which defines the norm $\|\cdot\|$ in the integrability condition (in $\log \|D\Phi^{\pm 1}\|$).

We introduce a measurable Riemannian metric associated to the pseudo Riemannian metric $\mathcal{J}$ by choosing a measurable field of positive subspaces $P(x)$, $x \in M$, and considering their $\mathcal{J}$–orthogonal complements, which we denote by $Q(x)$, $x \in M$. The Riemannian metric is equal to $\mathcal{J}$ on P and $-\mathcal{J}$ on Q, and the two subspaces are orthogonal. Let $\|\cdot\|_J$ denote the respective norm. We can now choose a measurable family of orthogonal frames in P and Q so that the derivative $D\Phi$ becomes a measurable matrix cocycle with $\mathcal{J}$–separated values. If we had the integrability of $\log \|D\Phi^{\pm 1}\|_J$, we could apply Theorem 3.6 directly. It turns out that we do not need to impose any further integrability conditions. Indeed, there is a measurable function $c(x) > 0$ such that

$$\frac{1}{c(x)}\|\cdot\| \leq \|\cdot\|_J \leq c(x)\|\cdot\|.$$

For every $\epsilon > 0$ consider the subset $M_\epsilon = \{x \in M \mid c(x) \geq \epsilon\}$ and the return map $\Phi_\epsilon : M_\epsilon \to M_\epsilon$. We clearly have the integrability of $\log \|D\Phi_\epsilon^{\pm 1}\|$ which leads to the integrability of $\log \|D\Phi_\epsilon^{\pm 1}\|_J$. Hence we can apply to $D\Phi_\epsilon$ the estimates in Theorem 3.6. The Lyapunov exponents for $D\Phi_\epsilon$ and $D\Phi$ differ by the average return time to M_ϵ which is close to 1 for small ϵ. The final observation is that the average of the coefficient $\sigma_d(D\Phi_\epsilon)$ is not smaller than the average of the coefficient $\sigma_d(D\Phi)$. It follows from the supermultiplicativity (1.1) of σ_d. Hence in the limit of $\epsilon \to 0$ we obtain the desired estimate. □

To develop the theory of $\mathcal{J}$–separated flows $\Phi^t : M^n \to M^n, t \in \mathbb{R}$, we need the velocity vector field $F(x)$ of the flow, $F(x) = \frac{d}{dt}\Phi^t(x)_{|t=0}$, to be nonzero everywhere and the form $\mathcal{J}$ to be degenerate. More precisely the positive index of inertia of $\mathcal{J}$ is equal to $p \geq 1$, the negative index of inertia is equal to $q \geq 1$ and $p+q = n-1$, and for the bilinear form $\mathcal{J}$ the linear form $\mathcal{J}(F, \cdot)$ vanishes identically.

Equivalently we consider the quotient tangent spaces

$$\widetilde{T}_x M^n = T_x M^n / span\{F(x)\}$$

and the nondegenerate forms $\mathcal{J}$ on them. The derivative $D\Phi^t$ projects naturally to the quotient spaces and we can apply the theory developed previously to the derivative cocycles on the quotient spaces.

In particular if for a smooth flow the derivative cocycle is strictly $\mathcal{J}$–separated then by Proposition 4.1 we get a splitting of the quotient space, but in general it does not lift to the splitting of the tangent space. We can lift the splitting in the case of strict $\mathcal{J}$–monotonicity by the classical proof of Anosov [**A**] (see also [**W8**]) and we obtain the counterpart of the Lewowicz's Theorem 4.2 for flows.

THEOREM 4.4. *If the derivative cocycle of a smooth flow is strictly $\mathcal{J}$–monotone then the flow is Anosov.*

Appendix. Boltzmann – Sinai Gas of Hard Spheres

In our discussion we will follow the approach developed in [**W2**] and [**W5**]. We consider the system of N spherical particles in a d-dimensional vessel, either a d-dimensional torus or a box. The particles have arbitrary masses: $m_1, m_2, \ldots, m_N$, and they collide elastically with each other (and the walls of the box). The Hamiltonian of the system is

$$H = \sum_{i=1}^{N} \frac{p_i^2}{2m_i}$$

where $p_i \in \mathbb{R}^d$ is the momentum of i-th particle, $i = 1, \ldots, N, d \geq 2$.

The Hamiltonian differential equations are linear

$$\dot{q}_i = \frac{p_i}{m_i}, \qquad \dot{p}_i = 0, \quad i = 1, ..., N.$$

Our phase space is a linear symplectic space. All its tangent spaces are naturally identified with it but for the sake of clarity we will distinguish between a point in the phase space and a tangent vector by placing δ in front of q and p. We choose the constant Lagrangian subspaces in the $(\delta q, \delta p)$-space $\mathbb{R}^{dN} \times \mathbb{R}^{dN}$

$$L_1 \equiv \mathbb{R}^{dN} \times \{0\} = \{dp_1 = \cdots = dp_N = 0\}$$

and

$$L_2 \equiv \{0\} \times \mathbb{R}^{dN} = \{dq_1 = \cdots = dq_N = 0\}.$$

This is the choice which makes the integrable system of noninteracting particles $\mathcal{Q}$–monotone.

We see that the velocity vector field is in L_1. The quadratic form $\mathcal{Q}$ is equal to

$$\mathcal{Q} = \sum_{k=1}^{N} \langle dq_k, dp_k \rangle,$$

where $\langle dq_k, dp_k \rangle = \sum_{i=1}^{d} dq_{k,i} dp_{k,i}$.

The linearized equations are

$$\frac{d}{dt}\delta q_i = \frac{\delta p_i}{m_i}, \qquad \frac{d}{dt}\delta p_i = 0.$$

We obtain immediately that between collisions

$$\frac{d}{dt}Q\left((\delta q, \delta p)\right) = \sum_{i=1}^{N} \frac{(\delta p_i)^2}{m_i} \geq 0,$$

which means that the system is $\mathcal{Q}$–monotone. Moreover it follows that L_2 enters $\mathcal{C}_+$ immediately and completely.

We proceed to investigate collisions of two particles in dimension $d \geq 2$. Mathematically speaking the collisions are described by a symplectomorphism defined on the boundary of the phase space. More precisely, not all positions of the balls are allowed since they cannot overlap. The configurations with at least two particles touching each other form the boundary of the phase space. The intersection of this boundary with a given total energy level, say $\{H = \frac{1}{2}\}$, has a canonical symplectic structure (its tangent can be identified with the factor space by the line spanned by the velocity vector). Collisions are then described by a symplectomorphism of this symplectic manifold (a gross simplification – this boundary has many discontinuities and so we only get piecewise differentiability, but it is still possible to talk about Lyapunov exponents). So we have here a combination of a Hamiltonian flow and a symplectomorphism. Such systems were discussed abstractly in [**W3**] and they are called there flows with collisions, the boundary being called the collision manifold.

Let the colliding particles have positions q_1, q_2, momenta p_1, p_2, masses m_1, m_2 and radii r_1, r_2. The collision manifold is given locally by $\|q_2 - q_1\| = r = r_1 + r_2$ and its tangent space by

$$\langle e, dq_2 - dq_1 \rangle = 0 \quad \text{where} \quad e = \frac{1}{r}(q_2 - q_1), \|e\| = 1.$$

To facilitate the calculations we introduce locally new symplectic coordinates (x, y), the center of mass, relative position and the respective momenta

$$x_1 = \frac{m_1}{m_1 + m_2} q_1 + \frac{m_2}{m_1 + m_2} q_2, \quad p_1 = \frac{m_1}{m_1 + m_2} y_1 - y_2$$

$$x_2 = q_2 - q_1, \qquad p_2 = \frac{m_2}{m_1 + m_2} y_1 + y_2.$$

In these coordinates the collision manifold is $\|x_2\| = r$, its tangent space $\langle x_2, dx_2 \rangle = 0$ and the collision map is given by the formulas

$$y_1^+ = y_1^-, \quad y_2^+ = y_2^- - \frac{2}{r^2}\langle y_2^-, x_2 \rangle x_2.$$

Differentiating we obtain

$$dy_2^+ = dy_2^- - \frac{2}{r^2}\langle dy_2^-, x_2 \rangle x_2 - \frac{2}{r^2}\langle y_2^-, dx_2 \rangle x_2 - \frac{2}{r^2}\langle y_2^-, x_2 \rangle dx_2 \tag{A.1}$$

Since we have that

$$Q = \langle dq_1, dp_1 \rangle + \langle dq_2, dp_2 \rangle + \cdots = \langle dx_1, dy_1 \rangle + \langle dx_2, dy_2 \rangle + \ldots,$$

we obtain on the collision manifold

$$
\begin{aligned}
Q^+ - Q^- &= \langle dx_2, dy_2^+\rangle - \langle dx_2, dy_2^-\rangle = -\frac{2}{r^2}\langle y_2^-, x_2\rangle\langle dx_2, dx_2\rangle \\
&= -\frac{2m_1m_2}{(m_1+m_2)r^2}\langle \frac{p_2^-}{m_2} - \frac{p_1^-}{m_1}, q_2 - q_1\rangle\langle dq_2 - dq_1, dq_2 - dq_1\rangle \geq 0.
\end{aligned} \tag{A.2}
$$

Thus monotonicity is verified also for the collisions. (The inequality $\langle \frac{p_2^-}{m_2} - \frac{p_1^-}{m_1}, q_2 - q_1\rangle > 1$ holds automatically for all nondegenerate collisions. It simply means that the particles are indeed colliding and not flying away from each other.)

Let us note that when approaching a collision, before we can apply (A.1) and (A.2) to a vector $(\delta q, \delta p)$ we have to project it onto the tangent space of the collision manifold, $\langle e, dq_2 - dq_1\rangle = 0$, along the velocity vector of the flow.

$$
(\delta q_i, \delta p_i) \mapsto \left(\delta q_i - \lambda\frac{p_i}{m_i}, \delta p_i\right), \quad \lambda = \frac{\langle \delta q_2 - \delta q_1, e\rangle}{\langle \frac{p_2}{m_2} - \frac{p_1}{m_1}, e\rangle}.
$$

This projection reflects the fact that nearby trajectories do not arrive at the collision at the same time. It does not change the value of the form $\mathcal{Q}$ but it does change the tangent vector.

The formula for the increase of the form $\mathcal{Q}$ shows that if a vector in L_1 does not enter $\mathcal{C}_+$ as a result of a nondegenerate collision of the first two particles then

$$
\delta q_2^- - \delta q_1^- = 0.
$$

If we take into account the necessary projection preceding the application of the derivative of the collision map it means that $\delta q_2 - \delta q_1$ is parallel to $\frac{p_2}{m_2} - \frac{p_1}{m_1}$ (then the projection makes the two components equal). Thus in every nondegenerate collision of two particles all vectors from L_1 enter $\mathcal{C}_+$ except for the subspace of codimension $d-1$ which stays in L_1.

It is clear that there are many special orbits and vectors in L_1 which do not enter $\mathcal{C}_=$ for a long time (ever).

Establishing strict monotonicity for almost all orbits in the system is difficult. This result was obtained by Simányi and Szász, [**S-Sz**]. One obstacle to strict monotonicity is the possibility that the particles are split into two families with all the collisions occurring inside one of the two families. Let us denote the union of orbits on which no such splitting is possible by Z. Simányi and Szász, [**S-Sz**], Theorem 5.1, discovered a short and transparent proof of the fact that for the system on the torus the set Z has full measure.

Using Proposition 2.5 and Theorem 3.10 we can easily prove the following

PROPOSITION A.1. *There are at least $(N-1)(d-1)$ positive Lyapunov exponents almost everywhere on the set Z.*

PROOF. We have established that each nondegenerate collision pushes vectors from L_1 into $\mathcal{C}_+$ except for a codimension $d-1$ subspace which stays in L_1. This exceptional subspace evolves then inside L_1 until the next collision. At that time some vectors from this evolved exceptional subspace are pushed into $\mathcal{C}_+$, unless the new exceptional subspace coincides with it. In general we will obtain a new exceptional subspace of higher codimension (containing all the vectors that stay in

L_1 longer). The question is what is the guaranteed codimension of this exceptional subspace after many collisions.

Each collision contributes $d-1$ linear conditions (equations) only in variables corresponding to the particles involved in the collision. During the evolution between collisions these equations change but no new variables may be introduced into them.

On orbits from Z after some time all the particles are connected by a chain of collisions. Let us inspect the equations determining the evolved exceptional subspaces generated by a minimal chain of $N-1$ collisions connecting all the particles (i.e., in every collision from this chain a particle collides for the first time, we skip the collisions between particles that collided earlier). We obtain $(N-1)(d-1)$ equations. We claim that they are linearly independent. Indeed, suppose k particles have collided generating $(k-1)(d-1)$ equations. These equations contain only the variables associated with the k particles. Note that in the forward evolution these equations are changed but no new variables may be added into them until a new particle collides with one of these k particles. A collision with a new particle generates $d-1$ equations linearly independent in the d variables associated with the new particle. Hence these new $d-1$ equations are linearly independent of the previous equations (because they do not depend on these new variables). We conclude that after the $N-1$ collisions in the minimal chain of collisions we will have $(N-1)(d-1)$ linearly independent equations.

It follows from the above discussion that on orbits from Z after sufficiently long time all the vectors from L_1 are pushed into $\mathcal{C}_+$ except possibly for a subspace of codimension at least $(N-1)(d-1)$. Our Propositions follows now from Proposition 2.5 and Theorem 3.10. □

Note that in the above proof no distinction is made between the cases of the torus and the cube as vessels. In the case of the torus the total momentum is the first integral of the system which leads to $2d$ zero Lyapunov exponents. Complete hyperbolicity has not yet been established for the case of particles in a box, although there is no doubt that it does hold. The special case of two particles was studied by Simányi, [**S2**].

References

[A] D. V. Anosov, *Geodesic flows on Riemannian manifolds with negative curvature*, Proc. Steklov Inst. Math. **90** (1967).

[B] P. Bougerol, *Kalman filtering with random coefficients and contractions*, Siam J. Control Optim. **31** (1993), 942 – 959.

[Ch-S] N. I. Chernov, Ya. G. Sinai, *Ergodic properties of some systems of 2-dimensional discs and 3-dimensional spheres*, Russ. Math. Surv. **42** (1987), 181-207.

[C] W. A. Coppel, *Dichotomies in stability theory*, Lecture Notes in Math. 629, Springer-Verlag (1978).

[H-Y] H. Y. Hu, L.-S. Young, *Nonexistence of SBR measures for some diffeomorphisms that are "almost Anosov"*, Erg. Th. Dyn. Sys. **15** (1995), 67 – 76.

[K-H] A. Katok, B. Hasselblatt, *Introduction to the modern theory of dynamical systems*, Cambridge UP, 1995.

[Ko] A. Korányi, *A Schwarz lemma for bounded symmetric domains*, Proc. AMS **17** (1966), 210 - 213.

[L] J. Lewowicz, *Lyapunov functions and topological stability*, J.Diff.Eq. **38** (1980), 192 – 209.

[L-W] C. Liverani, M. P. Wojtkowski, *Generalization of the Hilbert metric to the space of positive definite matrices*, Pacific Journal of Math. **166** (1994), 339 -355.

[M] R. Markarian, *Billiards with Pesin region of measure one*, Commun. Math. Phys. **118** (1988), 87 – 97.

[P1] V. P. Potapov, *The multiplicative structure of J–contractive matrix functions*, AMS Translations (2) **15** (1960), 131 – 243.

[P2] V. P. Potapov, *Linear fractional transformation of matrices*, AMS Translations (2) **138** (1988), 21 – 35.

[P3] V. P. Potapov, *A theorem on the modulus. I*, AMS Translations (2) **138** (1988), 55 – 65.

[Py] I. I. Pyatetskii-Shapiro, *Automorphic functions and the geometry of classical domains*, Gordon & Breach, 1969.

[Ra] M.-S. Raghunathan,, *A proof of Oseledec's multiplicative ergodic theorem*, Israel J. Math. **32** (1979), 356 - 362.

[Ro] C. Robinson, *Dynamical Systems: Stability, Symbolic Dynamics and Chaos*, CRC Press, 1995.

[Ru] D. Ruelle,, *Ergodic theory of differential dynamical systems*, Publ. Math. IHES **50** (1979), 275 - 306.

[S1] N. Simányi, *The Characteristic Exponents of the Falling Ball Model*, Commun. Math. Phys. **182** (1996), 457 – 468.

[S2] N. Simányi, *Ergodicity of Hard Spheres in a Box*, Ergodic theory and dynamical systems **19** (1999), 741-766.

[Su] M. Suzuki, *The intrinsic metrics on the circular domains in* $\mathbb{C}^n$, Pacific Journal of Math. **112** (1984), 249 -256.

[Sa-K] A.M. Samoilenko, V. L. Kulik, *Exponential dichotomy of an invariant torus of dynamical systems*, Differential Equations **15** (1979), 1019 - 1025.

[S-Sz] N. Simányi, D. Szász, *Hard Ball Systems Are Completely Hyperbolic*, Annals of Math. **149** (1999), 35-96.

[W1] M. P. Wojtkowski, *Invariant families of cones and Lyapunov exponents*, Ergodic Theory and Dynamical Systems **5** (1985), 145–161.

[W2] M. P. Wojtkowski, *Measure Theoretic Entropy of the system of hard spheres*, Ergodic Theory and Dynamical Systems **8** (1988), 133–153.

[W3] M. P. Wojtkowski, *A system of one dimensional balls with gravity*, Commun.Math.Phys. **126** (1990), 507 – 533.

[W4] M. P. Wojtkowski, *The system of one dimensional balls in an external field II*, Commun.Math.Phys. **127** (1990), 425 – 432.

[W5] M. P. Wojtkowski, *Systems of classical interacting particles with nonvanishing Lyapunov exponents*, Lyapunov Exponents, Proceedings, Oberwolfach 1990, L. Arnold, H. Crauel, J.-P. Eckmann (Eds), Lecture Notes in Math. 1486, Springer-Verlag (1991), 243 – 262.

[W6] M. P. Wojtkowski, *Hamiltonian systems with linear potential and elastic constraints*, Fundamenta Mathematicae **157** (1998), 305 – 341.

[W7] M. P. Wojtkowski, *Complete hyperbolicity in Hamiltonian systems with linear potential and elastic constraints*, Reports on Mathematical Physics **44** (1999).

[W8] M. P. Wojtkowski, *Magnetic flows and Gaussian thermostats*, Fundamenta Mathematicae **163** (2000), 177 – 191.

[W9] M. P. Wojtkowski, *W-flows on Weyl manifolds and Gaussian thermostats*, to appear in Jour. Math. Pur. Appl. (2000).

[W-L] M. P. Wojtkowski, C. Liverani, *Conformally symplectic dynamics and symmetry of the Lyapunov spectrum*, Commun. Math. Phys. **194** (1998), 47 – 60.

Department of Mathematics, University of Arizona, Tucson, Arizona 85 721 USA
E-mail address: maciejw@math.arizona.edu

Part IIb

Geodesic Flows

Proceedings of Symposia in Pure Mathematics
Volume **69**, 2001

Geodesic Flows in Manifolds of Nonpositive Curvature

Patrick Eberlein

Contents

Supported in part by NSF Grant DMS-9625452.

I. Introduction

In this section we give a very brief survey of geodesic flows on spaces of negative curvature. For a more complete account see the articles [Ano], [EHS] and [Hed] and the references in these articles.

From the beginning, at least from the 1920's, the properties of the geodesic flow on a space of strictly negative curvature have been studied with a variety of methods–geometry, analysis, ergodic theory and even brute force matrix manipulations that later became more sophisticated Lie theoretic methods. This has made the subject of geodesic flows both more interesting and more difficult for researchers who may have to digest unfamiliar points of view. These notes are an attempt to describe some results and methods from Riemannian geometry that will be useful in understanding the geometric approach.

It is useful to consider only unit speed geodesics in a Riemannian manifold M, which we always assume to be C^∞, and in this case we obtain a flow called the *geodesic flow* on the space T_1M of all unit vectors tangent to M. Given a unit vector v tangent to M and a real number t we define $g^t(v)$ to be the velocity at time t of the unique geodesic γ_v with initial velocity v. If M is compact, or more generally geodesically complete, then $g^t : T_1M \to T_1M$ is defined for every real number t, and the flow transformations $\{g^t\}$ are easily seen to obey the rule $g^{t+s} = g^t \circ g^s = g^s \circ g^t$ for all real numbers s and t. If M has dimension n, then T_1M has dimension $2n - 1$ since the natural projection $\pi : T_1M \to M$ has fibers that are unit spheres of dimension $n - 1$, the unit vectors at a fixed point of M.

In the 1920's it was suspected that geodesic flows on the unit tangent bundle T_1M of a compact 2-dimensional manifold with Gauss curvature $K \equiv -1$ should have very special properties: a dense set of vectors whose orbits under $\{g^t\}$ are periodic, a set of vectors of full measure whose $\{g^t\}$ orbits are dense in T_1M and a variety of even stronger properties including ergodicity and mixing. The leading researchers during this period were Nielsen, Koebe, Morse and others, and a good survey may be found in [Hed]. The early results exploited the algebraic structure of $SL(2,\mathbb{R})$, the group of 2×2 real matrices of determinant 1 and also, after equating elements A and $-A$, the identity component of the group of isometries of the hyperbolic plane with Gaussian curvature -1.

During the 1930's G. Hedlund led the transition to more geometric and analytic methods, and he succeeded in proving ergodicity and mixing properties for the geodesic flow in T_1M for a compact, 2-dimensional manifold with Gauss curvature $K \equiv -1$. During the period 1937 to 1942 E. Hopf [Ho 1, 2] succeeded in extending the new geometric methods to obtain ergodicity in the case where M is a compact, 2-dimensional manifold with strictly negative but nonconstant negative curvature. Hopf's method involved using what later came to be called the stable and unstable Anosov foliations in T_1M. In the case that $\dim M = 2$ these both are 1-dimensional foliations in the 3-dimensional space T_1M. Together with the 1-dimensional foliation Z tangent to the geodesic flow, one now has 3 foliations E^{ss}, E^{uu} and Z that are independent and span the tangent spaces of T_1M. Using clever ad hoc arguments that work only in dimension 2, Hopf showed that the foliations E^{ss}, and E^{uu} are C^1 in T_1M if M has dimension 2 and strictly negative Gauss curvature K. Hopf then devised a general method to prove ergodicity of the geodesic flow, which essentially has still not been improved upon. Moreover, Hopf realized that his method would generalize to arbitrary dimensions, provided that one could prove that the $(n-1)$-dimensional foliations E^{ss} and E^{uu} are still C^1 in T_1M if M is a compact Riemannian manifold with strictly negative sectional curvature. Hopf was able to prove that these foliations are C^1 only in special cases such as constant negative sectional curvature, when these foliations are in fact C^∞.

The study of geodesic flows on compact negatively curved spaces languished until the mid or late 60's when D. Anosov [Ano] overcame Hopf's technical problem with new methods and a new viewpoint. Putting the geodesic flow in the broader context of U-flows (later called Anosov), Anosov realized that one did not really need the foliations E^{ss} and E^{uu} to be C^1 to prove ergodicity of the flow, but only absolutely continuous. He succeeded in proving absolute continuity in the general Anosov setting, and he also showed that the foliations E^{ss} and E^{uu} are not always C^1. In fact, the foliations are rarely C^1, even in the special case of geodesic flows of compact negatively curved manifolds, so that Hopf's original method could not be generalized as planned. See sections VI and VII for further details.

During the 1970's Pesin [Pe 1-4] and others developed a theory of nonuniformly hyperbolic systems, and Pesin applied them to compact manifolds of nonpositive sectional curvature that today are called rank 1 manifolds (see section III G for a definition of rank). (In fact, Pesin considered an even more general class of manifolds.) In general, rank 1 manifolds have geodesic flows that in many ways are similar to those of manifolds of strictly negative sectional curvature, a special class of rank 1 manifolds. However, an arbitrary rank 1 manifold may contain a great deal of zero curvature, which introduces technical complications into proving ergodicity of the geodesic flow that have still not been completely overcome, even in dimension 2. The dynamical behavior of the geodesic flow in a compact, rank 1 manifold of nonpositive sectional curvature remains a major open problem.

Spurred by the Mostow Rigidity Theorem [Mo], the attention of researchers in geometry and dynamical systems turned during the 80's more toward rigidity phenomena such as the characterization of Riemannian symmetric spaces by various geometric and dynamical conditions. Some of these problems are described in section VII, along with many partial solutions that are outstanding even in the special cases considered. In section VIII we discuss the spectacular result of Besson-Courtois-Gallot [BCG 1, 2] that unified the earlier results and settled a number of the remaining open problems.

For a more detailed description of this article see the table of contents.

II. Preliminaries on Riemannian manifolds

A. Riemannian metric and Riemannian volume element. A C^∞ manifold X is a *Riemannian manifold* if for each point x of X there exists an inner product $<,>_x$ on the tangent space T_xX and the assignment $x \mapsto <,>_x$ is C^∞; that is, if V, W are C^∞ vector fields on X, then the function $x \mapsto <V(x), W(x)>_x$ is a C^∞ function on X. Define the norm $|v|$ of a vector v in T_xX in the usual way by $|v| = <v,v>^{1/2}$, and define the length of a C^1 curve $c : [a,b] \to X$ by $L(c) = \int_a^b |c'(t)|dt$. Define the distance $d(x,y)$ between points x and y of X to be the infimum of the lengths of all curves joining x to y. The distance function $d(x,y)$ is positive when $x \neq y$ and is called the *Riemannian metric* determined by the *Riemannian structure* $\{<,>_x : x \in X\}$.

If $\varphi = (\varphi_1, \ldots, \varphi_n) : U \to \varphi(U) \subseteq \mathbb{R}^n$ is a local coordinate system defined on an open set U of X, then we obtain for each point x of U a positive definite symmetric matrix $(g_{ij}(x))$ defined by $g_{ij}(x) = <\frac{\partial}{\partial x_i}, \frac{\partial}{\partial x_j}>(x)$. This is the classical description of the Riemannian structure $<,>$ in local coordinates.

A volume form for an n-dimensional, orientable C^∞ manifold X is a choice of a nowhere vanishing n-form ω on X, and the integral of a function $f : X \to \mathbb{R}$ with compact support is defined to be the integral of $f\omega$ over X. Now fix an orientation on an orientable Riemannian manifold X. If $\{e_1(x), \ldots, e_n(x)\}$ is an orthonormal basis for T_xX that is consistent with the fixed orientation of X, and if $\{\theta_1(x), \ldots, \theta_n(x)\}$ is the corresponding dual basis of 1-forms, then the assignment $x \mapsto \omega(x) = \theta_1(x) \wedge \cdots \wedge \theta_n(x)$ is the *Riemannian volume element* determined by the Riemannian structure $x \mapsto <,>_x$. The definition of $\omega(x)$ depends only on the orientation of X and is independent of the oriented orthonormal basis $\{e_1(x), \ldots, e_n(x)\}$ of T_xX.

B. The Levi Civita connection and covariant differentiation along curves. Given a C^∞ vector field W on a Riemannian manifold $X = \{X, <,>\}$ and a tangent vector v in T_xX we wish to define the directional derivative of W in the direction v as a vector $\nabla_v W$ in T_xX. If $X = \mathbb{R}^n$ with the usual inner product on each tangent space, and W is a vector field on $\mathbb{R}^n$ with component functions $W_1, \ldots, W_n$, then we define $\nabla_v W$ to be the vector in $T_x\mathbb{R}^n$ whose components are the standard directional derivatives $D_v W_1, \ldots, D_v W_n$. To define the directional derivative $\nabla_v W$ of a vector field W in an arbitrary Riemannian manifold X we follow the Koszul approach by stating as axioms certain properties that we wish to be satisfied and then showing that these axioms uniquely determine the definition of $\nabla_v W$. All of these properties are clearly satisfied by $\nabla_v W$ as defined above in $\mathbb{R}^n$ with the standard inner products on each tangent space. We then derive the classical formulas for ∇, the Levi Civita connection, in local coordinates using the Christoffel symbols. The Riemannian curvature tensor R can then be defined by ∇, and it will be, but the classical Christoffel symbol formulation of R will be avoided.

Axioms for the Levi Civita connection. Let there be given C^∞ vector fields V, W on a Riemannian manifold X and arbitrary vectors v, w in T_xX. Then there exists a unique vector $\nabla_v W$ in T_xX such that the following properties are satisfied.

1) (Leibnizian property) If $f : X \to \mathbb{R}$ is any C^∞ function, then

$$\nabla_v(fW) = v(f)W(x) + f(x)\nabla_v W\,,$$

where $v(f)$ denotes the derivative of f at x in the direction v.

2) ($\mathbb{R}$-linearity)
   a) $\nabla_{v+w}W = \nabla_v W + \nabla_w W$
      $\nabla_{av}W = a(\nabla_v W)$ for any real number a.
   b) $\nabla_v(V + W) = \nabla_v V + \nabla_v W$
      $\nabla_v(aW) = a(\nabla_v W)$ for any real number a.

3) $\nabla_V W - \nabla_W V = [V, W]$, where $[V, W]$ denotes the Lie bracket of V and W, and $\nabla_V W$ and $\nabla_W V$ denote the vector fields on X given by $(\nabla_V W)(x) = \nabla_{V(x)}W$ and $(\nabla_W V)(x) = \nabla_{W(x)}V$.

4) $v{<}V, W{>} = {<}\nabla_v V, W(x){>} + {<}V(x), \nabla_v W{>}$, where ${<}V, W{>}$ is the C^∞ function on X defined by ${<}V, W{>}(x) = {<}V(x), W(x){>}$ and $v{<}V, W{>}$ denotes the derivative of ${<}V, W{>}$ in the direction v.

Christoffel symbols and their formulas in local coordinates. Let $\varphi = (\varphi_1, \ldots, \varphi_n) : U \to \phi(U) \subseteq \mathbb{R}^n$ be a local coordinate system defined on an open set U of X. For $1 \le i, j, k \le n$ we define C^∞ functions $\Gamma^k_{ij} : U \to \mathbb{R}$ by

$$\nabla_{\partial/\partial x_i}\partial/\partial x_j = \sum_{k=1}^{n} \Gamma^k_{ij}\, \partial/\partial x_k\,.$$

The functions $\{\Gamma^k_{ij}\}$ are called the *Christoffel symbols* determined by the Levi Civita connection ∇. By 3) above and the fact that $[\partial/\partial x_i, \partial/\partial x_j] = 0$ it follows that the Christoffel symbols are symmetric in the two lower indices; that is, $\Gamma^k_{ij} = \Gamma^k_{ji}$ for all i, j, k.

Using the axioms for ∇ above and the symmetry of the Christoffel symbols $\{\Gamma^k_{ij}\}$ in the lower two indices it is routine to derive the classical formulas for the $\{\Gamma^k_{ij}\}$ in terms of the functions $g_{ij} : U \to \mathbb{R}$ and their partial derivatives; namely,

$$\Gamma^k_{ij} = \frac{1}{2}\sum_{s=1}^{n} \left(\partial g_{sj}/\partial x_i + \partial g_{is}/\partial x_j - \partial g_{ij}/\partial x_s\right) g^{sk}$$

where $g^{sk} = (g^{-1})_{sk}$.

Covariant differentiation of vector fields along curves. Let $c : [a, b] \to X$ be a C^∞ curve in a Riemannian manifold X, and let $Y : [a, b] \to TX$ be a C^∞ vector field along c; that is, $Y(t) \in T_{c(t)}X$ for every $t \in [a, b]$. We wish to define a new vector field $Y'(t)$ along c in an axiomatic way analogous to the definition of the Levi Civita connection above. This notion is needed to define parallel translation of vectors along arbitrary curves.

There exists for each C^∞ vector field $Y(t)$ along c a unique C^∞ vector field $Y'(t)$ along c such that the following properties are satisfied.

1) (Leibnizian property) If $f : [a, b] \to \mathbb{R}$ is any C^∞ function, then

$$(fY)'(t) = f'(t)Y(t) + f(t)Y'(t) \qquad \text{for all } t \in [a, b].$$

2) $(Y_1 + Y_2)'(t) = Y_1'(t) + Y_2'(t)$, for any two vector fields Y_1, Y_2 along c; $(aY)'(t) = aY'(t)$ for all $a \in \mathbb{R}$ and all $t \in [a, b]$.
3) If Y is the restriction to c of a vector field Y^* on X, then $Y'(t) = \nabla_{c'(t)} Y^*$ for all $t \in [a, b]$.
4) For any two vector fields Y_1, Y_2 along c the function $t \mapsto \langle Y_1(t), Y_2(t)\rangle$ satisfies the equation

$$\frac{d}{dt}\langle Y_1(t), Y_2(t)\rangle = \langle Y_1'(t), Y_2(t)\rangle + \langle Y_1(t), Y_2'(t)\rangle$$

Remarks.

(1) An arbitrary curve $c : [a, b] \to X$ may be highly singular or even constant so that not every vector field Y along c is the restriction to c of a vector field Y^* of X.

(2) For curves $c : [a, b] \to U$, where U is a coordinate neighborhood of X, one can define the covariant derivative $Y'(t)$ of a vector field $Y(t)$ along c in terms of local coordinate expressions involving the component functions of $Y(t)$ and $c(t)$ and the Christoffel functions restricted to c. We leave the derivation of such an expression to the diligent motivated reader.

C. Parallel translation of vectors along curves. Let $c : [a, b] \to X$ be an arbitrary C^∞ curve in a Riemannian manifold X. A C^∞ vector field $Y(t)$ along c is said to be *parallel* if the covariant derivative $Y'(t)$ is identically zero. Writing the equations for $Y'(t) \equiv 0$ in local coordinates, a standard differential equations argument shows that for every vector v in $T_{c(a)}X$ there exists a unique parallel vector field $Y(t)$ along c such that $Y(a) = v$. The vector $Y(b)$ is called the *parallel translate* of v along the curve c from $c(a)$ to $c(b)$.

Note that if $Y_1(t)$ and $Y_2(t)$ are any two parallel vector fields along c, then the function $t \mapsto \langle Y_1(t), Y_2(t)\rangle$ is constant in t by property 4) above. In particular the lengths of $Y_1(t)$ and $Y_2(t)$ and the angles between $Y_1(t)$ and $Y_2(t)$ are constant in t, justifying the name parallel translation.

D. Curvature.

Riemannian curvature tensor. Let X be a C^∞ Riemannian manifold. Given three C^∞ vector fields U, V and W on X we define a new vector field

$$R(U, V)W = \nabla_U(\nabla_V W) - \nabla_V(\nabla_U W) - \nabla_{[U,V]} W,$$

where ∇ denotes the covariant derivative of the Levi Civita connection defined above. One can show that the value of $R(U, V)W$ at a point x of X depends only on the values of U, V and W at x. Hence for each point x of X we obtain a trilinear map $R : (T_x X) \times (T_x X) \times (T_x X) \to T_x X$ given by $(u, v, w) \mapsto R(u, v)w = \{R(U, V)W\}(x)$, where U, V and W are C^∞ vector fields with values u, v and w at x.

This is the *Riemannian curvature tensor* of X. We make no attempt to satisfy the reader who wishes to see the classical tensor expression in terms of Christoffel symbols and their derivatives. This expression would illustrate all too clearly how simple geometric concepts can be made to look both ugly and deep, on the one hand discouraging mathematicians from learning geometry and on the other hand impressing nonmathematicians unreasonably [V].

We will need the Riemannian curvature tensor to define the second order differential equation whose solutions are the Jacobi vector fields on geodesics γ. Jacobi vector fields on a geodesic γ are basic objects that describe the first order behavior of the geodesics in a neighborhood of γ.

It is a straightforward exercise to verify from the definitions that the curvature tensor R satisfies the following symmetry equations for all vectors u, v, w and z at a fixed point of X:

$$
\begin{aligned}
&R(u,v)w = -R(v,u)w \\
&R(u,v)w + R(v,w)u + R(w,u)v = 0 \\
&<R(u,v)w, z> = -<R(u,v)z, w> \\
&<R(u,v)w, z> = <R(w,z)u, v>
\end{aligned}
$$

Curvature operators. Each vector v in T_xX defines a curvature operator $R_v : T_xX \to T_xX$ by $R_v(u) = R(u,v)v$. It follows from the symmetry equations above that each curvature operator R_v is a symmetric linear transformation on T_xX.

Sectional curvature. To each point x of a Riemannian manifold X and to each 2-dimensional subspace Π of the tangent space T_xX one may assign a number

$$K(\Pi) = <R(x,y)y, x>/|x \wedge y|^2 ,$$

where $\{x, y\}$ is any basis of Π and $|x \wedge y|^2 = |x|^2|y|^2 - <x,y>^2$. The number $K(\Pi)$ is called the *sectional curvature* of Π, and it is independent of the basis $\{x, y\}$ of Π.

If X has constant sectional curvature c (i.e. $K(\Pi) = c$ for all 2-planes Π), then the curvature tensor R has the following simple form: $R(u,v)w = c\{<v,w>u - <u,w>v\}$ for all vectors u, v and w in an arbitrary tangent space T_xX. It is not difficult to show that the Euclidean space $\mathbb{R}^n$ of any dimension with the standard inner product on each tangent space has a curvature tensor that is identically zero, and hence $\mathbb{R}^n$ has zero sectional curvature. The sphere in $\mathbb{R}^n$ of radius a with the Euclidean inner product on each tangent space has constant sectional curvature $1/a^2$. The hyperbolic space of any dimension (see section III D below) has constant negative sectional curvature. A basic theorem in Riemannian geometry asserts that any simply connected Riemannian manifold X with constant sectional curvature c that is complete as a metric space must be isometric to one of the three examples above, depending upon the magnitude and sign of c.

Sectional curvature may also be described in a simpler, if slightly imprecise, geometric fashion. Given a 2-plane Π in T_xX let Σ_Π denote the surface in X consisting of the union of all geodesics in X that pass through x and are tangent to Π. Then $K(\Pi)$ is the Gaussian curvature at x of the surface Σ_Π. (Of course, this requires knowing what the Gaussian curvature is in an abtract 2-dimensional Riemannian manifold.) If one knows all sectional curvatures of 2-planes at a given point x of X, then, in principle, one can recover the Riemannian curvature tensor at x by a complicated polarization formula.

Mean curvature of a hypersurface. Let X be a Riemannian manifold of arbitrary dimension n, and let Y be a hypersurface in X, that is, a submanifold of codimension 1. Let U be one of the two C^∞ unit vector fields on Y that is orthogonal to Y, and define a linear transformation $S : T_yY \to T_yX$ by $S(v) = -\nabla_v U$ for any vector v in T_yY. It is not difficult to show that the image of S is contained in T_yY and that $S : T_yY \to T_yY$ is a symmetric linear transformation. The trace of S is called the *mean curvature* of Y at y.

E. Geodesics and geodesic flow. A geodesic in a Riemannian manifold X is defined to be a C^∞ curve $c : [a, b] \to X$ whose velocity vector field $c'(t)$ is a parallel vector field along c. The fact that $c'(t)$ is parallel implies immediately that a geodesic $c(t)$ has constant speed since by the discussion above parallel vector fields have constant norm. This somewhat formal definition of geodesic also leads to a second order ordinary differential equation in local coordinates that involves the Christoffel symbols. The diligent reader may derive this equation as an exercise, and the less diligent reader may simply believe the statement above or look it up somewhere. The point is that the existence and uniqueness theory of ordinary differential equations shows that for every point x in X and every vector v in T_xX there exists a positive number $\varepsilon = \varepsilon(x, v)$ and a unique geodesic $\gamma_v : (-\varepsilon, \varepsilon) \to X$ such that $\gamma_v(0) = x$ and $\gamma_v'(0) = v$.

Geodesic completeness and the theorem of Hopf-Rinow. In general a geodesic $c(t)$ may not be defined on the whole real line. For example, remove the origin from the Euclidean plane with its standard inner product $<,>$ on each tangent space. The geodesics of the new space $X = \mathbb{R}^2-$origin still are straight lines, but those that used to run through the origin can no longer do so and are therefore not defined on all of $\mathbb{R}$. In most cases Riemannian geometers only consider Riemannian manifolds that are geodesically complete; that is, every geodesic is defined on all of $\mathbb{R}$. This is guaranteed if X is compact. In general the following theorem of Hopf and Rinow characterizes geodesic completeness. See for example [Mi] for a proof.

THEOREM. *Let X be a Riemannian manifold. Then the following conditions are equivalent :*

1) *All geodesics of X are defined on $\mathbb{R}$.*
2) *For some point x of X all geodesics of X that pass through x are defined on $\mathbb{R}$.*
3) *If d denotes the Riemannian metric of X, then the metric space (X, d) is complete; that is, every Cauchy sequence converges.*
4) *Every closed, bounded subset of X is compact.*

Geodesics as critical points of arc length. Next we give a geometric description of geodesics that conforms better to the usual notion of geodesics as shortest curves. Note that the parametrization of a geodesic is important in Riemannian geometry. Geodesics must always have constant speed and to be a geodesic it is not enough for a curve c to be the shortest connection between its endpoints. (However, any constant speed reparametrization of c will then be a geodesic by the discussion that follows.)

Given any two points p, q in a Riemannian manifold X let Ω_{pq} denote the collection of all piecewise C^∞ curves $c : [0, 1] \to X$, and let $L : \Omega_{pq} \to \mathbb{R}$ denote the length function. The geodesics from p to q can be described as the critical points of the length functional L on Ω_{pq} in a sense we now make precise.

A *fixed endpoint variation* of a piecewise C^∞ curve $c : [0, 1] \to X$ is a C^∞ family of piecewise C^∞ curves $c_t : [0, 1] \to X$, $-\varepsilon < t < \varepsilon$, such that $c_0 = c$ and $c_t(0) = c(0)$, $c_t(1) = c(1)$ for all t. (Formally, we require the map $F : [0, 1] \times (-\varepsilon, \varepsilon) \to X$ given by $F(s, t) = c_t(s)$ to be C^∞.) The curve c is said to be a *critical point* of the variation $\{c_t\}$ if $L'(0) = 0$, where $L(t) = L(c_t)$, the length of c_t, for all t. If $c : [0, 1] \to X$ has constant speed, necessary if c is to be a geodesic, then c is a geodesic of X if and only if c is a critical point of every variation $\{c_t\}$ of c. See for example [Mi] for a proof. In particular, if c has constant speed and is a shortest curve in Ω_{pq}, then c is a geodesic from p to q. However, not every geodesic from p to q is a shortest curve from p to q. For example, on the standard 2-sphere of radius 1 where all great circles have length 2π any great circle is a shortest curve until the antipode of the starting point is reached, but beyond that the great circle is no longer a shortest curve although it is still a geodesic.

One may think of Ω_{pq} as an infinite dimensional manifold in which the points of Ω_{pq} are curves from p to q, parametrized on $[0, 1]$. The tangent space $T_c\Omega_{pq}$ at the curve c consists of all initial velocities of piecewise C^∞ fixed endpoint variations of c; that is, $T_c\Omega_{pq}$ is the infinite dimensional real vector space of piecewise C^∞ vector fields $Y(t)$ on c such that $Y(0) = Y(1) = 0$. The nature of a geodesic c as a critical point of the arc length functional L can be investigated by studying the second derivative $L''(0)$ for all possible fixed endpoint variations of c. This second derivative information is encoded in a symmetric bilinear form $I : T_c\Omega_{pq} \times T_c\Omega_{pq} \to \mathbb{R}$ called the *index form*. The form I is positive definite if and only if c is shorter than all *nearby* curves of Ω_{pq}. If $I(Y, Y)$ is negative for some vector field Y in $T_c\Omega_{pq}$, then one may use Y to construct a fixed endpoint variation $\{c_t\}$ of c in which all nearby curves c_t are shorter than c; i.e. $L(c_t) < L(c)$ if $|t|$ is small. It turns out that the maximum dimension of a subspace W of $T_c\Omega_{pq}$ on which I is negative definite is always *finite*, and the dimension of W can be described precisely by the Morse Index Theorem in terms of *conjugate points* of c, which we define below. See for example [Mi] for details, including a definition and properties of the index form I.

Geodesic flow. Let X be a geodesically complete Riemannian manifold; that is, all geodesics of X are defined on $\mathbb{R}$. Let T_1X denote the collection of unit vectors tangent to X. We have seen that for every vector v in T_1X there exists a unique geodesic γ_v in X with initial velocity v. For each number t in $\mathbb{R}$ we define a diffeomorphism $g^t : T_1X \to T_1X$ by $g^t(v) = \gamma_v'(t)$, the velocity of γ_v at time t. Since X is geodesically complete the diffeomorphisms are defined for all t in $\mathbb{R}$, and the collection of diffeomorphisms $\{g^t\}$ is a 1-parameter group; that is, $g^t \circ g^s = g^s \circ g^t = g^{t+s}$ for all s, t in $\mathbb{R}$. The diffeomorphisms $\{g^t\}$ are called the *geodesic flow* in T_1X.

F. Riemannian exponential map and Jacobi vector fields. Let X be a geodesically complete Riemannian manifold. For each point x of X there exists an important map $\exp_x : T_xX \to X$ called the *Riemannian exponential map*. Given a vector v in T_xX, not necessarily of length 1, we define $\exp_x(v) = \gamma_v(1)$, the value of γ_v at time 1. It is not difficult to show that $\exp_x(tv) = \gamma_v(t)$ for all t in $\mathbb{R}$ so that $\exp_x$ maps the straight lines through the origin in T_xX into the geodesics of X that start at x. We shall see shortly that the critical points of $\exp_x$ have an important geometric meaning.

Let $c : [a, b] \to X$ be a geodesic of X, which for convenience we assume to have unit speed. A C^2 vector field $Y(t)$ on c is defined to be a *Jacobi vector field* if it satisfies the second order linear differential equation

$$Y''(t) + R(Y(t), c'(t))c'(t) = 0 \tag{J}$$

on $[a, b]$. Here $Y'(t)$ denotes the covariant derivative of $Y(t)$ along c, and $Y''(t)$ denotes the covariant derivative of $Y'(t)$ along c, or equivalently, the second covariant derivative of $Y(t)$ along c. R denotes the curvature tensor of X defined above. If we define $R(t)$ to be the symmetric curvature operator $v \mapsto R(v, c'(t))c'(t)$ determined by $c'(t)$, then we may rewrite the Jacobi equation (J) above as

$$Y''(t) + R(t)(Y(t)) = 0$$

By using an orthonormal family $\{E_i(t)\}$ of parallel vector fields along $c(t)$ the equation above may be converted into a second order matrix differential equation where $Y(t)$ is an $n \times 1$ column vector and $R(t)$ is a symmetric $n \times n$ matrix. The standard theory of ordinary differential equations says that given a vector v at a point x of X and any two vectors w, w^* in T_xX there exists a unique Jacobi vector field $Y(t)$ along the geodesic γ_v such that $Y(0) = w$ and $Y'(0) = w^*$, where as usual $Y'(t)$ denotes the covariant derivative of Y along γ_v. If we let $J(\gamma_v)$ denote the collection of all Jacobi vector fields on γ_v, then it is apparent from the linear equation (J) and the remarks just made that $J(\gamma_v)$ is a vector space of dimension $2n$, where n is the dimension of X.

Jacobi vector fields as geodesic variation vector fields. It is useful to describe an equivalent geometric definition of Jacobi vector fields. An (arbitrary) C^∞ variation of a C^∞ curve $c : [a, b] \to X$ is a C^∞ family of curves $\{c_t\}$, $-\varepsilon < t < \varepsilon$, such that $c_0 = c$. (We no longer require that the curves $\{c_t\}$ have the same endpoints as in the discussion of geodesics above.) A C^∞ variation of c determines a *variation vector field* $Y(s)$ on c; define $Y(s)$ to be the initial velocity of the curve $t \mapsto c_t(s)$ for each fixed s in $[a, b]$. If c is a geodesic of X, then a C^2 vector field $Y(t)$ on c is a Jacobi vector field if and only if Y is the variation vector field of a *geodesic* variation of c; that is, a C^∞ variation $\{c_t\}$ of c in which each of the curves c_t is a geodesic of X.

Conjugate points and critical points of the exponential map. Let $c : [a, b] \to X$ be a nonconstant geodesic in a (geodesically) complete Riemannian manifold. The points $c(a)$ and $c(b)$ are *conjugate* along c if there exists a nonzero Jacobi vector field $Y(t)$ along c such that $Y(a) = 0$ and $Y(b) = 0$. It is not difficult to show that $Y(t)$ can be represented as the variation vector field of a geodesic variation $\{c_t\}$ such that $c_t(a) = c(a)$ for all t. The fact that $Y(b) = 0$ means that the curve $\sigma(t) = c_t(b)$ has initial velocity zero and may be regarded as constant to first order. Geometrically one may think of the geodesics $\{c_t\}$ as a family of light rays emerging from the point $c(a)$ and focusing at the point $c(b)$. This behavior is typical of spaces with positive sectional curvature if one thinks of great circles on the sphere emerging from the north pole and meeting again at the south pole.

The existence of conjugate points along a geodesic is related to the critical points of the exponential map as follows. Let x be a point of a complete Riemannian manifold X, and let v be a vector in T_xX. Then v is a critical point of the Riemannian exponential map $\exp_x : T_xX \to X$ if and only if the points $\gamma_v(0) = x$ and $\gamma_v(1)$ are conjugate along γ_v. To see this, let ξ be a nonzero vector in T_xX such that $d\exp_x(\xi_v) = 0$, where ξ_v denotes the initial velocity of the curve $t \mapsto v + t\xi$ in T_xX. If c_t is the geodesic starting at x with initial velocity $v + t\xi$ and if Y is the variation vector field of the geodesic variation $\{c_t\}$, then $Y(t)$ is a nonzero Jacobi vector field such that $Y(0) = 0$ and $Y(1) = 0$.

If X is a complete Riemannian manifold with nonpositive sectional curvature (i.e. $K \leq 0$), then we shall see that any nonzero Jacobi vector field $Y(t)$ with $Y(0) = 0$ on a geodesic $c(t)$ grows in norm at least as fast as a linear function. It follows that no nonzero Jacobi vector field can vanish twice on a geodesic, and by the remarks above this means that every exponential map $\exp_x : T_xX \to X$ is nonsingular and, in fact, a covering map. In the case that X is simply connected it follows that each exponential map $\exp_x : T_xX \to X$ is a diffeomorphism, which means geometrically that for any two distinct points x and y in X there is a unique geodesic (up to constant speed reparametrizations) joining x to y. This is the theorem of Cartan-Hadamard.

G. Isometries and local isometries. Let X and Y be Riemannian manifolds. An *isometry* $p : X \to Y$ is a diffeomorphism such that $<dp_x(v), dp_x(w)> = <v, w>$ for all vectors v, w in T_xX and all points x of X. Clearly, $p^{-1} : Y \to X$ is also an isometry. It is not difficult to show that an isometry p carries geodesics to geodesics and Jacobi vector fields on a geodesic $c(t)$ to Jacobi vector fields on the geodesic $(p \circ c)(t)$. If Π is a 2-dimensional subspace of a tangent space T_xX, then the sectional curvature of Π equals that of $dp(\Pi) \subseteq T_{p(x)}Y$. In general, isometries preserve all possible geometric properties defined by the Riemannian structure $<,>$.

From the definition of the Riemannian metric d it follows that if $p : X \to Y$ is an isometry, then $d(p(x), p(x^*)) = d(x, x^*)$ for all points x, x^* in X. Conversely, if $p : X \to Y$ is a diffeomorphism such that $d(p(x), p(x^*)) = d(x, x^*)$ for all points x, x^* in X, then it is not difficult to show that p is an isometry. In summary, one may describe the isometries as the distance preserving diffeomorphisms.

A C^∞ map $p : X \to Y$ is called a *local isometry* if $<dp_x(v), dp_x(w)> = <v, w>$ for all vectors v, w in T_xX and all points x of X (i.e. we drop the condition that p be a diffeomorphism). Any local isometry is nonsingular, and hence by the inverse function theorem for any point x of X there exist open neighborhoods U of x in X and V of $p(x)$ in Y such that $p : U \to V$ is a diffeomorphism. The map $p : U \to V$ is therefore an isometry between the Riemannian manifolds U and V, which explains the term local isometry.

All of the geometric quantities discussed above (geodesics, sectional curvature, Jacobi vector fields etc.) are defined by equations that depend on local information in the Riemannian manifold. Hence local isometries also preserve all locally defined geometric quantities such as geodesics, sectional curvature and Jacobi vector fields. Moreover, the preimages of geodesics in Y under a local isometry $p : X \to Y$ are geodesics in X.

H. Geometry of the tangent bundle with the Sasaki metric. Let X be a C^∞ differentiable manifold and let TX denote the tangent bundle of X, which consists of the union of all tangent spaces T_xX, for $x \in X$. There is a natural projection map $\pi : TX \to X$ that sends a vector v in T_xX to its basepoint x in X. The fibers of π are the tangent spaces T_xX as x ranges over X, and hence TX is a manifold of dimension $2n$, where n is the dimension of X.

If X is in addition a Riemannian manifold, then the Riemannian metric on X induces a Riemannian metric on TX that is called the Sasaki metric. To describe it we first define the *connection map* $\mathcal{K} : TX \to X$, which unlike the projection map $\pi : TX \to X$ depends on the Riemannian structure of X. For further properties of the connection map see [E5] and [GKM].

Given a vector ξ in $T_v(TX)$ let $v(t)$ be a curve in TX such that $v(0) = v$ and $v(t)$ has initial velocity ξ. Let $c(t) = \pi(v(t))$ so that $v(t)$ may be regarded as a C^∞ vector field on the curve $c(t)$ in X. Now define $\mathcal{K}(\xi)$ to be the vector

$v'(0)$ in T_xX, where $x = \pi(v)$, the base point of v, and $v'(t)$ denotes the covariant derivative of $v(t)$ along $c(t)$. An explicit computation of $\mathcal{K}(\xi)$ in local coordinates shows that the definition of $\mathcal{K}(\xi)$ does not depend on the curve $v(t)$ chosen but only its initial velocity. Following our earlier practice, we leave the definition of $\mathcal{K}(\xi)$ in local coordinates as an exercise for the diligent reader. The subbundles $\mathrm{Ker}(\mathcal{K})$ and $\mathrm{Ker}(d\pi)$ of TX are called the *horizontal* and *vertical* subbundles respectively. It is not hard to see that TX is the direct sum of the horizontal and vertical subbundles.

We are now ready to define the Sasaki metric in TX. Given vectors ξ, η in $T_v(TX)$ we define

$$<\xi,\eta>_{TX} = <d\pi(\xi), d\pi(\eta)>_X + <\mathcal{K}(\xi), \mathcal{K}(\eta)>_X ,$$

where $\pi : TX \to X$ and $\mathcal{K} : TX \to X$ denote the projection and connection maps respectively. This bilinear form on TX is positive definite since $\mathrm{Ker}(\mathcal{K})$ and $\mathrm{Ker}(d\pi)$ are disjoint except for $\{0\}$.

Identification of $T_v(TX)$ with $J(\gamma_v)$. For each vector ξ in $T_v(TX)$ we let Y_ξ be the unique Jacobi vector field on γ_v such that $Y_\xi(0) = d\pi(\xi)$ and $Y_\xi'(0) = \mathcal{K}(\xi)$. It is not difficult to show that the map $\xi \mapsto Y_\xi$ is an isomorphism of $T_v(TX)$ onto $J(\gamma_v)$, the space of Jacobi vector fields on γ_v. Geometrically, one may also describe this situation as follows. Given a vector $\xi \in T_v(TX)$ let $v(t)$, $-\varepsilon < t < \varepsilon$, be a C^∞ curve in TX such that $v(0) = v$ and $v(t)$ has initial velocity ξ. If c_t denotes the geodesic with initial velocity $v(t)$, then $\{c_t\}$ is a geodesic variation of the geodesic $c_0 = \gamma_v$, and the corresponding variation vector field is a Jacobi vector field on γ_v that is easily seen to be the Jacobi vector field $Y_\xi(t)$ defined analytically above.

If T_1X denotes the bundle of unit tangent vectors of X, a codimension 1 submanifold of TX, then the isomorphism $\xi \mapsto Y_\xi$ carries $T_v(T_1X)$ onto $J^*(\gamma_v)$, the codimension 1 subspace of $J(\gamma_v)$ consisting of those Jacobi vector fields $Y(t)$ on γ_v such that $Y'(t)$ is orthogonal to $\gamma_v'(t)$ for all t.

Note that each map $g^t : TX \to TX$ leaves invariant T_1X since $g^t(v) = \gamma_v'(t)$ for any unit vector v and the velocity vector field $\gamma_v'(t)$ has constant norm since it is a parallel vector field on γ_v.

Now let $\{g^t\}$ denote the geodesic flow on TX or T_1X. If ξ is any vector in $T_v(TX)$, with corresponding Jacobi vector field Y_ξ on γ_v, then it follows from the definitions that $|dg^t(\xi)|^2 = |Y_\xi(t)|^2 + |Y_\xi'(t)|^2$ for all t. This fact allows one to estimate the growth rate of $|dg^t(\xi)|$ by using the Jacobi equation to estimate the growth rate of $|Y_\xi(t)|$, which, when the sectional curvature of X is strictly negative, is essentially the same as that of $|Y_\xi'(t)|$ for those Jacobi vector fields Y_ξ most important for the geodesic flow. When the sectional curvature is bounded above by a negative constant $-a^2$, then comparisons to Jacobi vector fields in a space of constant sectional curvature $-a^2$ yield sharp estimates for the growth rate of Y_ξ. See III B for further details.

We now combine the discussion of subsections F and G to obtain the following useful result that relates Jacobi vector fields to the differentials of the exponential maps.

PROPOSITION. *Let X be a complete Riemannian manifold, and let $\exp_x : T_xX \to X$ be the exponential map at a point x of X. Let v, ξ be any vectors in T_xX and let $\xi_v \in T_v(T_xX) \subseteq T_v(TX)$ denote the initial velocity of the curve $t \mapsto v(t) = v + t\xi$. Then $d\exp_x(\xi_v) = Y(1)$, where $Y(t)$ is the unique Jacobi vector field on the geodesic $c(t) = \exp_x(tv)$ such that $Y(0) = 0$ and $Y'(0) = \xi$.*

PROOF. For $|t| < \varepsilon$ and $s \in \mathbb{R}$ define $c_t(s) = \exp_x(s(v + t\xi))$. Each curve c_t is a geodesic, with initial velocity $v(t) = v + t\xi$ and the variation vector field is the Jacobi vector field $Y(s)$ on $c(s) = c_0(s) = \exp_x(sv)$ such that $Y(0) = d\pi(\xi)$ and $Y'(0) = \mathcal{K}(\xi)$, where $\mathcal{K} : TX \to X$ denotes the connection map. However, $d\pi(\xi) = 0$ since $d\pi(\xi)$ is the initial velocity of the constant curve $\pi(v(t))=\{x\}$. By definition $\mathcal{K}(\xi) = Z'(0)$, the covariant derivative at $t = 0$ of the vector field $Z(t) = v + t\xi$ on the constant curve $\pi(v(t)) = \{x\}$, but this is precisely ξ. □

The invariant volume element on TX and T_1X. There is a natural volume element on the tangent bundle TX of a Riemannian manifold X and a closely related natural volume element on the unit tangent bundle T_1X. In both cases these are the Riemannian volume elements as determined by the Sasaki metric defined above, but they also have other and better known descriptions. We explain briefly.

Given a differentiable manifold X there is a canonical 1-form θ on the cotangent bundle T^*X consisting of the union of all cotangent spaces $\{T_x^*X = \mathrm{Hom}(T_xX, \mathbb{R}), x \in X\}$ together with the natural projection $\pi^* : T^*X \to X$ that sends an element of T_x^*X to x. For $\omega \in T_x^*X$ and $\xi \in T_\omega(T^*X)$ we define $\theta(\xi) = \omega(d\pi^*(\xi))$. If $\Omega = d\theta$, then Ω is a symplectic 2-form on T^*X and in particular $\Omega \wedge \cdots \wedge \Omega$ (n times) is a nowhere vanishing volume element on T^*X, where $n = \dim X$.

If X has a Riemannian structure $<,>$, then there is a natural diffeomorphism $\varphi : TX \to T^*X$: given a vector v in T_xX let $\varphi(v) = \omega_v$ be that element in T^*X such that $\omega_v(\xi) = <\xi, v>$ for all vectors ξ in T_xX. Under this identification of TX and T^*X we let θ and Ω also denote the corresponding 1 and 2-forms in TX as well as their restrictions to the unit tangent bundle T_1X. One can show that the geodesic flow $\{g^t\}$ in either TX or T_1X leaves invariant the forms θ and Ω; that is, $(g^t)^*\theta = \theta$ and $(g^t)^*\Omega = \Omega$ for all $t \in \mathbb{R}$. The Riemannian volume elements in TX and T_1X with respect to the Sasaki metric may now be described respectively as $\Omega \wedge \cdots \wedge \Omega$ (n times) and $\theta \wedge (\Omega \wedge \cdots \wedge \Omega)$ ($n - 1$ times).

III. Manifolds of nonpositive sectional curvature

Notation: In the sequel we will use M to denote a complete Riemannian manifold (usually compact) and $\tilde{M}$ to denote its universal Riemannian cover; that is, $\tilde{M}$ is the unique simply connected differentiable manifold that covers M, and $\tilde{M}$ is equipped with the unique inner products on tangent spaces that make the differential maps of the covering map $p : \tilde{M} \to M$ linear isometries.

A. Definition of nonpositive curvature by triangle comparisons. Let Δ be a triangle in a complete Riemannian manifold X whose sides are geodesics and also shortest curves connecting the vertices A, B and C. Let the sides of Δ have lengths a, b and c as shown in the figure below. Now let Δ^* be a triangle in the Euclidean plane with the same side lengths a, b and c and opposite vertices A^*, B^* and C^*. Let P and P^* be points in Δ and Δ^* that belong to corresponding sides and have the same distances from the opposite vertices, as shown in the figure below.

Using Jacobi vector field arguments one can prove that if A and A^* are the vertices opposite P and P^*, then $d(A, P) \leq d(A^*, P^*)$ for all triangles Δ and Δ^* if and only the sectional curvature of X is nonpositive. Similarly, X has sectional

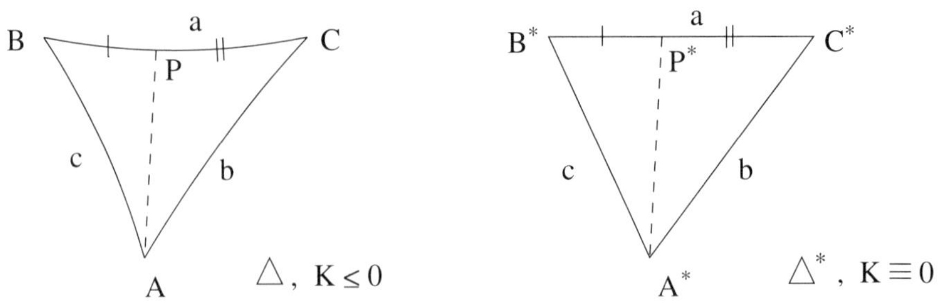

FIGURE 1.

curvature $\leq -c^2 < 0$ if and only if analogous triangle comparisons with a hyperbolic space of constant sectional curvature $-c^2 < 0$ are valid.

Conversely, one can use the triangle comparison definition above to define metric spaces X of curvature ≤ 0 or $\leq -c^2 < 0$, often when no differentiable structure for X exists, provided that the metric spaces X admit geodesics with reasonably nice properties. This generalization of Riemannian geometry to (possibly) nondifferentiable spaces of nonpositive or strictly negative curvature has been extremely fruitful for geometers, but it lies at some distance from smooth dynamical systems and will not be treated in these notes.

B. Growth of Jacobi vector fields. Let $Y(t)$ be a Jacobi vector field on a unit speed geodesic $c(t)$ in a complete Riemannian manifold X. If $Y^{\tan}(t)$ and $Y^{\perp}(t)$ denote the components of $Y(t)$ that are tangent to and orthogonal to $c'(t)$, respectively, then it follows from the Jacobi equation that $Y^{\tan}(t)$ and $Y^{\perp}(t)$ are also Jacobi vector fields on $c(t)$. If $Y(t) = f(t)c'(t)$ is a tangential Jacobi vector field on $c(t)$, then it follows easily from the Jacobi equation that $f(t)$ is a linear function $at + b$ for suitable constants a and b. Therefore, in estimating the growth of the norm of a Jacobi vector field, the only interesting case occurs when $Y(t)$ is orthogonal to $c'(t)$. Here the influence of the sign of the sectional curvature is significant.

Let $Y(t)$ be a Jacobi vector field on a unit speed geodesic $c(t)$ in a complete Riemannian manifold M with sectional curvature $K \leq 0$. If $f(t) = |Y(t)|^2 = <Y(t), Y(t)>$, then a direct computation shows that $f''(t) \geq 0$ for all t; that is, f is a convex function on $\mathbb{R}$. Specifically, we compute $f'(t) = 2<Y(t), Y'(t)>$ and by using the Jacobi equation we obtain

$$\begin{aligned} f''(t) &= 2\{|Y'(t)|^2 + <Y''(t), Y(t)>\} \\ &= 2\{|Y'(t)|^2 - <R(Y(t), c'(t))c'(t), Y(t)>\} \\ &= 2\{|Y'(t)|^2 - K(Y(t), c'(t))|Y(t) \wedge c'(t)|^2\} \geq 0\,, \end{aligned}$$

where $Y'(t)$ and $Y''(t)$ denote the first and second covariant derivatives of $Y(t)$ along $c(t)$ and $K(Y(t), c'(t))$ denotes the sectional curvature of the 2-plane spanned by $Y(t)$ and $c'(t)$. (If $Y(t)$ and $c'(t)$ are collinear then we adopt the convention that $K(Y(t), c'(t)) = 0$ since in this case $R(Y(t), c'(t))c'(t) = 0$ by symmetries of the curvature tensor described above.)

From the properties of convex functions from $\mathbb{R}$ to $\mathbb{R}$ and the arguments of the previous paragraph we may immediately conclude the following.

PROPOSITION. *Let M be a complete Riemannian manifold with sectional curvature $K \leq 0$, and let $Y(t)$ be a Jacobi vector field on a unit speed geodesic $c(t)$ of M. Then either*

1) *$|Y(t)|$ is nondecreasing on $\mathbb{R}$, and $|Y(t)| \to +\infty$ as $t \to \infty$, or*
2) *$|Y(t)|$ is nonincreasing on $\mathbb{R}$, and $|Y(t)| \to +\infty$ as $t \to -\infty$, or*
3) *$|Y(t)|$ is constant on $\mathbb{R}$ and $Y(t)$ is a parallel vector field on $c(t)$.*

The proposition above is a simple example of the important role that convex functions play in the geometry of complete manifolds M with sectional curvature $K \leq 0$. A similar but more difficult computation for the function $g(t) = |Y(t)|$, where $Y(t)$ is a Jacobi vector field on a unit speed geodesic $c(t)$, shows that $g''(t) \geq 0$ whenever $g(t) \neq 0$. If $Y(0) = 0$, then it is not difficult to show that $g'(0^+) = |Y'(0)|$. We obtain

(#) If $Y(t)$ is a Jacobi vector field on a unit speed geodesic $c(t)$ with $Y(0) = 0$, then $|Y(t)| \geq t|Y'(0)|$ for all $t > 0$.

In particular, Jacobi vector fields $Y(t)$ that start at zero in a manifold M of nonpositive sectional curvature grow in norm at least as fast as Euclidean Jacobi vector fields with the same initial rate of increase. The inequality (#) is the key fact needed to prove the triangle comparison result stated above, but we shall not pursue this.

Growth rates in manifolds with $K \leq -c^2 < 0$. Generally speaking, if one can solve equations like the Jacobi equation in spaces where the sectional curvature is identically $-c^2 < 0$, then one can obtain estimates for the solutions of the corresponding equations in manifolds with $K \leq -c^2 < 0$ or $K \geq -c^2$. In the case of the Jacobi equation , the mechanism for obtaining such estimates is the Rauch Comparison Theorem (cf [CE]), which is a generalization of the classical Sturm comparison theorem that applies to the growth of solutions to the scalar equation $y''(t) + k(t)y(t)$, where $k(t) \geq c$ for all t or $k(t) \leq c$ for all t and c is some real number.

To illustrate with a specific example, let $Y(t)$ be a nonzero Jacobi vector field orthogonal to $c'(t)$ on a unit speed geodesic $c(t)$ in a Riemannian manifold M with constant sectional curvature $K \equiv \lambda$. The norm $y(t) = |Y(t)|$ of a Jacobi vector field satisfies the scalar equation $y''(t) + \lambda y(t) = 0$, for which the solutions are easy to find. If $K \equiv -c^2$ and if $Y(0) = 0$, then $|Y(t)| = \sinh(ct)|Y'(0)|$ for all $t \geq 0$, where $\sinh(t) = (1/2)(e^t - e^{-t})$ is the hyperbolic sine.

In general the Rauch Comparison Theorem yields the following:

Let M be a complete Riemannian manifold with sectional curvature satisfying $-b^2 \leq K \leq -a^2$, where a and b are positive constants. If $Y(t)$ is a nonzero Jacobi vector field orthogonal to $c'(t)$ on a unit speed geodesic $c(t)$ of M with $Y(0) = 0$, then

$$(1/a)\sinh(at)|Y'(0)| \leq |Y(t)| \leq (1/b)\sinh(bt)|Y'(0)| \text{ for all } t \geq 0\,.$$

In other words, $|Y(t)|$ increases at least as fast as in the constant curvature case corresponding to the upper curvature bound and no faster than the constant curvature case corresponding to the lower curvature bound.

The discussion above of the geodesic flow states that

$$|dg^t(\xi)|^2 = |Y_\xi(t)|^2 + |Y'_\xi(t)|^2 \text{ for all } t\,,$$

where Y_ξ is the Jacobi vector field with $Y_\xi(0) = d\pi(\xi)$ and $Y_\xi'(0) = \mathcal{K}(\xi)$ and $\mathcal{K}$ is the connection mapping. In most important geometric situations $|Y_\xi'(t)|$ is at most a bounded multiple of $|Y_\xi(t)|$ for sufficiently large t so that $|dg^t(\xi)|$ is at most a bounded multiple of $|Y_\xi(t)|$ for sufficiently large t. For example, if M has constant sectional curvature $K \equiv -c^2$, then $y(t) = |Y_\xi(t)|$ satisfies the scalar equation $y''(t)-c^2y(t) = 0$. If $Y_\xi(0) = 0$, then

$$|Y_\xi'(t)| = y'(t) = c\{\cosh(ct)/\sinh(ct)\}y(t) \leq 2cy(t) = 2c|Y_\xi(t)|$$

if t is sufficiently large. If the sectional curvature of M satisfies $-b^2 \leq K \leq -c^2$, then the Rauch Comparison Theorem yields

$$c\{\cosh(ct)/\sinh(ct)\}|Y_\xi(t)| \leq |Y_\xi'(t)| \leq b\{\cosh(bt)/\sinh(bt)\}|Y_\xi(t)| \text{ for all } t.$$

See also the lemma at the end of VI A.

C. The Riemannian exponential map is a covering map. The theorem of Cartan-Hadamard. Let M be a complete Riemannian manifold with sectional curvature $K \leq 0$. Let m be any point of M, and let $\exp_m : T_mM \to M$ denote the Riemannian exponential map at m. We saw earlier at the end of II F that if v, ξ are any tangent vectors in T_mM and if $\xi_v \in T_v(T_mM)$ denotes the initial velocity of the curve $t \mapsto v + t\xi$, then $d\exp_m(\xi_v) = Y(1)$, where $Y(t)$ is the Jacobi vector field on $\gamma_v(t) = \exp_m(tv)$ such that $Y(0) = 0$ and $Y'(0) = \xi$. By the discussion above in III B we saw that $|Y(t)| \geq t|Y'(0)| = |t\xi|$ for all $t \geq 0$. Evaluating at $t = 1$ we obtain

$$(*) \qquad |d\exp_m(\xi_v)| \geq |\xi| \text{ for all } \xi_v \in T_v(T_mM) \text{ or equivalently:}$$
$$\text{If } \gamma(t) \text{ is any } C^1 \text{ curve in } T_mM, \text{ then } L(\gamma) \leq L(\exp_m \circ \gamma).$$

In particular, $\exp_m : T_mM \to M$ is a nonsingular map. In fact, we can say even more:

PROPOSITION. *Let M be a complete Riemannian manifold with sectional curvature $K \leq 0$. Let m be any point of M, and let $\exp_m : T_mM \to M$ denote the Riemannian exponential map at m. Then $\exp_m$ is a covering map. If M is simply connected, then $\exp_m$ is a diffeomorphism of M onto $\mathbb{R}^n$, $n = \dim M$.*

COROLLARY (Theorem of Hadamard-Cartan). *Let M be a complete Riemannian manifold with sectional curvature $K \leq 0$. For any two distinct points p and q there exists a unique (up to parametrization) geodesic γ_{pq} joining p to q.*

PROOF OF THE COROLLARY. Given distinct points p and q of M it suffices to show that there exists a unique geodesic $\gamma : [0,1] \to M$ such that $\gamma(0) = p$ and $\gamma(1) = q$. By the definition and discussion of the exponential map, every geodesic $\gamma(t)$ with $\gamma(0) = p$ has the form $\gamma(t) = \exp_p(tv)$ for some tangent vector v of T_pM. Let $\gamma(t)$ and $\gamma^*(t)$ be two geodesics parametrized on $[0,1]$ that join p to q, and choose vectors v and v^* in T_pM such that $\gamma(t) = \exp_p(tv)$ and $\gamma^*(t) = \exp_p(tv^*)$ for all t. Then $q = \gamma(1) = \exp_p(v)$, and the same argument shows that $q = \exp_p(v^*)$. Hence $v = v^*$ and $\gamma = \gamma^*$ since $\exp_p : T_pM \to M$ is a diffeomorphism. □

PROOF OF THE PROPOSITION. We shall need the following result from Riemannian geometry whose proof we omit.

LEMMA. *Let X and Y be complete Riemannian manifolds, and let $p : X \to Y$ be a local isometry; that is, $<dp_x(v), dp_x(w)> = <v, w>$ for all vectors v and w in T_xX and any point x of X. Then p is a covering map.*

We now prove the Proposition above. Fix a point m of M. By the inequality $(*)$ above the exponential map $\exp_m : T_mM \to M$ is nonsingular. (In particular, $\exp_m$ is a local diffeomorphism by the inverse function, which is a necessary condition for a C^∞ covering map.) Since $\exp_m : T_mM \to M$ is nonsingular, we may give the Euclidean space T_mM the unique Riemannian structure $<,>$ such that $<d\exp_m(v), d\exp_m(w)> = <v, w>$ for all vectors v and w in T_mM; i.e. the unique Riemannian structure $<,>$ that makes $\exp_m : T_mM \to M$ a local isometry. See II G. By hypothesis M is a complete Riemannian manifold, and hence by the lemma above it suffices to show that T_mM with the Riemannian metric $<,>$ is a complete Riemannian manifold.

By properties of the exponential map (cf. the discussion in II G) the preimages of the geodesics in M that start at m are precisely the straight lines in T_mM that pass through the origin in T_mM; that is the curves $t \mapsto tv$ for an arbitrary vector v in T_mM. It is a well known fact in Riemannian geometry that local isometries map geodesics to geodesics and that the preimages of geodesics in the range manifold are geodesics in the domain manifold. In the present case it means that the straight lines through the origin in T_mM are geodesics with respect to the Riemannian structure $<,>$ defined on T_mM. Since these straight line geodesics of $<,>$ are defined on $\mathbb{R}$ it follows by one of the equivalent formulations of the Hopf-Rinow theorem (see II E above) that T_mM with the Riemannian structure $<,>$ is complete. By the lemma above this completes the proof that $\exp_m : T_mM \to M$ is a covering map for every point m of M. Finally, any C^∞ covering map between simply connected spaces must be a diffeomorphism, which completes the proof of the Proposition. □

D. Examples: Riemannian symmetric spaces.

Definition of symmetric and locally symmetric space. A complete Riemannian manifold X is called a *Riemannian symmetric space* if for every point x of X there exists an isometry s_x of X that fixes x and reverses all the geodesics of X that start at x; that is,

$(**)$ If $\gamma(t)$ is any geodesic of X with $\gamma(0) = x$, then $s_x(\gamma(t)) = \gamma(-t)$ for all t.

The isometries $\{s_x : x \in X\}$ are called *geodesic symmetries*, and they are the analogues of reflection through the origin $(x \mapsto -x)$ in Euclidean space.

For any point x in a Riemannian manifold X the geodesic symmetry s_x may be defined in a neighborhood U of x by $(**)$, where now geodesics passing through x are defined in $(-\varepsilon, \varepsilon)$ for some small positive number ε. The manifold X is called *locally symmetric* if all of these locally defined geodesic symmetries $\{s_x : x \in X\}$ are local isometries. A theorem of E. Cartan says that any complete simply connected locally symmetric space must be a symmetric space.

Since isometries can also be described as diffeomorphisms that preserve the Riemannian metric (cf. the discussion in II G) one may also describe the locally symmetric spaces X as those Riemannian manifolds that satisfy a local SAS (side-angle-side) congruence axiom, as illustrated in the figure below.

If Δ is a triangle in X with vertices x, y and z whose sides are shortest geodesics, then the triangle $s_x(\Delta)$ has vertices x, $s_x(y)$ and $s_x(z)$. By the definition of the

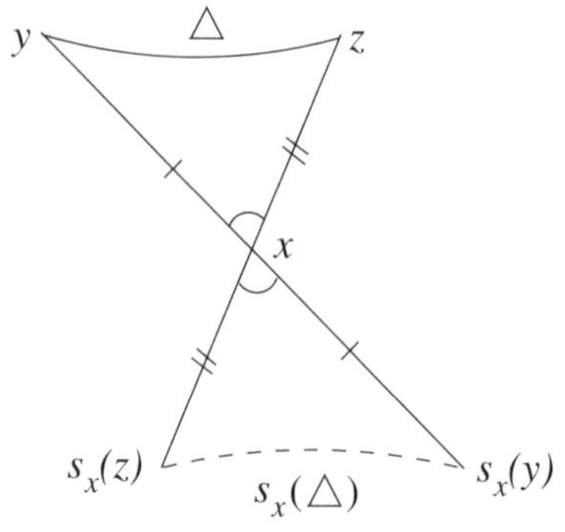

FIGURE 2.

geodesic symmetry, the sides of $s_x(\Delta)$ joining the vertices x to $s_x(y)$ and x to $s_x(z)$ are also geodesics with the same lengths as the corresponding geodesic sides in Δ that join x to y and x to z. The geodesic triangles Δ and $s_x(\Delta)$ therefore have the same side lengths and included angle at x. The condition that s_x be a local isometry is precisely the condition that $d(y, z) = d(s_x(y), s_x(z))$ for all points y, z in some neighborhood U of x, which in turn is equivalent to saying that X satisfies a SAS congruence condition in U for all geodesic triangles with vertex x.

Examples of Riemannian symmetric spaces. By the discussion above any complete, simply connected Riemannian manifold X that satisfies a local SAS congruence axiom is a Riemannian symmetric space. In particular spheres ($K \equiv c^2 > 0$) and Euclidean spaces ($K \equiv 0$) are Riemannian symmetric spaces. Riemannian symmetric spaces were completely classified by E. Cartan, and the irreducible examples are divided into the categories of compact type, noncompact type and Euclidean type. Spheres, not surprisingly, are examples of Riemannian symmetric spaces of compact type. Compact, connected Lie groups with finite center that are equipped with a biinvariant metric $<,>$ (i.e. left and right translations are isometries) are other examples of symmetric spaces of compact type. The symmetric spaces of Euclidean type are the Euclidean spaces with $K \equiv 0$ and the quotient spaces of these by lattices of translations (i.e. the flat tori).

Symmetric spaces of noncompact type. The symmetric spaces X of noncompact type are precisely the simply connected symmetric spaces with sectional curvature $K \leq 0$ that cannot be written as the Riemannian product of a flat Euclidean space with another space of nonpositive sectional curvature (i.e X has no Euclidean de Rham factor). If X is a Riemannian symmetric space of noncompact type, then X is diffeomorphic to a coset space G/K, where G is a connected, semisimple Lie group with no compact normal subgroups and K is a maximal compact subgroup of G. The group G is not unique but may be chosen so that every maximal compact subgroup K has finite center. If G/K is given a suitable Riemannian metric for which the left translations $L_g : g^*K \to gg^*K$ are isometries, then G becomes a Riemannian symmetric space and G is the identity component of the isometry group of G/K. By classifying first the complex simple Lie algebras and then the real simple Lie algebras, which are the simple summands of the Lie algebras of the groups G above, E. Cartan was able to carry out his classification of Riemannian symmetric spaces. See [Hel 2] for further details.

Rank of a symmetric space of noncompact type. Let X be a symmetric space of noncompact type. A *k-flat* in X is defined to be a complete, totally geodesic submanifold F of X that with the induced Riemannian metric on the tangent planes of F is isometric to a k-dimensional Euclidean space $\mathbb{R}^k$ with the standard flat Riemannian metric $<,>$ (i.e. $K \equiv 0$). A submanifold Y of X is *totally geodesic* if every geodesic of Y is also a geodesic of X. A 1-flat by convention is a geodesic of X defined on $\mathbb{R}$.

We define the *rank* of X to be the largest integer k such that every geodesic of X is contained in at least one k-flat F. The rank of X also has an equivalent algebraic definition, which we omit, in terms of the Lie algebra of the group G, where X is represented as a coset space G/K.

Example. If $G = SL(n, \mathbb{R})$, the group of $n \times n$ real matrices with determinant 1, and if $K = SO(n, \mathbb{R}) = \{A \in SL(n, \mathbb{R}) : AA^t = A^tA = I\}$, then G/K becomes a symmetric space of rank $n - 1$. It is diffeomorphic to the set P of all positive definite, symmetric $n \times n$ matrices with determinant 1. One can see this by defining a transitive action of G on P by setting $g(X) = gXg^t$ for $g \in G$ and $X \in P$. The subgroup of G that fixes the identity I in P is precisely K.

If F is the set of all diagonal matrices $\lambda = \operatorname{diag}(\lambda_1, \ldots, \lambda_n)$ in P, then the submanifold F of P has dimension $n - 1$ since $\det(\lambda) = 1$. In fact, F is an $(n-1)$-flat, and every $(n-1)$-flat of P has the form $g(F)$ for some $g \in G$.

Rank 1 symmetric spaces of noncompact type. The symmetric spaces X of noncompact type and rank 1 are precisely those symmetric spaces for which the sectional curvature satisfies $K \leq -c^2 < 0$ for some positive constant c. It follows from these curvature conditions that X is simply connected. In these notes we will be primarily interested in the rank 1 symmetric spaces of noncompact type. We begin by describing the real hyperbolic spaces with constant negative sectional curvature. There are three standard models, each with certain advantages.

Model I of real hyperbolic space (upper half space)

Let $H^n = \{(x_1, \ldots, x_n) \in \mathbb{R}^n : x_n > 0\}$ and equip H^n with the inner products $<,>_x$ on the tangent spaces T_xH^n, $x = (x_1, \ldots, x_n) \in H^n$, given by $<v_x, w_x>_x = (1/x_n^2)(v \cdot w)$, where for $\xi \in \mathbb{R}^n$, ξ_v denotes the initial velocity of $t \mapsto v + t\xi$, $v = (v_1, \ldots, v_n)$, $w = (w_1, \ldots, w_n)$ and $v \cdot w = v_1w_1 + \cdots + v_nw_n$. The sectional curvature in this model satisfies $K \equiv -1$. For $n = 2$ the identity component of the group of isometries is $SL(2, \mathbb{R})$, where an element $g = \begin{bmatrix} a & b \\ c & d \end{bmatrix} \in SL(2, \mathbb{R})$ acts on H^2 by the fractional linear transformation $z \mapsto (az + b)/(cz + d)$. For $n \geq 3$ the isometry group of H^n is more easily described by model III.

Model II of real hyperbolic space (open unit n-ball)

Let $H^n = \{(x_1, \ldots, x_n) \in \mathbb{R}^n : \sum_{k=1}^{n} x_k^2 < 1\}$, and equip H^n with the inner products $<,>_x$ on the tangent spaces T_xH^n, $x = (x_1, \ldots, x_n) \in H^n$, given by $<v_x, w_x>_x = 1/(1 - \sum_{k=1}^{n} x_k^2)^2(v \cdot w)$. The sectional curvature in this model satisfies $K \equiv -4$. For $n = 2$ the identity component of the isometry group is the subgroup of the fractional linear transformations of the complex plane that leaves invariant the unit disk.

Model III of real hyperbolic space (hyperboloid)

Let $H^n = \{(x_1, \ldots, x_{n+1}) \in \mathbb{R}^{n+1} : x_{n+1} > 0 \text{ and } \sum_{k=1}^{n} x_k^2 - x_{n+1}^2 = -1\}$, and equip H^n with the inner products $<,>_x$ on the tangent spaces T_xH^n, $x = (x_1, \ldots, x_{n+1}) \in H^n$, given by $<v_x, w_x>_x = L(v,w)$, where $v = (v_1, \ldots, v_{n+1})$, $w = (w_1, \ldots, w_{n+1})$ and $L(v,w) = v_1w_1 + \cdots + v_nw_n - v_{n+1}w_{n+1}$. The symmetric bilinear form L is nondegenerate but not positive definite on $\mathbb{R}^{n+1}$. However, when restricted to the tangent spaces of H^n, L is positive definite. The identity component of the isometry group of H^n is $SO(n,1)$, the subgroup of $GL(n+1,\mathbb{R})$ consisting of those invertible linear transformations of $\mathbb{R}^{n+1}$ that preserve the bilinear form L; that is, $T \in SO(n,1) \Leftrightarrow L(Tv,Tw) = L(v,w)$ for all vectors v, w in $\mathbb{R}^{n+1}$. The sectional curvature in this model satisfies $K \equiv -1$.

This model has the big advantage that it shows the hyperbolic functions very clearly in the trigonometric formulas, and these formulas are closely related in a strikingly obvious way to those of spherical trigonometry. The other two models of the real hyperbolic geometry, though better known, give few clues why the underlying geometry has anything to do with the hyperbolic trigonometric functions.

Other rank 1 symmetric spaces (cf. [Mo, pp. 134-141]). All of these spaces have strictly negative but variable sectional curvature. If the upper bound for the sectional curvature is normalized to be -1, then the lower bound for the sectional curvature is -4 in all cases. These rank 1 symmetric spaces of variable negative sectional curvature can be classified into two infinite families and one exceptional space as follows:

Complex hyperbolic spaces

For $n \geq 2$ let $\mathbb{C}H^n = \{z = (z_1, \ldots, z_n) \in \mathbb{C}^n : |z|^2 = z_1\bar{z}_1 + \cdots + z_n\bar{z}_n < 1\}$, and equip $\mathbb{C}H^n$ with the inner product $<,>_z$ on $T_z\mathbb{C}H^n$ obtained as in the unit ball model of real hyperbolic space by multiplying the standard inner product of $\mathbb{C}^n = \mathbb{R}^{2n}$ by $1/(1-|z|^2)^2$. The space $\mathbb{C}H^n$ has real dimension $2n$ and may be represented as a coset space by $SU(1,n)/U(n)$. The identity component of the isometry group is $SU(1,n)$.

Quaternionic hyperbolic spaces

For $n \geq 2$ let $\mathbb{H}H^n = \{\xi = (\xi_1, \ldots, \xi_n) \in \mathbb{H}^n : |\xi|^2 = \xi_1\bar{\xi}_1 + \cdots + \xi_n\bar{\xi}_n < 1\}$, and equip $\mathbb{H}H^n$ with the inner product $<,>_\xi$ on $T_\xi\mathbb{H}H^n$ obtained as in the unit ball model of real hyperbolic space by multiplying the standard inner product of $\mathbb{H}^n = \mathbb{R}^{4n}$ by $1/(1-|\xi|^2)^2$. The space $\mathbb{H}H^n$ has real dimension $4n$ and may be represented as a coset space by $Sp(1,n)/Sp(n)$. The identity component of the isometry group is $Sp(1,n)$.

Cayley hyperbolic "plane"

This is an exceptional space of real dimension 16 that may be represented as a coset space by $F_4/\text{Spin}(9)$. The identity component of the isometry group is F_4. For further details see [Mo, 136-141].

E. Convexity properties and the Cartan Fixed Point Theorem.

Convex functions and convex sets. A continuous function $f : \mathbb{R} \to \mathbb{R}$ is *convex* if $f((1-t)x + ty) \leq (1-t)f(x) + tf(y)$ for all $0 \leq t \leq 1$ and all x, y in $\mathbb{R}$. Convexity for continuous functions $f : I \to \mathbb{R}$, where I is an interval in $\mathbb{R}$, is defined in the same way. If $f(x)$ is C^2, then f is convex $\Leftrightarrow$ $f''(t) \geq 0$ for all t in the domain I of f.

If X is a complete Riemannian manifold, then a continuous function $f : X \to \mathbb{R}$ is convex if $f \circ \gamma : \mathbb{R} \to \mathbb{R}$ is convex for every geodesic $\gamma : \mathbb{R} \to X$. The existence of nonconstant convex functions $f : X \to \mathbb{R}$ has strong implications for the geometry of X. For example, if $X_c = \{x \in X : f(x) \leq c\}$, then it follows routinely from the definitions that X_c is strongly convex; that is for any two points x and x^* in X_c, *every* geodesic in X from x to x^* is contained in X_c.

If $\tilde{M}$ is a complete, simply connected Riemannian manifold with sectional curvature $K \leq 0$, then by the Cartan-Hadamard theorem there is a unique geodesic joining any two points of $\tilde{M}$, and hence strong convexity in this case is the same as the usual convexity in Euclidean space. In such a space $\tilde{M}$ there are many convex functions and convex sets as in the special case of a Euclidean space. The geometry of the spaces $\tilde{M}$ has very special properties due to this convexity, and these were first studied in a systematic way by R. Bishop and B. O'Neill in [BO]. We shall give only a few examples now. We already had a hint of this convexity earlier when we observed that the function $t \mapsto |Y(t)|^2$ is a C^∞ convex function on $\mathbb{R}$ for any Jacobi vector field $Y(t)$ on a geodesic $c(t)$ in a nonpositively curved Riemannian manifold.

Examples of convex functions in $\tilde{M}$.

1) If $C \subseteq \tilde{M}$ is a closed convex subset, then the function $d_C : \tilde{M} \to \mathbb{R}$ given by $d_C(x) = d(x, C) = \inf\{d(x, y) : y \in C\}$ is a continuous convex function. Here $d(\ ,\ )$ denotes the Riemannian distance function.
2) Fix a point m of $\tilde{M}$ and define a function $d_m : \tilde{M} \to \mathbb{R}$ by $d_m(x) = d(x, m)$. Then $d_m : \tilde{M} \to \mathbb{R}$ is a continuous convex function for every point m of $\tilde{M}$. The function $(d_m)^2 : \tilde{M} \to \mathbb{R}$ is a C^∞ convex function for all m. If r is any positive number, then $\tilde{M}_r = \{x \in \tilde{M} : d_m(x) \leq r\}$ is precisely the closed ball of radius r and center m. Hence all closed metric balls are convex subsets of $\tilde{M}$.
3) If φ is an isometry of $\tilde{M}$, then the displacement function $d_\varphi : \tilde{M} \to \mathbb{R}$ given by $d_\varphi(m) = d(m, \varphi(m))$ is a continuous convex function. Its square $(d_\varphi)^2 : \tilde{M} \to \mathbb{R}$ is a C^∞ convex function.

Cartan Fixed Point Theorem. Let $I(X)$ denote the group of isometries of a Riemannian manifold X. Equipped with the compact open topology it is a classical result that $I(X)$ is a Lie group, possibly zero dimensional (i.e. discrete). If X is complete and x is any point of X, then $I(X)_x$, the subgroup of $I(X)$ that fixes x, is always a compact subgroup of $I(X)$. If $\tilde{M}$ is any complete, simply connected Riemannian manifold with sectional curvature $K \leq 0$, then a theorem of E. Cartan says that the converse holds: if G is any compact subgroup of $I(\tilde{M})$, then $G \subseteq I(\tilde{M})_m$ for some point m of $\tilde{M}$; i.e. m is a common fixed point of G. It follows that the groups $\{I(\tilde{M})_m : m \in \tilde{M}\}$ are the maximal compact subgroups of $I(\tilde{M})$.

Cartan needed the fixed point theorem to prove that all maximal compact subgroups of a connected semisimple Lie group G are conjugate in G. However, he also used it to prove that every nonidentity element of the fundamental group of a compact manifold of nonpositive sectional curvature has infinite order. We take a few moments to describe the proof of the latter result.

Using the convexity of metric balls in $\tilde{M}$ and some trigonometric inequalities (law of cosines) that also arise from the convexity of $\tilde{M}$, one can show that for every bounded subset A of $\tilde{M}$ there is a unique closed metric ball B of smallest radius that contains A. One might call the center of B the spherical center of gravity of A. If G is a compact group of isometries of $\tilde{M}$ and if m is any point of $\tilde{M}$, then the orbit $A = G(m)$ is a bounded subset of $\tilde{M}$ that is invariant under G. If B is a smallest closed metric ball that contains A, then so is $g(B)$ for every element g of G. Hence $g(B) = B$ for all g in G, and the elements of G all fix the center of B, the spherical center of gravity of A.

F. Fundamental group of a complete nonpositively curved manifold.

The fundamental group of M acting as a discrete group of isometries of $\tilde{M}$. If $p : X \to Y$ is any covering projection, where X is simply connected, then the fundamental group of Y is isomorphic to the group Γ of deck transformations of p, the group of diffeomorphisms φ of X such that $p \circ \varphi = p$. By definition the group Γ leaves the fibers of p invariant, and since X is simply connected Γ acts transitively on each fiber of p. Moreover, no nonidentity element of Γ fixes any point of X. If p is a local isometry, then it follows immediately from the definition of Γ that the elements of Γ are isometries of X. We shall always think of the fundamental group of a complete Riemannian manifold M with sectional curvature $K \leq 0$ as a discrete group of isometries of the universal Riemannian cover $\tilde{M}$, which by definition is the simply connected cover $\tilde{M}$ equipped with the Riemannian structure that makes the covering projection a local isometry.

Let M be any complete Riemannian manifold with sectional curvature $K \leq 0$. The Cartan-Hadamard theorem says that the universal Riemannian cover $\tilde{M}$ is diffeomorphic to any one of its tangent spaces, a Euclidean space of the same dimension. Hence the homotopy groups $\pi_k(M)$ are zero for all $k \geq 2$ since a covering projection induces an isomorphism of the kth homotopy groups for all $k \geq 2$. This means that all of the homotopy of M is concentrated in $\pi_1(M)$, the fundamental group of M. Hence any two compact manifolds with nonpositive sectional curvature and isomorphic fundamental groups are homotopy equivalent. In fact, results of F. T. Farrell and L. Jones [FJ 1, 2] show that one can say even more.

THEOREM. *Let M_1 and M_2 be compact manifolds of nonpositive sectional curvature whose fundamental groups are isomorphic. Then*

1) *M_1 and M_2 are homeomorphic.*
2) *There exist examples where M_1 and M_2 are homeomorphic but not diffeomorphic.*

Although there are in principle infinitely many different Riemannian structures $<,>$ that one can put on a compact C^∞ manifold X, it is an article of faith among geometers that the geometry of a Riemannian structure $<,>$ on X that satisfies certain curvature restrictions is strongly influenced by the topology of X. This is certainly the case if $X = M$ has nonpositive sectional curvature. The topology of M is somehow encoded in the fundamental group of M by the remarks above, and hence it is perhaps not surprising that the algebraic properties of the fundamental group of M strongly reflect the geometric properties of M. This is discussed in some detail in [EHS]. Here we derive only one property of the fundamental group, valid even if M is noncompact.

PROPOSITION. *Let M be a complete Riemannian manifold with sectional curvature $K \leq 0$. Then every nonidentity element of the fundamental group of M has infinite order.*

PROOF. Suppose that $\Gamma = \pi_1(M)$ has a nonidentity element φ of finite order, and let G be the finite cyclic subgroup of Γ generated by φ. If we regard G as a group of isometries of the universal cover $\tilde{M}$, as explained above, then G is certainly compact since it is finite, and hence G fixes some point m of $\tilde{M}$ by the Cartan Fixed Point Theorem. However, this contradicts the fact that nonidentity elements of Γ cannot have fixed points in $\tilde{M}$. □

G. Rank of a nonpositively curved manifold. Let M be a complete Riemannian manifold with sectional curvature $K \leq 0$, and let v be any unit vector tangent to M. We define $r(v)$ to be the dimension of the space of Jacobi vector fields $Y(t)$ on the geodesic $\gamma_v : \mathbb{R} \to M$ that are parallel (i.e. $Y'(t) \equiv 0$). One knows that $r(v) \geq 1$ since the velocity vector field $Y(t) = \gamma_v'(t)$ is always a parallel Jacobi vector field by the definitions of geodesic and Jacobi vector field and the symmetries of the curvature tensor. Now define $\text{rank}(M)$ to be the infimum of the integers $r(v)$ as v ranges over all unit vectors tangent to M.

If $p : \tilde{M} \to M$ is the universal covering of M, then $\text{rank}(\tilde{M}) = \text{rank}(M)$ since the local isometry p maps geodesics $c(t)$ in $\tilde{M}$ to geodesics $(p \circ c)(t)$ in M and (parallel) Jacobi vector fields $Y(t)$ on $c(t)$ to (parallel) Jacobi vector fields $dp(Y(t))$ on $(p \circ c)(t)$. If $\tilde{M}$ is a Riemannian symmetric space, then one can show the definition of rank given here yields the same integer as the definition of rank of a symmetric space given above in III D.

If M is a Riemannian product $M_1 \times M_2$, then

$$\text{rank}(M) = \text{rank}(M_1) + \text{rank}(M_2) \geq 2\,.$$

However, negative sectional curvature implies rank 1 and even more.

LEMMA. *Let X be a complete Riemannian manifold, and let $Y(t)$ be a Jacobi vector field on a unit speed geodesic $c : \mathbb{R} \to X$. Let $Y^{tan}(t)$ and $Y^{\perp}(t)$ denote the decomposition of $Y(t)$ into Jacobi vector fields tangent and orthogonal to $c'(t)$ respectively (cf. III B). Then*

1) *If $Y(t)$ is parallel ($Y'(t) \equiv 0$), then $Y^{tan}(t) = ac'(t)$ for some real number a, and hence $Y^{tan}(t)$ and $Y^{\perp}(t)$ are both parallel on c.*
2) *If X has sectional curvature $K \leq 0$ and $Y(t)$ is parallel, then either $Y^{\perp}(t)$ is identically zero or $K(Y^{\perp}(t), c'(t)) \equiv 0$, where $K(Y^{\perp}(t), c'(t))$ denotes the sectional curvature of the 2-plane spanned by $Y^{\perp}(t)$ and $c'(t)$.*

As an immediate consequence of 2) of the lemma we obtain the following

COROLLARY 1. *Let M be a complete Riemannian manifold with strictly negative sectional curvature. Then $r(v) = 1$ for all $v \in T_1M$, and in particular M has rank 1.*

PROOF OF THE LEMMA.

1) Since $Y(t)$ is parallel it follows by differentiating $<Y(t), Y(t)>$ that $|Y(t)|$ is constant in t. Hence $|Y^{\tan}(t)| \leq |Y(t)| \equiv c$ is uniformly bounded above on $\mathbb{R}$.

However $Y^{\tan}(t) = (bt + a)c'(t)$ for some real numbers a and b by the discussion in III B. It follows that $b = 0$, which proves that $Y^{\tan}(t) = ac'(t)$ is parallel since $c'(t)$ is parallel. Since $Y^{\perp}(t) = Y(t) - Y^{\tan}(t)$, the difference of two parallel vector fields, it follows that $Y^{\perp}(t)$ is also parallel, which proves 1).

2) Suppose that X has sectional curvature $K \leq 0$, $Y(t)$ is a parallel vector field on some unit speed geodesic $c(t)$ and $Y^{\perp}(t)$ is not identically zero. Then $Y^{\perp}(t)$ is parallel by 1) and hence $|Y^{\perp}(t)| \equiv \alpha > 0$. However, by the discussion above in III B the function $f(t) = |Y^{\perp}(t)|^2$ is also a C^{∞} convex function whose second derivative is given by

$$\begin{aligned} f''(t) &= 2\left\{|Y^{\perp'}(t)|^2 - \langle R(Y^{\perp}(t), c'(t))c'(t), Y^{\perp}(t)\rangle\right\} \\ &= -2K(Y^{\perp}(t), c'(t))|Y^{\perp}(t) \wedge c'(t)|^2 \\ &= -2K(Y^{\perp}(t), c'(t))|Y^{\perp}(t)|^2 \\ &= -2K(Y^{\perp}(t), c'(t))\alpha^2 . \end{aligned}$$

Assertion 2) follows since the second derivative of a constant function is zero. □

From the lemma above, the proof of the lemma and the definition of rank we obtain

COROLLARY 2. *Let M be a complete Riemannian manifold with nonpositive sectional curvature. For each unit vector v tangent to M let $J_p^{\perp}(v)$ denote the vector space of parallel Jacobi vector fields $Y(t)$ on the unit speed geodesic γ_v such that $Y(t)$ is orthogonal to $\gamma_v'(t)$ for all $t \in \mathbb{R}$ and $K(Y(t), \gamma_v'(t)) = 0$ for all $t \in \mathbb{R}$. Then* $\dim(J_p^{\perp}(v)) = r(v) - 1$. *In particular, if* $\operatorname{rank}(M) = k \geq 2$, *then* $\dim(J_p^{\perp}(v)) \geq k - 1 \geq 1$ *for all vectors v tangent to M.*

Remark. In view of Corollary 2 one may think of a perpendicular, parallel Jacobi vector field $Y(t)$ as an "infinitesimal flat strip", and one may hope to piece together infinitesimal flat strips to form k-flats in $\tilde{M}$ when $k = \operatorname{rank}(M) \geq 2$. In fact this happens when M is compact [BBE], and this is an important step in the proof of the higher rank rigidity theorem stated below. One shows first that $r(v) = k$ for all unit vectors v in a dense open set $\mathfrak{R}$ of "regular" vectors in the unit tangent bundle $T_1\tilde{M}$ of the universal Riemannian cover $\tilde{M}$ of M. For each v in $\mathfrak{R}$ let $\mathfrak{P}(v)$ denote the k-dimensional subspace of $T_v(T_1\tilde{M})$ that corresponds to the k-dimensional subspace of parallel Jacobi vector fields on γ_v. In [BBE] it is shown that the distribution $\mathfrak{P}$ is integrable in $\mathfrak{R} \subseteq T_1\tilde{M}$ and the integral manifold through a vector v in $\mathfrak{R}$ is the set of all vectors tangent to a k-flat F_v and parallel to v.

Higher rank rigidity theorem. If one excludes Riemannian product manifolds $M = M_1 \times M_2$ and manifolds M whose universal Riemannian cover $\tilde{M}$ is a Riemannian product manifold or a symmetric space of noncompact type and rank $k \geq 2$, then it is hard to think of other examples of compact nonpositively curved manifolds of rank at least two. In fact there aren' t any as W. Ballmann and K. Burns-R. Spatzier showed independently in [Ba 2] and [BuSp].

THEOREM. *Let M be a compact Riemannian manifold with* $\operatorname{rank}(M) = k \geq 2$ *whose universal Riemannian cover $\tilde{M}$ is not a Riemannian product manifold. Then $\tilde{M}$ is a Riemannian symmetric space with* rank $= k$.

Remarks.

(1) The proof in [Ba 2] and [BuSp] is also valid if M has finite Riemannian volume and the sectional curvature satisfies $K \geq -a^2$ for some positive constant a. The restriction $K \geq -a^2$ was later removed in [EH] and [Ba 1], and the rigidity theorem was extended to the class of higher rank complete manifolds M whose unit tangent bundles consist entirely of vectors that are nonwandering with respect to the geodesic flow.

(2) U. Hamenstaedt [Ha] obtained an analogue to the rigidity theorem above for spaces with sectional curvature $K \leq -1$ by replacing the notion of rank with the notion of hyperbolic rank. For an extension of this result to homogeneous spaces see [Co 2].

THEOREM. *Let M be a compact Riemannian manifold with sectional curvature $K \leq -1$. For each unit vector v tangent to M define $r_h(v)$ to be the dimension of the space of Jacobi vector fields $Y(t)$ on γ_v such that $K(Y(t), \gamma_v'(t)) \equiv -1$ and $Y(t)$ is orthogonal to $\gamma_v'(t)$ for all $t \in \mathbb{R}$. If $r_h(v) \geq 1$ for all unit vectors v tangent to M, then the universal Riemannian cover $\tilde{M}$ is isometric to a symmetric space of rank 1.*

IV. Sphere at infinity of a complete simply connected manifold of nonpositive sectional curvature

A. Asymptotic geodesics and cone topology for $\tilde{M}(\infty)$. In this section we consider complete, simply connected manifolds of nonpositive sectional curvature.

Two unit speed geodesics γ and σ of $\tilde{M}$ are said to be *asymptotic* if there exists a positive number c such that $d(\gamma(t), \sigma(t)) \leq c$ for all $t \geq 0$. This generalizes the notion of parallel geodesics in flat Euclidean space, and in the unit disk model of the hyperbolic plane asymptotic geodesics "meet" at a single point on the boundary unit circle. We wish to generalize this model of the hyperbolic plane by adding an exterior "sphere at infinity" of dimension $n-1$, where n is the dimension of M.

The asymptote relation is an equivalence relation on the unit speed geodesics of $\tilde{M}$, and we define $\tilde{M}(\infty)$ to be the set of equivalence classes of unit speed geodesics of $\tilde{M}$. The set $\tilde{M}(\infty)$ will be called the *sphere at infinity* for $\tilde{M}$. Given a point p of $\tilde{M}$ and a unit speed geodesic γ of $\tilde{M}$ there exists a unique unit speed geodesic σ of $\tilde{M}$ such that $\sigma(0) = p$ and σ is asymptotic to γ (sketch of proof below). This defines a bijection ϕ_p between $\tilde{M}(\infty)$ and the $n-1$ sphere of unit vectors at p, and this bijection induces a topology $\mathfrak{T}_p$ on $\tilde{M}(\infty)$ that makes ϕ_p a homeomorphism. In fact, it is not difficult to show that this topology $\mathfrak{T}_p$ is independent of the point p, and we let $\mathfrak{T}$ denote this common topology.

The topology $\mathfrak{T}$ has an alternate description in terms of "truncated cones" with vertex at a fixed basepoint p of $\tilde{M}$. For this reason the topology $\mathfrak{T}$ is often called the *cone topology* for $\tilde{M}$. See [EO] for further details.

We now outline the proof that for each point p of $\tilde{M}$ and each unit speed geodesic γ of $\tilde{M}$ there is a unique unit speed geodesic σ of $\tilde{M}$ such that $\sigma(0) = p$ and σ is asymptotic to γ. This assertion may also be restated by saying that for each point p of $\tilde{M}$ and each point x of $\tilde{M}(\infty)$ there exists a unique unit speed geodesic γ "joining p to x"; that is, $\gamma(0) = p$ and γ belongs to the asymptotic equivalence class x.

The assertion above is a familiar fact for parallel geodesics in flat Euclidean space, and the result is not much more difficult to prove in the general case. First, to prove the existence of at least one asymptote σ to γ with $\sigma(0) = p$, one considers any sequence of real numbers $\{t_n\}$ diverging to $+\infty$ and defines $\{\sigma_n\}$ to be the sequence of unit speed geodesic segments such that $\sigma_n(0) = p$ and $\sigma_n(s_n) = \gamma(t_n)$, where $s_n = d(p, \gamma(t_n))$. If v is a cluster point of the sequence of unit vectors $\{\sigma_n'(0)\}$ at p, then it follows from convexity properties of $\tilde{M}$ that the geodesic σ with $\sigma(0) = p$ and $\sigma'(0) = v$ is an asymptote to γ. To prove the uniqueness of asymptotes let us assume that γ has two asymptotes σ_1 and σ_2 with $\sigma_1(0) = \sigma_2(0) = p$. Then $d(\sigma_1(t), \sigma_2(t)) \leq c$ for some positive number c and all $t \geq 0$. However, convexity properties of $\tilde{M}$ that follow from the growth rate of Jacobi vector fields $Y(t)$ with $Y(0) = 0$ (cf. III B) show that $d(\sigma_1(t), \sigma_2(t)) \geq d(c_1(t), c_2(t))$ if c_1 and c_2 are two unit speed geodesics in flat Euclidean space that issue from the same point and have the same vertex angle as the angle between σ_1 and σ_2 at p. If σ_1 and σ_2 were distinct, then this would imply that $d(\sigma_1(t), \sigma_2(t)) \to \infty$ as $t \to +\infty$, a contradiction. Therefore $\sigma_1 = \sigma_2$, and asymptotes to a geodesic γ from a fixed point p are unique.

B. Busemann functions and horospheres. Let v be a unit vector tangent to $\tilde{M}$, and define a *Busemann function* $B_v : \tilde{M} \to \mathbb{R}$ by $B_v(p) = \lim_{t\to\infty} d(p, \gamma_v(t)) - t$, where γ_v denotes the unique geodesic with initial velocity v. The limit defining the Busemann function exists by the triangle inequality, and each Busemann function is C^2 (sketch of proof below) but not in general C^∞. If v and w are unit vectors in $\tilde{M}$ such that the corresponding geodesics γ_v and γ_w are asymptotes, then the function $B_v - B_w$ is constant in $\tilde{M}$. This means that $\operatorname{grad}(B_v)$ and the level sets of B_v depend only on the asymptote class x in $\tilde{M}(\infty)$ of the geodesic γ_v. The level sets of a Busemann function B_v are called *horospheres* at the point x in $\tilde{M}(\infty)$ determined by γ_v. The discussion above shows that for each point p of $\tilde{M}$ and each point x of $\tilde{M}(\infty)$ there is a unique horosphere $H(p, x)$ at x that contains the point p. If $\tilde{M}$ has dimension 2, particularly when $\tilde{M}$ is the hyperbolic plane with Gaussian curvature $K \equiv -1$, horospheres are called horocycles.

We outline the proof that each Busemann function B_v is C^2. For further details see [HI]. Given a unit vector v tangent to $\tilde{M}$ and a positive real number t, we define $B_{v,t} : \tilde{M} \to \mathbb{R}$ by $B_{v,t}(p) = d(p, \gamma_v(t)) - t$. Then by definition $B_v = \lim_{t\to\infty} B_{v,t}$ and this convergence is uniform on compact subsets of $\tilde{M}$. For each fixed t the function $B_{v,t}$ is a continuous convex function and is C^∞ on $\tilde{M} - \{\gamma_v(t)\}$ (see III E). The gradients $\{\operatorname{grad}(B_{v,t})\}$ are unit vector fields that converge as $t \to \infty$, uniformly on compact subsets of $\tilde{M}$, to a unit vector field V on $\tilde{M}$. Moreover, the symmetric $(2,0)$ tensor fields $\{S_{v,t,p}\}$ given by $S_{v,t,p}(w^*, w) = \langle \nabla_{w^*} \operatorname{grad}(B_{v,t}), w \rangle$ for arbitrary vectors w^*, w in $T_p\tilde{M}$ converge as $t \to \infty$, uniformly on compact subsets of $\tilde{M}$, to a symmetric $(2,0)$ tensor field $S_{v,p}$. Finally, one shows that $\operatorname{grad}(B_v) = V$ and $S_{v,p}(w^*, w) = \langle \nabla_{w^*} \operatorname{grad}(B_v), w \rangle$ for arbitrary vectors w^*, w in $T_p\tilde{M}$. Hence B_v is a C^2 function on $\tilde{M}$ for all unit vectors v tangent to $\tilde{M}$.

C. Extensions of isometries to homeomorphisms of the sphere at infinity. Given an isometry φ of $\tilde{M}$ and a point x of $\tilde{M}(\infty)$ we define $\bar{\varphi}(x)$ to be the asymptote class of the geodesics $\{\varphi \circ \gamma\}$, where γ is any unit speed geodesic of $\tilde{M}$ that belongs to x. It is routine to check from the definitions that $\bar{\varphi}$ is a well defined homeomorphism of the boundary sphere $\tilde{M}(\infty)$ with the cone topology $\mathfrak{T}$.

In particular, if M is any complete Riemannian manifold with sectional curvature $K \leq 0$, then the fundamental group of M, regarded as a discrete group Γ of isometries of $\tilde{M}$, extends to a group Γ of homeomorphisms of $\tilde{M}(\infty)$. We shall see in the next subsection that the topological properties of the action of Γ on $\tilde{M}(\infty)$ strongly reflect the topological properties of the geodesic flow on the unit tangent bundle T_1M of M. To obtain information about the measurable properties of the geodesic flow (e.g. ergodicity) one must introduce measures on $\tilde{M}(\infty)$. This is described in section V.

Although $\tilde{M}(\infty)$ has a natural topological structure it supports no differentiable structure except in special cases.

D. Relating the action of the geodesic flow on T_1M to the action of $\pi_1(M)$ on $\tilde{M}(\infty)$. Let M be a manifold of nonpositive sectional curvature that is compact or more generally complete and with the property that every unit vector is nonwandering with respect to the geodesic flow $\{g^t\}$ in the unit tangent bundle T_1M. (This means that for every open set O in TM there exists a sequence of numbers $\{t_n\}$ diverging to $+\infty$ such that $g^{t_n}(O) \cap O$ is nonempty for every n.) Let Γ denote the fundamental group of M, and let $\tilde{M}$ denote the universal Riemannian cover of M. Recall that Γ acts by isometries on $\tilde{M}$ and by homeomorphisms on $\tilde{M}(\infty)$. The next two results show how one can relate the action of Γ on $\tilde{M}(\infty)$ to the action of the geodesic flow $\{g^t\}$ on T_1M. See [E 1, 3, 4] for the proofs.

THEOREM. *Let M be a complete Riemannian manifold with sectional curvature $K \leq 0$ such that every unit vector in the unit tangent bundle T_1M is nonwandering with respect to the geodesic flow $\{g^t\}$. Let Γ denote the fundamental group of M, and let $\tilde{M}$ denote the universal Riemannian cover of M. Then the following conditions are equivalent.*

1) *Some orbit of Γ in $\tilde{M}(\infty)$ is dense in $\tilde{M}(\infty)$.*
2) *Every orbit of Γ in $\tilde{M}(\infty)$ is dense in $\tilde{M}(\infty)$.*
3) *The geodesic flow $\{g^t\}$ has a dense orbit in T_1M.*

THEOREM. *Let M be a complete Riemannian manifold with sectional curvature $K \leq 0$ such that every unit vector in the unit tangent bundle T_1M is nonwandering with respect to the geodesic flow $\{g^t\}$. Let Γ denote the fundamental group of M, and let $\tilde{M}$ denote the universal Riemannian cover of M. Let $\tilde{M}$ be irreducible; that is, $\tilde{M}$ cannot be written as the Riemannian product of two manifolds of positive dimension. Then the following conditions are equivalent.*

1) *$\tilde{M}$ is isometric to a symmetric space of noncompact type and rank $k \geq 2$.*
2) *There exists a proper closed subset Y of $\tilde{M}(\infty)$ such that Γ leaves Y invariant.*
3) *The closure of every orbit of Γ in $\tilde{M}(\infty)$ is a proper closed subset of $\tilde{M}(\infty)$.*

V. Measures on the sphere at infinity ([Co 1], [Kai])

We now assume that $\tilde{M}$ is a complete, simply connected Riemannian manifold whose sectional curvature is bounded between two negative constants: $-b^2 \leq K \leq -a^2 < 0$. In this case there are various measures that one can define on the sphere at infinity $\tilde{M}(\infty)$, and these measures coincide if $\tilde{M}$ is a Riemannian symmetric space. These measures are actually families of measures indexed by the points of

$\tilde{M}$ and members of a family are absolutely continuous with respect to one another. We describe three of these families. Each family also induces an ergodic measure on $T_1 M$.

A. Harmonic measures $\{\nu_p : p \in \tilde{M}\}$**.** Under the curvature conditions $-b^2 \leq K \leq -a^2 < 0$ in $\tilde{M}$, M. Anderson–R. Schoen and D. Sullivan showed that the Dirichlet problem on $\tilde{M}(\infty)$ has an affirmative answer for continuous functions $f : \tilde{M}(\infty) \to \mathbb{R}$ (see [And], [AS] and [Su 2]).

THEOREM. *Let $\tilde{M}$ be a complete, simply connected Riemannian manifold with sectional curvature satisfying $-b^2 \leq K \leq -a^2 < 0$. Let $f : \tilde{M}(\infty) \to \mathbb{R}$ be a continuous function. Then there exists a unique harmonic function $\bar{f} : \tilde{M} \to \mathbb{R}$ whose limiting boundary values on $\tilde{M}(\infty)$ equal the function f.*

This result was later extended by W. Ballmann in [Ba 3] to the case that $\tilde{M}$ has rank 1 as defined in III G and is the universal Riemannian cover of a compact manifold M.

Under the conditions of the theorem, for each point p of $\tilde{M}$ one can define a linear functional $L_p : C(\tilde{M}(\infty)) \to \mathbb{R}$ by $L_p(f) = \bar{f}(p)$, where as above $\bar{f}$ is the unique harmonic function in $\tilde{M}$ that solves the Dirichlet problem with boundary values f. Here $C(\tilde{M}(\infty))$ denotes the space of continuous functions $f : \tilde{M}(\infty) \to \mathbb{R}$. We now define ν_p to be the unique measure on $\tilde{M}(\infty)$ such that $L_p(f) = \int_{\tilde{M}(\infty)} f d\nu_p$ for all f in $C(\tilde{M}(\infty))$. If $f \equiv 1$ on $\tilde{M}(\infty)$, then $\bar{f} \equiv 1$ on $\tilde{M}$, and hence each measure ν_p is a probability measure on $\tilde{M}(\infty)$. The measures $\{\nu_p : p \in \tilde{M}\}$ are all absolutely continuous with respect to each other, and their Radon-Nikodym derivatives are given by the following formula :

$$\frac{d\nu_p}{d\nu_q} = K(q, p, \xi) \qquad \text{for } (q, p, \xi) \in \tilde{M} \times \tilde{M} \times \tilde{M}(\infty)$$

where $K : \tilde{M} \times \tilde{M} \times \tilde{M}(\infty) \to \mathbb{R}$ satisfies the following conditions

a) $K(p, p, \xi) = 1$ for all $p \in \tilde{M}$ and all $\xi \in \tilde{M}(\infty)$.
b) For $(q, \xi) \in \tilde{M} \times \tilde{M}(\infty)$ the function $p \mapsto K(q, p, \xi)$ is harmonic in $\tilde{M}$.
c) $K(q, p, \xi) \to 0$ if $p \to \xi^* \neq \xi$, $\xi^* \in \tilde{M}(\infty)$.

Remark. The measures $\{\nu_p : p \in \tilde{M}\}$ are Γ-equivariant for any group Γ of isometries of $\tilde{M}$; that is, for each point $p \in \tilde{M}$, each $\gamma \in \Gamma$ and each ν–measurable set $A \subseteq \tilde{M}(\infty)$ one has $\nu_{\gamma p}(\gamma A) = \nu_p(A)$. This property is especially useful when Γ is a discrete cocompact group of isometries.

B. Patterson - Sullivan measures $\{\mu_p : p \in \tilde{M}\}$**.** Let $\tilde{M}$ be a complete, simply connected Riemannian manifold whose sectional curvature satisfies the inequalities $-b^2 \leq K \leq -a^2 < 0$, and let Γ be a discrete cocompact group of isometries of $\tilde{M}$. The Patterson-Sullivan measures on $\tilde{M}(\infty)$ determined by Γ are a family $\{\mu_p : p \in \tilde{M}\}$ of finite measures with the following properties :

1) The measures $\{\mu_p : p \in \tilde{M}\}$ are absolutely continuous with respect to each other and the Radon-Nikodym derivatives are given by

$$\frac{d\mu_p}{d\mu_q}(\theta) = e^{-hB_q(p,\theta)} \quad \text{for almost all } \theta \text{ in } \tilde{M}(\infty),$$

where h denotes the topological entropy of $\tilde{M}/\Gamma$ and $B_q(\cdot,\theta)$ denotes the Busemann function B_v, where γ_v is the unique geodesic of $\tilde{M}$ such that γ_v belongs to the asymptote class θ and $\gamma_v(0) = q$.

2) The measures $\{\mu_p : p \in \tilde{M}\}$ are Γ-equivariant; that is, for each point $p \in \tilde{M}$, each $\gamma \in \Gamma$ and each μ-measurable set $A \subseteq \tilde{M}(\infty)$ one has $\mu_{\gamma p}(\gamma A) = \mu_p(A)$.

Remarks.

1) ([Ma]) The topological entropy h for compact manifolds M of nonpositive sectional curvature can also be defined as the asymptotic volume growth as $r \to \infty$ of geodesic balls of radius r in the universal cover $\tilde{M}$; that is, $h = \lim_{r\to\infty} (1/r)\log B(r,p)$, where $B(r,p)$ is the volume of the closed geodesic ball with center p and radius r. If the sectional curvature of $\tilde{M}$ satisfies $-b^2 \le K \le -a^2 < 0$, then $a \le h \le b$.

2) Under the strictly negative curvature conditions above there is a unique family $\{\mu_p : p \in \tilde{M}\}$ of Patterson-Sullivan measures for each discrete cocompact group Γ of isometries of $\tilde{M}$. These were first constructed by Patterson in [Pa] in the case that $\tilde{M}$ is the hyperbolic plane, but the construction is valid for any discrete group Γ of isometries of $\tilde{M}$. This construction, which we outline below, was later extended by D. Sullivan [Su 1] to discrete groups of isometries Γ in a real hyperbolic space with constant negative sectional curvature and arbitrary dimension. In this context Sullivan proved uniqueness of the Patterson-Sullivan measures, but his arguments are valid in fact for any cocompact discrete group of isometries Γ in a space $\tilde{M}$ with sectional curvature $-b^2 \le K \le -a^2 < 0$. In [Kn 1] G. Knieper proved the existence and uniqueness of Patterson-Sullivan measures for a discrete cocompact group of isometries Γ of a space $\tilde{M}$ that has rank 1 as defined in III G.

Construction of the Patterson-Sullivan measures. This construction follows the original method of Patterson. For simplicity we assume that the sectional curvature satisfies $-b^2 \le K \le -a^2 < 0$, and that Γ is a discrete, cocompact group of isometries of $\tilde{M}$. As above, let h denote the topological entropy of $\tilde{M}/\Gamma$.

For points p, q in $\tilde{M}$ and a real number s define $P_\Gamma(s,p,q) = \sum_{\gamma\in\Gamma} e^{-sd(p,\gamma q)}$, the *Poincaré series* associated to Γ. One can show that this series converges for $s > h$ and diverges for $s \le h$. Now fix a number $s > h$ and a point q of $\tilde{M}$. For every point p of $\tilde{M}$ define a measure

$$\mu_{p,q,s} = (1/P_\Gamma(s,q,q)) \sum_{\gamma\in\Gamma} e^{-sd(p,\gamma q)} \delta_{\gamma q}\,,$$

where δ_m denotes the point mass concentrated at the point m of $\tilde{M}$. If $\{s_n\}$ is any sequence converging to h with $s_n > h$ for all n, then any weak limit μ_p of the sequence $\{\mu_{p,q,s_n}\}$ defines a family of measures $\{\mu_p : p \in \tilde{M}\}$ that satisfies properties 1) and 2) above.

C. Lebesgue measures $\{\lambda_p : p \in \tilde{M}\}$. These measures are the easiest and most obvious to construct on $\tilde{M}(\infty)$. For each point p of $\tilde{M}$ just use the natural bijection between $S_p\tilde{M}$, the sphere of unit vectors at p, and the sphere at infinity

$\tilde{M}(\infty)$ to transport the Riemannian Lebesgue measure λ_p on $S_p\tilde{M}$ to a measure, also denoted λ_p, on $\tilde{M}(\infty)$. Unfortunately, the Lebesgue measures also seem to satisfy some perverse conservation principle by being the most difficult to work with.

D. Barycenter map for probability measures. As above we assume that $\tilde{M}$ satisfies $-b^2 \le K \le -a^2 < 0$. Fix a basepoint p in $\tilde{M}$. If λ is a probability measure on $\tilde{M}(\infty)$, then we may define a function $\mathcal{B}_\lambda : \tilde{M} \to \mathbb{R}$ by

$$\mathcal{B}_\lambda(q) = \int_{\tilde{M}(\infty)} B_p(q,\theta)d\lambda(\theta)$$

where as before $B_p(\cdot,\theta)$ denotes the Busemann function B_v, where γ_v is the unique geodesic of $\tilde{M}$ such that γ_v belongs to the asymptote class θ and $\gamma_v(0) = p$. This function was earlier considered by A. Douady and C. Earle in [DE] in the case that $\tilde{M}$ is the hyperbolic plane, and by M. Burger and V. Schroeder [BuSc] in the case that $\tilde{M}$ is an arbitrary complete, simply connected manifold of nonpositive sectional curvature, with special applications in the case that $\tilde{M}$ is a Riemannian symmetric space.

LEMMA ([BCG 1, 2]). *For each probability measure λ on $\tilde{M}(\infty)$ the function $\mathcal{B}_\lambda$ is a convex function on $\tilde{M}$. If λ is a nonatomic probability measure on $\tilde{M}(\infty)$, then $\mathcal{B}_\lambda$ has a unique minimum point* $\text{bar}(\lambda)$ *in $\tilde{M}$.*

The result defines a real valued "barycenter" map $\lambda \mapsto \text{bar}(\lambda)$ on the space of all nonatomic probability measures λ on $\tilde{M}(\infty)$. It is an important tool in the result of Besson-Courtois-Gallot described in section VII.

We give an idea of the proof. The function $\mathcal{B}_\lambda$ is an average relative to λ of Busemann functions, $B_p(\cdot,\theta)$ as θ ranges over $\tilde{M}(\infty)$. Since each Busemann function is convex on $\tilde{M}$, so is $\mathcal{B}_\lambda$. The strictly negative sectional curvature implies that $\mathcal{B}_\lambda$ is strictly convex, and hence if $\mathcal{B}_\lambda$ has a minimum it is unique. To see that $\mathcal{B}_\lambda$ does have a minimum in $\tilde{M}$, suppose that it does not and let $\{p_n\}$ be a divergent sequence such that $\mathcal{B}_\lambda(p_n) \to \inf \mathcal{B}_\lambda$. If γ_n is the unique geodesic from the basepoint p to p_n, then the convexity of $\mathcal{B}_\lambda$ implies that $\mathcal{B}_\lambda$ is bounded on γ_n by $\max\{\mathcal{B}_\lambda(p), \mathcal{B}_\lambda(p_n)\}$. Since $\{p_n\}$ is an unbounded sequence in $\tilde{M}$, the geodesic segments $\{\gamma_n\}$ converge, passing to a subsequence if necessary, to a geodesic ray $\gamma : [0,\infty) \to \tilde{M}$, and by continuity the function $\mathcal{B}_\lambda$ is uniformly bounded above on $\gamma[0,\infty)$ by $\mathcal{B}_\lambda(p)$. However, the strictly negative curvature of $\tilde{M}$ implies that for any geodesic γ of $\tilde{M}$ the Busemann function $B_p(\gamma t,\theta)$ diverges to $+\infty$ as $t \to +\infty$, unless γ belongs to the asymptote class θ. From the definition of $\mathcal{B}_\lambda$ and the fact that the measure λ is nonatomic it follows that $\mathcal{B}_\lambda(\gamma t) \to +\infty$ as $t \to +\infty$ for any geodesic γ of $\tilde{M}$. This contradiction completes the proof of the Lemma. □

VI. Anosov foliations in the unit tangent bundle T_1M ([Ano])

Let $\{\varphi^t\}$ be a C^∞ flow on a compact C^∞ Riemannian manifold X. The flow is said to be an *Anosov flow* if the following conditions are satisfied :

1) The vector field V in X defined by the flow $\{\varphi^t\}$ never vanishes on X.
2) For each $x \in X$ the tangent space T_xX is a direct sum

$$T_xX = E^{ss}(x) \oplus E^{uu}(x) \oplus Z(x)$$

where $Z(x)$ is the 1-dimensional subspace spanned by $V(x)$ and $E^{ss}(x)$, $E^{uu}(x)$ define continuous subbundles of TX with positive dimension independent of x.

3) There exist positive numbers a, b and c such that
   i) For any $\xi \in E^{ss}(x)$

$$|d\varphi^t(\xi)| \leq a|\xi|e^{-ct} \quad \text{for all } t \geq 0$$
$$|d\varphi^t(\xi)| \geq b|\xi|e^{-ct} \quad \text{for all } t \leq 0$$

   ii) For any $\xi \in E^{uu}(x)$

$$|d\varphi^t(\xi)| \geq b|\xi|e^{ct} \quad \text{for all } t \geq 0$$
$$|d\varphi^t(\xi)| \leq a|\xi|e^{ct} \quad \text{for all } t \leq 0$$

The subbundles E^{ss} and E^{uu} are called the *strong stable* and *strong unstable* bundles. They are not in general C^1, but they always have integral submanifolds through each point x of X.

We now specialize to the case of a geodesic flow $\{g^t\}$ in the unit tangent bundle T_1M of a compact Riemannian manifold M of nonpositive sectional curvature. In this case one can use Jacobi vector fields to define continuous subbundles E^{ss} and E^{uu} in T_1M that have integral manifolds through every point v of T_1M. These subbundles E^{ss} and E^{uu} in general will have nonzero intersection at some points v of T_1M, and the geodesic flow is an Anosov flow precisely when $E^{ss}(v) \cap E^{uu}(v) = \{0\}$ for all v in T_1M. In particular, the geodesic flow is Anosov when the sectional curvature of M is strictly negative, but as we shall see, much weaker curvature conditions suffice for the geodesic flow to be Anosov.

A. Stable and unstable Jacobi vector fields. Let M be a complete Riemannian manifold with sectional curvature $K \leq 0$, and let $\gamma : \mathbb{R} \to M$ be a unit speed geodesic of M. A Jacobi vector field $Y(t)$ on $\gamma(t)$ is said to be *stable* (respectively *unstable*) if $|Y(t)|$ is uniformly bounded above for all $t \geq 0$ (respectively $|Y(t)|$ is uniformly bounded above for all $t \leq 0$).

The next result shows that the stable and unstable Jacobi vector fields on a geodesic γ of M are both n-dimensional vector subspaces of the $2n$-dimensional vector space $J(\gamma)$ of Jacobi vector fields on γ. Here $n = \dim M$.

PROPOSITION.

1) *For every vector $v \in T_{\gamma(0)}M$ there exists a unique stable Jacobi vector field $Y(t)$ on γ such that $Y(0) = v$.*
2) *For every vector $v \in T_{\gamma(0)}M$ there exists a unique unstable Jacobi vector field $Y(t)$ on γ such that $Y(0) = v$.*
3) *A Jacobi vector field $Y(t)$ is both stable and unstable $\Leftrightarrow$ $Y(t)$ is parallel on γ.*

PROOF. We provide only a sketch to give the general idea.

1) The fact that no Jacobi vector field $Y(t)$ vanishes twice in a manifold with nonpositive sectional (cf. III B) implies that given real numbers $a < b$ and arbitrary vectors $v \in T_{\gamma(a)}M$ and $w \in T_{\gamma(b)}M$ there exists a unique Jacobi vector field $Y(t)$ on γ such that $Y(a) = v$ and $Y(b) = w$. In particular, given a vector $v \in T_{\gamma(0)}M$ and a positive number t there exists a unique Jacobi vector field Y_t on γ such that $Y_t(0) = v$ and $Y_t(t) = 0$.

(Existence) One can show that for any sequence $\{t_n\}$ of positive real numbers that diverges to ∞ there exists a subsequence $\{s_n\}$ such that the covariant derivatives $\{Y'_{s_n}(0)\}$ converge to some vector w in $T_{\gamma(0)}M$. If $Y(t)$ is the Jacobi vector field such that $Y(0) = v$ and $Y'(0) = w$, then it is not difficult to see that $Y(t)$ is a stable Jacobi vector field. The fact that the function $u \mapsto |Z(u)|$ is convex for any Jacobi vector field Z (cf. III B) means that $u \mapsto |Y_{s_n}(u)|$ is bounded above by $|v| = |Y_{s_n}(0)|$ on the interval $[0, s_n]$ for every n. Hence $u \mapsto |Y(u)|$ is bounded above by $|v|$ on the interval $[0, \infty)$ since $Y_{s_n}(u) \to Y(u)$ for every real number u as $n \to \infty$.

(Uniqueness) If $Y_1(t)$ and $Y_2(t)$ are two stable Jacobi vector fields with $Y_1(0) = Y_2(0)$, then $Z(t) = Y_1(t) - Y_2(t)$ is a stable Jacobi vector field such that $Z(0) = 0$. However, if $Z(t)$ is nonzero, then $Z(t) \geq t|Z'(0)|$ for all $t \geq 0$ by the discussion in III B, which is impossible for a nonzero stable Jacobi vector field. Therefore $Z = 0$ and $Y_1 = Y_2$.

2) The same as 1) except for trivial modifications.

3) If the Jacobi vector field $Y(t)$ is both stable and unstable, then the function $f(t) = |Y(t)|^2$ is a convex function that is bounded on $\mathbb{R}$. Bounded convex functions on $\mathbb{R}$ must be constant, and hence

$$0 = f''(t) = 2\left\{|Y'(t)|^2 - K(Y(t), \gamma'(t))|Y(t) \wedge \gamma'(t)|^2\right\}$$

by the discussion in III B. Since both terms inside the curly brackets are ≥ 0 they must both be zero and in particular $Y'(t) \equiv 0$; that is $Y(t)$ is parallel along γ. Conversely, a parallel Jacobi vector field $Y(t)$ has constant norm since if $f(t) = |Y(t)|^2$, then $f'(t) = 2\langle Y(t), Y'(t)\rangle \equiv 0$. From the definitions, $Y(t)$ is both stable and unstable. □

Remark. The rate of growth or decay of the norm of a stable or unstable Jacobi vector field $Y(t)$ is essentially the same as that of its covariant derivative $Y'(t)$. More precisely one can show the following result whose proof (omitted) follows from an elementary compactness argument.

LEMMA. *Let M be a compact Riemannian manifold of nonpositive sectional curvature. Then there exists a positive constant a such that if $Y(t)$ is a stable or unstable Jacobi vector field on a geodesic γ of M, then $|Y'(t)| \leq a|Y(t)|$ for all $t \in \mathbb{R}$.*

Using an analytic argument one can achieve the same result if $-a^2$ is a lower bound for the sectional curvature on M.

B. The stable and unstable foliations E^s and E^u in $T(T_1M)$. Let M be a complete Riemannian manifold with sectional curvature $K \leq 0$, and let $\gamma : \mathbb{R} \to M$ be a unit speed geodesic of M. Let $J^s(\gamma)$ and $J^u(\gamma)$ denote the n-dimensional subspaces of stable and unstable Jacobi vector fields in the $2n$-dimensional vector space $J(\gamma)$ of all Jacobi vector fields on γ. Let $J^*(\gamma)$ be the codimension 1 subspace of $J(\gamma)$ that consists of all Jacobi vector fields $Y(t)$ on γ such that $\langle Y'(t), \gamma'(t)\rangle \equiv 0$. We recall from section II H that there is a natural isomorphism $\xi \mapsto Y_\xi$ between $T_v(T_1M)$ and $J^*(\gamma_v)$ such that $|dg^t(\xi)|^2 = |Y_\xi(t)|^2 + |Y'_\xi(t)|^2$ for all $t \in \mathbb{R}$. Using this isomorphism we define

$$E^s(v) = \{\xi \in T_v(T_1M) : Y_\xi \in J^s(\gamma)\}$$
$$E^u(v) = \{\xi \in T_v(T_1M) : Y_\xi \in J^u(\gamma)\}$$

Each subspace $E^s(v)$ and $E^u(v)$ has dimension $n = \dim M$. The subbundles E^s and E^u will be called the *stable* and *unstable* foliations of $T(T_1M)$.

Using the remarks above and the lemma from the previous subsection A one can also describe the subbundles E^s and E^u as follows:

$$E^s(v) = \{\xi \in T_v(T_1M) : |dg^t(\xi)| \text{ is uniformly bounded above for all } t \geq 0\}$$
$$E^u(v) = \{\xi \in T_v(T_1M) : |dg^t(\xi)| \text{ is uniformly bounded above for all } t \leq 0\}$$

By analogy with the Proposition of the previous subsection A one obtains the following

PROPOSITION. *Let M be a complete Riemannian manifold with sectional curvature $K \leq 0$. Let $\xi \in T_v(T_1M)$ be an arbitrary vector. Then the following statements are equivalent :*

1) $\xi \in E^s(v) \cap E^u(v)$.
2) *The vector field Y_ξ is a parallel Jacobi vector field on the geodesic γ_v.*
3) *The function $t \mapsto |dg^t(\xi)|$ is constant on $\mathbb{R}$.*
4) *The function $t \mapsto |dg^t(\xi)|$ is bounded above on $\mathbb{R}$.*

PROOF. We prove the assertions in cyclic order.

If $\xi \in E^s(v) \cap E^u(v)$, then by definition $Y_\xi \in J^s(\gamma) \cap J^u(\gamma)$. By the proposition of the previous subsection $J^s(\gamma) \cap J^u(\gamma)$ is the subspace of parallel Jacobi vector fields on γ. Hence 1) $\Rightarrow$ 2). If Y_ξ is a parallel vector field on γ, then $|Y_\xi(t)|^2$ is constant on $\mathbb{R}$ since this function has derivative zero, but by the discussion above $|Y_\xi(t)|^2 = |Y_\xi(t)|^2 + |Y_\xi'(t)|^2 = |dg^t(\xi)|^2$. Hence 2) $\Rightarrow$ 3), and 3) $\Rightarrow$ 4) is obvious. If the function $t \mapsto |dg^t(\xi)|^2 = |Y_\xi(t)|^2 + |Y_\xi'(t)|^2$ is bounded above on $\mathbb{R}$, then so is the function $t \mapsto |Y_\xi(t)|^2$. Hence the Jacobi vector field Y_ξ is both stable and unstable, which means precisely that $\xi \in E^s(v) \cap E^u(v)$. This proves that 4) $\Rightarrow$ 1) and completes the proof of the Proposition. □

Preservation of the foliations E^s and E^u under local isometries. If $p : M \to M^*$ is a local isometry between complete Riemannian manifolds M and M^* with sectional curvature $K \leq 0$ (i.e. a covering map), then $dp : T_1M \to T_1M^*$ is also a local isometry (and covering map) with respect to the Sasaki metrics in T_1M and T_1M^*. If $Y(t)$ is a Jacobi vector field on a geodesic γ of M, then $Y^*(t) = dp(Y(t))$ is a Jacobi vector field on the geodesic $\gamma^*(t) = (p \circ \gamma)(t)$ and $|Y(t)| = |Y^*(t)|$ for all t. Hence dp carries stable (unstable) Jacobi vector fields $Y(t)$ on γ to stable (unstable) Jacobi vector fields Y^* on γ^*, and we conclude that $dp(E^s(v)) = E^s(dp(v))$ and $dp(E^u(v)) = E^u(dp(v))$ for all vectors v in T_1M.

As we shall see next the subbundles E^s and E^u have integral manifolds through every point of T_1M, and by the discussion above the integral manifolds in M of these bundles project under p to the integral manifolds in M^* of these bundles.

Geometric meaning of the stable and unstable bundles E^s and E^u. The subbundles E^s and E^u are continuous but rarely C^1. Nevertheless, the subbundles admit integral manifolds through every point v of T_1M. These integral manifolds have a purely geometric description, and their existence requires only the completeness of the manifold M with nonpositive sectional curvature. By the preceding paragraph it suffices to describe these integral manifolds in the case that M is in fact $\tilde{M}$, a complete, simply connected manifold with sectional curvature $K \leq 0$. Note that the covering projection $p : \tilde{M} \to M$ is a local isometry by the definition of universal Riemannian covering.

Let $\tilde{M}$ be a complete, simply connected manifold with $K \leq 0$. For each unit vector v in $\tilde{M}$ and each point p of $\tilde{M}$ let $v^s(p)$ denote the initial velocity of the unique unit speed geodesic starting at p that is asymptotic to γ_v. Then the set of vectors $W^s(v) = \{v^s(p) : p \in \tilde{M}\}$ is a submanifold of $T_1\tilde{M}$ diffeomorphic to $\tilde{M}$ under the projection $\pi : T_1\tilde{M} \to \tilde{M}$. The submanifold $W^s(v)$ is an integral manifold of the bundle E^s, and $W^s(v)$ is called the *stable manifold* through v.

Similarly, for any unit vector v in $\tilde{M}$ and each point p of $\tilde{M}$ we let $v^u(p)$ denote the initial velocity of the unique unit speed geodesic σ that starts at p and is *backward* asymptotic to γ_v; that is, $\sigma(0) = p$ and $d(\sigma(t), \gamma_v(t)) \leq c$ for some positive constant c and all $t \leq 0$. Then $W^u(v) = \{v^u(p) : p \in \tilde{M}\}$ is also a submanifold of $T_1\tilde{M}$ diffeomorphic to $\tilde{M}$ under the projection $\pi : T_1\tilde{M} \to \tilde{M}$, and $W^u(v)$ is an integral manifold of E^u called the *unstable manifold* through v.

C. The strong stable and strong unstable foliations E^{ss} and E^{uu} in $T(T_1M)$. Given a unit speed geodesic $\gamma(t)$ in M we say that a Jacobi vector field $Y(t)$ is a *strong stable* Jacobi vector field on γ if $Y(t)$ is orthogonal to $\gamma'(t)$ for all t and $Y(t)$ is a stable Jacobi vector field on γ. Similarly, we say that a Jacobi vector field $Y(t)$ is a *strong unstable* Jacobi vector field on γ if $Y(t)$ is orthogonal to $\gamma'(t)$ for all t and $Y(t)$ is an unstable Jacobi vector field on γ. Let $J^{ss}(\gamma)$ and $J^{uu}(\gamma)$ denote the vector spaces of strong stable and strong unstable Jacobi vector fields on γ.

LEMMA. *Let γ be a unit speed geodesic of M, and let w be a vector in $T_{\gamma(0)}M$ orthogonal to $\gamma'(0)$. Then there exists a unique strong stable (respectively strong unstable) Jacobi vector field $Y(t)$ on γ such that $Y(0) = w$. In particular, for any unit speed geodesic γ of M each of the vector spaces $J^{ss}(\gamma)$ and $J^{uu}(\gamma)$ has dimension $n-1$, where $n = \dim M$.*

PROOF. We prove this only for $J^{ss}(\gamma)$ since the proof for $J^{uu}(\gamma)$ is essentially the same. Note first, that if $Y(t)$ is any Jacobi vector field on γ, and $f(t) = \langle Y(t), \gamma'(t)\rangle$ then

$$f''(t) = \langle Y''(t), \gamma'(t)\rangle = -\langle R(Y(t), \gamma'(t))\gamma'(t), \gamma'(t)\rangle = 0$$

for all t by the Jacobi equation, symmetries of the curvature tensor R and the fact that $\gamma'(t)$ is a parallel vector field on γ by the definition of geodesic. Hence there exist real numbers a and b such that $\langle Y(t), \gamma'(t)\rangle = at + b$ for all t. If $Y(t)$ is stable or unstable, then $|\langle Y(t), \gamma'(t)\rangle|$ is bounded for $t \geq 0$ or $t \leq 0$ respectively since $|\gamma'(t)| \equiv 1$. We conclude that $a = 0$ in either case, which proves

$(*)$ $\langle Y(t), \gamma'(t)\rangle$ is identically constant if $Y(t)$ is either a stable or unstable Jacobi vector field

Now let w be an arbitrary vector at $\gamma(0)$ that is orthogonal to $\gamma'(0)$. By the discussion above in subsection A there is a unique stable Jacobi vector field $Y(t)$ on γ such that $Y(0) = w$. Since $Y(t)$ is orthogonal to $\gamma'(t)$ at $t = 0$ by the choice of w it follows from $(*)$ that $Y(t)$ is orthogonal to $\gamma'(t)$ for all t. Hence $Y(t)$ is a strong stable Jacobi vector field, and the proof of the lemma is complete. □

We now define the strong stable and strong unstable foliations E^{ss} and E^{uu} in T_1M in the same way that we defined the stable and unstable foliations E^s and E^u in T_1M. Using the isomorphism $\xi \mapsto Y_\xi$ between $T_v(T_1M)$ and $J^*(\gamma_v)$ we define

$$E^{ss}(v) = \{\xi \in T_v(T_1M) : Y_\xi \in J^{ss}(\gamma)\}$$
$$E^{uu}(v) = \{\xi \in T_v(T_1M) : Y_\xi \in J^{uu}(\gamma)\}$$

Each subspace $E^{ss}(v)$ and $E^{uu}(v)$ has dimension $n-1$ by the lemma above, where $n = \dim M$.

As in the discussion of the stable and unstable foliations we may relate $E^{ss}(v)$ and $E^{uu}(v)$ as follows. We omit the proof, which is similar.

PROPOSITION. *Let M be a complete Riemannian manifold with sectional curvature $K \leq 0$. Let $\xi \in T_v(T_1M)$ be an arbitrary vector. Then the following statements are equivalent :*

1) $\xi \in E^{ss}(v) \cap E^{uu}(v)$.
2) *The vector field Y_ξ is a parallel Jacobi vector field on the geodesic γ_v, that is orthogonal to $\gamma'(t)$ for all t in $\mathbb{R}$.*
3) *The function $t \mapsto |dg^t(\xi)|$ is constant on $\mathbb{R}$ and $<d\pi(dg^t(\xi)), g^tv> = 0$ for all t in $\mathbb{R}$.*
4) *The function $t \mapsto |dg^t(\xi)|$ is bounded above on $\mathbb{R}$, and $<d\pi(\xi), v> = 0$.*

As a consequence of this result we obtain the following

COROLLARY. *Relative to the Sasaki metric in T_1M the bundles E^{ss} and E^{uu} are both orthogonal to the 1-dimensional foliation Z tangent to the geodesic flow.*

PROOF. If V is the unit vector field in T_1M whose flow transformations are the geodesic flow $\{g^t\}$, then $V(v)$ spans $Z(v)$ for all v in T_1M. Hence $\mathcal{K}(\xi) = 0$ for any vector $\xi = \alpha V(v)$ in $Z(v)$, where $\mathcal{K} : T(TM) \to TM$ is the connection map (cf. II H). Given a vector v in T_1M let ξ be any vector in $E^{ss}(v)$ and $\eta = \alpha V(v)$ any vector in $Z(v)$. Then

$$<\xi, \eta> = <d\pi(\xi), d\pi(\eta)> + <\mathcal{K}(\xi), \mathcal{K}(\eta)> = \alpha <d\pi(\xi), v> = 0$$

by 4) of the Proposition above. Hence $E^{ss}(v)$ and $Z(v)$ are orthogonal, and a similar argument shows that $E^{uu}(v)$ and $Z(v)$ are orthogonal. □

Geometric meaning of the strong stable and strong unstable bundles E^{ss} and E^{uu}. Like the stable and unstable bundles E^s and E^u, the bundles E^{ss} and E^{uu} also have integral manifolds $W^{ss}(v)$ and $W^{uu}(v)$ with nice geometric descriptions through each point v of T_1M. As before it also suffices to do this in the case that $M = \tilde{M}$, a complete, simply connected manifold of nonpositive curvature.

Given a vector v in $T_1\tilde{M}$ we let $B_v : \tilde{M} \to \mathbb{R}$ denote the Busemann function determined by v as defined above in IV B. Let $H(v) = B_v^{-1}(0)$ denote the horosphere determined by v that passes through the basepoint of v. Recall that for any point p of $\tilde{M}$ we let $v^s(p)$ denote the initial velocity of the unique unit speed geodesic starting at p that is asymptotic to γ_v, and we let $v^u(p)$ denote the initial velocity of the unique unit speed geodesic starting at p that is backward asymptotic to γ_v. Now define

$$W^{ss}(v) = \{v^s(p) : p \in H(v)\}$$
$$W^{uu}(v) = \{v^u(p) : p \in H(-v)\}$$

Then both $W^{ss}(v)$ and $W^{uu}(v)$ are $n-1$ manifolds in $T_1\tilde{M}$ that are diffeomorphic to the horospheres $H(v)$ and $H(-v)$ under the projection $\pi : T_1\tilde{M} \to \tilde{M}$. It is not difficult to show that $W^{ss}(v)$ and $W^{uu}(v)$ are integral manifolds of the bundles E^{ss} and E^{uu}. The manifolds $W^{ss}(v)$ and $W^{uu}(v)$ may also be described as the sets of inward unit normals to $H(v)$ and outward unit normals to $H(-v)$ respectively.

Relationship between the foliations E^{ss}, E^{uu}, E^s, E^u and their integral manifolds. Invariance under the geodesic flow. For each v in T_1M let $Z(v)$ denote the 1-dimensional subspace of $T_v(T_1M)$ that is tangent to the geodesic flow. From the definitions and the discussion above it is not difficult to obtain the following result whose proof we leave as an exercise.

PROPOSITION. *Let M be a complete Riemannian manifold with sectional curvature $K \leq 0$, and let v be any vector in T_1M. Then*

1) *$E^s(v) = E^{ss}(v) \oplus Z(v)$ and $E^u(v) = E^{uu}(v) \oplus Z(v)$.*
2) *$W^s(g^tv) = W^s(v) = g^tW^s(v)$ and $W^{ss}(g^tv) = g^tW^{ss}(v)$.*
   $W^u(g^tv) = W^u(v) = g^tW^u(v)$ and $W^{uu}(g^tv) = g^tW^{uu}(v)$.
3) *$W^s(v) = \bigcup_{t\in\mathbb{R}} g^tW^{ss}(v)$ and $W^u(v) = \bigcup_{t\in\mathbb{R}} g^tW^{uu}(v)$.*

D. Conditions for the foliations E^{ss} and E^{uu} to be Anosov. We must now assume that M is compact to get sufficient control over the subbundles E^{ss} and E^{uu} in T_1M to prove something reasonable. Note that if $\xi \in Z(v) \subseteq T_v(T_1M)$ (i.e. ξ is tangent to the geodesic flow), then $|dg^t(\xi)| \equiv |\xi|$.

THEOREM ([E 5]). *Let M be a compact Riemannian manifold with sectional curvature $K \leq 0$. Then the following conditions are equivalent :*

1) *The geodesic flow $\{g^t\}$ in T_1M is an Anosov flow.*
2) *$E^{ss}(v) \cap E^{uu}(v) = \{0\}$ for all $v \in T_1M$.*
3) *Let $\xi \in T_v(T_1M)$ be a vector such that the function $t \mapsto |dg^t(\xi)|$ is bounded on $\mathbb{R}$. Then $\xi \in Z(v)$.*
4) *There exists no nonzero parallel Jacobi vector field $Y(t)$ on a geodesic γ of M such that $Y(t)$ is orthogonal to $\gamma'(t)$ for all t in $\mathbb{R}$.*
5) *The rank $r(v) = 1$ for every vector v in T_1M (cf. III G).*

Moreover, if any of these equivalent statements hold, then the Anosov foliations are precisely the foliations E^{ss} and E^{uu}.

PROOF. We prove the statements in cyclic order, except for 5) $\Rightarrow$ 1), which is beyond the scope of these notes. See [E 5] for a proof of this statement.

1) $\Rightarrow$ 2). Let the geodesic flow be Anosov and let ξ be an element of $E^{ss}(v) \cap E^{uu}(v)$ for some vector v of T_1M. Let V denote the vector field in T_1M whose flow is the geodesic flow $\{g^t\}$. By the definition of Anosov flow we may write $\xi = \xi^{ss} + \xi^{uu} + \alpha V(v)$, where ξ^{ss} is tangent to the strong stable Anosov foliation, ξ^{uu} is tangent to the strong unstable Anosov foliation and α is some real number. By the first proposition in subsection C above we know that $d\pi(\xi)$ is orthogonal to v and $t \mapsto |dg^t(\xi)|$ is a constant function on $\mathbb{R}$. If $\xi^{ss} \neq 0$, then $|dg^t(\xi^{ss})|$ increases

exponentially as $t \to -\infty$, $|dg^t(\alpha V(v))| \equiv \alpha$ and $|dg^t(\xi^{uu})|$ goes to zero as $t \to -\infty$, which contradicts the fact that $t \mapsto |dg^t(\xi)|$ is a constant function on $\mathbb{R}$. Therefore $\xi^{ss} = 0$ and a similar argument shows that $\xi^{uu} = 0$. We conclude that $\xi = \alpha V(v)$. However, $0 = <d\pi(\xi), v> = \alpha$, which implies that $\xi = 0$.

2) $\Rightarrow$ 3). Suppose that 2) holds, and let $\xi \in T_v(T_1M)$ be a vector such that the function $t \mapsto |dg^t(\xi)|$ is bounded on $\mathbb{R}$. Then by the propositions in subsections B and C we conclude that

$$\xi \in E^s(v) \cap E^u(v) = \{E^{ss}(v) \oplus Z(v)\} \cap \{E^{uu}(v) \oplus Z(v)\} = Z(v).$$

3) $\Rightarrow$ 4). Let $Y(t)$ be a parallel Jacobi vector field on a geodesic γ such that $Y(t)$ is orthogonal to $\gamma'(t)$ for all t in $\mathbb{R}$. If $v = \gamma'(0)$ and ξ is that vector in $T_v(T_1M)$ such that $Y = Y_\xi$, then the discussion in subsection C shows that $<d\pi(\xi), v> = 0$ and $t \mapsto |dg^t(\xi)|$ is constant on $\mathbb{R}$. By 3) it follows that $\xi = \alpha V(v)$ for some real number α, which implies that $d\pi(\xi) = \alpha v$. Finally, $0 = <d\pi(\xi), v> = \alpha$, which shows that $\xi = 0$ and $Y = Y_\xi = 0$.

4) $\Leftrightarrow$ 5). It follows directly from the definition in III G of $r(v)$ for a vector v in T_1M that 4) and 5) are equivalent.

5) $\Rightarrow$ 1). This is the tough direction. See [E 5].

Next suppose that any of the 5 equivalent statements hold, and let E^{*ss} and E^{*uu} denote the Anosov foliations in T_1M. We show that $E^{*ss} = E^{ss}$ and $E^{*uu} = E^{uu}$, which will complete the proof of the theorem.

Note first that $T_v(T_1M) = E^{ss}(v) \oplus E^{uu}(v) \oplus Z(v)$ for all v in T_1M; by equivalence 2) $E^{ss}(v) + E^{uu}(v) + Z(v)$ is a direct sum in which $E^{ss}(v)$ and $E^{uu}(v)$ both have dimension $n-1$ and $T_v(T_1M)$ has dimension $2n-1$. Since $T_v(T_1M) = E^{*ss}(v) \oplus E^{*uu}(v) \oplus Z(v)$ by the definition of the Anosov conditions it suffices to prove that $E^{*ss}(v) \subset E^{ss}(v)$ and $E^{*uu}(v) \subseteq E^{uu}(v)$. We prove only the first inclusion since the proof of the second is very similar.

To prove that $E^{*ss}(v) \subseteq E^{ss}(v)$ we recall from subsection C above that the foliations E^{ss} and E^{uu} are both orthogonal to the 1-dimensional foliation Z tangent to the geodesic flow relative to the Sasaki metric on T_1M. If ξ is any vector in $Z(v)$, then it follows directly from the definition of the Sasaki metric that $|dg^t(\xi)| \equiv |\xi|$.

Now let ξ be any vector in $E^{*ss}(v)$ for some v in T_1M, where E^{*ss} denotes the strong stable Anosov foliation of T_1M. Since $T_v(T_1M) = E^{ss}(v) \oplus E^{uu}(v) \oplus Z(v)$ there exist vectors $\xi_1 \in E^{ss}(v)$, $\xi_2 \in E^{uu}(v)$ and $\xi_3 \in Z(v)$ such that $\xi = \xi_1 + \xi_2 + \xi_3$. In terms of the corresponding Jacobi vector fields on the geodesic γ_v we have $Y_\xi = Y_{\xi_1} + Y_{\xi_2} + Y_{\xi_3}$.

We show first that $\xi_2 = 0$. By the definitions of $E^{ss}(v)$ and $Z(v)$ and the discussion above the function $|Y_{\xi_1}(t)|^2$ is bounded above for $t \geq 0$ and the function $|Y_{\xi_3}(t)|^2$ is constant on $\mathbb{R}$. Hence $|Y_{\xi_1}(t) + Y_{\xi_3}(t)|$ is bounded above for $t \geq 0$. The convex function $|Y_{\xi_2}(t)|^2$ is bounded above for $t \leq 0$ by the definition of $E^{uu}(v)$, and hence $|Y_{\xi_2}(t)|^2$ is nondecreasing on $\mathbb{R}$. The function $|Y_{\xi_2}(t)|^2$ must also be bounded above for $t \geq 0$ since otherwise the function $|Y_\xi(t)| = |Y_{\xi_1}(t) + Y_{\xi_2}(t) + Y_{\xi_3}(t)|$ would be unbounded for $t \geq 0$, contradicting the fact that $|Y_\xi(t)|$ decreases exponentially to zero since ξ lies in the strong stable Anosov foliation $E^{*ss}(v)$. Hence the convex function $|Y_{\xi_2}(t)|^2$ is bounded on $\mathbb{R}$, and we conclude that $Y_{\xi_2}(t)$ is a parallel Jacobi vector field by earlier discussion. Since $Y_{\xi_2}(t)$ is orthogonal to $\gamma_v'(t)$ for all t by the definition of $E^{uu}(v)$ the fact that $Y_{\xi_2}(t)$ is parallel implies that $\xi_2 \in E^{ss}(v) \cap E^{uu}(v) = \{0\}$ by equivalence 2) of the theorem.

We have shown that $\xi_2 = 0$ and hence that $\xi = \xi_1 + \xi_3$, where $\xi_1 \in E^{ss}(v)$ and $\xi_3 \in Z(v)$. Since the foliations E^{ss} and Z are orthogonal relative to the Sasaki metric we obtain

$$|dg^t(\xi)|^2 = |dg^t(\xi_1)|^2 + |dg^t(\xi_3)|^2 = |dg^t(\xi_1)|^2 + |\xi_3|^2 \geq |\xi_3|^2$$

for all $t \in \mathbb{R}$. However, $|dg^t(\xi)|$ decreases exponentially to zero as $t \to +\infty$ since $\xi \in E^{*ss}(v)$, and this forces $\xi_3 = 0$ and $\xi = \xi_1 \in E^{ss}(v)$. This shows that $E^{*ss}(v) \subseteq E^{ss}(v)$ and completes the proof of the theorem. □

VII. Some outstanding problems of geometry and dynamics

Here we list some outstanding problems that researchers in differential geometry and dynamical systems began to consider seriously about 15 years ago. Each partial solution to these problems raised new questions, and it was obvious that the problems were closely related. All of the problems involved possible (even probable) characterizations of rank 1 symmetric spaces of negative curvature by various geometric or dynamic properties. There was a bewildering variety of tools and possible approaches to the resolution of these problems, and each new advance seemed to open up even more options.

A. The Katok entropy conjecture. In [Kat] A. Katok showed that on any compact surface M with genus $g \geq 2$ if one considered all Riemannian metrics on M with volume$(M) = 1$, then the metrics g with the smallest possible topological entropy were precisely the metrics g_0 with constant negative curvature. In the years that followed he focused attention on the conjecture that if (M, g_0) is any compact locally symmetric manifold with topological entropy $h(g_0)$ and volume $(M, g_0) = 1$, then for any other Riemannian metric g with volume $(M, g) = 1$ the topological entropy $h(g)$ should be $\geq h(g_0)$, with equality $\Leftrightarrow$ (M, g_0) is isometric to (M, g). See also [Gro 1, 2]. Katok's 2-dimensional methods in fact prove the conjecture for arbitrary dimensions whenever g is conformally equivalent to g_0.

B. Smoothness of Anosov foliations and Riemannian symmetric spaces. If $\{\varphi^t\}$ is an Anosov flow on a compact Riemannian manifold X, then Anosov proved that the foliations E^{ss} and E^{uu} are in general absolutely continuous but not C^1. The situation is not much better if one considers only those Anosov flows that are geodesic flows on the unit tangent bundles of compact Riemannian manifolds of strictly negative sectional curvature. As remarked earlier, E. Hopf in [Ho 1, 2] proved that the 1-dimensional foliations E^{ss} and E^{uu} in the unit tangent bundle of a compact negatively curved 2-dimensional Riemannian manifold are C^1. L. Green [Gre] and M. Hirsch-C. Pugh [HP] showed that the foliations are C^1 in arbitrary dimensions if the normalized sectional curvature satisfies $-4 < K \leq -1$. Progress stopped here, and only the Riemannian symmetric spaces were known to have C^∞ Anosov foliations. It became clear that C^∞ foliations were not to be expected and that this condition might even characterize Riemannian symmetric spaces.

C. The geodesic conjugacy problem. Let X and Y be compact Riemannian manifolds whose geodesic flows are "the same"; that is, there exists a C^k homeomorphism $F : T_1X \to T_1Y$ for some integer $k \geq 0$ such that $F \circ g_X^t = g_Y^t \circ F$ for all $t \in \mathbb{R}$. Under these conditions we say that X and Y are C^k *geodesically conjugate.* Any dynamical properties defined by the geodesic flow are the same for both manifolds, and it is tempting to ask if geodesically conjugate manifolds must be isometric.

In general the answer to the question above is no. See [CK] for a counterexample. However, evidence mounted steadily that if the sectional curvature of X and Y is nonpositive, then C^0 geodesically conjugate manifolds are isometric. For a long time steady progress was made from the late 80's through the mid 90's but the desired result remained just out of reach, even in special cases.

D. Harmonic and asymptotically harmonic spaces. A Riemannian manifold is said to be *harmonic* if all geodesic balls have constant mean curvature that depends only on the radius of the ball. Harmonicity is an extremely strong property that is satisfied by Riemannian symmetric spaces and their quotient manifolds, the locally symmetric spaces. Lichnerowicz many years ago conjectured in [Li] that harmonic spaces must be locally symmetric, but little progress was made toward a resolution of the conjecture.

For reasons that will become clearer below the concept of an *asymptotically harmonic* space of strictly negative sectional curvature arose during the 80's. A complete, simply connected Riemannian manifold $\tilde{M}$ of strictly negative sectional curvature is called *asymptotically harmonic* if all horospheres of $\tilde{M}$ have constant mean curvature. Every horosphere through a given point p of $\tilde{M}$ is a limit of geodesic balls that pass through p with radii going to $+\infty$. If $\tilde{M}$ is harmonic, then it follows that $\tilde{M}$ is asymptotically harmonic. Nevertheless, even with this weaker looking concept it didn't seem to be any easier to find asymptotically harmonic spaces of strictly negative sectional curvature that weren't already symmetric spaces.

E. Early partial solutions. We now describe briefly some of the research on the questions above that was published in the past 15 years. We apologize in advance for the incompleteness of this survey.

Katok entropy conjecture. Gromov in [Gro 1] considered the quantity $h^n(g)\mathrm{vol}(g)$ for a compact Riemannian manifold (X, g), where $n = \dim X$, $h(g)$ is the topological entropy of X and $\mathrm{vol}(g)$ is the volume of X with respect to the Riemannian volume determined by g. This quantity is invariant under multiplication of the metric g by positive constants, and if one considers only metrics g with $\mathrm{vol}(g) = 1$ then it is obviously equivalent to the topological entropy. Gromov showed that for any metric g this quantity is bounded below by $C_n|X|$, where C_n is an explicit constant that depends only on n and $|X|$ denotes the *simplicial volume* of X. The simplicial volume depends only on the topology of X and is positive if X admits a Riemannian metric with strictly negative curvature. If X admits a metric with $K \equiv -1$, then $|X|$ can be explicitly computed in terms of the volume of X. Although Gromov's result did not resolve the Katok entropy conjecture it focused attention on the scale invariant quantity $h^n(g)\mathrm{vol}(g)$, which eventually played a key role in the solution of the entropy conjecture, and others as well.

Smoothness of Anosov foliations and Riemannian symmetric spaces. Perhaps the first step in the resolution of the smoothness problem of the Anosov foliations was achieved by E. Ghys in [Gh]. He proved that if M is a compact 2-dimensional Riemannian manifold with $K < 0$ whose Anosov foliations are C^∞, then M has constant negative curvature.

Kanai [Ka] made a big leap forward when he proved that if M is a compact manifold of arbitrary dimension with sectional curvature satisfying $-9/4 < K < -1$ and C^∞ Anosov foliations, then M is C^∞ geodesically conjugate to a compact manifold of the same dimension with $K \equiv -1$. This spectacular result was convincing evidence that negatively curved spaces with C^∞ Anosov foliations must be Riemannian symmetric spaces, and it also focused new attention on the geodesic conjugacy problem, which it now seemed must also be solved if further progress were to be made.

A. Katok and R. Feres in [FK] improved the Kanai result by weakening the pinching condition to $-4 < K < -1$. This is the optimal pinching condition for a manifold geodesically conjugate to a space with constant negative sectional curvature. The other rank 1 symmetric spaces have normalized sectional curvatures satisfying $-4 \leq K \leq -1$.

Y. Benoist, P. Foulon and F. Labourie in [BFL] then dispensed with all rank 1 symmetric spaces in one shot by proving that if M is a compact manifold with strictly negative sectional curvature and C^∞ Anosov foliations, then M is C^∞ geodesically conjugate to a compact rank 1 locally symmetric manifold.

The geodesic conjugacy problem. Slightly after the result of Kanai stated above, the problem of geodesic conjugacy received new attention from an entirely different direction through the work of C. Croke [Cr] and J.-P. Otal [O 1, 2]. They showed independently that if M and M^* are compact 2-dimensional manifolds with sectional curvature $K \leq 0$ that are also C^0 geodesically conjugate, then M and M^* are isometric. The methods however were very definitely limited to dimension 2.

C. Croke and B. Kleiner in [CK] proved the useful result that if X and Y are any two compact Riemannian manifolds that are C^1 geodesically conjugate, then they have the same volume.

Using the higher rank rigidity theorem Croke, Kleiner and P. Eberlein were able to show in [CEK] that if M and M^* are two compact manifolds with sectional curvature $K \leq 0$ such that one of them has rank $k \geq 2$, then M and M^* are isometric if they are C^0 geodesically conjugate.

The geodesic conjugacy problem still remained open for manifolds with strictly negative sectional curvature. It is a known but apparently unpublished fact that if M and M^* are compact manifolds with strictly negative sectional curvature whose geodesic flows are C^0 conjugate, then there is a geodesic conjugacy $F : T_1M \to T_1M^*$ such that $F(-v) = -F(v)$ for all unit vectors v tangent to M. (Note that geodesic conjugacies are not unique since if F is one then $F^* = F \circ g^t$ is another for any real number t.)

Harmonic and asymptotically harmonic spaces. After many years of inactivity, the interest in the Lichnerowicz conjecture for harmonic spaces was revived when Z. Szabo proved it for compact simply connected harmonic spaces [Sz]. However, a short time later E. Damek and F. Ricci proved in [DR] that the Lichnerowicz conjecture was false for simply connected spaces of nonpositive curvature, more precisely for certain homogeneous, nonpositively curved, 1-dimensional solvable extensions of 2-step nilpotent Lie groups N of "Heisenberg type". The spaces of Heisenberg type arise from representations of the Clifford algebra on a real inner product space $\{V, <,>\}$. There are infinitely many examples of this kind, all of which are harmonic, including at least one for V of any dimension $k \geq 1$. The symmetric space examples occur only when $k = 1$, 3 or 7. However, there was still hope that the Lichnerowicz conjecture might be true if one required compactness of the nonpositively curved manifolds under consideration. A result of E. Heintze [He 1, 2] for strictly negative curvature that was later extended by R. Azencott and E. Wilson [AW 1, 2] to nonpositive sectional curvature says that if $\tilde{M}$ is a complete, simply connected homogeneous Riemannian manifold with sectional curvature $K \leq 0$, then $\tilde{M}$ is a symmetric space if it admits a quotient manifold that is compact or noncompact with finite volume.

Interest in asymptotically harmonic spaces of negative curvature was increased by three striking results in the early 90's.

P. Foulon and F. Labourie in [FL] showed that if M is a compact manifold of strictly negative sectional curvature whose universal Riemannian cover $\tilde{M}$ is asymptotically harmonic, then M is C^∞ geodesically conjugate to a compact locally symmetric manifold of strictly negative sectional curvature.

F. Ledrappier in [Le] showed that if M is any compact manifold of strictly negative sectional curvature, then the universal Riemannian cover $\tilde{M}$ is asymptotically harmonic $\Leftrightarrow$ the Patterson-Sullivan measures $\{\mu_p : p \in \tilde{M}\}$ coincide with the harmonic measures $\{\nu_p : p \in \tilde{M}\}$.

C. Yue in [Y 2] showed that if M is any compact manifold of strictly negative sectional curvature, then the universal Riemannian cover $\tilde{M}$ is asymptotically harmonic $\Leftrightarrow$ the Lebesgue measures $\{\lambda_p : p \in \tilde{M}\}$ coincide with either the Patterson-Sullivan measures or with the the harmonic measures.

Although the results described in this subsection obviously represented major progress, they also seemed to confuse the issue hopelessly by relating all possible relevant concepts to each other in a way that offered no clear path for the future. How should one disentangle the threads ?

VIII. The work of Besson - Courtois - Gallot

A. Statement of the main result.

THEOREM ([BCG 1, 2]). *Let (M, g_0) and (X, g) be compact Riemannian manifolds of dimension $n \geq 3$ such that M is a locally symmetric space with sectional curvature $K < 0$. Let $f : X \to M$ be a continuous map with nonzero degree. Then*

1) *$h^n(X,g)vol(X,g) \geq |deg(f)|h^n(M,g_0)vol(M,g_0)$, where h and vol denote topological entropy and volume respectively.*
2) *If equality holds in 1), then after rescaling the metric of M by a positive constant, f is homotopic to a Riemannian covering map.*

B. Corollaries of the main result. As we shall see most of the questions discussed in the previous section can now be answered, and the partial results fit together nicely into a larger framework.

COROLLARY 1 (Mostow Rigidity Theorem). *Let (M, g_0) and (X, g) be compact locally symmetric Riemannian manifolds of dimension ≥ 3 with sectional curvature $K < 0$. If the fundamental groups of M and X are isomorphic, then for some positive constant c, (M, cg_0) is isometric to (X, g).*

PROOF. Since both spaces have nonpositive curvature the homotopy groups π_k vanish for $k \geq 2$ for both M and X. Hence the isomorphism between fundamental groups induces homotopy equivalences $f : X \to M$ and $g : M \to X$ such that $f \circ g$ and $g \circ f$ are homotopic to the identity. In particular both maps f and g

have degree 1 and both manifolds have the same dimension $n \geq 3$. Applying the theorem to both f and g we obtain equality in the statement of the theorem, and the equality assertion of the theorem now completes the proof. □

COROLLARY 2. *Let (M, g_0) and (X, g) be compact Riemannian manifolds of arbitrary dimension with sectional curvature $K < 0$, and let (M, g_0) be locally symmetric. If (M, g_0) and (X, g) are C^1 geodesically conjugate, then they are isometric.*

PROOF. The fact that (M, g_0) and (X, g) are C^0 geodesically conjugate means that they have the same dimension and isomorphic fundamental groups by a standard argument. Arguing as in the proof of Corollary 1 there are homotopy equivalences $f : X \to M$ and $g : M \to X$ with degree 1. Clearly (M, g_0) and (X, g) have the same topological entropy since entropy (in its original dynamical definition) is defined by the geodesic flow. Moreover, both (M, g_0) and (X, g) have the same volume by the result of Croke and Kleiner since the geodesic conjugacy is C^1. If the manifolds have dimension 2, then the result follows by Katok's entropy result, and if the dimension is $n \geq 3$ then the result follows as in the proof of Corollary 1 by the equality assertion of the theorem. No rescaling of the metric by a positive constant c is necessary since the manifolds (M, g_0) and (X, g) have the same volume. □

COROLLARY 3. *Let M be a compact Riemannian manifold with sectional curvature $K < 0$ whose universal Riemannian covering $\tilde{M}$ is asymptotically harmonic. Then $\tilde{M}$ is a symmetric space.*

PROOF. By the result of Foulon-Labourie quoted in section VII the manifold M is C^∞ geodesically conjugate to a locally symmetric space with sectional curvature $K < 0$. Now apply Corollary 2. □

COROLLARY 4. *Let M be a compact Riemannian manifold with sectional curvature $K < 0$. Suppose that any two of the families of measures $\{\lambda_p : p \in \tilde{M}\}$ (Lebesgue), $\{\mu_p : p \in \tilde{M}\}$ (Patterson-Sullivan) or $\{\nu_p : p \in \tilde{M}\}$ (harmonic) coincide. Then M is locally symmetric.*

PROOF. The universal Riemannian cover $\tilde{M}$ is asymptotically harmonic by the results of Ledrappier and Yue stated in section VII. Now apply Corollary 3. □

COROLLARY 5. *Let M be a compact Riemannian manifold with sectional curvature $K < 0$ whose Anosov foliations for the geodesic flow in T_1M are C^∞. Then M is locally symmetric.*

PROOF. By the result of Benoist-Foulon-Labourie stated in section VII the manifold M is C^∞ geodesically conjugate to a locally symmetric space with sectional curvature $K < 0$. Now apply Corollary 2. □

The next result is an affirmative answer to the Katok entropy conjecture.

COROLLARY 6. *Let X be a compact C^∞ manifold with Riemannian metrics g and g_0 such that vol$(M, g) =$ vol$(M, g_0) = 1$ and (M, g_0) is locally symmetric with sectional curvature $K < 0$. Then $h(g) \geq h(g_0)$ with equality $\Leftrightarrow$ (M, g_0) is isometric to (M, g).*

PROOF. This is a direct application of the theorem in the case that f is the identity map. No rescaling of the metric is necessary since $\text{vol}(M, g) = \text{vol}(M, g_0) = 1$. □

The next result summarizes the preceding corollaries.

COROLLARY 7. *Let M be a compact Riemannian manifold with sectional curvature $K < 0$. Then the following conditions are equivalent:*

1) *M is locally symmetric.*
2) *The universal Riemannian cover $\tilde{M}$ of M is asymptotically harmonic.*
3) *The universal Riemannian cover $\tilde{M}$ of M is harmonic.*
4) *The Anosov foliations of the geodesic flow in T_1M are C^∞.*
5) *Any two of the Patterson-Sullivan $\{\mu_p : p \in \tilde{M}\}$, harmonic $\{\nu_p : p \in \tilde{M}\}$ or Lebesgue $\{\lambda_p : p \in \tilde{M}\}$ families of measures are equal.*

PROOF. The implications 1) $\Rightarrow$ 3) $\Rightarrow$ 2) are well known in the first case and obvious in the second. The implication 2) $\Rightarrow$ 1) is Corollary 3. The implications 1) $\Rightarrow$ 4) and 1) $\Rightarrow$ 5) are well known. Corollaries 4 and 5 are the implications 5) $\Rightarrow$ 1) and 4) $\Rightarrow$ 1), which completes the proof. □

C. Sketch of the proof of the main result ([BCG 2]). We now sketch the proof of the main result, but for simplicity only in the special case that (M, g_0) and (X, g) are compact manifolds with strictly negative sectional curvature $K < 0$, (M, g_0) is locally symmetric and the map $f : X \to M$ in the statement of the theorem is a homotopy equivalence. Note that this special case is enough to prove all the corollaries stated above.

Step 1

Let $(\tilde{M}, \tilde{g}_0)$ and $(\tilde{X}, \tilde{g})$ denote the universal Riemannian covers of the compact manifolds (M, g_0) and (X, g). Let Γ and Γ^* denote the fundamental groups of X and M acting as isometries on $(\tilde{X}, \tilde{g})$ and $(\tilde{M}, \tilde{g}_0)$. Let $f : X \to M$ be a homotopy equivalence, and let $\varphi : \Gamma \to \Gamma^*$ denote the induced isomorphism of fundamental groups. Then

1) There exists a lift $\tilde{f} : \tilde{X} \to \tilde{M}$ such that $\tilde{f} \circ \gamma = \varphi(\gamma) \circ \tilde{f}$ for any element γ of Γ.
2) The map $\tilde{f} : \tilde{X} \to \tilde{M}$ induces a Γ-equivariant homeomorphism $\bar{f} : \tilde{X}(\infty) \to \tilde{M}(\infty)$; that is, $\bar{f} \circ \gamma = \varphi(\gamma) \circ \bar{f}$ for any element γ of Γ, where the elements of Γ and Γ^* are now acting as homeomorphisms on the spheres at infinity $\tilde{X}(\infty)$ and $\tilde{M}(\infty)$.

Let $\mathfrak{M}_1(\tilde{X}(\infty))$ and $\mathfrak{M}_1(\tilde{M}(\infty))$ denote the space of nonatomic probability measures on the spheres at infinity $\tilde{X}(\infty)$ and $\tilde{M}(\infty)$. Let $\bar{f}_* : \mathfrak{M}_1(\tilde{X}(\infty)) \to \mathfrak{M}_1(\tilde{M}(\infty))$ denote the Γ-equivariant bijection induced by the Γ-equivariant homeomorphism $\bar{f} : \tilde{X}(\infty) \to \tilde{M}(\infty)$. One now uses the Patterson-Sullivan measures to construct an important Γ-equivariant map $\tilde{F} : \tilde{X} \to \tilde{M}$, which by virtue of the Γ-equivariance descends to a map $F : X \to M$. The map $F : X \to M$ will turn out to be an isometry when equality holds in the statement of the theorem.

Step 2

Let $\tilde{F} : \tilde{X} \to \tilde{M}$ be the composition of the maps

$$p \mapsto \mu_p \mapsto \bar{f}_*(\mu_p) \mapsto \operatorname{bar}(\bar{f}_*(\mu_p)),$$

where bar : $\mathfrak{M}_1(\tilde{M}(\infty)) \to \tilde{M}$ is the barycenter map defined in section V D. Then $\tilde{F}$ has the following properties:

1) $\tilde{F} \circ \gamma = \varphi(\gamma) \circ \tilde{F}$ for all $\gamma \in \Gamma$, where $\varphi : \Gamma \to \Gamma^*$ is the isomorphism of Step 1. Hence $\tilde{F} : \tilde{X} \to \tilde{M}$ descends to a map $F : X \to M$.
2) $F : X \to M$ is C^1 and $|Jac(F)| \leq \{h(X,g)/h(M,g_0)\}^n$. If equality holds at some point $p \in X$, then D_pF is a homothety by $h(X,g)/h(M,g_0)$.

Step 3 (Conclusion)

Let ω and ω_0 denote the volume elements of (X,g) and (M,g_0) respectively. Then

$$\begin{aligned}\mathrm{vol}(M,g_0) &= \int_M \omega_0 = \int_X F^*(\omega_0) = \left|\int_X Jac(F)\omega)\right| \\ &\leq \int_X |Jac(F)|\omega \leq \{h(X,g)/h(M,g_0)\}^n \int_M \omega \\ &= \{h(X,g)/h(M,g_0)\}^n \mathrm{vol}(X,g)\,.\end{aligned}$$

This gives the inequality of the theorem since $|\deg f| = 1$. Suppose now that equality holds and rescale the metric of M or X so that $\mathrm{vol}(X,g) = \mathrm{vol}(M,g_0)$. Then $h(X,g) = h(M,g_0)$ and $|Jac(F)| \equiv \{h(X,g)/h(M,g_0)\} = 1$. Hence by 2) of Step 2 D_pF is a linear isometry for all p in X, and we conclude that $F : (X,g) \to (M,g_0)$ is an isometry. □

References

[And] M. Anderson, "The Dirichlet problem at infinity for manifolds of negative curvature", J. Diff. Geom. 18 (1983), 701–721.

[Ano] D. V. Anosov, Geodesic flows on closed Riemann manifolds with negative curvature, Proc. Steklov. Instit. Math. 90 (1967).

[AS] M. Anderson and R. Schoen, "Positive harmonic functions on complete manifolds of negative curvature", Annals of Math (2) 121 (1985), 429–461.

[AW1] R. Azencott and E. Wilson, "Homogeneous manifolds with negative curvature, I", Trans. Amer. Math. Soc. 215 (1976), 323–362.

[AW2] R. Azencott and E. Wilson, "Homogeneous manifolds with negative curvature, II", Mem. Amer. Math. Soc. vol. 8, 178 (1976), 1–102.

[Ba 1] W. Ballmann, *Lectures on Spaces of Nonpositive Curvature*, DMV Seminar, vol. 25, Birkhäuser, Boston, 1995.

[Ba 2] ——, "Nonpositively curved manifolds of higher rank", Annals of Math. 122 (1985), 597–609.

[Ba 3] ——, "On the Dirichlet problem at infinity for manifolds of nonpositive curvature", Forum Math. 1 (1989), 201–213.

[BBE] W. Ballmann, M. Brin and P. Eberlein, "Structure of manifolds of nonpositive curvature, I", Annals of Math. (2) 122 (1985), 171–203.

[BCG1] G. Besson, G. Courtois and S. Gallot, "Entropies et Rigidités des Espaces Localement Symétriques de Courbure Strictement Négative", Geometric and Functional Analysis, vol. 5 (5) (1995), 731–799.

[BCG2] G. Besson, G. Courtois and S. Gallot, "Minimal entropy and Mostow's rigidity theorems", Ergod. Th. Dyn. Sys. 16 (1996), 623–649.

[BFL] Y. Benoist, P. Foulon and F. Labourie, "Flots d'Anosov a distributions stable et instable différentiables", J. Amer. Math. Soc. 5 (1) (1992), 33–74.

[BGS] W. Ballmann, M. Gromov and V. Schroeder, *Manifolds of Nonpositive Curvature*, Progress in Math. vol. 61, Birkhäuser, Boston, 1985.

[BO] R. Bishop and B. O' Neill, "Manifolds of negative curvature", Trans. Amer. Math. Soc. 145 (1969), 1–49.

[BuSc] M. Burger and V. Schroeder, "Amenable groups and stabilizers of measures on the boundary of a Hadamard manifold", Math. Ann. 276 (1987), 505–514.

[BuSp] K. Burns and R. Spatzier, "Manifolds of nonpositive curvature and their buildings", Publ. Math. IHES 65 (1987), 35–59.

[Ca1] E. Cartan, *Lecons sur la Géométrie des Espaces de Riemann*, Gauthier-Villars, Paris, 1946.

[Ca2] ——, "Sur une classe remarquable d'espaces de Riemann", Bull. Soc. Math. France 54 (1926), 214–264 ; 55 (1927), 114–134.

[Ca3] ——, "Sur certaines formes Riemanniennes remarquables des géométries a groupe fondamentale simple", Ann. Sci. École Normale Sup. 44 (1927), 345–467.

[Ca4] ——, "Les espaces Riemanniennes symétriques", Verh. Int. Math. Kongr. I, Zürich (1932), 152–161.

[CE] J. Cheeger and D. Ebin, *Comparison theorems in Riemannian Geometry*, North Holland, Amsterdam, 1975.

[CEK] C. Croke, P. Eberlein and B. Kleiner, "Conjugacy and rigidity for nonpositively curved manifolds of higher rank", Topology 35 (1996), 273–286.

[CK] C. Croke and B. Kleiner, "Conjugacy and rigidity for manifolds with a parallel vector field", J. Diff. Geom. 39 (3) (1994), 659–680.

[Co 1] C. Connell, "Asymptotic harmonicity of negatively curved homogeneous spaces and their measures at infinity", preprint, 1998.

[Co 2] ——, "A characterization of homogeneous spaces with positive hyperbolic rank", preprint, 1998.

[Cr] C. Croke, "Rigidity for surfaces of nonpositive curvature", Comm. Math. Helvet. , 65 (1) (1990), 150–169.

[DE] A. Douady and C. Earle, "Conformally natural extensions of homeomorphisms of the circle", Acta Math. Soc. 27 (1992), 139–142.

[DR] E. Damek and F. Ricci, "A class of non-symmetric harmonic Riemannian spaces", Bull. Amer. Math. Soc. 27 (1992), 139–142.

[E1] P. Eberlein, "Geodesic flows on negatively curved manifolds, II", Trans. Amer. Math. Soc. 178 (1973), 57–82.

[E2] ——, *Geometry of Nonpositively Curved Manifolds*, University of Chicago Press, Chicago, 1996.

[E3] ——, "Manifolds of nonpositive curvature", *Global Differential Geometry* (ed. S.-S. Chern), MAA Stud. in Math., vol. 27 (1989), 223–258.

[E4] ——, "Symmetry diffeomorphism group of a manifold of nonpositive curvature, II", Indiana Math Jour. 37 (1988), 735–752.

[E5] ——, "When is a geodesic flow of Anosov type, I", J. Diff. Geom. 8 (3) (1973), 437-463.

[EH] P. Eberlein and J. Heber, "A differential geometric characterization of symmetric spaces of higher rank", Publ. Math. IHES 71 (1990), 33–44.

[EHS] P. Eberlein, U. Hamenstädt and V. Schroeder, "Manifolds of nonpositive curvature", Proc. Symp. Pure Math., vol. 54 (part 3), Amer. Math. Soc., Providence (1993), 179–227.

[EO] P. Eberlein and B. O ' Neill, "Visibility manifolds", Pacific J. Math. 46 (1973), 45–109.

[FJ 1] F. T. Farrell and L. Jones, "Negatively curved manifolds with exotic smooth structures", J. Amer. Math. Soc. 2 (1989), 899–908.

[FJ 2] ——, "A topological analogue of Mostow's rigidity theorem", J. Amer. Math. Soc. 2 (1989), 257–370.

[FL] P. Foulon and F. Labourie, "Sur les variétés compactes asymptotiquement harmoniques", Invent. Math. 109 (1992), 97–111.

[Gh] E. Ghys, "Flots d' Anosov dont les feuilletages stables sont différentiables", Ann. Sci. École Normale Sup. 20 (1987), 251–270.

[GKM] D. Gromoll, W. Klingenberg and W. Meyer, *Riemannsche Geometrie im Grossen*, Lecture Notes in Math., vol. 55, Springer, 1968.

[Gre] L. Green, "Generalized geodesic flow", Duke Math J. 41 (1974), 115–126; correction 42 (1975), 381.

[Gro 1] M. Gromov, "Volume and bounded cohomology", Publ. Math. IHES 56 (1981), 213–307.

[Gro 2] ——, "Filling Riemannian manifolds", J. Diff. Geom. 18 (1983), 1–147.

[Ha] U. Hamenstädt, "A geometric characterization of compact locally symmetric spaces", J. Diff. Geom. 34 (1991), 193–221.

[Hed] G. Hedlund, "The dynamics of geodesic flows", Bull. Amer. Math. Soc. 45 (1939), 241-260.

[Hei 1] E. Heintze, "On homogeneous manifolds of negative curvature", Math. Ann. 211 (1974), 23–34.

[Hei 2] ——, "Compact quotients of homogeneous negatively curved Riemannian manifolds", Math. Zeit. 140 (1974), 79–80.

[Hel 1] S. Helgason, *Differential Geometry and Symmetric Spaces*, Academic Press, New York, 1962.

[Hel 2] ——, *Differential Geometry, Lie groups and Symmetric Spaces*, Academic Press, New York 1978.

[HI] E. Heintze and H.C. ImHof, "On the geometry of horospheres ", J. Diff. Geom. 12 (1977), 481–491.

[Ho 1] E. Hopf, "Statistik der geodätischen Linien in Mannigfaltigkeiten negativer Krümmung", Ber. Verh. Sachs. Akad. Wiss. Leipzig 91 (1939), 261–304.

[Ho 2] ——, "Statistik der Lösungen geodätischer Probleme vom unstabilen Typus , II" Math. Ann. 117 (1940), 590–608.

[HP] M. Hirsch and C. Pugh, "Smoothness of horocycle foliations", J. Diff. Geom. 10 (1975), 225–238.

[Kai] V. Kaimanovich, "Invariant measures of the geodesic flow and measures at infinity on negatively curved manifolds", Ann. Instit. Henri Poincaré 53 (4) (1990), 361–393.

[Kan] M. Kanai, "Geodesic flows on negatively curved manifolds with smooth stable and unstable foliations", Erg. Th. Dyn. Syst. 8 (1988), 215-239.

[Kat] A. Katok, "Entropy and closed geodesics", Erg. Th. Dyn. Syst. 2 (1982), 339–367.

[Kn 1] G. Knieper, "On the asymptotic geometry of nonpositively curved manifolds" , Geom. and Funct. Anal. 7 (1997), 755–782.

[Kn 2] ——, "Spherical means on compact manifolds of negative curvature", Diff. Geom. and Appl. 4 (1994), 361–390.

[Kn 3] ——, "The uniqueness of the measure of maximal entropy for geodesic flows on rank 1 manifolds", Annals of Math. 148 (1998), 291–314.

[Le] F. Ledrappier, "Harmonic measures and Bowen - Margulis measures", Israel J. Math. 71 (1990), 275–287.

[Li] A. Lichnerowicz, "Sur les espaces riemanniennes completement harmoniques", Bull. Soc. Math. France 72 (1944), 146–169.

[Ma] A. Manning, "Topological entropy for geodesic flows", Annals of Math. (2) 110 (1979), 567–573.

[Mi] J. Milnor, *Morse Theory*, Annals of Math. Stud., vol. 51, Princeton University Press, Princeton, 1963.

[Mo] G. Mostow, *Strong Rigidity of Locally Symmetric Spaces*, Annals of Math. Stud., vol. 78, Princeton University Press, Princeton, 1973.

[O 1] J. -P. Otal, "Le spectre marqué des surfaces a courbure négative", Annals of Math. (2) 131 (1990), 151–162.

[O 2] ——, "Sur les longueurs des géodésiques d' une métrique a courbure négative dans le disque", Comment. Math. Helv. 65 (1990), 334–347.

[Pa] S. J. Patterson, "The limit set of a Fuchsian group", Acta Math. 136 (1976), 241–273.

[Pe 1] Ya. Pesin, "Characteristic Lyapunov indicators and smooth ergodic theory", Russian Math Surveys 32 (1977), no. 4, 55–114.

[Pe 2] ——, "Formulas for the entropy of a geodesic flow on a compact Riemannian manifold without conjugate points", Math Notes 24 (1978), 796–805.

[Pe 3] ——, "Geodesic flows on closed Riemannian manifolds without focal points", Math. USSR - Izv. 11 (1977), 1195–1228.

[Pe 4] ——, "Geodesic flows with hyperbolic behavior of the trajectories and objects connected with them", Russian Math Surveys 36 (1981), no. 4, 1–59.

[Su 1] D. Sullivan," The density at infinity of a discrete group of hyperbolic motions", Publ. Math. IHES 50 (1979), 171–202.

[Su 2] ——, "The Dirichlet problem at infinity for a negatively curved manifold", J. Diff. Geom. 18 (1983), 723–732.

[Sz] Z. Szabo, "The Lichnerowicz conjecture on harmonic manifolds", J. Diff. Geom. 31 (1990), 1–28.

[V] S. S. Van Dine, *The Bishop Murder Case*, Charles Scribner's Sons, New York, 1929, pp. 123–125; 133; 144-146.

[Y 1] C. B. Yue, "Brownian motion on Anosov foliations and manifolds of negative curvature", J. Diff. Geom. 41 (1995), 159–183.

[Y 2] ——, "On Sullivan's Conjecture", Random and Computational Dynamics, (1992), no. 1, 131–145.

[Y 3] ——, "Rigidity and dynamics around manifolds of negative curvature", Math. Res. Lett. (1994), no. 1, 123–147.

Department of Mathematics, University of North Carolina, Chapel Hill, NC 27599

Proceedings of Symposia in Pure Mathematics
Volume **69**, 2001

Closed Geodesics and the Uniqueness of the Maximal Measure for Rank 1 Geodesic Flows

Gerhard Knieper

1. Introduction

In this lecture we provide a survey of some recent results on the asymptotic geometry and dynamics on compact nonpositively curved manifolds [**12, 13**]. It will be assumed that the geodesic flow has some hyperbolic behavior, i.e., at least for some orbits the only zero Lyapunov exponents are the trivial ones, corresponding to the flow direction (nonuniform hyperbolicity).

The results extend the work of Bowen [**4, 5**] and Margulis [**14**] on the growth rate of closed orbits for uniformly hyperbolic dynamical systems as the geodesic flow on a manifold of negative curvature. The main tool in their work is a construction of an invariant measure for which the closed trajectories are uniformly distributed. This measure is of maximal entropy (maximal measure), and as was shown by Bowen [**5**], it is uniquely determined by this fact. It turns out that some of those results hold also true in our nonuniform hyperbolic situation (see [**10**] for a reference on hyperbolic dynamics and entropy).

In this lecture, (M, g) will denote a compact n- dimensional Riemannian manifold of nonpositive curvature (see [**1**], [**3**], [**8**] for references on the theory of manifolds of nonpositive curvature). The theorem of Hadamard and Cartan implies that the universal cover X is diffeomorphic to $\mathbb{R}^n$ (X together with the lifted metric is called a Hadamard manifold). The manifold M can be represented as X/Γ, where Γ denotes the group of deck transformations which acts properly discontinuously and isometrically on the Hadamard manifold X. The group Γ is isomorphic to the fundamental group $\pi_1(M)$ of M, and the action of Γ can be extended to the ideal boundary

$$X(\infty) := \{\text{equivalence classes of asymptotic geodesic rays}\}.$$

With respect to the sphere topology $X(\infty)$ is homeomorphic to the $(n-1)$-dimensional sphere S^{n-1}. The topologies on X and $X(\infty)$ have natural extensions to $\bar{X} = X \cup X(\infty)$ (cone topology) such that $\bar{X}$ is homeomorphic to a closed n–ball. Moreover, Γ acts on $\bar{X}$ as a subgroup of homeomorphisms (see [**9**] for details).

2000 *Mathematics Subject Classification.* Primary 37C40, 53D25; Secondary 37D40.
Supported by a NSF conference Grant .

Let $c_v : \mathbb{R} \to M$ denote the geodesic with initial velocity $\dot{c}_v(0) = v \in SM$, and let $\phi^t : SM \to SM$ denote the geodesic flow on the unit tangent bundle SM given by $\phi^t(v) = \dot{c}_v(t)$. Furthermore,

$$\mathrm{rank}(v) := \dim\{\text{parallel Jacobi fields along } c_v\} \geq 1$$

is the dimension of the space of parallel Jacobi fields along c_v. In particular, if J is a parallel Jacobi field perpendicular to c_v, the Jacobi equation implies that the curvature of the plane spaned by $\{\phi^t(v), J(t)\}$ vanishes for all t. Therefore, $\mathrm{rank}(v) = 1$ if the trajectory $\phi^t(v)$ is hyperbolic. For periodic orbits the reverse statement is also true. Hence, the geometric rank (which generalizes the algebraic rank of symmetric spaces [**8**])

$$\mathrm{rank}M := \min_{v \in SM} \mathrm{rank}(v)$$

measures the amount of hyperbolicity of the geodesic flow. In particular, the nonuniform hyperbolicity of the geodesic flow implies that the rank of M equals one. The rank rigidity theorem, proved by Ballmann [**2**] and Burns/Spatzier [**6**] (using different methods), shows that most compact manifolds of nonpositive curvature are rank 1 metrics, in the sense that the universal covering of a higher rank manifold is either a product or a symmetric space.

Examples of compact rank 1 manifolds:

(1) Surfaces of genus bigger than one and nonpositive curvature are the only 2-dimensional examples of rank 1. By the uniformization theorem, they also carry metrics of constant negative curvature.

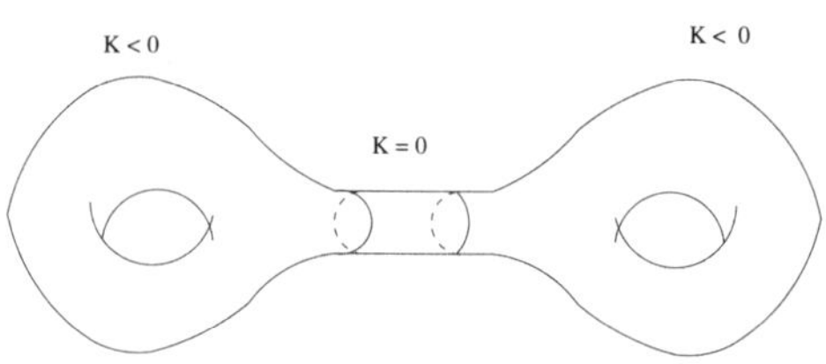

FIGURE 1. rank 1 surface

(2) In higher dimensions there are examples which do not allow negatively curved metrics. Consider a surface of nonpositive curvature, as indicated in Figure 1, and cut it off along a closed geodesic α on the flat part of the surface. Its end will be isometric to a flat cylinder $I \times \alpha$ (see Figure 2). Taking a product of this surface with a circle β of the same length as α, we obtain a 3-dimensional manifold of nonpositive curvature (rank 2), whose boundary is isometric to the flat torus $\alpha \times \beta$. In order to obtain a 3-manifold without boundary we consider a second copy and identify the boundaries by interchanging the factors α and β of the two tori as shown in the figure below. Although the resulting manifold M is locally a Riemannian product it is nevertheless a rank 1 manifold. While geodesics staying in one copy, i.e., geodesics which do not cross the torus $\alpha \times \beta$ are of rank 2, all other geodesics are of rank 1 (see the indicated geodesic $(c(t), q)$ in Figure 2). Since the 2–torus $T^2 = \alpha \times \beta$ is a flat totally geodesic submanifold of M, $\mathbb{Z}^2$ is a subgroup of the fundamental group of $\pi_1(M)$.

Therefore, a well-known theorem of Preissmann implies that M cannot carry a metric of negative curvature. Even the existence of a geodesic Anosov flow is impossible.

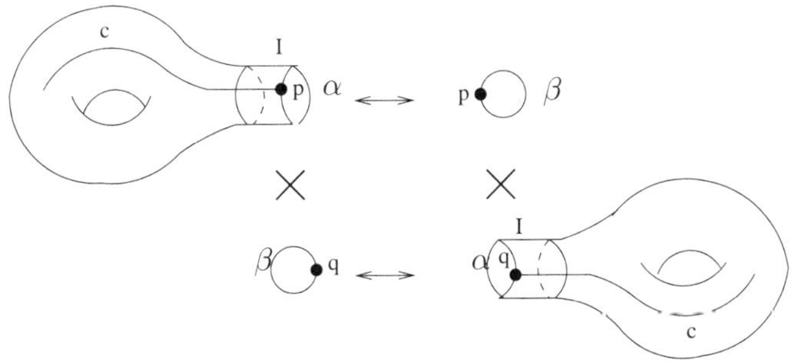

FIGURE 2. rank 1 manifold of $dim M = 3$

For a rank 1 manifold M the unit tangent bundle SM splits in two subsets invariant under the geodesic flow, namely, the regular set

$$\text{reg} := \{v \in SM | \text{ rank } (v) = 1\}$$

and the singular set

$$\text{sing} := SM \setminus \text{reg}.$$

It is easy to see that reg is an open subset of SM. One can also prove that this set is dense. However, it is an open question whether the regular set has full Liouville measure μ_L, in other words it is not known whether $\mu_L(\text{sing}) = 0$ holds in general. An affirmative answer to this question would yield the ergodicity of the geodesic flow with respect to the Liouville measure.
An important invariant which measures the orbit complexity of a dynamical system is the topological entropy (see e.g. [**10**]). In the case of nonpositive curvature this number is given by the exponential growth of volume of geodesic balls or spheres on the universal covering X, i.e.,

$$h = h_{\text{top}}(\phi) = \lim_{r \to \infty} \frac{\log \text{vol} B(p,r)}{r} = \lim_{r \to \infty} \frac{\log \text{vol} S(p,r)}{r},$$

where $B(p,r)$ and $S(p,r)$ are geodesic balls and spheres of radius r about $p \in X$.

The aim of this lecture is

(1) to give a precise estimate for the volume growth of geodesic spheres (balls) on X (see Section 5).
(2) to give an estimate on the growth rate of regular closed geodesics (see Section 6).
(3) to outline a proof of the uniqueness of the measure of maximal entropy for the geodesic flow. Furthermore, we will obtain an estimate for the growth rate of singular closed geodesics (see Sections 7-9).

2. Basics on rank 1 geometry

We will now discuss certain well-known properties which are crucial for the dynamics and geometry of rank 1 manifolds (see [**1**] for details). For each $k \in \mathbb{N}$

$$d_k(v,w) = \max\{d(c_v(t), c_w(t)) \mid 0 \leq t \leq k\}$$

defines a metric on the unit tangent bundle of a Riemannian manifold, which will be useful for our future considerations.

LEMMA 2.1. *Let X be a Hadamard manifold and c a rank 1 geodesic on X. For each $\epsilon > 0$ there are neighborhoods $U \subset X(\infty)$ of $c(-\infty)$ and $V \subset X(\infty)$ of $c(+\infty)$ such that for all $\xi \in U$ and $\eta \in V$ there exists a* rank 1 *geodesic h connecting ξ and η such that $d_1(\dot{c}(0), \dot{h}(0)) \leq \epsilon$. Furthermore, the flat strip theorem implies that h is uniquely determined up to parametrization.*

Elements γ in the group $\mathrm{Iso}(X)$ of isometries of X which translate a geodesic (axis) $c : \mathbb{R} \to X$, i.e., $\gamma(c(t)) = c(t + t_0)$ for some $t_0 > 0$ are called axial. If X/Γ is compact all isometries $\gamma \in \Gamma$ are axial. The associated axes on X correspond to closed geodesics on X/Γ. If c is a rank 1 axis, γ behaves like in negative curvature, i.e., the end points of c are attracting and repelling fixed points for γ. More precisely, one has:

LEMMA 2.2. *Let X be a Hadamard manifold and $c : \mathbb{R} \to X$ a rank 1 axis of $\gamma \in Iso(X)$. For all neighborhoods $U \subset \bar{X}$ of $c(-\infty)$ and $V \subset \bar{X}$ of $c(+\infty)$ there exists $n_0 \in \mathbb{N}$ such that for all $n \geq n_0$*

$$\gamma^n(\bar{X} \setminus U) \subset V \text{ and } \gamma^{-n}(\bar{X} \setminus V) \subset U$$

for all $n \geq n_0$.

REMARK Both lemmata imply that the end point of a rank 1 axis can be connected to any other point in $X(\infty)$ by a rank 1 geodesic.

There are many rank 1 axes if X/Γ has finite volume.

LEMMA 2.3. *Let X be a Hadamard manifold and $\Gamma \subset Iso(X)$ such that X/Γ is a manifold of finite volume. If $c : \mathbb{R} \to X$ is a rank 1 geodesic, then there exists a sequence of rank 1 axes c_n of deck tranformations $\gamma_n \in \Gamma$ such that $c_n \to c$.*

Proof. First one observes that in the case of finite volume the geodesic flow is non-wandering (Poincaré-recurrence). This implies that the end points of any geodesic $c : \mathbb{R} \to X$ are Γ-dual, i.e., there exists a sequence $\gamma_n \in \Gamma$ such that for one $p \in X$ (hence, for all $p \in X$) $\gamma_n p \to c(+\infty)$, $\gamma_n^{-1} p \to c(-\infty)$. A generalization of Lemma 2.2 shows that the closure of sufficiently small neighborhoods about $c(+\infty)$ and $c(-\infty)$ are invariant under γ_n and γ_n^{-1}. Since one can assume that those neighborhoods are homeomorphic to a closed ball the fixed point theorem of Brouwer yields the claim.

REMARK Under the above assumption one can even prove that the regular set, i.e., the set of rank 1 vectors is dense . Hence, the rank 1 axes are dense in SX and therefore, closed geodesics are dense in SX/Γ.

COROLLARY 2.4. Γ *acts minimally on $X(\infty)$, i.e., $\overline{\Gamma\xi} = X(\infty)$ for all $\xi \in X(\infty)$ (see Figure 3).*

Proof. Fix $\xi \in X(\infty)$ and consider for $\eta \in X(\infty)$ with $\eta \neq \xi$ a neighborhood $V \subset X(\infty)$ of η (see Figure 3). By the remark above, we can choose a rank 1 axis c of some element $\gamma \in \Gamma$ such that $c(+\infty) \in V$ and $c(-\infty) \neq \xi$. Choose a neighborhood $U \subset X(\infty)$ of $c(-\infty)$ which does not contain ξ. Then Lemma 2.2 implies the existence of $n \in \mathbb{N}$ such that $\gamma^n(X(\infty)\setminus U) \subset V$. In particular, $\gamma^n\xi \in V$.

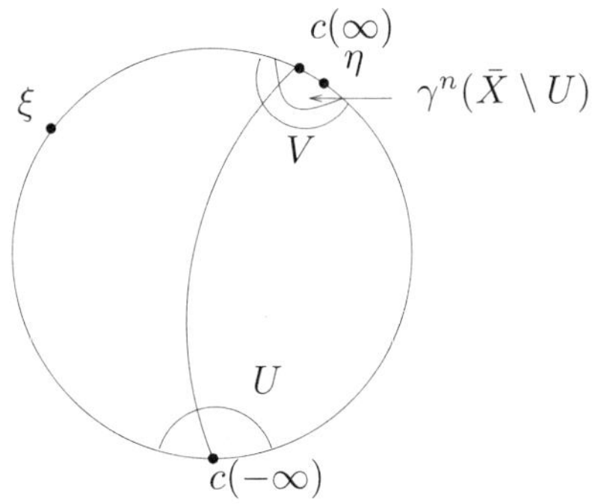

FIGURE 3

3. Busemann densities

A main tool in the ergodic theory of geodesic flows on nonpositively curved manifolds is the study of Busemann densities (conformal densities) supported on the ideal boundary $X(\infty)$.

Let X be a Hadamard manifold and let $\Gamma \subset \mathrm{Iso}(X)$ be a discrete subgroup. A family of finite Borel measures $\{\mu_p\}_{p\in X}$ on $X(\infty)$ is called an α-dimensional Busemann density if

a) For all $p \in X$
$$\mathrm{supp}\mu_p \subset \Lambda(\Gamma),$$
where $\Lambda(\Gamma)$ denotes the limit set of Γ.

b) For all $p, q \in X$
$$\frac{d\mu_q}{d\mu_p}(\xi) = e^{-\alpha b_p(q,\xi)}$$
for almost all $\xi \in X(\infty)$, where for the geodesic ray $c : [0,\infty) \to \mathbb{R}$ with $c(0) = p$ and $c(\infty) = \xi$ the limit
$$b_p(q,\xi) = \lim_{t\to\infty} d(q, c(t)) - t$$
defines the Busemann function based at ξ, normalized in such a way that $b_p(p,\xi) = 0$ and $b_p(q,\xi) \to -\infty$ if $q \to \xi$.

c) $\{\mu_p\}_{p\in X}$ is Γ-equivariant, i.e.,
$$\mu_{\gamma p}(\gamma A) = \mu_p(A)$$
for all Borel sets $A \subset X(\infty)$.

There is a standard way to construct such a measure. In the case of Fuchsian groups this is due to Patterson [**16**]. For general hyperbolic spaces it has been intensively studied by Sullivan [**17**]. The construction can be done as follows (see

[**12**] for further details). Assume that the group $\Gamma \subset \mathrm{Iso}(X)$ is of divergence type, i.e. the Poincaré series $P : \mathbb{R} \times X \times X \to \mathbb{R}$ given by

$$P(s,x,y) = \sum_{\gamma \in \Gamma} e^{-sd(x,\gamma y)}$$

diverges if s is equal to the critical exponent

$$\delta(\Gamma) = \inf\{s \geq 0 | \sum_{\gamma \in \Gamma} e^{-sd(x,\gamma y)} < \infty\}.$$

If not, one weights the Poincaré series $P(s,x,y)$ with a positive monotone increasing function $f : \mathbb{R}^+ \to \mathbb{R}^+$ such that for each $a > 0$ we have that $\frac{f(r+a)}{f(r)} \to 1$ for $r \to \infty$ and

$$\tilde{P}(s,x,y) = \sum_{\gamma \in \Gamma} f(d(x,\gamma y)) e^{-sd(x,\gamma y)}$$

diverges for $s = \delta(\Gamma)$.
Fix a point $x \in X$. Then, for $s > \delta$ and $p \in X$, consider the measure

$$\mu_{p,x,s} = \frac{\sum_{\gamma \in \Gamma} e^{-sd(p,\gamma x)} \delta_{\gamma x}}{P(s,x,x)},$$

where δ_y is the Dirac mass associated to $y \in X$. It has the following obvious properties.

(1) $e^{-sd(p,x)} \leq \mu_{p,x,s}(\bar{X}) \leq e^{sd(p,x)}$. In particular, the measures are finite.
(2) $\Gamma x \subset \mathrm{supp}\mu_{p,x,s} \subset \overline{\Gamma x}$.

Now choose for a fixed $p \in X$ a weak limit $\lim_{k \to \infty} \mu_{p,x,s_k} =: \mu_p$, where $s_k \downarrow \delta(\Gamma)$. The divergence of the series $P(s,x,x)$ for $s = \delta(\Gamma)$ and the discreteness of Γ yield that the support of μ_{p,x,s_k} moves into the limit set $\Lambda(\Gamma) = \overline{\Gamma x} \cap X(\infty)$ if $k \to \infty$. More precisely, one obtains:

PROPOSITION 3.1. *For each $q \in X$ (and not only for p) the weak limit*

$$\lim_{k \to \infty} \mu_{q,x,s_k} =: \mu_q$$

exists. The family of measures $\{\mu_q\}_{q \in X}$ is a $\delta(\Gamma)$-dimensional Busemann density.

REMARK A Busemann density constructed in such a way is also called a Patterson-Sullivan measure. In general, this construction depends on the various choices which have been made. However, if X/Γ is compact, there is only one Busemann density. Futhermore, $\delta(\Gamma)$ coincides with the topological entropy h (see also the remark at the end of Section 4).

4. Semilocal properties of Busemann densities

First we will show that on rank 1 manifolds X with compact quotient $M = X/\Gamma$ any Busemann density $\{\mu_p\}_{p \in X}$ has certain semilocal properties (see also Section 2 in [**13**]). For $\xi \in X(\infty)$ and $p \in X$ consider the projections

$$pr_\xi : X \to X(\infty)$$

and

$$pr_p : X \setminus \{p\} \to X(\infty)$$

along geodesics emanating from ξ and p, respectively. Hence, $pr_\xi(x) = c_{\xi,x}(\infty)$, where $c_{\xi,c}$ is the geodesic with $c_{\xi,x}(-\infty) = \xi$, $c_{\xi,x}(0) = x$ and $pr_p(x) = c_{p,x}(\infty)$, where $c_{p,x}$ is the geodesic with $c_{p,x}(0) = p$, $c_{p,x}(d(p,x)) = x$. If $A \subset X$ and $p \in X$, the image $pr_p(A)$ of A is always understood to be the set $pr_p(A \setminus \{p\})$.

PROPOSITION 4.1. *Let X be a Hadamard manifold with compact quotient and let $\{\mu_p\}_{p\in X}$ be an α-dimensional Busemann density. Then there exist constants $R > 0, \ell > 0$ such that*

a)

$$\mu_x(pr_p B(x,\rho)) \geq \ell$$

for all $x \in X$ and all $p \in X(\infty) \cup X = \bar{X}$, where $B(x,\rho)$ is the open geodesic ball of radius $\rho \geq R$ about x.

Furthermore, there is a constant $b = b(\rho) > 0$ such that for all $x \in X \setminus \{p\}$ and $\xi = c_{p,x}(-\infty)$,

b)

$$\frac{1}{b} e^{-\alpha d(p,x)} \leq \mu_p(pr_\xi(B(x,\rho))) \leq b e^{-\alpha d(p,x)}.$$

A similar estimate holds if we project from $p \in X$, namely there is a constant $a = a(\rho) > 0$ such that for all $x \in X$,

c)

$$\frac{1}{a} e^{-\alpha d(p,x)} \leq \mu_p(pr_p(B(x,\rho))) \leq a e^{-\alpha d(p,x)}.$$

Proof. The estimates in b) and c) follow from a) and the defining properties of μ_p. Namely, if $x \in X$ and $A \subset X(\infty)$ is any measurable set then

$$\mu_p(A) = \int_A e^{-\alpha b_x(p,\eta)} d\mu_x(\eta).$$

In particular, if A is of the form $pr_\xi B(x,\rho)$ or $pr_p B(x,\rho)$ then $|(b_x(p,\eta) - d(p,x))|$ is bounded by a constant for all $\eta \in A$. This yields b) and c).
The first estimate is a consequence of the following steps.

Step 1: supp$\mu_p = X(\infty)$ *for one and, hence, for all $p \in X$ since, as we showed in Corollary 2.4, Γ acts minimally on $X(\infty)$.*

Step 2: *For each $\xi \in X(\infty)$ and $x \in X$ there exists $r > 0$ such that $pr_\xi(B(x,r))$ contains some open set.*
To see this, choose a rank 1 axis c on X. By the remark following Lemma 2.2 the end points of such a geodesic can be connected to all other points of $X(\infty)$. For $\xi \in X(\infty)$ let $h_{\xi,\eta}$ be a geodesic with $h_{\xi,\eta}(+\infty) = \eta = c(+\infty)$ and $h_{\xi,\eta}(-\infty) = \xi$ (see Figure 4). Since $h_{\xi,\eta}$ is a rank 1 geodesic, Lemma 2.2 implies that for $\epsilon > 0$ and $x = x_0 = h_{\xi,\eta}(0)$ there exists an open neighborhood U of η such that $U \subset pr_\xi(B(x,\epsilon))$. If x is arbitrary one chooses $r = d(x_0,x) + \epsilon$.
For $v \in S_x X = \pi^{-1}(x)$ and $\epsilon > 0$ consider the open neighborhood

$$C_\epsilon(v) = \{c_w(\infty) \mid w \in S_x X \text{ and } \angle(v,w) < \epsilon\}.$$

Fix a compact set K such that $\cup_{\gamma\in\Gamma}\gamma(K) = X$ and a reference point $x_0 \in K$. Then the following uniform version of Step 2 holds.

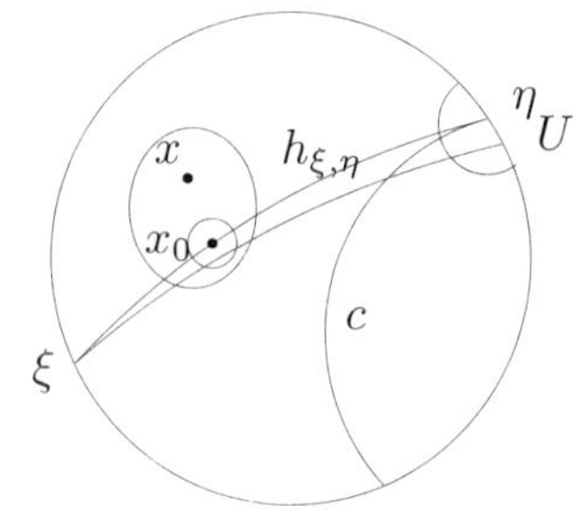

FIGURE 4

Step 3: *There exist $R > 0$ and $\epsilon > 0$ such that for all $p \in K$ and $x \in \bar{X}$*

$$C_\epsilon(v) \subset pr_x(B(p, R))$$

for some $v \in S_{x_0}X$.
Suppose Step 3 is false. Then there are sequences $p_n \in K$, $x_n \in \bar{X}$ such that

$$C_{1/n}(v) \not\subset pr_{x_n}(B(p_n, n))$$

for all $v \in S_{x_0}X$. Since $x_n \in \mathrm{Int}B(p_n, n)$ implies $pr_{x_n}(B(p_n, n)) = X(\infty)$, we can assume after choosing a subsequence that $x_n \to \xi$ and $p_n \to p \in K$. Then Step 2 implies the existence of $r > 0$, $\epsilon > 0$ and $v_0 \in S_{x_0}X$ such that $C_\epsilon(v_0) \subset pr_\xi(B(p, r))$. The continuity of the projection implies the existence of n_0 such that for all $n \geq n_0$ we have: $C_{\epsilon/2}(v_0) \subset pr_{x_n}(B(p_n, r))$. But this contradicts the choice of the sequence if $1/n < \epsilon/2$ and $n > r$.

The next step follows from the positivity of μ_p on open sets.

Step 4: *For all $\epsilon > 0$ there exists a constant $\ell = \ell(\epsilon) > 0$ such that*

$$\mu_p(C_\epsilon(v)) > \ell$$

for all $v \in S_{x_0}X$ and $p \in K$.

Now consider $x \in \bar{X}$ and $p \in X$. Choose $\gamma \in \Gamma$ such that $\gamma p \in K$. Since $\mu_p(pr_x(B(p, \rho)) = \mu_{\gamma p}(pr_{\gamma x}B(\gamma p, \rho))$ the estimate a) in the proposition is a consequence of Step 3 and Step 4.
REMARK Using Proposition 4.1, one can prove the uniqueness of the Busemann density (see [**12**]). Hence, this measure is given by the Patterson construction and therefore, its dimension coincides with the topological entropy h.

5. Volume growth of geodesic spheres

The local estimate of the Busemann density in 4.1 c) can be used to obtain some asymptotic properties of the geometry (see also Section 5 of [**12**]).

THEOREM 5.1. *Let $M = X/\Gamma$ be a compact rank one manifold and $p \in X$. Then there exists a constant $a > 1$ such that*

$$\frac{1}{a} \leq \frac{vol\, S(p, r)}{e^{hr}} \leq a$$

for all $r > 0$.

Proof. Choose $\rho \geq 2R$, where R is as in Proposition 4.1 and let $x_1, \ldots, x_m$ be a maximal ρ-separating set in $S(p,r)$, i.e., a set of maximal cardinality such that $d(x_i, x_j) \geq \rho$ for $i \neq j$. In particular, $\cup B(x_i, \rho) \supseteq S(p,r)$ and $B(x_i, \rho/2) \cap B(x_j, \rho/2) = \emptyset$ for $i \neq j$. The local property of the Busemann density implies

$$\frac{1}{b} e^{-hr} \leq \mu_p(pr_p(B(x_i, \tilde{\rho}) \cap S(p,r))) \leq b e^{-hr}$$

for all $\tilde{\rho}$ with $2R \geq \tilde{\rho} \geq R$ and a constant $b > 0$. In particular,

$$\mu_p(X(\infty)) \leq \sum_{i=1}^{m} \mu_p(pr_p(B(x_i, \rho) \cap S(p,r))) \leq mbe^{-hr}$$

and

$$\frac{1}{b} me^{-hr} \leq \mu_p(\bigcup_{i=1}^{m} pr_p(B(x_i, \rho/2) \cap S(p,r))) \leq \mu_p(X(\infty))$$

and, hence,

$$\frac{1}{\beta} e^{hr} \leq m \leq \beta\, e^{hr}$$

for a constant $\beta > 1$. It follows from the uniformity of the geometry that

$$\frac{1}{\ell} \leq \operatorname{vol}(B(x_i, \tilde{\rho}) \cap S(p,r)) \leq \ell$$

for $r \geq r_0$, $R \leq \tilde{\rho} \leq 2R$. Therefore, we obtain

$$\operatorname{vol} S(p,r) \leq \sum_{i=1}^{m} \operatorname{vol}(B(x_i, \rho) \cap S(p,r)) \leq \beta \ell e^{hr}$$

and

$$\operatorname{vol} S(p,r) \geq \sum_{i=1}^{m} \operatorname{vol}(B(x_i, \rho) \cap S(p,r)) \geq \frac{1}{\beta \ell} e^{hr}$$

which yields the claim.

6. Growth rate of regular (hyperbolic) closed geodesics

Let $P_{\text{reg}}(t)$ denote the set of regular (hyperbolic) primitive closed geodesics of period $\leq t$ in X/Γ. Unlike in the case of singular closed geodesics a free homotopy class which contains a regular closed geodesic cannot contain any other closed geodesic. This follows from the fact that a pair of homotopic closed geodesics would have lifts to the universal covering with bounded distance. Therefore, they would bound a flat strip by the flat strip theorem (see e.g.[**3**]) and hence both geodesics would be singular.

THEOREM 6.1. *There are constants $A, B > 0$ such that*

$$B \frac{e^{ht}}{t} \leq \# P_{reg}(t) \leq A \frac{e^{ht}}{t}.$$

The proof of the upper bound follows from a pre-trace formula technique (this will appear in our upcoming review in the Handbook on Dynamical Systems).

Sketch of the proof of the lower bound

Geometrically distinct closed geodesics modulo free homotopy are in one-to-one correspondence with equivalence classes in Γ. Two elements $\gamma_1, \gamma_2 \in \Gamma \setminus \{\mathrm{id}\}$ are called equivalent if and only if there exists $n, m \in \mathbb{Z}$ and $\eta \in \Gamma$ such that

$$\gamma_1^m = \eta\gamma_2^n\eta^{-1} .$$

Denote by $[\Gamma]$ the set of equivalence classes. We call $\gamma \in \Gamma \setminus \{\mathrm{id}\}$ primitive if it cannot be written in the form $\gamma = \beta^m$, $m \geq 2$ for some $\beta \in \Gamma$. Each $[\gamma] \in [\Gamma]$ can be represented as

$$[\gamma] = \{\eta\gamma_0^n\eta^{-1} | \gamma_0 \in \Gamma,\ \gamma_0 \text{ primitive },\ \eta \in \Gamma, n \in \mathbb{Z}\} .$$

For $\gamma \in \Gamma$ denote by

$$\ell(\gamma) = \inf\{d(p, \gamma p) | p \in X\}$$

the length of γ. Since our manifold is compact, there exists $p_0 \in X$ such that $d(p_0, \gamma p_0) = \ell(\gamma)$. The geodesic through $p_0, \gamma p_0$ is an axis of γ, and the projection onto M is a closed geodesic. Its prime period is given by

$$\ell([\gamma]) = \min\{\ell(\eta) | \eta \in [\gamma]\} = \ell(\gamma_0),$$

where γ_0 is a primitive element representing $[\gamma]$. If the axis of γ is regular, the corresponding closed geodesic on M is uniquely determined. For open subsets $U, V \subset X(\infty)$ define

$$\begin{aligned} A_{\mathrm{reg}}(U, V, t) = &\{\gamma \in \Gamma | \gamma \text{ has an regular axis } c \text{ with end points in } U, V \\ &\text{and } \ell(\gamma) \leq t\}. \end{aligned}$$

The proof of the lower bound follows immediately from the following proposition which in turn is a consequence of the estimates on volume growth given in Section 5 and the arguments given in [**11**].

PROPOSITION 6.2. *There exist open sets $U, V \subset X(\infty)$ and constants $b, b' > 0$ such that*

(1) $\#A_{reg}(U, V, t) \geq be^{ht}$
(2) *For all $\gamma \in A_{reg}(U, V, t)$, we have*

$$\#\{[\gamma] \cap A_{reg}(U, V, t)\} \leq b't.$$

7. Closed geodesics and the measure of maximal entropy

As the example in Section 1 (Figure 2) shows, the singular closed geodesics can grow exponentially. An interesting question is whether they can have the same exponential growth rate as the regular ones, i.e. is it possible that

$$\limsup_{t \to \infty} \frac{\log \#P_{\mathrm{sing}}(t)}{t} = h_{\mathrm{top}} \ ?$$

It turns out that the answer is no, due to the following positive answer to a conjecture of A. Katok [**7**]. The conjecture contained three parts stated in the following theorem (for more details see [**13**]).

THEOREM 7.1. *Let M be a compact Riemannian rank 1 manifold and ϕ^t : $SM \to SM$ be the geodesic flow on the unit tangent bundle SM. Then the following holds:*

(1) *The geodesic flow has a unique measure $\mu_{\max}$ of maximal entropy. This measure is ergodic and concentrated on the regular set, i.e., the singular set has measure zero.*

(2) *Let μ_T be the measure supported on the finite number of regular closed geodesics of length $\leq T$, defined by*

$$\int f d\mu_T = \frac{1}{\#P_{reg}(T)} \sum_{c \in P_{reg}(T)} \frac{1}{\ell(c)} \int_0^{\ell(c)} f(\dot{c}(s))ds \,,$$

for $f \in C^0(SM)$, where $\ell(c)$ denotes the period of the closed geodesic c. Then, in the weak-topology on the space of probability measures on SM,*

$$\mu_{\max} = \lim_{T \to \infty} \mu_T .$$

(3) *$h(\phi_{|sing}) < h(\phi)$ where $h(\phi_{|sing})$ denotes the topological entropy of the geodesic flow restricted to the singular set.*

The statements in 2. and 3. follow from the first statement. Since

$$\lim_{T \to \infty} \frac{\log \#P_{\text{reg}}(T)}{T} = h,$$

a computation shows (see [**13**]) that each weak limit of μ_T converges to a measure of maximal entropy and therefore, part two follows from part one.
The variational principle for entropy (see e.g. [**10**]) implies the existence of a sequence of invariant probability measures μ_n concentrated on the singular set such that

$$h(\phi_{|\text{sing}}) \geq h_{\mu_n}(\phi) \geq h(\phi_{|\text{sing}}) - \frac{1}{n}.$$

Let ν be a weak* limit of μ_n. The upper semicontinuity of entropy implies $h_\nu(\phi) = h(\phi_{|\text{sing}})$. Since sing is a closed subset of SM, the weak* limit ν is concentrated on the singular set. In particular, $\nu \neq \mu_{\max}$ and therefore, the uniqueness of the measure of maximal entropy implies that $h > h_\nu(\phi) = h(\phi_{|\text{sing}})$.

Corollary 7.2.

$$\limsup_{t \to \infty} \frac{\log \#P_{\text{sing}}(t)}{t} < h.$$

The proof follows from the second statement in Theorem 7.1 since

$$\limsup_{t \to \infty} \frac{\log \#P_{\text{sing}}(t)}{t} \leq h_{\text{top}}(\phi_{|\text{sing}}) < h_{\text{top}}(\phi).$$

The rest of the lecture is devoted to the construction and the proof of the uniqueness of the measure of maximal entropy.

8. Construction of a measure of maximal entropy

The Patterson-Sullivan measure will be crucial for the construction of the measure $\mu = \mu_{\max}$ of maximal entropy (see also Section 2 of [**13**]).

Denote by $P : SX \to X(\infty) \times X(\infty)$ the end point projection given by $P(v) = (c_v(-\infty), c_v(\infty))$, where c_v is the geodesic with $\dot{c}_v(0) = v$. Then

$$\mathcal{G}^E = P(SX)$$

is the set of points in $X(\infty) \times X(\infty)$ minus the diagonal which can be connected by a, in general not uniquely determined, geodesic. For a fixed point $p \in X$,

$$d\bar{\mu}(\xi, \eta) = e^{h\beta_p(\xi,\eta)} d\mu_p(\xi) d\mu_p(\eta)$$

defines a Γ-invariant measure on $\mathcal{G}^E$, where β_p is the Gromov product given by

$$\beta_p(\xi,\eta) = -(b_p(q,\xi) + b_p(q,\eta)) .$$

Here q is any point on a geodesic c connecting ξ and η. In geometrical terms, $\beta_p(\xi,\eta)$ is the length of the segment c which is cut out by the horoballs through (p,ξ) and (p,η). Since $\mathrm{grad}_q b_p(q,\xi) = -\mathrm{grad}_q b_p(q,\eta)$ for all points on geodesics connecting ξ and η, this number is independent of the choice of q. Now, $\bar\mu$ induces a ϕ^t-invariant measure μ on SX by setting for each Borel subset $A \subset SX$

$$\mu(A) = \int_{\mathcal{G}^E} \mathrm{vol}(\pi(P^{-1}(\xi,\eta) \cap A)) d\bar\mu(\xi,\eta),$$

where $\pi : SX \to X$ is the canonical projection and vol is the induced volume element on the submanifolds $\pi(P^{-1}(\xi,\eta))$. By the Γ-invariance of $\bar\mu$ this measure descends to a flow-invariant measure on SM. Finally, we normalize it so that it becomes a probability measure, i.e., $\mu(SM) = 1$.

Computation of the entropy of μ

Let $\mathcal{A} = \{A_1, \ldots, A_m\}$ be a measurable partition of SM such that the diameters of all elements in $\mathcal{A}$ are less than ϵ with respect to the metric d_1 introduced in Section 2. If $\alpha \in \mathcal{A}_\phi^{(n)}$, where $\mathcal{A}_\phi^{(n)} = \vee_{i=0}^{n-1} \phi^{-i}\mathcal{A}$, then, it follows that

$$\alpha \subset \bigcap_{k=0}^{n-1} \phi^{-k} B_{d_1}(\phi^k v, \epsilon)$$

provided $v \in \alpha$.

LEMMA 8.1. *Let $0 < \epsilon < \min\{R, \mathrm{inj}(M)\}$ where $\mathrm{inj}(M)$ is the injectivity radius of M. Then there is a constant $a > 0$ such that*

$$\mu(\alpha) \le e^{-hn} a$$

for all $\alpha \in \mathcal{A}_\phi^{(n)}$.

Proof. Since $\alpha \subset \bigcap_{k=0}^{n-1} \phi^{-k} B_{d_1}(\phi^k v, \epsilon)$ if $v \in \alpha$, we have for all $w \in \alpha$ and $t \in [0,n]$ that $d(c_v(t), c_w(t)) \le \epsilon$ for all. Let $p \in X$ be the reference point used in the definition of the measure μ and let $\tilde v \in SX$ be a lift of v such that $d(\pi\tilde v, p) \le \mathrm{diam} M = r_0$. By assumption ϵ is smaller than the injectivity radius of M and, consequently, we can lift α to $\tilde\alpha \subset SX$ such that $d(c_{\tilde w}(t), c_{\tilde v}(t)) \le \epsilon$ for all $\tilde w \in \tilde\alpha$ and $t \in [0,n]$. Let $c_{\tilde v}(n) = x$ and $\xi = c_{\tilde w}(-\infty)$. Consider the geodesic $c_{\xi,x}$ connecting ξ and x such that $c_{\xi,x}(n) = x$. Then,

$$d(c_{\xi,x}(0), p) \le d(c_{\xi,x}(0), \pi(\tilde w)) + d(\pi(\tilde w), p) \le 2\epsilon + r_0,$$

i.e., $\xi \in pr_x(B(p, r_0 + 2\epsilon))$. In particular,

$$P(\tilde\alpha) \subset \bigcup_{\eta \in pr_x(B(p,r_0+2\epsilon))} \{\eta\} \times pr_\eta(B(x,\epsilon)),$$

where $P : SX \to \mathcal{G}^E$ denotes the end point projection.

For each $\eta \in pr_x(B(p, r_0 + 2\epsilon))$ choose a point $q \in B(p, r_0 + 2\epsilon)$ that lies on the geodesic $c_{\eta,x}$. Then, using the transformation rule for the Busemann density, Proposition 4.1, and the estimate

$$d(q, x) \geq d(x, \pi\tilde{v}) - d(\pi\tilde{v}, p) - d(p, q) \geq n - r_0 - (2\,\epsilon + r_0),$$

we obtain

$$\mu_p(pr_\eta(B(x, \epsilon))) \leq e^{hd(p,q)} \mu_q(pr_\eta(B(x, \epsilon))) \leq e^{h(r_0+2\epsilon)} b\, e^{-hd(q,x)} \leq \tilde{b} e^{-hn}$$

for a constant $\tilde{b} > 0$. Since $\mu_p(pr_x(B(p, r_0 + 2\epsilon))) \leq \mu_p(X(\infty)) < \infty$ the proof follows from the definition of μ.

THEOREM 8.2. *The measure μ is a measure of maximal entropy for the geodesic flow, i.e.,*

$$h_\mu(\phi) = h(\phi) = h.$$

Proof. Choose a partition $\mathcal{A}$ as above. Then, by Lemma 8.1,

$$H(\mathcal{A}_\phi^{(n)}) = \sum_{\alpha \in \mathcal{A}_\phi^{(n)}} \mu(\alpha)\,(-\log \mu(\alpha)) \geq (hn - \log a) \sum_{\alpha \in \mathcal{A}_\phi^{(n)}} \mu(\alpha) = hn - \log a.$$

Hence, $h(\phi, \mathcal{A}) \geq h$, and the theorem follows from $h \geq h_\mu(\phi) \geq h(\phi, \mathcal{A})$.

9. Uniqueness of the maximal measure

In some sense, our proof of the uniqueness of the maximal measure can be viewed as a refinement of Bowen's original proof given in the uniform hyperbolic situation (see [**5**] and [**10**]). However, in the nonuniform hyperbolic case the dynamics of the singular geodesics can be quite complicated and a priori it could well be that the singular set carries another measure of maximal entropy.

Before we begin with the proof of the uniqueness, we need to establish the ergodicity of the maximal measure constructed in Section 8. It will be also of importance that partitions of a certain fixed size are enough for the computation of entropy.

PROPOSITION 9.1. (1) *The maximal measure μ is ergodic with respect to the geodesic flow and $\mu(reg) = 1$.*

(2) *Let ϕ^k be the time k-map of the geodesic flow and $\mathcal{A}$ be a partition such that, for each $A \in \mathcal{A}$, the diameter with respect to d_k is smaller than inj$(M)/3$, where inj(M) is the injectivity radius of M. Then*

$$h_\nu(\phi^k) = h_\nu(\phi^k, \mathcal{A})$$

for all invariant probability measures ν.

Proof. The first part follows from the structure of μ together with the Hopf argument. Since μ is positive on open sets it follows from the ergodicity that the regular set has full measure. The second part follows since ϕ^k is entropy expansive. For the metric d_k the expansivity constant is inj$(M)/3$ (for more details see Sections 3 and 4 of [**13**]).

Since the measure M of maximal entropy is ergodic, the proof of its uniqueness only requires to show that all invariant measures singular with respect to μ have less entropy. This follows since any measure ν can be decomposed into a singular ν_{sing} and absolutely continuous ν_{abs} part, i.e., there exists a constant $b \in [0, 1]$ such that $\nu = b\nu_{\text{sing}} + (1 - b)\nu_{\text{abs}}$. Hence, the ergodicity of μ implies $\nu_{\text{abs}} = \mu$ and, therefore, the entropy $h_\nu = bh_{\nu_{\text{sing}}} + (1-b)h < h$, if $h_{\nu_{\text{sing}}} < h$.

Steps of the proof of the uniqueness of μ (see also Section 5 of [**13**])

Step 1: Semilocal properties of μ
Let R be the constant introduced in Proposition 4.1 and $p \in \mathcal{F}$ be a fixed reference point in a fundamental domain $\mathcal{F}$. For $x \in X$ and $R > 0$ let

$$D(x, R', R) = \{v \in SB(p, R') \mid c_v(r) \in B(x, R) \text{ for some } r > 0\}$$

be the set of unit tangent vectors with foot points in $B(p, R')$, whose forward orbit meets the ball $B(x, R)$ (see Figure 5).

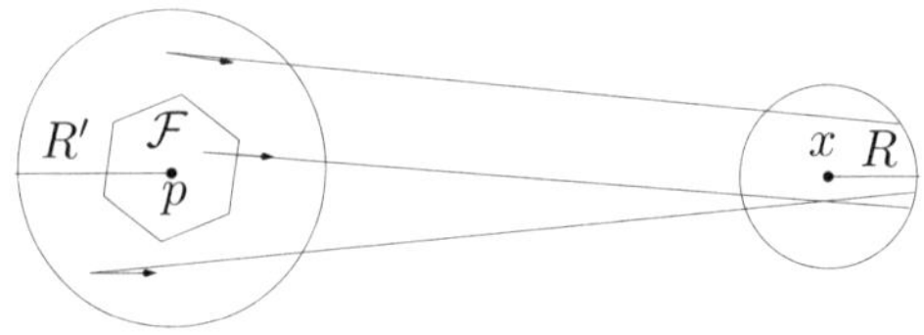

FIGURE 5

LEMMA 9.2. *For $R' > 2R + 1/2$ with $\mathcal{F} \subset B(p, R')$ and $d(p, x) \geq R' + R$ we obtain:*

$$\mu(D(x, R', R)) \geq c' e^{-hd(p,x)},$$

with a constant $c' > 0$ not depending on x.

Proof. If $P : SX \to X(\infty) \times X(\infty)$ is the end point projection map given by $P(v) = (c_v(-\infty), c_v(+\infty))$, one shows that $P(D(x, R', R))$ contains the set

$$A := \bigcup_{\eta \in pr_x(B(p,R))} \{\eta\} \times pr_\eta(B(x, R)).$$

From the local property of the measure μ_p (see 4.1) we obtain the estimates $\mu_p(pr_x(B(p, R))) > \ell$ and $\mu_p(pr_\eta(B(x, R))) \geq b' e^{-hd(p,x)}$. Hence, the lemma follows from the definition of μ and the fact that the intersection of the set $D(x, R', R)$ and a geodesic c with $(c(-\infty), c(+\infty)) \in A$ is a geodesic arc of length at least 1. The next lemma is a consequence of the convexity of the distance function.

LEMMA 9.3. *For all $x \in X$ with $d(x, p) \geq n + R' + R$ the cardinality of a (d_n, δ)-separated set of $D(x, R', R)$ is bounded from above by a constant $a = a(\delta, R', R)$*

Step 2: Construction of a suitable covering
Now consider a maximal $2R$-separated set $x_1, \ldots, x_{k(n)}$ of the geodesic sphere $S(p, r(n))$ of radius $r(n) = 2n + R' + R$ about p. Since this set is $2R$-separated the balls $B(x_i, R)$ are pairwise disjoint, and since it is maximal, the balls $B(x_i, 2R)$ cover $S(p, r(n))$. Therefore, the sets $D(x_i, R', R)$ are pairwise disjoint and the sets $D(x_i, R', 2R)$ cover the unit tangent bundle $SB(p, R')$ of the ball $B(p, R')$.
Thus we can choose a partition $F_1^{2n}, \ldots, F_{k(n)}^{2n}$ of $SB(p, R')$ such that

$$D(x_i, R', R) \subset F_i^{2n} \subset D(x_i, R', 2R)$$

for each i. Let $Q : SB(p, R') \to SM$ be the restriction to $SB(p, R')$ of the projection $SX \to SM$ and let $L_i^{2n} = Q(F_i^{2n})$. Note that the map Q is at most α-to-one for some constant $\alpha > 0$.

LEMMA 9.4. *$L^{2n} = \{L_1^{2n}, \ldots, L_{k(n)}^{2n}\}$ is a covering such that each $v \in SM$ lies in at most α sets from L^{2n}. Furthermore, there is a constant $\beta > 0$ such that $\mu(L_i^{2n}) \geq \beta e^{-2hn}$ for all $i \in \{1, \ldots, k(n)\}$.*

Proof. The first assertion follows since Q is at most α-to-one. The second assertion follows from Step 1 and the fact that $\mu(A) \leq \alpha\mu(Q(A))$ for all measurable sets $A \subset SB(p, R')$.

Step 3: Partitions of small diameter

To prove the uniqueness of the measure of maximal entropy we have to deal with partitions of small diameter, i.e., we need to pass from semilocal to a local scale. Let $V = \{v_1, \ldots, v_m\}$ be a maximal (d_{2n}, ϵ)-separated set of SM. Choose a partition $\mathcal{B}$ such that for each $B \in \mathcal{B}$ there exists $v_i \in V$ such that

$$B_{d_{2n}}(v_i, \epsilon/2) \subset B \subset B_{d_{2n}}(v_i, \epsilon).$$

LEMMA 9.5.

$$\#\{B \in \mathcal{B} \mid B \cap L_i^{2n} \neq \emptyset\} \leq a(\epsilon, R', 2R),$$

where $a(\epsilon, R', 2R)$ is the constant defined in Step 2.

Proof. Since $L_i^{2n} = Q(F_i^{2n})$ we can lift

$$\{v_j \in V \mid B_{d_{2n}}(v_j, \epsilon) \cap L_i^{2n} \neq \emptyset\}$$

to an (d_{2n}, ϵ)-separated set of $F_i^{2n} \subset D(x, R', 2R)$. Therefore, the result follows from 9.3.

Step 4: Separating measures using coverings

Now consider the push-forward $A^n = \{\phi^n L_i^{2n}\}$ of the covering L^{2n} by the time-n-map of the geodesic flow. It follows from the definition of L^{2n} that for each $A \in A^n$ and any two vectors $v, w \in A$ there exists a continuous curve $\alpha : [0,1] \to SM$ connecting v and w such that the length of $\pi \circ \phi^t \alpha$ is bounded by R' for all $t \in [-n, n]$. The reason is that the set F_i^{2n} is contained in $D(x_i, R', 2R)$. We will use the covering A^n to separate measures in the following way.

LEMMA 9.6. *Let ν, μ be mutually singular Borel probability measures on SM such that the singular set has measure 0 with respect to μ. Then there exists a union C_n of subsets of A^n such that*

$$\mu(C_n) \to 0, \quad \nu(C_n) \to 1$$

for $n \to \infty$.

Proof. Choose a set Ω containing the singular set which is of full measure w.r.t. ν but of zero measure w.r.t. μ. Now consider compact subsets K_1 of Ω respectively K_2 of $SM \setminus \Omega$ having almost full measure w.r.t ν respectively μ. Choose the union C_n of subsets of the covering A_n such that their elements intersect K_1. Since the elements of C_n correspond to geodesics staying in a fixed tubular neighborhood for the time intervall $[-n, n]$ and K_2 is part of the regular set it follows from the

flat strip theorem (see e.g. [**3**]) and a compactness argument that for large n the elements of C_n do not intersect K_2. Consequently, $\mu(C_n)$ is close to 0 and $\nu(C_n)$ is close to 1.
In the last step we will show that the measure of maximal entropy is unique. For that the following standard estimate is useful:

$$-\sum_{i=1}^{k} a_i \log a_i \leq \left(\sum_{i=1}^{k} a_i\right) \log k + \frac{1}{e} \tag{1}$$

provided $a_1, \ldots, a_k \geq 0$ and $\sum_{i=1}^{k} a_i \leq 1$.
Step 5: Proof of the uniqueness

THEOREM 9.7. *Let ν be an invariant Borel probability measure which is singular with repect to the maximal measure μ. Then*

$$h_\nu(\phi) < h = h_\mu(\phi).$$

Proof. Fix $\epsilon \leq \frac{\mathrm{inj}(M)}{3}$ and let $V = \{v_1, \ldots, v_m\} \in SM$ be a maximal (ϵ, d_n)-separated set. Choose a partition $\mathcal{B}^n$ such that for each $B \in \mathcal{B}^n$ we have

$$B_{d_{2n}}(v_i, \frac{\epsilon}{2}) \subset B \subset B_{d_{2n}}(v_i, \epsilon)$$

for some $v_i \in V$. Proposition 9.1(2) implies

$$2nh_\nu(\phi^1) = h_\nu(\phi^{2n}) = h_\nu(\phi^{2n}, \mathcal{B}^n) \leq H_\nu(\mathcal{B}^n) = -\sum_{B \in \mathcal{B}^n} \nu(B) \log \nu(B).$$

Consider the covering $L^{2n} = \{L_1^{2n}, \ldots, L_k^{2n}\}$ of SM constructed in Step 2 and its push-forward $\{\phi^n L_i^{2n}\} = A^n$. Since ν and μ are mutually singular and the singular set has measure 0 with respect to μ, Lemma 9.6 in Step 4 implies the existence of sets C_n consisting of a union of elements of A^n such that $\mu(C_n) \to 0$ and $\nu(C_n) \to 1$ as $n \to \infty$. Using the above estimate (1), we obtain:

$$\begin{aligned}
2nh_\nu(\phi) \leq & -\sum_{\{B \in \mathcal{B}^n \mid \phi^n B \cap C_n \neq \emptyset\}} \nu(B) \log \nu(B) \\
& - \sum_{\{B \in \mathcal{B}^n \mid \phi^n B \cap C_n = \emptyset\}} \nu(B) \log \nu(B) \\
\leq & \left(\sum_{\{B \in \mathcal{B}^n \mid \phi^n B \cap C_n \neq \emptyset\}} \nu(B)\right) \log \#\{B \in \mathcal{B}^n \mid \phi^n B \cap C_n \neq \emptyset\} \\
& + \left(\sum_{\{B \in \mathcal{B}^n \mid \phi^n B \cap C_n = \emptyset\}} \nu(B)\right) \log \#\{B \in \mathcal{B}^n \mid \phi^n B \cap C_n = \emptyset\} + \frac{2}{e} \\
\leq & \, b_n \log \#\{B \in \mathcal{B}^n \mid \phi^n B \cap C_n \neq \emptyset\} + (1 - b_n) \log \#\{\mathcal{B}^n\} + \frac{2}{e},
\end{aligned}$$

where

$$b_n = \sum_{\{B \in \mathcal{B}^n \mid \phi^n B \cap C_n \neq \emptyset\}} \nu(B).$$

Using Lemma 9.5, we estimate:

$$\#\{B \in \mathcal{B}^n \mid \phi^n B \cap C_n \neq \emptyset\}$$

$$\leq \sum_{\{L_i^{2n} \in L^{2n} \mid \phi^n L_i^{2n} \subset C_n\}} \#\{B \in \mathcal{B}^n \mid \phi^n B \cap \phi^n L_i^{2n} \neq \emptyset\}$$
$$\leq a(\epsilon, R', 2R) \#\{L_i^{2n} \in L^{2n} \mid \phi^n L_i^{2n} \subset C_n\}.$$

Furthermore, using Lemma 9.4

$$\#\{L_i^{2n} \mid \phi^n L_i^{2n} \subset C_n\} = \#\{L_i^{2n} \mid L_i^{2n} \subset \phi^{-n} C_n\}$$
$$\leq \alpha \frac{\mu(C_n)}{\min \mu(L_i^{2n})} \leq \alpha' \mu(C_n) e^{2hn}$$

for a constant $\alpha' > 0$. Moreover, replacing C_n by SM, the same argument yields

$$\#\{\mathcal{B}^n\} \leq a(\epsilon, R', 2R) \alpha \frac{\mu(SM)}{\min \mu(\phi^n L_i^{2n})} \leq \tilde{\alpha} e^{2hn}$$

for a suitable constant $\tilde{\alpha} > 0$. Hence,

$$(*) \qquad 2nh_\nu(\phi) - 2nh - \frac{2}{e} \leq b_n \log\left(\mu(C_n)\, \alpha' a(\epsilon, R', R)\right) + (1 - b_n) \log(\tilde{\alpha}).$$

Since

$$b_n = \sum_{\{B \in \mathcal{B}^n \mid \phi^n B \cap C_n \neq \emptyset\}} \nu(\phi^n B) \geq \nu(C_n) \to 1$$

and $\mu(C_n) \to 0$ the first term of the right hand side of $(*)$ tends to $-\infty$, as $n \to \infty$. Since the second term tends to 0, this implies $h_\nu(\phi) < h(\phi)$.

10. Some open Problems

(1) As we have seen above, for rank 1 manifolds the measure of maximal entropy is the unique equilibrium state for the constant function. Are there other functions which have a unique equilibrium state?
(2) For which rank 1 manifolds has the geodesic flow, restricted to the singular set, positive topological entropy? Are there examples beyond graph manifolds?
(3) If the singular set has positive topological entropy, does the geodesic flow restricted to the singular set have a unique measure of maximal entropy? Find a precise estimate for the growth rate of singular closed geodesics.
(4) Let (M, g_0) be a compact Riemannian manifold of negative curvature and g be any other metric. The set of minimal g–geodesics (minimal as geodesics on the universal covering) determines a closed invariant set for the geodesic flow. Determine the measure of maximal entropy of the geodesic flow restricted to the minimal set. Is this measure unique? [1]
(5) Find a criterion for nonuniform hyperbolic flows (beyond geodesic flows) which guarantees the existence and uniqueness of a maximal measure.

[1]This problem is another modern outgrowth of a 1924 paper by Marston Morse [**15**], whose fundamental idea played an important role in many developments in hyperbolic geometry of the 20th century including the rigidity theorem of Mostow and Gromov's generalization to the notion of quasiisometric equivalence

References

[1] W. Ballmann: *Lectures on spaces of nonpositive curvature.* DMV Seminar, Band 25 (1995).

[2] W. Ballmann: *Nonpositively curved manifolds of higher rank.* Annals of Math. **122** (1985), 597-609.

[3] W. Ballmann, M. Gromov and V. Schroeder: *Manifolds of nonpositive curvature.* Progress in Math., vol. 61, Birkhäuser, Boston 1985.

[4] R. Bowen: *Periodic Orbits for hyperbolic flows.* Amer. J. Math. **94** (1972), 1-30.

[5] R. Bowen: *Maximizing entropy for a hyperbolic flow.* Math. System Theory **7** (1973), 300-303.

[6] K. Burns and R. Spatzier: *Manifolds of nonpositive curvature and their buildings.* Publ. Math. IHES **65** (1987), 35-59.

[7] K. Burns and A. Katok: *Manifolds with non-positive curvature.* Ergod. Th. & Dynam. Sys. **5** (1985), 307-317.

[8] P. Eberlein: *Geometry of nonpositively curved manifolds.* Chicago Lectures in Mathematics. University of Chicago Press, 1996.

[9] P. Eberlein and B. ONeill: *Visibility manifolds.* Pacific J. Math. **46** (1973), 45-109.

[10] A. Katok and B. Hasselblatt: *Introduction to the modern theory of dynamical systems.* Encyclopedia of mathematics and its applications, vol. 54, Cambridge University Press, 1995.

[11] G. Knieper: *Das Wachstum der Äquivalenzklassen geschlossener Geodätischer in kompakten Riemannschen Mannigfaltigkeiten.* Archiv der Mathematik **40** (1983), 559-568.

[12] G. Knieper: *On the asymptotic geometry of nonpositively curved manifolds.* GAFA **7** (1997), 755-782.

[13] G. Knieper: *The uniqueness of the measure of maximal entropy for geodesic flows on rank 1 manifolds.* Annals of Math. **148** (1998), 291-314.

[14] G.A. Margulis: *Applications of ergodic theory to the investigation of manifolds of negative curvature.* Funct. Anal. Appl. **3** (1969), 335-336.

[15] M.Morse:*A fundamental class of geodesics on any closed surface of genus greater than one.* Trans. Amer. Math. Soc. **26** (1924), 25-60.

[16] S. Patterson: *The limit set of Fuchsian group.* Acta Math. **136** (1976), 241-273.

[17] D. Sullivan: *The density at infinity of a discrete group of hyperbolic motions.* Publ. Math. IHES **50** (1979), 171-202.

Fakultät für Mathematik, NA 5/33, Ruhr-Universität, 44780 Bochum, Germany
E-mail address: gknieper@math.ruhr-uni-bochum.de
URL: http://www.ruhr-uni-bochum.de/mathematik10

Part IIc

Algebraic Systems and Rigidity

Proceedings of Symposia in Pure Mathematics
Volume **69**, 2001

Invariant Measures for Actions of Higher Rank Abelian Groups

Boris Kalinin and Anatole Katok

ABSTRACT. The first part of the paper begins with an introduction into Anosov actions of $\mathbb{Z}^k$ and $\mathbb{R}^k$ and an overview of the method of studying invariant measures for such actions based on consideration of conditional measures along various invariant foliations. The main body of that part contains a detailed proof of a modified version of the main theorem from [**KS3**] for actions by toral automorphisms of with applications to rigidity of the measurable structure of such actions with respect to Lebesgue measure. In the second part principal technical tools for studying nonuniformly hyperbolic actions of $\mathbb{Z}^k$ and $\mathbb{R}^k$ are introduced and developed. These include Lyapunov characteristic exponents, nonstationary normal forms and Lyapunov Hoelder structures. At the end new rigidity results for $\mathbb{Z}^2$ actions on three-dimensional manifolds are outlined.

In this paper we discuss various results concerning invariant measures for actions of higher–rank abelian groups, i.e. $\mathbb{Z}^k$ and $\mathbb{R}^k$ for $k \geq 2$ on compact differentiable manifolds which display certain hyperbolic behavior. Similarly to the rank one case, hyperbolicity can be full or partial, and uniform or nonuniform. Full (corr. partial) uniform hyperbolicity appears for Anosov (corr. partially hyperbolic) actions. Nonuniform hyperbolicity appears for actions preserving measures for which all (for the full case) or some (for partial case) Lyapunov characteristic exponents do not vanish. Two parts of the paper deal with the uniform and nonuniform cases correspondingly.

While the final text of this paper is the product of a joint effort, the basic drafts of various parts were written separately: that for Section 3 was written by B. Kalinin and based on a part of his Ph. D. thesis; for the rest of the paper the draft was written by A. Katok.

We would like to thank Ralf Spatzier for carefully reading the paper and making a number of valuable suggestions and corrections.

The first author was partially supported by NSF grant DMS-9704776.
The second author was partially supported by NSF grants DMS-9704776 and DMS-0071339.

Part I. UNIFORMLY HYPERBOLIC ACTIONS

1. Definitions and examples

1.1. Anosov and partially hyperbolic actions.

1.1.1. *Anosov actions of discrete and continuous groups.* A smooth action α of a discrete group Γ on a compact manifold M is called *Anosov* if for some $\gamma \in \Gamma$, $\alpha(\gamma)$ is an Anosov diffeomorphism, *i.e.* the tangent bundle TM splits in an $\alpha(\gamma)$ invariant way into two subbundles (distributions) E^- and E^+ called the *stable* (or *contracting*) and *unstable* (or *expanding*) bundles (or distributions) correspondingly such that for some λ, $0 < \lambda < 1$ and $C \geq 1$ and for all positive integers m,

$$\|D(\alpha(\gamma^m))(v)\| \leq C\lambda^m, \tag{1.1.1}$$

if $v \in E^-$ and

$$\|D(\alpha(\gamma^m))(v)\| \geq C^{-1}\lambda^{-m}, \tag{1.1.2}$$

if $v \in E^+$.

Here $\|\cdot\|$ is the norm in TM generated by a Riemannian metric on M and D is the derivative map. The constant C (but not λ) depends on the choice of Riemannian metric.

More generally, a smooth locally free action of a Lie group G is called Anosov (or *normally hyperbolic*) if for a certain element conditions (1.1.1) and (1.1.2) hold for E^- and E^+ in the splitting $TM = E_o \oplus E^- \oplus E^+$, where E_o is the tangent bundle to the orbit foliation of the action. See [**KS1**] for a more detailed discussion. Such an element γ is then called *normally hyperbolic.* In this case the distributions E^- and E^+ are often called *the strong stable* and *the strong unstable* distributions.

The general theory of Anosov actions even for higher–rank abelian groups is not well developed. The first serious obstruction is the following open question.

Problem. *For an Anosov action of* $\mathbb{R}^k$, $k \geq 2$ *is the set* $\mathcal{N}$ *of normally hyperbolic elements dense?*

A similar question for $\mathbb{Z}^k$ actions is reduced to this one via suspension construction (Section 1.2.2). The set $\mathcal{N}$ is open and its connected components are convex cones. In all known examples $\mathbb{R}^k \setminus \mathcal{N}$ is the union of finitely many hyperplanes (*cf.* Lyapunov hyperplanes, Sections 1.2.3, 1.3, 5.2).

1.1.2. *Partially hyperbolic actions.* A more general class is formed by partially hyperbolic actions. A G action α is called *partially hyperbolic* if there is an $\gamma \in G$ and a $\alpha(\gamma)$–invariant splitting $TM = E^- \oplus E^+ \oplus E^0$, with E^- and E^+ satisfying (1.1.1) and (1.1.2) as before and E^0 the "slow" distribution with

$$(C')^{-1}\mu^{-m} \geq \|D(\alpha(\gamma^m))(v)\| \leq C'\mu^m \tag{1.1.3}$$

for some μ, $\lambda < \mu \leq 1$ and $C' \geq 1$.

1.1.3. *Contrast between rank one and higher rank.* Any discussion of Anosov actions of higher rank abelian groups requires a certain condition of being "genuinely higher rank" in order to avoid products of rank one actions and other situations which can be reduced to the rank one case. The structure of such actions in the higher–rank case looks quite different and much more rigid than in the classical rank one situation of diffeomorphisms ($\mathbb{Z}$–actions) and flows ($\mathbb{R}$-actions). All known examples of such actions satisfying the "genuine higher rank" assumptions are algebraic up to a differentiable conjugacy. In particular, most algebraic actions are known to be differentiably rigid [**KS2**]. For the algebraic actions a natural condition is absence of rank one algebraic factors. (See e.g. Section 3.1, condition ($\mathcal{R}'$)). Thus studying invariant measures for irreducible algebraic Anosov (and in certain cases, also partially hyperbolic) actions is quite natural.

One of the main differences between the rank one and higher rank situations may be highlighted as follows. In the former case the robust orbit structure is well–described by the Markov model via a Markov partition (see, e.g. [**KH**], Section 18.7). In particular, any invariant measure for the topological Markov chain associated with a Markov partition produces an invariant measure for the hyperbolic system in question. The variety of invariant measures, both well–structured and not, for a topological Markov chain, is huge; hence the same is true for the hyperbolic systems. In the higher–rank case the Markov model is not applicable, since natural topological Markov chains consist of maps with infinite topological entropy. Hence algebraic models should be understood more directly.

1.2. Higher Rank Actions by Toral Automorphisms. The most basic examples of smooth hyperbolic actions of higher rank abelian groups are actions by automorphisms of a torus and their suspensions.

1.2.1. *Preliminaries.* Denote by $GL(m,\mathbb{Z})$ the group of integral $m\times m$ matrices with determinant 1 or -1. Any matrix $A\in GL(m,\mathbb{Z})$ defines an automorphism of the torus $\mathbb{T}^m=\mathbb{R}^m/\mathbb{Z}^m$ which we denote by F_A.

The automorphism F_A is ergodic with respect to Lebesgue measure on $\mathbb{T}^m$ if and only if no eigenvalue of A is a root of unity.

This can be seen by considering the dual automorphism on the group of characters $\mathbb{Z}^m$. If A (and hence the transposed matrix A^t) has an rth root of unity as an eigenvalue there is a rational (and hence also an integer) invariant vector for $(A^t)^r$. Hence the operator $U_{F_A^r}$ induced by $(F_A)^r$ on L^2, has a nonconstant invariant function, the corresponding character χ, and $\sum_{i=0}^{r-1}\chi\circ(F_A^i)$ is an invariant nonconstant function for U_{F_A}.

On the other hand, if there are no roots of unity among the eigenvalues, all orbits of A^t acting on the characters, save the constants, are infinite and hence the operator $U_{F_A^r}$ has countable Lebesgue spectrum in the orthogonal complement to the constant and thus F_A is ergodic.

Furthermore, in the latter case A has an eigenvalue of absolute value greater then one (this can be seen by using a bit of linear algebra and a compactness argument), and F_A is a Bernoulli automorphism with respect to Lebesgue measure [**Kat**].

Any $\mathbb{Z}^k$ action α by automorphisms of $\mathbb{T}^m$ is given by an embedding $\rho_\alpha:\mathbb{Z}^k\to GL(m,\mathbb{Z})$ such that $\alpha(n)=F_{\rho_\alpha(n)}$, where $n=(n_1,...,n_k)\in\mathbb{Z}^k$.

Let α_1 and α_2 be two $\mathbb{Z}^k$ actions by automorphisms of $\mathbb{T}^{m_1}$ and $\mathbb{T}^{m_2}$ correspondingly. The action α_2 is called an *algebraic factor* of a_1 if there exists an epimorphism h: $\mathbb{T}^{m_1}\to\mathbb{T}^{m_2}$ such that $h\circ\alpha_1=\alpha_2\circ h$.

The action α is called *irreducible* if any algebraic factor has finite fibers, *i.e.* acts on the torus of the same dimension. Equivalently, α does not have any invariant subtorus of lower dimension, or ρ_α contains a matrix with irreducible characteristic polynomial [**B**]. In [**KS3, KS4**] irreducible actions are called completely irreducible.

1.2.2. *Suspension construction.* It will be more convenient for us to operate with $\mathbb{R}^k$ actions so we would like to pass from an action of $\mathbb{Z}^k$ to the corresponding action of $\mathbb{R}^k$. This is the so-called suspension construction.

Suppose $\mathbb{Z}^k$ acts on $\mathbb{T}^m$. Embed $\mathbb{Z}^k$ as a lattice in $\mathbb{R}^k$. Given a $\mathbb{Z}^k$ action α on the torus let $\mathbb{Z}^k$ act on $\mathbb{R}^k \times \mathbb{T}^m$ by

$$\bar{\alpha}(n)(t,x) = (t-n, \alpha(n)(x)) \tag{1.2.1}$$

and form the quotient

$$M = \mathbb{R}^k \times \mathbb{T}^m / \mathbb{Z}^k. \tag{1.2.2}$$

Note that the "vertical" action V of $\mathbb{R}^k$ on $\mathbb{R}^k \times \mathbb{T}^m$ by $V_s(t,x) = (t+s, x)$ commutes with the $\mathbb{Z}^k$-action and therefore descends to M. This $\mathbb{R}^k$-action is called the *suspension* of the $\mathbb{Z}^k$-action.

Note that any $\mathbb{Z}^k$-invariant measure on $\mathbb{T}^m$ lifts to a unique $\mathbb{R}^k$-invariant measure on M.

The manifold M is a fibration over the "time" torus $\mathbb{T}^k$ with the fiber $\mathbb{T}^m$. We note that TM splits into the direct sum $TM = T_f M \oplus T_o M$ where $T_f M$ is the subbundle tangent to the $\mathbb{T}^m$ fibers and $T_o M$ is the subbundle tangent to the orbit foliation.

Remark. The suspension construction is of very general nature. Namely, let α be an action of $\mathbb{Z}^k$ on a space X with a certain structure (measure, topology, differentiable, homogeneous, etc). Then (1.2.1) defines an action on the product and the factor (1.2.2) possesses the structure of the skew product over $\mathbb{T}^k$ with the fiber X and inherits the structure from the fiber. This structure is preserved by the suspension action of $\mathbb{R}^k$.

1.2.3. *Eigenvalues and Lyapunov Exponents.* Let us denote by $A_1, ..., A_k$ the generators of the action α, i.e. the images under ρ_α of the standard generators of $\mathbb{Z}^k$. For each A_i $\mathbb{R}^m$ splits as the direct sum of the root spaces of A_i:

$$\mathbb{R}^m = \bigoplus_{\lambda \in Sp A_i} Ker(A_i - \lambda)^m.$$

Since the matrices A_i commute there exists an invariant splitting $\mathbb{R}^m = \bigoplus V_j$ which is a common refinement of the above splittings. It is called the root decomposition for the action α. It follows that the tangent bundle $T\mathbb{T}^m = \mathbb{T}^m \times \mathbb{R}^m$ splits into the direct sum of the invariant subbundles corresponding to subspaces V_j. The Lyapunov exponent $\lambda(\alpha(n), v)$ exists for every element $\alpha(n) \in G_\alpha$ and every vector $v \in T\mathbb{T}^m$. If v lies in one of the subbundles then $\lambda(\alpha(n), v)$ equals to the logarithm of the absolute value of the corresponding eigenvalue of element $\alpha(n)$. Moreover, $\lambda(\alpha(\cdot), v)$ is an additive functional on $\mathbb{Z}^k$. Combining the subspaces V_j corresponding to the same Lyapunov exponent we obtain a more robust decomposition than the root decomposition which is called *the Lyapunov decomposition* for the action α. For the detailed discussion of Lyapunov exponents for $\mathbb{Z}^k$ and $\mathbb{R}^k$ actions by automorphisms of tori and solenoids including the non-Archimedean exponents appearing from p-adic valuations, see [**KS3**, Section 2 and Appendix]. We will return to the discussion of Lyapunov exponents in greater generality, first for algebraic actions in Section 1.3, and then in the general setting in Section 5.

The Lyapunov exponent for the suspension corresponding to $T_o M$ is always identically zero. To exclude this trivial case, when we speak of Lyapunov exponents we will always mean the Lyapunov exponents corresponding to $T_f M$. These Lyapunov exponents of the $\mathbb{R}^k$ action are the extensions of the Lyapunov exponents of the $\mathbb{Z}^k$ action to the linear functionals on $\mathbb{R}^k$. The kernels of non-zero Lyapunov exponents are called *Lyapunov hyperplanes*. Lyapunov exponents may be proportional to each other with positive or negative coefficients. In this case, they define the same Lyapunov hyperplane. The action is Anosov (Section 1.1) if there are no non-trivial identically zero Lyapunov exponents.

Definition. The action α is called *totally nonsymplectic (TNS)* if no Lyapunov exponents are proportional with a negative coefficient.

For a generalization of this concept see Section 5.2.

An element $a \in \mathbb{R}^k$ is called *regular* if it does not belong to any Lyapunov hyperplane. All other elements are called *singular*. We call a singular element *generic* if it belongs to only one Lyapunov hyperplane. A regular element for an Anosov action is called an *Anosov element*.

In the situation we are considering now, *i.e* for actions by automorphisms of the torus and for the suspensions of such actions, every Lyapunov subspace is uniquely integrable to a homogeneous foliation. This is not always the case for more general actions including homogeneous ones (Section 1.3); the *coarse Lyapunov decomposition* which combines positively proportional Lyapunov exponents is always integrable (Section 5.2).

For an element $a \in \mathbb{R}^k$ let us define the *stable, unstable* and *center* distributions E_a^-, E_a^+ and E_a^0 as the sum of the Lyapunov spaces for which the value of the corresponding Lyapunov exponent on a is negative, positive, and 0 respectively. For any singular element a the center distributions E_a^0 always contains an invariant subdistribution E_a^I, called *isometric* distribution. For an element a we will denote the integral foliations of the stable, unstable, center and isometric distributions E_a^-, E_a^+, E_a^0 and E_a^I by W_a^-, W_a^+, W_a^0 and W_a^I correspondingly.

1.3. General algebraic actions. Now we will consider "essentially algebraic" partially hyperbolic (in particular Anosov) actions of either $\mathbb{R}^k$ or $\mathbb{Z}^k$ more general than actions by toral automorphisms and their suspensions.

We recall that an action of a group G on a compact manifold is *Anosov* if some element $g \in G$ acts normally hyperbolically with respect to the orbit foliation (Section 1.1; see [**KS1**] for more details).

1.3.1. *Affine actions of discrete groups.* To clarify the notion of an algebraic action, let us first define affine algebraic actions of discrete groups. Let H be a connected Lie group with $\Lambda \subset H$ a cocompact lattice. Define $\mathrm{Aff}(H)$ as the set of diffeomorphisms of H which map right invariant vector-fields on H to right invariant vectorfields. Define $\mathrm{Aff}(H/\Lambda)$ to be the diffeomorphisms of H/Λ which lift to elements of $\mathrm{Aff}(H)$. Finally, define an action ρ of a discrete group G on H/Λ to be *affine algebraic* if $\rho(g)$ is given by some homomorphism $G \to \mathrm{Aff}(H/\Lambda)$. Let $\mathfrak{h}$ be the Lie algebra of H. Identifying $\mathfrak{h}$ with the right invariant vectorfields on H, any affine algebraic action determines a homomorphism $\sigma : G \to \mathrm{Aut}\ \mathfrak{h}$. Call σ the *linear part* of this action. We will also allow quotient actions of these on finite quotients of H/Λ, e.g. on infranilmanifolds. For any Anosov algebraic action of a discrete group G, H has to be nilpotent (cf. eg. [**GS**], Proposition 3.13).

1.3.2. *Algebraic $\mathbb{R}^k$-actions.* Suppose $\mathbb{R}^k \subset H$ is a subgroup of a connected Lie group H. Let $\mathbb{R}^k$ act on a compact quotient H/Λ by left translations where Λ is a lattice in H. Suppose C is a compact subgroup of H which commutes with $\mathbb{R}^k$. Then the $\mathbb{R}^k$-action on H/Λ descends to an action on $M = C \setminus H/\Lambda$. The general algebraic $\mathbb{R}^k$-action ρ is a finite factor of such an action. Let $\mathfrak{c}$ be the Lie algebra of

C. The *linear part* of ρ is the representation of $\mathbb{R}^k$ on $\mathfrak{c} \setminus \mathfrak{h}$ induced by the adjoint representation of $\mathbb{R}^k$ on the Lie algebra $\mathfrak{h}$ of H.

Let us note that the suspension (Section 1.2.2) of an algebraic $\mathbb{Z}^k$-action is an algebraic $\mathbb{R}^k$-action (cf. [**KS1**], Section 2.2).

Let ρ be an algebraic action of $\mathbb{R}^k$, not necessarily Anosov. For a given $a \in \mathbb{R}^k$ we denote the strong stable foliation of $\rho(a)$ by $\mathcal{W}_a^-$, the strong stable distribution by E_a^-, and the 0-Lyapunov space by E_a^0. Note that E_a^0 is always integrable for algebraic actions. Denote the corresponding foliation by $\mathcal{W}_a^0$. For a ρ–invariant Borel probability measure μ let us denote by ξ_a the partition into ergodic components of the element $\rho(a)$, by $h_\mu(a)$ the measure–theoretic entropy of the map $\rho(a)$ and by $\pi(a)$ the π–partition of $\rho(a)$. Recall that $\pi(a)$ is in fact equal to the measurable hull $\xi(\mathcal{W}_a^-)$ of the partition into the leaves of the strong stable foliation $\mathcal{W}_a^-$.

1.3.3. *Lyapunov exponents.* Let us describe the structure of the linear part of an algebraic $\mathbb{R}^k$ actions in more detail. In fact, we will first consider the case on the actions on H/Λ. The Lie algebra $\mathfrak{h}$ splits into root spaces $\mathfrak{h}^\alpha, \alpha \in \Sigma$ of the adjoint representation. The roots are linear functionals on $\mathbb{R}^k$ and their real parts in fact coincide with the Lyapunov exponents of the action ρ. The positive half–space $\Re\alpha \geq 0$ is called a *Lyapunov half–space* and its boundary a *Lyapunov hyperplane.* Let us denote the sum of root spaces with a given Lyapunov half–space L by $\mathfrak{h}^L$. Combining roots with the same real parts and taking corresponding right–invariant distribution we obtain *Lyapunov distributions*; Lyapunov distributions for different Lyapunov exponents form the *Lyapunov decomposition.* Combining Lyapunov distributions with the same Lyapunov half–space one obtains *coarse Lyapunov distributions* which form the *coarse Lyapunov decomposition.*

The strong stable distribution E_a^- for an element $a \in \mathbb{R}^k$ equals the sum of Lyapunov distributions E^α for those α for which $\alpha(a) < 0$. In particular, E_a^- is the sum of certain coarse Lyapunov distributions. Conversely, any coarse Lyapunov distribution equals to the intersection of strong stable distributions for certain elements of the action. Thus, any coarse Lyapunov distribution integrates to a homogeneous foliation whose leaves are in fact cosets of a nilpotent subgroup of H. For the Lyapunov half–space L let us denote this nilpotent subgroup by N_L. By picking an element a on the boundary of L which does not lie on any other Lyapunov hyperplane (such an element is called *generic singular*) and an element b inside L we see that $E^L = E_a^0 \cap E_b^-$ and hence E_b^- is the direct sum of $E_{a_i}^0 \cap E_b^-$ for several generic singular elements α_i.

Notice that unlike the case of actions by automorphisms of a torus Lyapunov distributions may not be integrable.

The case of an action on a double coset space is not too different. In fact the kernel of the projection lies in the 0-Lyapunov distributions for *all* elements $a \in \mathbb{R}^k$. Thus the picture on M is essentially the same as in H/Λ, only the zero Lyapunov exponent has lower multiplicity. This is why, in particular, the double coset construction appears in many standard examples of Anosov $\mathbb{R}^k$ actions.

1.4. Conditional measures. We will study the rigidity of invariant measures based on understanding of their conditional measures on some natural invariant foliations. Let us briefly recall how a probability measure ν on a manifold M determines a system of conditional measures on a foliation F. For a more detailed overview see [**KS3**], Section 4.

Denote by $\mathcal{B}$ the Borel σ-algebra on M. A *measurable partition* ξ of M is a partition of M such that, up to a set of measure 0, the quotient space M/ξ is separated by a countable number of measurable sets [**R**]. For every x in a set of full ν-measure there is a probability measure ν_x^ξ defined on $\xi(x)$, the element of ξ containing x, and satisfying the following properties: If $\mathcal{B}_\xi$ is the sub-σ-algebra of $\mathcal{B}$ whose elements are unions of elements of ξ, and $A \subset M$ is a measurable set, then $x \mapsto \nu_x^\xi(A)$ is $\mathcal{B}_\xi$-measurable and $\nu(A) = \int \nu_x^\xi(A)\nu(dx)$. These conditions determine the measures ν_x^ξ uniquely.

Let F be a foliation with smooth leaves and let us denote by $F(x)$ the leaf through x. Even though all foliations that we are going to consider in the next section will be linear, the partition into the leaves of F is not a measurable partition in general. Conditional measures on the leaves of the foliations that we will be working with are σ-finite locally finite measures ν_x^F defined up to a multiplicative constant. In other words, for almost every $x \in M$ and for open sets $A, B \subset F(x)$ with compact closures one can canonically define *the ratio* $\frac{\nu_x^F(A)}{\nu_x^F(B)}$. The conditional measures can be defined as follows.

Let us call a measurable partition ξ *subordinate* to F if for ν-a.e. x we have $\xi(x) \subset F(x)$ and $\xi(x)$ contains a neighborhood of x open in the submanifold topology of $F(x)$. Note that two different partitions subordinate to the same foliation determine conditional measures that are scalar multiples when restricted to the intersection of an element of one partition with an element of the other partition. Thus there is a locally finite measure ν_x^F on $F(x)$ uniquely defined up to scaling that restricts to a scalar multiple of a conditional measure for each partition subordinate to F. The measures ν_x^F form the system of conditional measures on the leaves of F.

Let $\sigma(F)$ denote the σ-algebra of all sets that consist a.e. of complete leaves of F. It corresponds to a unique measurable partition which is called the *measurable hull* of F, and is denoted by $\xi(F)$. It is the finest measurable partition whose elements consist a.e. of the entire leaves of F.

Notice that if the conditional measures on leaves of a foliation are δ-measures as will often be the case in the course of our considerations then *a posteriori* the partition into leaves of the foliation turns out to be measurable, indeed equivalent to the partition ϵ of the space into single points.

The next proposition is one of the expressions of fact that entropy measure uncertainly in the future given complete knowledge of the past. In particular, zero entropy implies that the past uniquely determines the future. This proposition is a particular case of [**KS3**, Proposition 4.1].

PROPOSITION 1.1. *Let f be a diffeomorphisms of a compact manifold, W be a an f-invariant expanding foliation, i.e. $\|Df^1\|_{TW}\| < 1$, and μ be an f-invariant Borel probability measure.*

If the entropy $h_\mu(f) = 0$ then conditional measures induced by μ on the leaves of the foliation W are δ-measures, i.e. for almost every leaf the conditional measure is concentrated at a single point.

If $h_\mu(f) > 0$ then conditional measures on almost all leaves are nonatomic.

1.5. Invariant measures for algebraic Anosov actions. All known ergodic invariant measures for algebraic Anosov actions of higher–rank abelian groups can be classified as follows

(1) Haar measures on homogeneous invariant submanifolds;
(2) Lifts of invariant measures from rank one factors on invariant homogeneous submanifolds.

The first case includes the measures concentrated on compact (periodic) orbits; these measures obviously have zero entropy for all elements of the action. Note that the union of such orbits is dense.

Appearance of the second case indicates certain reducibility. While it is reasonable to conjecture that no other ergodic measures exist ([**KS3**], Introduction) current methods allow to deal only with measures which somehow reflect the hyperbolic behavior of the system. The standard assumption would be that a certain element of the action has positive entropy with respect to the measure. This is crucial since our methods are based on considerations of *conditional* measures induced by an invariant measure on stable foliations of various elements of an action as well as on various invariant subfoliations of such foliations. As 1.1 shows such a measure must induce nontrivial (in fact, nonatomic) conditional measures on the leaves of the stable and unstable foliations for any element of the action.

A hyperbolic map has zero entropy with respect to an invariant measure if and only if the conditional measures on its contracting foliation are atomic which happens to be equivalent in this case that these are δ–measures. Under various assumptions we are able to show that an alternative to being atomic for the conditional measure is to be Haar on a certain homogeneous submanifold. This in turn in certain situations leads to the conclusion that the measure itself is of type (1) above.

This approach has been developed in [**KS3**]. The case of actions by automorphisms of the torus (as well as the more general non–invertible situation) is discussed in [**KS3**] in greater detail; the symmetric space actions (Weyl chamber flows and related actions) as well as twisted Weyl chamber flows are also considered. There are gaps in the proofs of the main results both for the toral case (Theorem 5.1) and for the symmetric space case (Theorem 7.1), which were partially corrected in [**KS4**]. In Section 3 we give a complete argument for the proof of the main theorem in the toral case with a properly modified formulation. This theorem yields the rigidity of measure–preserving conjugacies, centralizers and factors. These results are presented in Section 4; [**KKS**] contains applications of these results which use some subtle number–theoretic information. Apart from presenting complete proofs of the above theorem our arguments have an extra value because they elaborate an important method whose uses extend beyond this particular proof. For example, essentially the same method is used in the proof of rigidity of joinings in [**KaK**].

Before doing that in the next section we will explain some of the main features of the method by considering in a less formal way a special case of a $\mathbb{Z}^2$ action on the three–dimensional torus by hyperbolic automorphisms. In this example the underlying geometric structure is quite transparent and a number of complications which appear in more general situations do not show up. In more general cases in order to conclude the desired dichotomy (Haar on a rational subtorus or a homogeneous invariant submanifold, or zero entropy) one needs extra assumptions on the action (e.g. irreducibility in the toral case) or on the measure (such as K-property, or, in the symmetric space case, weak mixing). The conclusion in general is also somewhat weaker.

2. The simplest higher rank models

2.1. Cartan action on the three–dimensional torus. The following special case provides an excellent insight in the core features of the method used to study rigidity of invariant measures.

Let $A \in SL(3,\mathbb{Z})$ be any hyperbolic matrix with distinct real eigenvalues. By passing to a power, if necessary, we may assume that its eigenvalues λ_1, λ_2, λ_3 are all positive. By the Dirichlet unit theorem the centralizer of A in $SL(3,\mathbb{Z})$ is a finite extension of $\mathbb{Z}^2$. See [**KKS**] for a more detailed discussion and [**BS**] for the number–theoretic background. In particular, this centralizer contains a subgroup isomorphic to $\mathbb{Z}^2$ which consist of hyperbolic matrices. Such a subgroup σ determines an action of $\mathbb{Z}^2$ on the three–dimensional torus $\mathbb{T}^3$ by hyperbolic automorphisms which we will denote by ρ. (See Section 1.2) All elements of σ are simultaneously diagonalizable. The eigenvectors define three one–dimensional ρ–invariant linear foliations on the torus, W_1, W_2 and W_3. Each of these foliations possesses the natural affine parameter which is preserved by ρ. This is just another way of saying that the Euclidean length along each foliation is multiplied by each element of the action ρ by a constant which is equal to the absolute value of the corresponding eigenvalue. While no element of the action has an eigenvalue one or minus one there are elements for which the eigenvalue is arbitrary close to one. A useful technical devise is the *suspension* construction which by passing from the torus to a certain five–dimensional solvmanifold (an extension of $\mathbb{T}^2$ by $\mathbb{T}^3$) produces an $\mathbb{R}^2$ action α quite closely reflecting the features of the action ρ. (See Section 1.2.2) In particular, the "fiber" $\mathbb{T}^3$ direction splits into three invariant directions generating invariant foliations which we will still denote by W_1, W_2 and W_3. For each of these foliations there is an *irrational* direction in $\mathbb{R}^2$ such that the action along this direction preserves the length along the foliation. This direction is that of the kernel of the corresponding Lyapunov characteristic exponent, i.e. the Lyapunov hyperplanes defined in 1.2.3 which are of course just lines in our case. Call this direction the *critical* direction for the corresponding foliation. See Section 5.2 below for pertinent definitions in much greater generality.

2.2. Rigidity of invariant measures for Cartan actions. Now consider an ergodic invariant measure for the action ρ. We will show that

($\mathcal{D}$) *Either every element of the action has zero entropy with respect to this measure or the measure is Lebesgue measure on* $\mathbb{T}^3$.

First, extend the measure in a canonical way to an (also ergodic) invariant measure μ of the suspension action α. Let W be one of the three invariant foliations described above. The action preserves the system of conditional measures on the leaves on W. (See Section 1.4). Our argument works if the action in the critical direction is ergodic with respect to μ, or, more generally,

($\mathcal{E}$) *The ergodic components of the action in the critical direction consist of the whole leaves of* W.

This assumption will be later justified for any measure ergodic with respect to ρ. Under assumption ($\mathcal{E}$) we will show that

($\mathcal{T}$) *For almost every leaf* L *of the foliation* W *the conditional measure* μ_L *is invariant under the set of translations of full* μ_L *measure.*

Naturally, translations are defined with respect to the Euclidean length parameter. Here we mean invariance in an exact sense, not up to a scalar multiple. In other words, the measure μ_L is defined up to a scalar multiple but for any choice of normalization the measure is invariant with respect to almost every translation.

2.2.1. Thus L can be identified with the real line. Let us show that property $(\mathcal{T})$ implies that

The measure μ_L is either concentrated in a single point, or is a counting measure concentrated on a certain arithmetic progression, or is invariant under all translations, i.e. is simply Lebesgue measure on the line.

For, the measure is either atomic or continuous (translations cannot mix two parts). In the former case the atoms all have the same measure and hence must be isolated since otherwise the measure would not be locally finite. If there is more than one atom then the support is invariant by *every* translation taking one atom into another, which hence have to form a lattice.

If the measure is nonatomic than the set of translations under which it is invariant contains two rationally independent translations. Any locally finite measure on the line invariant under two rationally independent translations generates a finite measure on the circle (the factor by the first translation) invariant under an irrational translation and is hence Lebesgue.

Notice that ergodicity of the action implies that only one of the possibilities is realized for almost every leaf L.

The lattice case is, in fact, impossible due to the presence of the elements in the actions which expand the length parameter along W. For, any such element would map the set of leaves with a given value of the progression into the set with a strictly bigger value contradicting the Poincaré recurrence theorem.

Now assume that for almost every leaf L the measure μ_L is concentrated in a single point. The foliation W is the strong stable foliation of a certain elements of our action. But this implies that the entropy of those elements is equal to zero. These elements form a convex open set in $\mathbb{R}^2$ (a *Weyl chamber*, see Section 5.2. The inverses of these elements also have zero entropy. Since entropy is a sub–additive function on $\mathbb{R}^2$ ([**H**], Theorem B), it is equal to zero identically on $\mathbb{R}^2$.

If on the other hand, the measures μ_L are Lebesgue then the same is true for the conditional measures on the linear foliation on the torus. A fixed translation along the leaves of such a foliation is the linear flow on the torus which is in our case uniquely ergodic due to the fact that our action is irreducible over rationals (no invariant rational subtori). Since the conditional measures are invariant under translations so is the global measure which produces those conditionals. Hence by unique ergodicity the measure is itself Lebesgue.

An alternative argument does not use unique ergodicity of irrational translations. Instead one notice that any of the three invariant foliations W_1, W_2 and W_3 can play the role of W in the above argument. Since for any of these there is an element of the action for which it is the whole stable foliation, if at least one of the three systems of conditional measures are atomic the entropy of this element and hence of any element of the action vanishes. On the other hand, if all three systems of conditional measures are Lebesgue, then by Fubini Theorem the global measure is also Lebesgue.

2.2.2. Now we will explain why property $(\mathcal{T})$ holds. This is done in two steps.

Step 1. The measures are invariant in the affine sense: the translations are proportional up to a scalar multiple. To that end, let us fix the normalization at μ–almost every point x in such a way that the conditional measure of the interval $I(x)$ of length one on the leaf $L(x)$ centered at x is equal to one. Identifying $I(x)$ with the standard interval $[0, 1]$ via the length parameter we obtain a Borel map from our space to the set of probability measures on the unit interval provided with the weak* topology. On a compact subset C of measure arbitrary close to one this map is continuous (Luzin Theorem). A typical compact piece of a leaf intersects such a subset by a set of almost full conditional measure. Now use our ergodicity assumption $(\mathcal{E})$. Starting from a typical point x on a typical leaf L and moving in the critical direction the interval $I(x)$ comes arbitrary close to μ_L almost every point $y \in L$. In particular, if one assumes that both x and y are typical points of one of the continuity sets C described above, one may assert that the returns also appear on the set C. But this implies that in the limit the images of the conditional measures on $I(x)$ weakly converge to the conditional measure on $I(y)$. On the other hand, since we move along the critical direction the interval $I(x)$ is simply translated. Hence, the normalized conditional measure on $I(y)$ coincides with a translation of the normalized conditional measure on $I(x)$.

Step 2. The normalization constant which we will denote $c(x, y)$, is in fact equal to one. This again follows from the Poincaré recurrence. For, obviously, this constant is equivariant with respect to the action: for $a \in \mathbb{R}^2$, $c(\alpha(a)x, \alpha(a)y) = c(x, y)$. Secondly this is cocycle: $c(x, y)c(y, z) = c(x, z)$. The latter condition implies that along a typical leaf the density is exponential with respect to the natural length parameter. Taking an element a for which $\alpha(a)$ contracts the foliation W we see that the exponent must grow contradicting the Poincaré recurrence Theorem again, unless it is zero, *i.e.* the conditional measure is indeed Lebesgue.

2.2.3. Finally, we need to check assumption $(\mathcal{E})$ for any α–invariant ergodic measure, namely to show that the ergodic components for the action in the critical direction consist of the whole leaves of W. To that end we will use the structure of stable and unstable foliations for different elements or the action. In fact, for any of the three foliations there exists an element $a \in \mathbb{R}^2$ such that this foliation is the stable foliation of $\alpha(a)$ and the sum of the other two is the unstable one. On the other hand, the classical Hopf argument shows that ergodic decomposition for any element consists of complete leaves of its stable and unstable foliations: the positive (negative) time average of any continuous function is constant along the stable (unstable) leaves. The remaining ingredient is an important observation that for a normally hyperbolic (generic) element of the action the measurable hulls of partitions into leaves of the stable and unstable foliation coincide, because each of them generates the Pinsker σ–algebra (maximal σ–algebra with zero entropy).

Let us denote by ξ_a the partition into ergodic component of the element $\alpha(a)$ and by $\xi(W)$ the measurable hull of the partition into the leaves of the foliation W. Now let a be a non–zero element in the critical direction, W' be the one–dimensional stable foliation of a, W'' be the remaining foliation and $b \in \mathbb{R}^2$ be a regular (non–critical) element such that W' is the stable foliation of the element $\alpha(b)$. Thus we have the following inequalities:

$$\xi_a \leq \xi(W') = \pi(\alpha(b)) = \xi(W \oplus W'') \leq \xi(W). \tag{2.2.1}$$

2.3. $\times 2$, $\times 3$ and automorphisms of a solenoid. A very similar situation appears for commuting expanding maps of the circle. The basic example is the action of $\mathbb{Z}_+^2$ generated by multiplications by 2 and by 3 (mod 1). The question about invariant measures of this actions was posed by Furstenberg in 1967 [**F**]; it was solved for measures with positive entropy by D. Rudolph, [**R**]. It was an attempt to understand Rudolph's result in a geometric fashion that led the second author to the consideration of the model on the three–torus described above. Now in order to prove that the only ergodic positive entropy invariant measure for the multiplications by 2 and by 3 is Lebesgue we pass to the natural extension for this action. The phase space for this natural extension is a solenoid, the dual group to the discrete group $\mathbb{Z}(1/2, 1/3)$. It is locally modeled on the product of $\mathbb{R}$ with the groups of 2–adic and 3–adic numbers. Thus while topologically the solenoid is one–dimensional, there are three Lyapunov exponents, one for the real direction, and two for the non–Archimedean ones. Since the multiplication by 2 is an isometry in the 3–adic norm and vice versa the critical lines in this case are the two axis and the line $x \log 2 + y \log 3 = 0$ which does not intersect the first quadrant. All the above arguments work in this case verbatim with the real foliation playing the role of W. The unique ergodicity of the flow of translations along the real direction follows form the construction of the solenoid.

This argument of course extends to multiplications by p and q unless for some natural numbers k and l, $p^k = q^l$.

2.4. Other types of rigidity. The models discussed in this section are also very convenient for demonstrating a other types of rigidity phenomena which appear in actions of higher rank abelian groups.

The first type is rigidity of vector valued Hoelder and differentiable cocycles. Rigidity in these cases means that every cocycle from a given class is cohomologous to a constant coefficient cocycle, i.e. a homomorphism from the acting group to the vector space with the cohomology given by a transfer function from the same class or with moderate loss of regularity. See [**KS1**] for 1-cocycles over differentiable actions, [**KSch**] for 1-cocycles over actions by automorphisms of compact groups and [**KK**] for higher-order cocycles. A nice survey which also discusses other types of cocycles, such as those with values with compact abelian groups as well as nonabelian groups is [**NT**].

Another type of rigidity is local differentiable rigidity: any smooth action close to a given action in C^1 topology is differentiably conjugate (also maybe with a small loss of regularity) to the original action; in the continuous case up to an automorphism of $\mathbb{R}^k$ close to identity (see [**KS2**]). It is interesting to point that the proof of local differentiable rigidity involves a construction of a certain family of invariant geometric structures on certain invariant foliations of the perturbed action. In the simplest cases this structure is an affine connection. Then the structural stability implies existence of continuous conjugacy between the original and perturbed action (up to a time change in the continuous case). The principal idea in the proof that the conjugacy must be smooth involves showing that it intertwines a standard structure on the leaves of an invariant foliation for the algebraic action with the corresponding structure of the perturbed one. These latter structure plays a role very similar to that of conditional measures in the setup of the present paper.

There are also global rigidity results which assume that an action has certain topological (e.g. homotopy type) and dynamical (i.e. Anosov) properties and assert conjugacy with a standard algebraic action [**KL, MQ**].

The forthcoming notes [**K2**] discuss all these types of rigidity primarily in the simplest setting of actions of $\mathbb{Z}^2$ by automorphisms of $\mathbb{T}^3$.

3. Rigidity of invariant measures for actions by toral automorphisms

3.1. The Main Theorem and its Corollaries. In this section we discuss the main results on the rigidity of invariant measures for higher rank abelian actions by toral automorphisms. Theorem 3.1 is a modified version of the main result of A. Katok and R. Spatzier for the case of $\mathbb{Z}^k$ action by toral automorphisms (Theorem 5.1' of [**KS4**], which is in turn a modification of Theorem 5.1 of [**KS3**]). We give a complete self-contained proof of this theorem which is based on the original proof in [**KS3**] and [**KS4**].

The modification reflects the new version of Lemma 5.8' from [**KS4**] (see Lemma 3.11). The proof of Lemma 5.8' in [**KS4**] contains a gap since the ergodicity of the action may not imply that subspaces S_x are parallel. This in particular forces a slight modification in the formulation of the result, namely splitting of the measure into finitely many components. We give a more detailed proof of Lemma 3.2 (Lemma 5.4 from [**KS3**]) and a more elementary proof of Lemma 3.4 (Lemma 5.6 from [**KS3**]). Lemma 3.6 and Lemma 3.8 are similar to Lemma 5.9 and Lemma 5.10 in [**KS3**]. and the proofs employ the idea of the proofs in [**KS3**]. While the latter lemmas are correct their proofs in [**KS3**] lack some essential details. We give new arguments to complete the proofs.

Let α and α' be actions of $\mathbb{Z}^k$ by toral automorphisms and let α' be an algebraic factor of α. Then α' is called a *rank-one factor* of α if $\alpha'(\mathbb{Z}^k)$ has a subgroup of finite index which consists of powers of a single map.

For an action α of $\mathbb{Z}^k$ by automorphisms a torus the following two conditions are equivalent [**S**].

$(\mathcal{R})$: The action α contains a subgroup ρ, isomorphic to $\mathbb{Z}^2$, which consists of ergodic automorphisms.

$(\mathcal{R}')$: The action α does not possess non–trivial rank one algebraic factors.

THEOREM 3.1. *Let α be an $\mathbb{R}^k$-action with $k \geq 2$ induced from $\mathbb{Z}^k$ action by automorphisms of $\mathbb{T}^m$ satisfying condition $(\mathcal{R})$. Assume that μ is an ergodic invariant measure for α such that there are generic singular elements $a_1, \dots, a_k$ and a regular element $b \in \mathbb{R}^k$ such that*

C1: : $E_b^- = \sum_i (E_{a_i}^0 \cap E_b^-)$ *(where the sum need not be direct) and*
C2: : $\xi_{a_i} \leq \xi(E_{a_i}^0 \cap E_b^-)$.

Then either the invariant measure $\mu_{\mathbb{T}^m}$ for the $\mathbb{Z}^k$ action by automorphisms of $\mathbb{T}^m$ has zero entropy for all elements of the action or it decomposes as $\mu_{\mathbb{T}^m} = \frac{1}{N}(\mu_1 + ... + \mu_N)$, where measures μ_i, $i = 1, ..., N$, are invariant under $\alpha(\Gamma)$ for some finite index subgroup $\Gamma \subset \mathbb{Z}^k$. The actions $(\alpha(\Gamma), \mu_i)$ are algebraically isomorphic by the toral automorphisms from $\alpha(\mathbb{Z}^k)$. Each μ_i, $i = 1, ..., N$, is an extension of a zero entropy measure in an algebraic factor for $\alpha(\Gamma)$ of smaller dimension with Haar conditional measures in the fibers.

Condition **C2** (which of course depends on **C1** while the latter does not mean much by itself) is a generalization of the condition ($\mathcal{E}$) form Section 2.1. It will play a similar role in the proof. There are various situations where these conditions can be established. It is trivially satisfied if every one-parameter subgroup of the action α is ergodic, or, equivalently if the original $\mathbb{Z}^k$ action by toral automorphisms is weakly mixing.

COROLLARY 3.1. *Let α be a $\mathbb{Z}^k$-action with $k \geq 2$ by automorphisms of a torus satisfying condition ($\mathcal{R}$). Then any weakly mixing invariant measure for α is an extension of a zero entropy measure in an algebraic factor with Haar conditional measures in the fibers.*

The following corollary of Theorem 3.1 will be used in Section 4.

THEOREM 3.2. *Let α be an action of $\mathbb{Z}^2$ by ergodic toral automorphisms and let μ be an α–invariant weakly mixing measure such that for some $\mathbf{m} \in \mathbb{Z}^2$, $\alpha(\mathbf{m})$ is a K-automorphism. Then μ is a translate of Haar measure on an α–invariant rational subtorus.*

PROOF. By Theorem 3.1 the measure μ is an extension of a zero entropy measure for an algebraic factor of smaller dimension with Haar conditional measures in the fiber. Since α contains a K-automorphism it does not have nontrivial zero entropy factors. Hence the factor in question is the action on a single point and μ itself is a Haar measure on a rational subtorus. □

In the special case considered in Section 2.2 the ergodicity assumption ($\mathcal{E}$) for the critical direction was deduced from ergodicity. The deduction was based on the interlacing of stable and unstable foliations which allowed to produce (2.2.1). Now we will show how a similar argument allows to verify conditions of Theorem 3.1 in much greater generality.

THEOREM 3.3. *Any ergodic invariant measure μ for a totally nonsymplectic Anosov action of $\mathbb{Z}^k$ by automorphisms of a torus either has zero entropy for all elements of the action or decomposes as $\mu = \frac{1}{N}(\mu_1 + ... + \mu_N)$, where measures μ_i, $i = 1, ..., N$, are invariant under $\alpha(\Gamma)$ for some finite index subgroup $\Gamma \subset \mathbb{Z}^k$. The actions $(\alpha(\Gamma), \mu_i)$ are algebraically isomorphic by the toral automorphisms from $\alpha(\mathbb{Z}^k)$. Each μ_i, $i = 1, ..., N$, is an extension of a zero entropy measure in an algebraic factor for $\alpha(\Gamma)$ of smaller dimension with Haar conditional measures in the fibers.*

PROOF. First for any Anosov TNS action of $\mathbb{Z}^k$, $k \geq 2$, since for $k = 1$ any positive exponent is negatively proportional to any negative exponent. Thus we need to check conditions **C1, C2**. For that it is obviously sufficient to show that for any generic singular element $a \in \mathbb{R}^k$, $\xi_a \leq \xi(E_a^0)$.

For such an element a a single Lyapunov exponent vanishes and the corresponding Lyapunov distribution coincides with E_a^0. Take a regular element b nearby for which the corresponding Lyapunov exponent is positive and all other exponents have the same signs as for a. Thus $E_b^+ = E_a^+ \oplus E_a^0$ and $E_b^- = E_a^-$. Birkhoff averages with respect to $\alpha(a)$ of any continuous function are constant on the leaves of ξ_a. Since such averages generate the algebra of $\alpha(a)$ invariant functions we conclude that $\xi_a \leq \xi(E_a^-)$. On the other hand both and both $\xi(E_b^-)$ and $\xi(E_b^+)$ coincide with the Pinsker algebra $\pi(\alpha(b))$. Thus we conclude

$$\xi_a \leq \xi(E_a^-) = \xi(E_b^-) = \pi(\alpha(b)) = \xi(E_b^+) = \xi(E_a^+ \oplus E_a^0) \leq \xi(E_a^0).$$

□

COROLLARY 3.2. *Any ergodic invariant measure for an irreducible totally non-symplectic Anosov action of $\mathbb{Z}^k$ by automorphisms of a torus either has zero entropy for all elements of the action or is Lebesgue measure on the torus.*

3.2. Scheme of the Proof of Theorem 3.1. *Step 1.* Let us consider one of the generic singular elements a_i from the statement of Theorem 3.1 and a Lyapunov exponent λ such that $\lambda(a_i) = 0$. We denote by F the invariant foliation $W_{a_i}^I \cap W_b^- \cap W^\lambda$ where W^λ is the Lyapunov foliation corresponding to λ.

In Section 3.3 we use Lemmas 3.2, 3.3, 3.4 to establish the following dichotomy.

LEMMA 3.1. *For any foliation F described above either*

(1) *The conditional measure μ_x^F is atomic for μ-a.e. x, or*
(2) *The conditional measure μ_x^F is a Haar measure on an affine subspace S_x of positive dimension for μ-a.e. x.*

Step 2. If for all foliations F described in *Step 1* the first alternative of Lemma 3.1 takes place then we prove in Section 3.4 that the conditional measures on the foliation W_b^- are atomic. By Proposition 1.1 this implies that the entropy of $\alpha(b)$ is equal to 0. Since the stable foliation is the same for all elements in the same Weyl chamber we see that the entropy is 0 for all elements in the same Weyl chamber as b. Since the entropies of $\alpha(b)$ and $\alpha(-b)$ are the same we conclude the entropy is 0 for all elements in the Weyl chamber of b. Then it follows from sublinearity of entropy [**H**] that all elements of the action have 0 entropy.

This completes the proof of Theorem 3.1 in the case when the first alternative of Lemma 3.1 takes place for all foliations F described in Step 1.

Step 3. Suppose that for some foliation F described in *Step 1* the second alternative of Lemma 3.1 takes place, i.e. the conditional measure μ_x^F is a Haar measure on an affine subspace S_x for μ-a.e. x. We note that by ergodicity of the action the subspaces S_x have the same dimension for μ-a.e. x. We may assume that this dimension is positive. Then Lemma 3.11 shows that the measure on the torus decomposes as $\mu_{\mathbb{T}^m} = \frac{1}{N}(\mu_1 + ... + \mu_N)$, where measures μ_i, $i = 1, ..., N$, are invariant under $\alpha(\Gamma)$ for some finite index subgroup $\Gamma \subset \mathbb{Z}^k$. The actions $(\alpha(\Gamma), \mu_i)$ are algebraically isomorphic and each μ_i, $i = 1, ..., N$, is an extension of a zero entropy measure in an algebraic factor of smaller dimension for $\alpha(\Gamma)$ with Haar conditional measures in the fibers. We restrict the action α to the finite index subgroup Γ and consider invariant measure μ_1. If the factor–measure has zero entropy for all elements of this action we are done. Otherwise we may assume that in the factor some element still has positive entropy and repeat the argument. Note that condition $(\mathcal{R})$ is inherited by any algebraic factor as well as conditions **C1, C2**. We will arrive at a factor of the factor and so on. Since at every step the dimension of the factor drops this process has to stop, thus producing a factor with zero entropy. It is also clear by induction that the conditionals are in fact Haar measures on the fibers.

Notice that in the case of $\mathbb{Z}^2$ action on $\mathbb{T}^3$ considered in Section 2.2 Steps 2 and 3 are not needed.

3.3. Proof of the Lemma 3.1. In this section we prove the following three lemmas from [**KS3**] which establish the dichotomy of Lemma 3.1 from Section 3.2. Lemma 3.2 generalizes the argument from the Step 1 in Section 2.2.2, and Lemma 3.4 the argument from Step 2 in the same section.

The general setup for these lemmas is as follows. Suppose that $a \in \mathbb{R}^k$ is a generic singular element and $F \subset W^I_a$ is some $\alpha(a)$-invariant subfoliation of W^I_a with leaves of dimension d. Denote by $B^F_1(x)$ the closed unit ball in $F(x)$ about x with respect to the flat metric. Denote by μ^F_x the system of conditional measures on F normalized by the requirement $\mu_x(B^F_1(x)) = 1$ for all x in the support of μ and by G_x the subgroup of isometries of $F(x)$ which preserve μ^F_x up to a scalar multiple.

To simplify the notation from now on we will write a instead of $\alpha(a)$ when this does not cause any confusion.

Recall that we denote by ξ_a the partition into ergodic components of element a and by $\xi(F)$ the measurable hull of F.

LEMMA 3.2. *([**KS3**], Lemma 5.4) Suppose that $\xi_a \leq \xi(F)$. Then for μ-a.e. x, G_x is closed and the support of μ^F_x is the orbit of x under the group G_x . Furthermore, $\phi_* \mu^F_x = \mu^F_{\phi x}$ for any $\phi \in G_x$.*

PROOF. To prove the last statement of the lemma we note that the normalizations of the conditional measures $\phi_*\mu^F_x$ and μ^F_y coincide since $\phi_*\mu^F_x(B^F_1(y) = \mu^F_x(B^F_1(x)) = 1$ due to the fact that ϕ is an isometry.

Since G_x maps the support of μ^F_x to itself we only need to show that G_x is closed and acts transitively on the support of μ^F_x.

We first show that G_x is closed. Let $\{\phi_n\} \subset G_x$ be a sequence of isometries converging to an isometry ϕ. We need to show that $\phi \in G_x$. Let $y_n = \phi_n(x)$ and $y = \phi(x)$. We can choose a radius r such that the boundaries of balls $B^F_r(x)$ and $B^F_r(y)$ carry no conditional measure. Let us denote by $\mu^{F,r}_z$ the conditional measure normalized by $\mu^{F,r}_z(B^F_r(z)) = 1$. Since the balls $B^F_r(y_n)$ converge to $B^F_r(y)$ we conclude that $\mu^{F,r}_{y_n}$ converge to $\mu^{F,r}_y$ in weak* topology. Since $\phi_n \in G_x$ we see that $(\phi_n)_*\mu^{F,r}_x = \mu^{F,r}_{y_n}$ since their normalizations coincide. Since $(\phi)_*\mu^{F,r}_x$ is the weak* limit of $(\phi_n)_*\mu^{F,r}_x$ it follows that $(\phi)_*\mu^{F,r}_x = \mu^{F,r}_y$. Since $\mu^{F,r}_y$ is a scalar multiple of $\mu^{F,r}_x$, we conclude that ϕ belongs to G_x.

It remains to show that for μ-a.e. x and μ^F_x-a.e. $y \in F(x)$ there exists an isometry ϕ of $F(x)$ such that $y = \phi(x)$ and $\mu^F_y = \phi_*\mu^F_x$. Since G_x is closed and any set of full measure is dense in the support of μ^F_x, it would follow that G_x is, in fact, transitive on the support of μ^F_x.

Let X_1 be the set of all x such that the ergodic component E_x of a passing through x is well-defined and contains $F(x)$ (up to a set of μ^F_x-measure 0). By the assumption on a, the set X_1 has full μ-measure. Let μ_x be the induced measure on E_x. Recall that a acts isometrically on $F(x)$.

Let us fix $R > 0$ and introduce the following notation

$$\mu^F_{x,R} = \mu^F_x|_{B^F_R(x)}.$$

We can canonically identify each ball $B^F_R(x)$ with the standard ball B_R in $\mathbb{R}^{\dim F}$ of radius R. Thus we can consider the system of measures $\mu^F_{x,R}$ as a measurable function from M to the weak* compact set of Borel probability measures on B_R (compare with the proof of $(\mathcal{T})$, step 1 in Section 2.2). By Luzin's theorem, we can take an increasing sequence of closed sets K_i contained in the support of μ such that

(1) $\mu(K) = 1$, where $K = \bigcup_{i=1}^{\infty} K_i$

(2) $\mu^F_{x,R}$ depends continuously on $x \in K_i$ with respect to the weak* topology.

Set $X_2 = X_1 \cap K$. Since the transformation induced by a on $X_2 \cap E_x$ is ergodic, the transformation induced by a on $X_1 \cap K_i \cap E_x$ is also ergodic for any i. Hence the set X_3, which consists of points $x \in X_2$ whose orbit $\{a^n x\}_{n\in\mathbb{Z}}$ is dense in $X_1 \cap K_i \cap E_x$ for all i, has full μ-measure.

Let $x \in X_3$ and $y \in X_3 \cap F(x)$. Then $x, y \in X_1 \cap K_i \cap E_x$ for some i. Hence there exists a sequence $n_k \to \infty$ such that the points $y_k = a^{n_k} x \in X_1 \cap K_i \cap E_x$ converge to y. Since $\mu^F_{x,R}$ depends continuously on $x \in K_i$ with respect to the weak* topology the measures $\mu^F_{y_k,R}$ weak* converge to the measure $\mu^F_{y,R}$. Let us denote the isometry $a^{n_k}|_{F(x)}$ by ϕ_k. We note that $(\phi_k)_*\mu^F_{x,R} = \mu^F_{y_k,R}$ since they are conditional measures on the same leaf $F(y_k)$ and their normalizations coincide. Taking a subsequence if necessary, we may assume that ϕ_k converge uniformly on compact sets to some isometry ϕ of $F(x)$ with $\phi(x) = y$. Since $(\phi)_*\mu^F_{x,R}$ is the weak* limit of $(\phi_k)_*\mu^F_{x,R}$ it follows that $(\phi)_*\mu^F_{x,R} = \mu_{y,R}$.

We conclude that for any $R > 0$ there exists a set X_3 of full μ measure such that for any $x \in X_3$ and $y \in X_3 \cap F(x)$ there exists an isometry ϕ of $F(x)$ with the properties $\phi(x) = y$ and $\mu^F_{y,R} = \phi_*\,\mu^F_{x,R}$.

We can now take a sequence $R_i \to \infty$ and choose a set X of full μ-measure such that for any $x \in X$, $y \in X \cap F(x)$ and any i there exists an isometry ϕ_i of $F(x)$ with the property $\mu^F_{y,R_i} = (\phi_i)_*\,\mu^F_{x,R_i}$. Taking a converging subsequence we obtain that for any $x \in X$ and $y \in X \cap F(x)$ there exists an isometry ϕ of $F(x)$ with the properties $\phi(x) = y$ and $\mu^F_y = \phi_*\,\mu^F_x$. This completes the proof of the lemma since we may assume that the set X is chosen so that for $x \in X$ the set $X \cap F(x)$ has full μ^F_x-measure. □

LEMMA 3.3. *([**KS3**], Lemma 5.5) In addition to the assumptions in Lemma 3.2, let F be contained in the intersection of W^I_a with a Lyapunov subspace for a non-zero Lyapunov exponent λ. Then for μ-a.e. x, the support S_x of μ^F_x is an affine subspace of $F(x)$ whose dimension is constant for almost every x..*

PROOF. By Lemma 3.2, S_x is the orbit of a closed group of isometries. Therefore S_x is a submanifold, possibly disconnected. Note that the maximal principal curvature of S_x is constant along S_x. Let $\kappa(x)$ denote this constant.

Let b be any element such that $\lambda(b) < 0$. Note that b maps S_x to $S_{b\,x}$. Iterates of b exponentially contract the fibers of F. In particular, since the exponential contractions in all directions inside F are the same, any curve with positive principal curvature will be mapped to curves with exponentially increasing principal curvatures. Hence $\kappa(b^n x)$ goes to infinity for μ-a.e. x unless $\kappa(x) = 0$. This is impossible by Poincaré recurrence. Thus $\kappa(x) \equiv 0$, and hence the support of μ^F_x is a union of non-intersecting affine subspaces.

Let us now show that the support is connected. Suppose to the contrary that the support is a union $\cup A_i$ of at least two affine subspaces A_i. Let d_x denote the minimum of the distances from x to any A_i which does not contain x. Since the support is a closed subset, $d_x > 0$ for all x. Note that $d_{b^n x} \to 0$ as $n \to \infty$. This is again a contradiction to Poincaré recurrence. The dimension of S_x is α-invariant and hence is constant due to the ergodicity of the action α. □

LEMMA 3.4. *([**KS3**], Lemma 5.6) Under the assumptions of Lemma 3.3, μ_x^F is Haar measure on S_x.*

PROOF. By Lemma 3.2 the group G_x of isometries of $F(x)$ which map μ_x^F to its scalar multiple acts transitively on the support S_x of the conditional measure of μ_x^F. By Lemma 3.3, S_x is an affine space. For any x and $y \in S_x \subset F(x)$ let us define the scaling coefficient $c_x(y)$ by the equality $\phi\mu^F = c_x(y)\mu^F$, where μ^F is the conditional measure on $F(x) = F(y)$ and $\phi \in G_x$ is an isometry such that $\phi(x) = y$. In other words $c_x(y)$ can be calculated as

$$c_x(y) = \frac{\phi_*\mu^F(A)}{\mu^F(A)} = \frac{\mu^F(\phi^{-1}A)}{\mu^F(A)} = \frac{\mu^F(A)}{\mu^F(\phi A)}$$

for any set A of positive conditional measure. We note that since the conditional measures are defined up to a scalar multiple it is clear that the definition does not depend on a particular choice of μ^F. Since the image of the unit ball $\phi(B_1^F(x))$ is the same for all isometries $\phi \in G_x$ such that $\phi(x) = y$ we conclude that the definition does not depend on a particular choice of ϕ either.

Since we can take the test set A such that the conditional measure of the relative (to $F(x)$) boundary of ϕA is equal to zero, we conclude that that for a fixed x the coefficient $c_x(y)$ depends continuously on y.

We see that either for $\mu - a.e.\ x$ $c_x(y) = 1$ for all $y \in S_x$, hence μ_x^F is Haar on S_x, or there exists a set X of positive measure such that $c_x(y)$ is not identically equal to 1 for $x \in X$. In the latter case for some $\epsilon > 0$ we can define a finite positive measurable function

$$f_\epsilon(x) = \inf\{r : \exists\, y \in S_x\ s.t.\ d(x,y) < r\ and\ |c_x(y) - 1| > \epsilon\}$$

on some subset $Y \subset X$ of positive μ-measure. By measurability there exists N and a set Z of positive measure on which f_ϵ takes values in the interval $(1/N, N)$. We will show that

$$f_\epsilon(b^n x) \to 0 \text{ as } n \to \infty \tag{3.3.1}$$

uniformly on Z for an element b which contracts foliation F. This will prove the lemma since it contradicts to the recurrence of Z under b.

We will prove now that

$$c_x(y) = c_{bx}(by) \tag{3.3.2}$$

for μ-a.e. x and $y \in S_x$. Since the iterates of b exponentially contract the leaves of F this invariance property implies that $f_\epsilon(b^n x) \le C\lambda^n f_\epsilon(x)$, for some C,$\lambda > 0$, hence (3.3.2) implies (3.3.1)

To prove (3.3.2) we use Lemma 3.2. For any $y \in S_x$ there exists an isometry ϕ of $F(x)$ with the properties $\phi(x) = y$ and $\mu_y^F = (\phi)_*\mu_x^F$. Since such ϕ can be obtained as a limit of a sequence of some powers of a restricted to $F(x)$ we may assume that it commutes with b in the following sense. The map $\psi = b \circ \phi \circ b^{-1}$ is an isometry of $F(bx)$. Thus we have $b\circ\phi = \psi\circ b$. We note that since b preserves the family of conditional measures (up to a scalar multiple) ψ preserves the conditional measures on $F(bx)$ up to a scalar multiple. Hence for any set $A \subset F(x)$ of positive conditional measure we have

$$\begin{aligned} c_x(y) &= \frac{\mu^F(A)}{\mu^F(\phi A)} = \frac{(b_*\mu^F)(bA)}{(b_*\mu^F)(b \circ \phi A)} \\ &= \frac{(b_*\mu^F)(bA)}{(b_*\mu^F)(\psi \circ bA)} = \frac{(b_*\mu^F)(bA)}{(b_*\mu^F)(\psi(bA))} = c_{bx}(by). \end{aligned}$$

□

3.4. The Case of Atomic Conditional Measures. In this section we consider the case when for all foliations F, described in *Step 1* of the scheme of the proof, the first alternative of Lemma 3.1 takes place, i.e. the conditional measures are atomic. We use Lemmas 3.5, 3.6, and 3.8 to show that in this case the conditional measures on the foliation W_b^- are atomic. This by Proposition 1.1 implies that the entropy of $\alpha(b)$ is equal to 0 and allows us to proceed as in *Step 2* of the scheme of the proof.

REMARK 3.1. *This part of the argument is not needed in the case considered in Section 2.2 since the foliation F in that case coincides with the complete stable foliation of some regular element b and hence zero entropy follows right away.*

As in *Step 1* of the scheme of the proof of Theorem 3.1 we consider one of the generic singular elements a_i from the statement of Theorem 3.1 and a Lyapunov exponent λ such that $\lambda(a_i) = 0$. Let us denote by F the invariant foliation $W_{a_i}^I \cap W_b^- \cap W^\lambda$ where W^λ is the Lyapunov foliation corresponding to λ. We know that the conditional measure μ_x^F is atomic for μ-a.e. x.

First we use Lemma 3.5 to conclude that the conditional measures on the foliation $W_{a_i}^I \cap W_b^-$ are atomic for each a_i. We note that Lemma 3.2 implies that the conditional measures on the whole foliation $W_{a_i}^I \cap W_b^-$ are supported on smooth submanifold of the leaves.

Once we know that the conditional measures on foliations $W_{a_i}^I \cap W_b^-$ for all i are atomic Lemma 3.6 shows that the conditional measures on all foliations $W_{a_i}^0 \bigcap W_b^-$ are also atomic. Then to show that the conditional measures on the whole W_b^- are atomic we use the inductive process as in the end of the outline of the proof of Theorem 5.1 in [**KS3**].

We restrict the action to a 2-plane which contains b and intersects all the Lyapunov hyperplanes in generic lines. Then we can replace each a_i by an element c_i in the intersection of the 2-plane with the unique Lyapunov hyperplane that contains a_i. Since the elements a_i and c_i have the same center foliation for each i we see that

$$E_b^- = \sum_i (E_{a_i}^0 \cap E_b^-) = \sum_i (E_{c_i}^0 \cap E_b^-)$$

and the conditional measures on all $W_{c_i}^0 \cap W_b^-$ are atomic. We can now reorder the elements c_i in such a way that the Lyapunov exponent that correspond to $E_{c_j}^0 \cap E_b^-$ is negative on c_i for each $j < i$. Starting with $W_{c_1}^0 \cap W_b^-$ we apply Lemma 3.8 inductively to prove that the conditional measures are atomic on $(W_{c_1}^0 \oplus W_{c_2}^0) \cap W_b^-$, on $(W_{c_1}^0 \oplus W_{c_2}^0 \oplus W_{c_3}^0) \cap W_b^-$, and so on until we exhaust the whole W_b^-.

To complete the proof that the conditional measures on the foliation W_b^- are atomic we need to prove Lemmas 3.5, 3.6, and 3.8 below.

The proof of Lemma 3.5 follows the proof of Lemma 5.8 in [**KS3**]. Lemma 3.6 is identical to Lemma 5.9 in [**KS3**] and Lemma 3.8 is a slight modification of Lemma 5.10 from [**KS3**].

Notice that any α-invariant subfoliation of W_a^I splits into its intersections with Lyapunov foliations.

LEMMA 3.5. *Let F be an invariant subfoliation of $W_a^I \cap W_b^-$ and let $F = \sum_\lambda (F \cap W^\lambda)$ be the splitting into its intersections with the Lyapunov foliations. Assume that the conditional measures on all foliations $F \cap W^\lambda$ are atomic and the support S_x of measure μ_x^F is a smooth submanifold of $F(x)$. Then the conditional measures on F are also atomic.*

PROOF. The support S_x of measure μ_x^F is a smooth submanifold which intersects every $F \cap W^\lambda$ in at most one point. Let λ be the Lyapunov exponent smallest on b. Let D be the distribution of tangent spaces of S_x. It is measurable, and b-invariant and C^∞ on $F(x)$. Since D cannot intersect the component in the W_λ-direction in a subspace of positive dimension, D must be tangent to the sum $\sum_{\mu \neq \lambda} W_\mu$ by b-invariance. By taking the Lyapunov exponents inductively in increasing order, we see that D is trivial and μ_x^F are atomic. □

LEMMA 3.6. *Let $a \in \mathbb{R}^k$ be a generic singular element and $b \in \mathbb{R}^k$ be a regular element. If the conditional measures on the foliation $W_a^I \cap W_b^-$ are atomic, then the conditional measures on the foliation $W_a^0 \cap W_b^-$ are also atomic.*

PROOF. Let us introduce the following notations for the proof of the lemma:

$$E = E_a^0 \cap E_b^- \,, \qquad E^I = E_a^I \cap E_b^- .$$

E^I is the subspace of E spanned by the isometric directions of a, i.e. the eigendirections for real eigenvalues and invariant 2-planes for pairs of complex eigenvalues. Note that E and E^I are both a and b invariant. The corresponding foliations will be denoted by

$$F = W_a^0 \cap W_b^- \quad \text{and} \quad F^I = W_a^I \cap W_b^-$$

correspondingly.

We will prove the lemma by showing inductively that the conditional measures are atomic on the foliations that correspond to larger and larger subfoliations of F, until we exhaust the whole F.

The proof consists of two parts. One part is the basic inductive step which shows that if we add a certain one dimensional foliation to the previously constructed foliation than the conditional measures will be again atomic. This part is established in Lemma 3.7. The other part is the inductive process itself which we explain below. In this part we choose the one dimensional directions and the proper ordering of them to ensure our ability to make the inductive step. This choice is based on the relation between the algebraic properties of elements a and b as linear transformations of E.

The a and b invariant subspace E splits as the direct sum of its intersections with the root subspaces (the generalized eigenspaces) of a:

$$E = \bigoplus_{\lambda \in Sp(a)} Ker(a - \lambda)^m .$$

Each term of this direct sum is b invariant since b commutes with a. Hence it can be split into the direct sum of root subspaces for the restriction of b. Thus we obtain the splitting $E = \bigoplus U_i$ which is invariant under both elements a and b. Let us denote by λ_i (correspondingly ν_i) the eigenvalue of a (correspondingly b) on the subspace U_i. We have that $|\lambda_i| = 1$ and $|\nu_i| < 1$ for all i. Each U_i is filtered as

$$U_i \cap E^I = V_i^0 \subset ... \subset V_i^{d_i} = U_i \ , \text{ where } \quad V_i^k = Ker(a - \lambda_i)^{k+1} \cap U_i$$

and $(d_i + 1)$ is the maximal dimension of Jordan blocks of a on U_i. Note that since b commutes with a it leaves each subspace V_i^k invariant. So we see that each V_i^k is both a and b invariant. We assign the "weight" $|\nu_i|^{\frac{1}{j}}$ to each subspace V_i^j, we assign 0 weight to subspace V_i^0. Let us enumerate the weights in the increasing order: $0 = w_0 < w_1 < ... < w_{j_0}$. We denote by E_j the direct sum of all V_i^k whose weights equal to w_j. Now E filtrates as

$$E^I = E_0 \subset ... \subset E_{j_0} = E.$$

Even though the subspaces V_i^k that correspond to the complex eigenvalues are defined only over $\mathbb{C}$ all subspaces E_j are defined over $\mathbb{R}$. This follows from the structure of the Jordan normal form for a over reals. We note again that each E_j is a and b invariant.

We will show inductively that the conditional measures on the foliations corresponding to E_j are atomic.

Let us fix j and assume inductively that the conditional measures on the foliation corresponding to E_{j-1} are atomic. We will prove that the conditional measures on the foliation corresponding to E_j are also atomic by showing that the conditional measures are atomic on the foliations corresponding to larger and larger subspaces of E_j. The dimension of the subspace will be increased by one at each step until we exhaust the whole E_j.

When we would like to add one direction in some U_i for which λ_i or ν_i is not real, instead of elements a and b we will use their suitable powers a^{s_i} and b^{t_i} s.t. $\lambda_i^{s_i} = 1$ and $0 < \nu_i^{t_i} < 1$. By taking the square of the element, if necessary, we may assume that there exist real logarithms and hence real powers of a and b. The ergodicity of these elements is not needed since we only will be using the fact that they preserve the measure. Note that the splitting $E = \bigoplus U_i$ and filtrations $U_i \cap E^I = V_i^0 \subset ... \subset V_i^{d_i} = U_i$ are the same for all such powers. Hence the order of weights and filtration $E^I = E_0 \subset ... \subset E_{j_0} = E$ are also the same.

E_j splits into its intersections with U_i as $E_j = \bigoplus V_i^{k_i}$. Then the intersections of E_{j-1} with U_i are either $V_i^{k_i}$ or $V_i^{k_i-1}$. We will consider only the intersections of E_j with U_i of the second type since we do not need to add any directions along the other U_i's. The operator $(b^{t_i} - \nu_i^{t_i})$ is nilpotent on $V_i^{k_i}/V_i^{k_i-1}$ since it is nilpotent on U_i. Let $d_i^1, ..., d_{i,P_i}$ be the dimensions of the cyclic subspaces of $(b^{t_i} - \nu_i^{t_i})$ on $V_i^{k_i}/V_i^{k_i-1}$ and $e_{i,1}, ..., e_{i,P_i} \in V_i^{k_i}$ be some representatives of their generators. Then the vectors

$$e_{i,p}^l := (b^{t_i} - \nu_i^{t_i})^{d_i^p-(l+1)} e_{i,p} \quad for \quad p = 1, ..., P_i \quad and \quad l = 0, ..., d_i^p - 1$$

form a basis of $V_i^{k_i}$ relative to $V_i^{k_i-1}$. Finally we have

$$E_j = E_{j-1} \oplus \bigoplus_i \bigoplus_{p=1}^{P_i} \bigoplus_{l=0}^{d_i^p-1} < e_{i,p}^l >$$

We will add the directions $< e^l_{i,p} >$ to E_{j-1} one by one in the order of increasing of the value of $\frac{l}{k_i}$. The rest of the proof describes the procedure of adding.

Let $D =< e^{l_0}_{i_0,p_0} >$ be the new direction to be added at this step. We introduce the following notations for this step: $k_0 := k_{i_0}$, $e_l := e^l_{i_0,p_0}$ for $l = 0, ..., l_0$, $A := a^{s_{i_0}}$ and $B := b^{t_{i_0}}$. The eigenvalues of A and B on U_{i_0} are correspondingly 1 and ν, where $0 < \nu < 1$. The vectors $e^k_l := (A-1)^{k_0-k} e_l$ for $l = 0, ..., l_0$ and $k = 0, ..., k_0$ form a Jordan normal basis for A on some subspace $V \subset U_i$ which will be of particular interest for this step. Note that the subspaces U_{i_0} and V are invariant for A and B as subspaces over $\mathbb{R}$.

Let us denote by D_1 the A and B invariant subspace which is the sum of E_{j-1} and previously added directions. We denote the corresponding foliation by F_1. We assume inductively that the conditional measures on the foliation F_1 are atomic. It is easy to see that the subspace $D_2 = D_1 \oplus D$ is also defined over $\mathbb{R}$ and both A and B invariant. Let us denote the corresponding foliation by F_2.

To complete the proof Lemma 3.6 it remains to prove the following lemma which shows that the conditional measures on the foliation F_2 are atomic.

LEMMA 3.7. *If the conditional measures on the foliation F_1 are atomic then the conditional measures on the foliation F_2 are also atomic.*

PROOF. Let us consider a measurable partition, subordinate to F_2, which consists "mainly" of small "rectangles" of the same size with sides parallel to the basis directions and has the following property. The measure of the set Int_γ is at least 0.99 for some $\gamma > 0$, where Int_γ consists of points inside rectangles on the distance at least γ from the relative (to the leaf of F_2) boundary of the rectangle that contains the point.

By induction hypothesis there exists a set of full measure which intersects any fiber of F_1 in at most one point. Approximating this set from inside we can find a compact set K with the same property. We would like to consider only "good" part of the measure μ, so we introduce a new measure μ_X by $\mu_X(.) = \mu(. \cap X)$, where $X = K \cap Int_\gamma$ with $\mu(X) \geq 0.98$. Let us consider the system of the conditional measures of μ_X w.r.t. the measurable partition into the rectangles (the remaining elements have μ_X measure 0). These measures will be referred to as the conditional measures of rectangles. We will regard each rectangle as a direct product of its vertical (F_1) and horizontal (D) directions. We observe the following dichotomy:

(1) Either for every rectangle, in a set of positive measure, at least 1/3 of its conditional measure is concentrated on a single vertical leaf, hence at one point,
(2) Or there exists a lower bound $d > 0$ for the width of a vertical strip of a rectangle that can carry at least 1/3 of its conditional measure, for any rectangle in a set Y consisting of whole rectangles with $\mu_x(Y) > 0.97$

In the first case the existence of atoms for the conditional measures of rectangles implies the existence of atoms for the conditional measures on foliation F_2. Since this foliation is contracted by b the existence of atoms forces the conditional measures on F_2 be atomic. This can be seen as in the proof of Proposition 4.1 in [**KS3**]. If x is an atom of the conditional measure then there exists a small neighborhood U of x in the leaf $F_2(x)$ such that $\mu^{F_2}_x(U - \{x\}) < \varepsilon\, \mu^{F_2}_x(\{x\})$. Pushing $\mu^{F_2}_x$ backward and using Poincaré recurrence, we see that for a typical x, $\mu^{F_2}_x$ is concentrated at x.

In the latter case each rectangle in Y can be split into three vertical strips of width at least d so that both the left and the right ones have the conditional measure at least 1/3. We again may assume that the conditional measures do not have atoms since otherwise we could argue as in the first case.

The intersection of the compact set K with any rectangle is a graph of a continuous (not necessarily everywhere defined) function from the vertical direction to horizontal. Moreover the family of these functions is equicontinuous. Let us take some ϵ and find δ given by the equicontinuity. Combining contraction provided by B and shear provided by A we will find such an element $A^m B^n$ that the image of any rectangle from Y has the following properties:

(1) The image is δ-narrow in the horizontal direction
(2) The distance along the vertical direction between the images of the right and the left strips is at least ϵ
(3) The image is sufficiently small (its diameter less than γ) so that it can not intersect γ-interiors of two different rectangles simultaneously.

Under these conditions the images of the right and the left strips can not intersect $Z = X \cap Y$ simultaneously. But this implies that $\mu_X(A^m B^n(Z) \cap Z) \leq \frac{2}{3}\mu_X(Z)$ which is impossible since $\mu(Z) = \mu_X(Z) = \mu_X(Y) \geq 0.97$. To prove that the conditional measures on F_2 are atomic it remains to show that such an element $A^m B^n$ exists.

To satisfy the second condition we obtain the required shear along the direction of the eigenvector e_0^0. On subspace V element B provides uniform contraction by factor ν and possibly some shear. Since no shear accumulates in V along $D = < e_{l_0} >$ direction the size along D direction of the image under B^n of any rectangle is $\nu^n s$, where s is the size of the original rectangle. Since with respect to B vector e_{l_0} has height l_0 over e_0 the distance between the images under B^n of the right and left strips along e_0 direction is at least $c_0 n^{l_0} \nu^n d$ for some small $c_0 > 0$ and sufficiently large n. After this we apply A^m, where m will be exponential in n. We see that the size along D direction is still $\nu^n s$. Since vector e_0 has height k_0 (with respect to A) the leading term of the distance between the images of the right and left strips under $A^m B^n$ along the direction of the eigenvector e_0^0 is $c_1 m^{k_0} n^{l_0} \nu^n d$. Indeed, the other terms with the same power of m have smaller power of n while terms with smaller power of m are negligible since m will be exponential in n. We see that the other terms are bounded by $C_2 m^{k_0} n^{l_0 - 1} \nu^n s$, where C_2 is large (depending only on the structure of the subspace V for A and B), so the distance between the images of the right and left strips along e_0^0 direction is at least $c_3 m^{k_0} n^{l_0} \nu^n d$ for sufficiently large m and n and some small c_3 (depending only on k_0 and l_0). Similarly, the size of the image along all V directions can be estimated from above by $C_4\, m^{k_0} n^{l_0} \nu^n s$.

We would like to show that we can control the size of the image along other directions in E_1 as effectively as along the V directions. For each i the intersection $D_1 \cap U_i$ splits further into A and B invariant real subspaces that can be constructed similarly to V starting from some $e_{i,p}^l \in D_1$ such that $e_{i,p}^{l+1} \notin D_1$.

Let us consider any one of these A and B invariant real subspaces and denote it by $V' \subset D_1 \cap U_i$. As above, we can estimate the size of the image along V' by $C_5\, m^{k_i} n_*^l \nu_*^n s$, where ν_* is the corresponding eigenvalue of b.

We will now specify the choice of the element $A^m B^n$. First we note that the size of the image of any rectangle along D direction is $\nu^n s$ independently on m.

Hence the condition (1) will be automatically satisfied once n is chosen sufficiently large.

We observe that the ratio of the desired shear and the size along V directions is bounded. To make the shear and the size along V directions bounded for large m and n we take m and n so that $m \sim (n^{l_0}\nu^n)^{-\frac{1}{k_0}}$. In this case we will obtain the following estimate of the size of image along the subspace V':

$$C_6(\nu_*^{\frac{1}{k_i}}/\nu^{\frac{1}{k_0}})^{k_i n} n^{k_0(\frac{l}{k_i}-\frac{l_0}{k_0})} s.$$

We observe that this estimate is also bounded since $\nu_*^{\frac{1}{k_i}} = \nu^{\frac{1}{k_0}}$ and $\frac{l}{k_i} < \frac{l_0}{k_0}$ due to the the ordering of weights and the order of adding directions inside E_j. Note that if V' is constructed starting from some basis vector in E_{j-1} then $\nu_*^{\frac{1}{k_i}} < \nu^{\frac{1}{k_0}}$ due to the ordering of weights and the estimate will be in fact exponentially small.

We conclude that by taking m and n large and so that $m \sim (n^{l_0}\nu^n)^{-\frac{1}{k_0}}$ we can satisfy the condition (1) for any δ while producing the shear bounded below and the size of the images of rectangles bounded above. Then given the desired bound γ on the size of the image we can apply a bounded number of iterates of b to adjust the above estimate for the size to be γ. After that we choose ϵ smaller than the adjusted lower bound for the shear. For this ϵ there exists some δ by the equicontinuity. The above considerations show that we can now take n sufficiently large and $m \sim (n^{l_0}\nu^n)^{-\frac{1}{k_0}}$ to satisfy all three conditions. □

This finishes the proof of Lemma 3.6. □

LEMMA 3.8. *Let W be an invariant subfoliation of W_a^0 and F be an invariant subfoliation of W_a^- for some element a. Suppose that $F \oplus W \subset W_b^-$ for some element b and that the conditional measures of μ on both foliations F and W are atomic. Then the conditional measures on the foliation $F \oplus W$ are also atomic.*

PROOF. We will reduce the lemma to its special case, Lemma 3.9. the reduction goes as follows. Denote $W^I = W \cap W_a^I$. Then Lemma 3.9 shows that the conditional measures on the foliation $F \oplus W^I$ are also atomic.

After that we add root directions of W as in Lemma 3.6 and prove that the conditional measures on $F \oplus W$ are also atomic. The only difference is that in the basic step, Lemma 3.7, we need to control the size of the images of rectangles not only along W^I and previously added root directions but also along F directions. But such control is trivial since by the conditions of the lemma F is contracted by both b and a.

LEMMA 3.9. *Let W be an invariant subfoliation of W_a^I and F be an invariant subfoliation of W_a^- for some element a. Suppose that $F \oplus W \subset W_b^-$ for some element b and that the conditional measures of μ on both foliations F and W are atomic. Then the conditional measures on the foliation $F \oplus W$ are also atomic.*

PROOF. Similar to Lemma 3.6 we prove this lemma inductively by adding one dimensional subfoliations of W to the foliation F one until we exhaust the whole $F \oplus W$. Lemma 3.10 shows that on each step of this process we obtain a foliation with atomic conditional measures.

Let us consider the Jordan normal form of b on the subspace that corresponds to foliation W. This subspace splits over $\mathbb{C}$ into the sum $\bigoplus U_i$ of root subspaces of

b. Each U_i is an eigenspace of a corresponding to some eigenvalue λ_i with $|\lambda_i| = 1$. We denote by ν_i the corresponding eigenvalue of b.

In order to add a one dimensional direction inside some specific U_i for which λ_i or ν_i is complex we replace elements a and b by their proper powers a^{s_i} and b^{t_i} to make the corresponding eigenvalues real: $\lambda_i^{s_i} = 1$ and $0 < \nu_i^{t_i} < 1$. This allows us to consider U_i as a subspace over $\mathbb{R}$.

Unlike Lemma 3.6, for this proof we do not need to follow any particular order of adding. We need only to make sure that when we add a new direction in some U_i we again obtain a subspace which is both a and b invariant and defined over $\mathbb{R}$. The desired partial ordering of directions can be obtained by fixing some Jordan normal basis for the real operator b^{t_i} on U_i.

The next lemma proves the inductive step and completes the proof of Lemma 3.9.

LEMMA 3.10. *Let F_1 and F_2 be invariant subfoliation of W_a^I and F be an invariant subfoliation of W_a^- for some element a. Suppose that F_2 is one-dimensional, $(F \oplus F_1 \oplus F_2) \subset W_b^-$ for some element b, and that the conditional measures of μ on both foliations $F_1 \oplus F_2$ and $F \oplus F_1$ are atomic. Then the conditional measures on the foliation $F \oplus F_1 \oplus F_2$ are also atomic.*

PROOF. Since the conditional measures on the foliations $F_1 \oplus F_2$ and $F \oplus F_1$ are atomic we can find a set of full measure and its compact subset K of measure at least 0.99 which intersect any leaf of the foliations $F_1 \oplus F_2$ and $F \oplus F_1$ in at most one point.

Consider a measurable partition, subordinate to $F \oplus F_1 \oplus F_2$, which consists "mainly" of small "rectangles" of the same size with sides parallel to the foliations F, F_1 and F_2 and has the following property: the measure of the set Int_γ is at least 0.99 for some $\gamma > 0$, where Int_γ consists of points inside rectangles on the distance at least γ from the relative (to the leaf of $F \oplus F_1 \oplus F_2$) boundary of the rectangle that contains the point.

We would like to consider only "good" part of the measure μ, so we introduce a new measure μ_X by $\mu_X(.) = \mu(. \cap X)$, where $X = K \cap Int_\gamma$ with $\mu(X) \geq 0.98$. Let us consider the system of the conditional measures of μ_X w.r.t. the measurable partition into the rectangles (the remaining elements have μ_X measure 0). These measures will be referred to as the conditional measures of rectangles. We regard each rectangle as a direct product of its F, F_1 and F_2 directions. We observe the following dichotomy:

(1) Either for every rectangle, in a set of positive measure, at least 1/3 of its conditional measure is concentrated on a single $F \oplus F_1$ leaf, hence at one point,
(2) Or there exists a set Y with $\mu_X(Y) > 0.97$ which consists of whole rectangles and a number $d > 0$ such that for any rectangle in Y any subrectangle that carries at least 1/3 of the conditional measure has width at least d along F_2-direction.

In the first case the existence of atoms for the conditional measures of rectangles implies the existence of atoms for the conditional measures on foliation $F \oplus F_1 \oplus F_2$. As in the proof of Lemma 3.7 we see that since this foliation is contracted by b the existence of atoms forces the conditional measures on $F \oplus F_1 \oplus F_2$ be atomic.

In the latter case each rectangle in Y can be split into three subrectangles of width at least d along one-dimensional F_2 direction so that both the left and the right ones have the conditional measure at least $1/3$. We again may assume that the conditional measures do not have atoms since otherwise we could argue as in the first case.

Now we regard the intersection of the compact set K with any rectangle as a graph of a continuous (not necessarily everywhere defined) function from F-direction to $F_1 \oplus F_2$-direction. The family of these functions is equicontinuous. We would like to show that this equicontinuity contradicts to the recurrence under the action of a properly chosen element.

Since b contracts $F \oplus F_1 \oplus F_2$ we can find sufficiently large number n such that the size of the image of any rectangle under the action of b^n is γ-small. The distance along the F_2 direction between the images of the right and the left subrectangles, which was at least d, becomes at least d'. Let us fix some $\epsilon < d'$ and consider $\delta > 0$ given by the equicontinuity. Since a acts isometrically on $F_1 \oplus F_2$ and contracts F we can choose k so large that the image of any rectangle under $a^k b^n$ is δ-small in F direction. We see that the element $a^k b^n$ satisfies the following conditions:

(1) The image of any rectangle is δ-narrow in the F direction
(2) The distance along the F_2 direction between the images of the right and the left subrectangles is at lest ϵ
(3) The diameter of the image of any rectangle is less than γ, hence the image cannot intersect γ-interiors of two different rectangles simultaneously.

Under these conditions the images of the right and the left subrectangles can not intersect $Z = X \cap Y$ simultaneously. But this implies that $\mu_X(a^k b^n(Z) \cap Z) \leq \frac{2}{3}\mu_X(Z)$ which is impossible since $\mu(Z) = \mu_X(Z)\mu_X(Y) \geq 0.97$. This shows that the second alternative of the dichotomy is impossible and proves that the conditional measures on $F \oplus F_1 \oplus F_2$ are atomic. □

Thus Lemma 3.8 is proved. □

This finished the proof of Theorem 3.1 in the case of atomic conditional measures.

3.5. The case of Haar conditional measures.

LEMMA 3.11. *Let F be an invariant foliation. Suppose that for μ – a.e. x the conditional measure μ_x^F is a Haar measure on an affine subspace S_x with* $\dim S_x = l \geq 1$. *Then there exist a finite index subgroup Γ of $\mathbb{Z}^k$, rational subtori $T_i \subset \mathbb{T}^m$, $i = 1, ..., N$, of the same dimension, and Borel probability measures μ_i, $i = 1, ..., N$, on $\mathbb{T}^m$ such that:*

(1) *For μ – a.e. x the closure of S_x is a translation of T_i for some i,*
(2) *The original measure decomposes as $\mu_{\mathbb{T}^m} = \frac{1}{N}(\mu_1 + ... + \mu_N)$,*
(3) *μ_i, $i = 1, ..., N$ is invariant under the group of translations in the direction of T_i,*
(4) *T_i and μ_i, $i = 1, ..., N$, are invariant under $\alpha(\Gamma)$.*
(5) *All actions $(\alpha(\Gamma), \mu_i)$ are algebraically isomorphic by the toral automorphisms from $\alpha(\mathbb{Z}^k/\Gamma)$.*

PROOF. Let us denote the closure of S_x by $T(x)$. $T(x)$ is a rational subtorus which corresponds to the minimal rational subspace that contains S_x. Hence

$\dim T(x)$ is an invariant function. Since the original measure is ergodic we conclude that $\dim T(x)$ is constant μ-a.e. Let us call two points x and y equivalent if $T(x) = T(y)$. This equivalence relation gives rise to a measurable partition of $\mathbb{T}^m$ into rational subtori.

We first show that the conditional measures on these tori are Lebesgue. Indeed, let us consider a typical torus T with the conditional measure μ_T. The torus T is foliated by affine subspaces S_x in such a way that the conditional measures of μ_T on S_x are Haar for μ_T-a.e. x. One can assign an orthonormal basis to each subspace S_x in a measurable way. This produces a measurable $\mathbb{R}^l$ action on T which preserves μ_T. Let us take the ergodic decomposition of μ_T with respect to this $\mathbb{R}^l$ action and consider one ergodic component. The direction of S_x is invariant under the action since the subspace S_x is the same for all points on one trajectory. Hence almost all subspaces S_x within one ergodic component have the same direction, i.e. are parallel. It follows that the measure on this ergodic component is invariant under translations in this direction. By the construction of T the closure of S_x equals T for all x. This implies that the measure on any ergodic component is a Haar measure on T. Hence μ_T is the Lebesgue measure on T.

Since there can be at most countably many classes of parallel tori we conclude that there exists a torus T_1 such that the set E_1 consisting of all points x for which $T(x)$ is parallel to T_1 has positive measure. By recurrence, for any generator A_j of the action α there exists $n > 0$ such that $\mu_{\mathbb{T}^m}(E_1 \cap A_j^n(E_1)) > 0$. This means that for any point x in this intersection the element A_j^n maps $T(x)$ to a parallel torus. Since A_j^n is an affine map it follows that A_j^n preserves this class of parallel tori. Hence $A_j^n(E_1) = E_1$. In the same way it follows that the set E_1 is invariant with respect to the action of a finite index subgroup $\Gamma \subset \mathbb{Z}^k$. The orbit of the set E_1 consists of finitely many sets E_1, ..., E_N of equal measure which correspond to the elements of $\mathbb{Z}^k/\Gamma$. By ergodicity of the original measure the union of these sets has full measure. Hence $\mu_{\mathbb{T}^m} = \frac{1}{N}(\mu_1 + ... + \mu_N)$, where $\mu_i = \mu_{\mathbb{T}^m}|_{E_i}$. We note that $T(x)$ is parallel to the torus $T_i = a(T_1)$ for μ_i-a.e. x, where a is an element of the action α which maps E_1 to E_i. We conclude that T_i and μ_i are invariant under $\alpha(\Gamma)$ and that μ_i is invariant under the group of translations in the direction of T_i. □

This finishes the proof of Theorem 3.1 □

4. Measure–theoretic rigidity of conjugacies, centralizers, and factors

In this section we apply results about rigidity of invariant measures for actions by toral automorphisms to study fine measurable structure of such actions with respect to *Lebesgue* measure. The conclusions are rather striking: in total contrast to the rank one case (where the automorphisms are Bernoulli and hence the entropy is the only invariant of measurable isomorphism) in the higher rank case fine algebraic information is coded into measurable structure of the action. This section is taken almost *verbatim* from [**KKS**].

4.1. Conjugacies. Suppose α and α' are measurable actions of the same group G by measure–preserving transformations of the spaces (X, μ) and (Y, ν), respectively. If $H : (X, \mu) \to (Y, \nu)$ is a metric isomorphism (conjugacy) between the actions then the lift of the measure μ onto the graph $H \subset X \times Y$ coincides with the lift of ν to graph H^{-1}. The resulting measure η is a very special case of a *joining* of α and α': it is invariant under the diagonal (product) action $\alpha \times \alpha'$ and its projections to X and Y coincide with μ and ν, respectively. Obviously the projections establish metric isomorphism of the action $\alpha \times \alpha'$ on $(X \times Y, \eta)$ with α on (X, μ) and α' on (Y, ν) correspondingly.

Similarly, if an automorphism $H : (X, \mu) \to (X, \mu)$ commutes with the action α, the lift of μ to graph $H \subset X \times X$ is a self-joining of α, i.e. it is $\alpha \times \alpha$–invariant and both of its projections coincide with μ. Thus an information about invariant measures of the products of different actions as well as the product of an action with itself may give an information about isomorphisms and centralizers.

The use of this joining construction in order to deduce rigidity of isomorphisms and centralizers from properties of invariant measures of the product was first suggested in this context by J.-P. Thouvenot.

In both cases the ergodic properties of the joining would be known because of the isomorphism with the original actions. Very similar considerations apply to the actions of semi–groups by noninvertible measure–preserving transformations. We will use Theorem 3.2

Conclusion of Theorem 3.2 obviously holds for any action of $\mathbb{Z}^k$, $d \geq 2$ which contains a subgroup $\mathbb{Z}^2$ satisfying assumptions of Theorem 3.2. Thus we can deduce the following result which is central for our constructions.

THEOREM 4.1. *Let α and α' be two actions of $\mathbb{Z}^k$ by automorphisms of $\mathbb{T}^n$ and $\mathbb{T}^{n'}$ correspondingly and assume that α satisfies condition* $(\mathcal{R})$. *Suppose that $H : \mathbb{T}^n \to \mathbb{T}^{n'}$ is a measure–preserving isomorphism between (α, λ) and (α', λ), where λ is Haar measure. Then $n = n'$ and H coincides* (mod 0) *with an affine automorphism on the torus $\mathbb{T}^n$, and hence α and α' are algebraically isomorphic.*

PROOF. First of all, condition $(\mathcal{R})$ is invariant under metric isomorphism, hence α' also satisfies this condition. But ergodicity with respect to Haar measure can also be expressed in terms of the eigenvalues; hence $\alpha \times \alpha'$ also satisfies $(\mathcal{R})$. Now consider the joining measure η on graph $H \subset \mathbb{T}^{n+n'}$. The conditions of Theorem 3.2 are satisfied for the invariant measure η of the action $\alpha \times \alpha'$. Thus η is a translate of Haar measure on a rational $\alpha \times \alpha'$–invariant subtorus $\mathbb{T}' \subset \mathbb{T}^{n+n'} = \mathbb{T}^n \times \mathbb{T}^{n'}$. On the other hand we know that projections of $\mathbb{T}'$ to both $\mathbb{T}^n$ and $\mathbb{T}^{n'}$ preserve Haar measure and are one–to–one. The partitions of $\mathbb{T}'$ into pre–images of points for each of the projections are measurable partitions and Haar measures on elements are conditional measures. This implies that both projections are onto, both partitions are partitions into points, and hence $n = n'$ and $\mathbb{T}' = \text{graph}\, I$, where $I : \mathbb{T}^n \to \mathbb{T}^n$ is an affine automorphism which has to coincide (mod 0) with the measure–preserving isomorphism H. □

We will call two actions α and α' by automorphisms of a $\mathbb{T}^m$ measurably (algebraically) isomorphic up to a time change if for some $C \in GL(m, \mathbb{Z})$ $\alpha \circ C$ is measurably (algebraically) isomorphic to α'. Since a time change is in a sense a trivial modification of an action one is primarily interested in distinguishing actions up to a time change. The corresponding rigidity criterion follows immediately from Theorem 4.1.

COROLLARY 4.1. *Let α and α' be two actions of $\mathbb{Z}^k$ by automorphisms of $\mathbb{T}^n$ and $\mathbb{T}^{n'}$, respectively, and assume that α satisfies condition* $(\mathcal{R})$. *If α and α' are measurably isomorphic up to a time change then they are algebraically isomorphic up to a time change.*

4.2. Centralizers. Applying Theorem 4.1 to the case $\alpha = \alpha'$ we immediately obtain rigidity of the centralizers.

COROLLARY 4.2. *Let α be an action of $\mathbb{Z}^k$ by automorphisms of $\mathbb{T}^n$ satisfying condition $(\mathcal{R})$. Any invertible Lebesgue measure–preserving transformation commuting with α coincides* (mod 0) *with an affine automorphism of $\mathbb{T}^n$.*

Any affine transformation commuting with α preserves the finite set of fixed points of the action. Hence the centralizer of α in affine automorphisms has a finite index subgroups which consist of automorphisms and which corresponds to the centralizer of $\rho_\alpha(\mathbb{Z}^d)$ in $GL(n,\mathbb{Z})$.

Thus, in contrast with the case of a single automorphism, the centralizer of such an action α is not more than countable, and can be identified with a finite extension of a certain subgroup of $GL(n,\mathbb{Z})$. As an immediate consequence we obtain the following result.

PROPOSITION 4.1. *For any d and k, $2 \le d \le k$, there exists a $\mathbb{Z}^k$–action by hyperbolic toral automorphisms such that its centralizer in the group of Lebesgue measure–preserving transformations is isomorphic to $\{\pm 1\} \times \mathbb{Z}^k$.*

PROOF. Consider a hyperbolic matrix $A \in SL(k+1,\mathbb{Z})$ with irreducible characteristic polynomial and real eigenvalues such that the origin is the only fixed point of F_A. Consider a subgroup of $Z(A)$ isomorphic to $\mathbb{Z}^k$ and containing A as one of its generators. This subgroup determines an embedding $\rho : \mathbb{Z}^d \to SL(k+1,\mathbb{Z})$. It is not difficult to see that all matrices in $\rho(\mathbb{Z}^k)$ are hyperbolic and hence ergodic, condition $(\mathcal{R})$ is satisfied. Hence by Corollary 4.2, the measure–theoretic centralizer of the action α_ρ coincides with its algebraic centralizer, which, in turn, and obviously, coincides with centralizer of the single automorphism F_A isomorphic to $\{\pm 1\} \times \mathbb{Z}^k$. □

4.3. Factors, noninvertible centralizers and weak isomorphism. A small modification of the proof of Theorem 4.1 produces a result about rigidity of factors.

THEOREM 4.2. *Let α and α' be two actions of $\mathbb{Z}^k$ by automorphisms of $\mathbb{T}^n$ and $\mathbb{T}^{n'}$ respectively, and assume that α satisfies condition $(\mathcal{R})$. Suppose that $H : \mathbb{T}^n \to \mathbb{T}^{n'}$ is a Lebesgue measure–preserving transformation such that $H \circ \alpha = \alpha' \circ H$. Then α' also satisfies $(\mathcal{R})$ and H coincides* (mod 0) *with an epimorphism $h : \mathbb{T}^n \to \mathbb{T}^{n'}$ followed by translation. In particular, α' is an algebraic factor of α.*

PROOF. Since α' is a measurable factor of α, every element which is ergodic for α is also ergodic for α'. Hence α' also satisfies condition $(\mathcal{R})$. As before consider the product action $\alpha \times \alpha'$ which now by the same argument also satisfies $(\mathcal{R})$. Take the $\alpha \times \alpha'$ invariant measure $\eta = (\mathrm{Id} \times H)_*\lambda$ on graph H. This measure provides a joining of α and α'. Since $(\alpha \times \alpha', (\mathrm{Id} \times H)_*\lambda)$ is isomorphic to (α, λ) the conditions of Corollary 3.2 are satisfied and η is a translate of Haar measure on an invariant rational subtorus $\mathbb{T}'$. Since $\mathbb{T}'$ projects to the first coordinate one-to-one we deduce that H is an algebraic epimorphism (mod 0) followed by a translation. □

Similarly to the previous subsection the application of Theorem 4.2 to the case $\alpha = \alpha'$ gives a description of the centralizer of α in the group of all measure–preserving transformations.

COROLLARY 4.3. *Let α be an action of $\mathbb{Z}^k$ by automorphisms of $\mathbb{T}^n$ satisfying condition ($\mathcal{R}$). Any Lebesgue measure–preserving transformation commuting with α coincides* (mod 0) *with an affine map on $\mathbb{T}^n$.*

THEOREM 4.3. *Let α be an action of $\mathbb{Z}^d$ by automorphisms of $\mathbb{T}^n$ satisfying condition ($\mathcal{R}$) and α' another $\mathbb{Z}^d$-action by toral automorphisms. Then (α, λ) is weakly isomorphic to (α', λ') if and only if ρ_α and $\rho_{\alpha'}$ are isomorphic over $\mathbb{Q}$, i.e. if α and α' are finite algebraic factors of each other.*

PROOF. By Theorem 4.2, α and α' are algebraic factors of each other. This implies that α' acts on the torus of the same dimension n and hence both algebraic factor–maps have finite fibers. But this is equivalent to isomorphism over $\mathbb{Q}$ (See [**KKS**], Proposition 2.1). □

Part II. NONUNIFORMLY HYPERBOLIC ACTIONS

The second type of hyperbolicity is *non–uniform*. A discussion of this concept in full generality will appear in [**BKP**, Chapter 1]. A particular, and most important case of non–uniform hyperbolicity appears in the presence of a *hyperbolic measure*, i.e. an invariant measure with non–vanishing Lyapunov exponents (Sections 5 and 7.1).

In the rank one case this is the basis of a very fruitful structural theory at the root of which lie modified Bernoulli and Markov paradigms ([**P, KM, BP, BKP**]). Not surprisingly, the analogy with the uniform case extends to the actions of higher–rank groups and hyperbolic measures display certain rigidity properties not dissimilar to those for Anosov actions. In Sections 5, 6, and 7 we describe some of the principal technical tools and certain results concerning the nonuniformly hyperbolic situation. Some of these tools can be naturally defined in greater generality than the case of actions by diffeomorphisms of compact manifolds which is of principal interests for us and we develop appropriate contexts accordingly. In the last section we indicate how certain rigidity results appear in this context. The two most important ingredients of this approach is the concept of Lyapunov Hölder structure which is the proper framework for the cocycle rigidity results and the generalization of approach from Section 2.2 based on the consideration of conditional measures on contracting foliations. Detailed presentation of these results will appear in a separate paper.

5. Linear extensions of $\mathbb{Z}^k$ actions

We begin with a general treatment of Lyapunov characteristic exponents which can be defined for the linear extensions of actions by measure–preserving transformations of a finite measure space. To simplify notations a bit we will restrict the discussion to the discrete time case ($\mathbb{Z}^k$ actions); the results for the $\mathbb{R}^k$ case can actually be deduced from those for $\mathbb{Z}^k$ by taking the $\mathbb{Z}^k$ subgroup in an $\mathbb{R}^k$ action and showing that its Lyapunov decomposition is if fact $\mathbb{R}^k$ invariant.

5.1. Lyapunov exponents and Lyapunov metric. Let (X, μ) be a probability Lebesgue space $\alpha : \mathbb{Z}^k \times X \to X$ an ergodic action of the group $\mathbb{Z}^k$ by measure–preserving transformations of the space (X, μ), $A : \mathbb{Z}^k \times X \times \mathbb{R}^m \to X \times \mathbb{R}^m$

a *linear extension* of the action α. Such an extension is determined by a *matrix-valued cocycle* $\mathcal{A} : \mathbb{Z}^k \times X \to GL(m, \mathbb{R})$ as follows

$$A(n, x, t) = (\alpha(n, x), \mathcal{A}(n, x)(t)),$$

where by the group property

$$\mathcal{A}(n_1 + n_2, x) = \mathcal{A}(n_2, \alpha(n_1, x))\mathcal{A}(n_1, x). \tag{5.1.1}$$

THEOREM 5.1 (Multiplicative Ergodic Theorem for $\mathbb{Z}^k$ actions). *Suppose for each $n \in \mathbb{Z}^k$,*

$$\log \|\mathcal{A}(n, x)\| \in L^1(X, \mu). \tag{5.1.2}$$

Then there exist linear functionals $\chi_1, \ldots, \chi_l$ on $\mathbb{R}^k$ and for μ-a.e. point $x \in X$ a decomposition of the fiber $\mathbb{R}^m_x$ over x:

$$\mathbb{R}^m_x = E_{\chi_1}(x) \oplus \cdots \oplus E_{\chi_l}(x) \tag{5.1.3}$$

such that for $i = 1, \ldots, l$ and for any $v \in E_{\chi_i}(x)$ one has

$$\lim_{n\to\infty} \frac{\log \|\mathcal{A}(n, x)t)\| - \chi_i(n)}{\|n\|} = 0. \tag{5.1.4}$$

and

$$\lim_{n\to\infty} \frac{\log \det \mathcal{A}(n, x) - \sum_{i=1}^{l} m_i \chi_i(n)}{\|n\|} = 0, \tag{5.1.5}$$

where $m_i = \dim E_{\chi_i}(x)$. Moreover, the set of points where this decomposition is defined is α-invariant and decomposition itself is A-invariant.

Definitions. The functionals $\chi_1, \ldots, \chi_l$ are called the *Lyapunov characteristic exponents* of A. The dimension m_i of the space $E_{\chi_i}(x)$ is called the *multiplicity* of the exponent χ_i. The decomposition (5.1.3) is called the *(fine) Lyapunov decomposition* at the point x. Points where the assertions of the theorem are satisfied will be called *regular*. The set of all regular points will be usually denoted by the letter Λ, sometimes with extra indices. A basis $t_1, \ldots, t_m$ of $\mathbb{R}^m_x$ for a regular point x is called *regular* if it agrees with the Lyapunov decomposition (5.1.3) i.e. can be split into bases of the spaces $E_{\chi_1}(x), \ldots, E_{\chi_l}(x)$.

Remarks.

1. Due to the cocycle equation (5.1.1) it is sufficient to require the integrability condition (5.1.2) only for a subset of elements which generate $\mathbb{Z}^k$ as a semigroup, e.g. for the standard generators and their opposites.
2. For any $n \in \mathbb{Z}^k$ the values $\chi_i(n), i = 1, \ldots, l$ are equal to the Lyapunov characteristic exponent of the extension $A(n)$ of the map $\alpha(n)$ in the usual sense [**KM**],Section S2. Even though the map $\alpha(n)$ may not be ergodic with respect to μ its Lyapunov exponents are invariant with respect to the action of the whole group $\mathbb{Z}^k$ and hence constant almost everywhere. Naturally, for a particular n the values of $\chi_i(n)$ may coincide for some i so the map $F(n)$ may have fewer than l distinct exponents with higher multiplicities.
3. One may think of the group $\mathbb{Z}^k$ as embedded into $\mathbb{R}^k$ as the standard integer lattice. The values $\chi_i(t)$ for $t \in \mathbb{R}^k \setminus \mathbb{Z}^k$ do not make sense in the context of the action α. However they are easily interpreted as the Lyapunov characteristic exponents of the elements of the *suspension* action.

Theorem 5.1 can be easily deduced from the standard Oseledets Multiplicative ergodic theorem for linear extensions of a single measure–preserving transformation [**O**],[**BKP**] by a simple induction process. Namely, one first applies Oseledets theorem to the first generator of the $\mathbb{Z}^k$ action. The Lyapunov decomposition for it is invariant under the whole action so one can apply Oseledets theorem to the restriction of the extension of the second generator to each element of the Lyapunov decomposition and so on. Once one obtained a decomposition for which limits exist for multiples of all generators, existence of the limits (5.1.4), (5.1.5) and the linearity of exponents follow easily from the cocycle relation (5.1.1).

Denote the standard scalar product in $\mathbb{R}^m$ by $< \cdot, \cdot >$. Given $\epsilon > 0$ and a regular point x we will call *the standard ϵ–Lyapunov scalar product (or metric)* the scalar product $< \cdot, \cdot >_{x,\epsilon}$ in $\mathbb{R}^m_x$ defined as follows:

For $u, v \in E_{\chi_i}(x),\ i = 1, \dots, l$:

$$< u, v >_{x,\epsilon} = \sum_{n \in \mathbb{Z}^k} < \mathcal{A}(n,x)u, \mathcal{A}(n,x)v > \exp(-2\chi_i(n) - \epsilon\|n\|),$$

where the series converges due to (5.1.4).

For $u \in E_{\chi_i}(x),\ v \in E_{\chi_j}(x),\ i \neq j\ \ < u, v >_{x,\epsilon} = 0$.

We will usually omit the word "standard" and will call this scalar product *ϵ–Lyapunov scalar product*, or, if ϵ has been fixed and no confusion may appear, simply *Lyapunov scalar product.* The norm generated by this scalar product will be called the (standard ϵ–) Lyapunov norm and denoted by $\| \cdot \|_{x,\epsilon}$

By construction of the standard ϵ–Lyapunov norm one has for $u \in E_{\chi_i}(x)$,

$$\exp(\chi_i(n) - \epsilon\|n\|)\|u\|_{x,\epsilon} \le \|\mathcal{A}(n,x)u)\|_{x,\epsilon} \le \exp(\chi_i(n) + \epsilon\|n\|)\|u\|_{x,\epsilon}. \tag{5.1.6}$$

Notice that by definition $\|u\|_{x,\epsilon} \ge m^{-1/2}\|u\|$.

Denote for a $\delta > 0$ by $\Lambda_{\epsilon,\delta}$ the following subset of the set of regular points:

$$\Lambda_{\epsilon,\delta} = \{x \in X :\ \forall u \in \{x\} \oplus \mathbb{R}^m,\ \|u\|_{x,\epsilon} \le \delta^{-1}\|u\|\}. \tag{5.1.7}$$

A real–valued function ϕ on X is called *tempered at x* (with respect to the action α) if

$$\lim_{n \to \infty} \frac{\phi(\alpha(n,x))}{n} = 0.$$

It is easy to see that a tempered function can be estimated from above by a *slowly changing* function. Namely, let ϕ be tempered and let $\epsilon > 0$. We define

$$\phi_\epsilon(x) = \max_{n \in \mathbb{Z}^k}(|\phi(\alpha(n,x))| - \epsilon\|n\|).$$

Obviously $\phi_\epsilon(x) \ge |\phi(x)|$. A simple calculation shows that ϕ_ϵ indeed changes slowly:

$$\phi_\epsilon(x) - \epsilon\|n\| \le \phi_\epsilon(\alpha(n,x)) \le \phi_\epsilon(x) + \epsilon\|n\|. \tag{5.1.8}$$

The following fact is proved in the same way as for $k = 1$ in [**P**],[**BKP**].

Proposition 5.1 (Temperedness of the Lyapunov metrics). *For any regular point $x \in X$ any non-zero $u \in \mathbb{R}^m_x$ and any $\epsilon > 0$ the function* $\log \frac{\|u\|_{x,\epsilon}}{\|u\|}$ *is tempered.*

Two cocycles $\mathcal{A}$ and $\mathcal{B}$ over the same action α are called *equivalent* if there exists a function $C : X \to GL(m, \mathbb{R})$ such that

$$\max\{\log\|C\|, \log\|C^{-1}\|\}$$

is a tempered function and

$$\mathcal{B}(n,x) = C^{-1}(x)\mathcal{A}(n,x)C(\alpha(x,n)).$$

Equivalent cocycles have the same Lyapunov characteristic exponents with the same multiplicities. The above proposition implies that any measurable coordinate change which takes the standard scalar product to the standard ϵ–Lyapunov scalar product is tempered. Thus any cocycle $\mathcal{A}$ satisfying (5.1.2) is equivalent to a block cocycle with $m_i \times m_i$ diagonal blocks $\mathcal{A}_i(n,x)$, $i = 1,\dots,l$ such that

$$\|\mathcal{A}_i(n,x)\| \le \exp(\chi_i(n) + \epsilon\|n\|) \quad \text{and} \quad \|\mathcal{A}_i^{-1}(n,x)\| \ge \exp(-\chi_i(n) - \epsilon\|n\|).$$

We will call any measurable coordinate change of this kind an *ϵ–Lyapunov coordinate change* an the resulting equivalent cocycle an *ϵ–reduced* cocycle.

5.2. Lyapunov hyperplanes and Weyl chambers. Now we can generalize certain notions previously introduced in Sections 1.2.3 and 1.3 for special cases.

The hyperplane $\ker\chi \subset \mathbb{R}^k$, where χ is a non-zero Lyapunov exponent, is called a *Lyapunov hyperplane.* The subspace $\chi^{-1}(-\infty,0)$ (corr $\chi^{-1}(0,\infty)$) is called a *negative* (corr. *positive) Lyapunov half-space.* A Lyapunov hyperplane L is called *rational* if $L \cap \mathbb{Z}^k$ is a lattice in L, *totally irrational* if $L \cap \mathbb{Z}^k = \{0\}$, and *partially irrational* otherwise. For $k = 2$ we will call Lyapunov hyperplanes *Lyapunov lines.* Naturally , every Lyapunov line is either rational or totally irrational; in the latter case we will call it simply irrational.

A linear extension of an ergodic $\mathbb{Z}^k$ action is called *partially hyperbolic* if there is at least one non-zero Lyapunov exponent and *hyperbolic* if all Lyapunov exponents are different from zero.

A hyperbolic linear extension is called *totally non–symplectic (TNS)* if non of the Lyapunov exponents is proportional to another with the negative coefficient of proportionality.

An element $n \in \mathbb{Z}^k$ is called *regular* if n does not belong to any of the Lyapunov hyperplanes. A regular element for a hyperbolic linear extension of a $\mathbb{Z}^k$ action is called *hyperbolic.* A *Weyl chamber* is a connected component of the complement to the union of all Lyapunov hyperplanes.

Each Weyl chamber is an open convex polyhedral cone in $\mathbb{R}^k$. Inside a Weyl chamber every non-zero Lyapunov exponent has a constant sign. Conversely, the locus of points in $\mathbb{R}^k$ for which each non-zero Lyapunov exponent has a particular sign is either empty or is a Weyl chamber. Thus any Weyl chamber can be characterized as a minimal non-empty intersection of positive and negative Lyapunov half-spaces.

For a partially hyperbolic element $n \in \mathbb{Z}^k$ and a regular point x we set

$$E_n^+(x) = \bigoplus_{i:\chi_i(n)>0} E_{\chi_i}(x) \text{ and } E_n^-(x) = \bigoplus_{i:\chi_i(n)<0} E_{\chi_i}(x).$$

These subspaces are called correspondingly the *expanding* (or *unstable*) and the *contracting* (or *stable*) subspaces for n at the point x. Notice that for any $n \in \mathbb{Z}^k$ if one takes in the definition of the standard ϵ–Lyapunov metric $\epsilon < \epsilon_0(n) = \frac{\min|\chi_i(n)|}{\|n\|}$ where the minimum is taken over all exponents which do not vanish at n, then the extension $A(n)$ uniformly contracts the ϵ–standard Lyapunov metric in the contracting subspaces of all regular points and uniformly expands it in the expanding subspaces of those points. By making an ϵ–Lyapunov coordinate change we obtain an equivalent cocycle generating an extension B such that $B(n)$ uniformly contracts the standard Euclidean metric in the contracting subspaces of all regular points and uniformly expands it in the expanding subspaces of those points.

By definition contracting and expanding subspaces are the same for all elements in the same Weyl chamber. Any minimal non-zero intersection of expanding subspaces for various regular elements corresponds to a positive Lyapunov half-space H and will be denoted by $E_H(x)$. It is equal to the sum of the subspaces E_{χ_i} for those χ_i's which are positive on the subspace H. In other words, all these exponents are proportional with positive proportionality coefficients. Thus for any Lyapunov half–space H one can find a uniquely defined Lyapunov characteristic exponent $\chi(H)$ (called the *bottom exponent* for H) and positive numbers $1 = c_1(H) < c_2(H) <, \dots, < c_{m(H)}(H)$ such that

$$E_H(x) = \bigoplus_{i=1}^{m(H)} E_{c_i(H)\chi(H)}.$$

Let us denote

$$E_H^i(x) = \bigoplus_{j=i}^{m(H)} E_{c_j(H)\chi(H)}.$$

The nested sequence $E_H(x) = E_H^1(x) \supset E_H^2(x) \supset \cdots \supset E_H^{m(H)}(x) = E_{c_{m(H)}(H)\chi(H)}$ is called the *upper filtration* of the space $E_H(x)$.

Number all positive Lyapunov half–spaces by $H_1, \dots, H_s$. We will call the decomposition

$$\mathbb{R}_x^m = \bigoplus_{i=1}^{s} E_{H_i}(x)$$

the *coarse Lyapunov decomposition* at the point x. Extension A is called *multiplicity–free* if all Lyapunov exponents are simple and no two of them are positively proportional. In other words, in the multiplicity–free case the elements of the coarse Lyapunov decomposition are one–dimensional.

5.3. Strongly hyperbolic extensions. In the case of multiple exponents it is convenient to count each exponent the number of times equal to its multiplicity so that there are always exactly m exponents. Since this will not cause any confusion, from now on we will denote exponents listed this way by $\chi_1, \dots, \chi_m$, unless explicitly stated otherwise. This allows us to define the *Lyapunov map* $\Psi_A : \mathbb{R}^k \to \mathbb{R}^m$ by $\Psi_A = (\chi_1, \dots, \chi_m)$. The Lyapunov map is defined up to a permutation of coordinates.

Thus the linear extension μ is hyperbolic if and only if $\operatorname{Im}\Psi_A$ does not lie in any coordinate hyperplane. We will call $\dim\operatorname{Ker}\Psi_A$ the *defect* of μ and denote it by $d(A)$. Equivalently, $d(A)$ is equal to the dimension of the intersection of all Lyapunov hyperplanes. Furthermore, $\dim\operatorname{Im}\Psi_A$ is called the *rank* of A and is denoted by $r(A)$. Equivalently, $r(A)$ is equal to the maximal number of linearly independent Lyapunov exponents.

Definition A hyperbolic linear extension A is called *strongly hyperbolic* if $d(A) = 0$, i.e. if the intersection of all Lyapunov hyperplanes consists of the origin.

Obviously, $d(A) + r(A) = k$ and since $r(A) \le m$

$$k \le m + d(A).$$

In particular, for any strongly hyperbolic measure

$$k \le m. \tag{5.3.1}$$

5.4. Resonances. Let $\chi_1, \dots, \chi_l$ be the different Lyapunov exponents of an extension A as in Section 1.1.We will say that A possesses an *essential resonance* if the following two conditions hold:

($\mathcal{R}1$) There exist $i \in \{1, \dots, l\}$ and non-negative integers $s_1, \dots, s_l$ such that

$$\chi_i = \sum_{j \neq i} s_j \chi_j \tag{5.4.1}$$

($\mathcal{R}2$) There exists a Weyl chamber $\mathcal{W}$ such that in $\mathcal{W}$ $\chi_i < 0$ and $\chi_j < 0$ for all j such that $s_j \neq 0$.

If A does not possess any essential resonances we will call it *essentially non-resonant.* If in addition all exponents are simple (multiplicity one) we will say that A is *non-resonant.*

Let $x_1, \dots, x_l$ be negative numbers. We will call a relation of the form

$$x_i = \sum_{j \neq i} s_j x_j, \tag{5.4.2}$$

where $s_1, \dots, s_l$ are non-negative integers, a *resonance relation* between $x_1, \dots, x_l$. Notice that any given set of negative numbers may satisfy only finitely many resonance relations.

PROPOSITION 5.2. *There exists an element $n \in \mathbb{Z}^k \cap \mathcal{W}$ such that*

1. *The multiplicities of the Lyapunov exponents for the extension $A(n)$ are the same as for the extension A.*
2. *Any resonance relation between the negative Lyapunov characteristic exponents of $A(n)$ comes from an essential resonance for A.*

A proof of this statement is contained in the proof of [**K1**], Theorem 4.4.

6. Normal forms and linearization of non-linear extensions

6.1. Extensions by non-linear contractions. We will be interested later on in the way a smooth $\mathbb{Z}^k$ action with a partially hyperbolic invariant measure acts on the family of stable manifolds of its regular element. This naturally leads us to the study of measurable extensions of a measure–preserving transformation by diffeomorphisms preserving the zero section with negative Lyapunov exponents for the derivative at that section, as well as centralizers of such extensions.

Let as before (X, μ) be a probability Lebesgue space and $\alpha : \mathbb{Z}^k \times X \to X$ be an ergodic action of the group $\mathbb{Z}^k$ by measure–preserving transformations of the space (X, μ). We will assume that an extension Φ of α is defined in a neighborhood of the zero section, preserves the zero section and acts by C^∞ diffeomorphisms in the fibers. In other words, Φ can be written in coordinates $(x, t) \in X \times \mathbb{R}^m$ as

$$\Phi(n, x, t) = (\alpha(n, x), \mathcal{F}(n, x)(t)), \tag{6.1.1}$$

where $\mathcal{F}(n, 0) = 0$ and $\mathcal{F}$ is C^∞ in t.

The derivatives in the t variable at the zero section which we will denote simply by $D\mathcal{F}_0(n, x)$ define a linear extension of α. We assume that for some $n_0 \in \mathbb{Z}^k$, $D\mathcal{F}_0(n_0, x)$ has negative characteristic exponents. We will also assume that all partial derivatives of all orders are tempered functions.

We will consider the following natural notion of equivalence for extensions of the form (6.1.1). Namely, we will allow coordinate changes which are measurable in x, preserve each fiber $\mathbb{R}^m_x$, fix the origin, are C^∞ in each fiber and have tempered logarithm of all derivatives of all orders at the zero section. We will call such coordinate changes *admissible*. Our main purpose is to show that any extension can be brought to a certain normal form by an admissible coordinate change. We will do it by first considering the element $F(n_0)$. We will find a normal form for it and them describe the centralizer of that normal form. This description will show that the admissible coordinate change bringing $F(n_0)$ to its normal form also brings the whole extension F to a certain normal form. The simplest kind of such a normal form is linear. The obstacles to linearization of $F(n_0)$ via an admissible coordinate change are resonance relations between the Lyapunov characteristic exponents of the derivative at the zero section. Presence of such relations might force existence of certain non–linear resonance terms in the normal form.

6.2. Normal forms. The results of this sub–section represent one of several natural "non–stationary" generalizations of well-known facts about normal forms of smooth contractions near a fixed point or an invariant manifold which can be for example in [**Be**] and [**BK**]. Another version of the non–stationary normal form theory for contractions adapted to the study of uniformly hyperbolic dynamical systems is developed in [**GK**] and [**G**]. The latter version is a crucial in the proofs of local differentiable rigidity both for actions of abelian groups [**KS2**], as well as higher rank Lie groups and their lattices [**MQ**].

Let (X, μ) be a probability Lebesgue space, $f : X \to X$ be a measure–preserving transformation of (X, μ), not necessarily ergodic, U an open neighborhood of the origin in $\mathbb{R}^m$ and $\Phi : X \times U \to X \times \mathbb{R}^m$ an extension of f satisfying assumptions corresponding those for $\Phi(n_0)$ above. Namely, it preserves the zero section, is C^∞ along the fibers, the Lyapunov characteristic exponents for the linear extension defined by the derivatives at zero section are constant almost everywhere and negative and all higher derivatives are tempered functions. Let $\chi_i, \dots \chi_l$ be different Lyapunov characteristic exponents of the derivative extension and $m_1, \dots, m_l$ be their multiplicities. Consider all the resonance relations (5.4.2) between the numbers $\chi_i, \dots \chi_l$. Represent $\mathbb{R}^m$ as the direct sum of the spaces $\mathbb{R}^{m_i}, \dots, \mathbb{R}^{m_l}$ and let $(t_1, \dots, t_l)$ be the corresponding coordinate representation of a vector $t \in \mathbb{R}^m$. Let $P : \mathbb{R}^m \to \mathbb{R}^m$; $(t_1, \dots, t_l) \mapsto (P_1(t_1, \dots, t_l), \dots, P_l(t_1, \dots, t_l))$ be a polynomial map preserving the origin. We will say that the map P is of *resonance type* if it contains only such homogeneous terms in $P_i(t_1, \dots, t_l)$ with degree of homogeneity s_j in the coordinates of t_j, $i = 1, \dots, l$ for which the resonance relation $\chi_i = \sum_{j \neq i} s_j \chi_j$ holds. It is easy to see that polynomial maps of the resonance type with invertible derivative at the origin form a group. Since there are only finitely many resonance relations between $\chi_1, \dots, \chi_l$ this is a finite–dimensional Lie group. We will denote this group by G_χ. In particular, if there are no resonance relations between the numbers $\chi_1, \dots \chi_l$ then $G_\chi = GL(m, \mathbb{R})$, the group of linear automorphisms of $\mathbb{R}^m$.

THEOREM 6.1 (non–stationary normal form for contractions). *There exists an admissible coordinate change in $X \times \mathbb{R}^m$ which transforms Φ to an extension Ψ of the following form*

$$\Psi(x, t) = (f(x), \mathcal{P}_x(t)) \tag{6.2.1}$$

where for almost every $x \in X$, $\mathcal{P}_x \in G_\chi$.

The proof of Theorem 6.1 follows one of the usual schemes in the normal form theory. (see, e.g. [**KH**, Section 6.6], for the local case and [**GK**] for the uniformly hyperbolic nonstationary case). It includes three steps:

Step 1 Finding a tempered *formal* coordinate change, i.e. the Taylor series at the zero section for the desired coordinate change.

Step 2 Constructing a tempered smooth (C^∞) coordinate change for which the formal power series found at Step 1 is the Taylor series at the zero section

Step 3 The coordinate change constructed at Step 2 conjugates our extension with an extension which is C^∞ *tangent* to the derivative extension. We show that any two tempered C^∞ tangent contracting extensions are conjugate via a C^∞ tempered coordinate change C^∞ tangent to identity

We will outline Step 1 since at that step the role played by resonances becomes apparent. In particular it becomes clear that the nonuniform case is a more direct generalization of the local one and the Lyapunov exponents play exactly the same role as the eigenvalues. In the uniform case considered in [**GK**] an extra narrow band condition is required and not only resonance but also subresonance terms have to be allowed in the normal form. Of course the price paid for this simplification is that the dependence on the base point in the nonuniform nonstationary normal form is only measurable even if all the data are smooth.

Steps 2 and 3 are very similar to the uniform nonstationary arguments in [**GK**] since the use of Lyapunov metrics essentially reduces our situation to the case of an extension by uniform contractions.

First we make a linear coordinate which brings the derivative at the zero section to the block form as described at the end of Section 5.1. Denote the $m_i \times m_i$, $i = 1, \dots, l$ blocks thus obtained by $\Phi_1(x), \dots, \Phi_l(x)$. Denote the coordinates corresponding this block form by $t_1, \dots, t_l$, where $t_i \in \mathbb{R}^{m_i}$.

We will look inductively for successive jets of the desired coordinate change at the zero section to eliminate the non–resonance terms of higher and higher order.The base of the induction is the linear coordinate change described above. At the nth inductive step we start form the extension whose nth jets are polynomials of resonance type. We will find a coordinate change with the identity linear part and polynomial homogeneous non–linear part of degree $n+1$. Composing this coordinate change with the result from the previous steps would not change the lower order terms. The t_i component $C_x(t_1, \dots, t_l)$ of the coordinate change with degree of homogeneity s_j in the coordinates of t_j, $j = 1 \dots, l$ satisfies the following functional equation

$$\Phi_i(x) C_x(t_1, \dots, t_l) - C_{f(x)}(\Phi_1(x)t_1, \dots, \Phi_l(x)t_l) = H_x(t),$$

where $H_x(t)$ is a known function determined by the previous steps of the process. Let for $t = (t_1, \dots, t_l)$, $\Phi(t) = (\Phi_1(t_1), \dots, \Phi_l(t_l))$. In the non–resonance case we can solve this equation by one of the two "telescoping sums": if

(6.2.2)
$$\chi_i > \sum s_j \chi_j,$$

then

(6.2.3)
$$C_x(t) = \sum_{m=0}^{\infty} \Phi_i^{-1}(x) \dots \Phi_i^{-1}(f^m(x)) H_{f^m(x)}(\Phi(f^m(x)) \dots \Phi(f(x))(\Phi(x))(t)),$$

where the series in the right–hand part converges exponentially; if

$$\chi_i < \sum s_j \chi_j, \tag{6.2.4}$$

then

$$C_x(t) = \tag{6.2.5}$$

$$-\sum_{m=1}^{\infty} \Phi_i(f^{-1}(x)) \ldots \Phi_i(f^{-m}(x)) H_{f^{-m}(x)}(\Phi^{-1}(f^{-m}(x)) \ldots \Phi^{-1}(f^{-1}(x))(t)),$$

where again the series converges exponentially. Notice that the former case holds for all but finitely many homogeneous terms. The resonance terms of degree $n+1$ do not change, at this and successive steps. However that they were likely to have changed at the previous steps. Thus, while the type of normal form is determined by the first jet (the derivative extension), coefficients of its nth jet depend on the $(n-1)$st jet of the original extension.

THEOREM 6.2 (Centralizer for resonance maps). *Let g be a non–singular (not necessarily measure–preserving) transformation of the space (X, μ) commuting with f, and $\Gamma(x,t) = (g(x), \mathcal{G}_x(t))$ be an extension of g by C^∞ local diffeomorphisms preserving the zero section and commuting with an extension Ψ of the form (6.2.1). Then for almost every $x \in X$, $\mathcal{G}_x \in G_\chi$.*

For certain applications it is useful to define another class of polynomial maps associated with a given set of Lyapunov exponents. As before let $P : \mathbb{R}^m \to \mathbb{R}^m$; $(t_1, \ldots, t_l) \mapsto (P_1(t_1, \ldots, t_l), \ldots, P_l(t_1, \ldots, t_l))$ be a polynomial map preserving the origin. We will say that the map P is of *sub–resonance type* if it has homogeneous terms in $P_i(t_1, \ldots, t_l)$ with degree of homogeneity s_j in the coordinates of t_j, $i = 1, \ldots, l$ only if

$$\chi_i \le \sum_{j \ne i} s_j \chi_j.$$

In other words, in addition to resonance terms we also allow terms satisfying the strict inequality (6.2.4). The product of two maps of sub–resonance type is again a map of sub–resonance type; thus, as in the case of resonance maps, polynomial maps of sub–resonance type with invertible derivative at the origin form a group. It is larger than G_χ but still finite dimensional. We will denote this group by SR_χ. In particular, if $\min \chi_i > 2 \max \chi_i$, then $SR_\chi = GL(m, \mathbb{R})$. Theorem 6.2 has a counterpart for maps of sub–resonance type.

THEOREM 6.3 (Centralizer for sub–resonance maps). *Let Ψ be an extension of a measure–preserving transformation f; $(X, \mu \to (X, \mu)$ of the following form*

$$\Psi(x,t) = (f(x), \mathcal{P}_x(t)), \text{ where for almost every } x \in X,\ \mathcal{P}_x \in SR_\chi.$$

Let g be a non–singular transformation of the space (X, μ) commuting with f, and $\Gamma(x,t) = (g(x), \mathcal{G}_x(t))$ be an extension of g by C^∞ local diffeomorphisms preserving the zero section and commuting with an extension Ψ . Then for almost every $x \in X$, $\mathcal{G}_x \in SR_\chi$.

The proofs of theorems 6.2 and 6.3 are similar to the proof of Theorem 1.2 from [**GK**] with a proper allowance for non–uniformity of the situation. Details will appear in a separate paper.

7. Pesin theory for $\mathbb{Z}^k$ actions

7.1. Local $\mathbb{Z}^k$ actions and rank restrictions.

7.1.1. *Locally maximal sets.* The main situation when the general setup described in the previous two sections arises is the extension of a smooth $\mathbb{Z}^k$ action by derivatives. A somewhat more general semi-local setup is as follows.

Let M be a differentiable manifold, $U \subset M$ an open set and $\Lambda \subset U$ a compact subset. Let $f_1, \ldots, f_k : U \to M$ be commuting diffeomorphic embeddings of class $C^{1+\epsilon}$ for some $\epsilon > 0$ preserving the set Λ. Obviously, the maps $f_1, \ldots, f_k$ restricted to the set Λ generate an action of the group $\mathbb{Z}^k$ on Λ. Outside Λ the action of the whole group $\mathbb{Z}^k$ may not be defined. We will call this situation a *local $\mathbb{Z}^k$ action near* Λ. Our standard notation for a local action will be F so that for $n = (n_1, \ldots, n_k) \in \mathbb{Z}^k$ one has $F(n_1, \ldots, n_k) = f_1^{n_1} \circ \cdots \circ f_k^{n_k}$. The set Λ is called a *locally maximal set* for the local action F if for some open set $V \supset \Lambda$ the set Λ is the biggest invariant set contained in V, i.e.

$$\Lambda = \bigcap_{n \in \mathbb{Z}^k} F(n)V.$$

Any such set V will be called a *separating neighborhood* for Λ.

7.1.2. *Lyapunov exponents and hyperbolic measures for local actions.* Let μ be a Borel probability F-invariant ergodic measure such that $\operatorname{supp}\mu \subset \Lambda$. Even though the tangent bundle TM or its restriction to Λ, $T_\Lambda M$ may not be trivial, it is always trivial up to a set of measure zero. So the derivatives of the local action DF acting on $T_\Lambda M$ generate a linear extension of F as described in Section 5. Different trivializations define cohomologous linear extensions (see [**KM**]) so various properties of this extensions discussed in Section 5 do not depend on a particular choice of the trivialization. Accordingly we will attribute these properties to the action F itself. Thus we will speak of the Lyapunov characteristic exponents, hyperbolic measures, Weyl chambers, etc of a local action.

In the smooth situation various feature of the structure of the derivative extension are more of less directly reflected in the structure of the (non-linear) action itself. Accordingly we will associate the attributes of the derivative extension to the action and the measure. Notice that cocycles associated to the derivative extension depend on a choice of a measurable trivialization of the tangent bundle but cocycles corresponding to different trivializations are cohomologous. Thus Lyapunov exponents for the derivative cocycle with respect to an ergodic invariant measure are constant almost everywhere and thus are well defined as well as all notions derived form those exponents: Lyapunov hyperplanes, Weyl chambers, Lyapunov map, rank, defect, etc. We will thus speak about Lyapunov exponents of a local action with respect to an invariant measure, about partially hyperbolic, strongly hyperbolic and TNS (totally nonsymplectic) measures. In this case we will use the notation Ψ_μ for the Lyapunov map, $r(\mu)$ for the rank and so on.

7.1.3. *TNS actions.* Cartan actions ([**KKS**]) of which the $\mathbb{Z}^2$ actions of section 2.1 are the simplest examples are TNS. See [**NT**, Section 5] for a detailed discussion of TNS actions, in particular for specific examples of such actions on Tori and nilmanifolds. Among algebraic Anosov actions discussed in Section 1.3 TNS actions appear only among actions by automorphisms of tori and nilmanifolds, their finite affine extensions and suspensions. Most of the other interesting examples, are related in one way or another with semisimple Lie groups, and hence possess at least a partial simplectic structure in certain transverse directions. Hence some Lyapunov exponents appear in pairs which differ by the sign. See the description of standard Anosov actions in [**KS1**].

7.1.4. *The rank restriction.*

PROPOSITION 7.1 ([**K1**]). *If* $\operatorname{Im}\Psi_\mu$ *intersects the positive octant* $\mathbb{R}^n_+$ *(and hence also the negative octant* $-\mathbb{R}^n_+$*) then the measure* μ *is atomic.*

COROLLARY 7.1. *If* μ *is a hyperbolic measure for a local* $\mathbb{Z}^k$ *action and* $k = \dim M + d(\mu)$*, then* μ *is atomic. In particular, any strongly hyperbolic ergodic invariant measure for a local* $\mathbb{Z}^k$ *action on a* k*–dimensional manifold is atomic.*

PROOF. Obviously $r(\mu) = \dim M$ so $\operatorname{Im}\Psi_\mu = \mathbb{R}^{\dim M} \supset \mathbb{R}^{\dim M}_+$ □

Corollary 7.1 immediately implies that if a local $\mathbb{Z}^k$ action on an n–dimensional manifold possesses a strongly hyperbolic non–atomic invariant measure than

$$k \leq n - 1. \tag{7.1.1}$$

This inequality cannot be improved. In fact, for any $n \geq 2$ there is a $\mathbb{Z}^{n-1}$ action on the n–dimensional torus $\mathbb{T}^n$ by hyperbolic automorphisms for which Lebesgue measure is invariant and strongly hyperbolic. A "surgery" described in [**KL**] allows to produce actions of $\mathbb{Z}^{n-1}$ with strongly hyperbolic absolutely continuous invariant measures on certain n–dimensional manifolds other than torus. The rank of these measures is equal to $n - 1$.

A certain feature of these examples holds in general.

COROLLARY 7.2. *If* μ *is a non-atomic hyperbolic measure for a local* $\mathbb{Z}^k$ *action* F *and* $r(\mu) = \dim M - 1$ *then there exists a regular element* $n \in \mathbb{Z}^k$ *such that* $F(n)$ *has exactly one positive Lyapunov exponent and this exponent is simple.*

Further results of [**K1**] show that in the low–dimensional situations, *i.e.* for $n = 2$ or 3 the only way for a non–atomic hyperbolic measure to violate (7.1.1) is to have elements in the $\mathbb{Z}^k$ action with massive (in particular, full measure) sets of fixed points. In the real–analytic case this leads to the following conclusion:

PROPOSITION 7.2 ([**K1**]). *Any hyperbolic invariant measure for an effective real– analytic action of* $\mathbb{Z}^2$ *on a compact (real–analytic) surface or an effective real– analytic action of* $\mathbb{Z}^3$ *on a compact (real–analytic) three–manifold is atomic.*

7.2. Regular neighborhoods and regular sets. (See [**BP**] for a more detailed discussion in the rank one case.) Consider an ergodic invariant measure μ for a local $\mathbb{Z}^k$ action F. For μ–a.e. point $x \in \Lambda$ we can construct for any $\epsilon > 0$ the ϵ–Lyapunov metric for the derivative extension. Fix a sufficiently small $\epsilon > 0$; typically it is sufficient to assume that ϵ is less than a quarter of the minimal difference between two different Lyapunov exponents of the derivative extension. There are two types of regular neighborhoods for regular points in the sense of Section 5.1. The neighborhoods of the first kind are constructed by taking "boxes" i.e. products of balls in the Lyapunov metric of a fixed sufficiently small size in the spaces $E_{\chi_i}(x)$, $i = 1, \dots, l$ and exponentiating these boxes by the exponential map of the reference smooth Riemannian metric. The fine Lyapunov decomposition and the ϵ–Lyapunov metric for any $\epsilon > 0$ are continuous when restricted to any set where $\frac{\|u\|_{x,\epsilon}}{\|u\|}$ is bounded. In particular the sets $\Lambda_{\epsilon,\delta}$ defined by (5.1.7) are closed in this situation and the ϵ–Lyapunov metric and the regular neighborhoods are continuous on those sets. Such a neighborhood comes provided with a standard coordinate system, i.e. a smooth embedding of the standard box B in $\mathbb{R}^m$ into M. Continuity means continuous dependence of this map from the base. point in a proper topology, typically, since we mostly deal with C^∞ maps, the topology will be the C^∞ topology. Similarly to the rank one case [**Br**] it can be shown later that these structures are in fact Hoelder continuous.

The second type of regular neighborhoods is described in [**KM**], Theorem S.3.1. These neighborhoods are obtained by "truncating" the boxes in Lyapunov metric further to make their size changing slowly with respect to the reference metric. In most situation any of the two types of regular neighborhoods may be used.

7.3. Stable and unstable manifolds. (See [**BP**].) Let us assume that μ is a hyperbolic measure. Specifying the discussion from Section 5.2 to this case we obtain the coarse Lyapunov decomposition of $T_\Lambda M$:

$$T_\Lambda M = \bigoplus_{i=1}^{s} E_{H_i}(x).$$

Hadamard-Perron Theorem for $\mathbb{Z}^k$ actions asserts that each element of this decomposition as well as distributions obtained by the upper filtration in such an element can be uniquely integrated to a family of smooth submanifolds which possesses certain regularity properties away from sets of small measure. More specifically fix a sufficiently small ϵ as before and let $\Lambda_\epsilon = \cup_{\delta>0}\Lambda_{\epsilon,\delta}$.

Let E be a distribution from an upper filtration of the coarse Lyapunov decomposition and E' be the sum of the elements of the fine Lyapunov decomposition which do not belong to E and $x \in \Lambda_\epsilon$. In the regular neighborhood $\mathcal{R}(x)$ of x of the first kind, there is a natural product structure given by the exponential of the decomposition $T_xM = E(x) \oplus E'(x)$. Then *the local integral manifold* $\mathcal{E}(x) \subset \mathcal{R}(x)$ has the form $\exp_x \operatorname{graph} \phi_x$, where $\phi_x : B(E(x)) \to E'(x)$. Here $B(E(x))$ is the box around the origin in $E(x)$ of fixed size in the Lyapunov metric used in the construction of the regular neighborhoods and ϕ_x is a smooth function (in fact, its degree of regularity is the same as for the action F), which vanishes at the origin together with its first differential. Furthermore, if $\|D_xF(n)\|_{x,\epsilon} < 1-\epsilon$, then $F(n)(\mathcal{E}(x)) \subset \mathcal{E}(F(n)x)$ and $F(n)$ contracts $\mathcal{E}(x)$. In fact, these manifolds can be obtained by considering upper filtrations for individual regular elements of the action constructing the corresponding families of integral manifolds and taking their intersections. In the case $E = E_{H_i}$ one takes intersections of local stable manifolds for regular elements of the action.

The *global integral manifold* $\mathfrak{E}(x)$ is defined as the smallest invariant set containing the local integral manifold $\mathcal{E}(x)$, i.e.

$$\mathfrak{E}(x) = \cup_{n\in\mathbb{Z}^k} F(n)(\mathcal{E}(F^{-n}(x))).$$

Global integral manifolds are images of the Euclidean space under smooth injective immersions.

As we mentioned before in the description of the regular sets $\Lambda_{\epsilon,\delta}$ it is sufficient (and convenient) to fix a sufficiently small value of the parameter ϵ. This way we obtain a family of compact sets depending on a single parameter which we will denote by θ. This parameter roughly can be thought as the "guaranteed size" of integral manifolds for the elements of the coarse Lyapunov decomposition and their upper filtrations.

8. Rigidity of hyperbolic measures in dimension three: an outline

The crucial feature of the argument in Section 2.2 was presence of the critical direction such that the action along that direction was isometric along the foliation W and sufficiently ergodic. If we want to apply this type of arguments in more general situation we need to understand when a similar situation may appear.

In the case of algebraic actions discussed in Part I both the guidance and the restrictions are provided by linear algebra. Now we will consider a situation which looks much more general from dynamical point of view than our basic example of an action of $\mathbb{Z}^2$ on $\mathbb{T}^3$ by hyperbolic automorphisms but whose geometric features are very similar.

8.1. Lyapunov Hoelder cocycle rigidity for TNS measures. Let us consider an action α of $\mathbb{Z}^2$ by diffeomorphisms of a compact three-dimensional manifold M. The key regularity assumption is $C^{1+\epsilon}$, for some $\epsilon > 0$, but at the moment we can assume that the action is C^∞. Consider an ergodic α-invariant measure μ such that all three Lyapunov exponents are different from zero and none of them are proportional. In other words, the three Lyapunov lines (see Section 5.2) are all different. These lines divide $\mathbb{R}^2$ into six Weyl chambers. For a μ almost every point $x \in M$ there exists a measurable α-invariant splitting of the tangent space T_xM into three one-dimensional subspaces $E_1(x)$, $E_2(x)$, and $E_3(x)$ corresponding to three Lyapunov exponents. Each of these distributions as well as their pairwise sums are uniquely integrable to α-invariant families of smooth manifolds defined almost everywhere with respect to the measure μ (See Section 7.3). While these distributions in general are not continuous, they are Hölder with respect to ϵ-Lyapunov metrics defined in Section 5.1. Fix a smooth Riemannian metric on M and let $J_i(x,a) = \log ||D\alpha(a)|_{E_i(x)}||$, $i = 1, 2, 3$. The function J_i are additive one cocycles, i.e. $J_i(x, a_1 + a_2) = J_i(x, a_1) + J_i(\alpha(a_1)(x), a_2)$. Naturally these cocycles are also Hölder with respect to ϵ-Lyapunov metrics. For these class of cocycles we can prove cohomological rigidity: for such a cocycle J one can find a function H, also satisfying a Hölder condition in a Lyapunov metric, and a homomorphism $\sigma : \mathbb{Z}^2 \to \mathbb{R}$ such that

$$J(x, a) = \sigma(a) + H(\alpha(a)(x)) - H(x).$$

8.2. Invariant affine structure on contracting manifolds. Let W be one of the three families of one-dimensional invariant manifolds. It turns out that there is a uniquely defined family of α-invariant smooth affine parameters on these manifolds. This is a particular case of the normal form results discussed in Section 6.2 (Theorems 6.2 and 6.3 since in the case of a single exponent there are no resonance relations. the proof is in fact more direct and simple than in the general case. The cocycle trivialization result described in the previous subsection allows to normalize the affine parameters in such a way that they are transformed by the transformation $\alpha(a)$ with the constant expansion or contraction coefficient $\exp \sigma(a)$. Now pass to the suspension of the action α which we also denote by α. This action in the direction of the corresponding Lyapunov line preserves the normalized affine parameters. From here on we can closely intimate the arguments for the automorphisms of $\mathbb{T}^3$ from Section 2.2 exercising a certain care due to the fact that this structure is only measurable.

8.3. Conditional measures and rigidity. Consider the system of conditional measures on the leaves of W and identify a typical global leaf with $\mathbb{R}$ using a point in the support of the conditional measure as the origin and the length parameter normalized at that point. At this moment it is important to keep in mind that the normalizations at different (even typical) points of the same leaf need not agree. However by Luzin Theorem type argument they do not differ more than a fixed amount on a set of large measure. Assumption $(\mathcal{E})$ holds essentially by the same reasons as in Section 2.2. The first part of the proof of the key assertion $(\mathcal{T})$ which uses weak* compactness and Luzin theorem holds essentially *verbatim*. Thus, the conditional measure on a typical leaf of W is invariant (maybe up to a rescaling) with respect to an almost every translation. Again assuming that the conditionals are atomic implies that the every element of the action has zero entropy. In the case of continuous conditional measures one sees again that the support of it is a closed subgroup of $\mathbb{R}$, hence the whole $\mathbb{R}$. The argument showing that the rescaling cocycle is identically equal to one again holds. Thus, the conditionals are Lebesgue with respect to the smooth invariant affine parameter, hence, absolutely continuous. After that simple arguments using Ledrappier's converse [**L**] to the Pesin entropy formula [**P**] show that the measure μ itself must be absolutely continuous. Thus the conclusion is the following very general counterpart of the statement (D) from section 2.2.

THEOREM 8.1. *Let μ be an ergodic invariant measure for a smooth action of $\mathbb{Z}^2$ on a compact three-dimensional manifold M such that all three Lyapunov characteristic exponents are different from zero and none of them are proportional to each other. Then either μ is absolutely continuous or all elements of the action have zero entropy with respect to μ.*

A detailed proof of this theorem will appear in a separate paper. An important corollary is the following result which to the bet of our knowledge is the first instance in the differentiable dynamics in dimension greater than one when global data (the homotopy class in our case) force existence of an invariant geometric structure for an action of noncompact amenable group.

THEOREM 8.2. *Any $\mathbb{Z}^2$ action α on $\mathbb{T}^3$ which is homotopic to an action α_0 by hyperbolic automorphisms has an ergodic absolutely continuous invariant measure.*

PROOF. There is a surjective continuous map $\phi : \mathbb{T}^3 \to \mathbb{T}^3$ such that

$$\phi \circ \alpha = \alpha_0 \circ \phi.$$

This fact can be proven as follows. First, by a theorem of Franks ([**KH**, Theorem 2.6.1]) there is such a map homotopic to identity (call it $\phi(\mathbf{m})$) for each element $\alpha(\mathbf{m})$ for $\mathbf{m} \in \mathbb{Z}^2 \setminus \{0\}$ and it is unique. Uniqueness implies that the centralizer of any hyperbolic automorphism of a torus in the group of homeomorphisms of the torus coincides with its centralizer in the group of automorphisms; in particular, this centralizer contains no more that one element in each homotopy class. Since for any $\mathbf{m}_1, \mathbf{m}_2 \in\in \mathbb{Z}^2 \setminus \{0\}$ the map $\phi(\mathbf{m}_1)^{-1} \circ \alpha(\mathbf{m}_2) \circ \phi(\mathbf{m}_1)$ commutes with $\alpha_0(\mathbf{m}_1)$ and is homotopic to $\alpha_0(\mathbf{m}_2) = \phi(\mathbf{m}_2)^{-1} \circ \alpha(\mathbf{m}_2) \circ \phi(\mathbf{m}_2)$ we deduce that

$$\phi(\mathbf{m}_1) = \phi(\mathbf{m}_2).$$

In other words, the algebraic action α_0 is a factor of the action α, or, equivalently α is an extension of α_0.

Consider the set $\mathcal{M}$ of all Borel probability measures ν on $\mathbb{T}^3$ such that $(\phi)_*\nu = \lambda$, Lebesgue measure. The set $\mathcal{M}$ is convex, weak* compact and α invariant. Hence by Tychonoff theorem it contains an α invariant measure. Since α_0 is ergodic with respect to λ almost every ergodic component of an α-invariant measure $\nu \in \mathcal{M}$ also belongs to $\mathcal{M}$. Let μ be such an ergodic measure. We will show that it is absolutely continuous.

Since $(\phi)_*\mu = \lambda$, for any $\mathbf{m} \in \mathbb{Z}^2$, and since for each $\mathbf{m} \in \mathbb{Z}^2$ either the stable or unstable foliation for $\alpha_0(\mathbf{m})$ is one-dimensional, one has

$$h_\mu(\alpha(\mathbf{m})) \geq h_\lambda(\alpha_0(\mathbf{m})) = |\log \max |\chi_i(\mathbf{m})||,$$

where $\chi_1(\mathbf{m}), \chi_2(\mathbf{m})$ and $\chi_3(\mathbf{m})$ are eigenvalues of the matrix $\rho_\alpha(\mathbf{m})$.

This implies in particular that the entropies $h_\mu(\alpha(\mathbf{m}))$ for all $\mathbf{m} \in \mathbb{Z}^2 \setminus \{0\}$ are uniformly bounded away from zero.

In order to apply Theorem 8.1 we need to show that μ is a hyperbolic measure and that all three Lyapunov lines are different. Notice that this is of course true for the Lebesgue measure with respect to α_0; after all, this was the starting point of the argument in Section 2.2. If there are no more than two Lyapunov lines for α then along at least one of these lines at least two of the three Lyapunov exponents vanish. By Ruelle inequality [**KM**] this means that the entropy of the elements of the suspension along that line vanishes. Since there are elements of $\mathbb{Z}^2$ either on the line (if the latter is rational) or arbitrary close to it (if it is irrational), there are nonzero elements of α with arbitrary small entropy, a contradiction. □

References

[BKP] L. Barreira, A. Katok and Ya. Pesin, *Non–uniformly hyperbolic dynamical systems*, monograph in preparation

[BP] L. Barreira and Ya. Pesin, *Lectures on Lyapunov exponents and smooth ergodic theory*, in Smooth ergodic theory and its applications, Proc. Symp. Pure Math, AMS, 2001.

[Be] G. R. Belitskii, *Equivalence and normal forms of germs of smooth mappings*, Russian Math. Surveys, **31** (1978), 107–177

[B] D. Berend, *Multi-invariant sets on tori* Trans. Amer. Math. Soc **280**, (1983), 509–533.

[BS] Z.I. Borevich and I.R. Shafarevich, *Number Theory*, Academic Press, New York, 1966.

[Br] M. Brin, *Holder continuity of invariant distributions* Appendix a to *Lectures on Lyapunov exponents and smooth ergodic theory*, by L. Barreira and Ya. Pesin, in Smooth ergodic theory and its applications, Proc. Symp. Pure Math, AMS, 2001.

[BK] I. U. Bronstein and A. Ja. Kopanskii, *Smooth invariant manifolds and normal forms* World scientific, 1994.

[F] H. Furstenberg, *Disjointness in ergodic theory, minimal sets, and a problem in diophantine analysis*, Math. Syst. Theory **1** (1967), 1–49

[GS] E. Goetze and R. J. Spatzier, *Smooth classification of Cartan actions of higher rank semisimple Lie groups and their lattices*, Ann. Math. **150** (1999), 743–773

[GK] M. Guysinsky and A. Katok, *Normal forms and invariant geometric structures for dynamical systems with invariant contracting foliations*, Math. Res. Lett. **5** (1998), 149–163.

[G] M. Guysinsky *The theory of nonstationary normal forms*, Erg. Th. and Dynam. Syst. **21** (2001), to appear.

[H] H. Huyi, *Some ergodic properties of commuting diffeomorphisms*, Erg. Th. and Dynam. Syst. **13** (1993), 73–100

[KaK] B. Kalinin and A. Katok, *Measurable rigidity and disjointness for $\mathbb{Z}^k$ actions by toral automorphisms*, Erg. Th. and Dynam. Syst. , to appear.

[K1] A. Katok, *Hyperbolic measures and commuting maps in low dimension*, Discrete and cont. Dyn. Syst. **2** (1996), 397–411

[K2] A. Katok, *An introduction to rigidity of higher rank abelian group actions* in Ergodic theory, geometric rigidity and number theory , Proc of the Special year at the Newton Institute, Springer Verlag, to appear.

[KH] A. Katok and B. Hasselblatt, *Introduction to the modern theory of dynamical systems*, Cambridge University Press, 1995.

[KKS] A. Katok, S. Katok and K. Schmidt, *Rigidity of measurable structure for $\mathbb{Z}^d$ actions by automorphisms of a torus*, preprint, 2000.

[KK] A. Katok and S. Katok, *Higher cohomology for abelian groups of toral automorphisms*, Erg. Th. and Dynam. Syst. **15** (1995), 569–592

[KL] A. Katok and J. Lewis, *Global rigidity results for lattice actions on tori and new examples of volume-preserving actions*, Israel J. Math. **93** (1996), 253–280

[KM] A. Katok and L. Mendoza, *Dynamical systems with nonuniformly hyperbolic behavior*, Supplement to "Introduction to the modern theory of dynamical systems" (A. Katok and B. Hasselblatt) 1995, 659–700

[KNT] A. Katok, V. Niţică and A. Török, *Nonabelian Cohomology of Abelian Anosov Actions*,Erg. Th. and Dynam. Syst. **20** (2000), 259 288

[KSch] A. Katok and K. Schmidt, *The cohomology of expansive $\mathbb{Z}^d$ actions by automorphisms of compact abelian groups* Pacific J. Math **170** (1995) 105–142

[KS1] A. Katok and R. J. Spatzier, *First cohomology of Anosov actions of higher rank Abelian groups and applications to rigidity*, Publ. Math. IHES **79** (1994), 131–156

[KS2] A. Katok and R. J. Spatzier, *Differential Rigidity of Anosov Actions of Higher Rank Abelian Groups and Algebraic Lattice Actions*, Proc. Steklov Inst. Math **216**, (1997), 292–319

[KS3] A. Katok and R. J. Spatzier, *Invariant Measures for Higher Rank Hyperbolic Abelian Actions*, Erg. Th. and Dynam. Syst. **16** (1996), 751–778

[KS4] A. Katok and R. J. Spatzier, *Corrections to "Invariant measures for higher rank hyperbolic abelian actions"*, Erg. Th. and Dynam. Syst. **18**, (1998), 503-507.

[Kat] I. Katznelson, *Ergodic automorphisms of $\mathbb{T}^n$ are Bernoulli shifts*, Israel J. Math **10**, (1971), 186-195

[L] F. Ledrappier, *Propriétés ergodiques des measures de Sinai*, Publ. Math. IHES **59** (1984), 163–188

[MQ] G. Margulis and N. Qian, *Local rigidity of weakly hyperbolic actions of higher rank real Lie groups and their lattices*, Erg. Th. and Dynam. Syst. **21** (2001), to appear

[NT] V. Niţică and A. Török, *On the cohomology of Anosov actions* in Ergodic theory, geometric rigidity and number theory , Proc of the Special year at the Newton Institute, Springer Verlag, to appear.

[O] V. I. Oseledets, *A multiplicative ergodic theorem. Liapunov characteristic numbers for dynamical systems*, Trans. Moscow math. Soc **19** (1968), 197–221

[P] Y. Pesin, *Characteristic exponents and smooth ergodic theory*, Russian Math Surveys **32** (1977), 55–114

[R] V. A. Rohklin, *On the fundamental ideas of measure theory*, A.M.S. Transl. (1), **10** (1962), 1–52

[R] D. Rudolph, *$\times 2$ and $\times 3$ invariant measures and entropy*, Ergod. Th. and Dynam. Syst. **10** (1990), 395–406

[S] A. Starkov, *First cohomology group, mixing and minimal sets of commutative group of algebraic action on torus*, J. Math. Sci (New York) **95**, (1999), 2576–2582.

University of Michigan, Ann Arbor, MI
E-mail address: kalinin@math.lsa.umich.edu

The Pennsylvania State University, University Park, PA
E-mail address: katok_a@math.psu.edu

Proceedings of Symposia in Pure Mathematics
Volume **69**, 2001

Some Applications of Homogeneous Dynamics to Number Theory

Dmitry Kleinbock

ABSTRACT. This survey paper is not a complete reference guide to number-theoretical applications of ergodic theory. Instead, the plan is to consider an approach to a class of problems involving Diophantine properties of n-tuples of real numbers, namely, describe a specific dynamical system which is naturally connected with these problems.

1. A glimpse at Diophantine approximation

For motivation, let us start by looking at two (vaguely defined) Diophantine problems:

PROBLEM 1. *Given a nondegenerate indefinite quadratic form of signature (m,n), study the set of its values at integer points.*

Here is a precise statement along these lines, conjectured in 1929 by Oppenheim [**Op1**] and proved in 1986 by Margulis [**Ma3**]:

THEOREM 1.1. *Let B be a real nondegenerate indefinite quadratic form of signature (m,n), $k = m+n > 2$. Then either B is proportional to a rational form, or* $\inf_{\mathbf{x}\in\mathbb{Z}^k\smallsetminus\{0\}} |B(\mathbf{x})| = 0$.

One possible approach to the problem is to write $B(\mathbf{x}) = \lambda S_{m,n}(g\mathbf{x})$, where $\lambda\in\mathbb{R}$, $g\in SL_k(\mathbb{R})$ and

$$S_{m,n}(x_1,\dots,x_k) = x_1^2+\dots+x_m^2 - x_{m+1}^2 - \dots - x_k^2$$

(a linear unimodular change of variables). Then the problem reduces to studying values of the standard form $S_{m,n}$ of signature (m,n) applied to the collection of vectors of the form $\{g\mathbf{x}\mid \mathbf{x}\in\mathbb{Z}^k\}$. And the dynamical approach consists of studying the action of the stabilizer of the form $S_{m,n}$ on such collections.

PROBLEM 2. *Given m vectors $\mathbf{y}_1,\dots,\mathbf{y}_m\in\mathbb{R}^n$ (viewed as linear forms $\mathbf{x}\mapsto \mathbf{y}_i\cdot\mathbf{x}$, $\mathbf{x}\in\mathbb{R}^n$) how small (simultaneously) can be the values of $|\mathbf{y}_i\cdot\mathbf{q}+p_i|$, $p_i\in\mathbb{Z}$, when $\mathbf{q}=(q_1,\dots,q_n)\in\mathbb{Z}^n$ is far from 0?*

The author was supported in part by NSF Grants DMS-9704489 and DMS-0072565.

Let us also illustrate this by a conjecture, this time still open. Here we specialize to the case of just one linear form given by $\mathbf{y} \in \mathbb{R}^n$. The following is known as Littlewood's (1930) Conjecture:

CONJECTURE 1.2. *For every* $\mathbf{y} \in \mathbb{R}^n$, $n \geq 2$, *one has*

$$\inf_{\mathbf{q}\in\mathbb{Z}^n\smallsetminus\{0\},\, p\in\mathbb{Z}} |\mathbf{y}\cdot\mathbf{q}+p|\cdot\Pi_+(\mathbf{q}) = 0\,, \tag{1.1}$$

where $\Pi_+(\mathbf{q})$ *is defined to be equal to* $\prod_{i=1}^n \max(|q_i|,1)$ *or, equivalently,* $\prod_{q_i\neq 0}|q_i|$.

To approach Problem 2, one can put together

$$\mathbf{y}_1\cdot\mathbf{q}+p_1,\dots,\mathbf{y}_m\cdot\mathbf{q}+p_m \quad \text{and} \quad q_1,\dots,q_n,$$

and consider the *lattice*

$$\left\{\begin{pmatrix} Y\mathbf{q}+\mathbf{p} \\ \mathbf{q}\end{pmatrix}\middle|\, \mathbf{p}\in\mathbb{Z}^m,\, \mathbf{q}\in\mathbb{Z}^n\right\} = L_Y\mathbb{Z}^{m+n}\,, \tag{1.2}$$

where

$$L_Y \stackrel{\text{def}}{=} \begin{pmatrix} I_m & Y \\ 0 & I_n\end{pmatrix} \tag{1.3}$$

and Y is the matrix with rows $\mathbf{y}_1^T,\dots,\mathbf{y}_m^T$. In this case, the orbit of the lattice (1.2) under a certain group action provides a way to study Diophantine properties of $\mathbf{y}_1,\dots,\mathbf{y}_m$.

In both cases, we use the initial data of a number-theoretic problem to construct a lattice in a Euclidean space, and then work with the collection of all such objects (lattices). Our goal is to describe several principles responsible for a particular class of applications of flows in the space of lattices to number theory. For more details and a broader picture the reader is referred to a number of extensive reviews of homogeneous actions and interactions with number theory which have appeared during the last 10 years, such as: ICM talks of Margulis [**Ma5**], Ratner [**Ra3**], Dani [**D5**] and Eskin [**E**], books [**St2**] and [**BMa**], and survey papers [**D6, D7, KSS, Ma7, St1**].

The structure of this paper is as follows: in the next section we collect all basic facts about the space of lattices, and discuss a "lattice" approach to studying values of quadratic forms at integer points (Problem 1 and Theorem 1.1 in particular). Then in §§ 3 and 4 we take several sub-problems of Problem 2 and describe recent results obtained by means of homogeneous dynamics. The last section is devoted to Conjecture 1.2 and related issues, that is, so called "multiplicative Diophantine approximation".

2. The space of lattices

Phase space. Fix $k \in \mathbb{N}$ and consider

$$\Omega \stackrel{\text{def}}{=} \text{the set of unimodular lattices in } \mathbb{R}^k$$

(discrete subgroups with covolume 1). That is, any lattice $\Lambda \in \Omega$ is equal to $\mathbb{Z}\mathbf{v}_1 \oplus \cdots \oplus \mathbb{Z}\mathbf{v}_k$, where the set $\{\mathbf{v}_1,\dots,\mathbf{v}_k\}$ (called a *generating set* of the lattice) is linearly independent, and the volume of the parallelepiped spanned by $\mathbf{v}_1,\dots,\mathbf{v}_k$ is equal to 1.

An element of Ω which is easy to distinguish is $\mathbb{Z}^k$ (the standard lattice). In fact, any $\Lambda \in \Omega$ is equal to $g\mathbb{Z}^k$ for some $g \in G \stackrel{\text{def}}{=} SL_k(\mathbb{R})$. That is, G acts transitively on Ω, and, further, $\Gamma \stackrel{\text{def}}{=} SL_k(\mathbb{Z})$ is the stabilizer of $\mathbb{Z}^k$. In other words, Ω is isomorphic to the homogeneous space G/Γ.

Measure. One can consider a Haar measure on G (both left and right invariant) and the corresponding left-invariant measure on Ω. It is well known that the resulting measure happens to be finite. We denote by μ the normalized Haar measure on Ω.

Topology. Two lattices are said to be close if their generating sets are close. This defines a topology on Ω which coincides with the quotient topology on G/Γ. An important feature is that Ω is not compact (in other words, Γ is a *non-uniform* lattice in G). More precisely, one has

THEOREM 2.1 (Mahler's Compactness Criterion, see [**R**]). *A subset K of Ω is bounded iff there exists $\varepsilon > 0$ such that for any $\Lambda \in K$ one has $\inf_{\mathbf{x}\in\Lambda\smallsetminus\{0\}} \|\mathbf{x}\| \geq \varepsilon$. In other words, define*

$$\Omega_\varepsilon \stackrel{\text{def}}{=} \big\{\Lambda \in \Omega \;\big|\; \|\mathbf{x}\| < \varepsilon \text{ for some } \mathbf{x} \in \Lambda \smallsetminus \{0\}\big\}\,;$$

then $\Omega \smallsetminus \Omega_\varepsilon$ is compact.

Action. Ω is a topological G-space, with the (continuous) left action defined by

$$g\Lambda = \{g\mathbf{x} \mid \mathbf{x} \in \Lambda\} \quad \text{or} \quad g(h\Gamma) = (gh)\Gamma\,.$$

One can consider the action of various subgroups (one- or multi-parameter) or subsets of G. Thus one gets an interesting class of dynamical systems. Several important features of these systems are worth mentioning.

First, the geometry of the phase space is "uniform": a small enough neighborhood of every point of Ω is isometric to a neighborhood of identity in G. In other words, many geometric constructions can be reduced to algebraic manipulations in G.

Second, the very rich representation theory of G can be heavily used. Namely, the G-action on Ω can be studied via the regular representation of G on $L^2(\Omega)$

The two features above in fact apply for all *homogeneous actions*, that is, actions of subgroups of a Lie group G on the quotient space G/Γ where Γ is a lattice in G. There are also important features specifically for the space $\Omega = SL_k(\mathbb{R})/SL_k(\mathbb{Z})$: namely, combinatorial structure of the space of lattices, as well as intuition coming from the theory of Diophantine approximation.

In what follows we will focus our attention on the space Ω, but most of the results will be valid in much bigger generality of homogeneous actions, which will be indicated. The reader is referred to [**AGH, Ma6, R, St2, Z**] for general facts about Lie groups, discrete subgroups and homogeneous spaces.

Classification of actions. Let G be a Lie group and Γ a discrete subgroup. Since $g(h\Lambda) = (ghg^{-1})g\Lambda$ for every $\Lambda \in G/\Gamma$ and $g, h \in G$, local properties of the g-action are determined by the differential of the conjugation map, $\mathrm{Ad}_g(x) = \frac{d\left(g\exp(tx)g^{-1}\right)}{dt}\big|_{t=0}$ (here x belongs to the Lie algebra of G). An element $g \in G$ is said to be: *unipotent* if $(\mathrm{Ad}_g - \mathrm{Id})^j = 0$ for some $j \in \mathbb{N}$ (equivalently, all eigenvalues of

Ad_g are equal to 1); *quasi-unipotent* if all eigenvalues of Ad_g are of absolute value 1; *partially hyperbolic* if it is not quasi-unipotent.

Given $g \in G$, define

$$H^{\pm}(g) = \{h \in G \mid g^{-l}hg^l \to e \text{ as } l \to \pm\infty\}$$

(*expanding and contracting horospherical subgroups*). Then G is locally a direct product of $H^-(g)$, $H^+(g)$ and another subgroup $H^0(g)$, and g is quasiunipotent iff $H^0(g) = G$ (that is, $H^-(g)$ and $H^+(g)$ are trivial). Furthermore, for any $\Lambda \in G/\Gamma$ the orbits $H^-(g)\Lambda$, $H^+(g)\Lambda$ and $H^0(g)\Lambda$ are leaves of *stable, unstable and neutral* foliations on G/Γ.

We now specialize to the case $G = SL_k(\mathbb{R})$ and $\Gamma = SL_k(\mathbb{Z})$.

EXAMPLE. The simplest case is when $k = 2$: then $\Omega = G/\Gamma$ is isomorphic to the unit tangent bundle to the surface $\mathbb{H}^2/SL_2(\mathbb{Z})$. The geodesic flow is then given by the action of $\begin{pmatrix} e^t & 0 \\ 0 & e^{-t} \end{pmatrix}$, and the horocycle flow – by the action of $\begin{pmatrix} 1 & t \\ 0 & 1 \end{pmatrix}$ (the simplest example of a unipotent flow).

MORE EXAMPLES. Suppose that $g \in G$ is diagonalizable over $\mathbb{R}$, and take a basis of $\mathbb{R}^k$ in which $g = \mathrm{diag}(\underbrace{\lambda_1, \dots, \lambda_1}_{i_1 \text{ times}}, \dots\dots, \underbrace{\lambda_l, \dots, \lambda_l}_{i_l \text{ times}})$, $\lambda_1 > \dots > \lambda_l$.
Then $H^-(g)$ and $H^+(g)$ are subgroups of lower- and upper- triangular groups. An important special case occurs when g as above comes from "the most singular" direction in a Weyl chamber of the Lie algebra of G; that is, when it has only two distinct eigenvalues. In this case one can write $k = m + n$ and consider a one-parameter subgroup of $G = SL_k(\mathbb{R})$ given by

$$g_t = \mathrm{diag}(e^{t/m}, \dots, e^{t/m}, e^{-t/n}, \dots, e^{-t/n}) . \tag{2.1}$$

Then the expanding horospherical subgroup of G relative to g_1 is exactly $\{L_Y \mid Y \in M_{m\times n}(\mathbb{R})\}$, where L_Y is as defined in (1.3).

Ergodic properties. Here the main tool is the representation theory of semisimple Lie groups. By a theorem of Moore [**Mo1**], the action of any noncompact closed subgroup of G on $\Omega = G/\Gamma$ is ergodic and, moreover, mixing; in other words, matrix coefficients $(g\varphi, \psi)$ of square-integrable functions on Ω with mean value zero tend to 0 as $g \to \infty$ in G. (Here $(\cdot,\cdot)$ stands for the inner product in $L^2(\Omega)$.) In fact for smooth functions this decay is exponential, as shown in the following

THEOREM 2.2 (DECAY OF CORRELATIONS). *There exists $\beta > 0$ such that for any two functions $\varphi, \psi \in C^\infty_{comp}(G/\Gamma)$ with $\int \varphi\, d\mu = \int \psi\, d\mu = 0$ and any $g \in G$ one has*

$$\left| \int (g\varphi \cdot \psi)\, d\mu \right| \le \mathrm{const}(\varphi, \psi) e^{-\beta\|g\|} .$$

In particular, if g_t is partially hyperbolic, then

$$\left| \int (g_t\varphi \cdot \psi)\, d\mu \right| \le \mathrm{const}(\varphi, \psi, g_t) e^{-\gamma t} . \tag{2.2}$$

See [**Mo2, Ra1**] for $k = 2$, [**KS**] for $k > 2$.

The following result can be derived from the mixing property of partially hyperbolic actions on G/Γ:

THEOREM 2.3 (Uniform distribution of unstable leaves, [**KM1**]). *Let g_t be a partially hyperbolic one-parameter subgroup of G, $H = H^+(g_1)$, ν a Haar measure on H. Then for any open subset V of H, any $\varphi \in C^\infty_{comp}(G/\Gamma)$ and any compact subset Q of G/Γ, the average of φ over the g_t-image of $V\Lambda$, $\Lambda \in G/\Gamma$, tends to the integral of φ as $t \to \infty$ uniformly (in Λ) on compact subsets of G/Γ; that is,*

$$\frac{1}{\nu(g_tVg_{-t})}\int_{g_tVg_{-t}} \varphi(hg_t\Lambda)\,d\nu(h) \to \int_{G/\Gamma} \varphi\,d\mu\,. \tag{2.3}$$

REMARKS. Moore's theorem (that is, a criterion for mixing of subgroup actions) was proved under the assumption that

$$\begin{gathered} G \text{ is a connected semisimple Lie group with finite center} \\ \text{and no compact factors, and } \Gamma \text{ is an irreducible lattice in } G\,, \end{gathered} \tag{2.4}$$

(A lattice is irreducible if it is not, up to commensurability, a product of lattices in simple factors of G.) See also [**BM, Ma5**] for more general ergodicity and mixing criteria. Theorems 2.2 was proved in [**KS**] assuming (2.4) and in addition that

$$\text{all simple factors of } G \text{ have property (T).} \tag{2.5}$$

One also knows, see [**Bek,** Lemma 3], that Theorem 2.2 holds when the group G is simple. In [**KM3**] it was shown that one can remove condition (2.5) but instead assume that $\Gamma \subset G$ is a non-uniform lattice.

Theorem 2.3 is also proven in the generality of the assumption (2.4) (in fact, one only needs mixing of the g_t-action), and as long as Theorem 2.2 holds, the convergence in (2.3) is exponential in t. See [**KM1, K4**] for more details and generalizations, and [**Ma7,** Remark 3.10] for references to other related results and methods.

Recurrence of unipotent trajectories. In this subsection we are back to the case $G = SL_k(\mathbb{R})$ and $\Gamma = SL_k(\mathbb{Z})$. It is an elementary geometric observation that horocyclic trajectories on $SL_2(\mathbb{R})/SL_2(\mathbb{Z})$ do not run off to infinity. It is much harder to prove that the same holds for any unipotent flow on $\Omega = SL_k(\mathbb{R})/SL_k(\mathbb{Z})$ for $k \geq 3$ [**Ma2, D3**]. The theorem below, due to Dani (1985), is a quantitative strengthening; it shows that for any unipotent orbit one can find a compact subset of Ω such that the density of time that the orbit spends in this set is as close to 1 as one wishes:

THEOREM 2.4 [**D3**]. *For any $\Lambda \in \Omega$ and any $\delta > 0$ there exists $\varepsilon > 0$ such that for any unipotent subgroup $\{u_x | x \in \mathbb{R}\}$ of G and any $T > 0$ one has*

$$\big|\{x \in [0,T] \mid u_x\Lambda \in \Omega_\varepsilon\}\big| \leq \delta T\,.$$

The proof is based on the combinatorial structure of the space of lattices. We will obtain the theorem above as a corollary of a more general fact in §4.

Orbit closures of unipotent flows. It has been proved by Hedlund that any orbit of the horocycle flow on $SL_2(\mathbb{R})/SL_2(\mathbb{Z})$ is either periodic or dense. A far-reaching generalization has been conjectured by Raghunathan and proved in full generality by Ratner. In particular, one has the following

THEOREM 2.5. [**Ra2**] *Let G be a connected Lie group, Γ a lattice in G, and let U be a subgroup of G generated by unipotent one-parameter subgroups. Then for any $x \in G/\Gamma$ there exists a closed subgroup L containing U such that the closure of the orbit Ux coincides with Lx and there is an L-invariant probability measure supported on Lx.*

We emphasize that Theorem 2.5 allows one to understand all (not just almost all) orbits. It is this feature which is responsible for applications of this theorem to number theory. Let us illustrate it by sketching the reduction of the Oppenheim Conjecture (Theorem 1.1) to Ratner's theorem.

COROLLARY 2.6. *Let $S(x_1, x_2, x_3) = 2x_1x_3 - x_2^2$, and*

$$H_S = \{h \in SL_3(\mathbb{R}) \mid S(h\mathbf{x}) = S(\mathbf{x})\ \forall \mathbf{x} \in \mathbb{R}^3\} \cong SO(2,1)$$

(the stabilizer of S). Then any relatively compact orbit $H_S\Lambda$, Λ a lattice in $\mathbb{R}^3$, is compact.

PROOF. H_S is generated by its unipotent one-parameter subgroups, namely

$$u(t) = \begin{pmatrix} 1 & t & t^2/2 \\ 0 & 1 & t \\ 0 & 0 & 1 \end{pmatrix} \text{ and } u^T(t) = \begin{pmatrix} 1 & 0 & 0 \\ t & 1 & 0 \\ t^2/2 & t & 1 \end{pmatrix},$$

and there are no intermediate subgroups between H_S and $SL_3(\mathbb{R})$. Hence by Theorem 2.5 any H_S-orbit is either closed or dense. □

The following is crucial for the deduction of the Oppenheim conjecture from the above corollary (implicitly stated in [**CS**] and later observed by Raghunathan):

LEMMA 2.7. *Let B be a real nondegenerate indefinite quadratic form in 3 variables. Write $B(\mathbf{x}) = \lambda S(g\mathbf{x})$ for some $g \in SL_3(\mathbb{R})$. Then the orbit $H_S g\mathbb{Z}^3$ is relatively compact if and only if*

$$|B(\mathbf{x})| \geq \varepsilon \text{ for some } \varepsilon > 0 \text{ and all } \mathbf{x} \in \mathbb{Z}^3 \smallsetminus \{0\}. \tag{2.6}$$

PROOF. By transitivity of the action of the stabilizer H_B of the form B on the level sets of B in $\mathbb{R}^3 \smallsetminus \{0\}$ and by continuity of B at zero, assertion (2.6) is equivalent to the norm of $h\mathbf{x}$ being not less than ε for some $\varepsilon > 0$ and all $\mathbf{x} \in \mathbb{Z}^3 \smallsetminus \{0\}$ and $h \in H_B$. The latter, in view of Theorem 2.1, is equivalent to the orbit $H_B\mathbb{Z}^3$ being relatively compact in $\Omega = SL_3(\mathbb{R})/SL_3(\mathbb{Z})$. But $H_B = g^{-1}H_S g$, therefore the orbit $H_B\mathbb{Z}^3$ is relatively compact if and only if so is the orbit $H_S g\mathbb{Z}^3$. □

COROLLARY 2.8. *Let B be a real nondegenerate indefinite quadratic form in 3 variables. If (2.6) holds, then B is proportional to a rational form.*

PROOF. The previous corollary implies that $H_B\mathbb{Z}^3$ is compact; but since this orbit can be identified with $H_B/H_B \cap SL_3(\mathbb{Z})$, this shows that $H_B \cap SL_3(\mathbb{Z})$ is Zariski dense in H_B, which is equivalent to H_B being defined over $\mathbb{Q}$, hence the claim. □

To derive Theorem 1.1 from the above corollary, one can then observe that if B is a real irrational nondegenerate indefinite quadratic form in k variables and $l < k$ then $\mathbb{R}^k$ contains a rational subspace L of dimension l such that the restriction of B

to L is irrational nondegenerate and indefinite (the proof can be found in [**DM1**]). Hence the validity of Theorem 1.1 in the case $k = 3$ implies the general case.

It is worthwhile to mention that in the paper [**Op2**] Oppenheim modified his conjecture replacing the claim $\inf_{\mathbf{x}\in\mathbb{Z}^k\smallsetminus\{0\}} |B(\mathbf{x})| = 0$ with "0 is a non-isolated accumulation point of $B(\mathbb{Z}^k)$", which he showed to be equivalent to the density of $B(\mathbb{Z}^k)$ in $\mathbb{R}$. This stronger form of the conjecture was also proved by Margulis [**Ma4**]. It is not difficult to derive it from Theorem 2.5, first reducing to the case $k = 3$: if B is not proportional to a rational form, the orbit $H_B\mathbb{Z}^3$ is not closed, hence (by non-existence of intermediate subgroups between H_B and $SL_3(\mathbb{R})$) it is dense in Ω, and the density of $B(\mathbb{Z}^3)$ in $\mathbb{R}$ follows.

Finally let us briefly mention quantitative extensions of the above results. For B as above, an open interval $I \subset \mathbb{R}$ and a positive T one defines

$$V_{I,B}(T) \stackrel{\text{def}}{=} \{\mathbf{x} \in \mathbb{R}^k \mid B(\mathbf{x}) \in I, \|\mathbf{x}\| \le T\}$$

and

$$N_{I,B}(T) \stackrel{\text{def}}{=} \#\big(\mathbb{Z}^k \cap V_{I,B}(T)\big)\,,$$

then one has $N_{I,B}(T) \to \infty$ as $T \to \infty$ for every nonempty $I \subset \mathbb{R}$. The next theorem, a compilation of results from [**DM2**] and [**EMM**], describes the growth of this counting function comparing it to the volume of $V_{I,B}(T)$.

THEOREM 2.9. *Let B be a real nondegenerate indefinite quadratic form of signature (m, n), $m+n > 2$, $m \le n$, which is not proportional to a rational form, and let $I \subset \mathbb{R}$ be a nonempty open interval. Then*

(a) [**DM2**] $\displaystyle\liminf_{T\to\infty} \frac{N_{I,B}(T)}{|V_{I,B}(T)|} = 1\,;$

(b) [**EMM**] *if $n \ge 3$, then* $\displaystyle\lim_{T\to\infty} \frac{N_{I,B}(T)}{|V_{I,B}(T)|} = 1\,.$

In the exceptional cases, i.e. for forms of signature $(2, 1)$ and $(2, 2)$, there are counterexamples showing that $N_{I,B}(T)$ can grow like $\text{const}\cdot|V_{I,B}(T)|(\log T)^{1-\varepsilon}$, and it is proved in [**EMM**] that $\text{const}\cdot|V_{I,B}(T)|\log T$ is an asymptotically exact upper bound for $N_{I,B}(T)$.

For both parts the crucial step is to approximate the counting function $N_{I,B}(T)$ by values of integrals of certain functions along orbits in the space of lattices. Part (a) relies upon Ratner's uniform distribution theorem (a refinement of Theorem 2.5 for one-parameter unipotent subgroups), while the second part involves delicate estimates based on combinatorics of lattices. See [**DM1, DM2, EMM, Ma7**] for more details on the proofs and further refinements and generalizations.

3. Metric linear Diophantine approximation and lattices

The basic object to study in this section will be the set $M_{m\times n}(\mathbb{R})$ of $m \times n$ real matrices. The word "metric" refers to considering solution sets of Diophantine inequalities in terms of the Lebesgue measure, or, when the sets are of measure zero, for finer analysis, in terms of the Hausdorff dimension. We refer the reader to the books [**C, H, S3, Sp3**] for a detailed exposition.

In what follows, $\psi(\cdot)$ will be a positive non-increasing function $\mathbb{R}_+ \mapsto \mathbb{R}_+$. We are going to use it to measure the precision of approximation of a real number α by rational numbers as follows: we would like the fractional part of αq to be not bigger than $\psi(|q|)$ for infinitely many $q \in \mathbb{Z}$ in order to call α "sufficiently well approximable" (this notion being dependent on ψ). More precisely, let us say that α is *ψ-approximable* if there are infinitely many $q \in \mathbb{Z}$ such that

$$|\alpha q + p| \le \psi(|q|) \quad \text{for some } p \in \mathbb{Z}\,.$$

In order to consider a matrix analogue of this notion, one needs to choose a norm on $\mathbb{R}^k$ (we will do it by setting $\|\mathbf{x}\| = \max_{1\le i\le k}|x_i|$). Then one says that a matrix $Y \in M_{m\times n}(\mathbb{R})$ (viewed as a system of m linear forms in n variables) is *ψ-approximable* if there are infinitely many $\mathbf{q} \in \mathbb{Z}^n$ such that

$$\|Y\mathbf{q} + \mathbf{p}\|^m \le \psi(\|\mathbf{q}\|^n) \quad \text{for some } \mathbf{p} \in \mathbb{Z}^m\,. \tag{3.1}$$

The above normalization (raising norms in the power equal to the dimension of the space, instead of the traditional $\|Y\mathbf{q} + \mathbf{p}\| \le \psi(\|\mathbf{q}\|)$ as in [**Dod**] or [**BD**]) is convenient for many reasons: in our opinion it makes the structure more transparent and less dimension-dependent, and simplifies the connection with homogeneous flows.

The whole theory starts from a positive result of Dirichlet, namely

THEOREM 3.1. *Every $Y \in M_{m\times n}(\mathbb{R})$ is ψ_0-approximable, where $\psi_0(x) = \frac{1}{x}$.*

Clearly the faster ψ decays, the smaller is the set of ψ-approximable matrices. The next theorem, Groshev's [**Gr**] generalization of earlier results of Khintchine, provides the zero-one law for the Lebesgue measure of this set:

THEOREM 3.2 (The Khintchine-Groshev Theorem). *Almost every (resp. almost no) $Y \in M_{m\times n}(\mathbb{R})$ is ψ-approximable, provided the sum $\sum_{l=1}^{\infty}\psi(l)$ diverges (resp. converges).*

Now say that $Y \in M_{m\times n}(\mathbb{R})$ is *badly approximable* if it is not $c\psi_0$-approximable for some $c > 0$; that is, if there exists $c > 0$ such that $\|Y\mathbf{q} + \mathbf{p}\|^m\|\mathbf{q}\|^n \ge c$ for all $\mathbf{p} \in \mathbb{Z}^m$ and all but finitely many $\mathbf{q} \in \mathbb{Z}^n$ (equivalently: all $\mathbf{q} \in \mathbb{Z}^n \smallsetminus \{0\}$).

Note that in the case $m = n = 1$, $\alpha \in \mathbb{R}$ is badly approximable if and only if coefficients in the continued fraction expansion of α are bounded. Using continued fractions, Jarnik proved in 1928 that badly approximable numbers form a set of Hausdorff dimension one; for arbitrary m, n the corresponding fact, i.e. full Hausdorff dimension of badly approximable systems, was established by Schmidt in 1969 [**S2**].

The following interpretation of this property in terms of homogeneous dynamics is due to Dani. Throughout this section we will fix $m, n \in \mathbb{N}$ and put $k = m + n$.

THEOREM 3.3 [**D1**]. *$Y \in M_{m\times n}(\mathbb{R})$ is badly approximable iff the trajectory $\{g_t L_Y \mathbb{Z}^k \mid t \in \mathbb{R}_+\}$, with L_Y as in (1.3) and g_t as in (2.1), is bounded in the space Ω of unimodular lattices in $\mathbb{R}^k$.*

Instead of giving the proof (which, besides the original paper [**D1**] can be found in [**K2, K3**]) let us point out the similarity between the above theorem and Lemma 2.7. Indeed, denote by S the function on $\mathbb{R}^k$ given by

$$S(x_1, \dots, x_k) = \max(|x_1|, \dots, |x_m|)^m \max(|x_{m+1}|, \dots, |x_k|)^n\,.$$

Then Y is badly approximable iff for some $c > 0$ one has

$$S(L_Y\mathbf{x}) \ge c \text{ for all } \mathbf{x} = (\mathbf{p}, \mathbf{q}) \in \mathbb{Z}^m \times (\mathbb{Z}^n \smallsetminus \{0\})\,. \tag{3.2}$$

Furthermore, the one-parameter group $\{g_t\}$ as in (2.1) is essentially (up to the compact part) the stabilizer of S, and, as in the proof of Lemma 2.7, one can show that (3.2) is equivalent to the norm of $g_t L_Y \mathbf{x}$ being bounded away from zero for all $\mathbf{x} \in \mathbb{Z}^k \setminus \{0\}$ and $t \geq 0$, that is, to the statement that $\{g_t L_Y \mathbb{Z}^k \mid t \in \mathbb{R}_+\} \cap \Omega_\varepsilon = \varnothing$ for some $\varepsilon > 0$.

From the above theorem and the aforementioned result of Schmidt, Dani derived

COROLLARY 3.4. *The set*

$$\{\Lambda \in \Omega \mid \{g_t\Lambda \mid t \geq 0\} \textit{ is bounded}\},$$

with $\{g_t\}$ *as in* (2.1), *has full Hausdorff dimension.*

PROOF. Indeed, any $\Lambda \in \Omega$ can be written as $\begin{pmatrix} B & 0 \\ C & D \end{pmatrix} L_Y \mathbb{Z}^k$, therefore one has

$$g_t\Lambda = g_t \begin{pmatrix} B & 0 \\ C & D \end{pmatrix} g_{-t} \cdot g_t L_Y \mathbb{Z}^k .$$

But as we saw in one of the examples of §2, $\{L_Y \mid Y \in M_{m\times n}(\mathbb{R})\}$ is the expanding horospherical subgroup of G relative to g_1; thus the conjugation of the neutral and contracting parts plays no role and the trajectory $g_t\Lambda$ is bounded iff so is $g_t L_Y \mathbb{Z}^k$.□

A possibility to generalize the statement of the last corollary (to actions of other one-parameter groups on other homogeneous spaces) was mentioned by Dani in [**D2**] and later conjectured by Margulis [**Ma5,** Conjecture (A)]. The latter conjecture was settled by Margulis and the author in 1996. Let us state here the following weakened version.

THEOREM 3.5 [**KM1, K4**]. *Let G be a Lie group, Γ a lattice in G, $F = \{g_t \mid t \geq 0\}$ be a one-parameter subsemigroup of G consisting of semisimple*[1] *elements, and let $H = H_+(g_1)$ be the expanding horospherical subgroup corresponding to F. Assume in addition that the F-action on G/Γ is mixing. Then for any closed F-invariant null subset Z of G/Γ and any $x \in G/\Gamma$, the set*

$$\{h \in H \mid Fhx \textit{ is bounded and } \overline{Fhx} \cap Z = \varnothing\}$$

has full Hausdorff dimension. In particular, if $\{g_t\}$ is partially hyperbolic, then the set $\{x \in G/\Gamma \mid Fx$ is bounded and $\overline{Fx} \cap Z = \varnothing\}$ has full Hausdorff dimension.

Note that abundance of exceptional orbits is a feature of many *chaotic* dynamical systems. See e.g. [**AN1, AN2, D4, Dol1, U**]. In the situation of Theorem 3.5, the construction of bounded orbits (or, more generally, orbits staying away from a fixed part of the space) comes from uniform distribution of images of expanding leaves (Theorem 2.3). More precisely, first one reduces the problem to the case (2.4), and then considers natural "rectangular" partitions of H (called *tessellations* in [**KM1**] and [**K1**]) and studies their behavior under the automorphism $h \mapsto g_t h g_{-t}$ of H. Theorem 2.3 is used to show than one can cover the set of "bad" points by relatively small number of rectangles. Then those rectangles are used to

[1] $g \in G$ is called *semisimple* if the operator Ad_g is diagonalizable over $\mathbb{C}$

create a Cantor set consisting of points with orbits avoiding Z and staying within a compact subset of G/Γ. See [**KM1, K1, K4**] for details and generalizations[2]

So far we have illustrated the impact of ideas coming from Diophantine approximation to ergodic theory. On the other hand, Theorem 3.5 and Dani's correspondence (Theorem 3.3) can be used as an alternative proof of the aforementioned result of Schmidt on abundance of badly approximable systems of linear forms. What follows is another application to number theory, which produces a new result and demonstrates the power of ideas relating the two fields.

Let us consider an inhomogeneous twist of approximation of real numbers by rationals. Instead of just one real number α take a pair $\langle \alpha, \beta \rangle$, consider an *affine form* $x \mapsto \alpha x + \beta$ and look at fractional parts of its values at integers. Similarly, a system of m affine forms in n variables will be then given by a pair $\langle Y, \mathbf{b} \rangle$, where $Y \in M_{m\times n}(\mathbb{R})$ and $\mathbf{b} \in \mathbb{R}^m$. Let us denote by $\tilde{M}_{m\times n}(\mathbb{R})$ the direct product of $M_{m\times n}(\mathbb{R})$ and $\mathbb{R}^m$. Now say that a system of affine forms given by $\langle Y, \mathbf{b} \rangle \in \tilde{M}_{m\times n}(\mathbb{R})$ is *ψ-approximable* if there are infinitely many $\mathbf{q} \in \mathbb{Z}^n$ such that

$$\|Y\mathbf{q} + \mathbf{b} + \mathbf{p}\|^m \le \psi(\|\mathbf{q}\|^n) \quad \text{for some } \mathbf{p} \in \mathbb{Z}^m\,,$$

and *badly approximable* if it is not $c\psi_0$-approximable for some $c > 0$; that is, there exists a constant $\tilde{c} > 0$ such that for every $\mathbf{p} \in \mathbb{Z}^m$ and all but finitely many $\mathbf{q} \in \mathbb{Z}^n \smallsetminus \{0\}$ one has

$$\|Y\mathbf{q} + \mathbf{b} + \mathbf{p}\|^m \|\mathbf{q}\|^n > \tilde{c}\,.$$

It can be proved (and follows from an inhomogeneous version of the Khintchine-Groshev Theorem, see [**C**]) that the set of badly approximable $\langle Y, \mathbf{b} \rangle \in \tilde{M}_{m\times n}(\mathbb{R})$ is of measure zero. However, all known examples of badly approximable $\langle Y, \mathbf{b} \rangle \in \tilde{M}_{m\times n}(\mathbb{R})$ belong to a countable union of proper submanifolds of $\tilde{M}_{m\times n}(\mathbb{R})$, hence form a set of positive Hausdorff codimension. Yet a modification of the dynamical approach described above works in this case as well. Namely, one considers a collection of vectors

$$\left\{ \begin{pmatrix} Y\mathbf{q} + \mathbf{b} + \mathbf{p} \\ \mathbf{q} \end{pmatrix} \middle| \mathbf{p} \in \mathbb{Z}^m, \mathbf{q} \in \mathbb{Z}^n \right\} = L_Y \mathbb{Z}^k + \begin{pmatrix} \mathbf{b} \\ 0 \end{pmatrix},$$

which is an element of the space $\hat{\Omega} = \hat{G}/\hat{\Gamma}$ of *affine lattices* in $\mathbb{R}^k$, where

$$\hat{G} \stackrel{\text{def}}{=} \operatorname{Aff}(\mathbb{R}^k) = G \ltimes \mathbb{R}^k \text{ and } \hat{\Gamma} \stackrel{\text{def}}{=} \Gamma \ltimes \mathbb{Z}^k\,.$$

In other words,

$$\hat{\Omega} \cong \{\Lambda + \mathbf{w} \mid \Lambda \in \Omega,\ \mathbf{w} \in \mathbb{R}^k\}\,.$$

Note that the quotient topology on $\hat{\Omega}$ coincides with the natural topology on the space of affine lattices: that is, $\Lambda_1 + \mathbf{w}_1$ and $\Lambda_2 + \mathbf{w}_2$ are close to each other if so are $\mathbf{w}_i$ and the generating elements of Λ_i. Note also that $\hat{\Omega}$ is non-compact and has finite Haar measure, and that Ω (the set of *true* lattices) can be identified with a subset of $\hat{\Omega}$ (affine lattices containing the zero vector). Finally, g_t as in (2.1) acts on $\hat{\Omega}$, and it is not hard to show that the expanding horospherical subgroup

[2]In particular, it follows from the methods of [**KM1**] that one can remove the assumption of semisimplicity of elements of F, but then one needs the F-action to be exponentially mixing, that is, (2.2) must hold for any φ, ψ as in Theorem 2.2.

corresponding to g_1 is exactly the set of all elements of $\hat{G}$ with linear part L_Y and translation part $\begin{pmatrix} \mathbf{b} \\ 0 \end{pmatrix}$, $Y \in M_{m\times n}(\mathbb{R})$ and $\mathbf{b} \in \mathbb{R}^m$.

Now, for $\varepsilon > 0$, define

$$\hat{\Omega}_\varepsilon \stackrel{\text{def}}{=} \left\{ \Lambda \in \hat{\Omega} \mid \|\mathbf{x}\| < \varepsilon \text{ for some } \mathbf{x} \in \Lambda \right\}.$$

Then $\hat{\Omega} \setminus \hat{\Omega}_\varepsilon$ is a closed (non-compact) set disjoint from Ω.

THEOREM 3.6 [**K4**]. *Let $F = \{g_t \mid t \geq 0\}$ be as in* (2.1). *Then*

$$F\left(L_Y\mathbb{Z}^k + \begin{pmatrix} \mathbf{b} \\ 0 \end{pmatrix}\right) \quad \textit{is bounded and stays away from } \Omega$$

$$\Downarrow$$

$$F\left(L_Y\mathbb{Z}^k + \begin{pmatrix} \mathbf{b} \\ 0 \end{pmatrix}\right) \subset \hat{\Omega} \setminus \hat{\Omega}_\varepsilon \textit{ for some } \varepsilon > 0$$

$$\Downarrow$$

$$\langle Y, \mathbf{b}\rangle \textit{ is badly approximable}$$

The proof is basically a slight modification of ideas involved in the proof of Theorem 3.3. It follows from the results of [**BM**] (see also [**Ma5**]) that the F-action on $\hat{\Omega}$ is mixing. Since $\Omega \subset \hat{\Omega}$ is closed, null and g_t-invariant, Theorem 3.5 applies and one gets

COROLLARY 3.7. *The set of badly approximable $\langle Y, \mathbf{b}\rangle \in \tilde{M}_{m\times n}(\mathbb{R})$ has full Hausdorff dimension.*

See [**K4**] for details, remarks and extensions.

We close the section by stating a theorem generalizing Dani's correspondence (Theorem 3.3) to ψ-approximable systems. First we need a simple "change of variables" lemma.

LEMMA 3.8. *Fix $m, n \in \mathbb{N}$ and $x_0 > 0$, and let $\psi : [x_0, \infty) \mapsto (0, \infty)$ be a non-increasing continuous function. Then there exists a unique continuous function $\varepsilon : [t_0, \infty) \mapsto (0, \infty)$, where $e^{kt_0} = x_0^m/\psi(x_0)^n$, such that*

$$\text{(3.3a)} \qquad \textit{the function} \quad t \mapsto e^t\varepsilon(t)^n \quad \textit{is strictly increasing and unbounded},$$

$$\text{(3.3b)} \qquad \textit{the function} \quad t \mapsto e^{-t}\varepsilon(t)^m \quad \textit{is nonincreasing},$$

and

$$\text{(3.4)} \qquad \psi\big(e^t\varepsilon(t)^n\big) = e^{-t}\varepsilon(t)^m \quad \forall\, t \geq t_0\,.$$

Conversely, given $t_0 \in \mathbb{R}$ and a continuous function $\varepsilon : [t_0, \infty) \mapsto (0, \infty)$ such that (3.3ab) *hold, there exists a unique continuous non-increasing function $\psi : [x_0, \infty) \mapsto (0, \infty)$, with $x_0 = e^{t_0}\varepsilon(t_0)^n$, satisfying* (3.4).

See [**KM3**] for the proof. In many cases one can explicitly solve (3.4) to express $\varepsilon(\cdot)$ knowing $\psi(\cdot)$ and vice versa. For example if $\psi(x) = c\psi_0(x) = c/x$, the equation (3.4) gives $cc^{-t}\varepsilon(t)^{-n} = e^{-t}\varepsilon(t)^m$, and one sees that the corresponding function ε

is constant (more precisely, $\varepsilon(t) \equiv c^{1/k}$). Or one can take $\psi(x) = c\psi_\beta(x)$, where $\psi_\beta(x) = \frac{1}{x^{1+\beta}}$, $\beta > 0$; then $\varepsilon(t)$ decreases exponentially, namely

$$\varepsilon(t) = c^{\gamma/\beta} e^{-\gamma t}\,, \text{ where } \gamma = \frac{\beta}{(1+\beta)m + n}\,.$$

Now we can state a generalization of Theorem 3.3:

THEOREM 3.9. *$Y \in M_{m\times n}(\mathbb{R})$ is ψ-approximable iff there exist arbitrarily large positive t such that $g_t L_Y \mathbb{Z}^k \in \Omega_{\varepsilon(t)}$, where $\{g_t\}$ is as in (2.1), L_Y as in (1.3), and $\varepsilon(\cdot)$ is the function corresponding to ψ as in the previous lemma.*

Loosely speaking, good rational approximations for Y correspond to far excursions of the orbit into the "cusp neighborhoods" Ω_ε. In other words, one can measure the "growth rate" of the orbit in terms of hitting the sets $\Omega_{\varepsilon(t)}$ in time t for infinitely many $t \in \mathbb{N}$, and fast-growing orbits would correspond to systems approximable with a fast-decaying approximation function.

It is shown in [**KM3**] how the above correspondence provides an alternative (dynamical) proof of Theorem 3.2. More precisely, one can use ergodic properties of the g_t-action on Ω (exponential decay of correlations, see Theorem 2.2) to prove the following

THEOREM 3.10. *Let $\varepsilon(\cdot)$ be any positive function. Then for almost all (resp. almost no) $\Lambda \in \Omega$ one has $g_t\Lambda \in \Omega_{\varepsilon(t)}$ for infinitely many $t \in \mathbb{N}$, provided the sum*

$$\sum_{t=1}^{\infty} \varepsilon(t)^k \tag{3.5}$$

diverges (resp. converges).

We remark that the ratio $\mu(\Omega_\varepsilon)/\varepsilon^k$ is shown in [**KM3**] to be bounded from both sides; therefore the sum (3.5) is finite/infinite iff so is $\sum_{t=1}^\infty \mu(\Omega_{\varepsilon(t)})$. This places the above theorem in the rank of Borel-Cantelli type results. See [**KM3**] for generalizations and applications, and [**CK, CR, Dol2, Ph, Su**] for other results of similar flavor.

Another application of the correspondence of Theorem 3.9 is coming in the next section.

4. Diophantine approximation on manifolds

We start from the setting of the previous section but specialize to the case $m = 1$; that is, to Diophantine approximation of just one linear form given by $\mathbf{y} \in \mathbb{R}^n$. Recall that Theorem 3.2 says that whenever $\sum_{l=1}^\infty \psi(l)$ is finite, almost every $\mathbf{y}$ is not ψ-approximable; that is, the inequality

$$|\mathbf{q}\cdot\mathbf{y} + p| \le \psi(\|\mathbf{q}\|^n) \tag{4.1}$$

has at most finitely many solutions. It is instructive to sketch an elementary proof: for fixed $p, \mathbf{q}$, the set of $\mathbf{y}$ satisfying (4.1) is a $\frac{\psi(\|\mathbf{q}\|^n)}{\|\mathbf{q}\|}$-neighborhood of a hyperplane

$$\mathbf{q}\cdot\mathbf{y} + p = 0\,; \tag{4.2}$$

thus if one restricts $\mathbf{y}$ to lie in $[0,1]^n$ (or any other bounded subset of $\mathbb{R}^n$), the set of solutions will have measure at most const$\cdot\frac{\psi(\|\mathbf{q}\|^n)}{\|\mathbf{q}\|}$. Since there are at most const$\cdot\|\mathbf{q}\|$ admissible values of p, the sum of measures of all sets of solutions is at most

$$\sum_{\mathbf{q}\in\mathbb{Z}^n}\psi(\|\mathbf{q}\|^n)\asymp\sum_{l=1}^{\infty}l^{n-1}\psi(l^n)\asymp\sum_{l=1}^{\infty}\psi(l),$$

and the proof is finished by an application of the Borel-Cantelli Lemma.

Recall that $\psi_\beta(x) = x^{-(1+\beta)}$, $\beta > 0$, was one of the examples of functions realizing the convergence case in the Khintchine-Groshev Theorem. Say that $\mathbf{y}\in\mathbb{R}^n$ is *very well approximable (VWA)* if it is ψ_β-approximable for some $\beta > 0$. Thus almost all $\mathbf{y}\in\mathbb{R}^n$ are not VWA.

Now consider the following problem, raised by Mahler in 1932 [**M**]: is it true that for almost all $x\in\mathbb{R}$ the inequality

$$|p+q_1x+q_2x^2+\cdot+q_nx^n|\leq\|\mathbf{q}\|^{-n(1+\beta)}$$

has at most finitely many solutions? In other words, for a.e. $x\in\mathbb{R}$, the n-tuple

$$\mathbf{y}(x)=(x,x^2,\dots,x^n) \tag{4.3}$$

is not VWA. The proof presented above does not work, since this time one has to estimate the measure of intersection of the curve (4.3) with the sets of solutions of inequalities (4.1), and for some choices of $p,\mathbf{q}$ (namely for those which make the hyperplane (4.2) nearly tangent to the curve (4.3)) it is hard to produce a reasonable estimate.

This problem stood open for more than 30 years until it was solved in 1964 by Sprindžuk [**Sp1, Sp2**]. Earlier several special cases were considered, and, quoting Sprindžuk's survey paper, the problem rapidly revealed itself to be non-trivial and involving "deep and complicated phenomena in which arithmetical properties of numbers are closely entangled with combinatorial-topological properties" of the curve. The solution to Mahler's problem has eventually led to the development of a new branch of metric number theory, usually referred to as "Diophantine approximation with dependent quantities" or "Diophantine approximation on manifolds". We invite the reader to look at Sprindžuk's monographs [**Sp2, Sp3**] and a recent book [**BD**] for a systematic exposition of the field.

Mahler's problem and its generalizations have several motivations. The original motivation of Mahler comes from transcendental number theory. Indeed, the n-tuple (4.3) is not VWA if and only if for every $\beta > 0$ there are at most finitely many polynomials $P\in\mathbb{Z}[x]$ with degree at most n such that $|P(x)| < h(P)^{-n(1+\beta)}$, where $h(P)$ is the height of P; loosely speaking, x is "not very algebraic", and the affirmative solution to the problem shows that almost all x are such.

Another motivation comes from KAM theory: it is known that behavior of perturbation of solutions of ODEs is related to Diophantine properties of coefficients. If the latter are restricted to lie on a curve or submanifold of $\mathbb{R}^n$, it may be important to know that almost all values have certain approximation properties. See [de la Llave's lectures, this volume] and [**BD,** Chapter 7].

However, from the author's personal viewpoint, the appeal of this branch of number theory lies in its existing and potential generalizations. In a sense, the affirmative solution to Mahler's problem shows that a certain property of $\mathbf{y}\in\mathbb{R}^n$ (being not VWA) which holds for generic $\mathbf{y}\in\mathbb{R}^n$ in fact holds for generic points on the curve (4.3). In other words, the curve inherits the above Diophantine property from the ambient space, unlike, for example, a line $\mathbf{y}(x) = (x,\dots,x)$ – it is clear that every point on this line is VWA. This gives rise to studying other subsets of $\mathbb{R}^n$ and other Diophantine properties, and looking at whether this inheritance phenomenon takes place.

Note that the curve (4.3) is not contained in any affine subspace of $\mathbb{R}^n$ (in other words, constitutes an essentially n-dimensional object). The latter property, or, more precisely, its infinitesimal analogue, is formalized in the following way. Let V be an open subset of $\mathbb{R}^d$. Say that an n-tuple $\mathbf{f} = (f_1, \dots, f_n)$ of C^l functions $V \mapsto \mathbb{R}$ is *nondegenerate at* $x \in V$ if the space $\mathbb{R}^n$ is spanned by partial derivatives of $\mathbf{f}$ at x of order up to l. If $M \subset \mathbb{R}^n$ is a d-dimensional smooth submanifold, we will say that M is *nondegenerate at* $\mathbf{y} \in M$ if any (equivalently, some) diffeomorphism $\mathbf{f}$ between an open subset V of $\mathbb{R}^d$ and a neighborhood of $\mathbf{y}$ in M is nondegenerate at $\mathbf{f}^{-1}(\mathbf{y})$. We will say that $\mathbf{f} : V \to \mathbb{R}^n$ (resp. $M \subset \mathbb{R}^n$) is *nondegenerate* if it is nondegenerate at almost every point of V (resp. M, in the sense of the natural measure class on M). If the functions f_i are analytic, it is easy to see that the linear independence of $1, f_1, \dots, f_n$ over $\mathbb{R}$ in V is equivalent to all points of $M = \mathbf{f}(U)$ being nondegenerate. Thus the above nondegeneracy condition can be viewed as an infinitesimal version of not lying in any proper affine subspace of $\mathbb{R}^n$.

It appears that many known and anticipated results in the field fall in the framework of the following vague

META-CONJECTURE. *"Any" Diophantine property of vectors in an ambient space (e.g. $\mathbb{R}^n$) which holds for almost all points in this space should hold for generic points on a nondegenerate smooth submanifold M of the space.*

It was conjectured in 1980 by Sprindžuk [**Sp4,** Conjecture H_1] that almost all points on a nondegenerate analytic submanifold of $\mathbb{R}^n$ are not VWA. This conjecture was supported before and after 1980 by a number of partial results, one of the first being Schmidt's proof [**S1**] for nondegenerate planar curves. The general case was settled in 1998 by Margulis and the author using the dynamical approach. Namely, the following was proved:

THEOREM 4.1 [**KM2**]. *Let M be a nondegenerate smooth submanifold of $\mathbb{R}^n$. Then almost all points of M are not VWA.*

This is the result we will focus on later in this section.

In another direction, Sprindžuk's solution to Mahler's problem was improved in 1964 by Baker [**B1**] and later (1984) by Bernik [**Bern, BD**]; the latter proved that whenever $\sum_{l=1}^{\infty} \psi(l)$ is finite, almost all points of the curve (4.3) are not ψ-approximable. And several years ago Beresnevich [**Bere1**] proved the divergence counterpart, thus establishing a complete analogue of the Khintchine-Groshev Theorem for the curve (4.3).

It turned out that a modification of the methods from [**KM2**] allows one to prove the convergence part of the Khintchine-Groshev Theorem for any nondegenerate manifold. In other words, the following is true:

THEOREM 4.2. *Let M be a nondegenerate smooth submanifold of $\mathbb{R}^n$ and let ψ be such that $\sum_{l=1}^{\infty} \psi(l)$ is finite. Then almost all points of M are not ψ-approximable.*

This is proved in [**BKM**] and also independently in [**Bere2**]. A work on the divergence case is currently in progress. (We note that the main result of [**BKM**]

has a stronger "multiplicative" version (see Theorem 5.3) which is currently not doable by classical Sprindžuk-style methods developed in [**Bere2**].)

In this survey we will indicate a proof of Theorem 4.1 by first restating it in the language of flows on the space of lattices. For this we set $k = n + 1$ and look at the one-parameter group

$$g_t = \operatorname{diag}(e^t, e^{-t/n}, \dots, e^{-t/n}). \tag{4.4}$$

acting on $\Omega = SL_k(\mathbb{R})/SL_k(\mathbb{Z})$, and given $\mathbf{y} \in \mathbb{R}^n$, consider $L_{\mathbf{y}} \stackrel{\text{def}}{=} \begin{pmatrix} 1 & \mathbf{y}^T \\ 0 & I_n \end{pmatrix}$ (cf. (2.1) and (1.3)). It follows from Theorem 3.9 and the example discussed afterwards that $\mathbf{y} \in \mathbb{R}^n$ is VWA iff for some $\gamma > 0$ there exist arbitrarily large positive t such that

$$g_t L_{\mathbf{y}} \mathbb{Z}^k \in \Omega_{e^{-\gamma t}}. \tag{4.5}$$

Equivalently, for some $\gamma > 0$ there are infinitely many $t \in \mathbb{N}$ such that (4.5) holds.

With this in mind, let us turn to the setting of Theorem 4.1. Namely let V be an open subset of $\mathbb{R}^d$ and $\mathbf{f} = (f_1, \dots, f_n)$ an n-tuple of C^k functions $V \mapsto \mathbb{R}$ which is nondegenerate at almost every point of V. The theorem would be proved if we show that for any $\gamma > 0$ the set

$$\{x \in V \mid g_t L_{\mathbf{f}(x)} \mathbb{Z}^k \in \Omega_{e^{-\gamma t}} \text{ for infinitely many } t \in \mathbb{N}\}$$

has measure zero. In other words, a submanifold $\mathbf{f}(V)$ of $\mathbb{R}^n$ gives rise to a submanifold $L_{\mathbf{f}(V)}\mathbb{Z}^k$ of the space of lattices, and one needs to show that thegrowth rate of generic orbits originating from this submanifold is consistent with the growth rate of an orbit of a generic point of Ω (see Theorem 3.10 for an explanation of why lattices Λ such that $g_t\Lambda \in \Omega_{e^{-\gamma t}}$ for infinitely many $t \in \mathbb{N}$ form a null subset of Ω).

Now one can use the Borel-Cantelli Lemma to reduce Theorem 4.1 to the following statement:

THEOREM 4.3. *Let V be an open subset of $\mathbb{R}^d$ and $\mathbf{f} = (f_1, \dots, f_n)$ an n-tuple of C^k functions $V \mapsto \mathbb{R}$ which is nondegenerate at $x_0 \subset V$. Then there exists a neighborhood B of x_0 contained in V such that for any $\gamma > 0$ one has*

$$\sum_{t=1}^{\infty} |\{x \in B \mid g_t L_{\mathbf{f}(x)} \mathbb{Z}^k \in \Omega_{e^{-\gamma t}}\}| < \infty. \tag{4.6}$$

Here is the turning point of the argument: t-dynamics gives way to x-dynamics, namely, a natural way to demonstrate (4.6) is to fix t and think of the set $\{g_t L_{\mathbf{f}(x)} \mathbb{Z}^k\}$ as of an orbit of certain action (not a group action!), the goal being to prove that a substantial part of this "orbit" lies outside of "cusp neighborhoods" $\Omega_{e^{-\gamma t}}$ uniformly for all t. What immediately comes to mind is the recurrence property of unipotent orbits, that is, Theorem 2.4. And it turns out that a modification of the argument used to prove the latter theorem allows one to estimate the amount of "time" x that the "trajectory" $x \mapsto g_t L_{\mathbf{f}(x)} \mathbb{Z}^k$ spends "close to infinity" in Ω. More precisely, the following can be proved:

THEOREM 4.4. *Let V, $\mathbf{f}$ and x_0 be as in Theorem 4.3. Then there exists a neighborhood B of x_0 contained in V and constants $C, \rho > 0$ such that for any $\varepsilon > 0$ and any positive t one has*

$$|\{x \in B \mid g_t L_{\mathbf{f}(x)} \mathbb{Z}^k \in \Omega_\varepsilon\}| < C\varepsilon^{1/dn}|B|. \tag{4.7}$$

It is straightforward to verify that Theorem 4.3 follows from the above uniform estimate.

It remains to explain why the behavior of the curve $x \mapsto g_t L_{\mathbf{f}(x)}\mathbb{Z}^k$ is similar to that of the unipotent orbit. In fact it has been understood a long time ago that the main property of the unipotent actions on which the recurrence estimates are based is the polynomial dependence of u_x on x (see [**Ma2, Sh**]). One may wonder what is so special about polynomials – and it turns out that the crucial property is roughly "not making very sharp turns". More precisely, here is the definition motivated by the analysis of the proofs in [**Ma2**] (see also [**EMS**]): for $C, \alpha > 0$ say that a continuous function f on an open set $V \subset \mathbb{R}^d$ is *(C,α)-good on V* if for any open ball $B \subset V$ and any positive ε one has

$$\left|\{x \in B \mid |f(x)| < \varepsilon \cdot \sup_{x\in B} |f(x)|\}\right| \le C\varepsilon^\alpha |B|\,.$$

In other words, a good function which takes small values on a big part of a ball is not allowed to grow very fast on the remaining part of the ball. The main example is provided by polynomials:

LEMMA 4.5. *Any polynomial $f \in \mathbb{R}[x]$ of degree not greater than k is $(4k, 1/k)$-good on $\mathbb{R}$.*

This easily follows from Lagrange's interpolation formula, see [**DM2, KM2**]. The next theorem (the main result of [**KM2**]) therefore provides a generalization of Theorem 2.4 to polynomial trajectories on Ω. To state it we need to introduce some notation. If Δ is a discrete subgroup of $\mathbb{R}^k$ (not necessarily a lattice) generated by $\mathbf{v}_1, \dots, \mathbf{v}_l$, let us measure its norm, $\|\Delta\|$, by the norm of the exterior product $\mathbf{v}_1 \wedge \cdots \wedge \mathbf{v}_l$. For this one needs to extend the norm from $\mathbb{R}^k$ to its exterior algebra. If $\mathbf{e}_1, \dots, \mathbf{e}_k$ are standard base vectors of $\mathbb{R}^k$, the elements $\mathbf{e}_I \stackrel{\text{def}}{=} \mathbf{e}_{i_1} \wedge \cdots \wedge \mathbf{e}_{i_l}$, $I = \{i_1, \dots, i_l\} \subset \{1, \dots, k\}$ form a basis of $\bigwedge^l(\mathbb{R}^k)$. Since Diophantine applications call for the supremum norm, we will extend $\|\cdot\|$ to $\bigwedge(\mathbb{R}^k)$ by setting $\|\sum_I w_I \mathbf{e}_I\| = \max_I |w_I|$.

Now let us consider a curve in Ω given by $x \mapsto h(x)\mathbb{Z}^k$, where h is some function from $\mathbb{R}^d$ to $GL_k(\mathbb{R})$. It turns out that in order to understand its recurrence properties one has to keep an eye on norms of all discrete subgroups of $h(x)\mathbb{Z}^k$; in particular, it will be necessary to prove that all those norms (as functions of x) are (C,α)-good for some C, α. In fact, it will suffice to look at the coordinates of $h(x)(\mathbf{v}_1 \wedge \cdots \wedge \mathbf{v}_l)$ where $\mathbf{v}_1, \dots, \mathbf{v}_l$ form a basis of $\Delta \subset \mathbb{Z}^k$; one can easily show that if all components of a vector function are (C,α)-good, the norm of this function is also (C,α)-good.

THEOREM 4.6. *Let $d, k \in \mathbb{N}$, $C, \alpha > 0$, $0 < \rho \le 1/k$, and let a ball $B = B(x_0, r_0) \subset \mathbb{R}^d$ and a map $h : \tilde{B} \to GL_k(\mathbb{R})$ be given, where $\tilde{B}$ stands for $B(x_0, 3^k r_0)$. Assume that for any subgroup Δ of $\mathbb{Z}^k$,*

(i) *the function $x \mapsto \|h(x)\Delta\|$ is (C,α)-good on $\tilde{B}$;*

(ii) $\sup_{x\in B} \|h(x)\Delta\| \ge \rho$.

Then for any positive $\varepsilon \le \rho$ one has

$$\left|\{x \in B \mid h(x)\mathbb{Z}^k \in \Omega_\varepsilon\}\right| \le \text{const}(d,k) \cdot \left(\frac{\varepsilon}{\rho}\right)^\alpha |B|\,. \tag{4.8}$$

COROLLARY 4.7. *For any lattice Λ in $\mathbb{R}^k$ there exists a constant $\rho = \rho(\Lambda) > 0$ such that for any one-parameter unipotent subgroup $\{u_x\}_{x\in\mathbb{R}}$ of $SL_k(\mathbb{R})$, for any $T > 0$ and any $\varepsilon \le \rho$, one has*

$$|\{0 < x < T \mid u_x\Lambda \in \Omega_\varepsilon\}| \le \mathrm{const}(k)\left(\frac{\varepsilon}{\rho}\right)^{1/k^2} T. \tag{4.9}$$

This is clearly a quantitative strengthening of Theorem 2.4, with an explicit estimate of δ in terms of ε.

PROOF. Write Λ in the form $g\mathbb{Z}^k$ with $g \in GL_k(\mathbb{R})$, and denote by h the function $h(x) = u_x g$. For any $\Delta \subset \mathbb{Z}^k$ with basis $\mathbf{v}_1, \dots, \mathbf{v}_j$, the coordinates of $h(x)(\mathbf{v}_1 \wedge \cdots \wedge \mathbf{v}_j)$ will be polynomials in x of degree not exceeding k^2. Hence the functions $x \mapsto \|h(x)\Delta\|$ will be $(C, 1/k^2)$-good on $\mathbb{R}$, where C is a constant depending only on k. Now let $\rho \stackrel{\mathrm{def}}{=} \min\left(1/k, \inf_{\Delta\in\mathcal{L}(\mathbb{Z}^k)} \|g\Delta\|\right)$, positive by the discreteness of Λ in $\mathbb{R}^k$. Then $\|h(0)\Delta\| \ge \rho$ for any $\Delta \subset \mathbb{Z}^k$, therefore, with the the substitutions $B = (0, T)$, $\alpha = 1/k^2$ and $d = 1$ assumptions (i) and (ii) of Theorem 4.6 are satisfied, and one immediately gets (4.9) from (4.8). □

As was mentioned in §2, the proof of Theorem 4.6 is based on delicate combinatorial (partially ordered) structure of the space of lattices, and the reader is referred to [**KM2**] or [**BKM**] (most of the ideas are borrowed from [**Ma2**] and [**D3**]). Assuming the latter theorem, we conclude by presenting a

SKETCH OF PROOF OF THEOREM 4.4. Take a positive t and consider $h(x) \stackrel{\mathrm{def}}{=} g_t L_{\mathbf{f}(x)}$; clearly all one needs to prove (4.7) is to check conditions (i) and (ii) of Theorem 4.6 for every $\Delta \subset \mathbb{Z}^k$. An elementary computation shows that the coordinates of $h(x)(\mathbf{v}_1 \wedge \cdots \wedge \mathbf{v}_l)$ for any choice of vectors $\mathbf{v}_i$ are linear combinations of functions $f_1, \dots, f_n$ and 1. Consider first the case of original Mahler's conjecture, with $d = 1$ and $f_i(x) = x^i$. Then, as in the proof of Corollary 4.7, condition (i) is automatic due to Lemma 4.5. Further, a straightforward computation of the action of $h(x)$ on exterior products of vectors in $\mathbb{R}^k$ shows that at least one coefficient of at least one polynomial arising as a coordinate must have absolute value not less than 1. This implies that for every interval B there exists a constant ρ (independent of t) such that (ii) holds.

It remains to pass from this special case to the general situation of functions $f_1, \dots, f_n$ on $\mathbb{R}^d$ coordinatizing a nondegenerate submanifold of $\mathbb{R}^n$. Here one basically has to show that locally these functions behave like polynomials. Indeed, the following was proved in [**KM2**]:

LEMMA 4.8. *Let $\mathbf{f} = (f_1, \dots, f_n)$ be a C^l map from an open subset V of $\mathbb{R}^d$ to $\mathbb{R}^n$, and let $x_0 \in V$ be such that $\mathbb{R}^n$ is spanned by partial derivatives of $\mathbf{f}$ at x_0 of order up to l. Then there exists a neighborhood $U \subset V$ of x_0 and positive C such that any linear combination of $1, f_1, \dots, f_n$ is $(C, 1/dl)$-good on V.*

Now to finish the proof one simply has to choose U according to the above lemma, then pick a ball $\tilde{B}$ centered at x_0 and contained in U, and finally take B to be a concentric ball with radius 3^k times smaller. This implies condition (i) with $\alpha = 1/dn$ and some constant C independent of Δ and t, and (ii) follows as a result of a computation described above: one shows that at least one coordinate of $h(x)(\mathbf{v}_1 \wedge$

$\cdots \wedge \mathbf{v}_l)$ must have the form $c_0 + \sum_{i=1}^n c_i f_i(x)$ with $\max(|c_1|, \dots, |c_n|) \ge 1$, and therefore one gets a lower bound (again independent of t and Δ) for $\sup_{x\in B} \|h(x)\Delta\|$.□

5. Multiplicative approximation

We have already seen in Conjecture 1.2 how the magnitude of the integer vector $\mathbf{q}$ was measured by taking the *product* of coordinates rather than the maximal coordinate (that is the norm of the vector). Let us formalize it by saying, for ψ as before, that $Y \in M_{m\times n}(\mathbb{R})$ is *ψ-multiplicatively approximable* (ψ-MA) if there are infinitely many $\mathbf{q} \in \mathbb{Z}^n$ such that

$$\Pi(Y\mathbf{q} + \mathbf{p}) \le \psi\big(\Pi_+(\mathbf{q})\big) \quad \text{for some } \mathbf{p} \in \mathbb{Z}^m .$$

where for $\mathbf{x} = (x_1, \dots, x_k) \in \mathbb{R}^k$ one defines $\Pi(\mathbf{x}) = \prod_{i=1}^k |x_i|$ and $\Pi_+(\mathbf{x}) = \prod_{i=1}^k \max(|x_i|, 1)$. Clearly any ψ-approximable system of linear forms is automatically ψ-MA, but not necessarily other way around. Similarly to the standard setting, one can define *badly multiplicatively approximable (BMA)* and *very well multiplicatively approximable (VWMA)* systems. It can be easily shown that almost no $Y \in M_{m\times n}(\mathbb{R})$ are ψ-MA if the sum

$$\sum_{l=1}^{\infty} (\log l)^{k-2} \psi(l)$$

converges (here we again set $k = m + n$); in particular, VWMA systems form a set of measure zero. The converse (i.e. a multiplicative analogue of Theorem 1.2) can be proved using methods of Schmidt; the case $n = 1$ is contained in [**G**].

On the other hand, saying that a vector $\mathbf{y} \in \mathbb{R}^n$ (viewed as a linear form $\mathbf{q} \mapsto \mathbf{y}\cdot\mathbf{q}$) is not BMA is equivalent to (1.1); in other words, Conjecture 1.2 states that no $\mathbf{y} \in \mathbb{R}^n$, $n \ge 2$, is badly multiplicatively approximable. A more general statement that no $Y \in M_{m\times n}(\mathbb{R})$ is BMA unless $m = n = 1$ in fact reduces to this conjecture; moreover, as it is the case with Theorem 1.1, it is enough to prove Conjecture 1.2 for $n = 2$.

It seems natural to bring lattices into the game. In fact, one can observe the similarity between the statements of the Oppenheim and Littlewoods's conjectures. Indeed, to say that $Y \in M_{m\times n}(\mathbb{R})$ is not BMA amounts to saying that 0 is the infimum of absolute values of a certain homogeneous polynomial at integer points. Similarly to what was done for quadratic forms, a linear change of variables transforms this polynomial into the product of coordinates $\Pi(\mathbf{x})$, $\mathbf{x} \in \mathbb{R}^k$, and according to the scheme developed in the preceding sections, the dynamical system reflecting Diophantine properties of A must come from the group stabilizing Π, that is, the full diagonal subgroup D of $SL_k(\mathbb{R})$. Thus the problems rooted in multiplicative Diophantine approximation bring us to *higher rank actions* on Ω (in implicit form this was already noticed in the paper [**CS**] of Cassels and Swinnerton-Dyer). More precisely, one can state a multiplicative version of generalized Dani's correspondence (Theorem 3.8), relating multiplicative Diophantine properties of A to orbits of the form $\{gL_Y\mathbb{Z}^k \mid g \in D_+\}$ where D_+ is a certain open chamber in the group D. (For a version of such a correspondence see [**KM3,** Theorem 9.2].)

We illustrate this principle by two examples below, where for the sake of simplicity of exposition we specialize to the case $m = 1$ (one linear form $\mathbf{q} \mapsto \mathbf{y}\cdot\mathbf{q}$, $\mathbf{y} \in \mathbb{R}^n$), setting $k = n + 1$. We will need the following notation: for $\mathbf{t} = (t_1, \dots, t_n) \in \mathbb{R}^n$ let us denote $\sum_{i=1}^n t_i$ by t and define

$$g_{\mathbf{t}} = \operatorname{diag}(e^t, e^{-t_1}, \dots, e^{-t_n}) \in SL_k(\mathbb{R}). \tag{5.1}$$

Littlewoods's conjecture. One can show that $\mathbf{y} \in \mathbb{R}^n$ satisfies (1.1) (that is, it is BMA) iff the trajectory $\{g_{\mathbf{t}} L_{\mathbf{y}} \mathbb{Z}^k \mid \mathbf{t} \in \mathbb{R}_+^n\}$ is bounded in the space of lattices in $\mathbb{R}^k$. Thus Conjecture 1.2 is equivalent to the statement that every trajectory as above is unbounded. In fact, an argument rooted in the "Isolation Theorem" of Cassels and Swinnerton-Dyer [**CS, LW, Ma4**] shows that the latter statement can be reduced to the following

CONJECTURE 5.1. *Let D be the subgroup of diagonal matrices in $SL_k(\mathbb{R})$, $k \geq 3$. Then any relatively compact orbit $D\Lambda$, Λ a lattice in $\mathbb{R}^k$, is compact.*

Notice that Theorem 3.5 shows that the above statement does not hold if $k = 2$. This highlights the difference between rank-one and higher rank dynamics. Note also the similarity between Corollary 2.6 and Conjecture 5.1, showing that higher rank hyperbolic actions share some features with unipotent dynamics. In fact, Conjecture 5.1 is a special case of a more general hypothesis, see [**Ma8,** Conjecture 1], which, roughly speaking, says that for a connected Lie group G, a lattice $\Gamma \subset G$ and a closed subgroup H of G, any orbit closure $\overline{Hx}$, $x \in G/\Gamma$, is an orbit of an intermediate subgroup $L \supset H$ of G unless "it has a good reason not to" (the latter reasons must be coming from certain one-parameter quotient actions). See [**Ma8,** §1] for more detail.

Multiplicative approximation on manifolds. Since every VWA vector is VWMA (that is, ψ_ε-MA for some $\varepsilon > 0$) but not other way around, it is a more difficult problem to prove that a generic point on a nondegenerate manifold is not very well multiplicatively approximable. This has been known as Conjecture H_2 of Sprindžuk [**Sp4**]; the polynomial special case (that is, a multiplicative strengthening of Mahler's problem) was conjectured by Baker in [**B2**]; both conjectures stood open, except for low-dimensional special cases, until [**KM2**] where the following was proved:

THEOREM 5.2. *Let M be a nondegenerate smooth submanifold of $\mathbb{R}^n$. Then almost all points of M are not VWMA.*

The strategy of the proof of Theorem 4.1 applies with minor changes. One shows (see [**KM2,** Lemma 2.1 and Corollary 2.2] for a partial result) that $\mathbf{y} \in \mathbb{R}^n$ is VWMA iff for some $\gamma > 0$ there are infinitely many $\mathbf{t} \in \mathbb{Z}_+^n$ such that

$$g_{\mathbf{t}} L_{\mathbf{y}} \mathbb{Z}^k \in \Omega_{e^{-\gamma t}} \tag{5.2}$$

(here, as defined above, t stands for $\sum_{i=1}^n t_i$). Therefore it is enough to use Theorem 4.6 to prove a modification of the measure estimate of Theorem 4.4 with g_t as in (4.5) replaced by $g_{\mathbf{t}}$ as in (5.1).

Finally let us mention a multiplicative version of Theorem 4.2, proved in [**BKM**] by a modification of the method described above:

THEOREM 5.3. *Let M be a nondegenerate smooth submanifold of $\mathbb{R}^n$ and let ψ be such that $\sum_{l=1}^\infty (\log l)^{n-1} \psi(l)$ is finite. Then almost all points of M are not ψ-multiplicatively approximable.*

Acknowledgements

This survey is based on a minicourse of lectures given at the AMS Summer Research Institute in Smooth Ergodic Theory and applications (Seattle, 1999). The author is grateful to the organizers and participants of the workshop, and especially to Alex Eskin for sharing the responsibility and the fun of giving the course. Thanks are also due to participants/organizers of the workshop on Ergodic Theory, Rigidity and Number Theory (Cambridge UK, January 2000) where the presentation of material was tested one more time in a series of lectures, to Gregory Margulis, Alexander Starkov and Barak Weiss for helpful discussions, and to Stella for her endless patience and support.

References

[AGH] L. Auslander, L. Green and F. Hahn, *Flows on homogeneous spaces*, Annals of Mathematics Studies, No. 53, Princeton University Press, Princeton, N.J., 1963.

[AN1] A. G. Abercrombie and R. Nair, *An exceptional set in the ergodic theory of Markov maps of the interval*, Proc. London Math. Soc. **75** (1997), 221–240.

[AN2] ———, *An exceptional set in the ergodic theory of rational maps of the Riemann sphere*, Ergodic Theory Dynam. Systems **17** (1997), 253–267.

[B1] A. Baker, *On a theorem of Sprindzhuk*, Proc. Roy. Soc. London **A 292** (1966), 92–104.

[B2] ———, *Transcendental number theory*, Cambridge Univ. Press, Cambridge, 1975.

[Bek] M. Bekka, *On uniqueness of invariant means*, Proc. Amer. Math. Soc. **126** (1998), 507–514.

[Bere1] V. Beresnevich, *On approximation of real numbers by real algebraic numbers*, Acta Arith. **90** (1999), 97–112.

[Bere2] ———, *A Groshev type theorem for convergence on manifolds*, Acta Math. Hungar. (to appear).

[Bern] V. Bernik, *A proof of Baker's conjecture in the metric theory of transcendental numbers*, Doklady Akad. Nauk SSSR **277** (1984), 1036–1039. (Russian)

[BD] V. Bernik and M. M. Dodson, *Metric Diophantine approximation on manifolds*, Cambridge Univ. Press, Cambridge, 1999.

[BKM] V. Bernik, D. Kleinbock and G. A. Margulis, *Khintchine-type theorems on manifolds: the convergence case for standard and multiplicative versions*, Preprint (2000).

[BM] J. Brezin and C. C. Moore, *Flows on homogeneous spaces: a new look*, Amer. J. Math. **103** (1981), 571–613.

[BMa] B. Bekka and M. Mayer, *Ergodic theory and topological dynamics of group actions on homogeneous spaces*, Cambridge University Press, Cambridge, 2000.

[C] J. W. S. Cassels, *An introduction to Diophantine approximation*, Cambridge Tracts in Math., vol. 45, Cambridge Univ. Press, Cambridge, 1957.

[CK] N. Chernov and D. Kleinbock, *Dynamical Borel-Cantelli lemmas for Gibbs measures*, Israel J. Math. (to appear).

[CR] J.-P. Conze and A. Raugi, *Convergence des potentiels pour un opérateur de transfert, applications aux systèmes dynamiques et aux chaînes de Markov*, Preprint (1999).

[CS] J.W.S. Cassels and H.P.F. Swinnerton-Dyer, *On the product of three homogeneous forms and indefinite ternary quadratic forms*, Philos. Trans. Roy. Soc. London **248, Ser. A** (1955), 73–96.

[D1] S. G. Dani, *Divergent trajectories of flows on homogeneous spaces and Diophantine approximation*, J. Reine Angew. Math. **359** (1985), 55–89.

[D2] ———, *Bounded orbits of flows on homogeneous spaces*, Comment. Math. Helv. **61** (1986), 636–660.

[D3] ———, *On orbits of unipotent flows on homogeneous spaces*, II, Ergodic Theory Dynamical Systems **6** (1986), 167–182.

[D4] ———, *On orbits of endomorphisms of tori and the Schmidt game*, Ergodic Theory Dynam. Systems **8** (1988), 523–529.

[D5] ———, *Flows on homogeneous spaces and Diophantine approximation*, Proceedings of the International Congress of Mathematicians, Vol. 1, 2 (Zürich, 1994), Birkhäuser, Basel, 1995, pp. 780–789.

[D6] ———, *Flows on homogeneous spaces: a review*, Ergodic theory of Z^d actions (Warwick, 1993–1994), Cambridge Univ. Press, Cambridge, 1996, pp. 63–112.

[D7] ———, *Dynamical systems on homogeneous spaces*, in: Dynamical systems, ergodic theory and applications,, Springer-Verlag, Berlin, 2000.

[Dod] M. M. Dodson, *Geometric and probabilistic ideas in metric Diophantine approximation*, Russian Math. Surveys **48** (1993), 73–102.

[Dol1] D. Dolgopyat, *Bounded orbits of Anosov flows*, Duke Math. J **87** (1997), 87–114.

[Dol2] ———, *Limit theorems for partially hyperbolic systems*, Preprint (1998).

[DM1] S.G. Dani and G.A. Margulis, *Values of quadratic forms at primitive integral points*, Invent. Math. **98** (1989), 405–424.

[DM2] ———, *Limit distributions of orbits of unipotent flows and values of quadratic forms*, Adv. in Soviet Math., vol. 16, Amer. Math. Soc., Providence, R.I., 1993, pp. 91–137.

[E] A. Eskin, *Counting problems and semisimple groups*, Proceedings of the International Congress of Mathematicians, Vol. II (Berlin, 1998), Doc. Math., 1998, pp. 539–552.

[G] P. Gallagher, *Metric simultaneous diophantine approximation*, J. London Math. Soc. **37** (1962), 387–390.

[Gr] A. V. Groshev, *Une théorème sur les systèmes des formes linéaires*, Dokl. Akad. Nauk SSSR **9** (1938), 151–152.

[EMM] A. Eskin, G. A. Margulis and S. Mozes, *Upper bounds and asymptotics in a quantitative version of the Oppenheim conjecture*, Ann. Math. **147** (1998), 93–141.

[EMS] A. Eskin, S. Mozes and N. Shah, *Non-divergence of translates of certain algebraic measures*, Geom. Funct. Anal. **7** (1997), 48–80.

[H] G. Harman, *Metric number theory,* London Mathematical Society Monographs, New Series, vol. 18, The Clarendon Press, Oxford University Press, New York, 1998.

[K1] D. Kleinbock, *Nondense orbits of flows on homogeneous spaces*, Ergodic Theory Dynamical Systems **18** (1998), 373–396.

[K2] ———, *Flows on homogeneous spaces and Diophantine properties of matrices*, Duke Math. J. **95** (1998), 107–124.

[K3] ———, *Bounded orbit conjecture and diophantine approximation*, in: Proceedings of the International Colloquium on Lie Groups and Ergodic Theory, TIFR, Mumbai, 1998, pp. 119–130.

[K4] ———, *Badly approximable systems of affine forms*, J. Number Theory **79** (1999), 83–102.

[KM1] D. Kleinbock and G. A. Margulis, *Bounded orbits of nonquasiunipotent flows on homogeneous spaces*, Amer. Math. Soc. Transl. **171** (1996), 141–172.

[KM2] ———, *Flows on homogeneous spaces and Diophantine approximation on manifolds*, Ann. Math. **148** (1998), 339–360.

[KM3] ———, *Logarithm laws for flows on homogeneous spaces*, Inv. Math. **138** (1999), 451–494.

[KS] A. Katok and R. Spatzier, *First cohomology of Anosov actions of higher rank Abelian groups and applications to rigidity*, Inst. Hautes Études Sci. Publ. Math. **79** (1994), 131–156.

[KSS] D. Kleinbock, N. Shah, and A. Starkov, *Homogeneous flows, applications to number theory, and related topics*, in: Handbook on Dynamical Systems (Encyclopedia of Math Sciences Series), Elsevier (to appear).

[LW] E. Lindenstrauss and B. Weiss, *On sets invariant under the action of the diagonal group*, SUNY Stony Brook IMS Preprint # 1999/8.

[M] K. Mahler, *Über das Mass der Menge aller S-Zahlen*, Math. Ann. **106** (1932), 131–139.

[Ma1] G. A. Margulis, *On some problems in the theory of U-systems (in Russian)*, Thesis, Moscow University, 1970.

[Ma2] ———, *On the action of unipotent group in the space of lattices*, Proceedings of the Summer School on group representations, (Budapest 1971), Académiai Kiado, Budapest, 1975, pp. 365–370.

[Ma3] ———, *Formes quadratriques indéfinies et flots unipotents sur les espaces homogénes. C. R. Acad. Sci. Paris Sér. I Math.* **304** (1987), 249–253.

[Ma4] ———, *Discrete subgroups and ergodic theory*, Proc. of the conference "Number theory, trace formulas and discrete groups" in honour of A. Selberg (Oslo, 1987), Academic Press, Boston, MA 1989, 377–398.

[Ma5] ———, *Dynamical and ergodic properties of subgroup actions on homogeneous spaces with applications to number theory*, Proceedings of the International Congress of Mathematicians, Vol. I, II (Kyoto, 1990), Math. Soc. Japan, Tokyo, 1991, pp. 193–215.

[Ma6] ———, *Discrete subgroups of semisimple Lie groups*, Springer-Verlag, Berlin and New York, 1991.

[Ma7] ———, *Oppenheim conjecture*, Fields Medallists' lectures, World Sci. Publishing, River Edge, N.J., 1997, pp. 272–327.

[Ma8] ———, *Problems and conjectures in rigidity theory*, in: Mathematics: Frontiers and Perspectives, Amer. Math. Soc., Providence, R.I., 2000.

[Mo1] C. C. Moore, *Ergodicity of flows on homogeneous spaces*, Amer. J. Math. **88** (1966), 154–178.

[Mo2] ———, *Exponential decay of correlation coefficients for geodesic flows*, Group Representations, Ergodic Theory, Operator Algebras and Mathematical Physics, Math. Sci. Res. Inst. Publ., vol. 6, Springer-Verlag, Berlin and New York, 1987, pp. 163–181.

[Op1] A. Oppenheim, *The minima of indefinite quaternary quadratic forms*, Proc. Nat. Acad. Sci. USA **15** (1929), 724–727.

[Op2] ———, *Values of quadratic forms I*, Quart. J. Math. Oxford Ser. (2) **4** (1953), 54–59.

[Ph] W. Philipp, *Some metrical theorems in number theory*, Pacific J. Math. **20** (1967), 109–127.

[R] M. S. Raghunathan, *Discrete subgroups of Lie groups*, Springer-Verlag, Berlin and New York, 1972.

[Ra1] M. Ratner, *The rate of mixing for geodesic and horocycle flows*, Ergodic Theory Dynamical Systems **7** (1987), 267–288.

[Ra2] ———, *Raghunathan's topological conjecture and distribution of unipotent flows*, Duke Math. J. **63** (1991), 235–280.

[Ra3] ———, *Interactions between ergodic theory, Lie groups, and number theory*, Proceedings of the International Congress of Mathematicians, Vol. 1, 2 (Zürich, 1994), Birkhäuser, Basel, 1995, pp. 157–182.

[S1] W. M. Schmidt, *A metrical theorem in Diophantine approximation*, Canadian J. Math. **12** (1960), 619–631.

[S2] ———, *Badly approximable systems of linear forms*, J. Number Theory **1** (1969), 139–154.

[S3] ———, *Diophantine approximation*, Springer-Verlag, Berlin and New York, 1980.

[Sh] N. Shah, *Limit distribution of polynomial trajectories on homogeneous spaces*, Duke Math. J. **75** (1994), 711–732.

[Sp1] V. Sprindžuk, *More on Mahler's conjecture*, Doklady Akad. Nauk SSSR **155** (1964), 54–56 (Russian); English transl. in Soviet Math. Dokl **5** (1964), 361–363.

[Sp2] V. Sprindžuk, *Mahler's problem in metric number theory*, Translations of Mathematical Monographs, vol. 25, Amer. Math. Soc., Providence, R.I., 1969.

[Sp3] ———, *Metric theory of Diophantine approximations*, John Wiley & Sons, New York-Toronto-London, 1979.

[Sp4] ———, *Achievements and problems in Diophantine approximation theory*, Russian Math. Surveys **35** (1980), 1–80.

[St1] A. N. Starkov, *New progress in the theory of homogeneous flows*, Russian Math. Surveys **52** (1997), 721–818.

[St2] ———, *Dynamical systems in homogeneous spaces*, Translations of Mathematical Monographs, vol. 190, Amer. Math. Soc., Providence, R.I., 2000.

[Su] D. Sullivan, *Disjoint spheres, approximation by imaginary quadratic numbers, and the logarithm law for geodesics*, Acta Math. **149** (1982), 215–237.

[U] M. Urbanski, *The Hausdorff dimension of the set of points with nondense orbit under a hyperbolic dynamical system*, Nonlinearity **2** (1991), 385–397.

[Z] R. Zimmer, *Ergodic theory and semisimple groups*, Birkhäuser, Boston, 1984.

Department of Mathematics, Brandeis University, Waltham, MA 02454-9110
E-mail address: kleinboc@brandeis.edu

Proceedings of Symposia in Pure Mathematics
Volume **69**, 2001

Measurable Rigidity of Algebraic $\mathbb{Z}^d$-Actions

Klaus Schmidt

1. Introduction

Let X be a compact abelian group with identity element 0_X, Borel field $\mathcal{B}_X$ and normalized Haar measure λ_X. An *algebraic* $\mathbb{Z}^d$*-action* on X is a group homomorphism $\alpha\colon \mathbf{n} \mapsto \alpha^{\mathbf{n}}$ from $\mathbb{Z}^d$, $d \geq 1$, into the group $\mathrm{Aut}(X)$ of continuous automorphisms of X. An algebraic $\mathbb{Z}^d$-action α on X is *expansive* if there exists an open set $\mathcal{O} \subset X$ with $\bigcap_{\mathbf{n}\in\mathbb{Z}^d} \alpha^{-\mathbf{n}}(\mathcal{O}) = \{0_X\}$, where 0_X is the identity element of X, and *irreducible* if every closed, α-invariant subgroup $Y \subsetneq X$ is finite. The action α is *ergodic* or *mixing* if the Haar measure λ_X of X is ergodic or mixing under α.

If α is an algebraic $\mathbb{Z}^d$-action on X and $Y \subset X$ a closed, α-invariant subgroup, then we denote by α_Y and $\alpha_{X/Y}$ the $\mathbb{Z}^d$-action induced by α on Y and X/Y, respectively.

Algebraic $\mathbb{Z}^d$ actions can be classified algebraically, topologically or measurably. In order to introduce the appropriate notions of conjugacy we assume that α_i is an algebraic $\mathbb{Z}^d$-action on a compact abelian group X_i with Haar measure λ_{X_i}, where $i = 1, 2$. A surjective Borel map $\phi\colon X_1 \longrightarrow X_2$ is a *measurable factor map* of α_1 and α_2 if

$$\lambda_{X_1}\phi^{-1} = \lambda_{X_2}, \tag{1.1}$$

and if

$$\phi \circ \alpha_1^{\mathbf{n}}(x) = \alpha_2^{\mathbf{n}} \circ \phi(x) \tag{1.2}$$

for every $\mathbf{n} \in \mathbb{Z}^d$ and λ_{X_1}-a.e. $x \in X_1$. A bijective measurable factor map $\phi\colon X_1 \longrightarrow X_2$ is a *measurable conjugacy* of α_1 and α_2.

A continuous surjective group homomorphism $\phi\colon X_1 \longrightarrow X_2$ is an *algebraic factor map* of α_1 and α_2 if it satisfies (1.2) for every $\mathbf{n} \in \mathbb{Z}^d$ and $x \in X_1$. A bijective algebraic factor map $\phi\colon X_1 \longrightarrow X_2$ is an *algebraic conjugacy* of α_1 and α_2. Topological factor maps and topological conjugacy are defined similarly.

The action α_2 is a *measurable*, *algebraic* or *topological factor* of α_1 if there exists a measurable, algebraic or topological factor map $\phi\colon X_1 \longrightarrow X_2$ of α_1 and

2000 *Mathematics Subject Classification.* Primary 37A15, 37A25, 37A35, 37A45, 37B50; Secondary 13E05.

Key words and phrases. Automorphisms of compact groups, ergodic theory of $\mathbb{Z}^d$-actions, rigidity phenomena.

α_2. The actions α_1, α_2 are *measurably, algebraically* or *topologically conjugate* if there exists a measurable, algebraic or topological conjugacy $\phi\colon X_1 \longrightarrow X_2$ of α_1 and α_2, and they are *weakly measurably* (resp. *weakly algebraically*) *conjugate* if each of them is a measurable (resp. algebraic) factor of the other.

Finally we call a map $\phi\colon X_1 \longrightarrow X_2$ *affine* if it is of the form

$$\phi(x) = \psi(x) + x' \tag{1.3}$$

for every $x \in X_1$, where $\psi\colon X_1 \longrightarrow X_2$ is a continuous surjective group homomorphism and $x' \in X_2$.

For expansive algebraic $\mathbb{Z}^d$-actions on compact connected abelian groups the topological and algebraic classifications coincide (i.e. topological conjugacy implies algebraic conjugacy), by [**20**, (4.10) and Theorem 5.9]. For algebraic $\mathbb{Z}^d$-actions on disconnected groups this is no longer true: let A be a finite abelian group, and let $\Omega = A^{\mathbb{Z}^d}$ be the compact abelian group consisting of all maps $\omega\colon \mathbb{Z}^d \longrightarrow A$, furnished with the product topology and coordinate-wise addition. We write every $\omega \in \Omega$ as $\omega = (\omega_{\mathbf{n}})$ with $\omega_{\mathbf{n}} \in A$ for every $\mathbf{n} \in \mathbb{Z}^d$ and define the *shift-action* σ of $\mathbb{Z}^d$ on Ω by

$$(\sigma_{\mathbf{m}}\omega)_{\mathbf{n}} = \omega_{\mathbf{m}+\mathbf{n}} \tag{1.4}$$

for every $\omega \in \Omega$ and $\mathbf{m}, \mathbf{n} \in \mathbb{Z}^d$. Clearly, σ is an expansive algebraic $\mathbb{Z}^d$-action on Ω. If A' is a second abelian group with the same cardinality as A, then the resulting shift-actions σ and σ' of $\mathbb{Z}^d$ on $\Omega = A^{\mathbb{Z}^d}$ and $\Omega' = A'^{\mathbb{Z}^d}$ are obviously topologically, but not necessarily algebraically, conjugate.

For $d = 1$, any algebraic $\mathbb{Z}$-action is determined by the powers of a single group automorphism α. If α is ergodic, then it is Bernoulli (cf. e.g. [**1**], [**2**], [**7**], [**12**], [**14**]), which implies that two such actions with equal entropy are measurably conjugate even if they are algebraically non-conjugate. If $d > 1$ and α_1, α_2 are algebraic $\mathbb{Z}^d$-actions on compact abelian groups X_1, X_2 with completely positive entropy (with respect to Haar measure), then they are Bernoulli by [**17**], and can thus again be measurably conjugate without being algebraically conjugate. However, if these actions have zero entropy, measurable conjugacy may have much stronger implications.

THEOREM 1.1 ([**6**], [**11**]). *Let $d > 1$, and let α_1 and α_2 be mixing and expansive algebraic $\mathbb{Z}^d$-actions on compact abelian groups X_1 and X_2, respectively. If α_1 is irreducible, and if $\phi\colon X_1 \longrightarrow X_2$ is a measurable conjugacy of α_1 and α_2, then α_2 is irreducible and ϕ is λ_{X_1}-a.e. equal to an affine map.*

COROLLARY 1.2. *Let $d > 1$, and let α_1 and α_2 be measurably conjugate irreducible, mixing and expansive algebraic $\mathbb{Z}^d$-actions on compact abelian groups X_1 and X_2, respectively. Then α_1 and α_2 are algebraically conjugate.*

Actions with completely positive entropy and irreducible actions lie at opposite ends of the spectrum of ergodic algebraic $\mathbb{Z}^d$-actions. As mentioned above, the kind of isomorphism rigidity described in Theorem 1.1 is impossible for actions with completely positive entropy. However, it could conceivably hold for all mixing algebraic $\mathbb{Z}^d$-actions with zero entropy (i.e. without Bernoulli factors — cf. [**13**]). In this context it is worth pointing out the similarity between Theorem 1.1 and M. Ratner's rigidity theorem for horocycle flows (or their time-1-maps) in [**15**]: in both situations any measurable conjugacy carries with it the ambient algebraic structure. However, the underlying dynamical properties of the systems differ considerably: horocycle flows have zero entropy and are uniquely ergodic, whereas the algebraic $\mathbb{Z}^d$-actions under consideration here have a dense set of periodic orbits and their individual automorphisms are Bernoulli.

Theorem 1.1 implies that every Haar-measure-preserving invertible transformation T on a compact abelian group X which commutes with an irreducible, expansive and mixing $\mathbb{Z}^d$-action α on X is affine. This makes it possible — at least in principle — to determine the measurable centralizer of α. In order to provide the necessary notation we assume that α is an algebraic $\mathbb{Z}^d$-action on a compact abelian group X and write $\mathrm{Aff}(X)$ for the group of all continuous surjective affine bijections $\psi \colon X \longrightarrow X$. Put

$$(1.5)\qquad \begin{aligned} \bar{Z}(\alpha) &= \{\psi \in \mathrm{Aff}(X) : \phi \circ \alpha^{\mathbf{n}} = \alpha^{\mathbf{n}} \circ \psi \text{ for every } \mathbf{n} \in \mathbb{Z}^d\}, \\ Z(\alpha) &= \{\phi \in \mathrm{Aut}(X) : \phi \circ \alpha^{\mathbf{n}} = \alpha^{\mathbf{n}} \circ \phi \text{ for every } \mathbf{n} \in \mathbb{Z}^d\}. \end{aligned}$$

COROLLARY 1.3. *Let α be an irreducible, expansive and mixing algebraic $\mathbb{Z}^d$-action on a compact abelian group X, and let $Z_{\lambda_X}(\alpha)$ be the* measurable centralizer *of α, i.e. the group of all λ_X-preserving invertible transformations $T \colon X \longrightarrow X$ with $T \circ \alpha^{\mathbf{n}} = \alpha^{\mathbf{n}} \circ T$ λ_X-a.e., for every $\mathbf{n} \in \mathbb{Z}^d$. Then the following properties hold.*

(1) *Every $T \in Z_{\lambda_X}(\alpha)$ coincides λ_X-a.e. with an element $\psi \in \bar{Z}(\alpha)$;*
(2) *$\bar{Z}(\alpha)$ is countable and has an abelian subgroup of finite index.*

PROOF. The first assertion is clear from Theorem 1.1. For the second assertion we have to appeal to terminology and results from later sections. Proposition 2.2 shows that α is weakly algebraically equivalent to a $\mathbb{Z}^d$-action of the form $\alpha_{\mathfrak{R}_d/\mathfrak{p}}$ for some prime ideal $\mathfrak{p} \subset \mathfrak{R}_d$. Since α is irreducible, the algebraic factor maps $\chi \colon X \longrightarrow X_{\mathfrak{R}_d/\mathfrak{p}}$ and $\chi' \colon X_{\mathfrak{R}_d/\mathfrak{p}} \longrightarrow X$ between α and $\alpha_{\mathfrak{R}_d/\mathfrak{p}}$ are finite-to-one. Hence there exists a subgroup of finite index $Z' \subset \bar{Z}(\alpha)$ such that every $\phi \subset Z'$ acts trivially on $\ker(\chi)$ and thus induces an element $\phi' \in \bar{Z}(\alpha_{\mathfrak{R}_d/\mathfrak{p}})$. As $\bar{Z}(\alpha_{\mathfrak{R}_d/\mathfrak{p}})$ is abelian this implies the second assertion of this corollary (cf. [**6**] and [**11**]). Note that the abelian subgroup of finite index $A \subset \bar{Z}(\alpha)$ obtained in this manner is isomorphic to a subgroup of finite index in $\bar{Z}(\alpha_{\mathfrak{R}_d/\mathfrak{p}})$. □

This brief survey is organized as follows. In Section 2 we give a description of irreducible algebraic $\mathbb{Z}^d$-actions. Section 3 deduces Theorem 1.1 from a result about invariant measures of algebraic $\mathbb{Z}^d$-actions (Theorem 3.3). The Sections 4 and 5 contain brief sketches of the proofs of Theorem 3.3 in the two relevant cases of connected and zero-dimensional groups, respectively. The last section illustrates Theorem 1.1 with examples.

2. Irreducible $\mathbb{Z}^d$-actions

Following [**8**], [**18**] and [**20**] we denote by $\mathfrak{R}_d = \mathbb{Z}[u_1^{\pm 1}, \dots, u_d^{\pm 1}]$ the ring of Laurent polynomials with integral coefficients in the commuting variables $u_1, \dots, u_d$, and write every $f \in \mathfrak{R}_d$ as $f = \sum_{\mathbf{m} \in \mathbb{Z}^d} c_f(\mathbf{m}) u^{\mathbf{m}}$ with $u^{\mathbf{m}} = u_1^{m_1} \cdots u_d^{m_d}$ and $c_f(\mathbf{m}) \in \mathbb{Z}$ for every $\mathbf{m} = (m_1, \dots, m_d) \in \mathbb{Z}^d$.

Let α be an algebraic $\mathbb{Z}^d$-action on a compact abelian group X. We denote by $\widehat{X}$ the additively written dual group of X, write $\langle a, x \rangle$ the value of a character $a \in \widehat{X}$ at a point $x \in X$, and define the *dual action* $\hat{\alpha} \colon \mathbf{n} \mapsto \hat{\alpha}^{\mathbf{n}}$ of $\mathbb{Z}^d$ on $\widehat{X}$ by

$$\langle \hat{\alpha}^{\mathbf{n}} a, x \rangle = \langle a, \alpha^{\mathbf{n}} x \rangle$$

for every $\mathbf{n} \in \mathbb{Z}^d$, $x \in X$ and $a \in \widehat{X}$. For $f \in \mathfrak{R}_d$, $x \in X$ and $a \in \widehat{X}$, we set

$$f(\alpha)(x) = \sum_{\mathbf{n}\in\mathbb{Z}^d} c_f(\mathbf{n})\alpha^{\mathbf{n}}x, \qquad f(\hat{\alpha})(a) = \sum_{\mathbf{n}\in\mathbb{Z}^d} c_f(\mathbf{n})\hat{\alpha}^{\mathbf{n}}a, \tag{2.1}$$

and note that $f(\alpha)\colon X \longrightarrow X$ is a group homomorphism with dual homomorphism $\widehat{f(\alpha)} = f(\hat{\alpha})\colon \widehat{X} \longrightarrow \widehat{X}$,

The group $\widehat{X}$ is a module over the ring $\mathfrak{R}_d$ with operation

$$f \cdot a = f(\hat{\alpha})(a) \tag{2.2}$$

for $f \in \mathfrak{R}_d$ and $a \in \widehat{X}$. In particular,

$$u^{\mathbf{m}} \cdot a = \hat{\alpha}^{\mathbf{m}} a \tag{2.3}$$

for $\mathbf{m} \in \mathbb{Z}^d$ and $a \in \mathfrak{M}$. This *dual module* $\mathfrak{M} = \widehat{X}$ of α is Noetherian (and hence countable) whenever α is expansive (cf. [**18**, Proposition 5.4]).

Conversely, if $\mathfrak{M}$ is an $\mathfrak{R}_d$-module, we define an algebraic $\mathbb{Z}^d$-action $\alpha_{\mathfrak{M}}$ on the compact abelian group

$$X_{\mathfrak{M}} = \widehat{\mathfrak{M}} \tag{2.4}$$

by setting

$$\hat{\alpha}_{\mathfrak{M}}^{\mathbf{m}} a = u^{\mathbf{m}} \cdot a \tag{2.5}$$

for every $\mathbf{m} \in \mathbb{Z}^d$ and $a \in \mathfrak{M}$.

A prime ideal $\mathfrak{p} \subset \mathfrak{R}_d$ is *associated* with an $\mathfrak{R}_d$-module $\mathfrak{M}$ if there exists an $a \in \mathfrak{M}$ with $\mathfrak{p} = \operatorname{ann}(a) = \{f \in \mathfrak{R}_d : f \cdot a = 0\}$. The set $\operatorname{asc}(\mathfrak{M})$ of all prime ideals associated with $\mathfrak{M}$ has the property that

$$\bigcup_{\mathfrak{p}\in\operatorname{asc}(\mathfrak{M})} \mathfrak{p} = \{f \in \mathfrak{R}_d : \text{multiplication by } f \text{ on } \mathfrak{M} \text{ is not injective}\}. \tag{2.6}$$

According to [**18**], [**13**] and [**20**], many properties of an algebraic $\mathbb{Z}^d$-action α can be expressed in terms of the prime ideals associated with the dual module $\mathfrak{M} = \widehat{X}$ of α.

PROPOSITION 2.1. *Let $\mathfrak{M}$ a countable $\mathfrak{R}_d$-module.*

(1) *For any $\mathbf{n} \in \mathbb{Z}^d$, the following conditions are equivalent.*
  (a) *$\alpha_{\mathfrak{M}}^{\mathbf{n}}$ is ergodic;*
  (b) *$\alpha_{\mathfrak{R}_d/\mathfrak{p}}^{\mathbf{n}}$ is ergodic for every prime ideal $\mathfrak{p}$ associated with $\mathfrak{M}$;*
  (c) *No prime ideal $\mathfrak{p}$ associated with $\mathfrak{M}$ contains a polynomial of the form $u^{l\mathbf{n}} - 1$ with $l \geq 1$.*

(2) *The following conditions are equivalent.*
  (a) *$\alpha_{\mathfrak{M}}$ is ergodic;*
  (b) *$\alpha_{\mathfrak{R}_d/\mathfrak{p}}$ is ergodic for every prime ideal $\mathfrak{p}$ associated with $\mathfrak{M}$;*
  (c) *No prime ideal $\mathfrak{p}$ associated with $\mathfrak{M}$ contains a set of the form $\{u^{l\mathbf{n}} - 1 : \mathbf{n} \in \mathbb{Z}^d\}$ with $l \geq 1$.*

(3) *The following conditions are equivalent.*
  (a) *$\alpha_{\mathfrak{M}}$ is mixing;*
  (b) *$\alpha_{\mathfrak{M}}^{\mathbf{n}}$ is ergodic for every non-zero $\mathbf{n} \in \mathbb{Z}^d$;*
  (c) *$\alpha_{\mathfrak{M}}^{\mathbf{n}}$ is mixing for every non-zero $\mathbf{n} \in \mathbb{Z}^d$;*
  (d) *$\alpha_{\mathfrak{R}_d/\mathfrak{p}}$ is mixing for every prime ideal $\mathfrak{p}$ associated with $\mathfrak{M}$;*

(e) *None of the prime ideals associated with* $\mathfrak{M}$ *contains a polynomial of the form* $u^{\mathbf{n}} - 1$ *with* $\mathbf{0} \neq \mathbf{n} \in \mathfrak{R}_d$.

(4) *The following conditions are equivalent.*
   (a) α *is expansive;*
   (b) $\mathfrak{M}$ *is Noetherian and* $\alpha_{\mathfrak{R}_d/\mathfrak{p}}$ *is expansive for every prime ideal* $\mathfrak{p}$ *associated with* $\mathfrak{M}$*;*
   (c) *For every prime ideal* $\mathfrak{p}$ *associated with* $\mathfrak{M}$, $V_{\mathbb{C}}(\mathfrak{p}) \cap \mathbb{S}^d = \varnothing$, *where*

$$V_{\mathbb{C}}(\mathfrak{p}) = \{c \in (\mathbb{C} \smallsetminus \{0\})^d : f(c) = 0 \text{ for every } f \in \mathfrak{p}\}.$$

(5) *The following conditions are equivalent.*
   (a) $\alpha_{\mathfrak{M}}$ *is has positive entropy;*
   (b) $\alpha_{\mathfrak{R}_d/\mathfrak{p}}$ *has positive entropy for some prime ideal* $\mathfrak{p}$ *associated with* $\mathfrak{M}$*;*
   (c) *At least one prime ideal* $\mathfrak{p}$ *associated with* $\mathfrak{M}$ *has the property that* $\mathfrak{p}$ *is principal and* $\alpha_{\mathfrak{R}_d/\mathfrak{p}}$ *is mixing.*

For a proof we refer to [**20**, Propositions 6.6 and 19.4, and Theorem 6.5]. The next result is contained in [**20**, Proposition 6.6 and Theorem 29.2], or in [**6**] and [**11**].

PROPOSITION 2.2. *Let* α *be an irreducible and ergodic algebraic* $\mathbb{Z}^d$*-action on a compact zero-dimensional abelian group* X. *Then there exists a unique prime ideal* $\mathfrak{p} \subset \mathfrak{R}_d$ *with the following properties.*

(1) $\alpha_{\mathfrak{R}_d/\mathfrak{p}}$ *is ergodic;*
(2) *For every ideal* $I \supsetneq \mathfrak{p}$ *in* $\mathfrak{R}_d$, $\mathfrak{R}_d/I$ *is finite;*
(3) α *is weakly algebraically conjugate to* $\alpha_{\mathfrak{R}_d/\mathfrak{p}}$*;*
(4) *The group* X *is zero-dimensional if and only if* $\mathfrak{p}$ *contains a rational prime* $p > 1$, *and connected otherwise.*

Conversely, if $\mathfrak{p} \subset \mathfrak{R}_d$ *is a prime ideal satisfying the conditions* (1)–(2) *above, then the* $\mathbb{Z}^d$*-action* $\alpha = \alpha_{\mathfrak{R}_d/\mathfrak{p}}$ *on* $X_{\mathfrak{R}_d/\mathfrak{p}}$ *is irreducible and ergodic.*

PROPOSITION 2.3. *Let* α *be an irreducible and ergodic algebraic* $\mathbb{Z}^d$*-action on a compact abelian group* X. *Then* $h(\alpha^{\mathbf{n}}) < \infty$ *for every* $\mathbf{n} \in \mathbb{Z}^d$, *where* $h(\cdot)$ *denotes topological entropy.*

Conversely, if α *is an expansive, ergodic* (*and mixing*) *algebraic* $\mathbb{Z}^d$*-action on a compact abelian group* X *with* $h(\alpha^{\mathbf{n}}) < \infty$ *for every* $\mathbf{n} \in \mathbb{Z}^d$, *and if* $\mathfrak{p} \subset \mathfrak{R}_d$ *is a prime ideal associated with the module* $\mathfrak{M} = \widehat{X}$, *then the* $\mathbb{Z}^d$*-action* $\alpha_{\mathfrak{R}_d/\mathfrak{p}}$ *on* $X_{\mathfrak{R}_d/\mathfrak{p}}$ *is irreducible, ergodic* (*and mixing*).

The proof of Proposition 2.3 is similar to that of Corollary 2.3 in [**11**].

REMARKS 2.4. (1) Let α be an irreducible and ergodic algebraic $\mathbb{Z}^d$-action on a compact abelian group X, and let $\mathfrak{p} \subset \mathfrak{R}_d$ be the prime ideal satisfying the conditions (1)–(3) in Proposition 2.2. Proposition 2.1 shows that α is mixing if and only if $\alpha^{\mathbf{m}} \neq \mathrm{id}_X$ whenever $\mathbf{0} \neq \mathbf{m} \in \mathbb{Z}^d$, where id_X is the identity automorphism of X.

(2) If $\mathfrak{q} \subset \mathfrak{R}_d$ is a prime ideal satisfying the conditions (1)–(2) in Proposition 2.2, and if $\alpha = \alpha_{\mathfrak{R}_d/\mathfrak{q}}$, then the the prime ideal $\mathfrak{p} \subset \mathfrak{R}_d$ satisfying the conditions (1)–(3) in Proposition 2.2 is equal to $\mathfrak{q}$.

Motivated by Proposition 2.2 we take a closer look at $\mathbb{Z}^d$-actions of the form $\alpha_{\mathfrak{R}_d/\mathfrak{p}}$, where $\mathfrak{p} \subset \mathfrak{R}_d$ is a prime ideal. Let $\mathbb{T} = \mathbb{R}/\mathbb{Z}$, and let $\Omega = \mathbb{T}^{\mathbb{Z}^d}$, furnished with the product topology and component-wise addition. We write every $\omega \in \Omega$ as $\omega = (\omega_{\mathbf{n}})$ with $\omega_{\mathbf{n}} \in \mathbb{T}$ for every $\mathbf{n} \in \mathbb{Z}^d$ and define the shift-action σ of $\mathbb{Z}^d$ on Ω by (1.4). The pairing

$$\langle h, \omega \rangle = e^{2\pi i \sum_{\mathbf{n}\in\mathbb{Z}^d} c_h(\mathbf{n})\omega_{\mathbf{n}}}, \quad h \in \mathfrak{R}_d, \omega \in \Omega, \tag{2.7}$$

identifies the additive group $\mathfrak{R}_d$ with the dual group $\widehat{\Omega}$ of Ω. With this identification the automorphism $\hat{\sigma}^{\mathbf{n}}$ of $\mathfrak{R}_d$ dual to the shift $\sigma^{\mathbf{n}}$ on Ω consists of multiplication by $u^{\mathbf{n}}$. If $I \subset \mathfrak{R}_d$ is an ideal, then

$$I^{\perp} = \widehat{\mathfrak{R}_d/I} = X_{\mathfrak{R}_d/I} = \{\omega \in \Omega : \langle h, \omega \rangle = 1 \text{ for every } h \in I\} \tag{2.8}$$

is a closed, shift-invariant subgroup of Ω, and $\alpha_{\mathfrak{R}_d/I}$ is the restriction of σ to $X_{\mathfrak{R}_d/I}$.

Now consider the special case where $I = \mathfrak{p} \subset \mathfrak{R}_d$ is a prime ideal containing a rational prime $p > 1$. For any $k \geq 1$ we denote by F_{p^k} the field with p^k elements and write

$$\mathfrak{R}_d^{(p^k)} = F_{p^k}[u_1^{\pm 1}, \dots, u_d^{\pm 1}] \tag{2.9}$$

for the ring of Laurent polynomials in the variables $u_1, \dots, u_d$ with coefficients in F_{p^k}. Again we write every $h \in \mathfrak{R}_d^{(p^k)}$ as $h = \sum_{\mathbf{n}\in\mathbb{Z}^d} c_h(\mathbf{n})u^{\mathbf{n}}$ with $c_h(\mathbf{m}) \in F_{p^k}$ for every $\mathbf{m} \in \mathbb{Z}^d$.

We define a ring homomorphism $f \mapsto f_{/p}$ from $\mathfrak{R}_d$ to $\mathfrak{R}_d^{(p)}$ by reducing each coefficient of f modulo p. If

$$\bar{\mathfrak{p}} = \{f_{/p} : f \in \mathfrak{p}\}, \tag{2.10}$$

then $\bar{\mathfrak{p}} \subset \mathfrak{R}_d^{(p)}$ is again a prime ideal, and the map $f \mapsto f_{/p}$ induces an $\mathfrak{R}_d$-module isomorphism

$$\mathfrak{R}_p/\mathfrak{p} \cong \mathfrak{R}_d^{(p)}/\bar{\mathfrak{p}}. \tag{2.11}$$

We set $\Omega_p = F_p^{\mathbb{Z}^d}$, write every $\omega \in \Omega_p$ as $\omega = (\omega_{\mathbf{n}})$ with $\omega_{\mathbf{n}} \in F_p$ for every $\mathbf{n} \in \mathbb{Z}^d$, and use the pairing

$$\langle h, \omega \rangle = e^{2\pi i (\sum_{\mathbf{n}\in\mathbb{Z}^d} c_h(\mathbf{n})\omega_{\mathbf{n}})/p}, \quad h \in \mathfrak{R}_d^{(p)}, \omega \in \Omega_p,$$

to identify $\mathfrak{R}_d^{(p)}$ with the dual group $\widehat{\Omega}_p$ of Ω_p. Then

$$\widehat{\mathfrak{R}_d/\mathfrak{p}} = \widehat{\mathfrak{R}_d^{(p)}/\bar{\mathfrak{p}}} = X_{\mathfrak{R}_d^{(p)}/\bar{\mathfrak{p}}} = \{\omega \in \Omega_p : \langle h, \omega \rangle = 1 \text{ for every } h \in \mathfrak{p}\}, \tag{2.12}$$

and $\alpha_{\mathfrak{R}_d^{(p)}/\bar{\mathfrak{p}}}$ is the restriction to $X_{\mathfrak{R}_d^{(p)}/\bar{\mathfrak{p}}}$ of the shift-action σ of $\mathbb{Z}^d$ on Ω_p.

3. Mixing sets and invariant measures

Let α be an algebraic $\mathbb{Z}^d$-action on a compact abelian group X, and let μ be an α-invariant probability measure on the Borel field $\mathcal{B}_X$ of X. Following **[9]**–**[10]** we call a nonempty finite set $S \subset \mathbb{Z}^d$ *μ-mixing* under α if

$$\lim_{k\to\infty} \mu\Bigl(\bigcap_{\mathbf{n}\in S} \alpha^{-k\mathbf{n}}(B_{\mathbf{n}})\Bigr) = \prod_{\mathbf{n}\in S} \mu(B_{\mathbf{n}})$$

for every collection $(B_{\mathbf{n}},\ \mathbf{n} \in S)$ in $\mathcal{B}_X$, and *μ-nonmixing* otherwise. A nonempty finite set $S \subset \mathbb{Z}^d$ is *minimal μ-nonmixing* if it is μ-nonmixing, but every nonempty subset $S' \subsetneq S$ is μ-mixing. A λ_X-(non-)mixing set is called a (non-)mixing set of α.

Measurably conjugate algebraic $\mathbb{Z}^d$-actions obviously have the same nonmixing sets. Furthermore, the nonmixing sets of an algebraic $\mathbb{Z}^d$-actions on a compact abelian group X are determined by the prime ideals $\mathfrak{p} \subset \mathfrak{R}_d$ associated with the dual module $\mathfrak{M} = \widehat{X}$ of α.

PROPOSITION 3.1 ([**20**], [**10**]). *Let α be an algebraic $\mathbb{Z}^d$-actions on a compact abelian group X with dual module $\mathfrak{M} = \widehat{X}$. For every nonempty finite set $S \subset \mathbb{Z}^d$ the following conditions are equivalent.*

(1) *S is a mixing set of α;*

(2) *S is mixing for every $\alpha_{\mathfrak{R}/\mathfrak{p}}$ with $\mathfrak{p} \in \operatorname{asc}(\mathfrak{M})$.*

If the group X is connected and λ_X is mixing under α, then α is mixing of every order, and hence every nonempty finite set $S \subset \mathbb{Z}^d$ is mixing under α, by [**21**]. If X is disconnected, then a mixing algebraic $\mathbb{Z}^d$-action α on X has nonmixing sets if and only if λ_X does not have completely positive entropy (cf. [**11**], [**20**, Section 27] and Proposition 3.2 below). In particular, if α is an irreducible and mixing algebraic $\mathbb{Z}^d$-action on a compact zero-dimensional abelian group X, then α has nonmixing sets.

PROPOSITION 3.2 ([**11**]). *Let $d > 1$, α an irreducible and ergodic algebraic $\mathbb{Z}^d$-action on a compact zero-dimensional abelian group X, and let $\mathfrak{p} \subset \mathfrak{R}_d$ be the prime ideal and $1 < p \in \mathfrak{p}$ the rational prime described in Proposition 2.2. We define the prime ideal $\overline{\mathfrak{p}} \subset \mathfrak{R}_d^{(p)}$ by (2.10).*

(1) *The $\mathbb{Z}^d$-actions α and $\alpha_{\mathfrak{R}_d/\mathfrak{p}} = \alpha_{\mathfrak{R}_d^{(p)}/\overline{\mathfrak{p}}}$ have the same nonmixing sets;*

(2) *For every nonzero element $h \in \overline{\mathfrak{p}} \subset \mathfrak{R}_d^{(p)}$, the* support

$$\mathcal{S}(h) = \{\mathbf{n} \in \mathbb{Z}^d : c_h(\mathbf{n}) \neq 0\}$$

of h is a nonmixing set of $\alpha_{\mathfrak{R}_d/\mathfrak{p}}$ and hence of α.

In this terminology we can express Theorem 1.1 as a consequence of the following result.

THEOREM 3.3. *Let $d > 1$, α_1 and α_2 irreducible, mixing and expansive algebraic $\mathbb{Z}^d$-actions on compact abelian groups X_1 and X_2, respectively, and let $\phi\colon X_1 \longrightarrow X_2$ be a measurable factor map of α_1 and α_2. We set $X = X_1 \times X_2$, denote by $\alpha = \alpha_1 \times \alpha_2$ the product-action of $\mathbb{Z}^d$ on X, and assume that one of the following two conditions is satisfied.*

(1) *α_1 is mixing of every order;*

(2) *α_1 and α_2 share a minimal nonmixing set $S \subset \mathbb{Z}^d$.*

Then every joining ν of λ_{X_1} and λ_{X_2} such that the $\mathbb{Z}^d$-action α on (X, ν) is measurably conjugate to α_1 on (X_1, λ_{X_1}), is a translate of the Haar measure of a closed α-invariant subgroup $Y \subset X$.

PROOF OF THEOREM 1.1, USING THEOREM 3.3. Let $\phi\colon X_1 \longrightarrow X_2$ be a measurable conjugacy of α_1 and α_2. We choose a prime ideal $\mathfrak{q} \subset \mathfrak{R}_d$ which is associated with $\mathfrak{M}_2 = \widehat{X}_2$ and an element $a \in \mathfrak{M}_2$ with $\mathfrak{q} = \operatorname{ann}(a)$. The inclusion $\mathfrak{N} = \mathfrak{R}_d \cdot a \subset \mathfrak{M}_2$ determines a dual factor map $\psi\colon X_2 \longrightarrow X_2' = \widehat{\mathfrak{N}}$ of α_2 and $\alpha_2' = \alpha_{\mathfrak{N}}$. We set $\phi' = \psi \circ \phi\colon X_1 \longrightarrow X_2'$, $X' = X_1 \times X_2'$, $\alpha' = \alpha_1 \times \alpha_2'$, and denote by ν' the α'-invariant probability measure on $\Gamma(\phi') = \{(x, \phi'(x)) : x \in X_1\} \subset X'$ which satisfies that $\nu'\pi_1^{-1} = \lambda_{X_1}$ and $\nu'\pi_2'^{-1} = \lambda_{X_2'}$, where $\pi_1 \colon X \longrightarrow X_1$ and $\pi_2' \colon X \longrightarrow X_2'$ are the coordinate projections.

Since α_1 is irreducible, $h(\alpha_2^{\mathbf{n}}) = h(\alpha_1^{\mathbf{n}}) < \infty$ for every $\mathbf{n} \in \mathbb{Z}^d$ by Corollary 1.2. Hence α_2' is irreducible, mixing and expansive by the Propositions 2.1 and 2.2.

If X_1 is connected, then α_1 is mixing of every order by [**21**], and Theorem 3.3 implies that ν' is a translate of the Haar measure of a closed, α_2'-invariant subgroup of X'. It follows that ϕ' coincides λ_{X_1}-a.e. with an affine map.

We denote by $\psi' \colon X_1 \longrightarrow X_2'$ the homomorphism part of ϕ' (cf. (1.3)). As α_1 is irreducible, ψ' is finite-to-one. Hence $\phi' = \psi \circ \phi$ and ψ are both finite-to-one, and α_2 is irreducible by Proposition 2.2. We repeat the first part of the proof with α_2' and X_2' replaced by α_2 and X_2 and obtain that ϕ coincides λ_{X_1}-a.e. with an affine map.

If X_1 is not connected, it is zero-dimensional by irreducibility, and α_1 has nonmixing sets by Proposition 3.2. We fix a minimal nonmixing set $S \subset \mathbb{Z}^d$ for α_1 and α_2. Proposition 3.1 allows us to choose the associated prime ideal $\mathfrak{q} \subset \mathfrak{R}_d$ considered above such that S is minimal nonmixing for $\alpha_{\mathfrak{R}_d/\mathfrak{q}}$. Theorem 3.3 implies that ν' is a translate of the Haar measure of a closed, α_2'-invariant subgroup of X'. It follows that ϕ' coincides λ_{X_1}-a.e. with an affine map, and the proof is completed as in the connected case. □

4. The proof of Theorem 3.3 in the absence of nonmixing sets

If α_1 has no nonmixing sets, then it is mixing of every order, and the same is true for α_2, and X_1 and X_2 are connected (cf. the Propositions 2.1–2.3 and 3.1–3.2). In this case Theorem 3.3 is in essence a consequence of a result by Katok and Spatzier ([**5**]) on invariant measures for commuting toral automorphisms.

Definition 4.1. An algebraic $\mathbb{Z}^d$-action α on a compact abelian group X is *semi-irreducible* if there exist an integer $n \geq 1$, irreducible algebraic $\mathbb{Z}^d$-actions α_i on compact abelian groups X_i, $i = 1, \dots, n$, and a continuous, surjective, finite-to-one group homomorphism $\phi \colon X \longmapsto \bar{X} = X_1 \times \dots \times X_n$ such that $\phi \cdot \alpha^{\mathbf{n}} = \bar{\alpha}^{\mathbf{n}} \cdot \phi$ for every $\mathbf{n} \in \mathbb{Z}^d$, where $\bar{\alpha} = \alpha_1 \times \dots \times \alpha_n$ is the product-action of $\mathbb{Z}^d$ on $\bar{X}$.

The class of semi-irreducible actions is easily seen to be closed under taking direct products and algebraic factors. The following version of the result by Katok and Spatzier is suited for proving Theorem 3.3.

Theorem 4.2. *Let $d > 1$, and let α be an semi-irreducible expansive and mixing algebraic $\mathbb{Z}^d$-action on a compact connected abelian group X and μ an α-invariant and mixing probability measure on X. Then there exists a closed α-invariant subgroup $Y \subset X$ with the following properties:*

(1) *μ is invariant under translation by Y,*
(2) *The projection of μ onto X/Y has zero entropy under $\alpha_{X/Y}^{\mathbf{n}}$ for every $\mathbf{n} \in \mathbb{Z}^d$, where $\alpha_{X/Y}$ is the action induced by α on X/Y.*

Theorem 4.2 clearly implies Theorem 3.3 for connected groups, since μ has completely positive entropy under each $\alpha^{\mathbf{n}}$. By using an induction argument we may reduce the proof of Theorem 4.2 to the case where μ is not invariant under translation by any infinite subgroup of X.

In order to explain the proof of Theorem 4.2 in this special case we assume initially that $X = \mathbb{T}^n$ for some $n \geq 3$ (no smaller n is possible, since $d > 1$ and

α is mixing), and that the matrices $\alpha^{\mathbf{m}} \in \mathrm{GL}(n, \mathbb{Z})$, $\mathbf{m} \in \mathbb{Z}^d$, can be diagonalized simultaneously over $\mathbb{C}$. Then there exists a decomposition of $\mathbb{R}^n$ into eigenspaces $V_1, \dots, V_k$ of the linear action $\bar{\alpha}$ of α on $\mathbb{R}^n$ with eigenvalues

$$\eta_i \colon \mathbb{Z}^d \longrightarrow \mathbb{R}^{\times} \text{ if } V_i \cong \mathbb{R}, \;\; \eta_i \colon \mathbb{Z}^d \longrightarrow \mathbb{C}^{\times} \text{ if } V_i \cong \mathbb{C},$$

where $\bar{\alpha}^{\mathbf{n}} v = \eta_i(\mathbf{n}) v$ for every $i = 1, \dots, k$, $v \in V_i$ and $\mathbf{n} \in \mathbb{Z}^d$.

Fix $\mathbf{m} \in \mathbb{Z}^d$ such that $\alpha^{\mathbf{m}}$ is hyperbolic, set $P^-(\mathbf{n}) = \{i : |\eta_i(\mathbf{n})| < 1\}$, and assume without loss in generality that $P^-(\mathbf{n}) = \{1, \dots, l\}$, $l < k$.

We intersect the leaves of the foliation $\mathcal{F}$ of $X = \mathbb{T}^n = \mathbb{R}^n/\mathbb{Z}^n$ by the cosets of $V_1 \subset X$ with a finite partition $\mathcal{P}$ of X into 'nice' convex sets: the resulting measurable partition generates a countably generated Borel sigma-algebra $\mathcal{A} \subset \mathcal{B}_X$. The following illustration shows part of a typical leaf L of $\mathcal{F}$ (on the left), and some of the atoms of $\mathcal{A}$ into which L is decomposed by $\mathcal{P}$ (on the right).

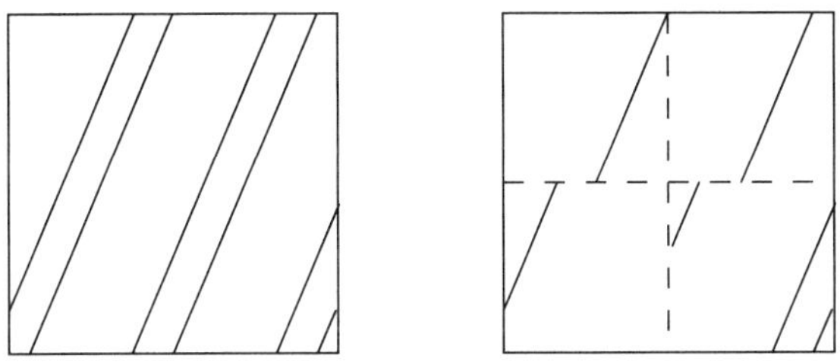

Let $\{\mu_x^{\mathcal{A}} : x \in X\}$ be the decomposition of μ with respect to $\mathcal{A}$, and let $J_\mu^{\mathcal{A}} \colon \mathbb{Z}^d \times X \longrightarrow \mathbb{R}$ be the *information cocycle* of $\mathcal{A}$, given by

$$J_\mu^{\mathcal{A}}(\mathbf{n}, x) = \log \frac{d\mu_x^{\mathcal{A}}}{d\mu_x^{\alpha^{-\mathbf{n}}(\mathcal{A})}}(x) = \log \frac{\mu_x^{\alpha^{-\mathbf{n}}(\mathcal{A})}([x]_{\mathcal{A} \vee \alpha^{-\mathbf{n}}(\mathcal{A})})}{\mu_x^{\mathcal{A}}([x]_{\mathcal{A} \vee \alpha^{-\mathbf{n}}(\mathcal{A})})}$$

for every $\mathbf{n} \in \mathbb{Z}^d$ and $x \in X$. The cocycle equation

$$J_\mu^{\mathcal{A}}(\mathbf{n}, \alpha^{\mathbf{n}'} x) + J_\mu^{\mathcal{A}}(\mathbf{n}', x) = J_\mu^{\mathcal{A}}(\mathbf{n} + \mathbf{n}', x)$$

is just the chain-rule of Radon-Nikodym derivatives. If

$$\mathbf{n} \in S = \{\mathbf{k} \in \mathbb{Z}^d : |\eta_1(\mathbf{k}) - 1| < 1/2\},$$

then each atom of $\mathcal{A}$ intersects only a bounded number of atoms of $\alpha^{-\mathbf{n}}(\mathcal{A})$ and vice versa. This implies that $\int |J_\mu^{\mathcal{A}}(\mathbf{n}, \cdot)| \, d\mu$ is bounded on the infinite set S. As μ is mixing under α this can be used to show that there exists an $\mathcal{A}$-measurable Borel map $b \colon X \longrightarrow \mathbb{R}$ with

$$b(\alpha^{\mathbf{n}} x) + J_\mu^{\mathcal{A}}(\mathbf{n}, x) - b(x) = \theta(\mathbf{n})$$

μ-a.e. for every $\mathbf{n} \in \mathbb{Z}^d$, where $\theta \colon \mathbb{Z}^d \longrightarrow \mathbb{R}$ is a group homomorphism.

We 'correct' the measures $\mu_x^{\mathcal{A}}$ by setting

$$\nu_x = e^{-b(x)} \mu_x^{\mathcal{A}}$$

for every $x \in X$. Then

$$\log \frac{d\nu_x}{d\nu_{\alpha^{\mathbf{n}} x} \alpha^{-\mathbf{n}}}(x) = \theta(\mathbf{n})$$

for every $\mathbf{n}, x$.

Fix $x \in X$ for the moment and combine the measures ν_y, $y \in x + V_1 \subset X$, to a sigma-finite measure $\bar{\rho}_x$ on $x + V_1$. The bijection $x + v \mapsto v$ from $x + V_1$ to V_1

sends $\bar{\rho}_x$ to a sigma-finite measure ρ_x on V_1. This map $x \mapsto \rho_x$ has the following properties:

(1) $\rho_x = e^{\theta(\mathbf{n})} \rho_{\alpha^{\mathbf{n}} x} M_{\eta_1(\mathbf{n})}$

(2) $\rho_{x+v} = \rho_v T_{-v}$,

where M_t is multiplication by t and T_v translation by v on V_1.

Ergodicity implies that there are only two possibilities: either each ρ_x is concentrated in a single atom, or the map $x \mapsto \rho_x$ is constant and each ρ_x is translation-invariant (some caution is needed in this part of the argument).

The latter possibility is excluded, since it would imply the invariance of μ under V_1 and hence under the closure of V_1 in X. In the former case the measure μ is concentrated on a set which intersects each leaf of $\mathcal{F}$ in at most one point.

We repeat the argument with V_2 replacing V_1. After l steps we obtain that μ is supported on a set which intersects each coset of $V_1 + \cdots + V_l$ in at most one point, which implies that $\alpha^{\mathbf{n}}$ has zero entropy.

In order to deal with the general case, where X is not necessarily a finite-dimensional torus, we have to give an alternative description of irreducible $\mathbb{Z}^d$-actions on compact connected abelian groups, taken from [**18**]–[**20**]; the algebraic background can be found in [**3**] and [**22**].

Let $d \geq 1$, $c = (c_1, \dots, c_d) \in (\overline{\mathbb{Q}}^{\times})^d$, and let $K = \mathbb{Q}(c)$ be the algebraic number field generated by $\{c_1, \dots, c_d\}$. We write P^K, P^K_f, and P^K_∞ for the sets of places (= equivalence classes of valuations), finite places and infinite places of K. For every $v \in P^K$ we write K_v for the completion of K at the place v, and choose a Haar measure λ_v on the locally compact additive group K_v. Since K_v is a topological ring, we can define a distinguished valuation $|\cdot|_v \colon K_v \longmapsto \mathbb{R}$ in v by setting

$$\lambda_v(aB) = |a|_v \lambda_v(B)$$

for every $a \in K_v$ and every compact set $B \subset K_v$. The set

$$R_v = \{r \in K_v : |r|_v \leq 1\}$$

is a compact subset of K_v. If $v \in P^K_f$, then R_v is, in addition, open, and is the unique maximal compact subring of K_v. Furthermore,

$$\mathfrak{o}_K = \bigcap_{v \in P^K_f} R_v \tag{4.1}$$

is the ring of integral elements in K. Let

$$F = P^K_\infty \cup \{v \in P^K_f : |c_i|_v \neq 1 \text{ for some } i \in \{1, \dots, d\}\}, \tag{4.2}$$

Then F is finite by Theorem III.3 in [**22**], and we denote by

$$\iota \colon K \longmapsto V = \prod_{v \in F} K_v \tag{4.3}$$

the diagonal embedding $r \mapsto (r, \dots, r)$, $r \in K$, and put

$$R = \{a \in K : |a|_v \leq 1 \text{ for every } v \in P^K_f \smallsetminus F\} \supset \mathfrak{o}_K. \tag{4.4}$$

Then $\iota(R)$ is a discrete, co-compact, additive subgroup of V, and we consider the compact quotient group

$$Y_c = \widehat{R} = V \,/\, \iota(R) \cong \Bigl(\prod_{v \in P^K_\infty} K_v \times \prod_{v \in F \cap P^K_f} R_v \Bigr) \Big/ \iota(\mathfrak{o}_K), \tag{4.5}$$

where $\widehat{R}$ is the Pontryagin dual of R, and write

$$\pi\colon V \longmapsto Y_c \tag{4.6}$$

for the quotient map. In the special case where $F = P_\infty^K$ (i.e. where each c_i, $i = 1, \dots, d$, is a unit in $\mathfrak{o}_K$),

$$R = \mathfrak{o}_K, \qquad V \cong \mathbb{R}^{r(K)}, \qquad Y_c = \widehat{\mathfrak{o}_K} \cong \mathbb{T}^{r(K)}, \tag{4.7}$$

with

$$r(K) = |\{v \in P_\infty^K : K_v = \mathbb{R}\}| + 2|\{v \in P_\infty^K : K_v = \mathbb{C}\}|. \tag{4.8}$$

For every $a \in R$ and $b = (b_v, v \in F) \in V$ we set $a \cdot b = (ab_v, v \in F)$. Since $a \cdot \iota(R) \subset \iota(R)$, multiplication by a defines a surjective group homomorphism $\theta_a\colon Y_c \longmapsto Y_c$ by

$$\theta_a(b + \iota(R)) = a \cdot b + \iota(R), \tag{4.9}$$

for every $b \in V_F$. Then $\theta_{aa'} = \theta_a \cdot \theta_{a'}$ for all $a, a' \in R_F$. In particular, if $R_F^\times \subset R_F$ is the group of units (i.e. invertible elements) in R_F, then θ_a is a continuous group automorphism of Y_F for every $a \in R_F^\times$. This allows us to define a $\mathbb{Z}^d$-action β_c by automorphisms of Y_c by setting

$$\beta_c^{\mathbf{n}} y = \theta_{c^{\mathbf{n}}}(y) \tag{4.10}$$

for every $\mathbf{n} = (n_1, \dots, n_d) \in \mathbb{Z}^d$, where

$$c^{\mathbf{n}} = c_1^{n_1} \cdots c_d^{n_d}.$$

PROPOSITION 4.3 (Proposition 7.2 in [**20**]). *Suppose that $d \geq 1$, and that α is a mixing and expansive algebraic $\mathbb{Z}^d$-action on a compact connected abelian group X. Then α is irreducible if and only if there exist a point $c = (c_1, \dots, c_d) \in (\overline{\mathbb{Q}}^\times)^d$ with the following properties.*

(i) $|c^{\mathbf{m}}| \neq 1$ *for every nonzero* $\mathbf{m} = (m_1, \dots, m_d) \in \mathbb{Z}^d$,

(ii) α *is weakly algebraically conjugate to the* $\mathbb{Z}^d$*-action* β_c *on* Y_c.

Furthermore,

$$h(\alpha^{\mathbf{n}}) = h(\beta_c^{\mathbf{n}}) = \sum_{v \in F} \log \max \{1, |c^{\mathbf{n}}|_v\} \tag{4.11}$$

for every $\mathbf{n} \in \mathbb{Z}^d$, *where* F *appears in* (4.2).

We return to the proof of Theorem 4.2 in the general case. Equation (4.5) allows us to assume the following (perhaps after modifying α and X by a finite-to-one algebraic factor map):

(1) $X = \prod_{i=1}^n K_i \,/\, D$, where each K_i is a locally compact abelian field with characteristic zero and $D \subset \prod_{i=1}^n K_i$ is a discrete co-compact subgroup;

(2) There exist homomorphisms $\gamma_i\colon \mathbb{Z}^d \longrightarrow K_i^\times$, $i = 1, \dots, d$, where $K_i^\times$ is the multiplicative group $K_i \smallsetminus \{0\}$, with the following property: if

$$\theta^{\mathbf{m}}(a, \dots, a_n) = (\gamma_1(\mathbf{m})a_1, \dots, \gamma_n(\mathbf{m})a_n)$$

for every $\mathbf{m} \in \mathbb{Z}^d$ and $(a_1, \dots, a_n) \in \prod_{i=1}^n K_i$, then $\theta^{\mathbf{m}}(D) = D$ and $\alpha^{\mathbf{m}}$ is the automorphism of X induced by $\theta^{\mathbf{m}}$.

This description of α and X is sufficiently similar to the toral case described earlier to allow essentially the same proof as before.

5. The proof of Theorem 3.3 in the presence of nonmixing sets

If α_1 and α_2 share a nonmixing set $S \subset \mathbb{Z}^d$ then the following lemma allows us to synchronize the 'nonmixing times' of α_1 and α_2.

LEMMA 5.1 ([**11**]). *Let α_1 and α_2 be algebraic $\mathbb{Z}^d$-actions on compact abelian groups X_1 and X_2, respectively. If α_2 is a measurable factor of α_1, and if $S \subset \mathbb{Z}^d$ is a nonempty finite set which is minimal nonmixing for both α_1 and α_2, then there exist nonzero elements $a_i(\mathbf{n}) \in \widehat{X}_i$, $\mathbf{n} \in S$, $i = 1, 2$, such that*

$$\sum_{\mathbf{n}\in S} \hat{\alpha}_1^{k\mathbf{n}}(a_1(\mathbf{n})) = \sum_{\mathbf{n}\in S} \hat{\alpha}_2^{k\mathbf{n}}(a_2(\mathbf{n})) = 0 \tag{5.1}$$

for every k in an infinite subset $K \subset \mathbb{N}$.

For the proof of Theorem 3.3 in the zero-dimensional case (i.e. in the presence of nonmixing sets) we use Lemma 5.1 to find nonzero elements $a_i(\mathbf{n}) \in \widehat{X}_i$, $\mathbf{n} \in S$, $i = 1, 2$, such that (5.1) holds for every k in an infinite subset $K \subset \mathbb{N}$. It follows that

$$\sum_{\mathbf{n}\in S} \hat{\alpha}_1^{k\mathbf{n}}(f_1(\hat{\alpha}_1)(a_1(\mathbf{n}))) = \sum_{\mathbf{n}\in S} \hat{\alpha}_2^{k\mathbf{n}}(f_2(\hat{\alpha}_2)(a_2(\mathbf{n}))) = 0 \tag{5.2}$$

for every $k \in K$ and $f_1, f_2 \in \mathfrak{R}_d$.

Put $X = X_1 \times X_2$ and denote by $\alpha = \alpha_1 \times \alpha_2$ the product-$\mathbb{Z}^d$-action on X. We fix a nonempty finite set S which is minimal nonmixing simultaneously for λ_{X_1}, λ_{X_2} and ν. As every proper subset of S is ν-mixing, the Fourier transform $\hat{\nu}\colon \widehat{X} = \widehat{X}_1 \times \widehat{X}_2 \longrightarrow \mathbb{C}$ satisfies that

$$\begin{aligned}
&\overline{\hat{\nu}\big(f_1(\hat{\alpha}_1)(a_1(\mathbf{m})), f_2(\hat{\alpha}_2)(a_2(\mathbf{m}))\big)} \\
&\quad = \hat{\nu}\big(-\hat{\alpha}_1^{k\mathbf{m}}(f_1(\hat{\alpha}_1)(a_1(\mathbf{m}))), -\hat{\alpha}_2^{k\mathbf{m}}(f_2(\hat{\alpha}_2)(a_2(\mathbf{m})))\big) \\
&\quad = \lim_{\substack{k\to\infty \\ k\in K}} \hat{\nu}\Big(\sum_{\mathbf{n}\in S\smallsetminus\{\mathbf{m}\}} \big(\hat{\alpha}_1^{k\mathbf{n}}(f_1(\hat{\alpha}_1)(a_1(\mathbf{n}))), \hat{\alpha}_2^{k\mathbf{n}}(f_2(\hat{\alpha}_2)(a_2(\mathbf{n})))\big)\Big) \\
&\quad = \prod_{\mathbf{n}\in S\smallsetminus\{\mathbf{m}\}} \hat{\nu}\big(f_1(\hat{\alpha}_1)(a_1(\mathbf{n})), f_2(\hat{\alpha}_2)(a_2(\mathbf{n}))\big)
\end{aligned}$$

for every $\mathbf{m} \in S$. By varying $\mathbf{m} \in S$ we see that

$$\big|\hat{\nu}\big(f_1(\hat{\alpha}_1)(a_1(\mathbf{m})), f_2(\hat{\alpha}_2)(a_2(\mathbf{m}))\big)\big| \in \{0, 1\}$$

for every $\mathbf{m} \in S$ and $f_1, f_2 \in \mathfrak{R}_d$.

We fix an element $\mathbf{n} \in S$, consider the $\hat{\alpha}$-invariant subgroup

$$\mathfrak{N} = \big\{\big(f_1(\hat{\alpha}_1)(a_1(\mathbf{n})), f_2(\hat{\alpha}_2)(a_2(\mathbf{n}))\big) : f_1, f_2 \in \mathfrak{R}_d\big\} \subset \widehat{X}.$$

Since $|\hat{\nu}(a)| \in \{0, 1\}$ for every $a \in \mathfrak{N}$, the projection of ν onto the quotient group $Z = X/\mathfrak{N}^{\perp} = \widehat{\mathfrak{N}}$ is a translate of the Haar measure of a closed, α_Z-invariant subgroup, where α_Z is the $\mathbb{Z}^d$-action induced by α on Z. Since the map $\pi\colon X \longrightarrow Z$ is finite-to-one by irreducibility and ν is ergodic, an elementary skew-product argument shows that ν is also a translate of the Haar measure of a closed, α-invariant subgroup $Y \subset X$. This proves Theorem 3.3 in the zero-dimensional case.

REMARK 5.2. There is a significant difference between the connected and zero-dimensional cases in Theorem 3.3. Assume for simplicity that $d > 1$, and that α is an irreducible, expansive and mixing algebraic $\mathbb{Z}^d$-action on a compact abelian group X. If X is connected, then it is not known whether there exists a nonatomic probability measure $\mu \neq \lambda_X$ on X which is invariant and ergodic under α; in some special cases it has even been conjectured that no such measures exist (cf. e.g. [**4**], [**16**]). Even the existence of nonatomic mixing measures $\mu \neq \lambda_X$ has not been ruled out. However, if μ is an α-invariant mixing measure with $h_\mu(\alpha^{\mathbf{n}}) > 0$ for some expansive $\alpha^{\mathbf{n}}$, then $\mu = \lambda_X$ by [**5**].

If X is zero-dimensional, then there always exist nonatomic α-invariant and ergodic probability measures $\mu \neq \lambda_X$ with $h_\mu(\alpha^{\mathbf{n}}) > 0$ for some nonzero $\mathbf{n} \in \mathbb{Z}^d$. One can also construct nonatomic α-invariant and ergodic probability measures μ with $h_\mu(\alpha^{\mathbf{n}}) = 0$ for every $\mathbf{n} \in \mathbb{Z}^d$. We refer to [**9**] and [**19**] for examples.

6. Examples

EXAMPLE 6.1. The matrix

$$A = \begin{pmatrix} 0 & 1 & 0 \\ 0 & 0 & 1 \\ 1 & 3 & 0 \end{pmatrix} \in \mathrm{GL}(3, \mathbb{Z})$$

has the irreducible characteristic polynomial

$$f = x^3 - 3x - 1$$

with roots

$$a_1 = -1.532\ldots,\ a_2 = -0.3473\ldots,\ a_3 = 1.879\ldots$$

Hence A has a two-dimensional expanding subspace, but the expanding subspace of A^{-1} has dimension 1. It follows that A and A^{-1} are not algebraically conjugate, although they are measurably conjugate. Since the characteristic polynomial of A is irreducible, A does not have an invariant subtorus and the $\mathbb{Z}$-action $n \mapsto A^n$ is irreducible.

EXAMPLE 6.2. Let

$$A = \begin{pmatrix} 3 & 4 \\ 1 & 1 \end{pmatrix},\quad B = \begin{pmatrix} 0 & 1 \\ 1 & 4 \end{pmatrix},\quad C = \begin{pmatrix} 3 & 2 \\ 2 & 1 \end{pmatrix}.$$

If

$$M = \begin{pmatrix} 1 & 3 \\ 0 & 1 \end{pmatrix},\quad M' = \begin{pmatrix} 1 & 3 \\ 0 & 2 \end{pmatrix},$$

then $AM = MB$ and $M \in \mathrm{SL}(2, \mathbb{Z})$, i.e. A and B are algebraically conjugate.

Also, $CM' = M'B$, but $M' \notin \mathrm{GL}(2, \mathbb{Z})$. Since there is no $M'' \in \mathrm{GL}(2, \mathbb{Z})$ with $CM'' = M''B$, B and C are not algebraically conjugate, but they are measurably conjugate: they are Bernoulli with equal entropy. The $\mathbb{Z}$-actions $n \mapsto A^n$, $n \mapsto B^n$, $n \mapsto C^n$ are all irreducible.

Theorem 1.1 shows that for irreducible, expansive and mixing algebraic $\mathbb{Z}^d$-actions with $d > 1$ topological and algebraic conjugacy are equivalent. The following elementary observations, taken from [**6**], are sometimes useful for checking that two given $\mathbb{Z}^d$-actions are algebraically nonconjugate.

Let α be an algebraic $\mathbb{Z}^d$-action on a compact abelian group X. We say that $\hat{\alpha}$ is *cyclic* if the dual module $\mathfrak{M} = \widehat{X}$ is singly generated, i.e. if $\mathfrak{M} = \mathfrak{R}_d \cdot a$ for some $a \in \mathfrak{M}$. In this case $\mathfrak{M} \cong \mathfrak{R}_d/I$ for some ideal $I \subset \mathfrak{R}_d$.

PROPOSITION 6.3. *Let α be an irreducible and mixing algebraic $\mathbb{Z}^d$-action on a compact connected abelian group X. Then $\hat{\alpha}$ is cyclic if and only if α is algebraically conjugate to $\alpha_{\mathfrak{R}_d/\mathfrak{p}}$ for some prime ideal $\mathfrak{p} \subset \mathfrak{R}_d$ (cf. Proposition 2.2).*

If $\hat{\alpha}$ is cyclic, and if β is an algebraic $\mathbb{Z}^d$-action on a compact abelian group Y which is weakly algebraically conjugate to α, then β is algebraically conjugate to α if and only if $\hat{\beta}$ is cyclic.

COROLLARY 6.4. *Let α be an irreducible algebraic $\mathbb{Z}^d$-action on a compact abelian group X with the property that $\hat{\alpha}^{\mathbf{n}}$ is cyclic for some $\mathbf{n} \in \mathbb{Z}^d$. If β is an algebraic $\mathbb{Z}^d$-action on a compact abelian group Y which is weakly algebraically conjugate to α, but for which $\hat{\beta}^{\mathbf{n}}$ is not cyclic, then $\hat{\beta}$ is not cyclic.*

COROLLARY 6.5. *Suppose that α and β are algebraic $\mathbb{Z}^d$-actions on compact abelian groups such that $\hat{\alpha}$ is cyclic, but not $\hat{\beta}$. Then α is not algebraically conjugate to any re-parametrization of β (i.e. to any $\mathbb{Z}^d$-action of the form $\mathbf{n} \mapsto \beta^{A\mathbf{n}}$ with $A \in GL(d, \mathbb{Z})$).*

We illustrate Proposition 6.3 and its corollaries with some examples.

EXAMPLE 6.6. Let α be the $\mathbb{Z}^2$-action on $\mathbb{T}^3$ given by $\alpha^{\mathbf{n}} = A^{n_1} B^{n_2}$ for every $\mathbf{n} = (n_1, n_2) \in \mathbb{Z}^2$, where

$$A = \begin{pmatrix} 1 & -1 & -1 \\ -1 & -2 & -1 \\ -1 & -4 & -2 \end{pmatrix}, \qquad B = \begin{pmatrix} 1 & 2 & -1 \\ -1 & -2 & 2 \\ 2 & 5 & -2 \end{pmatrix}.$$

The matrices A and B have the same irreducible characteristic polynomial

$$g = x^3 + 3x^2 - 6x + 1,$$

A is irreducible and hyperbolic, and α is irreducible, expansive and mixing.

Let

$$V = \begin{pmatrix} 2 & -2 & -1 \\ 0 & -3 & 0 \\ 1 & -4 & -2 \end{pmatrix},$$

$$A' = V^{-1}AV = \begin{pmatrix} 2 & -4 & -1 \\ 1 & -4 & -1 \\ 1 & -5 & -1 \end{pmatrix},$$

$$B' = V^{-1}BV = \begin{pmatrix} 0 & 1 & 0 \\ 0 & 0 & 1 \\ -1 & 6 & -3 \end{pmatrix},$$

and denote by α' the irreducible $\mathbb{Z}^2$-action $\mathbf{n} \mapsto A'^{n_1} B'^{n_2}$. It is clear that $\alpha^{\mathbf{n}}$ and $\alpha'^{\mathbf{n}}$ are measurably conjugate for every $\mathbf{n} \in \mathbb{Z}^d$, but α and α' are not algebraically conjugate, since $\hat{B}'$ is cyclic, but $\hat{B}$ is not. Hence they are measurably non-conjugate by Theorem 1.1.

EXAMPLE 6.7. Denote by σ the shift-action (1.4) of $\mathbb{Z}^d$ on $\bar{X} = (\mathbb{Z}/2\mathbb{Z})^{\mathbb{Z}^d}$. For every nonempty finite set $E \subset \mathbb{Z}^d$ we denote by $X_E \subset \bar{X}$ the closed shift-invariant subgroup consisting of all $x \in \bar{X}$ whose coordinates sum to 0 in every translate of E in $\mathbb{Z}^d$. If E has at least two points then X_E is uncountable and the restriction σ_E of σ to X_E is an expansive algebraic $\mathbb{Z}^d$-action.

For $d = 2$ and the subset

$$E = \{(0,0), (1,0), (0,1)\} \subset \mathbb{Z}^2,$$

the $\mathbb{Z}^2$-action σ_E on X_E is called *Ledrappier's example*: σ_E is mixing and expansive, but not mixing of order 3 (for every $n \geq 0$, $x_{(0,0)} + x_{(2^n,0)} + x_{(0,2^n)} = 0$).

We also consider the subsets

$$\begin{aligned} E_1 =& \{(0,0), (1,0), (2,0), (1,1), (0,2)\}, \\ E_2 =& \{(0,0), (2,0), (0,1), (1,1), (0,2)\}, \\ E_3 =& \{(0,0), (1,0), (2,0), (0,1), (1,1), (0,2)\}. \end{aligned}$$

of $\mathbb{Z}^2$. The shift-actions $\sigma_i = \sigma_{E_i}$ of $\mathbb{Z}^2$ on $X_i = X_{E_i}$ are mixing, irreducible and expansive.

For every $\mathbf{n} \in \mathbb{Z}^2$, the automorphisms $\sigma_i^{\mathbf{n}}$ are measurably conjugate, but for $i \neq j$ the $\mathbb{Z}^2$-actions σ_i and σ_j are measurably nonconjugate by Theorem 1.1.

The shift-action σ_4 of $\mathbb{Z}^2$ on X_{E_4} with

$$E_4 = \{(0,0), (2,0), (0,2)\}$$

is also mixing and expansive, but reducible: the map $\phi\colon X_{E_4} \to X_E$, given by

$$\phi(x)_{(m,n)} = x_{m,n} + x_{(m+1,n)} + x_{(m,n+1)},$$

is a shift-commuting surjective group homomorphism with kernel X_E.

EXAMPLE 6.8 (Conjugacy of $\mathbb{Z}^2$-actions with positive entropy). We define the sets $E_1, E_2, E_3 \subset \mathbb{Z}^2$ as in Example 6.7, set

$$Y_i = \left\{ x = (x_{\mathbf{n}}) \in (\mathbb{Z}/4\mathbb{Z})^2 : \sum_{\mathbf{m} \in E_i} x_{\mathbf{m}+\mathbf{n}} = 0 \ (\mathrm{mod}\ 2) \ \text{ for every } \ \mathbf{n} \in \mathbb{Z}^2 \right\},$$

and denote by τ_i the restriction to Y_i of the shift-action (1.4) of $\mathbb{Z}^2$ on $\Omega = (\mathbb{Z}/4\mathbb{Z})^{\mathbb{Z}^2}$. The entropy formula in [**13**, Theorem 4.2 and Lemma 4.2] (or a fairly straightforward direct calculation) shows that $h(\tau_i) = \log 2$ for $i = 1, 2, 3$. Theorem 6.5 in [**13**] implies that the Pinsker algebra $\pi(\tau_i)$ of τ_i is the sigma-algebra $\mathcal{B}_{Y_i/Z_i}$ of Z_i-invariant Borel sets in Y_i, where

$$Z_i = \{x = (x_{\mathbf{n}}) \in Y_i : x_{\mathbf{n}} = 0 \,(\mathrm{mod}\, 2) \ \text{ for every } \ \mathbf{n} \in \mathbb{Z}^2\}.$$

Then the $\mathbb{Z}^2$-action τ_i' induced by τ_i on Y_i/Z_i is algebraically conjugate to the shift-action σ_i on the group $X_i = X_{E_i}$ in Example 6.7.

Since any measurable conjugacy of τ_i and τ_j would map $\pi(\tau_i)$ to $\pi(\tau_j)$ and induce a conjugacy of τ_i' and τ_j' and hence of σ_i and σ_j, Example 6.7 implies that τ_i and τ_j are measurably nonconjugate for $1 \leq i < j \leq 3$.

EXAMPLE 6.9. Let $f = 1 + \omega u_1 + \omega^2 u_2 \in \mathfrak{R}_2^{(4)}$, where $\omega \in F_4$ and $1 + \omega + \omega^2 = 0$ (cf. (2.9)). The inclusion $\mathfrak{R}_2^{(2)} \subset \mathfrak{R}_2^{(4)}$ induces an injective $\mathfrak{R}_2$-module homomorphism $\jmath\colon \mathfrak{M} = \mathfrak{R}_2^{(2)}/\bar{\mathfrak{p}} \longrightarrow \mathfrak{M}' = \mathfrak{R}_2^{(4)}/\bar{\mathfrak{q}}$, where

$$\bar{\mathfrak{p}} = (1 + u_1 + u_2 + u_1^2 + u_1 u_2 + u_2^2)\mathfrak{R}_2^{(2)} \subset \mathfrak{R}_2^{(2)}$$

and

$$\bar{\mathfrak{q}} = f\mathfrak{R}_2^{(4)}.$$

Since $\jmath(\mathfrak{M})$ has index 2 in $\mathfrak{M}'$, there exists a two-to-one surjective dual homomorphism $\psi\colon X_{\mathfrak{M}'} = \widehat{\mathfrak{M}} \longrightarrow X' = \widehat{\mathfrak{M}'}$ with $\alpha_{\mathfrak{M}}^{\mathbf{n}} \cdot \psi = \psi \cdot \alpha_{\mathfrak{M}'}^{\mathbf{n}}$ for every $\mathbf{n} \in \mathbb{Z}^2$.

Since $\hat{\alpha}_{\mathfrak{M}}$ is cyclic, but $\hat{\alpha}_{\mathfrak{M}'}$ is not, Proposition 6.3 implies that the irreducible and mixing $\mathbb{Z}^2$-actions $\alpha_{\mathfrak{M}}$ and $\alpha_{\mathfrak{M}'}$ are not measurably conjugate, although they are weakly algebraically conjugate.

Further examples can be found in [**6**] and [**11**], where we also compute explicitly the measurable centralizers of such actions.

References

[1] N. Aoki, *A simple proof of the Bernoullicity of ergodic automorphisms of compact abelian groups*, Israel J. Math. **38** (1981), 189–198.

[2] N. Aoki and H. Totoki, *Ergodic automorphisms of T^∞ are Bernoulli transformations*, Publ. RIMS, Kyoto Univ. **10** (1975), 535–544.

[3] J.W.S. Cassels, *Local Fields*, Cambridge University Press, Cambridge, 1986.

[4] H. Furstenberg, *Disjointness in ergodic theory, minimal sets, and a problem in diophantine approximation*, Math. Systems Theory **1** (1967), 1–49.

[5] A. Katok and R.J. Spatzier, *Invariant measures for higher-rank hyperbolic abelian actions*, Ergod. Th. & Dynam. Sys. **16** (1996), 751–778; *Corrections*, **18** (1998), 507–507.

[6] A. Katok, S. Katok and K. Schmidt, *Rigidity of measurable structure for algebraic actions of higher-rank abelian groups*, ESI-Preprint: ftp://ftp.esi.ac.at/pub/Preprints/esi850.ps.

[7] Y. Katznelson, *Ergodic automorphisms of T^n are Bernoulli shifts*, Israel J. Math. **10** (1971), 186–195.

[8] B. Kitchens and K. Schmidt, *Automorphisms of compact groups*, Ergod. Th. & Dynam. Sys. **9** (1989), 691–735.

[9] B. Kitchens and K. Schmidt, *Markov subgroups of $(\mathbb{Z}_{/2})^{\mathbb{Z}^2}$*, Contemp. Math. **135** (1992), 265–283.

[10] B. Kitchens and K. Schmidt, *Mixing sets and relative entropies for higher dimensional Markov shifts*, Ergod. Th. & Dynam. Sys. **13** (1993), 705–735.

[11] B. Kitchens and K. Schmidt, *Isomorphism rigidity of irreducible algebraic* $\mathbf{Z}^d$*-actions*, Invent. Math. (to appear), ESI-Preprint: ftp://ftp.esi.ac.at/pub/Preprints/esi761.ps.

[12] D. Lind, *Ergodic automorphisms of the infinite torus are Bernoulli*, Israel J. Math. **17** (1974), 162–168.

[13] D. Lind, K. Schmidt and T. Ward, *Mahler measure and entropy for commuting automorphisms of compact groups*, Invent. Math. **101** (1990), 593–629.

[14] G. Miles and R.K. Thomas, *Generalized torus automorphisms are Bernoullian*, in: Studies in Probability and Ergodic Theory, Advances in Mathematics Supplementary Studies, vol. 2, Academic Press, New York-London, 1987, 231–249.

[15] M. Ratner, *Rigidity of horocycle flows*, Ann. of Math. **115** (1982), 597–614.

[16] D.J. Rudolph, $\times 2$ *and* $\times 3$ *invariant measures and entropy*, Ergod. Th. & Dynam. Sys. **10** (1990), 395–406.

[17] D.J. Rudolph and K. Schmidt, *Almost block independence and Bernoullicity of* $\mathbb{Z}^d$*-actions by automorphisms of compact groups*, Invent. Math. **120** (1995), 455–488.

[18] K. Schmidt, *Automorphisms of compact abelian groups and affine varieties*, Proc. London Math. Soc. **61** (1990), 480–496.

[19] K. Schmidt, *Invariant measures for certain expansive* Z^2*-actions*, Israel J. Math. **90** (1995), 295–300.

[20] K. Schmidt, *Dynamical systems of algebraic origin*, Birkhäuser Verlag, Basel-Berlin-Boston, 1995.

[21] K. Schmidt and T. Ward, *Mixing automorphisms of compact groups and a theorem of Schlickewei*, Invent. Math. **111** (1993), 69–76.

[22] A. Weil, *Basic Number Theory*, Springer Verlag, Berlin-Heidelberg-New York, 1974.

Mathematics Institute, University of Vienna, Strudlhofgasse 4, A-1090 Vienna, Austria

and

Erwin Schrödinger Institute for Mathematical Physics, Boltzmanngasse 9, A-1090 Vienna, Austria

E-mail address: klaus.schmidt@univie.ac.at

Part IId

KAM-theory

Proceedings of Symposia in Pure Mathematics
Volume **69**, 2001

Almost reducibility of linear quasi-periodic systems

L. H. Eliasson

ABSTRACT. We discuss reducibility and almost reducibility of linear quasi-periodic systems. We prove a result showing that when such a system is analytic and close to constant coefficients and has Diophantine quasi-periodic frequencies it is always almost reducible in the sense that it can be analytically conjugate to a system arbitrarily close to constant coefficients. We discuss when this approach converges and gives full reducibility, and when it does not. We discuss in particular the quasi-periodic Schrödinger equation and skew-products on $SO(3,\mathbb{R})$. We also discuss the relevance of this result for the perturbation theory of lower-dimensional tori in Hamiltonian systems.

1. Introduction

Let

$$f:\mathbb{R}\to gl(K,\mathbb{R})$$

be matrix-valued quasi-periodic function and consider the linear system

$$(*)\qquad x'(t) \;=\; f(t)x(t).$$

This system is said to be *Lyapunov-reducible* [**NS**] if it has a fundamental solution of the form

$$z(t)e^{tA},$$

where A is a constant matrix and $z(t)$ is an invertible differentiable (or at least absolutely continuous) matrix such that

$$z(t) \quad \text{and} \quad z(t)^{-1}$$

are bounded. In particular, the growth properties of solutions of a reducible system are the same as those of constant coefficient systems. Solutions of such systems are bounded or exponentially increasing/decreasing, or (exceptionally) polynomially increasing. Moreover, bounded solutions can never decrease to zero both in forward and in backward time.

It is clear that this concept says nothing about the "rotational" behavior of the solutions. For example, if all solutions of (*) are bounded — as happens when $f(t)$ is an anti-symmetric matrix — then the system is Lyapunov-reducible. For

1991 *Mathematics Subject Classification.* Primary 34C27, 37C55, 37H99, 37J40.
Key words and phrases. skew-products, quasi-periodic, KAM.

quasi-periodic systems it is natural to consider a stronger concept which takes into account the "rotational" behavior of the solutions. We say that the system is *(quasi-periodically) reducible* if the the matrix $z(t)$ is quasi-periodic. Reducibility in this paper will mean such quasi-periodic reducibility. In the next section we shall give an even more restricted formulation where we impose some conditions on the frequencies of $z(t)$.

We shall first review some results about reducibility. For a more detailed presentation from the point of view of time-discrete systems we refer to [**E1**]. Periodic solutions are always reducible [**NS**] but quasi-periodic systems are not — not even Lyapunov-reducible. The first examples had a function f that were only continuous [**Mi,G**], but later there were given examples with f smooth and even analytic [**J,JM**].

Associated to any quasi-periodic function $f(t)$ there is a *frequency vector* $\omega \in \mathbb{R}^d$ and a *hull function*

$$F : \mathbb{T}^d \to gl(K, \mathbb{R}), \qquad \mathbb{T} = \mathbb{R}/(2\pi\mathbb{Z}),$$

such that

$$f(t) \;=\; F(t\omega). \tag{**}$$

The frequency vector may be *Diophantine* — a property that holds for almost all vectors in $\mathbb{R}^d$ — or *Liouville* — a property that holds for a residual set of measure zero in $\mathbb{R}^d$. The hull function is always continuous, but it could fail to be any smoother than continuous even if f is analytic [**JM**].

In the non-reducible examples, referred to above, the hull function of f is only continuous and/or its frequency vector is Liouville. With KAM-theory it became possible to handle the small divisor problem occurring for systems with Diophantine frequencies and with smooth hull functions close to a constant function. A quasi-periodic system $(*) + (**)$ is reducible if

$$\begin{cases} \omega \text{ is Diophantine,} \\ F \text{ is sufficiently close — in a strong topology — to a constant system } B, \\ B \text{ is hyperbolic with simple spectrum.} \end{cases}$$

More generally, hyperbolic systems with simple spectrum are reducible under general conditions [**JS**].

The counter examples referred to above shows that this result is not true for non-hyperbolic matrices B. But a classical result from KAM, first described systematically in [**Mo**], states that many-parameter families of quasi-periodic systems $(*) + (**)$, with hull function F depending on the parameters, have many members that are reducible if

$$\begin{cases} \omega \text{ is Diophantine,} \\ F \text{ is sufficiently close — in a strong topology — to a constant system,} \\ \text{there are sufficiently many parameters.} \end{cases}$$

That in general only one parameter is needed became obvious with the ideas developed in [**E2**] — the case $F \in sp(K, \mathbb{R})$ follows as a corollary of the theorem of this paper — and was proven for general systems in [**JoS**]. The common feature of all these results is that the transformation matrix $z(t)$ is close to the identity matrix and that arithmetical conditions are required from the eigenvalues of the matrix A — the *Floquet exponents*. Such conditions can be fulfilled in one-parameter families, but only for a subset — typically a Cantor set of large measure — of parameter values.

It turns out that the arithmetical conditions on the Floquet exponents are imposed by the requirement that $z(t)$ should be close to the identity matrix, and by working with a more general class of matrices $z(t)$ one can relax very much these conditions. In this paper we shall discuss this mechanism and use it to prove a result about *almost reducibility* for which the arithmetical properties of the Floquet exponents are no restriction at all. We say that (*) is almost reducible if

$$\begin{cases} x(t) &= z(t)y(t) \\ y'(t) &= g(t)y(t) \end{cases}$$

with $g(t)$ arbitrarily close to a constant matrix. Theorem 8 gives a precise formulation of this statement. A similar result for compact groups is given in [**K1**].

This result suggests that reducibility in one-parameter families close to constant coefficients with Diophantine frequencies should occur, not only for a *large set*, but for *almost all* values in one-parameter families . This is indeed the case for the Schrödinger equation [**E3**], for certain one-parameter families in $SO(3,\mathbb{R})$ [**K1**] and in general compact groups [**K2,K3**], and there are strong reasons to believe that this is the general situation.

We cannot expect that all systems are reducible, not even with Diophantine frequencies and smooth hull function. The first examples was given for time-discrete systems — the almost Mathieu equation [**DS,Si**] — and the first time-continuous example occurred in [**FW**]. Later were even given smooth examples that were close to constant coefficients [**E3,H1,E4**].

Now we feel that the situation for linear quasi-periodic systems — both time-continuous and time-discrete — with Diophantine frequencies and analytic hull function is quite well understood close to constant coefficients. The typical system in measure sense is reducible while the typical system in topological sense is non-reducible. For non-compact systems the situation far from constant coefficients is different, and non-uniform hyperbolicity seems to be a prevailing property — see [**E1**] and references therein. For compact groups, however, it is an open problem whether this local structure also is the global one.

2. Reducibility

Let $f(t)$ be a *quasi-periodic function* with frequency vector ω and hull function F. Clearly, all the functions

$$f_\varphi(t) \;=\; F(\varphi + t\omega)$$

are quasi-periodic function with the same frequency vector and with hull functions that are translates of each other.

We can assume without restriction that the frequency vector ω is rationally independent, i.e. the *frequencies*

$$< n, \omega > =: \sum_{j=1}^{d} n_j\omega_j \neq 0, \qquad \forall n \in \mathbb{Z}^d \setminus 0.$$

This is the same as requiring that the straight line $\{t\omega\colon t \in \mathbb{R}\}$ projected onto $\mathbb{T}^d$ is dense.

Each quasi-periodic function gives rise to a family, parameterized by the angles $\varphi \in \mathbb{T}^d$, of systems of linear differential equations with quasi-periodic time-dependence

$$(1') \qquad x'(t) \;=\; F(\varphi + t\omega)x(t), \quad \varphi \in \mathbb{T}^d,$$

or equivalently to a *linear skew-product (flow)* on $\mathbb{T}^d \times GL(K,\mathbb{R})$ defined by

$$(1'') \qquad \begin{cases} \varphi' = \omega \\ x' = F(\varphi)x. \end{cases}$$

The flow of (1″) has the form

$$(\varphi_0 + t\omega, x(t;\varphi_0,x_0)) \;=\; (\varphi_0 + t\omega, x(t;\varphi_0,I)x_0),$$

where $x(t;\varphi,I)$ is the fundamental solution of (1′). The determinant of the matrix $x(t;\varphi,I)$ is always positive

If F belongs to some Lie algebra then $x(t)$ belongs to the corresponding Lie group: for example the orthogonal group $SO(K,\mathbb{R})$[1]

$$F^* = -F \quad \Rightarrow \quad x^* \;=\; x^{-1};$$

or the symplectic group $SP(K;\mathbb{R})$

$$F^* = -JFJ^* \quad \Rightarrow \quad x^* \;=\; Jx^{-1}J,$$

where J is the standard symplectic involution on R^K (K even).

Constant coefficients. The system (1′) is easy to analyze in the case of constant coefficients, i.e. when $F(\varphi) \;=\; A$ is independent of φ. Then

$$x(t;\varphi_0,x_0) = e^{At}x_0$$

and the behavior is completely determined by the set of eigenvalues $\sigma(A)$ of A (for generic A). The skew-product (1″) has an *invariant foliation* which is a product foliation

$$\mathbb{T}^d \times \overline{\{e^{tA}x_0 : t \in \mathbb{R}\}}, \quad x_0 \in GL(K,\mathbb{R}).$$

If $A \in so(3,\mathbb{R})$, then

$$\sigma(A) \;=\; \{\pm i\alpha, 0\}$$

and the leaves of the invariant foliation of $\mathbb{T}^d \times SO(3,\mathbb{R})$ are

$$\mathbb{T}^d \times \overline{\{e^{tA}x_0 : t \in \mathbb{R}\}} \;\approx\; \mathbb{T}^d \times \mathbb{T}.$$

Reducibility. There are three equivalent formulation of (quasi-periodic) reducibility. The first formulation is as *a representation of the fundamental solution of* (1′)

$$(2) \qquad x(t;\varphi,I) \;=\; Z(\varphi + t\omega)e^{tA}Z(\varphi)^{-1},$$

as a product of a quasi-periodic matrix and an exponential matrix.

[1] Star denotes transposition.

The second formulation of reducibility is more geometric. It says that there is *a change of variables*

$$(3) \qquad (\psi, y) = (\varphi, Z^{-1}(\varphi)x)$$

transforming the skew-product $(1'')$ to constant coefficients

$$(4) \qquad \begin{cases} \psi' = \omega \\ y' = Ay. \end{cases}$$

So if $(1'')$ is reducible then it has an invariant foliation. We can distinguish three different cases of reducible systems. If $Z = I$ then the system has constant coefficients and the invariant foliation is a product foliation. If $Z \sim I$, then the system is close to constant coefficients and the invariant foliation is close to a product foliation. If $Z \not\sim I$, then the invariant foliation is far from a product foliation even if the system itself is close to constant coefficients.

The third formulation is the one that usually is used in perturbation theory. It is as *a solution of the (non-linear) pde*

$$(5) \qquad \partial_\omega Z(\varphi) \;=\; F(\varphi)Z(\varphi) - Z(\varphi)A,$$

where ∂_ω denotes the directional derivative with respect to φ in the direction ω.

The matrix $Z(\varphi)$ will not be defined on $\mathbb{T}^d$ but on some covering — we shall assume that it is defined on the double torus[2]

$$Z : 2\mathbb{T}^d \to GL(K, R), \qquad 2\mathbb{T}^d = (\mathbb{R}/4\pi\mathbb{Z})^d.$$

Z must be differentiable in the direction ω, but otherwise one can consider different kinds of smoothness. For (quasi-periodic) reducibility the natural condition is that Z is continuous on $\mathbb{T}^d$, but one could just as well consider matrices that are bounded and measurable on $\mathbb{T}^d$. Nevertheless, in all cases treated by perturbation theory Z is quite smooth.

Floquet exponents. If (2) holds we refer to the elements of $\sigma(A)$, i.e the eigenvalues of A, as the *Floquet exponents* of (1).

LEMMA. *The real parts of $\sigma(A)$ are unique — these are the Lyapunov exponents of (1). The imaginary parts of $\sigma(A)$ are determined modulo*

$$\{\frac{1}{2} < n, \omega >: n \in \mathbb{Z}^d\}.$$

PROOF. Let $x(t;\varphi, I) \;=\; Z(\varphi + t\omega)e^{At}Z(\varphi)^{-1}$. Then

$$\lim_{T\to\infty} \frac{1}{2T}\int_{-T}^{T} x(t;\varphi, I)e^{-\gamma t}dt$$

has some finite and $\neq 0$ component only if $\Re(\gamma) \in \Re(\sigma(A))$ and

$$\Im(\gamma) \in \Im(\sigma(A)) + \{\frac{1}{2} < n, \omega >: n \in \mathbb{Z}^d\}.$$

□

The following example shows that the imaginary parts are not unique in general.

[2]This is the right covering for $GL(K, \mathbb{R})$, $SL(2, \mathbb{R})$ and $SO(3\mathbb{R})$. But for other groups other coverings may be relevant[**K2,K3**]

EXAMPLE. If

$$F(\varphi) = \begin{pmatrix} 0 & <n,\omega> \\ -<n,\omega> & 0 \end{pmatrix}$$

for some $n \in \mathbb{Z}^d$, then

$$Z_k(\varphi) = exp\begin{pmatrix} 0 & <k,\varphi> \\ -<k,\varphi> & 0 \end{pmatrix}$$

verifies

$$x(t;\varphi,I) = Z_k(\varphi+t\omega)exp\{t\begin{pmatrix} 0 & <n-k,\omega> \\ -<n-k,\omega> & 0 \end{pmatrix}\}Z_k(\varphi)^{-1}$$

for any $k \in \mathbb{Z}^d$.

There is no known way of characterizing the imaginary part of the Floquet exponents a priori, except in the case of $sl(2,\mathbb{R})$. If $F(\varphi) \in sl(2,\mathbb{R})$ then

$$\lim_{t\to\infty} \frac{1}{t} arg(x(t;\varphi,I)v) = \rho$$

— called the *rotation number* of (1) — exists $\forall\varphi \in \mathbb{T}^d, \forall v \in \mathbb{R}^2\backslash 0$ and is independent of φ and v [**JM,H2**]. If (2) holds then clearly

$$\rho \in \Im(\sigma(A)) + \{\frac{1}{2} <n,\omega>: n \in \mathbb{Z}^d\}.$$

The non-uniqueness of the Floquet exponents plays an important role in the perturbation theory. We shall discuss this issue in the next two sections.

3. Linear quasi-periodic systems with constant coefficients

In the perturbation theory of the equation (5)

$$\partial Z_\omega(\varphi) = (A+\varepsilon F(\varphi))Z(\varphi) - Z(\varphi)(A+\varepsilon B),$$

it is essential to be able to solve the "linearized equation"

$$\partial_\omega X(\varphi) = AX(\varphi) - X(\varphi)A + F(\varphi) - \hat{F}(0). \tag{6}$$

This is a small divisor problem the solution of which depends on smoothness properties of F and on arithmetical properties of both ω and the differences of the eigenvalues of A.

We say that a unit vector ω is *Diophantine* if for some $0 < \kappa < 1, 0 < \tau$

$$DC(\kappa,\tau): \qquad |<n,\omega>| > \frac{\kappa}{|n|^\tau} \qquad \forall n \in \mathbb{Z}^d \setminus 0.$$

Almost all $\omega \in S^{d-1}$ are Diophantine. A function F defined on $2\mathbb{T}^d$ with values in a normed space V belongs to $\underline{\mathcal{A}_r}$ if[3]

$$\begin{cases} F \text{ is analytic in } |\Im\varphi| \le r \\ |F|_r =: \sup_{|\Im\varphi|\le r} |F(\varphi)| < \infty. \end{cases}$$

[3]As norm we use, unless otherwise said, the operator norm.

The Fourier series of F will be written[4]

$$F(\varphi) \sim \sum_{n\in\frac{1}{2}\mathbb{Z}^d} \hat{F}(n)e^{i<n,\omega>t}.$$

LEMMA 1. *If $\frac{\omega}{2} \in DC(\kappa,\tau)$ and $F : 2\mathbb{T}^d \to \mathbb{R}^K$ is in $\mathcal{A}_r$, then there exists a unique function $X : 2\mathbb{T}^d \to \mathbb{R}^K$ such that*

$$\begin{cases} \partial_\omega X(\varphi) = F(\varphi) - \hat{F}(0) \\ \hat{X}(0) = 0. \end{cases}$$

The function X is in $\mathcal{A}_{r'}$ and

$$| X |_{r'} \leq cte. \frac{1}{\kappa(r-r')^\tau} | F |_r \qquad \forall r' < r. \tag{7}$$

(The constant depends on d,τ.)

It follows that, for all φ, the function $x(t) = X(\varphi + t\omega)$ is a quasi-periodic solution of

$$x'(t) = F(\varphi + t\omega)$$

if, and only, if $\hat{F}(0) = 0$.

PROOF. Formally we have

$$X(\varphi) \sim \sum_{n\neq 0} \frac{\hat{F}(n)}{i<n,\omega>} e^{i<n,\varphi>}$$

and we only need to estimate this sum.

Since $F \in \mathcal{A}_r$ and $\frac{\omega}{2} \in DC(\kappa,\tau)$ we have for all $n \neq 0$

$$| \hat{F}(n) | \leq e^{\;|n|r} | F |_r$$

and

$$| \frac{1}{i<n,\omega>} | \leq \frac{| 2n |^\tau}{\kappa}.$$

Therefore

$$| \sum_{n\neq 0} \frac{\hat{F}(n)}{i<n,\omega>} e^{i<n,\,\cdot\,>} |_{r'} \leq (\sum_{n\neq 0} | 2n |^\tau e^{-|n|(r-r')})\frac{1}{\kappa} | F |_r \leq$$

$$\leq c_1(\sum_{N=1}^{\infty} N^{\tau+d-1}e^{-N(r-r')})\frac{1}{\kappa} | F |_r \leq$$

$$\leq c_2(\tau+d-1)!\frac{1}{\kappa(r-r')^{\tau+d-1}} | F |_r,$$

which proves the lemma with the exponent $\tau + d - 1$. Rüssmann proved the result with the better exponent τ by using that for many n's, $|< n,\omega >|$ is indeed much larger that the lower bound $\frac{\kappa}{|n|^\tau}$. See for example [**R,E2**]. □

---

[4]The letter i will, unless otherwise said, always denote $\sqrt{-1}$.

We say that a number e is *Diophantine of order N with respect to ω* if for some $0<\kappa<1, 0<\tau$

$$DC_\omega^N(\kappa,\tau): \quad |\, e- <n,\omega>| \geq \frac{\kappa}{|\, n\,|^\tau} \qquad \forall n\in\mathbb{Z}^d\setminus 0,\ |\, n\,|\leq N.$$

We say that e is Diophantine with respect to ω if it is Diophantine of all orders, i.e $e\in DC_\omega(\kappa,\tau)=\cap_{N\geq 1}DC_\omega^N(\kappa,\tau)$. Almost all real numbers e are Diophantine with respect to ω.

LEMMA 2. *If $A\in gl(K,\mathbb{R})$ with $\sigma(i\cdot A)\subset DC_{\frac{\omega}{2}}(\kappa,\tau)$ and $F:2\mathbb{T}^d\to\mathbb{R}^K$ is in $\mathcal{A}_r$, then there exists a unique function $X:2\mathbb{T}^d\to\mathbb{R}^K$ such that*

$$\begin{cases} \partial_\omega X(\varphi)=AX(\varphi)+F(\varphi)-\hat{F}(0) \\ \hat{X}(0)=0. \end{cases}$$

The function X is in $\mathcal{A}_{r'}$ and

$$(8)\qquad |\, X\,|_{r'}\leq\ cte.(1+|\, A\,|)^{K-1}\left(\frac{1}{\kappa(r-r')^\tau}\right)^K |\, F\,|_r \qquad \forall r'\in]r-1,r[.$$

Moreover, if $\hat{F}(n)=0$ for all $|\, n\,|>N$, then the same result holds under the weaker condition that $\sigma(i\cdot A)\subset DC_{\frac{\omega}{2}}^{2N}(\kappa,\tau)$.

(The constant depends on d,τ,K.)

It follows that, for all φ, the function $x(t)=X(\varphi+t\omega)+v$ is a quasi-periodic solution of

$$x'(t)\ =\ Ax(t)+F(\varphi+t\omega)$$

if, and only if, $Av=\hat{F}(0)$.

PROOF. Clearly

$$X(\varphi)\ \sim\ \sum_{n\neq 0}(i<n,\omega>I-A)^{-1}\hat{F}(n)e^{i<n,\varphi>}$$

and we only need to estimate this sum.

Since we are only interested in norm estimates we can assume that A is upper triangular, so that

$$i<n,\omega>I-A\ =\ D-N$$

with D diagonal and N nilpotent. Then

$$(i<n,\omega>I-A)^{-1}\ =\ D^{-1}(I+(ND^{-1})+\cdots+(ND^{-1})^{K-1}).$$

The terms of $D^{-1}(ND^{-1})^{j-1}$ are of the form $\frac{a}{b}$ with

$$|\, a\,|\leq|\, 2A\,|^{j-1}$$

and $b\ =$ a j-fold product of eigenvalues of $(i<n,\omega>I-A)$. Each such term gives now a Fourier series which can be estimated as in the proof of Lemma 1.

The final statement is obvious. □

A special class of functions. Let $A \in gl(K,\mathbb{R})$. We define a $\mathbb{Z}_2^d$*-decomposition* Λ *of* A as a decomposition of $\mathbb{R}^K$ into a sum of A-invariant subspaces Λ_n, $n \in \mathbb{Z}_2^d$.[5]

If we choose a basis for each subspaces Λ_n, then the matrix

$$C_\Lambda = (\dots \Lambda_n \dots)_{n\in\mathbb{Z}_2^d}$$

will conjugate A to block diagonal form. We let

$$C_\Lambda^{-1} = \begin{pmatrix} \vdots \\ \Omega_n^* \\ \vdots \end{pmatrix}_{n\in\mathbb{Z}_2^d} .$$

Then we say that $F : 2\mathbb{T}^d \to gl(K,\mathbb{R})$ is *conjugate to* $\mathbb{T}^d$ *over* Λ if

$$\Omega_m^* F(\varphi)\Lambda_n e^{-i<\frac{m}{2}-\frac{n}{2},\varphi>}$$

is defined on $\mathbb{T}^d$ for all $m,n \in \mathbb{Z}_2^d$.

It is clear that this definition does not depend on the choice of basis for the subspaces Λ_n. The following properties are also obvious.

LEMMA 3. *Let* Λ *be a* $\mathbb{Z}_2^d$*-decomposition of* $A \in gl(K,\mathbb{R})$ *and let* $F, G : 2\mathbb{T}^d \to gl(K,\mathbb{R})$ *be conjugate to* $\mathbb{T}^d$ *over* Λ. *Then the following hold.*

1. *If* $B \in gl(K,\mathbb{R})$ *satisfies*

$$C_\Lambda^{-1} B C_\Lambda = (B_m^n)_{m,n\in\mathbb{Z}_2^d}$$

with $B_m^n = 0$ *when* $n \neq m$, *then* Λ *is a* $\mathbb{Z}_2^d$*-decomposition of* $A + B$.

2. *If*

$$C_\Lambda^{-1}\hat{F}(0)C_\Lambda = (B_m^n)_{m,n\in\mathbb{Z}_2^d},$$

then $B_m^n = 0$ *when* $n \neq m$.

3. $F + G$ *and* FG *are conjugate to* $\mathbb{T}^d$ *over* Λ.

In general we have no estimates of C_Λ^{-1}. We say that Λ is a δ*-decomposition of* A if there is a basis for the Λ_n's such that

$$\begin{cases} | C_\Lambda | \leq 1 \\ | C_\Lambda^{-1} | \leq \text{cte.}(\frac{1+|A|}{\delta})^{K(K+1)}. \end{cases} \tag{9}$$

For the constant we choose the constant in Lemma A (Appendix) which only depends on K. Then any $\mathbb{Z}_2^d$-decomposition, such that[6]

$$\underline{dist}(\sigma(A/\Lambda_n), \sigma(A/\Lambda_m)) > \delta$$

when $n \neq m$, is a δ-decomposition.

Arithmetical properties of A. For $A \in gl(K,\mathbb{R})$ (provided with the usual matrix norm) we define the linear operator

$$ad_A : B \to [A,B] = AB - BA.$$

The norm of this operator is

$$| ad_A | \leq 2 \, | A |$$

[5]This means that n is a vector in $\mathbb{Z}^d$ with components $0, 1$.

[6]By $\underline{dist}(X,Y)$ we understand the shortest distance between any two point x, y in X and Y.

and its eigenvalues are exactly the K^2 differences $e-f$ of the eigenvalues $e, f \in \sigma(A)$ (counted with multiplicity).

We would like to impose that $\sigma(i \cdot ad_A) \subset DC^{2N}_{\frac{\omega}{2}}(\kappa,\tau)$ but this condition cannot always be fulfilled if we want to remain on the torus $2\mathbb{T}^d$. Instead we shall consider the slightly weaker property. Let Λ be a $\mathbb{Z}_2^d$-decomposition of A and assume that

$$\begin{cases} \sigma(i\cdot A/\Lambda_m) - \sigma(i\cdot A/\Lambda_n) \subset DC^{2N}_{\frac{\omega}{2}}(\kappa,\tau) & \text{if } m \neq n \\ \sigma(i\cdot A/\Lambda_m) - \sigma(i\cdot A/\Lambda_n) \subset DC^{N}_{\omega}(\kappa,\tau) & \text{if } m = n \end{cases}$$

We shall denote this property, which is weaker than $\sigma(i \cdot ad_A) \subset DC^{2N}_{\frac{\omega}{2}}(\kappa,\tau)$ but stronger than $\sigma(i \cdot ad_A) \subset DC^{N}_{\omega}(\kappa,\tau)$, symbolically by

$$ad_A \in DC^{2N}_{\frac{\omega}{2},\Lambda}(\kappa,\tau).$$

PROPOSITION 4. *Let $A \in gl(K,\mathbb{R})$ with a δ-decomposition Λ, $\delta \leq 1$, and assume*

$$ad_A \in DC^{2N}_{\frac{\omega}{2},\Lambda}(\kappa,\tau), \qquad \forall N \geq 1.$$

Let $F : 2\mathbb{T}^d \to gl(K,\mathbb{R})$ be conjugate to $\mathbb{T}^d$ over Λ and in $\mathcal{A}_r$.

Then there exists a unique function $X : 2\mathbb{T}^d \to gl(K,\mathbb{R})$ such that

$$\begin{cases} \partial_\omega X(\varphi) = [A, X(\varphi)] + F(\varphi) - \hat{F}(0) \\ \hat{X}(0) = 0. \end{cases}$$

X is conjugate to $\mathbb{T}^d$ over Λ and in $\mathcal{A}_{r'}$ and

$$|X|_{r'} \leq cte. \left(\frac{1+|A|}{\delta}\right)^{2K^4} \left(\frac{1}{\kappa(r-r')^\tau}\right)^{K^2} |F|_r \tag{10}$$

for all $r' \in]r-1, r[$.

Moreover, if $\hat{F}(n) = 0$ for all $|n| > N$, then the same result holds under the weaker condition

$$ad_A \in DC^{2N}_{\frac{\omega}{2},\Lambda}(\kappa,\tau).$$

(The constant depends on d, τ, K.)

PROOF. Choose a basis for Λ such that C_Λ satisfies (9) and consider

$$\tilde{X} = C^{-1}XC, \quad \tilde{A} = C^{-1}AC, \quad \tilde{F} = C^{-1}FC.$$

Notice that

$$|\tilde{F}|_r \leq \text{cte.} \left(\frac{1+|A|}{\delta}\right)^{K(K+1)} |F|_r \,.$$

Then

$$\begin{cases} \partial_\omega \tilde{X}(\varphi) = [\tilde{A}, \tilde{X}(\varphi)] + \tilde{F}(\varphi) - \hat{\tilde{F}}(0) \\ \hat{\tilde{X}}(0) = 0. \end{cases}$$

Since $\tilde{A}$ is block diagonal this equation decouples into sub-block components parametrized by $m, n \in \mathbb{Z}_2^d$

$$\begin{cases} \partial_\omega \tilde{X}^n_m(\varphi) = \tilde{A}_m \tilde{X}^n_m(\varphi) - \tilde{X}^n_m(\varphi)\tilde{A}_n + \tilde{F}^n_m(\varphi) - \hat{\tilde{F}}{}^n_m(0) \\ \hat{\tilde{X}}{}^n_m(0) = 0. \end{cases}$$

The operator

$$X \to \tilde{A}_m \tilde{X} - X\tilde{A}_n$$

is a linear operator on a space of dimension $\leq K^2$ with norm $\leq \mid \tilde{A}_m \mid + \mid \tilde{A}_n \mid$.

If $n = m$ then $\tilde{F}^n_m$ is defined on $\mathbb{T}^d$ and we can apply Lemma 2 since

$$\sigma(i \cdot \tilde{A}_m) - \sigma(i \cdot \tilde{A}_n) \subset DC_\omega(\kappa, \tau).$$

On the other hand, if $n \neq m$ then $\tilde{F}^n_m$ is defined on $2\mathbb{T}^d$, and we can apply Lemma 2 since

$$\sigma(i \cdot \tilde{A}_m) - \sigma(i \cdot \tilde{A}_n) \subset DC_{\frac{\omega}{2}}(\kappa, \tau).$$

These are precisely the conditions guaranteed by $ad_A \in DC^{2N}_{\frac{\omega}{2},\Lambda}(\kappa, \tau)$ $\forall N \geq 1$. Lemma 2 therefore gives for all m, n

$$\mid \tilde{X}^n_m \mid_{r'} \leq \text{ cte.}(1+ \mid \tilde{A}_m \mid + \mid \tilde{A}_n \mid)^{K^2-1} \left(\frac{1}{\kappa(r-r')^\tau}\right)^{K^2} \mid \tilde{F}^n_m \mid_r$$

and, hence,

$$\mid \tilde{X} \mid_{r'} \leq \text{ cte.}(1+ \mid \tilde{A} \mid)^{K^2-1} \left(\frac{1}{\kappa(r-r')^\tau}\right)^{K^2} \mid \tilde{F} \mid_r .$$

Using the estimate (9) of C_Λ^{-1} we get (10).

That X is conjugate over Λ is clear from the construction. The final remark is obvious. □

It follows that, for all φ, the function $x(t) = X(\varphi + t\omega) + v$ is a quasi-periodic solution of

$$x'(t) = [A, x(t)] + F(\varphi + t\omega)$$

if, and only if, $[A, v] = \hat{F}(0)$.

4. Reduction of resonant eigenvalues

In the previous section we have seen that arithmetical properties of the eigenvalues of A play a role in the solution of the "linearized equation". In this section we shall see how we can conjugate a constant coefficient system, defined by a matrix A with a decomposition Λ such that

$$ad_A \notin DC^{2N}_{\frac{\omega}{2},\Lambda}(\kappa, \tau),$$

to a new constant coefficient system, defined by some matrix $\tilde{A}$ with the same decomposition Λ but with

$$ad_{\tilde{A}} \in DC^{2N'}_{\frac{\omega}{2},\Lambda}(\kappa', \tau)$$

for some $\kappa' \leq \kappa$, $N' \geq N$.

Reduction of eigenvalues. Let $A \in gl(K, \mathbb{R})$ and let $\alpha_1, \ldots, \alpha_K$ be its eigenvalues. Suppose we have a mapping

$$\sigma(A) \ni \alpha_j \to \tilde{\alpha}_j = \alpha_j - i < m_j, \omega > \in \mathbb{C} \tag{11}$$

with $m_j \in \frac{1}{2}\mathbb{Z}^d$. Then we define $\underline{A(\tilde{\alpha})}$ to be the same matrix as A but with each eigenvalue α replaced by the corresponding eigenvalue $\tilde{\alpha}$. The matrix $A(\tilde{\alpha})$ has the same spectral decomposition — i.e. decomposition into Jordan blocks — as A but with spectrum $\{\alpha_j - i < m_j, \omega >: j = 1, \ldots, K\}$. Moreover $A(\tilde{\alpha})$ is real if, and only if,

$$\alpha_j = \overline{\alpha}_k \quad \Rightarrow \quad m_j = -m_k, \tag{12}$$

i.e. the mapping commutes with complex conjugation. The matrix

$$\exp_{A,m}(\varphi) \;=\; \exp[A - A(\alpha - i < m, \varphi >)]$$

is an invertible matrix $2\mathbb{T}^d \to GL(K, \mathbb{R})$ satisfying

$$\begin{cases} \partial_\omega Y(\varphi) \;=\; AY(\varphi) - Y(\varphi)A(\tilde{\alpha}) \\ Y(0) = I. \end{cases}$$

$\exp_{A,m}(\varphi)$ is semi-simple (i.e. diagonalizable) and acts by multiplication by $e^{i<m_j,\varphi>}$ on each $\Lambda(\alpha_j)$ — the generalized eigenspace of A corresponding to eigenvalue α_j. Clearly $\exp_{A,m}$ is real if, and only if, (12) holds.

The estimates of the modified matrix $A(\alpha - i < m, \omega >)$ and of $\exp_{A,m}(\varphi)$ depend on the spacing of the eigenvalues of A and on the m_j's. Assume therefore

$$\mid \alpha_j - \alpha_k \mid \leq \delta \quad \Rightarrow \quad m_j = m_k. \tag{13}$$

Then we get from Lemma A (see Appendix) the following estimates of the matrices $A(\alpha - i < m, \omega >)$ and $\exp_{A,m}$.

LEMMA 5. *Assume that conditions (12) and (13) hold. Then the matrix* $A(\tilde{\alpha}) = A(\alpha - i < m, \omega >)$ *is real and satisfies*

$$\mid A(\tilde{\alpha}) - A \mid \leq \; cte. \left(\frac{1+ \mid A \mid}{\delta}\right)^{K(K+1)} \max_j |< m_j, \omega >|,$$

and the mapping $\exp_{A,m} : 2\mathbb{T}^d \to GL(K, \mathbb{R})$ *satisfies*

$$\mid \exp_{A,m} \mid_r \; \leq \; cte. \left(\frac{1+ \mid A \mid}{\delta}\right)^{K(K+1)} \max_j e^{d|m_j|r}.$$

(The constant only depends on K.)

Diophantine properties. Consider K complex numbers $\alpha_1, \ldots, \alpha_K$ (not necessarily different) such that for each number α_j also $\overline{\alpha}_j$[7] is one of these numbers. Suppose that the numbers are divided into groups $[\alpha_j]$ such that

$$\begin{cases} \operatorname{diam}\{\alpha_k \in \mathbb{C} : \alpha_k \in [\alpha_j]\} \; \leq \; \delta \\ \overline{[\alpha_j]} \; = \; [\overline{\alpha}_j]. \end{cases}$$

Two groups are equal — if, and only if, they contain the same α_j's — or different. Considered as a subset of $\mathbb{C}$ a group may be any set of cardinality at most K with diameter $\leq \delta$.

[7] Bar denotes complex conjugation.

LEMMA 6. *Assume* $\frac{\omega}{2} \in DC(\kappa, \tau)$ *and let*

$$N \in \mathbb{N} \setminus 0, \qquad \delta \leq \frac{\kappa}{K 2^{\tau+2} N^{\tau}}.$$

Then there exist $m_1, \dots, m_K \in \frac{1}{2}\mathbb{Z}^d$ *that satisfy for all* j, k

$$\alpha_j = \overline{\alpha}_k \quad \Rightarrow \quad m_j = -m_k, \tag{14}$$

$$| m_j | \leq \; KN, \tag{15}$$

$$\alpha_j \in [\alpha_k] \quad \Rightarrow \quad m_j = m_k, \tag{16}$$

and, for $\tilde{\alpha}_j \;=\; \alpha_j - i < m_j, \omega >$,

$$i(\tilde{\alpha}_j - \tilde{\alpha}_k) \in \begin{cases} DC^{2N}_{\frac{\omega}{2}}(\kappa', \tau) & \text{if } [\tilde{\alpha}_j] \neq [\tilde{\alpha}_k] \text{ and } [\overline{\tilde{\alpha}}_k] \\ DC^{N}_{\omega}(\kappa', \tau) & \text{if } [\tilde{\alpha}_j] = [\tilde{\alpha}_k] \text{ or } [\overline{\tilde{\alpha}}_k] \end{cases} \tag{17}$$

for any

$$\kappa' \leq \frac{\kappa}{K 2^{\tau+2} N^{\tau}}.$$

Moreover, if not all $m_j = 0$, *then there are at least two numbers* α_j, α_k *such that*

$$| \alpha_j - \alpha_k | \geq \frac{\kappa}{2^{\tau+1} N^{\tau}} \quad \text{and} \quad | \tilde{\alpha}_j - \tilde{\alpha}_k | \leq \kappa'. \tag{18}$$

PROOF. Suppose that there exists some $n = n_{j,k} \in \frac{1}{2}\mathbb{Z}^d$, $0 <| n |\leq N$, such that

$$| \alpha_j - \alpha_k - i < n, \omega >| \leq \frac{\kappa'}{| n |^{\tau}}$$

and with $n \in \mathbb{Z}^d$ if $[\alpha_j] = [\alpha_k]$ or $[\overline{\alpha}_k]$. (If no such n exists we are done with $m_j = 0$ for all j.) The existence of such a vector n implies in particular that

$$| \alpha_j - \alpha_k | \geq \frac{\kappa}{2^{\tau+1} N^{\tau}} \; \geq \; K\delta + K\kappa' \geq \delta$$

so $[\alpha_j] \neq [\alpha_k]$.

We then have two cases: 1) $[\alpha_k] = [\overline{\alpha}_j]$ (in which case $n \in \mathbb{Z}^d$); 2) $[\alpha_k] \neq [\alpha_j]$ and $\neq [\overline{\alpha}_j]$.

In the second case we define for all $\alpha \in [\alpha_k]$

$$\tilde{\alpha} = \alpha + i < n, \omega >, \qquad \overline{\tilde{\alpha}} = \overline{\alpha} - i < n, \omega >,$$

and $\tilde{\alpha} = \alpha$ for all other α's. In the first case we define for all $\alpha \in [\alpha_k] = [\overline{\alpha}_j]$

$$\tilde{\alpha} = \alpha + i < \frac{n}{2}, \omega >, \qquad \overline{\tilde{\alpha}} = \overline{\alpha} - i < \frac{n}{2}, \omega >,$$

and $\tilde{\alpha} = \alpha$ for all other α's. In both cases we have

$$| \alpha_j - \alpha_k | \geq \frac{\kappa}{2^{\tau+1} N^{\tau}} \quad \text{and} \quad | \tilde{\alpha}_j - \tilde{\alpha}_k | \leq \kappa'.$$

In this way we obtain reduced numbers $\tilde{\alpha}_j \;=\; \alpha_j - i < m_j, \omega >$ satisfying (14-16) and with all $| m_j | \leq N$. But the reduced system $\tilde{\sigma} = \{\tilde{\alpha}_1, \dots, \tilde{\alpha}_K\}$ may not fulfill (17). If not, we can repeat the same procedure with the numbers α_j replaced by $\tilde{\alpha}_j$. Each time we repeat this procedure a new (because $\kappa' \leq \frac{\kappa}{K2^{\tau+2}N^\tau}$) pair α_j, α_k will be shifted so that $|\tilde{\alpha}_j - \tilde{\alpha}_k| \leq \kappa'$. Hence, unless condition (17) is fulfilled, after at most $K-1$ many steps we have $|\tilde{\alpha}_j - \tilde{\alpha}_k| \leq K\kappa'$ for all j,k. But this implies that (17) holds.

The final statement is obvious. □

We shall need an even stronger statement.

LEMMA 6′. *Assume* $\frac{\omega}{2} \in DC(\kappa,\tau)$ *and let*

$$N \in \mathbb{N}\setminus 0, \qquad \delta \leq \frac{\kappa}{K(1+2K)^{\tau(K-2)}2^{\tau+2}N^\tau}.$$

Then there exist $m_1,\ldots,m_K \in \frac{1}{2}\mathbb{Z}^d$ *that satisfy for all* j,k

$$\alpha_j = \overline{\alpha}_k \quad \Rightarrow \quad m_j = -m_k, \tag{14′}$$

$$|m_j| \leq \ M \ \leq (1+2K)^K N, \tag{15′}$$

$$\alpha_j \in [\alpha_k] \quad \Rightarrow \quad m_j = m_k, \tag{16′}$$

and, for $\tilde{\alpha}_j = \alpha_j - i < m_j, \omega >$,

$$i(\tilde{\alpha}_j - \tilde{\alpha}_k) \in \begin{cases} DC_{\frac{\omega}{2}}^{2(N+2M)}(\kappa',\tau) & \text{if } [\tilde{\alpha}_j] \neq [\tilde{\alpha}_k] \text{ and } [\overline{\tilde{\alpha}}_k] \\ DC_{\omega}^{N+2M}(\kappa',\tau) & \text{if } [\tilde{\alpha}_j] = [\tilde{\alpha}_k] \text{ or } [\overline{\tilde{\alpha}}_k] \end{cases} \tag{17′}$$

for any

$$\kappa' \leq \frac{\kappa}{K(1+2K)^{\tau(K-1)}2^{\tau+2}N^\tau}.$$

Moreover, if not all $m_j = 0$, *then there are at least two numbers* α_j, α_k *such that*

$$|\alpha_j - \alpha_k| \geq \frac{\kappa}{2^{\tau+1}N^\tau} \quad \text{and} \quad |\tilde{\alpha}_j - \tilde{\alpha}_k| \leq \kappa'. \tag{18′}$$

PROOF. Since $\delta \leq \frac{\kappa}{K2^{\tau+2}N_1^\tau}$, $N_1 = N$, we can apply Lemma 6 to find $\tilde{\alpha}_j^1$ satisfying (14-17) for any $\kappa' \leq \frac{\kappa}{K2^{\tau+2}N_1^\tau}$. If it also satisfies (17′) we are done with $M_1 = KN_1$.

Otherwise we let $N_2 = N_1 + 2M_1$. Since $\delta \leq \frac{\kappa}{K2^{\tau+2}N_2^\tau}$ we can apply Lemma 6 to find $\tilde{\alpha}_j^2$ satisfying (14-17) for any $\kappa' \leq \frac{\kappa}{K2^{\tau+2}N_2^\tau}$. If it also satisfies (17′) we are done with $M_2 = KN_2$.

Otherwise we continue in the same way. This process will end after at most $K-1$ many steps because then, if not before, the difference between any pair of number $\tilde{\alpha}_j^{K-1}$, $\tilde{\alpha}_k^{K-1}$ will be less than $K\kappa'$ and this implies that (17′) is fulfilled. □

From Lemma 5 and 6′ we derive the following result.

PROPOSITION 7. *Let* $\frac{\omega}{2} \in DC(\kappa, \tau)$. *Let* $A \in gl(K, \mathbb{R})$ *have a* δ-*decomposition* Λ, $\delta \leq 1$, *and let* $F : 2\mathbb{T}^d \to gl(K, \mathbb{R})$ *be conjugate to* $\mathbb{T}^d$ *over* Λ. *Let* $N \in \mathbb{N} \setminus 0$ *and assume that*

$$\delta \leq \frac{\kappa}{K(1+2K)^{\tau(K-2)}2^{\tau+2}N^{\tau}}.$$

Then there exists vectors $m_1, m_2, \cdots \in \frac{1}{2}\mathbb{Z}^d$,

$$| m_j | \leq M \leq (1+2K)^K N,$$

such that $Y(\varphi) = \exp_{A,m}(\varphi)$ *is real valued and satisfies*

$$\begin{cases} \partial_\omega Y(\varphi) = AY(\varphi) - Y(\varphi)A(\tilde{\alpha}) \\ \tilde{\alpha}_j = \alpha_j - i < m_j, \omega > \end{cases}$$

with

(19)
$$| Y |_r \leq cte. \left(\frac{1+ | A |}{\delta}\right)^{2K(K+1)} e^{dMr},$$

(20)
$$| A(\tilde{\alpha}) - A | \leq cte. \left(\frac{1+ | A |}{\delta}\right)^{2K(K+1)} N,$$

(21) $Y^{-1}FY$ *is conjugate to* $\mathbb{T}^d$ *over some* δ-*decomposition* $\tilde{\Lambda}$ *of* $A(\tilde{\alpha})$,

and

(22)
$$ad_{A(\tilde{\alpha})} \in DC^{2(N+2M)}_{\frac{\omega}{2},\tilde{\Lambda}}(\kappa', \tau)$$

for any

(23)
$$\kappa' \leq \frac{\kappa}{K(1+2K)^{\tau(K-1)}2^{\tau+2}N^{\tau}}.$$

(The constant depends on K *and* τ.*)*

PROOF. We have

$$C_\Lambda^{-1} A C_\Lambda = \begin{pmatrix} \ddots & & 0 \\ & A_n & \\ 0 & & \ddots \end{pmatrix}_{n \in \mathbb{Z}_2^d}.$$

Moreover, by Lemma A (see Appendix) we can assume that

$$A_n = \begin{pmatrix} \ddots & & 0 \\ & A_{n,j} & \\ 0 & & \ddots \end{pmatrix}_j,$$

is a complex block matrix where the spectrum of each block $A_{n,j}$ has diameter $\leq \delta$. This decomposition is equivalent to a grouping of the eigenvalues $\alpha_1, \ldots, \alpha_K$ of A, and we can apply Lemma $6'$. This gives the m_j's and, hence, $Y(\varphi)$ and $A(\tilde{\alpha})$. The estimates (19) and (20) follows from Lemma 5.

Now $\Omega_k^*(Y^{-1}FY)\Lambda_n$ is

$$= \begin{pmatrix} \ddots & & 0 \\ & e^{-i<m_{k_j},\varphi>} & \\ 0 & & \ddots \end{pmatrix}_j \Omega_k^* F \Lambda_n \begin{pmatrix} \ddots & & 0 \\ & e^{i<m_{n_{j'}},\varphi>} & \\ 0 & & \ddots \end{pmatrix}_{j'} .$$

If we identify

$$m_{k_j} - \frac{k}{2} = m_{n_{j'}} - \frac{n}{2}, \qquad \text{mod } \mathbb{Z}_2^d$$

we get a new grouping of the eigenvalues $\tilde{\alpha}_1, \ldots, \tilde{\alpha}_K$ which defines a δ-decomposition $\tilde{\Lambda}$ of $A(\tilde{\alpha})$. Clearly $Y^{-1}FY$ is conjugate to $\mathbb{T}^d$ over $\tilde{\Lambda}$.

(22-23) holds by construction. □

5. Almost reducibility

THEOREM 8. *Let $A \in gl(K,\mathbb{R})$ and let $F : \mathbb{T}^d \to gl(K,\mathbb{R})$ be in $\mathcal{A}_r$. If $\omega \in DC(\kappa,\tau)$, then there is a constant ε_0 — that only depends on $d,\tau,K,|A|,\kappa,r$ — such that if*

$$| F |_r = \varepsilon \le \varepsilon_0,$$

then there exist an infinite sequence of real matrices $A = A^1, A^2, \ldots$ and an infinite sequence of mappings

$$\sigma(A^j) \ni \alpha_k^j \to \tilde{\alpha}_k^j = \alpha_k^j - i < m_k^j, \omega >, \qquad m_k^j \in \frac{1}{2}\mathbb{Z}^d$$

satisfying

$$(24) \qquad | A^{j+1} - A^j(\tilde{\alpha}^j) | \le \varepsilon_j = \begin{cases} \varepsilon & j = 1 \\ \varepsilon_{j-1}^{\frac{3}{2}} & 2 \le j, \end{cases}$$

and an infinite sequence of mappings

$$\begin{cases} Z^j : 2\mathbb{T}^d \to GL(K,\mathbb{R}) \\ F^j : 2\mathbb{T}^d \to gl(K,\mathbb{R}) \end{cases}$$

satisfying

$$\begin{cases} \partial_\omega Z^j(\varphi) = (A + F(\varphi))Z^j(\varphi) - Z^j(\varphi)(A^{j+1} + F^{j+1}(\varphi)) \\ Z^j(0) = I \end{cases}$$

with

$$(25) \qquad \begin{cases} | F^j |_{r_j} \le \varepsilon_j, & r_j \searrow \\ | Z^j - exp_{A^1,m^1} \cdot \ldots \cdot exp_{A^j,m^j} |_{r_j} \le \varepsilon^{\frac{1}{4}}. \end{cases}$$

Moreover, if

$$(26) \qquad \sigma(i \cdot ad_{A^j}) \subset DC_{\frac{\omega}{2}}^{2N_j}(\kappa_j,\tau), \qquad \begin{cases} \kappa_j \sim (\frac{r_j}{|\log(\varepsilon_j)|})^\tau \\ N_j \sim \frac{1}{r_j} | \log(\varepsilon_j) |^2, \end{cases}$$

then

$$m^j = 0 \quad \text{and} \quad r_{j+1} = \left(1 - \frac{1}{|\log(\varepsilon_j)|}\right) r_j.$$

REMARK. The theorem remains true if F is conjugate to $\mathbb{T}^d$ over a δ-decomposition of A and $\varepsilon \leq \varepsilon_0$ for some constant ε_0 that only depends on $d, \tau, K, | A |, \kappa, r$ and δ. Moreover, F^{j+1} will be conjugate to $\mathbb{T}^d$ over a decomposition of A^{j+1}.

From Theorem 8 we get for all $j \geq 1$

$$\begin{cases} \partial_\omega Z^j(\varphi) = (A + F(\varphi))Z^j(\varphi) - Z^j(\varphi)(A^{j+1} + F^{j+1}(\varphi)) \\ Z^j(0) = I \end{cases}$$

with

$$| F^{j+1} |_{r_{j+1}} \leq \varepsilon_{j+1}.$$

Clearly

$$\lim_{j\to\infty} | F^j |_0 \to 0.$$

Hence, the system

$$x'(t) = (A + F(\varphi + t\omega)x(t)$$

is reducible *if* the sequences — or at least some subsequences —

$$\lim_{j\to\infty} Z^j = Z \quad \text{and} \quad \lim_{j\to\infty} A^j = B, \tag{27}$$

because then

$$\partial_\omega Z = (A + F)Z - ZB.$$

For this it is sufficient that $r_j \to r_\infty > 0$ and $| Z^j |_{r_\infty}$ is bounded, which follows if (26) hold *for all* $j \geq 1$. In this case $Z \sim I$ and $B \sim A$, and the invariant foliation is close to a product foliation. But (27) also follows if (26) occurs *for all* $j \geq 1$ *except finitely many.* Then $Z \not\sim I$ and $B \not\sim A$, and the invariant foliation is not close to a product foliation, but may be quite wild.

REMARK. If $A_\lambda = A' + \lambda A''$, $detA'' \neq 0$, then it is not hard to see that

$$\sigma(i \cdot adA_\lambda^j) \subset DC_{\frac{\omega}{2}}^{2N_j}(\kappa_j, \tau), \qquad \begin{cases} \kappa_j \sim \left(\frac{r_j}{|\log(\varepsilon_j)|}\right)^\tau \\ N_j \sim \frac{1}{r_j} | \log(\varepsilon_j) |^2, \end{cases}$$

for all $j \geq 1$ outside a small set of λ-values. This was first proven for multi-parameter families [**Mo**] and then for one-parameter families [**E2,JoS**]. It is much harder to show that this occurs for all $j \geq 1$, except finitely many, outside a set of measure zero.

CONJECTURE. For any generic one-parameter family of analytic $F_\lambda(\varphi)$ in

$$| F_\lambda - A |_r \leq \varepsilon_0(d, \tau, K, | A |, \kappa, r)$$

the system

$$x'(t) = F_\lambda(\varphi + t\omega)x(t)$$

is reducible for a.e. λ.

6. Applications

This conjecture has been proven for certain families in $SL(2,\mathbb{R})$ $SO(3,\mathbb{R})$ and more general compact groups [**E3,K1,K2,K3**]. We discuss these results in this section.

Quasi-periodic Schrödinger. Let $v : \mathbb{T}^d \to \mathbb{R}$ be continuous and consider

$$H_\varphi y(t) \;=:\; -y''(t) + v(\varphi + t\omega)y(t) \;=\; Ey(t) \tag{28}$$

or, equivalently, the first order system

$$\begin{pmatrix} y \\ y' \end{pmatrix}'(t) \;=\; (A_E + F(\varphi + t\omega))\begin{pmatrix} y \\ y' \end{pmatrix}(t). \tag{29}$$

Let ρ_E be the rotation number of the system (29). We recall that ρ_E is *Diophantine with respect to* $\frac{\omega}{2}$ if there exist $\gamma, \sigma > 0$ such that

$$|\,\rho_E - < n, \frac{\omega}{2} > | \geq \frac{\gamma}{|\, n\,|^\sigma} \quad \forall n \in \mathbb{Z}^d \setminus 0,$$

i.e. $\rho_E \in DC_{\frac{\omega}{2}}(\gamma, \sigma)$, and we say that ρ_E is *rational with respect to* $\frac{\omega}{2}$ if

$$\rho_E \;= < n, \frac{\omega}{2} > \quad \text{some } n \in \mathbb{Z}^d \setminus 0.$$

THEOREM [**E3**]. *If $\omega \in DC(\kappa, \tau)$ and $v \in \mathcal{A}_r$, then there exists $\varepsilon_0(\tau, \kappa, r)$ such that if*

$$|\, v\,|_r \leq \varepsilon_0,$$

then (29) is reducible for all E such that ρ_E is rational or Diophantine with respect to $\frac{\omega}{2}$. Moreover, this happens for almost all E

This first part is a consequence of Theorem 8. A similar consequence for systems in $GL(K, \mathbb{R})$ would say:

T. *If $A \in gl(K, \mathbb{R})$ and $\omega \in DC(\kappa, \tau)$ and if $F : \mathbb{T}^d \to gl(K, \mathbb{R})$ is in $\mathcal{A}_r$, then there exists $\varepsilon_0(\kappa, \tau, r)$ such that if*

$$|\, F\,|_r \leq \varepsilon_0,$$

then (29) is reducible for all E such that the imaginary parts of the Floquet exponents are rational or Diophantine with respect to $\frac{\omega}{2}$.

And the conjecture would in this language be phrased:

C. *For any generic one-parameter family of analytic $F_\lambda(\varphi)$ in*

$$|\, F_\lambda - A\,|_r \leq \varepsilon_0(K, \kappa, \tau, r, |\, A\,|)$$

the imaginary parts of the Floquet exponents of the system

$$x'(t) \;=\; F_\lambda(\varphi + t\omega)x(t)$$

are rational or Diophantine with respect to $\frac{\omega}{2}$ for a.e. λ.

The only problem with this formulation is that for general systems there is no a priori characterization of the Floquet exponents. For systems in $SL(2, \mathbb{R})$, such as the Schrödinger equation, however, there is a characterization of the imaginary part of the Floquet exponent in terms of the rotation number and the formulation therefore does make sense for these systems.

The second part of the theorem follows from a result in spectral theory that identifies the rotation number with the integrated density of states. There are many other relations between dynamics and spectral theory [**E5**].

For large systems, reducibility is not an a.e.-phenomenon. Instead non-uniform hyperbolicity — which is inconsistent with reducibility in the way we have formulated this concept — seems to be a prevailing phenomena. This has been shown for time-continuous systems only in one case [**FW**]. For time-discrete systems the situation is more studied [**FW,S,E6,Y**] but still not completely understood.

Linear quasi-periodic skew-products on $SO(3,\mathbb{R})$. Let $F:\mathbb{T}^d\to so(3,\mathbb{R})$ and $A\in so(3,\mathbb{R})$, and consider the skew-product

$$\begin{cases} \varphi' = \omega \\ x' = (A+F(\varphi))x. \end{cases} \tag{30}$$

THEOREM [**K1**]. *Assume that $\omega\in DC(\kappa,\tau)$, $F\in\mathcal{A}_r$ and $A=A'+\lambda A''$, $\det A''\neq 0$. Then there exists an $\varepsilon_0(\kappa,\tau,r,A'')$ such that if*

$$|\, F\,|_r \leq \varepsilon_0,$$

then (30) is reducible for a.e. λ.

The invariant foliation is for a large set of λ close to a product foliation, but for a small set of λ, close to the exceptional measure-zero set, the invariant foliation may become quite wild. So what happen for the exceptional λ's? The following theorem gives an answer to that question. Non-reducible skew-products with different dynamical behavior have also been studied in [**H1**]

THEOREM [**E5**]. *Assume that $\omega\in DC(\kappa,\tau)$ and $F\in\mathcal{A}_r$. Then there exists an $\varepsilon_0(d,\tau,A,\kappa,r)$ such that in*

$$|\, F\,|_r \leq \varepsilon_0$$

the generic (for $|\ |_r$-topology) skew-product (30) is uniquely ergodic.

So the picture of the dynamics of systems close to constant coefficients is the following:

* almost all (in measure sense) systems are reducible

* generic (in topological sense) systems are uniquely ergodic.

Is this picture true also globally?

About lower-dimensional tori. Let

$$H(\varphi,r,z,\varepsilon) \;= <\omega,r> +\frac{1}{2}<Qr,r> +\frac{1}{2}<\Omega z,z> +\hat{H}(\varphi,r,z,\varepsilon)$$

with $\varphi\in\mathbb{T}^d$, $r\in\mathbb{R}^d$ and $z\in\mathbb{R}^{2\nu}$. The Hamiltonian equations are

$$\begin{cases} \varphi' = \frac{\partial H}{\partial r} \\ r' = -\frac{\partial H}{\partial \varphi} \\ z' = J\frac{\partial H}{\partial z}, \quad J=\begin{pmatrix} 0 & I_\nu \\ -I_\nu & 0 \end{pmatrix}. \end{cases}$$

When $\varepsilon=0$ there exists an invariant torus of dimension d,

$$\varphi\in\mathbb{T}^d,\ r=0,\ z=0,$$

with

$$\begin{cases} \text{basic frequencies } \omega \\ \text{normal frequencies } \sigma(J\Omega)\cup\{0\}. \end{cases}$$

In the perturbation theory of this torus one must assume

$$(*)\qquad \begin{cases} \omega \in DC(\kappa,\tau) \quad \& \quad \det Q \neq 0 \\ \sigma(i\cdot J\Omega) \subset DC_\omega(\kappa,\tau) \quad \& \quad \det J\Omega \neq 0, \end{cases}$$

because the first order correction $X(\varphi)$ of the torus must satisfy

$$\partial X(\varphi) \;=\; \begin{pmatrix} \begin{pmatrix} 0 & Q \\ 0 & 0 \end{pmatrix} & 0 \\ 0 & J\Omega \end{pmatrix} X(\varphi) + \varepsilon \begin{pmatrix} \frac{\partial \hat{H}}{\partial r} \\ -\frac{\partial \hat{H}}{\partial \varphi} \\ J\frac{\partial \hat{H}}{\partial z} \end{pmatrix} (\varphi, 0, 0, 0),$$

and solving this equation with constant coefficients requires (*).

In order to study higher order corrections one is lead to equations

$$(31)\qquad \partial X(\varphi) \;=\; (A+\varepsilon F(\varphi))X(\varphi) + \varepsilon' G(\varphi),$$

for $\varepsilon' << \varepsilon$. This equation with variable coefficients can be solved if the homogeneous part is reducible. Indeed, if

$$\partial_\omega Z(\varphi) \;=\; (A+\varepsilon F(\varphi))Z(\varphi) - Z(\varphi)B$$

then $X(\varphi) = Z(\varphi)Y(\varphi)$ is a solution of (31) if, and only if,

$$(32)\qquad \partial_\omega Y(\varphi) \;=\; BY(\varphi) + \varepsilon' Z(\varphi)^{-1}G(\varphi).$$

Solving (32) requires again *a condition on the normal frequencies*, i.e. of the eigenvalues of B. But reducibility requires *a condition on the difference of the normal frequencies* i.e. the eigenvalues of ad_B.

In the perturbation theory of lower dimensional tori [**E2,Me,Ku**] one therefore imposes a condition not only on the normal frequencies but also on their differences. However, noticing that it suffices to solve (31)

$$\text{mod } (\varepsilon')^{1+\sigma}$$

for some $\sigma > 0$, one sees that almost reducibility of the homogeneous equation is sufficient and that, therefore, the differences of the normal frequencies play no role.[8]

REMARK. In ∞-dimensional systems the differences of the normal frequencies is a serious problem and the result of this paper does not provide any solution. The reason is that when $K=\infty$ the only perturbation admitted by Theorem 8 is zero.

7. Proof of theorem 8

Let $\omega \in DC(\kappa,\tau)$ and let $A \in gl(K,\mathbb{R})$ with a δ-decomposition Λ, $\delta \leq 1$. Let $F : 2\mathbb{T}^d \to gl(K,\mathbb{R})$ be a mapping in $\mathcal{A}_r$ which is conjugate to $\mathbb{T}^d$ over Λ, and let $\varepsilon = \mid F \mid_r$. F has a Fourier series

$$F(\varphi) \;=\; \sum_{n\in\frac{1}{2}\mathbb{Z}^d} \hat{F}(n)e^{i<n,\varphi>}$$

[8] Bourgain [**B**] has constructed invariant tori, imposing no conditions on the differences of the normal frequencies, by solving equation (31) without using reducibility of the homogeneous equation at all.

and we define its truncation at order N by

$$\{F\}_N(\varphi) = \sum_{|n|\leq N} \hat{F}(n)e^{i<n,\varphi>}.$$

We have

$$| F - \{F\}_N |_{r'} \leq \text{ cte.}\frac{1}{(r-r')^d} N^{d-1}e^{-N(r-r')} | F |_r$$

and we assume that

$$\text{cte.}\frac{1}{(r-r')^d}N^{d-1}e^{-N(r-r')}\varepsilon \leq \frac{1}{4}\varepsilon^{\frac{3}{2}}. \tag{33}$$

We look for $Z(\varphi)$ of the form $I + X(\varphi)$. Then

$$\partial_\omega X(\varphi) = (A + F(\varphi))(I + X(\varphi)) - (I + X(\varphi))(A + A' + G(\varphi)),$$

and we think of I and A as of order 0, $\{F\}_N$, X and A' as of order 1, and $F - \{F\}_N$ and G as of order 2. We identify terms of equal order to get

$$\partial_\omega X(\varphi) = [A, X](\varphi) + \{F\}_N(\varphi) - A' \tag{34}$$

and

$$G(\varphi) = (I + X(\varphi))^{-1}(F(\varphi)X(\varphi) - X(\varphi)A' + (F - \{F\}_N)(\varphi)). \tag{35}$$

Now we shall solve equation (34) using Proposition 4 and then estimate G from (35). In order to do this we must consider two cases.

Case I. If

$$ad_A \in DC^{2N}_{\frac{\omega}{2},\Lambda}(\kappa', \tau)$$

we have, by Proposition 4, a unique solution $X(\varphi)$ of (34) with $\hat{X}(0) = 0$ and $A' = \hat{F}(0)$. This solution satisfies

$$| X |_{r'} \leq \text{ cte.}\left(\frac{a}{\delta}\right)^{2K^4}\left(\frac{1}{\kappa'(r-r')^\tau}\right)^{K^2}\varepsilon \qquad \forall r' \in]r-1, r[,$$

where $a =| A | +1$. We assume that

$$\text{cte.}\left(\frac{a}{\delta}\right)^{2K^4}\left(\frac{1}{\kappa'(r-r')^\tau}\right)^{K^2}\varepsilon \leq \frac{1}{4}\varepsilon^{\frac{1}{2}}. \tag{36}$$

Then

$$| (I + X)^{-1} - I |_{r'} \leq \frac{1}{4 - \varepsilon^{\frac{1}{2}}}\varepsilon^{\frac{1}{2}} \leq \frac{1}{3}\varepsilon^{\frac{1}{2}}$$

and

$$| G |_{r'} \leq \varepsilon^{\frac{3}{2}}.$$

By Proposition 4, X is conjugate to $\mathbb{T}^d$ over Λ and therefore, by Lemma 3(3), also G . By Lemma 3(1,2) it follows that Λ is a δ-decomposition of $A + A'$.

Now (33) will be fulfilled by taking N larger than some $f(r - r', \varepsilon)$, and (36) will be fulfilled by taking ε smaller than some $g(\delta, \kappa', r - r', a)$. For example if

$$N \geq \text{ cte.}\frac{1}{r-r'} | \log(\varepsilon) |, \tag{37}$$

and

$$\varepsilon \ \leq \ \text{cte.}\left(\frac{\delta}{a}\right)^{4K^4}(\kappa')^{2K^2}(r-r')^{2\tau K^2}. \tag{38}$$

Moreover, if $Z(\varphi) = (I + X(\varphi))(I + X(0))^{-1}$, then we get

$$|\, Z - I \,|_{r'} \ \leq \ \varepsilon^{\frac{1}{2}}.$$

Case II. In this case

$$ad_A \notin DC^{2N}_{\frac{\omega}{2},\Lambda}(\kappa', \tau).$$

Then we first apply Proposition 7 to find, for

$$N^\tau \ \leq \ \text{cte.}\frac{\kappa}{\delta}, \tag{39}$$

a transformation $Y(\varphi)$ such that

$$\partial_\omega Y(\varphi) \ = \ AY(\varphi) - Y(\varphi)\tilde{A}$$

with

$$|\, Y\,|_{\tilde{r}},\ |\, Y^{-1}\,|_{\tilde{r}} \leq \ \text{cte.}\left(\frac{1+|\,A\,|}{\delta}\right)^{2K(K+1)} e^{dM\tilde{r}},$$

for some $M \leq (2K+1)^K N$, and

$$|\, \tilde{A} - A \,| \leq \ \text{cte.}\left(\frac{1+|\,A\,|}{\delta}\right)^{2K(K+1)} N.$$

Moreover, by (21-23) there exists a δ-decomposition $\tilde{\Lambda}$ of $\tilde{A}$ such that $Y^{-1}FY$ is conjugate to $\mathbb{T}^d$ over $\tilde{\Lambda}$ and

$$ad_{\tilde{A}} \in DC^{2(N+2M)}_{\frac{\omega}{2},\tilde{\Lambda}}(\kappa', \tau)$$

for any

$$\kappa' \ \leq \ \text{cte.}\frac{\kappa}{N^\tau}. \tag{40}$$

This implies that

$$\partial_\omega Y(\varphi) \ = \ (A + F(\varphi))Y(\varphi) - Y(\varphi)(\tilde{A} + \tilde{F}(\varphi)),$$

where $\tilde{F}$ is an conjugate to $\mathbb{T}^d$ over $\tilde{\Lambda}$ and

$$|\, \tilde{F}\,|_{\tilde{r}} \leq \ \tilde{\varepsilon} \ = \ \text{cte.}\left(\frac{1+|\,A\,|}{\delta}\right)^{4K(K+1)} e^{2dM\tilde{r}}\varepsilon, \quad \forall \tilde{r} \leq r,$$

$$|\, \tilde{A} \,| \leq \ \tilde{a} - 1 = \text{cte.}\left(\frac{1+|\,A\,|}{\delta}\right)^{2K(K+1)}.$$

We can now apply the same procedure as in Case I. We start by truncating $\tilde{F}$ at order $2M + N$. This gives

$$\begin{aligned} |\, \tilde{F} - \{\tilde{F}\}_{2M+N} \,|_{r'} \ &\leq \ \text{cte.}\left(\frac{a}{\delta}\right)^{4K(K+1)} \sum_{|n|\geq N} |\, \hat{F}(n)\,|\, e^{(|n|+2dM)r'} \\ &\leq \ \text{cte.}\left(\frac{a}{\delta}\right)^{2K(K+1)} \frac{1}{(r-r')^d} N^{d-1} e^{2dMr'} e^{-N(r-r')}\varepsilon. \end{aligned}$$

If we assume

$$(\tilde{33})\qquad \text{cte.}\left(\frac{a}{\delta}\right)^{4K(K+1)}\frac{1}{(r-r')^d}N^{d-1}e^{2dMr'}e^{-N(r-r')}\varepsilon \;\le\; \frac{1}{4}\varepsilon^{\frac{3}{2}}.$$

we obtain, as before in Case I, estimates of

$$\partial_\omega\tilde{X}(\varphi) \;=\; [\tilde{A},\tilde{X}](\varphi)-\{\tilde{F}\}_{2M+N}(\varphi)-A'$$

and

$$\tilde{G}(\varphi) \;=\; (I+\tilde{X}(\varphi))^{-1}(\tilde{F}(\varphi)\tilde{X}(\varphi)-\tilde{X}(\varphi)A'+(\tilde{F}-\{\tilde{F}\}_{2M+N})(\varphi).$$

By Proposition 4 we get for all $r'\in]\tilde{r}-1,\tilde{r}[$

$$|\,\tilde{X}\,|_{r'} \le \text{cte.}\left(\frac{\tilde{a}}{\delta}\right)^{2K^4}\left(\frac{1}{\kappa'(\tilde{r}-r')^{\tau}}\right)^{K^2}|\,\tilde{F}\,|_{\tilde{r}},$$

and

$$|\,\tilde{G}\,|_{r'} \;\le\; \varepsilon^{\frac{3}{2}},$$

if we assume

$$(\tilde{36})\qquad \text{cte.}\left(\frac{\tilde{a}}{\delta}\right)^{2K^4}\left(\frac{1}{\kappa'(\tilde{r}-r')^{\tau}}\right)^{K^2}\tilde{\varepsilon} \;\le\; \frac{1}{4}\varepsilon^{\frac{1}{2}}.$$

In the same way as in Case I we see that $\tilde{\Lambda}$ is a δ-decomposition of $\tilde{A}+A'$ and that both $\tilde{X}$ and $\tilde{G}$ are conjugated over $\tilde{\Lambda}$.

Now we have to fulfill (39), (40), $(\tilde{33})$ and $(\tilde{36})$. If we choose δ according to (39) and κ' according to (40) we get

$$(\tilde{33})\qquad \text{cte.}a^{4K(K+1)}\frac{1}{(r-r')^d}N^{4\tau K(K+1)+d-1}e^{2dMr'}e^{-N(r-r')} \;\le\; \varepsilon^{\frac{1}{2}},$$

and

$$(\tilde{36})\qquad \text{cte.}a^{5K^5(K+1)}\left(\frac{1}{\tilde{r}-r'}\right)^{\tau K^2}e^{2dM\tilde{r}}N^{2K^4+5\tau K^5(K+1)}\varepsilon^{\frac{1}{2}} \;\le\; 1.$$

If we let $\tilde{r}=\min(r,\frac{1}{M})$ then the terms $e^{2dM\tilde{r}}$ and $e^{2dMr'}$ will be under control. Now $(\tilde{33})$ will be fulfilled by taking $N\ge f(r-r',\varepsilon,a)$ and $(\tilde{36})$ will be fulfilled by taking $\varepsilon\le g(N,\tilde{r}-r',a)$. Both these conditions can be satisfied simultaneously, for example if

$$(41)\qquad N \;=\; \text{cte.}\frac{1}{r-r'}\,|\log(\varepsilon)\,|,$$

and

$$(42)\qquad \varepsilon \;\le\; \text{cte.}\frac{(\tilde{r}-r')^{3\tau K^2}}{a^{15K^6}}.$$

Moreover, if $Z(\varphi)=\exp_{A,m}(\varphi)(I+\tilde{X}(\varphi))(I+\tilde{X}(0))^{-1}$, then we get

$$|\,Z-\exp_{A,m}\,|_{r'} \;\le\; \text{cte.}(aN^{\tau})^{2K(K+1)}\varepsilon^{\frac{1}{2}}.$$

which is $\le\varepsilon^{\frac{1}{4}}$ if (41-42) hold.

Notice that when δ and κ' are chosen according to (39) and (40), respectively, then (41-42) implies (37-38). Everything is now specified except r' and we decide to let

$$\tilde{r} - r' = \frac{1}{\log(\frac{1}{\varepsilon})} r. \tag{43}$$

Iteration. Choose now $\varepsilon_1 = \varepsilon,\ r_1 = r,\ \delta_1 = \delta,\ a_1 = a$ and

$$\begin{cases} \varepsilon_{j+1} = \varepsilon_j^{\frac{3}{2}} & a_{j+1} = \tilde{a}_j + \varepsilon_j \\ r_{j+1} = \tilde{r}_j - \frac{1}{|\log(\varepsilon_j)|}\tilde{r}_j & \delta_j = \text{cte.}\frac{\kappa}{N_j^\tau} \\ N_j = \text{cte.}\frac{1}{r_j - r_{j+1}} \mid \log(\varepsilon_j) \mid & \kappa_j = \text{cte.}\frac{\kappa}{N_j^\tau}. \end{cases}$$

Here we define

$$\begin{cases} \tilde{a}_j = a_j & \text{or} \quad = \text{cte.} a_j^{2K(K+1)} N_j^{\tau 2K(K+1)+1} \\ \tilde{r}_j = r_j & \text{or} \quad = \frac{r_j}{M_j}, \quad M_j = (2K+1)^K N_j \end{cases}$$

according to if $ad_{A^j} \in DC^{2(N_j+2M_j)}_{\frac{\omega}{2},\Lambda^j}(\kappa_j,\tau)$ or not.

Then (42) holds for all $j \geq 1$, i.e.

$$\varepsilon_j \leq \text{cte.}\frac{(\tilde{r}_j - r_{j+1})^{3\tau K^2}}{a_j^{15K^6}}$$

if ε_1 is sufficiently small.

So we can iterate the construction with $F^1 = F,\ A^1 = A$ and we find, for all $j \geq 1$, mappings Z_j and A^{j+1} such that

$$\partial_\omega Z_j(\varphi) = (A^j + F^j(\varphi))Z_j(\varphi) - Z_j(\varphi)(A^{j+1} + F^{j+1}(\varphi)).$$

These mappings satisfy the relations $A^{j+1} = \tilde{A}^j + \tilde{\hat{F}}^j(0)$ with

$$\tilde{A}^j = A^j(\tilde{\alpha}^j), \qquad \tilde{\alpha}^j = \alpha^j - i < m^j, \omega >,$$

and the estimates

$$\begin{cases} \mid F^j \mid_{r_j} \leq \varepsilon_j \\ \mid Z_j - exp_{A^j,m^j} \mid_{r_{j+1}} \leq \text{cte.}(a_j N_j^\tau)^{2K(K+1)} \varepsilon_j^{\frac{1}{2}} \\ \mid exp_{A^j,m^j} \mid_{r_{j+1}} \leq \text{cte.}(a_j N_j^\tau)^{2K(K+1)}. \end{cases}$$

Then it is easy to see that the A^j's and the m^j's and the functions

$$Z^j(\varphi) = Z_1(\varphi) \ldots Z_j(\varphi),$$

verify the statement of the theorem. The statement (26) follows since under this condition we are in Case I and therefore $\tilde{a}_j = a_j$ and $\tilde{r}_j = r_j$. This concludes the proof of Theorem 8. □

Appendix

Let $A \in gl(K, \mathbb{R})$. For all $\alpha \in \sigma(A)$, let $\Lambda(\alpha)$ be the generalized eigenspace of A associated to α. Since A is real we have $\Lambda(\overline{\alpha}) = \overline{\Lambda(\alpha)}$.

LEMMA A. *Let $\cup_1^s \sigma_j$ be a disjoint decomposition of $\sigma(A)$ such that*

σ_j and σ_k, $j \neq k$, are separated by at least δ

and

for any σ_j there is a unique $\sigma_{\tau_j} = \overline{\sigma_j}$.

Let Λ_j be the invariant space associated to σ_j — provided with a unitary basis such that $\Lambda_{\tau_j} = \overline{\Lambda_j}$. Then the matrix $C = (\Lambda_1, \ldots, \Lambda_s)$ satisfies

1.
$$C^{-1}AC = \begin{pmatrix} A_1 & & 0 \\ & \ddots & \\ 0 & & A_s \end{pmatrix};$$

2. *the matrix*
$$C\begin{pmatrix} B_1 & & 0 \\ & \ddots & \\ 0 & & B_s \end{pmatrix}C^{-1}$$
is real if $B_{\tau_j} = \overline{B_j}$ for all j;

3.
$$| C^{-1} | \leq \max\Big(\Big(cte.\frac{| A |}{\delta}\Big)^{K(K+1)}, 1\Big).$$

(The constant only depends on K.)

PROOF. 1. is clear by definition and 2. follows since we have chosen a basis that respects complex conjugation. We only need to consider 3. Fix a Λ_j and let Λ be a sum of all the other invariant subspaces — provided with a unitary basis. Then the spectrum of $\Lambda_j^* \Lambda \Lambda^* \Lambda_j$ is bounded away from 1 by at least

$$\gamma^2 = \Big(\text{cte.}\frac{1}{K}\Big)^{K+2}\Big(\frac{\delta}{| A |}\Big)^{K(K+1)}$$

(see [**E7**], Appendix A).

Let $\tilde{\Lambda}_j$ be the invariant subspace of the adjoint matrix A^* corresponding to σ_j — recall that $\sigma(A) = \sigma(A^*)$. Then

$$\tilde{\Lambda}_j = (\sum_{k \neq j} \Lambda_k)^{\perp}.$$

Now

$$C^{-1} = \begin{pmatrix} R_1^{-1} & & 0 \\ & \ddots & \\ 0 & & R_s^{-1} \end{pmatrix} \cdot \begin{pmatrix} \tilde{\Lambda}_1^* \\ \vdots \\ \tilde{\Lambda}_s^* \end{pmatrix}, \qquad R_j = (\tilde{\Lambda}_j)^* \Lambda_j.$$

Since $\Lambda_j^* \tilde{\Lambda}_j \tilde{\Lambda}_j^* \Lambda_j$ is Hermitian and equals

$$I - \Lambda_j^* \Lambda \Lambda^* \Lambda_j,$$

it follows that

$$| R_j^{-1} | = | (\Lambda_j^* \tilde{\Lambda}_j \tilde{\Lambda}_j^* \Lambda_j)^{-1} (\Lambda_j^* \tilde{\Lambda}_j | \leq \frac{1}{\gamma^2}.$$

□

References

[B] J. Bourgain, *On Melnikov's persistency problem*, Math. Res. Lett. **4** (1997), 445-458.

[DS] E. I. Dinaburg, Ya. G. Sinai, *The one-dimensional Schrödinger equation with quasi-periodic potential*, Funkt. Anal. i. Priloz. **9** (1975), 8–21.

[E1] L. H. Eliasson, *Reducibility and point spectrum for linear quasi-periodic skew-products*, Proceedings of the ICM, Berlin 1998. Documenta Mathematica **Extra Volume 2** (1998), http:-//www.matematik.uni-bielefeld.de/documenta, 779–788.

[E2] L. H. Eliasson, *Perturbations of stable invariant tori for Hamiltonian systems*, Ann. Sc. Norm. Sup. Pisa Cl. Sci. **15** (1988), 115–147.

[E3] L. H. Eliasson, *Floquet solutions for the one-dimensional quasi-periodic Schrödinger equation*, Commun. Math. Phys. **146** (1992), 447–482.

[E4] L. H. Eliasson, *Ergodic skew systems on* $SO(3,\mathbb{R})$, preprint ETH-Zürich (1991).

[E5] L. H. Eliasson, *One-dimensional quasi-periodic Schrödinger operators — Dynamical systems and spectral theory*, in A. Balog et. al. (eds.): Proceedings of the European Congress of Mathematics, Budapest 1996, Birkhäuser Verlag, Basel, 1998.

[E6] L. H. Eliasson, *Discrete one-dimensional quasi-periodic Schrödinger operators with pure point spectrum*, Acta Math **179** (1997), 153–196.

[E7] L. H. Eliasson, *Perturbations of linear quasi-periodic systems*, in Proceedings from CIME-Summer School on Small Divisors, Cetraro 1998.
(to be published by the organizers)

[FW] J. Fröhlich, T. Spencer, P. Wittver, *Loclization for a class of one-dimensional quasi-periodic Schrödinger operators*, Commun. Math. Phys. **132** (1990), 5–25.

[G] A. Gordon, *On the point spectrum of the one-dimensional Schrödinger operator*, Russ. Math. Surveys **31** (1976), 457–258.

[H1] M. Herman, *Non topological conjugacy of skew products on* $SU(2)$, manuscript (1989).

[H2] M. Herman, *Une méthode pour minorer les exposants de Lyapunov*, Comment. Math. Helvetici **58** (1983), 453–502.

[J] R. A. Johnson, *The recurrent Hill's equation*, J. Diff. Equations **46** (1982), 165–193.

[JM] R. A. Johnsson, J. Moser, *The rotation number for almost periodic potentials*, Commun. Math. Phys. **84** (1982), 403–438.

[JS] R. A. Johnsson, G. R. Sell, *Smoothness of spectral subbundles and reducibility of quasi-periodic linear differential systems*, J. Diff. Eq. **41** (1981), 262–288.

[JoS] A. Jorba, C Simo, *On the reducibility of linear differential equations with quasi-periodic coefficients*, J. Diff. Eq. **98** (1992), 111–124.

[K1] R. Krikorian, *Réductibilité presque partout des systèmes quasi-périodiques analytiques dans le cas* $SO(3)$, C. R. Acad. Sci. Paris **321, Série I** (1995), 1039–1044.

[K2] R. Krikorian, Astérisque **259** (1999).

[K3] R. Krikorian, *Reductibilité presque partout des flots fibrés quasi-périodiques á valeurs dans des groupes compacts*, Ann. Sci. École Norm. Sup. **32** (1999), 187-240.

[Ku] S. Kuksin, *Hamiltonian perturbations of* ∞*-dimensional linear systems with an imaginary spectrum*, Funct. Anal. Appl. **21** (1987), 192–205.

[Me] V. K. Melnikov, *On some cases of conservation of conditionally periodic motions under a small change of the Hamiltonian function*, Soviet Mathematics Doklady **6** (1985), 1592–1596.

[Mi] V.M. Millionščikov, *Proof of the existence of irregular systems of linear differential with almost periodic coefficients*, Diff. Equations **4** (1968), 203–205.

[Mo] J. Moser, *Convergent series expansions for quasiperiodic motions*, Math. Ann. **169** (1967), 136–176.

[NS] V.V. Nemytskii, V.V. Stepanov, *Qualitative theory of differential equations*, Princeton University Press, Princeton, New Jersey, 1960.

[R] H. Rüssmann, *On optimal estimates for the solutions of linear partial differential equations of the first order with constant coefficients on the torus*, Springer Lecture Notes in Physics **38** (1975), 598–624.

[Si] B. Simon, *Almost periodic Schrödinger operators: a review*, Adv. Appl. Math. **3** (1982), 463–490.

[S] Ya. G. Sinai, *Anderson localization for the one-dimensional difference Schrödinger operator with a quasi-periodic potential*, J. Stat. Phys. **46** (1987), 861–909.

[Y] L.S. Young, *Lyapunov exponents for some quasi-periodic cocycles*, Ergod. Th.& Dynam. Sys. **17** (1997), 483–504.

Department of Mathematics, Royal Institute of Technology, S-10044 Stockholm, Sweden

E-mail address: hakane@math.kth.se

Proceedings of Symposia in Pure Mathematics
Volume **69**, 2001

A Lecture on the Classical KAM Theorem

Jürgen Pöschel

1. The Classical KAM-Theorem

a. The purpose of this lecture is to describe the KAM theorem in its most basic form and to give a complete and detailed proof. This proof essentially follows the traditional lines laid out by the inventors of this theory, *K*olmogorov, *A*rnold and *M*oser (whence the acronym 'KAM'), and the emphasis is more on the underlying ideas than on the sharpness of the arguments. After all, KAM theory is not only a collection of specific theorems, but rather a methodology, a collection of ideas of how to approach certain problems in perturbation theory connected with "small divisors".

b. The classical KAM theorem is concerned with the stability of motions in hamiltonian systems, that are small perturbations of integrable hamiltonian systems. These integrable systems are characterized by the existence of action angle coordinates such that the hamiltonian depends on the action variable alone — see [**2,14**] for details. Thus we are going to consider hamiltonians of the form

$$H(p,q) = h(p) + f_\epsilon(p,q), \qquad f_\epsilon(p,q) = \epsilon f_*(p,q,\epsilon)$$

for small ϵ, where $p = (p_1, \dots, p_n)$ are the action variables varying over some domain $D \subset \mathbb{R}^n$, while $q = (q_1, \dots, q_n)$ are the conjugate angular variables, whose domain is the usual n-torus $\mathbb{T}^n$ obtained from $\mathbb{R}^n$ by identifying points whose components differ by integer multiples of 2π. Thus, f_ϵ has period 2π in each component of q. Moreover, all our hamiltonians are assumed to be *real analytic* in all arguments.

The equations of motion are, as usual,

$$\dot p = -H_q(p,q), \qquad \dot q = H_p(p,q)$$

in standard vector notation, where the dot indicates differentiation with respect to the time t, and the subscripts indicate partial derivatives. The underlying phase space is $D \times \mathbb{T}^n \subset \mathbb{R}^n \times \mathbb{T}^n$ with the standard symplectic structure

$$\upsilon = \sum_j dp_j \wedge dq_j.$$

2000 *Mathematics Subject Classification.* Primary 37K55, 58F07, 70H05.

The hamiltonian vectorfield X_H associated with the equations of motions then satisfies $\upsilon \lrcorner X_H = -dH$.

We assume that the number n of degrees of freedom is at least 2, since one degree of freedom systems are always integrable.

c. For $\epsilon = 0$ the system is governed by the unperturbed, integrable hamiltonian h, and the equations of motion reduce to

$$\dot{p} = 0, \qquad \dot{q} = \omega$$

with

$$\omega = h_p(p).$$

They are easily integrated – hence the name *integrable system* – and their general solution is

$$p(t) = p_0, \qquad q(t) = q_0 + \omega(p_0)t.$$

Hence, every solution curve is a straight line, which, due to the identification of the q-coordinates modulo 2π, is winding around the *invariant torus*

$$\mathcal{T}_{p_0} = \{p_0\} \times \mathbb{T}^n$$

with constant frequencies, or winding numbers, $\omega(p_0) = (\omega_1(p_0), \ldots, \omega_n(p_0))$. Such tori with linear flow are also called *Kronecker tori*.

In addition these tori are *Lagrangian*. That is, the restriction of the symplectic form υ to their tangent space vanishes, and their dimension is maximal with respect to this property.

Thus, the whole phase space is *foliated* into an n-parameter family of invariant Lagrangian tori $\mathcal{T}_{p_0}$, on which the flow is linear with constant frequencies $\omega(p_0)$. – This is the geometric picture of an integrable hamiltonian system.

It should be kept in mind that due to the introduction of action angle coordinates these solutions are related to "real world solutions" by some coordinate transformation, which is periodic in $q_1, \ldots, q_n$. Expanding such a transformation into Fourier series and inserting the linear solutions obtained above, the "real world solutions" are represented by series of the form

$$\sum_{k \in \mathbb{Z}^n} a_k(p_0)\, e^{\langle k, q_0 \rangle + t \langle k, \omega(p_0) \rangle}, \qquad a_k \in \mathbb{R}^{2n},$$

where $\langle \cdot, \cdot \rangle$ denotes the usual scalar product. Thus, every solution is now *quasi-periodic* in t: its frequency spectrum in general does not consist of integer multiples of a single frequency – as is the case with periodic solutions –, but rather of integer combinations of a finite number of different frequencies. In essence, the "real world solutions" are superpositions of n oscillations, each with its own frequency. Moreover, these quasi-periodic solutions occur in families, depending on the parameter q_0, which together fill an invariant embedded n-torus.

Let us return to action angle coordinates. We observe that the topological nature of the flow on each Kronecker torus crucially depends on the arithmetical properties of its frequencies ω. There are essentially two cases.

1. – The frequencies ω are *nonresonant*, or *rationally independent*:

$$\langle k, \omega \rangle \neq 0 \qquad \text{for all } 0 \neq k \in \mathbb{Z}^n.$$

Then, on this torus, each orbit is dense, the flow is ergodic, and the torus itself is minimal.

2. – The frequencies ω are *resonant*, or *rationally dependent*: that is, there exist integer relations

$$\langle k, \omega \rangle = 0 \qquad \text{for some } 0 \neq k \in \mathbb{Z}^n.$$

The prototype is $\omega = (\omega_1, \dots, \omega_{n-m}, 0 \dots, 0)$, with $1 \leq m \leq n-1$ trailing zeroes and nonresonant $(\omega_1, \dots, \omega_{n-m})$. In this case the torus decomposes into an m-parameter family of invariant $n-m$-tori. Each orbit is dense on such a lower dimensional torus, but not in $\mathbb{T}^n$.

A special case arises when there exist $m = n - 1$ independent resonant relations. Then each frequency $\omega_1, \dots, \omega_n$ is an integer multiple of a fixed non-zero frequency ω_*, and the whole torus is filled by *periodic* orbits with one and the same period $2\pi/\omega_*$.

In an integrable system the frequencies on the tori may or may not vary with the torus, depending on the nature of the *frequency map*

$$h_p\colon\ D \to \Omega, \quad p \mapsto \omega(p) = h_p(p),$$

where $\Omega \subset \mathbb{R}^n$. We now make the assumption that this system is *nondegenerate* in the sense that

$$\det h_{pp} = \det \frac{\partial \omega}{\partial p} \neq 0$$

on D. Then h_p is an open map, even a local diffeomorphism between D and some open *frequency domain* $\Omega \subset \mathbb{R}^n$, and "the frequencies ω effectively depend on the amplitudes p", as a physicist would say. It follows that nonresonant tori and resonant tori of all types each form dense subsets in phase space. Indeed, the resonant ones sit among the nonresonant ones like the rational numbers among the irrational numbers.

This "frequency-amplitude-modulation" is a genuinely nonlinear phenomenon. By contrast, in a linear system the frequencies are the same all over the phase space. As we will see, this is essential for the stability results of the KAM theory. As it is said, "the nonlinearities have a stabilizing effect".

d. Now we consider the perturbed hamiltonian. The objective is to prove the persistence of invariant tori for small $\epsilon \neq 0$.

The first result in this direction goes back to Poincaré and is of a negative nature. He observed that the *resonant* tori are in general *destroyed* by an arbitrarily small perturbation. In particular, out of a torus with an $n-1$-parameter family of periodic orbits, usually only *finitely* many periodic orbits survive a perturbation, while the others disintegrate and give way to *chaotic behavior*. – So in a nondegenerate system a *dense* set of tori is usually destroyed. This, in particular, implies that a generic hamiltonian system is *not integrable* [**6,9,27**].

Incidentally, it would not help to drop the nondegeneracy assumption to avoid resonant tori. If h is too degenerate, the motion may even become ergodic on each energy surface, thus destroying all tori [**7**].

A dense set of tori being destroyed there seems to be no hope for other tori to survive. Indeed, until the fifties it was a common belief that arbitrarily small perturbations can turn an integrable system into an ergodic one on each energy surface. In the twenties there even appeared an – erroneous – proof of this "ergodic hypothesis" by Fermi.

But in 1954 Kolmogorov observed that the converse is true – the *majority* of tori survives. He proved the persistence of those Kronecker systems, whose frequencies ω are not only nonresonant, but are *strongly nonresonant* in the sense that there exist constants $\alpha > 0$ and $\tau > 0$ such that

$$|\langle k, \omega \rangle| \geq \frac{\alpha}{|k|^\tau} \qquad \text{for all } 0 \neq k \in \mathbb{Z}^n, \tag{1}$$

where $|k| = |k_1| + \cdots + |k_n|$. Such a condition is called a *diophantine* or *small divisor condition.*

The existence of such frequencies is easy to see. Let Δ_α^τ denote the set of all $\omega \in \mathbb{R}^n$ satisfying these infinitely many conditions with fixed α and τ. Then Δ_α^τ is the complement of the open dense set $R_\alpha^\tau = \bigcup_{0 \neq k \in \mathbb{Z}^n} R_{\alpha,k}^\tau$, where

$$R_{\alpha,k}^\tau = \{ \omega \in \mathbb{R}^n : |\langle k, \omega \rangle| < \alpha / |k|^\tau \} .$$

Obviously, for any bounded domain $\Omega \subset \mathbb{R}^n$, we have the Lebesgue measure estimate $m(R_{\alpha,k}^\tau \cap \Omega) = O(\alpha / |k|^{\tau+1})$, and thus

$$m(R_a^\tau \cap \Omega) \leq \sum_k m(R_{\alpha,k}^\tau \cap \Omega) = O(\alpha) ,$$

provided that $\tau > n - 1$. Hence, $R^\tau = \bigcap_{\alpha > 0} R_\alpha^\tau$ is a set of measure zero, and its complement

$$\Delta^\tau = \bigcup_{\alpha > 0} \Delta_\alpha^\tau$$

is a set of *full measure* in $\mathbb{R}^n$, for any $\tau > n - 1$. In other words, *almost every* ω in $\mathbb{R}^n$ belongs to Δ^τ, $\tau > n - 1$, which is the set of all ω in $\mathbb{R}^n$ satisfying (1) for some $\alpha > 0$ while τ is fixed..

As an aside we remark that $\Delta^\tau = \varnothing$ for $\tau < n - 1$, because for every nonresonant ω,

$$\min_{0 \neq |k|_\infty \leq K} |\langle k, \omega \rangle| \leq \frac{|\omega|}{K^{n-1}} \tag{2}$$

by Dirichlet's pigeon hole argument. And for $\tau = n - 1$, the set Δ^{n-1} has measure zero, but Hausdorff dimension n – see [**17**] for references. So there are continuum many diophantine frequencies to the exponent $n - 1$, although they form a set of measure zero. – But here we will fix $\tau > n - 1$ and drop it from the notation, letting $\Delta_\alpha = \Delta_\alpha^\tau$.

e. But although almost all frequencies are strongly nonresonant for any fixed $\tau > n - 1$, it is *not true* that almost all tori survive a given perturbation f_ϵ, no matter how small ϵ. The reason is that the parameter α in the nonresonance condition limits the size of the perturbation through the condition

$$\epsilon \ll \alpha^2.$$

Conversely, under a given small perturbation of size ϵ, only those Kronecker tori with frequencies ω in Δ_α with

$$\alpha \gg \sqrt{\epsilon},$$

do survive. Thus, we can not allow α to vary, but have to *fix* it in advance.

To state the KAM theorem, we therefore single out the subsets

$$\Omega_\alpha \subset \Omega, \qquad \alpha > 0,$$

whose frequencies belong to Δ_α and also have at least distance α to the boundary of Ω. These, like Δ_α, are *Cantor sets*: they are closed, perfect and *nowhere dense*, hence of first Baire category. But they also have *large Lebesgue measure*:

$$m(\Omega - \Omega_\alpha) = O(\alpha),$$

provided the boundary of Ω is piecewise smooth, or at least of dimension $n-1$ so that the measure of a boundary layer of size α is $O(\alpha)$.

The main theorem of Kolmogorov, Arnold and Moser can now be stated as follows.

THE CLASSICAL KAM THEOREM [**1,8,10**]. *Suppose the integrable hamiltonian h is nondegenerate, the frequency map h_p is a diffeomorphism $D \to \Omega$, and $H = h + f_\epsilon$ is real analytic on $\bar{D} \times \mathbb{T}^n$. Then there exists a constant $\delta > 0$ such that for*

$$|\epsilon| < \delta\alpha^2$$

all Kronecker tori $(\mathbb{T}^n, \omega)$ of the unperturbed system with $\omega \in \Omega_\alpha$ persist as Lagrangian tori, being only slightly deformed. Moreover, they depend in a Lipschitz continuous way on ω and fill the phase $D \times \mathbb{T}^n$ up to a set of measure $O(\alpha)$.

Here, "real analytic on $\bar{D} \times \mathbb{T}^n$" means that the analyticity extends to a uniform neighborhood of D.

It is an immediate and important consequence of the KAM theorem that small perturbations of nondegenerate hamiltonians are *not ergodic*, as the Kronecker tori form an invariant set, which is neither of full nor of zero measure. Thus the ergodic hypothesis of the twenties was wrong.

It has to be stated again, however, that this invariant set, although of large measure, is a *Cantor set* and thus has no interior points. It is therefore impossible to tell with finite precision whether a given initial position falls onto an invariant torus or into a gap between such tori. From a physical point of view the KAM theorem rather makes a *probabilistic* statement: with probability $1-O(\alpha)$ a randomly chosen orbit lies on an invariant torus and is thus perpetually stable.

f. We conclude with some remarks about the necessity of the assumptions of the KAM theorem.

First, neither the perturbation nor the integrable hamiltonian need to be real analytic. It suffices that they are differentiable of class C^l with

$$l > 2\tau + 2 > 2n$$

to prove the persistence of individual tori [**12,15,20**]. For their Lipschitz dependence some more regularity is required [**16**].

The nondegeneracy condition may also be relaxed. It is not necessary that the frequency map is open. Roughly speaking, it suffices that the intersection of its range with any hyperplane in $\mathbb{R}^n$ has measure zero. For example, if it happens that

$$h_p(p) = \big(\omega_1(p_1), \ldots, \omega_n(p_1)\big)$$

is a function of p_1 only (and thus completely degenerate), it suffices to require that

$$\det\left(\frac{\partial^j \omega_i}{\partial p_1^j}\right)_{1\le i,j\le n} \neq 0.$$

For a more general statement see [**18**], and [**19,22,26**] for proofs.

Finally, the hamiltonian nature of the equations is almost indispensable. Analogous result are true for reversible systems [**13,16**]. But in any event the system has to be *conservative*. Any kind of dissipation immediately destroys the Cantor family of tori, although isolated ones may persist as attractors.

2. The KAM Theorem with parameters

a. Instead of proving the classical KAM theorem directly, we are going to deduce it from another KAM theorem, which is concerned with perturbations of a family of linear hamiltonians. This is accomplished by introducing the frequencies of the Kronecker tori as independent parameters. This approach was first taken in [**11**].

To this end we write $p = p_0 + I$ and expand h around p_0 so that

$$h(p) = h(p_0) + \langle h_p(p_0), I\rangle + \int_0^1 (1-t)\,\langle h_{pp}(p_t)I, I\rangle\, dt,$$

where $p_t = p_0 + tI$. By assumption, the frequency map is a *diffeomorphism*

$$h_p\colon\ D \to \Omega, \quad p_0 \mapsto \omega = h_p(p_0).$$

Hence, instead of $p_0 \in D$ we may introduce the frequencies $\omega \in \Omega$ as independent parameters, determining p_0 uniquely. – Incidentally, the inverse map is given as

$$g_\omega\colon\ \Omega \to D, \quad \omega \mapsto p_0 = g_\omega(\omega),$$

where g is the *Legendre transform* of h, defined by $g(\omega) = \sup_p(\langle p, \omega\rangle - h(p))$. See [**2**] for more details on Legendre transforms.

Thus we write $h(p) = e(\omega) + \langle\omega, I\rangle + P_h(I;\omega)$ and $f_\epsilon(p,q) = P_\epsilon(I,q;\omega)$ with

$$P_h = \int_0^1 (1-t)\,\langle h_{pp}(p_t)I, I\rangle\, dt, \quad P_\epsilon = \epsilon f_*(p_0 + I, q, \epsilon).$$

Writing also θ instead of q for the angular variables, we obtain the family of hamiltonians $H = N + P$ with

$$N = e(\omega) + \langle\omega, I\rangle\,, \quad P = P_h(I;\omega) + P_\epsilon(I,\theta;\omega).$$

They are real analytic in the coordinates (I, θ) in $B \times \mathbb{T}^n$, B some sufficiently small ball around the origin in $\mathbb{R}^n$, as well as in the parameters ω in some *uniform* neighborhood of Ω.

This family is our new starting point. For $P = 0$ it reduces to the *normal form* $N = e(\omega) + \langle \omega, I \rangle$. There is an invariant Kronecker torus $\mathcal{T}_\omega = \{0\} \times \mathbb{T}^n$ with constant vectorfield

$$X_N = \sum_j \omega_j \frac{\partial}{\partial \theta_j}$$

for each $\omega \in \Omega$, and all these tori are given by the family

$$\Phi_0\colon \ \mathbb{T}^n \times \Omega \to B \times \mathbb{T}^n, \quad (\theta, \omega) \mapsto (0, \theta)$$

of trivial embeddings of $\mathbb{T}^n$ over Ω into phase space. Moreover, each such torus clearly is Lagrangian. Our aim is to prove the persistence of a 'subfamily' of such Lagrangian Kronecker tori under sufficiently small perturbations $P \neq 0$ over the Cantor set $\Omega_\alpha \subset \Omega$ of frequency parameters ω. – Thus, instead of proving the existence of a Cantor family of invariant tori in *one* hamiltonian system, we first prove the existence of *one* invariant torus within a Cantor family of hamiltonian systems.

This change of perspective has several advantages. — The unperturbed hamiltonian is as simple as possible, namely linear. This simplifies the KAM proof. — The frequencies are separated from the actions. This makes their rôle more transparent. For example, the Lipschitz dependence of the tori on ω is easily established. — Generalizations such as weaker nondegeneracy conditions and extension to infinite dimensional systems are easier. Also, this approach lends itself to applications in bifurcation theory, where systems naturally depend on parameters.

b. To state the basic result quantitatively we need to introduce a few notations. Let

$$D_{r,s} = \{I : |I| < r\} \times \{\theta : |\mathrm{Im}\, \theta| < s\} \subset \mathbb{C}^n \times \mathbb{C}^n$$

and

$$O_h = \{\omega : |\omega - \Omega_\alpha| < h\} \subset \mathbb{C}^n$$

denote complex neighborhoods of the torus $\{0\} \times \mathbb{T}^n$ and Ω_α, respectively, where $|\cdot|$ stands for the sup-norm of real vectors. The sup-norm of functions on $D_{r,s} \times O_h$ is denoted by $|\cdot|_{r,s,h}$.

We will also consider the Lipschitz constants of mappings with respect to ω. We define

$$|\varphi|_{\mathrm{L}} = \sup_{v \neq \omega} \frac{|\varphi(v) - \varphi(\omega)|}{|v - \omega|},$$

where the underlying domain will be clear from the context or indicated by a subscript.

THEOREM A. *Let $H = N + P$. Suppose P is real analytic on $D_{r,s} \times O_h$ with*

$$|P|_{r,s,h} \leq \gamma \alpha r s^\nu, \qquad \alpha s^\nu \leq h,$$

where $\nu = \tau + 1$ and γ is a small constant depending only on n and τ. Suppose also that $r, s, h \leq 1$. Then there exists a Lipschitz continuous map $\varphi\colon\ \Omega_\alpha \to \Omega$ close *to the identity and a Lipschitz continuous family of real analytic torus embeddings* $\Phi\colon\ \mathbb{T}^n \times \Omega_\alpha \to B \times \mathbb{T}^n$ *close to* Φ_0, *such that for each* $\omega \in \Omega_\alpha$ *the embedded tori are Lagrangian and*

$$X_H|_{\varphi(\omega)} \circ \Phi = D\Phi \cdot X_N.$$

Moreover, Φ *is real analytic on* $T_* = \{\theta : |\mathrm{Im}\,\theta| < s/2\}$ *for each* ω, *and*

$$|W(\Phi - \Phi_0)|\,, \alpha s^\nu\, |W(\Phi - \Phi_0)|_{\mathrm{L}} \le \frac{c}{\alpha r s^\nu}\, |P|_{r,s,h}\,,$$
$$|\varphi - id|\,, \alpha s^\nu\, |\varphi - id|_{\mathrm{L}} \le \frac{c}{r}\, |P|_{r,s,h}\,,$$

uniformly on $T_* \times \Omega_\alpha$ *and* Ω_α, *respectively, where* c *is a large constant depending only on* n *and* τ, *and* $W = \mathrm{diag}(r^{-1}Id, s^{-1}Id)$.

By slight abuse of notation we wrote X_N also for the analogous constant vectorfield on $\mathbb{T}^n$ alone. – The theorem states that for each ω in Ω_α there is an embedded invariant Kronecker torus $\mathcal{T}_\omega = \Phi(\mathbb{T}^n;\omega)$ with frequencies ω for the hamiltonian vectorfield X_H at the slightly *shifted* parameter value $\tilde{\omega} = \varphi(\omega)$. Conversely, for each $\tilde{\omega}$ in the slightly deformed Cantor set

$$\tilde{\Omega}_\alpha = \varphi(\Omega_\alpha) \subset \Omega,$$

the vectorfield $X_H|_{\tilde{\omega}}$ admits an invariant Kronecker torus $\mathcal{T}_\omega$ with frequencies $\omega = \varphi^{-1}(\tilde{\omega})$. Each such torus is Lagrangian and close to the corresponding unperturbed torus.

c. The Lipschitz estimates allow us to control the measure of $\tilde{\Omega}_\alpha$ and its complement. To this end we first extend φ to a lipeomorphism $\bar{\varphi}$ of Ω.

PROPOSITION 1. *There exists an extension of* φ *to a lipeomorphism* $\bar{\varphi}\colon \Omega \to \Omega$ *with*

$$|\bar{\varphi} - id|_{\mathrm{L},\Omega} \le \max\left(|\varphi - id|_{\mathrm{L},\Omega_\alpha}\,, \alpha^{-1}\,|\varphi - id|_{\Omega_\alpha}\right).$$

PROOF. Let ψ be any coordinate function of $\varphi - id$ defined on Ω_α. Define ψ on $\Omega^c = \mathbb{R}^n - \Omega$ to be zero. Then

$$|\psi|_{\mathrm{L},\Omega_\alpha \cup \Omega^c} \le \max\left(|\varphi - id|_{\mathrm{L},\Omega_\alpha}\,, \alpha^{-1}\,|\varphi - id|_{\Omega_\alpha}\right) < 1.$$

According to Appendix B we can extend ψ to a function $\bar{\psi}$ on all of $\mathbb{R}^n$ preserving its Lipschitz constant . Doing this with every coordinate of φ we obtain an extension $\bar{\varphi}$ of φ such that $\bar{\varphi} = id$ on Ω^c and

$$|\bar{\varphi} - id|_{\mathrm{L},\mathbb{R}^n} = |\psi|_{\mathrm{L},\Omega_\alpha \cup \Omega^c} < 1.$$

Hence $\bar{\varphi}$ is a lipeomorphism on $\mathbb{R}^n$. Since it is the identity outside of Ω, it is also a lipeomorphism of Ω, extending φ. □

PROPOSITION 2. *For* $\tilde{\Omega}_\alpha = \varphi(\Omega_\alpha)$ *the estimate*

$$m(\Omega - \tilde{\Omega}_\alpha) = O(\alpha)$$

holds, where the implicit constant depends only on Ω.

PROOF. Let $\bar{\varphi}$ be the extension of Proposition 1. Then

$$\begin{aligned} m(\Omega - \tilde{\Omega}_\alpha) &= m(\Omega - \varphi(\Omega_\alpha)) \\ &= m(\Omega - \bar{\varphi}(\Omega_\alpha)) \\ &= m(\bar{\varphi}(\Omega - \Omega_\alpha)) \\ &\leq |\bar{\varphi}|_{\mathrm{L},\Omega}\, m(\Omega - \Omega_\alpha) \\ &= O(\alpha). \quad \square \end{aligned}$$

The Cantor family of torus embeddings may also be extended to a family over Ω. This extension can even be chosen so that the additional embedded tori are still Lagrangian, though of course not invariant.

PROPOSITION 3. *There exists an extension of Φ to a Lipschitz continuous family of real analytic torus embeddings $\bar{\Phi}$: $\mathbb{T}^n \times \Omega \to B \times \mathbb{T}^n$ such that each embedded torus is Lagrangian, and the estimates for $\bar{\Phi}$ are the same as for Φ, though with a different constant c.*

d. We thus arrive at the following conclusion.

THEOREM B. *Suppose the assumptions of Theorem A are satisfied. Then there exist a lipeomorphism $\bar{\varphi}: \Omega \to \Omega$ close to the identity and a family of torus embeddings*

$$\bar{\Phi}\colon\ \mathbb{T}^n \times \Omega \to \mathbb{R}^n \times \mathbb{T}^n$$

close to Φ_0 such that for every parameter value

$$\tilde{\omega} \in \tilde{\Omega}_\alpha = \bar{\varphi}(\Omega_\alpha)$$

the hamiltonian vectorfield $X_H|_{\tilde{\omega}}$ admits an invariant Lagrangian Kronecker torus $\mathcal{T}_\omega = \Phi(\mathbb{T}^n; \omega)$, where $\omega = \bar{\varphi}^{-1}(\tilde{\omega})$. Moreover, the estimates for $\bar{\varphi}$ and $\bar{\Phi}$ are the same as for φ and Φ in Theorem A, though with a different constant c, and

$$m(\Omega - \tilde{\Omega}_\alpha) = O(\alpha),$$

where the implicit constant depends only on Ω.

We will see at the end of section 5 that the map φ actually can be assigned ω-derivatives of every order on the Cantor set Ω_α. This may be formalized by introducing the intrinsically defined notion of a differentiable function on an arbitrary closed set [**24,25**]. The point is that – due to the Whitney extension theorem – such functions can be extended to functions on the whole space with the same differentiability properties. The upshot is that there even exists an extension of φ to a C^∞-function $\bar{\varphi}$ on Ω. The same applies to Φ and leads to the notion of smooth foliations of invariant tori over Cantor sets [**16**].

e. We now prove the classical KAM theorem. Introducing the frequencies as parameters we wrote the hamiltonian as $H = N + P$, where

$$P = P_h + P_\epsilon$$

is real analytic on $B \times \mathbb{T}^n \times \bar{\Omega}$, B some small ball around the origin in $\mathbb{R}^n$. Thus we can fix some small h and s, with $s^\nu \leq h$, so that P is real analytic on the complex domain $D_{r,s} \times O_h$ for all small r, and so that

$$|P|_{r,s,h} \leq |P_h|_{r,s,h} + |P_{f_\epsilon}|_{r,s,h} \leq Mr^2 + F\epsilon,$$

where M is a bound on the hessian of h and $F = \sup_{p,q,\epsilon} |f_*(p,q,\epsilon)|$.

To meet the smallness condition of Theorems A and B, we choose r by $Mr^2 = F\epsilon$ and arrive at the condition

$$2F\epsilon \leq \gamma\alpha r s^\nu = \gamma\alpha s^\nu \sqrt{\frac{F\epsilon}{M}},$$

or

$$\epsilon \leq \frac{\gamma^2\alpha^2 s^{2\nu}}{4FM} = \delta\alpha^2, \qquad \delta = \frac{\gamma^2 s^{2\nu}}{4FM}.$$

So there is a δ depending on n, τ and H such that Theorems A and B apply for $\epsilon \leq \delta\alpha^2$.

By construction, an orbit $(I(t), \theta(t))$ for the hamiltonian H at the parameter value $\tilde{\omega}$ translates into an orbit $(p(t), q(t)) = (p_0(\tilde{\omega}) + I(t), \theta(t))$ for this hamiltonian in p, q-coordinates. It therefore follows with Theorem B that the mapping

$$\Psi\colon\ \mathbb{T}^n \times \Omega \to D \times \mathbb{T}^n,$$

which is a composition of $\bar{\Phi}$ and $\bar{\varphi}$ with

$$\Xi\colon\ B \times \mathbb{T}^n \times \Omega \to D \times \mathbb{T}^n, \qquad (I, \theta; \omega) \mapsto (h_p^{-1}(\omega) + I, \theta),$$

is an embedding of an invariant Lagrangian Kronecker torus $(\mathbb{T}^n, \omega)$ for every frequency $\omega \in \Omega_\alpha$. Moreover, Ψ is *Lipschitz close* to the real analytic unperturbed embedding

$$\Psi_0\colon\ \mathbb{T}^n \times \Omega \to D \times \mathbb{T}^n, \qquad (\theta; \omega) \mapsto (h_p^{-1}(\omega), \theta).$$

It follows that the measure of the complement of all those tori in the phase space is bounded by a constant times the measure of $\Omega_\alpha \times \mathbb{T}^n$, hence is $O(\alpha)$. This finishes the proof of the classical KAM theorem.

3. Outline of the proof of Theorem A

a. We prove Theorem A by a rapidly converging iteration procedure that was proposed by Kolmogorov [**8**]. At each step of this scheme a hamiltonian

$$H_j = N_j + P_j$$

is considered, which is a small perturbation of $N_j = e_j + \langle \omega, I \rangle$. A coordinate and parameter transformation $\mathcal{F}_j$ is constructed such that

$$H_j \circ \mathcal{F}_j = N_{j+1} + P_{j+1}$$

with another normal form N_{j+1} and a much smaller error term P_{j+1}. Namely,

$$|P_{j+1}| \leq C\, |P_j|^\kappa$$

with some exponent $\kappa > 1$. Repetition of this process leads to a sequence of transformations $\mathcal{F}_0, \mathcal{F}_1, \ldots$, whose products $\mathcal{F}^j = \mathcal{F}_0 \circ \mathcal{F}_1 \circ \cdots \circ \mathcal{F}_{j-1}$ converge to an embedding of an invariant Kronecker torus.

In the meantime a number of other proofs have been given, for example by formulating some generalized implicit function theorem suited for small divisor problems [**28**], or by referring to an implicit function theorem in tame Frechet spaces [**4**]. Recently, Salamon and Zehnder [**21**] gave a proof that avoids coordinate transformations altogether and works in configuration space. Also, Eliasson [**5**] described a way of using power series expansions and majorant techniques in a very tricky way.

But here we stick to the traditional method of proof, as it probably is the most transparent way to get to know the basic techniques. They are indeed quite flexible and robust, and not at all restricted to perturbations of integrable hamiltonian systems. As we mentioned in the beginning, these techniques rather amount to a *strategy* of how to approach a large class of perturbation problems.

b. To describe one cycle of this iterative scheme in more detail we now drop the subscript j.

First, the perturbation P is approximated by some hamiltonian R by truncating its Taylor series in I at first order and its Fourier series in θ at some suitable high order K. The approximation error $P - R$ will be small, and we now consider the hamiltonian $\bar{H} = N + R$ instead of $H = N + P$. The purpose of this approximation will become clear later.

The transformation $\mathcal{F}$ consists of a parameter dependent symplectic change of coordinates Φ and a change φ of the parameters alone:

$$\mathcal{F} = (\Phi, \varphi)\colon\ (I, \theta; \omega) \mapsto (\Phi(I, \theta; \omega), \varphi(\omega)).$$

Moreover, this coordinate transformation is of the form

$$\Phi\colon\ (I, \theta; \omega) \to (U(I, \theta; \omega), V(\theta; \omega)),$$

where U is *affine in* I, and V is independent of I. Such transformations $\mathcal{F}$ form a group $\mathcal{G}$ under composition.

We obtain Φ as the time-1-map of the flow X_F^t of a hamiltonian F, which is affine in I. That is, Φ is the time-1-shift determined by the equations of motion

$$\dot{I} = -F_\theta(I, \theta; \omega), \quad \dot{\theta} = F_I(\theta; \omega).$$

Then Φ is symplectic for each ω.

To describe the transformed hamiltonian $H \circ \Phi$ we recall that for a function K,

$$\frac{d}{dt} K \circ X_F^t = \{K, F\} \circ X_F^t,$$

the *Poisson bracket* of K and F evaluated at X_F^t. Indeed,

$$\begin{aligned} \frac{d}{dt} K \circ X_F^t \Big|_{t=0} &= \sum_{1 \le j \le n} K_{\theta_j} \dot{\theta}_j + K_{I_j} \dot{I}_j \\ &= \sum_{1 \le j \le n} K_{\theta_j} F_{I_j} - K_{I_j} F_{\theta_j} = \{K, F\}, \end{aligned}$$

and the general formula follows.

So we can use Taylor's formula to expand $\bar{H}\circ\Phi = \bar{H}\circ X_F^t\big|_{t=1}$ with respect to t at 0 and write

$$\begin{aligned}\bar{H}\circ\Phi &= N\circ X_F^t\big|_{t=1} + R\circ X_F^t\big|_{t=1}\\ &= N + \{N,F\} + \int_0^1 (1-t)\,\{\{N,F\},F\}\circ X_F^t\,dt\\ &\quad + R + \int_0^1 \{R,F\}\circ X_F^t\,dt\\ &= N + \{N,F\} + R + \int_0^1 \{(1-t)\,\{N,F\}+R,F\}\circ X_F^t\,dt.\end{aligned}$$

This is a *linear* expression in R and F – the *linearization* of $\bar{H}\circ\Phi$ – plus a *quadratic* integral remainder. That is, if R and F are both roughly of order ϵ, then the integral will roughly be of order $\epsilon^2 \ll \epsilon$ and may be ascribed to the next perturbation $P_+ = P_{j+1}$.

The point is to find F such that

$$N + \{N,F\} + R = N_+$$

is again a normal form. Equivalently, we want to solve

$$\{F,N\} + \hat{N} = R, \qquad \hat{N} = N_+ - N \tag{3}$$

for F and $\hat{N}$ when R is given. Suppose for a moment that such a solution exists. Then $(1-t)\,\{N,F\}+R = (1-t)\hat{N}+tR$, and altogether we obtain

$$H\circ\Phi = \bar{H}\circ\Phi + (P-R)\circ\Phi = N_+ + P_+$$

with $N_+ = N + \hat{N}$ and

$$P_+ = \int_0^1 \{(1-t)\hat{N}+tR,F\}\circ X_F^t\,dt + (P-R)\circ X_F^1 \tag{4}$$

as the new error term.

c. Let us consider equation (3) first on a formal level. Clearly,

$$\partial_\omega F \overset{\text{def}}{=} \{F,N\} = \sum_j F_{\theta_j} N_{I_j} = \sum_j \omega_j F_{\theta_j}$$

is a first order partial differential operator on the torus $\mathbb{T}^n$ with constant coefficients ω. Expanding F into a Fourier series,

$$F = \sum_{k\in\mathbb{Z}^n} F_k e^{i\langle k,\theta\rangle},$$

with coefficients depending on I and ω, we find

$$\partial_\omega F = \sum_{k\in\mathbb{Z}^n} i\,\langle k,\omega\rangle\, F_k e^{i\langle k,\theta\rangle}.$$

Thus, ∂_ω admits a basis of eigenfunctions $e^{i\langle k,\theta\rangle}$ with eigenvalues $i\langle k,\omega\rangle$, $k\in\mathbb{Z}^n$. In other words, ∂_ω diagonalizes with respect to this basis.

If ω is now *nonresonant*, then these eigenvalues are all different from zero except when $k=0$. We then can solve for all Fourier coefficients R_k of the given function R, except for R_0, which is given by the mean value of R over $\mathbb{T}^n$,

$$R_0=[R]=\frac{1}{\operatorname{vol}\mathbb{T}^n}\int_{\mathbb{T}^n}R\,d\theta.$$

Hence, if R is given, then we can always solve the equation $\partial_\omega F=R-[R]$ formally by setting

$$F=\sum_{0\neq k\in\mathbb{Z}^n}\frac{R_k}{i\langle k,\omega\rangle}e^{i\langle k,\theta\rangle}. \tag{5}$$

We are still free to add a θ-independent function to F, but we choose to normalize F so that $[F]=0$.

Finally, equation (3) is completely solved by setting

$$\hat{N}=[R].$$

Of course, this choice of $\hat{N}$ is in no way uniquely determined, but this is in some sense the simplest one.

d. There is a more systematic interpretation of the preceding construction. For irrational ω, the domain of ∂_ω, consisting of all formal Fourier series in θ (ignoring the other coordinates here), *splits* into two *invariant subspaces*, its nullspace $\mathcal{N}$ consisting of all constant functions, and its range $\mathcal{R}$, consisting of all series with vanishing constant term. Moreover, ∂_ω is invertible on $\mathcal{R}$.

Decompose R into its respective components in $\mathcal{N}$ and $\mathcal{R}$,

$$R=R_{\mathcal{N}}+R_{\mathcal{R}}.$$

The projection onto $\mathcal{N}$ is given by taking the mean value, so

$$R_{\mathcal{N}}=[R],\qquad R_{\mathcal{R}}=R-[R].$$

The equation

$$\partial_\omega F+\hat{N}=R=R_{\mathcal{R}}+R_{\mathcal{N}}$$

is then solved by "solving componentwise",

$$\hat{N}=R_{\mathcal{N}}=[R],\qquad \partial_\omega F=R_{\mathcal{R}}=R-[R],$$

where the latter can be solved uniquely for F in $\mathcal{R}$, since ∂_ω is invertible on $\mathcal{R}$.

This general procedure – "solve for all the terms you can solve for, and keep the rest" – is at the basis of all normal form theory. It just happens to take a particularly simple form in our case.

e. So far our considerations were formal. But in estimating the series representation (5) of F, we are confronted with the well known and notorious problem of "small divisors". Even if ω is nonresonant, infinitely many of the divisors $\langle k, \omega\rangle$ become arbitrarily small in view of (2), threatening to make the series (5) divergent.

This divergence is avoided, if ω is required to be strongly nonresonant. To formulate this key lemma, let $\mathcal{A}^s$ denote the space of all analytic functions u defined in the complex strip $\{\theta : \sup_j |\operatorname{Im}\theta_j| < s\} \subset \mathbb{C}^n$ with bounded sup-norm $|u|_s$ over that strip. Let

$$\mathcal{A}_0^s = \{\, u \in \mathcal{A}^s : [u] = 0\,\},$$

and recall that $\omega \in \Delta_a^\tau$ satisfies $|\langle k, \omega\rangle| \geq \alpha / |k|^\tau$ for all $0 \neq k \in \mathbb{Z}^n$.

LEMMA 1. *Suppose that* $\omega \in \Delta_\alpha^\tau$. *Then the equation*

$$\partial_\omega u = v, \qquad v \in \mathcal{A}_0^s,$$

has a unique solution u *in* $\bigcup_{0<\sigma<s} \mathcal{A}_0^{s-\sigma}$, *with*

$$|u|_{s-\sigma} \leq \frac{c}{\alpha\sigma^\tau} |v|_s\,, \qquad 0 < \sigma < s,$$

where the constant c *depends only on* n *and* τ.

PROOF. We prove the lemma with $\sigma^{\tau+n}$ in place of σ^τ to avoid lengthy technicalities. The interested reader is referred to **[17]** or **[15]** for a proof of the sharper result. – Expanding u and v into Fourier series, the unique formal solution u with $[u] = 0$ is

$$u = \sum_{0 \neq k \in \mathbb{Z}^n} \frac{v_k}{i\,\langle k, \omega\rangle}\, e^{i\langle k,\theta\rangle}.$$

As to the estimate we recall that the Fourier coefficients of an analytic function on $\mathbb{T}^n$ decay exponentially fast:

$$|v_k| \leq |v|_s\, e^{-|k|s},$$

where $|k| = |k_1| + \dots |k_n|$. See Lemma A.1 for a reminder. Together with the small divisor estimate for ω we obtain

$$|u|_{s-\sigma} \leq \sum_{k\neq 0} \frac{|v_k|}{|\langle k, \omega\rangle|} e^{|k|(s-\sigma)} \leq \frac{|v|_s}{\alpha} \sum_{k \neq 0} |k|^\tau\, e^{-|k|\sigma}.$$

The infinite sum is now easily estimated by constant times $\sigma^{-\tau-n}$. □

We observe that ∂_ω^{-1} is *unbounded* as an operator in $\mathcal{A}_0^s$. It is *bounded* only as an operator from $\mathcal{A}_0^s$ into the larger spaces $\mathcal{A}_0^{s-\sigma}$, with its bound tending to infinity as σ tends to zero. This phenomenon is known as "loss of smoothness" affected by the solution operator ∂_ω^{-1}, and is the main culprit why small divisor problems are technically so involved. During the iteration we have to let $\sigma \to 0$ in order to stay in the classes $\mathcal{A}_0^s$. But then the operator ∂_ω^{-1} is getting unbounded. By the rapid convergence of the Newton scheme, however, the error term converges to zero even faster, thus allowing to overcome this effect of the small divisors.

It is absolutely essential for Lemma 1 to be true that ω satisfies infinitely many small divisor conditions, thus restricting ω to a *Cantor* set with no interior points. On the other hand, we will also need to transform the frequencies and thus want them to live in *open* domains. This conflict is resolved by approximating P by a trigonometric polynomial R. Then only finitely many Fourier coefficients need to be considered at each step, and only *finitely many* small divisor conditions need to be required, which are easily satisfied on some open ω-domain. Of course, during the iteration more and more conditions have to be satisfied, and in the end these domains will shrink to some Cantor set.

f. We still have to finish one cycle of the iteration. Solving (3) we arrive at $H \circ \Phi = N_+ + P_+$, where

$$N_+ = N + [R] = e_+(\omega) + \langle \omega + v(\omega), I \rangle ,$$

since $[R]$ is affine in I and independent of θ. To write N_+ again in normal form, we have to introduce

$$\omega_+ = \omega + v(\omega)$$

as new frequencies. Since v is small, there exists an inverse map $\varphi \colon \omega_+ \mapsto \omega$ by the implicit function theorem – see Appendix A. With this change of parameters,

$$N_+ = e_+(\omega_+) + \langle \omega_+, I \rangle$$

is again in normal form. This finishes one cycle of the iteration.

g. The next section describes the quantitative details, and the final section its iteration.

4. The KAM Step

a. To avoid a flood of constants we will write

$$u \lessdot v, \qquad u \cdot\!< v,$$

if there exists a positive constant $c \geq 1$, which depends *only on n and τ* and could be made explicit, such that $u \leq cv$ and $cu \leq v$, respectively. – Now let P be a real analytic perturbation of some normal form N.

THE KAM STEP. *Suppose that $|P|_{r,s,h} \leq \epsilon$ with*

$$\text{(a)} \qquad \epsilon \cdot\!< \alpha \eta r \sigma^{\nu},$$

$$\text{(b)} \qquad \epsilon \cdot\!< hr,$$

$$\text{(c)} \qquad h \leq \frac{\alpha}{2K^{\nu}},$$

for some $0 < \eta < 1/8$, $0 < \sigma < s/5$ and $K \geq 1$, where $\nu = \tau + 1$. Then there exists a real analytic transformation

$$\mathcal{F} = (\Phi, \varphi) \colon \; D_{\eta r, s-5\sigma} \times O_{h/4} \to D_{r,s} \times O_h$$

in the group $\mathcal{G}$ such that $H \circ \mathcal{F} = N_+ + P_+$ with

$$|P_+|_{\eta r, s-5\sigma, h/4} \lessdot \frac{\epsilon^2}{\alpha r \sigma^{\nu}} + \left(\eta^2 + K^n e^{-K\sigma} \right) \epsilon.$$

Moreover,

$$\begin{aligned} |W(\Phi - id)|\,, \left|W(D\Phi - Id)W^{-1}\right| &\lessdot \frac{\epsilon}{\alpha r \sigma^\nu}\,, \\ |\varphi - id|\,, h\,|D\varphi - Id| &\lessdot \frac{\epsilon}{r}\,, \end{aligned}$$

uniformly on $D_{\eta r, s-5\sigma} \times O_h$ *and* $O_{h/4}$*, respectively, with the weight matrix* $W = \operatorname{diag}(r^{-1}Id, \sigma^{-1}Id)$.

b. The proof of the KAM Step follows the lines of the preceding section and consists of six small steps. Except for the last step everything is uniform in O_h, whence we write $|\cdot|_{r,s}$ for $|\cdot|_{r,s,h}$ throughout.

1. Truncation. We approximate P by a hamiltonian R, which is affine in I and a trigonometric polynomial in θ. To this end, let Q be the linearization of P in I at $I = 0$. By Taylor's formula with remainder and Cauchy's estimate – see Appendix A for a reminder –, we have

$$|Q|_{r,s} \lessdot \epsilon, \qquad |P - Q|_{2\eta r, s} \lessdot \eta^2 \epsilon.$$

Then we simply truncate the Fourier series of Q at order K to obtain R. By Lemma A.2,

$$|R - Q|_{r,s-\sigma} \lessdot K^n e^{-K\sigma} \epsilon.$$

Since the factor $K^n e^{-K\sigma}$ will be made small later on, we also have

$$|R|_{r,s-\sigma} \lessdot \epsilon.$$

See Appendix A for some remarks about this truncation of Fourier series.

2. Extending the small divisor estimate. The nonresonance conditions (6) are assumed to hold on the real set Ω_α only. But assumption (c) implies that

$$|\langle k, \omega \rangle| \geq \frac{\alpha}{2\,|k|^\tau} \qquad \text{for all } 0 \neq |k| \leq K \tag{7}$$

for all ω in the neighborhood O_h of Ω_α. Indeed, for $\omega \in O_h$ there is some $\omega_* \in \Omega_\alpha$ with $|\omega - \omega_*| < h$, hence

$$|\langle k, \omega - \omega_* \rangle| \leq |k| \cdot |\omega - \omega_*| \leq Kh \leq \frac{\alpha}{2K^\tau} \leq \frac{\alpha}{2\,|k|^\tau}$$

for $|k| \leq K$. Together with the estimate (6) for $\langle k, \omega_* \rangle$ this proves the claim.

3. Solving the linearized equation $\{F, N\} + \hat{N} = R$. We solve this equation as described in the preceding section. We have $\hat{N} = [R]$ and thus

$$|\hat{N}|_r \leq |R|_{r,s-\sigma} \lessdot \epsilon.$$

We can solve for F uniformly for all ω in O_h because of (7) and the fact that R contains only Fourier coefficients up to order K, by truncation. Hence the estimate of Lemma 1 applies as well, and we obtain a real analytic function F with

$$|F|_{r,s-2\sigma} \lessdot \frac{|R|_{r,s-\sigma}}{\alpha\sigma^\tau} \lessdot \frac{\epsilon}{\alpha\sigma^\tau} .$$

With Cauchy we get $|F_\theta|_{r,s-3\sigma} \lessdot \epsilon/\alpha\sigma^{\tau+1}$ and $|F_I|_{r/2,s-2\sigma} \lessdot \epsilon/\alpha r\sigma^\tau$, hence

$$\frac{1}{r}|F_\theta|, \frac{1}{\sigma}|F_I| \lessdot \frac{\epsilon}{\alpha r\sigma^\nu}$$

uniformly on $D_{r/2,s-3\sigma}$ with $\nu = \tau + 1$.

4. Transforming the coordinates. The coordinate transformation Φ is obtained as the real analytic time-1-map of the flow X_F^t of the hamiltonian vectorfield X_F, with equations of motions

$$\dot{I} = -F_\theta, \qquad \dot{\theta} = F_I.$$

With assumption (a) and the preceding estimates we can assure that we have $|F_\theta| \le \eta r \le r/8$ and $|F_I| \le \sigma$ on $D_{r/2,s-3\sigma}$ uniformly in ω. Therefore, the time-1-map is well defined on $D_{r/4,s-4\sigma}$, with

$$\Phi = X_F^t\big|_{t=1} \colon \; D_{r/4,s-4\sigma} \to D_{r/2,s-3\sigma}, \tag{8}$$

and

$$|U - id| \le |F_\theta| \lessdot \frac{\epsilon}{\alpha\sigma^\nu}, \qquad |V - id| \le |F_I| \lessdot \frac{\epsilon}{\alpha r\sigma^{\nu-1}}$$

on that domain for $\Phi = (U, V)$. The Jacobian of Φ is

$$D\Phi = \begin{pmatrix} U_I & U_\theta \\ 0 & V_\theta \end{pmatrix},$$

since F is *linear* in I, hence F_I and V are *independent* of I. By the preceding estimates and Cauchy,

$$|U_I - Id| \lessdot \frac{\epsilon}{\alpha r\sigma^\nu}, \quad |U_\theta| \lessdot \frac{\epsilon}{\alpha\sigma^{\nu+1}}, \quad |V_\theta - Id| \lessdot \frac{\epsilon}{\alpha r\sigma^\nu}$$

on the domain $D_{r/8,s-5\sigma} \supseteq D_{\eta r,s-5\sigma}$. This proves all the estimates for Φ. Finally, we observe that $|U - id| \le |F_\theta| \le \eta r$ implies that also Φ: $D_{\eta r,s-5\sigma} \to D_{2\eta r,s-4\sigma}$.

5. New error term. To estimate P_+ as given in (4) we first consider the term $\{R, F\}$. Again, by Cauchy's estimate,

$$\begin{aligned} |\{R,F\}|_{r/2,s-3\sigma} &\lessdot |R_I|\,|F_\theta| + |R_\theta|\,|F_I| \\ &\lessdot \frac{\epsilon}{r}\cdot\frac{\epsilon}{\alpha\sigma^\nu} + \frac{\epsilon}{\sigma}\cdot\frac{\epsilon}{\alpha r\sigma^{\nu-1}} \lessdot \frac{\epsilon^2}{\alpha r\sigma^\nu} . \end{aligned}$$

The same holds for $|\{\hat{N}, F\}|_{r/2,s-3\sigma}$. Together with (8) and $\eta < \frac{1}{8}$ we get

$$\left| \int_0^1 \{(1-t)\hat{N} + tR, F\} \circ X_F^t \, dt \right|_{\eta r, s-5\sigma}$$
$$\leq \left| \{(1-t)\hat{N} + tR, F\} \right|_{r/2, s-4\sigma} \lessdot \frac{\epsilon^2}{\alpha r \sigma^\nu} .$$

The other term in (4) is bounded by

$$\begin{aligned} |(P-R)\circ\Phi|_{\eta r,s-5\sigma} &\leq |P-R|_{2\eta r,s-4\sigma} \\ &\leq |P-Q|_{2\eta r,s-4\sigma} + |Q-R|_{2\eta r,s-4\sigma} \\ &\lessdot \left(\eta^2 + K^n e^{-K\sigma}\right)\epsilon . \end{aligned}$$

These two estimates together give the bound for $|P_+|$.

6. Transforming the frequencies. Finally, we have to invert the map

$$\omega \mapsto \omega_+ = \omega + v(\omega), \qquad v = \hat{N}_I = [R_I]$$

to put $N + \hat{N}$ back into a normal form N_+. With assumption (b) and Cauchy's estimate we can assure that

$$|v|_{h/2} = |\hat{N}_I|_{h/2} \lessdot \frac{\epsilon}{r} \leq \frac{h}{4}.$$

The implicit function theorem of Appendix A applies, and there exists a real analytic inverse map $\varphi: O_{h/4} \to O_{h/2}$, $\omega_+ \mapsto \omega$, with the estimates

$$|\varphi - id|\,, h\,|D\varphi - Id| \lessdot \frac{\epsilon}{r}$$

on $O_{h/4}$. Setting $N_+ = (N + \hat{N}) \circ \varphi$ the proof of the KAM Step is completed.

5. Iteration and proof of Theorem A

a. We are going to iterate the KAM Step infinitely often, choosing appropriate sequences for the parameters σ, η and so on. To motivate our choices, let us start by fixing a geometric sequence for σ, say, $\sigma_+ = \sigma/2$, where the plus sign indicates the corresponding parameter value for the next step. Let $r_+ = \eta r$, and let us consider the *weighted error terms*

$$E = \frac{\epsilon}{\alpha r \sigma^\nu}, \qquad E_+ = \frac{\epsilon_+}{\alpha r_+ \sigma_+^\nu} .$$

Then we have

$$E_+ \lessdot \frac{E^2}{\eta} + \left(\eta^2 + K^n e^{-K\sigma}\right)\frac{E}{\eta} .$$

Suppose we can choose η and K so that $\eta^2 = E$ and $K^n e^{-K\sigma} \leq E$. Then

$$E_+ \lessdot \eta^{-1} E^2 = E^\kappa, \qquad \kappa = \frac{3}{2} .$$

That is, $E_+ \le c^{\kappa-1}E^\kappa$ for some constant c determined by the KAM Step and depending only on n and τ. Consequently,

$$cE_+ \le c^\kappa E^\kappa,$$

and this scheme converges exponentially fast for $E < c^{-1}$.

It remains to discuss our assumptions

$$\text{(d)} \qquad \eta^2 = E$$

$$\text{(e)} \qquad K^n e^{-K\sigma} \le E$$

as well as assumptions (a–c) of the KAM Step. There is no obstacle to take (d) as the *definition* of η, as this *implies* $E \cdot\!< \eta$ and thus (a) for E sufficiently small. The other three conditions amount to

$$\frac{\epsilon}{r} \cdot\!< h \le \frac{\alpha}{2K^\nu}, \qquad K^n e^{-K\sigma} \le E = \frac{\epsilon}{\alpha r \sigma^\nu}.$$

They are combined into the sufficient condition

$$\text{(f)} \qquad K^\nu \sigma^\nu e^{-K\sigma} \le E\sigma^\nu = \frac{\epsilon}{\alpha r} \cdot\!< \frac{h}{\alpha} \le \frac{1}{2K^\nu}.$$

Now it suffices to set up geometric sequences for K and h such that, say,

$$K_+\sigma_+ = 2K\sigma, \qquad K_+^\nu h_+ \le K^\nu h.$$

Then these inequalities hold inductively, provided they hold initially, with $K_0\sigma_0$ sufficiently large. In particular, after fixing σ_0, K_0 and E_0, we may set

$$h_0 = \alpha c_0 E_0 \sigma_0^\nu$$

with a suitable constant c_0.

b. We are now ready to set up our parameter sequences. Let

$$\sigma_{j+1} = \frac{\sigma_j}{2}, \quad K_{j+1} = 4K_j, \quad h_{j+1} = \frac{h_j}{4^\nu},$$

where $\sigma_0 = s_0/20$, and K_0 is chosen so large that the left hand side of (f) is smaller than its right hand side, and small enough so that E_0 can be fixed to meet (a) and (d). In addition, $K_0\sigma_0$ has to be so large that the left hand side of (f) decreases at least at an exponential rate κ. Thus we need $1 \cdot\!< K_0\sigma_0$. Subsequently, we fix h_0 as above.

Next, let

$$E_{j+1} = c^{\kappa-1}E_j^\kappa, \quad r_{j+1} = \eta_j r_j, \quad \eta_j^2 = E_j,$$

where r_0 is still free and c given by the KAM Step. Finally, we define $s_{j+1} = s_j - 5\sigma_j$ and the complex domains

$$D_j = \{|I| < r_j\} \times \{|\operatorname{Im}\theta| < s_j\}, \quad O_j = \{|\omega - \Omega_\alpha| < h_j\}.$$

Note that $s_j \to s/2$ and $r_j \to 0$. – Now let $H = N + P_0$.

Iterative Lemma. *Suppose P_0 is real analytic on $D_0 \times O_0$ with*

$$|P_0|_{r_0,s_0,h_0} \le \epsilon_0 \stackrel{\text{def}}{=} \alpha E_0 r_0 \sigma_0^\nu,$$

Then for each $j \ge 0$ there exists a normal form N_j and a real analytic transformation $\mathcal{F}^j = \mathcal{F}_0 \circ \cdots \circ \mathcal{F}_{j-1}$: $D_j \times O_j \to D_0 \times O_0$ in the group $\mathcal{G}$ such that $H \circ \mathcal{F}^j = N_j + P_j$ with

$$|P_j|_{r_j,s_j,h_j} \le \epsilon_j \stackrel{\text{def}}{=} \alpha E_j r_j \sigma_j^\nu.$$

Moreover,

$$\left|\bar{W}_0(\mathcal{F}^{j+1} - \mathcal{F}^j)\right| \lessdot \frac{\epsilon_j}{r_j h_j}$$

on $D_{j+1} \times O_{j+1}$ with the weight matrix $\bar{W}_0 = \operatorname{diag}(r_0^{-1} Id, \sigma_0^{-1} Id, h_0^{-1} Id)$.

c. *Proof.* Letting $\mathcal{F}_0 = id$, there is nothing to do for $j = 0$. To proceed by induction, we have to check the assumptions of the KAM Step for each $j \ge 0$. But (a) is satisfied by the definition of η_j and the sufficiently small choice of E_0, and (b–c) hold by the definition of h_j and K_j and the choice of their initial values.

We obtain a transformation $\mathcal{F}_j$: $D_{j+1} \times O_{j+1} \to D_j \times O_j$ taking $H_j = N_j + P_j$ into $H_j \circ \mathcal{F}_j = N_{j+1} + P_{j+1}$ with

$$\begin{aligned} |P_{j+1}| &\lessdot \epsilon_j E_j + \left(\eta_j^2 + K_j^n e^{-K_j \sigma_j}\right) \epsilon_j \\ &\lessdot \epsilon_j E_j = \alpha E_j^2 r_j \sigma_j^\nu \lessdot \alpha \eta_j^{-1} E_j^2 r_{j+1} \sigma_{j+1}^\nu. \end{aligned}$$

Since $\eta_j^{-1} E_j^2 = E_j^\kappa = c^{1-\kappa} E_{j+1}$, we obtain $|P_{j+1}| \le \epsilon_{j+1}$ by an appropriate choice of c as required. Thus, the transformation $\mathcal{F}^{j+1} = \mathcal{F}^j \circ \mathcal{F}_j = \mathcal{F}_0 \circ \cdots \circ \mathcal{F}_j$ takes H into $N_{j+1} + P_{j+1}$ with the proper estimate for P_{j+1}.

The estimate of $\mathcal{F}^j$ requires a bit more, though elementary work. We observe that the estimates of the KAM Step and Cauchy imply

$$\left|\bar{W}_j(\mathcal{F}_j - id)\right|, \left|\bar{W}_j(\bar{D}\mathcal{F}_j - Id)\bar{W}_j^{-1}\right| \lessdot \max\left(\frac{\epsilon_j}{\alpha r_j \sigma_j^\nu}, \frac{\epsilon_j}{r_j h_j}\right) \lessdot \frac{\epsilon_j}{r_j h_j}$$

on $D_{j+1} \times O_{j+1}$, where $\bar{D}$ denotes the Jacobian with respect to I, θ and ω, and $\bar{W}_j = \operatorname{diag}(r_j^{-1} Id, \sigma_j^{-1} Id, h_j^{-1} Id)$. We then have

$$\begin{aligned} \left|\bar{W}_0(\mathcal{F}^{j+1} - \mathcal{F}^j)\right| &= \left|\bar{W}_0(\mathcal{F}^j \circ \mathcal{F}_j - \mathcal{F}^j)\right| \\ &\le \left|\bar{W}_0 \bar{D}\mathcal{F}^j \bar{W}_j^{-1}\right| \left|\bar{W}_j(\mathcal{F}_j - id)\right| \\ &\lessdot \left|\bar{W}_j(\mathcal{F}_j - id)\right| \\ &\lessdot \frac{\epsilon_j}{r_j h_j} \end{aligned}$$

provided we can uniformly bound the first factor in the second row on $D_j \times O_j$.

But by induction we have $\bar{D}\mathcal{F}^j = \bar{D}\mathcal{F}_0 \circ \cdots \circ \bar{D}\mathcal{F}_{j-1}$, with the Jacobians evaluated at different points, which we do not indicate. Since $\left|\bar{W}_i \bar{W}_{i+1}^{-1}\right| \le 1$ for all i, we can use a telescoping argument and the inductive estimates for the $\mathcal{F}_i$ to obtain

$$\begin{aligned}\left|\bar{W}_0\bar{D}\mathcal{F}^j\bar{W}_j^{-1}\right| &\le \left|\bar{W}_0\bar{D}\mathcal{F}_0\circ\cdots\circ\mathcal{F}_{j-1}\bar{W}_j^{-1}\right| \\ &\le \left|\bar{W}_0\bar{D}\mathcal{F}_0\bar{W}_0^{-1}\right|\left|\bar{W}_0\bar{W}_1^{-1}\right|\times\cdots \\ &\qquad\times\left|\bar{W}_{j-1}\bar{D}\mathcal{F}_{j-1}\bar{W}_{j-1}^{-1}\right|\left|\bar{W}_{j-1}\bar{W}_j^{-1}\right| \\ &\le \prod_j\left(1+\frac{c_1\epsilon_j}{r_jh_j}\right),\end{aligned}$$

which is uniformly bounded and small, since ϵ_j/r_jh_j converges rapidly to zero. □

d. We can now prove Theorem A by applying the Iterative Lemma to $H = N+P$, letting $P_0 = P$ and $r_0 = r$, $s_0 = s$. We have $h_0 \cdot\!< \alpha\sigma_0^\nu \le \alpha s^\nu \le h$ by construction and assumption, and

$$|P_0|_{r_0,s_0,h_0} \le |P|_{r,s,h} \le \epsilon \le \gamma\alpha rs \le \epsilon_0 = \alpha E_0 r_0\sigma_0^\nu,$$

by fixing the constant γ in Theorem A sufficiently small.

By the estimates of the Iterative Lemma the maps $\mathcal{F}^j$ converge uniformly on

$$\bigcap_{j\ge 0} D_j\times O_j = T_*\times\Omega_\alpha, \qquad T_* = \{0\}\times\{|\operatorname{Im}\theta| < s/2\},$$

to a map $\mathcal{F}$ consisting of a family of embeddings $\Phi\colon \mathbb{T}^n\times\Omega_\alpha \to D\times\mathbb{T}^n$ and a parameter transformation $\varphi\colon\Omega_\alpha\to\Omega$, which are real analytic on $\mathbb{T}^n$ and uniformly continuous on Ω_α. Moreover,

$$\left|\bar{W}_0(\mathcal{F}-id)\right| \cdot\!< \frac{\epsilon_0}{r_0h_0}$$

uniformly on $T_*\times\Omega_\alpha$ by the usual telescoping argument.

From the estimate $\left|H\circ\mathcal{F}^j - N_j\right| \cdot\!< \epsilon_j$ on $D_j\times O_j$ we obtain

$$\left|W_j(J(D\Phi^j)^t\nabla H\circ\mathcal{F}^j - J\nabla N_j)\right| \cdot\!< \frac{\epsilon_j}{r_j\sigma_j}$$

on $T_*\times O_j$ with $W_j = \operatorname{diag}(r_j^{-1}Id, \sigma_j^{-1}Id)$. The symplectic nature of the map Φ^j and the uniform estimate of $\bar{W}_0\bar{D}\mathcal{F}^j\bar{W}_j^{-1}$ above then imply

$$\left|X_H\circ\mathcal{F}^j - D\Phi^j\cdot X_N\right| \cdot\!< \frac{\epsilon_j}{r_j\sigma_j}$$

on $T_*\times\Omega_\alpha$ for all j, where X_N is the hamiltonian vectorfield of $N = \langle\omega, I\rangle$. Going to the limit we obtain

$$X_H\circ\mathcal{F} = D\Phi\cdot X_N.$$

Thus, Φ is an embedding of the Kronecker torus $(\mathbb{T}^n,\omega)$ as an invariant torus of the hamiltonian vectorfield X_H at the parameter $\varphi(\omega)$. Moreover, this torus is

Lagrangian, since

$$\Phi^* \upsilon = \lim_{j\to\infty} (\Phi^j)^* \upsilon\big|_{I=0} = \lim_{j\to\infty} \upsilon|_{I=0} = 0$$

by the symplecticity of the Φ^j.

e. Let us now look at the ω-derivatives of the $\mathcal{F}^j$. Since E_j converges to zero at an *exponential* rate, we have

$$\frac{\epsilon_j}{r_j h_j^l} \to 0 \qquad \text{for all } l \geq 0.$$

Hence, *all* ω-derivatives of the $\mathcal{F}^j$ converge uniformly on $T_* \times \Omega_\alpha$, and we could assign ω-derivatives of any order to the limit map $\mathcal{F}$ on the Cantor set Ω_α [**24**]. Without making this concept precise, however, we can at least conclude that $\mathcal{F}$ is *Lipschitz continuous* in ω. Its Lipschitz norm is bounded by the limit of the bounds on the first ω-derivatives of the $\mathcal{F}^j$. The usual Cauchy estimate yields

$$\left|\bar{W}_0(\mathcal{F} - id)\right|_{\mathrm{L}} \lessdot \frac{\epsilon_0}{r_0 h_0^2}$$

on $T_* \times \Omega_\alpha$.

f. We finally look at the estimate of $\mathcal{F}$. So far it does not reflect the actual size ϵ of the perturbation, since we fixed E_0 and thus ϵ_0 independently of ϵ. But we observe that everything is still all right if in all the estimates for P_j, $\mathcal{F}_j$ and $\mathcal{F}^j$, the ϵ_j are scaled down by the linear factor

$$\frac{\epsilon}{\epsilon_0} = \alpha E_0 r_0 \sigma_0^\nu .$$

Scaling down our estimates of $\mathcal{F}$ by this factor we can finally extract our estimates of Φ and φ as stated in Theorem A, since $\alpha \sigma_0^\nu \lessdot h_0$. This finishes the proof.

A. Some facts about analytic functions

a. First we recall a variant of the Cauchy estimate, which is used over and over. Let D be an open domain in $\mathbb{C}^n$, let $D_r = \{\, z : |z - D| < r \,\}$ be the neighborhood of radius r around D, and let F be an analytic function on D_r with bounded sup-norm $|f|_r$. Then

$$\left|f_{z_j}\right|_{r-\rho} \leq \frac{1}{\rho} |f|_r$$

for all $0 < \rho < r$ and $1 \leq j \leq n$. This follows immediately from the Cauchy estimate for one complex variable.

b. Next we give the estimate for the Fourier coefficients of an analytic function v on $\mathbb{T}^n$ used in the proof of Lemma 1. Recall that $\mathcal{A}^s$ denotes the space of all functions on $\mathbb{T}^n$ bounded and analytic in the strip $\{\theta : |\operatorname{Im} \theta| < s\}$.

LEMMA A.1. *If $v \in \mathcal{A}^s$, then $v = \sum_k v_k e^{i\langle k,\theta\rangle}$ with*

$$|v_k| \leq |v|_s \, e^{-|k|s}$$

for all $k \in \mathbb{Z}^n$.

PROOF. The Fourier coefficients v_k of v are given by

$$v_k = \frac{1}{(2\pi)^n} \int_{\mathbb{T}^n} v(\theta) e^{-i\langle k,\theta\rangle}\, d\theta.$$

Since the integral of an analytic function over a closed contractible loop in any of the coordinate planes is zero, and since v is 2π-periodic in each argument also in the complex neighborhood, the path of integration may be shifted into the complex, so that

$$v_k = \frac{1}{(2\pi)^n} \int_{\mathbb{T}^n} v(\theta - i\varphi) e^{-i\langle k,\theta - i\varphi\rangle}\, d\theta$$

for any constant real vector φ with $|\varphi| < s$. Choosing $\varphi = (s-\sigma)(e_1, \dots, e_n)$ with $0 < \sigma < s$ and $e_j = \operatorname{sgn} k_j$, $1 \le j \le n$, we obtain

$$|v_k| \le |v|_s\, e^{-|k|(s-\sigma)}$$

for all $\sigma > 0$. Letting $\sigma \to 0$ the lemma follows. □

We can now also estimate very roughly the remainder, when we truncate the Fourier series of v at order K to obtain $T_K v = \sum_{|k|\le K} v_k e^{i\langle k,\theta\rangle}$.

LEMMA A.2. *If* $v \in \mathcal{A}^s$, *then*

$$|v - T_K v|_{s-\sigma} \le CK^n e^{-K\sigma} |v|_s\,, \qquad 0 \le \sigma \le s,$$

where the constant C *only depends on* n.

PROOF. With Lemma A.1,

$$\begin{aligned} |v - T_K v|_{s-\sigma} &\le \sum_{|k|>K} |v_k|\, e^{|k|(s-\sigma)} \\ &\le |v|_s \sum_{|k|>K} e^{-|k|\sigma} \le |v|_s \sum_{l>K} 4^n l^{n-1} e^{-l\sigma}, \end{aligned}$$

by summing first over all k with $|k| = l$, whose number is bounded by $4^n l^{n-1}$. The last sum is then easily bounded by a constant times $K^n e^{-K\sigma}$. □

There are much more efficient ways to approximate a periodic function v by trigonometric polynomials. The above crude way amounts to multiplying the Fourier transform $\hat{v}$ of v with a *discontinuous* cut off function. Instead, one should multiply $\hat{v}$ with a *smooth* cut off function ψ_K. For instance, one could take $\psi_K(x) = \psi(x/K)$, where ψ is a fixed function, which is 1 on the ball $|x| \le \frac{1}{2}$, 0 outside the ball $|x| \ge 1$, and between 0 and 1 otherwise. Transforming back,

$$(\hat{v}\psi_K)\hat{} = v * \hat{\psi}_K$$

amounts to a convolution of v with a real analytic approximation of the identity $\hat{\psi}_K$, as $K \to \infty$. Such smoothing operators have many interesting properties. For more details, see for example [**28**].

c. We finally formulate a special version of the implicit function theorem for analytic maps, which we need to invert the frequency map during the KAM Step. Recall that O_h is an open complex neighborhood of radius h of some subset Ω of $\mathbb{R}^n$. In the following, $|\cdot|$ denotes the sup-norm for vectors and maps, and the induced operator-norm for Jacobians.

LEMMA A.3. *Suppose f is a real analytic map from O_h into $\mathbb{C}^n$. If $|f - id| \le \delta \le h/4$ on O_h, then f has a real analytic inverse φ on $O_{h/4}$. Moreover,*

$$|\varphi - id|\,, \quad \frac{h}{4}\,|D\varphi - Id| \le \delta$$

on this domain.

PROOF. Let $\eta = h/4$. Let u, v be two points in $O_{2\eta}$ with $f(u) = f(v)$. Then we have

$$u - v = (u - f(u)) - (v - f(v))\,,$$

hence $|u - v| \le 2\delta \le 2\eta$. It follows that the segment $(1-s)u + sv$, $0 \le s \le 1$, is strictly contained in $O_{3\eta}$. Along this segment,

$$\theta = \max |Df - I| < \delta/\eta \le 1$$

by Cauchy's inequality and so

$$|u - v| \le |Df - I|\,|u - v| \le \theta\,|u - v|$$

by the mean value theorem. It follows that $u = v$. Thus, f is one-to-one on the domain $O_{2\eta}$.

By elementary arguments from degree theory the image of $O_{2\eta}$ under f covers O_η, since $|f - id| \le \delta$. So f has a real analytic inverse φ on O_η, which clearly satisfies $|\varphi - id| \le \delta$. Finally,

$$\begin{aligned} |D\varphi - I|_\eta &\le \left|(Df)^{-1} - I\right|_{2\eta} \\ &\le \frac{1}{1 - |Df - I|_{2\eta}} - 1 \le \frac{1}{1 - \delta/2\eta} - 1 \le \frac{\delta}{\eta} \end{aligned}$$

by applying Cauchy to the domain $O_{2\eta}$. □

B. Lipschitz functions

Let $B \subset \mathbb{R}^n$ be a closed set. We prove the basic fact – used in section 2 – that a Lipschitz continuous function $u\colon B \to \mathbb{R}$ can be extended to a Lipschitz continuous function $U\colon \mathbb{R}^n \to \mathbb{R}$ without affecting its Lipschitz constant

$$|u|_{\mathrm{L},B} = \sup_{\substack{x,y\in B\\ x\ne y}} \frac{|u(x) - u(y)|}{|x - y|}\,,$$

where on $\mathbb{R}^n$ we may take any norm $|\cdot|$. That is, we have

$$U|_B = u, \qquad |U|_{\mathrm{L},\mathbb{R}^n} = |u|_{\mathrm{L},B}\,,$$

Incidentally, for this extension B could be *any* point set.

Indeed, U is given by

$$U(x) = \sup_{z \in B}(u(z) - \lambda |z - x|), \qquad x \in \mathbb{R}^n,$$

where $\lambda = |u|_{\mathrm{L},B}$. By the triangle inequality,

$$(u(z) - \lambda |z - y|) \geq (u(z) - \lambda |z - x|) - \lambda |x - y| .$$

Taking suprema over z we obtain $U(y) \geq U(x) - \lambda |x - y|$, or equivalently $U(x) - U(y) \leq \lambda |x - y|$. Interchanging x and y we obtain

$$|U(x) - U(y)| \leq \lambda |x - y| ,$$

whence $|U|_{\mathrm{L},\mathbb{R}^n} \leq |u|_{\mathrm{L},B}$. We leave it to the reader to check that $U = u$ on B.

References

[1] V. I. Arnol'd, *Proof of a theorem by A.N. Kolmogorov on the invariance of quasi-periodic motions under small perturbations of the Hamiltonian*, Russian Math. Surveys **18** (1963), 9–36.

[2] V. I. Arnol'd, *Mathematical Methods of Classical Mechanics*, Springer, 1978.

[3] V. I. Arnol'd (ed.), *Dynamical Systems III, Encyclopaedia of Mathematical Sciences Volume 3*, Springer, 1988.

[4] J.-B. Bost, *Tore invariants des systèmes dynamiques Hamiltoniens (d'apres Kolmogorov, Arnold, Moser, Rüssmann, Zehnder, Herman, Pöschel, ...)*, Séminaire Bourbaki, Astérisque **133–134** (1986), 113–157.

[5] L. H. Eliasson, *Absolutely convergent series expansions for quasi periodic motions*, Math. Phys. Electron. J. **2** (1996), Paper 4, 33 p..

[6] C. Genecand, *Transversal homoclinic points near elliptic fixed points of area-preserving diffeomorphisms of the plane*, Dynamics Reported, New Series, Volume 2, Springer-Verlag, 1993, pp. 1–30.

[7] A. B. Katok, *Ergodic properties of degenerate integrable Hamiltonian systems*, Math. USSR Izv. **7** (1973), 185–214.

[8] A. N. Kolmogorov, *On the conservation of conditionally periodic motions for a small change in Hamilton's function*, Dokl. Akad. Nauk SSSR **98** (1954), 527–530, (Russian). English translation in: Lectures Notes in Physics 93, Springer, 1979.

[9] L. Markus and K. Meyer, *Generic Hamiltonian systems are neither integrable nor ergodic*, Mem. Am. Math. soc. **144** (1974).

[10] J. Moser, *On invariant curves of area preserving mappings of an annulus*, Nachr. Akad. Wiss. Gött., Math. Phys. Kl. (1962), 1–20.

[11] J. Moser, *Convergent series expansions for quasi-periodic motions*, Math. Ann. **169** (1967), 136–176.

[12] J. Moser, *On the continuation of almost periodic solutions for ordinary differential equations*, Proc. Int. Conf. Func. Anal. and Rel. Topics, Tokyo, 1969, pp. 60–67.

[13] J. Moser, *Stable and Random Motions in Dynamical Systems*, Princeton University Press, 1973.

[14] J. Moser and E. Zehnder, *Lectures on Hamiltonian Systems*, Lecture Notes, ITCP, 1991.

[15] J. Pöschel, *Über invariante Tori in differenzierbaren Hamiltonschen Systemen*, Bonn. Math. Schr. **120** (1980), 1–103.

[16] J. Pöschel, *Integrability of Hamiltonian systems on Cantor sets*, Comm. Pure Appl. Math. **35** (1982), 653–695.

[17] H. Rüssmann, *On optimal estimates for the solution of linear partial differential equations of first order with constant coefficients on the torus*, Dynamical Systems, Theory and Applications, J. Moser (ed.), Lecture Notes in Physics 38, Springer, 1975, pp. 598–624.

[18] H. Rüssmann, *Non-degeneracy in the perturbation theory of integrable dynamical systems*, Number Theory and Dynamical Systems, M. M. Dodson, J. A. G. Vickers (eds.), London Math. Soc. Lecture Note Series 134, Cambridge University Press, 1989, pp. 5–18.

[19] H. Rüssmann, *Invariant tori in the perturbation theory of weakly nondegenerate integrable Hamiltonian systems*, Preprint, Mainz (1998).

[20] D. Salamon, *The Kolmogorov-Arnold-Moser theorem*, Preprint, ETH-Zürich, 1986.

[21] D. Salamon and E. Zehnder, *KAM theory in configuration space*, Comment. Math. Helv. **64** (1989), 84–132.

[22] M. Sevryuk, *KAM-stable Hamiltonians*, J. Dyn. Control Syst. **1** (1995), 351–366.

[23] C. L. Siegel and J. K. Moser, *Lectures on Celestial Mechanics*, Springer, 1971.

[24] E. Stein, *Singular Integrals and Differentiability Properties of Functions*, Princeton, 1970.

[25] H. Whitney, *Analytic extensions of differentiable functions defined in closed sets*, Trans. A.M.S. **36** (1934), 63–89.

[26] Junxiang Xu, Jiangong You and Qingjiu Qiu, *Invariant tori for nearly integrable Hamiltonian systems with degeneracy*, Math. Z. **226** (1997), 375–387.

[27] E. Zehnder, *Homoclinic points near elliptic fixed points*, Comm. Pure Appl. Math. **26** (1973), 131–182.

[28] E. Zehnder, *Generalized implicit function theorems with applications to some small divisor problemss I*, Comm. Pure Appl. Math. **28** (1975), 91–140, *II,* **29,** 1976, 49–111.

Mathematisches Insitut A, Universität Stuttgart Pfaffenwaldring 57, 70569 Stuttgart, Germany

E-mail address: poschel@mathematik.uni-stuttgart.de

Proceedings of Symposia in Pure Mathematics
Volume **69**, 2001

A Lagrangian Proof of the Invariant Curve Theorem for Twist Mappings

M. Levi and J. Moser

1. Background on generating functions

This proof is contained in unpublished lecture notes of a course held by J. Moser at the ETH-Zürich in June 1986.

In contrast to the original "Hamiltonian" approach, where the invariant curve is sought in the phase space, the present Lagrangian approach is based on reducing the problem of finding an invariant curve to solving a nonlinear second difference equation in the configuration space. This approach is modeled after the proof of an analogue theorem for elliptic partial differential equations [**9**] where it was needed since the transformation technique fails for partial differential equations. In our situation the invariance of the desired curve under a mapping leads to a nonlinear second order difference equation which will be derived in the next section and will be solved in the subsequent sections.

Let us consider an area–preserving twist mapping $\varphi : \mathbb{R}^2 \to \mathbb{R}^2$ of the covering plane of the cylinder $\mathbb{R}(\text{mod } 1) \times \mathbb{R}$. Assume that $\varphi : (x_1, y_1) \to (x_2, y_2)$ has a generating function h:

$$\begin{aligned} h_1(x_1, x_2) &= -y_1, \\ h_2(x_1, x_2) &= \ \ y_2. \end{aligned} \tag{1}$$

Here h_1, h_2 are the derivatives with respect to the first or second argument of h. Similarly we will denote the second derivatives by h_{11}, h_{12}, h_{22}. The following theorem establishes a sufficient condition under which a mapping is given by (1).

THEOREM 1. *Any smooth twist cylinder map φ satisfying the monotone twist condition $\partial x_2/\partial y_1 > 0$ possesses a generating function h such that the map is given by (1) implicitly. Moreover, the map is exact: $\int_{\varphi\gamma} y dx = \int_\gamma y dx$, where γ is an arbitrary smooth noncontractible circle on the cylinder, iff $h(x_1 + 1, x_2 + 1) = h(x_1, x_2)$ and $h_{12} < 0$.*

2. Reduction to a difference equation

The existence of an invariant curve is equivalent to a second order differential equation; this is the main point of this section.

Consider a closed curve wrapping around the cylinder, with parametric equations $x = u(\theta)$, $y = v(\theta)$ in the covering (x, y)–plane of the cylinder, with $u(\theta) - \theta$ and $v(\theta)$ periodic of period 1, Figure 1.

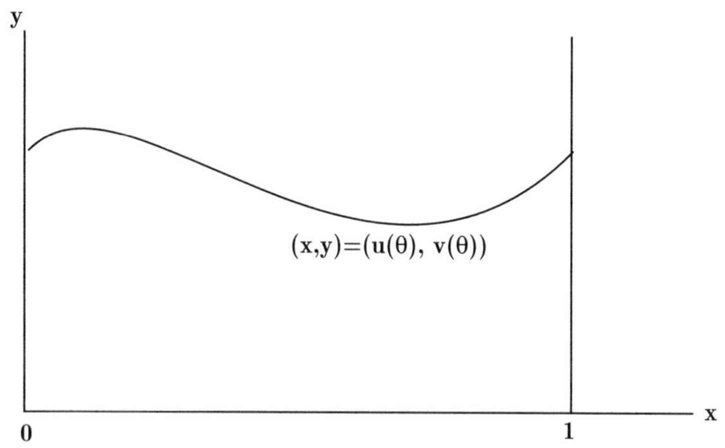

FIGURE 1. The invariant curve

We will find an invariant curve which satisfies the invariance condition

$$\varphi(w(\theta)) = w(\theta + \omega), \tag{2}$$

with a prescribed rotation number ω. Moreover, u is assumed to be strictly monotone in θ. According to (2) the restriction of φ to the curve is conjugate to a rotation by ω.

The following is a Lagrangian formulation of the invariance condition (2), as it occurs in Mather's theory.

THEOREM 2. *The curve $w(\theta) = (u(\theta), v(\theta))$ satisfies the invariance condition (2) with φ given by (1) iff the horizontal function $u(\theta)$ satisfies the second order difference equation*

$$E(u(\theta)) \equiv h_1(u(\theta), u(\theta + \omega)) + h_2(u(\theta - \omega), u(\theta)) = 0. \tag{3}$$

To prove the theorem we simply shift θ by $-\omega$ in the second equation in (1) and add the two equations in (1).

The Hamiltonian problem of finding $w(\theta)$ has thus been reduced to the Lagrangian equation for a single function $u(\theta)$. If $u(\theta)$ has been so determined one finds $v = v(\theta)$ from the formula $v = -h_1(u, u^+)$; here and in the following we use the abbreviation $u^+ = u(\theta + \omega)$, $u^- = u(\theta - \omega)$.

REMARK 3. The equation (3) is the Euler–Lagrange variational equation for the variational problem $\delta \int_0^1 h(u, u^+) d\theta = 0$ (Percival's variational principle).

REMARK 4. The mean value of $u_\theta E(u)$ is zero:

$$\int_0^1 u_\theta E(u) d\theta = 0 \tag{4}$$

as follows from the invariance of the Lagrangian $\int_0^1 h(u,u^+)d\theta$ under θ-translations $u(\theta) \to u(\theta + c)$ or from the identity

$$u_\theta E(u) = \frac{\partial}{\partial \theta} h(u, u^+) - \nabla(u_\theta h_2(u^-, u))$$

where $\nabla f = f(\theta + \omega) - f(\theta)$. Integration gives the claim.

EXAMPLE 5. Consider the standard map

$$\begin{aligned} x_2 &= x_1 + y_1 + S'(x_1) \\ y_2 &= \qquad\; y_1 + S'(x_1) \end{aligned}$$

the generating function is

$$h(x_1, x_2) = \frac{1}{2}(x_1 - x_2)^2 + S(x_1)$$

with a periodic function S of period 1 the equation (3) takes the form

$$u(\theta + \omega) - 2u(\theta) + u(\theta - \omega) = S'(u(\theta)).$$

3. Main theorem

The invariant curve theorem stated here is in effect an implicit function theorem stating that in the vicinity of a near-solution: $E(u_0) \approx 0$ there exists an exact solution: $E(u) = 0$ provided, in particular, that ω is Diophantine.

Notations. Let W_r be the set of 1-periodic real analytic functions of θ bounded on the strip $|\text{Im}\theta| \leq r$. Introduce the maximum norm

$$|f|_r \equiv \sup_{|Im\theta| \leq r} |f(\theta)|.$$

Assumptions. We assume that h is analytic for $(x_1, x_2) \in \mathcal{D} \subset \mathbf{C}^2$ real for (x_1, x_2) and invariant under translations mentioned in Theorem 1. For $R > 0$, let $\mathcal{D}_R \subset \mathcal{D}$ denote the largest set whose R-neighborhood lies in $\mathcal{D}$, Figure 2.

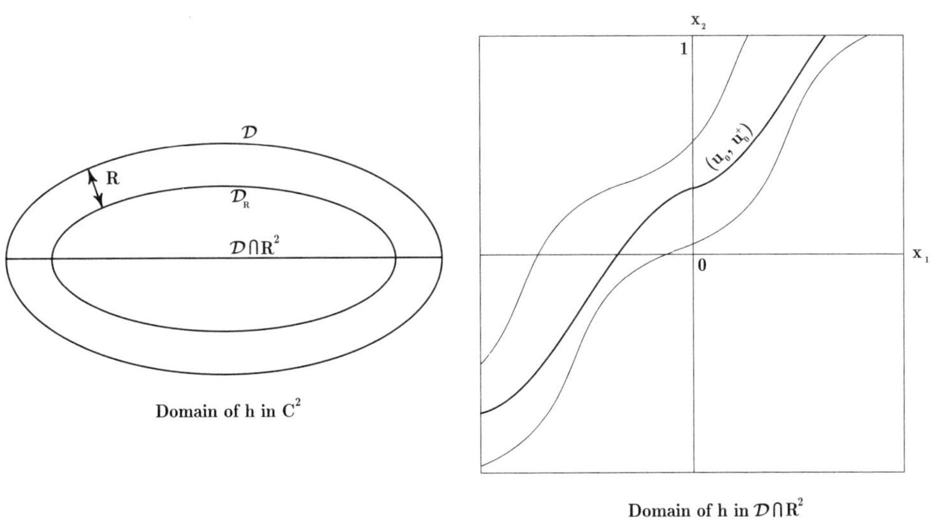

FIGURE 2. Schematic image of the domain of h

Thus the points in $\mathcal{D}_R$ are "safely" inside $\mathcal{D}(h)$. We assume that

$$\min_{\mathcal{D}}|h_{12}| > \kappa, \quad \kappa > 0, \tag{5}$$

and that there exists $M > 0$ such that

$$|h|_{C^3(\mathcal{D})} < M. \tag{6}$$

Let $u_0(\theta) - \theta \in W_r$ for some $0 < r < 1$. Assume, moreover, that

$$(u_0, u_0^+) \in \mathcal{D}_R \text{ for } |\mathrm{Im}\theta| < r, \tag{7}$$

so that (u_0, u_0^+) is "safely" inside the domain of h, Figure 2, and that for some $N_0 > 0$ (possibly large)

$$|(u_0)_\theta|_r < N_0, \ |(u_0)_\theta^{-1}|_r < N_0. \tag{8}$$

THEOREM 6 (Main theorem). *Assume that ω is Diophantine: there exist $K > 0$ and $\sigma > 0$ such that*

$$|\omega - \frac{p}{q}| \geq \frac{K}{q^{2+\sigma}} \quad \forall \text{ integer } p, q \neq 0. \tag{9}$$

Assume that h and u_0 satisfy the above conditions. There exists $\delta = \delta(r, h, M, N_0, K, \sigma, \kappa)$ such that if $|E(u_0)|_r < \delta$, then there exists a unique solution u near u_0 of $E(u) = 0$ with $u(\theta) - \theta \in W_{r/2}$ and mean value of $u(\theta) - \theta$ equal to zero.

Application to the near-integrable twist map. Consider a small perturbation of the linear twist map:

$$\begin{aligned} x_2 &= x_1 + y_1 + \varepsilon f(x_1, y_1, \varepsilon) \\ y_2 &= y_1 + \varepsilon g(x_1, y_1, \varepsilon), \end{aligned} \tag{10}$$

assumed to be area-preserving and exact. The map is given by a generating function $h(x_1, x_2) = \frac{1}{2}(x_2 - x_1)^2 + \varepsilon H(x_1, x_2, \varepsilon)$ which we assume to be defined on $\mathcal{D} = \{(x_1, x_2) : \ a < \mathrm{Re}(x_2 - x_1) < b, \ |\mathrm{Im}\ x_1| < 1, \ |\mathrm{Im}\ x_2| < 1\}$. To apply the main theorem, choose a Diophantine $a < \omega < b$ and take $u_0(\theta) = \theta$. It is a simple exercise to verify the conditions (5)-(8), along with $|E(u_0)|_r = O(\varepsilon)$. Theorem 6 yields an invariant curve in the annulus $a < y < b$ provided ε is small enough.

REMARK 7. It is important to observe that the smallness assumption $|E(u_0)|_r < \delta$ in the main theorem is more general than the standard near-integrability assumption requiring the map to be of the form (10).

4. The homological equation

We will solve the problem $E(u) = 0$ by a modification of Newton's iteration, starting with u_0. Seeking an improvement $\tilde{u} = u + v$ of u, we write

$$E(u + v) = E(u) + E'(u)v + Q,$$

where Q is the remainder and where

$$E'(u)v = (h_{11} + h_{22}^-)v + h_{12}v^+ + h_{12}^- v^-. \tag{11}$$

The first idea would be to kill linear terms:

$$E'(u)v = -E(u), \tag{12}$$

in order to make $E(u+v)$ quadratically small compared with $E(u)$. This is the basis of the standard Newton iteration. The equation (12) is, however, difficult to solve. Fortunately, it is not necessary to kill linear terms: making them quadratically small would suffice. To determine precisely how to modify the equation (12), we multiply both sides by u_θ and subtract the quadratically small term $v\frac{d}{d\theta}E(u) = vE'(u)u_\theta$ from the left–hand side. The result is the homological equation, no longer equivalent to (12):

$$u_\theta E'(u)v - vE'(u)u_\theta = -u_\theta E(u). \tag{13}$$

When v is determined from this equation, $E(u+v)$ turns out to be still quadratically small compared to $E(u)$ as we shall see later, but first we show how to solve (13). Note that

$$u_\theta E'(u)v - vE'(u)u_\theta = h_{12}(u_\theta v^+ - u_\theta^+ v) + h_{12}^-(u_\theta v^- - u_\theta^- v);$$

introducing now the new unknown $w = \frac{v}{u_\theta}$ we rewrite the above as

$$h_{12}u_\theta u_\theta^+(w^+ - w) - h_{12}^- u_\theta^- u_\theta(w - w^-) = \nabla^*(h_{12}u_\theta u_\theta^+ \nabla w)$$

where

$$\nabla f(\theta) \equiv f(\theta + \omega) - f(\theta), \quad \nabla^* f(\theta) \equiv f(\theta) - f(\theta - \omega).$$

To summarize, we have transformed (13) into

$$\nabla^*(h_{12}u_\theta u_\theta^+ \nabla w) = -u_\theta E(u). \tag{14}$$

The Newton step. is thus defined as follows. Given u, find w from the homological equation (14) and set the updated

$$\tilde{u} = u + v, \; v = u_\theta w. \tag{15}$$

In the sections below we will show that with appropriate choices of spaces and constants the repeated iterations of this step converge to the solution of our problem $E(u) = 0$.

An outline of the proof of the main theorem. We will show that one step (15) produces a new error that is quadratic in the preceding one, as one would expect from a Newton-type step: $|E(\tilde{u})|_\rho \sim |E(u)|_r^2/(r-\rho)^{4\gamma}$. This is done in section 6. The improvement, however, is counteracted by the small denominator, due to small divisors. In section 7 we show that the step can be iterated infinitely many times and that convergence wins if appropriate choices are made. In the next section 5 we study the homological equation.

A heuristic motivation. The above trick of writing the correction term in the form $u_\theta w$ is essential for this approach. It can be motivated by noticing that, in the case that u is already a solution of $E(u) = 0$, the operator $E'(u)$ has as zero solutions the multiples of u_θ. Thus by setting $v = u_\theta w$ the corresponding operator has the constants as zero solutions ("variation of constants"). Moreover, since the operator E' is symmetric also $u_\theta E'(u)u_\theta$ is symmetric, so that also its adjoint has the constants in its zero space, as is borne out by the formula $u_\theta E'(u)u_\theta w = (\nabla^* a \nabla)w$ (with $a = h_{12}u_\theta u_\theta^+$). Therefore the range of this operator contains only functions of mean value zero and Equation (4) appears as a necessary compatibility condition for the solvability of equation (14).

5. Solving the homological equation

The following lemma gives an estimate on the solution of (14). We treat u as given and w as the unknown.

LEMMA 8. *Assume that* $u(\theta)$ *satisfies*

$$(u, u^+) \in \mathcal{D}_R \ \text{for}\ |Im\,\theta| < r, \tag{16}$$

and

$$|u_\theta|_r < N,\ |u_\theta^{-1}|_r < N \ \text{for}\ |Im\,\theta| < r. \tag{17}$$

Then (14) has a unique solution $w \in W_\rho$, *with* $w_0 \equiv \int_0^1 w d\theta = 0$ *for any* $0 < \rho < r$, *and the correction* $v = u_\theta w$ *satisfies the estimates*

$$|v|_\rho \le \frac{c}{(r-\rho)^{2\tau}}|E(u)|_r,\quad |v_\theta|_\rho \le \frac{c}{(r-\rho)^{2\tau+1}}|E(u)|_r, \tag{18}$$

where $c = c(M, N, K, \sigma)$ *and* $\tau = 2 + \sigma$.

We will prove this lemma by a two-fold application of the simpler lemma:

LEMMA 9 (Decrease of domain of analyticity). *Let* ω *satisfy the Diophantine condition (9), and let* $g \in W_r$ *with mean value* $[g] = 0$. *The equation*

$$\nabla\psi = g, \tag{19}$$

has a unique solution $\psi \in W_{r'}$ *with* $[\psi] = 0$ *for any* $0 < r' < r$. *Moreover,*

$$|\psi|_{r'} < c(K, \sigma)\frac{|g|_r}{(r-r')^\tau}. \tag{20}$$

PROOF OF LEMMA 9. The Fourier coefficients of g and ψ are related via

$$\psi_n = \frac{g_n}{e^{2\pi i n\omega} - 1} \ne 0,\ \psi_0 = 0. \tag{21}$$

The Diophantine condition (9) gives the lower bound on the denominators:

$$|e^{2\pi i n\omega} - 1| > \frac{c(K)}{|n|^{1+\sigma}}, \tag{22}$$

while $g \in W_r$ implies

$$|g_n| \le |g|_r e^{-2\pi|n|r}. \tag{23}$$

Using (23) and (22) in (21) we obtain, for $0 < r' < s < r$:

$$|\psi_n| \le |g|_r c^{-1}(K)e^{-2\pi r|n|}|n|^{1+\sigma} = |g|_r c^{-1}(K)e^{-2\pi s|n|}e^{-2\pi(r-s)|n|}|n|^{1+\sigma}. \tag{24}$$

We estimate the right hand side by using the estimate $xe^{-x} \le e^{-1}$ for positive x, which for $x = \frac{a|n|}{b}$ gives the inequality $e^{-a|n|}|n|^b \le e^{-b}(b/a)^b$ for all n; with $a = 2\pi(r-s)$ and $b = 1+\sigma$ we obtain the result

$$|\psi_n| \le c_1(K,\sigma)|g|_r\frac{1}{(r-s)^{1+\sigma}}e^{-|n|2\pi s}.$$

From this estimate on ψ_n we obtain

$$|\psi|_{r'} \le \sum |\psi_n|e^{|n|2\pi r'} \le \frac{2c_1|g|_r}{(r-s)^{1+\sigma}}\left(1 - e^{-2\pi(s-r')}\right)^{-1} \le \frac{2c_1|g|_r}{(r-s)^{1+\sigma}(s-r')}$$

which for $s = \frac{r+r'}{2}$ gives [1] the desired estimate (20) for ψ. This completes the proof of Lemma 9. □

[1]In the last step we used the inequality $(1 - e^{-2\pi q})^{-1} < q^{-1}$, $0 < q < \frac{1}{2}$.

PROOF OF LEMMA 8. Introducing the abbreviations $(h_{12}u_\theta u_\theta^+)^{-1} = p$ and $-u_\theta E(u) = g$, we rewrite (14) as a system, where μ is a constant to be chosen:

$$\left\{ \begin{array}{l} \nabla^* \psi = g \\ p^{-1} \nabla w = \psi + \mu. \end{array} \right. \tag{25}$$

By Lemma 9 (recall that the mean value $[g] = 0$ by (4)) we get a unique ψ with zero mean and satisfying (20):

$$|\psi|_{r'} \leq \frac{c(K,\sigma)}{(r-r')^\tau} |g|_r. \tag{26}$$

for any $0 < r' < r$. We now determine w from the second equation in (25): $\nabla w = p(\psi + \mu)$. Since the left–hand side has mean value zero, we must require the same of the right–hand side:

$$\mu = -\frac{\int p\psi d\theta}{\int p d\theta}. \tag{27}$$

From the above estimates on h_{12} and on u_θ we obtain $|\mu| < \kappa M N^4 c \frac{|g|_r}{(r-r')^\tau}$. We conclude that there exists a unique solution w satisfying

$$|w|_\rho \leq c_1 \frac{|p(\psi + \mu)|_{r'}}{(r'-\rho)^\tau} \leq \frac{c_2}{(r'-\rho)^\tau (r-r')^\tau} |g|_r;$$

here (26) and (27) were used in the second inequality. Setting $r' = (r+\rho)/2$ we obtain, recalling that $|u_\theta| < N$:

$$|w|_\rho \leq \frac{c_3}{(r-\rho)^{2\tau}} |E(u)|_r, \quad c_3 = c_3(M, N, K, \kappa, \sigma). \tag{28}$$

Using $|u_\theta| \leq N$ again we get the first estimate of (18). Using the Cauchy estimate $|v_\theta|_\rho \leq \frac{|v|_r}{s-\rho}$ (with appropriate s) one obtains also the second inequality (18). This completes the proof of Lemma 8 and the analysis of the homological equation. □

6. Quadratic dependence of the error

The updated $\tilde{u} = u + v$ $(v = u_\theta w)$ obtained by Newton's iteration is expected, according to our discussion of section 2, to give a quadratic improvement of the error. The following lemma makes this precise:

LEMMA 10. *Let u satisfy the assumptions of Lemma 8 and, moreover, let $\tilde{u} = u + v$ satisfy $(\tilde{u}, \tilde{u}^+) \in \mathcal{D}_R$ for $|Im\ \theta| < \rho$, for some $0 < \rho < r$. Then*

$$|E(\tilde{u})|_\rho \leq \frac{c_6}{(r-\rho)^{4\tau}} |E(u)|_r^2, \tag{29}$$

where $c_6 = c_6(M, N, K, \kappa, \sigma)$.

PROOF. With u as specified, we estimate the error

$$|E(\tilde{u})|_\rho \equiv |E(u+v)|_\rho = |E(u) + E'(u)v + Q|_\rho$$

with Q denoting the remainder. From (13) we obtain

$$u_\theta E(u) + u_\theta E'(u)v = vE'(u)u_\theta,$$

or $E(u) + E'(u)v = wE'(u)u_\theta = w\frac{d}{d\theta}E(u)$. Using (28) for w, together with the Cauchy estimate $|\frac{d}{d\theta}E(u)|_\rho \le c\frac{|E(u)|_r}{r-\rho}$, we obtain

$$|E(u) + E'(u)v|_\rho \le c_4\frac{|E(u)|_r^2}{(r-\rho)^{2\tau+1}} < c_4\frac{|E(u)|_r^2}{(r-\rho)^{4\tau}}, \tag{30}$$

where $c_4 = c_4(M, N, K, \kappa, \sigma)$. Moreover, the error Q is quadratically small in terms of $|E(u)|_r$. Indeed, by Taylor's formula we have for some $0 \le t \le 1$:

$$Q = \frac{1}{2}\frac{d^2}{dt^2}E(u+tv).$$

From (3) we have

$$|Q|_\rho \le c_5|v|_\rho^2, \tag{31}$$

where c_5 depends only on $|h|_{C^3}$. From this, from (18) and from (30) we obtain the desired estimate (29). □

7. The limiting process

In this section we specify the details of iteration of the Newton step (15) and show that it converges to the solution of $E(u) = 0$.

We choose the sequence $r = r_0 > r_1 > \dots$, with $r_n \to r_\infty > 0$, according to $r_n = r_\infty + 2^{-n}(r_0 - r_\infty)$, and construct a sequence $u_0, u_1, \dots, u_n, \dots$ via $u_n = u_{n-1} + (u_{n-1})_\theta w_{n-1}$, where $w_{n-1} \in W_{r_n}$ is the solution of the homological equation (14) with $u = u_{n-1}$.

We show first that the solution of $E(u) = 0$ exists (provided δ is small enough) by repeated application of Lemmas 8 and 10; their applicability will be justified below.

Applying (18) of Lemma 8 to $u = u_n$, $r = r_n$, $N = 2N_0$, we obtain the estimates for corrections $v_n = u_{n+1} - u_n$:

$$|v_n|_{r_{n+1}} \le \frac{c_1}{(r_n - r_{n+1})^{2\tau}}\epsilon_n, \tag{32}$$

where $\epsilon_n = |E(u_n)|_{r_n}$ and where $c_1 = c(M, 2N_0, K, \sigma)$ is the constant from Lemma 8, and

$$|(v_n)_\theta|_{r_{n+1}} \le \frac{c_1}{(r_n - r_{n+1})^{2\tau+1}}\epsilon_n. \tag{33}$$

By Lemma 10, see (29), ϵ_n satisfy

$$\varepsilon_{n+1} \le c_6\frac{\varepsilon_n^2}{(r_n - r_{n+1})^{4\tau}} = c_7 a^n \epsilon_n^2, \tag{34}$$

where $a = 2^{4\tau}$ and $c_7 = c_6\frac{2^{4\tau}}{(r_0 - r_\infty)^{4\tau}}$. Thus the geometric growth a^n competes with quadratic decay ε_n^2; with $\varepsilon_0 \equiv \delta$ chosen sufficiently small – it is here that the choice of δ in Theorem 6 comes in – the quadratic decay wins: $\varepsilon_n \to 0$. Indeed, rescaling the sequence by introducing

$$\eta_n = a^{n+1}c_7\varepsilon_n \tag{35}$$

we obtain $\eta_{n+1} \le \eta_n^2$. Therefore, if $\eta_0 = ac_7\varepsilon_0 < 1$ we have $\eta_n \le \eta_0^{2^n} \to 0$ and $\varepsilon_n \to 0$. By (32) we then have $|v_n|_{r_\infty} \to 0$ faster than exponentially, so that $u_\infty = \lim_{n\to\infty} u_n = u_0 + \sum_0^{n-1} v_k$ is well defined in $|\mathrm{Im}\ \theta| < r_\infty$. We conclude: $E(u_\infty) = \lim_{n\to\infty} E(u_n) = 0$.

We prove now that the assumptions (16) and (17) of Lemmas 8 and 10 hold with $u = u_n$, $N = 2N_0$, $r = r_n$ at nth step for all $n \geq 1$. It suffices to show that for all $n \geq 1$

$$|u_n - u_0|_{r_n} < \frac{R}{2} \tag{36}$$

and

$$|(u_n - u_0)_\theta|_{r_n} < \frac{1}{2N_0}, \tag{37}$$

for all $n = 1, 2, \ldots$. Indeed, one verifies that (36) together with (7) implies (16) with $u = u_n$, $r = r_n$, while (37) with (8) imply (17) with $u = u_n$, $N = 2N_0$ and $r = r_n$. Thus the constant c_1 in (32) and (33) is given by $c_1 = c(M, 2N_0, K, \kappa, \sigma)$, where c is from Lemma 8. Similarly, $c_6 = c_6(M, 2N_0, K, \kappa, \sigma)$. It remains thus to prove (36) and (37).

For this purpose we first verify, for $\lambda \leq 4\tau$, hence $2^\lambda \leq a$, the estimate

$$\sum_0^\infty \frac{\varepsilon_n}{(r_n - r_{n+1})^\lambda} \leq c_7^{-1}(r_0 - r_\infty)^{-\lambda} \sum_0^\infty \eta_n \leq \frac{2\eta_0}{c_7(r_0 - r_\infty)^\lambda} = \frac{2a\varepsilon_0}{(r_0 - r_\infty)^\lambda}, \tag{38}$$

for $\eta_0 < 1/2$, where we used $\eta_n < \eta_0^{n+1}$. This sum can be made small by choice of $\varepsilon_0 = \delta$.

To prove (36) and (37) we assume that $n \geq 1$ is the smallest integer for which at least one these inequalities is violated. Then, for $\nu < n$ these inequalities hold and we can use the estimate (32), and with (38) we get

$$|u_n - u_0|_{r_n} < \sum_{\nu=0}^{n-1} |v_\nu|_{r_\nu} \leq \sum_{\nu=0}^{n-1} \frac{c_1 \varepsilon_\nu}{(r_\nu - r_{\nu+1})^{2\tau}} < c_1 2a \frac{\varepsilon_0}{(r_0 - r_\infty)^{2\tau}}.$$

Similarly, we find from (33)

$$|(u_n - u_0)_\theta|_{r_n} < c_1 2a \frac{\varepsilon_0}{(r_0 - r_\infty)^{2\tau+1}}.$$

By further decreasing $\delta = \varepsilon_0$, if necessary, we assure that (36) and (37) hold for all n. This completes the proof of the Main Theorem.

REMARK 11. We have proven along the way that if $\frac{|E(u_0)|_r}{(r-\rho)^{4\tau}} < \delta'$ (with a δ' independent of r, ρ) then there exists a solution $u \in W_\rho$ of $E(u) = 0$ satisfying the estimates

$$|u - u_0|_\rho < c_1 2a \frac{|E(u_0)|_r}{(r-\rho)^{2\tau}}, \quad |(u - u_0)_\theta|_\rho < c_1 2a \frac{|E(u_0)|_r}{(r-\rho)^{2\tau+1}}.$$

8. Appendix

A. Small Twist.

1) Statement of the result. For several applications it is important to treat the case where the "twist" $\frac{\partial x_2}{\partial y_1}$ in formula (10) is small. This degenerate case gives difficulties in the representation of the mapping by a generating function, since this requires solving the equation $x_2 = f(x_1, y_1)$ for y_1. On the other hand this situation is unavoidable as it arises, for example, in the stability problem of an elliptic fixed point of an area-preserving mapping. We will show how these difficulties can be overcome, essentially by proper rescaling and introducing a more suitable generating

function. In the subsequent section we will apply the resulting theorem to the above mentioned stability problem.

We assume that the area-preserving real analytic map φ depends on a parameter $\gamma \in (0,1)$ and has the form

$$x_2 = x_1 + a(\gamma) + \gamma y_1 + f(x_1, y_1, \gamma)$$
$$y_2 = y_1 + g(x_1, y_1, \gamma) \tag{39}$$

where the functions on the right hand side are analytic in the complex disc $|y_1| \leq 1$ and in $|Im\ x_1| \leq r$. We denote $W_{r,s}$ the space of real-analytic functions, bounded analytic in $|y_1| < s$, $|Im\ x_1| < r$ of period 1 in x_1, with the norm $|f|_{r,s} = sup|f|$ over this domain.

THEOREM 12. *There exists a positive constant δ, independent of $\gamma \in (0,1)$ such that if*

$$|f|_{r,1} + |g|_{r,1} < \gamma\delta$$

then there exists an invariant curve $x_1 = u(\theta), y_1 = v(\theta)$ in $-1 < v(\theta) < 1$ with both $u(\theta) - \theta = \hat{u}(\theta)$ and $v(\theta)$ of period 1, real analytic and with $u_\theta > 0$.

This theorem differs from the previous one as the rotation number ω is not prescribed but has to be constructed in an interval of length γ. We need not impose any smoothness conditions on $a(\gamma)$. As the rule this theorem is used in the following way: If $|f|_{r.1} + |g|_{r,1} = o(\gamma)$ then there exists an invariant curve for sufficiently small γ. (Here we use the Landau notation: $f = o(\gamma^k)$ means $|f|\gamma^{-k} \to 0$ for $\gamma \to 0$, while $f = O(\gamma^k)$ means $|f|\gamma^{-k}$is bounded).

2) Generating function. To construct a generating function for this mapping we do not use x_1, x_2 as independent variables but rather x_1 and the rescaled variable $p = \gamma^{-1}(x_2 - x_1 - a(\gamma))$ so that

$$p = y_1 + \gamma^{-1} f(x_1, y_1, \gamma).$$

Since $\gamma^{-1} f = O(\delta)$ is assumed to be small we can solve this equation for $y_1 = F(x_1, p, \gamma) = p + O(\delta)$ and, inserting this into the second equation of (39), we obtain $y_2 = G(x_1, p, \gamma) = p + O(\delta)$. These functions can be defined for $|p| < \vartheta$, for any $0 < \vartheta < 1$, e.g. for $\vartheta = \frac{3}{4}$ if δ is sufficiently small.

Now we use that $y_2 dx_2 - y_1 dx_1$ is exact, hence also

$$\gamma^{-1}(y_2 dx_2 - y_1 dx_1) = \gamma^{-1}(y_2 - y_1)dx_1 + y_2 dp := dl(x_1, p, \gamma)$$

which defines the generating function $l = l(x_1, p, \gamma)$ up to a constant. Thus we have $y_2 - y_1 = G - F = \gamma l_{x_1}, y_2 = G = l_p$, so that $l(x_1, p) = \frac{1}{2}p^2 + O(\delta)$, where the estimate is understood as being uniform in γ.

If we compare this generating function $l(x_1, p)$ (we suppress the γ) with $h = h(x_1, x_2)$ we find the relation $l(x_1, p) = \gamma^{-1} h(x_1, x_1 + a + \gamma p)$. Note that the domain of definition of h depends on γ while that of l is fixed, say $|p| < \vartheta$.

The function $l(x, p)$ can be viewed as the discrete analogue of the Lagrange function $L(x, \dot{x})$ in mechanics.

3) The difference equation. We express the condition that $x_1 = u(\theta) = \theta + \hat{u}(\theta)$, $y_1 = v(\theta)$ represents an invariant curve with rotation number ω -which is yet to be determined: Notice that the invariance requires that $x_2 = u(\theta + \omega)$, hence

$$p = \gamma^{-1}(u^+ - u - a) = \nabla \hat{u} + \alpha$$

where we defined the rescaled ∇ and α by $\nabla f = \gamma^{-1}(f(\theta+\omega)-f(\theta))$, $\alpha = \gamma^{-1}(\omega - a)$. Similarly we define the rescaled functional $E(u)(=\gamma^{-1}(h_1(u,u^+)+h_2(u^-,u)))$ by

$$E(u) = \gamma^{-1}(l_p - l_p^-) - l_{x_1} = \nabla^* l_p(x_1, \nabla\hat{u}+\alpha) - l_{x_1}(x_1, \nabla\hat{u}+\alpha).$$

One recognizes the equation $E(u) = 0$ as the analogue of the Euler equation in mechanics.

4) Determination of ω. To find a solution of $E(u) = 0$ we have to determine ω so that α lies in the domain of definition of l. It suffices that

$$|\omega - a| < \gamma/2 \text{ so that } -\frac{1}{2} < \alpha < \frac{1}{2}.$$

If we want to find a number ω in this interval satisfying a Diophantine condition we have to replace the factor K in (9) by γK and require that

$$|\omega - \frac{p}{q}| \geq \gamma K q^{-\sigma-2}. \tag{40}$$

One can show by a simple measure theoretical argument that for $\gamma > 0$ there exists a number $\omega = \omega(\gamma)$ satisfying these two requirements (see Siegel-Moser [**15**]).

Having fixed ω we have with $u_0 = \theta$ that $E(u_0) = \nabla^*(\alpha) + O(\delta) = O(\delta)$.

Now we may proceed like in the proof of the main theorem:

THEOREM 13. *With ω chosen as above there exists a $\delta' > 0$ depending on r, σ, K but not on γ such that if $|E(u_0)|_r < \delta'$ there exists a unique solution $u(\theta)$ of $E(u) = 0$ with $\hat{u}(\theta) = u(\theta) - \theta \in W_{r/2}$ and with mean value zero.*

5) Proof. The proof proceeds as before; we just have to verify that the estimates of lemma 9 still hold, uniformly in γ, if we use the rescaled ∇ and the modified Diophantine condition (40). Indeed the small divisors can be estimated independently of γ, as follows. If m is chosen as an integer satisfying $|n\omega - m| \leq \frac{1}{2}$ we have

$$\gamma^{-1}|e^{2\pi i n\omega} - 1| = \gamma^{-1} 2\sin\pi|n\omega - m| \geq \gamma^{-1}4|n\omega - m| \geq 4Kn^{-\sigma-1}.$$

The rest of the proof is straightforward.

B. Stability of elliptic fixed points. An area-preserving real-analytic mapping φ can, in the neighborhood of a fixed point, which we take to be the origin, be written in the form $\varphi : (u,v) \to (u\cos\alpha - v\sin\alpha, u\sin\alpha + v\cos\alpha) + O(u^2+v^2)$, or in complex notation with $w = u + iv$

$$w \to we^{i\alpha} + O(|w|^2)$$

If $q\alpha/2\pi$ is not an integer for $q = 1, 2, ..., k$ then one can find symplectic coordinates such that the mapping is in Birkhoff normal form up to order $k-1$, i.e.

$$w \to we^{i\psi} + O(|w|^k) \tag{41}$$

where ψ is real polynomial of degree $\leq \frac{k}{2} - 1$ in $|w|^2$. Its coefficients are invariant under symplectic transformations; they are the so-called Birkhoff invariants.

THEOREM 14. *If in (41) ψ is not a constant then the origin is a stable fixed point under φ.*

PROOF. We may assume that $\psi = \alpha + \beta |w|^{2m}$ where $\beta \neq 0$ and $2 \leq 2m \leq k-1$. Truncating at the order $2m+2$ we may assume that the mapping has the form

$$w \to we^{i\psi} + O(|w|^{2m+2}), \ \psi = \alpha + \beta |w|^{2m}.$$

Since for φ^{-1} the coefficient β is replaced by $-\beta$ we may also assume $\beta > 0$. The stability will be established by exhibiting invariant curves surrounding the origin in any of its neighborhoods. More specifically we will construct such invariant curves in any annulus

$$\varepsilon^2(1-\varepsilon^\nu) \leq |w|^2 \leq \varepsilon^2(1+\varepsilon^\nu), \ \nu = \frac{1}{3}$$

for sufficiently small $\varepsilon > 0$. Notice that this annulus is rather narrow since its width is of the order $\varepsilon^{1+\nu} = \varepsilon^{4/3}$ which is small compared to the radius ε.

The existence of the invariant curves will be deduced from the theorem of the previous section, by using suitable "polar coordinates" x, y which are defined by

$$w = \varepsilon \sqrt{1 + \varepsilon^\nu y}\, e^{2\pi i x}$$

so that the above annulus is given by $-1 \leq y \leq 1$. One verifies that the Jacobian of this coordinate change is a constant, namely $-\pi\varepsilon^{2+\nu}$. Therefore the transformed mapping $(x_1, y_1) \to (x_2, y_2)$ will preserve the area element $dx \wedge dy$ and, in fact be exact symplectic.

To express the mapping in these coordinates we observe that for the image point $w_2 = \varphi(w)$ we have $|w_2|^2 = |w|^2 + O(|w|^{2m+3})$ hence $\varepsilon^2(1+\varepsilon^\nu y_2) = \varepsilon^2(1+\varepsilon^\nu y_1) + O(\varepsilon^{2m+3})$ or $y_2 = y_1 + O(\varepsilon^{2m+1-\nu})$. Similarly we find $x_2 = x_1 + \frac{1}{2\pi}(\alpha + \beta\varepsilon^{2m}(1+\varepsilon^\nu y_1)^m) + O(\varepsilon^{2m+1})$ or, expanding the mth power, we have

$$x_2 = x_1 + a + \gamma y_1 + O(\varepsilon^{2m+2\nu}), \ y_2 = y_1 + O(\varepsilon^{2m+1-\nu})$$

where $a = \frac{1}{2\pi}(\alpha + \beta\varepsilon^{2m}), \gamma = \frac{m\beta}{2\pi}\varepsilon^{2m+\nu} > 0$. Notice that the two error terms have the same order by our choice of ν.

This mapping has the form required for the theorem of the previous section, and since $\varepsilon^{2m+2\nu} = \varepsilon^{2m+1-\nu} = O(\gamma^{1+\mu})$ with $\mu = \frac{1}{6m+1} > 0$ the error term is indeed small compared to γ. Moreover, the mapping is real analytic in $|y_1| < 1, |Im\ x_1| < r$. Thus, for sufficiently small ε, every such annulus contains an invariant curve, proving our claim. □

C. Comments on the literature. The original invariant curve theorem [**8**] was proven under different assumptions: The mapping there was neither real analytic nor exact symplectic but only a sufficiently smooth mapping satisfying a certain curve intersection property (which entered also in G. D. Birkhoff's proof of his extension of the "last geometric theorem of Poincaré" [**1**]. In this sense the present statement is more restrictive but still sufficient for many applications. It should be pointed out that our proof can be extended to cover the case of smooth area-preserving mappings, by employing the approximation technique described in [**11**], [**14**]. The case of real analytic mappings satisfying the curve intersection property has been treated in [**15**], but using transformation theory.

As mentioned in the introduction the present proof originated in the study of analogue theorems for partial differential equations, where a transformation theory is not available. This was carried out in [**9**], where also the trick of the "variation of constants" can be found. For the Hamiltonian case the corresponding proof was carried out by Salamon and Zehnder in [**14**].

There is a large literature on the more delicate aspects of the invariant curve theorem. For example, the minimal smoothness assumption for the validity of the theorem have been studied by Rüssmann [**12**] and in the extensive papers by M. Herman [**3**], [**4**], where one finds a wealth of further results on this problem, e.g on the break-down of invariant curves. In the analytic case the Diophantine condition can be replaced by a weaker one, the so-called Brjuno condition. This situation has been studied with great precision by Rüssmann, see [**13**]. In particular, we want to point out that his arguments in [**12**] and [**10**] lead to an improvement of the estimates in Lemma 9: one can replace the exponent $\sigma + 2$ by $\sigma + 1$ in (20), and consequently one can replace $\tau = \sigma + 2$ by $\tau = \sigma + 1$ in the text following this Lemma.

This list of references is certainly not complete, and the reader can find more information in the papers quoted. In particular, the Section 16 of the survey of Mather and Forni [**7**] contains a discussion of various results on annulus mappings.

We also want to point out that generically analytic maps possess near an elliptic fixed point besides the invariant curves also invariant Mather sets, which are Cantor sets. More precisely, if one fixes a germ of an elliptic fixed point of general type, one can, by arbitrarily small change of terms of arbitrary higher order, achieve that the so modified map possesses such nondegenerate Mather sets in every neighborhood of the fixed point. This was proven by C. Genecand [**2**], sharpening the earlier paper by Zehnder [**16**]. All the above mentioned proofs are based on functional analytic methods and iteration technique. An interesting proof avoiding these iterative approximations entirely is the paper by Y. Katznelson and Ornstein [**5**], [**6**]. It is based on delicate estimates on the orbits of the mappings. It gives sharper results, but is not easy to read.

Applying (18) of Lemma 8 to u_0 with $r = r_0$ and $N = N_0$, we obtain

$$|u_1 - u_0|_{r_1} < \frac{R}{4}$$

and

$$|(u_1 - u_0)_\theta| < \frac{1}{2U_0},$$

for small enough δ. The first of these two estimates implies $(u_1, u_1^+) \in D_R$, while the second, together with the bounds (8) on $(u_0)_\theta$, implies $|(u_1)_\theta|\ |(u_1)_\theta^{-1}| < 2U_0$.

References

[1] G. D. Birkhoff. An extension of Poincaré's Last Geometric Theorem. *Acta Math. Vol.* **47**, pages 297–311, 1925.

[2] C. Genecand. Transversal homoclinic points near elliptic fixed points of area-preserving diffeomorphisms in the plane. *Dynamics reported, 1-30, Dynamics Reported Expositions Dynamical Systems, (New Series)* **2**, 1993.

[3] M. R. Herman. Sur les courbes invariantes par les difféomorphismes de l'anneau I. *Asterisque Vol.* **103/104**, 1983.

[4] M. R. Herman. Sur les courbes invariantes par les difféomorphismes de l'anneau II. *Asterisque Vol.* **144**, 1986.

[5] Y. Katznelson and D. Ornstein. A new method for twist theorems. *Jour. d'Analyse Math.,* **60**, pages 157–208, 1993.

[6] Y. Katznelson and D. Ornstein. The smoothness of invariant curves. *Fourier Analysis and Applications, Special Issue*, pages 283–310, 1995.

[7] J. Mather and G. Forni. Action minimizing orbits in hamiltonian systems. *Lecture Notes in Mathematics* **1589**, *Spinger Verlag*, pages 92–186, 1994.

[8] J. Moser. On invariant curves of area-preserving mappings of an annulus. *Nachrichten der Akademie der Wissenschaften, math.-phys, Klasse IIa*, pages 1–20, 1962.

[9] J. Moser. A stability theorem for minimal foliations on a torus. *Ergodic Theory and Dynamical Systems* **8***, pages 251–281, 1988.

[10] H. R. Rüssmann. Notes on sums containing small divisors. *Communications Pure Appl. Mathematics* **39**, pages 755–758, 1970.

[11] H. R. Rüssmann. Über invariante Kurven differenzierbarer Abbildungen eines Kreisringes. *Nachrichten der Akademie der Wissenschaften, math.-phys, Klasse IIa*, pages 67–105, 1970.

[12] H. R. Rüssmann. On optimal estimates for the solutions of linear partial differential equations of first order with constant coefficients on a torus. *Lecture Notes in Physics* **138**, pages 598–624, 1975.

[13] H. R. Rüssmann. On the frequencies of quasi-periodic solutions of nearly integrable hamiltonian systems. *Seminar on Dynamical Systems, Progress in Nonlinear Differential equations and Their Application* **12**, pages 160–183, 1994.

[14] D. Salamon and E. Zehnder. KAM theory in configuration space. *Commentarii Math. Helv.* **64**, pages 84–132, 1989.

[15] C.L. Siegel and J.K. Moser. *Lectures on Celestial Mechanics.* Grundlehren der mathematischen Wissenschaften, **87**. Springer, 1971.

[16] E. Zehnder. Homoclinic points near elliptic fixed points. *Comm. Pure Appl. Math.* **26**, pages 131–182, 1973.

DEPARTMENT OF MATHEMATICS, THE PENNSYLVANIA STATE UNIVERSITY, UNIVERSITY PARK, PA 16802

Part III

Research Articles

Proceedings of Symposia in Pure Mathematics
Volume **69**, 2001

Thermodynamical Formalism for Piecewise Invertible Maps: Absolutely Continuous Invariant Measures as Equilibrium States

Jérôme Buzzi

ABSTRACT. We characterize, for piecewise invertible dynamical systems in arbitrary dimension, the invariant probability measures which are absolutely continuous w.r.t. a given measure m with Jacobian J as the equilibrium states w.r.t. $-\log J$. We assume mainly bounded distortion of $\log J$ and negativity of the **pressure of the boundary** of the pieces of the map.

The proof uses **shadowing** (defined in [**5,7**] for Markov extensions) to apply an abstract proposition of F. Ledrappier [**29**] (introduced to characterize a.c.i.m.'s on the interval) using topological pressure estimates. Contrarily to [**17,26**], we do not use Markov extensions or the more usual transfer operator techniques.

We present several applications, first to Lebesgue measure and multi-dimensional piecewise expanding maps with Jacobian *not necessarily Hölder* and satisfying a *generic* condition which even holds for *all* piecewise expanding and affine mappings in the plane. As a corollary, we get existence of finitely many ergodic a.c.i.m.'s. Second, we consider a.c.i.m.'s with positive exponent for interval maps with finitely many discontinuities and non-flat critical points, strengthening [**29**]. Finally we give an application to more general equilibrium states in a simple multi-dimensional setting, building on [**16,17**].

RÉSUMÉ. Nous caractérisons, pour des systèmes dynamiques inversibles par morceaux en dimension quelconque, les mesures de probabilité invariantes absolument continues par rapport à une mesure m de jacobien J comme les mesures d'équilibre pour $-\log J$. Nous supposons essentiellement que $\log J$ est à distortion bornée et que la **pression des bords** des morceaux est < 0.

La preuve utilise le **pistage** (défini dans [**5,7**] pour l'étude des extensions markoviennes) pour appliquer une proposition abstraite de F. Ledrappier [**29**] (introduite pour caractériser les a.c.i.m. sur l'intervalle) grâce à des estimations de pression topologique. Contrairement à [**17,26**], nous n'utilisons pas d'extension markovienne ou les techniques plus classiques d'opérateurs de transfert.

Nous présentons plusieurs applications, d'abord à la mesure de Lebesgue et aux systèmes multi-dimensionnels dilatants par morceaux, de jacobien *non nécessairement höldérien* et satisfaisant une condition *générique*, remplie par *toute* application dilatante et affine par morceaux du plan. On en déduit l'existence d'un nombre fini d'a.c.i.m.

2000 *Mathematics Subject Classification.* Primary 37Axx.

Key words and phrases. Absolutely continuous invariant measures, equilibrium states, pressure, thermodynamical formalism, multidimensional piecewise invertible dynamical systems, shadowing, boundary pressure.

ergodiques. Nous considérons ensuite les a.c.i.m. d'exposant > 0 pour des applications de l'intervalle ayant un nombre fini de discontinuités et de points critiques non-plats, complétant [**29**]. Finalement, nous donnons une application, basée sur [**16,17**], à des potentiels plus généraux dans un cadre multi-dimensionnel simple.

0. Introduction

The thermodynamical formalism is the study of **equilibrium states**, i.e., invariant probability measures μ, the **pressure** of which satisfies:

$$P(\mu, T) \stackrel{\text{def}}{=} h(\mu, T) - \int_X \log J \, d\mu = \sup_\nu P(\nu, T) = 0$$

where h is the measure-theoretic entropy (see, e.g., [**44**]) and J is some function, or weight.

For smooth expanding maps $T : X \to X$, it is well-known that, if J is the Jacobian of T w.r.t. some measure m, then a probability measure μ is an equilibrium state iff it is an absolutely continuous invariant probability measure (**a.c.i.m.** for short), a much more "geometric" property, which allows further study. The most interesting case is probably when m is Lebesgue measure.

This has been generalized to diffeomorphisms, uniformly hyperbolic [**40,2,35**], non-uniform hyperbolic [**30**] and arbitrary [**32**]. It is also known for piecewise monotonic maps on the interval in the expanding case [**43,19,17**] and in the case of continuous and piecewise $C^{1+\alpha}$ maps with non-flat critical points [**29**].

In this paper, we extend the above characterization to *piecewise* invertible dynamical systems in *arbitrary dimension*. We assume that the pressure of the set of discontinuities is negative (a smallness condition), that the Jacobian has bounded distortion and that the conformal meaure (denoted by m) is given (see [**12**]).

We give three applications of this characterization of equilibrium states. To begin with, we study a.c.i.m. w.r.t. Lebesgue measure in two quite different settings:

A. piecewise expanding maps in higher dimensions assuming negative boundary pressure, a condition which we show to hold generically (i.e., for "*almost all*" maps w.r.t. Lebesgue measure or Baire category). Perhaps more strikingly, this condition is *satisfied by all* piecewise affine and expanding maps in the plane —in fact, as M. Tsujii pointed out to us, this generalizes to piecewise real-analytic maps using [**10**] (see also [**41**]) and to all piecewise affine and expanding maps [**42**];

B. interval maps with discontinuities and (non-flat) critical points, generalizing the characterization already obtained by F. Ledrappier [**29**] to all measures, including the "degenerate" ones.

In case (A), we obtain, as a corollary of the above characterization, a **thermodynamical construction of the a.c.i.m.'s**. Thus we get existence and finite multiplicity of the a.c.i.m. for piecewise expanding maps under a *very weak smoothness* hypothesis (namely $C^{1+\Sigma}$, i.e., summable variation of the logarithm of the Jacobian) introduced by B. Schmitt [**39**] in a related setting. This assumption is weaker than the usual $C^{1+\alpha}$, for any $\alpha > 0$ (see [**18,1,3,38**]).

Finally we substitute to Lebesgue measure the conformal measure defined by a general potential and follow the general strategy of [**16,17**]. This allows the study of the:

C. existence of a unique equilibrium state in a simple multi-dimensional setting.

It appears that the condition "$\sup \phi < P(\phi)$" of [**16,17**] is the one-dimensional case of our small boundary pressure condition introduced below.

To state our results precisely we need some definitions.

DEFINITION 0.1. *A* **piecewise invertible dynamical system** *is* (X, P, T) *with:*

1. *X a locally connected metric space.*
2. *P a finite or countable collection of non-empty, open and relatively compact subsets of X which are pairwise disjoint. We write: $Y \stackrel{\text{def}}{=} \bigcup_{A \in P} A$.*
3. *$T : Y \to X$ such that every restriction $T : A \to f(A)$ for $A \in P$ extends to a homeomorphism T_A between neighborhoods of the closures.*

The **boundary** *of P is $\partial P = \bigcup_{A \in P} \partial A$.*

Let m be some Borel measure on X together with a weight $J : Y \to \mathbb{R}_+$ which is a **Jacobian** for (P, T, m): i.e., a measurable and m-integrable function such that:

$$m(TE) = \int_E J \, dm$$

if E is included in some element of P. (In particular: $T_* m << m$.) Remark that m determines J only up to a m-negligible set: $J = \sum_{A \in P} \frac{d(T_A{}^{-1})_* m}{dm}$, m-a.e.

Because our system is piecewise defined, we slightly modify the usual definition of topological pressure [**44**]:

DEFINITION 0.2. *The* **topological pressure** *of an arbitrary subset S of X is:*

$$P_{\text{top}}(S, T) \stackrel{\text{def}}{=} \limsup_{n \to \infty} \frac{1}{n} \log \sum_{\substack{A_0 \ldots A_{n-1} \in P^n \\ [A_0 \ldots A_{n-1}] \cap S \neq \emptyset}} J^n(A_0 \ldots A_{n-1})^{-1}$$

*where $P^n = P \times \cdots \times P$, $J^n(A_0 \ldots A_{n-1})$ is the supremum of $J \cdot J \circ T \cdots J \circ T^{n-1}$ over the n-***cylinder** *$[A_0 \ldots A_{n-1}] \stackrel{\text{def}}{=} A_0 \cap \cdots \cap T^{-n+1} A_{n-1}$.*

The **total pressure** *of (X, T) is:*

$$P_{\text{top}}(T) \stackrel{\text{def}}{=} P_{\text{top}}(X, T).$$

The **measure-theoretical pressure** *of an invariant probability measure μ of (X, T) is:*

$$P(\mu, T) \stackrel{\text{def}}{=} h(\mu, T) - \int_X \log J \, d\mu.$$

Remark that $P_{\text{top}}(T)$ is the standard topological pressure [**44**] of the symbolic dynamics of (X, P, T) w.r.t. the lifted potential $-\log J$.

The **small boundary pressure** condition is:

$$P_{\text{top}}(\partial P, T) < P_{\text{top}}(T).$$

REMARK 0.3. This condition turns out to be a fundamental assumption, which we believe to be not only technical. It is the **multi-dimensional version of the well-known condition** "$\sup \varphi < P_{\text{top}}(T)$" of [**16**] (see also [**33**]). □

Our main result is:

Main Theorem. *Let (X, T, P) be a piecewise invertible dynamical system together with a probability measure m and a Jacobian J. Assume the following:*

(*) *X compact and $m(X) < \infty$.*

(**) *$\log J : Y \to \mathbb{R}$ is continuous and:* $\sup_{A \in P^n} \sup_{x,y \in A} \log \frac{J^n(x)}{J^n(y)} \overset{n\to\infty}{\longrightarrow} 0$.

(***) *P is finite, $P_{\text{top}}(\partial P, T) < P_{\text{top}}(T|\operatorname{supp}(m))$ and:* $\operatorname{diam}(P^n) \overset{n\to\infty}{\longrightarrow} 0$.

Then:

1. *the variational principle holds:*

$$P_{\text{top}}(T|\operatorname{supp}(m)) = \sup_{\mu} P(\mu, T|\operatorname{supp}(m)) \le 0.$$

2. *if $m(X \setminus Y) = 0$ then $P_{\text{top}}(T|\operatorname{supp}(m)) = 0$.*
3. *there exist a.c.i.m.'s iff $P_{\text{top}}(T|\operatorname{supp}(m)) = 0$.*
4. *the a.c.i.m.'s make a finite dimensional simplex.*
5. *each a.c.i.m. has open support in* $\operatorname{supp}(m)$.

We give the three applications.

0.1. Multi-dimensional uniformly expanding maps with discontinuities.

Theorem A. *Let (X, P, T) be a piecewise invertible dynamical system with X a compact subset of $\mathbb{R}^d$ and P finite. Let m be Lebesgue measure and J a Jacobian w.r.t. it. Assume additionally that:*

(A0) *$m(X \setminus Y) = 0$.*

(A1) *T is uniformly expanding: there exists $\Lambda > 1$ such that for all $x, y \in Y$ with $P(x) = P(y)$ (i.e., in the same element of P), $\|T(x) - T(y)\| \ge \Lambda \|x - y\|$.*

(A2) *the modulus of continuity of the Jacobian $J : Y \to \mathbb{R}$:*

$$M_J(t) \overset{\text{def}}{=} \sup\{|J(x) - J(y)| : \|x - y\| \le t \text{ and } P(x) = P(y)\}$$

satisfies:

$$\int_0^1 \frac{M_J(t)}{t}\, dt < \infty.$$

(A3) *the boundary is small w.r.t. pressure:*

$$P_{\text{top}}(\partial P, T) < \max(P_{\text{top}}(T), 0).$$

Then:

1. *the* **a.c.i.m.***'s, i.e., the absolutely continuous invariant probability measures, are exactly the equilibrium states. Moreover, the variational principle holds, despite the discontinuities:*

$$P_{\text{top}}(T) = \sup_{\nu} P(\nu, T) = 0.$$

2. *the a.c.i.m.'s make up a non-empty, finite-dimensional simplex.*
3. *each a.c.i.m. μ has* **open support***, i.e., there is an open set Ω such that $\mu(\Omega) = 1$ and every non-empty open subset of Ω has positive μ-measure.*

$T : Y \to \mathbb{R}$ admitting a Jacobian satisfying condition (A2) is said to be **piecewise** $C^{1+\Sigma}$ (remark that such a T is not necessarily C^1).

Condition (A3) might look discouraging, but it holds true quite generally. Indeed, we have the following bound in terms of the (box) codimension of the boundary, the minimal expansion exponent and the exponential rate of accumulation of discontinuities:

$$P_{\text{top}}(\partial P, T) \leq -\operatorname{codim}(\partial P) \cdot \lambda_-(T) + h_{\text{mult}}(P, T).$$

(Proposition 5.2). Hence maps with reasonable partitions, uniform expansion and subexponential accumulation of discontinuites have negative boundary pressure. It turns out that this last property is almost always true in a sense to be made precise. Thus we can analyze (see section 5):

1. *all* multi-dimensional β-transformations [**6**], i.e., $x \mapsto B.x \pmod{Z^d}$ where B is an expanding affine map of $\mathbb{R}^d$.
2. *almost all* piecewise affine expanding maps in arbitrary dimensions [**9**].

and also the new case of:

3. *almost all* piecewise $C^{1+\Sigma}$ expanding maps in arbitrary dimensions.

Perhaps more strikingly, the pressure turns out to be negative even in case of fast exponential accumulation of discontinuities at least in the following case:

4. *all* piecewise expanding and affine maps of the plane (section 6).

REMARK 0.4.

• One can replace (A1)–(A2) by the weaker assumptions: both $\operatorname{diam}(P^n)$ and $\max_{A \in P^n} \operatorname{diam}(\log J^n([A]))$ converge to zero as $n \to \infty$. □

• We shall in fact get an almost explicit expression for the invariant density —see remark 1.3(iii).

□

0.2. **Interval maps with critical points and discontinuities.** We apply our ideas back to the problem considered by F. Ledrappier [**29**]. The natural class of interval maps w.r.t. our techniques is the following:

DEFINITION 0.5. *$T : [0,1] \to [0,1]$ is a* **regular $C^{1+\Sigma}$ interval map** *if there exist finitely many* **singular** *points $a_0 = 0 < a_1 < \cdots < a_N = 1$ and numbers $1 < \tau, \tau', K < \infty$ such that T is $C^{1+\Sigma}$ with non-vanishing derivative on each interval $]a_0, a_1[, \ldots,]a_{N-1}, a_N[$ and:*

(R1) *if $\tau^{-1} < \frac{y - a_i}{x - a_i} < \tau$ ($\forall i = 0, \ldots, N$), then $|T'(x)/T'(y)| \leq K$.*

(R2) *for each* **critical** *point $c\pm$, (i.e., $T'(c\pm) = 0$):*

$$\sum_{n \geq 0} -|T'(c \pm {\tau'}^{-n})| \log |T'(c \pm {\tau'}^{-n})| < \infty.$$

(R3) *for all $\lambda > 0$, for all $\delta > 0$ sufficiently small w.r.t. λ: if $|x_n - y_n| \leq e^{-\lambda n}$ and $\inf_i |x_n - a_i| \geq e^{-\delta n}$ for all large enough n, then:*

$$\sum_{n \geq 0} |\log |T'(x_n)/T'(y_n)|| < \infty.$$

As we shall check (see Lemmas 7.7 and 7.8), this class contains the PC-maps of [**29**] and also the piecewise $C^{1+\Sigma}$ interval maps with **non-flat critical points** defined at the end of part III (in particular, the non-PC example of P. COLLET).

Theorem B. *Let T be a regular $C^{1+\Sigma}$ interval map. Let μ be an invariant and ergodic probability measure for T with positive Lyapunov exponent. Then:*

μ is an a.c.i.m. if and only if it is an equilibrium state ($P(\mu, T) = 0$).

The big difference with [**29**] is that **shadowing allows us to control the "degenerate measures"** that had to be excluded from consideration in that paper.

For applications of the result of Ledrappier, see the paper [**4**] of H. Bruin and G. Keller.

0.3. **General equilibrium states for multi-dimensional piecewise affine maps.**

The previous results allow the study of thermodynamical formalism but with a weight specialized to the Jacobian induced by Lebesgue measure. In fact our methods apply more generally: we can study equilibrium states w.r.t. arbitrary potentials φ, *as soon as we can build the related conformal measure m*, i.e., which have $e^{-\varphi}$ as Jacobian. We give a very simple example of this approach:

Theorem C. *Let (X, P, T) be a piecewise affine and expanding dynamical system, i.e., (X, P, T) is a piecewise invertible system with P a finite collection of polytopes in $\mathbb{R}^d$ (i.e., intersections of finitely many half-spaces) and affine and expanding maps T_A.*

Assume that T is locally eventually onto (**l.e.o.**)*: for every non-empty open set $U \subset X$,*

$$\bigcup_{n\geq 0} T^n(U) \supset \bigcap_{n\geq 0} T^{-n}Y. \tag{0.6}$$

Let $\varphi : Y \to \mathbb{R}$ be C^Σ. (The topological and measure-theoretic pressures are understood to be w.r.t. φ.) Assume that:

$$P_{\text{top}}(\partial P, T) < P_{\text{top}}(T).$$

Then the variational principle holds: $P_{\text{top}}(T) = \sup_{\mu\in \mathrm{M}(T)} P(\mu, T)$, and this supremum is realized by a unique equilibrium state.

Let us give a completely explicit setting where the above hypothesis are fulfilled:

Corollary 0.7. *Let T be a multi-dimensional β-transformation, i.e., $X = [0,1]^d$, $T : x \mapsto B.x \mod \mathbb{Z}^d$ for some affine and expanding map B and P is the collection of maximal, open in $\mathbb{R}^d$, connected sets of continuity of T. Let $\varphi : Y \to \mathbb{R}$ be C^Σ. Assume that:*

1. *the minimal rate of expansion Λ of T is $> 1 + \sqrt{d}$.*
2. $\sup\varphi - \inf\varphi < \log\Lambda$.

Then the previous theorem applies.

Remark 0.8. As the proof will show, Theorem C follows from a careful reading of M. Denker and M. Urbański's mainly one-dimensional work [**16**], in combination with our equivalence result (proposition D below) and due respect for the small boundary pressure condition.

A much more general, non-linear version of this, using a different approach, is presented in [**12**]. □

0.4. **Outline of the proof.** We shall deduce the Main Theorem from the following abstract result. We consider (X, P, T, m, J) a piecewise invertible dynamical system with a positive, σ-finite Borel measure m on X admitting a Jacobian $J : Y \to \mathbb{R}$. P may be countably infinite.

We first prove equivalence of equilibrium states and a.c.i.m.'s in a rather general (weakly expanding) situation. This is essentially Ledrappier's proposition *plus* shadowing (the definition of which is recalled in section 2).

PROPOSITION D. *Let (X, P, T, m, J) be as above. Let μ be an invariant and ergodic measure for (X, T). Let $p : (Z, \nu, S) \to (X, \mu, T)$ be the natural extension. Extend J to Z by setting $J|Z = J \circ p$. Assume that:*

(i) *$m(A) < \infty$ for each $A \in P$ and $\mu(\mathrm{supp}(m)) = 1$.*

(ii) *$\log J \in L^1(\mu)$ and there exists Z_0 with full measure in Z such that for all $z, z_0 \in Z_0$ with $p(S^{-k}z)$ and $p(S^{-k}z_0)$ in the same element of P for all $k \geq 0$:*

$$\prod_{k\geq 1} \frac{J(S^{-k}z_0)}{J(S^{-k}z)} \text{ converges in }]0, \infty[$$

and is bounded independently of z.

(iii) *μ is not shadowed by ∂P, $H_\mu(P) < \infty$ and, for a.a. $x \in X$:*

$$\mathrm{diam}(P^n(x)) \overset{n\to\infty}{\longrightarrow} 0.$$

where $P^n(x)$ is the n-cylinder $[A_0 \dots A_{n-1}]$, containing x, or X if there is no such cylinder.

Then the following holds:

(D1) *$P(\mu, T) \leq 0$.*

(D2) *μ is an a.c.i.m. (w.r.t. m) iff $P(\mu, T) = 0$.*

(D3) *if μ is an a.c.i.m., then it has open support in $\mathrm{supp}(m)$, i.e., there is a set Ω, open in $\mathrm{supp}(m)$, such that $\mu(\Omega) = 1$ and every non-empty open subset of Ω has positive μ-measure.*

REMARK 0.9.

(i) Conclusion (D1) above is an "abstract" Ruelle-Margulis inequality: if $m =$ Lebesgue measure and T is expanding then $P(\mu, T) \leq 0$ is equivalent to: $h(\mu, T) \leq \int_X \sum_{i=1}^d \lambda_i \, d\mu$, where the λ_i's are the (necessarily positive) Lyapunov exponents. Also, *if there is an everywhere positive invariant density*, then (D1) has a very short and general proof (by normalizing J into a g-function, see [**28, p.187**]).

(ii) As in [**29**], one could prove that the natural extensions ν of a.c.i.m.'s are Bernoulli. □

To check that a measure μ with $P(\mu, T) > P_{\mathrm{top}}(\partial P, T)$ is not shadowed by the boundary, we shall use the following bound (a weighted variant of a result of [**5,7**]):

THEOREM S. *Let (X, P, T) be a piecewise invertible dynamical system such that X is totally bounded. Let $J : Y \to]0, \infty[$ be continuous and have bounded oscillations in the sense that $\sup_{P(x)=P(y)} |J(x)/J(y)| < \infty$.*

If μ is shadowed by $E \subset X$, $\log J \in L^1(\mu)$ and $\mathrm{diam}\, P^n(\cdot) \overset{n\to\infty}{\to} 0$ μ-a.e, then:

$$P(\mu, T) \leq P_{\mathrm{top}}(E, T).$$

In particular, if μ is shadowed by the boundary, then:

$$P(\mu,T) \leq P_{\text{top}}(\bigcup_{A\in P} \partial T(A),T) \leq P_{\text{top}}(\partial P,T).$$

Remark that without some continuity assumption on $\log J$, the topological entropy cannot be defined in a useful way.

Making uniform assumptions, one gets the Main Theorem.

Comments.

The direct use of **shadowing** introduced in our work [**5,7,8**] about Markov extensions is the main novelty in our approach. Observe also that it *avoids the use of Markov extensions*, contrary to [**17,26**] (see remark 1.3(iv)). We hope that this paper illustrates convincingly the naturalness and interest of shadowing, at least for (possibly non-uniformly) piecewise expanding systems. □

A significant difference with our previous works [**5,6,9**] is that, following the advice of G. KELLER about a related problem, we work with the pressure rather than the entropy. This results in gains in simplicity (no need for extension by random Lasota-Yorke maps like in [**9**]) and generality (e.g., the ability to treat maps with exponential accumulation of discontinuities). □

We do *not use the usually very powerful transfer operator techniques.* Let us give some justification. As B. SCHMITT pointed out to us, the transfer operator acting on the natural Banach space C^Σ defined below, has bad spectral properties. This forbids the usual approach of proving a Lasota-Yorke inequality, etc. (see however [**27**]). Also this would give us a lot of information about existence, multiplicity and statistical properties of the a.c.i.m.'s but *not* the fact that all equilibrium states are a.c.i.m.'s (the main point of this paper). A transfer operator proof of this, as in [**28,43,20**], seems quite difficult in a multi-dimensional setting:

First, one would probably need to use a notion of multi-dimensional variation defined by oscillation norms [**24,3,38**] losing control of sets of zero measure (with the Lebesgue measure replaced by the appropriate conformal measure). Second, general C^Σ functions do not have bounded variation, as multi-dimensional bounded variation corresponds more or less to "hölder plus discontinuities". Third, one would have to rule out *a priori* the existence of equilibrium states living outside the support of the a.c.i.m.'s. But the complement of this support can be quite large and quite complex even in the "simple" case of an onto, 2-dimensional β-transformation [**6**]. □

The $C^{1+\Sigma}$ condition was introduced in a related setting by B. Schmitt [**39**] who pointed out to us the following facts. (A2) naturally defines a Banach space:

$$C^\Sigma(Y) = \left\{ f \in C^0(Y,\mathbb{C}) : \sup|f| + \int_0^1 \frac{M_f(t)}{t}\,dt < \infty \right\},$$

this last quantity being the norm. As (A2) is weaker than C^α, for all $\alpha > 0$, the transfer operator has no spectral gap on this space: in fact the whole open unit disk is made of eigenvalues of infinite multiplicity (see, e.g., [**25,14**]). In particular, one can easily see that there are C^Σ functions with arbitrarily slow decay of auto-correlation. One could try to work on a smaller space but it is easy to build examples for which $T_*^n(m)$ *converges arbitrarily slowly* in $L^1(m)$ to the invariant density as $n \to \infty$. □

The study of the thermodynamical formalism done in this paper is completed in the following works:

[12] (with F. PACCAUT and B. SCHMITT) generalizes Theorem C (building of the conformal measure for an arbitrary potential) to the case of non-linear maps;

[11] (with V. MAUME-DESCHAMPS) studies the speed of mixing of the equibrium states built here;

There one assumes slightly stronger hypothesis than we do (but gets more general or more precise results).

Acknowledgements. I am happy to thank B. SCHMITT for many useful discussions especially about the summability condition (A2). I am also much indebted to G. KELLER for pointing out to me, in a related setting, the advantage of working explicitly with the pressure instead of separating the entropy from the weight.

A preliminary version of these results was presented to the *Séminaire de Théorie ergodique* du Laboratoire de Probabilités de l'Université de Paris Jussieu (March 3, 1998). It is a pleasure to thank the participants (especially F. LEDRAPPIER) for their interest and comments which improved this paper.

PART I: ABSTRACT RESULTS

1. Proof of Proposition D

We first review the ingredients of the proof: Ledrappier's abstract result and Rohlin's formula.

1.1. Ledrappier's Proposition.

PROPOSITION 1.1 (F. LEDRAPPIER **[29, Prop. 3.6]**). *Let S be some ergodic automorphism of a Lebesgue probability space $(Z,\mathcal{Z},\nu)$. Let η be a measurable partition of Z together with a conditional system of measures m_E, $E\in\eta$, and a measurable function $J:Z\to\mathbb{R}_+$ with $(\log J)^+$ or $(\log J)^-$ ν-integrable.*

Assume the following:

(E1) *$S^{-1}\eta$ is finer than η and $\bigvee_{n\geq 0}S^{-n}\eta$ is the partition into points.*

(E2) *$S:(S^{-1}(\eta(z)),m_{\eta(S^{-1}z)}|S^{-1}\eta(z))\to(\eta(z),m_{\eta(z)})$ is non-singular with Jacobian $J|S^{-1}\eta(z)$, for ν-a.a. $z\in Z$ ($\eta(z)$ is the element of η containing z).*

(E3) *$h(\nu,S)=H_\nu(S^{-1}\eta|\eta)\overset{\text{def}}{=}-\int_Z\log E_\nu(1_{S^{-1}\eta(z)}|\eta)\,\nu(dz)$.*

(E4) *for a.a. $z_0\in Z$, for all $z\in\eta(z_0)$,*

$$q_{z_0}(z)\overset{\text{def}}{=}\prod_{k\geq 1}\frac{J(S^{-k}z_0)}{J(S^{-k}z)}\in\,]0,\infty[.$$

is well-defined.

(E5) *$0<\int_{\eta(z_0)}q_{z_0}(z)\,m_{z_0}(dz)<\infty$ for ν-a.e. $z_0\in Z$ (m_{z_0} stands for $m_{\eta(z_0)}$).*

Then $P(\nu,S)\leq 0$ and:

$$P(\nu,S)=0\iff\forall S\in\mathcal{Z}\ \nu(S)=\int_Z d\nu(z_0)\int_{\eta(z_0)\cap S}q_{z_0}(z)\,m_{\eta(z_0)}(dz).$$

(1.2)

REMARK 1.3.

(i) We thank F. LEDRAPPIER for pointing out to us that his proof in [**29**] gives also the inequality $P(\nu, S) \leq 0$ above, an abstract version of the *Ruelle-Margulis inequality* [**36**]. (We had a geometrical proof (based on the usual one for a globally smooth map) for T a piecewise uniformly C^1 map and m = Lebesgue measure, without assuming $P_{\text{top}}(\partial P, T) < 0$.)

(ii) One of the strengths of this proposition of Ledrappier is that it needs *no uniformity* in condition (E5) above: compare with [**17**]. On the other hand, it seems that to have $|\log J^n(S^{-n}z_0)/J^n(S^{-n}z)|$ uniformly bounded for all z, z_0, n as in (E4) would not be enough to apply Ledrappier's approach, contrary to the situation in that paper.

(iii) In fact we get an almost explicit expression for the invariant density. In particular, under sufficiently uniform hypothesis, one can establish the regularity of the invariant density. Indeed, if their convergence is uniform, the infinite products $q_{z_0}(z)$ are as smooth as J and the boundary of the pieces $T^n[A_{-n} \ldots A_0]$ are as smooth as T and P. For instance, if T is uniformly expanding and $C^{1+\alpha}$ and ∂P has finite $d-1$-dimensional volume, then the invariant density is quasi-hölder in the sense of [**3,38**].

(iv) We could have proceeded as in [**17**] by using our isomorphism theorem for *connected Markov extensions* [**8**] and then projecting to the usual Markov extension. (This would give slightly different conditions and would require more work).

□

We recall Ledrappier's short and elegant proof for the convenience of the reader:

PROOF (LEDRAPPIER [**29**]). Let ρ be the probability measure defined by its decomposition over η as: $\rho = \int_Z \rho_z \, d\nu(z)$ with:

$$\rho_z(B) = Q(z)^{-1} \int_{B \cap \eta(z)} q_z \, dm_z \text{ with } Q(z) \stackrel{\text{def}}{=} \int_{\eta(z)} q_z \, dm_z.$$

We compute:

$$\begin{aligned}
\rho_z([S^{-n}\eta](z)) &= Q(z)^{-1} \int_{[S^{-n}\eta](z)} q_z \, dm_z \\
&= Q(z)^{-1} \int_{\eta(S^n z)} q_z(S^{-n}v) \frac{dm_{S^n z}(v)}{J^n(S^{-n}v)} \\
&= Q(z)^{-1} \int_{\eta(S^n z)} q_{S^n z}(v) \frac{J^n(S^{-n}v)}{J^n(z)} \frac{dm_{S^n z}(v)}{J^n(S^{-n}v)} \\
&= \frac{Q(S^n z)}{Q(z) J^n(z)}.
\end{aligned}$$

Hence,

$$-\int_Z \log \rho_z([S^{-n}\eta](z)) \, \nu(z) = \int_Z \log J^n \, d\nu = n \int_Z \log J \, d\nu$$

where the first equality follows from the following remarks: $\log \rho_z([S^{-n}\eta](z))$ has an integrable positive part (it is non-positive!) and therefore this first integral can be evaluated by using the ergodic theorem. But the ergodic averages, which must converge a.e., are easily seen to converge in measure towards $-\int_Z \log J^n \, d\nu$ using the computation above. Now,

$$nh(\nu, T) = H_\nu(S^{-n}\eta|\eta) = -\int \log \nu_z([S^{-n}\eta](z)) \, d\nu(z).$$

Therefore:

$$n\left(h(\nu, T) - \int_Z \log J \, d\nu\right) = \int_Z \log \frac{\rho_z([S^{-n}\eta](z))}{\nu_z([S^{-n}\eta](z))} \, d\nu(z)$$
$$= \int_Z \log \frac{d\bar{\rho}}{d\bar{\nu}} \, d\nu,$$

with $\bar{\nu}, \bar{\rho}$ the measures restricted to the σ-algebra given by $S^{-n}\eta$. By the concavity of the log, this is expression is bounded by $\log \int_Z \frac{d\bar{\rho}}{d\bar{\nu}} \, d\nu = 0$, proving that $P(\nu, T) \leq 0$. The strict concavity shows moreover that equality (i.e., Rohlin's formula) can hold only if the above derivative is constantly equal to 1. Hence $\bar{\nu} = \bar{\rho}$. But as $n \to \infty$, $S^{-n}\eta$ increases to the total σ-algebra, so that $\nu = \rho$. This proves eq. (1.2). □

1.2. **Rohlin formula.** a.c.i.m.'s have zero pressure by the following classical result:

LEMMA 1.4 (V.A. ROHLIN [**37**]). *Assume that $(X, \mathcal{B}, \mu, T)$ is an endomorphism of a Lebesgue space together with a countable measurable partition P such that $H_\mu(P) < \infty$, $\bigvee_{n\geq 0} T^{-n}P = \mathcal{B}$ and, for every $A \in P$, $T(A)$ is measurable and $T : A \to T(A)$ is invertible with Jacobian J w.r.t. a reference measure m such that: 1) $(\log J)^+$ or $(\log J)^-$ is μ-integrable; 2) for μ-a.e. x, $m(P^n(x)) < \infty$ for large n.*

If $\mu << m$, then $h(\mu, T) = \int \log J \, d\mu$.

REMARK 1.5. This case (possibly non-uniform expansion) is easy. Compare with [**31**] which addresses the case of a non-uniformly hyperbolic map and needs additional assumptions on μ. □

We reproduce a proof for the convenience of the reader:

PROOF [**29**]. We have:

$$h(\mu, T) = H_\mu\left(P \,\middle|\, \bigvee_{n\geq 1} T^{-n}P\right) = -\int_X \log \sum_{A\in P} 1_A \cdot E_\mu\left(1_A \,\middle|\, \bigvee_{n\geq 1} T^{-n}P\right) d\mu$$
$$= -\int_X \log \lim_{n\to\infty} \frac{\mu(P(x) \cap T^{-1}P^n(Tx))}{\mu(T^{-1}P^n(Tx))} \, d\mu(x).$$

Writing ρ for the density $\frac{d\mu}{dm}$, we compute, for $A = P(x)$:

$$\mu(A \cap T^{-1}P^n(Tx)) = \int_{A\cap T^{-1}P^n(Tx)} \rho \, dm = \int_{TA\cap P^n(Tx)} \frac{\rho}{J} \circ (T|A)^{-1} \, dm$$
$$= \int_{P^n(Tx)} 1_{TA} \cdot \frac{\rho}{J} \circ (T|A)^{-1} \, dm$$
$$\overset{n\to\infty}{\sim} m(P^n(Tx)) \times 1_{TA}(Tx) \cdot \frac{\rho}{J} \circ (T|A)^{-1}(Tx)$$
$$= m(P^n(Tx)) \times \frac{\rho(x)}{J(x)}$$

for m-a.e. x, using the Martingale Convergence Theorem. Hence, if $A = P(x)$,

$$E_\mu\left(1_A \,\middle|\, \bigvee_{n\geq 1} T^{-n}P\right)(x) = \lim_{n\to\infty} \frac{\mu(A\cap T^{-1}P^n(Tx))}{\mu(T^{-1}P^n(Tx))}$$
$$= \frac{\rho(x)}{J(x)\sum_{A\in P}(\rho/J)\circ(T|A)^{-1}(Tx)} = \frac{\rho(x)}{J(x)\rho(Tx)}$$

using the invariance of $\rho\cdot m$ under T. Thus:

$$h(\mu,T) = \int_X (\log J - \log\rho + \log\rho\circ T)\,d\mu = \int_X \log J\,d\mu,$$

the first integral being well-defined by the above computation and the last equality following from the remark that this first integral can be computed (looking at the ergodic decomposition of μ) as a Birkhoff average, which must converge a.e. by the ergodic theorem, but it clearly converges in measure towards $\int_X \log J\,d\mu$.

□

1.3. **Proof of (D1)-(D2).** Let (X,P,T,m,J,μ) be as in the statement of Proposition D, in particular, μ is not shadowed by ∂P.

The implication μ a.c.i.m. $\implies P(\mu,T) = 0$ is Rohlin's formula, which we just recalled. We now prove the (crucial) reverse implication, together with the inequality: $P(\mu,T)\leq 0$, in all cases, by applying Ledrappier's result. This will conclude the proof of (D1)-(D2). We define:

1. (Z,S,ν) to be the natural extension of $(\bigcap_{n\geq 0} T^{-n}Y, T,\mu)$ realized as:

$$\begin{aligned} Z = \{(x, A_0A_{-1}\dots)\in Y\times P^{\mathbb{Z}_-} : \exists (x_n)_{n\leq 0}\in Y^{\mathbb{Z}_-} \text{ s.t.}\\ x_0 = x \text{ and for all } n\leq 0\text{: } x_n\in A_n \text{ and } f(x_{n-1}) = x_n\}\\ S : (x, A_0A_{-1}\dots)\mapsto (T(x), P(T(x))A_0A_{-1}\dots)\end{aligned}$$

where $P(y)$ is the element of P containing y.

2. $\eta = \bigvee_{n\geq 0} S^n\xi$ with $\xi = \{\pi_1^{-1}(A) : A\in P\}$.
3. $m_E = m\circ\pi_1|E$ for all $E\in\eta$ and $J|Z = J\circ\pi_1$.

Remark that the element of η containing $(x,A)\in Z$ is:

$$\eta(x,A) = \left(\bigcap_{n\geq 0} T^n([A_{-n}\dots A_0])\right)\times\{A\}\subset Z.$$

It is the set of points in Z having the same past w.r.t. P as (x,A). It depends only on A so that we shall write, when convenient: $\eta(A)$. We shall also write m_z,ν_z instead of $m_{\eta(z)},\nu_{\eta(z)}$.

If we can apply Ledrappier's Proposition 1.1, then we shall have that $P(\mu,T) = P(\nu,T)\leq 0$ and that, in case of equality, almost each component ν_z is absolutely continuous w.r.t. m_z (integration then giving $\mu << m$ as was to be proved). Hence, (D1) and (D2) will follow from the:

Claim. *Conditions (E1)-(E5) of Proposition 1.1 are satisfied.*

Conditions (E1), (E2) and (E4) are trivially satisfied. (E3) follows from the computation:

$$H_\nu\left(S^{-1}\eta|\eta\right) = H_\nu\Big(\bigvee_{n\leq 1} S^{-n}\xi \Big| \bigvee_{n\leq 0} S^{-n}\xi\Big) = H_\nu\Big(S^{-1}\xi \Big| \bigvee_{n\leq 0} S^{-n}\xi\Big).$$

$H_\nu(\xi) = H_\mu(P) < \infty$ and ξ is a generating partition, so the last expression is well-known to be equal to $h(\nu, S^{-1}) = h(\nu, S)$ (see, e.g., [**44**]).

We turn to (E5). Obviously $\int_{\eta(z)} q_z\, dm_z < \infty$ as $\sup_y q_z(y) \leq C(z) < \infty$ and $m_z(\eta(z)) < \infty$ as $\pi_1\eta(z)$ is included in some element of P. On the other hand $q_z(\cdot) > 0$, hence it remains to prove $m_z(\eta(z)) > 0$ for a.a. $z \in Z$. As $\mu(\operatorname{supp} m) = 1$, it is enough to prove the following:

Sub-claim. *For a.a. $z \in Z$, $\pi_1\eta(z)$ contains a neighborhood of z.*

In particular, it will imply that one can make the following:

Assumption. *Each $\pi_1\eta(z)$ is an open set.*

Indeed, replace each element of η by its intersection with $\{z \in Z : \pi_1(z)$ is in the interior of $\pi_1\eta(z)\}$.

We now prove the sub-claim. Let $(x, A) = z$. Obviously,

$$x \in \bigcap_{n\geq 0} c(T^n[A_{-n}\dots A_0], x) \subset \bigcap_{n\geq 0} T^n([A_{-n}\dots A_0])$$

where $c(S, x)$ is the connected component of S containing x. The connected components of an open set are open because X is locally connected. Thus, $n \mapsto c(T^n[A_{-n}\dots A_0], x)$ is a non increasing sequence of non-empty open subsets containing x. Therefore, it is enough to prove that this nested sequence is eventually constant, i.e., for ν-a.e. (x, A), there exists $N < \infty$ such that for all $n \geq N$:

$$c(T^n[A_{-n}\dots A_0], x) = c(T^N[A_{-N}\dots A_0], x).$$

Consider a bad (x, A): there exists a sequence $n_1 < n_2 < \dots$ such that:

$$c(T^{n_i}[A_{-n_i}\dots A_0], x) \subsetneq c(T^{n_i-1}[A_{-n_i+1}\dots A_0], x). \tag{1.6}$$

We remark that this bad set is similar to the one which appears in the construction of the Markov Diagram of F. Hofbauer [**19**]. It is exactly the one in our multi-dimensional generalization [**8**]. We can therefore control it as we did there. We reproduce the reasoning for the convenience of the reader.

(1.6) is equivalent to:

$$T(A_{-n_i}) \not\supset c([A_{-n_i+1}\dots A_0], y_i) \text{ with } y_i \stackrel{\text{def}}{=} p(S^{-n_i}x).$$

We see in particular, that if (x, A) is bad then $S^{-1}(x, A)$ is also bad: consider the sequence $n_1 - 1 < n_2 - 1 < \dots$. Hence the set of bad (x, A)'s is S-invariant and, by the ergodicity of ν, must have zero or full measure.

Assume by contradiction that the bad set has full measure. As

$$c([A_{-n_i+1}\dots A_0], y_i)$$

is *connected* (by definition!), (1.6) is equivalent to:

$$\partial T(A_{-n_i}) \cap c([A_{-n_i+1}\dots A_0], y_i) \neq \emptyset$$

which implies that, for a.a. $z = (x, A) \in Z$, there exist integers $n_1 < n_2 < \dots$ such that:

$$P(T^k(S^{-n_i+1}z)) = P(T^k(s_i)) \quad \text{for all } k = 0, \dots, n_i - 1.$$

for some points $s_i \in T_{A_{-n_i}} \partial A_{-n_i}$. This says exactly that μ is shadowed by the boundary (see the formal definition below), contrary to assumption. □

1.4. **Proof of Conclusion (D3), open support.** Let μ be some a.c.i.m. of (X, T, m). By Rohlin's formula, $P(\mu, T) = 0$. Thus, Ledrappier's Proposition 1.1 implies:

$$\begin{aligned}\mu(E) = \nu(\pi_1^{-1}E) &= \int_Z d\nu(z_0) \int_{(\pi_1^{-1}E)\cap\eta(z_0)} q_{z_0}(z)\, dm_{z_0}(z)\\ &= \int_Z d\nu(z_0) \int_{E\cap\pi_1\eta(z_0)} q_{z_0}(x, A)\, dm(x)\end{aligned}$$

where $(x_0, A) := z_0 \in X \times P^{\mathbb{Z}_-}$. Hence μ is equivalent to $\tilde{\mu} = h \cdot m$ with:

$$h(x) = \int_Z 1_{\pi_1\eta(z_0)}(x)\, d\nu(z_0),$$

where $1_{\pi_1\eta(z_0)}$ is the characteristic function of $\pi_1\eta(z_0) \subset X$.

By the sub-claim, each $\pi_1\eta(z_0)$ is an open set. Hence, each $1_{\pi_1\eta(z_0)}$ is an upper semi-continuous function. This immediately implies the same for h. Thus $\Omega = \{h > 0\} \cap \operatorname{supp}(m)$ is an open set in $\operatorname{supp}(m)$ which satisfies the definition of an "open support". □

2. Proof of Theorem S

The crucial hypothesis in Proposition D is the non-shadowing of the measure by the boundary. To establish this, the key tool is Theorem S, which we now prove.

First recall the:

DEFINITION 2.1 [**5**]. *Let (X, P, T) be a piecewise invertible dynamical system with an invariant probability measure μ and a subset $E \subset X$. Let $p : (Z, S, \nu) \to (X, T, \mu)$ be the natural extension.*

μ is said to be **shadowed by** *E if for ν-a.e. $z \in Z$, there exist integers $n_1 < n_2 < \dots$ and points $e_1, e_2, \dots \in E$ such that:*

$$\forall 0 \le k < n_i \quad P(S^{-n_i+k}z) = P(T^k e_i)$$

(we implicitely assume that $e_i \in \bigcap_{k=0}^{n_i-1} T^{-k}Y$).

We say that μ **is shadowed by the boundary** *if it is shadowed by the set $\bigcup_{A\in P} \partial T(A)$.*

Let $\epsilon > 0$. Fix $E_1, E_2, \dots \subset E$ such that E_n contains exactly one point in each P, n-cylinder meeting E. By definition of $P_{\text{top}}(E,T)$, for all large n (say, $n \geq L$):

$$\sum_{e \in E_n} J^n(x)^{-1} \leq e^{(P_{\text{top}}(E,T)+\epsilon)n}.$$

We assume (without lost of generality) μ to be ergodic. Recall that the (r,n)-**ball** centered at $x \in X$ is:

$$B(x,r,n) = \{y \in X : \forall k = 0, \dots, n-1 \; d(T^k x, T^k y) < r\}.$$

By Katok's entropy formula [**22**], there exists $r > 0$ such that for all n large enough, one needs at last $e^{(h(\mu,T)-\epsilon)n}$ (r,n)-balls to cover a set with μ-measure at least $1/2$ (i.e., to get a $(r,n,1/2)$-**covering**).

Fix $1 \leq J_* < \infty$ so large that: $\int_{\{J>J_*\}} |\log J| \, d\mu < \epsilon$. Set $\delta = \min(\mu(J > J_*)/2, \epsilon/3 \log R, \epsilon/\log K)$ with $R < \infty$ the minimal number of $r/6$-balls necessary to cover the totally bounded set X.

Fix $M < \infty$ so large that $\operatorname{diam}(P^M(x)) < r/6$ and $\operatorname{diam} \log J(P^M(x)) < \epsilon$ except for a set of points x of measure $< \delta$.

By the ergodic theorem, one can find $G \subset X$ with measure $> 3/4$ and $N < \infty$ such that, for all $x \in G$, all $n \geq N$:

1. $J^n(x)^{-1} \geq e^{-(\lambda+\epsilon)n}$, where $\lambda = \int \log J \, d\mu$;
2. $\#\{k : \operatorname{diam}(P^M(T^k x)) < r/6 \text{ and } \operatorname{diam}(\log J(P^M(T^k x))) < \epsilon\} \geq (1-\delta)n$.
3. $\prod_{k: J(T^k x) > J_*} J(T^k x) \leq e^{\epsilon n}$ and $\#\{k : J(T^k x) > J_*\} \geq \frac{3}{2}\delta n$.

(k above is implicitely restricted to $[0, n[$.)

The shadowing assumption implies that the itineraries of typical orbits are made of very long beginnings of itineraries of boundary points. We sketch the proof already given in [**5, ???**] and [**7, ???**].

Consider the natural extension of $T : X \to X$ (i.e., that system made invertible). Define $n(x)$ to be the smallest integer larger than $\max(M\delta^{-1}, L)$ such that $P^{n(x)}(T^{-n(x)}x) = P^{n(x)}(e)$ for some $e \in E$. Set $N < \infty$ so large that the measure of points x such that $n(x) > N$ is less than δ. Finally, given a point x and a large integer n, set $a_0 = n$ and, for $i \geq 0$, define b_{i+1} to be the largest integer less than a_i such that $n(T^{b_i}) \leq N$, $\ell_{i+1} := n(T^{b_i})$ and $a_{i+1} = b_{i+1} - \ell_{i+1}$. Let r be the largest integer such that $a_r \geq 0$.

Thus, after a change of indices and maybe reducing slightly G and increasing N, we have obtained that for all $x \in G$, all $n \geq N$, one can find disjoint integer intervals $I_1, \dots, I_r \subset [0, n[$, $r \geq 1$, such that: 1) for each $I_i = [a_i, a_i + \ell_i[$, there exists $e_i \in E_{\ell_i}$ such that $P^{\ell_i}(e_i) = P^{\ell_i}(T^{a_i}x)$; 2) the length ℓ_i of each interval I_i is $\geq \max(M\delta^{-1}, L)$; 3) $\# \bigcup_i I_i \geq (1-\delta)n$. We call $D(x) \stackrel{\text{def}}{=} (r, I_1, \dots, I_r, e_1, \dots, e_r)$ a **description** of x.

Consider S_n a $(r/3, n, 3/4)$-covering. Discarding balls not meeting G (the union of which has measure $< 1/4$) and moving each remaining $x \in S_n$ within its $(r/3, n)$-ball, we can assume that $S_n \subset G$ and that it is still a $(r, n, 1/2)$-covering. Perhaps after removing some points, we can also assume S_n to be $(r/3, n)$-separated.

By Katok's formula $\#S_n \geq e^{(h(\mu,T)-\epsilon)n}$ for large n, so we get:

$$\sum_{x \in S_n} J^n(x)^{-1} \geq \#S_n \times e^{-(\lambda+\epsilon)n} \geq e^{(P(\mu,T)-2\epsilon)n}. \tag{2.2}$$

Thus to bound $P(\mu, T)$ it is enough to bound the sum on the left hand side. We first bound $J^n(x)^{-1}$ using the description $D(x)$:

$$\begin{aligned} J^n(x)^{-1} &\le \prod_{i=1}^{r} J^{\ell_i}(e_i)^{-1} e^{\epsilon \ell_i} K^M \times \prod_{k:\mathrm{diam}\log P^M(T^k x)>\epsilon} K \times \prod_{k\notin\bigcup_i I_i} J(T^k x) \\ &\le \prod_{i=1}^{r} J^{\ell_i}(e_i)^{-1} \times K^{rM} \times e^{\epsilon n} \times K^{\delta n} \times e^{\epsilon n} \\ &\le \prod_{i=1}^{r} J^{\ell_i}(e_i)^{-1} \times e^{4\epsilon n} =: J^n(D(x))^{-1}. \end{aligned}$$

(where K is the bound on oscillations of J inside elements of P) using $r \le \epsilon/M \log K$ and the following fact: $\prod_{k\in\mathcal{K}} J(T^k x) \le e^{\epsilon n}$ if $\mathcal{K} \subset [0, n[$ satisfies $\#\mathcal{K} \le \delta n$. Indeed, as $\#\mathcal{K} < \#\{k : J(T^k x) > J_*\}$, one can send every element of $\mathcal{K}$ into $\{k : J(T^k x) > J_*\}$, by a permutation on $[0, n[$, without decreasing any factor of the product. But the new product is manifestly bounded by $\prod_{k:J(T^k x)>J_*} J(T^k x) \le e^{\epsilon n}$, proving the bound.

Now, we claim that at most $e^{\epsilon n}$ points in S_n can have a given description. To see this, recall that any two points in S_n are $(r/3, n)$-separated. Consider $x \in S_n$. For all $0 \le k < n$ except $2\delta n + rM \le 3\delta n$, $\mathrm{diam}(P^M(T^k x)) < r/6$ and $P^M(T^k x)$ is determined by $D(x)$. Therefore if points in S_n have the same description and have the same positions up to a precision $r/6$ at all of these $3\delta n$ bad times, then they must be the same. Now there are only $R^{3\delta n} \le e^{\epsilon n}$ choices for these positions (up to precision $r/6$). This proves the claim.

Therefore:

$$\begin{aligned} \sum_{x\in S_n} J^n(x)^{-1} &= \sum_D \sum_{\substack{x\in S_n\\ D(x)=D}} J^n(x)^{-1} \le \sum_D e^{\epsilon n} \times J^n(D)^{-1} \\ &\le e^{5\epsilon n} \sum_{r,I_1,\dots,I_r} \sum_{e_1,\dots,e_r} \prod_{i=1}^{r} J^{\ell_i}(e_i)^{-1} = e^{5\epsilon n} \sum_{r,I_1,\dots,I_r} \prod_{i=1}^{r} \sum_{e\in E_{\ell_i}} J^{\ell_i}(e)^{-1} \\ &\le e^{5\epsilon n} \sum_{r,I_1,\dots,I_r} \prod_{i=1}^{r} e^{(P_{\mathrm{top}}(E,T)+\epsilon)\ell_i} = e^{(P_{\mathrm{top}}(E,T)+o(1))n} \end{aligned}$$

Indeed, $\ell_i \ge L$ so that $\sum_{e\in E_{\ell_i}} J^{\ell_i}(e)^{-1} \le e^{(P_{\mathrm{top}}(E,T)+\epsilon)\ell_i}$ and as $\min_i \ell_i \ge \epsilon^{-1}$, there are less than $\epsilon n \cdot C_n^{2r} \le \epsilon n \cdot C_n^{2\epsilon n} = e^{o(1)n}$ choices for $1 \le r \le \epsilon n$ and $I_1, \dots, I_r$ ($o(1)$ stands for a function of ϵ which tends to zero).

Theorem S is proved. □

3. Proof of the Main Theorem

(X, P, T, m, J) is a piecewise invertible dynamical system with a positive, σ-finite measure and a Jacobian.

3.1. **Preliminaries.** We shall use the following facts.

LEMMA 3.1. *For $\nu \in \mathrm{M_{erg}}(T)$ and $S \subset X$, $\nu(S) > 0$ implies that $P(\nu, \sigma) \le P_{\mathrm{top}}(S, T)$.*

(a corollary of the "pressurized" version of Katok's entropy formula [**22**].)

This lemma immediately gives: $P(\nu, T) \leq P_{\text{top}}(T)$ for any invariant probability measure ν.

Recall that the **symbolic dynamics** defined by (X, P, T) is (see, e.g., [**6**]), the left-shift σ acting on:

$$\Sigma_+ \stackrel{\text{def}}{=} \{A \in P^{\mathbb{N}} : \forall n \geq 0\ [A_0 \ldots A_n] \neq \emptyset\}.$$

It is straightforward to observe:

LEMMA 3.2. *Assume that* $\operatorname{diam}(P^n) \to 0$. *Let* $\Pi : \Sigma_+ \to X$ *be defined by* $\bigcap_{n\geq 0} \overline{[A_0 \ldots A_n]} = \{\Pi(A)\}$.
Then Π *is well-defined and its restriction:*

$$\Pi : \left(\sigma, \Sigma_+ \setminus \bigcup_{k\geq 0} \sigma^{-k}\Pi^{-1}(\partial P)\right) \to \left(T, X \setminus \bigcup_{k\geq 0} T^{-k}(\partial P)\right)$$

is a measurable isomorphism.

We now assume conditions (*)-(***) of the Main Theorem. By Theorem S, Proposition D applies to measures μ with $P(\mu, T) > P_{\text{top}}(\partial P, T)$.

3.2. **Proof of the variational principle.** Let $\tilde{X} = \operatorname{supp}(m)$ and $\tilde{T} = T|\tilde{X}$. Set $\tilde{\Sigma}_+ = \{A \in \Sigma_+ : \Pi(A) \in \tilde{X}\}$. By definition, $P_{\text{top}}(\tilde{T}) = P_{\text{top}}(\sigma|\tilde{\Sigma}_+, \varphi)$ where $\varphi(A) \stackrel{\text{def}}{=} \lim_{n\to\infty} \sup \log J([A_0 \ldots A_n])$. φ is continuous and well-defined by the assumptions: $\log J$ continuous and $\operatorname{diam}(P^n) \to 0$. By definition, $P_{\text{top}}(\tilde{T}) = P_{\text{top}}(\tilde{\Sigma}_+)$.

As $\sigma|\tilde{\Sigma}_+$ is a continuous self-map of a compact space and φ is continuous, we have Walters' variational principle [**44**]:

$$P_{\text{top}}(\tilde{T}) = \sup_{\mu \subset \text{M}_{\text{erg}}(\tilde{\Sigma}_+)} P(\mu, \sigma).$$

$P_{\text{top}}(\tilde{T}) > P_{\text{top}}(\partial P, T)$ so that we can restrict μ in the above supremum to the ergodic measures with $P(\mu, \sigma) > P_{\text{top}}(\partial P, T)$. But the ergodic and invariant probability measures of $(\tilde{X}, \tilde{T})$ and $(\tilde{\Sigma}_+, \sigma)$, which have pressure $> P_{\text{top}}(\partial P, T)$, are in a one-to-one, isomorphic correspondence (see Lemmas 3.2 and 3.1). Hence:

$$P_{\text{top}}(\tilde{T}) = \sup_{\mu \in \text{M}_{\text{erg}}(\tilde{T})} P(\mu, \tilde{T}) = \sup_{\mu \in \text{M}(\tilde{T})} P(\mu, \tilde{T}). \tag{3.3}$$

- $P_{\text{top}}(\tilde{T}) \leq 0$.

We have just seen that $P_{\text{top}}(\tilde{T})$ is the supremum of $P(\mu, \tilde{T})$ over the ergodic measures μ with $P(\mu, \tilde{T}) > P_{\text{top}}(\partial P, T)$. But one can apply Ledrappier's Proposition 1.1 to the natural extensions ν of these μ. One finds: $P(\mu, \tilde{T}) = P(\nu, \tilde{T}) \leq 0$ and finally: $P_{\text{top}}(\tilde{T}) \leq 0$, as claimed. □

- $P_{\text{top}}(S, T) \geq 0$ if $m(S) > 0$ and $m(X \setminus Y) = 0$.

By assumption (**), there exists $C > 0$ such that, for all $n \geq 0$, all non-empty cylinders $A \in P^n$:

$$J^n(x) \geq C \cdot \sup_{x\in[A]} J^n = C \cdot J^n(A).$$

Remark that $m(X \setminus Y) = 0$ implies, as $J > 0$, that $m(X \setminus \bigcap_{k\geq 0} T^{-k}Y) = 0$. Hence:

$$\begin{aligned} m(S) = m\left(\bigcap_{k=0}^{n-1} T^{-k}Y \cap S\right) &\leq \sum_{\substack{A\in P^n\\ [A]\cap S\neq\emptyset}} m([A]) \\ &\leq \sum_{\substack{A\in P^n\\ [A]\cap S\neq\emptyset}} C^{-1}\cdot J^n(A)^{-1}\cdot m(T^n[A]) \\ &\leq \sum_{\substack{A\in P^n\\ [A]\cap S\neq\emptyset}} C^{-1}\cdot J^n(A)^{-1}\cdot m(X). \end{aligned} \tag{3.4}$$

Using $m(S) > 0$ and $m(X) < \infty$, one sees that $P_{\text{top}}(S,T) \geq 0$. □

3.3. **Proof of existence.** By (D2), the a.c.i.m.'s and the equilibrium states (μ with $P(\mu,T) = 0$) coincide.

First recall that $P_{\text{top}}(\tilde{T}) \leq 0$ and $P(\mu,\tilde{T}) \leq P_{\text{top}}(\tilde{T})$. Thus, if $P_{\text{top}}(\tilde{T}) \neq 0$, then $P_{\text{top}}(\tilde{T}) < 0$ and there is no a.c.i.m.

Now, assume $P_{\text{top}}(\tilde{T}) = 0$. As in the proof of the variational principle, the lift φ of $\log J$ to the symbolic dynamics Σ_+ is continuous. Moreover, the shift is continuous and expansive on the compact space $\tilde{\Sigma}_+$, hence there exist equilibrium states for the symbolic dynamics (see, e.g., [**44**]). As $P_{\text{top}}(\partial P,T) < P_{\text{top}}(\tilde{T}) = 0$, we see that equilibrium states can be identified with invariant measures of (X,T) with zero pressure and these are a.c.i.m.'s by Proposition D. □

3.4. **Proof of finite multiplicity.** Assume by contradiction that there exist infinitely many ergodic a.c.i.m.'s: one can find an infinite sequence $\mu_1, \mu_2, \ldots$ of mutually singular a.c.i.m.'s.

Remark first that the existence of a single a.c.i.m. implies, by (D1) and (D2), that $P_{\text{top}}(T) = 0$.

Identifying these measures with the Π-isomorphic measures on the symbolic dynamics Σ_+ with maximal, zero pressure, one can assume that they converge vaguely to some $\mu \in \mathrm{M}(\Sigma_+)$. By the upper semicontinuity of the pressure on Σ_+, μ must have non-negative pressure. As $P_{\text{top}}(T) = 0$, it must have zero pressure. As $P_{\text{top}}(\partial P,T) < 0$, Lemmas 3.1 and 3.2 imply that one can identify μ with a Π-isomorphic invariant measure of (X,T), having also zero pressure. By conclusion (D2) again, this measure μ is an a.c.i.m. (possibly not ergodic).

Being a.c.i.m.'s, μ, resp. μ_n, has open support: there exists an open set Ω, resp. Ω_n, such that $\mu(\Omega) = 1$, resp. $\mu_n(\Omega_n) = 1$, and every non empty open subset has positive measure.

By the definition of vague convergence, $1 = \mu(\Omega) \leq \liminf_{k\to\infty} \mu_k(\Omega)$ (Ω being open): for large k, $\mu_k(\Omega) > 0$, hence $\Omega \cap \Omega_k$ is a non empty open set. Fix such a k. Repeating the same reasoning with $\mu(\Omega \cap \Omega_k) > 0$ instead of $\mu(\Omega) = 1$, we get that for some $n > k$, $\Omega \cap \Omega_k \cap \Omega_n$ is a non empty open set. But this implies that μ_k and μ_n are not mutually singular: a contradiction, which proves the finiteness. □

The Main Theorem is proved. □

PART II: Piecewise Expanding Maps & Lebesgue Measure

This part is devoted to the deduction of Theorem A from the abstract results and its application to various classes of piecewise expanding multi-dimensional maps. We first recover some results of ours about cases where discontinuities only accumulate subexponentially, without having to use Markov extensions or random Lasota-Yorke maps as in [**26,9**].

We then carry these results over to the weakly smooth case: expanding piecewise $C^{1+\Sigma}$ maps satisfying a generic condition. All this follows from a bound on the topological pressure of a set in terms of its codimension, the expansion rate and the speed of accumulation of discontinuities.

We finally establish a key estimate for the geometry of piecewise affine and expanding maps of the plane which allows the analysis of every such map, despite possible fast exponential accumulation of discontinuities.

4. Proof of Theorem A

Let (X, P, T) be a piecewise expanding piecewise invertible dynamical system. Write m for the Lebesgue measure and J for the unique continuous jacobian on Y defined by T and m.

We assume conditions (A0)-(A3). We simply have to check conditions (*)-(***) of Proposition E, the conclusions of which are the same as those of Theorem A as $\text{supp}(m) = X$. The only points which are not completely obvious are that 1) the uniform expansion and the $C^{1+\Sigma}$ condition together imply that:

$$\sup_{A\in P^n} \text{diam}(\log J^n([A])) \overset{n\to\infty}{\longrightarrow} 0; \tag{4.1}$$

and that 2) $P_{\text{top}}(\partial P, T) < P_{\text{top}}(T)$.

We first take care of point 1). (A1) implies that $\text{diam}(P^n) \overset{\text{def}}{=} \max_{A\in P^n} \text{diam}(A)$ is at most $\text{diam}(X) \cdot \Lambda^{-n}$. Now, the left hand side in eq. (4.1) is smaller than:

$$\sum_{n\geq 1} \max_{A\in P^n} \text{diam}(\log J([A])) \leq \sum_{n\geq 1} \frac{1}{\Lambda} \cdot M_J(\text{diam}(P^n)) \leq \sum_{n\geq 1} \frac{1}{\Lambda} M_J(C \cdot \Lambda^{-n})$$

where $C = \text{diam}(X)$ and M_J is the modulus of continuity of J.

Using that $t \mapsto M_J(t)$ is a non-negative, non-decreasing function and the change of variable $t = C\Lambda^{-u}$, we get:

$$\int_{C\Lambda^{-n-1}}^{C\Lambda^{-n}} \frac{M_J(t)}{t}\, dt = \int_n^{n+1} \frac{M_J(t)}{t} \log\Lambda \cdot t\, du \geq \log\Lambda \cdot M_J(C\Lambda^{-n}).$$

Hence, (A2) implies $\sum_{n\geq 1} M_J(C\Lambda^{-n}) < \infty$ and in particular eq. (4.1).

We now check 2) by remarking that: $P_{\text{top}}(\partial P, T) < 0$ implies $P_{\text{top}}(\partial P, T) < P_{\text{top}}(T)$. But this is simply because $P_{\text{top}}(T) \geq 0$: see the computation (3.4) in the proof of Proposition E. □

5. Subexponential Accumulation of Discontinuities

In this section we bound the boundary pressure in terms of the expansion rate and the rate of accumulation of discontinuities under iteration of P and deduce from this three applications of Theorem A to cases where discontinuities accumulate subexponentially.

The third case (generic expanding piecewise $C^{1+\Sigma}$ maps) is new.

5.1. **Boundary Pressure Estimate.** To state our estimate we need the following definitions. First the **minimal uniform exponent** is:

$$\lambda_-(T) \stackrel{\text{def}}{=} \lim_{n\to\infty} \frac{1}{n} \log \Lambda_-(T^n) \text{ with } \Lambda_-(S) \stackrel{\text{def}}{=} \inf_{(x,v)} \|S'(x).v\|.$$

(the above infimum is restricted to the x's where S' makes sense and to the v's with norm 1).

The exponential rate of accumulation of the discontinuities is measured by the following quantity (the **entropy multiplicity**, see, e.g., [**6**]):

$$h_{\text{mult}}(P,T) = \lim_{n\to\infty} \frac{1}{n} \log \operatorname{mult}(P^n) \text{ where } \operatorname{mult}(P) = \max_{x\in X} \#\{A \in P : \overline{A} \ni x\}.$$

The **codimension** of $E \subset \mathbb{R}^d$, $\operatorname{codim}(E)$, is the supremum of the numbers c such that, for all small enough $\epsilon > 0$,

$$m(B(E,\epsilon)) \leq \text{const} \cdot \epsilon^c$$

(i.e., $\operatorname{codim}(E) \stackrel{\text{def}}{=} \liminf_{\epsilon\to 0} \log m(B(E,\epsilon))/\log\epsilon$) where $B(E,\epsilon)$ is the set of points of $\mathbb{R}^d$ at a distance $< \epsilon$ from E and m is the Lebesgue measure on $\mathbb{R}^d$.

DEFINITION 5.1. *Let r be a number > 1 or $r = 1+\Sigma$. $(X,P,\{T\}_A)$ is a* **piecewise uniformly C^r map** *if it is a piecewise invertible dynamical system with finite P and such that, for each $A \in P$, $T_A : U_A \to V_A$ is a C^r-diffeomorphism between neighborhoods of $\overline{A}$ and $\overline{T(A)}$ in $\mathbb{R}^d$.*

PROPOSITION 5.2. *Assume that (X,P,T) is a piecewise expanding, piecewise uniformly $C^{1+\Sigma}$. Then:*

$$P_{\text{top}}(\partial P,T) \leq -\operatorname{codim}\partial P \cdot \lambda_-(T) + h_{\text{mult}}(P,T).$$

The interest of this bound is that, as we shall see: $h_{\text{mult}}(P,f) = 0$ *almost always holds true.* Hence, expansion is (generically) enough to ensure the crucial condition $P_{\text{top}}(\partial P,T) < 0$ appearing in Theorem A.

PROOF. Remark first that under the uniform expansion and uniform $C^{1+\Sigma}$ assumption, a computation like in the proof of Theorem A gives:

$$\sup_{A\in P^n} \operatorname{diam}(\log J^n(B(A_0,\epsilon_0)\cap T_{A_0}^{-1}(B(A_1,\epsilon_0))\cap \dots \\ \dots \cap T_{A_0\dots A_{n-2}}^{-n+1} B(A_{n-1},\epsilon_0))) \xrightarrow{n\to\infty} 0. \tag{5.3}$$

where $\epsilon_0 > 0$ is so small that $B(A,\epsilon_0) \subset U_A$, the domain of T_A.

It is convenient to consider the following map $\mathcal{T}$, a "one-to-one, multi-valued extension" of T. Reducing $\epsilon_0 > 0$ if necessary, we assume that $\#\{A \in P : A \cap B(x,\epsilon_0) \neq \emptyset\} \leq \operatorname{mult}(P)$ for all $x \in X$.

Let P^* be the set of finite words on P, including the empty word ε. Let $\mathcal{X}$ be the collection of all subsets of $X \times P^*$. Define $\mathcal{T} : \mathcal{X} \to \mathcal{X}$ by:

$$\mathcal{T}(E) = \{(f_C(x), AC) \in X \times P^* : (x,A) \in E \text{ and } C \in P \text{ and } x \in B(C,\epsilon_0)\}.$$

Let $K \subset X$ and $n = 1,2,\dots$. Set:

$$K' = B(K, \Lambda_-(T)^{-n}\epsilon_0) \times \{\varepsilon\}.$$

It is easy to check that for all $B = B_0 \ldots B_{n-1} \in P^n$:

$$\mathcal{T}^n(K') \supset B(T_B^n(K \cap \overline{[B_0 \ldots B_{n-1}]}), \epsilon_0) \times \{B\}.$$

Thus, if $K \cap \overline{[B_0 \ldots B_{n-1}]} \neq \emptyset$, then:

$$m(\pi_B(\mathcal{T}^n(K))) \geq \text{const} \cdot \epsilon_0^d. \tag{5.4}$$

where $\pi_B(E) = \{x \in X : (x, B) \in E\}$. Now,

$$K' = \bigcup_{B \in P^n} \left((T_B^n)^{-1} \pi_B(\mathcal{T}^n(E))\right) \times \{\varepsilon\}.$$

This union is not disjoint but each point of K' is obtained at most $\text{mult}(P, \epsilon_0)^n$ times. Hence, using eqs. (5.3) and (5.4):

$$\begin{aligned} m(K') \cdot \text{mult}(P, \epsilon_0)^n &\geq \sum_{B \in P^n} \int_{\pi_B(\mathcal{T}^n(E))} \text{jac}(T_B^n)^{-1}\, dm \\ &\geq \sum_{B \in P^n} m(\pi_B(\mathcal{T}^n(E))) \cdot C^{-1} J^n(B)^{-1} \\ &\geq (\text{const} \cdot \epsilon_0^d) \cdot C^{-1} \sum_{\substack{B \in P^n \\ K \cap \overline{[B_0 \ldots B_{n-1}]} \neq \emptyset}} J^n(B)^{-1}. \end{aligned}$$

Now, for any $\alpha > 0$, $m(K') \leq \text{const} \cdot (\epsilon_0 \Lambda_-(T)^{-n})^{\text{codim}(K) - \alpha}$. Thus, we get:

$$P_{\text{top}}(K, T) \leq -(\text{codim}(K) - \alpha) \log \Lambda_-(T) + \log \text{mult}(P).$$

Going to a sufficiently high order iterate and letting $\alpha \to 0$, we get the Proposition.□

5.2. **Multi-dimensional β-transformations.** Recall that these transformations are the self-maps T of $[0, 1[^d$ given by $x \mapsto B.x \pmod{\mathbb{Z}^d}$ where B is an affine and expanding transformation of $\mathbb{R}^d$. We have proved [**6**] that $h_{\text{mult}}(P, T) = 0$ for these T. Clearly, $\text{codim}(\partial P) = 1$ and $\lambda_-(T) > 0$ so that $P_{\text{top}}(\partial P, T) < 0$. The other conditions are trivially satisfied:

COROLLARY 5.5. *Theorem A applies to all multi-dimensional β-transformations.*

(This was first proved in [**6**] using Markov extensions.)

5.3. **Generic piecewise affine and expanding maps.**

Let (X, P, T) be a piecewise expanding and affine map, i.e., a piecewise invertible dynamical system with $X \subset \mathbb{R}^d$, P a finite collection of open, bounded polytopes (i.e., finite intersections of half-spaces) and such that each map T_A, $A \in P$, is affine and expanding.

Again all conditions of Theorem A except (A3) are trivially satisfied by such maps.

It is not true that all these maps satisfy $h_{\text{mult}}(P, T) = 0$ (see [**6**]), but most of them do, "most" in the following sense. Remark that the set of piecewise affine maps is in an obvious way, a discrete union of finite-dimensional vector spaces, hence it is endowed with natural notions both of Baire category and of negligible sets w.r.t. Lebesgue measure. In both these senses, the piecewise affine maps with $h_{\text{mult}}(P, T) > 0$ are negligible [**9**]. Thus:

Corollary 5.6. *Theorem A applies to all piecewise affine and expanding maps, except for a set negligible both w.r.t. Baire category and Lebesgue measure.*

(This was first proved in [**9**] using Markov extensions, the equivalence result of [**17**] and random Lasota-Yorke maps.)

5.4. **Generic piecewise smooth and expanding maps.** We consider piecewise uniformly C^r maps, for $r > 1$, possibly non-integer, or $r = 1 + \Sigma$ (see the beginning of this section). Let P be a finite partition together with an extension, i.e., the choice of a neighborhood of the closure of each $A \in P$. Recall that a map is said to be piecewise uniformly C^r if for each $A \in P$, we are given a C^r extension of T_A to the chosen neigborhood of A. We can then equip the set of maps thus obtained (for each degree of smoothness $1 < r < \infty$ and $1 + \Sigma$) with an obvious Banach space structure. One can then prove the genericity of $h_{\text{mult}}(P, T) = 0$ [**15**] in this setting. Thus we get:

Corollary 5.7. *Assume that P is a finite partition, satisfies* $\operatorname{codim}(\partial P) > 0$ *and that some extension of P is chosen.*

The piecewise uniformly C^r and expanding maps (X, P, T) which do not satisfy $P_{\text{top}}(\partial P, T) < 0$ are a negligible set in the sense of Baire category.

6. Piecewise Affine Expanding Maps of the Plane

This section is devoted to the key estimate opening up the study of arbitrary expanding plane piecewise affine maps: though the discontinuities accumulate exponentially fast in some cases, the boundary pressure is *always negative.* Hence Theorem A applies.

This illustrates the big advantage of working with the pressure instead of considering separately the entropy and the weight.

Recall that a **piecewise affine map** is a piecewise invertible dynamical system (X, P, T) with $X \subset \mathbb{R}^d$, P finite and such that each map T_A, $A \in P$, is affine and each $A \in P$ is a polytope (a finite intersection of half-spaces). It is **plane** if $X \subset \mathbb{R}^2$.

The **minimal uniform exponent** of T is: $\lambda_-(T) \stackrel{\text{def}}{=} \lim_{n\to\infty} \frac{1}{n} \log \Lambda_-(T^n)$ with $\Lambda_-(S) \stackrel{\text{def}}{=} \inf_{(x,v)} \|S'(x).v\|$. (the above infimum is restricted to the x's where S' makes sense and to the v's with norm 1).

T is **eventually expanding** if $\lambda_-(T) > 0$.

Theorem 6.1. *Let (X, P, T) be a piecewise affine and eventually expanding plane map.*

For any **convex curve** *γ, i.e., a connected compact subset of $X \cap \partial C$ for some convex set C:*

$$P_{\text{top}}(\gamma, T) \leq -\lambda_-(T) < 0.$$

Corollary 6.2. *Theorem A applies to arbitrary expanding piecewise affine plane maps.*

Remark 6.3. This is a partial generalization of the fact (due to G. Keller [**23**]) that there is no exponential accumulation of singularities for expanding piecewise holomorphic maps (see [**10**]). □

For convenience we shall write $\mathcal{C}$ for the set of convex curves as defined above.

The proof splits into two lemmas: the first one deals with the pressure of itineraries above single points and the second one deduces from this the result about curves.

6.1. **Pressure of points.**

The weighted local complexity of the iterates of (X, P, T) is measured by the following quantity:

DEFINITION 6.4. *The* **weighted multiplicity** *is:*

$$\text{mult}_J(P) \stackrel{\text{def}}{=} \sup_x \text{mult}_J(P, x) \text{ where } \text{mult}_J(P, x) \stackrel{\text{def}}{=} \sum_{\substack{A \in P \\ \overline{A} \ni x}} J(A)^{-1}$$

and the **multiplicity pressure***:*

$$P_{\text{mult}}(P, T) = \lim_{n \to \infty} \frac{1}{n} \log \text{mult}_J(P^n).$$

The previous limit exists by sub-multiplicativity: $\text{mult}_J(P^{n+m}) \leq \text{mult}_J(P^n) \cdot \text{mult}_J(P^m)$.

LEMMA 6.5. *Let (X, P, T) be a piecewise affine and expanding plane map. Then:*

$$P_{\text{mult}}(T, P) \leq -2\lambda_-(T).$$

To prove this lemma, it is convenient to define an **angular sector** $\mathcal{S}_x$ at some point $x \in \mathbb{R}^2$ to be the germ at x of $S \subset \mathbb{R}^2$ such that either $S = \emptyset$ or $\mathbb{R}^2$, or ∂S is the union of two half straight lines at x. We shall confuse the germ S_x with this "canonical representative". We write $\angle(S)$ for the non-oriented measure of the angle S.

We shall also use:

LEMMA 6.6. *Let S be some angular sector and A be some affine plane transformation. Then:*

$$\angle(A(S)) \leq \frac{|\det(A)|}{\Lambda_-(A)^2} \angle(S).$$

To see this, remark that $A(S \cap B(0, r/\Lambda_-(A))) \supset B(0, r) \cap A(S)$ and that $\angle(S) = area(S \cap B(0, r)) \cdot 2/r^2$ for any $r > 0$ small enough.

PROOF OF LEMMA 6.5. It is enough to show that, for some $C_* < \infty$, for all $n = 1, 2, \ldots$:

$$\text{mult}_J(P^n) \leq C_* \cdot \left(\frac{4}{\Lambda_-(T)^2}\right)^n,$$

The statement of the lemma will then follow from this estimate applied to the iterate (X, P^N, T^N) using $P_{\text{mult}}(T, P) = P_{\text{mult}}(T^N, P^N)/N$ for $N \to \infty$.

Pick $\theta_- > 0$ strictly smaller than the smallest angle defined at any point $x \in \overline{A}$ by any element $A \in P$. Set:

$$C_* \stackrel{\text{def}}{=} \#P \cdot \left(\frac{2\pi}{\theta_-} + 1\right).$$

We prove the following claim by induction on $n = 0, 1, 2, \ldots$: for any $x \in X$, there exists a partition $\mathcal{S}_n$ of the germ of the neighborhood of x, into angular sectors σ such that:

1. σ is contained in some element of P^{n+1}.

2. $\angle(T^n\sigma) \leq \theta_-$.
3. $\sum_{\sigma\in\mathcal{S}_n} J^n(A)^{-1} \leq C_*\cdot(4/\Lambda_-(T)^2)^n$, if A is the element of P^n containing σ.

For $n=0$, this claim is obvious, considering the choice of C_*.

We now assume this for some $n \geq 0$ and prove it for $n+1$.

Let $\mathcal{S}_n$ be given by the induction hypothesis. Consider $\sigma \in \mathcal{S}_n$. By Lemma 6.6,

$$\angle(T^{n+1}(\sigma)) \leq \frac{|\det(T'_B)|}{\Lambda_-(T'_B)^2}\theta_-,$$

if B is the element of P containing $T^n(\sigma)$. Hence we can divide σ into at most: $|\det(T'_B)|/\Lambda_-(T'_B)^2+1$ subsectors τ, each with angle $\angle(T^{n+1}\tau) \leq \theta_-$.

By the choice of θ_-, dividing each subsector $T^{n+1}\tau$ according to P gives at most two subsectors at $T^{n+1}_A x$ for each τ (with A the element of P^n containing τ). The resulting subsectors σ' satisfy the conditions (1) and (2) above. Let $\mathcal{S}_{n+1}$ be the collection of subsectors thus obtained. We check condition (3):

$$\begin{aligned}\sum_{\sigma'\in\mathcal{S}_{n+1}} J^{n+1}(AB)^{-1} &= \sum_{\sigma\in\mathcal{S}_n} J^n(A)^{-1} \sum_{\sigma'\subset\sigma} J(B)^{-1}\\ &\leq C_*\cdot\left(\frac{4}{\Lambda_-(T)^2}\right)^n \times \max_{B\in P}\left\{2\left(\frac{|\det(T'_B)|}{\Lambda_-(T'_B)^2}+1\right)\times\frac{1}{|\det(T'_B)|}\right\}\\ &\leq C_*\cdot\left(\frac{4}{\Lambda_-(T)^2}\right)^{n+1},\end{aligned}$$

using $|\det(T'_B)| \geq \Lambda_-(T)^2$. □

6.2. Pressure of curves.

Theorem 6.1 now follows from:

Lemma 6.7. *Let (X,P,T) be a piecewise affine and expanding plane map. If γ is a convex curve:*

$$P_{\mathrm{top}}(\gamma,T) \leq \max(-\lambda_-(T), P_{\mathrm{mult}}(T,P)).$$

Proof. We claim that there exists $\epsilon_0>0$ such that any ball $B(x,\epsilon_0)$ meets only line segments of ∂P which start from a common point. To see this, remark that ∂P has only a finite number of vertices, i.e., points where several line segments of ∂P meet.

Hence P divides any ball $B(x,\epsilon_0)$ by half straight lines, all originating from the same point $y\in X$. In particular,

$$\mathrm{mult}_J(P,x,\epsilon_0) \stackrel{\mathrm{def}}{=} \sum_{\substack{A\in P\\ A\cap B(x,\epsilon_0)\neq\emptyset}} J(A)^{-1} \leq \mathrm{mult}_J(P,y) \leq \mathrm{mult}_J(P) \qquad \forall x\in X.$$

We claim that for each $n=0,1,\dots$, for each $\gamma_*\in\mathcal{C}$, there exists a collection $\mathcal{P}_n \subset \mathcal{C}\times P^n$ such that:

1. for all $A\in P$, $\bigcup_{(\gamma,A)\in\mathcal{P}_n}\gamma \supset \gamma_*\cap\overline{A}$.
2. if $(\gamma,A)\in\mathcal{P}_n$, the length $|T^n_A\gamma|$ is $<\epsilon_0$.
3.

$$\sum_{(\gamma,A)\in\mathcal{P}_n} J^n(A)^{-1} \leq C_*\Big(8\max(1/\Lambda_-(T),\mathrm{mult}_J(P))\Big)^n. \tag{6.8}$$

As usual, this claim, applied to arbitrarily high order iterates of (X, P, T) will give the Lemma.

We first fix $C_* < \infty$ so large that this claim holds for $n = 0$ and all $\gamma_* \in \mathcal{C}$. We prove it by induction for $n \geq 1$. Thus assume that it holds for some $n \geq 0$ and let us prove it for $n + 1$.

Let $(\gamma, A) \in \mathcal{P}_n$.

Let $Q_1 \stackrel{\text{def}}{=} \{q_0, q_1, \dots, q_{N+1}\}$ be successive points on γ such that: q_0 and q_{N+1} are the endpoints of γ (if they exists, otherwise just assume $q_0 = q_{N+1}$) and each sub-curve $[q_i q_{i+1}] \subset \gamma$, $i = 0, \dots, N-1$, resp. N, is mapped by T_A^{n+1} to a curve of length ϵ_0, resp. at most ϵ_0.

We define the successors of (γ, A) to be $(\gamma' \cap T_A^{-n}(\overline{B}), A \cap T^{-n}B)$, γ' being a sub-curve of γ of the form $[q_i q_{i+1}]$ for $i = 0, \dots, N$ and B ranging over the elements of P such that these two intersections are non-empty. We define $\mathcal{P}_{n+1}$ to be the set of all successors of all elements of $\mathcal{P}_n$.

$\mathcal{P}_{n+1}$ clearly satisfies conditions (1) and (2) above (use: $\bar{A} \cap T^{-n}(\bar{B})$ is convex and also contains $\overline{A \cap T^{-n}B}$). We have to check condition (3). We fix some $(\gamma, A) \in \mathcal{P}_n$ and focus on its successors $(\gamma', AB) \stackrel{\text{def}}{=} (\gamma', A \cap T^{-n}B)$. We have to bound:

$$\sum_{(\gamma', AB) \leftarrow (\gamma, A)} J^{n+1}(AB)^{-1} = J^n(A)^{-1} \underbrace{\sum_{(\gamma', AB) \leftarrow (\gamma, A)} J(B)^{-1}}_{S}$$

We first consider the sum S' over the successors $(\gamma', A \cap T^{-n}B) \in \mathcal{C} \times P^{n+1}$ such that at least one endpoint of γ' is on ∂P: this endpoint is in $\partial_1 (T_A^n(\gamma) \cap \partial B)$ where ∂_1 designates the boundary w.r.t. the one-dimensional topology (e.g., $\partial_1 \gamma$ is the set of endpoints of γ). ∂B is locally the union of two half straight lines. Hence, $T_A^n(\gamma) \cap \partial B$ is the union of at most two intervals (which can be reduced to points), and $\partial_1(T_A^n(\gamma) \cap \partial B)$ contains at most 4 points. Therefore, each $B \in P$ can appear in at most 4 terms in the sum S'. Remark also that $T_A^n \gamma$ is included in a ball with radius ϵ_0. Hence:

$$S' \leq 4 \sum_B J^n(B)^{-1} \leq 4 \cdot \operatorname{mult}_J(P).$$

We consider now the rest $S - S'$ of the sum, i.e., the sum of the terms in S indexed by (γ', AB) such that $\partial_1 \gamma' \subset Q_1 \setminus \partial P$, i.e., $\gamma' = [q_i, q_{i+1}]$ for some i. Each $[q_i q_{i+1}]$ is a non-trivial interval and therefore can be contained in the closure of at most 2 elements of P^{n+1}. Also, except if $\gamma' = [q_N q_{N+1}]$, $T^{n+1}\gamma'$ has length ϵ_0. These two facts imply the following bound on the sum $S - S'$ (using $|\det(T'_B)| = \Lambda_-(T_B)\Lambda_+(T_B)$ and $|T^n \gamma| \leq \epsilon_0$):

$$\begin{aligned} S - S' &\leq \underbrace{\frac{2}{\inf |\det T'|}}_{\gamma' = [q_N q_{N+1}]} + \sum_{\substack{(\gamma', B) \\ \gamma' \neq [q_N q_{N+1}]}} \frac{1}{|\det(T'_B)|} \\ &\leq \frac{2}{\Lambda_-(T)} \left(1 + \frac{1}{\epsilon_0} \sum_{\gamma'} \frac{\epsilon_0}{\Lambda_+(T_B)} \right) \\ &\leq \frac{2}{\Lambda_-(T)} \left(1 + \frac{1}{\epsilon_0} \sum_{\gamma'} |T^n \gamma'| \right) \leq \frac{4}{\Lambda_-(T)}. \end{aligned}$$

Therefore, the sum in condition (3) for $\mathcal{P}_{n+1}$ is at most $8\max(\Lambda_-(T)^{-1}, \text{mult}_J(P))$ times the one for $\mathcal{P}_n$. The bound in condition (3) for $\mathcal{P}_{n+1}$ follows. □

REMARK 6.9. The bound (6.8) is uniform w.r.t. $\gamma_* \in \mathcal{C}$. Moreover, one can of course take $\epsilon_0 > 0$ arbitrarily small. This increases C_*, but an important fact is that the dependence of C_* on ϵ_0 can be removed if we only consider $\gamma_* \in \mathcal{C}$ with diameter $< \epsilon_0$. □

Part III : Non-uniform expansion

7. Interval maps with critical points

Let $T : [0,1] \to [0,1]$ be a regular $C^{1+\Sigma}$ map of the interval. Let μ be an invariant and ergodic probability measure with exponent $\lambda(\mu, T) > 0$. We study the equivalence of a.c.i.m.'s and equilibrium states and prove Theorem B.

Recall that the implication (μ a.c.i.m. $\implies P(\mu, T) = 0$) is the very general and classical Rohlin's formula, a proof of which was recalled in Lemma 1.4. Now we shall check that Proposition D applies, so that $P(\mu, T) = 0$ implies that μ is an a.c.i.m., concluding the proof of Theorem B.

Take $c_0 = 0 < c_1 < \cdots < c_N = 1$, numbers $\tau > 1$ and $K < \infty$ as in the definition of a regular $C^{1+\Sigma}$ map. We define P to be the countable set of connected components of $]0,1[\setminus C$ where C is the set of:

1. all singular points $c_0, \ldots, c_N$.
2. $c \pm \tau^{-n}$ for all n large enough, for each critical point $c\pm$.
3. k/N_P for $0 < k < N_P$.

with N_P a sufficiently large integer.

REMARK 7.1.

(i) A similar partition was used by F. Hofbauer and P. Raith [**21**].

(ii) The motivation for this is quite simple, we need 1) the elements of P to be small enough so that T' will be constant, up to a uniform factor, on each one of them; 2) the endpoints of P to accumulate quickly enough on a finite set so that ∂P will be effectively finite (from the point of view of entropy). □

We establish some basic facts about P to apply Proposition D.

7.1. P has finite entropy.

LEMMA 7.2. *Let $\mu \in \mathrm{M}(T)$ be such that $\lambda(\mu, T) > -\infty$. Then the $C^{1+\Sigma}$ regularity of T implies:*

$$H_\mu(P) < \infty.$$

PROOF. We compute: $H_\mu(P) \stackrel{\text{def}}{=} -\sum_{A\in P} \mu(A) \log \mu(A)$ by splitting $P = P' \cup P''$ with:

$$P' = \{A \in P : \mu(A) \geq |T'(A)| \stackrel{\text{def}}{=} \sup_{x\in A} |T'(x)|\} \text{ and } P'' = P \setminus P'.$$

First, as $t \mapsto -\log t$ is non-increasing and $\mu(A) \geq |T'(A)| \geq |T'(x)|$ for $A \in P'$ and $x \in A$:

$$-\sum_{A\in P'} \mu(A) \log \mu(A) \leq -\sum_{A\in P'} \mu(A) \log |T'(A)| \leq -\int_{\bigcup P'} \log |T'|\, d\mu < \infty$$

using $\log|T'| \in L^1(\mu)$ (as $\lambda > -\infty$). Second, as $t \mapsto -t\log t$ is non-decreasing on $[0, 1/2]$:

$$-\sum_{A\in P''} \mu(A)\log\mu(A) \leq -\sum_{A\in P''} |T'(A)|\log|T'(A)| + \text{const}$$
$$\leq -\sum_{c\pm}\sum_{n\geq 1} K|T'(c\pm\tau^{-n})|\log(K|T'(c\pm\tau^{-n})|) + \text{const} < \infty,$$

using (R1) and (R2). □

7.2. **μ is not shadowed by ∂P.** First, the natural partition Q generates w.r.t. any ergodic invariant probability measure of non-zero entropy. Hence, by Theorem S, if μ is shadowed by ∂P: then $h(\mu, T) \leq h_{\text{top}}(\partial P, T)$, where:

$$h_{\text{top}}(\partial P, T) = \limsup_{n\to\infty} \frac{1}{n}\#\{A \in Q^n : \overline{[A]} \cap \partial P \neq \emptyset\}$$

(we consider T as the piecewise invertible system $([0,1], Q, T)$, not $([0,1], P, T)$).

Now, the assumptions: $P(\mu, T) \geq 0$ and $\lambda(\mu, T) > 0$ immediately imply: $h(\mu, T) > 0$. Hence, the:

Claim. *μ of $([0,1], Q, T)$ is not shadowed by ∂P.*

will follow from:

Lemma 7.3. $h_{\text{top}}(\partial P, T) = 0$.

Proof. We prove the statement for $C = \{c, c+\tau^{-1}, c+\tau^{-2}, \dots\}$ where $c+$ is a critical point. The proof of the statement for ∂P is only more of the same.

Fix $\epsilon > 0$ smaller than the distance between any two singular points of T.

For each $n \geq 1$, let $\mathcal{R}_n$ be the set of elements $A \in Q^n$ such that:

1. $\overline{A}$ contains $c+\tau^{-m}$ for some $m = 1, 2, \dots, m_* = [(\log\Lambda\cdot n - \log\epsilon)/\log\gamma] + 1$,
2. A meets $]c, c+\epsilon\Lambda^{-n}[$.

Obviously, $h_{\text{top}}(\partial P, T) \leq \limsup_{n\to\infty} \frac{1}{n}\log\#\mathcal{R}_n$. Moreover we can forget the A's given by (1) above, as there are only $\mathcal{O}(n)$ of them. We redefine $\mathcal{R}_n$ accordingly and we prove that $\#\mathcal{R}_n \leq e^{o(1)n}$.

It is convenient to consider the graph G with:

1. vertices: (A, k) with $1 \leq k \leq n$, $A \in Q^k$ containing an element of $\mathcal{R}_n$.
2. arrows: $(A, k) \to (B, k+1)$ iff $B \subset A$.

G is a tree, with $\#\mathcal{R}_n$ leafs and with all branches of length n.

For each $(A, k) \in G$, $T^k(A)$ is an interval of the form $]T^q d, T^{q'}d'[$, for some integers q, q' and endpoints d, d' of Q. Moreover this interval has length $< \epsilon$.

Remark that if $(A, k) \to (B, k+1)$ then either:

$$T^{k+1}(B) = T(T^k(A)) =]T^{q+1}d, T^{q'+1}d'[$$

(in which case $(B, k+1)$ is the sole successor of (A, k)) or (agreeing that $]a, b[=]b, a[$):

$$T^{k+1}(B) = T(T^k(A)) \cap D =]T^{q+1}d, d''[$$

for some $D \in Q$ and some endpoint d'' of D (in which case there are two successors, and no more, because $|T^{k+1}(A)| \leq \epsilon$).

Set: $q(A, k) \stackrel{\text{def}}{=} \min(q, q')$. Define:

$$q(\epsilon) \stackrel{\text{def}}{=} \min\{k > 0 : 0 < |T^k(d\pm) - e| < \epsilon \text{ for some } d, e \in \partial Q, \text{ and some sign } \pm\}.$$

We have:

1. G has a unique root $(]c,d[,1)$ for some $d \in \partial Q$, because of the choice of $\epsilon > 0$.
2. if (A,k) has more than one successor, then $q(A,k) \geq q(\epsilon)$, (A,k) has two successors and each one satisfies: $q(B,k+1) = 0$;
3. if $(B,k) \to (B',k+1)$ then $q(B',k+1) \leq q(B,k)+1$.

From this it follows that the number of elements of order n is $\#\mathcal{R}_n \leq 2^{n/q(\epsilon)}$. But $q(\epsilon) \to \infty$ as $\epsilon \to 0$. This proves the lemma. □

7.3. A.E. exponential decay of $\operatorname{diam}(P^n(S^{-n}z))$.

LEMMA 7.4. *Let $0 < \alpha < 1$, $n = 1,2,\dots$ and I_0 be a subinterval of $[0,1]$ such that I_0 is included in some element of P^n. Then, for all $x \in I_0$:*

$$\frac{|I_n|}{|I_0|} \geq \frac{|(T^n)'(x)| \times \prod_{k\in B_n} |T'(T^k x)|}{\exp\left(\frac{K}{\alpha}\sum_{k=0}^{n-1} M_J(|I_k|)\right)(1+KM_J(1))^{\#B_n}}.$$

where $B_n = \{0 \leq k < n : |T'(T^k x)| < \alpha\}$, $I_k = T^k(I_0)$, K is from (R1) in the definition of regularity and M_J is the modulus of continuity of $J = |T'|$.

PROOF. We start from $|I_n|/|I_0| = |(T^n)'(y)|$ for some $y \in I_0$. Now,

$$\begin{aligned}
|(T^n)'(y)| &= \frac{|(T^n)'(x)|}{\prod_{k=0}^{n-1}\left|\frac{T'(T^k x)}{T'(T^k y)}\right|} = \frac{|(T^n)'(x)|}{\prod_{k=0}^{n-1}\left|\frac{T'(T^k x)-T'(T^k y)+T'(T^k y)}{T'(T^k y)}\right|}\\
&\geq \frac{|(T^n)'(x)|}{\prod_{k=0}^{n-1}\left(\frac{M_J(|I_k|)}{|T'(T^k y)|}+1\right)} \geq \frac{|(T^n)'(x)|}{\prod_{k=0}^{n-1}\left(\frac{KM_J(|I_k|)}{|T'(T^k x)|}+1\right)}\\
&\geq \frac{|(T^n)'(x)|}{\prod_{k\notin B_n}\left(\frac{KM_J(|I_k|)}{\alpha}+1\right)\times\prod_{k\in B_n}\left(\frac{KM_J(1)}{|T'(T^k x)|}+1\right)}\\
&\geq \frac{|(T^n)'(x)| \times \prod_{k\in B_n}|T'(T^k x)|}{\prod_{k\notin B_n}\left(\frac{K}{\alpha}M_J(|I_k|)+1\right)\times\prod_{k\in B_n}\left(KM_J(1)+|T'(T^k x)|\right)}\\
&\geq \frac{|(T^n)'(x)| \times \prod_{k\in B_n}|T'(T^k x)|}{\prod_{k\notin B_n}\left(\frac{K}{\alpha}M_J(|I_k|)+1\right)\times(KM_J(1)+\alpha)^{\#B_n}}.
\end{aligned}$$

Finally, use $1+x \leq e^x$ and $\alpha \leq 1$ to get the claim. □

COROLLARY 7.5. *Let $p : (Z,\nu,S) \to (X,\mu,T)$ be the natural extension as in section 1. For ν-a.e. $z \in Z$:*

$$\limsup_{n\to\infty}\frac{1}{n}\log\operatorname{diam}P^n(p(S^{-n}z)) = -\lambda$$

where λ is the exponent of μ, assumed to be > 0.

PROOF. Fix $\epsilon > 0$. Let $\alpha > 0$ be so small that $B \stackrel{\text{def}}{=} \{x : |T'(x)| < \alpha\}$ satisfies:

$$\mu(B) < \frac{\epsilon}{\log(1+KM_J(1))} \text{ and } \int_B |\log|T'||\,d\mu < \epsilon.$$

By Birkhoff's theorem, one can find $Z_1 \subset Z$ with $\nu(Z_1) > 0$ and $N_1 < \infty$ such that for all $z \in Z_1$ and $n \geq N_1$:

$$|(T^n)'(S^{-n}z)| \geq e^{(\lambda-\epsilon)n}$$
$$\prod_{k\in B_n} |T'(S^{-k}z)|| \geq e^{-\epsilon n}$$
$$(1+KM_J(1))^{\#B_n} \leq e^{\epsilon n}$$

where we have abbreviated $p(S^{-n}z)$ to $S^{-n}z$ and defined $B_n = \{0 \leq k < n : |T'(S^{-k}z)| < \alpha\}$.

If necessary increase N_1 so that:

$$\exp \frac{K}{\alpha}\left(M_J(1) + \sum_{k\geq 1} M_J(e^{-(\lambda-4\epsilon)k})\right) \leq e^{\epsilon N_1}$$

(recall that the left hand side is finite by the $C^{1+\Sigma}$ assumption —see section 3).

Now, fix the parameter N_P in the definition of P so large that:

$$N_P > e^{(\lambda-4\epsilon)N_1}.$$

We claim that for all $z \in Z_1$ and all $n \geq 0$:

$$|P^n(S^{-n}z)| \leq e^{-(\lambda-4\epsilon)n}. \tag{7.6}$$

First, this holds for $0 \leq n \leq N_1$ because of the small diameter of P, i.e., the above choice of N_P. We prove it for $n > N_1$ by induction. We assume that, for some $m \geq N_1$, eq. (7.6) holds for all $0 \leq n < m$ and prove it for $n = m$. The previous Lemma gives:

$$|P^m(S^{-m}z)| \leq \frac{\exp\left(\frac{K}{\alpha}\sum_{k=1}^{m} M_J(|P^k(S^{-k}z)|)\right) \times (1+KM_J(1))^{\#B_m}}{|(T^m)'(S^{-m}z)| \times \prod_{k\in B_m} |T'(S^{-m}z)|}$$

using $T^k(P^m(S^{-m}z)) \subset P^{m-k}(S^{-m+k}z)$. The induction hypothesis and the above ergodic estimates give:

$$|P^m(S^{-m}z)| \leq \frac{\exp\frac{K}{\alpha}\left(\sum_{k=1}^{m-1} M_J(e^{-(\lambda-4\epsilon)k}) + M_J(1)\right) \times e^{\epsilon m}}{e^{(\lambda-\epsilon)m} \times e^{-\epsilon m}}$$
$$\leq \exp\frac{K}{\alpha}\left(M_J(1) + \sum_{k\geq 1} M_J(e^{-(\lambda-4\epsilon)k})\right) \cdot e^{-(\lambda-3\epsilon)m} \leq e^{-(\lambda-4\epsilon)m}.$$

This completes the induction and deals with $z \in Z_1$.

To prove the statement for general $z \in Z$, remark that, as $\mu(Z_1) > 0$, for almost all $z \in Z$, there exists $m < \infty$ such that $S^{-m}z \in Z_1$ and, obviously, for all $n \geq m$:

$$P^n(S^{-n}z) \subset P^{-n+m}(S^{-n+m}(S^{-m}z)).$$

This deals with $z \in Z$. To get λ instead of $\lambda - 4\epsilon$, just recall that $\epsilon > 0$ was arbitrarily small. $\square$

7.4. **Conclusion of the Proof of Theorem B.** We gather the above facts to check the hypothesis of Proposition D.

Condition (i) ($m([0,1]) < \infty$ and $\mu(\mathrm{supp}(m)) = 1$) is obviously satisfied with m the Lebesgue measure on $[0,1]$.

Condition (iii) is: μ not shadowed by ∂P, $H_\mu(P) < \infty$ and $\operatorname{diam} P^n(\cdot) \to 0$ a.e. when $n \to \infty$. The first two points follow from Lemma 7.2 and the Claim before Lemma 7.3. Consider the third point. We have seen that: $\operatorname{diam}(P^n(S^{-n}z)) \to 0$ when $n \to \infty$ a.e. This implies that $\operatorname{diam}(P^n(x)) \to 0$ in probability. Hence, there exists a subsequence $n_1 < n_2 < \dots$ such that $\operatorname{diam}(P^{n_i}(x)) \to 0$ a.e. But $n \mapsto \operatorname{diam}(P^n(x))$ is non-increasing and thus convergence along a subsequence implies convergence of the full sequence. So condition (iii) is fulfilled.

For condition (ii), notice that $\log J \in L^1(\mu)$ as $J = |T'|$ is bounded and $\lambda \neq -\infty$. It remains to check the convergence of the infinite product

$$\prod_{k \geq 1} J(S^{-k}z_0)/J(S^{-k}z)$$

for a.a. $z_0 \in Z$, all $z \in Z$ such that: $P(S^{-k}z) = P(S^{-k}z_0)$ for all $k \geq 0$.

Notice that, because of $\log J \in L^1(\mu)$ (as $\lambda > -\infty$), we have, for every $\delta > 0$, for a.a. $z_0 \in Z$ and all large n:

$$|J(S^{-k}z_0)| \geq e^{-\delta k}.$$

Thus the convergence of the infinite product is precisely the regularity assumption (R3) and condition (ii) is fulfilled.

Hence we can apply Proposition D and we see that $P(\mu, T) = 0$ and that μ is absolutely continuous. Theorem B is proved. □

We finally check that, as claimed in the introduction, the following two classes of maps are $C^{1+\Sigma}$ regular.

LEMMA 7.7. *A PC-map T [**29**] is always $C^{1+\Sigma}$ regular.*

PROOF. First T is piecewise $C^{1+\beta}$ hence it is piecewise $C^{1+\Sigma}$.

Second, for each critical point $c\pm$ (i.e., $T'(c\pm) = 0$), there exist $C < \infty$ and $\alpha = \alpha(c, \pm) > 0$ such that: $C^{-1} \leq |T'(c \pm h)|/h^\alpha \leq C$ for all small enough $h > 0$. Hence $|T'(x)/T'(y)| \leq C^2 2^{2\alpha}$ for x, y near $c\pm$ and $\tau^{-1} < (x-c)/(y-c) < \tau$ with $\tau = 2$. Away from the critical points a similar inequality holds by the continuity of T'. Thus (R1) is satisfied for some large $K < \infty$.

(R2) is satisfied because, if $c\pm$ is a critical point, then: $|T'(c \pm 2^{-n})| \leq C 2^{-\alpha n}$. Hence:

$$\sum_n -|T'(c \pm 2^{-n})| \log |T'(c \pm 2^{-n})| \leq \sum_n C 2^{-\alpha n}(\alpha n \log 2 - \log C) < \infty.$$

(R3) is satisfied because, as $\log 1 + x \leq x$:

$$\sum_n |\log |T'(x_n)/T'(y_n)|| \leq \sum_n \frac{|T'(y_n) - T'(x_n)|}{|T'(x_n)|} \leq \sum_n \mathrm{const} \frac{|x_n - y_n|^\beta}{|x_n|^\alpha} < \infty$$

if $0 < \delta < \lambda\beta/\alpha$. □

LEMMA 7.8. *A piecewise $C^{1+\Sigma}$ map T with non-flat critical points is $C^{1+\Sigma}$ regular.*

PROOF. Such a T is piecewise $C^{1+\Sigma}$ by definition.

For x, y close to a critical point $c+$, $\tau^{-1} < (x-c)/(y-c) < \tau$ implies:

$$|T'(x)/T'(y)| \leq |C(x)/C(y)|\tau^{2\alpha}\left(1 + \frac{\log\tau}{|\log|x-c||}\right)^{\beta} \leq 2\tau^{2\alpha}(1+\log\tau)^{\beta}$$

Continuity of T' away from critical points is now enough to prove Property (R1).

To prove (R2) compute:

$$-\sum_n |T'(c\pm\tau^{-n})|\log|T'(c\pm\tau^{-n})|$$
$$\leq \sum_n C(\tau^{-n})(\log\tau)^{-\beta}\cdot n^{-\beta}\tau^{-\alpha n}(\alpha\log\tau\cdot n + |\beta|\log n + \log C(\tau^{-n}) + |\beta|\log\tau)$$

and remark that this series converges for $\alpha > 0$ and whatever β, and for $\alpha = 0$ and $\beta > 1$.

To prove (R3), remark first that, for $x, y > 0$:

$$\begin{aligned}\left|x^{\alpha}|\log x|^{-\beta} - y^{\alpha}|\log y|^{-\beta}\right| &\leq \text{const}\cdot x^{\alpha-1}|\log x|^{-\beta}(\alpha + \beta/|\log x|)\times|x-y| \\ &\leq \text{const}\cdot x^{\alpha-1}|\log x|^{-\beta}\times|x-y|\end{aligned}$$

if $x/2 \leq y \leq 2x$. Then compute:

$$\begin{aligned}\sum_n |\log|T'(x_n)/T'(y_n)|| &\leq \sum_n \frac{|T'(y_n)-T'(x_n)|}{|T'(x_n)|} \\ &\leq \sum_n \frac{C(y_n)\text{const}\cdot\|x_n\|^{\alpha-1}|\log\|x_n\||^{-\beta}\times|x_n-y_n|}{\text{const}\cdot\|x_n\|^{\alpha}|\log\|x_n\||^{-\beta}} \\ &\quad + \frac{M_C(|x_n-y_n|)|T'(x_n)|}{|T'(x_n)|} \\ &< \sum_n \text{const}\frac{|x_n-y_n|}{\|x_n\|} + M_C(|x_n-y_n|) \\ &\leq \sum_n \text{const}\cdot e^{-(\lambda-\delta)n} + M_C(e^{-\lambda n}) < \infty.\end{aligned}$$

where $\|x\| = \min_c |x-c|$ and M_C is a common modulus of continuity of the functions C appearing in the definition of non-flat critical points. Thus, (R3) is proved. □

Part IV: Equilibrium States

8. Locally eventually onto piecewise affine and expanding maps

We prove Theorem C following the strategy developped in a one-dimensional setting by M. Denker, G. Keller and M. Urbański [**16,17**], but using our abstract results instead of their Theorem 1 [**17**] and also the boundary pressure estimates from [**6**]. Let (X, P, T) and $\varphi : Y \to \mathbb{R}$ be as in the theorem. By adding a constant, we assume that $P_{\text{top}}(T) = 0$ (here and below, the pressures are understood to be w.r.t. φ).

We build an $e^{-\varphi}$**-conformal measure** m on $\bigcap_{n\geq 0} T^{-n}Y$ to be able to apply our proposition E. Conformality means that m is a probability measure concentrated on $\bigcap_{n\geq 0} T^{-n}Y$ such that for every **special set**, i.e., every subset $S \subset X$

such that S and $T(S)$ are measurable and the restriction $T : S \to T(S)$ is a measurable isomorphism, we have:

$$m(T(S)) = \int_S e^{-\varphi}\, dm.$$

T being locally eventually onto (l.e.o., eq. (0.6)), such an m will have full support, allowing the application of Proposition E from which the conclusions of Theorem C will follow immediately (indeed, l.e.o.+open support of the equilibrium state give the uniqueness).

We follow [**16**] for the construction of m. First, we need to consider a continuous map and a continuous potential on a compact space: let (Σ_+, σ) be the symbolic dynamics together with the lift of the potential $\varphi : \Sigma_+ \to \mathbb{R}$. As T is expanding, we have a well-defined projection $\Pi : \Sigma_+ \to X$ defined by $\{\Pi(A)\} = \bigcap_{n\geq 0} \overline{[A_0 \dots A_n]}$. It gives an isomorphism between $(\Sigma_+ \setminus \bigcup_{n\geq 0} \sigma^{-n}\Pi^{-1}(\partial P), \sigma)$ and $(\bigcap_{n\geq 0} T^{-n}Y, T)$.

We make the following easy observations (in the language of [**16**]):

1. $\sigma|\Sigma_+$ has **no critical points**: every point admits a neighborhood U which is a special set.
2. the **singular points** of $\sigma|\Sigma_+$ are in $\Pi^{-1}(\partial P)$: outside this set the map σ is open.

We define sets E_n, $n \geq 1$ to apply the construction principle of [**16, p.372**].

Assume that the distance on Σ_+ is $d(u, v) = 2^{-n}$ with $n = \inf\{k \geq 0 : u_k \neq v_k\}$. Fix $\epsilon = 1$. Set $E_0 = \emptyset$ and let $E_{n+1} \supset \sigma^{-1}E_n$ be some maximal $(\epsilon, n+1)$-separated set in Σ_+ (i.e., E_{n+1} contains a point in every $n+1$-cylinder in Σ_+). The construction principle then gives a probability measure m which we shall show to be a conformal measure.

We are going to prove condition (3.6) of [**16**] with $D = X$. Inspection shows that this condition is fulfilled if:

$$\limsup_{n\to\infty} \frac{1}{n} \log \sum_{u\in E_{n+1}\setminus\sigma^{-1}E_n} e^{S_n\varphi(u)} \leq P_{\text{top}}(\partial P, T)$$

where $S_n\varphi = \sum_{k=0}^{n-1} \varphi \circ \sigma^k$. (Recall that $P_{\text{top}}(\partial P, T) < P_{\text{top}}(T)$.)

This inequality is an immediate consequence of the following claim: let $u \in E_{n+1} \setminus \sigma^{-1}E_n$. u is then contained in a $n+1$-cylinder meeting $\Pi^{-1}(\partial P)$.

Let $A = u_0$. $\sigma(u)$ is in the n-cylinder of some $v \in E_n$, as this set (ϵ, n)-spans by construction. If $\Pi(v) \in T(A)$, then $\Pi(Av)$ would be in A, hence $Av \in \Sigma_+$, so that $Av \in \sigma^{-1}E_n$. But, clearly, u would then be in the $n+1$-cylinder defined by Av and yet distinct from it as $u \notin \sigma^{-1}E_n$: a contradiction to the fact that E_{n+1} is $(\epsilon, n+1)$-separated. Thus $\Pi(v) \notin T(A)$.

We now use the assumption that (X, P, T) *is piecewise affine* and more precisely the following consequence of it: the Π-image of every cylinder is convex, therefore connected. Therefore, the n-cylinder defined by v must project by Π accross $\partial T(A)$, as its image contains points both inside $(\Pi(\sigma(u)))$ and outside $(\Pi(v))$ of $T(A)$: there exists $w \in \Pi^{-1}(\partial TA)$ such that w is in the n-cylinder of v. Of course, the $n+1$-cylinder of $Aw \in \Pi^{-1}(\partial P)$ contains u. But this was our claim. Thus, condition (3.6) of [**16**] with $D = X$ holds.

By Lemma 3.3 of [**16**], this implies that for every special set $A \subset \Sigma_+ \setminus \Pi^{-1}(\partial P)$:

$$m(\sigma A) = \int_A e^{-\varphi}\, dm.$$

It remains to check this for special sets included in $\Pi^{-1}(\partial P)$. This will be clear if we prove that:

$$m\left(\bigcup_{k\geq 0} \sigma^k(\Pi^{-1}(\partial P))\right) = 0. \tag{8.1}$$

Now the proof of Lemma 3.4 of [**16**] gives that, for every compact subset K contained in some n-cylinder:

$$\int_K e^{-S_n\varphi}\, dm \leq m(\sigma^n K) \leq 1.$$

Therefore, a computation similar to eq. (3.4) implies that $m(K) = 0$ for every compact subset K of Σ_+ with topological pressure < 0. But $P_{\text{top}}(\sigma^k(\Pi^{-1}(\partial P)), \sigma) = P_{\text{top}}(\Pi^{-1}(\partial P), \sigma)$ which is equal to $P_{\text{top}}(\partial P, T) < P_{\text{top}}(T) = 0$. Hence eq. (8.1) holds: we have proved that m is indeed a $e^{-\varphi}$-conformal measure on Σ_+.

As $e^{-\varphi} > 0$, eq. (8.1) implies: $m\left(\bigcup_{k\geq 0} \sigma^{-k}(\Pi^{-1}(\partial P))\right) = 0$. Hence we can identify m with a measure on $\bigcap_{k\geq 0} T^{-k}Y$, which is clearly again an $e^{-\varphi}$-conformal measure. This finishes the construction of m and the proof of Theorem C. □

PROOF OF COROLLARY 0.7. We prove that the hypothesis of Theorem C are fulfilled, namely that such a T is l.e.o. and satisfies $P_{\text{top}}(\partial P, T) < P_{\text{top}}(T)$.

Clearly, $P_{\text{top}}(\partial P, T) \leq h_{\text{top}}(\partial P, T) + \sup \varphi$. By [**6, Prop. 4**]:

$$h_{\text{top}}(\partial P, T) \leq \sup_{(x,V)} \log J_{d-1}(x, V) + h_{\text{mult}}(T, P)$$

where $h_{\text{mult}}(T, P)$ is the entropy multiplicity defined in Section 3 and $J_{d-1}(x, V)$ is the Jacobian of $T'(x) : V \to T'(x).V$ w.r.t. $d-1$-dimensional Euclidean volume, x ranging over Y and V over the $d-1$-dimensional subspaces of $\mathbb{R}^d$. Now, $|\det B| = |\det T'(x)| \geq J_{d-1}(x, V) \cdot \Lambda$ and $h_{\text{mult}}(T, P) = 0$ for a β-transformation according to [**6, Lemma 1**]. Hence:

$$P_{\text{top}}(\partial P, T) \leq \log|\det B| - \log\Lambda + \sup\varphi.$$

Similarly,

$$P_{\text{top}}(T) \geq h_{\text{top}}(T) + \inf\varphi = \log|\det B| + \inf\varphi.$$

Hence, $\sup\varphi - \inf\varphi < \log\Lambda$ implies $P_{\text{top}}(\partial P, T) < P_{\text{top}}(T)$.

It remains to see that the β-transformation T is l.e.o. under the condition: $\Lambda > 1 + \sqrt{d}$. But this is [**6, Lemma 5**] and the remark that follows. The corollary is proved. □

References

[1] K. Adl-Zarabi, *Absolutely continuous invariant measures for piecewise expanding C^2 transformations in R^n on domains with cusps on the boundaries*, Ergodic Theory Dynam. Systems **16** (1996), 1–18.

[2] R. Bowen, *Equilibrium states and the ergodic theory of Anosov diffeomorphisms*, LNM 470, Springer, 1975.

[3] M. Blank, *Discreteness and continuity in problems of chaotic dynamics*, Translations of Mathematical Monographs, 161, American Mathematical Society, Providence, RI, 1997.

[4] H. Bruin, G. Keller, *Equilibrium states for S-unimodal maps*, Ergod. Th. & Dynam. Syst. **18** (1998), 765–789.

[5] J. Buzzi, *Entropies et représentation markovienne des applications régulières de l'intervalle*, Thèse, Université Paris-Sud, Orsay, 1995.

[6] ———, *Intrinsic Ergodicity of Affine Maps on* $[0,1]^d$, Monat. Math. **124** (1997), 97–118.

[7] ———, *Intrinsic Ergodicity of Smooth Interval Maps*, Israel J. Math. **100** (1998), 125–161.

[8] ———, *Markov extensions for multi-dimensional dynamical systems*, Israel J. Math. **112** (1999), 357–380.

[9] ———, *Absolutely continuous invariant measures for generic piecewise affine and expanding maps*, Int. J. Chaos & Bif. **9** (1999), 1743–1750.

[10] ———, *Absolutely continuous invariant measures for real-analytic piecewise expanding maps of the plane*, Ergod. Th. & Dynam. Syst. **20** (2000), 697–708.

[11] J. Buzzi, V. Maume-Deschamps, *Decay of correlations for multi-dimensional piecewise invertible dynamical systems*, preprint (1999).

[12] J. Buzzi, F. Paccaut, B. Schmitt, *Conformal measures for multi-dimensional piecewise invertible dynamical systems*, Ergod. Th. & Dynam. Syst. (to appear).

[13] P. Collet, *Some ergodic properties of maps of the interval*, Dynamical systems (Temuco, 1991/1992), Travaux en Cours, 52, Hermann, Paris, 1996, pp. 55–91.

[14] P. Collet, S. Isola, *On the essential spectrum of the transfer operator for expanding maps*, Commun. Math. Phys. **139** (1991), 551–557.

[15] B. Cowieson, *Piecewise expanding maps in R^d*, Ph. D. thesis, U.C.L.A., Berkeley, 1998.

[16] M. Denker, M. Urbański, *On the existence of conformal measures*, Trans. Amer. Math. Soc. **328** (1991), 563–587.

[17] M. Denker, G. Keller, M. Urbański, *On the uniqueness of equilibrium states for piecewise monotone mappings*, Studia math. **97** (1990), 27–36.

[18] P. Góra, A. Boyarsky, *Absolutely continuous invariant measures for piecewise expanding C^2 transformation in R^N*, Israel J. Math. **67** (1989), 272–286.

[19] F. Hofbauer, *β-shifts have unique maximal measure*, Mh. Math. **85** (1978), 189–198.

[20] F. Hofbauer, G. Keller, *Ergodic properties of invariant measures for piecewise monotonic transformations.*, Math. Z. **180** (1982), 119–140.

[21] F. Hofbauer, P. Raith, *The Hausdorff dimension of an ergodic invariant measure for a piecewise monotonic map of the interval*, Canad. Math. Bull. **35** (1992), 84–98.

[22] A. Katok, *Lyapunov exponents, entropy and periodic orbits for diffeomorphisms*, Publ. Math. I.H.E.S. **51** (1980), 137–173.

[23] G. Keller, *Propriétés ergodiques des endomorphismes dilatants, C^2 par morceaux, des régions bornées du plan*, Thèse, Université de Rennes, 1979.

[24] ———, *Generalized bounded variation and applications to piecewise monotonic transformations*, Z. Wahrsch. Verw. Gebiete **69** (1985), 461–478.

[25] ———, *On the rate of convergence to equilibrium in one-dimensional systems*, Comm. Math. Phys. **96** (1984), 181–193.

[26] ———, *Lifting measures to Markov extensions*, Mh. Math. **108** (1989), 183–200.

[27] A. Kondah, V. Maume, B. Schmitt, *Vitesse de convergence vers l'état d'équilibre pour des dynamiques markoviennes non höldériennes*, Ann. Inst. H. Poincaré Probab. Statist. **33** (1997), 675–695.

[28] F. Ledrappier, *Principe variationnel et systèmes dynamiques symboliques*, Z. Wahrscheinlichkeitstheorie und Verw. Gebiete **30** (1974), 185–202.

[29] ———, *Some properties of absolutely continuous invariant measures of an interval*, Ergod. Th. & Dynam. Sys. **1** (1981), 77–93.

[30] ———, *Propriétés ergodiques des mesures de Sinaï*, Publ. Math. Inst. Hautes Etudes Sci. **59** (1984), 163–188.

[31] F. Ledrappier, J.-M. Strelcyn, *A proof of the estimation from below in Pesin's entropy formula*, Ergod. Th. Dynam. Syst. **2** (1982), 203–219.

[32] F. Ledrappier, L.-S. Young, *The metric entropy of diffeomorphisms. Part I: Characterization of measures satisfying Pesin's entropy formula*, Annals Math. **122** (1985), 509–539.

[33] C. Liverani, B. Saussol, S. Vaienti, *Conformal measure and decay of correlations for covering weighted systems*, Ergod. Th. Dynam. Syst. **18** (1998), 1399–1420.

[34] D.J. Rudolph, *Fundamentals of measurable dynamics*, Clarendon Press, Oxford, 1990.

[35] D. Ruelle, *A measure associated with Axiom A attractors*, Amer. J. Math. **98** (1976), 619–654.

[36] ———, *An inequality for the entropy of differentiable maps*, Bol. Soc. Bras. Mat. **9** (1978), 83–87.

[37] V.A. Rohlin, *Exact endomorphism of a Lebesgue space*, Amer. Math. Soc. Transl. (2) **39** (1964), 1–36.

[38] B. Saussol, *Absolutely continuous invariant measures for multidimensional expanding maps*, preprint 1998 (http://www.cpt.univ-mrs.fr).

[39] B. Schmitt, *Contributions à l'étude de systèmes dynamiques unidimensionnels en théorie ergodique*, Thèse, Dijon, 1986.

[40] Y. Sinai, *Gibbs measure in ergodic theory*, Russ. Math. Surveys **27** (1972), 21–69.

[41] M. Tsujii, *Absolutely continuous invariant measures for piecewise real-analytic expanding maps on the plane*, Commun. Math. Phys. **208** (2000), 605–622.

[42] ———, *Absolutely continuous invariant measures for piecewise affine and expanding maps*, Preprint Hokkaido University (1998).

[43] P. Walters, *Equilibrium states for β-transformations and related transformations*, Math. Z. **159** (1978), 65–88.

[44] ———, *An introduction to ergodic theory*, Springer, 1981.

CMAT / U.M.R. 7640 du C.N.R.S. / École Polytechnique, 91128 Palaiseau Cedex, France

E-mail address: buzzi@math.polytechnique.fr

Proceedings of Symposia in Pure Mathematics
Volume **69**, 2001

Smoothness of Holonomy Maps Derived from Unstable Foliation

M. Guysinsky

ABSTRACT. We prove that the unstable foliation holonomy on an area preserving Anosov diffeomorphism on a two-dimensional torus has the derivative with bounded variation only if the diffeomorphism is smoothly conjugate to a linear map.

1. Historic background

In 1962 V. I. Arnold and Ya. G. Sinai published a paper [**Arnold, Sinai**] where they proved existence of what we now call stable and unstable foliations for small perturbations of hyperbolic linear map on a torus. For a two-dimensional torus they used these foliations to prove that the perturbed map is conjugate to the original linear map. A little later D. V. Anosov [**Anosov**] proved that it is true for any uniformly hyperbolic (called now Anosov) map and J. Franks [**Frank**] proved that any Anosov map on the n dimensional torus is conjugate to a linear map. But in early sixties people just started to work in this area, and paper by V. I. Arnold and Ya. G. Sinai was one of the first steps there. The idea was quite elegant. We can choose a closed curve transversal to the unstable foliation and homotopic to one of the two generators of the fundamental group. Using the unstable foliation we can define the first return map on this curve. Take any point on this curve and follow the unstable curve passing through the point till the next intersection with the curve. It is proved in [**Arnold, Sinai**] that this map is well defined. Moreover, we can think about this curve as a circle, therefore we get a map from a circle to itself. By Denjoy theorem this map is conjugate to a rotation. After that it is not difficult to find a change of coordinates conjugating the linear map on the torus and the perturbed one. The only flaw in this proof is that Denjoy theorem could be applied only if the map is smooth enough. In [**Arnold, Sinai**] authors claim that Oceledec (who was an undergraduate student at the Moscow State University in early sixties) proved in his semester paper that the first return map described above has the first derivative of bounded variation. This is exactly what you need in order to apply the Denjoy theorem. But very soon after publishing the paper V. I. Arnold and Ya. G. Sinai published a correction [**Arnold, Sinai1**] where they admitted that the smoothness that they can prove is not enough to finish the

2000 *Mathematics Subject Classification.* Primary 37E25, 37C15.

proof. Later D. V. Anosov showed that even for an area–preserving map on the two dimensional torus the angle between the stable and unstable foliations is not C^2 if some data in periodic points is not 0(see [**Anosov**]). Then S. Hurder and A. Katok [**Hurder, Katok**] studied the smoothness of the angle between stable and unstable directions for area preserving Anosov maps on the two–dimensional torus or volume preserving Anosov flows on the three–dimensional closed manifold. They generalized Anosov's construction and introduced a special cocycle (they called it Anosov cocycle) and proved that if this cocycle is not a coboundary then the first derivative of the angle along stable or unstable curves belongs to Zigmund class and not differentiable, and if the first derivative has higher smoothness then in fact Anosov cocycle is a coboundary and the angle function is C^∞. This does not necessarily imply that variation is not bounded, but it gives serious doubts about it. Below (see **Corollary**) I show that, in fact, if Anosov cocycle is not a coboundary, then the first derivative for the first return map can not have a bounded variation for any segment. The only case when variation is bounded is when Anosov cocycle is a coboundary and then the first return map is C^∞ and the map is C^∞ conjugate to a linear map (see [**Hurder, Katok**])

Does it mean that the approach was wrong? The answer is not necessarily. At least for one special case, if a map is area–preserving, now, almost 40 years later this gap in the proof can be fixed. From the results published in 1990 by S. Hurder and A. Katok [**Hurder, Katok**] follows that the first derivative of the first return map belongs to Zigmund class and in the survey written by M. Jacobson and G. Swiatek (to be published, see [**Jacobson, Swiatek**]) the Denjoy theorem is proved under exactly these conditions.

2. Definitions and results

Let T^2 be a two–dimensional torus and $f : T^2 \to T^2$ be an Anosov, area–preserving diffeomorphism. Let E^s_p,E^u_p and L^s_p,L^u_p be stable,unstable directions and manifolds for f at the point p. We can define (see [**DeLatte**]) for every point $p \in T^2$ new local coordinates $(\xi_p(x,y), \psi_p(x,y))$ with p at the origin, which have the following properties:

(1) They are C^∞ in x, y and functions $\xi_p(x,y), \psi_p(x,y)$ and all their derivatives depend C^1 on p.
(2) The change of coordinates is area–preserving for any p.
(3) After the change of coordinates at p and $f(p)$ the map f has the following form

$$f(\tilde{x},\tilde{y}) = \left(\lambda_p \tilde{x} \Phi_p(\tilde{x}\tilde{y}), \lambda_p^{-1} \tilde{y} \Phi_p^{-1}(\tilde{x}\tilde{y})\right) \tag{1}$$

in the r–neighborhood of the point p (r doesn't depend on p), λ_p is the contraction rate for the E^s, $\Phi_p(0) = 1$, $\Phi_p(z)$ is C^∞ in z and $\Phi_p(z)$ and all its derivatives C^1 depend on p.
(4) In new coordinates, curves L^s_p and L^u_p have the form $(\tilde{x}, 0)$ and $(0, \tilde{y})$ in the r-neighborhood of the point p .

Definition 1. $\phi(p) = \Phi'_p(0)$

It is easy to see that $\phi(p)$ is a C^1 cocycle. It is called Anosov cocycle.

Assume that we have two C^2–curves $l(t)$ and $k(t)$ $(0 \le t \le 1)$ transversal to the unstable foliation. These two curves and unstable foliation define a holonomy map which we denote(by slightly abusing notations) as $H(l(t))$ or $H(t)$ or $H_{l,k}$.

Theorem 1. Let $p = l(t_0)$ and $q = H(p)$ where $0 < t_0 < 1$. If $\sum_{i=1}^{n} \phi(f^{-i}(q))$ is not bounded over n, then $H''(t_0)$ doesn't exist.

Remark. $\sum_{i=1}^{n} \phi(f^{-i}(q))$ is bounded if and only if $\sum_{i=1}^{n} \phi(f^{-i}(p))$ is bounded, where p is any point that belongs to L^u_q.

Corollary. If $\phi(p)$ is not a coboundary, then $H''(t)$ doesn't exist a.e.. Therefore $H'(t)$ is not absolutely continuous and variation of $H'(t)$ is unbounded for any segment.

3. Proofs

Proof (Corollary):

If $H''(t)$ is defined on a set of non–zero measure, then from the Theorem 1 and Remark we see that $\sum_{i=1}^{n} \phi(f^{-i}(p))$ is bounded on a set of non–zero area. Set of points where this sum is bounded is invariant under the map f. Thus from the ergodicity of $f(x)$ it follows that $\sum_{i=1}^{\infty} \phi(f^{-i}(p))$ is bounded a.e.. Therefore we can find a transitive point such that at this point the sum is bounded and conclude that the sum is zero for all periodic points. From the Livshitz Theorem follows that ϕ is a coboundary.

Proof (Theorem 1):

Case 1: First we consider the case when $l(t)$ and $k(t)$ are parts of L^s_p and L^s_q and $dist(p, q) < r$ (see properties 3 and 4). Let $(x, y)_p$ be a point with coordinates (x, y) in the local coordinates for the point p. Then, $p = (0, 0)_p$ and $q = (0, y)_p$ for some value y. We can also parameterize curves $l()$ and $k()$ as $(x, 0)_p$ and $(x, y)_p$. We denote as $(a_p(x, y), b_p(x, y))_p$ the coordinates for the point of intersection of $L^s_{(0,y)_p}$ and $L^u_{(x,0)_p}$. We see that $H(x) = H((x, 0)_p) = a_p(x, y)$.

Let $c_p(x, y) = a_p(x, y)b_p(x, y)$. From the definition of a_p, b_p and c_p and properties of stable and unstable foliations follows that

(1) The functions $a'_{p,x}$, $a'_{p,y}$, $a''_{p,xy} = a''_{p,yx}$, $b'_{p,x}$,$b'_{p,y}$, $b''_{p,xy} = b''_{p,yx}$, $c'_{p,x}$, $c'_{p,y}$, $c''_{p,xy} = c''_{p,yx}$ exist and depend continuously on p.

(2)

$$\begin{array}{lll} a(x,0) = x & b(x,0) = 0 & c(x,0) = 0 \\ a'_x(x,0) = 1 & b'_x(x,0) = 0 & c'_x(x,0) = 0 \\ a(0,y) = 0 & b(0,y) = y & c(0,y) = 0 \\ a'_y(0,y) = 0 & b'_y(0,y) = 1 & c'_y(0,y) = 0\,. \end{array} \tag{2}$$

From the formula (1) it follows that

$$f((x, 0)_p) = (\lambda_p x, 0)_{f(p)} \quad \text{and} \quad f((0, y)_p) = (0, \lambda_p^{-1} y)_{f(p)}$$

and hence

$$f(L^s_{(0,y)_p} \cap L^u_{(x,0)_p}) = f(L^s_{(0,y)_p}) \cap f(L^u_{(x,0)_p}) = L^s_{(0,\lambda_p^{-1}y)_{f(p)}} \cap L^u_{(\lambda_p x,0)_{f(p)}}$$

or

$$f(a_p(x, y), b_p(x, y)) = (a_{f(p)}(\lambda_p x, \lambda_p^{-1} y), b_{f(p)}(\lambda_p x, \lambda_p^{-1} y))\,.$$

On the other hand if we apply the formula (1) directly to the point $(a_p(x, y), b_p(x, y))$ we see that

$$f(a_p(x, y), b_p(x, y)) = (\lambda_p a_p(x, y)\Phi_p(c_p(x, y), \lambda_p^{-1} b_p(x, y)\Phi_p^{-1}(c_p(x, y))$$

therefore

$$a_{f(p)}(\lambda_p x, \lambda_p^{-1} y) = \lambda_p a_p(x, y)\Phi_p(c_p(x, y)) \tag{3}$$

and

$$c_{f(p)}(\lambda_p x, \lambda_p^{-1} y) = c_p(x,y)\,.$$

Let us iterate formula (3) n times to obtain the general formula

$$(4)\qquad a_{f^n(p)}\Big(\prod_{i=0}^{n-1}\lambda_{f^i(p)}x, \prod_{i=0}^{n-1}\lambda^{-1}_{f^i(p)}y\Big) = \prod_{i=0}^{n-1}\lambda_{f^i(p)}a_p(x,y)\prod_{i=0}^{n-1}\Phi_{f^i(p)}(c_p(x,y))\,.$$

If we apply the last formula to the point $f^{-n}(p)$ instead of p, and to $(x, \Lambda_n y)$ instead of (x,y) and take $\Lambda_n = \prod_{i=1}^{n}\lambda_{f^{-i}(p)}$ then instead of formula (4) we obtain

$$(5)\qquad a_p(\Lambda_n x, y) = \Lambda_n a_{f^{-n}(p)}(x, \Lambda_n y)\prod_{i=1}^{n}\Phi_{f^{-i}(p)}(c_{f^{-n}(p)}(x,\Lambda_n y))\,.$$

After differentiating both sides of the formula (5) with respect to x we see that

$$\Lambda_n a'_{p,x}(\Lambda_n x, y) =$$

$$\left[\Lambda_n(a'_{f^{-n}(p),x} + a_{f^{-n}(p)}\sum_{i=1}^{n}\frac{\Phi'_{f^{-i}(p)}c_{f^{-n}(p)}c'_{f^{-n}(p),x}}{\Phi_{f^{-i}(p)}})\prod_{i=1}^{n}\Phi_{f^{-i}(p)}(c_{f^{-n}(p)})\right](x,\Lambda_n y)$$

or if we denote $f^{-n}(p)$ as p_i and cancel Λ_n

$$(6)\qquad a'_{p,x}(\Lambda_n x, y) = \left[\left(a'_{p_n,x} + a_{p_n}\sum_{i=1}^{n}\frac{\Phi'_{p_i}(c_{p_n})c'_{p_n,x}}{\Phi_{p_i}}\right)\prod_{i=1}^{n}\Phi_{p_i}(c_{p_n})\right](x,\Lambda_n y)\,.$$

We are going to expand functions from the righthandside of (6) assuming that Λ_n is very small. We shall use formulas (2)

$$(7)\qquad \begin{aligned}
a_{p_n}(x,\Lambda_n y) &= a_{p_n}(x,0) + a'_{p_n,y}(x,0)\Lambda_n y + o(\Lambda_n)\\
&= x + a'_{p_n,y}(x,0)\Lambda_n y + o(\Lambda_n)\\
c_{p_n}(x,\Lambda_n y) &= c_{p_n}(x,0) + c'_{p_n,y}(x,0)\Lambda_n y + o(\Lambda_n)\\
&= c'_{p_n,y}(x,0)\Lambda_n y + o(\Lambda_n)\\
a'_{p_n,x}(x,\Lambda_n y) &= a'_{p_n,x}(x,0) + a''_{p_n,xy}(x,0)\Lambda_n y + o(\Lambda_n)\\
&= 1 + a''_{p_n,xy}(x,0)\Lambda_n y + o(\Lambda_n)\\
c'_{p_n,x}(x,\Lambda_n y) &= c'_{p_n,x}(x,0) + c''_{p_n,xy}(x,0)\Lambda_n y + o(\Lambda_n)\\
&= c''_{p_n,xy}(x,0)\Lambda_n y + o(\Lambda_n)\\
\Phi_{p_i}(c_{p_n}(x,\Lambda_n y)) &= \Phi_{p_i}(c'_{p_n,y}(x,0)\Lambda_n y + o(\Lambda_n))\\
&= 1 + \phi(p_i)c'_{p_n,y}(x,0)\Lambda_n y + o(\Lambda_n)\\
\Phi'_{p_i}(c_{p_n}(x,\Lambda_n y)) &= \phi(p_i) + O(\Lambda_n)\,.
\end{aligned}$$

All $o()$ and $O()$ in the formulas above can be chosen uniformly over p. If we put $x = 0$ in the equation (6) then we have that

$$a'_{p,x}(0,y) = a'_{p_n,x}(0,\Lambda_n y) + a_{p_n}(0,\Lambda_n y)\sum_{i=1}^{n}c'_{p_n,x}(0,\Lambda_n y)$$

or

$$a'_{p,x}(0,y) = 1 + O(n\Lambda_n)\,.$$

Using the fact that Λ_n is exponentially decreasing with n we see that

$$a'_{p,x}(0,y) = 1\,.$$

In the same way we can prove that

$$b'_{p,y}(x,0) = 1\,. \tag{8}$$

Therefore

$$\begin{aligned} c'_{p,y}(x,0) &= a'_{p,y}(x,0)b_p(x,0) + a_p(x,0)b'_{p,y}(x,0) = x \\ c''_{p,xy}(x,0) &= c''_{p,yx}(x,0) = 1\,. \end{aligned} \tag{9}$$

From the expansions (7) and exponential decreasing of Λ_n follows that

$$\sum_{i=1}^{n} \frac{\Phi'_{p_i}(c_{p_n})c'_{p_n,x}}{\Phi_{p_i}}(x,\Lambda_n y) = \sum_{i=1}^{n} \phi(p_i)c''_{p_n,xy}(x,0)\Lambda_n y + o(\Lambda_n)$$

or

$$\sum_{i=1}^{n} \frac{\Phi'_{p_i}(c_{p_n})c'_{p_n,x}}{\Phi_{p_i}}(x,\Lambda_n y) = \sum_{i=1}^{n} \phi(p_i)\Lambda_n y + o(\Lambda_n) \tag{10}$$

and

$$\prod_{i=1}^{n} \Phi_{p_i}(c_{p_n}(x,\Lambda_n y)) = 1 + \sum_{i=1}^{n} \phi(p_i)c'_{p_n,y}(x,0)\Lambda_n y + o(\Lambda_n)$$

or

$$\prod_{i=1}^{n} \Phi_{p_i}(c_{p_n}(x,\Lambda_n y)) = 1 + \sum_{i=1}^{n} \phi(p_i)x\Lambda_n y + o(\Lambda_n)\,. \tag{11}$$

Finally, putting formulas (7), (10) and (11) in (6) we have that

$$a'_{p,x}(\Lambda_n x, y) = 1 + a''_{p_n,xy}(x,0)\Lambda_n y + 2xy\Lambda_n \sum_{i=1}^{n} \phi(p_i) + o(\Lambda_n)\,. \tag{12}$$

Last formula allows us to prove Theorem 1,

$$\frac{a'_{p,x}(\Lambda_n, y) - a'_{p,x}(0,y)}{\Lambda_n x} = a''_{p_n,xy}(x,0)\frac{y}{x} + 2x\sum_{i=1}^{n} \phi(p_i) + \frac{o(\Lambda_n)}{\Lambda_n x}\,. \tag{13}$$

It follows immediately from the last equality that $a''_{p,xx}(0,y)$ does not exist if $\sum_{i=1}^{n} \phi(p_i)$ is not bounded.

Case 2: Assume now that curves $l()$ and $k()$ are transversal at the points p and q to the unstable foliation, but still $dist(p,q) < r$. We also assume that the curve l is parameterized as $(x, l(x))_p$ in the neighborhood of the point p.

Let us denote as $H_{L^s_p,l}$ the holonomy map from L^s_p to l. Then $H_{L^s_p,l}(0) = 0$, hence $|H_{L^s_p,l}(x)| < c_0 x$ for small x. We are going to prove that $H_{L^s_p,l}(x) = x + O(x^2)$. Indeed, for the function $l(x)$ we have that

$$l(H_{L^s_p,l}(x)) = c_1 x + o(x)\,.$$

We denote as $Angle(x,y)_p$ the angle between the stable and unstable manifolds at the point $(x,y)_p$ in new coordinates at the point p. Then $Angle(0,0)_p = \frac{\pi}{2}$ The angle function is continuously differentiable with respect to x and y. Hence if we know that $|y| < c_1 x + o(x)$ then

$$Angle(x,y)_p = \frac{\pi}{2} + c_2 x + c_3 y + o(\sqrt{x^2+y^2}) = \frac{\pi}{2} + c_4 x + o(x)$$

$$|\cos(Angle(x,y)_p)| < c_5 x\,.$$

In order to find $H_{L_p^s,l}(x) - x$ we have to integrate the function $\cos(Angle(x,y)_p)$ along unstable foliation between points $(x,0)_p$ and $(x, H_{L_p^s,l}(x))$. The part of the unstable curve between these two points has the length bounded by $c_6 x$ hence

$$H_{L_p^s,l}(x) - x = O(x^2).$$

The same way we can prove that

$$H_{L_q^s,k} = x + O(x^2)$$

and hence

$$H_{k,L_q^s} = H^{-1}_{L_q^s,k} = x + O(x^2).$$

Finally, the holonomy map $H_{L_p^s,L_q^s}$ from *Case 1* can be written as

$$H_{L_p^s,L_q^s} = H^{-1}_{L_q^s,k} \circ H_{l,k} \circ H_{L_p^s,l}$$

and if $H''_{l,k}$ exists, then

$$H_{L_p^s,L_q^s} = x + O(x^2).$$

The last equality contradicts formula (13), if $\sum_{i=1}^n \phi(p_i)$ is not bounded.

Case 3: Now we consider the general case, when curves $l()$ and $k()$ are transversal at the points p and q to the unstable foliation, but nothing is known about the distance between p and q.

Let H_{l_n,k_n} be a holonomy map from the curve $f^{-n}(l)$ to the curve $f^{-n}(k)$ along unstable foliation. It is obvious that $H''_{l,k}(p)$ exists if and only if $H''_{l_n,k_n}(p_n)$ exists. The distance between p_n and q_n is small for n big enough, hence this case is equivalent to the *Case 2*.

References

[Arnold, Sinai] V. I. Arnold, Ya. G. Sinai, On small perturbations of the automorphisms of a torus. (Russian) Dokl. Akad. Nauk SSSR 144 1962 695–698

[Arnold, Sinai1] V. I. Arnold, Ya. G. Sinai, Correction. (Russian) Dokl. Akad. Nauk SSSR, 150, 1963, 958

[Anosov] D. V. Anosov, Geodesic flows on closed Riemann manifolds with negative curvature. Proceedings of the Steklov Institute of Mathematics, No. 90 (1967)

[DeLatte] D. DeLatte, Nonstationary normal forms and cocycle invariants. Random Comput. Dynamics 1 (1992/93), no. 2, 229–259.

[Frank] J. Franks Anosov diffeomorphisms on tori. Trans. Amer. Math. Soc. 145 1969 117–124

[Hurder, Katok] S. Hurder, A. Katok Differentiability, rigidity and Godbillon-Vey classes for Anosov flows. Inst. Hautes Études Sci. Publ. Math. No. 72, (1990)

[Jacobson, Swiatek] M. Jacobson, G. Swiatek, One-Dimensional Maps, in Handbook in Dynamical Systems, to be published, Elsevier, 2001

[Katok, Hasselblat] A. Katok, B. Hasselblat Introduction to the modern theory of dynamical systems. Cambridge University Press, 1995

Department of Mathematics, Tufts University
E-mail address: mguysi01@tufts.edu

Proceedings of Symposia in Pure Mathematics
Volume **69**, 2001

Schrödinger Operators and Topological Pressure on Manifolds of Negative Curvature

Viorel Niţică and Frederico Xavier

Abstract. The bilaplacians of the Busemann functions on the universal cover of a compact manifold M of negative curvature induce a function $\mathcal{B}^{(2)}$ on the unit tangent bundle of M. It is conjectured that $\mathcal{B}^{(2)}$ is cohomologous to a non-positive continuous function if and only if M is locally symmetric. It is shown here that $\mathcal{B}^{(2)}$ is not cohomologous to a negative continuous function. In particular, its topological pressure is non-negative. The proof is based on the construction of certain Schrödinger operators.

1. Introduction

Let N be a complete simply-connected manifold of negative curvature, Γ a co-compact lattice of the group of isometries of N and $\pi : UN \to N$ the natural projection of the unit tangent bundle of N. To every $v \in UN$ one can associate the Busemann function $B_v : N \to \mathbb{R}$ given by $B_v(x) = \lim_{t\to\infty}(d(\gamma_v(t), x) - t)$, where γ_v is the geodesic of N satisfying $\gamma_v(0) = \pi(v), \gamma'(0) = v$. If g is an isometry of N one has $B_{gv} \circ g = B_v$. In particular, $\Delta^j B_v = \Delta^j B_{gv} \circ g$, $j \geq 1$, where Δ is the Laplace-Beltrami operator of N. Hence $UN \ni v \to (\Delta^j B_v)(\pi v)$ is Γ-invariant and as such it descends to a (Hölder) function $\mathcal{B}^{(j)}$ on the unit tangent bundle UM of $M := N/\Gamma$. The function $\mathcal{B}^{(1)}$, which assigns to v the mean curvature at πv of the v-horosphere, is especially important because its integral with respect to the Liouville measure gives the entropy (the Pesin formula-see [**Y**] and [**KH**]). By contrast, no relation between $\mathcal{B}^{(j)}$ and the geodesic flow of M seems to have been known in the case $j \geq 2$. The object of this note is to prove the following result about the bilaplacian cocycle $\mathcal{B}^{(2)}$. First recall that $\phi_1, \phi_2 : UM \to \mathbb{R}$ are called cohomologous if there exists $\phi_3 : UM \to \mathbb{R}$ such that $(\phi_2 - \phi_1)(x) = \frac{d}{dt}(\phi_3(g_t(x)))|_{t=0}$; here g_t stands for the geodesic flow.

Theorem. *$\mathcal{B}^{(2)}$ is not cohomologous to a negative continuous function. In particular, the topological pressure of $\mathcal{B}^{(2)}$ is non-negative.*

A current theme in differential geometry and dynamical systems is the question of finding dynamical properties associated to geometrically defined objects that

---

2000 *Mathematics Subject Classification.* Primary 53C20.

The second author was partially supported by NSF Grant 9625392

might allow for a characterization of locally symmetric spaces of negative curvature. This is the subject of extensive research. A major rigidity result of G. Besson, G. Courtois and S. Gallot [B-C-G] states that locally symmetric spaces minimize entropy when the volume is prescribed (for the case $n = 2$ this was proved by A. Katok). Combining this with results of P. Foulon and F. Laburie [F-L], and F. Ledrappier [F], one has that a compact orientable Riemannian manifold whose Bowen-Margulis measure and harmonic measure coincide is locally symmetric. Let us mention also the entropy conjecture (due to Katok and proved in [K] for surfaces) that a compact orientable Riemannian manifold whose Bowen-Margulis measure and Liouville measure coincide must be locally symmetric.

In light of the remarks above and motivated by our theorem we propose the following characterization of locally symmetric spaces of negative curvature.

Conjecture. *A compact manifold M of negative curvature is locally symmetric if and only if its bilaplacian cocycle is cohomologous to a non-positive function.*

The "only if" part of the conjecture follows from the fact that symmetric spaces with negative curvature are asymptotically harmonic, i.e., their horospheres have constant mean curvature. From the results of [FL] and [BCG] it is known that asymptotically harmonic spaces with compact quotients are necessarily symmetric. (The conclusion may fail if the manifold in question does not admit a compact quotient-see [DR]).

There is evidence to suggest that the techniques of the present paper can be pushed to prove that if $\mathcal{B}^{(2)}$ is cohomologous to a non-positive function, then $\mathcal{B}^{(2)}$ must in fact be a trivial cocycle. Suppose that one can go further and prove that the Busemann functions are actually biharmonic. It appears that a simple argument involving ergodicity and Fatou's theorem on non-tangential limits of bounded harmonic functions (see [M], for instance) would show then that $\widetilde{M}$ is asymptotically harmonic. As observed above, this would imply that M is locally symmetric.

The statement in the theorem about the topological pressure can be established as follows. It follows directly from the definition that if the Hölder continuous function $\psi : UM \to \mathbb{R}$ has negative pressure then, for sufficiently large n, the function $\psi_n(x) := \frac{1}{n}\int_0^n \psi(g_t(x))dt$ has a negative upper bound (see [**KH**], page 623). On the other hand, by the Livsic theorem (see [**L**]), the functions ψ and ψ_n are cohomologous. Putting these two facts together one sees that if the pressure of the the function $\psi = \mathcal{B}^{(2)}$ is negative, the function must be cohomologous to a negative cocycle.

The proof of the above theorem is based on the examination of the spectral properties of certain Schrödinger operators defined on the universal cover of M. If the conclusion of the theorem were false one could construct a Schrödinger operator with empty spectrum, an impossibility for self-adjoint operators.

2. Commutators of Hilbert space operators

An important task in the spectral theory of self-adjoint operators is to formulate conditions to ensure that the spectral measures of a given operator H are absolutely continuous with respect to the Lebesgue measure. One such technique is Putnam's theorem. It involves producing an auxiliary operator A for which the commutator $i[H, A]$ is positive in the sense of quadratic forms. As pointed out in [**CFKS**], page

60, a byproduct of the proof of Putnam's theorem is the fact that the estimate

$$\langle i[H, A]x, x\rangle = 2\mathrm{Im}\langle Hx, Ax\rangle \geq c|x|^2, \ c > 0,$$

is impossible for a pair of *bounded* self-adjoint operators H and A. Below we formulate an extension of this observation which is valid for unbounded self-adjoint operators.

Lemma 1. *Let H be a self-adjoint operator on a Hilbert space $\mathcal{H} \neq 0$, and let $\mathcal{C} \subset D(H)$ be a core of H, i.e., $\mathcal{C}$ is dense in $D(H)$ in the graph norm $|x|_H^2 = |Hx|^2 + |x|^2$ of H. Let A be a symmetric operator with domain $\mathcal{C}$. It is impossible for H and A to satisfy simultaneously the two properties below:*
1. $|Im\langle Hx, Ax\rangle| \geq c|x|^2, \ c > 0, \ x \in \mathcal{C}$;
2. $|Ax| \leq C(|\langle Hx, x\rangle|^{\frac{1}{2}} + |x|), \ x \in \mathcal{C}$.

Proof. The proof consists in showing that if H and A satisfy (i) and (ii) the spectrum of H has to be empty. This is impossible since the Hilbert space has non-zero dimension. We observe that the spectrum of H is the union of the set of eigenvalues of finite multiplicity and the essential spectrum. A point $\lambda \in \mathbb{R}$ belongs to the essential spectrum if there is a sequence of unit vectors $x_n \in D(H)$ which converge weakly to 0 and satisfy $|(H - \lambda I)x_n| \to 0$. Since $\mathcal{C}$ is dense in $D(H)$ in the graph norm, we may actually take x_n to belong to $\mathcal{C}, |x_n| \geq \delta > 0$. If, on the other hand, λ is an eigenvalue (of finite or infinite multiplicity), $Hx = \lambda x, |x| = 1$, we set $x_n = x$. In either case we may compute, using (i),(ii) and $\mathrm{Im}\langle(H - \lambda)x, Ax\rangle = \mathrm{Im}\langle Hx, Ax\rangle$:

$$c\delta^2 \leq c|x_n|^2 \leq |(H - \lambda I)x_n||Ax_n| \leq$$
$$C|(H - \lambda I)x_n|\Big(|x_n| + \big(|\langle(H - \lambda I)x_n, x_n\rangle| + |\lambda||x_n|^2\big)^{\frac{1}{2}}\Big).$$

Clearly the quantity on the right side goes to zero as $n \to \infty$. This contradiction establishes the lemma. □

3. An integral identity involving the bilaplacian

The object of this section is to establish the integral identity (1) below. The case $q = 0$ was the starting point in **[X]** of an investigation of the absolute continuity of the spectral measure of the Laplacian on complete manifolds. In the present sign convention, Δu is the trace of the Hessian of u.

Lemma 2. *Let M be a Riemannian manifold and f, u smooth real-valued functions on M, at least one of which has compact support. Let also q be smooth on M. Then*

$$\int(-\Delta u + qu)\left(\nabla f \cdot \nabla u + \frac{\Delta f}{2}u\right) =$$

$$\int Hessf(\nabla u, \nabla u) - \frac{1}{4}\int(\Delta^2 f + 2\nabla f \cdot \nabla q)u^2. \tag{1}$$

PROOF. We treat the case $q = 0$ first. Consider the following identity ([**KW**], page 32):

$$2\Delta u(\nabla f \cdot \nabla u) = \mathrm{div}(2(\nabla f \cdot \nabla u)\nabla u - |\nabla u|^2 \nabla f) - 2\mathrm{Hess}f(\nabla u, \nabla u) + (\Delta f)|\nabla u|^2.$$

After integration we have

$$2\int (\Delta u)(\nabla f \cdot \nabla u) = -2\int \mathrm{Hess}f(\nabla u, \nabla u) + \int (\Delta f)|\nabla u|^2.$$

Using

$$\int (\Delta f)|\nabla u|^2 = \frac{1}{2}\int (\Delta f)\Delta(u^2) - \int u(\Delta u)\Delta f$$

we obtain

$$\int (\Delta u)(\nabla f \cdot \nabla u) = -\int \mathrm{Hess}f(\nabla u, \nabla u) + \frac{1}{4}\int (\Delta f)\Delta(u^2) - \frac{1}{2}\int u\Delta u\Delta f$$

so that

$$\int (\Delta u)(\nabla f \cdot \nabla u + \frac{1}{2}u\Delta f) = -\int \mathrm{Hess}f(\nabla u, \nabla u) + \frac{1}{4}\int (\Delta f)\Delta(u^2).$$

The result in the case $q = 0$ then follows by transforming the second integral on the right hand side.

To deal with the case $q \neq 0$, we start by computing the (formal) L^2 adjoint X^* of a vector field X on M. Since $\int \mathrm{div}Y = 0$ whenever Y has compact support, we have with $u, v \in C_0^\infty(M)$:

$$0 = \int \mathrm{div}(uvX) = \int u(\mathrm{div}X)v + \int uXv + \int vXu,$$

so that $X^* = -X - \mathrm{div}X$. In particular, $\langle Xu, u\rangle + \frac{1}{2}\langle(\mathrm{div}X)u, u\rangle = 0$. Taking $v = qu$, one sees that $\langle qXu, u\rangle + \langle uXq, u\rangle = \langle X(uq), u\rangle = \langle uq, X^*u\rangle$. Using the formula for X^* one then has

$$\langle Xu + \frac{1}{2}(\mathrm{div}X)u, qu\rangle = -\frac{1}{2}\langle(Xq)u, u\rangle$$

and (1) follows by taking $X = \nabla f$. □

4. Proof of the Theorem

We argue by contradiction and write $\mathcal{B}^{(2)} = W_1 + W_2$, with $W_1 < 0$ and

$$W_2(v) = \frac{d}{dt}W_3(g_t(v))|_{t=0}, \ v \in UM.$$

Fix $z \in N(\infty)$ and let $\alpha : UN \to UM$ be the natural map. Next, we define $q : N \to \mathbb{R}$ by

$$q(x) = -\frac{1}{2}W_3(\alpha(\nabla B_z(x))).$$

Note that q is bounded. Since $-\Delta$ is essentially self-adjoint on $C_0^\infty(N, \mathbb{C})$ (recall that N is complete), the same is true of $-\Delta + q$ (see [**Ka**], page 288). Taking $f = B_z$, the quantity $\Delta^2 f + 2\nabla f \nabla q$ which figures in Lemma 2 becomes

$$\begin{aligned}(\Delta^2 B_z + 2\nabla B_z \cdot \nabla q)(x) &= \mathcal{B}^{(2)}(\alpha(\nabla B_z(x))) - \nabla B_z(x) \cdot \nabla (W_3(\alpha(\nabla B_z(x))))\\ &= \mathcal{B}^{(2)}(\alpha(\nabla B_z(x))) - \frac{d}{dt}(W_3(g_t(\alpha(\nabla B_z(x)))))|_{t=0}\\ &= \mathcal{B}^{(2)}(\alpha(\nabla B_z(x))) - W_2(\alpha(\nabla B_z(x)))\\ &= W_1(\alpha(\nabla B_z(x))) < 0.\end{aligned}$$

Since W_1 is continuous and B_z is convex, Lemma 2 implies for every $u \in C_0^\infty(N, \mathbb{R})$, and some $\delta > 0$:

$$\int (-\Delta u + qu)(\nabla f \cdot \nabla u + \frac{\Delta f}{2} u) \geq \delta \int u^2. \tag{2}$$

Next, let H be the unique self-adjoint extension of $(-\Delta + q)|_{C_0^\infty(N,\mathbb{C})}$. Let also A be defined on $C_0^\infty(N, \mathbb{C})$ by $Au = i(\nabla f \cdot \nabla u + \frac{\Delta f}{2} u)$. It is easy to see that A so defined is symmetric. Indeed, the skew-symmetry of iA amounts to the identity

$$\int (\nabla f \nabla u + \frac{\Delta f}{2} u) u = 0, \; u \in C_0^\infty(N, \mathbb{R}),$$

which was essentially established at the end of the proof of Lemma 1. Applying (2) to u_1, u_2 and adding we have

$$\mathrm{Im}\langle Hu, Au\rangle \geq \delta |u|^2, \; u = u_1 + iu_2 \in C_0^\infty(N, \mathbb{C}).$$

In other words, the pair H, A obeys (i) of Lemma 1. To reach a contradiction we now show that (ii) of Lemma 1 is also satisfied. The key here is the fact that q is bounded from below. Since the curvature is bounded, so is ΔB_z. Hence, for $u \in C_0^\infty(N, \mathbb{C})$,

$$\begin{aligned}|Au|^2 = \int |\nabla f \cdot \nabla u + \frac{\Delta f}{2} u|^2 &\leq C \int (|\nabla u|^2 + |u|^2)\\ &\leq C \int (-\Delta u + qu)\bar{u} + C \int (1-q)|u|^2\\ &\leq C'(|\langle Hu, u\rangle| + |u|^2).\end{aligned}$$

Taking square roots one obtains (ii) of lemma 1. Hence the pair H, A violates the conclusion of Lemma 1, and this represents a contradiction to the assumption that $\mathcal{B}^{(2)}$ is cohomologous to a negative function. □

References

[BCG] G. Besson, G. Courtois, G. Gallot, *Entropies et rigidities des espaces localement symetriques des curbure strictement negative*, Geom. Funct. Anal. **5** (1995), 731–799.

[CFKS] H. L. Cycon, R. G. Froese, W. Kirsch, B. Simon, *Schrödinger Operators*, Springer-Verlag, 1987.

[DR] E. Damek, F. Ricci, *A class of non-symmetric harmonic Riemannian spaces*, Bull. Amer. Math. Soc. **27** (1992), 139–142.

[FL] P. Foulon, F. Labourie, *Sur les varietes compacts asymptotiquement harmoniques*, Invent. Math. **109** (1992), 97–111.

[Ka] T. Kato, *Perturbation theory for linear operators*, Springer-Verlag, 1966.

[K] A. Katok, *Entropy and closed geodesics*, Ergod. Theory Dynam. Systems **2** (1982), 339–365.

[KH] A. Katok, B. Hasselblat, *Introduction to the modern theory of dynamical systems*, Cambridge University Press, 1995.

[KW] J. Kazdan, F. Warner, *Curvature functions for compact 2-manifolds*, Ann. Math. **99** (1974), 14–47.

[Le] F. Ledrappier, *Harmonic measures and Bowen-Margulis measures*, Israel J. Math. **3** (1990), 275–287.

[L] A. Livsic, *Cohomology of dynamical systems*, Math. USSR, Izvestia **6** (1972), 1278–1301.

[M] F. Mouton, *Comportement asymptotique des fonctions harmoniques en courbure negative*, Comment. Math. Helv. **70** (1995), 475–505.

[X] F. Xavier, *Convexity and absolute continuity of the Laplace-Beltrami operator*, Math. Annalen **282** (1988), 579–585.

[Y] C. Yue, *Rigidity and dynamics around manifolds of negative curvature*, Math. Research Letters **1** (1994), 123–147.

Department of Mathematics, Notre Dame University, Notre Dame, IN 46556-5683

E-mail address: nitica.1@nd.edu, fxavier@nd.edu

Proceedings of Symposia in Pure Mathematics
Volume **69**, 2001

Isoperimetric and Ergodic Properties of Horospheres in Symmetric Spaces

Norbert Peyerimhoff

In this article we are mainly concerned with a lower estimate of the Cheeger constant of a horosphere $\mathcal{H}(v)$ in a higher rank symmetric space $\tilde{V}$ of noncompact type (depending on the unit normal direction $v \in S\tilde{V}$ of the horosphere). The Cheeger constant $h(M)$ of a complete manifold M is defined as

$$h(M) = \inf_{K \subset M} \frac{\text{area}(\partial K)}{\text{vol}(K)},$$

where K runs through all compact, connected subsets with smooth boundary. $h(M)$ may be considered as an intrinsic growth measure of M. Our estimate suggests the following possible growth relation: The bigger the growth rate of a horosphere, the smaller the asymptotic volume density of increasing geodesic spheres in normal direction, and vice versa. We also like to mention that the asymptotic volume density in direction $v \in S\tilde{V}$ agrees with the topological entropy of the restriction of the geodesic flow $\Phi_t : SV \to SV$ to the Φ_t-invariant subspace $d\pi(Gv) \subset SV$ (with $V = \Gamma \backslash \tilde{V}$ any compact Riemannian quotient, $\pi : \tilde{V} \to V$ the canonical projection, and G the connected component of the identity of the isometry group of $\tilde{V}$). This topological entropy was calculated by R. J. Spatzier [**Sp**, Ch. 11].

We will also show that the horosphere normal to a direction of maximal asymptotic volume density admits an exhaustion $\bigcup_j K_j$ by compact sets K_j with

$$\lim_{j \to \infty} \frac{\text{area}(\partial K_j)}{\text{vol}(K_j)} = 0. \tag{1}$$

We refer to horospheres admitting such an exhaustion as "good horospheres", while the others are called "bad horospheres". The question, which horospheres admit exhaustions satisfying (1) in general spaces of non-positive curvature was also raised in [**Gr**, §5].

Finally, we discuss some implications. As an immediate consequence we obtain an explicit exponential lower estimate for the volume growth of intrinsic geodesic balls in bad horospheres. Interestingly, good horospheres have exponential volume growth of intrinsic geodesic balls, as well. Hence, for good horospheres, intrinsic geodesic balls are not the right choice for an exhaustion satisfying (1). In contrast

---

2000 *Mathematics Subject Classification.* Primary 53C35, 37D40; Secondary 37C40, 53C12.
Supported by an NSF conference grant.

to the bad horospheres, however, we don't have informations about an explicit exponential lower bound.

In another application we consider a compact Riemannian quotient $V = \Gamma\backslash\tilde{V}$. It follows that good horospheres in $\tilde{V}$ distribute uniformly under projection onto this compact quotient.

Added in proof: After completion of this article, we learned from P. Pansu that H. F. Hoke III (see [**Ho**]) proved in 1989 that a simply connected Lie group G has vanishing Cheeger constant if and only if G is amenable and unimodular. This covers our Theorem 2.3 since good horospheres are unimodular solvable groups. C. Druţu remarked that good horospheres are not quasi–isometric to bad ones, since vanishing of the Cheeger constant is a quasi–isometry invariant (see [**Ka**]). It would be interesting to know whether a more general statement is true, namely, that any two non–isometric horospheres of a higher rank symmetric space are also not quasi–isometric.

1. Metric and Riemannian volume element of a horosphere

Let us fix some notation at the beginning. Let $\tilde{V}$ be a symmetric space of noncompact type of rank $k \geq 1$ and $p_0 \in \tilde{V}$. Any unit vector $v_0 \in S_{p_0}\tilde{V}$ ($S\tilde{V}$ denotes the unit tangent bundle of $\tilde{V}$) determines a Busemann function $b(q) = \lim_{t\to\infty} d(q, c(t)) - t$ and a horosphere $\mathcal{H}(v_0) = b^{-1}(0) \subset \tilde{V}$. Here c is the geodesic with $\dot{c}(0) = v_0$. Geometrically, $\mathcal{H}(v_0)$ can be considered as the limit set of increasing geodesic spheres with center $c(t)$ and radius t.

Let G be the connected component of the identity of the isometry group of $\tilde{V}$, $K \subset G$ be the isotropy group of p_0, and $\mathfrak{g}$ and $\mathfrak{k}$ be the corresponding Lie algebras. The Cartan involution $\theta_{p_0} : \mathfrak{g} \to \mathfrak{g}$ induces a decomposition $\mathfrak{g} = \mathfrak{k} \oplus \mathfrak{p}$ and one obtains a positive definite inner product ϕ_{p_0} on $\mathfrak{g}$ by

$$\phi_p(X, Y) = -B(\theta_{p_0} X, Y),$$

where B denotes the Killing form. Let $\mathfrak{a} \subset \mathfrak{p}$ be a maximal abelian subspace and $\mathfrak{g} = \mathfrak{g}_0 + \sum_{\alpha\in\Lambda} \mathfrak{g}_\alpha$ be the corresponding root space decomposition. We choose a positive Weyl chamber $\mathfrak{a}^+ \subset \mathfrak{a} - \bigcup_{\alpha\in\Lambda} \ker(\alpha)$ and we denote the set of positive roots by Λ^+. This yields an Iwasawa decomposition $G = NAK$ with $A = \exp(\mathfrak{a})$, $N = \exp(\mathfrak{n})$, and $\mathfrak{n} = \sum_{\alpha\in\Lambda^+} \mathfrak{g}_\alpha$. $\mathfrak{n}$ is a nilpotent Lie algebra.

Henceforth, we consider $\tilde{V}$ as the solvable group NA via the identification $g \mapsto gp_0$. The symmetric space metric of $\tilde{V}$ coincides then with a left invariant metric on NA and $A \cong \mathbb{R}^k$ coincides with a totally geodesic flat submanifold of $\tilde{V}$, tangent to v_0.

Our next aim is to describe the pullback of the metric of $\tilde{V}$ via the diffeomorphism

$$\varphi : \mathfrak{a} \times N \to \tilde{V}, \quad \varphi(H, x) = x \exp(H)p_0.$$

This diffeomorphism induces an identification $T_{p_0}\tilde{V} \cong \mathfrak{a} \times \mathfrak{n}$. Let $\langle\cdot,\cdot\rangle_0$ and $\langle\cdot,\cdot\rangle_\alpha$ denote the restrictions of the symmetric space metric of $T_{p_0}\tilde{V}$ to the subspaces $\mathfrak{a}$ and $\mathfrak{g}_\alpha$. The decomposition $\mathfrak{n} = \sum_{\alpha\in\Lambda^+} \mathfrak{g}_\alpha$ extends to a left–invariant decomposition $TN = \sum_{\alpha\in\Lambda^+} \mathcal{D}_\alpha$. The left–invariant extension of the metric $\langle\cdot,\cdot\rangle_\alpha$ to the distribution $\mathcal{D}_\alpha \subset TN$ is again denoted by $\langle\cdot,\cdot\rangle_\alpha$.

LEMMA 1.1. *Let $q = \varphi(H, x) \in \tilde{V}$. The pullback of the symmetric space metric in $T_q\tilde{V}$ (w.r.t. φ), applied to the vectors*

$$v_j = H_j + \sum_{\alpha\in\Lambda^+} w_j^\alpha \ \in\ \mathfrak{a}\times T_xN \ \cong\ T_{(H,x)}(\mathfrak{a}\times N), \quad j = 1, 2,$$

is given by

$$\langle H_1, H_2\rangle_0 + \sum_{\alpha\in\Lambda^+} e^{-2\alpha(H)}\langle w_1^\alpha, w_2^\alpha\rangle_\alpha. \tag{2}$$

PROOF. We use the following imbedding theorem (see [**Eb**, Thm. 2.6.5, Prop. 2.19.5] and [**Fe**, Prop. 7.5.1]): Let $n = \dim G$. Choosing a suitable orthonormal base $E_1, \ldots, E_n$ of the space $(\mathfrak{g}, \phi_{p_0})$, we obtain an identification $\mathfrak{g} \cong \mathbb{R}^n$ with the following properties:

i) $\mathrm{Ad} : G \to SL(n, \mathbb{R})$ is a faithful representation and $\mathrm{Ad}(G) = \mathrm{Ad}(G)^\top$.

ii) The map

$$F : \tilde{V} \to \tilde{V}_n = SL(n, \mathbb{R})/SO(n), \quad F(gp_0) = [\mathrm{Ad}(g)]_{SO(n)} \in \tilde{V}_n$$

is an isometry onto a complete, totally geodesic submanifold of $\tilde{V}_n$ with the restricted Riemannian metric of the ambient space $\tilde{V}_n$ (after, possibly, rescaling the metrics of the irreducible factors of $\tilde{V}$).

iii) Let A_n, resp., $N_n \subset SL(n, \mathbb{R})$ denote the group of diagonal matrices, resp., upper triangular matrices with 1's on the diagonal. Then we have

$$\mathrm{Ad}(K) \subset SO(n), \quad \mathrm{Ad}(A) \subset A_n, \quad \mathrm{Ad}(N) \subset N_n.$$

iv) Let E_{ij} denote the $n \times n$ matrix with a 1 in position (i, j) and zeros elsewhere, $\mathfrak{g}_{ij} = \mathbb{R}\cdot E_{ij}$, and $\alpha_{ij}(\mathrm{diag}(t_1, \ldots, t_n)) = t_i - t_j$. Then we have, for all roots $\alpha \in \Lambda$:

$$\mathfrak{g}_\alpha = \mathfrak{g} \cap \bigoplus_{\substack{1\le i,j\le n\\ \alpha_{ij}|\mathfrak{a}=\alpha}} \mathfrak{g}_{ij},$$

where we tacitly used the inclusions $\mathfrak{g} \subset \mathfrak{sl}(n, \mathbb{R})$ and $G \subset SL(n, \mathbb{R})$ induced by Ad.

At first, we prove (2) for $v_1 \in \mathfrak{a}$ and $v_2 \in (\mathcal{D}_\alpha)_x$. Let $L_x : N \to N$ denote left–translation by $x \in N$, and $w_2^\alpha = dL_x(X_2^\alpha)$ with $X_2^\alpha \in \mathfrak{n}$. Then v_1 and v_2 are represented by the curves

$$c_1(t) = \varphi(H + tH_1, x) = x\exp(H)\exp(tH_1),$$
$$c_2(t) = \varphi(H, x\exp(tX_2^\alpha)) = x\exp(tX_2^\alpha)\exp(H) = x\exp(H)\exp(te^{-\alpha(H)}X_2^\alpha).$$

Left–invariance of the metric implies

$$\langle \dot{c}_1(0), \dot{c}_2(0)\rangle = \langle \dot{c}_3(0), \dot{c}_4(0)\rangle,$$

with $c_3(t) = \exp(tH_1)$ and $c_4(t) = \exp(te^{-\alpha(H)}X_1^\alpha)$. Let $\widehat{F}(gp_0) = \mathrm{Ad}(g)$ for $gp_0 \in \tilde{V}$, $g \in NA$, $\gamma_j = F \circ c_j : \mathbb{R} \to \tilde{V}_n$ and $\widehat{\gamma}_j = \widehat{F} \circ c_j : \mathbb{R} \to SL(n, \mathbb{R})$ for $j = 3, 4$. Then we obtain, e.g., by [**Mo**, §3]:

$$\begin{aligned}\langle \dot{c}_3(0), \dot{c}_4(0)\rangle &= \langle \dot{\gamma}_3(0), \dot{\gamma}_4(0)\rangle_{F(p_0)}\\ &= \mathrm{trace}((\dot{\widehat{\gamma}}_3(0) + \dot{\widehat{\gamma}}_3(0)^\top)(\dot{\widehat{\gamma}}_4(0) + \dot{\widehat{\gamma}}_4(0)^\top)).\end{aligned}$$

The trace vanishes, since $\dot{\widehat{\gamma}}_3(0) \in \mathfrak{a}_n$ is a diagonal matrix and $\dot{\widehat{\gamma}}_4(0) \in \mathfrak{n}_n$ has vanishing diagonal entries.

Next, we prove (2) for $v_1 \in (\mathcal{D}_\alpha)_x, v_2 \in (\mathcal{D}_\beta)_x$, $\alpha, \beta \in \Lambda^+$. As above, v_1, v_2 are represented by curves

$$c_1(t) = x \exp(H) \exp(te^{-\alpha(H)} X_1^\alpha), \quad c_2(t) = x \exp(H) \exp(te^{-\beta(H)} X_2^\beta),$$

and we obtain

$$\begin{aligned} (3) \quad \langle \dot{c}_1(0), \dot{c}_2(0) \rangle &= e^{-\alpha(H)-\beta(H)} \langle \dot{c}_3(0), \dot{c}_4(0) \rangle \\ (4) \quad &= e^{-\alpha(H)-\beta(H)} \operatorname{trace}((\dot{\widehat{\gamma}}_3(0) + \dot{\widehat{\gamma}}_3(0)^\top)(\dot{\widehat{\gamma}}_4(0) + \dot{\widehat{\gamma}}_4(0)^\top)), \end{aligned}$$

where $c_3(t) = \exp(tX_1^\alpha)$, $c_4(t) = \exp(tX_2^\beta)$, and $\widehat{\gamma}_3, \widehat{\gamma}_4$ are defined as above. For every root $\alpha \in \Lambda^+$, we introduce the following set of indices

$$I_\alpha = \{(i,j) \mid \alpha_{ij}|_{\mathfrak{a}} = \alpha\}.$$

For $\alpha \neq \beta$, we obviously have $I_\alpha \cap I_\beta = \emptyset$. Since $\dot{\widehat{\gamma}}_3(0) \in \sum_{(i,j)\in I_\alpha} \mathbb{R}E_{ij}$ and $\dot{\widehat{\gamma}}_4(0) \in \sum_{(i,j)\in I_\beta} \mathbb{R}E_{ij}$, the trace in (4) vanishes if $\alpha \neq \beta$. In the case $\alpha = \beta$ we conclude from (3):

$$\langle \dot{c}_1(0), \dot{c}_2(0) \rangle = e^{-2\alpha(H)} \langle X_1^\alpha, X_2^\alpha \rangle_\alpha = e^{-2\alpha(H)} \langle w_1^\alpha, w_2^\alpha \rangle_\alpha.$$

The case $v_1, v_2 \in \mathfrak{a}$ is trivial. This finishes the proof. □

The following lemma gives an algebraic description of any horosphere through p_0 whose normal vector v_0 lies in the closure of the positive Weyl chamber $\mathfrak{a}^+$.

LEMMA 1.2. *In the statements below, we assume that the vector $v_0 \in S_{p_0}\tilde{V}$ corresponds, via the identification $T_{p_0}\tilde{V} \cong \mathfrak{p}$, to an element in $\overline{\mathfrak{a}^+}$, which we denote by H_0.*

a) *Let 0_x denote the zero vector of T_xN. The pushforward $Z = \varphi_* Z_0 : \tilde{V} \to S\tilde{V}$ of the smooth vector field*

$$Z_0 : \mathfrak{a} \times N \to T(\mathfrak{a} \times N),$$
$$Z_0(H,x) = H_0 + 0_x \in \mathfrak{a} \times T_xN \cong T_{(H,x)}(\mathfrak{a} \times N)$$

enjoys the following properties: $Z(p_0) = v_0$, and $Z(q)$ is asymptotic to v_0 for all $q \in \tilde{V}$.

b) *Let $\mathfrak{a}'$ denote the orthogonal complement of H_0 in $(\mathfrak{a}, \langle\cdot,\cdot\rangle_0)$. The horosphere $\mathcal{H}_{v_0} \subset \tilde{V}$ is given by*

$$\mathcal{H}_{v_0} = \varphi(\mathfrak{a}' \times N).$$

PROOF. Part a) follows immediately from the lemma of [**BGS**, p. 245]. Let $\mathcal{D} \subset T\tilde{V}$ denote the distribution orthogonal to the vector field Z. The horospheres normal to Z are the maximal integral manifolds of $\mathcal{D}$. Thus, part b) is a consequence of Lemma 1.1. □

We conclude from the lemmas above

PROPOSITION 1.3. *Let $v_0 \in S_{p_0}\tilde{V}$, $H_0 \in \overline{\mathfrak{a}^+}$, and $\mathfrak{a} = \mathbb{R}H_0 \perp \mathfrak{a}'$ be defined as in the previous lemma. The pullback of the metric of the horosphere $\mathcal{H} = \mathcal{H}_{v_0} \subset \tilde{V}$ at $q = \varphi(H,x) \in \mathcal{H}$ (via φ) is given by*

$$\langle H_1, H_2 \rangle_0 + \sum_{\alpha \in \Lambda^+} e^{-2\alpha(H)} \langle w_1^\alpha, w_2^\alpha \rangle_\alpha,$$

for the vectors $v_j = H_j + \sum_\alpha w_j^\alpha \in \mathfrak{a}' \times T_x N$, $j = 1, 2$. *Accordingly, the Riemannian volume element* $\mathrm{dvol}_\mathcal{H}$ *of the horosphere at* $q \in \mathcal{H}$ *takes the form*

$$\varphi^* \mathrm{dvol}_\mathcal{H}(q) = e^{-\rho(H)} d(\lambda^{k-1} \otimes \nu)(H, x),$$

where λ^{k-1} *is the standard Lebesgue measure of* $(\mathfrak{a}', \langle \cdot, \cdot \rangle_0) \cong \mathbb{R}^{k-1}$ *and* $d\nu$ *is the Riemannian volume element of the space* $(N, \bigoplus_{\alpha \in \Lambda^+} \langle \cdot, \cdot \rangle_\alpha)$, *and*

$$\rho = \sum_{\alpha \in \Lambda^+} m_\alpha \alpha : \mathfrak{a} \to \mathbb{R}, \tag{5}$$

with $m_\alpha = \dim \mathfrak{g}_\alpha$.

For any linear map $\alpha : \mathfrak{a} \to \mathbb{R}$, let $H_\alpha \in \mathfrak{a}$ denote the unique vector satisfying $\alpha = \langle H_\alpha, \cdot \rangle_0$. Then H_ρ is a vector of the positive Weyl chamber $\mathfrak{a}^+$ (see Lemma 1.4 below). The corresponding direction $v_\rho = \frac{H_\rho}{\|H_\rho\|_0} \in S_{p_0} \tilde{V}$ (w.r.t. the identification $\mathfrak{p} = T_{p_0} \tilde{V}$) can be characterized geometrically as follows: For any direction v of the closed spherical Weyl chamber $\mathcal{W} = \overline{\mathfrak{a}^+} \cap S_{p_0} \tilde{V}$, let c_v denote the geodesic with $\dot{c}_v(0) = v$, and $f(v, r)$ denote the volume density of the geodesic sphere about p_0 at $c_v(r)$. Then the function $d_{\mathrm{as}}(v) = \lim_{r \to \infty} \frac{\ln(f(v,r))}{r}$, defined on the closed spherical Weyl chamber $\mathcal{W}$, attains a unique maximum at $v = v_\rho$. For that reason we refer to $v_\rho \in S\tilde{V}$ as a "maximal direction". Calculations as in [**Kn**] yield $d_{\mathrm{as}}(v) = \langle H_\rho, v \rangle_0$ for $v \in \mathcal{W}$. Comparison with [**Sp**, Ch. II] shows that this agrees with the topological entropy of the geodesic flow on the invariant subset $d\pi(Gv) \subset SV$, where $\pi : \tilde{V} \to V$ is the canonical projection to a compact Riemannian quotient V.

LEMMA 1.4. *Let* $\rho : \mathfrak{a} \to \mathbb{R}$ *be defined as in* (5). *Then we have* $H_\rho \in \mathfrak{a}^+$.

REMARK. The proof of Lemma 1.4 was communicated to me by D. Akhiezer. The corresponding statement for complex semisimple Lie groups is a well known fact.

PROOF. Let $\Delta \subset \Lambda^+$ denote the set of simple roots. We have to show $\alpha(H_\rho) > 0$ for all $\alpha \in \Delta$. For $\alpha \in \Delta$, let $s_\alpha : \mathfrak{a} \to \mathfrak{a}$ denote the reflection at the hyperplane $\ker \alpha \subset \mathfrak{a}$, i.e.,

$$s_\alpha(H) = H - \frac{2\alpha(H)}{\|H_\alpha\|_0^2} H_\alpha. \tag{6}$$

s_α permutes the root vectors $\{H_\beta \mid \beta \in \Lambda\}$, as well as the corresponding root spaces (see [**Eb**, Prop. 2.9.10]). This implies $m_{\beta'} = m_\beta$ for $H_{\beta'} = s_\alpha(H_\beta)$. For any root $\gamma \in \Lambda$, there is a unique representation $H_\gamma = \sum_{\beta \in \Delta} c_\beta H_\beta$, and the coefficients c_β are either all non–negative, or non–positive. In particular, $\gamma \in \Lambda^+$ if and only if $c_\beta \geq 0$ for all $\beta \in \Delta$. We conclude from (6) that

$$s_\alpha(H_\gamma) = \sum_{\substack{\beta \in \Delta \\ \beta \neq \alpha}} c_\beta H_\beta - (c_\alpha + \sum_{\substack{\beta \in \Delta \\ \beta \neq \alpha}} \frac{2 c_\beta \alpha(H_\beta)}{\|H_\alpha\|_0^2}) H_\alpha.$$

Since, for $\gamma \in \Lambda^+ - \{\alpha, 2\alpha\}$, the representation of $s_\alpha(H_\gamma)$ has some strictly positive coefficients, we conclude in this case $\gamma' \in \Lambda^+$ for $H_{\gamma'} = s_\alpha(H_\gamma)$. Consequently, s_α

also permutes the elements of $\{H_\beta \mid \beta \in \Lambda^+ - \{\alpha, 2\alpha\}\}$. This implies

$$\begin{aligned} s_\alpha(H_\rho) = \sum_{\beta\in\Lambda^+} m_\beta s_\alpha(H_\beta) &= \sum_{\beta\in\Lambda^+-\{\alpha,2\alpha\}} m_\beta H_\beta - (m_\alpha + 2m_{2\alpha})H_\alpha \\ &= H_\rho - 2(m_\alpha + 2m_{2\alpha})H_\alpha \end{aligned} \tag{7}$$

Combining (6) and (7), we obtain

$$\alpha(H_\rho) = (m_\alpha + 2m_{2\alpha})\|H_\alpha\|_0{}^2 > 0.$$

□

2. Isoperimetric properties of horospheres

The Cheeger constant of a noncompact Riemannian manifold M is defined as

$$h(M) = \inf_{K\subset M} \frac{\mathrm{area}(\partial K)}{\mathrm{vol}(K)},$$

where K runs through all compact, connected subsets with smooth boundary. In the first part of this section we prove the following lower estimate of the Cheeger constant of horospheres:

THEOREM 2.1. *Let $v_0 \in S_{p_0}\tilde{V}$ be a vector corresponding to an element $H_0 \in \overline{\mathfrak{a}^+}$ via the identification $T_{p_0}\tilde{V} \cong \mathfrak{p}$, and $\mathfrak{a} = \mathbb{R}H_0 \perp \mathfrak{a}'$. Then we have for the horosphere $\mathcal{H} = \mathcal{H}_{v_0}$:*

$$h(\mathcal{H}) \geq m_\rho(v_0) = \max_{\substack{H\in\mathfrak{a}' \\ \|H\|_0=1}} \rho(H).$$

PROOF. Our considerations refer to the space $\mathfrak{a}' \times N$, which can be identified with $\mathcal{H}$ via the isometry $\varphi : \mathfrak{a}' \times N \to \mathcal{H}$. Let $H_1 \in \mathfrak{a}'$ be a unit vector with $\rho(H_1) > 0$. (Note that if $\rho(H) = 0$ for all $H \in \mathfrak{a}'$, there is nothing to show.) Let $\mathfrak{a}'' \subset \mathfrak{a}'$ be the subspace orthogonal to $\mathbb{R}H_1$. We refer to the H_1-direction as the vertical direction. $\pi : \mathfrak{a}' \times N \mapsto \mathfrak{a}'' \times N$ denotes the orthogonal projection, and λ^{k-2} denotes the standard Lebesgue measure of $(\mathfrak{a}'', \langle\cdot,\cdot\rangle_0) \cong \mathbb{R}^{k-2}$. Let $K \subset \varphi^{-1}(\mathcal{H})$ be a compact, connected subset with smooth boundary, $\pi_0 : \partial K \to \mathfrak{a}'' \times N$ be a restriction of π, and $K_0 = \pi_0(\partial K)$. The set

$$C = \{q = (H, x) \in \partial K \mid \mathbb{R}H_1 + 0_x \subset T_q\partial K \subset \mathfrak{a}' \times T_x N\}$$

of points of ∂K with vertical tangent space coincides with the set of critical points of π_0. $U = K_0 - \pi(C)$ is an open set and $(\lambda^{k-2} \otimes \nu)(\pi(C)) = 0$, by Sard's Theorem. For each $(H, x) \in U$, the set $\pi^{-1}(H, x) \cap K \subset (H + \mathbb{R}H_1, x) \cong \mathbb{R}$ consists of finitely many closed disjoint intervals, i.e.

$$\pi^{-1}(H, x) \cap K \cong \bigcup_{i=1}^{p} [a_{2i-1}, a_{2i}]$$

with $a_1 < a_2 < \cdots < a_{2p}$ (p is constant on connected components of U). Let $f_{\mathrm{bot}}(H, x) = a_1$ and $f_{\mathrm{top}}(H, x) = a_{2p}$. Then we obviously have

$$\mathrm{area}(\partial K) \geq \mathrm{area}(K_{\mathrm{bot}}) + \mathrm{area}(K_{\mathrm{top}}) = \int_{K_{\mathrm{bot}}} dS_{\mathrm{bot}} + \int_{K_{\mathrm{top}}} dS_{\mathrm{top}},$$

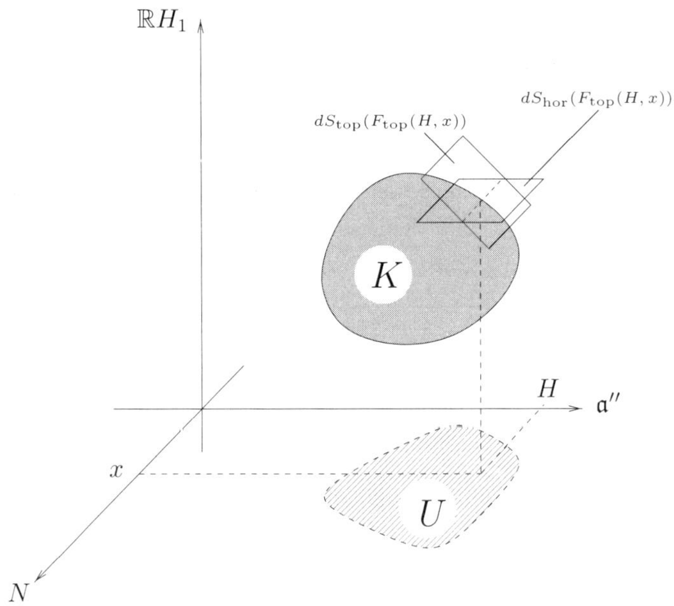

FIGURE 1. Replacing the area element of $K_{\mathrm{top}} \subset \partial K$ by the horizontal area element yields the lower area estimate (8)

where dS_{bot} and dS_{top} are the Riemannian area elements of the bottom and top component

$$K_{\mathrm{bot}} = F_{\mathrm{bot}}(U), \qquad F_{\mathrm{bot}}(H,x) = (H + f_{\mathrm{bot}}(H,x)H_1, x),$$
$$K_{\mathrm{top}} = F_{\mathrm{top}}(U), \qquad F_{\mathrm{top}}(H,x) = (H + f_{\mathrm{top}}(H,x)H_1, x)$$

of ∂K. Let dS_{hor} denote the horizontal area element, i.e., $dS_{\mathrm{hor}}(v_1, \ldots, v_n) = 1$ for all oriented orthonormal bases $v_1, \ldots, v_n \in T((\mathfrak{a}'' + tH_1) \times N)$, $t \in \mathbb{R}$ fixed, and $dS_{\mathrm{hor}}(v_1, \ldots, v_n) = 0$ for $v_1 \in \mathbb{R}H_1$. A straightforward calculation yields

$$\begin{aligned} \operatorname{area}(K_{\mathrm{top}}) &= \int_{K_{\mathrm{top}}} dS_{\mathrm{top}} = \int_U F^*_{\mathrm{top}} dS_{\mathrm{top}} \geq \int_U F^*_{\mathrm{top}} dS_{\mathrm{hor}} \\ &= \int_U e^{-\rho(H + f_{\mathrm{top}} H_1)} d(\lambda^{k-2} \otimes \nu)(H,x). \end{aligned} \tag{8}$$

We obtain an analogous inequality for the bottom component K_{bot} of ∂K. Let

$$K_U = \{\, (H + tH_1, x) \mid (H,x) \in U,\, t \in [f_{\mathrm{bot}}(H,x), f_{\mathrm{top}}(H,x)] \,\}.$$

The inequality

$$\begin{aligned} \operatorname{vol}(K) \leq \operatorname{vol}(K_U) &= \int_U \int_{f_{\mathrm{bot}}(H,x)}^{f_{\mathrm{top}}(H,x)} e^{-\rho(H+tH_1)} dt \; d(\lambda^{k-2} \otimes \nu)(H,x) \\ &= \frac{1}{\rho(H_1)} \int_U e^{-\rho(H + f_{\mathrm{bot}} H_1)} - e^{-\rho(H + f_{\mathrm{top}} H_1)} d(\lambda^{k-2} \otimes \nu)(H,x), \end{aligned}$$

implies, together with (8), that

$$\frac{\operatorname{area}(\partial K)}{\operatorname{vol}(K)} \geq \frac{\operatorname{area}(K_{\mathrm{top}}) + \operatorname{area}(K_{\mathrm{bot}})}{\operatorname{vol}(K_U)} \geq \rho(H_1).$$

This finishes the proof. □

Next, we investigate the particular horosphere $\mathcal{H} = \mathcal{H}_{v_\rho}$, where $v_\rho \in S_{p_0}\tilde{V}$ corresponds to $\frac{H_\rho}{\|H_\rho\|_0} \in \mathfrak{a}^+$ via the identification $T_{p_0}\tilde{V} \cong \mathfrak{p}$. Since $\rho(H) = \langle H_\rho, H\rangle_0$, the volume element of this horosphere is simply $d(\lambda^{k-1} \otimes \nu)$, and the solvable group NA' ($A' = \exp(\mathfrak{a}')$) is unimodular in this case. Let $\Lambda^+ = \{\alpha_1, \ldots, \alpha_m\}$. For any vector $\mathfrak{u} = (u_1, \ldots, u_m) \in (\mathbb{R}^+)^m$, let $N_\mathfrak{u}$ denote the Riemannian manifold N equipped with the metric $\langle\cdot,\cdot\rangle_\mathfrak{u} = \bigoplus_{j=1}^m u_j \langle\cdot,\cdot\rangle_{\alpha_j}$. $\mathfrak{e}$ denotes the vector $(1, \ldots, 1) \in \mathbb{R}^m$. The following lemma will become useful below:

Lemma 2.2. *Let $W \subset N$ be a closed, smooth manifold of dimension r, $\mathfrak{u} \in \mathbb{R}^m$, and $d, D > 0$ be constants satisfying $d\mathfrak{e} \leq \mathfrak{u} \leq D\mathfrak{e}$. Then we have*

$$d^{r/2}\operatorname{area}_{N_\mathfrak{e}}(W) \leq \operatorname{area}_{N_\mathfrak{u}}(W) \leq D^{r/2}\operatorname{area}_{N_\mathfrak{e}}(W). \tag{9}$$

Proof. At any point $x \in W$, the area element of $W \subset N_\mathfrak{u}$, in terms of an arbitrary base $v_1, \ldots, v_r$ of T_xW, is given by $\sqrt{\det(\langle v_i, v_j\rangle_\mathfrak{u})}\, v_1^* \wedge \cdots \wedge v_r^*$, where $v_1^*, \ldots, v_r^* \in (T_xW)^*$ is the dual base of $v_1, \ldots, v_r$. Consequently, (9) follows from the pointwise inequality

$$d^r \det(\langle v_i, v_j\rangle_\mathfrak{e}) \leq \det(\langle v_i, v_j\rangle_\mathfrak{u}) \leq D^r \det(\langle v_i, v_j\rangle_\mathfrak{e}) \tag{10}$$

for one particularly chosen base of T_xW. Let $A_x : T_xN \to T_xN$ denote the symmetric operator (w.r.t. $\langle\cdot,\cdot\rangle_\mathfrak{e}$) defined by $\langle v, w\rangle_\mathfrak{u} = \langle A_x v, w\rangle_\mathfrak{e}$. Obviously, $d\,\mathrm{Id} \leq A_x \leq D\,\mathrm{Id}$. Let $P_x : T_xN \to T_xW$ denote orthogonal projection with respect to $\langle\cdot,\cdot\rangle_\mathfrak{e}$. One easily checks that $P_xA_x : T_xW \to T_xW$ is, again, a symmetric operator w.r.t. $\langle\cdot,\cdot\rangle_\mathfrak{e}$ and satisfies $d\,\mathrm{Id} \leq P_xA_x \leq D\,\mathrm{Id}$. Then, (10) holds for a choice of eigenvectors $v_1, \ldots, v_r$ of P_xA_x. □

Theorem 2.3. *Let $v_\rho \in S_{p_0}\tilde{V}$ be the unique maximal direction of a spherical Weyl chamber $\mathfrak{a}^+ \cap S_{p_0}\tilde{V}$, and $\mathcal{H} = \mathcal{H}_{v_\rho}$ be the corresponding horosphere. Then $\mathcal{H}$ can be exhausted by a monotone increasing sequence $\{K_j\}$ of compact, connected subsets with smooth boundaries ∂K_j satisfying*

$$\lim_{j\to\infty} \frac{\operatorname{area}(\partial K_j)}{\operatorname{vol}(K_j)} = 0.$$

In particular, we have $h(\mathcal{H}_{v_\rho}) = 0$.

Similarly, there exists, for a given $\epsilon > 0$, a monotone increasing sequence $\{K_j'\}$ of compact, connected subsets satisfying

$$\lim_{j\to\infty} \frac{\operatorname{vol}((\partial K_j')_\epsilon)}{\operatorname{vol}(K_j')} = 0,$$

where $(\partial K_j')_\epsilon$ denotes the closed ϵ-tube around $\partial K_j'$ in $\mathcal{H}$.

Proof. We identify again the horosphere $\mathcal{H}$ with $\mathfrak{a}' \times N$. Let $K = B_r \times K_0 \subset \mathcal{H}$, where B_r is a $(k-1)$-dimensional closed Euclidean ball in $\mathfrak{a}' \cong \mathbb{R}^{k-1}$ of radius $r > 0$, and K_0 is a compact subset of N. Then we obtain immediately

$$\operatorname{vol}(K) = \nu(K_0)\operatorname{vol}_{\mathfrak{a}'}(B_r) = C_1\nu(K_0)r^{k-1}.$$

Next we investigate the two components of the boundary $\partial K = (\partial B_r \times K_0) \cup (B_r \times \partial K_0)$. Obviously, we have

$$\operatorname{area}(\partial B_r \times K_0) = \nu(K_0)\operatorname{area}_{\mathfrak{a}'}(\partial B_r) = C_2 \nu(K_0) r^{k-2}.$$

We define $\mathfrak{u}(H) = (e^{-2\alpha_1(H)}, \ldots, e^{-2\alpha_m(H)})$ for $H \in \mathfrak{a}'$. Choosing $D > 0$ such that $e^{-2\alpha_i(H)} \leq D$ for all $H \in B_r$ and $1 \leq i \leq m$, we conclude from Lemma 2.2 that

$$\operatorname{area}_{N_{\mathfrak{u}(H)}}(\partial K_0) \leq D^{(l-1)/2}\operatorname{area}_{N_{\mathfrak{e}}}(\partial K_0),$$

where $l = \dim\,(N)$. This implies

$$\begin{aligned}\operatorname{area}(B_r \times \partial K_0) &= \int_{B_r} \operatorname{area}_{N_{\mathfrak{u}(H)}}(\partial K_0)\, d\lambda^{k-1}(H) \\ &\leq C_1 D^{(l-1)/2}\operatorname{area}_{N_{\mathfrak{e}}}(\partial K_0) r^{k-1}.\end{aligned}$$

Putting everything together, we obtain

$$\frac{\operatorname{area}(\partial K)}{\operatorname{vol}(K)} \leq D^{(l-1)/2}\frac{\operatorname{area}_{N_{\mathfrak{e}}}(\partial K_0)}{\nu(K_0)} + \frac{C_2}{C_1 r}.$$

Now the construction of a sequence $K_j \subset \mathcal{H}$ satisfying $\frac{\operatorname{area}(\partial K_j)}{\operatorname{vol}(K_j)} \to 0$ is easy. For a given $\delta > 0$, we can choose $r > 0$ large enough such that $\frac{C_2}{C_1 r} < \frac{\delta}{2}$. This choice yields an upper bound $D > 0$, as above. Since N is nilpotent, it has polynomial growth, and we can find a large enough set $K_0 \subset N$ such that $D^{(l-1)/2}\frac{\operatorname{area}_{N_{\mathfrak{e}}}(\partial K_0)}{\nu(K_0)} < \frac{\delta}{2}$. ($K_0$ is essentially a large enough geodesic ball with respect to a left invariant metric of N. By "essentially" we mean that we have to use a smooth approximation of the distance function if the left invariant metric has conjugate points. The details are carried out in [**KP**].) Then $K = B_r \times K_0$ satisfies $\frac{\operatorname{area}(\partial K)}{\operatorname{vol}(K)} < \delta$. ∂K fails to be smooth in the set $\partial B_r \times \partial K_0$. This problem can be solved by applying a standard approximation argument, which doesn't change $\operatorname{area}(\partial K)$ and $\operatorname{vol}(K)$ too much.

We outline the similar construction of the sequence $K_j' \subset \mathcal{H}$: Choose a fixed left invariant metric on N, and denote by $(K_0)_C \subset N$, resp., $(\partial K_0)_C \subset N$ the closed C-tube about K_0, resp., ∂K_0 with respect to this metric. Given $B_r \subset \mathfrak{a}'$ and $K = B_r \times K_0$ as above, there exists a $C > 0$ such that $(\partial K)_\epsilon$ is contained in

$$(B_{r+\epsilon} - \overset{\circ}{B}_{r-\epsilon}) \times (K_0)_C \cup B_{r+\epsilon} \times (\partial K_0)_C.$$

In this case we obtain the estimate

$$\frac{\operatorname{vol}((\partial K)_\epsilon)}{\operatorname{vol}(K)} \leq \left(1 + \frac{\epsilon}{r}\right)^{k-1} \frac{\nu((\partial K_0)_C)}{\nu(K_0)} + \frac{(r+\epsilon)^{k-1} - (r-\epsilon)^{k-1}}{r^{k-1}} \frac{\nu((K_0)_C)}{\nu(K_0)}.$$

Again, $\frac{\operatorname{vol}((\partial K)_\epsilon)}{\operatorname{vol}(K)}$ can be made arbitrarily small by first taking $r > 0$ appropriately, and then choosing $K_0 \subset N$ to be a suitably large geodesic ball with respect to the initially chosen left invariant metric. $\square$

3. Applications

We first discuss two consequences of Theorem 2.1. We remind the reader of the definition $m_\rho(v_0) = \max_{\substack{H \in \mathfrak{a}' \\ \|H\|_0 = 1}} \rho(H)$, where $\mathfrak{a} = \mathbb{R}H_0 \perp \mathfrak{a}'$ and $H_0 \in \mathfrak{p}$ is the vector (in the closed positive Weyl chamber) corresponding to v_0.

Corollary 3.1. *Let $v_0 \in S_{p_0}\tilde{V}$ be a vector with $m_\rho(v_0) > 0$. Let $B(p,r)$ denote the intrinsic geodesic ball (centered at p and of radius $r > 0$) of the horosphere $\mathcal{H}_{v_0}$. Then there are constants $r_0, C > 0$ such that*

$$\mathrm{vol}(B(p,r)) \geq Ce^{m_\rho(v_0)r} \quad \textit{for all } r \geq r_0. \tag{11}$$

Proof. Apply [**Cha2**, Prop. 6.3]. □

Remark. It is very likely that the estimate (11) is very crude. It would be very interesting to get a more precise understanding of the asymptotic behavior of $r \mapsto \mathrm{vol}_{\mathcal{H}}(B(p,r))$, and its dependance of the normal direction v_0.

In good horospheres we also have exponential volume growth of intrinsic geodesic balls. However, our argument doesn't give an explicit inequality.

Proposition 3.2. *Let $\mathcal{H} = NA'$ be a good horosphere. We consider $\mathcal{H}$ as a Riemannian manifold with left invariant Riemannian metric (induced from the metric of the ambient symmetric space). Then there exists a constant $c > 0$, such that*

$$\mathrm{vol}(B(p,r)) \geq e^{cr} \quad \textit{for all } r \geq r_0,$$

where $B(p,r)$ denotes the intrinsic geodesic ball of $\mathcal{H}$, as in Corollary 3.1.

Proof. Note that $H = N^+A'$ is a connected unimodular group, equipped with a left invariant Riemannian metric. Let λ denote the induced Riemannian volume, which coincides with the biinvariant Haar measure on H. We recall from [**Pa**, Ch. 6] that the following properties are equivalent:

i) For any compact neighborhood $C \subset H$ of the identity we have

$$\lambda(C^n)^{1/n} \to 1, \quad \text{as } n \to \infty.$$

ii) H is of type R, i.e., for all $X \in \mathfrak{h}$, the spectrum of $\mathrm{ad}X : \mathfrak{h} \to \mathfrak{h}$ is purely imaginary.

Obviously, our group $H = N^+A'$ violates property ii), since, for any nonsingular vector $Z \in \mathfrak{a}'$, the spectrum of $\mathrm{ad}Z : \mathfrak{h} \to \mathfrak{h}$ contains the values $\alpha(Z) \in \mathbb{R} - \{0\}$ for all $\alpha \in \Lambda^+$. Consequently, there exists a compact neighborhood $C \subset H$ of the identity with $\lambda(C^n)^{1/n} \to 1 + \epsilon > 1$, as $n \to \infty$. Let $r_0 > 0$ be large enough such that $C \subset B(p, r_0)$. We conclude $C^n \subset B(p, nr_0)$, because any $g \in C^n$ can be written in the form $g = g_1 \dots g_n$, with elements $g_j \in C$, and we obtain, by left invariance of the metric:

$$\begin{aligned} d(g,e) &\leq d(g_1 \cdots g_n, g_1 \cdots g_{n-1}) + d(g_1 \cdots g_{n-1}, g_1 \cdots g_{n-2}) + \cdots + d(g_1, e) \\ &= d(g_n, e) + d(g_{n-1}, e) + \cdots + d(g_1, e) \leq nr_0. \end{aligned}$$

This implies

$$\liminf_{n\to\infty} \frac{\log \mathrm{vol}(B(p, nr_0))}{n} = 2\delta > 0,$$

and, consequently, for n sufficiently large,

$$\mathrm{vol}(B(p, nr_0)) \geq e^{\delta n}.$$

The proof of the proposition is complete. □

COROLLARY 3.3. *Let $K \subset \mathcal{H}_{v_0}$ be any compact domain with piecewise smooth boundary. Then the first positive Dirichlet eigenvalue $\lambda_1(K)$ of the equation $\Delta f = \lambda f$ on K satisfies*

$$\lambda_1(K) \geq \frac{m_\rho(v_0)^2}{4}.$$

PROOF. Apply Cheeger's inequality (see, e.g., [**Cha1**, Thm. 3]). □

Theorems 2.1 and 2.3 imply that each spherical Weyl chamber contains precisely one unit vector v_ρ such that H_{v_ρ} is a *good horosphere* as defined in the introduction. (Note that v_ρ coincides with $\frac{H_\rho}{\|H_\rho\|_0}$, where H_ρ is given in terms of the root vectors as $H_\rho = \sum_{\alpha \in \Lambda^+} H_\alpha$.) We finally conclude the following

COROLLARY 3.4. *Let $\tilde{V}$ be a symmetric space of noncompact type, $\mathcal{H} = \mathcal{H}_{v_\rho} \subset \tilde{V}$ be a good horosphere, and $V = \Gamma \backslash \tilde{V}$ be a compact quotient. Let K_j be an exhaustion of $\mathcal{H}$ satisfying $\frac{\text{area}(\partial K_j)}{\text{vol}(K_j)} \to 0$. Then we have, for any Γ-periodic continuous function f on $\tilde{V}$,*

$$\frac{1}{\text{vol}(K_j)} \int_{K_j} f(q) dq \to \frac{1}{\text{vol}(V)} \int_V f dz, \quad \text{as } j \to \infty.$$

A geometric interpretation of Corollary 3.4 is that $\mathcal{H}_{v_\rho}$ distributes uniformly on the compact quotient V.

PROOF. We check the conditions stated in [**KP**, Thm. 1]: The mean curvature of the horosphere $\mathcal{H}$ is obviously constant and equal to the sum $\sum_j \sqrt{-\lambda_j}$, where the λ_j's are the eigenvalues of the curvature operator $w \mapsto R(w, v_\rho)v_\rho$. The mean curvature is strictly positive, since at least one of the λ_j's is negative (see [**Eb**, 2.15.]). Moving $\mathcal{H}$ in normal direction, we obtain a foliation of $\tilde{V}$ by isometric horospheres. The corollary follows now immediately from [**KP**, Thm. 1]. □

Corollary 3.4 is related to unique ergodicity of the strong unstable foliation in the rank one case (see [**Fü**] and [**BM**]), and to unique ergodicity of the N-action on $\Gamma \backslash G$ in the higher rank case (see [**Ve**]). The author intends to discuss these relations in his Habilitation thesis.

References

[BGS] W. Ballmann, M. Gromov, and V. Schroeder, *Manifolds of nonpositive curvature*, Birkhäuser, Boston, Basel, Berlin, 1985.

[BM] R. Bowen and Brian Marcus, *Unique ergodicity for horocycle foliations*, Isr. J. Math. **26** (1977), 43–67.

[Cha1] I. Chavel, *Eigenvalues in Riemannian Geometry*, Academic Press, Orlando, 1984.

[Cha2] I. Chavel, *Riemannian Geometry: A modern introduction*, Cambridge Univ. Press, Cambridge, 1993.

[Eb] P. B. Eberlein, *Geometry of nonpositively curves manifolds*, University of Chicago Press, Chicago, 1996.

[Fe] R. Feres, *Dynamical systems and semisimple groups – an introduction*, Cambridge Univ. Press, Cambridge, 1998.

[Fü] H. Fürstenberg, *The unique ergodicity of the horocyclic flow*, Recent Advances in Topological Dynamics (A. Beck, ed.) Springer, Berlin (1972), 95–115.

[Gr] M. Gromov, *Geometric group theory. Volume 2: Asymptotic invariants of infinite groups. Proceedings of the symposium held at the Sussex University, Brighton, July 14–19, 1991*, Cambridge Univ. Press, Cambridge, 1993.

[Ho] H. F. Hoke III, *Lie groups that are closed at infinity*, Trans. Amer. Math. Soc. **313** (1989), 721–735.

[Ka] M. Kanai, *Rough isometries, and combinatorial approximations of geometries of noncompact Riemannian manifolds*, J. Math. Soc. Japan **37** (1985), 391-413.

[Kn] G. Knieper, *On the asymptotic geometry of nonpositively curved manifolds*, GAFA **7** (1997), 755–782.

[KP] L. Karp, N. Peyerimhoff, *Horospherical means and uniform distribution of curves of constant negative curvature*, Math. Z. **231** (1999), 655–677.

[Mo] G.D. Mostow, *Strong rigidity of locally symmetric spaces*, Ann. of Math. Stud. **78**, Princeton University Press, Princeton, NJ, 1973.

[Pa] A.L.T. Paterson, *Amenability*, Mathematical surveys and monographs **28**, Amer. Math. Society, Providence, Rhode Island, 1988.

[Sp] R.J. Spatzier, *Dynamical properties of algebraic systems*, Dissertation, Warwick and Maryland, 1983.

[Ve] W. A. Veech, *Unique ergodicity of horospherical flows*, Amer. J. Math. **99** (1977), 827–859.

Fakultät für Mathematik NA 5/32 Ruhr-Universität Bochum 44780 Bochum Germany

E-mail address: peyerim@math.ruhr-uni-bochum.de
URL: http://www.ruhr-uni-bochum.de/mathematik10

Proceedings of Symposia in Pure Mathematics
Volume **69**, 2001

Minimal but not Uniquely Ergodic Diffeomorphisms

Alistair Windsor

Abstract. This paper provides a method of constructing smooth minimal diffeomorphisms whose set of ergodic measures has a given cardinality. The construction applies to compact manifolds admitting free S^1-actions. The diffeomorphisms constructed all lie in the closure of the space of periodic diffeomorphisms. However they cannot be obtained by Baire Category arguments.

1. Introduction

If f is a uniquely ergodic transformation on a separable metric space preserving a Borel measure μ then $f|_{\text{supp}(\mu)}$ is minimal, that is every orbit is dense [**7**, Proposition 4.1.18]. It was therefore a natural question whether minimality is sufficient for unique ergodicity. This is one in a series of questions which ask which topological properties are sufficient for more abstract metric properties. The unfortunate answer has been that while metric properties have strong topological consequences the converse is false. One reason for this is that there exist smooth cocycles which are measurable coboundaries but for which the transfer functions behave wildly topologically. This was known by A. N. Kolmogorov [**1**] who used this method to construct a time change of a linear flow on $\mathbb{T}^2$ with pure point spectrum but with discontinuous eigenfunctions. The same observation was independently made by H. Furstenberg [**3**] who used it to construct diffeomorphisms which are minimal but not uniquely ergodic (see [**7**, Corollary 12.6.4] for the essence of the construction). These diffeomorphisms are skew-products, are analytic, and all admit uncountably many ergodic measures.

A construction of minimal transformations with any given finite number, a countable number, or a continuum of ergodic invariant measures was provided by S. Williams [**6**] in the case of symbolic dynamics. This serves as very general counterexample to our earlier question in the case of symbolic systems. No such construction has been published for the smooth category. A construction of topologically transitive diffeomorphisms preserving a smooth measure and with a given number of ergodic components was announced by D. V. Anosov and A. B. Katok [**4**] without proof. We use a variation of their methods to construct minimal diffeomorphisms with a given number of ergodic measures. The idea of transversal

2000 *Mathematics Subject Classification.* Primary 37C40.
The author was partially supported by the NSF grants DMS97-04776 and DMS 98-03129.

cutting which is one of the keys for avoiding the measure zero exceptional set which occurs in [**4**] was communicated to the author by A.B Katok in lectures on smooth ergodic theory given at The Pennsylvania State University during 1998.

2. The Theorem

THEOREM. *Let $\alpha \in \mathbb{Q}$, M be a smooth manifold admitting a free C^∞ action of S^1, $\tau : S^1 \times M \to M$, and μ be a smooth positive measure invariant under τ. Let $d \in \mathbb{N}$.*

Then there exists a diffeomorphism T, arbitrarily close to $\tau_\alpha(\cdot) := \tau(\alpha, \cdot)$ which is minimal but has exactly d ergodic invariant measures which are absolutely continuous with respect to μ.

3. Structure of the Construction

The required diffeomorphism is constructed using a variant of the fast approximation - conjugation method pioneered by Anosov and Katok [**4**]. The required diffeomorphism T is obtained as the limit of a sequence of periodic diffeomorphisms, $\{T_n\}$, each of which preserve μ. Each periodic diffeomorphism T_n is smoothly conjugate to τ_{α_n} for some $\alpha_n \in \mathbb{Q}$,

$$T_n = h_n \circ \tau_{\alpha_n} \circ h_n^{-1}.$$

Let

$$\mathfrak{B}(M) := \{h \circ \tau_\alpha \circ h^{-1} : \alpha \in S^1,\, h \in \mathrm{Diff}^\infty(M)\} \subset \mathrm{Diff}^\infty(M).$$

By construction $T \in \overline{\mathfrak{B}(M)}$. Fathi and Herman showed in [**5**] that the collection of uniquely ergodic maps forms a dense G_δ in $\overline{\mathfrak{B}(M)}$. Thus unique ergodicity is generic in $\mathfrak{B}(M)$. This remains true if we restrict ourselves to $\mathfrak{B}(M,\mu) \subset \mathfrak{B}(M)$ where the conjugacy h is taken to preserve μ. Thus our construction is not generic.

The sequence of conjugacies, $\{h_n\}$, and "rotation numbers", $\{\alpha_n\}$ are constructed inductively.

$$h_n = A_1 \circ \cdots \circ A_n$$

where each A_i preserves μ. The "rotation numbers" satisfy

$$\alpha_{n+1} = \alpha_n + \frac{1}{m_n Q_n q_n}$$

where q_n is the denominator of α_n and m_n and Q_n are parameters defined later.

We take a partition of our space into d Borel sets of positive measure. These naturally support d absolutely continuous probability measures $\{\mu_i\}$, given by the normalized restrictions of μ. The required d ergodic invariant measures are constructed from the μ_i as the limits of the sequences

$$\xi_i^n := h_n^* \mu_i.$$

Notice that ξ_i^n is preserved by T_n.

4. Topological Preliminaries

The approach of Anosov and Katok applies to trivial bundles. For constructions in the measurable category this is not really a restriction since every smooth bundle is trivial in the measurable category. Furthermore any manifold admitting a smooth periodic flow is measurably a trivial bundle. Unfortunately our construction is not a measurable category construction and we must perforce deal with non-trivial bundles. These topological lemmas are part of our extension of the Anosov and Katok approach to the case of non-trivial bundles.

The approach of Anosov and Katok allowed for manifolds with boundary. Unfortunately our approach cannot deal with these. Obviously there cannot exist minimal diffeomorphisms of the disk since the boundary must be preserved and there is always a fixed point. However it might be supposed that the orbit of any other point could be made dense. That this is not possible was shown by Le Calvez and Yoccoz [**8**]. The ergodic properties of such a diffeomorphism may be more tractable.

LEMMA 1. *If N is a connected compact Riemannian manifold then there exist a constant K such that for all $\epsilon > 0$ there exists K embedded disks $\{D_i\}$ such that*

(1) $\operatorname{vol}(D_i) > 1 - \epsilon$ *where* vol *is the normalized Riemannian volume on N.*
(2) $\{D_i^\circ\}$ *is an open cover of N.*

PROOF. As N is a smooth compact Riemannian manifold there exists a smooth triangulation of N [**2**]. Let $\{\mathcal{S}_i\}_{i=0}^m$ be the set of simplices from the triangulation with the same dimension as N. For each simplex $\mathcal{S}_i$ construct a smoothly embedded closed disk K_i by taking a slightly smaller simplex within $\mathcal{S}_i$ and smoothing the corners. K_i can be constructed such that $\operatorname{vol}(K_i) > (1-\epsilon)\operatorname{vol}(\mathcal{S}_i)$.

Construct a graph with vertices $\{\mathcal{S}_i\}$ by putting an edge between $\mathcal{S}_i$ and $\mathcal{S}_j$ if they share a face. This is a connected graph which we will call the adjacency graph. Construct a maximal tree in the adjacency graph. We will call this the joining tree.

We describe the step by step procedure for constructing the disk D_j. Each disk D_j is constructed by joining the disks K_i described above but enlarging one disk so that the family forms a cover.

(1) Using coordinate charts construct a smoothly embedded disk κ_j such that $K_i \cap \kappa_j = \varnothing$ for all $i \neq j$ and κ_j° is a neighborhood of $\mathcal{S}_j$.
(2) For each edge in the joining tree not involving $\mathcal{S}_j$ construct a smooth tube which joins the appropriate K_i and K_k. Let U be a chart neighborhood about the center of their common face. Using the smooth coordinates in the interior of each simplex construct a smooth tube joining the appropriate K to the neighborhood U. Use the smooth coordinates in U to complete the tube.
(3) For each edge in the joining tree joining a K_i to κ_j we can construct a smooth tube using the smooth coordinates in the interior of $\mathcal{S}_i$.

This yields a smoothly embedded disk since we can show that the addition of a tube and disk to an embedded disk constructed by joining K_i and K_k is again an embedded disk and then proceed by induction. Let us imagine joining $K_i \subset \mathcal{S}_i$ to a smooth embedded disk via a tube which is contained in $K_i \cup K_k$ using the process outlined in step 2 above. Using the smooth coordinates in the interior of $\mathcal{S}_k$ we can show that the combination of the tube and K_k is diffeomorphic to a short tube with a spherical cap contained within the neighborhood U and this diffeomorphism

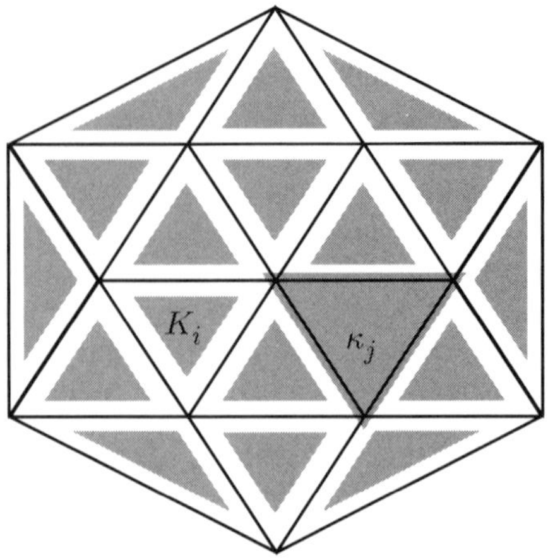

FIGURE 1. An example of the a triangulation with disks $\{K_i\}$ and disk κ_j.

extends via the identity. Now use the smooth coordinates in U to contract the tube and cap to a tube and cap contained in the interior of $\mathcal{S}_i$. Finally we use the smooth coordinates in the interior of $\mathcal{S}_i$ to show that the smooth embedded disk plus a tube and cap is diffeomorphic to the original embedded disk. This argument covers the joining of K_i and K_j but we can show using coordinate charts that connecting K_i and κ_j is equivalent to connecting K_i and K_j whereupon the earlier argument applies.

Thus we get a smoothly embedded disk with

$$\operatorname{vol}(D_j) > \sum \operatorname{vol}(K_i) > (1-\epsilon) \sum \operatorname{vol}(\mathcal{S}_i) = 1 - \epsilon.$$

Since D_i° contains a neighborhood of the simplex $\mathcal{S}_i$ the family $\{D_i\}$ covers N. The constant K is precisely the number of highest dimensional simplices in the triangulation of N. □

Let $B(r)$ denote the standard closed disk of radius r in $\mathbb{R}^{\dim N}$.

LEMMA 2 (Legerdemain Lemma). *Let N be a smooth connected compact manifold without boundary endowed with a smooth positive measure μ. Given any positive integer d there exists a cover of N by d sets $\{N_i\}$ such that*

(1) *each set N_i is closed and connected.*
(2) *$N_i^\circ \cap N_j = \varnothing$ for all $i \neq j$.*
(3) *N_i is a smoothly embedded disk for $1 \leq i \leq d-1$. In particular ∂N_i is an embedded sphere.*
(4) *there exists K such that for all $\epsilon > 0$ there is a cover of N by K smoothly embedded closed disks $\{D_i\}$, such that $\mu(D_i) > 1-\epsilon$ and $N_i \subset D_j^\circ$ for $1 \leq i \leq d-1$ and for all j.*

PROOF. Construct a triangulation, adjacency graph, and joining tree as in Lemma 1. Let $\{\mathcal{S}_i\}$ be the set of simplices from the triangulation with the same dimension as N. Fix $\epsilon_1 > 0$ and consider the collection of smooth embedded closed disks L_i constructed by taking a smaller simplex in $\mathcal{S}_i$ and smoothing the corners such that $\mu(L_i) > (1-\epsilon_1)\mu(\mathcal{S}_i)$. Now in any cover constructed using Lemma 1 (using μ in place of vol) with $\epsilon > \epsilon_1$ we will have $L_i \subset K_i^\circ$. Let $\{d_j\}$ be the family of embeddings, $d_j : B(1) \to N$, given by Lemma 1. There are only finitely many of these so, by continuity, there exists $\delta_0 > 0$ such that for all $0 < \delta < \delta_O$ and each j the embedded disk $d_j\big(B(1-\delta)\big)$ is connected and contains $\cup L_i$. Taking any $d-1$ disjoint smoothly embedded disks N_i in $\cup L_i$ and adding $N_d = N \setminus \bigcup_{i=1}^{d-1} N_i^\circ$ we obtain a cover with the required property. □

5. Preliminary Construction on a Trivial Sub-Bundle

The Legerdemain lemma reduces each step of the construction to several simpler sub-steps. Each sub-step is free of topological obstructions. Each step in our iterative construction is built up of a finite number of sub-steps each of which consists of a construction on a trivial sub-bundle.

Let $B(1)\times S^1$ be a trivial bundle endowed with normalized Lebesgue measure, λ. Let $\{R_1, \dots, R_{n-1}\}$ be a collection of disjoint smoothly embedded closed disks contained in $B(1)^\circ$ and let $R_n = \overline{B(1)\setminus \bigcup_{i=1}^{n-1} R_i}$. Define $\tilde{C} := B(1) \times S^1$ and $\tilde{C}_j := R_j \times S^1$. Each $\tilde{C}_j$ supports a measure given by

$$\nu_j(A) := \frac{\lambda(\tilde{C}_j \cap A)}{\lambda(\tilde{C}_j)}.$$

Lemma 3 (Sub-Bundle Lemma). *Let $\alpha = \frac{p}{q} \in \mathbb{Q}^+$, with $(p,q) = 1$, be given, and $\Phi = \{\varphi_1, \dots, \varphi_n\}$ be a given finite set of continuous functions on $\tilde{C}$.*

Then for all $\epsilon > 0$ there exists $\delta_0 > 0$ such that for all $0 < \delta < \delta_0$ there exists $Q \in \mathbb{N}$, and a measure-preserving diffeomorphism A such that

(1) *there exists a τ_α-invariant set Σ with $B(1-\delta) \times S^1 \subset \Sigma \subset B(1-\frac{\delta}{2}) \times S^1$ such that A is the identity off Σ.*
(2) *$A \circ \tau_\alpha = \tau_\alpha \circ A$.*
(3) *$\forall \varphi \in \Phi$, $\forall x \in \Sigma$, and $\forall \alpha' = \frac{p'}{q'}$ with $q' = mQq$ and $(p', q') = 1$*

$$\inf_{\nu \in \mathfrak{M}} \left| \frac{1}{q'} \sum_{\Pi^x} \varphi\big(A \circ \tau_{\alpha'}^i x\big) - \frac{\operatorname{Card}(\Pi^x)}{q'} \int \varphi \, d\nu \right| < \epsilon$$

where $\mathfrak{M}$ is the simplex of measures generated by $\{\nu_1, \dots, \nu_d\}$, $m \in \mathbb{N}$ is arbitrary, and $\Pi^x := \{i : \tau_{\alpha'}^i x \in \Sigma\}$.
(4) *$\forall \alpha' = \frac{p'}{q'}$, with $q' = mQq$ and $(p', q') = 1$, the map $A \circ \tau_{\alpha'} \circ A^{-1}$ is ϵ-minimal, that is every orbit intersects every ball of radius ϵ.*
(5) *$\forall \varphi \in \Phi$ and $1 \le i \le d$*

$$\left| \int \varphi \, d\nu_i - \int \varphi \circ A \, d\nu_i \right| < \epsilon.$$

(6) *$\lambda(A\tilde{C}_i \Delta \tilde{C}_i) < \epsilon$ for $1 \le i \le d$.*

5.1. Outline of the Proof of the Sub-Bundle Lemma. This lemma is the key step in proving Theorem 1. It is proved by constructing appropriate combinatorics on the bundle and then constructing a smooth realization of these combinatorics.

5.1.1. *Combinatorics.* In order to define combinatorics on the bundle it must be discretized. Cover the disk $B(1-\delta)$ by closed cubes of side length s, $\{C_i\}$, such that $C_i^\circ \cap C_j = \varnothing$ and $C_i^\circ \cap B(1-\delta) \neq \varnothing$. Choose s small enough so that $B(1-\delta) \subset \cup C_i \subset B(1-\delta/2)$. Later there with be further restrictions on s.

Associate each cube C_i with a region R_j. A cube C_i can be associated with a region R_j provided $C_i^\circ \cap R_j \neq \varnothing$. Let the sets of cubes associated to each R_j be connected. Let N_i denote the number of cubes associated with R_i, $N = \sum N_i$, and $L = \operatorname{lcm}(N_1, \dots N_n)$. Renumber the cubes C_i so that the first N_1 are associated with R_1, the next N_2 with R_2, and so on.

This produces a partial partition of $\Delta_{lq} = B(1) \times [0, \frac{1}{lq})$ into cubes, $Q_{i,j}$ where $1 \le i \le N$, $1 \le j \le N + kL$, by

$$Q_{i,j} = \{(x,t) : x \in C_i, \frac{j-1}{ql(N+kL)} \le t < \frac{j}{ql(N+kL)}\}.$$

This can be envisioned as a tower with the base $\cup C_i \subset B(1)$ and $N + kL$ floors of equal height.

Define A on Δ_{lq} and extend it to all of $\tilde{C}$ by requiring

$$A \circ \tau_\alpha = \tau_\alpha \circ A.$$

Outside of the union of the cubes define the map to be the identity. On the union of the cubes, $\cup Q_{i,j} \subset \Delta_{lq}$, choose A so that it induces the following permutation

$$A(Q_{i,j}) = \begin{cases} Q_{\alpha^{j-1}(i),j} & j \le N \\ Q_{\beta^{j-N-1}(i),j} & \text{otherwise} \end{cases} \tag{1}$$

where

$$\begin{aligned} \alpha &= (1 \ \dots \ N) \\ \beta &= (1 \ \dots \ N_1)(N_1 + 1 \ \dots \ N_1 + N_2) \ \dots \ (N - N_n + 1 \ \dots \ N_n) \end{aligned}$$

are permutations in the symmetric group on N elements. There is no restriction on the action of A within the cubes. Further properties, such as weak mixing, would require permuting the cubes between the levels of Δ_{lq}.

The cycle α is applied on the first N levels of $N + kL$ total levels in Δ_{lq}. Though the entire cycle is traversed only once in Δ_{lq}, there are lq such regions in $\tilde{C}$. Hence the frequency of these regions is determined by l. This region will give the ϵ-minimality. Since the levels all have equal measure, the measure of the area on which α is applied goes to 0 as $k \to \infty$. The idea behind the construction is to make this region occur frequently but have small measure, thus making this region topologically significant but measure theoretically insignificant.

On the remaining levels in the tower the permutation β is applied. This cyclically permutes the cubes corresponding to each region, R_j. This ensures that on some scale the orbits are uniformly distributed on $\tilde{C}_j$ with the small measure exception corresponding to the region on which α is applied. This is the key in proving conclusion 3 of Lemma 3 and is embodied in the main estimate (7).

5.1.2. *Smooth Realization.* As intuitively appealing as the above construction is, it is clearly not smooth. To smooth the construction it is necessary to relinquish control over part of the space. The cubes $Q_{i,j}$ are replaced by smaller cores which can be permuted as above by a C^∞-diffeomorphism of M. In order to retain sufficient control to be able to produce conclusion 3 it is necessary to control a large proportion of *every* orbit. This is where the construction leaves the measurable

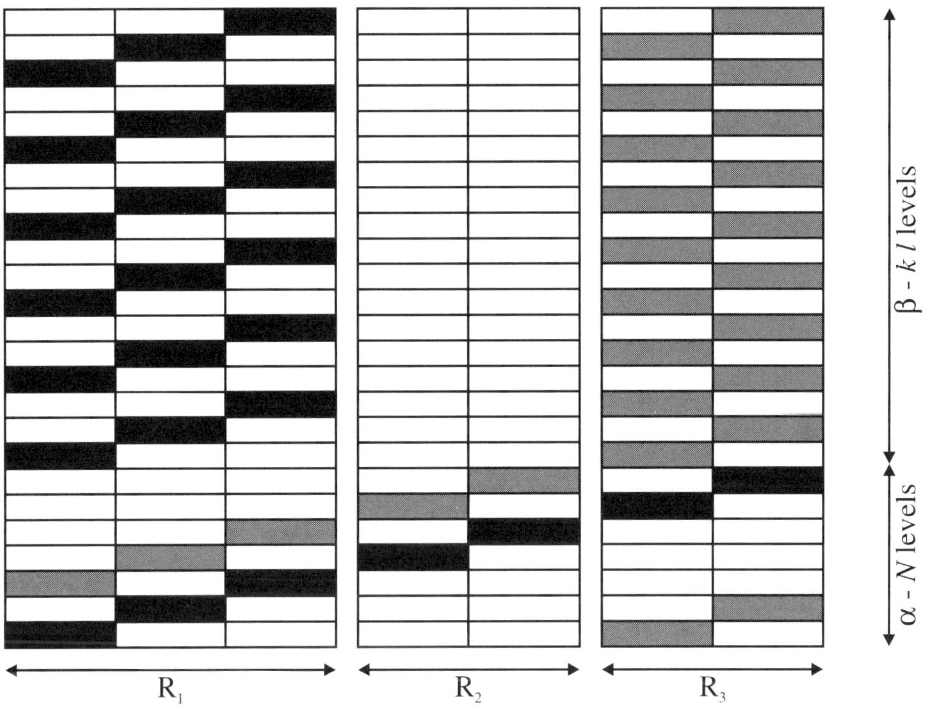

FIGURE 2. An example of the action of A where $d = 3$, $N_1 = 3$, $N_2 = N_3 = 2$, and $k = 3$.

category since for such a construction it would suffice to control most of the orbits at each step. This leads to control of *almost all* orbits of the limiting transformation. A measurable construction would be able to neglect a measure zero set of orbits. However for our construction this is fatal. A single orbit could support an ergodic invariant measure. A construction in the measurable category is sufficient to control the number of *absolutely continuous* ergodic invariant measures.

Let s be the side length of our decomposition and v be a diagonal vector of one of our cubes of side length s. Take a collection of cubes of side length s, $\{C_i\}$, such that $\cup_{t\in[0,1/2)}(B(1-\delta)-tv)\cap C_i \neq \emptyset$ and $\cup_{t\in[0,1/2)}(B(1-\delta)-tv) \subset \cup_i C_i \subset B(1-\delta/2)$.

As before, A is defined on Δ_{lq} and extended to $\tilde{C}$ by requiring it commute with $\tau_{\frac{1}{lq}}$. Partially partition Δ_{lq} by parallelepipeds

$$P_{i,j} := \Big\{(x,t) : x - \frac{lq(N+kL)t+1-j}{2}v \in C_i,\ \frac{j-1}{lq(N+kL)} \leq t < \frac{j}{lq(N+kL)}\Big\}.$$

Scale each parallelepiped about its center by a factor γ. This gives a family of identical disjoint parallelepipeds $\{\tilde{Q}_{i,j}\}$. These are called cores and play the role of the $Q_{i,j}$ in the previous section.

Define the map A to be a diffeomorphism inducing the same permutations as before (1). Any two adjacent parallelepipeds can be enclosed in a region diffeomorphic to a disk such that the images of the two parallelepipeds are related via a rotation about the center. Since any permutation of cores can be written as a

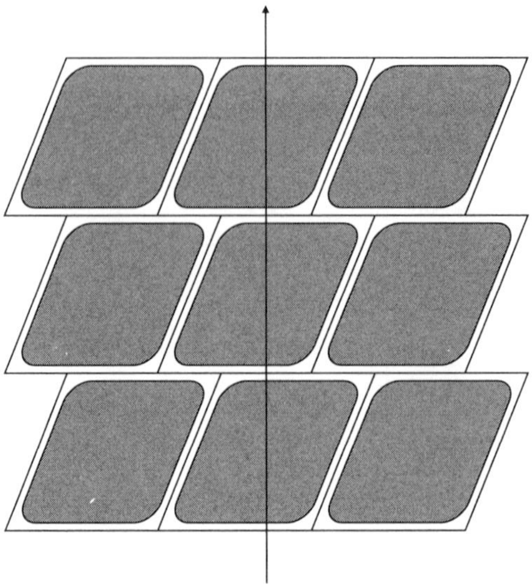

FIGURE 3. Example of cores. Arrow denotes the fiber direction. Notice the gaps are transverse to the fiber direction.

product of such transpositions A can be taken to be a measure preserving C^∞-diffeomorphism. Defined in the this manner the diffeomorphism A automatically satisfies conclusions 1 and 2 of Lemma 3.

5.2. Proof of the Sub-Bundle Lemma. There are several parameters of the construction: δ_0, s, γ, k, P, and l. The parameters δ_0, s, P, and γ are related to the geometry. The parameters k and l govern the combinatorics. Let $\kappa \geq 1$ be a global bound for $|\varphi|$ for all $\varphi \in \Phi$.

5.2.1. *Geometric Estimates.* Fix δ_0 so that for $1 \leq j \leq n$

$$\nu_j\big(B(1-\delta_0)\big) < \frac{\epsilon}{20\kappa}$$

and for $1 \leq i \leq n-1$ we have

$$R_i \subset B(1-\delta_0).$$

As the boundary of R_j is a smooth compact sub-manifold we can choose s sufficiently small such that the union of the parallelepipeds associated to $\tilde{C}_j$, denoted $\tilde{V}_j$, approximates $\tilde{C}_j$ in the sense that

$$\frac{\lambda(\tilde{C}_j \Delta \tilde{V}_j)}{\lambda(\tilde{C}_j)} < \frac{\epsilon}{18\kappa} \tag{2}$$

Consideration of the action of A leads naturally to the following partial partition of Δ_{lq} into "columns",

$$K_{a,b} := \begin{cases} \bigcup_{m=1}^{N} P_{a,m} & b = 1 \\ \bigcup_{m=1}^{L} P_{a,N+(b-2)L+m} & b \geq 2 \end{cases}$$

This is extended to a partial partition of $B(1) \times S^1$ by requiring it to commute with $\tau_{\frac{1}{lq}}$. Set

$$K_{a,b,c} := \tau_{\frac{1}{lq}}^{c} K_{a,b}.$$

The columns $K_{a,b,c}$ with $b = 1$ correspond to levels on which the cycle α-cycle is applied. The orbit of $A \circ \tau_\alpha \circ A^{-1}$ is almost equidistributed among the $\{K_{a,b,c} \subset \tilde{V}_i : b \neq 1\}$. This observation together with the following local estimate (3) are the key to proving the main estimate (7).

From the geometry of the parallelepipeds there exists a constant $C > 0$ which depends solely on the dimension of M such that $\operatorname{diam}(K_{a,b,c}) < Cs^{\dim M}$. Uniform continuity guarantees that $\operatorname{var}_{K_{a,b,c}} \varphi < \epsilon/9$ for all $\varphi \in \Phi$ given a sufficiently small s. Thus for all $\varphi \in \Phi$ and $y \in K_{a,b,c} \subset \tilde{V}_j$ the following local estimate holds

$$\left| \frac{1}{\nu_j(K_{a,b,c})} \int_{K_{a,b,c}} \varphi \, d\nu_j - \varphi(y) \right| < \frac{\epsilon}{9}. \tag{3}$$

Choosing P, the minimum number of points of $\tau_{\alpha'}$ within a level of our tower (corresponding to $m = 1$), to be large enough we can make the proportion of the orbit contained within cores arbitrarily close to the proportion of the fiber contained within the cores. This can be made arbitrarily close to 1 by choosing γ sufficiently close to 1.

Let $Q := l(N + kL)P$.

5.2.2. *Combinatorial Estimates.* Let

$$\Gamma := \bigcup_{i=0}^{lq-1} \tau^i_{\frac{1}{lq}} \left(B(1) \times \left[\frac{N}{lq(N+kL}, \frac{1}{lq} \right) \right).$$

Choose k sufficiently large that the proportion of a τ-orbit which lies in the levels of the tower which correspond to the cycle α satisfies

$$\lambda(\tilde{C} \backslash \Gamma) = \frac{N}{N+kL} < \frac{\epsilon}{9\kappa}. \tag{4}$$

Given the orbit of x under $\tau_{\alpha'}$ denote by π_1^x the collection of points on the orbit which lie in levels on which the cycle α is applied, π_2^x the collection of points which are not contained in the cores, and π_3^x the collection of points in cores in levels which correspond to the permutation β. Let $\Pi_j^x = \{i : A \circ \tau^i_{\alpha'}(x) \in \pi_j^x\}$. By choosing γ sufficiently close to 1, P large enough, and k large enough we can guarantee for all l that

$$\frac{\operatorname{Card}(\Pi_1^x \cup \Pi_2^x)}{q'} < \frac{\epsilon}{3\kappa} \qquad \frac{\operatorname{Card}(\Pi_3^x)}{q'} \geq \frac{\operatorname{Card}(\Pi^x)}{q'} - \frac{\epsilon}{3\kappa}. \tag{5}$$

Let the proportion of π_3^x that lies within $\tilde{V}_j$ be ω_j. The proportion of π_3^x contained in a column $K_{a,b,c} \subset \tilde{V}_j$ is, for $b \neq 1$

$$\rho_j := \frac{L}{lq(N+kL)} \frac{\omega_j}{N_j}.$$

Taking the sum over $\Pi_1^x \cup \Pi_2^x$ yields

$$\left| \sum_{i \in \Pi_1^x \cup \Pi_2^x} \varphi(A \circ \tau^i_{\alpha'} x) \right| < \kappa \operatorname{Card}(\Pi_1^x \cup \Pi_2^x). \tag{6}$$

Taking a similar sum over Π_3^x and using the local estimate (3) yields

$$\left| \sum_{\Pi_3^x} \varphi(A \circ \tau^i_{\alpha'} x) - \operatorname{Card}(\Pi_3^x) \sum_{j=1}^{d} \frac{\rho_j}{\nu_j(K)} \int_{\tilde{V}_j \cap \Gamma} \varphi \, d\nu_j \right| < \frac{\epsilon}{9} \operatorname{Card}(\Pi_3^x). \tag{7}$$

The geometry of the construction means that the measure of a column on which the permutation β is applied, $K = K_{a,b,c}$ with $b \neq 1$, is given by

$$\nu_j(K) = \frac{\nu_j(\tilde{V}_j)L}{lq(N+kL)N_j}$$

where $\tilde{V}_j$ is the "cylinder" containing the column.

This reduces the main estimate (7) to

$$\Big|\sum_{i\in\Pi_3^x} \varphi(A\circ\tau_{\alpha'}^i x) - \mathrm{Card}(\Pi_3^x)\sum_{i=1}^{s}\frac{\omega_i}{\nu_i(\tilde{V}_i)}\int_{\tilde{V}_i\cap\Gamma}\varphi\,d\nu_i\Big| < \frac{\epsilon}{9}\,\mathrm{Card}(\Pi_3^x). \tag{8}$$

Use (2) and (4) to get

$$\Big|\frac{1}{\nu_i(\tilde{V}_i)}\int_{\tilde{V}_i\cap\Gamma}\varphi d\nu_i - \int\varphi d\nu_i\Big| < \frac{2\epsilon}{9}.$$

Define $\nu := \sum_{i=1}^n \omega_i\nu_i$ and apply (8) to get

$$\Big|\sum_{i\in\Pi_3^x}\varphi(A\circ\tau_{\alpha'}^i x) - \mathrm{Card}(\Pi_3^x)\int\varphi d\nu\Big| < \frac{\epsilon}{3}\,\mathrm{Card}(\Pi_3^x). \tag{9}$$

The cardinality estimate for Π_3^x , (5), implies that

$$\Big|\Big(\frac{\mathrm{Card}(\Pi^x)}{q'} - \frac{\mathrm{Card}(\Pi_3^x)}{q'}\Big)\int\varphi\,d\nu\Big| < \frac{\epsilon}{3}.$$

Taking the estimates for the sums over $\Pi_1^x\cup\Pi_2^x$, (6), and Π_3^x, (9), dividing by q', and applying the previous estimate yields the desired property

$$\Big|\frac{1}{q'}\sum_{i\in\Pi^x}\varphi(A\circ\tau_{\alpha'}^i x) - \frac{\mathrm{Card}(\Pi^x)}{q'}\int\varphi\,d\nu\Big| < \epsilon.$$

To prove ϵ-minimality it is convenient to use the metric on $\tilde{C}$ given by

$$d((x,t),(x',t')) := \max\{d(x,x'),|t-t'|\}.$$

In this metric, which is equivalent to the usual metric, the ϵ-ball is a cylinder. Choosing $l > 2/\epsilon$ guarantees that the cylinder passes through a collection of levels on which the entire α-cycle is traversed. If s is chosen small enough then the cylinder must contain an entire column from this collection. Since every orbit meets every column, every orbit meets every ϵ-ball.

Observe that A preserves the decomposition into $\{\tilde{V}_i\}$ except on the levels corresponding to the cycle α. From (4) it follows that

$$\frac{\lambda(\tilde{V}_i\Delta A\tilde{V}_i)}{\lambda(\tilde{C}_i)} < \frac{2\epsilon}{9\kappa}.$$

By the triangle inequality and (2) we have

$$\frac{\lambda(\tilde{C}_i\Delta A\tilde{C}_i)}{\lambda(\tilde{C}_i)} < \frac{1\epsilon}{3\kappa}.$$

This implies conclusions 5 and 6.

This completes the proof of the lemma. □

6. Extension to the Entire Bundle

In this section we state and prove a proposition which extends the lemma above to the entire bundle. The proof is an induction over the disks provided by Lemma 2 applying Lemma 3 at each step.

Choose a cover of N by a d sets, $\{N_i\}$, using Lemma 2.

PROPOSITION 4. *Let $\alpha = \frac{p}{q} \in \mathbb{Q}^+$, with $(p,q) = 1$, be given, and $\Phi = \{\varphi_1, \dots, \varphi_n\}$ be a given finite set of continuous functions on M.*

Then for all $\epsilon > 0$ there exists $Q \in \mathbb{N}$, and a diffeomorphism A of M such that

(1) *$\forall \alpha' = \frac{p'}{q'}$, with $q' = mQq$ and $(p',q') = 1$, the map $A \circ \tau_{\alpha'} \circ A^{-1}$ is ϵ-minimal, that is every orbit intersects every ball of radius ϵ.*

(2) *$\forall \varphi \in \Phi$, $\forall x \in M$, and $\forall \alpha' = \frac{p'}{q'}$ with $q' = mQq$ and $(p',q') = 1$*

$$\inf_{u \in \mathfrak{M}'} \left| \frac{1}{q'} \sum_{i=0}^{q'-1} \varphi(A \circ \tau_{\alpha'}^i x) - \int \varphi \, du \right| < \epsilon$$

where $\mathfrak{M}'$ is the simplex of measures generated by $\{\mu_1, \dots, \mu_n\}$.

(3) *$A \circ \tau_\alpha = \tau_\alpha \circ A$.*

(4) *$\mu(A\tilde{N}_i \Delta \tilde{N}_i) < \epsilon$ for $1 \le i \le d$.*

6.1. Preliminaries. Construct D_j using Lemma 2 so that for all i, j

$$1 - \tilde{\mu}_i(D_j) < \min\left\{\frac{\epsilon}{6K\kappa}, \operatorname{vol}\left(\frac{\epsilon}{2}\right)\right\} \tag{10}$$

where $\operatorname{vol}\left(\frac{\epsilon}{2}\right)$ is the least measure of a $\frac{\epsilon}{2}$-ball in M and K is the number of disks required for N. Consider the normalized pull-back of the measure $\tilde{\mu}$ via the embedding d_i. This gives a smooth positive measure on $B(1)$ which, by Moser's Theorem [**7**, Theorem 5.1.27], is the push-forward of the Lebesgue measure on $B(1)$ by a diffeomorphism m_i. Let $\rho_i = d_i \circ m_i$ be the new measure-preserving embedding. Consider its extension to a map $\tilde{\rho}_i : \tilde{C} \to M$ defined by $\tilde{\rho}_i(x,t) = \tau_t \rho_i(x)$. By uniform continuity of $\tilde{\rho}_i$ there exists an ϵ' such that the pull-back of any $\epsilon/2$-ball centered at $x \in \tilde{D}_i$ contains a ball of radius ϵ'.

Let ζ_i^j denote the push-forward of ν_i on via $\tilde{\rho}_j$. The measures $\zeta_1^j, \dots, \zeta_d^j$ on the cylinder $\tilde{D}_j = \pi^{-1}(D_j) \subset M$ are related to the measures $\mu_i, \dots, \mu_d$ by

$$\zeta_i^j(A) = \frac{\mu_i(\tilde{D}_j \cap A)}{\mu_i(\tilde{D}_j)}$$

From (10) it is immediate that

$$\left| \int_{\tilde{D}_j} \varphi \, d\zeta_i^j - \int_{\tilde{D}_j} \varphi \, d\mu_i \right| < \frac{\epsilon}{6K} \qquad \left| \int_{\tilde{D}_j} \varphi \, d\mu_i - \int_M \varphi \, d\mu_i \right| < \frac{\epsilon}{6K}$$

and thus we have for each $\tilde{D}_j$

$$\left| \int_{\tilde{D}_j} \varphi \, d\zeta_i - \int_M \varphi \, d\mu_i \right| < \frac{\epsilon}{3K} \tag{11}$$

Henceforth Lemma 3 will be treated as applying to the $\tilde{D}_j$ and giving diffeomorphisms on M. The diffeomorphisms given by Lemma 2 are identity on a neighborhood of the boundary and hence can be extended by the identity to diffeomorphisms on M. Since $\{D_j^\circ\}$ is an open cover of N there exists $\delta_1 > 0$ so that

$\{D'_j := d_j(B(1-\delta_1))\}$ still cover N. For each application of Lemma 3 we will take $\delta < \min(\delta_0, \delta_1)$. We adopt our earlier convention and denote the associated trivial bundles by $\tilde{D}'_j$.

6.2. Details. Denote the α given in the Proposition by α_0.

6.2.1. *Induction Hypotheses.* Assume there are diffeomorphisms $A_1, \ldots, A_j$ with their associated sets $\Sigma_1, \ldots, \Sigma_j$ from Lemma 3 such that

(1) for all $x \in \bigcup_{i=1}^j \tilde{D}'_i$ and for all $\varphi \in \Phi$

$$\inf_{u \in \mathfrak{M}'} \left| \frac{1}{q_j} \sum_{i=0}^{q_j - 1} \varphi(A_1 \circ \cdots \circ A_j \circ \tau^i_{\alpha_j} x) - \int_M \varphi \, du \right| < j \frac{\epsilon}{K} \tag{12}$$

where $q_j = m_j Q_j q_{j-1}$ with $m_j \in \mathbb{N}$ arbitrary, $\mathfrak{M}'$ is the simplex generated by $\{\mu_1, \ldots \mu_n\}$, and K is the number of disks required to cover N.

(2) for all $x \in \bigcup_{i=1}^j D'_i$ the orbit of x under $A_1 \circ \cdots \circ A_j \circ \tau^i_{\alpha_j}$ is ϵ-dense.

(3) for all $\varphi \in \Phi$ and all i

$$\left| \int \varphi \circ A_1 \circ \cdots \circ A_j \, d\mu_i - \int \varphi \, d\mu_i \right| < j \frac{\epsilon}{3K^2} \tag{13}$$

(4) $\mu(A_1 \circ \cdots \circ A_1 \tilde{N}_i \Delta \tilde{N}_i) < j \frac{\epsilon}{K}$ for $1 \le i \le d$.

6.2.2. *Induction Step.* We need to construct a diffeomorphism A_{j+1} such that the above properties hold with j replaced by $j+1$. Fix m_j, let $\bar{A} := A_1 \circ \ldots A_j$, and let $\tilde{\epsilon} := \min\{\epsilon', \frac{\epsilon}{3K^2}\}$.

Apply Lemma 3 to $\tilde{D}_{j+1}$ with $\alpha = \alpha_j$, $\epsilon = \tilde{\epsilon}$, and $\Phi_{j+1} = \{\varphi \circ \bar{A} : \varphi \in \Phi\}$. For all $x \in \tilde{D}_{j+1}$ and $\varphi \in \Phi$ this gives

(14)
$$\inf_{\zeta \in \mathfrak{M}_{j+1}} \left| \frac{1}{q_{j+1}} \sum_{\Pi^x_{j+1}} \varphi(\bar{A} \circ A_{j+1} \circ \tau^i_{\alpha_{j+1}} x) - \frac{\operatorname{Card}(\Pi^x_{j+1})}{q_{j+1}} \int_{\tilde{D}_{j+1}} \varphi \circ \bar{A} \, d\zeta \right| < \tilde{\epsilon} < \frac{\epsilon}{3K^2}$$

where $\mathfrak{M}_{j+1}$ is the simplex of measures generated by $\{\zeta_1^{j+1}, \ldots, \zeta_n^{j+1}\}$. By Property 5 of Lemma 2

$$\left| \int \varphi \circ \bar{A} \circ A_{j+1} \, d\mu_i - \int \varphi \circ \bar{A} \, d\mu_i \right| < \frac{\epsilon}{3K^2}.$$

This together with the induction hypothesis, (13), yields

$$\left| \int \varphi \circ \bar{A} \circ A_{j+1} \, d\mu_i - \int \varphi \, d\mu_i \right| < (j+1) \frac{\epsilon}{3K^2} < \frac{\epsilon}{3K} \tag{15}$$

as required. Combining (11), (14), and (15) yields

$$\inf_{u \in \mathfrak{M}'} \left| \frac{1}{q_{j+1}} \sum_{\Pi^x_{j+1}} \varphi(\bar{A} \circ A_{j+1} \circ \tau^i_{\alpha_{j+1}} x)) - \frac{\operatorname{Card}(\Pi^x_{j+1})}{q_{j+1}} \int_M \varphi \, du \right| < \frac{\epsilon}{K}. \tag{16}$$

For orbits contained in Σ_{j+1}, which includes the orbit of any $x \in D'_{j+1}$, $\operatorname{Card}(\Pi_{j+1}) = q_{j+1}$ the previous estimate reduces to

$$\inf_{u \in \mathfrak{M}'} \left| \frac{1}{q_{j+1}} \sum_{i=0}^{q_{j+1}-1} \varphi(\bar{A} \circ A_{j+1} \circ \tau^i_{\alpha_{j+1}} x)) - \int_M \varphi \, du \right| < (j+1) \frac{\epsilon}{K} \tag{17}$$

which is the next step for (12). It remains to prove (17) for points in $\bigcup_{i=1}^j \tilde{D}'_i \backslash \tilde{D}'_{j+1}$. Orbits in $\bigcup_{i=1}^j \tilde{D}'_i$ but outside of Σ_{j+1} are unaffected by A_{j+1} and hence (17) follows from (12) since m_j was arbitrary. Consider an orbit in $\bigcup_{i=1}^{j} \tilde{D}_i'$ which intersects but is not contained in Σ_{j+1}. By Property 1 of Lemma 3 it follows that the portion of the orbit outside Σ_{j+1} can be decomposed into N τ_{α_j}-orbits where

$$N := \frac{C}{q_j} \qquad\qquad C := \mathrm{Card}\big((\Pi_{j+1}^x)^c\big).$$

Applying (12) to one of the τ_{α_j}-orbits produces

$$\inf_{u\in\mathfrak{M}'}\Big|\frac{1}{q_j}\sum_{i=0}^{q_j}\varphi(\bar{A}\circ A_{j+1}\circ\tau_{\alpha_j}^i x_l) - \int_M \varphi\, du\Big| < j\frac{\epsilon}{K}.$$

Multiplying by q_j and dividing by q_{j+1} yields

$$\inf_{u\in\mathfrak{M}'}\Big|\frac{1}{q_{j+1}}\sum_{i=0}^{q_j}\varphi(\bar{A}\circ A_{j+1}\circ\tau_{\alpha_j}^i x_l) - \frac{C}{q_{j+1}}\int_M \varphi\, d\Big(\frac{q_j}{C}u\Big)\Big| < j\frac{\epsilon}{K}\frac{q_j}{q_{j+1}}.$$

Summing over the N orbits and using the convexity of $\mathfrak{M}'$ gives

$$\inf_{u\in\mathfrak{M}'}\Big|\frac{1}{q_{j+1}}\sum_{(\Sigma_{j+1}^x)^c}\varphi(\bar{A}\circ A_{j+1}\circ\tau_{\alpha_{j+1}}^i x) - \frac{C}{q_{j+1}}\int_M \varphi\, du\Big| < j\frac{\epsilon}{K}.$$

Adding (16) to this gives (17) as required.

Since A_{j+1} only affects $\tilde{D}_{j+1}$ Lemma 3 gives

$$\mu(A_{j+1}\tilde{N}_i\Delta\tilde{N}_i) < \lambda(A_{j+1}\tilde{C}_i\Delta\tilde{C}_i) < \frac{\epsilon}{K}.$$

By induction hypothesis 4 and the triangle inequality the result follows.

Lemma 3 proves that any orbit contained in Σ_{j+1} is $\epsilon/2$-dense on D_{j+1}; the orbit of any point in $\tilde{D}_{j+1}'$ meets any $\epsilon/2$-ball centered at a point in $\tilde{D}_{j+1}$. By (10) any $\epsilon/2$-ball in M must intersect $\tilde{D}_{j+1}$. Thus any ϵ-ball in M contains an $\epsilon/2$-ball centered at a point in $\tilde{D}_{j+1}$. By the argument used above any other orbit in $\bigcup_{i=1}^{j}\tilde{D}_i'$ contains a τ_{α_j}-orbit which is unaffected by A_{j+1} and ϵ-dense by induction hypothesis 2.

This completes the proof of the proposition. □

7. The Proof of the Theorem

The proof follows from induction using Proposition 4.

LEMMA 5. *Let $T = h\circ\tau_\alpha\circ h^{-1}$ where h is a smooth measure preserving diffeomorphism of M and $\alpha\in\mathbb{Q}$. Let Φ be a finite set of continuous functions on M. For all $\epsilon>0$ there exists a smooth measure preserving diffeomorphism of M, A, and $\alpha' = p'/q'$ with $(p',q')=1$ such that*

$$T' := h\circ A\circ\tau_{\alpha'}\circ A^{-1}\circ h^{-1}$$

satisfies

(1) $d_{C^\infty}(T,T') < \epsilon$
(2) $d_q(T,T') := \max_{x\in M}\max_{0\le i\le q} d(T^i x,(T')^i x) < \epsilon$
(3) $\mu(h\circ A\,\tilde{N}_i\Delta h\tilde{N}_i) < \epsilon$

(4) *for all* $\varphi \in \Phi$

$$\min_{\xi\in\Xi}\left|\frac{1}{q'}\sum_{i=0}^{q'-1}\varphi((T')^i x) - \int \varphi d\xi\right| < \epsilon$$

where Ξ *is the simplex generated by* $\{\xi_1, \dots, \xi_n\}$ *where* $\xi_i = h^*\mu_i$.

PROOF. Apply Proposition 4 to τ_α with the set of functions $\{\varphi \circ h : \varphi \in \Phi\}$ to get A. Define α' by

$$\alpha' := \alpha + \beta \qquad \beta := \frac{1}{mQq}$$

where$m$ is arbitrary, Q is given by Proposition 4, and q is the denominator of α. Observe

$$\begin{aligned} T' &= h \circ A \circ \tau_{\alpha'} \circ A^{-1} \circ h^{-1} \\ &= h \circ A \circ \tau_\alpha \circ \tau_\beta \circ A^{-1} \circ h^{-1} \\ &= h \circ \tau_\alpha \circ A \circ \tau_\beta \circ A^{-1} \circ h^{-1} \end{aligned}$$

using the fact that A commutes with τ_α. Now as $m \to \infty$ by definition $\beta \to 0$ and hence $A \circ \tau_\beta \circ A^{-1} \to \mathrm{Id}$ smoothly. Therefore we can find m large enough that

$$d_{C^\infty}(T, T') < \epsilon \quad \text{and} \quad d_q(T, T') < \epsilon.$$

Furthermore, directly from Proposition 4

$$\mu(A\tilde{N}_i \Delta \tilde{N}_i) = \mu(h \circ A\tilde{N}_i \Delta h\tilde{N}_i) < \epsilon.$$

Similarly Proposition 4 generates the following inequality

$$\min_{u\in\mathfrak{M}'}\left|\frac{1}{q'}\sum_{i=0}^{q'-1}\varphi(h \circ A \circ \tau_{\alpha'}^i x) - \int \varphi \circ h\, du\right| < \epsilon.$$

Replacing x by $A^{-1} \circ h^{-1}x$ we get the desired property. □

PROOF OF THE THEOREM. Let $\{\epsilon_n\}$ be a summable monotone decreasing sequence and $\{\varphi_n\}$ a dense set of continuous functions. Applying the previous lemma inductively we can produce a sequence of numbers $\{\alpha_n\}$ and transformations $\{h_n\}$ such that the sequence of induced transformations $\{T_n\}$ satisfies

(1) $d_{C^\infty}(T_n, T_{n+1}) < \epsilon_{n+1}$
(2) $d_{q_n}(T_n, T_{n+1}) < \epsilon_{n+1}$ where q_n is the period of T_n
(3) $\mu(h_{n+1}\tilde{N}_i \Delta h_n \tilde{N}_i) < \epsilon_{n+1}$
(4) for $\varphi \in \{\varphi_1, \dots, \varphi_n\}$

$$\min_{\xi^n\in\Xi_n}\left|\frac{1}{q_{n+1}}\sum_{i=0}^{q_{n+1}-1}\varphi(T_{n+1}^i x) - \int \varphi\, d\xi^n\right| < \epsilon_{n+1}$$

where Ξ_n is the simplex generated by $\{\xi_1^n, \dots \xi_d^n\}$

The first condition guarantees that the sequence $\{T_n\}$ has a limit T. T is a C^∞ diffeomorphism.

The second condition means that the sequence $\{h_n\tilde{N}_i\}_{n=1}^\infty$ is a Cauchy sequence in the metric on the associated measure algebra. Since this space is complete there is a limit in the measure algebra.

$$\lim_{n\to\infty} h_n\tilde{N}_i = B_i$$

As a consequence we have weak convergence of the measures $\{\xi_i^n\}$,

$$\lim_{n\to\infty} \xi_i^n = \xi_i$$

where

$$\xi_i(A) = \frac{\mu(A \cap B_i)}{\mu(B_i)}.$$

Notice that, as h_n is measure-preserving, $\mu(B_i) = \mu(\tilde{N}_i)$. Notice also that convergence in the measure algebra is not adapted to the original topology. Although the support of ξ_i^n is $h_n \tilde{N}_i$ the limiting measure ξ_i is supported on the whole space. This is another manifestation of the disjunction between the topological setting and the measure-theoretic. Although measure-theoretically B_i is the limit of $h_n \tilde{N}_i$ the first is a dense set while the sets in the sequence are all closed proper subsets. Since the sets $\tilde{N}_i$ and $\tilde{N}_j$ are measurably disjoint so are the limits B_i and B_j.

The measures ξ_i are invariant under the limiting diffeomorphism T. Using the triangle inequality and the invariance of ξ_i^j under T_n

$$\xi_i = \lim_{n\to\infty} \xi_i^n = \lim_{n\to\infty} T_n^* \xi_i^n = T^* \xi_i.$$

It remains to establish that the measures ξ_i exhaust all invariant ergodic probability measures for T. Suppose there exist an invariant measure ζ which is not in the simplex determined by $\{\xi_i\}$. Without loss of generality we may assume that ζ is extremal, that is ζ is an ergodic invariant measure. By the Birkhoff Ergodic Theorem for $\varphi \in L^1(\zeta)$ and for ζ-almost every x

$$\lim_{n\to\infty} \frac{1}{n} \sum_{i=0}^{n-1} \varphi(T^i x) = \int \varphi \, d\zeta.$$

Since $\{\varphi_i\}$ is dense an approximation argument shows that for any continuous function φ and any x in a set of ζ-full measure

$$\int \varphi \, d\zeta = \lim_{n\to\infty} \frac{1}{n} \sum_{i=0}^{n-1} \varphi(T^i x) = \lim_{n\to\infty} \frac{1}{q_n} \sum_{i=0}^{q_n - 1} \varphi(T_n^i x) = \lim_{n\to\infty} \int \varphi \, d\xi^n$$

where ξ^n is in the simplex given by $\{\xi_i^n\}$. Furthermore the sequence of measures $\{\xi^n\}$ depends only on the point x and not on the function φ. Thus

$$\lim_{n\to\infty} \int \varphi \, d\xi^n = \int \varphi \, d\zeta$$

but this means that ζ is in the simplex generated by $\{\xi_i\}$ since this simplex is weakly closed. Contradiction.

This proves that the only possible ergodic measures are $\{\xi_i\}$. Since the sets $\{B_i\}_{i=1}^d$ are measurably disjoint the measures ξ_i are linearly independent. However any non-ergodic measure can be written as a linear combination of ergodic measures so all the ξ_i are ergodic. □

8. Further Results

The method used here to establish the result where d is finite also suffices to produce a diffeomorphism with a countable number of ergodic components. The only differences in the argument occur in the construction of the domains $\{N_i\}$ and in the induction step. We define the sets $\{N_i\}$ given by the Legerdemain Lemma by embedding one disk $d_1 : B(1) \to N$ and defining

$$N_i = \begin{cases} d_1\big(B(1)\big)^c & i = 1 \\ d_1\big(A(\frac{1}{j+1}, \frac{1}{j})\big) & i > 1 \end{cases}$$

where $A(r_1, r_2) = \{x : r_1 \leq \|x\| \leq r_2\}$. On the n-th step of the induction we will take the decomposition of N by $\{N_i'\}_{i=1}^n$ given by

$$N_i' = \begin{cases} N_i & i < n \\ \bigcup_{j=n}^{\infty} N_j & i = n. \end{cases}$$

The permutation α is defined and applied as before but the permutation β is taken to be the identity permutation on parallelepipeds associated to N_n'.

The argument to produce a diffeomorphism which is minimal but which has power of continuum ergodic measures is the simplest of the arguments. In this case we can dispense with the permutation β altogether. Instead of β we apply the identity. The singular measures on the fibers converge to the required ergodic measures.

Though the theorem here produces ergodic measures which are absolutely continuous it is possible to modify the construction to give a mixture of absolutely continuous and singular measures. To do this the diffeomorphisms A_n must be taken non-measure-preserving. Constructing the required A_n for this case can be envisioned as a two step process whereby we first construct a measure-preserving diffeomorphism exactly as was done in this paper and then compose it with a diffeomorphism which takes the cylinder $\tilde{N}_i$ to a smaller cylinder contained within it. This smaller cylinder will serve as the $\tilde{N}_i$ for the next step. This diffeomorphism can be chosen to be the identity on $\tilde{N}_j$ for $1 \leq j \leq d-1$ with $i \neq j$. If the measure of the cylinder $\tilde{N}_i$ becomes 0 in the limit then the limiting measure ξ_i will be singular.

References

[1] A. N. Kolmogorov, *On dynamical systems with integral invariant on the torus.* Doklady Akad. Nauk SSSR, vol 93, 1953, 763 – 766

[2] H. Whitney *Geometric integration theory.* Princeton University Press, Princeton, N. J., 1957

[3] H. Furstenberg, *Strict ergodicity and transformation of the torus.* Amer. J. Math. 83 1961 573–601.

[4] D. V. Anosov and A. B. Katok *New examples in smooth ergodic theory. Ergodic diffeomorphisms* Trudy Moskov. Mat. Obšč. 23 (1970), 3–36

[5] A. Fathi and M. R. Herman *Existence de difféomorphismes minimaux.* Dynamical systems, Vol. I—Warsaw, pp. 37–59. Asterisque, No. 49, Soc. Math. France, Paris, 1977 (French)

[6] S. Williams *Toeplitz minimal flows which are not uniquely ergodic.* Z. Wahrsch. Verw. Gebiete 67 (1984), no. 1, 95–107.

[7] A. Katok and B. Hasselblatt *Introduction to the modern theory of dynamical systems.* With a supplementary chapter by A. Katok and L. Mendoza. Encyclopedia of Mathematics and its Applications, 54. Cambridge University Press, Cambridge, 1995

[8] P. Le Calvez and J.-C. Yoccoz *Un théorème d'indice pour les homéomorphismes du plan au voisinage d'un point fixe* Ann. of Math. (2) 146 (1997), no. 2, 241–293 (French)

Department of Mathematics, University Park, The Pennsylvania State University
E-mail address: windsor@math.psu.edu
URL: http://www.math.psu.edu/windsor

Proceedings of Symposia in Pure Mathematics
Volume **69**, 2001

Piecewise Smooth Maps with Absolutely Continuous Invariant Measures and Uniformly Scaled Markov Partitions

Michael Jakobson

ABSTRACT. We consider families of piecewise smooth one-dimensional maps f_t and give an estimate of the measure of parameter values t with absolutely continuous invariant measures. Estimate depends on the relation between expansion, distortion and several other characteristics of the family. Construction uses " uniformly scaled " Markov partitions and can be adjusted to two-dimensional transformations.

1. Introduction

We consider one-parameter families of piecewise C^2-maps of the interval $f_t : I \to I$. The first proof of the existence of absolutely continuous invariant measures for positive measure set of parameter values t appeared in [**3**]. Since then several other proofs appeared which modified the approach of [**3**] in several directions, see in particular [**2**], [**4**], [**7**], [**9**], [**10**], [**11**], [**12**], [**13**].

The method of [**3**] can be called *induced hyperbolicity*. For a positive measure set of parameters $\mathcal{M}$ we associate with f a piecewise smooth "power map" F defined on a countable collection I_i of disjoint subintervals of I such that $F \mid I_i = f^{n_i}$, $\bigcup_i I_i$ has the full Lebesgue measure in I and the power map F is expanding. Moreover F satisfies conditions of the so called Folklore theorem ([**1**]) which implies the existence of an absolutely continuous invariant measure.

Results mentioned above are in a sense asymptotic. They prove that parameter values corresponding to maps with preperiodic critical points or to Misiurewicz maps are density points of $\mathcal{M}$. In this paper we estimate measure of $\mathcal{M}$ for general families f_t without assuming that t belong to a small neighborhood of a specific parameter value. We give an estimate of the measure of "good parameter values " based on expansion, distortion and other properties of the given family. Our proof here is a modification of [**3**] motivated in part by two-dimensional construction developed in [**6**]. We proceed by induction. At step n we construct domains $\Delta_i^{(n)}$ of the power map with sizes bounded between two exponents

$$c_1\alpha^n < \mid \Delta_i^{(n)} \mid < c_2\beta^n \tag{1}$$

2000 *Mathematics Subject Classification.* Primary 37D45, 37E05, 37C40.
Work supported in part by NSF grant DMS - 9803635.

We call such partitions *uniformly scaled.* One example of such construction was considered in [**5**].

Uniform scaling allows to generalize construction to two-dimensional case. The choice of β is motivated by 2-dimensional distortion estimates. Assuming β is given, we find maximal possible α compatible with inductive construction. Then α determines how small must be the jacobian in 2-dimensional construction.

In Section 2 we review properties of the first return map for a general unimodal family.

In Section 3 we use a special preliminary construction in order to create in the phase space a set of large measure, where power map has strong expansion. Families obtained by that construction provide examples, to which our technique applies. Then we formulate our main theorem for families of piecewise smooth maps.

In Section 4 we describe first non-standard steps of induction, $1 \leq n \leq z_0 - 1$, where z_0 depends on the family and on the choice of scaling.

In Section 5 we state inductive assumptions and describe the algorithm.

In Sections 6 we prove inductive estimates (1) and measure estimates in the phase space.

In Section 7 we prove distortion estimates, and parameter estimates. Then we transfer estimates from the phase space into the parameter space. Finally we discuss decay of correlations for absolutely continuous invariant measures obtained by that construction.

2. The first return map for unimodal families

2.1. Initial assumptions 1–4. Let us consider a C^2- family of unimodal maps $x \to f_t(x)$ such that for a specific parameter value $\bar{t}$ the transformation $f_{\bar{t}}$ maps some interval $[p, p^{-1}]$ twice onto itself. Here $p(t)$ is a repelling fixed point, $p^{-1}(t)$ its preimage. Parameter t varies within an interval $\mathcal{T} = [\bar{t} - \epsilon, \bar{t}]$. For $t = \bar{t}$ the critical value $w(\bar{t})$ coincides with $p^{-1}(\bar{t})$ and the following transversality condition holds

$$\frac{\partial w(t)}{\partial t} - \frac{\partial p^{-1}(t)}{\partial t} \Big|_{t=\bar{t}} \neq 0$$

Then using a C^2 change of variables x and t we can renormalize our family so that it satisfies the following assumptions:

1) As a function of two variables $f_t(x) = f(t, x)$ is twice continuously differentiable in some domain $\mathcal{D} \supset \mathcal{T} \times [p, p^{-1}]$.

2) For all $t \in \mathcal{T}$ maps f_t have the following properties:

(a) $f_t(p) = f_t(p^{-1}) = p$
(b) $\frac{\partial f}{\partial x}(t, 0) = 0$
(c) $f_t(x) > x$ when $x \in (p, 0]$
(d) $\frac{\partial f}{\partial t}(t, 0) = 1$

3) $f_{\bar{t}}$ has the following properties.

(a) $f(\bar{t}, 0) = p^{-1}$.
(b) $\frac{\partial^2 f}{\partial x^2}(\bar{t}, 0) = -2\bar{k} < 0$.
(c) $\frac{\partial f}{\partial x}(\bar{t}, p) = A_0 > 1$.

We assume that parameter varies within a small interval, so that for all t under consideration the fixed point p is repelling, $f_t(0)$ is close to p^{-1} and 0 is a non-degenerate critical point. Let us consider a related first return map. For each

t the map f_t has a fixed point $q_t \in (0, p^{-1})$. We denote q_t^{-1} the preimage of q_t which belongs to $(p, 0)$. Let $I_t = [q_t^{-1}, q_t]$ be the corresponding interval. By another change of variables we can make I_t independent of t and of length one. We formulate it as

4) For all $t \in \mathcal{T}$ $q_t = q$, $q_t^{-1} = q^{-1}$, $\mid [q^{-1}, q] \mid = 1$. Let us denote $[q^{-1}, q]$ by I and the first return map on I by T_t.

2.2. Properties of the first return map. T_t has an even number $2(N-1)$ of monotone branches $T_{\pm n}$, $n = 1, 2, \dots N-1$ located symmetrically with respect to the critical point, and a central parabolic branch T_N, where $N = N(t)$. The return time for T_n equals $n+1$. When $t = \bar{t}$ there is no parabolic branch and there is an infinite number of monotone branches T_n. When $t \to \bar{t}$ the number of monotone branches tends to ∞. The n-th branch of T_t is obtained as the following composition. First f_t is applied near to the critical point, then f_t is applied near to p^{-1} and after that f_t is iterated consecutively $n-1$ times starting near p and ending up in I. We denote the domain of T_n by $\Delta T_n = [x_n, x_{n+1}]$. The boundary points of ΔT_n are preimages of q, and we denote by x_n the "outer" point, so that $x_{n+1} \in (x_n, 0)$.

The following Proposition 1 describes properties of T_t, when parameter is sufficiently close to the limit value $\bar{t}$. Before formulating Proposition 1 we introduce several useful quantities.

For an arbitrary C^2-diffeomorphism $F : \Delta \to J$ the following quantity is used to estimate distortions of lengths.

$$\Theta(F, \Delta) = \max_{y \in \Delta} \left| \frac{\frac{\partial^2 F}{\partial x^2}(y)}{\frac{\partial F}{\partial x}(y)} \right| \times \mid \Delta \mid .$$

We call $\Theta(F, \Delta)$ - *distortions*. For $y_1, y_2 \in \Delta$ ratios of derivatives are bounded by

$$\left| \frac{\frac{\partial F}{\partial x}(y_1)}{\frac{\partial F}{\partial x}(y_2)} \right| < \exp(\Theta(F, \Delta)) . \tag{2}$$

So lengths are distorted less than $\exp \Theta(F, \Delta)$.

The following quantity measures dependence on the parameter

$$\left| \frac{\frac{\partial T_n}{\partial t}}{\frac{\partial T_n}{\partial x}}(t, x) \right| .$$

PROPOSITION 1. *I. For a sufficiently large n_0 there exists a sequence of parameter values t_n, $n > n_0$, such that, when $t = t_n$ the central branch T_{n-1} splits into two monotone branches and a new central branch T_n is born. When t changes between t_n and t_{n+1}, the critical value of T_n moves monotonically between q and q^{-1}. There exist $c_0 > 0$ and small $\epsilon_0 > 0$ such that t_n satisfy*

$$c_0 A_0^{-n}(1 - \epsilon_0) < \bar{t} - t_n < c_0 A_0^{-n}(1 + \epsilon_0) . \tag{3}$$

II. There exist constants $c_2 > c_1 > 0$ such that for $n > n_0$ and for all t the branches T_n satisfy :

$$c_1 A_0^n \mid x \mid < \left| \frac{\partial T_n}{\partial x}(t, x) \right| < c_2 A_0^n \mid x \mid . \tag{4}$$

$$c_1 A_0^n < \left| \frac{\partial T_n}{\partial t}(t,x) \right| < c_2 A_0^n\,. \tag{5}$$

For $t > t_{n+2}$ distortions of T_n are uniformly bounded, namely there exists a constant $\bar{l}_0$ such that

$$\Theta(T_n) < \bar{l}_0\,. \tag{6}$$

For $t > t_{n+2}$

$$c_1 A_0^{n/2} < \left| \frac{\frac{\partial T_n}{\partial t}}{\frac{\partial T_n}{\partial x}}(t,x) \right| < c_2 A_0^{n/2}\,. \tag{7}$$

III. Let $x_n(t)$ be the outer boundary point of ΔT_{nt}. Then for $n > n_0$ and $t > t_n$ holds

$$(1-\epsilon_0)\sqrt{\frac{t-t_n}{\bar{k}}} <\mid x_n(t) \mid< (1+\epsilon_0)\sqrt{\frac{t-t_n}{\bar{k}}}\,. \tag{8}$$

Proof of Proposition 1.

(a) *Expansion and distortion of staircase maps.*
p is a repelling fixed point for all t under consideration. Therefore there exists a sequence of preimages of the second fixed point q which form a staircase $q_t^{-1}, q_t^{-2}, \ldots, q_t^{-n}, \ldots$ converging to p. According to one of the initial assumptions $\frac{\partial f}{\partial x}(\bar{t}, p) = A_0 > 1$. Let $A(t) = \frac{\partial f}{\partial x}(t,p)$. As $f(t,x) \in C^2$, we get

$$A(t) = A_0 + \frac{\partial^2 f}{\partial x \partial t}(\bar{t}, p)(t-\bar{t}) + o(t-\bar{t})\,. \tag{9}$$

Let us fix a sufficiently large n_0. By continuity for every $\epsilon > 0$ there exists $\delta > 0$ such that for all $t \in [\bar{t}-\delta, \bar{t}]$ and for all $n > n_0$ holds

$$\mid q_t^{-n} - p \mid< \epsilon\,. \tag{10}$$

Let us consider the *staircase maps*

$$f^n(t,x) : [p, q_t^{-n}] \to [p,q]\,. \tag{11}$$

In a small neighborhood of p, where we are taking the first $n-n_0$ iterates of f, expansion is larger, than some $A_1 > 1$ close to A_0. In order to estimate derivatives we notice that f^n is a composition of the form $f^n = f^{n_0} \circ f^{n-n_0}$, where f^{n-n_0} is an iterate of an expanding C^2-map and f^{n_0} is a fixed C^2-map. Applying inductively distortion formula (291) (see a similar estimate (62) below), we get that distortions of f^n are uniformly bounded by some constant $\bar{k}_0$ independent of n and t

$$\Theta(f^n) < \bar{k}_0\,. \tag{12}$$

REMARK 1. *We call constants, which are determined only by properties of f and do not depend on the construction, uniform constants.*

From (12) we get for $x, y \in [p, q_t^{-n}]$ the ratios of expansions are bounded

$$\exp(-\bar{k}_0) < \left| \frac{\frac{\partial f^n}{\partial x}(t,x)}{\frac{\partial f^n}{\partial x}(t,y)} \right| < \exp(\bar{k}_0)\,. \tag{13}$$

Using the mean value theorem we get that for all n and for $x \in [p, q_t^{-n}]$ holds

$$\frac{|p-q|}{q_t^{-n}-p}\exp(-\bar{k}_0) < \left|\frac{\partial f^n}{\partial x}(t,x)\right| < \frac{|p-q|}{q_t^{-n}-p}\exp(\bar{k}_0)\,. \tag{14}$$

Similarly for all $n \geq 0$ holds

$$|q_t^{-n}-p| < C_0 A_1^{-n} \tag{15}$$

where C_0 is a uniform constant.

(b) *Dependence of staircase maps on the parameter.*
Next we estimate how points q_t^{-n} depend on t. From

$$f^n(t, q_t^{-n}) = q \tag{16}$$

we get

$$\frac{\partial f^n}{\partial t} + \frac{\partial f^n}{\partial x}\frac{d}{dt}(q_t^{-n}) = \frac{d}{dt}(q) = 0\,. \tag{17}$$

Therefore

$$\left|\frac{d}{dt}(q_t^{-n})\right| < \left|\frac{\frac{\partial f^n}{\partial t}(q_t^{-n})}{\frac{\partial f^n}{\partial x}(q_t^{-n})}\right|. \tag{18}$$

LEMMA 1. *There exists a uniform constant C_1 such that for all t under consideration, for any $n \in N$ and for any $x \in [p, q_t^{-n}]$ holds*

$$\left|\frac{\partial f^n}{\partial t}(x)\right| < C_1 \times n \tag{19}$$

PROOF. We have

$$\begin{aligned}
\frac{\partial f^n}{\partial t}(t,x) &= \frac{\partial f}{\partial t}(t, f(t,\ldots f(t,x))\ldots) \\
&= \frac{\partial f}{\partial t}(t, f^{n-1}x) + \frac{\partial f}{\partial x}(t, f^{n-1}x)\Big[\frac{\partial f}{\partial t}(t, f^{n-2}x) \\
&+ \frac{\partial f}{\partial x}(t, f^{n-2}x)\Big[\ldots\Big[\frac{\partial f}{\partial t}(t, fx) \\
&+ \frac{\partial f}{\partial x}(t, fx)\frac{\partial f}{\partial t}(t,x)\Big]\Big]\ldots\Big]\,.
\end{aligned} \tag{20}$$

As f is a C^2-map and $\frac{\partial f}{\partial t}(p) \equiv 0$ we have

$$\left|\frac{\partial f}{\partial t}(f^i x)\right| < K_2(q_t^{-(n-i)} - p) \tag{21}$$

where $K_2 = max_{x,t} \left| \frac{\partial^2 f}{\partial x \partial t} \right|$ is a uniform constant. We substitute estimates (14) and (21) into (20) and get

$$\begin{aligned}
\left|\frac{\partial f^n}{\partial t}(t,x)\right| &< |p-q|\exp(\bar{k}_0)K_2[(1 + \frac{1}{q_t^{-1}-p} \times (q_t^{-2}-p) \\
&+ \frac{1}{q_t^{-2}-p} \times (q_t^{-3}-p) + \ldots + \frac{1}{q_t^{-(n-1)}-p} \\
&\times (q_t^{-n}-p)] < C_1 \times n\,.
\end{aligned} \tag{22}$$

That proves (19) with

(23) $$C_1 =| p - q | \, max_{x,t} \, | \frac{\partial^2 f}{\partial x \partial t} | \exp \bar{k}_0 \, .$$

□

We substitute (15), (14) and (19) into (18) and get an estimate of the speed of boundary points

(24) $$| \frac{d}{dt}(q_t^{-n}) |< C_2 \frac{n}{A_1^n}$$

where C_2 is a uniform constant.

(c) *Right staircase and definition of* t_n.

To the sequence of left preimages q_t^{-n} converging to p it corresponds a sequence of right preimages $q_{rt}^{-(n+1)} = f_t^{-1}(q_t^{-n})$ converging to p^{-1}. As above we get that $q_{rt}^{-(n+1)}$ satisfy conditions (15), (22) and (24), and after changing C_0, C_1, C_2 if necessary, we can assume that these conditions are satisfied with the same constants.

If n_0 used above is sufficiently large, then (24) implies for all $n \geq n_0$

$$| \frac{d}{dt}(q_t^{-n}) | \ll 1, \; | \frac{d}{dt}(q_{rt}^{-(n+1)}) | \ll 1 \, .$$

As the critical value is moving with the unit speed we get that for every $n \geq n_0$ there exists a unique parameter value t_n such that

$$f_{t_n}(0) = q_{rt}^{-n} \, .$$

By definition the n-th critical branch is born at $t = t_n$. From (15) and Lemma 2 below we get

(25) $$| t_n - \bar{t} |< \frac{C_0(1+\epsilon)}{A_1^n}$$

(d) *Precise rate of accumulation of* t_n *toward* $\bar{t}$. Let us fix a sufficiently large n_1, which is at the same time sufficiently small comparatively to n_0. We define c_0 in (3) from the equation

(26) $$| p^{-1} - q_{r\bar{t}}^{-n_1} |= c_0 A_0^{-n_1} \, .$$

Then

(27) $$c_0 =| p - q | \frac{A_0^{n_1}}{| \frac{\partial f^{n_1}}{\partial x}(x_0) |}$$

where $x_0 \in [q_{r\bar{t}}^{-n_1}, p^{-1}]$ is defined from the mean value theorem. By continuity we get from (25) that there exists $\epsilon_1 = \epsilon_1(n_0)$, such that $\lim_{n_0 \to \infty} \epsilon_1 = 0$ and for any $t > t_{n_0}$ holds

(28) $$c_0(1-\epsilon_1)A_0^{-n_1} <| p^{-1} - q_{rt}^{-n_1} |< c_0(1+\epsilon_1)A_0^{-n_1} \, .$$

Let us consider f^{n-n_1} restricted to the interval $[q_{rt}^{-n}, p^{-1}]$. Notice that $f[q_{rt}^{-n}, p^{-1}] = [p, q_t^{-(n-1)}]$. As $f(t,x)$ is a C^2 map, we get from (15) that for any $x \in [p, q_t^{-(n-1)}]$ holds

(29) $$| f_x(t,x) - A(t) |< C_3 A_1^{-(n-1)} \, .$$

Let us consider $f_x^{n-1-n_1}|[p, q_t^{-(n-1)}]$. As we assumed n_1 to be large derivatives remain exponentially close to $A(t)$ for all iterates f^i with $i \leq n - n_1$. Then we get that for all t and all $n > n_1$ holds

(30) $$(1-\epsilon_2)A(t)^{-(n-n_1)} < \frac{| p^{-1} - q_{rt}^{-n} |}{| p^{-1} - q_{rt}^{-n_1} |} < (1+\epsilon_2)A(t)^{-(n-n_1)}.$$

Using (9) and (25) we get that for all $n \geq n_0$ and all $t \in [t_n, \bar{t}]$ there exist constants $c_{11}, c_{12} > 0$ such that

(31) $$\left(A_0 - \frac{c_{11}}{A_1^n}\right)^{(n-n_1)} < A(t)^{(n-n_1)} < \left(A_0 + \frac{c_{12}}{A_1^n}\right)^{(n-n_1)}.$$

If $n - n_1$ is large, (31) can be rewritten as

(32) $$(1-\epsilon_3)A_0^{(n-n_1)} < A(t)^{(n-n_1)} < (1+\epsilon_3)A_0^{(n-n_1)}$$

where $\epsilon_3 > 0$ is small. We combine (28), (30) and (32) and get that there exists $\epsilon_0 = \epsilon_0(\epsilon_1, \epsilon_2, \epsilon_3)$ such that for all $n \geq n_0$ and for all $t \in [t_n, \bar{t}]$ holds

(33) $$(1-\epsilon_0)c_0 A_0^{-n} <| p^{-1} - q_{rt}^{-n} |< (1+\epsilon_0)c_0 A_0^{-n}.$$

As the critical value moves with the unit speed, (33) is equivalent to

(34) $$| \bar{t} - t_n | \in c_0 A_0^{-n}(1 \pm \epsilon_0)$$

which proves (3) of Proposition 1.

Simultaneously previous arguments improve estimate (15). Namely for $t \in [t_n, \bar{t}]$ holds

(35) $$(1-\epsilon_0)c_0' A_0^{-n} <| q_t^{-n} - p |< (1+\epsilon_0)c_0' A_0^{-n}$$

where $c_0' = c_0 \frac{\partial f}{\partial x}(p^{-1}, \bar{t})$.

Similarly, (24) can be improved and we get

(36) $$| \frac{d}{dt}(q_t^{-n}) |< d_1 \frac{n}{A_0^n}$$

where

(37) $$d_1 = \exp \bar{k}_0 (1+\epsilon_0) K_2 \frac{| \frac{\partial f^{n_0}}{\partial x}(\theta_0) |}{A_0^{n_0}}$$

and θ_0 is defined from the mean value theorem.

(e) *Pull-back by parabolic branch.*
We shall use $h(t,x)$ to denote the restriction of the original unimodal map $f(t,x)$ to a small neighborhood of zero. Let us denote by $g_n(t,x)$ the n-th iterate of f, which maps $[q_{rt}^{-n}, q_{rt}^{-(n+1)}]$ onto $[q^{-1}, q]$. The branches T_n of the induced map are compositions of the form $T_n = g_n \circ h$.

First we estimate derivatives of g_n. From (33) and (13) we get that for all $t \in [t_n, \bar{t}]$ the maps g_n satisfy

(38) $$(1-\epsilon_0)(\exp -\bar{k}_0)c_1' A_0^n <| \frac{\partial g_n}{\partial x} |< (1+\epsilon_0)(\exp \bar{k}_0)c_1' A_0^n$$

where $c_1' = \frac{|p-q|}{c_0}$.

We estimate derivatives of $T_n(t,x)$ using

(39) $$\frac{\partial T_n}{\partial x} = \frac{\partial g_n \circ h}{\partial x} = \frac{\partial g_n}{\partial x}(t, h(t,x)) \times \frac{\partial h}{\partial x}(t,x)$$

and

$$\frac{\partial T_n}{\partial t} = \frac{\partial g_n \circ h}{\partial t} = \frac{\partial g_n}{\partial t}(t, h(t,x)) + \frac{\partial g_n}{\partial x}(t, h(t,x)) \times \frac{\partial h}{\partial t}(t,x)\,. \tag{40}$$

As $h(t,x)$ is a C^2-map and $\frac{\partial^2 h}{\partial x^2}(\bar{t},0) = -2\bar{k}$, we get

$$2\bar{k}(1-\epsilon)\times \mid x \mid < \left|\frac{\partial h}{\partial x}(t,x)\right| < 2\bar{k}(1+\epsilon)\times \mid x \mid \tag{41}$$

where $-2\bar{k} = \frac{\partial^2 h}{\partial x^2}(\bar{t},0)$.

Estimates (39), (38) and (41) prove estimate (4) of Proposition 1 with

$$c_1 = 2\bar{k}(1-\epsilon)\frac{\mid p-q \mid}{c_0}(\exp -k_0);\ c_2 = 2\bar{k}(1+\epsilon)\frac{\mid p-q \mid}{c_0}(\exp k_0)\,. \tag{42}$$

Because of the condition $\frac{\partial h}{\partial t}(0) \equiv 1$ we have for small x

$$1-\epsilon < \frac{\partial h}{\partial t} < 1+\epsilon$$

Similarly to Lemma 1 we get the following estimate

$$\mid \frac{\partial g_n}{\partial t}(t,x) \mid < C_1 \times n\,. \tag{43}$$

We substitute (38), (43) and the above estimate of $\frac{\partial h}{\partial t}$ into (40) and get (5) of Proposition 1.

According to the estimate (33) sizes of Δg_n equal

$$\mid \Delta g_n \mid = (1 \pm \epsilon)c_0(A_0^{-n} - A_0^{-(n+1)})\,. \tag{44}$$

When $\bar{t} > t > t_{n+2}$ the distance H between the interval $[q_{rt}^{-(n+1)}, q_{rt}^{-n}]$ and the critical value satisfies

$$H > (1-\epsilon)c_0(A_0^{-(n+1)} - A_0^{-(n+2)})\,. \tag{45}$$

We use formula (293) for distortions of compositions $T_n = g_n \circ h$ and get

$$\Theta(T_n) < \bar{l}_0 \tag{46}$$

where $\bar{l}_0$ can be expressed through $\bar{k}_0$ and A_0. That proves (6) of Proposition 1.

Similarly to (45) we estimate H from above

$$H < (1+\epsilon)c_0 A_0^{-(n+1)}(1 + A_0^{-1} + A_0^{-2} + \ldots)\,. \tag{47}$$

Using that in local coordinates $H = (1 \pm \epsilon)\bar{k}x^2$ we estimate $x \in \Delta T_n$ and get from (4)

$$c_4' A_0^{n/2} > \left|\frac{\partial T_n}{\partial x}\right| > c_4'' A_0^{n/2}\,. \tag{48}$$

The estimates (5) and (48) imply

$$\left|\frac{\frac{\partial T_n}{\partial t}}{\frac{\partial T_n}{\partial x}}\right| < c_5' A_0^{n/2}\,. \tag{49}$$

The above constants c_4', c_4'', c_5' can be expressed through $c_0, A_0, \bar{k}, \bar{k}_0$ and $\mid p-q \mid$. That proves (7) of Proposition 1. The Taylor expansion of $h(t,x)$

at 0 and $\frac{\partial h}{\partial t}(0) \equiv 1$ imply that the boundary points $x_n(t)$ of the central branch $h_n(t,x)$ satisfy

$$(1-\epsilon)\sqrt{\bar{k}^{-1}(t-t_n)} < \mid x_n(t) \mid < (1+\epsilon)\sqrt{\bar{k}^{-1}(t-t_n)} \tag{50}$$

which is (8) of Proposition 1. That concludes the proof of Proposition 1. □

REMARK 2. *The proof shows that c_1, c_2 are uniform constants.*

REMARK 3. *If $\bar{t}-t$ is small comparatively to $\bar{t}-t_n$, then the distance between the critical value $w(t)$ and q_{rt}^{-n} can be estimated as $(1\pm\epsilon) \mid p^{-1} - q_{rt}^{-n} \mid$. Using (34) we get for such t*

$$\mid x_n(t) \mid = \sqrt{\frac{c_0}{\bar{k}}} A_0^{-n/2}(1\pm\epsilon)\,. \tag{51}$$

2.3. Initial assumption 5 - uniform expansion. When t is greater than t_{n+2}, expansion of T_n is of the order $A_0^{n/2}$. We can express it in different terms. When t varies between t_n and t_{n+1} there are only two branches, which may have small derivatives. One of them is the parabolic branch. Also in the case when the height of the parabolic branch is small, two monotone branches adjacent to the central one are almost parabolic near to the respective end points. On these branches the derivatives change from very large to arbitrary small, when we approach the critical point. For all other branches derivatives grow exponentially with n.

When $t=\bar{t}$ all branches satisfy estimate (48). Notice that (48) does not imply expansion for branches with small numbers. For example T_i with small i might have an attracting fixed point. For quadratic polynomials with parameter values close to Chebyshev they are expanding. In that case $\mid \frac{\partial T_1}{\partial x} \mid > 3$ and $\mid \frac{\partial T_i}{\partial x} \mid$ grow with i as 2^i. Generally it is enough to assume that that some power of the first return map is expanding, but in order to simplify exposition, we assume that the first return map itself is expanding.

Initial assumption 5. *For $t=\bar{t}$ there exists $\rho_0 > 1$ such that any branch of $T_{\bar{t}}$ satisfies*

$$\left| \frac{\partial T_i}{\partial x}(\bar{t},x) \right| > \rho_0 \tag{52}$$

Initial assumption 5 may follow from some special properties of maps, such as negative Schwarzian derivative. It can be also verified numerically, because the derivatives $\frac{\partial T_n}{\partial x}$ grow exponentially, and we only need to check a finite number of branches T_n.

COROLLARY 1. *From the assumption 5 we get that all periodic points of $f_{\bar{t}}$ are repelling. In particular the fixed point q is repelling.*

REMARK 4. *If f_t is a family of unimodal maps satisfying initial assumptions 1 - 5 and g_t is a family sufficiently close to f_t in C^2-topology, then there exists a change of variable such that g_t satisfies 1 - 5.*

Proposition 1 describes properties of a piecewise smooth family obtained by the first return map. In that family domains of good maps ΔT_n are scaled as $A_0^{-n/2}$ for $n \le N_0 - 2$. Domains, which are constructed later inside ΔT_{N_0}, are scaled in an unrelated way. In the next section we change our family to a new one, which has large expansion on a set of large measure, and is adjusted to the beginning of induction.

3. Constructing a special piecewise smooth family

3.1. Getting a large expansion. We assume the first return map has $2N_0-1$ branches, and the critical value $w(t) = h(t, 0)$ moves between $q_{rt}^{-N_0}$ and $q_{rt}^{-(N_0+1)}$. The parameter interval under consideration is $\mathcal{T} = \mathcal{T}_{N_0} = [t_{N_0}, t_{N_0+1}]$. We fix a sufficiently large number

$$R > 1 \tag{53}$$

specified below and use powers of R to define various scales in our construction.

Let us consider branches of the first return map at the moment $\bar{t}$. Let T_i be one of them. If

$$\min \left| \frac{\partial T_i}{\partial x} \right| > R \tag{54}$$

then we do not change T_i.

We fix a small interval δ in the middle of I which is a union of ΔT_n, $p_0 \leq n$, where

$$p_0 < N_0 - 1\,. \tag{55}$$

The size of δ will be specified below. From (3) and (8) of Proposition 1 we get for all $t \in \mathcal{T}$

$$(1-\epsilon_0)\sqrt{\frac{c_0}{\bar{k}}}A_0^{-p_0/2} <| \delta |< (1+\epsilon_0)\sqrt{\frac{c_0}{\bar{k}}}A_0^{-p_0/2}\,. \tag{56}$$

Although δ may contain many good branches, at this stage it is considered as one element. We call such elements "undecided domains". A more detailed structure inside these domains appears at later steps of induction.

In our preliminary construction we use the following algorithm. Fix $t = \bar{t}_0$. Assume T_{i_0} with $i_0 < p_0$ does not satisfy (54). Then consider all compositions of the form $T_i \circ T_{i_0}$ where $i < p_0$. That leaves in the middle of the domain of T_{i_0} a preimage δ^{-1}. If for some $T_{i_1} \circ T_{i_0}$

$$\left| \frac{\partial (T_{i_1} \circ T_{i_0})}{\partial x} \right| > R$$

then we do not change that composition. For other branches we proceed similarly and consider triple compositions. Again that leaves in the middle of the domain $\Delta T_{i_1} \circ T_{i_0}$ a preimage δ^{-2}, because we do not consider compositions of the form $T_i \circ T_{i_1} \circ T_{i_0}$ with $i \geq p_0$. Initial assumption (5) implies that after no more than $M_1 = log_{\rho_0} R$ steps of the above construction we get

$$\left| \frac{\partial}{\partial x}(T_{i_s} \circ \ldots \circ T_{i_1} \circ T_{i_0}(\bar{t}, x)) \right| > R \tag{57}$$

for all new branches. As the number of branches involved in the above compositions $T_{i_s} \circ \ldots \circ T_{i_1} \circ T_{i_0}$ is finite, we get for any t close enough to $\bar{t}$

$$\left| \frac{\partial}{\partial x}(T_{i_s} \circ \ldots \circ T_{i_1} \circ T_{i_0}(t, x)) \right| > R\,. \tag{58}$$

According to (3) intervals $\mathcal{T}_N = [t_N, t_{N+1}]$ decrease exponentially as well as distances from $\mathcal{T}_N$ to the limit value $\bar{t}$. Therefore we can take N_0 sufficiently large so that (58) holds for all $t \in \mathcal{T}_{N_0}$. We denote new "good" branches by $f_0 = T_{i_s} \circ \ldots \circ T_{i_1} \circ T_{i_0}$ and their domains by Δf_0. By construction f_0 maps Δf_0 onto the whole interval I. The union of the intervals Δf_0 constitutes the original good set which we denote as $\mathcal{I}_0$. The complementary set $\mathcal{C}_0$ consists of δ and its preimages δ^{-l}. We call δ and δ^{-l} *undecided domains* in order to emphasize that they are partitioned later in the inductive process. We get the following partition of I.

$$(\xi) \quad I = \mathcal{I}_0 \cup \mathcal{C}_0 = \left(\bigcup \Delta f_0\right) \cup \delta \cup \left(\bigcup \delta^{-l}\right). \tag{59}$$

By construction elements of ξ depend continuously on t. We denote by G_0 maps from preimages δ^{-j} onto δ, and by ΔG_0 domains of these maps. If $\delta^{-1} \subset \Delta T_i$ then $G_0|\delta^{-1}$ coincides with T_i. If $\delta^{-2} \subset \Delta T_i \circ T_j$, then $G_0|\delta^{-2}$ coincides with $T_i \circ T_j$ and so on. As G_0 are defined where we stop iterating T_i, (58) may be not true for G_0. But in any case from initial assumption 5, we know that expansion of any T_k is greater than ρ_0. Therefore

$$\left| \frac{\partial G_0}{\partial x} \right| > \rho_0 \tag{60}$$

3.2. Distortion estimates. According to Proposition 1 distortions of such branches are uniformly bounded by some constant $\bar{l}_0$

$$\Theta(T_i, \Delta_i) < \bar{l}_0, \quad 1 \le i < N_0 - 1\,. \tag{61}$$

Notice that in the proof of (61) we used that distortions of staircase maps satisfy

$$\Theta(g_i) < \bar{k}_0 < \bar{l}_0\,.$$

Next we consider good branches $f_0 = T_{i_s} \circ \ldots \circ T_{i_1} \circ T_{i_0}$, which are compositions of initial branches T_i. Applying inductively estimate (291) for compositions of maps with expansion greater than ρ_0 we get

$$\Theta(f_0, \Delta f_0) \le \bar{l}_0 \left(1 + (\exp \bar{l}_0) \sum_1^\infty \rho_0^{-i}\right) \exp\left[\bar{l}_0 (\exp l_0) \sum_1^\infty \rho_0^{-i}\right]. \tag{62}$$

Moreover, the same estimate holds for compositions of f_0 with arbitrary staircase maps g_i. We denote the rhs of (62) by l_0. Then

$$\Theta(f_0),\ \Theta(f_0 \circ g_i) < l_0\,. \tag{63}$$

G_0 are compositions of the same type as f_0, but they are restricted to the preimages of δ_0. From the definition of distortion we get

$$\Theta(G_0, \delta^{-j}) \le l_0 \exp l_0 \mid \delta \mid\,. \tag{64}$$

3.3. Estimates of sizes. Distortion estimates imply some bounds on derivatives of G_0 and respective bounds on sizes of undecided domains that we shall use later. By construction we are taking compositions if expansion at some point is less than R. As distortions are bounded by l_0, we get that for all $t \in \mathcal{T}$ and for any x in the domain of a branch, which we compose, expansions cannot exceed $R \exp l_0$. Thus we have

$$R \exp l_0 > \left| \frac{\partial G_0}{\partial x} \right| > \rho_0 > 1\,. \tag{65}$$

We denote

$$K_0 = \exp(-l_0) R^{-1}\,. \tag{66}$$

Then undecided domains satisfy

$$K_0 \mid \delta \mid < \mid \delta^{-l} \mid < \mid \delta \mid \rho_0^{-1}. \tag{67}$$

Let us estimate sizes of Δf_0. As $\mid f_{0x} \mid > R$, we get

$$\mid \Delta f_0 \mid < R^{-1}.$$

In order to estimate sizes from below we estimate maximal expansion. It arises when we compose a branch with expansion already close to R with a branch T_i with maximal possible index $i = p_0 - 1$. From the definition of δ we get

$$\mid \frac{\partial f_0}{\partial x} \mid < a_1(\exp l_0) R \mid \delta \mid^{-1} \tag{68}$$

where $a_1 \geq 1$ is a uniform constant which can be expressed through A_0. Respectively

$$R^{-1} > \mid \Delta f_0 \mid > a_1^{-1}(\exp - l_0) R^{-1} \mid \delta \mid . \tag{69}$$

3.4. Measure estimates in the phase space. At each step of the preliminary construction we deleted in the middle of certain domains preimages of δ. As the upper bound of distortions is l_0, we get that a domain δ^{-k}deleted from Δ occupies at most $(\exp l_0) \mid \delta \mid$ fraction of $\mid \Delta \mid$. As the number of steps is at most $M_1 = 1 + \log_{\rho_0} R$ we get that the measure of the union of Δf_0 in (59) is at least

$$(1 - (\exp l_0) \mid \delta \mid)^{M_1}. \tag{70}$$

The constants *exp* l_0 and M_1 do not depend on δ. If δ is a small interval, then (70) is close to one, the measure of undecided domains is small and is estimated as

$$mes\ \mathcal{C}_0 < 1 - (1 - (\exp l_0) \mid \delta \mid)^{M_1} < c_1 \mid \delta \mid \exp l_0 \log_{\rho_0} R \tag{71}$$

where c_1 is a uniform constant.

3.5. Speed estimates. We estimate how elements of the initial partition move depending on the parameter. Initial branches T_i are of the form $T_i = g_i \circ h$, where g_i are " staircase " maps defined in the proof of Proposition 1.

For all t boundary points $x_i(t)$ of $\Delta(T_i)$ satisfy

$$g_i(t, h(t, x(t))) = q. \tag{72}$$

We use (38) and (40) - (43) and get

$$\frac{1-\epsilon}{2}\sqrt{\bar{k}^{-1}(t-t_i)^{-1}} < \left|\frac{dx_i(t)}{dt}\right| < \frac{1+\epsilon}{2}\sqrt{\bar{k}^{-1}(t-t_i)^{-1}}. \tag{73}$$

Estimate (73) is valid for all $t > t_i$. For domains ΔT_i which are elements of the initial partition ξ indices i satisfy

$$i < p_0.$$

We assume $\mid t - \bar{t} \mid$ is small comparatively to $\mid t - t_{p_0} \mid$, see condition (105) below. Then for $t \in \mathcal{T}_{N_0}$ the quotient

$$\frac{t - t_i}{\bar{t} - t_i}$$

is close to one. Using (3) we get that the boundary points of $\Delta(T_i)$ are moving with the speed

$$\left|\frac{dx_i(t)}{dt}\right| = \frac{1 \pm \epsilon}{2}\sqrt{\bar{k}^{-1} c_0^{-1}} A_0^{i/2}. \tag{74}$$

Maximal speed corresponds to $i = p_0$. Boundary points $x_{p_0 \pm}(t)$ of ΔT_{p_0} are simultaneously boundary points of $\delta = \delta_{p_0}$. Maximal size of $\delta(t)$ corresponds to the maximal parameter value $t_0 = t_{N_0+1}$ - the end point of $t \in \mathcal{T}_{N_0}$. Let us denote

(75) $$\delta_0 = \mid \delta(t_0) \mid .$$

From (74) and (51) we get that for any $t \in \mathcal{T}_{N_0}$ the end points of $\delta(t)$ are moving with the speed

(76) $$\left| \frac{dx_{p_0}(t)}{dt} \right| = \frac{1 \pm \epsilon}{\bar{k}} \delta_0^{-1} .$$

The length of $\mathcal{T}_{N_0}$ depends on the length of Δg_0. We assume in (105) below

$$\mid \Delta g_0 \mid < \frac{1}{2} R^{-6s} \bar{k} \delta_0^2 .$$

Then (76) implies that for $t \in \mathcal{T}_{N_0}$ variation of $\delta(t)$ does not exceed

$$\frac{1+\epsilon}{2} R^{-6s} \delta_0 .$$

We get for all $t \in \mathcal{T}_{N_0}$

(77) $$(1-\epsilon)\delta_0 < \mid \delta(t) \mid < \delta_0 .$$

Other elements of ξ_0 are domains of compositions $T_{i_s} \circ T_{i_{s-1}} \circ \ldots \circ T_{i_0}$, $i_k < p_0$, and domains $\delta^{-k} = T_{i_0}^{-1} \circ \ldots \circ T_{i_{s-1}}^{-1} \delta$. By construction, we are taking next composition, if expansion at some point is less than R. In particular, any map T_i which enters into composition satisfies (74), where $A_0^{i/2} < constR$.

We use chain rule formula (333) below, (74), (76) and get the following estimate. Let $G = T_{i_{s-1}} \circ \ldots \circ T_{i_0}$ be one of the above compositions, and let $\delta^{-k} = G^{-1}\delta$. Let G_{xmin} be the minimal expansion of G. Then speeds of the end points of δ^{-k} are bounded by

(78) $$b_1 R + \frac{1+\epsilon}{\bar{k}} \mid \delta \mid^{-1} \frac{1}{\mid G_{xmin} \mid}$$

where b_1 is a uniform constant.

As expansions of all branches are greater than $\rho_0 > 1$ speeds are bounded by

(79) $$b_1 R + \rho_0^{-1} \frac{1+\epsilon}{\bar{k}} \delta_0^{-1}$$

According to relations between parameters, see below, $\delta_0 = R^{-3}$, $b_1 < R$. Thus, the last expression is less than $\frac{1+\epsilon}{\bar{k}} \delta_0^{-1}$. So in all cases speeds are bounded by

(80) $$\frac{1+\epsilon}{\bar{k}} \delta_0^{-1}$$

3.6. Enlargements. As $\delta_0 \ll R^{-1}$ interval δ is in the middle of a much larger interval - a union of branches T_i which have derivatives larger than R. Respectively, these branches are not changed along previous construction. We define an *enlargement* $\hat{\delta} \supset \delta$

$$\hat{\delta} = \bigcup_{i \geq i_1} \Delta T_i$$

as a minimal union such that

$$\mid \hat{\delta} \mid > 2R^s \delta_0$$

As ΔT_i are scaled as $A_0^{-i/2}$

$$\mid \hat{\delta} \mid < 2(1+\epsilon) A_0^{1/2} R^s \delta_0 \, .$$

We assume

$$R^{2s} > 17 A_0 \, . \tag{81}$$

Then we get

$$2R^s \delta_0 < \mid \hat{\delta} \mid < \frac{1}{2} R^{2s} \delta_0 \, . \tag{82}$$

By construction preimages $\hat{\delta}^{-s}$ are defined, mapped onto $\hat{\delta}$ by the same branches G_0 which map δ^{-s} onto δ. Distortions of $G_0 | \hat{\delta}^{-s}$ satisfy

$$\Theta(G_0, \hat{\delta}^{-s}) < l_0 \exp l_0 R^{2s} \delta_0$$

and are small if δ_0 is smali enough. Then ratios of sizes are almost preserved and we get

$$2(1-\epsilon) R^s \mid \delta^{-s} \mid < \mid \hat{\delta}^{-s} \mid < \frac{1+\epsilon}{2} R^{2s} \mid \delta^{-s} \mid \, . \tag{83}$$

Enlargements $\hat{\delta}^{-s}$ and Δf_0 which do not belong to any enlargement form a partition $\hat{\xi}_0$ similar to (59). Below, when we choose admissible parameter values we put the iterates of the critical value outside of enlargements. Let $\hat{\mathcal{C}}_0$ be the measure of the union of enlargements. From (71) we get

$$mes \hat{\mathcal{C}}_0 < c_1 R^{2s} \delta_0 \exp l_0 \log R \, . \tag{84}$$

The following proposition describes analytic properties of maps obtained after preliminary construction.
We remind that $\bar{k}, \rho_0$ are parameters of the initial map, R, δ_0 are variable parameters which determine our preliminary construction, distortions are bounded by l_0, which does not depend of R, $\mid \delta \mid$.

PROPOSITION 2. *Let ξ be the partition (59). Let $\Delta f_0, \delta, \delta^{-s}$ be elements of ξ and let f_0, G_0 be the respective maps. Let $\mathcal{C}_0$ be the union of undecided domains in ξ. Let $\hat{\delta}$, $\hat{\delta}^{-s}$ be enlargements of δ, δ^{-s} and $\hat{\mathcal{C}}_0$ their union. Let $\mathcal{T} = \mathcal{T}_{N_0}$ be the parameter interval under consideration. There exist uniform constants $a_1 \geq 1, b_1 \geq 1, c_1 \geq 1$ and a small constant ϵ_0 such that for all $t \in \mathcal{T}$ the mappings constructed at the preliminary construction satisfy the following estimates:*

$$a_1 (\exp l_0) R \delta_0^{-1} > \mid \frac{\partial f_0}{\partial x} \mid > R \tag{85}$$

$$(\exp l_0) R > \mid \frac{\partial G_0}{\partial x} \mid > \rho_0 \tag{86}$$

$$\Theta(f_0, \Delta f_0) < l_0 \tag{87}$$

$$\delta_0 (1-\epsilon_0) < \mid \delta(t) \mid \leq \delta_0 \tag{88}$$

$$\delta_0 R^s < \mid \hat{\delta}(t) \mid < \frac{1}{2} \delta_0 R^{2s} \tag{89}$$

$$\Theta(G_0, \delta^{-s}) < l_0 \exp l_0 \delta_0, \;\; \Theta(G_0, \hat{\delta}^{-s}) < l_0 \exp l_0 \delta_0 R^{2s} \tag{90}$$

$$\left|\frac{f_{0t}}{f_{0x}}\right|, \left|\frac{G_{0t}}{G_{0x}}\right| < b_1 R + \frac{1+\epsilon_0}{\bar{k}}\delta_0^{-1}\frac{1}{\mid G_{0x}\mid} < \frac{1+\epsilon_0}{\bar{k}}\delta_0^{-1} \tag{91}$$

$$2R^s\delta_0 <\mid \hat{\delta} \mid< \frac{1}{2}R^{2s}\delta_0 \tag{92}$$

$$mes\ \mathcal{C}_0 < c_1\delta_0(\exp l_0)\log_{\rho_0} R\,. \tag{93}$$

Let us call $\delta(t)$ *parabolic zone* and $I\backslash\delta(t)$ *hyperbolic zone*. The above proposition describes properties of maps insidc hyperbolic zone.

REMARK 5. *Here and below we do not specify how small should be "small constants ". In any case we assume* $\epsilon < \frac{1}{10}$.

3.7. Preliminary construction inside parabolic zone. Parabolic zone δ is a union of ΔT_i, $p_0 \le i \le N_0$. Let us denote the critical branch T_{N_0} as $h_0 = g_0 \circ h$, where g_0 is the staircase map from N_0-th step onto I, and h is the original parabolic map.

It is convenient to consider partitions located on x-axis near the critical point as pull-backs of respective partitions located on y-axis near the critical value. In inductive construction at step n we consider neighborhoods of zero scaled as $\delta_0 R^{-sn}$ where $0 < s < 1$ is a parameter. In order to make estimates simpler at the first steps of induction, we assume that the domain of the parabolic branch ΔT_{N_0} is significantly smaller than δ_0. It is convenient to compare sizes on y- axis, where the respective scaling is $\bar{k}\delta_0^2 R^{-2sn}$. We assume that for all $t \in \mathcal{T}$

$$\mid \Delta g_0 \mid< \frac{1}{2}R^{-6s}\bar{k}\delta_0^2\,. \tag{94}$$

Maps inside parabolic zone are compositions $T_i = g_i \circ h$, where g_i are staircase maps. Notice that distortions of compositions $f \circ h$ are determined by distortions of f and relative distances of their domains from the critical value. Therefore, these quantities make part of our inductive estimates. Distortions, expansions and dependence on the parameter of g_i were estimated above. According to (63) we have

$$\Theta(g_i, \Delta g_i) < l_0\,. \tag{95}$$

As we consider domains $T_i = g_i \circ h$ inside parabolic zone we get using (95)

$$\left|\frac{\partial g_i}{\partial x}\right| > \frac{1-\epsilon}{\bar{k}}\delta_0^{-2}\exp(-l_0)\,. \tag{96}$$

Notice that $\mid \Delta g_i \mid$ is of the order A_0^{-i}. Respectively i is of the order $\log \mid \Delta g_i \mid$. We use (37) and get for $i \ge p_0$

$$\left|\frac{g_{it}}{g_{ix}}\right| < c\delta_0^2 \mid \log\delta_0 \mid\,. \tag{97}$$

Here

$$c = \Gamma \exp \bar{k}_0(1+\epsilon_0)K_2\frac{\mid \frac{\partial f^{n_0}}{\partial x}(\theta_0)\mid}{A_0^{n_0}}\,. \tag{98}$$

where Γ expresses conversion from A_0^{-n} scale to $\delta_0 R^{-ks}$ scale. When δ_0 is small, $c\delta_0^2 \mid \log\delta_0 \mid$ is also small.

Let us generalize situation and assume that a partition ζ on y-axis is given. One element of ζ is Δg_0 which contains the critical value. Other elements are domains of maps $\phi_0 : \Delta\phi_0 \to I$ and $\mathcal{G}_0 : \delta^{-s} \to \delta$ with distortions satisfying

$$\Theta(\phi_0, \Delta\phi_0) < l_0 \tag{99}$$

and

$$\Theta(\mathcal{G}_0, \Delta\mathcal{G}_0) < l_0 \exp l_0\delta_0 \tag{100}$$

Speeds of elements on y-axis are small. We assume

$$\left|\frac{\phi_{0t}}{\phi_{0x}}\right|, \left|\frac{\mathcal{G}_{0t}}{\mathcal{G}_{0x}}\right| < \alpha_0 \tag{101}$$

where α_0 is a small constant. From (97) speeds on y-axis decrease as sizes of domains. Based on (94) we shall assume

$$\left|\frac{g_{0t}}{g_{0x}}\right| < \alpha_0 R^{-6s} . \tag{102}$$

Similarly to the properties of partitions on x-axis we assume that for each element δ^{-s} of ζ its enlargement $\hat{\delta}^{-s}$ is defined and satisfies (89).

Next conditions imply that at the first three steps of induction the union of undecided domains constitutes a small fraction of the space near to the critical value. Let $y_0 = y_0(t_0)$ be the image of both boundary points of $\delta(t_0)$ and let w_0 be the end point of Δg_0 such that $\Delta g_0 \subset [y_0, w_0]$. Notice that w_0 coincides with the critical value at the moment t_0. Let us define

$$y_i = w_0 - (w_0 - y_0)R^{-2si}, \ i = 1, 2, 3 . \tag{103}$$

Let $\mathcal{C}(y_i)$ be the union of undecided domains which intersect the interval $[y_i, w]$. Then we assume

$$\frac{mes\mathcal{C}(y_i)}{\mid y_i - w_0 \mid} < R^{-3s} .$$

Let us call the image of the parabolic zone $[y_0, w_0] = h\bar{\delta}_0$ - *parabolic range* at t_0. When t varies in $\mathcal{T}$ $y_0(t)$ and the critical value $w(t)$ change, and $w(t)$ moves inside Δg_0. We denote by $w_0(t)$ the end point of $\Delta g_0(t)$ such that $w_0(t_0) = w_0 = w(t_0)$ and call $[y_0(t), w_0(t)]$ - *parabolic range* at t.

Our next proposition determines properties of a family of piecewise smooth maps inside parabolic zone through properties of the respective maps which domains belong to the parabolic range. Let α_0 be a small constant of the order $\delta_0^2 \log \delta_0$. Let R, s, l_0, $\bar{k}$, δ_0 be constants from Proposition 2, and let $\mathcal{T}$ parameter interval, such that the critical value $w(t) = h(t, 0)$ belongs to Δg_0 when $t \in \mathcal{T}$.

PROPOSITION 3. *For all $t \in \mathcal{T}$ parabolic range is a union*

$$[y_0, w_0] = \bigcup \Delta\phi_0 \cup \bigcup \Delta\mathcal{G}_0 \cup \Delta g_0 \tag{104}$$

with the following properties:

$$\mid \Delta g_0 \mid < \frac{1}{2} R^{-6s} \bar{k} \delta_0^2 . \tag{105}$$

Distortion estimates:

$$\Theta(\phi_0, \Delta\phi_0) < l_0 \tag{106}$$

$$\Theta(\mathcal{G}_0, \Delta\mathcal{G}_0) < l_0 \exp l_0\delta_0 \tag{107}$$

Expansion estimates:

$$\left|\frac{\partial \phi_0}{\partial x}\right|, \left|\frac{\partial \mathcal{G}_0}{\partial x}\right| > \frac{1}{\bar{k}} \delta_0^{-2} \exp(-l_0) \tag{108}$$

Parameter estimates:

$$\left|\frac{\phi_{0t}}{\phi_{0x}}\right|, \left|\frac{\mathcal{G}_{0t}}{\mathcal{G}_{0x}}\right| < \alpha_0 \tag{109}$$

$$\left|\frac{g_{0t}}{g_{0x}}\right| < \alpha_0 R^{-6s} . \tag{110}$$

Relative measure estimates: let y_i be the fixed points defined by (103) *and $\mathcal{C}(y_i)$ unions of respective undecided domains, then*

$$\frac{| \mathcal{C}(y_i) |}{| y_i - w_0 |} < R^{-3s} \tag{111}$$

REMARK 6. *In the inductive process we shall only use properties of a piecewise smooth family satisfying propositions* (2) *and* (3).
First return maps are examples of such families.

3.8. Choice of scales. At the n-th step of induction we shall construct partitions inside a neighborhood δ_n of the origin of the size $(1 \pm \epsilon) R^{-sn} \delta_0$, where $1 > s > 0$ is a parameter of our construction specified below. Motivated by $2 - dim$ construction we require that for some $\alpha > 1$ domains Δf of good branches located at the distance d from the origin satisfy

$$| \Delta f | < const\ d^{\alpha} .$$

It is convenient to measure sizes with respect to $R^{-sn}\delta_0$ scale. For that purpose we denote $r_1 = \alpha - 1$, and rewrite the upper bound at step n as

$$| \Delta f | < const\ \delta_0 R^{-s(r_1+1)n} . \tag{112}$$

Estimates from below are expressed similarly. We require that Δf inside δ_n satisfy

$$| \Delta f | > const\ \delta_0 R^{-s(r_2+1)n} . \tag{113}$$

As discussed in the introduction, we are looking for a minimal r_2 compatible with inductive construction.

3.9. Relations between parameters. We assume a piecewise family satisfies properties listed in Propositions 2 and (3) and use constants from these propositions. At the first steps of induction we construct new maps on y-axis by pulling back consecutively partition ξ_0. By assumption, distortions of maps that we compose are bounded by l_0. Then distortions of compositions that we obtain are bounded by

$$l_0^* = l_0 \left(1 + \sum_1^{\infty} \frac{\exp l_0}{R^i} \right) \exp \left(\sum_1^{\infty} \frac{l_0 \exp l_0}{R^i} \right) \tag{114}$$

see (157) below. A priori l_0^* can be a small number. We define

$$\tilde{l}_0 = max\{l_0^*, 1\}$$

At the n-th step of induction in order to satisfy upper bound (112) we pull-back partitions of the step $[nx]$, where $x < 1$ is a parameter of our construction. We use the following expressions for distortion estimates at step k of induction.

$$l_k = \tilde{l}_0(\sum_0^k R^{-xsi})(\exp \sum_1^k R^{-xsi}) . \tag{115}$$

Then we define

$$l_\infty = \tilde{l}_0(\sum_0^\infty R^{-xsi})(\exp \sum_1^\infty R^{-xsi}) . \tag{116}$$

The following relations between parameters of our construction are sufficient to make inductive estimates at different steps compatible. There is a lot of flexibility here, and many ways to improve construction in order to get more admissible parameters.

We assume

$$r_1 \geq 1 \tag{117}$$

is given. With given r_1 we define

$$s = min\{\frac{1}{3}, \frac{1 - 1/10}{r_1}\} . \tag{118}$$

Then we define

$$r_2 = \frac{4}{s} + 6 + r_1 . \tag{119}$$

Notice that (118) and (119) imply

$$r_2 \geq 19 . \tag{120}$$

We define

$$x = \frac{1}{8} . \tag{121}$$

Then we define z_0 as the minimal number satisfying

$$z_0 + 1 \geq \frac{6}{x} \tag{122}$$

and

$$s(2r_2 + 3) + 5 < s[r_2 - r_1 - (r_2 + 1)x]z_0 . \tag{123}$$

We assume δ_0 and R are related by

$$\delta_0 = R^{-3} . \tag{124}$$

We define

$$\chi_0 = 2R^{-s} . \tag{125}$$

After that we choose R as a minimal number satisfying the following conditions:

$$\frac{1 + 1/10}{1 - R^{-2s}} \chi_0^3 < \frac{1}{2} R^{-2s} \tag{126}$$

$$max\{l_\infty \exp l_\infty, \ 2 \exp l_\infty, \ 2\sqrt{a_1}\} < R^s \tag{127}$$

$$\frac{1 + 1/10}{R^s} + \frac{1}{\rho_0} < 1 \tag{128}$$

$$(1+\frac{1}{10})c_1R^{2s}\delta_0\exp(2l_0^*)\log_{\rho_0}R<1 \tag{129}$$

$$b_1<R \tag{130}$$

$$\exp(R^{-6s})<\frac{3}{2} \tag{131}$$

$$\left(1-c_1\delta_0\exp(l_0+l_0^*)\log_{\rho_0}R\right)^{[sr_1(z_0-1)+1]}>1-\chi_0^3\,. \tag{132}$$

Notice that (126) implies

$$R^s>16\,. \tag{133}$$

We leave to the reader to check that the above conditions are compatible for R large enough.

3.10. Main theorem. In order to make elements of our partitions properly sized at the first z_0-1 nonstandard steps of induction we do $k_0=[z_0s(r_2+3)]+1$ parameter choices, see (159). After that the measure of remaining parameter values is greater than

$$1-\mu_0=\left[1-(1+\frac{1}{10})c_1R^{2s}\delta_0\exp(2l_0^*)\log R\right]^{k_0} \tag{134}$$

see (160). From (129) it follows that $1-\mu_0>0$. We denote

$$m_0=[s(r_2+3)]+2 \tag{135}$$

the number of parameter choices at a given step of induction which is sufficient to make parameter intervals properly sized, see (212).

THEOREM 1. *Assume maps which constitute a piecewise smooth family F_t satisfy conditions of Propositions 2 and 3. Let parameters of our construction satisfy the above relations* (117) - (132). *Then there exists a set of parameter values $\mathcal{M}\subset\mathcal{T}$ such that for $t\in\mathcal{M}$ F_t have absolutely continuous invariant measures and domains of the respective power map are uniformly scaled. The relative measure of $\mathcal{M}$ in $\mathcal{T}$ satisfies*

$$mes\mathcal{M}>\mid\mathcal{T}\mid(1-\mu_0)\prod_{k=0}^{\infty}\left(1-(mes\mathcal{C}_0)exp(l_\infty)(2R^{-s})^{[(z_0+k)x-1]}\right)^{m_0}\,. \tag{136}$$

4. First steps of induction

4.1. First parameter choice. In order to make fluctuation of sizes small comparatively to sizes, we organize the general inductive construction in such a way that at step n intervals of variation of parameter $t\in\mathcal{T}$ satisfy

$$\mid\mathcal{T}\mid<\bar{k}\delta_0^2R^{-s(r_2+3)n}\,. \tag{137}$$

Regular inductive construction starts at step z_0. We begin the preliminary construction by reducing initial parameter intervals in order to satisfy (137) with $n=z_0$. Initial parameter interval $\mathcal{T}=[t_0,t_1]$ corresponds to the movement of the critical point $w(t)$ from one end point of $\Delta g_0(t)$ to the other end point. In order to make parameter choice we pull-back on Δg_0 partition $\hat{\xi}_0$, which consists of Δf_0 and enlarged undecided domains. Let us denote $g_0^{-1}\hat{\xi}_0=\hat{\zeta}_*$. From formula (333) it follows that elements of $\hat{\zeta}_*$ are moving with small speeds, see (180) below. At the same time the critical point $w(t)$ moves with the unit speed. Therefore for the whole domain Δg_0, as well as for each element of $\hat{\zeta}_*$, two moments are defined: when the critical value first touches that element, and when it leaves it. These moments determine the respective subintervals of $\mathcal{T}_0$. We consider parameter intervals corresponding to good branches as admissible, and intervals corresponding to enlarged undecided domains as non-admissible. Let $\mathcal{M}_1 = \bigcup \mathcal{T}(\phi_0)$ be the union of admissible parameter intervals. We need to estimate relative measure of $\mathcal{M}_1$.

First we prove one general estimate. Let us consider a point $w(t)$ moving with the unit speed across an interval, which varies slowly depending on the parameter. Lemma 2 below implies that up to a factor $1 \pm \epsilon$ initial parameter interval $\mathcal{T} = [t_0, t_1]$ has the same size as $\mid \Delta g_0(t_0) \mid$. If $\mid \Delta g_0(t_0) \mid$ is not small enough, then we do several parameter choices.

LEMMA 2. *Let $\Delta(t) = [x_1(t), x_2(t)]$ be an interval on the x-axis. Assume*

$$\left| \frac{\partial x_1}{\partial t} \right|, \left| \frac{\partial x_2}{\partial t} \right| < \epsilon .$$

Let $\tau(\Delta) = [t_1, t_2]$ be the respective interval on the parameter axis, where t_1 is the time when $w(t)$ enters $\Delta(t)$ and t_2 is the time when $w(t)$ leaves $\Delta(t)$. Then

$$\frac{1}{1+\epsilon} < \frac{\mid \tau(\Delta) \mid}{\Delta(t_1)} < \frac{1}{1-\epsilon} \tag{138}$$

and for any t', $t'' \in \tau(\Delta)$

$$\frac{1}{1+2\epsilon} < \frac{\mid \Delta(t') \mid}{\mid \Delta(t'') \mid} < \frac{1}{1-2\epsilon} . \tag{139}$$

PROOF. Elementary. For (138) consider the graphs of $x_2(t)$ and $x = t$. For (139) consider also the graph of $x_1(t)$. □

PROPOSITION 4.

$$\frac{mes(\mathcal{M}_1)}{\mid \mathcal{T}_0 \mid} > 1 - \hat{c}\delta_0 \log R \tag{140}$$

where $\hat{c}$ is a uniform constant.

PROOF. (a) From (102) we have

$$\left| \frac{g_{0t}}{g_{0x}} \right| < \alpha_0 R^{-6s} . \tag{141}$$

Then (105) and Lemma 2 imply

$$\mid \mathcal{T}_0 \mid < \frac{1+\epsilon}{2} R^{-6s} \bar{k} \delta_0^2 . \tag{142}$$

Let us estimate a relative change of lengths of undecided domains in the partition $\xi(t)$ when t varies within $\mathcal{T}_0$. We already know the change is small for $\delta(t)$. For preimages $\delta^{-s}(t)$ maximal relative change occurs, when lengths are minimal and changes are maximal, so we estimate sizes from below.

Let G be one of the maps that we used in the preliminary construction for pull-backs, G_{xmax} (respectively G_{xmin}) maximal (respectively minimal) expansion of G. Let $\delta^{-s} = G^{-1}\delta$. For any $t \in \mathcal{T}_0$

(143) $$\mid \delta^{-s}(t) \mid > \frac{\mid \delta(t) \mid}{G_{xmax}(t)}.$$

Let us estimate a shift $sh(t)$ of some boundary point of $\delta^{-s}(t)$ from its initial position at t_0. From (91) speeds of such points do not exceed

(144) $$v_{max} = b_1 R + \frac{1+\epsilon}{\bar{k}}\delta_0^{-1}\frac{1}{\mid G_{xmin} \mid}.$$

As distortion of G is small we get

(145) $$\frac{1}{\mid G_{xmin} \mid} < (1+\epsilon) \mid \delta^{-s}(t) \mid \delta_0^{-1}.$$

Then

(146) $$v_{max}(t) < b_1 R + \frac{1+\epsilon}{\bar{k}}\delta_0^{-2} \mid \delta^{-s}(t) \mid .$$

As

$$\mid \delta^{-s}(t) \mid \leq \mid \delta^{-s}(t_0) \mid + 2sh(t)$$

we get

(147) $$\mid \delta^{-s}(t) \mid < \mid \delta^{-s}(t_0) \mid + 2\int_0^{|\mathcal{T}_0|} b_1 R + \frac{1+\epsilon}{\bar{k}}\delta_0^{-2} \mid \delta^{-s}(u) \mid du.$$

By Gronwall inequality

(148) $$\mid \delta^{-s}(t) \mid < \big(\mid \delta^{-s}(t_0) \mid + 2 \mid \mathcal{T}_0 \mid b_1 R\big) \exp[\frac{1+\epsilon}{\bar{k}}\delta_0^{-2} \mid \mathcal{T}_0 \mid].$$

From (86)

$$\delta^{-s}(t_0) > (1-\epsilon)\delta_0(\exp -l_0)R^{-1}$$

Using (124), (127) and (130) we get

(149) $$\mid \delta^{-s}(t_0) \mid + 2 \mid \mathcal{T}_0 \mid b_1 R \leq \mid \delta^{-s}(t_0) \mid (1 + \frac{1}{10}).$$

From (105)

(150) $$\frac{1+\epsilon}{\bar{k}}\delta_0^{-2} \mid \mathcal{T}_0 \mid < (1+\epsilon)R^{-6s}.$$

From (131) exponent in (148) is less than $\frac{3}{2}$ and we get

(151) $$\mid \delta^{-s}(t) \mid < 2 \mid \delta^{-s}(t_0) \mid .$$

In the same way we estimate $\delta^{-s}(t)$ from below and get

(152) $$\frac{1}{2} \mid \delta^{-s}(t_0) \mid < \mid \delta^{-s}(t) \mid < 2 \mid \delta^{-s}(t_0) \mid .$$

(b) We conclude the proof of Proposition 4 by the arguments, which are also used at the general step of induction. Lemma 2 implies, that when $t \in \mathcal{T}_0$, the length of $\Delta g_0(t)$ varies by a factor close to 1.

As distortions of $g_0(t)$ are bounded by $\exp l_0$ we get for all t

(153) $$\exp(-l_0) < \frac{\frac{\partial g_0}{\partial x}(x_1,t)}{\frac{\partial g_0}{\partial x}(x_2,t)} < \exp l_0$$

for all $x_1,\ x_2 \in \Delta g_0(t)$.

Let $D(t) = \delta^{-s}(t)$ be any enlarged undecided domain - an element of the partition $\hat{\xi}(t)$. For enlarged undecided domains shifts of boundary points are estimated as above, but their sizes are R^s times larger. Therefore we get similarly to (152) a better estimate

(154) $$(1-\epsilon) \mid \hat{\delta}^{-s}(t_0) \mid \leq \mid \hat{\delta}^{-s}(t) \mid \leq (1+\epsilon) \mid \hat{\delta}^{-s}(t_0) \mid .$$

Using (153) we conclude that the relative measure of $g_0^{-1}(D(t))$ in $\Delta g_0(t)$ cannot exceed the measure of $D(t_0)$ by more than $(1+\epsilon)\exp l_0$. Applying Lemma 2 we get that the relative measure of respective parameter interval does not exceed $(1+\epsilon)D(t_0)\exp l_0$. Let $\hat{\mathcal{C}}_0$ be the set of enlarged undecided domains in ξ_0 at the moment t_0. Let $\mathcal{N}_1$ be the set of non-admissible parameters. We get

(155) $$\frac{mes(\mathcal{N}_1)}{\mid \mathcal{T}_0 \mid} \leq (1+\epsilon)(\exp l_0) mes\ \hat{\mathcal{C}}_0 .$$

From (93)

$$mes\ \hat{\mathcal{C}}_0 < c_1 R^{2s} \delta_0 (\exp l_0) \log_{\rho_0} R .$$

Therefore

(156) $$\frac{mes(\mathcal{M}_1)}{\mid \mathcal{T}_0 \mid} > 1 - \hat{c}\delta_0 \log_{\rho_0} R .$$

where $\hat{c} = (1+\epsilon)R^{2s} c_1 \exp 2l_0$ is a uniform constant.

That proves Proposition 4.

□

4.2. Subsequent parameter choices. As a result of the previous parameter choice the set of admissible parameter values is a union of $\mathcal{T}_i = \{t : w(t) \in \Delta_i = \Delta(\phi_i)\}$, where $\phi_i = f_i \circ g_0$. For $t \in \mathcal{T}_{i_1}$ we define the new central branch $h_1 = \phi_{i_1} \circ h = f_{i_1} \circ g_0 \circ h$.

We shall use notation F_k for branches, which contain critical value. So we denote $F_1 = \phi_{i_1}$ and $h_1 = F_1 \circ h$. Let t_0 be the parameter value, corresponding to the full height of h_1. From Lemma 2 the size of $\mathcal{T}_i$ is close to $\Delta F_1(t_0)$. As sizes of good branches in ξ are less than R^{-1}, we get taking into account distortion

$$\mid \Delta F_1(t_0) \mid < \mid \Delta g_0 \mid \exp l_0 R^{-1}$$

If needed we repeat the above construction. Then we obtain compositions of the form

$$h_k = f_{i_k} \circ \ldots \circ f_{i_2} \circ f_{i_1} \circ g_0 \circ h = F_k \circ h$$

where all f_{i_s} are hyperbolic maps defined in the preliminary construction. Distortions of f_i are bounded by l_0 and expansions are at least R. Using formula (291) for distortion of compositions we get $\Theta(F_k) < l_0^*$ where

(157) $$l_0^* := l_0 \left(1 + \sum_1^\infty \frac{\exp l_0}{R^i}\right) \exp\left(\sum_1^\infty \frac{l_0 \exp l_0}{R^i}\right) .$$

After k compositions we get using (105)

(158) $$\mid \Delta F_k \mid < \frac{1}{2} R^{-6s} \bar{k} \delta_0^2 R^{-k} \exp l_0^* .$$

Let

(159) $$k_0 = [z_0 s(r_2+3)] + 1 .$$

Then (137) will be satisfied after no more than k_0 parameter choices. Let $\epsilon_2 = (1+\epsilon)R^{2s}c_1\delta_0 \log R \exp(2l_0^*)$. From (129) we get $\epsilon_2 < 1$. Similarly to (156) we get that after k_0 steps of our construction the measure of admissible parameters is greater than

$$(1-\epsilon_2)^{k_0} = 1 - \mu_0 \,. \tag{160}$$

Clearly, $\mu_0 \to 0$ when $\delta_0 \to 0$.

4.3. Critical pull-back. When doing consecutive parameter choices we define simultaneously a nested sequence of parameter intervals $\mathcal{T}_{i_1} \supset \mathcal{T}_{i_1 i_2} \supset \ldots \supset \mathcal{T}_{i_1 i_2 \ldots i_m} = \mathcal{T}_*$ and an increasing sequence $\zeta_{i_1} \prec \zeta_{i_1 i_2} \prec \ldots \prec \zeta_{i_1 i_2 \ldots i_m} = \zeta_*$ of partitions on y axis.

When choosing parameter we pull-back $\hat{\xi}_0$. However, for constructions in the phase space we use pull-backs of ξ_0. Therefore $\zeta_{i_1} = g_0^{-1}\xi_0$, $\zeta_{i_1 i_2} = \phi_{i_1}^{-1}\xi_0$ where $\Delta\phi_{i_1}$ is the element of ζ_{i_1} containing $w(t)$ and so on. The last parameter choice fixes an element $\Delta_* = \Delta\phi_*$ of ζ_* which contains the critical value $w(t)$. Fixing a value $t_* \in \mathcal{T}_*$ is equivalent to fixing a position $w(t_*) \in \Delta_*$.

When we pull-back ζ_* by the initial parabolic branch h we create a partition ξ_* of some neighborhood of the critical point. Element $h^{-1}\Delta_*$ is a branched preimage of Δ_*. Other elements E of ζ, which are covered by the image of h have two diffeomorphic preimages $h_1^{-1}(E)$ and $h_2^{-1}(E)$, which belong to ξ_*. The pull-back by h^{-1} is called *critical pull-back.* When $w(t)$ moves through Δ_* elements of ξ_* vary continuously. Critical pull-backs create first "rough" partitions of the domains of consecutive parabolic branches. Many elements of these partitions are " oversized" and will be additionally partitioned at the respective steps of induction. Sizes of elements in partitions ζ_* and ξ_* are related in a following way. Let $w = h(0, t_0)$ be the critical value. In a small neighborhood of zero $h(x, t_0)$ is close to a quadratic map $w - \bar{k}x^2$. That implies the following estimate. If an element E is small comparatively to its distance $d(E)$ from the critical value, then

$$| h^{-1}E | = \frac{| E |}{2\sqrt{\bar{k}d(E)}}(1 \pm \epsilon_1) \,. \tag{161}$$

Let $d(h^{-1}E)$ be the distance from $h^{-1}E$ to the origin. Then

$$d(h^{-1}E) = \sqrt{\frac{d(E)}{\bar{k}}}(1 \pm \epsilon_2) \tag{162}$$

and

$$\frac{| h^{-1}E |}{d(h^{-1}E)} = \frac{E}{2d(E)}(1 \pm \epsilon) \,. \tag{163}$$

4.4. Partitions ζ_1, ξ_1, and respective inductive estimates at the first step of induction.

(a) *Partitions ζ_{11} and ξ_{11}.*
After several parameter choices we got admissible parameter intervals $\mathcal{T}$ small enough to make variation of sizes insignificant at the first z_0 steps of our construction. Let $\mathcal{T}_0$ be one of these intervals and let $t_0 \in \mathcal{T}_0$ be the parameter value such that the image of the parabolic branch covers the respective interval Δ_*. We use an algorithm based on sizes of elements at $t = t_0$. Then we prove that sizes satisfy inductive estimates for all $t \in \mathcal{T}_0$. Let us denote by ζ_{11} a partition on y- axis which restriction to

Δg_0 coincides with ζ_* and outside Δg_0 it coincides with partition ζ of Proposition 3. Because of (105) we are not dealing with the refined part of ζ_{11} at the first three steps of induction.

(b) *Refining monotone branches.*
For monotone branches inductive estimate (196) requires at step one

$$8R^{-r_2 s} < \frac{|E|}{d(E)} < \frac{1}{2}R^{-r_1 s}\,. \tag{164}$$

If for a given $\Delta\phi_0$ the upper bound in (164) is not satisfied then we pull-back ξ_0 and check (164) for new monotone branches. If needed we pull-back ξ_0 again and so on. The number of pull-backs grows with the step of induction. The related loss of measure of good branches is estimated below. We define partitions obtained by the above pull-backs ζ_{12} and ξ_{12}. Elements of ζ_{12} are domains of ϕ_0 undecided domains δ^{-s} obtained by the above pull-backs, and undecided domains δ^{-s} which already existed on y-axis and satisfied estimates of Proposition 3. Each time we pull-back ξ_0 onto some $\Delta\phi_0$ we consider a simultaneous pull-back of $\hat{\xi}_0$. Then each domain δ^{-s} is contained in a respective enlargement $\hat{\delta}^{-s}$. By construction two enlargements do not intersect or one of them contains another one. Then maximal enlargements and remaining branches $\Delta\phi_0$ form a partition $\hat{\zeta}_{12}$ such that $\zeta_{12} \succ \hat{\zeta}_{12}$.

(c) *Choice of* r_2 *and lower bounds at step one.*
Assume a pull-back of ξ_0 onto some element E is required. That means

$$\frac{|E|}{d(E)} > \frac{1}{2}R^{-sr_1}\,. \tag{165}$$

Minimal sizes of monotone branches in ξ_0 follow from (85). Taking into account distortion we get for a domain E_1 of a new branch obtained by the above pull-back

$$\frac{|E_1|}{d(E_1)} > \frac{1}{2}R^{-sr_1}\exp(-2l_0^*)a_1^{-1}R^{-1}\delta_0\,. \tag{166}$$

Then the lower bound in (164) follows from (119), (124) and (127).

By the same arguments the lower bound in (197) follows from the choice of r_2.

REMARK 7. *The gap between upper and lower bounds increases with the number of steps. Therefore, when we pull-back* ξ_0 *on oversized domains at steps* $2 \le n \le z_0 - 1$ *lower bounds will be satisfied.*

When checking the upper bound in (197) we notice that there are two kinds of undecided domains. First, there are a priori given δ^{-s} satisfying (111). As $x = \frac{1}{8}$ the upper bound R^{-3s} which follows from (111) is better than (197) at the first three steps of induction.

Second, there are δ^{-s} constructed as a result of the above refining of monotone branches. For such domains the ratio of their sizes to the distance to the critical value is bounded by $2(1+\epsilon)\delta_0\exp(l_0+l_0^*) < R^{-2} \ll R^{-3s}$, see a similar estimate (259) below. Thus, (197) follows in this case as well.

REMARK 8. *In a specific problem it might be preferable to find lower bounds at the first step by direct computation and define the corresponding* r_2.

(d) *Definition of* δ_1.
We remind that the end points of $\delta(t)$ are two critical preimages $h_i^{-1}y_0(t)$, $i = 1, 2$. We define $\delta_1(t)$ in a similar way.

Let $[y_1(t_0), w(t_0)]$ be the maximal union of elements in $\hat{\zeta}_{12}(t_0)$ such that

$$| \, [y_1(t_0), w(t_0)] \, | < \bar{k}\delta_0^2 R^{-2s} \, .$$

From (164) sizes of elements $\Delta\phi_0$ in $\hat{\zeta}_{12}$ located near $y_1(t_0)$ are small comparatively to their distances to the critical value. Similarly from (111) enlargements $\hat{\delta}^{-s}$ have the same property. That implies

$$(1-\epsilon)\bar{k}\delta_0^2 R^{-2s} < | \, [y_1(t_0), w(t_0)] \, | < (1+\epsilon)\bar{k}\delta_0^2 R^{-2s}$$

Let $\delta_1(t_0)$ be the interval bounded by two preimages $h^{-1}y_1(t_0)$. Then $\delta_1(t_0)$ satisfies

$$(1-\epsilon)\delta_0 R^{-s} < | \, \delta_1(t_0) \, | < (1+\epsilon)\delta_0 R^{-s}$$

For $t \in \mathcal{T}_0$ we define $\delta_1(t)$ by continuity so that it consists of the same elements of $\xi_{12}(t)$. Notice that by construction using enlargements we got the following property. If $\delta^{-s} \subset \delta_1$ then $\hat{\delta}^{-s} \subset \delta_1$. Similarly if $\delta^{-s} \subset \delta \setminus \delta_1$ then $\hat{\delta}^{-s} \subset \delta \setminus \delta_1$. We can say that our construction "respects enlargements". Let us estimate how the size of $\delta_1(t)$ can change. As in Proposition 2 we get that the boundary points x_i of $\delta_1(t)$ are moving with the speed

$$\left| \frac{dx_i}{dt} \right| < \frac{(1+\epsilon)}{\bar{k}} \delta_0^{-1} R^s \, . \tag{167}$$

On the other hand interval of parameter $\mathcal{T}_0$ satisfies (137) with $n = z_0$. Thus when t varies in $\mathcal{T}_0$ points $x_i(t)$ shift by an amount of the order $\delta_0 R^{-s[(r_2+3)z_0-1]}$, which is much smaller than the size of $\delta_1(t_0)$. So we get for all $t \in \mathcal{T}_0$

$$\delta_0 R^{-s}(1-\epsilon) < | \, \delta_1(t) \, | < \delta_0 R^{-s}(1+\epsilon) \, . \tag{168}$$

(e) *Canonical partition and collars.*
We define canonical partition

$$\xi_1 = \xi_{12} \mid \delta \setminus \delta_1 \, . \tag{169}$$

We define a *collar* $S(\delta_1)$ as a union of all good branches of ξ_1 inside $\delta \setminus \delta_1$.

$$S(\delta_1) = \delta \setminus \left(\delta_1 \bigcup \delta^{-k} \right) \, . \tag{170}$$

(f) *Final filling-in outside* δ.
We finish the construction of ξ_1 by pulling-back $\xi_1 \mid \delta \setminus \delta_1$ onto all domains δ^{-k} which lie outside δ. The collar $S(\delta_1^{-k}) \subset S(\delta^{-k})$ is defined as a pull - back of $S(\delta_1)$.

4.5. Partitions ξ_i, $2 \le i \le 3$ and related estimates.

(a) *Estimates of sizes*

Partitions ξ_i, $i = 2,3$ are constructed similarly to ξ_1. In particular, enlargements are respected. Using (105) and the definition (103) of points y_i we get that enlargements near y_i are small comparatively to their distances to the critical value and for all $t \in \mathcal{T}$

$$| \delta_i |= \delta_0 R^{-si}(1 \pm \epsilon) .$$

Notice that all undecided domains elements of $\xi_i|\delta_i \setminus \delta_{i+1}$ are preimages δ^{-k}. Similarly to (170) we define $S(\delta_{i+1})$ as a union of good branches in $\delta_i \setminus \delta_{i+1}$. Bounds (196) and (197) are satisfied by the same argument as at step one. That proves inductive estimates of sizes for elements of canonical partitions ξ_i, $i \le 3$. Non-canonical elements outside δ_i are obtained by filling-in of respective domains δ^{-k}. As derivatives outside δ are greater than $\rho_0 > 1$, sizes decrease, which implies upper estimates (201), (202), (200). Lower estimates are determined by lower estimates of elements δ^{-k} in ξ_0. From (86) sizes decrease less than by $R^{-1}\exp(-l_0)$. We define K_0 in (201), (200) as

(171) $$K_0 = \frac{1}{2} R^{-1} \exp(-l_0)$$

As distortions of maps G_0 are small, we get estimates from below which are $\frac{1-\epsilon}{2}$ times better than required in (201), (200). That proves size estimates.

(b) *One property of partitions ξ_i, $i = 1,2,3$ used at inductive steps $4 \le m \le z_0 - 1$.*

Let δ^{-k} be an element of a canonical partition of δ_m $m \le 2$. Its distance to zero is less than $\frac{(1+\epsilon)|\delta_m|}{2}$. We use (111) and (163) and get

(172) $$| \delta^{-k} |< R^{-3s}\frac{1+\epsilon}{4} | \delta_m | .$$

At steps $4 \le m \le z_0 - 1$ we use pull-backs of ξ_3 onto δ^{-k}. Notice that $\xi_3|\delta$ does not coincide with $\xi_i|\delta$, $i = 1,2$. The difference is related to the filling-in of domains δ^{-k} at step 2 and δ^{-k}, δ_1^{-k} at step 3. As a result of these fillings after step 2 we get from (172)

(173) $$| \delta_1^{-k} |< \frac{1+\epsilon}{4}\delta_0 R^{-4s} .$$

After step 3 we get for all elements δ_i^{-k}, $i = 0,1,2$ of the partition $\xi_3|\delta$

(174) $$| \delta_i^{-k} |< \frac{1+\epsilon}{4}\delta_0 R^{-5s} .$$

(c) *Distortion estimates.*

By definition of l_0^*

(175) $$\Theta(\phi_0) < l_0^* .$$

Respectively distortions of maps $\mathcal{G}_0 : \delta^{-k} \to \delta$ satisfy

(176) $$\Theta(G_0), \Theta(\mathcal{G}_0) < l_0^* \exp l_0^* \delta_0 .$$

From (124) and (127) the above estimate is less than R^{-2}. That is much smaller than (207) required at step $i = 0$. Notice that distortions of a priori given maps ϕ_0 and $\mathcal{G}_0$ from Proposition 3 satisfy estimates (106) and (107), which are better than respectively (175) and (176).

For canonical maps $f_i = \phi_0 \circ h$ and $G_i = \mathcal{G}_0 \circ h$ $i = 1, 2, 3$ distortion estimates follow from (291), (293) and respective ratios of sizes.

(177) $$\Theta(f_i) < l_0^* \exp \frac{1+\epsilon}{4} R^{-r_1 si} + \frac{1+\epsilon}{4} R^{-r_1 si}$$

$r_1 > 1$ implies that (177) is less than l_1 as required by (206).

For a priori given undecided domains we use (111) and get from (293)

(178) $$\Theta(G_i) < l_0^* \exp l_0^* \delta_0 \exp \left[\frac{1+\epsilon}{2} R^{-3s}\right] + \frac{1+\epsilon}{2} R^{-3s} .$$

The above bound is much smaller than the required estimate (207). Undecided domains obtained by monotone pull-backs satisfy better estimates because δ_0 is small.

(d) *Dependence on the parameter.*

When estimating dependence on the parameter we use formula (333). According to our construction we get properly sized domains on y-axis by taking compositions $F_0 = f_{01} \circ f_{02} \circ \ldots \circ f_{0k}$ and after that consider $\phi_0 \circ F_0$.

As speeds of elements of ξ_0 are less than $\frac{1+\epsilon}{k}\delta_0^{-1}$ and expansions of f_0 are greater than R we get from (333)

(179) $$\left|\frac{F_{0t}}{F_{0x}}\right| < \frac{1+\epsilon}{\bar{k}} \delta_0^{-1}(1 + R^{-1} + R^{-2} + \ldots) .$$

For compositions $F_0 \circ \phi_0$ we use $\left|\frac{\phi_{0t}}{\phi_{0x}}\right| < \alpha_0$ and (108) and get the following estimate of speeds for elements of partitions ζ_i, $1 \le i \le 3$ on y-axis.

(180) $$\left|\frac{\phi_t}{\phi_x}\right|, \left|\frac{\mathcal{G}_t}{\mathcal{G}_x}\right| < \alpha_0 + (1+\epsilon)\delta_0(\exp l_0)(1 + R^{-1} + R^{-2} + \ldots) .$$

As (180) is less than $\alpha_0 + R^{-3+s}$ it satisfies (332). For elements of ζ_i on x-axis we get similarly to (91)

(181) $$\left|\frac{f_t}{f_x}\right|, \left|\frac{G_t}{G_x}\right| < \frac{1+\epsilon}{\bar{k}} \mid \delta_i \mid^{-1} .$$

4.6. Partitions ξ_m, $4 \le m < z_0$.

(a) *Size estimates.*

For monotone branches which belong to partitions ξ_m, $m < z_0$, construction remains the same. We pull-back consecutively initial partition ξ_0 until sizes of monotone branches at the distance $R^{-2sm}y_0$ from the critical value satisfy the upper bound (196). After that we do the critical pull-back. Then the upper bound (198) follows from (163).

After several pull-backs we get monotone branches of the form

(182) $$\phi_k \circ h = f_{0i_1} \circ \ldots \circ f_{0i_k} \circ g_i \circ h$$

where f_{0i_l} are maps which belong to ξ_0, g_i are staircase maps and h is the parabolic map. Notice that each pull-back of ξ_0 produces several preimages of δ. As discussed above, δ^{-k} satisfy (197) at the first three steps. However, at the next steps some of them may be too big. Let us consider $\delta^{-k} = [a, b]$ where $b \in (a, 0)$ and $\delta_0 R^{-si} >\mid [a, 0] \mid> \delta_0 R^{-s(i+1)}$.

If δ^{-k} does not satisfy the upper bound (197) of step $i+1$, we do a new operation : *filling-in* of δ^{-k}. Let $G : \delta^{-k} \to \delta$ be the respective map. We pull-back $\xi_3 \mid \delta \setminus \delta_3$ by G^{-1} into δ^{-k}. Then we get a partition of δ^{-k} into central subdomain δ_3^{-k} and secondary preimages $\delta_m^{-(k+l)}$, $m \leq 2$. By construction δ^{-k} are separated from the critical value by respective enlargements. That implies the following estimate

(183)
$$\frac{\mid \delta_3^{-k} \mid}{d(\delta_3^{-k})} < 2(1+\epsilon)R^{-4s} .$$

From (174) it follows that ratios are much smaller for secondary preimages $\delta_i^{-(k+l)}$, $i = 0, 1, 2$. We use (183) in Section 7 when doing distortion and measure estimates.

Notice that the upper bound (197) may still be not satisfied after the above filling-in, but we do not require it at steps $m < z_0$. Together with δ_m^{-k} their enlargements $\hat{\delta}_m^{-k}$ are defined by construction. For $m \geq 1$ $\hat{\delta}_m^{-k} = \delta_{m-1}^{-k}$ satisfy

(184)
$$(1-\epsilon)R^s \mid \delta_m^{-k} \mid < \mid \hat{\delta}_m^{-k} \mid < (1+\epsilon)R^s \mid \delta_m^{-k} \mid$$

and for $m = 0$ sizes of enlargements follow from (92). From (174) and (183) it follows that enlargements are R^{-3s} times smaller than their distances to the critical value. Respectively their critical preimages are small comparatively to their distances to zero. That allows to construct a sequence of intervals δ_m, $m < z_0$, satisfying

(185)
$$\mid \delta_m \mid = (1 \pm \epsilon)R^{-sm}\delta_0$$

and a sequence of canonical partitions $\xi_m | \delta_{m-1} \setminus \delta_m$ respecting enlargements. Elements of these partitions are good branches and preimages δ_m^{-k}, $m \leq 2$.

Lower bounds (198) for monotone branches at step m are exponentially smaller than upper bounds because $r_2 > r_1$. The same is true about lower bounds (199) for undecided domains. The narrowest gap between estimates occurs for preimages of δ, which are inside $\delta \setminus \delta_1$. These are satisfied by the choice of r_2. For preimages $\delta^{-k} \subset \delta_i \setminus \delta_{i+1}$ the gap between actual sizes and lower bounds increases with i. Starting from $n = 4$ canonical partitions ξ_n may contain elements δ_3^{-k} obtained by filling-in of oversized elements δ^{-k}. Then

(186)
$$\frac{\mid \delta_3^{-k} \mid}{d(\delta_3^{-k})} > \frac{R^{-s}}{2} R^{-xsn} R^{-3s}(1-\epsilon) .$$

Comparing the rhs with

$$8R^{3r_2 s}R^{-sr_2 n}$$

in the inductive lower bound (197), we see that the minimal gap occurs for $n = 4$. Then the lower bound follows from (119). Collars $S(\delta_m) \subset \delta_{m-1}$ are defined as unions of monotone branches Δf which belong to $\delta_{m-1} \setminus \delta_m$.

After canonical partition is fixed, we complete the construction of ξ_m outside δ_{m-1}. That is done as at the general step of induction. Every preimage δ_{i-1}^{-k} is filled-in by the pull-back of respective canonical partitions $\xi_i | \delta_{i-1} \setminus \delta_i$. If some monotone branches obtained by such filling are oversized, we pull-back ξ_0 onto such branches until they satisfy upper bounds. Then lower bounds are satisfied by the choice of r_2 in (119) and K_0 in (171).

(b) *Measure estimates.*
In order to satisfy upper bounds on monotone branches at steps $m \leq z_0 - 1$ we need to pull-back ξ_0 no more than $[r_1 sm] + 1$ times. From estimate of $mes\ \mathcal{C}_0$ in Proposition 2 we estimate the maximal loss of measure as

(187) $$(1 - c_1\delta_0 \exp(l_0 + l_0^*) \log R)^{[r_1 s(z_0-1)]+1} = 1 - \mu_1 .$$

From (132)

(188) $$\mu_1 < \chi_0^3 .$$

In order to complete measure estimates we need to estimate the relative measure of undecided domains, which remain after filling-in of oversized δ^{-k}. That is a particular case of an estimate done at the general step of induction. After filling-in measures of undecided domains which remain at the place of δ^{-k} decrease by a factor $(1+\epsilon)R^{-3s}$. Without considering mutual position of good branches and undecided domains we can assume that all undecided domains are near to the end points of δ_i. Then after the critical pull-back their relative measure is multiplied by $(1+\epsilon)R^s$. It does not exceed $(1+\epsilon)R^{-2s}$. It is small comparatively to the relative measure of the central domain which is $(1 \pm \epsilon)R^{-s}$. Thus, at first steps, the total measure of undecided domains decreases by a factor $(1+\epsilon)R^{-s}$ per step. At a general step of induction we prove that measure of undecided domains decreases at least by $\chi_0 = 2R^{-s}$ per step. We define a collar

(189) $$S(\delta_i) = \delta \setminus \left(\delta_i \bigcup \delta^{-k}\right) .$$

Then we get

(190) $$\frac{| S(\delta_i) |}{| \delta_{i-1} |} > 1 - R^{-s}(1+\epsilon) .$$

(c) *Distortion and speed estimates.*
At the first three steps they were proved above. At the following steps they are proved in Section 7.

4.7. Dependence of sizes on the parameter. Let $\mathcal{T}$ be the parameter interval under consideration. By construction it satisfies (137). Estimates of Section 7 show that elements on y-axis are moving with a small speed. That implies

(191) $$\left| \frac{dx_i}{dt} \right| < \frac{(1+\epsilon)}{\bar{k}} \delta_0^{-1} R^{sn} .$$

From (137) and (191) we get by the arguments used in (168) that for all $t \in \mathcal{T}$

(192) $$\delta_0 R^{-si}(1 - \epsilon_0) <| \delta(t) |< \delta_0 R^{-si}(1 + \epsilon_0) .$$

Sizes of other elements in ξ_m satisfy inductive lower bounds. Therefore inductive arguments of a general step can be applied and we get that when t varies in $\mathcal{T}_0$, variation of size for any element of ξ_m, $m \leq z_0$ is small comparatively to its size. That finishes the construction of ξ_m, $m < z_0$.

5. Basic objects and operations. Inductive assumptions

In this section we assume the construction of the first $n \geq z_0$ steps completed and describe the construction at step $n+1$.

5.1. Basic objects.

(a) *Admissible parameters, good branches, parabolic branches.*
A set M_n of admissible parameter values, consists of a finite number of intervals $\mathcal{T} = \mathcal{T}_{i_1 i_2 \dots i_m}$. By construction $\mathcal{T}_{i_1 i_2 \dots i_m} \subset \mathcal{T}_{i_1 i_2 \dots i_{m-1}} \subset \dots \subset \mathcal{T}_0$. The nested sequence of intervals $\mathcal{T} = \mathcal{T}_{i_1 i_2 \dots i_k}$ characterizes the trajectory of the critical value $w(t) = h(t,0)$ under consecutive compositions. When $t \in \mathcal{T}_{i_1 i_2 \dots i_k}$, $w(t)$ moves through a domain of some branch $F_k = f_{i_k} \circ \dots \circ f_{i_1} \circ g_0$. The respective parabolic branch is $h_k = F_k \circ h$. When t varies from one end of $\mathcal{T}_{i_1 i_2 \dots i_k}$ to another, the height of the parabolic branch h_k varies from zero to one.

(b) *Partitions ξ_l.*
For each $t \in \mathcal{T} = \mathcal{T}_{i_1 i_2 \dots i_m}$ a sequence of partitions $\xi_l(t)$ of the interval I (up to a finite number of common end-points) is given, $z_0 \leq l \leq n$. Elements of ξ_l vary continuously with t.

$$(\xi_l) \qquad I = \bigcup_k \Delta f_k \cup \bigcup_{i,m} \delta_i^{-m} \cup \delta_l \,. \tag{193}$$

In the above partitions elements Δf_k are domains of good branches f_k. Elements δ_l are *central domains.* They contain zero, and satisfy the following conditions

$$(1-\epsilon)\delta_0 R^{-sl} < \mid \delta_l \mid < (1+\epsilon)\delta_0 R^{-sl} \,. \tag{194}$$

Elements δ_i^{-m} are preimages of central domains δ_i. They have indices $0 \leq i < l$ and $m > 0$.
δ_i^{-m} are mapped onto δ_i by diffeomorphisms G_i satisfying some expansion and distortion properties formulated below. δ_i and δ_i^{-m} are called *undecided domains.*

(c) *Canonical partitions.*
Let us consider a partition ξ_{l+1}. Its central element containing zero is δ_{l+1}. We call the restriction of ξ_{l+1} onto δ_l *canonical partition* of δ_l. The branches Δf_k, which are elements of $\xi_{l+1} \mid \delta_l$, are called *canonical branches.* The preimages δ_j^{-m}, which are elements of $\xi_{l+1} \mid \delta_l$, are called *canonical undecided domains.* Respectively elements Δf_k and δ_j^{-m} of ξ_{l+1}, which are outside δ_l, are called noncanonical.

(d) *Boundary points of δ_l and points y_l.*
Let $\delta_l = (x_{l1}, x_{l2})$. By construction $x_{l1}, x_{l2} = h^{-1} y_l$, where y_l is a common boundary end point of two adjacent elements $\Delta\phi_k$ of ζ_l. So we have a sequence of points on y-axis: $y_1 < y_2 < \dots < y_l < w(t)$, which determine intervals δ_l, $z_0 \leq l \leq n$.

(e) *Partitions ζ_l.*
Another sequence of partitions ζ_l closely related to ξ_l is considered on y-axis. ζ_l is a partition of some interval $[y_l, z_l]$ containing critical value. The restriction of ζ_l onto $[y_l, y_{l+1}]$ is the canonical partition on y-axis. Elements of ζ_l are domains of diffeomorphisms ϕ_k and $\mathcal{G}_i$, where ϕ_k map $\Delta\phi_k$ onto I, $\mathcal{G}_i$ map δ_i^{-m} onto central domains δ_i. These diffeomorphisms

satisfy certain expansion and distortion estimates formulated below. The map ϕ_k, such that $w(t) \in \Delta\phi_k$, has special notation F_k. Elements of canonical partition at step l are critical preimages of the respective elements of ζ_l.

(f) *Enlargements.*
For every undecided domain $\delta_i^{-s} = \mathcal{G}^{-1}\delta_i$ (or $\delta_i^{-s} = G^{-1}\delta_i$) the respective map $\mathcal{G}^{-1}$ (or G^{-1}) is defined on δ_{i-1} and has distortion less than R^{-s}. δ_{i-1}^{-s} is called *enlargement* of δ_i^{-s}. For δ^{-s} their enlargements δ_{-1}^{-s} coincide with $\hat{\delta}^{-s}$ defined in Section 4. Any two enlargements are either disjoint or one of them contains another one. In particular, if δ_i^{-s} is an element of a canonical partition $\xi_{k+1}|\delta_k$, then its enlargement is a subset of $\delta_k \setminus \delta_{k+1}$.

(g) *Collars.*
In order to control the measure of undecided domains, we assign to every domain δ_l a union of good branches Δf_k - elements of canonical partition $\xi_l \mid \delta_{l-1}$. We denote it $S(\delta_l)$ and call *collar* of δ_l.

5.2. Basic operations.

(a) Let $f : \Delta f \to I$ be a diffeomorphism, and let ξ_k be a partition of I. Then $f^{-1}(\xi_k)$ is a partition of Δf. That procedure is called *monotone pull-back.*
(b) Let ζ_k be a partition on y-axis, $w(t) = h(t, 0)$ - critical value of the original parabola. Admissible parameter values are $t \in \mathcal{T}$, such that $w(t)$ belongs to domains $\Delta\phi_i$ of good branches of the partition ζ_k. When we fix a domain $\Delta\phi_*$, which contains the critical value, it determines the next parameter interval $\mathcal{T}_* \subset \mathcal{T}$, and is called *parameter choice.*
(c) Let $w(t) \in \Delta\phi_*$, where $\Delta\phi_*$ is an element of the partition ζ_k. Then a partition $h^{-1}\zeta_k$ is defined, which has in the middle a parabolic branch $h_* = \phi_* \circ h$. Other elements of $h^{-1}\zeta_k$ are diffeomorphic preimages of elements $E \in \zeta_k$ covered by the image of h. They are of the form $E_{1,2} = h^{-1}(E)$. That procedure is called *critical pull-back.*
(d) Let δ_i^{-s} be an undecided domain - element of some partition ξ_k and let $G_i : \delta_i^{-s} \to \delta_i$ be a respective diffeomorphism. Let $\xi_{i+1} \mid \delta_i$ be the canonical partition of δ_i. Then $G_i^{-1}(\xi_{i+1} \mid \delta_i)$ is a partition of δ_i^{-s}. We call it *filling-in.*

More generally we define a *k-step filling-in* as a pull-back $G_i^{-1}(\xi_{i+k} \mid \delta_i)$ onto δ_i^{-s}.

5.3. Properly sized properties. Let $\mathcal{T}$ be an admissible parameter interval under consideration at step n. Let $h_n = F_n \circ h$ be the respective parabolic branch. When t varies within $\mathcal{T}$ the critical value $w(t) = h(t, 0)$ moves through ΔF_n. Let t_0 be the parameter value corresponding to the full height of h_n.

ΔF_n is called *properly sized,* if

$$\mid \Delta F_n(t_0) \mid < \bar{k}\delta_0^2 R^{-(r_2+3)sn} . \tag{195}$$

Let $E = [a, b]$ be an element of some partition on y-axis. Assume E is contained in the image of the parabolic branch h, and assume b lies between a and w.
Definition.

$$H(E) = \mid a - w \mid .$$

A domain of monotone branch $\Delta\phi = [a, b]$ is called *n-properly sized* if

$$\frac{1}{2} \times R^{-r_1 sn} > \frac{|\Delta\phi|}{H(\Delta\phi)} > 8 \times R^{-sr_2 n}. \tag{196}$$

Before formulating the respective property for undecided domains, we notice that indices i of undecided domains δ_i^{-m} encountered in our construction at step n, satisfy (252) below. That leaves a gap between upper and lower bounds in the following estimate. The gap grows with n. A domain $\delta_i^{-m} = [a, b]$ is called *n-properly sized* if

$$\frac{R^{-s}}{2} \times R^{-xsn} > \frac{|\delta_i^{-m}|}{H(\delta_i^{-m})} > 8 \times R^{r_2 si} R^{-sr_2 n}. \tag{197}$$

If only one part of the above inequalities hold, the respective element is called *n-upper properly sized* or respectively *n-lower properly sized.*

5.4. Inductive assumptions. Let r_1, r_2, R, δ_0, s, x, $\chi_0 = 2R^{-s}$, $K_0 = \frac{1}{2}R^{-1}(\exp -l_0)$ be constants defined in the preliminary construction.

We assume that for all $t \in \mathcal{T}$ partitions ξ_n, ζ_n and related objects and structures are defined, vary continuously with t, and satisfy the following estimates:

(a) Domains Δf of the canonical partition $\xi_n|\delta_{n-1}$ satisfy

$$\frac{1}{2} \times \delta_0 R^{-(r_1+1)sn} > |\Delta f| > 2 \times \delta_0 R^{-(r_2+1)sn}. \tag{198}$$

(b) Domains δ_i^{-m} of the canonical partition $\xi_n|\delta_{n-1}$ satisfy

$$\frac{1}{2} \times \delta_0 R^{-(1+x)sn} > |\delta_i^{-m}| > 2 \times \delta_0 R^{sr_2 i} R^{-(r_2+1)sn}. \tag{199}$$

(c) Non-canonical undecided domains δ_i^{-m} of ξ_n satisfy

$$\delta_0 R^{-sn} > |\delta_i^{-m}| > K_0 \delta_0 R^{r_2 si} R^{-s(r_2+1)n}. \tag{200}$$

(d) Non-canonical branches f of ξ_n constructed at step n satisfy

$$\delta_0 R^{-sn} > |\Delta f| > K_0 \delta_0 R^{-s(r_2+1)n}. \tag{201}$$

(e) Noncanonical branches f such that $\Delta f \subset \delta_i \setminus \delta_{i+1}$, $i < n$ satisfy

$$|\Delta f| < \delta_0 R^{-s(r_1+1)(i+1)}. \tag{202}$$

(f) Central domains δ_n satisfy

$$\delta_0 R^{-sn}(1 - \epsilon_n) < |\delta_n| < \delta_0 R^{-sn}(1 + \epsilon_n) \tag{203}$$

where ϵ_n decrease exponentially with n.

(g) Let $S(\delta_i) \subset \delta_{i-1}$ be the collar corresponding to δ_i, $1 \le i \le n$. Then

$$\frac{mes S(\delta_i)}{mes \delta_{i-1}} > 1 - \frac{3}{2} R^{-s}. \tag{204}$$

(h) Let $\mathcal{C}_n$ be the union of undecided domains - elements of ξ_n, then

$$mes\ \mathcal{C}_n < \chi_0^n mes\ \mathcal{C}_0. \tag{205}$$

(i) Good maps f_n and ϕ_n constructed at step n satisfy

$$\Theta(f_n), \Theta(\phi_n) < l_{[\frac{n}{4}]+1}. \tag{206}$$

(j) Maps $G_i, \mathcal{G}_i : \delta_i^{-m} \to \delta_i$ constructed at step n satisfy

(207) $$\Theta(G_i), \Theta(\mathcal{G}_i) < l_{[\frac{n}{4}]} R^{-(sxi+s)} .$$

(k) Maps f, G defined on elements of ξ_n satisfy

(208) $$\left| \frac{f_t}{f_x}(x(t), t) \right|, \left| \frac{G_t}{G_x}(x(t), t) \right| < \frac{2}{\bar{k}} \mid \delta_n \mid^{-1} .$$

(l) Maps $\phi, \mathcal{G}$ defined on elements of ζ_n satisfy

(209) $$\left| \frac{\phi_t}{\phi_x}(x(t), t) \right|, \left| \frac{\mathcal{G}_t}{\mathcal{G}_x}(x(t), t) \right| < \sigma_n$$

where σ_n are small constants defined in the proof.

(m) Element ΔF_n, which contains the critical value is properly sized at step n in the sense of (195), which means

(210) $$\mid \Delta F_n(t_0) \mid < \bar{k} \delta_0^2 R^{-(r_2+3)sn} .$$

5.5. Algorithm of step $n+1$. We outline the main substeps of step $n+1$. Detailed description and justification are given below in the proof of Theorem 2. We assume objects of step n satisfy inductive assumptions. They vary continuously with $t \in \mathcal{T}_n = \mathcal{T}_{i_1 i_2 \ldots i_l}$. The value $t_0 \in \mathcal{T}_n$ corresponds to the full height of the central branch $h_n = F_n \circ h$.

(a) *Substep 1 - construction of $\zeta_{n+1,*}$ and parameter choice.*
If

(211) $$\mid \Delta F_n(t_0) \mid < \bar{k} \delta_0^2 R^{-(r_2+3)s(n+1)}$$

then keep the same $\mathcal{T}_{n+1} = \mathcal{T}_n$, keep the same t_0, set $F_{n+1} = F_n$, set $\zeta_{n+1,1} = \zeta_n$ and go to Substep 2. Otherwise, define

$$\zeta^1_{n+1,*} = F_n^{-1} \xi_{[xn]} .$$

It is a partition of ΔF_n. Simultaneously construct

$$\hat{\zeta}^1_{n+1,*} = F_n^{-1} \xi_{[xn]-1} .$$

As good branches of $\xi_{[xn]-1}$ are also good branches of $\xi_{[xn]}$ domains $\Delta\phi$ which belong to $\hat{\zeta}^1_{n+1,*}$ also belong to $\zeta^1_{n+1,*}$. Consider the movement of $w(t)$ across $\hat{\zeta}^1_{n+1,*}$ and do parameter choice. Parameter choice fixes $F^1_{n+1} = f_{i_1} \circ F_n$ where f_{i_1} is a good branch of $\xi_{[xn]}$. Repeat such steps k_{n+1} times until $F^{k_{n+1}}_{n+1}$ satisfies (211). (212) below shows that $k_{n+1} \leq m_0 = [(r_2+3)s] + 2$.

Set $F_{n+1} = F^{k_{n+1}}_{n+1}$. A new parameter interval $\mathcal{T} = \mathcal{T}_{n+1}$ corresponds to the movement of $w(t)$ through ΔF_{n+1}. Fix the value $t_0 \in \mathcal{T}$, such that $h(t_0, 0)$ covers the whole interval $\Delta F_{n+1}(t_0)$. Let $\zeta_{n+1,*}$ be a partition of ΔF_n which elements were obtained by the above pull-backs. Define a partition $\zeta_{n+1,1}$ which coincides with ζ_n outside ΔF_n and with $\zeta_{n+1,*}$ on ΔF_n. Then go to Substep 2.

(b) *Substep 2 - construction of ζ_{n+1}.*
Consider n-located elements of $\zeta_{n+1,1}$. If a domain $\Delta\phi$ is not n-upper properly sized, then pull-back $\xi_{[nx]}$ by ϕ^{-1} onto $\Delta\phi$.

If an undecided domain δ_i^{-s} is n-properly sized, then do a three-step filling-in of that domain. If an undecided domain δ_i^{-s} is not n-upper properly sized, then do an l-step filling-in of that domain, where $3 \leq l \leq \frac{3}{2}(n+1)x$, is specified in the proof. The above pull-backs and fillings create a new partition $\zeta_{n+1,2} \succ \zeta_{n+1,1}$. Apply the above procedure to all oversized elements of $\zeta_{n+1,2}$ and so on. Repeat these steps until all n-located elements are n properly sized. Define y_{n+1}. Define canonical partition ζ_{n+1} as the resulting partition restricted to $[y_n, y_{n+1}]$. Then go to Substep 3.

(c) *Substep 3 - critical pull-back of ζ_{n+1} and canonical partition .* Construct the critical pull-back $\eta_{n+1} = h^{-1}\zeta_{n+1}|[y_n, y_{n+1}]$. Define δ_{n+1}. Define canonical partition $\xi_{n+1} \mid \delta_n \setminus \delta_{n+1}$. Define collar $S(\delta_{n+1})$. Then go to Substep 4.

(d) *Substep 4 - partition ξ_{n+1}.* For all undecided domains δ_i^{-s}, which are outside δ_n, do one-step filling-in. Consider oversized monotone branches $\Delta f \subset \delta_i \setminus \delta_{i+1}$ which appeared after filling-in. Pull-back $\xi_{[nx]}$ on such branches consecutively until they satisfy (202). That defines ξ_{n+1} outside δ_n.

5.6. Main technical result. In the last sections we prove

THEOREM 2. *If inductive assumptions are satisfied at steps $z_0 \leq k \leq n$ then applying algorithm of step $n+1$ we get partitions ζ_{n+1}, ξ_{n+1}, satisfying inductive assumptions of step $n+1$.*

6. Estimates of sizes at step $n+1$

6.1. Parameter choice. Assume ΔF_n is too large, so that it does not satisfy (211). Then we define the pull-back $\zeta_{n+1,*} = F_n^{-1}\xi_{[xn]}$. After that we do parameter choice.

According to (209) speeds of elements of ζ_n are less than some small ϵ_n. Then Lemma 2 implies that intervals of parameter $\mathcal{T} = \mathcal{T}_{n+1}$ are uniquely defined. Each interval $\mathcal{T}$ corresponds to the movement of $w(t)$ through some element $\Delta F_{n+1} \subset \Delta F_n$ of $\zeta_{n+1,*}$.

As ΔF_n is properly sized with respect to step n, it satisfies (210). So in order to satisfy properly sized property, it is enough to make new elements $R^{-(r_2+3)s}$ times smaller, than ΔF_n. As expansion of good branches is greater than R, we get using inductive assumption about distortions

$$| \Delta F_{n+1} | < \exp l_\infty R^{-m_0} | \Delta F_n | . \tag{212}$$

Then (127) implies that ΔF_{n+1} is properly sized.

REMARK 9. *Derivatives of good branches constructed at later steps grow exponentially. When $h_k(t,0)$ belongs to such elements, the size of $\Delta(F_{k+1})$ will be strongly reduced comparatively to the size of $\Delta(F_k)$. Then we skip many steps before the next parameter choice.*

6.2. Lower properly sized property for elements of $\zeta_{n+1,*}$. Let $\mathcal{T}$ be a given parameter interval $\mathcal{T}$ and $t_0 \in \mathcal{T}$ parameter value corresponding to the maximal height of the parabolic branch. At the next steps we subdivide $\mathcal{T}$ into smaller intervals $\mathcal{T}'$ and fix respective values of parameter $t_0' \in \mathcal{T}'$, $t_0' < t_0$. When t decreases $w(t)$ moves down with the unit speed 1 and elements on y-axis are moving with small speeds. Thus the distances H in (196), (197) can only decrease and lower estimates improve. Therefore we have only to verify lower estimates for elements created at the last step. Next notice that lower bounds increase when location grows. By construction locations m of elements of $\zeta_{n+1,*} = F_n^{-1}\xi_{[xn]}$ satisfy $m \geq \frac{r_2+3}{2}n$, because ΔF_n was properly sized at step n. Therefore it is enough to check that all new elements are $\frac{r_2+3}{2}n$ lower properly sized. As $\frac{r_2+3}{2} > 1$ lower properly sized estimates at step $\frac{r_2+3}{2}n$ are weaker than respective estimates at step $n+1$. Below in Proposition 6 we prove that $n+1$ lower properly sized estimates are satisfied. Respectively, elements of $\zeta_{n+1,*}$ are lower properly sized.

6.3. Monotone pull-back and resulting elements of ζ_{n+1}.

(a) Next we go to the Substep 2 of our algorithm. We consider elements of partition $\zeta_{n+1,1}$ on y-axis which are located above y_n. Notice that y_n is n-located and close to the boundary between $n-1$ and n-located elements.

Let us consider an element $\Delta\phi = [a,b], a < b < w(t_0)$, of $\zeta_{n+1,1}$. Assume $\Delta\phi$ is $n+1$-oversized, which means

$$\frac{\mid \Delta\phi \mid}{H(\Delta\phi)} > \frac{1}{2}R^{-r_1 s(n+1)}. \tag{213}$$

We pull-back $\xi_{[nx]}$ onto $\Delta\phi$ and get a new element $\Delta\phi' \subset \Delta\phi$. Then $H(\Delta\phi') \leq H(\Delta\phi)$ and respectively

$$\frac{\mid \Delta\phi' \mid}{H(\Delta\phi')} \geq \frac{\mid \Delta\phi' \mid}{H(\Delta\phi)}. \tag{214}$$

Let us check, that elements $\Delta\phi'$ resulting out of that pull-back are $n+1$-lower properly sized. According to (201) domains of good branches in $\xi_{[nx]}$ are greater than $K_0\delta_0 R^{-s(r_2+1)xn}$. On the other hand distortions of ϕ do not exceed l_n. Therefore

$$\frac{\mid \Delta\phi' \mid}{\mid \Delta\phi \mid} \geq K_0\delta_0 R^{-s(r_2+1)xn}\exp(-l_n). \tag{215}$$

We use (213) - (215) and get

$$\frac{\mid \Delta\phi' \mid}{H(\Delta\phi')} > \frac{K_0}{2}\exp(-l_n)\delta_0 R^{-s[(r_2+1)xn+r_1(n+1)]}. \tag{216}$$

Then we get the following condition sufficient for lower properly sized property:

$$\frac{K_0}{2}\exp(-l_\infty)\delta_0 R^{-s[(r_2+1)xn+r_1(n+1)]} > 8R^{-sr_2(n+1)} \tag{217}$$

(217) holds, if

$$\frac{K_0}{16}R^{s(r_2-r_1)}\exp(-l_\infty)\delta_0 > R^{-sn(r_2-r_1-(r_2+1)x)}. \tag{218}$$

The minimal n that we encounter in our construction is $n = z_0$. So we get lower properly sized elements if

$$\frac{K_0}{16}R^{s(r_2-r_1)}\exp(-l_\infty)\delta_0 > R^{-sz_0(r_2-r_1-(r_2+1)x)}. \tag{219}$$

That follows from (119), (123) and the definition of K_0.

(b) If some of the new elements are still oversized, then we pull-back $\xi_{[nx]}$ again onto these elements, and so on. We check how many times do we need to pull-back in order to make monotone branches properly sized. Assume we construct a new element $\Delta\phi' \subset \Delta\phi$. As at step $n+1$ we are considering n-located elements $\Delta\phi$ and $\Delta\phi'$, we get

$$\frac{H(\Delta\phi')}{H(\Delta\phi)} \geq R^{-s}(1-\epsilon)\,. \tag{220}$$

Respectively

$$\frac{|\,\Delta\phi'\,|}{H(\Delta\phi')} \leq \frac{|\,\Delta\phi'\,|}{H(\Delta\phi)} R^{s}(1+\epsilon)\,. \tag{221}$$

Notice that

$$|\,\Delta\phi\,| < H(\Delta\phi)\,.$$

As derivatives of good branches are greater than R, sizes of domains decrease after l compositions by a factor greater than R^l. Taking into account distortion, we get after l compositions

$$\frac{|\,\Delta\phi'\,|}{|\,\Delta\phi\,|} \leq \exp l_\infty R^{-l}\,. \tag{222}$$

The upper properly sized estimate at step $n+1$ is

$$\frac{|\,\Delta\phi'\,|}{H(\Delta\phi')} \leq \frac{R^{-s}}{2} R^{-r_1 s(n+1)}\,. \tag{223}$$

According to (127)

$$2\exp l_\infty < R^s\,. \tag{224}$$

Combining (221) - (224) we get a sufficient condition

$$R^{2s-l} < R^{-sr_1(n+1)}\,. \tag{225}$$

Hence, in order to get properly sized elements, it is enough to take l compositions, where l satisfies $l \geq sr_1(n+1)+2s$. So after no more than

$$l = [sr_1(n+1)+2s]+1 \tag{226}$$

compositions all n-located elements $\Delta\phi'$ become properly sized.

(c) When we pull-back $\xi_{[nx]}$ onto $\Delta\phi$, we also create undecided domains. Estimates for undecided domains are similar to those for domains of good branches. However, using the fact that undecided domains in ξ_m decrease exponentially with m, we show that their pull-backs are not only lower properly sized, but also upper properly sized.

First we prove some general properties of a filling-in operation. Assume we are doing a one-step filling-in of some domain δ_i^{-k} located on y-axis. Let E be an element obtained by that filling-in. Notice that $H(E) < H(\delta_i^{-k})$.

PROPOSITION 5. *For a given $c > 0$ let*

$$\frac{|\,\delta_i^{-k}\,|}{H(\delta_i^{-k})} > cR^{sr_2 i}\,. \tag{227}$$

Then elements δ_{i+1}^{-k}, δ_j^{-l} and Δf obtained by the one-step filling-in of δ_i^{-k} satisfy

$$\frac{|\delta_j^{-l}|}{H(\delta_j^{-l})} > 2(1-\epsilon)cR^{-(r_2+1)s}R^{sr_2 j} \tag{228}$$

$$\frac{|\Delta f|}{H(\Delta f)} > 2(1-\epsilon)cR^{-(r_2+1)s} \tag{229}$$

$$\frac{|\delta_{i+1}^{-k}|}{H(\delta_{i+1}^{-k})} > (1-\epsilon)cR^{-(r_2+1)s}R^{sr_2(i+1)}\,. \tag{230}$$

PROOF. The canonical partition of δ_i consists of the following elements: the next central domain δ_{i+1}, secondary preimages δ_j^{-q} and branches Δf. Respectively $\delta_i^{-k} = G_i^{-1}\delta_i$ is a union of $G_i^{-1}\delta_{i+1}$, $G_i^{-1}\delta_j^{-q}$, and $G_i^{-1}\Delta f$.

By inductive assumption (199) of step $i+1$ we have

$$|\delta_j^{-q}| > 2\times\delta_0 R^{sr_2 j}R^{-(1+r_2)s(i+1)}\,. \tag{231}$$

As δ_i satisfies

$$\delta_0 R^{-si}(1-\epsilon_i) < |\delta_i| < \delta_0 R^{-si}(1+\epsilon_i)\,. \tag{232}$$

we can rewrite (231) as

$$|\delta_j^{-q}| > |\delta_i|\, 2(1-\epsilon_i)\times R^{sr_2 j}R^{-sr_2 i}R^{-s(r_2+1)}\,. \tag{233}$$

Distortions of G_i^{-1} are small by inductive assumptions. Therefore preimages $G_i^{-1}\delta_j^{-q}$ satisfy

$$|G_i^{-1}\delta_j^{-q}| > |\delta_i^{-k}|\, 2(1-\epsilon)\times R^{sr_2 j}R^{-sr_2 i}R^{-s(r_2+1)}\,. \tag{234}$$

Then (228) follows from (234) and (227).

Similar estimates prove (229). Namely, for preimages $G_i^{-1}\Delta f$ we use (198) at step $i+1$, instead of (199), and after that repeat the above arguments.

Finally, (230) follows from (227) and (232). □

REMARK 10. *We got lower properly sized estimates for central subdomains which are two times worse than lower properly sized estimates for secondary preimages. Therefore if lower bound for central subdomain is satisfied, then it is also satisfied for secondary preimages. On the other hand, upper properly sized estimates do not involve indices of undecided domains. As central subdomains are much larger than secondary preimages we get a similar property for upper bounds.*

(d) We use Proposition 5 in the proof of

PROPOSITION 6. *After pull-back of $\xi_{[nx]}$ every undecided domain δ_i^{-r} is properly sized. Moreover, undecided domains that we obtain after a 3-step filling-in of δ_i^{-r}, are lower properly sized with respect to their locations.*

PROOF. Let δ_i^{-m} be an element of $\zeta_{n+1,1}$. Using (200), with $[nx]$ instead of n, and distortion assumptions we get

$$K_0(\exp -l_\infty)\delta_0 R^{r_2 si} R^{-s(r_2+1)xn} < \frac{\mid \delta_i^{-m} \mid}{\mid \Delta\phi \mid} < (\exp l_\infty)\delta_0 R^{-s[xn]}. \tag{235}$$

The lower properly sized property is satisfied if

$$\frac{\mid \delta_i^{-m} \mid}{H} > 8 \times R^{sr_2 i} R^{-sr_2(n+1)} \tag{236}$$

(236) follows from (235) by repeating arguments of (213) - (219).

Some undecided domains created by the above pull-back may have locations greater than n. Then they are considered at later steps of induction. If such a domain is properly sized at a respective step, then we do a 3-step filling-in. Therefore we need to check that after an additional 3-step filling-in all undecided domains remain lower properly sized. As such domains will be considered at steps $n+2$ and larger, the minimal location to be considered is $n+1$.

Let us apply Proposition 5 to the filling-in of δ_i^{-m}. From (235) it follows that we can take

$$c = \frac{K_0}{2}(\exp -l_\infty)\delta_0 R^{-s(r_2+1)xn}.$$

The worst estimate corresponds to the central subdomain. Applying (230) three times we get as above

(237)
$$\frac{\mid \delta_{i+3}^{-m} \mid}{H} > \frac{K_0}{2}(1-\epsilon)(\exp -l_\infty)\delta_0 R^{sr_2(i+3)} R^{-(r_2+1)sxn} R^{-3(r_2+1)s} R^{-(r_1 s(n+1)}.$$

Lower properly sized estimate for an undecided domain δ_{i+3}^{-l} at step $n+2$ holds, if

$$\frac{\mid \delta_{i+3}^{-l} \mid}{H} > 8 \times R^{sr_2(i+3)} R^{-r_2 s(n+2)} \tag{238}$$

and similar condition for monotone branches.

Therefore, a sufficient condition for lower properly sized property is

$$\frac{K_0}{2} R^{-(r_1 s(n+1)} R^{-3(r_2+1)s}(1-\epsilon)(\exp -l_\infty)\delta_0 R^{-(r_2+1)sxn} > 8 \times R^{-r_2 s(n+2)}. \tag{239}$$

The estimate improves when n grows. The most restrictive estimate below corresponds to the minimal $n = z_0$.

$$\frac{K_0}{16} R^{-s(2r_2+3)}\delta_0 \exp(-l_\infty)(1-\epsilon) > R^{-s[r_2-r_1-(r_2+1)x]z_0}. \tag{240}$$

Condition (240) follows from the definition of K_0 and inductive assumption (123).

Finally, we consider the upper estimates. The upper properly sized estimate for δ_i^{-m} will be satisfied, if

$$\frac{\mid \delta_i^{-m} \mid}{H(\delta_i^{-m})} < \frac{R^{-s}}{2} R^{-sx(n+1)}$$

From the rhs of (235) we get similarly to (221) that the required upper estimate, holds if

(241) $$R^{2s}(\exp l_\infty)\delta_0 R^{-s[xn]} < 1/2R^{-s(n+1)x}\,.$$

Inequality (241) holds because of (124) and (127). That proves Proposition 6. □

The above estimates can be summarized in a following way.

PROPOSITION 7. *Let $\Delta\phi$ be an n-located oversized element of $\zeta_{n+1,1}$. Then by using a sequence of consecutive pull-backs of $\xi_{[nx]}$ we make all monotone branches properly sized. All undecided domains constructed in that way are properly sized and they admit an additional 3- step filling-in. The number of pull-backs l applied until a particular monotone branch $\Delta\phi' \subset \Delta\phi$ becomes properly sized, satisfies*

(242) $$1 \le l \le [sr_1(n+1)+s]+2\,.$$

6.4. Filling-in and resulting elements of ζ_{n+1}.

(a) We turn to n-located undecided domains δ_i^{-m} - elements of $\zeta_{n+1,1}$. Let us first consider an element δ_i^{-m}, which is already $n+1$-properly sized. δ_i^{-m} was constructed at some previous step $k \le n$ of induction. In that case inductive estimates of step k prove that δ_i^{-m} admits an extra 3-step filling-in, such that all elements obtained by that filling-in are $n+1$-lower properly sized. So we do that filling and get $n+1$ properly sized elements.

(b) Next we consider the main case, when δ_i^{-m} is $n+1$- oversized. Respectively

(243) $$\frac{|\,\delta_i^{-m}\,|}{H} > 1/2R^{-xs(n+1)}\,.$$

By construction any δ_i^{-m} is separated from the critical value by its enlargement δ_{i-1}^{-m}. That implies

(244) $$\frac{|\,\delta_i^{-m}\,|}{H} > 2(1+\epsilon)R^{-s}\,.$$

We take maximal $2 \le j \le (n+1)x$, such that

(245) $$\frac{|\,\delta_i^{-m}\,|}{H} < 2(1+\epsilon)R^{-s(j-1)}\,.$$

Then

(246) $$2(1+\epsilon)R^{-s(j-1)} > \frac{|\,\delta_i^{-m}\,|}{H} > 1/2R^{-sj}\,.$$

In this situation we do an l-step filling-in of δ_i^{-m} where

(247) $$l = \left[\frac{3}{2}(n+1)x - j\right] + 2\,.$$

We prove below in Proposition 9 that with such choice of l new undecided domains, which appear after the filling-in, are properly sized.

(c) For several reasons we need to estimate from above indices i of undecided domains δ_i^{-m} depending on their location. In particular, the following estimates will be used several times. If we are doing parameter choice at step $n+1$, then $\Delta F_n(t_0)$ was properly sized at step n, but is not at step $n+1$. So we have

(248) $$\bar{k}\delta_0^2 R^{-(r_2+3)sn} >| \Delta F_n(t_0) |> \bar{k}\delta_0^2 R^{-(r_2+3)s(n+1)} .$$

From the lhs of (248) locations p of new elements, which partition ΔF_n, satisfy

(249) $$p \geq \frac{r_2+3}{2} n .$$

(249) can be written as

(250) $$n \leq \frac{2p}{(r_2+3)} .$$

At step n we pull-back $\xi_{[nx]}$. As the maximal index i for domains δ_i^{-m} in $\xi_{[nx]}$ equals $[nx]$, we get the following

LEMMA 3. *Let δ_i^{-m} be a p-located element of $\zeta_{n+1,*}$. Then*

(251) $$i \leq \frac{2px}{r_2+3} .$$

For canonical partitions we get the following upper bound on indices

PROPOSITION 8. *Elements δ_q^{-m} of canonical partition ζ_{n+1} on y-axis and elements of canonical partition $\xi_{n+1}|\delta_n$ have indices q satisfying*

(252) $$q < \frac{5}{2} nx - 3 .$$

From (121) it follows that

(253) $$\frac{5}{2} x < \frac{1}{3} .$$

Respectively we get

(254) $$q < \frac{n}{3} - 3 .$$

Proof of Proposition 8.

(i) In our construction there are "primary" domains δ_i^{-m} constructed either as elements of some ζ_{k*} when we are doing parameter choice, or in a result of a monotone pull-back of some ξ_{kx}. Other domains are constructed by filling-in of "primary" domains.
Let δ_i^{-m} be a primary domain constructed at some step k as an element of ζ_{k*}. Note that we are considering n located elements. In that case we get from (251)

(255) $$i \leq \frac{2nx}{r_2+3} .$$

(ii) Next let us consider a primary domain δ_i^{-m} constructed when at some step $k \leq n$ we pulled-back $\xi_{[kx]}$ onto an k -oversized element $\Delta\phi$. Notice that δ_i^{-m} is separated from the critical value at least by the distance between δ_i^{-m} and the respective end point of $\Delta\phi$. The image $\mathcal{G}_i\delta_i^{-m} = \delta_i^{-r} \subset I$ is separated from the end point of I by some distance d. We get the worst estimate by considering $\delta_i^{-r} \subset \delta^{-r}$, where δ^{-r} was constructed when we did a last subdivision of some monotone branch adjacent to the end point of I. Taking into account that δ_i is located in the middle of I we get

(256) $$\frac{\mid \delta_i^{-r} \mid}{d} < 2(1+\epsilon)\exp(l_0) \mid \delta_i \mid .$$

Taking into account distortion we get after pull-back onto y-axis

(257) $$\frac{\mid \delta_i^{-m} \mid}{H} < 2(1+\epsilon)\exp(l_0 + l_\infty) \mid \delta_i \mid .$$

As $\mid \delta_i \mid < (1+\epsilon)\delta_0 R^{-si}$ we get an estimate

(258) $$\frac{\mid \delta_i^{-m} \mid}{H} < 2(1+\epsilon)\exp(l_0 + l_\infty)\delta_0 R^{-si} .$$

From (127) and (124)

(259) $$\delta_0 2(1+\epsilon)\exp(l_0 + l_\infty) < R^{-2} .$$

As $s < \frac{1}{3}$ (258) implies that j in (246) satisfies

(260) $$j > i + 6 .$$

We can rewrite it as

(261) $$i < j - 6 < (n+1)x - 6 .$$

(iii) Let us consider indices after filling-in. Let us assume δ_i^{-m} is a non-central undecided subdomain constructed by a one-step filling-in of δ_p^{-r}, where p satisfies (255) or (261). Let n be the location of δ_i^{-m} and let k be the location of δ_p^{-r}. Then $n \geq k$. Notice that for $p < z_0$ we have by construction $i \leq 3$. Let us consider $p \geq z_0$. From (121) and (122) it follows that

(262) $$z_0 > 40 .$$

Applying inductively (252) and using (254) we get

(263) $$i < \frac{p}{3} < p - 26 .$$

Thus for any secondary preimage constructed by a filling-in index is smaller and (252) is satisfied.

(iv) Now we consider the case when δ_p^{-r} satisfies (255) or (260) and $\delta_i^{-m} = \delta_{p+q}^{-r}$ is the central subdomain. After an l-step filling-in indices increase by l. With l defined by (247) we get for i satisfying (255)

(264) $$i + l \leq \frac{2nx}{r_2 + 3} + l .$$

As $n \geq z_0$ and $z_0 x > 4$ we get in this case

(265) $$i + l \leq \frac{2nx}{r_2 + 3} + \frac{3}{2}(n+1)x .$$

If i satisfies (260) then using (247) we get

(266) $$i + l < j - 6 + l < \frac{5}{2}(n+1)x - 6 .$$

which is greater than (265). That proves Proposition 8 for domains constructed at step $n+1$.

(v) It remains to consider indices of domains which were constructed by a filling-in at a certain step, but are considered at a later step because of their location. Let δ_i^{-m} be a primary domain constructed at some step $k \leq n$ and considered at step $n+1$. By enlargement property it is small comparatively to its distance from the critical value. Then it is close to y_{n+1} and elements constructed by the filling-in are considered at step $n+2$. In that case we are doing a 3-step filling-in. That increases index of the central subdomain by 3 and we add 3 to the rhs of (266). Then (252) is satisfied with respect to the step $n+2$. The next proposition shows that new elements are properly sized with respect to the step $n+2$. That finishes the proof of Proposition 8. □

(d) Next we prove

PROPOSITION 9. *After an l-step filling-in, where l satisfies (247) we get $n+1$ properly sized undecided domains. Moreover they admit an additional 3-step filling-in.*

PROOF. (i) *Lower bounds.*
It is enough to check lower bounds at step $n+2$ after a 3-step filling-in. With j defined in (246) and l satisfying (247) we get for the central subdomain

$$\frac{|\delta_{i+l}^{-m}|}{H} > \frac{1}{2}R^{-s[\frac{3}{2}x(n+1)+2]}. \tag{267}$$

The lower bound (197) at step $n+2$ is satisfied if

$$\frac{1}{2}R^{-s[\frac{3}{2}x(n+1)+2]} > 8R^{-sr_2(n+2-q)} \tag{268}$$

where q satisfies (252). We use (133), (120) and (121) and get that (268) is satisfied for $n \geq 1$. That proves lower bound.
We checked lower bounds for central subdomains. A corollary from Proposition 5 implies that lower bounds for secondary undecided domains are also satisfied.

REMARK 11. *For measure estimates we require $l \geq 3$. From (246) and (247) we get*

$$l \geq \left[\frac{1}{2}(n+1)x\right]$$

Then (122) implies $l \geq 3$.

(ii) *Upper bounds.*
We prove upper bounds at step $n+1$. As we are doing a 3-step filling-in at the next step, upper bounds of step $n+2$ follow. With j from (246) and $l = \left[\frac{3}{2}x(n+1) - j\right] + 2$ we get

$$\frac{|\delta_{i+l}^{-m}|}{H} < 2(1+\epsilon)R^{-s(\frac{3}{2}x(n+1))}. \tag{269}$$

That results in a condition

$$2(1+\epsilon)R^{-s(\frac{3}{2}x(n+1)-2)} < \frac{1}{2}R^{-s(x(n+1)+1)}. \tag{270}$$

As $R^{-s} < \frac{1}{16}$ a sufficient condition for the respective upper properly sized bound (197) is

$$\frac{3}{2}x(n+1) > x(n+1) + 2 \tag{271}$$

or

$$z_0 + 1 \geq \frac{4}{x}$$

which follows from (122).

□

(e) Many monotone branches can be oversized after filling-in operations. Then we pull-back partition ξ_{xn} onto such branches again. By Proposition 7 all new undecided domains are properly sized, but some monotone branches can be still oversized We repeat monotone pull-back onto such branches and so on. In this way we construct a sequence of partitions $\zeta_{n+1,k}$. Let us call ζ_{n+1} the final partition with all elements properly sized. As undecided domains created at each step are properly sized, the maximal number of operations needed to construct ζ_{n+1} is given by (242). Respectively we need no more than $[sr_1(n+1)+s]+2$ pull-backs to make all elements properly sized.

That finishes the construction of partition ζ_{n+1}. Notice, that in particular we proved properties (196), (197) and (210) of step $n+1$.

6.5. Critical pull-back and sizes of elements. Let $\eta_{n+1} = h^{-1}\zeta_{n+1}$ be the critical pull-back of ζ_{n+1}. Let $\Delta\phi$ be an n-located element of ζ_{n+1} on y-axis. Let $[a,b] = \Delta f = h^{-1}\Delta\phi$ be its preimage on x-axis -element of η_{n+1}. Assume $b \in (a,0)$, and let d be the distance from a to zero. From the definition of n-located domain, we get

$$\delta_0(1-\epsilon)R^{-s(n+1)} < d < \delta_0(1+\epsilon)R^{-sn}\,. \tag{272}$$

Using (163), we get from (196)

$$R^{-s}\frac{1+\epsilon}{4}R^{-r_1 s(n+1)} > \frac{|\Delta f|}{d} > 4(1-\epsilon)R^{-sr_2(n+1)}\,. \tag{273}$$

(272) and (273) imply (198) at step $n+1$ for domains Δf under consideration. Similarly if δ_i^{-m} is an n-located element of ζ_{n+1}, and $\delta_i^{-(m+1)} = h^{-1}\delta_i^{-m}$ is its preimage, then using (163), we get from (197)

$$R^{-s}\frac{1+\epsilon}{4}R^{-xs(n+1)} > \frac{|\delta_i^{-(m+1)}|}{d} > 4(1-\epsilon)R^{sr_2 i}R^{-sr_2(n+1)}\,. \tag{274}$$

(272) and (274) imply (199) at step $n+1$.

6.6. Construction of y_{n+1}, δ_{n+1} and canonical partition at step $n+1$. (273) and (274) imply that elements of η_{n+1} located inside δ_n, but outside some $(1-\epsilon)\delta_0 R^{-s(n+1)}$- neighborhood of zero, satisfy estimates of canonical partition of step $n+1$. In particular these elements are exponentially smaller than $|\delta_0| R^{-s(n+1)}$. They are preimages of elements of ζ_{n+1} which are exponentially small comparatively to their distances from the critical value. By construction each undecided domain δ_i^{-m} in ζ_{n+1} has an enlargement δ_{i-1}^{-m} which is $(1\pm\epsilon)R^s$ times larger. Maximal (in the sense of inclusion) enlargements and domains $\delta\phi_n$ which do not belong to enlargements form a partition $\hat{\zeta}_{n+1}$. Let $\hat{y}_{n+1}$ be a point located at the distance

$\bar{k}\delta_0^2 R^{-2s(n+1)}$ from w. Then $\hat{y}_{n+1}$ belongs to some element E of $\hat{\zeta}_{n+1}$. By definition y_{n+1} is the end point of E closest to w. Next we define canonical partition on y-axis as $\zeta_{n+1}|[y_n, y_{n+1}]$. Finally we define δ_{n+1} as an interval between two critical preimages of y_{n+1}.

$$\delta_{n+1} = (h_1^{-1} y_{n+1}, h_2^{-1} y_{n+1}) \,. \tag{275}$$

We get

$$(1-\epsilon_n) \mid \delta_0 \mid R^{-s(n+1)} <\mid \delta_{n+1} \mid< (1+\epsilon_n) \mid \delta_0 \mid R^{-s(n+1)} \tag{276}$$

where ϵ_n is exponentially small. That proves inductive property (203) of step $n+1$. We define canonical partition as

$$\xi_{n+1}|\delta_n \setminus \delta_{n+1} = h^{-1}\zeta_{n+1}|(y_n, y_{n+1}) \,. \tag{277}$$

By construction inductive properties (198) and (199) are satisfied and enlargement property is respected. We define $S(\delta_{n+1})$ as the union of elements of canonical partition $\Delta f \subset \delta_n \setminus \delta_{n+1}$.

6.7. Final filling-in and refinement outside δ_n and sizes of elements. In order to finish the construction of ξ_{n+1} we consider preimages δ_j^{-r} outside δ_n and do one-step filling-in for each preimage. That creates a new central subdomain δ_{j+1}^{-r}, new monotone branches and new preimages δ_i^{-m}. Some monotone branches can be oversized in the sense of (202). Then we pull-back $\xi_{[nx]}$ onto such branches until they are properly sized. We have to check that elements, obtained by these operations, are properly sized. We have already checked size estimates of steps $n \le z_0$ in Section 4. By the choice of K_0 in (171) they are twice better than required. At a general step $n+1$ we begin by lower properly sized estimates. They are similar to the above estimates of Proposition 5. First we consider undecided domains. Notice that estimate for noncanonical domains (200) at step n is $K_0/2$ times weaker than canonical estimate (199). So, when a certain domain constructed at step $n \ge z_0$, as an element of some canonical partition, becomes noncanonical, it is much bigger than the respective lower estimate.

Let us consider consecutive central subdomains $\delta_n^{-k} \subset \delta_{n-1}^{-k}$. At each step the central domain decreases by R^{-s}, its index increases by 1 and n increases by one. An $1 \pm \epsilon$ variation due to distortion and variation of sizes of δ_n is compensated by initial difference between estimates. Therefore central subdomains remain lower properly sized at all steps. Arguments of Proposition 5 prove that at the next step the lower estimates for secondary preimages and for monotone branches improve by a factor $2(1-\epsilon)$. Finally, when doing refinement of oversized monotone branches, which appear after filling-in, we pull-back $\xi_{[nx]}$. Then at steps $n \ge z_0$ elements remain properly sized as proved in Proposition 7. At steps $n < z_0$ it was assured by the choice of r_2, and K_0 as discussed in Section 4. That proves lower estimates (201) and (200).

Next we check upper estimates. All monotone branches are upper properly sized by construction. Upper estimates for δ_j^{-r} constructed at steps $n < z_0$ may be not satisfied. However, by construction, if $\delta_j^{-r} \subset \delta_i \setminus \delta_{i+1}$, $i < z_0$, then

$$\mid \delta_j^{-r} \mid< (1+\epsilon) R^{-3s} \mid \delta_i \mid \,.$$

Thus, when δ_j^{-r} becomes noncanonical at step $i+1$ it satisfies inductive estimate, which is $(1+\epsilon)R^{-2s}$ times better than required for noncanonical domains. Therefore, at step z_0 upper estimates for noncanonical domains are satisfied. Let us check upper estimates at a general step of induction. We use that upper estimates for canonical domains δ_j^{-r} constructed at steps $n \geq z_0$, are $\frac{1}{2}R^{-snx}$ times smaller than respective estimates for noncanonical domains. As central subdomains decrease at a rate $(1 \pm \epsilon)R^{-s}$ per step we get that upper estimates for central subdomains are better than required by the respective factor $\frac{1+\epsilon}{2}R^{-snx}$ or $(1+\epsilon)R^{-2s}$. Secondary undecided domains, which appear after the filling-in of δ_j^{-r}, with $j < z_0$ are by construction at least R^{-3s} times smaller than δ_j^{-r}, which implies an upper bound R^{-2s} times better than required. Secondary undecided domains, which appear after the filling-in of δ_j^{-r}, with $j \geq z_0$ are at least $\frac{1+\epsilon}{2}R^{-xsj}$ times smaller than δ_j^{-r}. As $z_0 > \frac{1}{x}$ that implies the upper estimate (200).

6.8. Dependence of sizes on time. We proved properly sized estimates for $t = t_0$, which were significantly better, than required (respectively larger or smaller). Speed estimates in Section 7 and the size of parameter interval imply that for other $t \in \mathcal{T}$ sizes can only change by a factor close to one, see (347) - (350). So estimates of sizes hold for all $t \in \mathcal{T}$.

6.9. Relative measure of good branches. We need to check property (204) for good branches constructed at step $n+1$. In particular we need to construct a collar of δ_{n+1} with relative measure greater than $1 - \frac{3}{2}R^{-s} \mid \delta_n \mid$. First we estimate relative measures in partition ζ_{n+1} on y-axis, and after that pull-back onto x-axis.

In the original partition $\zeta_{n+1,1}$ the holes δ_i^{-m} constitute some subset $\mathcal{A}_1$ and the remaining subset $\mathcal{A}_2$ is occupied by monotone branches. By inductive assumption each filling on x-axis decreases the measure of holes at least by χ_0. When we do an l-step filling-in we reduce the measure of holes at least by by χ_0^l.

By construction $l \geq 3$. Therefore the measure of old holes is reduced at least by

$$(1+\epsilon)\chi_0^3\,. \tag{278}$$

Then we estimate the measure of new holes which appear at the place occupied by oversized monotone branches $\hat{f}_n$.

According to (218) we pull-back onto such branches partition $\xi_{[nx]}$. Each pull-back has distortion bounded by $\exp l_\infty$.

Let $mes\ \mathcal{C}_0$ be the measure of undecided domains in ξ_0. Condition (242) implies that after the construction of ξ_{n+1} is completed, the total relative measure of good branches remaining at the place of any domain $\Delta\hat{f}_n$ is greater than

$$\left(1 - (mes\ \mathcal{C}_0)\chi_0^{[nx]}\right)^{[sr_1(n+1)+s+2]} \tag{279}$$

As $\chi_0^{[nx]}$ decrease exponentially with n it is enough to estimate the above expression for $n = z_0$. Notice that it is greater than in (132) because in (132), $mes\mathcal{C}_0$ is not multiplied by $\chi_0^{[nx]}$. Therefore we get

$$\left(1 - (mes\ \mathcal{C}_0)\chi_0^{[z_0x]}\right)^{[sr_1(z_0+1)+s+2]} > 1 - \chi_0^3\,. \tag{280}$$

We constructed partition ζ_{n+1} on some interval $[y_n, a_n] \subset [y_n, w]$. From previous estimates we get that the total measure of holes between y_n and a_n does non exceed

$$(1+\epsilon)\chi_0^3 \tag{281}$$

Notice that generally $[y_n, a_n] \supset [y_n, y_{n+1}]$, because we do construction of step $n+1$ on n-located domains which can extend beyond y_{n+1}. But in any case a_n is less than the critical value. As

$$\mid [y_n, y_{n+1}] \mid > (1 - R^{-2s}(1+\epsilon_n)) \mid [y_n, w] \mid \tag{282}$$

the relative measure of holes inside $[y_n, y_{n+1}]$ does not exceed

$$\frac{(1+\epsilon)\chi_0^3}{(1 - R^{-2s}(1+\epsilon_n))} . \tag{283}$$

Let us denote by $\mathcal{E}$ the union of δ_i^{-m} inside $\delta_n \setminus \delta_{n+1}$. A priory after a critical pull-back estimate may worsen. If we assume that all domains δ_i^{-m} are located near y_{n+1}, then the relative measure of their preimages inside $\delta_n \setminus \delta_{n+1}$ is multiplied by R^s. Therefore

$$mes\ \mathcal{E} < \left(\frac{(1+\epsilon)\chi_0^3}{(1 - R^{-2s}(1-\epsilon_n))} R^s \right) mes(\delta_n \setminus \delta_{n+1}) . \tag{284}$$

From (126) we get

$$mes\ \mathcal{E} < \frac{1}{2} R^{-s} mes(\delta_n \setminus \delta_{n+1}) . \tag{285}$$

We define

$$S(\delta_{n+1}) = \delta_n \setminus (\delta_{n+1} \cup \mathcal{E}) . \tag{286}$$

From (276),(284) and (285) we get that the relative measure of $S(\delta_{n+1})$ inside δ_n is greater than $1 - \frac{3}{2}R^{-s}$. That proves (204). After filling-in the relative measure of holes inside δ_i^{-k} can increase by a distortion factor $1+\epsilon$. An additional increase of holes due to the final refinement is less than the one given by (280) and so much smaller than $\frac{1}{2}R^{-s}$. Therefore at each step, every hole δ is substituted by a union of new holes with total measure less than $\chi_0 \mid \delta \mid$. That proves (205) and finishes the proof of size and measure estimates at step $n+1$.

7. Distortion estimates, parameter dependence and measure of admissible parameter values

7.1. Distortion estimates. Let l_k, l_∞ be the numbers defined by (115), (116). We use f_k, ϕ_k to denote good maps constructed at step k. Domains Δf_k are elements of ξ_k located on x-axis, and $\Delta\phi_k$ are elements of ζ_k located on y-axis. Similarly $G_{ik}, \mathcal{G}_{ik}$ are mappings from undecided domains δ_i^{-m} to δ_i constructed at step k. Domains of G_{ik} are located on x-axis and domains of $\mathcal{G}_{ik}$ are located on y-axis.

Let $\Theta(F) = \Theta(F, \Delta)$ be the measure of distortion of a diffeomorphism F introduced in Section 1. We prove by induction

PROPOSITION 10. *For all* $n \geq 0$

$$\Theta(f_n) \leq l_{[\frac{n}{4}]+1} \tag{287}$$

$$\Theta(\phi_n) \leq l_{[\frac{n}{4}]+1} \tag{288}$$

(289) $$\Theta(G_{in}) \le l_{[\frac{n}{4}]} R^{-(sxi+s)}$$

(290) $$\Theta(\mathcal{G}_{in}) \le l_{[\frac{n}{4}]} R^{-(sxi+s)} .$$

PROOF. (a) *Distortion formulas.*
The following estimates for distortions of compositions (see [**3**]) are used in the proof. Let $f_1 : \Delta_1 \to I_1$ and $f_2 : \Delta_2 \to I_2$ be two monotone maps such that $\Delta_1 \subset I_2$. Then $f_1 \circ f_2$ is defined on $\Delta_{12} = f_2^{-1}\Delta_1$. Then

(291) $$\Theta(f_1 \circ f_2, \Delta_{12}) \le \Theta(f_1, \Delta_1) \exp \Theta(f_2, \Delta_{12}) + \Theta(f_2, \Delta_{12})$$

where

(292) $$\Theta(f_2, \Delta_{12}) \le \Theta(f_2, \Delta_2) \exp \Theta(f_2, \Delta_2) \frac{|\Delta_1|}{|I_2|} .$$

Let $f : \Delta \to I$ be a monotone map, let h be a parabolic map defined in a neighborhood of the origin and assume image of h contains an interval Δ. Let $\Delta^{-1} = h^{-1}\Delta$ be one of diffeomorphic preimages. Let us denote by $d(\Delta)$ be the distance between Δ and the critical value of h. If a neighborhood of 0 is sufficiently small, then h is close to quadratic. A straightforward calculation shows

(293) $$\Theta(f \circ h, \Delta^{-1}) \le \Theta(f) \exp\left(\frac{|\Delta|}{(2-\epsilon_0)d(\Delta)}\right) + \frac{|\Delta|}{(2-\epsilon_0)d(\Delta)}$$

where ϵ_0 depends on the neighborhood of zero. If distortion is increased, as in the above formulas (291), (293), due to multiplication by a factor $\exp\epsilon$ and a simultaneous addition of ϵ, then we call ϵ an *increment.* When l_k in (115) is increased due to an increment $\epsilon < R^{-sx(k+1)}$ the result is less than l_{k+1}.

(b) *Estimates at the first steps of induction $n < z_0$.*
Estimates of the first three steps were checked in Section 4. Some of the maps constructed at steps $n < z_0 - 1$ on y-axis are compositions

(294) $$\phi_k = f_{0i_1} \circ \ldots \circ f_{0i_k} \circ g_i$$

and

(295) $$\mathcal{G}_k = \phi_k | \delta^{-m}$$

where f_{0i_l} are maps which belong to ξ_0 and g_i are staircase maps. On x-axis we get compositions of the above maps with the parabolic map h. By the definition of l_0^*, see (114)

(296) $$\Theta(\phi_k) \le l_0^*$$

which is better than (288). Then

(297) $$\Theta(\mathcal{G}_k) \le l_0^* \exp l_0^* \delta_0$$

From (124) and (127) the above estimate is less than R^{-2}, which is better than (290). We remind that by construction elements $\Delta\phi_k$ are properly sized in the sense of (196). Using (295) and (291) we get for $f_k = \phi_k \circ g$ an estimate better than (287) with $n = 0$. If an element δ^{-k} is upper properly sized in the sense of (197) then using (293) we get an estimate which is at least $\frac{1+\epsilon}{4} R^{-sx}$ times better than required by (289). Thus distortions of maps constructed without filling-in are better than required.

At steps $n > 3$ some δ^{-k} on y-axis may be not upper properly sized. Then we pull-back $\xi_3 \mid \delta$. That creates inside δ^{-k} several new domains. Let us first consider a central preimage δ_3^{-k}, which is R^{-3s} times smaller than δ^{-k}. From (183)

$$\frac{\mid \delta^{-k} \mid}{d(\delta^{-k})} < 2(1+\epsilon)R^{-4s}\,. \tag{298}$$

Applying (293) we get that distortions of $G_1 = G \circ h : \delta_3^{-(k+1)} \to \delta_3$ satisfy

$$\Theta(G_1, \delta_3^{-(k+1)}) < (1+\epsilon)R^{-4s}\,. \tag{299}$$

The above estimate is better than (289) for $i = 3$. The above pull-back also creates inside δ^{-k} secondary preimages $\delta^{-(k+l)}$. From (174) we get much better estimates for such preimages.

Some of monotone branches obtained by a pull-back of $\xi_3 \mid \delta$ into δ^{-k} may be oversized. We pull-back ξ on such elements until they are properly sized. Distortion estimates for these branches are a particular case of estimates at a general step of induction with l_0 used instead of $l_{[nx]}$. The same is true for elements of ξ_i constructed outside of δ_i by the final filling-in and refinement. That proves inductive estimates of distortions at steps $i < z_0$.

(c) *Types of maps under consideration.*

Let us consider elements of canonical partition. Each element is obtained according to our algorithm through a sequence of pull-backs and/or fillings.

Let ϕ_k, $\mathcal{G}_k$ be the maps constructed before step $n+1$. By inductive assumption they satisfy estimates of step n. According to Proposition 6 each time when we pull-back the partition $\xi_{[nx]}$ by some monotone branch we create properly sized undecided domains which do not need further filling-in. Similarly according to Proposition 9 each time when we fill-in an element δ_i^{-m} up to δ_{i+l}^{-m} we create properly sized undecided domains which do not need further filling-in.

Each monotone pull-back on y-axis generates new compositions of the form

$$f_i \circ \phi_k$$

$$G_i \circ \phi_k$$

Each filling-in on y-axis generates new compositions of the form

$$f_j \circ \mathcal{G}_i$$

$$G_j \circ \mathcal{G}_i$$

After that we compose with h. Therefore we have to estimate distortions of the following types of compositions.

(i) $f_{i_s} \circ \ldots \circ f_{i_1} \circ \phi_k \circ h$
(ii) $\mathcal{G}_i \circ \phi_k \circ h$
(iii) $G_i \circ f_{i_s} \circ \ldots \circ f_{i_1} \circ \phi_k \circ h$
(iv) $G_j \circ \mathcal{G}_i \circ h$
(v) $f_{j_1} \circ \ldots \circ f_{j_p} \circ \mathcal{G}_i \circ h$
(vi) $G_j \circ f_{j_1} \circ \ldots \circ f_{j_p} \circ \mathcal{G}_i \circ h$

We split estimates into two steps. First we estimate distortions of elements on y-axis and after that estimate contribution of the critical pull-back. We prove that distortions of short compositions on y-axis

$$f_i \circ \phi_k$$
$$G_i \circ \phi_k$$
$$G_j \circ \mathcal{G}_i$$
$$f_j \circ \mathcal{G}_i$$

are smaller than respective inductive estimates for maps ϕ_i, $\mathcal{G}_i$ of step n. That makes unnecessary to consider multiple compositions f_{i_s} and reduces estimates in cases $(a) - (f)$ to estimates for compositions of the above four types with h.

(d) *Compositions* $f_i \circ \phi_k$.
At step $n+1$ we compose ϕ_k with various branches f_i constructed at steps $i \leq [nx]$. Thus, we estimate distortions of compositions

$$\phi_{k+1} = f_i \circ \phi_k$$

where by inductive assumption

(300) $$\Theta(f_i) < l_{[\frac{i}{4}+1]}, \quad \Theta(\phi_k) < l_{[\frac{n}{4}+1]} \, .$$

Applying (291) we get

(301) $$\begin{aligned} \Theta(\phi_{k+1}) &< \Theta(f_i)\exp[\Theta(\phi_k)(\exp\Theta(\phi_k)) \mid \Delta(f_i) \mid] \\ &+ \Theta(\phi_k)(\exp\Theta(\phi_k)) \mid \Delta(f_i) \mid \, . \end{aligned}$$

By inductive assumptions for $i \geq 1$

(302) $$\mid \Delta(f_i) \mid < \mid \delta_0 \mid R^{-s(r_1+1)i}$$

and for $i = 0$

(303) $$\mid \Delta(f_0) \mid < R^{-1} \, .$$

When $i \geq 1$, (127) and (124) imply

(304) $$\Theta(\phi_k)(\exp\Theta(\phi_k)) \mid \delta_0 \mid R^{-s(r_1+1)i} < R^{-sx(i+1)} \, .$$

When $i = 0$, (127) implies

(305) $$\Theta(\phi_k)(\exp\Theta(\phi_k))R^{-1} < R^{-(1-s)} \, .$$

We get that in any case the increment is less than $R^{-sx(\frac{i}{4}+1)}$. Therefore

(306) $$\Theta(f_i \circ \phi_k) = \Theta(\phi_{k+1}) < l_{[\frac{i}{4}]+1} \leq l_{[\frac{nx}{4}]+1} \, .$$

REMARK 12. *Actually we proved*

(307) $$\Theta(f_i \circ \phi_\infty) < l_{[\frac{i}{4}]+1} \, .$$

(e) *Compositions* $G_i \circ \phi_k$.
Next we estimate $\Theta(G_i \circ \phi_k)$, where ϕ_k is as above and G_i maps an element δ_i^{-m} of the partition $\xi_{[nx]}$ onto δ_i. We use (291) for composition $G_i \circ \phi_k$. As δ_i^{-m} is a properly sized element of $\xi_{[nx]}$ we have

(308) $$\mid \delta_i^{-m} \mid < \mid \delta_0 \mid R^{-xsn} \, .$$

Then a respective increment in distortion formula is less than

(309) $$\Theta(\phi_k)\exp\Theta(\phi_k) \mid \delta_0 \mid R^{-xsn} < R^{-(s+snx)} \, .$$

Respectively

(310) $$\Theta(G_i \circ \phi_k) < \Theta(G_i)\exp\Big(R^{-(s+snx)}\Big) + R^{-(s+snx)}\,.$$

As δ_i^{-m} is an element of $\xi_{[nx]}$ we get using inductive assumption

(311) $$\Theta(G_i) < l_{[\frac{nx}{4}]}R^{-s(xi+1)}\,.$$

We factor $R^{-s(xi+1)}$ out of relation (310) and get an increment less than $R^{-sx(n-nx)}$. As $n - nx \gg \frac{nx}{4}$ we get a better estimate than (290) at step $[nx]+1$

(312) $$\Theta(G_i \circ \phi_k) < l_{[\frac{[nx]+1}{4}]+1}R^{-s(xi+1)}\,.$$

(f) *Compositions $G_j \circ \mathcal{G}_i$ and $f_j \circ \mathcal{G}_i$.*
Next we estimate distortions of maps obtained by a k-step filling-in, $k \geq 3$. That operation creates three types of elements.

a) the new central domain δ_{i+k}^{-m}.

b) secondary preimages $\mathcal{G}_{in}^{-1}(\delta_j^{-l})$.

c) domains of monotone branches $\mathcal{G}_{in}^{-1}(\Delta f)$.

Taking into account

$$\mid \delta_{i+k} \mid = (1 \pm \epsilon_i) \mid \delta_i \mid R^{-sk}$$

we get in case a) $\mathcal{G}_{i+kn+1} = \mathcal{G}_{in} \mid \delta_{i+k}^{-m}$,

(313) $$\Theta(\mathcal{G}_{in} \mid \delta_{i+k}^{-m}) < \Theta(\mathcal{G}_{in})\exp\Theta(\mathcal{G}_{in})(1+\epsilon_i)R^{-sk}\,.$$

As $R^{-sk} \ll R^{-sxk}$ we get that (313) is exponentially small comparatively to (290) at zero step of induction.

In case b) we use inductive assumptions

(314) $$\Theta(G_{ji+k}) < l_{[\frac{i+k}{4}]}R^{-(sxj+s)}$$

(315) $$\Theta(\mathcal{G}_{in}) < l_{[\frac{n}{4}]}R^{-(sxi+s)}\,.$$

Let $\Theta_1 = \Theta(G_{ji+k} \circ \mathcal{G}_{in}, \delta_j^{-(m+l)})$ be the distortion that we estimate.

From inductive assumption

(316) $$\Theta(\mathcal{G}_{ji+k}) < l_{[\frac{i+k}{4}]}R^{-(sxj+s)}\,.$$

Using (291) we get

(317) $$\begin{aligned}\Theta_1 &< \Theta(G_{ji+k})\exp\Big(\Theta(\mathcal{G}_{in})(\exp\Theta(\mathcal{G}_{in}))\frac{\mid \delta_j^{-l} \mid}{\mid \delta_i \mid}\Big)\\ &+ \Theta(\mathcal{G}_{in})\Big(\exp\Theta(\mathcal{G}_{in})\Big)\frac{\mid \delta_j^{-l} \mid}{\mid \delta_i \mid}\,.\end{aligned}$$

Hence the increment does not exceed

(318) $$incr = \Theta(\mathcal{G}_{in})\Big(\exp\Theta(\mathcal{G}_{in})\Big)\frac{\mid \delta_j^{-l} \mid}{\mid \delta_i \mid}\,.$$

Size estimates imply that after a one-step filling-in the new central subdomain is much larger than secondary undecided domains. In particular, the following lemma holds

LEMMA 4. *Non-central subdomains in* $\xi_{i+k}|\delta_i$ *satisfy*

(319) $$\frac{|\delta_j^{-l}|}{|\delta_i|} < \frac{1+\epsilon}{2} R^{-sk(1-x)} R^{-sx(i+k)} .$$

PROOF. Let δ_m^{-l} be a domain constructed at step $i+p$, $p<k$. Then by inductive assumption (199)

(320) $$|\delta_m^{-l}| < \frac{1+\epsilon}{2} \times R^{-sx(i+p)} R^{-sp} |\delta_i| .$$

At the next steps its subdomains decrease at least at the rate R^{-s} per step. Respectively, at step $i+k$ we get for $\delta_j^{-l} \subset \delta_m^{-l}$

(321) $$|\delta_j^{-l}| < \frac{1+\epsilon}{2} |\delta_i| R^{-sx(i+p)} R^{-sp} R^{-s(k-p)}$$

which implies (319). □

Then using inductive assumptions for $\Theta(\mathcal{G}_{in})$ we get

(322) $$incr < \frac{1+\epsilon}{2} l_{[\frac{n}{4}]} R^{-(sxi+s)} R^{-sx(i+k)} R^{-sk(1-x)} .$$

We factor $l_{[\frac{i+k}{4}]} R^{-(sxj+s)}$ out of (317), use $j<i+k$ and (322) and get an estimate

(323) $$\Theta_1 < l_{[\frac{i+k}{4}]+1} R^{-(sxj+s)} .$$

From (254)

$$i+k < \left[\frac{n}{3}\right] .$$

Hence we get an estimate better than (290) with coefficient $l_{[\frac{n}{12}]+1}$.

In case c) the filling-in produces a new monotone map $f_i^{new} = f_{i+p}^{old} \circ \mathcal{G}_{in}$. Arguing as above we get that $\Theta(f_{i+p}^{old})$ is increased by

(324) $$\Theta(\mathcal{G}_{in})(\exp\Theta(\mathcal{G}_{in})) \frac{|\Delta f_{i+p}^{old}|}{|\delta_i|} .$$

Using $\frac{|\Delta f_{i+p}^{old}|}{|\delta_i|} < R^{-sr_1(i+p)}$. we get an estimate better than (288) at step $i+p+1$.

(g) *Compositions* $\phi_n \circ h$.
We proved that distortions of ϕ_n constructed above are less than inductive estimates of some previous step $k < [\frac{n}{3}]+1 < n$. Hence it is enough to consider compositions $\phi_n \circ h$ assuming inductive estimate of step n holds for ϕ_n and the domain Δ of ϕ_n is properly sized.

As Δ is properly sized we have

(325) $$\frac{|\Delta|}{H} < R^{-r_1 sn} .$$

We use inductive assumption and (293) and get

(326) $$\Theta(\phi_n \circ h) < l_{[\frac{n}{4}]+1} exp(R^{-r_1 sn}) + R^{-(r_1 sn)} .$$

As $r_1 \geq 1$ and $x = \frac{1}{8}$ we get an increment exponentially smaller than (288) at step $n+1$.

(h) *Compositions* $\mathcal{G}_j \circ h$.
We use properly sized property of undecided domains and (293) and get

(327) $$\Theta(\mathcal{G}_j \circ h) < \Theta(\mathcal{G}_j) \exp(R^{-xsn}) + R^{-xsn}\,.$$

We proved above that fillings on y-axis resulted in distortion estimate

(328) $$\Theta(\mathcal{G}_j) < l_{[\frac{n}{12}]+1} R^{-(sxj+s)}\,.$$

Hence

(329) $$\Theta(\mathcal{G}_j \circ h) < l_{[\frac{n}{12}]+1} R^{-(sxj+s)} \left(1 + R^{-sx(n-j-\frac{1}{x})}\right).$$

Secondary preimages δ_j^{-m} have indices $j < \frac{n}{3}$. Using (122) we get that (329) is less than (289).

(i) *Estimates for maps constructed at the final filling-in and refinement.*
By construction we are doing a one-step filling-in of undecided domains $\delta_i^{-m} \subset I \setminus \delta_n$. These estimates are a particular case of estimates for a k-step filling-in considered above. For monotone branches obtained by the final refinement outside δ_n we get the same estimate as (307) by the same arguments. Similarly for undecided domains obtained by the final refinement outside δ_n we get the same estimate as (312). That proves inductive estimates outside δ_n.

□

REMARK 13. *The choice of index* $\left[\frac{n}{4}\right]$ *in inductive assumptions of Proposition 10 is not essential. It can be* n*, which gives worse estimates, or it can be* $\left[\frac{n}{z_0}\right]$ *as well.*

7.2. Dependence on the parameter. We estimate dependence on the parameter for elements constructed at step $n+1$. Let Δf, ΔG be elements of ξ_n on $x-$axis, $\delta\phi$, $\Delta\mathcal{G}$ elements of ζ_n on y-axis. Let

$$\frac{\phi_t}{\phi_x}(x(t),t),\ \frac{\mathcal{G}_t}{\mathcal{G}_x}(x(t),t),\ \frac{f_t}{f_x}(x(t),t),\ \frac{G_t}{G_x}(x(t),t)$$

be the respective speeds. Let $\beta_0 = \frac{2}{k}$ and let α_0 be the constant from (109). Let

(330) $$\sigma_i = \alpha_0 + R^{-2+s}, i = 1,2,3,\ \ \sigma_i = \sigma_{i-1} + R^{-si}, i \geq 4\,.$$

PROPOSITION 11. *At any step* $n = 0, 1, \ldots$ *the maps under consideration satisfy*

(331) $$\frac{f_t}{f_x}, \frac{G_t}{G_x} < \beta_0 \mid \delta_n \mid^{-1}$$

(332) $$\frac{\phi_t}{\phi_x}, \frac{\mathcal{G}_t}{\mathcal{G}_x} < \sigma_n\,.$$

PROOF. (a) For the first three steps (331) follows from (181) and (332) follows from (180). Let F_k be a composition of good maps of the form

$$F_k = f_j \circ \ldots \circ f_1 \circ f_*$$

Using the chain rule we get

$$\begin{aligned}\left|\frac{\frac{\partial F_k}{\partial t}}{\frac{\partial F_k}{\partial x}}\right| &\leq \left|\frac{\frac{\partial f_*}{\partial t}}{\frac{\partial f_*}{\partial x}}\right| + \\ &+ \left|\frac{\frac{\partial f_1}{\partial t}}{\frac{\partial f_1}{\partial x}}\right| \times \left|\frac{\partial f_*^{-1}}{\partial x}\right| + \dots \\ &+ \left|\frac{\frac{\partial f_j}{\partial t}}{\frac{\partial f_j}{\partial x}}\right| \times \left|\frac{\partial (f_{j-1}\circ \dots \circ f_*)^{-1}}{\partial x}\right| . \end{aligned} \tag{333}$$

At step $n+1$ we consider compositions $f_{i_1} \circ \dots \circ f_{i_s} \circ \phi_n$, where $\Delta\phi_n$ is an element of ζ_n. By construction $\Delta\phi_n$ belongs to $[y_n, w]$. Thus

$$\mid \Delta\phi_n \mid < \mid [y_n, w] \mid < \bar{k}\delta_0^2 R^{-2sn} . \tag{334}$$

As distortion of ϕ_n is bounded by l_∞ we get

$$\mid \frac{\partial \phi_n^{-1}}{\partial x} \mid < \exp(l_\infty)\bar{k}\delta_0^2 R^{-2sn} . \tag{335}$$

By construction of step $n+1$ indices of maps f_{i_j} satisfy condition $i_j \leq [nx]$. Let $\phi_{n+1} = f_{i_1} \circ \dots \circ f_{i_s} \circ \phi_n$. Using inductive assumptions for speeds, (333) and the fact that expansions of f_{i_j} are greater than R we get

$$\begin{aligned}\frac{\phi_{n+1t}}{\phi_{n+1x}} &< \frac{\phi_{nt}}{\phi_{nx}} \\ &+ \exp(l_\infty)\delta_0 R^{-2sn+s[nx]} \\ &\times \left(1 + R^{-1} + \dots\right) . \end{aligned} \tag{336}$$

Using (127) and (124) we conclude that in this case estimate is better than (332).

Other elements are obtained by filling-in. Assume $\delta_i^{-k} \subset [y_n, w]$ is filled-in. As distortion of $\mathcal{G}_i$ is small we get similarly to (334) that expansion of $\mathcal{G}_i : \delta_i^{-k} \to \delta_i$ satisfies

$$(1-\epsilon)\frac{\delta_0 R^{-si}}{\bar{k}\delta_0^2 R^{-2sn}} . \tag{337}$$

By inductive assumption speeds in δ_i are less than $\frac{2}{k}\delta_0^{-1}R^{si}$. Applying (333) we get that after filling-in speeds do not exceed

$$\sigma_{n-1} + (1+\epsilon)2R^{-s(2n-2i)} \tag{338}$$

From (254) $i < \frac{n}{3} - 3$. Therefore the second term in (338) is small comparatively to $R^{-\frac{4}{3}sn}$.

Steps $3 < n < z_0$ are considered separately. For monotone branches constructed at these steps by pull-backs of ξ_0 (180) is valid and (332) follows. Then (331) follows similarly to (91). For elements obtained by filling-in at steps $3 < n < z_0$ we use that indices i of δ_i^{-k} satisfy $i \leq 3$. We use (331) and (108) and get an estimate

$$\left|\frac{\phi_t}{\phi_x}\right|, \left|\frac{\mathcal{G}_t}{\mathcal{G}_x}\right| < \alpha_0 + \frac{2}{\bar{k}}\delta_0^{-1}R^{3s}\exp l_0 \bar{k}\delta_0^2 < \alpha_0 + 2\delta_0 R^{3s}\exp l_0 \tag{339}$$

which is less than σ_3. At all steps we need to take into account additional pull-backs of $\xi_{[nx]}$ done at the final refinement, when after filling-in we obtain oversized monotone branches. Applying (333) we add consecutive terms to (338). The third term is obtained from the second by dividing additionally by expansion of some map f_1 where $\Delta f_1 \subset \delta_i$. The minimal expansion of such maps corresponds to $i = 1$ and satisfies

(340)
$$\mid \frac{\partial f_1}{\partial x} \mid > (1-\epsilon)\delta_0^{-1} R^{sr_1} .$$

Similarly to (336) consecutive pull-backs contribute $1 + R^{-1} + \ldots$ and we get a bound less than

(341)
$$\sigma_{n-1} + R^{-\frac{4}{3}sn}(1 + R^{-1} + R^{-2} + \ldots)) < \sigma_{n-1} + R^{-sn} .$$

That proves (332).

(b) When doing critical pull-back we consider a composition $f = \phi \circ h$ where ϕ is as above and the point x where we apply h lies outside δ_{n+1}. By the chain rule formula

(342)
$$\mid \frac{(\phi \circ h)_t}{(\phi \circ h)_x} \mid \leq \mid \frac{h_t}{h_x} \mid + \mid \frac{\phi_t}{\phi_x} \mid \times (\mid \frac{\partial h}{\partial x} \mid)^{-1} .$$

Outside δ_{n+1} the expression $(\mid \frac{\partial h}{\partial x} \mid)^{-1}$ is bounded by $\frac{(1+\epsilon)}{k} \times (\mid \delta_{n+1} \mid)^{-1}$. At the same time h_t is close to 1 and $\left|\frac{\phi t}{\phi x}\right|$ estimated above is small. That proves (331) for elements of canonical partition of step $n+1$.

(c) It remains to consider the final filling-in of undecided domains. Let f be a map defined on an element of the canonical partition $\xi_{i+1} \mid \delta_i,\ i \leq n$, and let $\delta_i^{-k} = G_i^{-1}\delta_i$ be a preimage of δ_i. We use $\mid \frac{\partial G_i}{\partial x} \mid > \rho_0$ and (128) which implies

(343)
$$\frac{1+\epsilon}{R^s} + \frac{1}{\rho_0} < 1 .$$

Applying (333) and (331) and using (343) we get

(344)
$$\mid \frac{(f \circ G_i)_t}{(f \circ G_i)_x} \mid < \beta_0 \mid \delta_n \mid^{-1} + \frac{\beta_0 \mid \delta_{n+1} \mid^{-1}}{\rho_0} < \beta_0 \mid \delta_{n+1} \mid^{-1}$$

That finishes the proof.

□

7.3. Measure of admissible parameter values. Finally, we show that the measure of inadmissible parameter values decreases proportionally to the measure of undecided domains in the phase space. According to (93) initial set of undecided domains $\mathcal{C}_0$ and initial set of enlarged undecided domains $\hat{\mathcal{C}}_0$ have small measures. In the phase space the measure of undecided domains decreases at least at a rate χ_0 per step. Thus, the measure of inadmissible domains in the phase space at step $z_0 + k$ is less than $(mes\ \mathcal{C}_0)\chi_0^{z_0+k}$. Assume we are doing parameter choice at step $n = z_0 + k$. The critical value $w(t) = h(0,t)$ is moving with the unit speed across some element of ζ_n - domain of a good branch $\Delta^*(t) = \Delta\phi(y,t)$. We pull-back onto $\Delta^*(t)$ elements of $\xi_{[nx]}$. By Proposition 11 the end points of these elements as well as the end points of $\Delta^*(t)$ are moving with a small speed less than ϵ. Let t_0 be the moment when the image of $h(t,0)$ covers the whole domain $\Delta^*(t)$. From Lemma 2 we get that when $w(t)$ goes through $\Delta^*(t)$ the length of $\Delta^*(t)$ satisfies

$$1-\epsilon < \frac{|\Delta^*(t)|}{|\Delta^*(t_0)|} < 1+\epsilon\,. \tag{345}$$

and the time $\mathcal{T}$ that $w(t)$ spends inside $\Delta^*(t)$ satisfies

$$1-\epsilon' < \frac{|\mathcal{T}|}{|\Delta^*(t_0)|} < 1+\epsilon' \tag{346}$$

Similar estimates are valid for elements of $\phi^{-1}\xi_{[nx]}$. By construction at step n

$$|\Delta^*(t_0)| < \bar{k} \mid \delta_0 \mid^2 R^{-(r_2+3)sn}\,. \tag{347}$$

Then we get from (346)

$$|\mathcal{T}| < (1+\epsilon)\bar{k} \mid \delta_0 \mid^2 R^{-(r_2+3)sn} \tag{348}$$

Now we are able to estimate how the lengths of the elements of ξ_n change during the time when $w(t)$ crosses $\Delta^*(t)$. As the elements are properly sized they are greater than

$$|\delta_0| R^{-(r_2+1)sn}\,. \tag{349}$$

Speeds of the boundary points of these elements are less than $\frac{2}{k} \mid \delta_0 \mid^{-1} R^{sn}$. Using (348) we get that during the time interval $\mathcal{T}$ the lengths cannot change more than by

$$2 \mid \delta_0 \mid R^{-(r_2+2)sn}\,. \tag{350}$$

That is exponentially small comparatively to (349). We get that the variation of length for any element of ξ_n is exponentially small comparatively to its length. That proves that size estimates of Section 6 are valid for all $t \subset \mathcal{T}$. When estimating measure of admissible parameter values we pull-back elements of $\xi_{[nx]}$ which are larger than elements of ξ_n. Respectively relative variation of sizes is smaller. As in the proof of Proposition 4, see (155), it implies that the relative time that $h(t,0)$ spends in the union of good branches is up to a uniform factor $(1 \pm \epsilon)\exp(l_\infty)$ proportional to the measure of these branches at a fixed moment $t_0 \in \mathcal{T}$. When doing parameter choice at step z_0+k we pull-back $\xi_{[(z_0+k)x-1]}$. The measure of inadmissible domains in that partition is less than $(mes\ \mathcal{C}_0)\chi_0^{[(z_0+k)x]-1}$. As we pull-back no more than m_0 times the measure remaining in the parameter space is greater than

$$\left(1-(\exp l_\infty)(mes\ \mathcal{C}_0)\chi_0^{[(z_0+k)x]-1}\right)^{m_0}\,. \tag{351}$$

We used μ_0 to denote the measure of inadmissible parameter values after step z_0. Respectively, after z_0+k steps of induction the remaining measure is greater than

$$(1-\mu_0)\prod_{k=0}^{\infty}\left(1-(\exp l_\infty)(mes\ \mathcal{C}_0)\chi_0^{[(z_0+k)x]-1}\right)^{m_0}\,. \tag{352}$$

That finishes the proof of Theorem 1.

7.4. f-invariant measure and decay of correlations. Folklore Theorem [**1**] implies that for an admissible parameter value t the power map $F_t = f^{n_i}|\Delta_i$ has an absolutely continuous invariant measure. The original map f_t has an absolutely continuous invariant measure if $\sum_i n_i \mid \Delta_i \mid$ is finite.
Let us denote by $N(n)$ the maximal number of iterates of the original map in all branches which belong to partitions ζ_n and ξ_n. From construction we get

$$N(n+1) < N(n) + c_1 n N([xn]) \tag{353}$$

where c_1 is a uniform constant. As $x < 1$ it implies by induction that for any $\epsilon > 0$ there exists $c_2 = c_2(\epsilon)$ such that for sufficiently large n

$$N(n) < c_2 R^{\epsilon n}\,. \tag{354}$$

We proved that the measure of undecided domains at step n decreases faster than $(2R^{-s})^n$. As all elements constructed at step $n+1$ are inside undecided domains of step n we get

$$\sum_i n_i \mid \Delta_i \mid < c_2 \sum_{n=0}^{\infty} (2R^{-(s-\epsilon)})^n\,. \tag{355}$$

According to (133) $R^{-s} < \frac{1}{16}$. The sum converges and we get an f-invariant absolutely continuous measure μ. Let $T(n)$ be the set of elements Δ_i satisfying $n_i > n$. The same arguments show that

$$\sum_{n=N}^{\infty} n \mid T(n) \mid \tag{356}$$

decrease faster than any power N^{-p}. Then the results of [**15**] imply for (f,μ) a decay of correlations faster than any power. On the other hand it is easy to check that there are admissible parameter values in our construction such that

$$\lim_{n\to\infty} \inf \log \mid \frac{\partial f^n}{\partial x}(w) \mid = 0\,. \tag{357}$$

From [**8**] it follows that the decay of correlations for such parameter values is slower than exponential. As proved in [**14**] systems (f,μ) from [**2**] have exponential decay of correlations. We conclude that some parameter values obtained by the above construction do not belong to the set of parameters from [**2**].

References

[1] R. Adler. Afterword to R. Bowen, Invariant measures for Markov maps of the interval. *Comm. Math. Phys.*, 69(1): 1–17, 1979.

[2] M. Benedicks and L. Carleson. On iterations of $1 - ax^2$ on $(-1,1)$. *Ann. Math.*, 122: 1–25, 1985.

[3] M. Jakobson. Absolutely continuous invariant measures for one-parameter families of one-dimensional maps. *Comm. Math. Phys.*, 81: 39–88, 1981.

[4] M. Jakobson. Families of one-dimensional maps and nearby diffeomorphisms. *Proceedings of ICM Berkeley, 1986*, 1150–1160, 1987.

[5] M. Jakobson. Uniformly scaled Markov partitions for unimodal maps. *J. of Math Sciences*, 95, 5, 2583–2608, 1999.

[6] M. Jakobson and S. Newhouse. SRB measures for non-hyperbolic attractors. *in preparation.*

[7] S. Luzzatto and M. Viana. Positive Lyapunov exponents for Lorenz-like families with criticalities. *Asterisque*, 261, 201–237, 2000.

[8] T. Nowicki and D. Sands. Non-uniform hyperbolicity and universal bounds for S-unimodal maps. *Inventiones Math.*, 132: 633–680, 1998.

[9] M. Pacifico, A. Rovella and M. Viana. Infinite-modal maps with global chaotic behavior. *Ann. Math.*, 148: 441–484, 1998.
[10] M. Rychlik. Another proof of Jakobson's theorem and related results. *Ergodic Theory and Dynamical Systems*, 8: 93–109, 1988.
[11] S. Senti. Dimension de Hausdorff de l'ensemble exceptionnel dans le theoreme de Jakobson. *These, L' Universite de Paris-Sud*, 2000.
[12] Ph. Thieullen, C. Tresser, L.-S. Young. Positive Lyapunov exponents for generic one-parameter families of unimodal maps. *J. Anal. Math.*, 64: 121–172, 1994.
[13] J.-C. Yoccoz. Jakobson's theorem. *Manuscript of course at College de France*, 1997.
[14] L.-S. Young. Decay of correlations for certain quadratic maps. *Comm. Math. Phys.*, 146, 1: 123–138, 1992.
[15] L.-S. Young. Ergodic theory of chaotic dynamical systems. *ICMP' 97, Brisbane, Intern. Press, Cambridge, MA* , 1999.

DEPARTMENT OF MATHEMATICS, UNIVERSITY OF MARYLAND, COLLEGE PARK, MD, 20742

Selected Titles in This Series

(Continued from the front of this publication)

47 **Paul Fong, Editor,** The Arcata conference on representations of finite groups (Humboldt State University, Arcata, California, July 1986)

46 **Spencer J. Bloch, Editor,** Algebraic geometry – Bowdoin 1985 (Bowdoin College, Brunswick, Maine, July 1985)

45 **Felix E. Browder, Editor,** Nonlinear functional analysis and its applications (University of California, Berkeley, July 1983)

44 **William K. Allard and Frederick J. Almgren, Jr., Editors,** Geometric measure theory and the calculus of variations (Humboldt State University, Arcata, California, July/August 1984)

43 **François Trèves, Editor,** Pseudodifferential operators and applications (University of Notre Dame, Notre Dame, Indiana, April 1984)

42 **Anil Nerode and Richard A. Shore, Editors,** Recursion theory (Cornell University, Ithaca, New York, June/July 1982)

41 **Yum-Tong Siu, Editor,** Complex analysis of several variables (Madison, Wisconsin, April 1982)

40 **Peter Orlik, Editor,** Singularities (Humboldt State University, Arcata, California, July/August 1981)

39 **Felix E. Browder, Editor,** The mathematical heritage of Henri Poincaré (Indiana University, Bloomington, April 1980)

38 **Richard V. Kadison, Editor,** Operator algebras and applications (Queens University, Kingston, Ontario, July/August 1980)

37 **Bruce Cooperstein and Geoffrey Mason, Editors,** The Santa Cruz conference on finite groups (University of California, Santa Cruz, June/July 1979)

36 **Robert Osserman and Alan Weinstein, Editors,** Geometry of the Laplace operator (University of Hawaii, Honolulu, March 1979)

35 **Guido Weiss and Stephen Wainger, Editors,** Harmonic analysis in Euclidean spaces (Williams College, Williamstown, Massachusetts, July 1978)

34 **D. K. Ray-Chaudhuri, Editor,** Relations between combinatorics and other parts of mathematics (Ohio State University, Columbus, March 1978)

33 **A. Borel and W. Casselman, Editors,** Automorphic forms, representations and L-functions (Oregon State University, Corvallis, July/August 1977)

32 **R. James Milgram, Editor,** Algebraic and geometric topology (Stanford University, Stanford, California, August 1976)

31 **Joseph L. Doob, Editor,** Probability (University of Illinois at Urbana-Champaign, Urbana, March 1976)

30 **R. O. Wells, Jr., Editor,** Several complex variables (Williams College, Williamstown, Massachusetts, July/August 1975)

29 **Robin Hartshorne, Editor,** Algebraic geometry – Arcata 1974 (Humboldt State University, Arcata, California, July/August 1974)

28 **Felix E. Browder, Editor,** Mathematical developments arising from Hilbert problems (Northern Illinois University, Dekalb, May 1974)

27 **S. S. Chern and R. Osserman, Editors,** Differential geometry (Stanford University, Stanford, California, July/August 1973)

26 **Calvin C. Moore, Editor,** Harmonic analysis on homogeneous spaces (Williams College, Williamstown, Massachusetts, July/August 1972)

ISBN 0-8218-2682-4

9 780821 826829

PSPUM/69